NOISE AND FLUCTUATIONS

To learn more about the AIP Conference Proceedings, including the
Conference Proceedings Series, please visit the webpage **http://proceedings.aip.org/proceedings**

NOISE AND FLUCTUATIONS

18th International Conference on
Noise and Fluctuations - ICNF 2005

Salamanca, Spain   19 – 23 September 2005

EDITORS
Tomás González
Javier Mateos
Daniel Pardo
University of Salamanca
Salamanca, Spain

SPONSORING ORGANIZATIONS
University of Salamanca
IEEE Electron Devices Society
Spanish Ministry of Education and Science
Education Council—Community of Castilla y León
Caja Duero

◎ **CD-ROM INCLUDED**

Melville, New York, 2005
AIP CONFERENCE PROCEEDINGS ■ VOLUME 780

Editors:

Tomás González
Javier Mateos
Daniel Pardo

Departamento de Física Aplicada
Universidad de Salamanca
Plaza de la Merced s/n
37008 Salamanca, SPAIN

E-mail: tomasg@usal.es
         javierm@usal.es
         dpardo@usal.es

Authorization to photocopy items for internal or personal use, beyond the free copying permitted under the 1978 U.S. Copyright Law (see statement below), is granted by the American Institute of Physics for users registered with the Copyright Clearance Center (CCC) Transactional Reporting Service, provided that the base fee of $22.50 per copy is paid directly to CCC, 222 Rosewood Drive, Danvers, MA 01923, USA. For those organizations that have been granted a photocopy license by CCC, a separate system of payment has been arranged. The fee code for users of the Transactional Reporting Services is: ISBN/0-7354-0267-1/05/ $22.50.

© 2005 American Institute of Physics

Permission is granted to quote from the AIP Conference Proceedings with the customary acknowledgment of the source. Republication of an article or portions thereof (e.g., extensive excerpts, figures, tables, etc.) in original form or in translation, as well as other types of reuse (e.g., in course packs) require formal permission from AIP and may be subject to fees. As a courtesy, the author of the original proceedings article should be informed of any request for republication/reuse. Permission may be obtained online using Rightslink. Locate the article online at http://proceedings.aip.org, then simply click on the Rightslink icon/"Permission for Reuse" link found in the article abstract. You may also address requests to: AIP Office of Rights and Permissions, Suite 1NO1, 2 Huntington Quadrangle, Melville, NY 11747-4502, USA; Fax: 516-576-2450; Tel.: 516-576-2268; E-mail: rights@aip.org.

L.C. Catalog Card No. 2005929442
ISBN 0-7354-0267-1
ISSN 0094-243X
Printed in the United States of America

Contents

Preface... xix
ICNF 2005 Organization, Committees, and Sponsors xxi

PLENARY LECTURES

Noise in Advanced Electronic Devices and Circuits........................... 3
   M. J. Deen and O. Marinov

THEORY

Control and Rectification in Collective Stochastic Systems (*invited*)............ 15
   J. M. R. Parrondo and L. Dinís
A Form of Active Brownian Motor-like on a (Nonlinear) Toda Lattice 21
   E. del Rio, W. Ebeling, and M. G. Velarde
Diffusion Acceleration in Randomly Switching Sawtooth Potential 25
   A. A. Dubkov and B. Spagnolo
Current Fluctuations in Degenerate Non-equilibrium Systems 29
   R. Katilius, S. Reggiani, and M. Rudan
Quantum Noise in Long Josephson Junctions 33
   V. V. Kurin and I. V. Pimenov
Noise-induced Patterns in Semiconductor Nanostructures and
Time-delayed Feedback Control.. 37
   E. Schöll, G. Stegemann, A. Amann, and A. G. Balanov
Control of Noise-induced Oscillations in Superlattices by Delayed
Feedback 41
   J. Hizanidis, A. G. Balanov, A. Amann, and E. Schöll
Current and Noise Suppression in AC-driven Coherent Transport
(*invited*).. 45
   S. Kohler, M. Strass, P. Hänggi, M. Rey, and F. Sols
High-order Current Correlation Functions in Kondo Systems 51
   A. Golub
System Size Stochastic Resonance from the Viewpoint of the
Nonequilibrium Potential.. 55
   H. S. Wio
Extensions of the Stochastic Model of the Overdamped Oscillator
Applied to AC Ionic Conductivity in Solids 59
   J. Bisquert
Fluctuation-dissipation Dispersion Relation for Systems with Slowly
Varying Parameters 63
   V. Belyi
A New Version of the Fluctuation-dissipation Relations 67
   S. F. Timashev and G. V. Vstovsky

Bispectrum Theory for Brownian Motion of Electrical Charge in
Non-linear RC-circuit... 71
    B. M. Grafov

$1/f$ Temperature Fluctuations in Solids 75
    H. Higuchi, S. Nakamura, and S. Ochi

Properties of the Sum-lengths of the Adjacent Level Crossing
Intervals of the Chaotic Process Generated by the Logistic Map 79
    T. Munakata and W. Schwarz

First Passage Time Algorithm for Signal and Filtered Multi-level
Noise... 83
    R. M. Howard

$1/f$-type Noise in View of Phonons Interface Percolation Dynamics 87
    S. V. Melkonyan, F. V. Gasparyan, V. M. Aroutiounian, and H. V. Asriyan

$1/f$ Noise in Fractal Quaternionic Structures 91
    T. Meškauskas and B. Kaulakys

A Fast Stochastic Digital Signal Generator Based on Chaotic Iteration........ 95
    F. Chong, W. Pei-Rong, X. Zhe, and Z. Wei-Yong

Estimating the Variance of Multiplicative Noise 99
    D. Evans

MATERIALS

Microwave Noise in Biased AlGaN/GaN and AlGaN/AlN/GaN
Channels... 105
    A. Matulionis, J. Liberis, and M. Ramonas

Generation of Interface States Due to Quantum-dot Growth in
Au/GaAs Schottky Diode Structures 109
    W. Choi, H. Nam, J. Lee, B. Yu, J. Song, H. Yang, and A. Chovet

$1/f$ Noise Enhancement in GaAs... 113
    J. I. Izpura and J. Malo

Unusual Fluctuations of Flux Flow at Low Temperature in
Superconducting Films.. 117
    S. Okuma, K. Kainuma, and T. Kishimoto

Study of Dendritic Avalanches by Current Noise Measurements in
High T_c Superconductors... 121
    E. Celasco, R. Eggenhöffner, G. Tolotto, and M. Celasco

Non-Gaussian Fluctuations in Biased Resistor Networks: Size Effects
versus Universal Behavior (*invited*)...................................... 125
    C. Pennetta, E. Alfinito, L. Reggiani, and S. Ruffo

Low-frequency Current Fluctuations in Post-hard Breakdown Thin
Silicon Oxide Films.. 131
    Y. Omura and K. Komiya

Noise and Charge Storage in Nb_2O_5 Thin Films 135
    V. Sedlakova, J. Sikula, L. Grmela, P. Hoeschl, Z. Sita, S. Hashiguchi, and
    M. Tacano

$1/f$ Noise in Low Density Two-dimensional Hole Systems in GaAs........... 139
    G. Deville, R. Leturcq, D. L'Hôte, R. Tourbot, C. J. Mellor, and M. Henini

Low-frequency Noise Measurements in $La_{0.7}Sr_{0.3}MnO_3$ Thin Films on (100) $SrTiO_3$ 143
    L. Méchin, F. Yang, S. Mercone, J. M. Routoure, S. Flament, C. Simon, and R. A. Chakalov

Acoustic Noise Spectrum of the Liquid Helium Boiling Process on Superconducting Bolometer Surface 147
    O. V. Pakhomov and I. A. Khrebtov

Monte Carlo Calculation of Diffusion Coefficient, Noise Spectral Density and Noise Temperature in HgCdTe 151
    C. Palermo, L. Varani, J. C. Vaissière, J. F. Millithaler, E. Starikov, P. Shiktorov, V. Gružinskis, and B. Azaïs

Noise Properties of High Resistivity Cl-doped Cadmium Telluride 155
    I. S. Virt, A. Kolek, V. D. Popovych, and I. S. Bilyk

Hot Electron Noise in N-type Semiconductors in Crossed Electric and Magnetic Fields 159
    F. Ciccarello and M. Zarcone

Low-frequency Noise Resolution of Ru-based Low-temperature Thick Film Sensors 163
    A. Kolek, P. Ptak, Z. Zawislak, A. W. Standler, and K. Mleczko

Current-driven Large Density Fluctuations of Vortices and Antivortices in the Corbino Disk 167
    S. Okuma and S. Morishima

Microwave Induced Effects on the Random Telegraph Signal in a MOSFET 171
    E. Prati, M. Fanciulli, G. Ferrari, M. Sampietro, and P. Fantini

Low-frequency Noise of the CdTe Crystals 175
    L. Grmela, J. Sikula, J. Zajacek, and P. Moravec

DEVICES

Low-frequency Noise in Si-based MOS Devices (*invited*) 181
    J. Jomaah and G. Ghibaudo

The Low-frequency Noise of Strained Silicon n-MOSFETs 187
    E. Simoen, G. Eneman, P. Verheyen, R. Delhougne, R. Rooyackers, R. Loo, W. Vandervorst, K. De Meyer, and C. Claeys

Low-frequency Noise Sensitivity to Technology Induced Mechanical Stress in MOSFETs 191
    P. Fantini and G. Ferrari

Can $1/f$ Noise in MOSFETs Be Reduced by Gate Oxide and Channel Optimization? 195
    M. Marin, J. C. Vildeuil, B. Tavel, B. Duriez, F. Arnaud, P. Stolk, and M. Woo

Impact of Interface Micro-roughness on Low-frequency Noise in (110) and (100) pMOSFETs 199
    P. Gaubert, A. Teramoto, T. Hamada, M. Yamamoto, K. Nii, H. Akahori, K. Kotani, and T. Ohmi

Modeling of Suppressed Shot Noise in Stress-induced Leakage Currents (*invited*) 203
    G. Iannaccone

LF Noise and Tunneling Current in Nanometric SiO_2 Layers 209
    J. Gurgul, C. Leroux, G. Ghibaudo, and J. A. Chroboczek

Evolution of RTS Source Activities in Saturation Range in n-MOSFETs for Different Oxidation Temperatures 213
    C. Leyris, A. Hoffmann, M. Valenza, J. C. Vildeuil, and F. Roy

Zero Cross Analysis of RTS Noise 217
    J. Pavelka, M. Tacano, M. Toita, J. Sikula, and T. Musha

High Magnetic Field Dependence of Capture/Emission Fluctuations of a Single Defect in Silicon MOSFETs 221
    E. Prati, M. Fanciulli, G. Ferrari, and M. Sampietro

Noise in Si and SiGe MOSFETs with High-k Gate Dielectrics (*invited*) 225
    M. von Haartman, B. G. Malm, P. E. Hellström, and M. Östling

Impact of Gate Material on Low-frequency Noise of nMOSFETs with 1.5 nm SiON Gate Dielectric: Testing the Limits of the Number Fluctuations Theory 231
    P. Srinivasan, E. Simoen, L. Pantisano, C. Claeys, and D. Misra

Gate and Drain Low-frequency Noise in HfO_2 nMOSFETs 235
    T. Nguyen, M. Valenza, F. Martinez, G. Neau, J. C. Vildeuil, G. Ribes,
    V. Cosnier, T. Skotnicki, and M. Müller

Intrinsic Fluctuations Induced by a High-k Gate Dielectric in Sub-100 nm Si MOSFETs 239
    A. J. García-Loureiro, K. Kalna, and A. Asenov

Contributions of Channel Gate and Overlap Gate Currents on $1/f$ Gate Current Noise for Thin Oxide Gate p-MOSFETs 243
    F. Martinez, A. Laigle, A. Hoffmann, M. Valenza, A. Veloso, and
    M. Jurczak

RF Noise Modeling in SiGe HBTs (*invited*) 247
    G. Niu, K. Xia, D. Sheridan, and S. Sweeney

Impact of Lateral Scaling on Low-frequency Noise of 200 GHz SiGe:C HBTs 253
    R. Venegas, N. Ouassif, A. Piontek, S. Van Huylenbroeck, and
    S. Decoutere

Influence of Carbon Incorporation on the Low-frequency Noise of Si/SiGe:C HBTs Based on 0.25 μm BiCMOS Technology 257
    P. Benoit, J. Raoult, C. Delseny, F. Pascal, J. C. Vildeuil, B. Szelag, and
    A. Monroy

Effect of Base/Collector Implant and Emitter-poly Overlap on Low-frequency Noise in SiGe HBTs 261
    M. M. U. Hoque, Z. Çelik-Butler, S. Martin, C. Knorr, and C. Bulucea

Low-frequency Noise in SOI SiGe HBTs Made by Selective Growth of the Si Collector and Non-selective Growth of SiGe Base 265
    N. Lukyanchikova, N. Garbar, A. Smolanka, M. Lokshin, S. Hall, O. Buiu,
    I. Z. Mitrovic, H. A. W. Mubarek, and P. Ashburn

Noise in SOI MOSFETs and Gate-all Around Transistors (*invited*) 269
  B. Iñiguez, A. Lázaro, H. A. Hamid, G. Pailloncy, G. Dambrine, and
  F. Danneville
Noise in Nanometric s-Si MOSFET for Low-power Applications 275
  K. Fobelets and J. E. Velázquez
Geometry Dependence of $1/f$ Noise in N- and P- Channel MuGFETs 279
  V. Subramanian, A. Mercha, A. Dixit, K. G. Anil, M. Jurczak,
  K. De Meyer, S. Decoutere, H. Maes, G. Groeseneken, and W. Sansen
Low-frequency Noise in Si/SiGe HFET 283
  M. Rodriguez, N. Zerounian, H. J. Herzog, T. Hackbarth, and F. Aniel
HF Noise Performance and Modelling of SiGe HFETs 287
  M. Enciso, N. Zerounian, P. Crozat, T. Hackbarth, J.-H. Herzog, and
  F. Aniel
**Model of the $1/f$ Noise in GaN/AlGaN Heterojunction Field Effect
Transistors** 291
  M. E. Levinshtein, A. P. Dmitriev, S. L. Rumyantsev, and M. S. Shur
**Low-frequency Noise Characterization of Hot-electron Degradation in
GaN-based HEMTs** 295
  S. Jha, J. Gao, C. F. Zhu, E. Jelenkovic, K. Y. Tong, M. Pilkuhn, C. Surya,
  and H. Schweizer
**Low-frequency Noise of AlGaN/GaN HEMT Grown on Al_2O_3, Si and
SiC Substrates** 299
  J. G. Tartarin, G. Soubercaze-Pun, A. Rennane, L. Bary, S. Delage,
  R. Plana, and J. Graffeuil
**Investigation of Shot Noise Reduction in InGaP HBTs with Different
Base Thickness** 303
  P. Sakalas, M. Schroter, and P. Zampardi
**Low-frequency Noise in SiGe Channel pMOSFETs on Ultra-thin
Body SOI with Ni-silicided Source/Drain** 307
  M. von Haartman, J. Hållstedt, J. Seger, B. G. Malm, P.-E. Hellström, and
  M. Östling
**Low-frequency Noise Characteristics of TaSiN/HfO$_2$/SRPO SiO$_2$
MOSFETs** 311
  S. P. Devireddy, Z. Çelik-Butler, H. H. Tseng, P. J. Tobin, F. Wang, and
  A. Zlotnicka
**Drain and Gate Current LF Noise in Advanced CMOS Devices with
Ultrathin Gate Oxides** 315
  T. Contaret, K. Romanjek, G. Ghibaudo, J. Chroboczek, F. Bœuf, and
  T. Skotnicki
**Apparent Noise Parameter Behavior in n-MOS Transistors Operating
from Subthreshold to Above-threshold Regions** 319
  A. Boukhenoufa, L. Pichon, C. Cordier, B. Cretu, L. Ding, R. Carin,
  J.-F. Michaud, and T. Mohammed-Brahim
**Effect of Oxide Thickness and Nitridation Process on PMOS Gate
and Drain Low-frequency Noise** 323
  F. Martinez, C. Leyris, M. Valenza, A. Hoffmann, F. Bœuf, T. Skotnicki,
  M. Bidaud, D. Barge, and B. Tavel

Characterization of Low-frequency Noise Sources in Planar Devices
Using Cross-shaped 4-terminal Devices.................................327
    V. Mosser and A. Kerlain
Low-frequency Noise Characterization of 90 nm Multiple Gate Oxide
CMOS Transistors 331
    N. Lukyanchikova, N. Garbar, A. Smolanka, M. Lokshin, S.-C. Lee,
    E. Simoen, and C. Claeys
TeraHertz Emission from Nanometric HEMTs Analyzed by Noise
Spectra..335
    J.-F. Millithaler, L. Varani, C. Palermo, J. Mateos, T. González, S. Pérez,
    D. Pardo, W. Knap, J. Lusakowski, N. Dyakonova, S. Bollaert, and
    A. Cappy
RTS in Submicron MOSFETs: High Field Effects........................339
    J. Pavelka, V. Sedlakova, J. Sikula, J. Havranek, M. Tacano, S. Hashiguchi,
    and M. Toita
Hooge Noise Parameter of GaN HFETs on SiC..........................343
    N. Tanuma, J. Pavelka, S. Yagi, H. Okumura, T. Uemura, M. Tacano,
    S. Hashiguchi and J. Sikula
Fundamental Effects in the Dependence of the $1/f$ Noise Spectrum on
the Bias Current in Semiconductor Diodes.............................347
    A. V. Yakimov

OPTOELECTRONICS AND PHOTONICS

Photonic Propagation Noise on Long Atmospheric Paths..................353
    P. J. Edwards, J. Gleeson, T. Smallhorn, A. P. Whichello, and D. Woodgate
Light Transport through Photonic Liquids.............................357
    L. S. Froufe-Pérez, S. Albaladejo, E. Sahagún, P. García-Mochales,
    M. Reufer, F. Scheffold, and J. J. Sáenz
About the Physical Origin of Pixel Flickering in Cooled $Hg_{0.7}Cd_{0.3}Te$
Infrared Photodetectors...361
    B. Orsal, J. P. Perez, M. Myara, R. Alabedra, C. Leyris, J.-P. Tourrenc, and
    P. Signoret
$1/f$ Optical Fluctuations in Quantum Well Laser Diodes 365
    A. V. Belyakov, L. K. J. Vandamme, and A. V. Yakimov
Investigation of the External Feedback Effects on the Relative
Intensity Noise Characteristics of AlGaInN Blue Laser Diodes...............369
    J. C. Yi, J. Y. Kim, and T. K. Yoo
Noise Modelling of Absolute High-T_c Superconducting Measuring
Instrument for X-ray Synchrotron Radiation............................373
    I. A. Khrebtov, K. V. Ivanov, and D. A. Khokhlov
Low and Medium Frequency Noise Levels of Fibered Very-High-
Power 1.460nm-pump Laser Diode Designed for Raman Amplification........377
    C. Chluda, M. Myara, J.-P. Perez, P. Signoret, and B. Orsal
Transmittances Distributions at the Diffusive-localized Crossover in
Disordered Wave-guides with Absorption..............................381
    L. S. Froufe-Pérez, P. García-Mochales, P. A. Serena, and J. J. Sáenz

Rice Representation of Noise Processes in Optical PLL's............385
    A. Arvizu M. and F. J. Mendieta J.
Acoustic Emission, Electrical and Light Fluctuations in
Optoelectronic Devices..389
    O. I. Vlasenko, V. P. Veleschuk, and O. V. Lyashenko
Local Instabilities in GaAsP Diode PN Junctions 393
    P. Koktavy and B. Koktavy
Photocurrent Noise in Quantum Dot Infrared Photodetectors 397
    A. Carbone, R. Introzzi, and H.C. Liu

MESOSCOPICS

Shot Noise in Mesoscopic Conductors: from Schottky to Bell (*invited*) 403
    M. Büttiker, P. Samuelsson, and E. V. Sukhorukov
Entanglement in a Noninteracting Mesoscopic Structure............409
    A.V. Lebedev, G. B. Lesovik, and G. Blatter
Current Noise Spectrum of Open Charge Qubits 413
    R. Aguado, N. Lambert, and T. Brandes
Fano Factor Reduction on the 0.7 Structure 417
    P. Roche, J. Ségala, D. C. Glattli, J. T. Nicholls, M. Pepper, A. C. Graham,
    K. J. Thomas, M. Y. Simmons, and D. A. Ritchie
Current Noise in Non-chiral Luttinger Liquids: Appearance of
Fractional Charge...421
    F. Dolcini, B. Trauzettel, I. Safi, and H. Grabert
Comparative Analysis of Sequential and Coherent Tunneling Models
of Shot Noise in Resonant Diodes (*invited*) 425
    V. Y. Aleshkin and L. Reggiani
Shot Noise Experiments in Multi-barrier Semiconductor
Heterostructures..431
    E. E. Mendez, W. Song, A. K. M. Newaz, Y. Lin, and J. Nitta
Transition between Pauli Exclusion and Coulomb Interaction in the
Noise Behavior of Resonant Tunneling Devices 435
    I. A. Maione, G. Basso, M. Macucci, G. Iannaccone, and B. Pellegrini
Decoherence and Current Fluctuations in Tunneling through Coupled
Quantum Dots...439
    G. Kießlich, P. Samuelsson, A. Wacker, and E. Schöll
Noise and Bistabilities in Quantum Shuttles......................442
    C. Flindt, T. Novotný, and A. P. Jauho
Noise Minimization in Quantum Transistors 446
    U. Gavish, B. Yurke, and Y. Imry
Length Dependence of the Fano Factor in Mesoscopic Cavities 450
    P. Marconcini and M. Macucci
Chaotic-to-Regular Crossover of Shot Noise in Mesoscopic
Conductors...454
    S. Rotter, F. Aigner, and J. Burgdörfer
Excess Noise in Carbon Nanotubes 458
    S. Reza, Q. T. Huynh, G. Bosman, J. Sippel, and A. G. Rinzler

Low-frequency Noise in Contacted Single-wall Carbon Nanotube 462
    S. Soliveres, A. Hoffmann, F. Pascal, C. Delseny, A. Salesse, M. S. Kabir,
    S. Bengtsson, O. Nur, M. Willander, and J. Deen

Hanbury Brown and Twiss Noise Correlations to Probe the Statistics
of GHz Photons Emitted by Quantum Conductors (*invited*) 466
    D. C. Glattli, J. Gabelli, L.-H. Reydellet, G. Fève, J. M. Berroir, B. Plaçais,
    and P. Roche

Measurements of Correlated Conductances and Noise Fluctuations
from 3-Lead Quantum Dots 472
    R. C. Toonen, M. Prada, H. Qin, A. K. Huettel, S. Goswami,
    M. A. Eriksson, D. W. van der Weide, K. Eberl, and R. H. Blick

Observation of Giant Thermal Noise Due to Multiple Andreev
Reflection in a Ballistic SNS Junction with an InGaAs-based
Heterostructure 476
    T. Akazaki, H. Nakano, J. Nitta, and H. Takayanagi

Full Counting Statistics of Mesoscopic Electron Transport 480
    W. Belzig

Chaotic Transport: from Quantum to Classical 484
    R. S. Whitney and P. Jacquod

Photo-assisted Shot Noise in the Fractional Quantum Hall Regime 488
    A. Crépieux, P. Devillard, and T. Martin

Observation of $1/f^\alpha$ Noise of GaInP/GaAs Triple Barrier Resonant
Tunneling Diodes 492
    N. Asaoka, M. Fukumitsu, M. Suhara, and T. Okumura

CIRCUITS AND SYSTEMS

Oscillator Noise Analysis (*invited*) 499
    A. Demir

Noise Simulation of Continuous-time $\Sigma\Delta$ Modulators 505
    J. Arias, L. Quintanilla, D. Bisbal, J. San Pablo, L. Enriquez, J. Vicente,
    and J. Barbolla

Noise Performance at Cryogenic Temperature of Microwave SiGeC
Low Noise Amplifier Using BiCMOS Technology 509
    S. Pruvost, S. Delcourt, F. Danneville, I. Telliez, G. Dambrine, M. Laurens,
    and A. Monroy

Nonlinear Noise in SiGe Bipolar Devices and Its Impact on
Radio-frequency Amplifier Phase Noise 513
    S. Gribaldo, G. Cibiel, O. Llopis, and J. Graffeuil

A New Method of Minimizing Noise Figure of CMOS LNAs 517
    D. Pienkowski and G. Boeck

Analysis and Design of a Frequency Synthesizer with Internal and
External Noise Sources 521
    G. S. Sangha and M. H. W. Hoffmann

Analytical Calculations for the Influence of Carrier's Phase Noise on
Modulating Signals 525
    D. Kondis, A. Birbas, and M. Birbas

Observation of Noise-induced Transitions in Radar Range Tracking Systems 529
    E. H. Abed and S. I. Wolk

BIOLOGICAL AND BIOMEDICAL SYSTEMS

Functional Roles of Noise and Fluctuations in the Human Brain (*invited*) 535
    Y. Yamamoto, R. Soma, K. Kitajo, L. A. Safonov, K. Yamanaka, I. Hidaka, K. Ohashi, D. Nozaki, Z. R. Struzik, L. M. Ward, and S. Kwak

Mapping of Synaptic-neuronal Impairment on the Brain Surface through Fluctuation Analysis 541
    T. Musha, T. Kurachi, N. Suzuki, and Y. Kosugi

Sleep Stage Dependence of Invariance Characteristics in Fluctuations of Healthy Human Heart Rate 545
    F. Togo, K. Kiyono, Z. R. Struzik, and Y. Yamamoto

Criticality and Universality in Healthy Heart Rate Dynamics 549
    Z. R. Struzik, K. Kiyono, J. Hayano, S. Sakata, S. Kwak, and Y. Yamamoto

Fluctuations of the Single Photon Response in Visual Transduction 553
    L. Shen, D. Andreucci, H. E. Hamm, and E. DiBenedetto

Spontaneous Movements of Mechanosensory Hair Bundles (*invited*) 557
    B. Nadrowski, P. Martin, and F. Jülicher

Brownian Dynamics Simulations of Ionic Current through an Open Channel 563
    R. Tindjong, R. S. Eisenberg, I. Kaufman, D. G. Luchinsky, and P. V. E. McClintock

Stochastic Resonance of Artificial Ion Channels Inserted in Small Membrane Patches 567
    R. H. Blick, S. Y. Choi, H. S. Kim, S. Ramachandran, and D. W. van der Weide

Starting and Stopping a Bistable Pacemaker: Stochastic Stimulation Identifies Critical Perturbations 571
    D. Paydarfar, D. B. Forger, and J. R. Clay

Nanoscale Electronic Noise Measurements 575
    L. Fumagalli, I. Casuso, G. Ferrari, G. Gomila, M. Sampietro, and J. Samitier

What Information Is Hidden in Chaotic Signals of Biological Systems? 579
    S. F. Timashev, G. V. Vstovsky, A. Y. Kaplan, and A. B. Solovieva

Information Transfer Analysis of Spontaneous Low-frequency Fluctuations in Cerebral Hemodynamics and Cardiovascular Dynamics 583
    T. Katura, N. Tanaka, A. Obata, H. Sato, and A. Maki

Stochastic Resonance of Na, K-ion Pumps on the Red Cell Membrane 587
    C.-H. Chang and T. Y. Tsong

Contraction of Information on Brain Wave Fluctuations by
Information Geometrical Approach 591
    H. Konno
Fluctuations in Neuronal Activity: Clues to Brain Function 595
    J. L. Pérez-Velazquez, R. Guevara, J. Belkas, R. Wennberg, G. Senjanoviè,
    and L. García-Dominguez
Changes in the Hurst Exponent of Heart Rate Variability during
Physical Activity 599
    N. Aoyagi, K. Kiyono, Z. R. Struzik, and Y. Yamamoto
Spatial Asymmetric Retrieval States in Binary Attractor Neural
Network 603
    K. Koroutchev and E. Korutcheva
Relationships among One-minute Oscillations in Oxygen Saturation
Level of Blood and Hemoglobin Volume in Calf Muscular Tissue and
One-minute Wave in Body Fluid Volume Change during Upright
Standing in Humans... 607
    K. Inamura, T. Mano, and S. Iwase
Modelization of Thermal Fluctuations in G Protein-coupled Receptors........ 611
    C. Pennetta, V. Akimov, E. Alfinito, L. Reggiani, G. Gomila, G. Ferrrari,
    L. Fumagalli, and M. Sampietro
Noise-driven Switching between Limit Cycles and Adaptability in a
Small-dimensional Excitable Network with Balanced Coupling 615
    L. A. Safonov and Y. Yamamoto
Stochastic Nonlinear Evolutional Model of the Large-scaled Neuronal
Population and Dynamic Neural Coding Subject to Stimulation.............. 619
    R. Wang and W. Yu

CHEMISTRY AND ELECTROCHEMISTRY

The Electrochemical Noise Technique—Applications in Corrosion
Research (*invited*) 625
    F. Mansfeld
Chaos-order Transitions during Electrochemical Corrosion of Silicon......... 631
    V. Parkhutik and E. Matveyeva
Chaos and Resonance in the Model for Current Oscillations at the
Si/Electrolyte Contact.. 635
    J. Grzanna, H. Jungblut, and H. J. Lewerenz
Stochastic Resonances and Highly Selective Separation Methods:
Application to the Detection of DNA Mutations........................... 639
    A. Estévez-Torres, L. Jullien, and A. Lemarchand
Baseline Noise in High-performance Liquid Chromatography with
Electrochemical Detection 643
    A. Kotani, Y. Hayashi, R. Matsuda, and F. Kusu
The Noise Diagnostics of Organic Electrolytes for Rechargeable
Lithium Batteries 647
    L. S. Kanevskii, B. M. Grafov, and M. G. Astafiev

MEASUREMENT TECHNIQUES

Novel Transimpedance Amplifier for Noise Measurements on Bio-electronic Devices 653
    G. Ferrari and M. Sampietro

Suppression of Offset and Drift in a DC Amplifier by Combination of Multiple Amplifiers 657
    S. Hashiguchi, Y. Takemoto, M. Ohki, M. Tacano, and J. Sikula

Experimental Study of Hysteretic Josephson Junctions as Threshold Detectors of Shot Noise 661
    M. Meschke, T. E. Nieminen, and J. P. Pekola

Accuracy of $1/f$ Noise Parameter Extraction in the Presence of Background Noise 665
    I. Slaidins and M. Zeltins

RTS Noise in Optoelectronic Coupled Devices 669
    A. Konczakowska, J. Cichosz, and B. Stawarz

Noise Scattering Patterns Method for Recognition of RTS Noise in Semiconductor Components 673
    J. A. Cichosz and A. Szatkowski

A Method of Two-terminal Excess Noise Measurement with a Reduction of Measurement System and Contact Noise 677
    A. Szewczyk, L. Spiralski, and L. Hasse

Progress Toward "Optical Beam Induced Noise" Measurement Set-up 681
    J. M. Routoure, L. Méchin, and S. Flament

Finite Bandwidth Related Errors in Noise Parameter Determination of PHEMTs 685
    W. Wiatr and D. Adamson

Uncertainty in Measuring Noise Parameters of a Communication Receiver 689
    K. Korcz, B. Palczynska, and L. Spiralski

-190 dBV^2/Hz Preamplifier for Low-frequency Noise Measurements 693
    S. Yokokura, N. Tanuma, M. Tacano, S. Hashiguchi, J. Sikula, Y. Kajiwara, and M. Hirasita

Analysis of Error Sources in On-wafer Noise Characterization of RF CMOS Transistors 697
    W. Wiatr

RELIABILITY

Low-frequency Noise Considerations for CMOS Analog Circuit Design (*invited*) 703
    R. Brederlow, J. Koh, G. I. Wirth, R. da Silva, M. Tiebout, and R. Thewes

Accelerated Aging of GaN Light Emitting Diodes Studied by $1/f$ and RTS Noise 709
    S. Bychikhin, L. K. J. Vandamme, J. Kuzmik, G. Meneghesso, S. Levada, E. Zanoni, and D. Pogany

Influence of Small Doses of Gamma Irradiation on Transport and
Noise Properties of SiC MESFETs 713
    S. A. Vitusevich, M. V. Petrychuk, A. M. Kurakin, S. V. Danylyuk,
    A. E. Belyaev, H.-Y. Cha, M. G. Spencer, L. F. Eastman, and N. Klein

Effects of Oxygen-related Traps in Silicon on the Generation-
Recombination Noise 717
    J. A. Jiménez Tejada, J. A. López-Villanueva, A. Godoy, J. E. Carceller,
    F. M. Gómez-Campos, and S. Rodríguez-Bolívar

Noise and I-V Characteristic as Characterization Tools for GaSb
Based Laser Diodes 721
    Z. Chobola, J. Vaněk, J. Kazelle, E. Hulicius, and T. Simecek

Foil Capacitors Reliability Prediction Based on Fluctuation and
Non-linear Phenomena 725
    L. Hasse, L. Spiralski, and J. Turczyński

Morphology of Nanostructured Semiconductors Studied Using Atomic
Force Microscopy Combined with Stochastic Signal Spectroscopy 729
    V. Parkhutik, Y. Makushok, and E. Metveyeva

MODELING

Modeling and Characterization of Noise in 90-nm RF CMOS
Technology (*invited*) 735
    A. J. Scholten, L. F. Tiemeijer, A. T. A. Zegers-van Duijnhoven,
    R. J. Havens, R. de Kort, R. van Langevelde, D. B. M. Klaassen,
    W. Jeamsaksiri, and R. D. M. A. Velghe

An Analytical Thermal Noise Model of the MOS Transistor Valid in
All Modes of Operation 741
    A. S. Roy and C. C. Enz

On the High-Frequency Noise Figures of Merit and Microscopic
Channel Noise Sources in Fabricated 90 nm PD SOI MOSFETs 745
    R. Rengel, M. J. Martín, G. Pailloncy, G. Dambrine, and F. Danneville

3D Monte Carlo Study of Thermal Noise in DG-MOSFET 749
    P. Dollfus, A. Bournel, and J.E. Velázquez

Non-linear Noise in Nanometric Schottky-barrier Diodes (*invited*) 753
    S. Pérez, T. González, P. Shiktorov, E. Starikov, V. Gružinskis, L. Reggiani,
    L. Varani, and J. C. Vaissière

Analytical Model of Noise Spectrum in Schottky-barrier Diodes 759
    P. Shiktorov, E. Starikov, V. Gružinskis, L. Reggiani, L. Varani, and
    J. C. Vaissière

Monte Carlo Simulation of Quantum Noise (*invited*) 763
    X. Oriols

Noise and Diffusion in Superlattices within the Wannier-Stark
Approach 769
    M. Rosini and L. Reggiani

Coulomb Suppression of Avalanche Noise in Double-drift IMPATT
Diodes 773
    A. Reklaitis and L. Reggiani

A Frequency Domain Spherical Harmonics Solver for the Langevin Boltzmann Equation (*invited*) 777
    C. Jungemann and B. Meinerzhagen

Two-dimensional Physics-based Low-frequency Noise Modeling of Bipolar Semiconductor Devices in Small- and Large- Signal Operation 783
    F. Bertazzi, G. Conte, S. Donati Guerrieri, F. Bonani, and G. Ghione

Governing Equations of the Green's Functions for the Short-circuit Terminal Noise Currents in Semiconductor Devices 787
    S.-M. Hong, C. H. Park, H. S. Min, and Y.-J. Park

Noise Enhancement as Indicator of Instability Onset in Semiconductor Structures 791
    E. Starikov, P. Shiktorov, V. Gružinskis, L. Reggiani, L. Varani, and J. C. Vaissière

Fluctuation of the Electron Scattering Probability in One-dimensional Atomic Chain Having Plural Degree of Freedom 795
    H. Akabane and M. Agu

On the High Frequency Limit of the Impedance Field Method for Si 799
    C. Jungemann and B. Meinerzhagen

Properties of Some Deterministic Noises 803
    J. Kumicák

Phase Transition and $1/f$ Noise in a Modified Bak-Tang-Wiesenfeld Sand Pile Model with Time-dependent Avalanche Propagation 807
    K. Kiyono, Z. R. Struzik, and Y. Yamamoto

Effect of Noise in the Estimation of Magnitudes with Spatial Dependence: A Spatial Statistics Technique Based on Kriging 811
    L. M. Sanchez-Brea and E. Bernabeu

MISCELLANEOUS

Density Correlations in Ultracold Atomic Fermi Gases 817
    W. Belzig, C. Schroll, and C. Bruder

Spatial and Angular Intensity Correlations of Waves in Disordered Media 821
    L. S. Froufe-Pérez, A. García-Martín, P. García-Mochales, G. Cwilich, and J. J. Sáenz

Seismic Pattern Recognition by Wavelet Based-higher Order Statistics 825
    S. S. Kharintsev and M. Kh. Salakhov

Author Index 829

PREFACE

This volume constitutes the proceedings of the 18th International Conference on Noise and Fluctuations (ICNF 2005), held in Salamanca (Spain) from 19 to 23 September 2005. The present conference is the continuation of the series of *International Conferences on Fluctuation Phenomena*, started in 1968 in Nottingham (England), and the series of *International Symposiums on 1/f Fluctuations*, initiated in 1977 in Tokyo, which combined in 1983 in Montpellier into the *International Conference on Noise in Physical Systems and 1/f Fluctuations* (ICNF). Since then ICNF has been organized bi-annually in Rome in 1985, Montreal in 1987, Budapest in 1989, Kyoto in 1991, St. Louis in 1993, Palanga in 1995, Leuven in 1997, Hong Kong in 1999, Gainesville, Florida, in 2001 (where the title was changed to *International Conference on Noise and Fluctuations*) and Prague in 2003.

The activity in the area of noise and fluctuations has intensified during the last few years. Noise and fluctuations are now recognized as extremely important in a variety of fields. Telecommunications, nanoelectronics, mesoscopic structures and biological systems are some spotlight examples. Noise is one of the ultimate factors in determining performance of information transfer and data manipulation systems, communications and computing. In contrast, noise can also be a key source of information about the properties of a particular system, as mentioned by R. Landauer: *The noise is the signal* [Nature **392**, 658 (1998)]. In some specific cases, mainly related to biological systems, apart from a source of information, noise can even perform a valuable (constructive) role.

Research work on noise and fluctuations involves quite different disciplines (physics, engineering, mathematics, biology, chemistry, signal theory, etc.) and requires both fundamental and technological scientific efforts. In this framework, events like ICNF, where researchers on noise working in different fields (and coming both from academic and industrial environments) can meet, present their last advances and exchange their ideas, become a basic catalyst for the progress in the field.

$1/f$ noise, its influence on a variety of systems, and the different models developed to explain it, continues being a priority subject at ICNF 2005. Noise in electronic devices and circuits also receives a lot of attention, with advanced MOS technologies being the main topic. Noise in mesoscopic structures (quantum noise, shot-noise suppression and enhancement) and noise in biological and biomedical systems are two subjects of increasing activity in recent years that are widely represented in this edition of ICNF. The theory underlying the different noise processes and the models to predict the noise behaviour also have a significant presence.

The conference features 2 plenary and 20 invited papers by internationally recognized experts in the field, reviewing the progress made and outlining recent trends. After a peer review process by the International and Scientific Program Committees, 175 contributed papers were selected from more than 220 submissions. Contributions originate from scientists in more than 35 different countries, which reflects the international character of the conference. Papers are presented in two parallel oral sessions and two poster sessions.

We thank the members of the International and Scientific Program Committees for their valuable and qualified assistance in the selection of invited speakers and review of the submitted contributions. We are also indebted to the Organizing Committee for their support.

We hope the conference will contribute to the worldwide progress in the field by promoting fluid communications and collaborations between researchers and groups from different backgrounds and disciplines.

Salamanca, June 6, 2005　　　　　　　　　　　　　　　　　Tomás González
　　　　　　　　　　　　　　　　　　　　　　　　　　　　　Javier Mateos
　　　　　　　　　　　　　　　　　　　　　　　　　　　　　Daniel Pardo

18th International Conference on Noise and Fluctuations

ICNF 2005

organized by

**Electronics Group
Applied Physics Department
University of Salamanca, Spain**

Conference Chair:

Tomás González, Applied Physics Department, University of Salamanca, Spain

ICNF International Advisory Committee

G. Bosman	University of Florida, Gainesville	USA
Z. Celik-Butler	University of Texas, Arlington	USA
C. L. Claeys	IMEC, Leuven	Belgium
J. Deen	McMaster University, Hamilton	Canada
T. González	University of Salamanca	Spain
P. H. Handel	University of Missouri, Saint Louis	USA
M. Macucci	University of Pisa	Italy
A. Matulionis	Semiconductor Physics Institute, Vilnius	Lithuania
T. Musha	Brain Functions Laboratory, Kawasaki	Japan
L. Reggiani	University of Lecce	Italy
J. Sikula	Brno University of Technology	Czech Republic
C. Surya	Hong Kong Polytechnic University	Hong Kong
M. Tacano	Meisei University, Tokyo	Japan
L. Varani	University of Montpellier II	France
A. V. Yakimov	Nizhny Novgorod State University	Russia
Y. Yamamoto	University of Tokyo	Japan

Sponsored by:

University of Salamanca
IEEE Electron Devices Society
Spanish Ministry of Education and Science
Education Council - Community of Castilla y León
Caja Duero

ICNF 2005 Scientific Program Committee

C. W. J. Beenakker	University of Leiden	The Netherlands
S. Bezrukov	NIH, Bethesda	USA
F. Bonani	Politechnic of Torino	Italy
G. Bosman	University of Florida, Gainesville	USA
M. Buttiker	University of Geneva	Switzerland
F. Danneville	IEMN, University of Lille	France
J. Deen	McMaster University, Hamilton	Canada
G. Ghibaudo	ENSERG, Grenoble	France
T. González	University of Salamanca	Spain
F. Green	University of New South Wales, Sidney	Australia
P. Hänggi	University of Augsburg	Germany
J. Mateos	University of Salamanca	Spain
V. Parkhutik	Politechnic University of Valencia	Spain
J. M. R. Parrondo	University Complutense of Madrid	Spain
F. Pascal	University of Montpellier II	France
D. Pavlidis	University of Michigan, Ann Arbor	USA
M. Sampietro	Politechnic of Milano	Italy
E. Simoen	IMEC, Leuven	Belgium
J. Starikov	Semiconductor Physics Institute, Vilnius	Lithuania
A. V. Yakimov	Nizhny Novgorod State University	Russia
Y. Yamamoto	University of Tokyo	Japan

ICNF 2005 Organizing Committee

T. González, D. Pardo, J. Mateos, S. Pérez, M. J. Martín,
R. Rengel, B. G. Vasallo, I. Iñiguez de la Torre
University of Salamanca, Spain

G. Gomila
University of Barcelona, Spain

PLENARY LECTURES

Noise in Advanced Electronic Devices and Circuits

M. Jamal Deen and O. Marinov

Dept. of Electrical and Computer Eng., McMaster University, Hamilton, ON L8S 4K1 Canada.

Abstract. State-of-the-art low-frequency and high-frequency noise performance and modeling in modern semiconductor devices and circuits are discussed. The increase of noise-to-DC current ratio may compromise the circuit applications in near future. The low-frequency noise (LFN) tends to a log-normal distribution. Since the random-telegraph-signal (RTS) noise is pronounced in submicron devices, then new techniques being used to characterize of multilevel RTS are discussed. High-frequency noise modeling and sample experimental results are presented, including the important effect of gate-tunneling current for future devices. For the RF circuits, we discuss the phase noise in voltage-controlled oscillators (VCO) based on ring oscillators and LC-tank VCOs with and without automatic amplitude control. Finally, the effects of hot-carrier stress on the performance of a VCO is presented and discussed.

Keywords: Noise in devices, RF integrated circuits, flicker noise, 1/f noise, RTS and phase noise, electrical stress, degradation and reliability, VCO, VCO with AAC

INTRODUCTION

The recent explosion in wireless and information technology has been one of the most dramatic applications of semiconductor technology in the past decade. This revolution is apparent in the ever increasing use of wireless products such as cell-phones, personal digital assistants, digital cameras and electronic entertainment systems. Key ingredients of this technological revolution have been the rapid advances in the quality and processing of materials, and in device, circuit and system design and integration. In the area of materials, the science and technology of dielectrics occupies a prominent place in providing the dominant technology, CMOS, with its important characteristics of negligible standby power dissipation, good input-output isolation and surface potential control for switching operations.

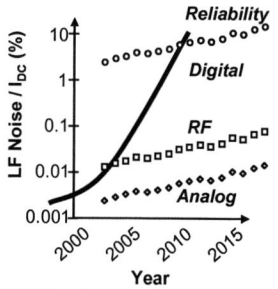

○ *Digital:* W×L=3×1 L_{min}^2, p-p Noise
f_{max}/f_{min}=1/error rate=10^{12}, t×σ=12dB

□ *RF:* W×L=1000×1 L_{min}^2, RMS Noise
f_{max}/f_{min}=1MHz/1Hz, t×σ=10dB

◇ *Analog:* W×L=500×20 L_{min}^2, RMS Noise
f_{max}/f_{min}=100kHz/1Hz, t×σ=6dB

FIGURE 1. Low-frequency noise is important for analog and RF applications. It is expected to also become a critical issue for digital circuits and for the reliability of the electronic systems in future.

However, with this explosion in semiconductor technology, noise has become a major concern. This is clearly underscored in the recent update of ITRS 2004 roadmap

[1] which highlights the importance of both low-frequency noise (LFN) as well as radio-frequency (RF) noise as silicon-based devices are scaled to deep-submicron dimensions in the nm range for the nanoscale transistors and circuits [2].

In the period before 2000, LFN was of concern mostly for analog applications. However, currently, the performance of RF integrated circuits (ICs) is a serious concern in applications affected by LFN. In future, it is expected that LFN will also affect the operation of digital circuits, and after ~2010, the reliability of electronic systems can decrease because the noise will be more than 10% of the DC biasing. This is schematically shown in Figure 1.

LOW-FREQUENCY NOISE

MOSFETs: The models for 1/f or flicker noise in MOSFETs are based on carrier mobility ($\Delta\mu$), carrier number (Δn), or correlated carrier number-mobility (Δn-$\Delta\mu$) fluctuations, and are typically represented by the following three expressions.

$$(\Delta\mu): \quad \frac{S_{I_D}}{I_D^2} = \frac{\alpha_H}{fnWL} \propto \frac{1}{Area \times (V_{GS}-V_T)}, \quad n = \frac{C_{ox}(V_{GS}-V_T)}{q}. \tag{1}$$

$$(\Delta n): \quad \frac{S_{I_D}}{I_D^2} = S_{FB}\left(\frac{g_m}{I_D}\right)^2 \propto \frac{1}{Area \times (V_{GS}-V_T)^2}, \quad S_{FB} = \frac{q^2 k T \lambda_T N_T}{f C_{ox}^2 WL}. \tag{2}$$

$$(\Delta n\text{-}\Delta\mu): \quad \frac{S_{I_D}}{I_D^2} = S_{FB}\left(1 + \sigma \, \mu_{eff} C_{ox} \frac{I_D}{g_m}\right)^2 \left(\frac{g_m}{I_D}\right)^2 \propto \frac{[1+\theta(V_{GS}-V_T)]^2}{Area \times (V_{GS}-V_T)^2}. \tag{3}$$

In these expressions, S_{ID} is noise spectral density in the drain current I_D, α_H is Hooge parameter, f is frequency, W and L are width and length of the transistor's gate, q is the electronic charge, C_{ox} is gate insulator capacitance per unit area, k is Boltzmann's constant, T is the absolute temperature, λ_T is the tunneling attenuation distance (see Table 1 for typical values), N_T is the oxide trap density per eV, μ_{eff} is the carrier mobility, σ is the scattering parameter and θ (~0.3-1V^{-1}) is the mobility degradation factor. In addition, another model, the BISIM3 (Δn-$\Delta\mu$) model can be used for predicting LFN in MOSFETs.

TABLE 1. Tunneling Attenuation Distances [3].

Gate Insulator	SiO$_2$	Al$_2$O$_3$	HfAlO$_x$	HfO$_2$
λ_T, nm	0.10	0.11	0.15	0.21

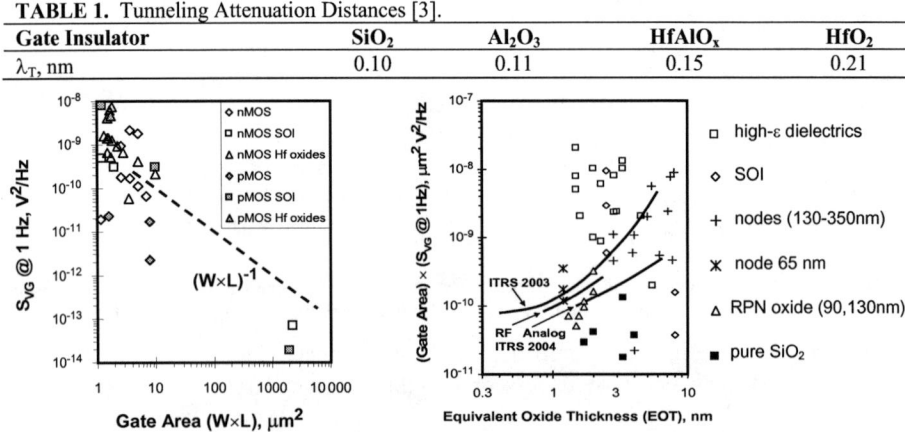

FIGURE 2. Recent data for gate referred voltage S_{VG} in MOSFETs from [3, 4, 5, 6, 7, 8, 9, 10].

However, the three theories above and the BSIM3 model cannot explain many experimental observations. Therefore, 1/f noise modeling in MOSFETs continues to be an active research area [11]. All models predict a noise increase when the device area (W×L) is reduced, which is also confirmed from recent experiments, as shown in Figure 2. For comparison of devices, the gate voltage referred noise $S_{VG} \propto S_{FB}$ at low gate overdrive voltage $|V_{GS}-V_T|<0.2V$ is usually used for nMOSFETs, for example in the ITRS [1], and according to Δn model for 1/f noise. However, the deviation from the prediction of this model is apparent in Figure 2, especially for SOI MOSFETs and when using gate dielectrics with high permittivity, such as HfO_2 or Al_2O_3, and λ_T in these dielectrics cannot explain the higher values for S_{VG}.

At present, much effort is devoted for developing advanced MOS devices by using gate dielectrics of high permittivity and by using germanium and carbon in heterostructure SiGe or strained silicon MOSFETs. As seen, however, in Figure 2 and Figure 3, the 1/f noise is increased in these devices by orders of magnitude when compared to $Si-SiO_2$ MOSFETs, and the extracted N_T values are very large. Thus, there are significant efforts to fabricate devices with higher C_{ox} and μ_{eff} without degrading too much the low-frequency noise performance of the transistors.

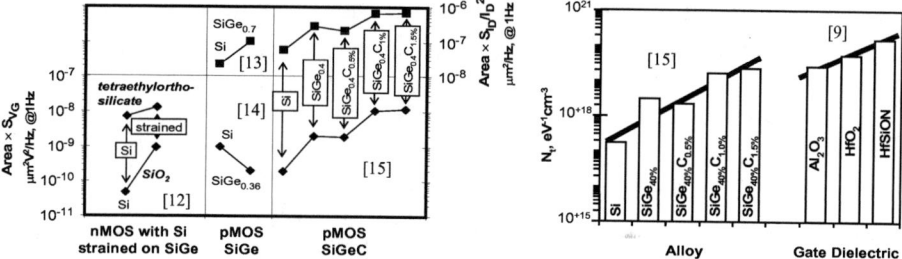

FIGURE 3. Noise and dielectric trap density in MOSFETs made with advanced fabrication techniques and novel gate dielectrics or semiconductor materials.

The data presented above are for the 1/f noise in MOSFETs. However, RTS noise and other kinds of LFN with Lorentzian spectra are routinely observed in transistors with submicron-areas [16, 17], especially in those with oxide thickness less than 3 nm [7, 18] and in SOI FETs [5]. In these devices, existing LFN models are empirically extended [5, 17] to account for the bias dependence of time constants and the large variations in the measured noise levels.

BJTs: The 1/f noise voltage, referred to the input base-emitter terminals in BJTs is much lower than the corresponding S_{VG} in MOSFETS, as shown in Figure 4. However, the normalized noise magnitude K_F, given by $K_F = S_{IC}/I_C^2 \approx S_{IB}/I_B^2$ in the collector and base currents, is of the same order of magnitude as the corresponding S_{ID}/I_D^2 in MOSFETs. This implies that the input noise current in BJTs is large compared to MOSFETs, and the BJT models, therefore, consider the noise in the base current as the dominant noise source.

In μm-sized BJTs with mono-silicon emitters, the 1/f noise is attributed to the fluctuations in diffusion at the base-emitter junction and in the base resistance, which result in S_{IB} being proportional to I_B and I_B^2, respectively. The modern sub-μm scale BJTs, however, use poly-silicon emitters (PEs), and it is found that the interfacial oxide (IFO) between polycrystalline and mono-silicon emitter is usually where the noise originates. The effective thickness (t_{IFO}) of the IFO ranges from 0.2 nm to less than 2 nm, and in contrast to gate oxide in MOSFETs, this IFO is not a uniformly continuous layer.

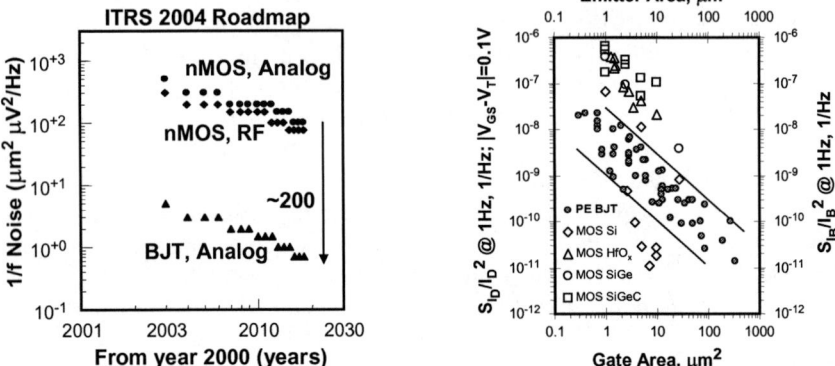

FIGURE 4. Comparison of 1/f noise in PE-BJTs and MOSFETs. The left figure shows the input normalized noise voltage and the right figure shows the output (MOSFET) and input (BJT) normalized noise current. Note that for many BJTs, the input and output normalized noise currents are equal.

Some of the popular models for the 1/f noise in PE BJTs assume trapping [19], two-step tunneling [20] or direct tunneling [21] of carriers in IFO, and they are respectively given by (meanings of symbols in [22]):

$$S_{I_B} = \frac{C(T) N_T}{f A_E \ln(\tau_2/\tau_1)} I_B^2, \qquad (4)$$

$$S_{I_B} = \frac{q^4 N_T t_{IFO}^2}{2 kT \pi \varepsilon_{ox}^2 \ln[\tau(x_0)/\tau(0)] f A_E} I_B^2, \qquad (5)$$

$$S_{I_B} = \left[1 + \sqrt{\frac{kT}{2\pi m^*}} T_{holes} \left(\frac{1}{S_m} + \frac{W_m}{D_m} + \frac{W_p}{D_p}\right)\right]^{-2} \frac{m^* q kT \tan\delta \, t_{IFO}^3}{3\hbar^2 \pi \varepsilon_{ox} f A_E V_0}. \qquad (6)$$

These models predict that the 1/f noise is inversely proportional to the emitter area A_E, but have different dependences on temperature and t_{IFO}. The noise variations studied by several research groups are shown in Figure 5. The $1/A_E$ dependence is apparent, with an average of $A_E \times K_F = 5.5 \times 10^{-9}$ μm² by a log-normal distribution with standard deviation of $\sigma \approx 3.7$ dB. Regarding IFO, the separate data series suggest two-step tunneling mechanism for the 1/f noise according to (5), since $A_E \times K_F \propto t_{IFO}^2$. However, the overall trend in the data is closer to $A_E \times K_F \propto t_{IFO}^3$ for direct tunneling according to (6). The range of values for t_{IFO} is narrow with an error of 0.1–0.3 nm, and the different theories cannot be distinguished reliably with existing experimental results. An exponential function, proposed in [28] empirically, can also fit the data.

Another approach to study the origin of 1/f noise in PE BJTs is through its temperature dependence. To date, either a weak increase or decrease, or no change in the noise is observed for temperatures from less than 200K to above 350K. The temperature variation of 1/f noise is small compared to variations due to biasing, input and output impedance loading [22, 23, 24], irradiation or electrical stress [21, 22], or the occurrence of noise with Lorentzian spectra in sub-μm² emitter area BJTs [19].

The individual Lorentzian components usually exhibit temperature dependences typical for generation-recombination processes [19, 23, 29], but these dependences are smeared out when averaging the noise of several devices, and in the 1/f noise in devices with emitters of ~μm^2 areas.

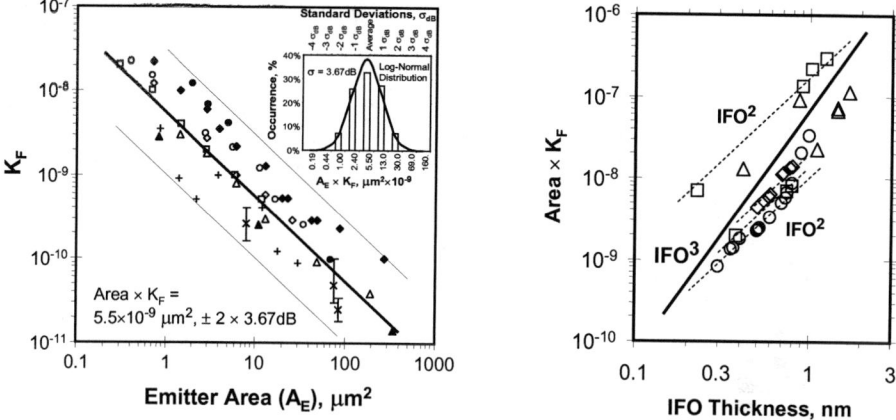

FIGURE 5. Dependence of normalized 1/f noise $K_F = f \times S_{IB}/I_B^2$ in PE-BJTs on emitter area (A_E) [21, 22, 23, 24, 25, 26] and interfacial oxide (IFO) thickness [21, 26, 27, 28]. The product ($A_E \times K_F$) has log-normal distribution shown in insert.

The 1/f noise models given by equations (4) to (6) above are applicable when the exponent x in the $S_{IB} \propto I_B^x$ dependence is x=2. However, sometimes x varies between 1 and 2 depending on the bias and between samples. A suggestion for splitting the noise into two components $S_{IB} = (K_{F1} \times I_B + K_{F2} \times I_B^2)/(f \times A_E)$ is presented in [26, 27], by using the crossover between the noise due to injection, mobility and tunneling in PE BJTs. Nevertheless, the electrical model [22, 23, 24] with noise sources in the base, emitter and collector is a good approach for characterization of the 1/f noise in BJTs.

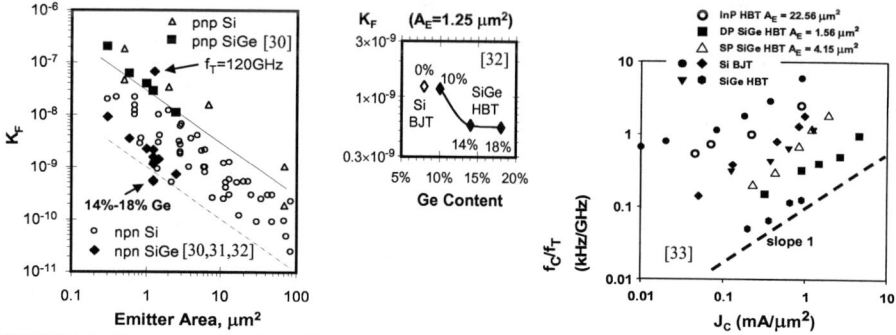

FIGURE 6. Current results for 1/f noise in HBTs (left figure), dependence on germanium content (middle figure) and for the ratio of corner frequency f_C between 1/f and shot noise to transit frequency f_T in BJTs (right figure).

Much current research in heterostructure BJTs (HBTs) for high-speed/high-frequency applications showed that the relations and numbers obtained for noise in Si BJTs are also valid for these HBTs. For example, the solid symbols in the first plot of Figure 6, which are for K_F in HBTs, are in the same range of the open symbols for Si

BJTs. The second plot in this figure implies that the noise in SiGe HBTs is almost independent of the germanium content in base. An important finding for phase noise in RF applications of BJTs is given in the third plot of the figure, providing that the ratio f_C/f_T of the corner frequency f_C between 1/f and shot noise and the transit frequency f_T is proportional to the current density in BJTs [33, 34], given by

$$\frac{f_c}{f_T} \approx \frac{K_f A_E \pi}{\beta q}(\tau_f J_c + V_t C_t) \, . \tag{7}$$

Noise Measurement and Statistics: The amplitude of RTS in sub-µm dimension MOSFETs may originate to flatband voltage fluctuation due to single electron trapping-detrapping and follows the relation [35, 38]

$$\frac{\Delta I}{I_D} = \frac{g_m}{I_D} \frac{q}{WLC_{ox}} \, . \tag{8}$$

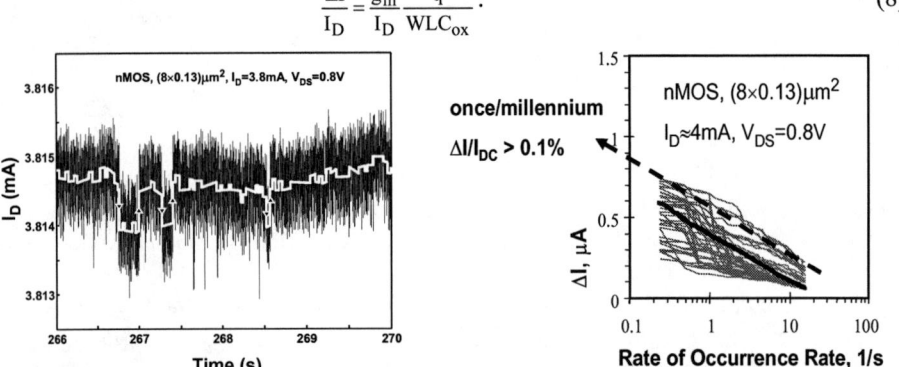

FIGURE 7. Multilevel RTS (left figure) and statistics for the rate of occurrence of RTS amplitudes (right figure).

It is also found that RTS can originate from gate tunneling [39], in SOI devices [5, 40], and after electrical stress [41]. The capture and emission time constants vary between samples and with the bias and temperature, causing difficulties for prediction of RTS noise, especially when switching between multiple levels due to several traps [16]. As illustrated in Figure 7, by inspecting for step discontinuities in the drain current, we obtain multilevel RTS, and we get that the RTS transition amplitudes are a logarithmic function of the rate of occurrence of these RTS transitions.

From several attempts to model the LFN dispersion, it is found that the dispersion increases inversely with the square root of gate or emitter area. The worst case for noise can be obtained from the standard deviation of the noise spectral density [17, 19, 29, 35]. It is noted, however, that the standard deviation of noise can be larger than its average [36], and that the noise has log-normal distribution [15], as shown in the insert of Figure 5. The log-normal distribution implies that the noise should be averaged geometrically [37] and the noise with its variation should be given by

$$S_{VAR} = S_{AVERAGE} \, 10^{\pm t \times \sigma_{dB}/10 dB}, \tag{9}$$

where, t=1...3 is chosen by desired confidence probability (0.6...0.99) for S_{VAR}, and $S_{AVERAGE}$ and σ_{dB} are obtained by calculations with noise levels in dB. The geometric averaging centers the mean values symmetrically in the range of noise variation, while the arithmetic mean is biased toward data with high values of noise.

HIGH-FREQUENCY NOISE

CMOS technology is being increasingly used in RF applications because of the higher unity current gain (f_T) or maximum oscillation (f_{max}) frequencies in scaled-down MOS devices. Several MOSFET noise models have been developed with the focus on the channel thermal noise [42]. For the typical saturation-region operation of MOSFETs in analog and RF ICs, the recent analytical model for the channel noise current [43, 44] is

$$S_{iD} = 4kT \frac{4V_{GT}^2 + V_0^2 - 2V_0 V_{GT}}{3V_{GT}^2 (V_{GT} - V_0)} \alpha I \tag{10}$$

where $V_0 = I/(WC_{ox}v_{sat})$, V_{GT} is the effective gate voltage and $\alpha \approx 1$. is the bulk-charge effect coefficient. As shown in Figure 8, using this simple, but relatively accurate small-signal model, we get excellent agreement to experiments for the normal noise parameters, given by

$$G_{opt} = \frac{|Y_{11}||Y_{21}|\sqrt{\beta I R_g}}{\beta I + R_g |Y_{21}|^2} \approx \left(\frac{f}{f_T}\right)\frac{\sqrt{\beta I R_g}}{R_n}, \quad B_{opt} = \frac{-\omega C_{gg}\beta I}{\beta I + R_g |Y_{21}|^2} \approx -\left(\frac{f}{f_T}\right)\frac{\beta I}{g_m R_n} \tag{11}$$

$$R_n = R_g + \frac{\beta I}{|Y_{21}|^2} \approx R_g + \frac{\beta I}{g_m^2} \quad \text{and} \quad NF_{min} \approx 1 + 2\left(\frac{f}{f_T}\right)\sqrt{\beta I R_g}\left(1 + \left(\frac{f}{f_T}\right)\sqrt{\beta I R_g}\right)$$

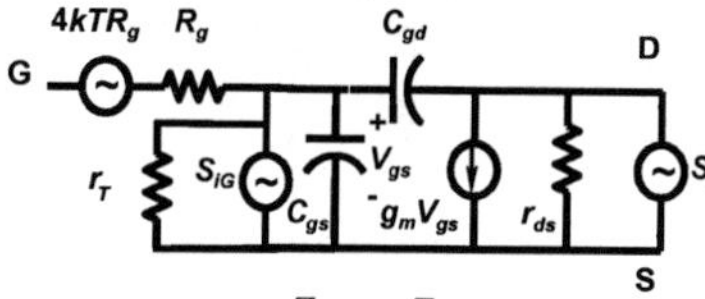

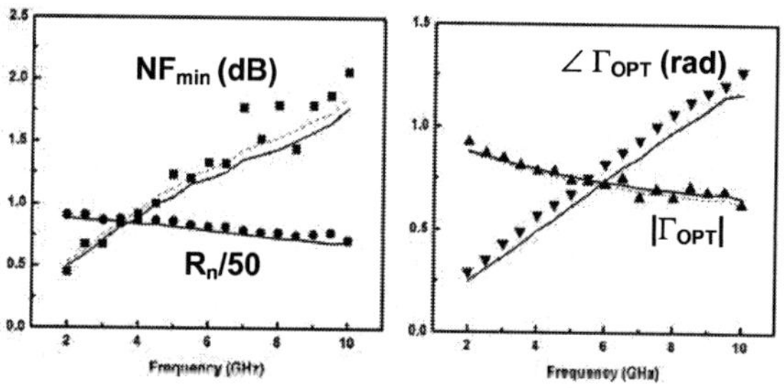

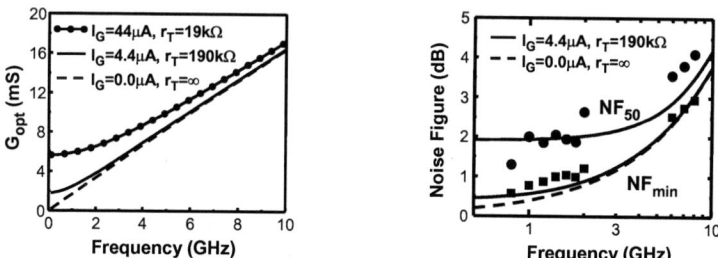

FIGURE 8. RF noise model and comparison between model predictions and experiments.

FIGURE 9. Increase of G_{opt} and NF_{min} due to gate tunneling current (GTC) given by the analytical model [45] (solid lines) and compared to experimental data from [46] (symbols) for a MOSFET with 1.5 nm gate oxide. The dashed lines represents models without GTC ($r_T = \infty$).

The expressions for these four RF noise parameters are changed by the gate tunneling current (GTC) in MOSFETs with very thin gate oxides [45]. This is in addition to the significant contribution of the GTC to LFN. Increasing the GTC, the gate oxide conductance $1/r_T$ increases and f_T decreases as $(2\pi C_{gg} f_T)^2 = g_m^2 - r_T^{-2}$. Then, G_{opt} and NF_{min} are increased as illustrated in Figure 9 with solid lines and compared to the case of $r_T=\infty$ and with experimental data for a MOSFET with thin gate oxide. From this figure, the effect of GTC is more pronounced at low frequencies, at which LFN most likely will dominate over thermal noise. However, the GTC RF noise can be important up to 10GHz for MOSFETs with ultra-thin gate oxides, because of the significant increase in the tunneling current and conductance for these oxides.

NOISE IN OSCILLATORS

For low-power, fully-integrated CMOS ring VCOs, phase noise is regarded as a key performance parameter [48]. Among many models which have been developed for phase noise in different types of oscillators, the time-variant model discussed in [50] identifies the mechanism by which the intrinsic device noise sources contribute to the total phase noise. This model demonstrated that the waveform asymmetry degrades phase noise by increasing the corner frequency between $1/f^3$ and $1/f^2$ components [50].

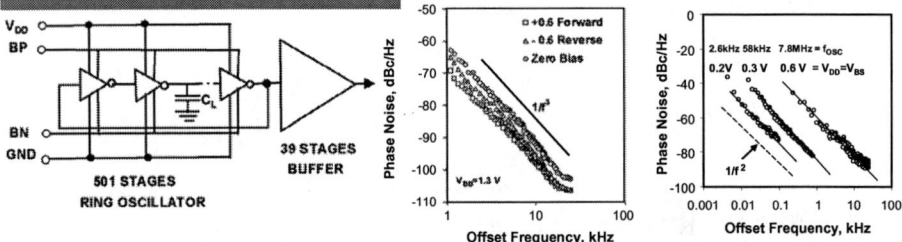

FIGURE 10. Circuit of 501-stage ring VCO with carefully designed buffer for accurate measurements (left figure). Variation of phase noise with body biasing (center figure) and in weak inversion with bodies tied to supply (right figure).

The circuit of a VCO with a ring of 501 CMOS inverters and a 3-stage buffer is shown in Figure 10. This VCO has two extra terminals - BP and BN from the bodies of the PMOS and NMOS transistors [49]. The oscillator is fabricated in a 0.18μm technology with the ratio of channel widths of $W_{PMOS}/W_{NMOS}=3.39$. This size adjustment provided a symmetrical voltage transfer characteristic at a supply voltage $V_{DD}=1.8V$. The frequency is tuned by forward body biasing up to 0.6V applied both to BP and BN, and also by the supply voltage V_{DD} from 1.8V down to 80mV.

The 1/f noise in MOSFETs decreases with forward body biasing by ~8dB/V [47]. When BP is tied to GND and BN is tied to V_{DD}, the 1/f noise is suppressed both by the low supply voltage and the forward body biasing, and the phase noise becomes −61dBc and −81dBc at offset frequencies of 1kHz and 10kHz, respectively, as shown in Figure 10 for $V_{DD}=0.6V$, indicating up-conversion of only white noise.

In modern RFICs, fully integrated low power VCOs [51] with low phase noise are desired along with automatic amplitude control (AAC) [52]. In Figure 11, the phase noise decreases in VCO with increasing the oscillation amplitude up to 0.5V, because the 1/f noise in the cross-coupled MOS transistors M1–M2 decreases when their operation changes from weak- to strong-inversion. At larger amplitudes, however, the 1/f noise from the gate bias (V_G) of M3 appears significant and the VCO phase noise increases. Without AAC, V_G is provided by a low-noise battery supply, and the increase in the phase noise is small. In contrast, the 1/f noise in the output of the AAC is larger, since the transistors in AAC are small in area, and the phase noise is ~30 times higher than for the case of battery biasing of M3 at oscillation amplitudes above 0.6V. Thus, the 1/f noise in AAC circuits is an important design issue for RFIC VCOs.

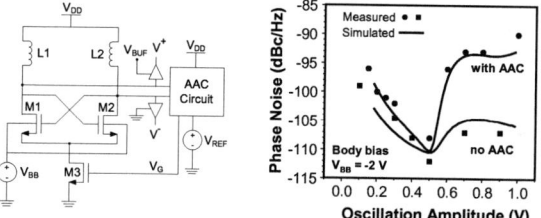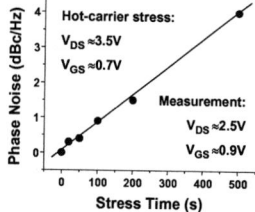

FIGURE 11. Schematic and phase noise at 1MHz offset for the VCO with AAC and without AAC. Increase in phase noise after 500s stress of cross-coupled transistors in specially designed LC oscillator.

With gate length reduction, problems associated with short channel MOS transistors become important. One issue is the generation of hot carriers in the channel owing to strong electric fields inside the device. These high–energy carriers impact ionize to create interface and oxide traps, which cause device performance degradation. Many studies on the hot carriers demonstrated degradation in the DC parameters and increase of low–frequency noise, which in turn increases the phase noise in LC oscillators. In the right plot of Figure 11, a sample from our experiments [53] on hot–carrier effects on a fully integrated LC tank CMOS oscillator is presented. We found that the oscillation amplitude decreases after stress, resulting in an increase of the phase noise.

CONCLUSION

The noise in devices is limiting the performance of analog and RF systems currently, and soon it might raise reliability issues in digital circuits. The input-referred voltage noise in BJTs is less than that in MOSFETs, but the normalized output referred noise currents are of similar magnitudes. The use of heterostructures for the semiconductors and novel dielectrics for the gate insulators can result in an increase in the noise in advanced silicon-based device structures. The LFN has log-normal distribution, which implies geometric averaging (arithmetic in dB). The increased levels of RTS noise in submicron devices require methods for analyses which use multilevel RTS. At high frequencies, it is not only important to have an accurate noise model, but the model should also be simple enough so that it can be used to guide the design of RFICs. A simple, but accurate noise model for the four noise parameters was discussed and verified experimentally. This model is suitable for use by RF and analog IC designers. Finally, the impact of low frequency noise on oscillators operating at various frequencies and the effects of high-field stress on the oscillator's phase noise was discussed. As technology advances, new issues related to noise in devices and circuits appear and the role of noise in the performance of both digital and analog circuits is expected to become increasingly important.

Acknowledgements: We are very grateful to several of our colleagues for their collaborations over the years in our noise research and many researchers from whose published work we have learnt much.

REFERENCES

1. SEMATECH, *ITRS 2004 Update*, http://public.itrs.net/.
2. Deen, M. J., and Pascal, F., *IEE Proc. Circuits Devices and Systems*, **151**(2), 125 (2004).
3. B. Min, B., et al, *IEEE Trans. Electron Devices*, **51**(10), 1679 (2004).
4. Chew. K., et al, *IEE Proc. Circuits Devices and Systems*, **151**(5), 415 (2004).
5. Simoen, E., et al, *IEEE Trans. Electron Devices*, **51**(6), 1008 (2004).
6. Binkley, D., et al, *IEEE Trans. Nuclear Science*, **51**(6), 3788 (2004).
7. Simoen, E., et al, *IEEE Trans. Electron Devices*, **51**(5), 780 (2004).
8. Ahsan, A., et al, *IEEE Electron Device Letters*, **25**(4), 211 (2004).
9. Min, B., et al, *Applied Physics Letters*, **86**(8), 082102 (2005).
10. Mercha, A., et al, *Proc. SPIE*, **5470** (Danneville, Bonani, Deen, Levinshtein; Eds.), 193 (2004).
11. Vandamme, L., et al, *Physica B*, **357**(3-4), 507 (2005).
12. Hua, W.-C., et al, *IEEE Electron Device Letters*, **25**(10), 693 (2004).
13. Myronov, M., et al, *Applied Physics Letters*, **84**(4), 610 (2004).
14. Prest, M., et al, *Applied Physics Letters*, **85**(24), 6019 (2004).
15. Deen, M. J., et al, *Proc. SPIE*, **5470** (Danneville, Bonani, Deen, Levinshtein; Eds.), 215 (2004).
16. Kramer, T., et al, *Physica E*, **19**(1-2), 13 (2003).
17. Brederlow, R., et al, *IEEE-IEDM*, 159 (1999).
18. Wu, J., et al, *Applied Physics Letters*, **85**(21), 5076 (2004).
19. Sanden, M., et al, *IEEE Trans. Electron Devices*, **49**(3), 514 (2002).
20. Kumar, V., et al, *IEEE Trans. Electron Devices*, **24**, 146 (1977).
21. Markus, H., et al, *IEEE Trans. Electron Devices*, **42**(4), 720 (1995).
22. Deen, M. J., et all, *IEE Proc. Circuits Devices and Systems*, **149**(1), 40 (2002).
23. Deen, M. J., et al, *Journal of Applied Physics*, **77**(12), 6278 (1995).
24. Deen, M. J., et al, *Journal of Applied Physics*, **84**(1), 625 (1998).
25. Delseny, C., et al., *Microelectronics Reliability*, **40**(11), 1869 (2000).
26. Hoque, M., et al, *IEEE Trans. Electron Devices*, **51**(9), 1504 (2004).
27. Hoque, M., et al, *Journal of Applied Physics*, **97**(8), 084501 (2005).
28. Simoen, E., et al, *Solid-State Electronics*, **42**(9), 1679 (1998).
29. Sanden, M., et al, *IEEE Electron Device Letters*, **22**(5), 242 (2001).
30. Zhao, E., et al, *IEEE Trans. Nuclear Science*, **51**(6), 3243 (2004).
31. Niu, G., et al, *IEEE Trans. Microwave Theory Techniques*, **53**(2), 506 (2005).
32. Tang, J., et al, *IEEE Trans. Electron Devices*, **51**(9), 1475 (2004).
33. Pascal, F., et al, *Proc. SPIE*, **5113** (Deen, Celik-Butler, Levinshtein; Eds.), 133 (2003).
34. Pascal, F., et al, , *IEE Proc. Circuits Devices and Systems*, **151**(2), 138 (2004).
35. Boutchacha, T., et al, *Physica Status Solidi. (a)* , **167**(1), 261 (1998).
36. Wirth, G., et al, *http://arxiv.org/ftp/cond-mat/papers/0409/0409030.pdf*, (2005).
37. Pintelon, R., et al, *IEEE Trans. Instrumentation and Measurement*, **37**(2), 213 (1988).
38. Buisson, O., et al, *Solid-State Electronics*, **35**(9), 1273 (1992).
39. Wu, J., et al, *Applied Physics Letters*, 85(21), 261 (2004).
40. Lukyanchikova, N., et al, *IEEE Electron Device Letters*, **25**(6), 433 (2004).
41. Petit, C., et al, *24th Int. Conference Microelectronics*, **2**, 641 (2004).
42. Asgaran, S., and Deen, M. J., *Workshop on Compact Modeling*, Boston, MA, 259 (2004).
43. Asgaran, S., et al, *IEEE Trans. Electron Devices*, **51**(12), 2109 (2004).
44. Chen, C-H., et al, *Proc. SPIE*, **5470** (Danneville, Bonani, Deen, Levinshtein; Eds.), 49 (2004).
45. Deen, M. J., et al, *Workshop on Compact Modeling*, Anaheim, CA, 35 (2005)
46. Momose, H. S., et al, , *Symp. VLSI Technology, Digest of Technical Papers*, 96 (1998).
47. Deen, M. J. and Marinov, O., *IEEE Trans. Electron Devices*, **49**(3), 409 (2002).
48. Deen, M. J., et al, *Proc. ICNF 2003*, 525 (2003).
49. Deen, M. J., et al, *Proc. IEEE-ISCAS 2003*, I697 (2003).
50. Kazemeini, M., et al, *Proc. IEEE-ISCAS 2003*, I704 (2003).
51. Deen, M. J., et al, *IEEE-NEWCAS Workshop*, (2005) 4 pages.
52. Murji, R., et al, *IEEE-RFIC Symposium*, (2005) 4 pages.
53. Naseh, S. , et al, *IEEE Trans. Electron Devices*, **50**(5), 1334 (2003).

THEORY

Control and Rectification in Collective Stochastic Systems

Juan M.R. Parrondo* and Luis Dinís*

*Dep. de Física Atómica, Molecular y Nuclear
and GISC (Grupo Interdisciplinar de Sistemas Complejos),
Universidad Complutense de Madrid, 28040-Madrid, Spain.*

Abstract. In this paper we review two basic and related mechanisms rectifying fluctuations: Brownian ratchets and paradoxical games. We focus our study on the effect of control in both ratchets and games.

Keywords: Brownian ratchets, Parrondo paradox, stochastic control
PACS: 05.40.a

INTRODUCTION

One of the most celebrated examples of the constructive role of noise is the rectification of thermal fluctuations by *Brownian ratchets*. In 1992, Magnasco [1] and Ajdari and Prost [2] introduced very simple models of ratchets capable to induce a systematic motion of Brownian particles. Since then, rectification of thermal fluctuations is becoming a major research topic in non equilibrium statistical mechanics [3], with potential applications in biology, condensed matter, and nanotechnology. On the other hand, Brownian ratchets have inspired some paradoxical gambling games known as *Parrondo paradox* [4, 5]: two losing games, which turn out to be winning when alternated.

Most Brownian ratchets, as well as Parrondo paradox, work by introducing an external time-dependent perturbation in an asymmetric equilibrium system. The models and applications studied so far have focused on periodic or random time-dependent perturbations [3]. Only very recently a *feedback controlled perturbation*, depending on the state of the system, has been considered [5, 6, 7, 8].

Introducing control in Brownian ratchets is relevant for the aforementioned applications. In particular, control theory is of extreme relevance in biology [9], and most protein motors would probably operate as control systems. In the context of Parrondo paradox, collective paradoxical games are suitable models to study frustration effects in stochastic control and collective decision making [5, 6, 7].

Here we will review the basic features of the flashing ratchet, one of the simplest models of a Brownian ratchet, and the paradoxical games. In the last part of the paper we will briefly address some of the most recent results on control of games and ratchets.

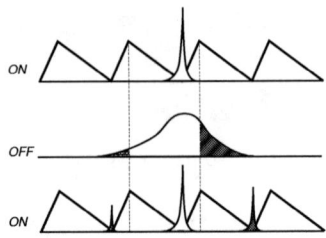

FIGURE 1. The flashing ratchet at work. The figure represent three snapshots of the potential and the density of particles.

THE FLASHING RATCHET

The flashing ratchet introduced by Ajdari and Prost [2] consists of a one-dimensional Brownian particle or, equivalently, an ensemble of independent particles, in the asymmetric sawtooth potential depicted in Fig. 1. It is not difficult to show that, if the potential is switched on and off periodically, the particles exhibit an average motion to the right. Let us assume that the temperature T is low enough to ensure that kT is much smaller than the maxima of the potential, and that we start with the potential switched on and with all the particles around one of its minima, as shown in the upper plot of figure 1. When the potential is switched off, the particles diffuse freely, and the density of particles spreads as depicted in the central plot of the figure. If the potential is then switched on again, each particle will move back to the initial minimum or to one of the nearest neighboring minima, depending on its position. As is apparent from the figure and due to the asymmetry of the potential, more particles fall into the right hand minimum, and thus a net motion of particles to the right is induced.

This motion can be seen as a rectification of the thermal noise associated with free diffusion. The diffusion is symmetric: some particles move to the right and some to the left, but their average position does not change. However, when the potential is switched on again, most of the particles that moved to the left are driven back to the starting position, whereas many particles that moved to the right are pushed further to the right hand minimum. The asymmetric potential acts as a rectifier: it "kills" most of the negative fluctuations and "promotes" most of the positive ones.

The effect remains if we add a small force toward the left, i.e., in a direction opposite to the induced motion. In this case, the ratchet exhibits a curious property: when the potential is permanently on or off, the Brownian particles move in the same direction as the force, whereas they move in the opposite direction when the potential flashes on and off. This effect is even more striking in the context of gambling games.

PARADOXICAL GAMES

Consider now the two following gambling games, A and B, in which a player can make a bet of 1 euro. $X(t)$ denotes the capital of the player, where $t = 0, 1, 2\ldots$ stands for the number of turns played. Game A consists of tossing a slightly biased coin so that the

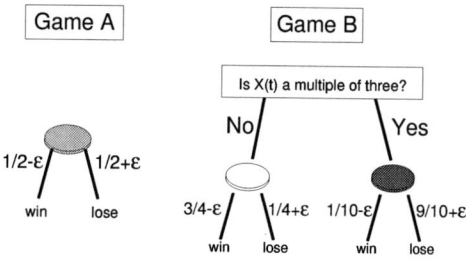

FIGURE 2. Rules of the paradoxical games.

player has a probability p_A of winning which is less than a half. That is, $p_A = 1/2 - \varepsilon$, where the bias ε is a small positive number.

The second game, B, is played with two biased coins, a "bad coin" and a "good coin". The player must toss the bad coin if her capital $X(t)$ is a multiple of 3, the probability of winning being $p_{\text{bad}} = 1/10 - \varepsilon$. Otherwise, the good coin is tossed and the probability of winning is $p_{\text{good}} = 3/4 - \varepsilon$. The rules of games A and B are sketched in Fig. 2, in which the darkness represents the "badness" of each coin.

For these choices of p_A, p_{good} and p_{bad}, both games are fair if $\varepsilon = 0$, in the sense that $\langle X(t) \rangle$ is constant. This is evident for game A, since the probabilities to win and lose are equal. The analysis of game B is more involved, but one can prove that the effect of the good and the bad coins cancel each other for $\varepsilon = 0$. On the other hand, both games have a tendency to lose if $\varepsilon > 0$, i.e., $\langle X(t) \rangle$ decreases with the number of turns t. Surprisingly enough, if the player randomly chooses the game to play in each turn, or plays them following some predefined periodic sequence such as ABBABB..., then her average capital $\langle X(t) \rangle$ is an increasing function of t, as can be seen in Fig. 3.

The paradox is closely related to the flashing ratchet. If we visualize the capital $X(t)$ as the position of a Brownian particle in a one dimensional lattice, game A, for

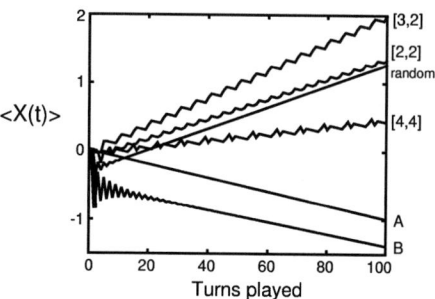

FIGURE 3. Average capital for 5000 players as a function of the number of turns for game A, B and their periodic and random combinations. $\varepsilon = 0.005$ and $[a,b]$ stands for periodic sequences where A (B) is played a (b) consecutive turns.

FIGURE 4. A random walk picture of game B compared with the ratchet potential.

$\varepsilon = 0$, is a discretization of free diffusion, whereas game B resembles the motion of the particle under the action of an asymmetric sawtooth potential. Fig. 4 shows this spatial representation for game B compared with the ratchet potential. The sawtooth potential has a short spatial interval in which the force is negative and a long interval with a positive force. Equivalently, game B uses a bad coin on a "short interval", i.e., on one site of every three on the lattice, and a good coin on a "long interval" corresponding to two consecutive sites which are not multiple of three.

COLLECTIVE DECISIONS IN PARADOXICAL GAMES

Consider now a set of N players who play game A or B against a casino and such that, in each turn, *all* of them must play the *same* game. They have to make a collective decision, choosing between game A or B in each turn. Among the different possible ways of achieving such a decision, it is reasonable to first consider a *majority rule* to select the game, that is, the game which receives more votes is the one to be played by all the players simultaneously. The vote of each individual is rather clear from the rules of game B: players with capital multiple of three vote for A, whereas the rest vote for B.

This strategy is optimal for a single player, but turns out to be losing if the number of players is large enough [7]. This can be seen in figure 5 where we show numerical results of the average capital per player for an increasing number of players ranging from 10 to 1000. One can observe that, the larger the number of players, the worse the results for the majority rule, becoming losing for a number of players between 50 and 100.

The reason is quite simple. Either if we play game A, B or a combination of both, the

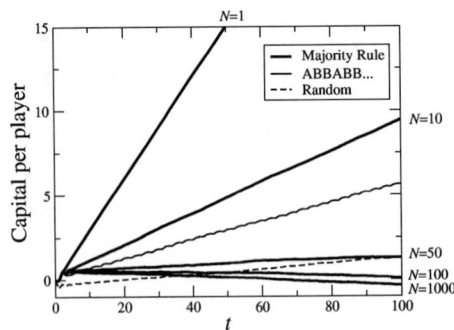

FIGURE 5. Simulation results for the average capital per player for $N = 10$, 50, 100, and 1000 players, $\varepsilon = 0.005$, and the three different strategies.

fraction π_0 of players with capital multiple of three lies, after a given number of turns, around 1/3, i.e., far below the majority 1/2. Consequently, the majority rule for N large selects B with probability close to 1 and we recall that selecting B in every turn yields a decreasing average capital (see figure 3). On the other hand, if, instead of using the majority rule, we select the game at random or following a periodic sequence, game A will be chosen even though $\pi_0 < 1/2$. This is a bad choice for the majority of the players, since playing B would make them toss the good coin. That is, the random or periodic selection contradicts from time to time the will of the majority. Nevertheless, it results on a steady gain for the whole ensemble.

The same effect can be observed (with some slightly modified games) even if the ensemble selects the best game in each turn, i.e., the game yielding the highest profit in this specific turn. The reason is that this strategy is only optimal for one turn but can be losing in the long run [6].

CONTROL IN RATCHETS

Similar phenomena can be observed in the flashing ratchet [8]. Consider an ensemble of N overdamped Brownian particles at temperature T in an external asymmetric periodic potential $V(x)$, that can be switched on and off. The dynamics of our system is described by the Langevin equations:

$$\gamma \dot{x}_i(t) = -\alpha(t) F(x_i(t)) + \xi_i(t); \qquad i = 1 \ldots N, \qquad (1)$$

where $x_i(t)$ the position of particle i, γ is the friction coefficient, $\xi_i(t)$ are thermal white noises, and $F(x)$ is the force derived from a potential similar to the one depicted in figure 1. Finally, $\alpha(t)$ is a control parameter which can take on the values 1 and 0, i.e., the only allowed operations on the Brownian motor consist of switching on and off the potential $V(x)$. We consider the following two switching strategies:

- *Periodic switching*: $\alpha(t+\tau) = \alpha(t)$, with $\alpha(t) = 1$ for $t \in [0, \tau/2)$, and $\alpha(t) = 0$ for $t \in [\tau/2, \tau)$. This case is equivalent to the periodic flashing ratchet, since particles are independent.
- *Controlled switching*:

$$\alpha(t) = \Theta(f(t)) \quad \text{with} \quad f(t) = \frac{1}{N} \sum_{i=1}^{N} F(x_i(t)), \qquad (2)$$

where $f(t)$ is the net force per particle and $\Theta(y)$ is the Heaviside function, $\Theta(y) = 1$ if $y \geq 0$ and 0 otherwise. As can be deduced form Eq. (1), this strategy maximizes the instant velocity of the center of mass, $\dot{x}_{cm}(t) = \frac{1}{N} \sum_{i=1}^{N} \dot{x}_i(t)$. Particles are not longer independent, due to the feedback control.

In figure 6, the stationary velocity of the center of mass is plotted as a function of the number of particles for the periodic switching with the optimal period [2, 3], and for the controlled switching [8]. The controlled switching yields a higher velocity than the periodic strategy only up to $N \simeq 1300$. We also see from simulations that the velocity goes to zero when $N \to \infty$, in the controlled case. In the figure, two approximations valid for small and large N respectively are also depicted (see [8] for details). By using one of these approximations, one can prove that the decay of the velocity is proportional to $1/\ln N$. This is surprising at first sight, because the controlled strategy given by (2) maximizes the instant velocity $\langle \dot{x}_{cm}(t) \rangle$. However, this local maximization does not ensure good results in the long term, exactly as happens in the games [6].

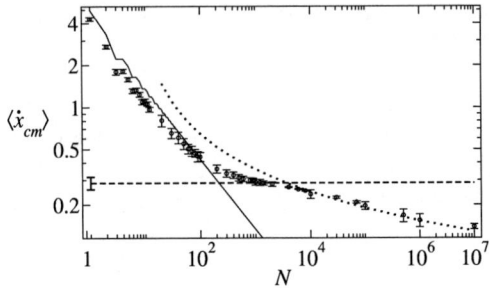

FIGURE 6. Expectation value of the speed of the center of mass, $\langle \dot{x}_{cm} \rangle$, for simulations of the periodic switching with optimal period (*dashed line*), simulations of the controlled ratchet (*circles with error bars*), an analytical approximation for small (*solid line*) and large N (*dotted line*).

CONCLUSIONS

Both in the flashing ratchet and in the paradoxical games, we have seen that greedy-like strategies only work in the presence of fluctuations, i.e., for a small number of particles or individuals, whereas yield systematic losses in large ensembles. This is just one of the interesting effects that one finds when applying control to ratchets and games, but we believe that the combination of control and rectification will be relevant in many applications in biology and nanotechnology, and could also shed some light on the old problem of the relationship between entropy and information.

ACKNOWLEDGMENTS

This work has been financially supported by grant FIS2004-0271 from MCYT (Spain).

REFERENCES

1. M. O. Magnasco, *Phys. Rev. Lett.* **71**, 1477 (1993).
2. A. Ajdari and J. Prost, *C.R. Acad. Sci. Paris II* **315**, 1635 (1993).
3. P. Reimann, *Phys. Rep.* **361**, 57 (2002).
4. G. P. Harmer and D. Abbott, *Nature* **402**, 846 (1999).
5. J.M.R. Parrondo and L. Dinís, *Contemp. Phys.* **45**, 147 (2004).
6. L. Dinis and J.M.R. Parrondo, *Europhys. Lett.* **63**, 319 (2003).
7. L. Dinis and J.M.R Parrondo, *Physica A* **343**, 701 (2004).
8. F. Cao, L. Dinís, and J.M.R. Parrondo, *Phys. Rev. Lett.* **93**, 040603 (2004).
9. H. Kitano, *Foundations of Systems Biology*, MIT Press, Cambridge, 2001.

A Form of Active Brownian motor-like on a (nonlinear) Toda lattice

E. del Rio*, W. Ebeling[†] and M.G. Velarde**

*Dpto. Fisica Aplicada, E.T.S.I. Aeronauticos. Universidad Politecnica de Madrid, Plaza Cardenal Cisneros 3, 28040 Madrid
[†]Institut für Physik, Humboldt-Universität Berlin, Newtonstraβe 15, 12489-Berlin, Germany
**Instituto Pluridisciplinar, Universidad Complutense de Madrid, Paseo Juan XXIII 1, 28040-Madrid, Spain.

Abstract. Results are provided about the evolution of a charged Brownian particle (an "electron") interacting with with particles or "units" placed on a one-dimensional Toda lattice. The thermal bath is a Gaussian white noise obeying Einstein's fluctuation-dissipation theorem. The electron-lattice interaction is modeled by a Coulomb pseudo-potential. Lattice compressions create soliton excitations (dissipative solitons) that may or may not bind the electron. The electron's eventual trajectory depends on the (noise) temperature and on the value of the Brownian damping coefficient. It also depends on the landscape displayed by the Coulomb pseudo-potential that allows waves traveling in either direction. Hence the system operates as a drifting ratchet, a kind of active Brownian motor for the transport of particles (or charges) along or against the solitonic motion.

Keywords: Brownian motor, drifting ratchet, noise, stochasticity, nonlinearity, dissipative solitons.
PACS: 05.40.Jc, 05.45.Yv, 05.70.Ln.

Earlier we have investigated the oscillation modes of a 1D Toda-lattice with N identical particles [1, 2, 3]. If the lattice particles (mean separation σ) are coupled to electric charges, nonlinear oscillations may play a key role in conduction and transport. In Ref.[?], the position of the lattice particles was approximated by $x_j = \sigma j + \Delta x_j$ where Δx_j is about a period-one ratchet function

$$\psi(z) = \begin{cases} \frac{M}{M-1}z & \text{if } z \in S_1 = [0, 1 - \frac{1}{M}) \\ M(1-z) & \text{if } z \in S_2 = [1 - \frac{1}{M}, 1] \end{cases}, \qquad \psi(z) = \psi(z+1). \tag{1}$$

By scaling amplitude and time to tune wave velocity along the ring, v_w, and period, τ, we have:

$$\Delta x_j = A\psi(\frac{\frac{\sigma}{v_w}j - t}{\tau}). \tag{2}$$

The electron-lattice particle (placed at y) interaction is approximated by a Coulomb-like pseudopotential

$$U_e(y - x_k) = \frac{-U_0}{\sqrt{(y-x_k)^2 + h^2}} \tag{3}$$

U_0 and h are parameters. Note that the results to be discussed below would be valid for almost any *repulsive* Toda-like interaction between lattice units and almost any

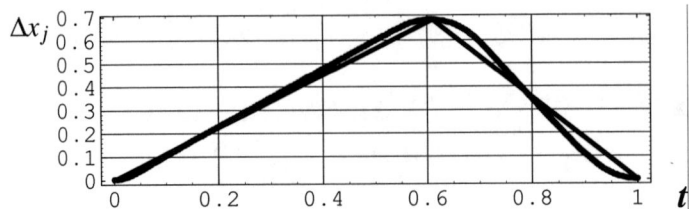

FIGURE 1. Piece-wise linear approximation, Eq. (2) with $v_w = -4.8$, $A = 0.68$, $\tau = 1$ and $M = 1.638$, together with the rescaled experimental time realization for a Toda ring.

reasonable electron-lattice interaction. The total potential applied to the electron is

$$U(t,y) = \sum_{j=n_y-n}^{n_y+n} U_e(\sigma j + A\psi(\frac{\frac{\sigma}{v_w}j - t}{\tau}) - y) \quad (4)$$

In Eq. (4) we assume that $y = n_y\sigma + \Delta y$ with $|\Delta y| < \sigma$ and we neglect the contributions from lattice particles beyond $n\sigma$. For a single electron or several if the interaction between the electrons is neglected, for the electron dynamics we can take

$$\frac{d^2y}{dt^2} + \gamma\frac{dy}{dt} + \frac{\partial U(t,y)}{\partial y} = \gamma\sqrt{2D}\xi(t). \quad (5)$$

The quantity $\gamma\sqrt{2D}\xi(t)$ (γ denotes positive damping and D the *spatial* diffusion constant) models a surrounding heat bath (Gaussian white noise), obeying a fluctuation-dissipation theorem. With $\frac{dU(t,y)}{dt} = 0$ we get the (local) velocity extrema. Here we have

$$\frac{dy}{dt} = \frac{A}{\tau}\frac{M}{M-1}\frac{(M-1)\sum_{j\in j_2}U'_j - \sum_{j\in j_1}U'_j}{\sum_{j\in j_2}U'_j + \sum_{j\in j_1}U'_j}, \quad U_j \equiv U_e(\sigma j + A\psi(\frac{\frac{\sigma}{v_w}j - t}{\tau}) - y) \quad (6)$$

where the indices j_1 and j_2 correspond, respectively, to S_1 and S_2 in Eq. (1).

Let us consider two cases. First assume that j_1 is empty. A point corresponding to this case is P_1 (any point along the line with positive slope far enough from the negative slope part) in Fig. 2(a). Then, according to Eq. (6) we get, e.g., $v_1 = M\frac{A}{\tau}$. In the opposite case, when the other set defined by j_2 is empty, (Fig. 2(a), P_2) the corresponding velocity extremum is, e.g., $v_2 = -\frac{M}{M-1}\frac{A}{\tau} < 0$. This mean that for any fixed time, there are intervals in the positions of the lattice particles where the wave-potential runs clockwise, as shown in Fig. 2(b), and intervals where the wave-potential runs in opposite direction. Note the crucial role played by the cut-off h in (3). The form, however, plays no significant role for the time being. Since $M \neq 2$ the ratchet function is asymmetric and we have $|v_1| \neq |v_2|$. In the noise-free case, for values of γ not too high, but larger than a given $\gamma_r \approx 80$, the particle moves clockwise with velocity, $v_r = M\frac{\sigma}{\tau} > 0$ slightly different from v_2 as the particle jumps periodically when it meets a minimum in opposite direction. Similarly, for $\gamma < \gamma_l$ with $\gamma_l(\approx 68)$ there is an interval of γ-values where the particle moves counterclockwise with velocity $v_l = v_w < 0$.

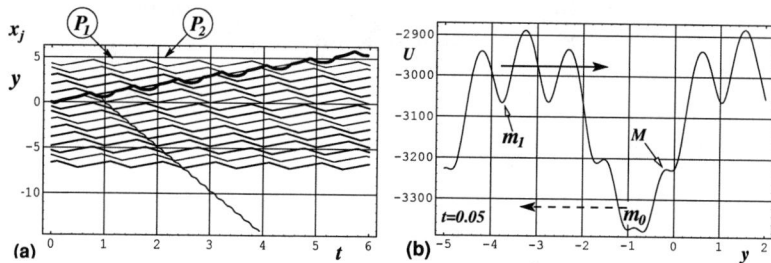

FIGURE 2. a) Positions of fifteen lattice "units", x_k, for $k = -11, \ldots, 3$, together with the trajectories for two noise-free particles with two different values of γ: $\gamma = 81.25$ with positive velocity and $\gamma = 70$ with negative velocity. b) Potential, Eq. (4), for $t=0.05$ with $h = 0.35$, $\sigma = 0.8$ and $U_0 = 239$. Arrows indicate the direction of motions along the potential-waveform. The solid arrow indicates the direction of motion for $\gamma = 81.25$ and the dashed arrow corresponds to $\gamma = 70$

Let us now consider a the role played by the noise with $\gamma > \gamma_r$. For low noise levels, the particle moves inside a potential minimum (Fig. 2(b), m_1) with velocity $v_r > 0$ traveling *against* the solitonic motion. After a short time, $\Delta t < \tau$, it finds a deeper minimum (Fig. 2(b), m_0) traveling with higher (in absolute value) negative velocity *along* the solitonic motion. At this time instant and due to the noise, the particle may change direction by changing its location hence the minimum where it seats, with probability $P_{r \to l}$. If the particle changes its direction of motion, a fraction of period later, it meets a new minimum traveling with $v_r > 0$ and has probability $P_{l \to r}$ to recover its previous antisolitonic motion direction. For a Brownian particle, the mean velocity depends on both probabilities

$$v_B = \left(1 - \frac{P_{r \to l}}{P_{l \to r}}\right) \bigg/ \left(1 - \frac{V_r}{V_l}\frac{P_{r \to l}}{P_{l \to r}}\right) v_r \quad (7)$$

Note that as $v_l < 0$ and $v_r > 0$, from Eq. (7) we have $v_B > 0$ if $P_{r \to l} < P_{l \to r}$ and $v_B < 0$, otherwise. As we have $\gamma > \gamma_r$ for very low noise level (temperature close to zero) we have $P_{r \to l} \approx 0$ and $P_{l \to r} \approx 1$ and then $v_B \approx v_r$.

Furthermore, $P_{r \to l}$ incorporates diffusion relative to the deterministic trajectory, $x_{det}(t)$. This is a form of Brownian motor [4], as here a new potential maximum appears (Fig. 2(b), M) as time proceeds. It diffuses according to

$$P(x_{det}|x;t) = \frac{1}{2\sqrt{\pi Dt}} e^{-\frac{(x_{det}-x)^2}{4Dt}}. \quad (8)$$

Hence the probability to be to the left of the new local maximum is

$$P_{r \to l} = \int_{-\infty}^{-x_0} P(x_{det}|x;t_0)dx = \frac{1}{2}\text{erfc}\left(\frac{x_0}{2\sqrt{Dt_0}}\right) \quad (9)$$

where $t_0 \approx 0.1$ is the time taken for the particle to go beyond the minimum m_0 in Fig. 2(b), and $x_0 = 0.05$ is the least distance from the maximum M for the deterministic trajectory. Note that $\lim_{D \to \infty} P_{r \to l} = 1/2$. On the other hand, $P_{l \to r}$ is an escape probability from the local potential barrier, P_{esc} as $P_{l \to r} = 1 - P_{esc}$. Obtaining an analytical expression of P_{esc} from the Langevin equation (5) is difficult, but in first approximation we can use Kramers theory for noise-induced escape from a single potential well. Hence

$$P_{l \to r} = e^{-\alpha t_1}, \quad \text{and} \quad \alpha = \frac{\omega}{2\pi} e^{\frac{-\Delta U \gamma}{D}} \quad (10)$$

The escape rate α is the product of an Arrhenius factor and a preexponential factor. $\omega \approx 50$ is the oscillation frequency in the metastable minimum of the potential, $\Delta U \approx 80$ is the mean value for the pseudopotential barrier and $t_1 \approx 0.2$ is the time while the barrier grows. We assume that in region II, for the maximum value of the pseudopotential barrier $P_{esc} \approx 0$, as the numerical integration shows. Using Eqs.(9),(10) together with Eq. (7) gives the velocity of the particle as a function of the random force. In region I of Fig.

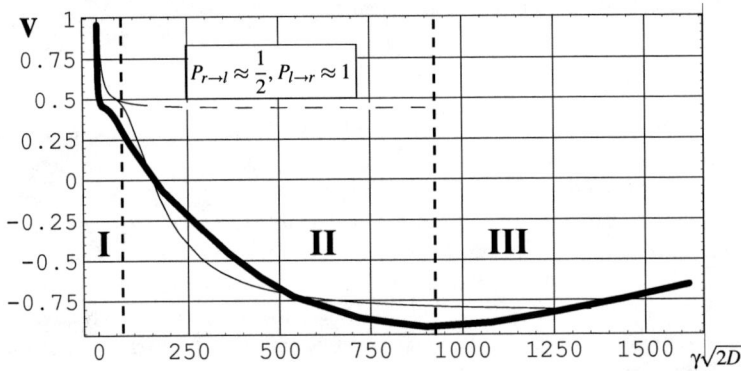

FIGURE 3. Velocity of a Brownian particles as a function of $\gamma\sqrt{2D}$. The heavy solid line comes from direct numerical integration averaging over trajectories of 100 particles during evolutions lasting 2000s each. The light solid line is the approximation using (7), (9) and (10).

3, from Eq. (10) we have $P_{l \to r} \approx 1$ and $P_{r \to l}$ approaches very quickly the value $\frac{1}{2}$, so v_B tends to 0.4. In region II, $P_{r \to l} \approx \frac{1}{2}$ but $P_{l \to r} < 1$ that slowly approaches $\exp(-\frac{\omega t_1}{2\pi})$ with $v_B \approx -0.8 < 0$. Note that in this region the particle changes sign according to the temperature value. In region III the noise level is so high that the particle starts flying away (with zero mean velocity) and the potential value, hence the barrier, is negligible relative to $k_B T$. Finally, let us conclude by saying that if we consider the electric current our system would be a kind of dynamo (in the sense of having current without electrostatic field).

REFERENCES

1. V. A. Makarov, E. del Río, W., Ebeling and M. G. Velarde. *Phys. Rev. E* **64**, 036601 (2001)
2. E. del Río, V. A. Makarov, W., Ebeling and M. G. Velarde. *Phys. Rev. E* **67**, 056208 (2003)
3. A. Chetverikov, W. Ebeling and M.G. Velarde. *Contr. Plasma Phys.*, **45** (2005)
4. R. Dan Astumian and M. Bier. *Phys. Rev. Lett.*, **72**, 1766–1769, (1994).

Diffusion Acceleration in Randomly Switching Sawtooth Potential

Alexander A. Dubkov* and Bernardo Spagnolo[†]

Radiophysics Department, Nizhni Novgorod State University, 23 Gagarin ave., 603950 Nizhni Novgorod, Russia
[†]*INFM-CNR, and Dipartimento di Fisica e Tecnologie Relative, Group of Interdisciplinary Physics, Università di Palermo, Viale delle Scienze, pad.18, I-90128 Palermo, Italy*

Abstract. We investigate an overdamped Brownian motion in symmetric sawtooth periodic potential switched by Markovian dichotomous noise between two configurations. The two configurations differ each other by a translation of half of period. The calculation of the effective diffusion coefficient is reduced to the mean first-passage time problem, and we obtain the exact expression valid for arbitrary mean rate of switchings and arbitrary intensity of white Gaussian noise. We find the area at parameters plane where acceleration of diffusion in comparison with the free diffusion case takes place.

INTRODUCTION

Brownian diffusion in fluctuating periodic potentials was investigated in the framework of molecular motors problem, i.e. unidirectional motion of Brownian particles along one-dimensional periodic structures (see, for example, review [1]). As a rule, a two-state model in which asymmetric (ratchet-like) potential randomly switches between two different configurations was considered [2], and so-called *on-off* ratchet scheme [3] in particular. But in the supersymmetric case [4] the mean flow of Brownian particles is zero, and the rate of diffusion is of interest.

In this paper we generalize the exact results for effective diffusion coefficient obtained in the overdamped limit for arbitrary fixed periodic potential [5] and for symmetric periodic potential modulated by external white Gaussian noise [6], on the case of randomly switching symmetric potential. The general equations obtained for arbitrary symmetric periodic potential are solved for the sawtooth potential, and the exact expression for the effective diffusion coefficient is derived without any assumptions on the intensity of driving white Gaussian noise and mean rate of switchings.

GENERAL EQUATIONS

We consider one-dimensional overdamped Brownian motion in randomly switching symmetric periodic potential $U(x)$ with the spatial period L

$$\dot{x} = -U'(x)\eta(t) + \xi(t), \qquad (1)$$

where $x(t)$ is the Brownian particle displacement in time t, $\xi(t)$ is the white Gaussian noise with zero mean and intensity $2D$, $\eta(t)$ is a Markovian dichotomous noise taking the values ± 1 with mean rate of switchings v. Thus, we investigate Brownian diffusion in a periodic potential flipping between two configurations $U(x)$ and $-U(x)$, i.e. in the "overturned" configuration the maxima of potential become the minima and vice versa. The symmetric potential satisfies the supersymmetry criterion: $-U(x) = U(x-L/2)$, and, as a result, the mean flow of Brownian particles $\langle \dot{x} \rangle$ in such a situation is zero [4].

We place for convenience all Brownian particles at the origin at $t=0$, and will calculate the effective diffusion coefficient as the following limit

$$D_{eff} = \lim_{t \to \infty} \langle x^2(t) \rangle / (2t). \tag{2}$$

Because of periodic potential, the diffusion process can be coarsely conceived as consecutive transitions of Brownian particle from points of potential minima $x_m = mL$ to nearest neighboring points $x_{m\pm 1}$. Thus, to obtain the asymptotic characteristic (2) we can use the equivalent "jumped diffusion" model [7] in which the effective diffusion coefficient D_{eff} is expressed in the following form

$$D_{eff} = L^2 / [\tau_+(0) + \tau_-(0)], \tag{3}$$

where $\tau_+(0)$ and $\tau_-(0)$ are the mean first times to reach one of absorbing boundaries $x = \pm L$ from the origin for the configuration of potential with $\eta(t) = 1$ and for "overturned" configuration ($\eta(t) = -1$). This is because the non-Markovian random process $x(t)$ has Markovian dynamics between flippings, i.e. involves an alternating pieces of two Markovian random processes $x_1(t)$ and $x_2(t)$ which are governed by Langevin equations (1) with $\eta(t) = 1$ and $\eta(t) = -1$ respectively.

Using the exact equations for the mean first passage times for Brownian diffusion in randomly switching potentials, derived in [8] (see also [9]), for our system (1), we find finally from Eq. (3)

$$D_{eff} = D \left[1 - \frac{2}{L} \int_0^L \left(1 - \frac{x}{L}\right) U'(x) \theta'(x) dx \right]^{-1}, \tag{4}$$

where function $\theta(x)$ is the solution of the following integro-differential equation

$$\theta''(x) - \frac{U'(x)}{D^2} \int_0^x U'(y) \theta'(y) dy - \frac{2v}{D} \theta(x) = -\frac{xU'(x)}{D^2} \tag{5}$$

with the boundary conditions: $\theta'(0) = 0, \theta(L) = 0$.

Unfortunately, Eqs. (4) and (5) cannot be solved for arbitrary symmetric periodic potential $U(x)$.

SAWTOOTH POTENTIAL

For the symmetric sawtooth periodic potential (see Fig. 1)

$$U(x) = \begin{cases} 2Ex/L, & 0 \leq x \leq L/2, \\ 2E(1-x/L), & L/2 \leq x \leq L \end{cases} \tag{6}$$

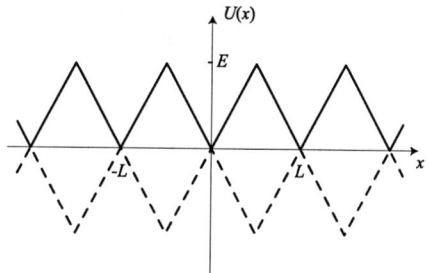

FIGURE 1. Switching sawtooth periodic potential.

from Eqs. (4) and (5) after complex rearrangements we obtain the following exact result for effective diffusion coefficient

$$D_{eff}/D = \left[2\alpha^2(1+\mu)\left(1-2\mu+4\mu\cosh\alpha+\mu^2\cosh2\alpha\right)\right]/[2\alpha^2\mu^2(1+\mu) \\ +2\mu\left(7-\mu+2\alpha^2\mu^2\right)\sinh^2\alpha+4\alpha\mu\left(1-3\mu+4\mu\cosh\alpha\right)\sinh\alpha \\ +8\left(1-6\mu+\mu^2\right)\sinh^2(\alpha/2)], \quad (7)$$

where $\alpha = \sqrt{E^2/D^2 + vL^2/(2D)}$ and $\mu = vL^2D/(2E^2)$ are two dimensionless parameters. Formula (7) is valid for arbitrary intensity of white Gaussian noise, mean rate of switchings and values of potential profile parameters.

Further we introduce for convenience two new dimensionless parameters $\beta = E/D$ and $\omega = vL^2/(2D)$ having a clear physical meaning. The parameter β is the ratio between the height of potential barriers and the intensity of white Gaussian noise. The parameter ω is the ratio between the time of a free diffusion through the distance L and the mean time interval between switchings. The dimensionless parameters α and μ, involved in Eq. (7), can be expressed in terms of β and ω as $\alpha = \sqrt{\beta^2 + \omega}$, $\mu = \omega/\beta^2$.

Let us analyze some limiting cases. At very rare flippings ($\omega \to 0$) we have $\alpha \simeq \beta$, $\mu \to 0$, and Eq. (7) gives

$$D_{eff}/D \simeq \beta^2/[2(\cosh\beta - 1)]. \quad (8)$$

Because of $(\cosh\beta - 1) > \beta^2/2$, the diffusion slows down in fixed periodic potential in comparison with the case when Brownian particles diffuse freely (see [5]). In the opposite case of very fast switchings ($\omega \to \infty$) we can predict the result. In such a situation Brownian particles "see" the average potential, i.e. $[U(x) + (-U(x))]/2 = 0$, and we arrive to diffusion in the absence of potential: $D_{eff} \simeq D$.

In the limiting case of very small barriers ($\beta \ll 1$) we find from Eq. (7)

$$\frac{D_{eff}}{D} \simeq 1 + \frac{\beta^2[(1+2\omega)\cosh 2\sqrt{\omega} - (4\cosh\sqrt{\omega} - 3)(1 + 4\sqrt{\omega}\sinh\sqrt{\omega} - 2\omega)]}{2\omega^2\cosh 2\sqrt{\omega}}. \quad (9)$$

As analysis of Eq. (9) shows, we obtain the enhancement of diffusion just at relatively fast switchings: $\omega > 9.195$. For very high potential barriers ($\beta \to \infty$) and fixed mean

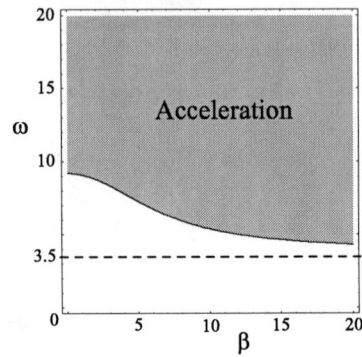

FIGURE 2. The area of diffusion acceleration.

rate of flippings ν we have $\alpha \simeq \beta \to \infty$, $\mu \to 0$, $\alpha^2 \mu \to \omega$. In such a case from Eq. (7) we obtain very interesting result: $D_{eff} = \nu L^2/7$.

Let us explain the result obtained. A diffusion is practically absent at very rare switchings because Brownian particles are not able to cross so high potential barriers. They can move in both directions by flippings only (mechanical diffusion). After new switching particles fall rapidly in the nearest potential wells and then wait next switching of a potential. As a result, the rate of mechanical diffusion increases with increasing mean rate of switchings and becomes greater than rate of free diffusion at $\omega > 3.5$.

The total area of diffusion acceleration, obtained by Eq. (7), is shown on the plane (β, ω) in Fig. 2 as shaded area. As it is seen from Fig. 2, this area lies inside the rectangular $\beta > 0$, $\omega > 3.5$.

ACKNOWLEDGMENTS

This work has been supported by MIUR, INFM, by Russian Foundation for Basic Research (project 05-02-16405), and by Federal Program "Scientific Schools of Russia" (project 1729.2003.2).

REFERENCES

1. P. Reimann, *Phys. Rep.*, **361**, 57–265 (2002).
2. R. D. Astumian, and M. Bier, *Phys. Rev. Lett.*, **72**, 1766–1769 (1994).
3. J-F. Chauwin, A. Ajdari, and J. Prost, *Europhys. Lett.*, **32**, 373–378 (1995).
4. P. Reimann, *Phys. Rev. Lett.*, **86**, 4992–4995 (2001).
5. R. Festa, and E. G. d'Agliano, *Physica A*, **90**, 229–244 (1978).
6. B. Spagnolo, A. A. Dubkov, and N. V. Agudov, *Physica A*, **340**, 265–273 (2004).
7. B. Lindner, M. Kostur, and L. Schimansky-Geier, *Fluct. and Noise Lett.*, **1**, R25–R39 (2001).
8. V. Balakrishnan, C. Van den Broeck, and P. Hänggi, *Phys. Rev. A*, **38**, 4213–4222 (1988).
9. B. Spagnolo, A. A. Dubkov, and N. V. Agudov, *Eur. Phys. J. B*, **40**, 273–281 (2004).

Current Fluctuations in Degenerate Non-Equilibrium Systems

R. Katilius*, S. Reggiani✧, and M. Rudan✧

*Semiconductor Physics Institute, 01108 Vilnius, Lithuania
✧ARCES and DEIS, University of Bologna, 40136 Bologna, Italy

Abstract. Noise characteristics of degenerate electron gas in two- or three-dimensional channels at moderate/high electric fields are investigated theoreticaly at densities sufficient for the electron temperature approximation to work. Current fluctuations are sensitive to additional correlation of occupancies of one-electron states created by inter-electron (e–e) collisions in non-equilibrium. Specificity of degenerate systems lies in the fact that the most non-trivial part of correlation is *intrinsically* of a *two-particle* nature. The degeneracy effects violate the macroscopic noise–response relations valid for Boltzmann's statistics. The method is worked out for calculating the source of correlation of the electron energies as a function of the applied electric field for different electron–lattice scattering mechanisms.

Keywords: degenerate non-equilibrium systems, current fluctuations, inter-particle collisions, electron-temperature approach
PACS: 05.40.-a, 72.70.+m, 72.20.Ht

INTRODUCTION

Having in mind the advances of modern micro- and nano-electronics, a generalization of the noise theory in order to include degenerate systems is called for. Recently [1] the theory of noise for a degenerate 2D or 3D electron gas was formulated in the electron-temperature approach (electron density being high enough). The generalization of the analytic relation between the intensities of current fluctuations (noise) and small-signal conductivity (I–V characteristics) was achieved, but for not-too-high electric fields only (for so-called "*warm electrons*"). At higher fields, the degeneracy violates the macroscopic noise–response relations valid in the electron-temperature approximation for Boltzmann's statistics. Noise properties of a degenerate gas beyond the "warm electron" region become, in a sense, independent of transport properties, and deserve a special study. The aim of the paper is the theoretical investigation and prediction of noise characteristics of a degenerate electron gas at moderate/high electric fields, at electron densities sufficient for the electron temperature approximation to work, i.e., the rate at which inter-electron (e–e) collisions redistribute energy *within* the electron system to be larger than the rate at which the electron system transfers the energy to the lattice.

ENERGY-DISTRIBUTION CORRECTION

In the electron temperature approximation, current fluctuations are sensitive to *correlation of electron energies* [1, 2]. The *"source"* of the correlation enters the expression for the *spectral intensity of longitudinal current fluctuations*. To complete the theory of fluctuations in degenerate systems in the electron-temperature approximation, it was necessary to work out the method for calculating the energy-correlation source as a function of the applied electric field for different electron-lattice scattering mechanisms.

The *"source"* term differs from zero so far as, *due to electron–lattice interaction*, the electron distribution does not take a *strictly Fermi–Dirac shape*. Due to overwhelming frequency of *e–e* collisions, the *deviation* $\Delta n(\varepsilon) = n(\varepsilon) - n_T(\varepsilon)$ of the electron-energy distribution $n(\varepsilon)$ from the Fermi–Dirac distribution $n_T(\varepsilon)$ is *small*. However, again due to rather high frequency of *e–e* collisions, the contribution of the correction $\Delta n(\varepsilon)$ to the macroscopic fluctuations (e.g., those of current) is remarkable (comparable with the standard contributions).

The importance of the correction $\Delta n(\varepsilon)$ is not relevant to the degenerate statistics only. However, for non-degenerate statistics, it has been possible to bypass the direct calculation of $\Delta n(\varepsilon)$ by means of a specific manipulation using the Boltzmann equation [2]. Such a manipulation does not work for degenerate systems. Thus, we are confronted with the situation where one *should* discriminate between the Fermi–Dirac and the actual energy distribution. Let us note that such a situation is met for the first time during the several scores of years of use of the Electron-Temperature Approach.

We see that the kinetic (transport) equation should be solved with a higher accuracy. Since $\Delta n(\varepsilon)$ is small, the *e–e* collision term in the transport equation can be *linearized* with respect to $\Delta n(\varepsilon)$. On the other hand, in the *electron–lattice collision term*, only the Fermi–Dirac part of the distribution function should be kept. Thus, the correction term $\Delta n(\varepsilon)$ satisfies a *linear integral equation*, with *the kernel* depending on the (high) frequency of the *e–e* collisions, while the right hand side, for nearly elastic scattering of electrons by lattice, takes the form of a differential operator containing the details of the scattering. For a 3D system, the equation reads:

$$I^\varepsilon[\Delta n] = -T^{-1}\varepsilon^{-1/2}\frac{d}{d\varepsilon}\left\{\left[\frac{2e^2E^2\tau_p(\varepsilon)}{3m} - \frac{T-T_0}{\tau_{en}(\varepsilon)}\right]\varepsilon^{3/2}n_T(\varepsilon)(1-n_T(\varepsilon))\right\} \quad (1)$$

where T and T_0 are the electron and lattice temperatures (in energy units), E is the applied electric field, the momentum- and energy-relaxation times are denoted as $\tau_p(\varepsilon)$ and $\tau_{en}(\varepsilon)$. $I^\varepsilon[\Delta n]$ is the *e–e* collision term *averaged over the constant-energy surface* and *linearized* with respect to $\Delta n(\varepsilon)$.

Equation (1) needs to be solved numerically for different values of the applied electric field, for different scattering-on-lattice mechanisms. The obtained correction term $\Delta n(\varepsilon)$ is to be inserted into the expressions for fluctuation characteristics.

ADDITIONAL CORRELATION

Kinetic theory of fluctuations in a degenerate gas was developed from the first principles by Kagan [3] and Muradov [4]. The equations contain sources of *additional* (kinetic) correlation among electrons, arising in the non-equilibrium state due to *e–e* collisions. The *source of correlation of occupancies of one-particle states* in a degenerate gas — the net simultaneous flow from (into) the *pair* of *single-electron states* due to *e–e* collision — is given by the expression

$$I^{pp'}\{n\} = (1 - n_p - n_{p'}) \sum_q I^{pp'}_{p-q, p'+q} + (n_{p'} - n_p) \sum_q (I^{p', p-q}_{p'-q, p} + I^{p', q-p'}_{p, q-p}) \quad (2)$$

The correlation-creating term is the ordinary *e–e* collision term entering the kinetic equation, but *without* one of the summations:

$$\sum_q I^{pp'}_{p-q, p'+q} \equiv \sum_q W^{pp'}_{p-q, p'+q} [n_p n_{p'} (1 - n_{p-q})(1 - n_{p'+q}) - n_{p-q} n_{p'+q} (1 - n_p)(1 - n_{p'})]. \quad (3)$$

Here n_p is the electron distribution function, $W^{pp'}_{p-q, p'+q}$ is the probability of the *e–e* collision changing the momenta of the two electrons before the collision, **p** and **p'**, into **p–q** and **p'+q** after the collision.

The *e–e* collision term entering the kinetic equation is $I^p = \sum_{p'q} I^{pp'}_{p-q, p'+q}$. The term $I^{pp'}_{p-q, p'+q}$, the source $I^{pp'}$, and the collision term I^p, due to conservation of energy in a collision event, *vanish after the insertion of any Fermi distribution* (including *equilibrium* distribution).

Specific for the degeneracy is a possibility for electrons with momenta **p** and **p'** to participate in *three different types of collisions* (in three types of processes corresponding to three topologically different diagrams [3, 4]). We conclude that the additional correlation is created by *three different types of fluxes in quasi-momentum space*. For non-degenerate statistics only one of the fluxes contributes, the remaining items in Eq. (2) being of higher order in the electron density.

In the electron-temperature approximation, the source of correlation of electron energies, $\Lambda = \sum_{pp'} \varepsilon_p \varepsilon_{p'} I^{pp'}$, enters the expression for the observable quantity — the spectral intensity of *longitudinal current fluctuations*. The source consists of two parts,

$$\Lambda = \Lambda^{(1)} + \Lambda^{(2)}. \quad (4)$$

The contribution $\Lambda^{(1)}$ is of the same form as in the case of Botzmann's statistics, and *is expressible in terms of transport characteristics* in the same manner as in the non-degenerate case (cf. Eq. (29) in [1]),

$$\Lambda^{(1)} \equiv \sum_{\mathbf{p}} \varepsilon_{\mathbf{p}}^2 \sum_{\mathbf{p'q}} I_{\mathbf{p-q,p'+q}}^{\mathbf{pp'}} = 2T^2 [E^2 \sigma /(T-T_0) - c_e / \tau_T]. \quad (5)$$

Here $c_e \equiv \sum_{\mathbf{p}} \varepsilon_{\mathbf{p}} dn_T(\varepsilon_{\mathbf{p}})/dT$ is the *specific heat* of the degenerate electron gas, and τ_T is the *the electron-temperature relaxation time* near the steady state.

Specific for a degenerate non-equilibrium system, the contribution $\Lambda^{(2)}$ is intrinsically a *two-particle* one. It reads

$$\Lambda^{(2)} \equiv 2 \sum_{\mathbf{pp'}} \varepsilon_{\mathbf{p}} \varepsilon_{\mathbf{p'}} n_{\mathbf{p}} \sum_{\mathbf{q}} (I_{\mathbf{p-q,p'+q}}^{\mathbf{pp'}} + I_{\mathbf{p'-q,p}}^{\mathbf{p',p-q}} + I_{\mathbf{p,q-p}}^{\mathbf{p',q-p'}}). \quad (6)$$

It's in the nature of things that $\Lambda^{(2)}$ is unexpressible in terms of the one-particle response. For the computation of $\Lambda^{(2)}$, the knowledge of the correction term $\Delta n(\varepsilon) = n(\varepsilon) - n_T(\varepsilon)$ — of the solution of the Eq. (1) — is mandatory.

The solution is to be inserted into the expression (6) for $\Lambda^{(2)}$. The results of the computation of $\Lambda^{(2)}$ for different scattering-on-lattice mechanisms should be compared with the contribution of $\Lambda^{(1)}$, in the non-equilibrium case existing for the Boltzmann statistics as well. The impact of degeneracy on the macroscopic current fluctuations is characterized by the ratio $\Lambda^{(2)} / \Lambda^{(1)}$.

The numerical solution determining the correction $\Delta n(\varepsilon)$ from Eq. (1) and, subsequently, the ratio $\Lambda^{(2)} / \Lambda^{(1)}$ for different practically important operating conditions is currently carried out on the lines described here.

ACKNOWLEDGMENT

This work has been partly funded by the European Commission under the frame of the Network of Excellence "SINANO" (Silicon-based Nanodevices, IST–506844).

REFERENCES

1. R. Katilius, *Phys. Rev.* **B69**, 245315 (2004). See also: *Proc. 17th ICNF* (Prague, 2003), pp. 25–30.
2. H. L. Hartnagel, R. Katilius, A. Matulionis. *Microwave Noise in Semiconductor Devices* (Wiley, New York, 2001). Chapter 12.
3. V. D. Kagan, *Sov. Phys. – Solid State* **17**, 1289 (1975).
4. M. I. Muradov, *Phys. Rev.* **B58**, 12883 (1998).

Quantum Noise in Long Josephson Junctions

Vladislav V. Kurin* and Igor V. Pimenov*

Institute for Physics of Microstructures of RAS, Nizhny Novgorod, 603950, Russia

Abstract. In the framework of tunneling Hamiltonian the Langevin equation describing quasiclassical dynamics affected by quantum noise in Extended Josephson Junctions at voltages comparable to superconducting gap was derived. Voltage-Current Characteristics and the shape of radiation spectrum produced was calculated in the high magnetic field approximation. Contributions both from tunneling, both quasiparticle and superconducting ones, and surface current were taken into account selfconsistently. It is found that fluctuations due to surface losses in electrodes contributes considerably to the linewidth.

Recently Extended Josephson Junctions (EJJ) have been successfully used as so called Flux-Flow Oscillator (FFO) for generating radiation of submm range, for pumping SIS mixers in integrated superconducting recievers and spectrometers [1] for radio astronomical applications. The need to improve the spectral resolution of such receivers makes investigation of the noise properties of EJJ extremely important. Another area of possible application of EJJ - flux logical devices for classical and quantum computing (see, for example [2]) also requires a thorough investigation of fluctuations since their level plays a crucial role in their ability to process information.

In this report we propose a theory that provides a self-consistent description of the dynamics and effects of quantum fluctuations in EJJ. Though the theory for the lumped junction taking into account the quantum behavior was developed long ago [3, 4] it is not directly applicable for describing the situation in the extended junction. Unlike the lumped junction, where only tunnel currents, quasiparticle and superconducting, play roles, in the extended Josephson junction placed into an external magnetic field there is additional current flowing along the electrodes which results in spatial inhomogeneity of magnetic field. The dissipative component of this current according to the fluctuation-dissipation theorem will serve as a source of additional, with respect to lumped junction, fluctuations. These high frequency fluctuations being converted by nonlinearity of the tunnel current can result in considerable modification of the noise properties of Josephson devices. Here the developed theory is applied to the problem of spectral properties of radiation produced by FFO.

Strictly speaking, the quantum dynamics of a Josephson junction (as of any other system) should be described by the reduced density matrix functional $\rho(\varphi, \varphi', t)$ which obeys the evolution equation [5]

$$\rho(\varphi, \varphi', t) = \int J(\varphi, \varphi'; \varphi_i, \varphi'_i; t) \rho(\varphi_i, \varphi'_i, 0) D\varphi_i D\varphi'_i,$$

where evolution operator $J(\varphi, \varphi'; \varphi_i, \varphi'_i; t)$ is the average of the evolution operator for the full density matrix with respect to initial quantum state of environment, usually called thermal bath, such as electronic degree of freedom and power supply circuits.

But if Josephson Junction is macroscopic, then its resistive state, even affected by quantum fluctuation, will be quasiclassical and can be described by a much simpler means of the so called Langevin equation which is equation for a classical variable, the Josephson phase difference $\varphi(x,t)$. For lumped junction the equation of this kind was derived in [3]. For an extended junction this equation should be modified to take into account the surface current along the electrodes and provided the magnetic field is smooth with respect to London penetration depth λ it can be represented in the following dimensionless form

$$\ddot{\varphi} + \int_0^\infty d\tau \left\{ \alpha_I(\tau) \sin \frac{\varphi(t,x) - \varphi(t-\tau,x)}{2} + \beta_I(\tau) \sin \frac{\varphi(t,x) + \varphi(t-\tau,x)}{2} \right\}$$

$$- \int_0^\infty \gamma_I(\tau) \varphi''(t-\tau,x) d\tau - j_{ext} = \Xi \quad (1)$$

where length is normalized to the static Josephson length, time- to the inverse plasma frequency, current - to the static critical current density $j_c = \int_0^\infty \beta_I(\tau) d\tau$, functions $\alpha_I(t)$ and $\beta_I(t)$, first obtained in [6] are expressed via products of normal and anomalous Green functions of different Josephson electrodes. They determine quasiparticle and superconducting components of the tunnel current through the junction. As usual the dot above the letter denotes the time derivative, the prime - the spatial one. The response function $\gamma_I(t)$ represents the inductance density of JJ and it is related to the surface impedance of Josephson electrodes and is defined by complex London penetration depth by

$$\gamma_I(t) = (d + 2\lambda(\omega = 0)) \int_{-\infty}^\infty \frac{1}{d + 2\lambda(\omega)} \exp(-i\omega t) \frac{d\omega}{2\pi}. \quad (2)$$

It can also be expressed via Green functions of electrodes, $j_{ext}(x)$ is the current distribution injected into the Junction. The L.H.S. of Eq.1, being equating to zero yields the equation averaged over quantum state of the junction and environment. For the lumped junction the equation of this kind was first obtained by Werthamer [6]. The expression for surface impedance and for function γ is determined by the well-known expression for penetration depth of transverse field into superconductor [7]. The R.H.S. of Eq. 1 $\Xi(x,t)$ is a sum of fluctuating currents

$$\Xi = \xi^{(1)}(t,x) \cos \frac{\varphi}{2} + \xi^{(2)}(t,x) \sin \frac{\varphi}{2} + \zeta(t,x), \quad (3)$$

where $\xi_1(x,t), \xi_2(x,t)$ describe current fluctuations related to tunnel current, the function $\zeta(x,t)$ represents noise current flowing along the junction. All these functions are Gaussian stochastic fields with zero mean values and correlation functions, defined by

$$\left\langle \xi^{(1,2)}(0,0) \xi^{(1,2)}(x,t) \right\rangle = \delta(x)(\alpha_R(t) \pm \beta_R(t)), \quad (4)$$

$$\left\langle \xi^{(1)}(0,0) \xi^{(2)}(x,t) \right\rangle = 0, \quad \langle \zeta(0,0)\zeta(x,t) \rangle = \delta''(x)\gamma_R(t) \quad (5)$$

where functions $\alpha_R(t), \beta_R(t), \gamma_R(t)$ defining the correlation properties of current fluctuations are related to response functions $\alpha_I(t), \beta_I(t), \gamma_I(t)$ by the fluctuation-dissipation theorem, which for spectral functions $\alpha_{R,I}(\omega), \beta_{R,I}(\omega), \gamma_{R,I}(\omega)$ looks in a standard way

$$(\alpha, \beta, \gamma)_R(\omega) = \coth\frac{\hbar\omega}{2T}\text{Im}(\alpha, \beta, \gamma)_I(\omega) \qquad (6)$$

If we neglect the terms describing the currents along the junction assuming $\gamma_{I,R} = 0$, we come to an equation for lumped junction derived in [3]. In the limiting case of slow with respect gap frequency Josephson phase $\varphi(t,x)$ the integral operators in Eq.1 turn into the local ones and the equation appear to be the well-known Sine-Gordon Equation with classical current sources describing thermal fluctuations. Such classical equation was used in [8] earlier. Eq. 1 should be supplemented by boundary conditions. Here we choose them as the conditions of absence of current at the junction ends and write down in the form $\int_{-\infty}^{t}\gamma_I(t-\tau)\varphi_x(\tau)d\tau\bigg|_{x=0,L} = h$ where h is the value of the dimensionless external magnetic field at the junction ends.

Calculations of the voltage current characteristic and the shape of the radiation spectral line can be carried out for a sufficiently high magnetic field and small noise intensity, close to work [9]. We seek the solution in the form $\varphi = \Omega t - hx + \psi + \vartheta$, where high frequency perturbation $\psi \ll 1$ is small and slow function θ describes phase diffusion due to noise, Ω is a Josephson frequency related to voltage. Then, assuming the tunneling supercurrent $\sim \beta_I \sin(\varphi + \varphi')/2$ as a small perturbation and equating the terms with equal order of magnitude, in the zero approximation we find unperturbed IV curve and explicit expression for noise source

$$g(\Omega) = \text{Im}\alpha(\Omega/2) = \int_0^\infty \alpha_I(\tau)\sin\frac{\Omega\tau}{2}d\tau = j_{ext},$$

$$\Xi_0 = \xi_1(t,x)\cos\frac{\Omega t - hx + \theta}{2} + \xi_2(t,x)\sin\frac{\Omega t - hx + \theta}{2} + \zeta(t,x)$$

In the first approximation we find linear integro-differential equation for high frequency phase ψ which can easily be solved by expansion $\psi(x,t) = \sum c_n(t)\psi_n(x)$ in eigenfunction $\psi_n(x) = \sqrt{L^{-1}(2-\delta_{0,n})}\cos L^{-1}\pi nx$. In the second approximation we find the equation for the slow phase, which reads

$$\ddot\theta + \frac{\dot\theta}{2}\int_0^\infty d\tau\tau\alpha_I(\tau)\cos\frac{\Omega\tau}{2} - \theta_{xx}\int_0^\infty \gamma_I(\tau)d\tau - \dot\theta_{xx}\int_0^\infty \gamma_I(\tau)\tau d\tau =$$

$$j_{ext} - I(\Omega + \dot\theta, h - \theta_x, x) + \Xi_{eff}(x,t). \qquad (7)$$

Here R.H.S. describes the distortion of IV curve due to the down conversion of Josephson oscillations on nonlinearities of quasiparticle and superconducting tunneling currents. Effective noise $\Xi_{eff}(x,t)$ is also determined by the same processes of noise conversion from the vicinity of the Josephson frequency to the low frequency region.

Averaging Eq.7 over the ensemble of random force and coordinate and stipulating that $\langle \dot{\theta} \rangle$ be limited in time we find IV curve $j_{ext} = I(\Omega, h)$ where

$$I(\Omega, h) = g(\Omega) + \frac{1}{2} \sum_n \text{Re} \beta_I(\Omega/2) \text{Im} \{\beta_I(\Omega/2) D_n^*(\Omega)\} |c_n|^2 |D_n(\Omega)|^{-2}$$

$$- \frac{1}{2} \text{Im} \{2\alpha_I(\Omega/2) - \alpha_I(3\Omega/2) - \alpha_I(-\Omega/2)\} \sum_n |\beta(\Omega/2) c_n|^2 |D_n(\Omega)|^{-2}. \quad (8)$$

Here are $c_n(h) = \int_0^L e^{-ihx} \psi_n(x) dx$ and $D_n(\Omega) = \gamma(\Omega) k_n^2 - [\alpha_I(3\Omega/2) - \alpha_I(-\Omega/2)]/4 - \Omega^2$. The first term in the sum describes the effects of the IV curve distortion due to nonlinearity of tunnel supercurrent such as Fiske resonances. The second one is due to quasiparticle nonlinearity and describes the effects of quasiparticle tunneling induced by Josephson radiation. We also find an equation for the slow fluctuating component $r_d^{-1} \dot{\theta} = \Xi_{eff}(x,t)$, where $r_d^{-1} = \partial I(\Omega, h)/\partial \Omega$ is the inverse differential resistance of Josephson junction. The shape of the radiation spectral line is defined by statistics of stochastic process $\psi_r \propto e^{i\Omega t + i\theta}$. On a natural assumption of a wide band of average noise Ξ_{eff} we find that the spectral line has a Lorentz shape with width $\Gamma = r_d^2 S_\Xi(0)$, where $S_\Xi(\omega)$ is the power spectrum of stationary component of current fluctuations expressed by

$$S_\Xi(0) = \alpha_R(\Omega/2) + \frac{1}{4} \sum_n \left\{ 2 \left(\text{Re} \beta_I(\Omega/2) \right)^2 |D_n(\Omega)|^{-2} + |\beta_I(\Omega/2)|^2 |A(\Omega)|^2 |D_n(\Omega)|^{-4} \right\}$$

$$|c_n|^2 [k_n^2 \gamma_R(\Omega) + \frac{1}{2}(\alpha_R(3\Omega/2) + \alpha_R(\Omega/2))],$$

where $A(\Omega) = \alpha_I(\Omega/2) - 3\alpha_I(-\Omega/2) + 3\alpha_I(-3\Omega/2) - \alpha_I(-5\Omega/2)$. The first term gives direct contribution of quasiparticle tunnel current, the second and the third one are related to conversion of high frequency noise resulting both from the tunnel and the surface losses on nonlinearities of quasiparticle and superconducting tunnel currents to vicinity of the zero frequency. Effectiveness of the down conversion depends on the response functions and external magnetic field.

Work is supported by Russian Foundation for Basic Research, grant N03-02-16533 and partly by ISTC N2445 and INTAS N01-0367.

REFERENCES

1. V.P. Koshelets at al, *Superconducting Science and Technology*, **17**, S127–S131, (2004).
2. A. Wallraff at al, *Nature*, **425**, 133, (2003)
3. U. Eckern, G. Schon and V. Ambegaokar, *Phys. Rev. B*, **30**, 6419, (1984)
4. A. I. Larkin, Yu. N. Ovchinnikov, *Phys. Rev. B*, **28**, 6281, (1983)
5. R.P. Feynman, A.R. Hibbs, *Quantum mechanics and path integrals*, McGraw-Hill, NY 1965
6. N. R. Werthamer, *Phys. Rev.*, **147**, 255 (1966)
7. J. Bardeen, D.C. Mattis, *Phys. Rev.*, **111**, 412 (1958)
8. N. Groenbech-Jensen, M. Salerno, M.R. Samuelsen, *Phys. Rev. B*, **46**, 308, (1992)
9. M. Salerno, M. R. Samuelsen, A. V. Yulin, *Phys. Rev. Lett.* **86**, 5397-5400 (2001)

Noise-induced patterns in semiconductor nanostructures and time-delayed feedback control

E. Schöll*, G. Stegemann*, A. Amann* and A. G. Balanov*

Institut für Theoretische Physik, Technische Universität Berlin, Hardenbergstr. 36, 10623 Berlin, Germany

Abstract. We study the constructive influence of noise upon the nonlinear dynamics of current density patterns in semiconductor nanostructures, and its control by time delayed feedback methods. In particular, we investigate noise-induced pattern formation in a double barrier resonant tunnelling diode described by a nonlinear reaction-diffusion model. For this purpose the parameters of the system are fixed at values below the Hopf bifurcation where the only stable state of the *deterministic* system is a spatially inhomogeneous "filamentary" steady state, and oscillating space-time patterns do not occur. We show that the addition of weak Gaussian white noise to the system gives rise to spatially inhomogeneous oscillations. As the noise intensity grows, the oscillations tend to become more and more spatially homogeneous, while simultaneously the temporal coherence of the oscillations decreases. We demonstrate that the application of a time delayed feedback loop, similar to that used in deterministic chaos control, allows one to control the temporal coherence and the time scales of the space-time patterns. Furthermore, with increasing control strength, a transition from spatially inhomogeneous, spiky oscillations to spatially homogeneous oscillations can be induced.

Keywords: nanostructures, resonant tunneling, noise-induced patterns, feedback control
PACS: 05.45.-a,05.40.-a,72.20.Ht

Semiconductor nanostructures exhibit a rich variety of nonlinear spatio-temporal dynamic behavior in charge transport [1]. A prominent example is the double barrier resonant tunnelling diode (DBRT). The characteristic feature of the DBRT is intrinsic bistability associated with a Z-shaped current-voltage characteristic [2]. It is the basis for a variety of interesting phenomena, including lateral spatio–temporal pattern formation of the current density [3, 4, 5, 6, 7, 8]. The general importance of noise in these nanostructures is well known [9].

For a long time it was widely believed that the effect of noise is only destructive, smearing–out any deterministic dynamics. However, recently noise has been found to evoke nontrivial, constructive response in nonlinear systems. It was shown, for instance, that random fluctuations are able to induce quite coherent patterns in extended media [10]. Moreover, time-delayed feedback which was earlier proposed to stabilize unstable states of a deterministic chaotic system [11], has been shown to allow also for control of noise-induced oscillations [12]. In the present work we show how noise can influence current density patterns in a semiconductor nanostructure, and how these can be controlled by time-delayed feedback.

We use the following model for the DBRT [6, 8]:

$$\frac{\partial a(x,t)}{\partial t} = f(a,u) + \frac{\partial}{\partial x}\left(D(a)\frac{\partial a}{\partial x}\right) + D_a\xi(x,t)$$
$$\frac{\partial u(t)}{\partial t} = \frac{1}{\varepsilon}(U_0 - u - rJ) + D_u\eta(t)$$
(1)

All quantities are dimensionless. The dynamical variable $a(x,t)$ describes the space- and time-dependent charge carrier density inside the quantum well, and $u(t)$ is the voltage drop across the device. The nonlinear function f models the net tunneling rate of the electrons through the two energy barriers into and out of the quantum well, and $D(a)$ is the effective diffusion coefficient, describing the diffusion of the electrons within the quantum well along the x–direction perpendicular to the current flow, $j(a,u) = \frac{1}{2}(f(a,u)+2a)$ is the local current density, and $J = \frac{1}{L}\int_0^L j\,dx$ gives the total current through the device of lateral width $L = 30$ (assuming Neumann boundary conditions). The two equations (1) are the local balance equation of the charge in the quantum well, and Kirchhoff's law of the circuit in which the device is operated, respectively. The external bias voltage is U_0. The dimensionless bifurcation parameter $\varepsilon = RC/\tau_a$ is related to the load resistance R (dimensionless: r), and the parallel capacitance C of the attached circuit, normalized by the tunneling time τ_a. For details of the model see [6, 7].

The two Gaussian white noise sources $\xi(x,t)$ and $\eta(t)$

$$\langle\xi(x,t)\rangle = \langle\eta(t)\rangle = 0 \quad (x \in [0,L]),$$
$$\langle\xi(x,t)\xi(x',t')\rangle = \delta(x-x')\delta(t-t'),$$
$$\langle\eta(t)\eta(t')\rangle = \delta(t-t').$$
(2)

of intensities D_a and D_u, respectively, describe effective contributions, e. g., of thermal fluctuations and shot noise [13].

The spatially homogeneous states of the system (1) yield a Z-shaped current–voltage characteristic. Additionally, there exist spatially inhomogeneous current density distributions or *current filaments*, which correspond to an additional branch in the current-voltage characteristic. The intersections of the different branches of the current-voltage characteristics with the load line determine the respective fixed points H (homogeneous) and I (inhomogeneous) for given U_0 and r.

In the noise–free case ($D_u = D_a = 0$) the inhomogeneous fixed point I is a stable focus for $\varepsilon < 6.4$. It corresponds to a stationary current filament. At $\varepsilon_{\text{Hopf}} \approx 6.4$ a Hopf bifurcation of I occurs which generates a limit cycle of spatio-temporal breathing oscillations of the filament (frequency $f_{\text{Hopf}} = 0.1674$). In the following we fix $\varepsilon = 6.2$ slightly below the Hopf bifurcation. The homogeneous fixed point H is stable with respect to homogeneous perturbations for $\varepsilon < 16.6$, but unstable with respect to spatially modulated perturbations. Thus, after some rapid transient, the deterministic system tends to the inhomogeneous filamentary fixed point I.

Next, we fix $D_a = 0.001$ and vary the noise intensity D_u (Fig. 1). While for small noise the system exhibits rather small oscillations around the inhomogeneous fixed point (a), with increasing noise intensity a transition to completely homogeneous oscillations occurs (c). For intermediate values of D_u one can see very nicely the competition

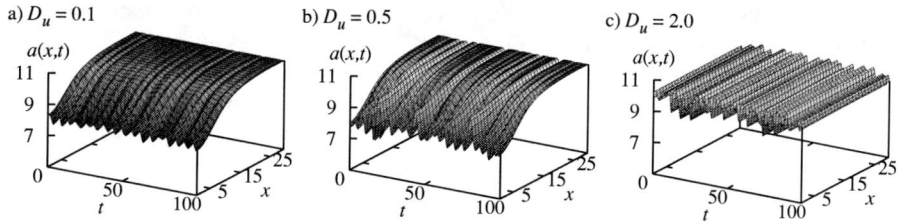

FIGURE 1. Noise-induced spatio–temporal dynamics of the charge carrier density $a(x,t)$. Simulation with a) $D_u = 0.1$, b) $D_u = 0.5$, c) $D_u = 2.0$. ($D_a = 0.001$, $\varepsilon = 6.2$, $U_0 = -84.2895$, $r = -35$.)

between the inhomogeneous and the spatially homogeneous modes, the former one dominating in Fig. 1(b).

The temporal coherence of the system can be measured by the correlation time $t_{\text{cor}} \equiv \frac{1}{\sigma^2} \int_0^\infty |\Psi(s)| ds$, where $\Psi(s) \equiv \langle (u(t) - \langle u \rangle)(u(t+s) - \langle u \rangle) \rangle_t$ is the autocorrelation function of the variable $u(t)$ and $\sigma^2 = \Psi(0)$ is its variance.

The correlation time versus noise intensity in Fig. 2(a) shows that the temporal coherence of the system in contrast to the spatial ordering decreases rapidly with increasing noise.

We shall now apply the method of time-delayed feedback control [12] to the noise-induced current density patterns in the DBRT. Since the voltage u, in contrast to $a(x,t)$, is easily accessible in a real experiment, we will add the control term only to the second eq. (1):

$$\frac{\partial u(t)}{\partial t} = \frac{1}{\varepsilon}(U_0 - u - rJ) + D_u \eta(t) - K[u(t) - u(t-\tau)]. \quad (3)$$

K is the control amplitude and τ is the time delay of the feedback loop.

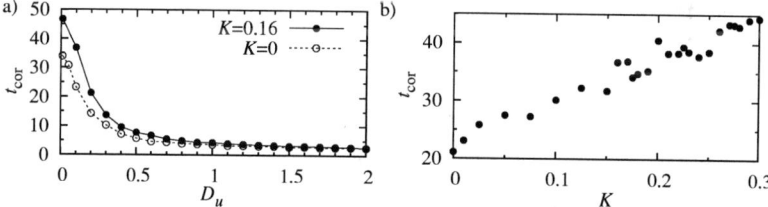

FIGURE 2. Enhancement of temporal correlation with time delayed feedback control ($\tau = 5.848$). a) Correlation time in dependence of noise amplitude with and without control. b) Correlation time in dependence of control amplitude for $D_u = 0.1$. Parameters as in Fig. 1.

We choose $\tau = 5.848$ close to the inverse intrinsic frequency f_{Hopf} of the noise-induced oscillations for low noise intensity. With this value and $K = 0.16$ the correlation time of the oscillations can be increased significantly for noise intensities ranging from $D_u = 0$ to 0.5 (Fig. 2(a)). At fixed noise intensity $D_u = 0.1$ the temporal correlation can be enhanced by increasing the control strength K (Fig. 2(b)). From a linear stability analysis around the inhomogeneous fixed point it follows that the spatially inhomogeneous fixed point undergoes a Hopf bifurcation around $K = 0.5$. Thus for sufficiently high K

the control loop changes already the deterministic dynamics of our system: The stable fixed points becomes unstable and a stable limit cycle is induced by the feedback control. With increasing K the shape of this induced limit cycle changes from small spatially inhomogeneous oscillations around the previously stable fixed point into a spatio-temporal spiking pattern (Fig. 3(a)) and further into a completely spatially homogeneous oscillation (Fig. 3(b)).

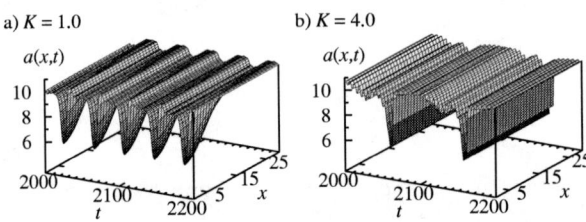

FIGURE 3. Spatio-temporal patterns induced by the control loop, $\tau = 5.848, D_u = 0.1$. a) For $K = 1.0$ the control loop induces a stable limit cycle corresponding to a spatio-temporal spiking pattern. b) For $K = 4.0$ the induced limit cycle is located within the homogeneous manifold of the system. Parameters as in Fig. 1.

In conclusion, we have shown that complex spatio–temporal current density patterns can be induced by noise in the double barrier resonant tunneling diode in a regime where the deterministic system is in a stable stationary state. By applying a simple time-delayed feedback scheme, the coherence and the time-scales of the patterns can be controlled.

We acknowledge helpful discussions with J. Pomplun and N. Janson. This work was supported by DFG in the framework of Sfb 555.

REFERENCES

1. E. Schöll, *Nonlinear spatio-temporal dynamics and chaos in semiconductors*, Cambridge University Press, Cambridge, 2001.
2. V. J. Goldman, D. C. Tsui, and J. E. Cunningham, *Phys. Rev. Lett.*, **58**, 1256 (1987).
3. A. Wacker, and E. Schöll, *J. Appl. Phys.*, **78**, 7352 (1995).
4. B. A. Glavin, V. A. Kochelap, and V. V. Mitin, *Phys. Rev. B*, **56**, 13346 (1997).
5. M. Meixner, P. Rodin, E. Schöll, and A. Wacker, *Eur. Phys. J. B*, **13**, 157 (2000).
6. E. Schöll, A. Amann, M. Rudolf, and J. Unkelbach, *Physica B*, **314**, 113 (2002).
7. J. Unkelbach, A. Amann, W. Just, and E. Schöll, *Phys. Rev. E*, **68**, 026204 (2003).
8. G. Stegemann, A. G. Balanov, and E. Schöll, *Phys. Rev. E*, **71**, 016221 (2005).
9. Y. M. Blanter, and M. Büttiker, *Phys. Rep.*, **336**, 1 (2000).
10. J. García-Ojalvo, A. Hernández-Machado, and J. M. Sancho, *Phys. Rev. Lett.*, **71**, 1542 (1993).
11. K. Pyragas, *Phys. Lett. A*, **170**, 421 (1992).
12. N. B. Janson, A. G. Balanov, and E. Schöll, *Phys. Rev. Lett.*, **93**, 010601 (2004).
13. L. L. Bonilla, O. Sánchez, and J. Soler, *Phys. Rev. B*, **65**, 195308 (2002).

Control of noise-induced oscillations in superlattices by delayed feedback

J. Hizanidis*, A. G. Balanov*, A. Amann* and E. Schöll*

*Institut für Theoretische Physik, Technische Universität Berlin
Hardenbergstr. 36, 10623 Berlin, Germany*

Abstract.
The dominant noise source, which effects the electron dynamics in semiconductor nanostructures, is shot noise, which is associated with the tunneling of individual carriers across a potential barrier. We consider noise-induced dynamics of electrons in a superlattice, which consists of alternating layers of two semiconductor materials with different band gaps. The parameters are fixed in the regime below the Hopf bifurcation of spatio-temporal oscillations, where in the absence of noise the system rests in a fixed point. It is shown that in this case noise applied to the superlattice can induce quite coherent oscillations of the current through the device. While the regularity of these oscillations depends on the noise intensity, their dominant frequency remains almost constant with variation of the noise level in the system. Further, we demonstrate that a time-delayed feedback scheme that was previously used to control purely temporal oscillations induced by noise, can not only enhance or deteriorate the regularity of stochastic spatio-temporal patterns but also allows for the manipulation of the system's timescales with varying time delay.

Keywords: superlattices, noise-induced patterns, time-delayed feedback
PACS: 05.45.-a, 05.40.-a, 72.70.+m

INTRODUCTION

Superlattices are semiconductor nanostructures which consist of alternating layers of two semiconductor materials with different band gaps. This leads to (periodic) spatial modulations of the conduction and valence band of the material, thus forming an energy band scheme consisting of a periodic sequence of potential barriers and quantum wells (see Fig. 1).

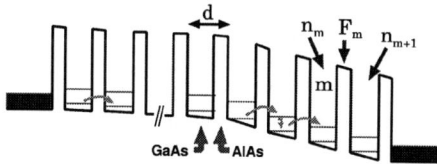

FIGURE 1. Superlattice energy band structure of alternating GaAs and AlAs layers under bias.

A sequential tunneling model is considered [1], which assumes that electrons are localized in one particular well and only weakly coupled to the adjacent wells. If a dc voltage is applied across the superlattice, sequential resonant tunneling of electrons between different wells leads to strongly nonlinear charge transport phenomena. This nonlinearity gives rise to complex spatiotemporal dynamics of the charge density and the field distribution within the device, including the formation of charge accumulation and depletion fronts associated with current oscillations [2, 3, 4].

The resulting tunneling current density $J_{m \to m+1}(F_m, n_m, n_{m+1})$ from well m to well $m+1$ depends only on the electric field F_m between both wells and the electron densities n_m and n_{m+1} in the respective wells (in units of cm^{-2}). The densities of electrons in each well are taken as dynamic variables of the system. The dynamic equations are then given by the continuity equations:

$$e\frac{dn_m}{dt} = J_{m-1 \to m} - J_{m \to m+1} \quad \text{for } m = 1, \ldots N, \tag{1}$$

where N is the number of wells in the superlattice. The applied voltage between emitter and collector gives rise to a global constraint (d is the superlattice period):

$$U = -\sum_{m=0}^{N} F_m d \tag{2}$$

NOISE-INDUCED PATTERNS

Now we extend the deterministic model including stochastic influences. The dominant noise source, which effects the electron dynamics in semiconductor nanostructures, is shot noise [5], which is associated with the fluctuations of the times between tunneling of electrons across a potential barrier. If the tunneling times are much smaller than any characteristic time scale of our system, we can roughly approximate those fluctuations by Gaussian white noise sources in the continuity equations for the electron densities. Charge conservation is satisfied by adding a noise term ξ_m to each current density $J_{m \to m+1}$:

$$e\frac{dn_m}{dt} = J_{m-1 \to m} + D\xi_m(t) - J_{m \to m+1} - D\xi_{m+1}(t), \tag{3}$$

where $\xi_m(t)$ is Gaussian white noise with

$$\langle \xi_m(t) \rangle = 0, \tag{4}$$
$$\langle \xi_m(t)\xi_{m'}(t') \rangle = \delta(t-t')\delta_{mm'}, \tag{5}$$

and D is the noise intensity. We next fix the parameters of the system slightly below a Hopf bifurcation, where without noise the only stationary solution is a stable fixed point that corresponds to a stationary depletion front localized over a small number of wells near the emitter. With increase of the noise intensity ($D > 0$), the current density starts oscillating in a rather regular way around the steady state (left panel of Fig. 2). (In Fig. 2 upper plots depict the charge density spatiotemporal dynamics and lower plots the current density time series). This oscillation corresponds to a, more or less, periodic motion of the depletion front as a whole (see inset of charge density plot). However, at larger noise intensities, the dynamics changes significantly. (Fig. 2, right panel). Current oscillations become sharply peaked and spiky, and the average current is shifted towards larger values. This is reflected in a more asymmetric motion of the depletion front (see inset). In particular we occasionally observe the onset of a tripole oscillation, where in addition to the existing depletion front, a dipole of an accumulation and a depletion front is generated close to the emitter.

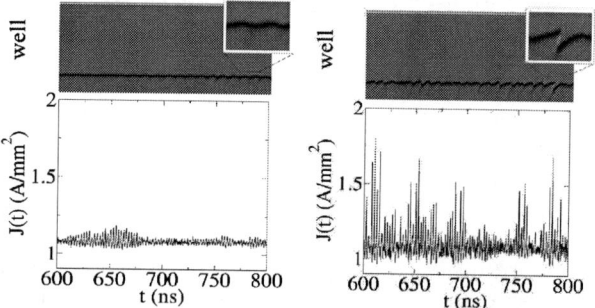

FIGURE 2. Dynamical behavior of the superlattice for different noise intensities $D = 0.1 As^{1/2}/mm^2$ (left panel) and $D = 0.5 As^{1/2}/mm^2$ (right panel).

To qualitatively understand how the noise level in the supperlattice affects the essential characteristics of noise-induced oscillations we consider spectra for different values of noise intensities D (Fig. 3, left plot). We see that an increase of the noise level broadens the spectral peak and suppresses secondary ones. At the same time the position of the main spectral peak, corresponding to the basic frequency of the oscillations, is almost unchanged. This is confirmed by the right plot of Fig. 3, where the dependence of the basic period T_0 (the inverse of the frequency at which the spectral peak is centered) of the noisy oscillations versus the noise intensity is presented. This basic period is close to the period of self-oscillations above the Hopf bifurcation.

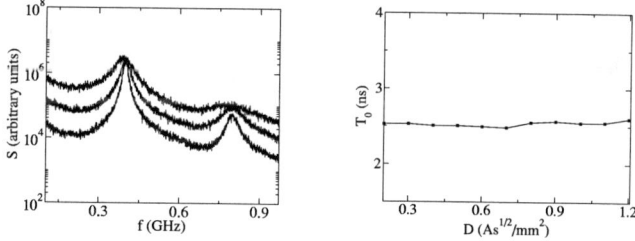

FIGURE 3. Main and secondary spectral peak of the power spectral density $S(2\pi f)$ vs. frequency f for increasing noise (from top to bottom: $D = 0.3, D = 0.5$ and $D = 1.0 As^{1/2}/mm^2$). Basic period T_0 vs noise intensity D.

CONTROL OF NOISE-INDUCED DYNAMICS BY DELAYED FEEDBACK.

Now we will try to control these noisy oscillations via time-delayed feedback [6, 7]. An easy way to implement control in the superlattice model is to simply add the control force, which is formed by the difference between $J(t)$ and its delayed version, to the

external voltage U, i.e.
$$U = U_0 - K(J(t) - J(t-\tau)), \quad (6)$$
where U_0 is the time-independent external voltage bias. Since both voltage and total current density are externally accessible global variables, such a control scheme is easy to implement experimentally.

Next we are interested in the effect of control on the coherence and timescales of the noise-induced oscillations. A natural choice for τ is the basic period of the Hopf oscillation (or integer multiples of it). As seen in the left plot of Fig. 4 the application of control indeed improves the coherence of the current signal, since the main peak in the power spectrum becomes narrower. In order to study the influence of control on the timescales of the system, the parameter to be varied will be the time delay τ. Calculating spectra for increasing τ it was found that additional peaks appear, while the main (most pronounced) peak moves towards lower frequencies. Plotting the period of the resulting main peak as a function of τ, we see that $T_0(\tau)$ has an almost piecewise linear, oscillatory character (right plot, Fig 4).

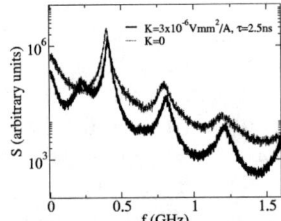 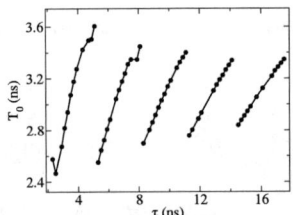

FIGURE 4. Power spectral density of noise-induced oscillations with and without control and basic period as a function of the time delay τ. $K = 3 \times 10^{-6} V mm^2/A$ and $D = 0.5$.

We therefore conclude that while the position of the main peak of the spectrum does not depend on the noise level in the case without control, it is indeed possible to shift its position by the proposed time delayed feedback scheme.

ACKNOWLEDGMENTS

This work was supported by DFG in the framework of Sfb 555. The authors would like to thank Grischa Stegemann and Natalia Janson for fruitful discussions.

REFERENCES

1. A. Wacker, *Phys. Rep.*, **357**, 1 (2002).
2. E. Schöll, *Nonlinear spatio-temporal dynamics and chaos in semiconductors*, Cambridge University Press, Cambridge, 2001.
3. L. L. Bonilla and H. T. Grahn, *Rep. Prog. Phys.* **68**, 577 (2005)
4. A. Amann and E. Schöll, *J. Stat. Phys.*, in print (2005).
5. Y. M. Blanter, and M. Büttiker, *Phys. Rep.*, **336**, 1 (2000).
6. K. Pyragas, *Phys. Lett. A*, **170**, 421 (1992).
7. N. B. Janson, and A. G. Balanov and E. Schöll, *Phys. Rev. Lett.*, **93**, 010601 (2004).

Current and noise suppression in ac-driven coherent transport

Sigmund Kohler*, Michael Strass*, Peter Hänggi*, Miguel Rey[†] and Fernando Sols**

*Institut für Physik, Universität Augsburg, Universitätsstraße 1, D-86135 Augsburg, Germany
[†]Departamento de Física Teórica de la Materia Condensada, Universidad Autónoma de Madrid, E-28049 Madrid, Spain
**Departamento de Física de Materiales, Universidad Complutense de Madrid, E-28040, Spain

Abstract.
We investigate the possibility to manipulate for the transport through heterostructures the dc current and its noise properties by an ac gate voltage. For a computation of the noise strength, we map the system to a tight-binding model for which noise suppression by ac fields has been predicted recently. The quality of this description is tested by comparing the transmission of the tight-binding system with a transfer-matrix approach.

Keywords: coherent transport, driven systems, noise
PACS: 05.60.Gg, 05.40.-a, 72.40.+w 73.63.-b,

INTRODUCTION AND MODELING

Semiconductor heterostructures represent a popular physical system for the investigation of mesoscopic transport and tunneling phenomena [1, 2]. In particular, these setups open various ways to study tunneling in time-dependent systems [3–7]. A straightforward possibility for introducing a time-dependence is the application of an ac transport voltage which only modulates the energies of the electrons in the leads while the potential inside the mesoscopic region remains time-independent. This kind of driving allows for a description within the Tien-Gordon theory [8] expressing the dc current in terms of the static transmission and an effective distribution function for the lead electrons. If the time-dependence enters via an external microwave field or an ac gate voltage, however, such an approach is generally insufficient [9]. A remarkable difference with respect to the static situation is the emergence of inelastic transport channels stemming from the emission or absorption of quanta of the driving field. This follows indeed from a recently presented Floquet theory for the transport through driven tight-binding systems [7, 9]. For the computation of the dc current, the latter approach justifies the applicability of a Landauer-like current formula where the static transmission is replaced by the time-averaged transmission of the time-dependent system.

The transmission of the transport channels can depend sensitively on the driving parameters; the contribution of certain channels can even vanish. For the transport across two barriers which enclose an oscillating potential well, Wagner [10] showed that it is possible to suppress the contribution of individual inelastic scattering channels. The total current, however, is given by the sum over all channels, and thus it is not possible

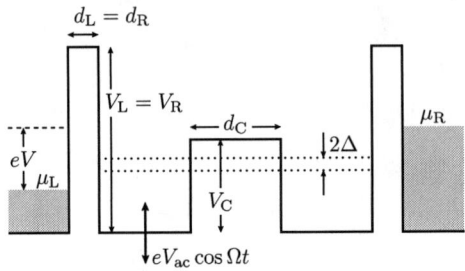

FIGURE 1. Model potential for the double-well heterostructure. In the numerical calculations, we employ barriers with the heights $V_L = V_R = 90\,\text{meV}$, $V_C = 40\,\text{meV}$ and the widths $d_L = d_R = 5\,\text{nm}$, $d_C = 15\,\text{nm}$. The dotted lines mark the energy of a metastable tunnel doublet with splitting energy 2Δ described by the Hamiltonian (2). The on-site energy of the left well is subject to an oscillating gate voltage with amplitude V_{ac}.

to isolate the contribution of a single channel in a current measurement. By contrast, in the case of transport through a two-level system with attached leads, driving with a dipole field has directly observable consequences: There, the driving not only affects the contribution of individual transport channels, but the dc current can be suppressed almost entirely [11, 12]. For the appearance of this *coherent current suppression*, it is essential that the central region consists of at least two weakly coupled wells which oscillate relative to each other [9].

Here, we explore the possibility of coherent current suppression in double-well heterostructures. Thereby, we compare two theoretical approaches to describe coherent transport in quantum-well structures: The transfer-matrix method and a tight-binding approach. As a model we consider the triple-barrier structure sketched in Fig. 1 where the driving enters via an oscillating gate voltage which modulates the bottom of the left well. The applied transport voltage is assumed to shift the Fermi energy of the left lead by $-eV$, $-e$ being the electron charge. Having a tight-binding approximation for the double-well system at hand, we are able to compute within the recently developed Floquet approach [7, 13] also the noise strength.

Transfer-matrix method

Following Landauer [14], we consider the coherent mesoscopic transport as a quantum mechanical scattering process. The central idea of this approach is the assumption that sufficiently far from the scattering region, the electronic single-particle states are plane waves and that their occupation probability is given by the Fermi function with the chemical potential depending on the applied voltage. The unitarity of evolution under coherent ac driving allows us to write the resulting currents as [15]

$$I = \frac{e}{h}\int dE\,[T_{RL}(E)f_L(E) - T_{LR}(E)f_R(E)], \qquad (1)$$

where $T_{RL}(E)$ denotes the total transmission probability of an electron with energy E from the left to the right lead. The term $T_{LR}(E)$ describes the respective scattering from the right to the left lead. These transmission probabilities comprise the sum over transverse modes and outgoing inelastic channels and can be computed with the transfer matrix method developed in Ref. [16]. The latter method relies on the fact that for a time-dependent spatially constant potential, the eigenfunctions are plane waves with an additional time-dependent phase factor. The propagator for a piecewise constant potential, in turn, can be constructed by imposing proper matching conditions at the boundaries of the adjacent layers of the heterostructure.

Tight-binding approximation

A different approach to study resonant tunneling in a driven double-well structure is based on the adoption of a tight-binding approximation where each well is represented by a single localized electron orbital. Then, the Hamiltonian of the transport setup reads $H(t) = H_{\text{wells}}(t) + H_{\text{leads}} + H_{\text{contacts}}$, where

$$H_{\text{wells}}(t) = -\Delta(c_L^\dagger c_R + c_R^\dagger c_L) + eV_{\text{ac}}\cos(\Omega t) c_L^\dagger c_L \qquad (2)$$

describes the electrons in the wells. The second term of the Hamiltonian (2) accounts for the harmonic driving of the traversing electrons in the left well via an oscillating gate voltage with amplitude V_{ac} and period $\mathcal{T} = 2\pi/\Omega$. The leads are modeled as ideal Fermi gases with the Hamiltonian $H_{\text{leads}} = \sum_{\ell,q} \varepsilon_{\ell q} c_{\ell q}^\dagger c_{\ell q}$, where $c_{\ell q}$ annihilates an electron in the lead with energy $\varepsilon_{\ell q}$ with $\ell = L, R$. As an initial condition, we employ the grand-canonical ensembles of electrons in the leads at inverse temperature $\beta = 1/k_B T$. Therefore, the lead electrons are characterized by the equilibrium Fermi distribution $f_\ell(\varepsilon_{\ell q}) = \{1 + \exp[-\beta(\varepsilon_{\ell q} - \mu_\ell)]\}^{-1}$. The localized state in each well couples via the tunneling matrix element $V_{\ell q}$ to the state $|\ell q\rangle$ in the respective lead. The Hamiltonian which describes this interaction has the form $H_{\text{contacts}} = \sum_{\ell,q} V_{\ell q} c_{\ell q}^\dagger c_\ell + \text{H.c.}$ The lead–well coupling is entirely specified by the spectral density $\Gamma_\ell(\varepsilon) = 2\pi \sum_q |V_{\ell q}|^2 \delta(\varepsilon - \varepsilon_{\ell q})$. Since, for the system at hand, the bandwidth of the conduction band of the leads is much larger than the energy regime where transport happens, the spectral densities are practically constant, i.e. $\Gamma_\ell(\varepsilon) = \Gamma_\ell$, which defines the so-called wide-band limit. By matching for the static case the transmissions of the transfer-matrix and the tight-binding approach, we find the tight-binding parameters $\Gamma_L = \Gamma_R = 0.16$meV and $\Delta = 0.23$meV.

Floquet transport theory

Starting from the Heisenberg equations of motion for the annihilation operators, one eliminates the lead operators and thereby obtains for the electrons on the dots a reduced set of equations. These are solved with the help of the retarded Green function obeying $[i\hbar d/dt - \mathcal{H}(t) + i\Gamma/2]G(t,t') = \delta(t-t')$, where $\mathcal{H}(t)$ is the single-particle Hamiltonian corresponding to double well Hamiltonian (2) and $\Gamma = \Gamma_L = \Gamma_R$.

The coefficients of the equation of motion for $G(t,t')$ are $2\pi/\Omega$-periodic and, consequently, its solution can be constructed with the help of the Floquet ansatz $|\psi_\alpha(t)\rangle = \exp[(-i\varepsilon_\alpha/\hbar - \gamma_\alpha)t]|\phi_\alpha(t)\rangle$. The Floquet states $|\phi_\alpha(t)\rangle$ obey the time-periodicity of the Hamiltonian and fulfill the eigenvalue equation

$$\left[\mathcal{H}(t) - \frac{i}{2}\Gamma - i\hbar\frac{d}{dt}\right]|\phi_\alpha(t)\rangle = (\varepsilon_\alpha - i\hbar\gamma_\alpha)|\phi_\alpha(t)\rangle. \tag{3}$$

This yields the retarded Green function $G(t,t') = -(i/\hbar)\sum_\alpha |\psi_\alpha(t)\rangle\langle\psi_\alpha^+(t')|\Theta(t-t')$.

In particular, one finds for the current a convenient Landauer-like expression with an additional sum over the Fourier index k, i.e. $T_{LR}(E) = \sum_k T_{LR}^{(k)}(E)$ [7, 13]. Since the symmetrized noise correlation function $S(t,t') = \langle [I(t), I(t')]_+ \rangle$ depends explicitly on both times, we characterize the noise by the time-average of its zero-frequency component, $S = (\Omega/2\pi)\int_0^{2\pi/\Omega} dt \int_{-\infty}^{+\infty} d\tau S(t, t - \tau)$.

High-frequency approximation

The Floquet treatment of the present transport problem allows for the implementation of a stationary perturbation scheme for driving frequencies much larger than all other frequency scales of the system [17]. This approach has recently been extended to transport situations which are characterized by the presence of leads [9, 12]; here we only mention the cornerstones of this approach and refer the reader to Ref. [7]. The starting point is the unitary transformation

$$U_0(t) = \exp\left\{-\frac{ie}{\hbar\Omega}V_{ac}\sin(\Omega t)c_L^\dagger c_L\right\}, \tag{4}$$

which is first applied to the quantum-well Hamiltonian (2). For sufficiently large driving frequencies $\Omega \gg \Delta/\hbar$, a separation of time scales is performed. Thereby, fast oscillations of the transformed Hamiltonian are neglected by averaging over a driving period. Finally, we arrive at the effective Hamiltonian for the quantum wells

$$\bar{H}_{eff} = \frac{1}{\mathcal{T}}\int_0^\mathcal{T} dt\left(U_0^\dagger H_{wells}(t)U_0 - i\hbar U_0^\dagger \dot{U}_0\right) = -\Delta_{eff}(c_L^\dagger c_R + c_R^\dagger c_L), \tag{5}$$

which is of the same form as in the static case but with the effective tunneling matrix element $\Delta_{eff} = J_0(eV_{ac}/\hbar\Omega)\Delta$, J_0 being the zeroth order Bessel function of the first kind.

The transformation (4) also affects the lead–well coupling. Applying $U_0(t)$ to $H_{contacts}$ and solving the Heisenberg equations for the lead and quantum-well operators in the wide-band limit, we find that the influence of the left lead is no longer determined by the Fermi function $f_L(\varepsilon)$ but rather by the effective electron distribution [7, 13]

$$f_{L,eff}(\varepsilon) = \sum_{n=-\infty}^{\infty} J_n^2\left(\frac{eV_{ac}}{\hbar\Omega}\right)f_L(\varepsilon + n\hbar\Omega). \tag{6}$$

The squares of the nth-order Bessel function of the first kind J_n weight the processes where an electron with energy ε is transmitted under the emission ($n < 0$) or absorption ($n > 0$) of $|n|$ photons.

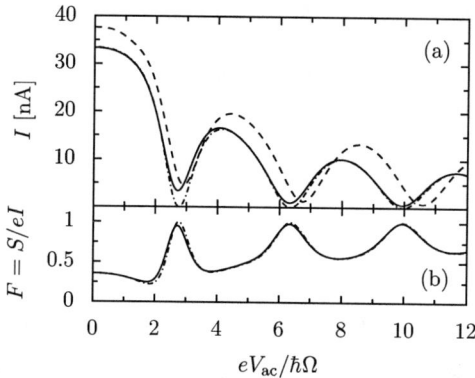

FIGURE 2. (a) Average current *vs.* driving amplitude obtained numerically from tight-binding (solid line) and transfer-matrix (dashed) methods compared to the high-frequency approximation (dashed-dotted). The driving parameters are $\hbar\Omega = 1.15\,\text{meV}$ and $V = 6.0\,\text{mV}$. (b) Corresponding Fano Factor.

COHERENT TRANSPORT SUPPRESSION

We now turn our attention to the coherent control of current. Tunneling suppression in a closed, driven system is known for more than a decade. For example for a driven bistable potential, tunneling breaks down at exact crossings of the quasi-energy spectrum. Then, one observes the so-called *coherent destruction of tunneling* [18] which has been studied in a number of cases [16, 19, 20], but in the context of transport between two leads has received attention only recently [9, 12, 21].

Surveying the time-averaged current calculated numerically from the transfer-matrix and the tight-binding method plotted in Fig. 2(a), we observe current minima for distinct values of $eV_{ac}/\hbar\Omega$ for frequencies in the microwave regime. The reason for the current suppressions becomes apparent from the effective tunnel matrix element $\Delta_{\text{eff}} = J_0(eV_{ac}/\hbar\Omega)\Delta$. This expression implies that the tunneling between the two wells and consequently the current vanishes whenever the ratio $eV_{ac}/\hbar\Omega$ assumes a zero of the Bessel function J_0, i.e. for the values 2.405, 5.520, 8.654, ... By varying the ratio between driving amplitude and frequency, we can thus tune the tunneling between the two wells and thereby control the current [12]. For a frequency $\Omega = 5\Delta/\hbar$, the analytically obtained current in Fig. 2 shows a remarkable agreement with the tight-binding result.

Figure 2(b) depicts the noise strength of the current for the tight-binding approximation characterized by the Fano factor $F = S/eI$. For zero driving amplitude, we find $F \approx 1/2$ which is characteristic for the transport through a double barrier [22]. Note that the central barrier is considerably lower and, thus, the outer barriers determine the transport. At the current suppression, the central barrier becomes the bottleneck. Then, the setup corresponds to a tunneling point contact with $F \approx 1$. In the crossover region, the noise can be even lower than in the static case, i.e. $F < 1/2$.

CONCLUSIONS

We have demonstrated that the current across a double-well heterostructure is strongly affected by the purely coherent influence of an oscillating gate voltage. We have used a transfer-matrix method as an exact approach to compute tunneling currents through such a system. We compared these results to those obtained from a tight-binding Floquet description. In particular, we find that the current suppression is controlled by the ratio of the driving frequency and amplitude. This can be understood by exploring the high-frequency limit within the tight-binding formalism. Since the effective inter-well coupling depends sensitively on the driving parameters, the transport properties of the double well can be controlled. The effective behavior ranging from transport through an almost open channel to a regime of rare tunnel events. The qualitative difference between these transport regimes is also reflected in the behavior of the Fano factor. In the crossover region, the driving can even reduce the noise level.

ACKNOWLEDGMENTS

We thank S. Camalet, G.-L. Ingold, and J. Lehmann for helpful discussions. This work has been supported through Acción Integrada no. HA2003-0091. Financial support is also acknowledged from MEC (Spain), Grant no. BFM2001-0172, Fundación Ramón Areces, and DFG (Germany), Graduiertenkolleg 283 and SFB 486.

REFERENCES

1. L. Esaki, and R. Tsu, *IBM J. Res. Dev.*, **14**, 61 (1970).
2. T. C. L. G. Sollner, W. D. Goodhue, P. E. Tannenwald, C. D. Parker, and D. D. Peck, *Appl. Phys. Lett.*, **43**, 588–590 (1983).
3. M. Grifoni, and P. Hänggi, *Phys. Rep.*, **304**, 229 (1998).
4. H. Qin, A. W. Holleitner, K. Eberl, and R. H. Blick, *Phys. Rev. B*, **64**, 241302 (2001).
5. W. G. van der Wiel, S. De Franceschi, J. M. Elzerman, T. Fujisawa, S. Tarucha, and L. P. Kouwenhoven, *Rev. Mod. Phys.*, **75**, 1 (2003).
6. G. Platero, and R. Aguado, *Phys. Rep.*, **395**, 1 (2004).
7. S. Kohler, J. Lehmann, and P. Hänggi, *Phys. Rep.*, **406**, 379 (2005).
8. P. K. Tien, and J. P. Gordon, *Phys. Rev.*, **129**, 647 (1963).
9. S. Camalet, S. Kohler, and P. Hänggi, *Phys. Rev. B*, **70**, 155326 (2004).
10. M. Wagner, *Phys. Rev. B*, **49**, 16544 (1994).
11. J. Lehmann, S. Camalet, S. Kohler, and P. Hänggi, *Chem. Phys. Lett.*, **368**, 282 (2003).
12. S. Kohler, S. Camalet, M. Strass, J. Lehmann, G.-L. Ingold, and P. Hänggi, *Chem. Phys.*, **296**, 243 (2004).
13. S. Camalet, J. Lehmann, S. Kohler, and P. Hänggi, *Phys. Rev. Lett.*, **90**, 210602 (2003).
14. R. Landauer, *IBM J. Res. Dev.*, **1**, 223 (1957).
15. M. Wagner, and F. Sols, *Phys. Rev. Lett.*, **83**, 4377 (1999); *Ann. Phys. (Leipzig)*, **9**, 776 (2000).
16. M. Wagner, *Phys. Rev. A*, **51**, 798 (1995).
17. J. H. Shirley, *Phys. Rev.*, **138**, B979 (1965).
18. F. Grossmann, T. Dittrich, P. Jung, and P. Hänggi, *Phys. Rev. Lett.*, **67**, 516 (1991).
19. M. Holthaus, *Phys. Rev. Lett.*, **69**, 351 (1992).
20. C. E. Creffield, and G. Platero, *Phys. Rev. B*, **65**, 113304 (2002).
21. M. Rey, M. Strass, S. Kohler, F. Sols, and P. Hänggi, *Chem. Phys.* (in press), cond-mat/0412221.
22. Ya. M. Blanter, and M. Büttiker, *Phys. Rep.*, **336**, 1 (2000).

High-Order Current Correlation Functions in Kondo Systems

A. Golub

Department of Physics, Ben-Gurion University of the Negev, Beer-Sheva, Israel

Abstract. We examine the statistics of current fluctuations in a junction with a quantum dot described by Kondo Hamiltonian. With the help of modified Keldysh technique we calculate the third current cumulant S_3. As a function of applied bias S_3 was obtained in three different regimes: Fermi liquid regime, crossover region and RG limit. Third cumulant shows strong non-linear voltage dependence. Only in the asymptotical limit its linear dependence on V is recovered.

Keywords: Third cumulant, Kondo effect, NCA
PACS: 72.10.Fk, 72.15.Qm, 73.63.Kv

THIRD CUMULANT

The direct electron transport is an important tool to study small junctions and quantum dots. The investigation of the current correlation functions helps to get additional information about physical properties of such systems. Interest in the third and higher moments has occurred, first, because its characteristics differ significantly from that of the second moment. In particular, [1] in non-interacting systems the third moment is insensitive to sample thermal noise, yet is more sensitive to the environment. Recently the first experimental study of the third current cumulant in mesoscopic tunnel junctions was reported [2]. The theory of full counting statistic is a theoretical framework which is used to analyze statistics of charge transfer in experiment. However, when the interactions are included it is necessary to go beyond the theory of full counting statistics. Recently for interacting systems third current cumulant was defined [3, 4] as a time ordered product of three current operators on standard Keldysh 2-lines contour. This correlation function is equal to a sum of partially time ordered products on the usual one line contour. Therefore, it does not represent non-time ordered third current cumulant. The relation of this function to the full counting statistics which, in fact, considers cumulants of charge operators, is also not clear. Here with the help of a multiple time contour (see Fig.1) [5] we provide a derivation of non-time ordered third momentum for the Kondo sistem. Hamiltonian for quantum dot in a junction is $H = H_L + H_R + H_J$ where

$$H_J = \sum_{\alpha\alpha'\sigma\sigma'} J_{\alpha\alpha'} c^\dagger_{\alpha\sigma}(0)(\frac{1}{4}\delta_{\sigma,\sigma'} + \vec{S}\vec{s}_{\sigma\sigma'})c_{\alpha'\sigma'}(0) \quad (1)$$

is Kondo hamiltonian. The first two terms correspond to non-interacting electrons in the two leads and $c_{\alpha\sigma}$ is the electron field operator of a lead. Index $\alpha = L, R$ indicates left (right) lead. We assume that the leads are dc-biased by applied voltage V. Here s is the one half spin matrix which acts on spin index of electron operators and S is the spin

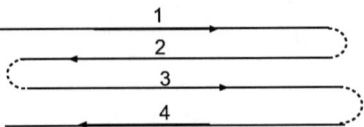

FIGURE 1. The four-axis time contour

operator of the dot. The bare coupling constants in (1) can be obtained by Schrieffer-Wolff transformation from the parent Anderson Hamiltonian and are related to the parameters of this Hamiltonian as $J_{\alpha\alpha'} = \sqrt{\Gamma_\alpha \Gamma_{\alpha'}}/(\pi v |\tilde{\varepsilon}|)$ and $\tilde{\varepsilon} \equiv (U - |\varepsilon|)\varepsilon/U$. Here U is the repulsive Hubbard coupling, ε denotes bare level energy of the dot and v is the density of states in a lead. The widths $\Gamma_{L,R} = 2\pi v |v_{L,R}|^2$ are expressed in terms of tunnelling matrix elements and density of electron states. We define the third momentum as a symmetric combination of non-time ordered correlation functions of three current operators. In a stationary situation this function can be written as

$$\tilde{S}_3(t_1 - t_2, t_2 - t_3) = \frac{1}{6} \sum_{P(\alpha\beta\gamma)} <I(t_\alpha)I(t_\beta)I(t_\gamma)> \quad (2)$$

where $P(\alpha\beta\gamma)$ is the permutation of 1,2,3. The Fourier-transformed value of (2) $\tilde{S}_3(\omega_1, \omega_2)$ is a function of two energy variables. Below we consider the zero frequency limit $\tilde{S}_3(\omega_1 = 0, \omega_2 = 0) \equiv \tilde{S}_3$ and introduce the third order cumulant S_3, that is, the irreducible part of correlation function (2). This cumulant yields the equation $S_3 = \tilde{S}_3 - 3\bar{I}S_2 + 2\bar{I}^3$. Here $\bar{I}$ is the averaged current and S_2 stands for the pair current correlation function (shot noise). In the Kondo regime this function was calculated [6].

To apply perturbation theory to a product of non-time ordered Heisenberg operators like S_3 we need to order these operators on some multiple time contour. For two operators (shot noise) the standard 2-lines Keldysh contour is sufficient. However, to arrange more then two current operators additional time axis must be included. S_3 is described by 4 lines contour (see Fig.1). [5]. The action consists of the integration over this four lines path. Alternatively, we can use one infinite time path, though, with a four independent field operators, corresponding to the operators on each four branches (Fig.1). Thus we have:

$$S = \frac{i}{\hbar} \int_{-\infty}^{\infty} dt [\sum_{k\alpha\sigma} \hat{c}^\dagger_{\alpha\sigma,k} \hat{G}_{k\alpha}^{-1} \hat{c}_{\alpha\sigma,k} - \sum_{j=1}^{4} \hat{\sigma}_z^j \hat{H}_J^j] \quad (3)$$

Here $\hat{c}^\dagger_{\alpha\sigma,k}, \hat{c}_{\alpha\sigma,k}$ are the operators in $4 \otimes 4$ Keldysh space, $\hat{G}_{k\alpha}$ denotes $4 \otimes 4$ Green's function of noninteracting leads; $\sigma_z^j = 1$ if j=1,3 and $\sigma_z^j = -1$ if j=2,4. The interaction (1) now acquires Keldysh index j. On this four fold contour the correlation function is represented by time ordered product of three current operators

$$\tilde{S}_3(t_1 - t_2, t_2 - t_3) = \frac{1}{6} \sum_{P(ijk)} <TI^i(t_1)I^j(t_2)I^k(t_3)>$$

Here superscripts i,j,k are corresponding Keldysh indices of the Heisenberg operators (electrons and spin $\vec{S}$) related to the first three paths in Fig.1 (each index runs from 1 to

3). Summation over all permutations P of these indices is performed. For temperatures $T \gg T_K$ ($T_K \simeq D_0 \exp[-1/(g_{LL}+g_{RR})]$, D_0 are the Kondo temperature and the effective bandwidth, correspondingly) logarithmic corrections which appear in perturbation theory can be disregarded. In the lowest (fourth) non-vanishing order of perturbation theory in $J_{\alpha,\alpha'}$ S_3 as a function of source-drain voltage is presented by inset in Fig.2 where $\underline{S} = S_3 e/T \sigma_B^2$ and $\sigma_B = \pi e^2 g_{LR}^2/\hbar$ is the conductance in the Born approximation.

The logarithmic divergences appear in the next (fifth) order expansion in couplings $g_{\alpha\alpha'}$. We compute S_3 to this order only for $T = 0$. In this case if voltages $eV \gg T_K$ the perturbation theory still can be used. On the level of 'poor man's scaling technique in the zero temperature limit we get

$$S_3(T=0) = \pi V [\frac{1}{8}\sigma_B^2 + 3\sigma_B \sigma(V) + \frac{2}{3}\sigma^2(V)] \tag{4}$$

where $\sigma(V) = 3\pi e^2 \Gamma_L \Gamma_R / [2\sqrt{\hbar}(\Gamma_L + \Gamma_R) \ln(eV/T_K)]^2$. The potential scattering contribution to conductance σ_B is not changed under RG transformations and renders a small correction to S_3.

FERMI LIQUID REGIME

To study regime where $T, eV < T_K$ we apply the mean field slave boson approximation (MFSB) to the Anderson hamiltonian [7]. In the slave-boson approach, the localized electron operator $d_\sigma^\dagger$ is represented by $\hat{f}_\sigma^\dagger \hat{b}$ with $\hat{b}$ and $\hat{f}_\sigma^\dagger$ being the standard boson and fermion operators. For a symmetric tunnelling the effective action is similar to (3) and can be written as $S_{eff}(\gamma) = S_b + S_f$ where the first term is nonoperator bosonic part of the action, while the last one explicitly depends on source fields $\gamma^k(t)$ and is given by

$$\begin{aligned} S_f(\gamma) &= \sum_\sigma \int dt \int dt' \hat{f}_\sigma^\dagger [G_{f\sigma}^{-1}(\gamma)] \hat{f}_\sigma \\ G_{f\sigma}^{-1}(\gamma) &= G_{0f\sigma}^{-1} - T_k(Q^+ G_{L\sigma} Q^- + Q^- G_{R\sigma} Q^+) \\ G_{0f\sigma}^{-1}(\omega) &= (\omega - \tilde{\varepsilon})\hat{\sigma}_z, \quad Q_k^\pm = \hat{\sigma}_z^k \pm \frac{ie}{2\hbar}\gamma^k(t) \end{aligned} \tag{5}$$

Here $G_{L,R}$ represents $4 \otimes 4$ matrix of the electrons propagators in the leads, $G_{0f\sigma}(\omega)$ is the Fourier transform of zero order slave fermions Green's function and $\hat{\sigma}_z$ is diagonal $4 \otimes 4$ matrix with elements $\hat{\sigma}_z^j$ (3). We also define $T_k = \Gamma b^2$ as an effective Kondo temperature. The $G_{f\sigma}(\omega)$ includes the Lagrange multiplier which shifts the localized level position ε to $\tilde{\varepsilon} = \varepsilon + \lambda$. Both free parameters T_k and renormalized level $\tilde{\varepsilon}$ are self-consistently determined by two equations that define the extremum of S_{eff} relative to b and $\tilde{\varepsilon}$ when $U \to \infty$ [7].

We can trace out the slave fermions and get a closed form for generating functional: $Z = \exp[-\frac{i}{2\hbar} Tr \ln G_f^{-1}(\gamma)]$. Then S_3 is obtained by taking the third variation of the logarithm of this functional on source fields. If $T \to 0$ then expression for S_3 is reduced to

$$S_3 = \frac{2}{\pi} e^3 \int_0^{eV} d\omega T^2(\omega)(1 - T(\omega)) \tag{6}$$

where $T(\omega) = \pi T_k \rho(\omega)$ and the spectral density of interacting level $\pi \rho(\omega) \equiv -ImG_d^r(\omega) = T_k/[(\omega - \tilde{\varepsilon})^2 + T_k^2]$ is introduced. Thus, we obtain third cumulant S_3 as a function of two dimensionless parameters eV/T_k and $\tilde{\varepsilon}/T_k$. A simple approximate expression for S_3 follows when these parameters are small $S_3 \approx 2e^6V^3/(3\pi T_k^2)$. When this parameters are not small we solve the mean field equations of MFSB theory) (see Fig2 for numerical result). The crossover regime $(T, eV \sim T_k)$ actually is extended to a several threshold values T_K. This region can also be studied within the present theory by using non-crossing approximation in non-equilibrium (NCA). A satisfied estimation of the 3rd noise in this case can be achieved by calculating equation (6) with a dot density of states which is derived in NCA : $T_K \rho(\omega) \to \Gamma \rho_{NCA}(\omega)$ (see Fig.2).

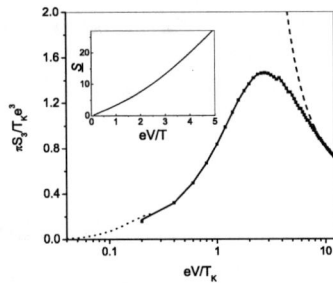

FIGURE 2. The third cumulant S_3 versus applied voltage eV/T_K. Dot curve represents MFSB calculations valid for small voltage $eV < T_K$. Solid line with black squares is obtained in NCA approximation. Dash line is the result of direct computation of Eq.(4). The curves which were calculated for two T_K differing by a factor of 10 practically coincide. This reflects the universality of 3rd cumulant in considered regime. (*inset*): Normalized 3rd cumulant $\underline{S}$ as function of eV/T in the high temperature limit $T \gg T_K$.

In conclusions, we have represented a general theory which allows to describe current fluctuations in quantum dot in the presence of interactions. Restricting ourselves to zero frequency we were able by our approach to cover three important regions: one is the in Fermi liquid limit, crossover regime, and weak coupling Kondo limit. The measurement of third momentum can be important for point junctions in the regime which exhibits 0.7 conductance plateau. Third cumulant is less sensitive to thermal fluctuations and therefore can be important tool for detecting the Kondo effect which may be relevant for 0.7 anomaly [8].

REFERENCES

1. L. S. Levitov and M. Reznikov, cond-mat/0111057.
2. B.Reulet, J. Senzier, and D.E. Prober, PRL., **91**, 196601 (2003).
3. D.B. Gutman and Yu. Gefen, Phys. Rev. B **68**, 035302 (2003).
4. A.V. Galaktionov, D.S. Golubev, and A.D. Zaikin, cond-mat/0403464; Phys. Rev. B **68**, 235333 (2003).
5. S.M. Kogan, Phys. Rev. A **44**, 8072 (1991).
6. Y. Meir, A. Golub, Phys. Rev. Lett. 88, 116802 (2002).
7. A.C. Hewson, The Kondo Problem to Heavy Fermions (Cambridge University Press, 1993).
8. Y. Meir *et al*, Phys. Rev. Lett. 89, 196802, 2002

System Size Stochastic Resonance from the Viewpoint of the Nonequilibrium Potential

Horacio S. Wio[*,†]

Instituto de Física de Cantabria, Universidad de Cantabria-CSIC, E-39005 Santander, Spain and Centro Atómico Bariloche and Instituto Balseiro, S.C. Bariloche, Argentina

Abstract. We study the phenomenon of system size stochastic resonance within the nonequilibrium potential's (NEP) framework. We analyze a simple spatially extended system and show that through the analysis of the NEP we can obtain a clear physical interpretation of this phenomenon in a wide class of extended systems.

Keywords: Stochastic resonance, nonequilibrium potential, extended systems
PACS: 05.40.+j,02.50.-r,87.10.+3

In almost all studies of *stochastic resonance* (SR) [1] the control variable was the noise intensity, while system's size didn't play a relevant role. However, some recent studies on biological models [2, 3] have shown that ion concentrations along cell membranes show intrinsic SR-like phenomena when varying the number of ion channels, with related results in sets of coupled excitable FitzHugh-Nagumo units [4]. This phenomenon, called *system size stochastic resonance* (SSSR), has also been found in globally coupled units described by a ϕ^4 theory [5], and in opinion formation models [6].

In a recent series of papers [7, 8, 9, 10] SR in extended systems was studied exploiting the concept of *nonequilibrium potential* (NEP). This is a Lyapunov functional of the associated deterministic system which, for nonequilibrium systems, plays a role similar to that of a thermodynamic potential in equilibrium thermodynamics [11, 12]. Such a NEP characterizes the global properties of the dynamics. A *mini-review* on the application of this approach to the study of SR in reaction-diffusion systems could be found in [10].

As SSSR occurs in extended systems, it is of extreme interest to describe this phenomenon within the NEP framework. With this goal in mind, we analyze the SSSR phenomenon exploiting the notion of NEP within the context of a simple reaction–diffusion model. The specific model we shall focus on, with a known form of the NEP, corresponds to a one-dimensional, one-component model [13, 14]. We emphasize that this model, with a piecewise linear form, **mimics** general bistable reaction–diffusion models [13] with a **general cubic like** nonlinear reaction term. In particular we will exploit some of the results for the influence of general boundary conditions [15] as well as previous studies of the NEP [12] and of SR [7, 8].

The particular non-dimensional form of the model that we work with is [15, 7, 8]

$$\frac{\partial}{\partial t}\phi = \frac{\partial^2}{\partial y^2}\phi + \Omega(\phi). \tag{1}$$

Where $\Omega(\phi) = -\phi + \phi_h \theta(\phi - \phi_c)$. We consider here a class of stationary structures $\phi(y)$

in $y \in [-y_L, y_L]$ with albedo boundary conditions (abc): $\frac{d\phi}{dy}|_{y=\pm y_L} = \mp k\phi(\pm y_L)$ ($k > 0$ is the albedo parameter). These are the spatially symmetric solutions to Eq.(1) already studied in [15]. The explicit forms of these stationary patterns are given by Eq. (9) of [15] and typical forms are shown in Fig. 4 in [15]. In the present case the NEP reads [12]

$$\mathscr{F}[\phi, k, y_L] = \int_{-y_L}^{y_L} \left\{ -\int_0^{\phi(y,t)} \Omega(\phi')\, d\phi' + \frac{1}{2}\left(\frac{\partial \phi(y,t)}{\partial y}\right)^2 \right\} dy + \frac{k}{2}\phi(y,t)^2 \bigg|_{\pm y_L}, \quad (2)$$

with $\gamma(k,y) = \sinh(y) + k\cosh(y)$, and $z = 1 - 2\phi_c/\phi_h$ ($-1 < z < 1$). Replacing the forms of the stationary nonhomogeneous solutions (see Eq.(9) in [15]), the explicit expression for $\mathscr{F}^{u,1} = \mathscr{F}[\phi_{u,1}, k, y_L]$ was obtained [7], while for $\phi_0 = 0$, we have $\mathscr{F}[\phi_0, k, y_L] = \mathscr{F}^0 = 0$ [7]. The expression for the double-valued coordinate y_c, at which $\phi = \phi_c$, is given by Eq. (11) in [15]. When $y_c^\pm$ exists and $y_c^\pm < y_L$, this pair of solutions represents a structure with a central *excited* zone ($\phi > \phi_c$) and two lateral *resting* regions ($\phi < \phi_c$). For each parameter set, there are two stationary solutions. It has been shown [15] that the structure with the smallest *excited* region ($\phi_u(y)$) is unstable, whereas the other one ($\phi_1(y)$) is linearly stable. The linearly stable homogeneous solution $\phi_0(y) = \phi_0 = 0$ always exists. These are the only stable stationary structures under the given abc.

Figure 1, part (a) depicts $\mathscr{F}[\phi, k, y_L]$ as a function of y_L, for fixed values of k and z (or ϕ_c/ϕ_h). The curves correspond to $\mathscr{F}^{u,1}$, whereas the horizontal line stands for $\mathscr{F}^0 = 0$. We focused on the bistable zone, the upper branch being the NEP of $\phi = \phi_u$, where $\mathscr{F}$ attains a maximum, while in the lower branches (for $\phi = \phi_0$, $\phi = \phi_1$), the NEP has local minima. We see that when y_L becomes small, the difference between the NEP for the states $\phi_u(y)$ and $\phi_1(y)$ reduces until, for $y_L \approx 0.72$, they coalesce and disappear (inverse saddle-node bifurcation). In part (b) of Fig. 2 we show $\mathscr{F}[\phi, k, y_L]$ but now as a function of k, for a fixed value of y_L and the same value of z. The initial difference between the NEP for $\phi_u(y)$ and $\phi_1(y)$ reduces for increasing k until, for $k \to \infty$, the values for Dirichlet bc are asymptotically reached. It is worth noting that, since the NEP for ϕ_u is $\mathscr{F}^u > 0$ and $\mathscr{F}^1 < 0$ for y_L large enough, and $\mathscr{F}^1 > 0$ for small values of y_L, $\mathscr{F}^1 = \mathscr{F}^0 = 0$ for an intermediate value $y_L = y_L^*$, point where the stable structures $\phi_1(y)$ and $\phi_0(y)$ exchange their relative stability.

In order to account for the effect of fluctuations, we include into Eq. (1) a fluctuation term $\xi(y,t)$ modeled as an additive noise source [10, 16], yielding a stochastic partial differential equation for the random field $\phi(y,t)$. We assume that the fluctuation term $\xi(y,t)$ is a Gaussian white noise with zero mean and a correlation function $\langle \xi(y,t)\xi(y',t')\rangle = 2\gamma\delta(t-t')\delta(y-y')$, γ being the noise strength.

As was discussed in [7, 8, 9, 10], using known results for activation processes in multidimensional systems [17], we estimate the activation rate using a Kramers' like result for the first-passage-time between attractors, $\langle \tau_i \rangle = \tau_0 \exp\{\Delta\mathscr{F}^i[\phi, k]/\gamma\}$, where $\Delta\mathscr{F}^i[\phi, k, y_L] = \mathscr{F}[\phi_u(y), k, y_L] - \mathscr{F}[\phi_i(y), k, y_L]$ ($i = 0, 1$). τ_0 is usually determined by the curvature of $\mathscr{F}[\phi, k, y_L]$ at its extreme and typically is, in one hand, several orders of magnitude smaller than the average time $\langle \tau \rangle$, while on the other does not change significantly when changing the system's parameters. Hence, in order to simplify the analysis, we assume that τ_0 is constant and scale it out of our results.

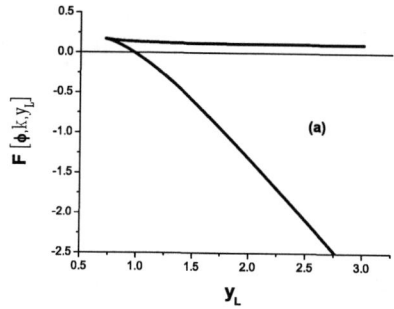 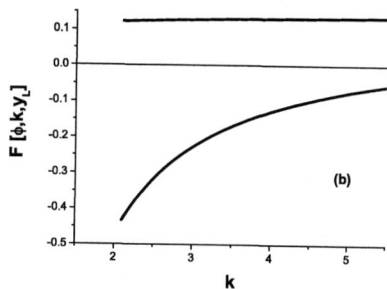

FIGURE 1. NEP evaluated at the stationary solutions $\phi_0(y)$, $\phi_1(y)$ and $\phi_u(y)$: (a) $\mathscr{F}[\phi,k,y_L]$ vs. y_L, with $k=3$; (b) $\mathscr{F}[\phi,k,y_L]$ vs. k, with $y_L=1.2$. In both cases $\phi_c/\phi_h=0.193$.

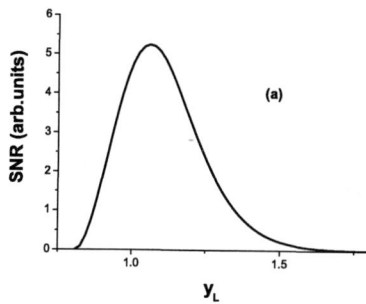 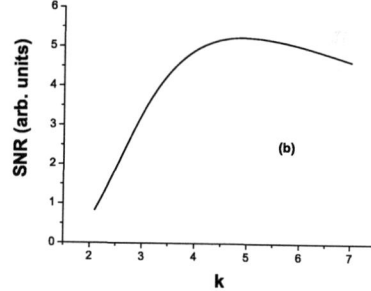

FIGURE 2. SNR, vs.: (a) y_L (for $k=3$.), (b) k (for $y_L=1.2$). In both cases $\gamma=0.1$, $\phi_c/\phi_h=0.193$.

We assume that the system is (adiabatically) subject to an external harmonic variation $\phi_c(t) = \phi_c + \delta\phi_c \cos(\omega t)$ [8, 10], and exploit the *two-state approximation* [1] as in [8, 9, 10], for details we refer to [9]. Up to first-order in the amplitude $\delta\phi_c$ (assumed to be "sub-threshold") the transition rates adopt the form $W_i = \tau_0^{-1} \exp[-\Delta\mathscr{F}^i[\phi,k,y_L]/\gamma]$, $\Delta\mathscr{F}^i[\phi,k,y_L] = \Delta\mathscr{F}^i[\phi,k,y_L] + \delta\phi_c[\partial\Delta\mathscr{F}^i[\phi,k,y_L]/\partial\phi_c]_{\phi_c=\phi_c^*} \cos(\omega t)$. This yields $W_i \simeq (1/2)[\mu_i \mp \alpha_i(\delta\phi_c/\gamma)\cos(\omega t)]$, where $\mu_i \approx \exp[-\Delta\mathscr{F}^i[\phi,k,y_L]/\gamma]$ and $\alpha_i \approx \pm\mu_i[d\Delta\mathscr{F}^i/d\phi_c]_{\phi_c}$ ($i=1,2$). These results allows us to get all the information to obtain the SNR, the details of the calculation can be found in [9]. We have now all the elements required to analyze the problem of SSSR.

Figure 2.a shows the typical behavior of SR as function of the system length y_L, for fixed values of k, γ and ϕ_c/ϕ_h. This is the typical response for a system exhibiting SSSR. Within the context of NEP, it results clear that the phenomenon arises due to the breaking of the NEP's symmetry: both attractors exchange their relative stability due to the variation of y_L as shown in Fig. 1.a. Hence, in this case, SSSR arises as a particular case of the more general discussion in [9]. In Fig. 2.b we show the curves of the SNR as a function of k, a parameter that indicates the degree of coupling with the environment, while keeping fixed values of y_L, and z. When k is not too large (high degree of reflectiveness at the boundary or reduced exchange with the environment), we see that the SNR changes for k varying from small to large values (a large k indicates that the boundaries become absorbent). The *robustness* of the systems' response when changing k is apparent. From Fig. 1.b, and according to the previous argument about the breaking of NEP's symmetry, this is the expected result.

It was also shown that the models discussed in [5] can be cast into the same NEP approach. The above results show that the NEP (even if not known in detail [18]) offers an extremely useful framework to analyze a wide spectrum of aspects associated to extended systems. Within this framework the phenomenon of SSSR and other aspects of SR in extended systems [9], can be clearly understood. In addition, we have seen a new form of resonant behavior through the variation of its coupling with the surroundings. In this case the system's response to an external signal becomes more *robust* or less sensitive to the precise value of the albedo parameter. All this aspects, together with a detailed analysis of cases discussed in [5] and of situations where the patterns (attractors) are noise-induced, will be discussed elsewhere [19].

ACKNOWLEDGMENTS

The author thanks G. Izús, B. von Haeften, R. Toral, C. Tessone for fruitful discussions, and thanks the European Commission for the award of a *Marie Curie Chair*.

REFERENCES

1. L. Gammaitoni, P. Hänggi, P. Jung and F. Marchesoni, Rev. Mod. Phys. **70**, 223 (1998).
2. G. Schmid, I. Goychuk, P. Hänggi, Europhys. Lett. **56**, 22 (2001); and Phys. Biol. **1**, 61 (2004).
3. P. Jung, J.W. Shuai, Europhys. Lett. **56**, 29 (2001); and Phys. Rev. Lett. **88**, 068102 (2003).
4. R. Toral, C. Mirasso, J. Gunton, Europhys. Lett. **61**, 162 (2003).
5. A. Pikovsky, A. Zaikin, M.A. de la Casa, Rev. Lett. **88**, 050601 (2002).
6. C. J. Tessone, R. Toral, Physica A **351**, 106 (2005).
7. H. S. Wio, Phys. Rev. E **54**, R3075 (1996).
8. F. Castelpoggi and H. S. Wio, Europhys. Lett. **38**, 91 (1997); and Phys. Rev. E **57**, 5112 (1998).
9. S. Bouzat and H. S. Wio, Phys. Rev. E **59**, 5142 (1999).
10. H. S. Wio, S. Bouzat and B. von Haeften, Proc. STATPHYS-21, Physica A **306C** 140 (2002).
11. R.Graham, in *Instabilities and Nonequilibrium Structures*, Eds.E.Tirapegui and D.Villaroel (D.Reidel, Dordrecht, 1987); H. S. Wio, in *4th. Granada Seminar in Computational Physics*, Eds. P. Garrido and J. Marro (Springer, Berlin, 1997).
12. G. Izús *et al*, Phys. Rev. E **52**, 129 (1995).
13. H. S. Wio, *An Introduction to Stochastic Processes and Nonequilibrium Statistical Physics* (World Scientific, 1994), chp.5.
14. W.J. Skocpol, M.R. Beasley, M. Tinkham, J. Appl. Phys. **45**, 4054 (1974); B. Ross, J.D. Lister, Phys. Rev. A **15**, 1246 (1977); D. Bedeaux, P. Mazur, Physica A **105**, 1 (1981).
15. C.L. Schat and H.S. Wio, Physica A **180**, 295 (1992).
16. J. García-Ojalvo and J. M. Sancho, *Noise in Spatially Extended Systems* (Springer, New York, 1999).
17. P. Hänggi, P. Talkner and M. Borkovec, Rev. Mod. Phys, **62**, 251 (1990).
18. H.S. Wio, M.Kuperman, F. Castelpoggi, G. Izús and R. Deza; Physica A **257**, 275 (1998)
19. B. von Haeften, G. Izús and H.S. Wio, *System Size Stochastic Resonance: General Nonequilibrium Potential Framework*, submitted to Phys. Rev. E (2005), cond-mat/0504131.

Extensions of the Stochastic Model of the Overdamped Oscillator Applied to AC Ionic Conductivity in Solids

Juan Bisquert

Departament de Ciències Experimentals, Universitat Jaume I, 12071 Castelló, Spain
bisquert@uji.es

Abstract. Several extensions of the overdamped stochastic oscillator applied to ac ionic conduction in solids are discussed and compared. A shift of the equilibrium position towards the particle position introduces long range displacement (dc conductivity) absent in the simple oscillator. Time correlations in the friction modify the high frequency conductivity. Time correlations in the elastic force produce a negative conductivity at low frequency.

Keywords: ac conduction, disordered solids, Langevin equation, memory effect.
PACS: 72.80.Ng, 78.55.Qr

INTRODUCTION

In many different kinds of solids with disordered structures such as glasses, structurally disordered crystals, and polymers the ionic ac conductivity displays the same qualitative features. At low frequencies, the long range displacement of carriers gives the frequency independent (dc) conduction σ_0. At increasing frequencies, the conductivity increases approximately as a power law.[1]

In ionic conduction the free carrier motion is limited by impacts with the lattice atoms in thermal motion that is represented by stochastic force and damping term in the Langevin equation. The Coulomb interactions of a carrier with the other carriers forms an "ionic atmosphere" around the central carrier. Under the influence of an external force, the central carrier may be pictured as experiencing a restoring force and the motion is described as an overdamped stochastic oscillator.[2] Here we consider the effect of the relaxation of the ionic atmosphere by stating an additional equation for the displacement of the potential towards the particle position. The result of memory terms both in the friction force and elastic force will be considered as well.

The stochastic oscillator is defined by the Langevin equation for a Brownian particle with spring constant k. If the inertia term can be neglected, the equation of motion for the displacement x of a particle of charge q in an electric field E is

$$-\zeta \frac{dx}{dt} - kx + qE + f = 0 \qquad (1)$$

where f is the stochastic force related to the friction coefficient ζ through the fluctuation-dissipation theorem, $<f(t)f(0)>=k_BT\zeta$. The conductivity can be obtained from the Kubo formula, and also from the deterministic equation ($f=0$) as $\sigma(\omega)=Nq\tilde{v}/\tilde{E}$, considering the velocity $\tilde{v}=i\omega\tilde{x}$ produced by a small field $E=\tilde{E}\exp(i\omega t)$ of pulsation ω. Taking the Laplace transform of Eq (1), one obtains the result

$$\sigma(\omega)=\sigma_\zeta \frac{i\omega}{\omega_\zeta+i\omega} \qquad (2)$$

where $\omega_\zeta=k/\zeta=\tau_\zeta^{-1}$ is the inverse of the decay time for return to the equilibrium position. Eq. (2) is shown in Fig. 1(a) and corresponds to the standard Debye relaxation. At low frequencies the displacement is slowed down by the restoring force and σ' decreases to zero as ω^2. At high frequencies ($\omega >> \omega_\zeta$) the ac conductivity σ' shows a plateau corresponding to the value $\sigma_\zeta=Nq^2/\zeta$. In this high frequency regime the particle behaves as if it were moving freely around the potential minimum under the influence of the viscous force only.

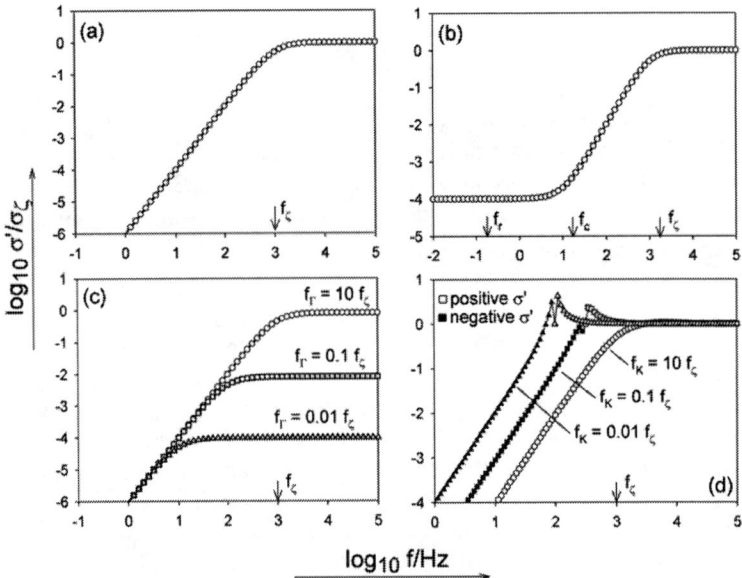

FIGURE 1. Representation of the real part of ac conductivity $\sigma'(\omega)$ vs. frequency for an overdamped oscillator, indicating the frequency $f_\zeta=\omega_\zeta/2\pi=10^3$ s^{-1}. (a) Ordinary overdamped oscillator. (b) Model with shifting equilibrium position. (c) Model with memory in the friction, representation parametric in $f_\Gamma=\omega_\Gamma/2\pi$. (c) Model with memory in the elastic force, representation of absolute value of $\sigma'(\omega)$ parametric in $f_K=\omega_K/2\pi$.

MODELS

Let us consider some extensions of the model of Eq. (2) that describe additional effects of interaction in the process of ionic conduction.

Relaxation of the equilibrium position

We consider the rearrangement of the potential surface during the motion of the particle.[3] When the hopping carrier is displaced from the equilibrium position in the ionic atmosphere, either it may return to the equilibrium position pulled by the restoring force, or a rearrangement of the surrounding charge towards the new position of the carrier may occur. This last effect has been amply discussed in the jump relaxation model by Funke and his co-workers.[4] If we denote by x_0 the equilibrium position of the potential well, the elastic force in Eq. (1) becomes $-k(x-x_0)$. The equation of motion for the centre of the surrounding charge, is

$$\frac{dx_0}{dt} = -\frac{1}{\tau_r}(x_0 - x) \qquad (3)$$

where $\tau_r = \omega_r^{-1}$ is the characteristic time for rearrangement of the potential well to the new position. The ac conductivity in this model, shown in Fig. 1(b), is[3]

$$\sigma(\omega) = \sigma_\zeta \frac{\omega_r + i\omega}{\omega_\zeta + \omega_r + i\omega} \qquad (4)$$

The general behaviour can be understood in terms of two types of motions: free carrier motion at high frequencies determined by microscopic friction, and the much slower long range motion of a carrier accompanied by the confinement cloud at low frequencies, with a dc conductivity $\sigma_0 = \sigma_\zeta/(1 + \omega_\zeta/\omega_r)$. The onset of ac conduction is at the frequency $\omega_c = (\omega_\zeta \omega_r)^{1/2}$.

Relaxation with memory in the friction

The dynamic effects of many-particle interactions are often taken into account with a memory function in the viscous drag, through the generalized Langevin equation (GLE)[5] that describes temporal correlations in the interaction between the particle and the medium. The GLE

$$-\int_0^\infty \Gamma(t-t')\frac{dx}{dt}(t')dt' - kx + qE + f = 0 \qquad (5)$$

has been applied to describe the microscopic mechanism of fast ionic conduction in solids.[6,7] The memory $\Gamma(t)$ is related to the stochastic force through the fluctuation-dissipation theorem as $<f(t)f(0)> = k_B T \Gamma(t)$. For an exponentially decaying kernel $\Gamma(t) = \zeta e^{-\omega_\Gamma t}$, the ac conductivity in this model, shown in Fig. 1(c), is

$$\sigma(\omega) = \sigma_\zeta \frac{i\omega}{\omega_\zeta + i\omega/(1+i\omega/\omega_\Gamma)} \qquad (6)$$

The time correlation in the friction modifies the high frequency limit of the conductivity provided that $\omega_\Gamma < \omega_\zeta$.

Relaxation with memory in the elastic force

Let us consider a fluctuation of the elastic potential that depends on the previous position of the particle, as described by the equation

$$-\zeta \frac{dx}{dt} - \int_0^\infty K(t-t')x(t')dt' + qE = 0 \qquad (7)$$

This model is similar to that based on GLE, Eq. (6), however we do not have a stochastic foundation for the model of Eq. (7), which is presented as a phenomenological ansatz. For an exponentially decaying kernel $K(t) = ke^{-\omega_K t}$, the ac conductivity in this model, shown in Fig. 1(d), is

$$\sigma(\omega) = \sigma_\zeta \frac{i\omega}{i\omega + \omega_\zeta/(1+i\omega/\omega_K)} \qquad (8)$$

At low frequencies the relaxation of the force introduces a *negative friction* which overcomes the positive component provided that $\omega_K < \omega_\zeta$. This negative friction supplies power to the external circuit and causes a negative conductivity. The relationship of this model with brownian motors, and also with the stochastic models with random elastic force,[8,9] needs to be further investigated.

REFERENCES

1. J. C. Dyre and T. B. Schroder, *Rev. Mod. Phys.* **72,** 873 (2000).
2. F. Vauchot, F. Henn, and J. C. Giuntini, *J. Non-Cryst. Solids* **307-310**, 1060 (2002).
3. J. Bisquert, V. Halpern, and F. Henn, *J. Chem. Phys.* **122**, 151101 (2005).
4. K. Funke and R. Hoppe, *Solid State Ionics* **40-41,** 200 (1990).
5. R. Zwanzig, *J. Stat. Phys.* **9**, 215 (1973).
6. M. A. Olson and S. A. Adelman, *J. Chem. Phys.* **83**, 1865 (1985).
7. P. Brüesch, L. Pietronero, S. Strässle, and H. R. Zeller, *Phys. Rev. B* **15**, 4631 (1977).
8. M. Gitterman, *Phys. Rev. E.* **67**, 057103 (2003).
9. P. Chovsta, and F. Slanina, *J. Phys. A: Math. Gen.* **35**, L277 (2002).

Fluctuation-Dissipation Dispersion Relation for Systems with Slowly Varying Parameters

V. Belyi

IZMIRAN, Russian Academy of Sciences, Troitsk, Moscow region, 142190, Russia and International Solvay Institutes for Physics and Chemistry, ULB-CP231, 1050, Brussels, Belgium
sbelyi@izmiran.ru

Abstract. A generalization of the fluctuation-dissipation formula for systems with slowly varying parameters is given using the Langevin approach and momentum method. It is shown that spectral function of the fluctuations in these systems is determined not only by the dissipation but also by the derivations of the dispersion. The non Joule dispersion contribution is characterized by a new nonlocal effect originating from an additional phase shift between the force and response of the system. That phase shift results from the parametric control to the system. The general formalism is illustrated for an oscillating electrical circuit. It is shown that in that systems the dispersive contributions strongly affect the quality factor.

Keywords: Fluctuation-dissipation theorem; Non-Marcovian processes.
PACS: 05.40.Hp; 05.40.-a; 05.10.Gg; 07.50.Hp

Fluctuations play an important role in the formation of the dissipative structures [1-3], in the sensitivity of the devices; find applications in diagnostic procedures [4-5]. In thermodynamic equilibrium, fluctuations are determined by the system temperature and the dissipation. The first fluctuation-dissipation relation between the diffusion coefficient and the dissipative friction coefficient was derived in early report independently by Einstein and Smoluchowski [6] in the theory of Brownian motion. Later, this relation was established by Nyquist [7] for electric circuits and experimentally confirmed by Johnson [8]. The Nyquist-Johnson relation was extended by Callen and Welton [9] to a general class of dissipative thermodynamic equilibrium systems. In the classical case the spectral function of the fluctuations of the variable x has the form:

$$\left(x^2\right)_\omega = \frac{2T}{\omega}\operatorname{Im}\alpha(\omega), \qquad (1)$$

where $\alpha(\omega)$ - is the response function, and T- is the temperature in energy units. For electrostatic field fluctuations in plasma, with $x = \delta\mathbf{E}$ and $\alpha(\omega) = -4\pi/\varepsilon(\omega,\mathbf{k})$, where $\varepsilon(\omega,\mathbf{k})$ is dielectric function. The corresponding fluctuation-dissipation relation is

$$\left(\delta\mathbf{E}\,\delta\mathbf{E}\right)_{\omega,\mathbf{k}} = \frac{8\pi T \operatorname{Im}\varepsilon(\omega,\mathbf{k})}{\omega|\varepsilon(\omega,\mathbf{k})|^2}. \qquad (2)$$

The matter becomes more delicate even in the local equilibrium case. We have indeed shown, that in the *collisional regime* the Callen-Welton formula should be revised [10]. There then appear new terms explicitly displaying dissipative nonequilibrium contributions and containing the interparticle collision frequency, the differences in the temperatures and the velocities, and also functions of the *real* parts of the dielectric susceptibilities. However, it is not evident that the plasma parameters can be kept *constant* in both space and time. Inhomogeneities in space and time of these quantities will certainly also contribute to the fluctuations. In the context of plasma physics using the Langevin approach and the time-space multiscale technique, it has been shown that the amplitude and the width of the spectral lines of the electrostatic field fluctuations are determined not only by the imaginary part of the dielectric susceptibility but also by the derivatives of its real part [11]:

$$(\delta E \delta E)_{\omega,\mathbf{k}} = \frac{8\pi T}{\omega} \frac{\operatorname{Im}\varepsilon(\omega,\mathbf{k}) + \frac{\partial^2}{\partial t \partial \omega}\operatorname{Re}\varepsilon(\omega,\mathbf{k}) - \frac{\partial^2}{\partial \mathbf{r} \partial \mathbf{k}}\operatorname{Re}\varepsilon(\omega,\mathbf{k})}{\operatorname{Re}^2\varepsilon(\omega,\mathbf{k}) + [\operatorname{Im}\varepsilon(\omega,\mathbf{k}) + \frac{\partial^2}{\partial t \partial \omega}\operatorname{Re}\varepsilon(\omega,\mathbf{k}) - \frac{\partial^2}{\partial \mathbf{r} \partial \mathbf{k}}\operatorname{Re}\varepsilon(\omega,\mathbf{k})]^2} \quad (3)$$

Similar expression is obtained for the electron form factor near the Langmiur resonance:

$$(\delta n_e \delta n_e)_{\omega,\mathbf{k}} =$$

$$\frac{k^2 T}{2\pi\omega} \frac{\operatorname{Im}\varepsilon(\omega,\mathbf{k}) + \frac{\partial^2}{\partial t \partial \omega}\operatorname{Re}\varepsilon(\omega,\mathbf{k}) - \frac{1}{k^2}\frac{\partial}{\partial r_i}k_j \frac{\partial}{\partial k_i}k_j \operatorname{Re}\varepsilon(\omega,\mathbf{k})}{\operatorname{Re}^2\varepsilon(\omega,\mathbf{k}) + [\operatorname{Im}\varepsilon(\omega,\mathbf{k}) + \frac{\partial^2}{\partial t \partial \omega}\operatorname{Re}\varepsilon(\omega,\mathbf{k}) - \frac{1}{k^2}\frac{\partial}{\partial r_i}k_j \frac{\partial}{\partial k_i}k_j \operatorname{Re}\varepsilon(\omega,\mathbf{k})]^2}. \quad (4)$$

As a result of the inhomogeneity, these properties become asymmetric with respect to the inversion of the sign of the frequency. In the kinetic regime the form factor is more sensitive to space gradients than the spectral function of the electrostatic field fluctuations. This asymmetry of lines can be used as a diagnostic tool to measure local gradients in the plasma.

For generalization of fluctuation-dissipation theorem to a general class system with slowly varying parameters we use the so called momentum method, which is alternative to the Langevin method. This method is based on the equation for two-time correlation function [12]. In general case both methods are equivalent, if one correctly determines the intensity of the Langevin source. Let us consider an arbitrary system whose evolution is described by the following equation:

$$\left(\frac{\partial}{\partial t} + \underline{\underline{L}}\right) \cdot \underline{\underline{G}}(t,t') = 0, \quad t > t' \quad (5)$$

where $L(t)-$ is generally a non self-conjugate, linear operator in the Hilbert space. This operator varies slowly in time. The term "slowly" means that the control parameter undergoes only a small change during the period of the system motion. $\underline{\underline{G}}(t,t')$ may be the Heisenberg operator. Then $\underline{\underline{L}} \cdot \underline{\underline{G}}(t,t')$ will be the commutator with the Hamiltonian. In other cases $\underline{\underline{G}}(t,t')$ could be a density matrix, and $\underline{\underline{L}}(t)$ would appear as the Liouville operator. Finally, for $\underline{\underline{G}}(t,t')$ we can take the two-time correlator $\underline{\underline{G}}(t,t') = \langle \delta f_{nm}(t) \delta f^*_{n'm'}(t') \rangle$ of the deviation from the referent state $f_n(t)$ of the density matrix in the energy representation $\delta f_{nm}(t)$. In such a case $\underline{\underline{L}}(t)$ takes into account the self-consistent field and collisions. The slow scale is much larger than the characteristic fluctuation time. We can therefore introduce a small parameter μ, which allows us to describe fluctuations on the basis of a multiple time scale analysis. Obviously, fluctuations vary on both "fast" and "slow" time scales. At first order with respect to μ the expression for the spectral function of the fluctuations in the classical limit $\hbar \to 0$ takes the form [13]:

$$(\delta A \, \delta B)_\omega = \frac{2T}{\omega} [\text{Im}\, \alpha_{AB}(\omega) + \frac{d^2}{dt d\omega} \text{Re}\, \alpha_{AB}(\omega)]. \tag{6}$$

For sake of convenience, we omit μ from that equation and throughout this communication, keeping in mind that the time derivatives are taken with respect to the slowly varying variables. When expanding the Green's function of Eq. (5) in terms of the small parameter μ, there appears an additional term at first order. It is important to note that the imaginary part of the response function is now replaced by the real part. If the quality factor of the system is of the order of 1 (it can be a broad-band system or a process near the zero frequency), the real and imaginary parts of the response function are of the same order and the correction is negligibly small. But in the case of systems with a high quality factor, for which the real part of the response function is greater than the imaginary part, the second small parameter appears to be inversely proportional to the quality factor. An example of such system with a high quality factor could be plasma fluctuations near the Langmuir frequency when the quality factor is inversely proportional to the small plasma parameter [11]. When this small parameter is comparable with μ, the second term in Eqs.(6) may have an effect comparable to the first term. This will be shown in the next example. At the second order in the expansion in μ, the corrections appear only in the imaginary part of the response function, and they can reasonably be neglected. It is therefore sufficient to retain the first order corrections to solve the problem.

As an example we consider the electrical oscillation circuit which can be used to model many oscillation processes in nature. We assume that all the circuit elements R, inductance L, and capacity C have the same temperature, which can change adiabatically. Therefore the system parameters R, L, and C will vary slowly in time. Moreover the change of these parameters may also be mechanical, due to the action of external forces, by "hand". It is this case that we will consider when evaluating the quality factor of LC-circuit. Applying the procedure above, we obtain the following expression for the spectral function of the current in the circuit.

$$\left(J^2\right)_\omega = \frac{2[\operatorname{Re} Z(\omega) + \frac{d^2}{dtd\omega}\operatorname{Im} Z(\omega)]T}{\operatorname{Im}^2 Z(\omega) + [\operatorname{Re} Z(\omega) + \frac{d^2}{dtd\omega}\operatorname{Im} Z(\omega)]^2}, \qquad (7)$$

where $Z(\omega) = R - i(L\omega - 1/C\omega)$ is the complex impedance. One can see that in the general case the spectral density of the e.m.f. for slow processes

$$(\tilde{E}^2)_\omega = 2T[\operatorname{Re} Z(\omega) + \frac{d^2}{dtd\omega}\operatorname{Im} Z(\omega)] \qquad (8)$$

depends on the frequency and is not always white noise. As the time derivative can have different signs, the dispersion corrections in Eq. (7) may both decrease and increase the line width. Therefore, at finite time intervals one can increase drastically the oscillation system quality factor by simultaneously increasing the inductance and decreasing the capacity. Similar situations can appear in other oscillating systems. These results are applicable to other systems and are important for the understanding of various behaviors observed in different field of physics, communication, chemistry and biophysics.

ACKNOWLEDGMENTS

This research was supported by the Russian Foundation for Basic Research (Grant No. 03-02-16345).

REFERENCES

1. Prigogine, I., *Science* **201**, 777 (1978).
2. Lefever, R., in *Fluctuations, Instabilities and Phase Transitions,* New York: Plenum, 1975.
3. Horsthemke, W., and Lefever, R., *Noise-Induced Transitions,* Berlin: Springer-Verlag, 1984.
4. Dougherty, J.P., and Farley, D.T., Proc. Roy. Soc. **A259**, 79 (1960).
5. Sheffield, J., *Plasma Scattering of electromagnetic Radiation,* New York: Academic, 1975.
6. Einstein, A., *Ann.Phys.* (Leipzig) **17**, 549 (1905); Smoluchowski, M.v., *Ann.Phys.* (Leipzig) **21**, 756 (1906)
7. Nyquist, H., Phys. Rev. **32**, 110 (1928).
8. Johnson, J.B., Phys. Rev. **32**, 97 (1928).
9. Callen, H.B., and Welton, T.A., Phys. Rev. **83**, 34 (1951).
10. Belyi, V.V., and Paiva-Veretennicoff, I., J. Plasma Physics **43**, 1 (1990).
11. Belyi, V.V., Phys. Rev. Lett. **88**, 255001 (2002).
12. Balescu, R., *Equilibrium and Nonequilibrium Statistical Mechanics*, New York: Wiley, 1975.
13. Belyi, V.V., Phys. Rev. E. **69**, 017104 (2004).

A New Version of the Fluctuation-Dissipation Relations

S.F. Timashev, *G.V. Vstovsky

L.Ya. Karpov Institute of Physical Chemistry, Moscow, Russia
*N.N. Semenov Institute of Chemical Physics, Russian Academy of Science, Moscow

Abstract. Flicker-Noise Spectroscopy as a new phenomenological approach for extracting information hidden in chaotic signals is presented. According to FNS, the information is provided by sequences of distinguishing types of irregularities – spikes, jumps, and discontinuities of derivatives of different orders, at all space-time hierarchical levels of systems. In this case, the power spectra $S(f)$ and difference moments $\Phi^{(2)}(\tau)$ of the 2nd order can be considered as generalization of fluctuation-dissipation relations (FDRs) introduced usually for stationary processes near equilibrium state, because $S(f)$ and $\Phi^{(2)}(\tau)$ are used in the case of strong non-equilibrium processes also. In this paper we present examples of the FDRs for non-equilibrium stationary processes (the Lèvy diffusion, isotropic fully developed turbulence).

Keywords: Flicker-Noise Spectroscopy, Fluctuation-dissipation relations, Lèvy diffusion, Fully developed turbulence.

THE MAIN EQUATIONS

A new type of information contained in chaotic time series $V(t)$ (t is a time), is introduced in the frame of Flicker-Noise Spectroscopy (FNS) [1, 2]. According to this phenomenological approach, the main information hidden in chaotic signals at an interval T is provided by sequences of distinguishing types of irregularities – spikes, jumps, and discontinuities of derivatives of different orders, at all space-time hierarchical levels of systems. FNS approach classifies the irregularities of different types by the generalized functions with zero carrier (the Dirac δ-functions and their derivatives), and the Heaviside θ-functions and functions with discontinuities of different order derivatives for potential singularities. In this case it is possible to introduce different types of information. The ability to distinguish irregularities means that the parameters or patterns characterizing the totality of properties of the irregularity sequences, are extracted from the power spectra $S(f)$ (f – frequency):

$$S(f) = \left| \int_{-T/2}^{T/2} \langle V(t)V(t+t_1) \rangle \cdot \exp(2\pi i f t_1) dt_1 \right|, \quad \langle (...) \rangle = \frac{1}{T} \int_{-T/2}^{T/2} (...) dt, \qquad (1)$$

and the difference moments $\Phi^{(p)}(\tau)$ of the pth order ($p = 1, 2, 3, ...$):

$$\Phi^{(p)}(\tau) = \left\langle [V(t) - V(t+\tau)]^p \right\rangle, \qquad (2)$$

where τ is a time delay parameter.

In this case, $\Phi^{(p)}(\tau)$ is formed exclusively by jumps of the dynamic variable at different space-time hierarchical levels of the system under consideration, and $S(f)$ is formed by spikes and jumps. In other words, the power spectra and difference moments of the 2nd order carry out different information, which complement each other. (It is different from the standard point of view, which could be adequate only for smooth signals). The characteristic information extracted from the $S(f)$ and $\Phi^{(p)}(\tau)$ dependencies are the "passport parameters" which are the correlation times, the parameters characterizing the loss of "memory" for this correlation time, characterizing the sequences of "spikes", "jumps" and discontinuities of derivatives of different orders (in the latter case time series for "quasi-derivatives" are formed).

In the case of "stationary" processes the "passport parameters" do not depend on the position of the T interval at the time axis as well as on the digital frequency. The latter is valid if a set of "invariants" K_i for every type of irregularities is introduced. These invariants K_i are characterized by a dimension of frequency $[s^{-1}]$ and have a sense of kinetic factors. Therefore, they could be introduced for exposition of the relaxation processes stipulating an establishment in a system of a stationary condition at presence of outer actions. It means that we can consider the $S(f)$ and $\Phi^{(2)}(\tau)$ dependences as a generalization of fluctuation-dissipation relations (FDRs) introduced usually for stationary processes near equilibrium state, because $S(f)$ and $\Phi^{(2)}(\tau)$ are used in the case of strong non-equilibrium processes also. In this case, the specific frequencies K_0 and K_1 are introduced to describe correlation links in the sequences of spikes and jumps correspondingly.

The simplest natural process is the establishment of a stationary condition in a system embedded in a thermostat approaching a thermodynamic equilibrium after a weak perturbation. It is possible to speak about a linear relaxation if the relaxation process could be described by one parameter, which is the Maxwellian time of a relaxation τ_M. In this case all frequency invariants K_j degenerate in an uniform Maxwellian frequency τ_M^{-1}, and are equal each other, that is $K_j \equiv K_M = \tau_M^{-1}$. The corresponding kinetic factor in the case of an electro-conducting system is $K_M^{el} = 4\zeta_{el}\pi\sigma_{el}$, where σ_{el} is the specific electro-conductivity (note, $K_M^{el} \sim 10^{13} Hz$ if $\sigma_{el}^{-1} \sim 1\ Ohm\cdot cm$) and $\zeta_{el} \sim 1$ is a dimensionless factor. For Brownian particles having a mass M, which diffuse in a medium with diffusion coefficient D, we have $K_M^D = \zeta_D \dfrac{k_B T_K}{MD}$, where k_B is the Boltzmann constant, T_K is the absolute temperature, and $\zeta_D \sim 1$ is a dimensionless factor.

In accordance with [1, 2], the contributions $S_S(f)$ and $S_R(f)$ to $S(f)$ due only to the spike-type and jump-type irregularities correspondingly are presented as:

$$S_S(f) \approx \frac{S_S(0)}{1+(2\pi f T_0)^{n_0}}, \quad S_R(f) = \int_0^\infty Cos(2\pi f\tau)\left[\Phi^{(2)}(\infty)-\Phi^{(2)}(\tau)\right]d\tau. \qquad (3)$$

Here $S_S(0)$ is an effective phenomenological parameter; n_0 characterizes the rate of "memory (correlation) loss" in a sequence of spikes within time intervals shorter than the correlation time T_0, $T_0^{-1} \equiv K_0$. The contribution $S_R(f)$ can be found using the expression for $\Phi^{(2)}(\tau)$ [1, 2]:

$$\Phi^{(2)}(\tau) = 2\cdot\sigma^p\cdot\left[1-\Gamma^{-1}(H)\cdot\Gamma(H,\tau/T_1)\right]^2, \quad \Gamma(s,x) = \int_x^\infty \exp(-t)\cdot t^{s-1}dt, \quad \Gamma(s)=\Gamma(s,0). \quad (4)$$

Here σ is the variance of the measured dynamic variable; the parameter H has the sense of the Hurst constant, which characterizes the rate at which the dynamic variable "forgets" its value within time intervals shorter than $T_1 \equiv (K_1)^{-1}$; $\Gamma(s)$ and $\Gamma(s, x)$ are the gamma- and incomplete gamma functions ($x \geq 0$ and $s > 0$), respectively. The introduced parameter $S_S(0)$, n_0, T_0, H, T_1 and σ can be considered as "passport" parameters of the stationary process under consideration.

In this case the term $S_R(f)$ can be expressed by the interpolation equation:

$$S_R(f) \approx S_R(0) \frac{1}{1+(2\pi f T_1)^{2H+1}},$$

$$S_R(0) = 4\sigma^2 T_{01} H \cdot \left\{1 - \frac{1}{2H\,\Gamma^2(H)} \int_0^\infty \Gamma^2(H,\xi)d\xi\right\}. \qquad (5)$$

It is important to note that the both contributions $S_S(f)$ and $S_R(f)$ to the power spectrum $S(f)$ are similar, although the corresponding parameters in (3) and (5) generally differ from each other: $S_R(0) \neq S_S(0)$, $T_1 \neq T_0$ and $2H \neq n_0 - 1$. It is possible to separate both contributions (3) and (5) in the "experimental" power spectra using known parameters H and T_1. Below we present $S(f)$ and $\Phi^{(2)}(\tau)$ as FDRs for two non-equilibrium stationary processes.

EXAMPLES OF FLUCTUATION-DISSIPATION RELATIONS

1. *Lèvy diffusion.* The anomalous diffusion, or Lèvy diffusion, is a stochastic process for which the root-mean-square displacement can be defined as

$$\langle (\Delta x)^2 \rangle_{pdf} = 2Dt_0 (\tau/t_0)^{2H}. \qquad (6)$$

Here D is a diffusion coefficient, t_0 is a characteristic time, and H is the Hurst index. The value $H = 1/2$ corresponds to the Fickian ("normal") diffusion. Case with $H > 1/2$ corresponds to the "enhanced" diffusion while $H_1 < 1/2$ corresponds to the diffusion with "geometric constraints". In accordance with the FNS logics, we introduce $K_0 = K_1 \equiv K_M^D = (T_1)^{-1}$ to Eq. (4), and obtain in the limit $\tau \ll T_1$ (the case of Lèvy diffusion corresponds to the small times):

$$\Phi^{(2)}(\tau) = \langle [x(t) - x(t+\tau)]^2 \rangle \equiv \langle (\Delta x)^2 \rangle \approx 2\sigma^2 \Gamma^{-2}(H+1) \cdot (\tau/T_1)^{2H}. \qquad (7)$$

Therefore: $t_0 = T_1$ and $D = \sigma^2 \Gamma^{-2}(H+1)/T_1$. The later relation can be tested if we analyze a relaxing process in the interval $T_{tot} > T_1$, when the D and T_1 values can be extracted separately. The description of the "experimental" power spectrum $S(f)$ with using Eq. (3) and Eq. (5) permits to find the parameters: n_0, T_0, $S_S(0)$.

2. *Fully developed turbulence.* Let us consider a case of a fully developed turbulence in a stream with average velocity U_0. We consider fluctuations of a local velocity $u(t)$ [cm/s] as dynamic variables to get its power spectra in the wave number space, $k = f/U_0$, and structural functions. The main feature of fluid dynamics in fully developed turbulent fluxes is the existence of the so-called inertial interval for variation of wave numbers (characteristic frequencies) of turbulent pulsations. In such range of wave number variation, energy dissipation is negligibly small. In addition, the energy flux appears independent of characteristic scales of pulsations l and it is equal to an average specific rate ε_K [cm^2/s^3] of energy dissipation. The inertial interval is

limited in the range of smaller wave numbers to the inverse magnitude of outer scale l_0. It is also limited on the large wave number range by the inverse magnitude of Kolmogorov dissipative scale $l_K \sim (v^3/\varepsilon_K)^{1/4}$, where v is the kinematical viscosity of the medium. At first, consider the Eq. (3) for the power spectrum $S_{Sh}(f)$ [cm^2/s] related to spike-irregularities of the dynamic variable $u(t)$. Introduce a new parameter, w_S [cm^2/s^2], characterizing the energetic content of the turbulent flux pulsations, and then define the specific frequency $K_0 = \varepsilon_K/w_S$ [1/s], assuming that only the value $\tau_K = (\varepsilon_K/w_S)^{-1}$ represents a characteristic relaxation time in a fully developed turbulent flux, stipulating its stationary state. Using dimension arguments, Eq. (3) can be presented as:

$$S_{Sh}(f) = \frac{S_{Sh}(0)}{1+(2\pi f T_0)^{n_0}} = A_{Sh} \frac{\varepsilon_K^{(3-n_0)/2} \cdot K_0^{(3n_0-5)/2} \cdot U_0^{n-1}}{f_0^{n_0} + f^{n_0}}; \quad f_0 = \frac{K_0}{2\pi} \qquad (8)$$

Here A_{Sh} is a dimensionless parameter. Note that the Eq. (8) is valid if the exponent $r \equiv [(3n_0 - 5)/2] > 0$, as it should be fulfilled $S_{Sh}(f) \to 0$ at $(K_0/f_{min}) \to 0$.

The inertial interval frequency boundaries $\{f_{min}, f_{max}\}$ can be defined as: $f_{min} \sim \varepsilon_K/u_0^2$, and $f_{max} = f_K \sim \varepsilon_K/w_D$. Here we introduce parameters: u_0 [cm/s] which is a standard deviation of local velocity fluctuations for the pulsations of the "outer" scale; and $w_D = (\varepsilon_K v)^{1/2}$ – the same parameter for the turbulent pulsations appropriate to the Kolmogorov dissipative scale, on which dissipation becomes significant and the extent of correlation links in dynamics of fluctuations sharply falls down. To reproduce the well-known Kolmogorov-Obukhov (KO) law on the base of Eq. (8), it requires the fulfillment of an additional condition: $K_0 \ll f_{min}$ or $u_0^2 \ll w_S$. In that case, the $S_{Sh}(0)$ magnitude has a maximum value at $n_0 = 5/3$, because $S_{Sh}(0) \sim (K_0/f_{min})^r \to 0$ at $K_0/f_{min} \ll 1$ and $r > 0$. Therefore, we must choose $n_0 \to 5/3$ out of other n_0 values only:

$$S_{Sh}(f) \to A_{Sh} \frac{\varepsilon_K^{2/3} \cdot U_0^{2/3}}{f^{5/3}} \qquad (9)$$

Eq. (9) corresponds to an "infinity memory" in correlation links among the sharpest fluctuations in turbulent pulsation sequence. Analysis of the $\Phi^{(2)}(\tau)$ and $S_{Rh}(f)$ dependences in this case shows that the $S_{Rh}(0)$ magnitude has a maximum value at $H_1 = 1/3$, and we reproduce the KO law (9) for the $S_{Rh}(f)$ dependence. However the corresponding dimensionless parameter A_{Rh} is small compared to A_{Sh}: $A_{Rh}/A_{Sh} \sim \eta$, where $\eta \equiv \{\sigma_1^2/[w_S^{2/3} u_0^{2/3}]\} \ll 1$, as $\sigma_1 \sim u_0$. In other words, the KO dependence is formed by spike-irregularities of the dynamic variable $u(t)$.

ACKNOWLEDGMENTS

The work was supported by Russ. Fond Bas. Res. (grants 05-02-17079, 04-02-16850).

REFERENCES

1. Timashev S.F., Vstovsky G.V., and Belyaev V.E. *Informative essence of chaos*, in: J. Sikula, ed. Noise and Fluctuations – ICNF 2003, Prague. Brno Univ. Techn., 2003. pp. 77-80.
2. Timashev S.F. Science of Complexity: phenomenological basis and possibility of application to problems of chemical engineering, *Theor. Found. Chem. Eng.* **34,** 301-312 (2000).

Bispectrum Theory for Brownian Motion of Electrical Charge in Nonlinear RC-circuit

Boris M. Grafov

A.N. Frumkin Institute of Electrochemistry of Russian Academy of Sciences
31 Leninskii prospekt, Moscow 119071, Russia

Abstract. The purpose of present report is to provide the bispectrum theory for Brownian motion of charge in the equilibrium non-linear RC-circuit (both resistance R and capacity C are nonlinear). The theory is based on the dual Langevin linear equations and on the Stratonovich fluctuation-dissipation relation for the voltage noise bispectrum of resistor. The equation for the asymmetry of the equilibrium electrical charge fluctuations in the nonlinear capacity is derived. This equation differs essentially from the Gibbs thermodynamics equation in several aspects. In line with the bispectrum theory, the asymmetry of equilibrium charge fluctuations depends on the resistance nonlinearity and is independent of the capacitance nonlinearity. We arrive at the fundamental conclusion that the Gibbs statistical method is inapplicable to the description of nonlinear thermodynamic

INTRODUCTION

There are two general theoretical ways to analyze the thermodynamic fluctuations. The classic way is to use the Gibbs statistical method. The second one is the noise approach. Let consider the charge fluctuations in the capacity. In line with noise approach the variance $<q^2>$ and asymmetry (skewnees) $<q^3>$ of thermodynamic charge fluctuations can be found by integrating single spectrum $<q_\omega q_\omega^*>$ and bispectrum $<q_\omega q_\nu q_{\omega+\nu}^*>$ over all frequencies ω and ν:

$$<q^2> = \frac{1}{2\pi} \int_{-\infty}^{\infty} d\omega <q_\omega q_\omega^*> \qquad (1)$$

$$<q^3> = \frac{1}{(2\pi)^2} \int_{-\infty}^{\infty} d\omega \int_{-\infty}^{\infty} d\nu <q_\omega q_\nu q_{\omega+\nu}^*> \qquad (2)$$

It is well known that, in the case of Brownian movement of electrical charge in the equilibrium RC-circuit, both ways lead to the same expression for the charge variance $<q^2>$ [1]:

$$<q^2> = kTC \qquad (3)$$

where k is Boltzmann's constant, T is the temperature, C is the capacity. For asymmetry of charge fluctuations in capacity the Gibbs statistical method yields [1]:

$$<q^3> = (kT)^2 dC/dE \qquad (4)$$

where dC/dE is the derivative of capacity with respect to the voltage.

The purpose of this paper is to consider the charge fluctuations asymmetry in the nonlinear RC-circuit (Fig.1) from the viewpoint of noise theory.

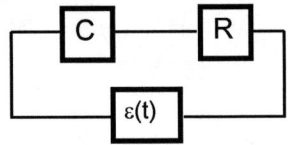

Fig.1. The RC-circuit: ε(t) is the Thevenin voltage noise generator

NYQUIST'S THEOREM

It is well known that the Nyquist fluctuation-dissipation theorem is dual and given by equations [1]:

$$<\varepsilon^*_\omega \varepsilon_\omega> = 4kT \operatorname{Re} Z_\omega \qquad (5)$$

$$<i^*_\omega i_\omega> = 4kT \operatorname{Re} G_\omega . \qquad (6)$$

We use the following notations: Z_ω is a small-signal impedance; $G_\omega = Z_\omega^{-1}$ is a small-signal admittance; $<\varepsilon^*_\omega \varepsilon_\omega>$ is a single spectrum of voltage noise $\varepsilon(t)$; $<i^*_\omega i_\omega>$ is a single spectrum of current noise $i(t)$; t is time. The voltage noise $\varepsilon(t)$ characterizes the Thevenin noise circuit. The current noise $i(t)$ characterizes the Norton noise circuit.

DUAL LANGEVIN EQUATIONS

The Nyquist fluctuation-dissipation theorem (5)-(6) is applicable both to linear and nonlinear equilibrium systems [1-2]. We have used [3] this property of the Nyquist theorem to generalize the Langevin stochastic equation to the nonlinear Brownian movement in the form of dual stochastic equations:

$$\varepsilon(t) = \int_{-\infty}^{\infty} dt_1 H^E(t-t_1) i(t_1) \qquad (7)$$

$$i(t) = \int_{-\infty}^{\infty} dt_1 H^I(t-t_1) \varepsilon(t_1), \qquad (8)$$

where $H^E(t)$ is a system function characterizing the small-signal voltage response and $H^I(t)$ is a system function characterizing the small-signal current response. The Langevin stochastic equation (7) is the map from the Norton current noise to the Tevenin voltage noise. The Langevin stochastic equation (8) is the map from the Tevenin voltage noise to the Norton current noise.

BISPECTRUM OF CURRENT FLUCTUATIONS

The dual Langevin stochastic equations (7)-(8) can be considered as the linear filter equations. One can conclude on the basis of (8) that bispectrum for current $i(t)$ of the circuit in Fig.1 is

$$<i_\omega i_\nu i^*_{\omega+\nu}> = G_\omega G_\nu G^*_{\omega+\nu} <\varepsilon_\omega \varepsilon_\nu \varepsilon^*_{\omega+\nu}> \qquad (9)$$

where the star denotes the complex conjugate quantity. Admittance G_ω of circuit in Fig.1 is given by equation:

$$G_\omega = \frac{1}{R+j\omega C} \qquad (10)$$

where j is the imaginary unit, R is the small signal resistance, and C is the small signal capacitance. Fluctuating current $i(t)$ and fluctuating charge $q(t)$ are interrelated by equation:

$$i(t) = dq(t)/dt. \qquad (11)$$

Equation (11) is linear. Therefore, we can use the following equation of linear filter theory:

$$<i_\omega i_\nu i^*_{\omega+\nu}> = (j\omega)(j\nu)(-j\omega-j\nu) <q_\omega q_\nu q^*_{\omega+\nu}>. \qquad (12)$$

It is known [4] that the bicovariance function of nonlinear resistor coincides with that for the nonlinear white noise. The corresponding voltage bispectrum is:

$$<\varepsilon_\omega \varepsilon_\nu \varepsilon^*_{\omega+\nu}> = B(kT)^2 dR(I)/dI\big|_{I=0} \qquad (13)$$

where B is the number and $dR(I)/dI\big|_{I=0}$ is the derivative of small signal resistance $R(I)$ with respect to steady-state current I at the equilibrium point. According to Stratonovich [4]

$$B = 24. \qquad (14)$$

CHARGE FLUCTUATION ASYMMETRY

Combining (9),(12),(13), we obtain for the charge fluctuations bispectrum:

$$<q_\omega q_\nu q^*_{\omega+\nu}> = \frac{B(kT)^2 dR(I)/dI|_{I=0}}{(j\omega)(j\nu)(-j\omega-j\nu)(R+j\omega C)(R+j\nu C)(R-j\omega C-j\nu C)} \quad (15)$$

After substituting (15) for $<q_\omega q_\nu q^*_{\omega+\nu}>$ in (2) and integrating over both frequencies ω and ν, we obtain for the charge fluctuation asymmetry the following expression:

$$<q^3> = \frac{B}{12}(kT)^2 CR^{-2} dR(I)/dI|_{I=0} . \quad (16)$$

It is seen that equations (4) and (16) are strongly different. According to the Gibbs statistical method, the charge fluctuation asymmetry is independent of linear or nonlinear resistor properties and is absent for the linear capacity. According to the noise approach, the charge fluctuation asymmetry is independent of nonlinear capacity properties and is absent for the linear resistor only. We arrive at the fundamental conclusion. The Gibbs statistical method does not work to calculate the asymmetry and higher cumulants of thermodynamic fluctuations. Therefore, a tremendous task arises - to create the statistical kinetics that could describe the nonlinear thermodynamic fluctuations.

ACKNOWLEDGMENTS

Author is grateful to L.B.Kish, A.M.Kuznetsov, S.F.Timashev, and R.M.Yulimetiev for useful discussions of nonlinear noise problems.

This work was supported by the Russian Foundation for Basic Research, project no. 02-03-32114.

REFERENCES

1. W.Bernard and H.B.Callen, *Rev. Mod. Phys.* **31**, 1017-1044 (1959).
2. Sh.Kogan, *Electronic Noise and Fluctuations in Solids*, Cambridge University Press, New York, 1996, pp.57-62.
3. B.M.Grafov, *Fluctuation and Noise Letters* **4**, L617-L622 (2004)
4. R.L.Stratonovich, *Izv. VUZ Radiofizika* **13**, 1512-1522 (1970)

1/f Temperature Fluctuations in Solids

Hisayuki Higuchi*, Shuichi Nakamura**, and Shikayuki Ochi***

*Department of Information Engineering, Faculty of Engineering, Maebashi Institute of Technology
420-1 Kamisadori, Maebashi-shi, Gunma 371-0816, Japan
**Former Affiliate is Device Development Center, Hitachi Ltd., Ohme-shi, Tokyo, Japan
***Former Affiliate is ULSI System Engineering, Ohme-shi, Tokyo, Japan

Abstract. We propose a 1/f temperature fluctuation model, based on the random walk of a phonon in solids. When the phonon reaches the boundary of the solid, it is absorbed by the boundary. Thus, the statistical characteristics of the absorbed phonon follow the Poisson counting process, i.e., the flow variance of the absorbed phonon is given by the average phonon flow. This relationship gives a 1/f fluctuation spectrum on the phonon density fluctuation that generates a 1/f temperature fluctuation. Using this relationship, we calculated a 1/f fluctuation in the resistors. The resulting power spectrum was about one tenth of the value reported by Voss, and the calculated Hooge's constant was around one fourth of the reported value. These quantitative and qualitative coincidences support our 1/f temperature fluctuation model.

Keywords: 1/f, fluctuation, temperature, noise.
PACS: 72.70.

INTRODUCTION

Fluctuations whose power spectra are inversely proportional to the frequencies commonly occur in resistors and various electronic devices as electrical noise. Also, 1/f spectra have been observed in traffic flows and music, [1] and [2]. These fluctuations, analyses, and models were reported and reviewed in the special issue on fluctuations [3]. However, 1/f fluctuation spectra have not been explained well. In this article, we analyzed the random walks of phonons using a statistical technique. During the analysis, we found a model that can successfully generate 1/f fluctuation spectra.

PROPOSED MODEL

A 1-dimensional resistor with two electrodes is shown in Fig. 1. It is assumed that the resistor is in thermal equilibrium with the reservoir electrodes. The phonons in the resistor are scattered randomly and move from x=0 at t=0 toward x=L. The trajectory is shown schematically. The phonon trajectory is analyzed from the viewpoint of statistical results, namely, by neglecting the trajectory details. This treatment leads us to the conclusion that the phonon flow is directed outwards, (i.e., from x=0 to x=L). The time when the phonon reaches the boundary at x=L, deviates randomly, but the average time can be given by $L^2/4D$, where D is the phonon diffusivity.

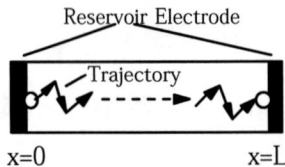

Figure 1. The trajectory of a phonon in a resistor with two electrodes.

The time inverse gives the number of phonons absorbed in the boundary at x=L in 1 second. After a given time t, the total number of the absorbed phonons is given as $(4D/L^2)t$. This phenomenon occurs randomly, following the Poisson statistics model. Therefore, the variance σ^2 of the absorbed phonon after t seconds is given by the total number of the absorbed phonons $(4D/L^2)t$. The standard deviation is given as equation (1).

$$\sigma = \frac{2\sqrt{Dt}}{L} \quad (1)$$

This relation shows that σ depends on $\sqrt{t}$, where t is the elapsed time after beginning to observe the phenomenon. Further, σ is a function of the distribution tempérée on t. Therefore, the Fourier transform can be executed on σ. The obtained σ in the frequency domain is

$$\sigma(\omega) = \frac{\sqrt{D}}{L\omega^{3/2}}. \quad (2)$$

Now, we get

$$\frac{d\sigma(\omega)}{d\omega} = \frac{\sqrt{D}}{L\,\omega^{1/2}}. \quad (3)$$

Equation (3) shows that the flow rate fluctuation of the absorbed phonon gives a 1/f power spectrum.

If the number of phonons in the sample is N, the power spectrum of the total phonon flow rate in the resistor is given by multiplying N by the value of a single phonon. The flow rate power spectrum is given by $N(D/L^2)/\omega$.

$$N\left(\frac{d\sigma(\omega)}{d\omega}\right)^2 = \frac{DN}{L^2\omega} \quad (4)$$

Next, the phonon flow rate fluctuations are converted to temperature fluctuations. In this process, we assumed that the phonon density fluctuation was given by the ratio of the phonon flow rate fluctuations to the average of their flow rate. The assumption is valid under the condition that the flow velocity is strong enough to replace all the phonons in the resistor, while measuring the inverse frequency. The condition is usually satisfied at a measuring frequency of 100 Hz, or less. This gives the 1/f fluctuation spectrum equation

$$\frac{\Delta T(\omega)^2}{T^2} = \frac{\Delta N(\omega)^2}{N^2}, \text{ where } \Delta N(\omega)^2 \text{ is given by equation (4).} \quad (5)$$

Then,

$$\Delta T(\omega)^2 = \frac{T^2 L^2}{16 N D \omega} = \frac{7 \times 10^{-11}}{\omega}. \quad (6)$$

We obtained $\Delta T(\omega)^2$ using the physical parameters of the resistor [4], which are summarized in Table 1.

TABLE 1. Physical parameters of the Au thin film resistor [4]

Parameter	
Temperature T	300 K
Length of resistor L	600 μm
Number of phonons in resistor N	6×10^{12}
Phonon diffusivity D	0.05 cm^2/s

DISCUSSIONS

We compared the value obtained from equation (6) with the measured values. The measured values are the values of thin metal films reported by Voss [4]. Therefore, we have to modify equation (6), because in metals, heat is mainly carried in the electrons, although thermal energy is stored in the phonons. Thus, both electrons and phonons, cause heat flow fluctuations, and these values must be converted into temperature fluctuations. We assumed that the temperature fluctuations increase due to the electrons, as shown in the first term on the left side of equation (7), and that temperature fluctuations were obtained by multiplying the phonon flow rate by the heat capacity, as shown on the right side terms of equation (7).

$$\left(\frac{kT}{2}\right)^2 \frac{D_n N}{L^2 \omega} + (kT)^2 \frac{D N}{L^2 \omega} = \left(\frac{4D}{L^2} C_{Vp}\right)^2 \Delta T^2 \quad (7)$$

where, k, D_n, D, and C_{Vp} are the Boltzmann constants, the diffusivity of the electrons and phonons, and the heat capacity of the phonons, respectively. In equation (7), we ignored the electron heat capacity. Using $C_{Vp} = kN$ in a 1-dimensional analysis, we obtained equation (8).

$$\Delta T(\omega)^2 = \frac{T^2 L^2}{16 D N \omega} \left(\frac{D_n}{4D} + 1\right) \quad (8)$$

$$= \frac{3.5 \times 10^{-10}}{\omega}$$

We deduced from equation (8) that the temperature fluctuations increase around 5 times when the electron effect is considered.

By inserting the equation parameters and measuring conditions used in Voss's article, we obtained a voltage fluctuation at 10 Hz. This value is around one tenth of the value reported by Voss. The calculated Hooge's constant was around one fourth of the value reported by Hooge [5]. These quantitative and qualitative coincidences support our 1/f temperature fluctuation model.

REFERENCES

1. T. Musha and H. Higuchi, Japan. J. Applied Physics, **15**, pp.1271-1275 (1976).
2. R. F. Voss, Proceedings of the Symposium on 1/f Fluctuations, pp. 199-205 (1977).
3. IEEE Transactions **41**, no. 11 (1994).
4. R. F. Voss and J. Clarke, Physical Review B, **13**, pp. 556-576 (1976).
5. N. F. Hooge, IEEE Transactions **41**, pp. 1926-1935 (1994).

APPENDIX

In our article, we used the Poisson counting process to generate a 1/f fluctuation spectrum. Here, we will show another process that can generate a 1/f fluctuation spectrum.

We consider the resistor of length L, shown in Figure A1 (a). The phonon located at x=0 when t=0 starts to move toward x=L. When it reaches x=L, the phonon is absorbed into the electrode. In the next moment, a new phonon is injected from the electrodes to maintain the thermal equilibrium. The injected phonon from the electrode at x=L is considered that the absorbed phonon has returned. The injected phonon from the electrode at x=0 starts to move toward the electrode at x=L. However, the phonon has already moved from x=0 to x=L. Therefore, it is plotted on the right side x=L domain. When it reaches x=2L, it is plotted in the same way. This process generates the phonon trajectory shown in Fig. A1 (b). The phonon distribution follows the Gaussian distribution function. The standard deviation of the distribution increases in proportion to $2(Dt)^{1/2}$. Thus the standard deviation of the velocity dx/dt of the phonon is given as $(D/t)^{1/2}$. This function can be Fourier transformed to $(D/\omega)^{1/2}$. This model also gives a 1/f fluctuation power spectrum in flow rate fluctuations of the phonon which is randomly walking.

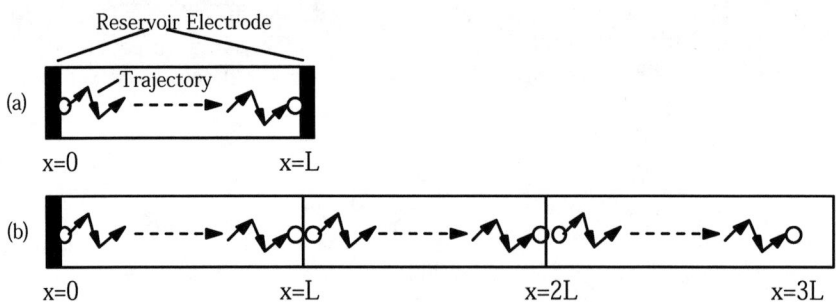

Figure A1. (a) A resistor with two reservoir electrodes at x=0 and x=L. Arrows indicate the phonon trajectory in the resistor. Figure A1 (b) shows a phonon trajectory plotted using the distance it travels.

Properties of the Sum-lengths of the Adjacent Level Crossing Intervals of the Chaotic Process Generated by the Logistic Map

Tsutomu Munakata[*] and Wolfgang Schwarz[**]

[*]Faculty of Engineering, Tamagawa University, 6-1-1 Tamagawa-Gakuenn, Machida-shi, Tokyo, Japan
[**]Faculty of Electrical Engineering and Information Technology, Dresden University of Technology
Mommsenstrasse 13, D-01062 Dresden, Germany,
Email: kreuz@eng.tamagawa.ac.jp, schwarz@iee1.et.tu-dresden.de

Abstract. In this paper we analyzed the statistical properties of the sum-lengths of the adjacent level crossing intervals of the chaotic process generated by the logistic map. We found a lot of new interesting phenomena and properties, which have never been observed in the past research of stationary random processes

Keywords: Level crossing interval, Chaotic process, Stochastic process, Logistic map.
PACS: 02.50.-r

INTRODUCTION AND THEORY

Since the work of S.O. Rice [1] many authors have been studied the statistical properties of level crossing intervals for stationary random processes such as Gaussian and Rayleigh processes. On the other hand the statistical properties of level crossing intervals of the chaotic process generated by the logistic map have been studied in the literature [2]. In this paper we analyzed the statistical properties of the sum-lengths of the adjacent level crossing intervals of the chaotic process generated by the logistic map. .

The chaotic process X(i) is generated by the logistic map of eq. $X_i = F(X_{i-1})$, where $F(x) = 4x(1-x)$. The chaotic sequence is converted into a binary signal by comparing every signal value with the level value L. The time moment where the discrete signal first-time exceeds the crossing level is called crossing point. Depending on the direction of crossing we distinguish up-crossing and down-crossing. The intervals between the edges of the binary signal are an integer multiple of the unit interval T between the discrete signal samples. T is set to 1 throughout this paper. The statistical properties of mean, variance, probability distribution of the sum-lengths of the adjacent crossing intervals, and the correlation between the crossing intervals are studied both theoretically and experimentally. The sum-lengths of k adjacent crossing intervals are denoted as $\tau_{k-1,\pm}$, where the index + means, that the interval begins with up-crossing, and - means, that the interval begins with down-crossing. For these level crossing intervals the probability distributions are defined as follows: $P_{k+}(n)$ is the probability that the crossing interval τ_{k+} has the length n, and $P_{k-}(n)$ is the probability that the crossing interval τ_{k-} has the length n. The probability distributions $P_{k+}(n)$ and $P_{k-}(n)$ can be given recursively from the eqs. (1) to (13), and they were newly formulated in this paper. Function $g(n,x)$ is the $(n+1)$th iterate of $f(x)$, and $g_{cr}(n,x)$ is the clipped function of $g(n,x)$ by level L. The value 1 or 0 of $g_{cr}(n,x)$ indicates, whether the process, which started at initial value x, stays above or below the level L at step n. The function $ng_{cr}(n,x)$ is the complement of $g_{cr}(n,x)$.

$$f(x) = 4x(1-x), \quad (1)$$

$$\begin{cases} g(0,x) = f(x), & n=0 \\ g(n,x) = f^{(n+1)}(x) = f(g(n-1,x)), & n \geq 1 \end{cases} \quad (2)$$

$$g_{cr}(n,x) \equiv \begin{cases} 1 & g(n,x) > L, \text{ for } n \geq 0, \\ 0 & g(n,x) \leq L, \text{ for } n \geq 0, \end{cases} \quad (3)$$

$$ng_{cr}(n,x) \equiv \begin{cases} 0 & g(n,x) > L, \text{ for } n \geq 0, \\ 1 & g(n,x) \leq L, \text{ for } n \geq 0. \end{cases} \quad (4)$$

$$R_1 = [f_1(L), min(f_2(L),L)], \quad \text{where} \quad f_1(x) = (1-\sqrt{1-x})/2, \text{ and } f_2(x) = (1+\sqrt{1-x})/2. \quad (5)$$

The function $h_0(0,1,x)$ in (6) and the region R_1 show the region of initial value x of the process, which cross the level L upward at step $n = 1$ as an start up-crossing. The start crossing at step $n=1$ is denoted as 0-th crossing. The crossing interval between 0 th and $k+1$ th crossing is described by the change of the function $h_k(n,i,x)$ for the step n. For the derivation of $P_{0+}(n)$, i.e. $k=0$ for the interval between 0-th and $1st$ crossing, $h_0(n,i,x)$ of eq.(6) and (7) are used. For the derivation of $P_{k+}(n)$, i.e. $k \geq 1$, the function $h_k(n,i,x)$ is used. For the odd number of k, the function $h_k(n,0,x)$ shows the region of the initial value x, that the process remains below the level L until step n, after having the k th crossing before step n, without including the k th crossing at step n. The function $h_k(n,1,x)$ shows the region of the initial value x, that the process remains below the level L until step n, after having the k th crossing until step n, including the k th crossing at step n. In the same manner, for the even number of k, the function $h_k(n,0,x)$ shows the region of the initial value x, that the process remains upper the level L until step n, after having the k th crossing before step n, without including the k th crossing at step n. The function $h_k(n,1,x)$ shows the region of the initial value x, that the process remains upper the level L until step n, after having the k th crossing until step n, including the k th crossing at step n.

for $k = 0$,
$$\begin{cases} h_0(0,1,x) \equiv f_{ar}(x) \equiv \begin{cases} 1, & f_1(L) \leq x < min(f_2(L),L), \\ 0, & \text{others}, \end{cases} & (6) \\ h_0(n,1,x) \equiv g_{cr}(n-1,x) \cdot h_0(n-1,1,x), \text{ for } n \geq 1. & (7) \end{cases}$$

for $k \geq 1$, $n = 0$,
$$h_k(0,0,x) \equiv 0, \text{ and } h_k(0,1,x) \equiv 0, \quad (8)$$

for $k \geq 1$ odd and $n \geq 1$,
$$h_k(n,0,x) \equiv ng_{cr}(n-1,x)h_k(n-1,1,x), \quad (9)$$
$$h_k(n,1,x) \equiv cr(h_k(n,0,x) + ng_{cr}(n-1,x)h_{k-1}(n-1,1,x)).$$

where $cr(x) \equiv \begin{cases} 1 & x > 0, \\ 0 & x \leq 0. \end{cases} \quad (10)$

for $k \geq 2$ even and $n \geq 1$,
$$h_k(n,0,x) \equiv g_{cr}(n-1,x)h_k(n-1,1,x), \quad (11)$$
$$h_k(n,1,x) \equiv cr(h_k(n,0,x) + g_{cr}(n-1,x)h_{k-1}(n-1,1,x)).$$

Eq. (8) shows the initial values of the functions $h_k(0,i,x)$ at step $n=0$. As shown in Eq.(9) and (11), if the step n increases, the function $h_k(n,i,x)$ can be calculated step by step from the function $h_k(n-1,i,x)$ and $h_{k-1}(n-1,i,x)$ recursively. The probability distribution $P_{k+}(n)$ is calculated from the difference of the probability $S_k(n,i)$ as follows:

$$S_k(n,i) = \int_0^1 h_k(n,i,x)f_p(x)dx, \text{ for } n \geq 0, i = 0,1, \quad \text{where} \quad f_p = 1/\left(\pi\sqrt{x(1-x)}\right). \quad (12)$$
$$P_{k+}(n) = \{S_k(n,1) - S_k(n+1,0)\}/S_0(0,1). \quad (13)$$

In the same manner, $P_{k-}(n)$ can be derived from only changing the polarity of crossings in the equations, but they are ommitted here because of the small space.

THEORETICAL AND EXPERIMENTAL RESULTS

The probability distributions $P_{k+}(n)$ and $P_{k-}(n)$ for $k=0$ to 5, i.e. the sum of up to 6 crossing interval-lengths, are studied in both theoretical and experimental ways. The value of crossing level L is varried between $L=0.05$ to 0.99. Fig.1 shows the typical results of $P_k(n)$ for $L=0.56$ and 0.25. The properties of $P_k(n)$s are summerized in Table 1. As well known, in case of stationary random process, the probability densities $P_{k+}(n)$ and $P_{k-}(n)$ are always equal with each other for the odd number of index k. Unlike to the case of stationary random processes, the probability distribution $P_{k+}(n)$ and $P_{k-}(n)$ of the chaotic process are not equal with each other for the odd number of index k. They show the different distributions in general. However for the level $L \geq 0.75$, or $L=0.5$, and $L=0.25$, $P_{k+}(n) = P_{k-}(n)$ is valid for odd k. Furthermore, at the level $L=0.5$, $P_{k+}(n) = P_{k-}(n)$ is valid for all k. For the level $L \geq 0.75$, $P_{k-}(n) = P_{k+1,\pm}(n+1) = P_{k+2,+}(n+2)$ is valid for even k. For $L=0.75, 0.5,$ and 0.25, $P_{k-}(n)=P_{0-}(n)*P_{k-1,+}(n)$, and $P_{k+}(n)=P_{0+}(n)*P_{k-1,-}(n)$ are valid. The correlations between successive crossing intervals have been also studied in both experimental and theoretical ways, and the correlation coefficients κ_is are given in Fig.2. If the level value is $L < 0.75$, $\kappa_{1+}, \kappa_{2+},$ and κ_{3+} show relative big values, and if the level L changes, they alternate plus to minus. At some points they show the value 0. On the other hand, if the first interval begins with down crossing, κ_{1-} and κ_{3-} have the value 0 for all value of L. But for the level $L > 0.75$, κ_{2-} is not 0. These results show that the property of κ strongly depends on the polarity of start crossing. Such propertits of polarity dependence are quite interesting, and they have never been seen in case of stationary random processes like Gaussian process.

TABLE 1. Typical properties of probability distribution $P_{k+}(n)$ and $P_{k-}(n)$.

L	$P_{0+}(n)$	$P_{0-}(n)$	$P_{1+}(n)$ $P_{1-}(n)$	$P_{k+}(n)$ $P_{k-}(n)$
0.75~1.0	$P_{0+}(n)=1$ (n=1) $=0$ (n:others)		$P_{1+}(n) = P_{1-}(n)$ $= P_{0-}(n-1)$	$P_{k+}(n) = P_{k-}(n)$, for k:odd $P_{k-}(n) = P_{k+1,\pm}(n+1) = P_{k+2,+}(n+2)$, for k:even
0.75	$P_{0+}(n)=1$ (n=1) $=0$ (n:others)	$P_{0-}(n) = (1/2)^n$	$P_{1+}(n) = P_{1-}(n)$ $= (1/2)^{n-1}$, $n \geq 2$	$P_{k+}(n) = P_{k-}(n)$, for k:odd $P_{k-}(n) = P_{k+1,\pm}(n+1) = P_{k+2,+}(n+2)$, for k:even $P_{k-}(n) = P_{0-}(n) * P_{k-1,+}(n)$, $P_{k+}(n) = P_{0+}(n) * P_{k-1,-}(n)$
0.5~0.75		$P_{0-}(n) = (1/2)^n$	$P_{1+}(n) \neq P_{1-}(n)$	$P_{k+}(n) \neq P_{k-}(n)$, for k:odd
0.5	$P_{0+}(n) = (1/2)^n$	$P_{0-}(n) = (1/2)^n$	$P_{1+}(n) = P_{1-}(n)$ $= (n-1)/2^n$, $n \geq 2$	$P_{k+}(n) = P_{k-}(n)$, for all k $P_{k-}(n) = P_{0-}(n) * P_{k-1,+}(n)$, $P_{k+}(n) = P_{0+}(n) * P_{k-1,-}(n)$
0.25~0.5		$P_{0-}(n) = (1/2)^n$	$P_{1+}(n) \neq P_{1-}(n)$	$P_{k+}(n) \neq P_{k-}(n)$, for k:odd
0.25	$P_{0+}(n) = (1/2)^{n/2}$ (n:even) $= 0$ (n:odd)	$P_{0-}(n) = (1/2)^n$	$P_{1+}(n) = P_{1-}(n)$ $= [2^{(n-2)/2}-1]/2^{n-1}$ (n:even) $= [2^{(n-1)/2}-1]/2^{n-1}$ (n:odd)	$P_{k+}(n) = P_{k-}(n)$, for k:odd $P_{k-}(n) = P_{0-}(n) * P_{k-1,+}(n)$, $P_{k+}(n) = P_{0+}(n) * P_{k-1,-}(n)$
0.0~0.25		$P_{0-}(n) = (1/2)^n$	$P_{1+}(n) \neq P_{1-}(n)$	$P_{k+}(n) \neq P_{k-}(n)$, for k:odd

CONCLUSION

1) The sum-lengths of adjacent level crossing intervals of the chaotic process derived from the logistic map were studied, and they showed a lot of interesting properties.

2) The most remarkable thing was the properties of polarity dependence of the crossing intervals. The statistical characteristics of the crossing intervals like probability distributions and correlations are strongly determined by the direction of start crossing, whether the intervals start with up crossing or down crossing.

3) The properties of the level crossing of the chaotic process are quite different from those of stationary random processes like the Gaussian process.

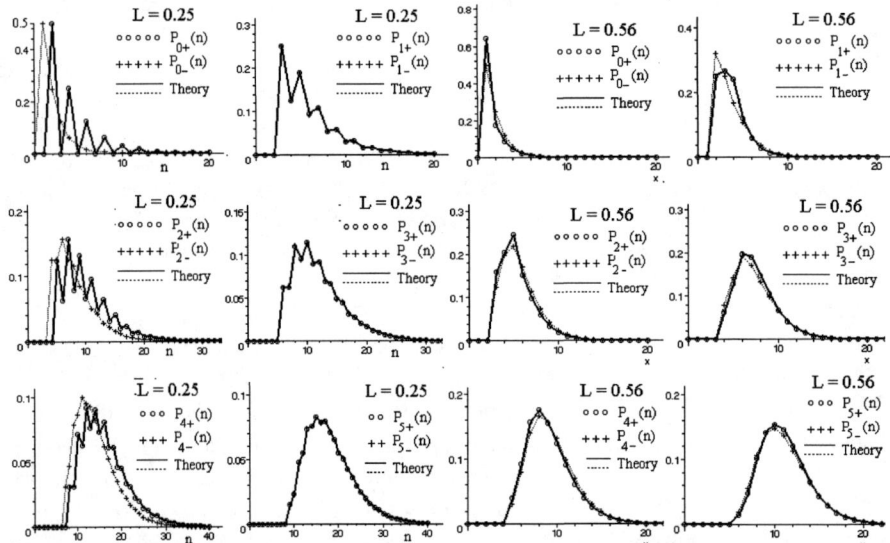

FIGURE 1. Probability distrubution $P_k(n)$ of the sum of adjacent level crossing intervals.

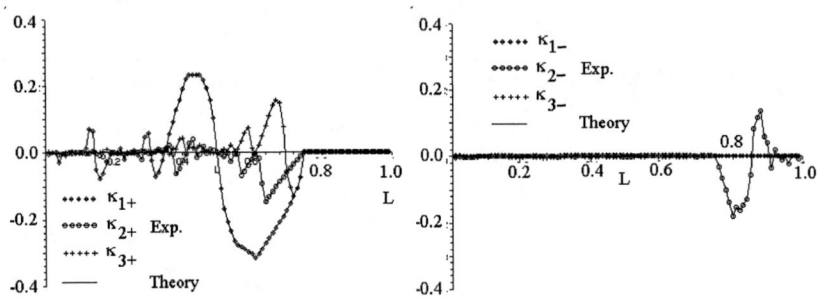

FIGURE 2. Correlation Coefficients κ_i between subsequent intervals.

REFERENCES

1. S. O. Rice, "The Mathematical Analysis of Random Noise," *B.S.T.J.*, 24, 1945, 46-156.
2. Tsutomu Munakata, and Wolfgang Schwarz, Properties of Level Crossing Intervals of the Chaotic Process Generated by the Logistic Map, , proc. 2004 International Symposium 0n Nonlinear Theory and its Applications, NOLTA2004, pp.617-620.
3. T.Munakata and D.Wolf, On the distribution of the level-crossing time intervals of random processes, Noise in Physical Systems and 1/f Noise Proc. 7th Int. Conf. on Noise in Physical Systems North-Holland Publ.Co., Editor M.Savelli Montpellier, 1983 , 49-52.

First Passage Time Algorithm For Signal And Filtered Multi-level Noise

Roy M. Howard

Department of Electrical and Computer Engineering,
Curtin University of Technology, GPO Box U 1987, Perth, 6845, Australia.

Abstract. An algorithm for determining the level crossing probabilities of a bounded random walk associated with filtering of dichotomous, or multi-level, noise is detailed. The algorithm is applied to the first passage time of a noise plus linear signal across a level by utilizing an orthogonal decomposition of the filtered noise. The algorithm is particularly suited to the transient noise case and good agreement with simulation results is demonstrated.

Keywords: First passage time, level crossing, dichotomous noise, bounded random walk
PACS: 02.50.Ey, 05.40.Fb

INTRODUCTION

Level crossing problems - from clock jitter in communication systems, to neuron firing characteristics in neurobiology and to option pricing in financial markets - have received significant attention over several decades, e.g. [1-3], due to their practical importance. Recent first passage time research includes [3-5]. Generally, Markov processes and stationarity are assumed; less attention has been paid to the non-Markovian and the non-stationary case.

In this paper the prototypical situation of the level crossing of a signal plus filtered noise, as shown schematically in Figure 1, is considered. The output of the linear filter for the interval $[0, N\Delta t]$, and for evaluation at the times $\{\Delta t, 2\Delta t, ..., N\Delta t\}$, yields a generalized form of a random walk with an ensemble:

$$\left\{ z(x_1, ..., x_N) = \left(x_1 h_1, ..., \sum_{i=1}^{N} x_i h_{N-i+1} \right) : \quad x_i \in S_X, h_q = \int_{(q-1)\Delta t}^{q\Delta t} h(\lambda) d\lambda \right\} \quad (1)$$

Here x_i is an outcome of a random variable X_i with a sample space S_X which is assumed to be independent of X_j for $i \neq j$. For the case where h is the unit step function and $S_X = \{\pm 1\}$ a standard random walk is defined and for this case the level crossing probability mass function is well known, e.g. [1], and has been solved, for example, for the case of boundaries of various shapes [4]. For the general case - where h is continuous and integrable - a bounded random walk is defined. This problem is a

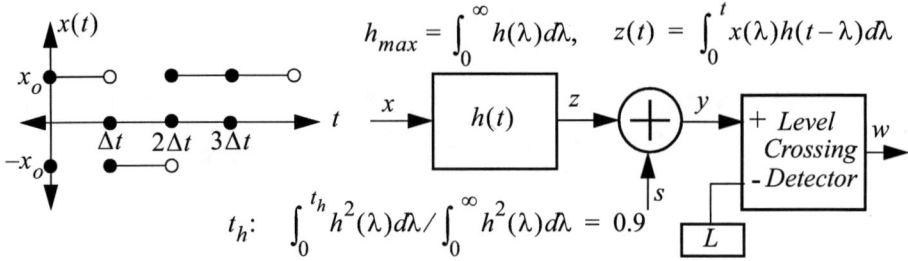

FIGURE 1. Schematic diagram of system being considered and parameter definitions. The input signal is illustrated for the dichotomous noise case.

variation on that considered in [6], for example, as the nature of the impulse response, assumed arbitrary here, implies, in general, a higher order (potentially infinite) DE than the first order DE that is typically assumed. The process, in general, is non-Markovian. The approach taken is to, first, convert the problem from a first passage time for the noise signal across a time dependent absorbing barrier (as defined by the signal) to a level crossing problem. This is achieved by decomposing the noise signal into orthogonal components based on the geometry of the time dependent barrier. Second, a level crossing algorithm, based on a tree structure associated with a bounded random walk, is implemented. The algorithm is particularly suited to the transient noise case and good agreement with simulation results is demonstrated.

LEVEL CROSSING - SIGNAL PLUS FILTERED NOISE

First, consider the dichotomous case where $S_X = \{\pm x_o\}$ and the filter output takes on the values: $\{x_1 h_1, x_1 h_2 + x_2 h_1, x_1 h_3 + x_2 h_2 + x_3 h_1, \ldots\}$ where $x_i \in S_X$. By considering all possible values of each random variable a tree structure of values can be formed for the filtered output. The rule for generating the two new values from each existing branch of length $i-1$ is $I_{i-1} \bullet H_{i-1} \pm x_o h_1$ where $I_{i-1} = (x_{i-1}, x_{i-2}, \ldots, x_1)$ represents values defining existing branches, $H_{i-1} = (h_2, h_3, \ldots, h_i)$ and $\bullet$ is the dot product operator. An algorithm for ascertaining the level crossing probability of a level L in $[0, N\Delta t]$ seconds then is:
- Add branches to tree according to $I_{i-1} \bullet H_{i-1} \pm x_o h_1$
- Sort tree branch values according to level
- If branch value $\geq L$ then delete branch from tree. Augment level crossing probability for $t = i\Delta t$ by $1/2^i$
- iterate

To avoid the exponential increase in branch numbers a representative ensemble [7] can be used after $[0, M\Delta t]$ according to the equivalence relationship:

$$Z_j R Z_r \quad \text{iff} \quad \sum_{k=1}^{M} |x_{jk} - x_{rk}| = 0, \quad \begin{cases} Z_j = z(x_{j1}, \ldots, x_{jN}) \\ Z_r = z(x_{r1}, \ldots, x_{rN}) \end{cases} \quad (2)$$

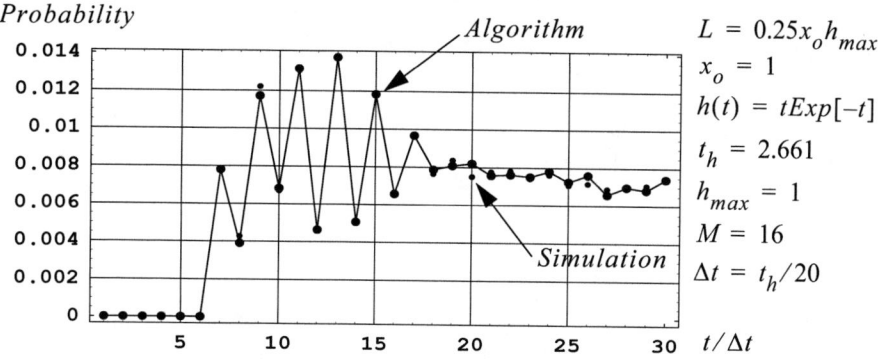

FIGURE 2. Level crossing probabilities. For the simulation 200000 trajectories were used.

for the case of $N > M$. For $[0, M\Delta t]$ the exact tree structure is used. For each time increment, $t \in \{(M+1)\Delta t, ..., N\Delta t\}$, one of the two possible branch values can be chosen, at random, to keep the branch number at a maximum of 2^M. Results are shown in Figure 2 for the case of $h(t) = tExp[-t]$ and a level crossing level, L, of $0.25 x_o h_{max}$. The parameters h_{max} and t_h are defined in Figure 1.

Second, consider the case where the signal can be modelled around the level crossing level L, which occurs at t_S, by an affine approximation $s(t) \approx L + k_S(t - t_S)$. Clearly, level crossing of this signal plus noise occurs when $z(t) = -k_S(t - t_S)$. The zero crossing time, t_{ZC}, is illustrated in Figure 3. As is evident from this Figure the first passage time across the level L occurs when the noise component, in the direction perpendicular to the time dependent barrier, reaches the level L_p defined by $L_p = t_S \cos[\theta]$ where $\theta = \pi/2 - \operatorname{atan}(k_S)$. Accordingly, by considering the geometry of the noise signal at the kth step, as illustrated in Figure 3, the level crossing takes place at the time $r\Delta t$ (assuming appropriate quantization) where

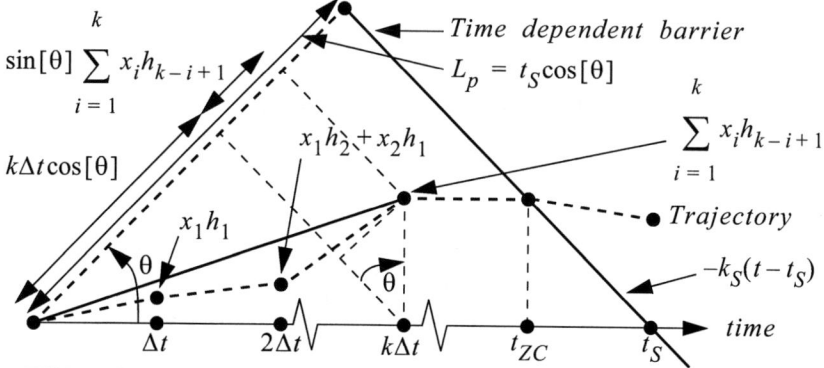

FIGURE 3. Components of noise trajectory that are perpendicular to time dependent barrier.

$$r\Delta t \cos[\theta] + \sin[\theta] \sum_{i=1}^{r} x_i h_{r-i+1} = t_S \cos[\theta] \qquad (3)$$

This decomposition has transformed a first passage problem across a time dependent barrier into a level crossing problem for a bounded random walk with linear drift. Results using the above specified algorithm, but modified for the case of a linear signal plus noise, yield the results shown in Figure 4 for the case of $\theta = \pi/3$.

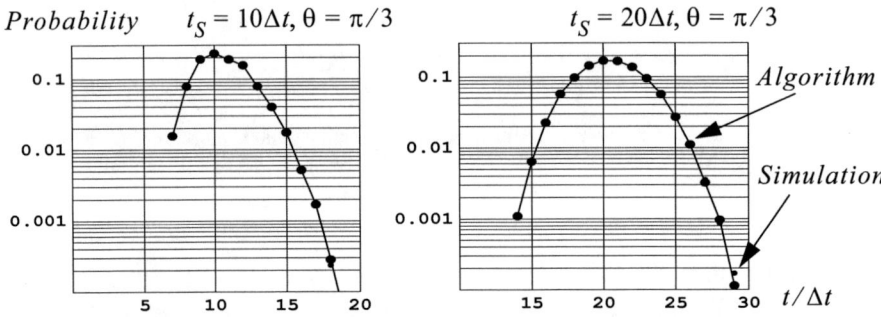

FIGURE 4. Level crossing probabilities for filtered dichotomous noise plus a linear signal $k_S(t-t_S)$. The parameters of Figure 2 have been used (except that $M=17$).

CONCLUSION

An algorithm for establishing the level crossing probability of filtered dichotomous noise has been proposed. The algorithm has been successfully applied to the transient case of the first passage time of a linear signal plus filtered dichotomous noise to a set level by decomposing the filtered noise into orthogonal components; orthogonality being based on the signal. Good agreement with simulation results has been demonstrated. Whilst the results have been presented for the dichotomous noise case they can be readily extended to the multi-level case.

REFERENCES

1. Chandrasekhar, S., 'Stochastic Problems in Physics and Astronomy', in *Selected Papers on Noise & Stochastic Processes*, edited by N. Wax, Dover, 1954, pp. 3-91.
2. Blake, I. F., and Lindsey, W. C., IEEE Transactions on Information Theory **19**, 295-315 (1973).
3. Romero, A. H., Sancho, J. M., and Lindenberg, K., Fluctuation and Noise Letters **2**, L79-100 (2002).
4. Nagar, A., and Pradhan, P., Physica A **320**, 141-148 (2003).
5. Basak, G. K., and Ho, K. R., Advances in Applied Probability **36**, 643-666 (2004).
6. Pawula, R. F., Porra, J. M., and Masoliver, J., Physical Review E **47**, 189-201 (1993).
7. Howard, R. M., 'The representative ensemble and its application to 1/f type random processes' in *Unsolved Problems of Noise and Fluctuations*, edited by D. Abbot et al., AIP Conference Proceedings 511, New York: American Institute of Physics, 2000, pp. 124-129.

1/f - type Noise in View of Phonons Interface Percolation Dynamics

S.V. Melkonyan, F.V. Gasparyan, V.M. Aroutiounian, H.V. Asriyan

*Dept. of Physics of Semiconductors & Microelectronics,
Yerevan State University, 1 Alex Manoogian, Yerevan 375025, Armenia
E-mails: smelkonyan@ysu.am ; fgaspar@ysu.am ; kisahar@ysu.am ; hasriyan@ysu.am*

Abstract. The influence of long-wave acoustic longitudinal-phonon percolation dynamics on 1/f -type noise level is modeled for homogeneous, non-degenerated and bounded semiconductors. Phonons percolation from semiconductor media to environment regions via so-called «refraction points» of phonons' wave vector phase space is modeled within framework of the bulk mechanism of electron lattice mobility fluctuation. On the base of this mechanism it is shown, that semiconductor surface is the source of suppression of 1/f-noise. It is indicated that in some certain applications of the Fluctuation Theory it is physically correct to use Schönfeld model to consider 1/f noises in semiconductors.

Keywords: 1/f noise, refraction points, percolation dynamics.
PACS: 72.70.+m:

INTRODUCTION

The main proposed possible mechanisms of origin of 1/f noise are well presented in many review papers [1-3]. The mechanisms can be divided on two main groups: surface and bulk related. The main disadvantages of these two groups are the absence of possibility to explain for example 1/f noise dependence on surface physical prosperities within the context of bulk models of 1/f noise and vice versa. And in our opinion any new proposed model has to try to overcome that disadvantage. In our opinion that disadvantage can be overcome in models focused on the role of electron phonon scattering processes related with bulk and surface aspects of that scattering.

PHONON INTERFACE PERCOLATION AND 1/F NOISE

During last years we proposed several ideas concerning 1/f noise new model based on electron-phonon scattering processes and in this paper we made an attempt to outline those studies in one approach [4-6]. These two works were fostered by early theoretical considerations of influence of the electron-phonon interactions on the level of 1/f noise [7], which at present can be considered as transitional results.

Electron lattice mobility fluctuation

In Ref.[4] the two main causes of origin of the mobility fluctuation of the electrons in homogeneous and non-degenerated semiconductors are discussed. The characteristic peculiarities of the current carrier mobility fluctuations, coming into existence in the bulk of the semiconductors, can be established basing on the following quasi-classical argumentation.

The density of the electron current **j** can be determined and represented as

$$\mathbf{j}(t) = -\frac{e}{4\pi^3} \int_{BZ} \mathbf{v_k} f_{k,c}(t) d\mathbf{k} \equiv en\mu_n \mathbf{E}, \qquad (1)$$

where $f_{k,c}$ is the distribution function (DF) of electrons in the conduction-band (c-band), $\mathbf{v_k}$ is velocity of c-band electron with $\mathbf{k}$ wave vector, $\mathbf{E}$ is the external electric field, t is the time; note that in Eq.(1) the integration via $\mathbf{k}$ is done within the Brillouin first zone (BZ).

It should be taken into account that the electron velocity $\mathbf{v_k}$ in the given $\mathbf{k}$ state of the conduction band is a constant (not fluctuating) quantity, while the population, i.e. number of electrons $f_{k,c} d\mathbf{k}$ in that state may undergo certain fluctuation. Note, that if in case of intra-band or inter-band (band to band or c-band to impurity state electron transitions) scatterings electron did pass (or leaved) into another quasi energy level of c-band, then such a transition in Eq.(1) will get reflected in form of $f_{k,c}$ DF fluctuation.

The analysis of the expression (1) shows that the mobility of an electron μ_n is a fluctuating quantity [4]. So, any change of electrons' distribution by the c-band levels, as it can be seen from the Eq.(1), leads to the change of the quantity μ_n. In general, the electron mobility fluctuations do origin due to the random character of the intra-band, so and the inter-band electron transitions (generation-recombination transitions). In particularly, at the intra-band scattering processes the electron quantum states before and after scattering stay in the same band (in our case in c-band). That is why at such scattering the electron concentration n does not change ($\tilde{n} = 0$). So from Eq.(1)

$$\tilde{\mu}_n = -\frac{1}{4\pi^3 \overline{n} \overline{E}^2} \int_{BZ} \mathbf{E} \mathbf{v_k} \tilde{f}_{k,c}(t) d\mathbf{k}. \qquad (2)$$

Further analyses of Eq.(2) is shown in Ref. [4], one can see that the mobility fluctuation is conditioned only by symmetric component of the electron DF fluctuation, i.e. by the fluctuations of the conduction electrons energy. That is the main and most important result obtained in Ref [4].

Electron energy fluctuations

Processes of origin and relaxation of fluctuations of the conduction electrons' energy fluctuations in space-homogenous and non-degenerate thermal equilibrium semiconductors, in their turn, are discussed and elaborated in detail in Refs. [5,6].

Let us consider the fluctuations of the lattice mobility of electrons $\tilde{\mu}_l$ at very important case, when the scatterings in semiconductor are conditioned by electron-phonon, phonon-electron and phonon-phonon interactions only.

A model of microscopic mechanism of origin and subsequent damping of the electrons' DF symmetric fluctuations is suggested in Ref. [6]. The initial random perturbation, characterizes the random redistribution of the energy between the branches or the parts of the branches of the phonons' energetic spectrum. These fluctuations, which origin as a result of random phonon-phonon scatterings, can be representing as a symmetric fluctuation of the equilibrium phonons' DF.

Discussing the processes of transition of the initial random perturbation of phonons to electrons in the paper [6] we established that a dynamic equilibrium state between the more long wave phonons and electrons does origin. In this long wavelength region the system of the interrelated Boltzmann kinetic equations for electrons and phonons has stationary solution. As a mechanism of damping of the energy of the dynamic equilibrium state the phonons quasi-momentum absolute value diffusion in quasi-momentum space is suggested. It is shown that in low frequency region the Fourier-component of the long-wavelength electrons DF's symmetric fluctuating component $\tilde{f}^s_{c,k}(\omega)$ is proportional to $\omega^{-1/2}$. Character or slow (diffusion) damping of the dynamic equilibrium state energy indicated, that the concept of "Thermal Equilibrium Semiconductor" can be reconsidered when one studies fluctuation phenomena. The electron-phonon system always stays in non-equilibrium condition from point of view of the fluctuation theory.

The law $\tilde{f}^s_{c,k}(\omega) \sim \omega^{-1/2}$ shows that spectral density of the electron mobility (or current) fluctuations $(S_I \sim \tilde{f}^s_{c,k}(\omega)\tilde{f}^{s,*}_{c,k}(\omega))$ has a $1/\omega$ (or $1/f$) form. The $\tilde{f}^s_{c,k}(\omega) \sim \omega^{-1/2}$ law (which was obtained by us in Ref.[6]) well fits to Schönfeld model and we showed that in some applications of the Fluctuation Theory it is physically correct to use Schönfeld model to consider 1/f noises in semiconductors.

Refraction points

From the fact of slow damping of the above-mentioned non-equilibrium dynamic equilibrium state (please consider the entire phrase as a unified term, for more details see [6]) follows that in the energy relaxation processes the electron and phonon surface scatterings processes must be taken into account.

As it is known in finite size semiconductor-contacting substrate media the collection (package) of phonon wave vectors **q** is discrete. This collection depends on the form and the sizes of the semiconductor-contacting substrate media. Following the well-known laws of waves refraction for phonon percolation by e.g. z axis it is easy to come to the

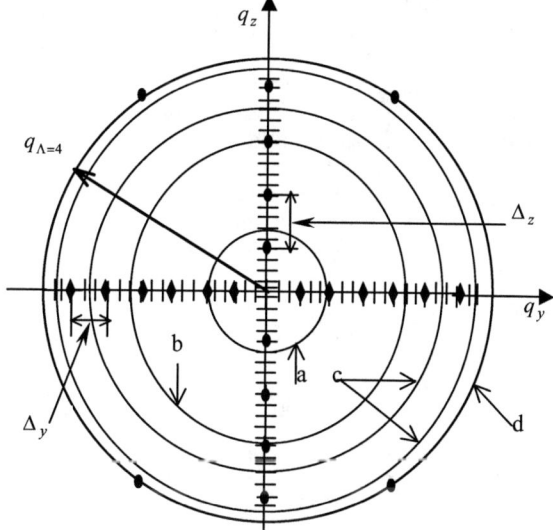

FIGURE 1. Cross-sections of iso-energetic surfaces of acoustic phonons. Dots show "refraction points" [5], Λ is the number of the refraction dots for each iso-energetic surface, Δ_z and Δ_y are the distances between the nearby magnitudes of refraction points of phonons, falling on directions **z** and **y**. Surfaces a,b, Λ= 2;c, Λ= 0; d, Λ= 4.

key equation, which yields the selections rules outlining those phonons in quasi-momentum space which are able to pass from semiconductor to contacting media (for details see Ref. [5])

$$(\frac{v_0 n_x}{L_x})^2 + (\frac{v_0 n_y}{L_y})^2 + (\frac{v_0 n_z}{L_z})^2 = (\frac{v_s n_x}{L_x})^2 + (\frac{v_s n_y}{L_y})^2 + (\frac{v_s m_z}{\Gamma l_z})^2. \qquad (3)$$

where $n_i = 0, \pm1, \pm2...$ and $m_i = 0, \pm1, \pm2...$ are the whole numbers, $i = x, y, z$, L_i are the linear sizes of the crystal; l_i are the sizes of environmental area, for non-crystalline environments $\Gamma = 2$ (continuum approximation), for crystalline environments $\Gamma = 1$, v_0 and v_s are the phonons velocities in the semiconductor and the environment, correspondingly. Solution of Eq. (3) is rather complex task, it shows that (acoustic) phonons refraction on the flat interface processes for certain discrete magnitudes of phonons wave vectors, which can be named. Solutions of Eq.(3) is given in details in Ref.[5] (see Eqs.(8)-(12)). The results are given on Fig.1, where «refraction points» are marked by black dots, Λ shows number of the refraction dots for each iso-energetic surface. Only those phonons having right combination of $\{q_x; q_y\}$ matching to those dots can refract on the surface perpendicular to the direction z and percolate to the contacting media thus subsequently helping to diminish 1/f noise level.

CONCLUSION

As it logically follows from the outlined approach if one can regulate the temp of the surface and interface losses of phonons: phonons percolation dynamics from semiconductor media to the contacting media, then the removal of L-phonon energy fluctuation form a semiconductor may cause significant 1/f noise level drop in the entire structure. The influence of the boundary scattering leak on 1/f noise level was experimentally verified by us for Al/n$^+$Si-nSi/Al and Ag/n$^+$Si-nSi/Ag structures having various contacting media topology and type [8]. To study that influence in details the consideration of the following problems were needed:

1. What are the conditions of phonons energy transfer from one media to another: phonon's percolation conditions? How that process dynamics affects on 1/f noise level in semiconductors? (see Ref. [4, 5])

2. How to interconnect bulk and surface models to avoid the disadvantages of 1/f noise models mentioned at the beginning of this paper? (see Ref. [4, 6])

As it follows from the above sited considerations and through theory developed in Refs. [4-6], by controlling percolation rate of phonons from semiconductor bulk region to the contacting media it turns possible to influence 1/f noise level of semiconductor devices. The mentioned preliminary experimental studies over Al/n$^+$Si-nSi/Al and Ag/n$^+$Si-nSi/Ag structures [8], which still require additional experimental studies over other materials having various contacting medias, where contacting media is selected in accordance to certain theoretical criteria developed in Refs. [4-6], show promising results for applications in MOS and MOSFET devices and certainly may significantly help to further Gas noise recognition technique in semiconductor-based sensors so and Noise spectroscopy and diagnostics after all.

ACKNOWLEDGMENTS

This work is supported by the Ministry of Education and Science of the Republic of Armenia (Base funding frames). Authors are also thankful to Prof. Can E. Korman, from the George Washington University.

REFERENCES

1. Hooge, F.N., *IEEE Trans.* **ED41**, 1926-1935 (1994); *Physica* **B311**, 238-249 (I), **B336**, 236-251 (II), (2003).
2. Weissman, M.B., *Rev. Mod. Phys.* **60**, Parts I-II, pp.538-571, (1988).
3. Kogan, Sh. M., *Electronic Noise and Fluctuations in Solids* (Cambridge Univ. Press, 1996).
4. Melkonyan, S.V., Gasparyan, F.V., Aroutiounian, V.M., Korman, C.E., *Noise and Information in Nanoelectronics, Sensors, and Standards III, Proc. of* **SPIE 5115-50**, 412-420 (2003).
5. Melkonyan, S.V., *Noise and Information in Nanoelectronics, Sensors, and Standards II, Proc. of* **SPIE 5472-50**, 401-408 (2004).
6. Melkonyan, S.V., Aroutiounian, V.M., Gasparyan, F.V., and Korman, C.E., *Physica* **B357** (398-407) 2005.
7. Melkonyan, S.V., et al, *Mod. Phys. Letters* **B12**, 1245-1254 (1998); Gasparyan, F.V., Asriyan H.V., *Mod. Phys. Letters* **B18**, 427-442 (2004); Asriyan H.V., et al, *Sensors & Actuators* **A113**, 338-343 (2004).
8. Asriyan, H.V., Shatveryan, A.A., Aroutiounian, V.M., Gasparyan, F.V., Melkonyan, S.V., Mkhitharyan, Z.H., Ayvazyan G., *Noise and Information in Nanoelectronics, Sensors, and Standards III, Proc. of* **SPIE 5846-25**, (2005).

$1/f$ Noise in Fractal Quaternionic Structures

T. Meškauskas[†][*] and B. Kaulakys[†]

[*]*Vilnius University, Naugarduko 24, LT-03225 Vilnius, Lithuania*
[†]*Institute of Theoretical Physics and Astronomy of Vilnius University, Goštauto 12, LT-01108 Vilnius, Lithuania*

Abstract. We consider the *logistic map* over *quaternions* $\mathbb{H} \sim \mathbb{R}^4$ and different 2D projections of Mandelbrot set in 4D quaternionic space. The approximations (for finite number of iterations) of such 2D projections are fractal circles. We show that a *point process* defined by radiuses R_j of fractal circles exhibits $1/f$ noise.

Keywords: $1/f$ noise, point process, logistic map, Mandelbrot set, quaternions, hypercomplex numbers
PACS: 05.40.–a, 05.45.Df, 02.50.Ey

INTRODUCTION

$1/f$ noise is observed in large diversity of real life and artificial systems, which behavior is usually defined by a complex interaction of many components. Complexity of the system usually assumes that long-term correlations are observed. Examples are processes and experimental data in condensed matter, traffic flow, quasar emissions, music, biological and medical systems, economic and financial data, human cognition and even distribution of prime numbers (see [1] and references herein).

Fluctuations of signals defined by time series obtained from such systems are found to be characterized by a *power spectral density* $S(f)$ diverging at low frequencies f like $1/f^\alpha$, here α is some real parameter. $1/f$ ($\alpha \approx 1$) noise is an intermediate between the white noise ($\alpha = 0$) with no correlation in time and the random walk (Brownian motion) noise ($\alpha = 2$) with no correlation between increments. Note that Brownian motion can be obtained integrating white noise and that taking the integral of the signal increases the exponent α by 2 while the inverse operation of differentiation decreases it by 2.

Parameter α is closely related to the Hurst exponent H. It is known that fluctuations which are fractionally homogeneous, i.e. unifractal or uniscaling, can be quantified by a single coefficient H and a single exponent α [2].

Possible generalization leads to multiscaling or multifractals, with the exponent H dependant on time. Therefore multifractal processes are characterized by a set of scaling relations or power laws with correspondingly many exponents α [3].

POINT PROCESSES AND $1/f$ NOISE

In many cases, the intensity of some current can be represented by a sequence of random (however, as a rule, mutually correlated) or pseudo-periodic pulses. It is known (see [4] and references herein) that only the transit times t_j of these pulses (and not the shapes of the pulses) are responsible for appearance of $1/f$ noise. The current $I(t)$ (see Fig. 1, left) is then expressed as $I(t) = \sum_j \delta(t-t_j)$, here $\delta(t)$ is the Dirac delta function.

Hence, instead of current $I(t)$, we further deal with *point process*, defined by the sequence $t_1, t_2, \ldots, t_N, \ldots$. The *power spectral density* of the current $I(t)$ is defined as

$$S(f) = \lim_{N \to \infty} \frac{2}{t_N - t_1} \left| \sum_{j=1}^{N} e^{-i2\pi f t_j} \right|^2 \qquad (1)$$

where $[t_1, t_N]$ is assumed to be the interval of observation.

In this approach the power spectral density of the signal depends on the statistics and correlations of *point process* (the transit times t_j) only. In [4] we proposed simple analytically solvable model for producing *point process* resulting in $S(f) \sim 1/f$ ($\alpha = 1$) noise. Discussion on the origin and universality of $1/f$ noise was continued in [5, 6]. Some further work, related to the applications of the theory of *point processes* and $1/f$ noise to econophysics, was done in [7, 8].

$1/f$ NOISE IN QUATERNIONIC MANDELBROT SET

Complex numbers $\mathbb{C} \sim \mathbb{R}^2$, along to their *real* predecessors $\mathbb{R}$, are widely used in nowadays mathematical modeling and scientific computing. Beside others, they have important applications in theories of complex systems, fractals and signal processing: famous Mandelbrot and Julia fractal sets are defined in $\mathbb{C}$, spectrum (Fourier transform) is defined as integral of complex function etc.

There are some clues that we should not stop with the computations in $\mathbb{R}$ and $\mathbb{C}$, and that further generalization to *quaternions* $\mathbb{H} \sim \mathbb{R}^4$ (introduced by Hamilton) or even *octonions* $\mathbb{G} \sim \mathbb{R}^8$ (introduced by Graves) are particulary interesting and valuable, even though the role of these *hypercomplex* numbers is not widely understood yet.

In order to define *hypercomplex* algebras, one has to consider not only two algebraic operations $+$ and $\times$, but also one geometric map: $x \mapsto \bar{x}$, where $\bar{x}$ denotes the conjugate vector of x.

The three operations are defined recursively as we define the algebras, in the following manner. Let A_k be the real *hypercomplex* algebra of dimension 2^k, $k \geq 1$. It is constructed recursively as $A_k = A_{k-1} \times A_{k-1}$ by means of the three following operations:

$$\begin{aligned}
\text{addition:} \quad & (a,b) + (c,d) = (a+c, b+d), \\
\text{conjugacy:} \quad & \overline{(a,b)} = (\bar{a}, -b), \\
\text{multiplication:} \quad & (a,b) \times (c,d) = (ac - \bar{d}b, da + b\bar{c}),
\end{aligned}$$

where ac denotes $a \times c$ in A_{k-1}. For $k = 0$, A_0 is taken to be the field $\mathbb{R}$ with the arithmetic operations $+$ and $\times$, the conjugacy map being the identity on $\mathbb{R}$: $a \mapsto \bar{a} = a \in \mathbb{R}$. This construction is known to algebraists as the Cayley-Dickson doubling process.

About computations with *hypercomplex* numbers, and why only *real numbers, complex numbers, quaternions* and *octions* are suitable for computations see [9, 10] and references herein.

Explicitly multiplication in $\mathbb{H}$ can be expressed as $(a,b,c,d) \times (a',b',c',d') = (a'',b'',c'',d'')$, with

$$a'' = aa' - bb' - cc' - dd'$$
$$b'' = ab' + ba' + cd' - dc'$$
$$c'' = ac' + ca' + db' - bd'$$
$$d'' = ad' + da' + bc' - cb'$$

We consider the *logistic map* over *quaternions* $\mathbb{H} \sim \mathbb{R}^4$

$$z_{k+1} = rz_k(1-z_k), \quad r, z_k \in \mathbb{H}, \quad k = 0, 1, \ldots. \qquad (2)$$

with given initial value z_0, for example $z_0 = (0.5, 0, 0, 0)$. The logistic map (2) has been extensively studied over $\mathbb{R}$ (real numbers) and $\mathbb{C}$ (complex numbers). Despite its great simplicity this map exhibits an extremely complex behaviour. The study of (2) on $\mathbb{R}$ gives birth to the Feigenbaum tree while the analysis of (1) on $\mathbb{C}$ leads to the famous Mandelbrot and Julia fractal sets.

We further deal with 2D projections of Mandelbrot set in 4D quaternionic space. Any two components of r are set to zero, while the remaining two vary. For example,

$$\mathcal{M}_{12} = \left\{ (r_1, r_2) : r = (r_1, r_2, 0, 0), \lim_{k \to \infty} |z_k| < \infty \right\},$$

$$\mathcal{M}_{24} = \left\{ (r_2, r_4) : r = (0, r_2, 0, r_4), \lim_{k \to \infty} |z_k| < \infty \right\}.$$

Note that $\mathcal{M}_{12}$ is just the famous Mandelbrot set in $\mathbb{C}$. We also get that $\mathcal{M}_{12} = \mathcal{M}_{13} = \mathcal{M}_{14}$ and $\mathcal{M}_{23} = \mathcal{M}_{24} = \mathcal{M}_{34}$.

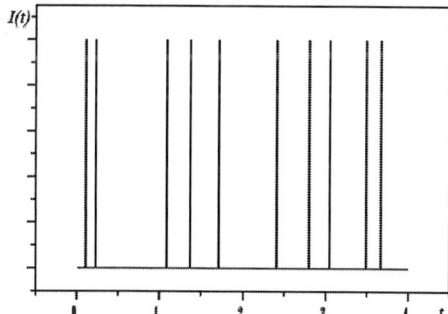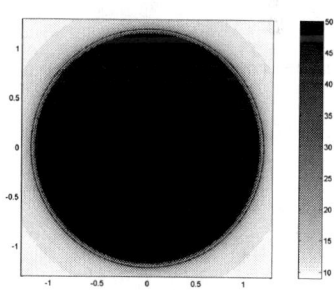

FIGURE 1. (Left) Current $I(t)$ vs time t. Such dependences appear when registering the consecutives heart beats, cars on a highway passing through the reference point, transactions in financial markets etc.; (Right) Approximation (after 50 iterations) of Mandelbrot set $\mathcal{M}_{23}$ (one gets exactly the same for $\mathcal{M}_{24}$ or $\mathcal{M}_{34}$).

The approximations (for finite number of iterations) of Mandelbrot set $\mathcal{M}_{23} = \mathcal{M}_{24} = \mathcal{M}_{34}$ (near its boundary) are fractal circles (see Fig. 1, right), dependant only on radius $R = \sqrt{r_2^2 + r_3^2}$.

Define the *point process* R_j as the values of radius of each circle – mathematically they are the values of R, small change of which result in significant change of number of iterations needed for $|z_k|$ to reach "infinity" (10^{10} for example). The values R_j correspond to peaks in Fig. 2, left.

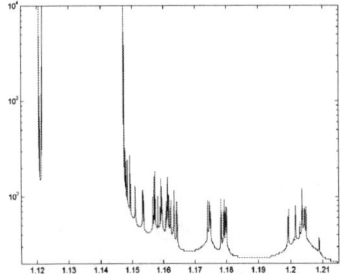 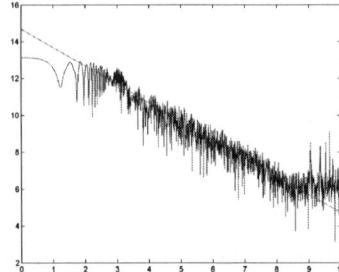

FIGURE 2. (Left) The number of iterations needed to reach $|z_k| > 10^{10}$ vs radius R when computing $\mathcal{M}_{23}$; (Right) $\log_{10} S(f)$ vs $\log_{10} f$ with $N = 796474$. The plot is compared to the function $1/f$.

According to (1), the *power spectral density* of such *point process* is defined as

$$S(f) \approx \frac{2}{R_N - R_1} \left| \sum_{j=1}^{N} e^{-i2\pi f R_j} \right|^2,$$

here N is the volume of point process data ($N \to \infty$, as R_j recording resolution increases).

We obtain (see Fig. 2, right) that $S(f) \sim 1/f$, i.e. radiuses R_j of fractal circles in Mandelbrot set $\mathcal{M}_{23}$ exhibit pure $1/f$ noise ($\alpha = 0$) or unifractal noise.

ACKNOWLEDGMENTS

We acknowledge support by the Lithuanian State and Studies Foundation.

REFERENCES

1. B. Pilgram, and D. T. Kaplan, *Physica D*, **114**, 108–122 (1998).
2. C.-K. Peng, S. Havlin, H. E. Stanley, and A. L. Goldberger, *Chaos*, **5**, 82–87 (1995).
3. B. B. Mandelbrot, *Multifractals and 1/f Noise*, Springer, New York, 1999.
4. B. Kaulakys, and T. Meškauskas, *Phys. Rev. E*, **58**(6), 7013–7019 (1998).
5. B. Kaulakys, and T. Meškauskas, *Microelectronics Reliability*, **11**(40), 1781–1785 (2000).
6. B. Kaulakys, and T. Meškauskas, "Models for generation of $1/f$ noise," in *Noise in Physical Systems and 1/f Fluctuations*, edited by C. Surya, ICNF 1999 Conference Proceedings 15th International Conference on Noise in Physical Systems and 1/f Fluctuations, HongKong, China, 1999, pp. 375–378.
7. V. Gontis, and B. Kaulakys, *Physica A*, **343**, 505–514 (2004).
8. V. Gontis, and B. Kaulakys, *Physica A*, **344**(1-2), 128–133 (2004).
9. F. Chaitin-Chatelin, T. Meškauskas, and A. N. Zaoui, *CERFACS Technical Report TR/PA/00/74*, available at http://www.cerfacs.fr/algor/reports/2000, (2000).
10. F. Chaitin-Chatelin, and T. Meškauskas, *Nonlinear Analysis*, **47**, 3391–3400 (2001).

A Fast Stochastic Digital Signal Generator Based on Chaotic Iteration

Fu Chong, Wang Pei-Rong, Xu Zhe and Zhu Wei-Yong

School of Information Science and Engineering, Northeastern University, Shenyang 110004, China

Abstract. In this paper, an improved random bit recycle multi-bit quantification algorithm based on Logistic mapping was proposed to resolve the high calculation complexity problem in generating stochastic sequence by using chaotic iteration and to improve the performance of the sequence. The balance and correlation property of the generated sequence were analyzed. The sequence was proved to obey the binary Bernoulli stochastic distribution and the auto and cross correlation was proved to obey normal distribution $N(0,1/N)$. The simulation results indicate that the algorithm has excellent balance and correlation property.

Keywords: stochastic signal, chaotic iteration, Logistic mapping, auto/cross correlation.
PACS: 05.45.Gg

INTRODUCTION

A chaotic system is a pseudo stochastic system generated by determinative iteration. It is widely used in information security fields such as data encryption, image encryption and spread spectrum communication system, etc. nowadays due to its unpredictable, initial sensitive and Gauss like statistical characteristics. The primitive generated chaotic sequence is analog and cannot be used in digital system directly. The binary quantification process is the key step in generating stochastic digital signals and it has a great influence on system performance. The quantification method widely used now is defining a binary quantification threshold function:

$$\sigma_c(x) = \begin{cases} 0 & x < c \\ 1 & x \geq c \end{cases}, \qquad (1)$$

then we can get the binary stochastic sequence $\{\sigma_c(f^n(x))\}_{n=0}^{N}$ by quantifying $f^n(x)$ with a properly selected threshold c in which $f^n(x)$ is the analog value of n times iteration and c is usually selected as the average of $\{f^n(x)\}_{n=0}^{N}$ [1]. By using this method, we can get only one bit through each iterating. For applications with large data, the computation amount is tremendous and the long sequence would come be periodic on the effect of computation precision. Some scholars had proposed a multi-bit quantification algorithm to reduce the calculation complication and improve the performance of the sequence [2]. In this paper, an improved random bit recycle multi-bit quantification algorithm based on Logistic mapping was proposed and its performance was analyzed.

CP780, *Noise and Fluctuations: 18th International Conference on Noise and Fluctuations-ICNF 2005*,
edited by T. González, J. Mateos, and D. Pardo
© 2005 American Institute of Physics 0-7354-0267-1/05/$22.50

STATISTIC PROPERTIES OF LOGISTIC MAPPING

Logistic mapping is defined as:

$$x_{n+1} = \mu x_n (1-x_n), \quad x_n \in [0,1], \text{ usually } \mu = 4. \tag{2}$$

The probability density of chaotic sequence generated by Eq.2 is:

$$\rho(x) = \begin{cases} \dfrac{1}{\pi\sqrt{x(1-x)}} & 0 < x < 1 \\ 0 & \text{others} \end{cases} \tag{3}$$

$\rho(x)$ is independent of initial value x_0 [3]. We can get the following properties by using the probability density defined by Eq.3 [4].

Property 1 The average of chaotic sequence generated by Eq.2 is:

$$\bar{x} = \lim_{N \to \infty} \frac{1}{N} \sum_{i=0}^{N-1} x_i = \int_0^1 x\rho(x)dx = 0.5. \tag{4}$$

Property 2 The auto correlation function of chaotic sequence $\{x_n\}$ is:

$$AC(\tau) = \lim_{N \to \infty} \frac{1}{N} \sum_{i=0}^{N-1} (x_i - \bar{x})(x_{i+\tau} - \bar{x}) = \int_0^1 xf^\tau(x)\rho(x)dx - \bar{x}^2 = \begin{cases} 0.125 & \tau = 0 \\ 0 & \tau \neq 0 \end{cases}. \tag{5}$$

Property 3 For two different initial value x_{01} and x_{02}, the cross correlation function of the two sequences is:

$$CC12(\tau) = \lim_{N \to \infty} \frac{1}{N} \sum_{i=0}^{N-1} (x_{i1}-\bar{x})(x_{(i+\tau)2}-\bar{x}) = \int\int x_1 f^\tau(x_2)\rho(x_1)\rho(x_2)dx_1 dx_2 - \bar{x}^2 = 0. \tag{6}$$

RANDOM BIT RECYCLE MULTI-BIT QUANTIFICATION ALGORITHM

The random bit recycle multi-bit quantification algorithm is constructed as follows:

(I) Selecting an initial value x_0, generating a real value x_i from x_{i-1} by Eq.2, $x_i \in [0,1]$, usually i must be large enough to skip some preceding generated values in order to enhance the security;

(II) Transforming x_i to binary form: $(0.a_0 a_1 a_2 ...)_2$, $a_i = 0$ or 1, selecting its first n bit $a_0 a_1 a_2 ... a_n$;

(III) Selecting another initial value x_0', generating another sequence $\{x_n\}$ by Eq.2, quantifying its $x_{m \times i}$ to $x_{m \times (i+1)}$ real values to sequence $\{r_n\}$ that composed of different integers from 1 to n with random orders, m may difference for each quantification;

(IV) Shifting the sub sequence $a_0 a_1 a_2 ... a_n$ according to $\{r_n\}$, generating other n subsequences with length $n+1$;

(V) Concatenating all the generated subsequences together. The rest may be deduced by analogy, thus getting the binary sequence with the length needed.

PERFORMANCE ANALYZING OF THE RANDOM BIT RECYCLE MULTI-BIT QUANTIFICATION ALGORITHM

Balance Property Analyzing

Lemma 1 The sequence generated by above quantification algorithm is a binary Bernoulli stochastic sequence. For sequence with length N, the distribution of the differences between 0 and 1 is shown in Table 1.

TABLE 1. The distribution of the differences between 0 and 1 for sequence with length N.

Number of differences	N	$N-2$	$N-4$	...	$-N+4$	$-N+2$	$-N$
Probability	$(\frac{1}{2})^N$	$C_N^1(\frac{1}{2})^N$	$C_N^2(\frac{1}{2})^N$	...	$C_N^{N-2}(\frac{1}{2})^N$	$C_N^{N-1}(\frac{1}{2})^N$	$(\frac{1}{2})^N$

Proof. The average of chaotic sequence generated by Eq.2 is 0.5, the correspondence binary form is 0.1000 ... ..., so the probability of a_0 is 0 or 1 are equal for each iteration. x_{i+1} satisfies:

$$\begin{cases} x_{i+1} \geq 0.5 & 0.5 - \frac{\sqrt{2}}{4} \leq x_i \leq 0.5 + \frac{\sqrt{2}}{4} \\ x_{i+1} < 0.5 & others \end{cases}, \quad (7)$$

because $\int_{0.5-\frac{\sqrt{2}}{4}}^{0.5+\frac{\sqrt{2}}{4}} \rho(x)dx = -\frac{\arcsin(1-2x)}{\pi} \Big|_{0.5-\frac{\sqrt{2}}{4}}^{0.5+\frac{\sqrt{2}}{4}} = 0.5$, thus the probability of a_0 is 0 or 1 for next iteration are all 0.5 no matter what the previous x_i is. The balance and independence properties of $a_1, a_2, \ldots$ can be proved similarly. The distribution of the differences between 0 and 1 in the generated sequence can be get through further analyzing, the average is 0 and the variance is N. The simulation result is shown in Fig. 1, it shows that the quantification algorithm has an excellent balance property.

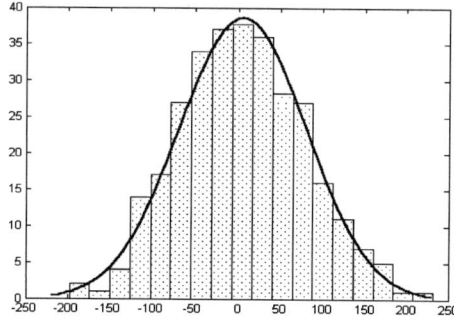

FIGURE 1. Frequency distribution of the differences between 0 and 1 of the sequence generated by the random bit recycle multi-bit quantification algorithm for $x_0 = 0.3$ and $x_0' = 0.4$.

Correlations Property Analyzing

The auto and cross correlation properties are import measurements to identify the performance of a stochastic sequence.

Lemma 2 The sequences $\{x_i\}_{i=1}^N$ and $\{y_i\}_{i=1}^N$ that quantified by above algorithm satisfies: When N is large enough, the auto correlation side lobe of $\{x_i\}_{i=1}^N$ and the cross correlation of $\{x_i\}_{i=1}^N$ and $\{y_i\}_{i=1}^N$ obey the normal distribution $N(0,1/N)$.

The simulation result of a 5000 length sequence is shown in Fig.2 for $x_0 = 0.3$ and $y_0 = 0.30001$. The maximum absolute values of auto correlation side lobe and cross correlation are 0.0166 and 0.0148, which indicate that the algorithm has an excellent correlation property.

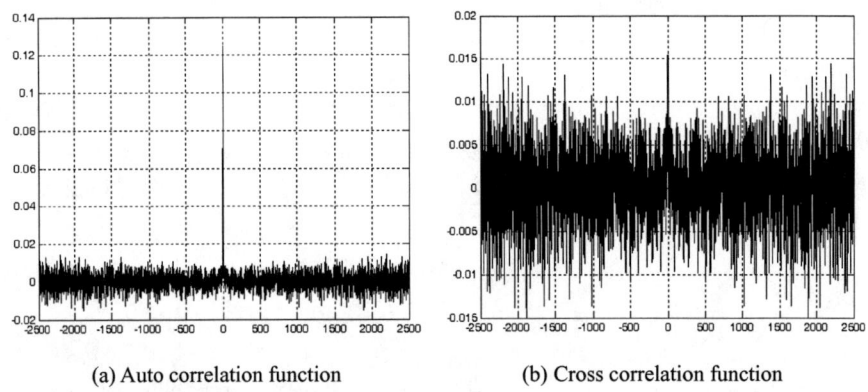

(a) Auto correlation function   (b) Cross correlation function

FIGURE 2. Auto and cross correlation functions of sequence generated by random bit recycle multi-bit quantification algorithm.

CONCLUSION

By using this quantification algorithm, we can get $n \times n$ bits through each iterating. The generating speed was largely increased and the performance of the sequence was guaranteed. This algorithm is general and can be used on other chaotic mappings such as Chebyshev, Lorenz, etc. similarly.

REFERENCES

1. Heidari-Bateni G, McGillem C D, *A Chaotic Direct-Sequnce Spread-Spectrum Communication System*. IEEE Trans. Commun **42**, 1524-1527 (1994).
2. Lipton J M, Dabke K P, *Spread Spectrum Communications Based on Chaotic Systems*. International J. Bifurcation and Chaos **6**, 2361-2374 (1996).
3. Parlitz U, Ergeinger S, *Robust Communication Based on Chaotic Spreading Sequences*. Physics Letters A **5**, 146-150 (1994).
4. Ling Cong, Wu Xiaofu, Sun Songgeng, *A General Efficient Method for Chaotic Signal Estimation*. IEEE Trans. On Signal Processing **47**, 1424-1428 (1999).

Estimating the variance of multiplicative noise

Dafydd Evans

School of Computer Science, Cardiff University, 5 The Parade, Cardiff CF24 3AA, Wales, UK

Abstract. When constructing non-parametric models from noisy data, it is useful to have information regarding the statistical properties of the noise distribution. In many cases, such information is not explicitly available, and must be estimated directly from the data. Under the hypothesis of *additive* noise, algorithms for estimating the variance of the noise distribution have appeared in the literature. In this paper we present a novel algorithm for estimating the noise variance under a *multiplicative* hypothesis.

Keywords: nonlinear modelling, multiplicative noise, near-neighbours
PACS: 02.50.Cw, 02.60.Gf, 02.70.Rr, 05.45.Tp

INTRODUCTION

Let $f : \mathbb{R}^m \to \mathbb{R}^n$ be a smooth function, mapping an input $x \in \mathbb{R}^m$ to an output $y \in \mathbb{R}^n$. The goal of statistical modelling is to estimate the function f from a finite set of observations $S = \{(x_1, y_1), \ldots, (x_M, y_M)\}$, a task made considerably more difficult by the presence of noise in the data. In this paper, *noise* is defined to be that part of the output which cannot be accounted for by any smooth transformation of the input. Noise is usually represented by a *random variable*, and simply knowing its *variance* σ^2 is important for many function estimation techniques. For example, σ^2 is used to define a stopping criterion for neural network training [1], and also to determine a threshold for wavelet de-noising [2]. Noise is invariably assumed to be *additive*:

$$y = f(x) + r \quad \text{where} \quad E(r) = 0 \qquad (1)$$

The Gamma test [3] is a non-parametric algorithm for estimating the variance of additive noise, using only the available data (x_i, y_i). A useful overview of the method can be found in [4]. In [5], the estimate computed by the algorithm is shown to be (weakly) consistent as the number of data points increases. In this paper, we present a related algorithm for estimating the variance of *multiplicative* noise:

$$y = f(x) r \quad \text{where} \quad E(r) = 1 \qquad (2)$$

NOISE ESTIMATION

To illustrate the ideas underpinning our algorithm, we first describe a new algorithm for the estimating the variance of additive noise. Our algorithms exploit the *near-neigbour* structure of the input points $x_1, \ldots, x_M$, which may be computed in time $O(M \log M)$ using kd-trees [6]. For any input point x, choose x' and x'' to be any two points from among its first p nearest neighbours in the set $\{x_1, \ldots, x_M\}$; let y' and y'' be the outputs corresponding to x' and x'', and let r' and r'' represent the noise measured on the outputs y' and y'' respectively. We assume that the noise values r, r' and r'' are independent and identically distributed, and also independent of the associated input points x, x' and x''; the input points are also assumed to be independent and identically distributed.

Our algorithm for additive noise is based on the product of *differences* $(y-y')(y-y'')$. Starting with (1) and applying the above conditions, it is easily shown that

$$E\big((y-y')(y-y'')\big) = \sigma^2 + E\big((f(x)-f(x'))(f(x)-f(x''))\big) \qquad (3)$$

where σ^2 is the variance of the noise distribution, and the expected value on the right is taken with respect to the distribution of the input points. We think of $E\big((y-y')(y-y'')\big)$ as an estimate of σ^2, while $E\big((f(x)-f(x'))(f(x)-f(x''))\big)$ quantifies the associated estimation error. Because f is approximately locally linear[1] at x, this error term satisfies

$$E\big((f(x)-f(x'))(f(x)-f(x''))\big) \approx G_f E\big(|x-x'||x-x''|\big) \qquad (4)$$

where G_f is related to the expected gradient of f (with respect to the input distribution).

For $k \in \mathbb{N}$, let x_{i_k} denote the kth nearest neighbour of x_i among the input points $x_1, \ldots, x_M$, and let y_{i_k} denote the output corresponding to x_{i_k}. For every $k \in \{1, \ldots, p\}$, we estimate the expected values $E\big((y-y')(y-y'')\big)$ and $E\big(|x-x'||x-x''|\big)$ by the sample means

$$\Gamma_k = \frac{1}{M}\sum_{i=1}^{M}(y_i - y_{i_k})(y_i - y_{i_{k+1}}) \quad \text{and} \quad \Delta_k = \frac{1}{M}\sum_{i=1}^{M}|x_i - x_{i_k}||x_i - x_{i_{k+1}}| \qquad (5)$$

respectively. In [5], is shown that sample means such as Γ_k and Δ_k satisfy a weak law of large numbers, in the sense that they converge to their expected values in probability as the number of data points increases. In view of this, we substitute the sample means of (5) for the distribution means of (3) and (4), which yields

$$\Gamma_k \approx \sigma^2 + G_f \Delta_k \qquad (6)$$

where we think of G_f as a (finite) *constant*, in the sense that it is independent of any particular realisation of the sample data $\{(x_i, y_i)\}$. Finally, we exploit this (approximate) linear relation between Γ_k and Δ_k, by performing linear regression on the pairs (Δ_k, Γ_k) to estimate the value of Γ_k in the limit as $\Delta_k \to 0$, which provides an estimate for σ^2.

THE ALGORITHM

To estimate the variance of multiplicative noise, rather than look at the product of differences $(y-y')(y-y'')$ we now consider the product of *ratios* $(y/y')(y/y'')$, or

---

[1] The unknown function f is assumed to be smooth, and is therefore approximately linear in sufficiently small regions around x. Here, the region of interest is the p-*nearest neighbour ball*, centred at x and having the pth nearest neighbour of x on its boundary. Whether this ball is 'sufficiently small' to ensure local linearity depends on the density of the input distribution near x.

equivalently $y^2/y'y''$. Because the noise values r, r' and r'' are independent of the inputs x, x' and x'', starting from (2) it easily follows that

$$\frac{E(y^2)}{E(y'y'')} = (\sigma^2+1)\left(\frac{E(f(x)^2)}{E(f(x')f(x''))}\right) \tag{7}$$

where we have used the fact that $E(r'r'') = E(r')E(r'') = 1$ and $E(r^2) = \sigma^2 - 1$. Furthermore, because f is smooth, by Taylor's theorem

$$f(x)^2 \approx f(x')f(x'') - \left((x-x')f(x'') + (x-x'')f(x')\right)\nabla f(x) \tag{8}$$

Thus by (7), and using the fact that x, x' and x'' are identically distributed,

$$\frac{E(y^2)}{E(y'y'')} \approx (\sigma^2+1) + G_f E\left((x-x') + (x-x'')\right) \tag{9}$$

where

$$G_f = -(\sigma^2+1)E\left(f(x)\nabla f(x)\right)/E(f(x')f(x'')) \tag{10}$$

Because f is smooth, G_f is bounded provided the noise variance σ^2 is finite, and also provided f is bounded (and not identically zero) over the set of possible inputs. Following the discussion leading to (5), for every $k \in \{1,\ldots,p\}$ we estimate the ratio $E(y^2)/E(y'y'')$ and the expected value $E\left((x-x') + (x-x'')\right)$ by the empirical values

$$\Gamma_k = \frac{\sum_{i=1}^M y_i^2}{\sum_{i=1}^M y_{i_k}y_{i_{k+1}}} \quad \text{and} \quad \Delta_k = \frac{1}{M}\sum_{i=1}^M \left((x_i - x_{i_k}) + (x_i - x_{i_{k+1}})\right) \tag{11}$$

respectively. As in (6), we then replace the expected values of (9) by the empirical values of (11), leading to

$$\Gamma_k \approx (\sigma^2+1) + G_f \Delta_k \tag{12}$$

Finally, we compute our estimate for σ^2 by exploiting this approximate linear relation between Γ_k and Δ_k, using simple linear regression to estimate the value of Γ_k in the limit as $\Delta_k \to 0$ (note that because the intercept estimates $\sigma^2 + 1$, we must subtract one from this to get the final estimate). It is interesting to note that the gradient of the regression line can be interpreted as an estimate of G_f, and might therefore represent some useful information regarding the unknown function f.

Our algorithm is explicitly stated as Algorithm 1.

Algorithm 1

1. Compute the p-nearest neighbour structure of the input points $\{x_1,\ldots,x_M\}$.
for $k \in \{1,\ldots,p\}$ **do**
   2. Compute the pair (Δ_k, Γ_k) as defined in (11).
end for
3. Perform linear regression on the pairs $\{(\Delta_k, \Gamma_k) : k = 1,\ldots,p\}$.
4. Return the intercept of the regression line with the $\Delta_k = 0$ axis (minus one).

EXPERIMENTAL RESULTS

We generated a set of 2000 points x_i, each selected uniformly at random from the unit interval $[0,1]$, and a set of 2000 noise values r_i, each selected according to a Gaussian distribution of unit mean and variance 0.2. The output points y_i were then constructed according to the rule $y_i = f(x_i)r_i$ where $f(x) = 3\sin(8\pi x) + \cos(23\pi x)$.

The left-hand plot of Figure 1 shows the output points y_i plotted against the input points x_i. The multiplicative nature of the noise is evident here – the noise becomes more pronounced as the distance between the underlying curve $f(x)$ and the $y = 0$ axis increases.

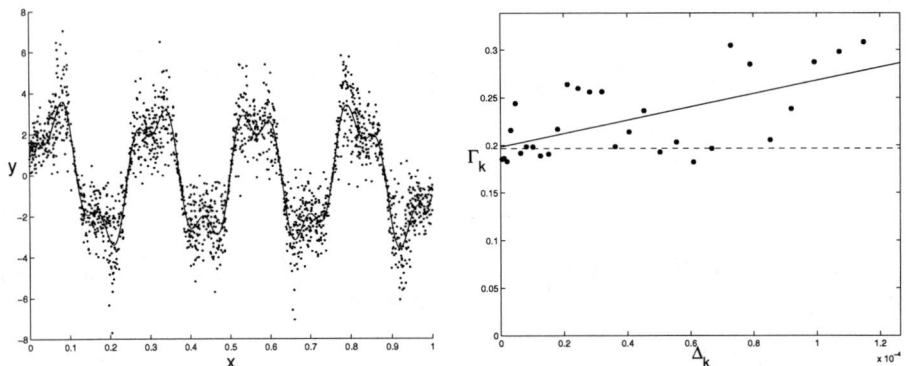

FIGURE 1. The noisy data (x_i, y_i) and the regression plot produced by the algorithm.

The data set $\{(x_i, y_i)\}$ was processed according to Algorithm 1, with $p = 30$. The resulting regression plot is shown on the right of Figure 1, where it can be seen that the intercept of the regression line (solid line) is in close agreement with the variance of the noise (dashed line). It can also be seen that the point estimates Γ_k become increasingly inaccurate as Δ_k increases. Experimental evidence suggests that our algorithm can successfully estimate the variance of multiplicative noise, provided there are sufficient data available. We have observed that more data points are required as the compexity of the function f increases, a fact that will be addressed in a future study.

ACKNOWLEDGMENTS

The author acknowledges the support of the Royal Society (URF 516002.K5736).

REFERENCES

1. S. Haykin, *Neural Networks*, Prentice Hall, 1998.
2. D. L. Donoho, and I. M. Johnstone, *Biometrika*, **81**, 425–455 (1994).
3. A. Stefánsson, N. Končar, and A. J. Jones, *Neural. Comput. Appl.*, **5**, 131–133 (1997).
4. A. J. Jones, *Comp. Manage. Sci.*, **1**, 109–149 (2004).
5. D. Evans, and A. J. Jones, *Proc. R. Soc. Lond. A*, **458**, 2759–2799 (2002).
6. J. L. Bentley, *Comm. ACM*, **18**, 309–517 (1975).

MATERIALS

Microwave Noise In Biased AlGaN/GaN And AlGaN/AlN/GaN Channels

A. Matulionis, J. Liberis, M. Ramonas

Semiconductor Physics Institute, A. Goštauto 11, Vilnius 01108, Lithuania
E-mail: matulionis@pfi.lt

Abstract. Noise temperature is measured at 10 GHz at room temperature for biased AlGaN/GaN and AlGaN/AlN/GaN two-dimensional channels. Interpretation of the experimental results, through Monte Carlo simulation, takes into account interaction of hot electrons with phonons. The calculated longitudinal noise temperature exceeds the transverse one, but the resultant anisotropy of noise becomes substantially weaker when accumulation of nonequilibrium (hot) optical phonons is taken into account. The considered intense interaction of hot electrons with hot phonons and weak coupling with the thermal bath shifts the hot-electron–hot-phonon subsystem closer to the equilibrium at the elevated temperature as compared with the hot-electron subsystem interacting with the equilibrium phonons.

Keywords: Microwave noise, GaN heterostructures, two-dimensional electron gas, Monte Carlo simulation, noise anisotropy, hot electrons, hot phonons
PACS: 02.70.Uu, 63.20.Kr, 72.20.Ht, 72.70.+m, 73.63.Hs

INTRODUCTION

Anisotropy of noise in an isotropic semiconductor is an indicator of displacement of electron subsystem from the thermal equilibrium. The anisotropy will be considered for a degenerate two-dimensional electron gas (2DEG) subjected to a high electric field in nitride channels, where the hot electrons and hot phonons are known to form the subsystem that is weakly coupled with the thermal bath [1]. Monte Carlo simulation shows that the hot-electron temperature can be introduced [2]. Our goal is to demonstrate that the noise anisotropy depends on accumulation of hot phonons.

EXPERIMENTAL DETAILS AND MONTE CARLO MODEL

The noise temperature was measured in the current direction for two-terminal samples supplied with coplanar ohmic electrodes and subjected to electric field applied in the 2DEG plane. The gated 10 GHz radiometric setup was used for noise temperature measurements [3,4]. Short 150 ns voltage pulses were applied to avoid channel self-heating [5]. AlN layer was inserted in AlGaN/AlN/GaN heterostructure in order to suppress the hot-electron real-space transfer noise [2]; this source dominated in AlGaN/GaN channels at high electric fields [1]. Electron density and low-field mobility data are given in Table 1 together with sample dimensions.

TABLE 1. Parameters of nitride 2DEG channels

Heterostructure	Electron Sheet Density, cm^{-2}	Electron Mobility, cm^2/V·s	Channel Length, µm	Electrode Width, µm
Al$_{.15}$Ga$_{.85}$N/GaN	4·10^{12}	1100	7	120
Al$_{.33}$Ga$_{.67}$N/AlN/GaN	1·10^{12}	1152	5	100

Monte Carlo simulation was carried out for AlGaN/AlN/GaN within a many-subband parabolic model. Inelastic scattering of electrons by longitudinal-optical and acoustic phonons was taken into account. Elastic scattering mechanisms (impurities, dislocations, etc.) were ignored. The acoustic phonons were assumed to be in equilibrium with the thermal bath, the displacement of optical phonons from the equilibrium was treated in the hot-phonon lifetime approximation (for details see [2]).

RESULTS

The experimental results on longitudinal noise temperature show (Fig. 1, stars) that the hot-electron real-space transfer is the dominant source of noise in the AlGaN/AlN channel at the supplied electric power P_s exceeding 10 nW/electron. This source is heavily suppressed when the thin AlN barrier is inserted. Thus, an almost linear dependence of the noise temperature on the supplied power is obtained for the AlGaN/AlN/GaN channel at $P_s > 10$ nW/electron (circles). In this channel, a steep increase in the noise temperature tends to appear when the power exceeds $P_s > 100$ nW/electron (this corresponds to the electric field over 60 kV/cm). An approximately square-root dependence is observed at a low supplied power. The change of behavior

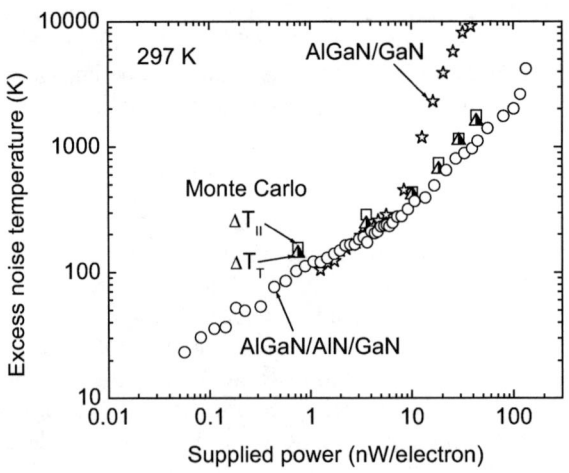

FIGURE 1. Dependence of excess noise temperature on supplied electric power. Experimental data: AlGaN/GaN (stars), and AlGaN/AlN/GaN (circles). Monte Carlo simulation: longitudinal noise (squares), transverse noise (triangles).

indicates that interaction of hot electrons with optical phonons becomes the dominant dissipation mechanism at a high supplied power [4,6].

The main features of noise in the AlGaN/AlN/GaN channel are reproduced through Monte Carlo simulation. When hot phonons are taken into account, the simulated longitudinal noise temperature (Fig.1, squares) is close to the transverse one (triangles). At a high density of 2DEG, the transverse noise temperature equals the electron temperature [3,7]. Consequently, the hot electron temperature in AlGaN/AlN/GaN can be estimated from the measured longitudinal noise temperature in the range of bias where the interaction of hot electrons with hot phonons is responsible for dissipation of the supplied power.

NOISE ANISOTROPY

The results of Monte Carlo simulation at a high bias are sensitive to the way we treat the interaction of hot electrons with optical phonons. The simulated noise anisotropy increases as the supplied power increases if the optical phonon distribution is assumed to remain at equilibrium with the thermal bath even at a high bias (Fig. 2, closed squares). Thus, the noise anisotropy is an indicator of displacement from the equilibrium. Note, that the anisotropy remains close to unity at a high supplied power if the hot-phonon accumulation is taken into account (Fig. 2, open squares) despite of an essential increase in electron temperature as the noise data show (Fig. 1).

Let us apply the electron temperature approximation to the hot electrons in the 2DEG channel in order to compare the simulated noise anisotropy with the anisotropy calculated in the electron temperature approximation.

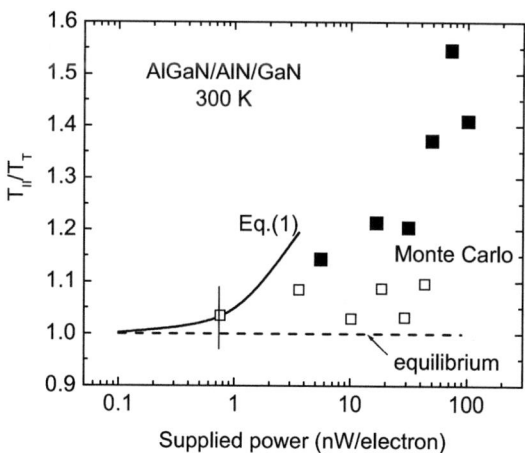

FIGURE 2. Dependence of noise temperature anisotropy on supplied electric power. Symbols: Monte Carlo simulation with hot phonon accumulation taken into account (open squares) and hot phonons neglected (closed squares). Solid line is Eq. (1). Dashed line stands for equilibrium.

In the electron temperature approximation under the dominant quasi-elastic scattering with the termal bath, the anisotropy of noise is given by [7]:

$$\frac{T_\parallel}{T_\perp} = 1 + \frac{T_\perp}{4(T_\perp - T_L)} \left(\frac{\sigma_\parallel}{\sigma_\perp} - 1 \right)^2 \frac{\sigma_\perp}{\sigma_\parallel} \qquad (1)$$

where $\sigma_\parallel$ is the differential conductivity, $\sigma_\perp$ is the transverse conductivity, and T_L is the lattice temperature. This expression is valid for warm electrons, it is compared (Fig. 2, solid line) with the results of simulation (squares).

Open squares are closer to dashed line and farther away from solid line as compared with closed squares (Fig. 2). This result can be interpreted as follows. High-energy electrons easily emit optical phonons. The emitted optical phonons stay in the channel until they disintegrate into acoustic phonons. They support a high probability for phonon absorption by low-energy electrons. Thus, accumulation of hot phonons supports intense exchange of energy between the electrons and the phonons. On the other hand, the hot electrons and hot phonons are weakly coupled with the thermal bath. As a result, one has a quasi-isolated hot-electron–hot-phonon subsystem. The obtained low anisotropy of noise indicates that, due to the re-absorption, the hot electrons and the hot phonons are almost at equilibrium at the elevated (hot-electron) temperature.

In conclusion, the noise data suggest that the hot-electron–hot-phonon subsystem is closer to the thermal equilibrium (at the elevated temperature) as compared with the hot-electron subsystem interacting with the equilibrium phonons.

ACKNOWLEDGMENTS

This work has been partly funded by the US Office of Naval Research (ONR Award No. N00014-03-1-0558 monitored by Dr. Colin Wood), the Lithuanian National Foundation for Science and Education (Contract No. T-24/05), and the European Commission within the Network of Excellence "SINANO" (IST–506844).

REFERENCES

1. A. Matulionis, J. Liberis, I. Matulionienė, M. Ramonas, L. F. Eastman, J. R. Shealy, V. Tilak, and A. Vertiatchikh, *Phys. Rev.* **B 68,** 035338 (2003).
2. M. Ramonas, A. Matulionis, J. Liberis, L. F. Eastman, X. Chen, Y.-J. Sun, *Phys. Rev.* **B 71,** 075324 (2005).
3. H. L. Hartnagel, R. Katilius, A. Matulionis, *Microwave Noise in Semiconductor Devices*, John Wiley & Sons, New York, 2001, 312 p.
4. A. Matulionis, J. Liberis, L. Ardaravičius, M. Ramonas, and J. Smart, *Semicond. Sci. Technol.* **17,** L9–L14 (2002).
5. L. Ardaravičius, J. Liberis, A. Matulionis, L. F. Eastman, J. R. Shealy, and A. Vertiatchikh, *physica status solidi* **(a) 201,** 203–206 (2004).
6. A. Matulionis, J. Liberis, M. Ramonas, I. Matulionienė, L. F. Eastman, A. Vertiatchikh, X. Chen, Y.-J. Sun, *physica status solidi* **(c) 2,** 2585–2588 (2005).
7. R. Katilius, *Phys. Rev.* **B 69,** 245315 (2004).

Generation of Interface States Due to Quantum-Dot Growth in Au/GaAs Schottky Diode Structures

W. Choi[1], H. Nam[1,2], J. Lee[1], B. Yu[1], J. Song[1], H. Yang[2], A. Chovet[3]

[1]*Nano Device Research Center, Korea Institute of Science and Technology, Seoul 136-791, Korea*
[2]*Department of Physics, Chung-Ang University, Seoul 156-756, Korea*
[3]*IMEP. CNRS/INPG UMR 5130, 38016 Grenoble, France*

Abstract. We investigated the low-frequency excess electrical noise characteristics of Au/GaAs Shottky diodes with and without self-assembled InAs quantum-dot layer grown by molecular beam epitaxy. The noise intensity shows $1/f$ behavior and non-quadratic current dependences. The current dependence is explained by the generation of interface states increasing toward the conduction band edge, in the diodes with quantum-dot layer, utilizing the model of random walk of electrons involving interface states. The extracted energy distributions of the interface states for the diodes with and without quantum-dot layer, are presented.

Keywords: InAs, GaAs, Quantum-dots, Schottky diodes, Low-frequency noise, Interface states, Random walk.
PACS: 73.20.At, 73.61.Ey, 73.80.Ey, 74.40.+k, 79.40.+z, 85.30.H

INTRODUCTION

Self-assemble semiconductor quantum-dots (QD's) provide revolutionary opportunity for new physical phenomena and also improvement of performance in electronic and nanophotonics devices [1]. Since a lot of stress is involved during the growth of the three-dimensional QDs, lattice defects are likely to be generated and diffused during the growth of structures containing QD layers. Among different electrical and optical characterization techniques, low-frequency current noise is known to be quite sensitive to the lattice defects. For low-frequency noise measurements, the Schottky barrier structure provides a unique test tool, since the band bending at the metal-semiconductor interface exposes the energetic and spatial distribution of the defects to the Fermi level and those traps located at the Fermi level contribute to the noise generation most. Low frequency current noise measurements can give useful information on the energetic and/or spatial location and the nature of the noise sources in both the depletion layer and the interface [2]-[5].

In this work we analyzed the low-frequency noise characteristics of Au/n-GaAs Schottky diode structures with and without InAs QD layers (preliminary results of which have been reported in ref. [2]) extracted energy distribution of interface states in the diodes. We find the growth of QD layers generates interface states with their density increasing toward the conduction band edge.

DEVICE STRUCTURE AND MEASUREMENTS

Samples used in this study were grown on (001) silicon doped n-typed GaAs substrates in a V80 MBE system. After growth of 1 μm-thick n-typed GaAs buffer layer at 580 °C with silicon dopant, all other structures on the buffer were grown at 480 °C. For samples type A, a single InAs QD layer was sandwiched by 100 Å-thick GaAs layers and capped by a 0.4 μm-thick n-typed GaAs layer. For samples type B (reference), the structure was the same without QD layer. For the QD layer, total coverage of InAs was 3 MLs. Growth rate of InAs and GaAs were ~ 0.07 ML/s and ~ 0.5 ML/s, respectively. Typical background pressure during growth is ~ 4×10^{-8} torr. Average width, height and density of the InAs QDs of ~ 40 nm, ~ 7 nm, and $4.1 \times 10^{10}/cm^2$, respectively, were identified from the atomic force microscopy (AFM) profile of an uncapped QD structure. After the epitaxial growth, Au contacts were fabricated by thermal evaporation through metal mask with 0.3 mm diameter holes.

The current-voltage (I-V) characteristics of the diodes were measured by HP4140B Semiconductor parameter Analyzer. Bias-dependent series resistance was extracted from the measured I-V curves to estimate the bias applied to the metal-semiconductor interface utilizing the analytic expression [6]. Most of the diodes showed linear part at low bias and ideality factors and saturation currents (and hence the barrier height) were determined from those regions utilizing conventional method.

For low-frequency current noise measurements, battery-operated SR570 Transimpedance Amplifier and Agilent E4440A Spectrum Analyzer were utilized at room temperature. All the equipments were PC controlled. The current dependence of the normalized current noise density is shown for diodes types A and B, respectively, in Fig. 1.

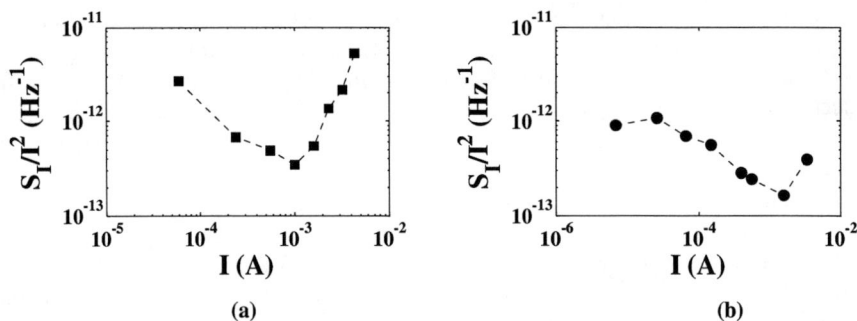

FIGURE 1. Current dependence of the normalized current noise intensity, $S_I/I^2(I)$. (a) The diode with a InAs QD layer (squares, type A), (b) the diode without QD layer (dots, type B, reference).

RESULTS AND DISCUSSION

In general, the current noise density follows the empirical relationship,

$$S_I = K \frac{I^\beta}{f^\gamma}. \tag{1}$$

In Schottky diode, the power index of the frequency, γ, is related to the bulk trap distribution in the depletion (or space charge) region, and the observed values of γ in these structures, between 0.95 and 1.05, indicate rather uniform distribution of bulk traps in the diodes [7]. The deviation of the value of β from 2 can only be explained by the random walk of electrons model [8]-[9], where the noise density is determined by the density of interface states, D_{it}, at the energy level in the band gap where Fermi level is located. The normalized noise density due to random walk of electrons at the metal-semiconductor interface, is given by [8],

$$\frac{S_I}{I^2} = \frac{G}{f}\frac{q}{kT}\left(\frac{q}{4\varepsilon_s}\right)^2 \frac{qD_{it}}{\pi N_d WA}, \tag{2}$$

where $G \approx 0.1$ and W is the depletion width. If the density of interface states is uniform, the noise density will follow quadratic current dependence with some modification due to the bias dependent depletion layer width. As we increase the bias we are probing the interface states density in the direction towards the conduction band edge. So β larger than 2 means that the interface states density is increasing towards the conduction band edge, which were the cases for type A. The extracted interface states density was on the order of 10^{11} to 10^{13} cm^{-2}eV^{-1}. For both types of samples, β was smaller than 2, indicating that the density increased towards the midgap. Existence of midgap interface states in GaAs is a well-documented fact.

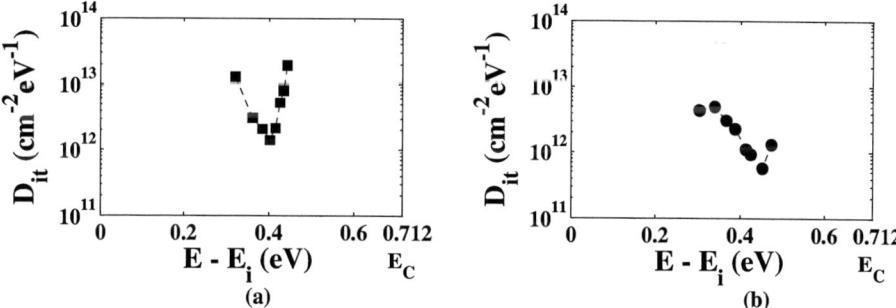

FIGURE 2. Energy distribution of interface states, $D_{it}(E)$, for (a) the diode with QD layer (squares, type A) and (b) the diode without QD layer (dots, type B). The zero energy level is the intrinsic Fermi level, E_i, and the conduction band edge, E_c, is indicated.

This is quite contrary to what Hastas *et al.* [3]-[5] observed with a similar structure. They observed $1/f^\gamma$ noise with $\gamma \sim 0.6$ in Shottky diode with QDs and attributed to the generation of energetically non-uniform bulk states. The difference may stem from the smaller dot density in our sample or different crystal quality of the epitaxial layers. Our results is somewhat similar to those of Hastas *el al.*'s sample with InAs quantum layer [3]-[5], where they observed *1/f* noise and the current dependence steeper than the quadratic and explained by the generation of interface states via the random walk model. However, the deduced amount of interface states was much larger than the normal interface state density by several orders of magnitude and the interface inhomogeneity at the Au/GaAs interface had to be considered. In our case, the derived interface state density was on the order of $10^{11} - 10^{13}/\text{eVcm}^2$ without considering any interface inhomogeneity. In some devices, random telegraph signal noise was observed at certain current levels (bias conditions), which might be related to the discrete energy levels in the QDs.

CONCLUSION

In summary, we extracted the energy distributions of the interface states from the low-frequency noise characteristics of Au/GaAs Schottky diodes with and without InAs QD layer. From the frequency dependence and the current dependence of the current noise density, the major noise generation mechanism in these diodes is found to be the random walk of electrons involving interface states via barrier height modulation. The growth of QD layer is found to generate trap states at the metal-semiconductor interface with its density increasing towards the conduction band edge up to 10^{13} cm^{-2}eV^{-1}, within the measurement range.

ACKNOWLEDGMENTS

This work was partially supported by Ministry of Science and Technology (and now Ministry of Commerce, Industry and Energy) through Nano R&D Program, and by French Embassy in Korea and KISTEP (now KOSEF) through STAR Program.

REFERENCES

1. M. S. Skolnick and D. J. Mowbray, *Annu. Rev. Mater. Res.* **34**, 181-218 (2004).
2. J. D. Song, W. J. Choi, I. K. Han, W. J. Cho, J. I. Lee, Y. B. Yu, C. H. Pyun, J. H. Kim, J. I. Song, and A. Chovet, SPIE Proceedings 5472, Maspalomas, Gran Canaria, Spain, 2004, pp. 432-439.
3. N. A. Hastas, C. A. Dimitriadis, L. Dozsa, E. Gombia, S. Amighetti, and P. Frigeri, *J. Appl. Phys.*, **93**, 3990-3994 (2003).
4. N. A. Hastas, C. A. Dimitriadis, L. Dozsa, and R. Mosca, *J. Appl. Phys.*, **93**, 5833-5835 (2003).
5. N. A. Hastas, D. H. Tassis, C. A. Dimitriadis, L. Dozsa, S. Franchi, and P. Frigeri, *Semiconduc. Sci. Technol.*, **19**, 461-467 (2004).
6. J. I. Lee, J. Brini, and C. A. Dimitriadis, *Electron. Lett.*, **34**, 1268-1269 (1998).
7. J. I. Lee, J. Brini, A. Chovet, and C. A. Dimitriadis, *Solid-St. Electron.*, **43**, 2181-2183 (1999).
8. J. I. Lee, J. Brini, A. Chovet, and C. A. Dimitriadis, *Solid-St. Electron.*, **43**, 2185-2189 (1999).
9. J. I. Lee, I. K. Han, D. C. Heo, J. Brini, A. Chovet, C. A. Dimitriadis and J. C. Jeong, *J. Korean Phys. Soc.*, **37**, 966-970 (2000).

1/f Noise Enhancement In GaAs

J.I. Izpura♣, J. Malo♦

♣*Departamento de Ingeniería Electrónica, ETSIT-UPM*
♦*Departamento de Sistemas Electrónicos y de Control, EUITT-UPM*
Universidad Politecnica de Madrid. 28040-Madrid, SPAIN.

Abstract. This paper presents a way to induce 1/f-like resistance noise in channels from their surface. Random fluctuations of surface charge are created optically in a small area of a GaAs surface by weak and localized optical pulses. The non null surface conductivity of n-GaAs and its surface space charge region, form a slow transmission line that allows the relaxation of such fluctuations over the surface. Their Field Effect on the GaAs channel as they diffuse, leads to a resistance noise with 1/f character well below the Hz range for the usual size of samples.

Keywords: 1/f noise, surface charge fluctuations, surface charge diffusion.
PACS: 73.50.Td, 85.30.De, 73.25.+i, 85.30.Tv

INTRODUCTION

The high gain of GaAs photoconductors at low illumination levels was explained recently by a photoconductance gain based on the geometrical variation of conductive volume in thin epitaxial layers, due to transversal photovoltages being developed at bordering space charge regions (BSCR) [1-2]. Thus, the above gain is due to changes in the conductance of devices not due to changes in their conductivity (inner material), but to changes in their conductive volume. This fact led us to consider if the geometry of GaAs conductors can be taken as perfectly constant in the dark, because very often they are bound by BSCR as that of the substrate-epilayer interface of n-GaAs layers grown on semi-insulating substrates. In this case, a resistance noise would exist in the above devices reflecting the "trembly" character of the BSCR enclosing them.

In addition to the above BSCR, naked n-GaAs surfaces also hold a BSCR, but their weak surface conductivity [3,4] do not allow to have on top a highly conductive "plate" like a FET gate, and the corresponding noise model for this surface BSCR has to consider the existence of random local charge fluctuations over the surface, that will relax with time in the way allowed by the system. Studying this case, we predict some time ago the existence of a new type of resistance noise in conductive channels bound by BSCR: an Interface-Induced Thermal noise (IIT noise) [5] due to the Field-Effect action on the channel of thermal voltage noises of capacitors (kT/C noise) associated with BSCR. Such proposal however, seems hard to be accepted without experimental proofs and here we present an approach to such a proof for the effect of the random appearance of local charge fluctuations in the surface BSCR of an n-GaAs conductive channel. It is shown that 1/f-like resistance noise can be induced in an n-GaAs channel from its surface, being a particular case of IIT noise.

Propagation of Charge Fluctuations Over the Surface

The ability of the GaAs surfaces to hold a surface charge density, together with their finite surface sheet resistance R_Θ [3-4] allow the spread of any charge or voltage fluctuation disturbing the mean value of the electrostatic energy stored under them. Such fluctuations require associated modulations of the channel width underneath, because the BSCR width below the surface will vary as the surface band bending or the electrostatic energy of the BSCR is disturbed [2]. As these fluctuations propagate over the surface, the accompanying channel thickness modulation will track them until they will "recombine" through the conductance shunting the BSCR or they will reach the limits of the surface in the edges of the sample if losses during their propagation are low. This will produce a channel width perturbation that leads to a resistance noise in the channel. Although this is a two-dimensional process, a one dimensional model for samples with a high L/W ratio is presented for clarity purposes. The exponential solutions of the differential equation for the 1-D transmission line depicted in Figure 1 are much more familiar than the solutions for the Bessel-related equation that appears in the 2-D treatment, although the final results are similar [5].

Figure 1-a shows the geometry and electrical circuit used for the above 1_D model. It shows the electrical circuit existing between an ohmic contact on the GaAs channel (terminal B) and an hypotetical ohmic contact on the surface, placed just over the former, thus being a lateral contact to the surface "skin" responsible for the finite surface conductivity of naked n-GaAs surfaces. With the surface capacitance defined by $C_\Theta=\varepsilon/u$ (in F/cm^2), the sum of all ∂C will be the low frequency BSCR capacitance $C_{LF}=W \times L \times C_\Theta$, and the sum of all ∂R will be $Rs=R_\Theta \times L/W$, the series resistance of the surface "skin" viewed as a parallel channel to the conductor one. The transmission line of Figure 1 has been solved to study frequency effects in planar capacitors [6], but it will be repeated briefly here for convenience of the reader. The voltage v(x) along the structure is given by:

$$\frac{\partial^2 v}{\partial x^2} = \frac{v(x)}{\frac{1}{j\omega R_\Theta C_\Theta}} = \frac{v(x)}{L_T^2(j\omega)} \quad (1)$$

where $j=\sqrt{-1}$ is the imaginary unit. Therefore, R_Θ, C_Θ and $\omega=2\pi f$, lead to a frequency-dependent penetration depth of the ac signal in the transmission line $L_T(\omega)$ defined as:

$$L_T(\omega) = \frac{1}{\sqrt{\omega R_\Theta C_\Theta}} = \frac{1}{\sqrt{2\pi f R_\Theta C_\Theta}} \quad (2)$$

For a long enough sample, (sample length L$\to\infty$ or $L_T(\omega) \ll L$), the impedance $Z(j\omega)$ between terminals A and B is:

$$Z(j\omega) = \frac{L_T(\omega)}{\sqrt{j} \times \sigma W H} = \frac{R_\Theta \times L_T(\omega)}{W \times \sqrt{j}} \quad (3)$$

From equation (3) the impulse response h(t) of our planar structure is

$$h(t) = \frac{1}{\sigma W H \times \sqrt{R_\Theta C_\Theta}} \times \frac{1}{\sqrt{\pi t}} \quad (4)$$

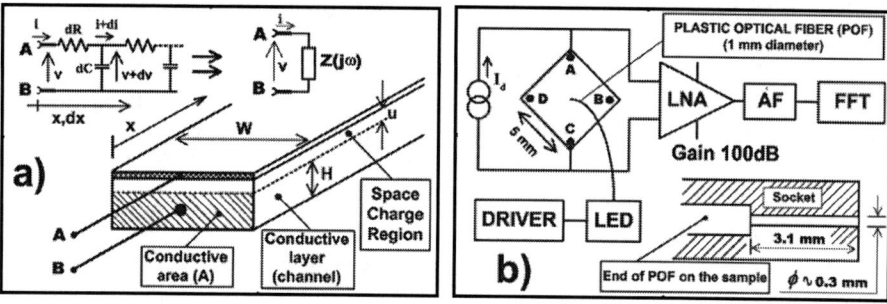

FIGURE 1. a) Distributed circuit proposed to study charge relaxations over the surface of n-GaAs layers and b) experimental setup used to enhance the 1/f noise of a GaAs Hall sample.

Then, any sudden "wire-like" charge fluctuation parallel to electrode A, appearing between terminals A and B (voltage fluctuation Δv) will decrease with the square root of time due to their propagation towards the inside of the structure (two times faster if we consider forward and backward propagations along x axis). The power spectrum of such voltage decays or diffusions of charge over the surface, is the 1/f one stated by equation (3). This diffusing surface charge will have an opposite charge density (as a channel thickness "bump") tracking it, thus producing a 1/f-like resistance noise in the channel. For samples of length L, this will be true for those spectral components such that $L_T(\omega) \ll L$. A transition frequency $f_T = 1/(2\pi R_\Theta C_\Theta L^2)$ sets the frequency limit for which $L_T(2\pi f_T)$ equals the sample length L. For the $R_\Theta = 63 G\Omega$ per square obtained from the I-V curve of a multifinger GaAs MESFET [3] and $C_\Theta \approx 10^{-7} F/cm^2$ coming from a typical BSCR width (u=0.1μm) in GaAs material doped in the $10^{17} cm^{-3}$ range, an $f_T = 0.25 Hz$ is obtained for L=100μm.

1/f-like Resistance Noise Enhancement in GaAs

Therefore, usual n-GaAs samples (L≥100μm) will generate quite pure 1/f resistance noise if charge fluctuations are created on their surfaces. Figure 1-b shows the setup to do it optically and to measure (simultaneously) voltage noise on a squared Hall planar GaAs sample (5x5mm^2) made from a 1μm thick n-type GaAs epitaxial layer grown by MBE on a semi-insulating substrate. A doping level $N_d = 7.7 \times 10^{16} cm^{-3}$ and an electron mobility $\mu_n = 3400 cm^2/Vs$ were measured in this sample, whose electrical resistance between ohmic contacts A and C (small alloyed In balls) was $R_{AC} = 480\Omega$. Pointing to the center of the sample and placed very close to the upper GaAs surface (≈0.1mm), the socketed end of a Plastic Optical Fiber (POF) allows to shine weakly just a "small area" (∅≈300μm) of the GaAs surface. This highly attenuated optical signal allows to vary locally the space charge region width under the surface, thus producing an initial surface charge "bump" that will start to diffuse. A red Light Emitting Diode (LED) injects short optical pulses (100μs) in the POF that produce ΔR_{AC} resistance transients giving transient voltage amplitudes ΔV_{AC} at the output of the low noise amplifier (LNA) roughly equal to the peak-to-peak noise signal of the sample. A 4th-order, low-pass anti-aliasing filter (AF) with 25KHz cut-off frequency and a PC-based sampling system and Fast Fourier Transform analyzer were used. The 16-bits sampling card was driven to take sets of 240Ksamples at 200Ksamples per second. The power spectra of 64 of the above sets are averaged and presented in Figures 2. For $I_d=0$, a flat spectrum at -166dB V^2/Hz appears due to the addition in power of $e_n \approx 4$nV/$\sqrt{Hz}$ due to the LNA and the thermal noise of R_{AC} ($e_{nR} \approx 2.8$nV/$\sqrt{Hz}$) (Figure 2-a). For $I_d=280\mu A$, however, a clear 1/f noise spectrum (1/f native) is measured well over the above one.

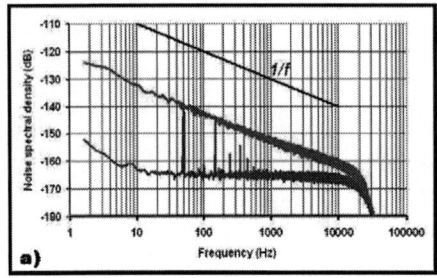 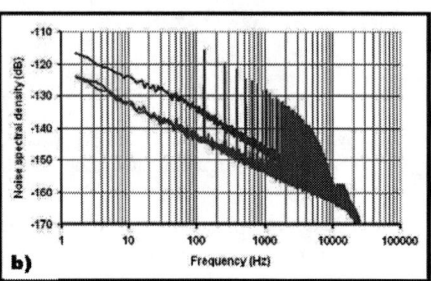

FIGURE 2. a) Thermal ($I_d=0$) and 1/f ($I_d=280\mu A$) noise power spectral densities measured in the dark b) noise power spectral densities measured under optical pumping at 130 pulsed per second.

Figure 2-b shows the noise spectra obtained with the same I_d, but driving the LED at 130 pulses per second (pps) fixed rate (line spectrum) and at 130 pps (on average) random-rate. The line spectrum has a background 1/f baseline that exactly matches the 1/f native noise of Figure 2-a, but it becomes the 1/f-like spectrum that our model predicts when charge fluctuations are created randomly. This $1/f^\beta$ spectrum with $\beta \approx 1.3$ in our case, reflects a lack of high-frequency components (e.g. over 1KHz) that is due, first, to the non null pulse width used as impulse function, whose spectral content drops at such frequencies. The zero at 10KHz of its corresponding Fourier Transform is apparent in Figure 2-b. The second reason for $\beta \neq 1$ is the rather wide initial area ($\phi \approx 300\mu m$) of each surface charge "bump" that starts to diffuse over the surface. Smaller initial "bumps" on the surface, as those of thermal origin due to trapping and emission of single electrons in surface states, would be by far smaller, leading to much faster initial diffusions making β closer to unity.

Hence, 1/f-like resistance noise can be induced in n-GaAs channels from a special interface: their naked surface. This kind of resistance noise induced in the channel by a Field Effect action from planar interfaces, leads to Lorentzian noise terms in general, often overlooked, able to synthesize 1/f noise over a wide frequency range [7].

REFERENCES

1. Izpura, J.I., and Muñoz, E., *Proc. IEEE-WOFE'97* , pp. 73-80 (1996).
2. Izpura, I., Valtueña, J.F., and Muñoz, E., *Semicond. Sci. Technol.* **12**, 678-686 (1997).
3. Choi, K.J., and Lee, J.L., *IEEE Trans. Electron Devices* **48**, 190-195 (2001).
4. Kim, J.Y., Lee, J., Kim, J., Kang, B., and Kwon, O'D., *Appl. Phys. Lett.* **25**, 4504-4506 (2003).
5. Izpura, J.I., (unpublished) (2004).
6. Izpura, J.I., *Semicond. Sci. Technol.* **16**, 243-249 (2001).
7. Izpura, J.I., Submitted to *IEEE Electron Device Lett.* (2005).

Unusual Fluctuations of Flux Flow at Low Temperature in Superconducting Films

S. Okuma, K. Kainuma, and T. Kishimoto

Research Center for Low Temperature Physics, Tokyo Institute of Technology
2-12-1, Ohokayama, Meguro-ku, Tokyo 152-8551, Japan

Abstract. We present the measurements of the fluctuating component of the flux(vortex)-flow voltage $\delta V(t)$ about the average in the low-temperature(T) vortex-liquid phase of amorphous Mo_xSi_{1-x} films. For the thick film $\delta V(t)$ and broad-band noise originating from the vortex motion is clearly visible in the quantum-vortex-liquid phase, where the amplitude of $\delta V(t)$ is pronounced and the distribution of $\delta V(t)$ is anomalously asymmetric, implying the unstational vortex motion. For the thin film the similar unusual vortex motion is observed in nearly the same reduced-T regime. Physical origin responsible for the phenomena is discussed.

Keywords: Vortices, Flux flow, Quantum fluctuations, Noise, Amorphous films
PACS: 74.40.+k, 74.25.Dw, 74.78.Db

Since the discovery of high-T_c oxide superconductors, the vortex phase diagram in the mixed state of type-II superconductors has been actively studied. It has been generally accepted that there is a vortex-liquid phase just below the upper-critical field. In the vortex-liquid phase the vortex motion causes nonzero dc resistivity ρ or voltage V in the presence of an arbitrarily small current I, while in the vortex-solid phase ρ is truly zero (as long as I is small enough to pin the solid). At high temperatures T properties of vortex lines are dominated by thermal fluctuations, while at sufficiently low T they are subject to quantum fluctuations. Analogously to liquid helium, quantum fluctuations are able to melt the vortex solid into the quantum vortex liquid (QVL) at $T=0$, which has been actually reported for several low-T_c [1-5] and high-T_c [6] superconductors. For thick (100 nm) amorphous (a-)Mo_xSi_{1-x} films, for which we present data in this paper, we have recently obtained convincing evidence for the QVL phase at low T [7,8]. However, dynamic properties of vortices in the QVL phase have not yet been studied in any system. Here we study the change in vortex dynamics associated with the change in vortex states from the thermal to quantum liquid by measuring the time(t)-dependent component of flux-flow voltage, $\delta V(t)$, about the average voltage V_0. The detailed data related to present work have been published elsewhere [9,10].

We prepared thick (100 nm; $x=47$ at.%) and thin (6 nm; $x=65$ at.%) films of a-Mo_xSi_{1-x} by coevaporation of pure Mo and Si. The mean-field transition temperatures T_{c0} for the thick and thin films are 1.13 and 1.96 K, respectively. The linear dc resistivity ρ and time-dependent voltage $V(t)$ induced by dc current I were measured using a four-terminal method. The voltage $V(t)$ enhanced with a preamplifier was recorded using a fast-Fourier transform spectrum analyzer, where we selected a time resolution of 39 or 390 μs to detect the asymmetric distribution of voltage fluctuations clearly, as described below. The magnetic field B was applied perpendicular to the plane of the film using a superconducting magnet in a persistent-current mode.

We mainly present the results of the thick film. For the thick film we can define a characteristic temperature $T_Q (\approx 0.1$ K) at which curvature of the Arrhenius plots of ρ in B changes from downward to upward upon cooling. This temperature marks the crossover from temperature dominated to quantum driven fluctuations in the liquid phase. Figures 1(a) and 1(c) show the fluctuating component of the (flux-flow) voltage, $\delta V(t)$, about the average voltage V_0, which is defined as $\delta V(t) \equiv V(t) - V_0$, for the thick film measured at $T = 0.15$ K ($>T_Q$) in $B = 3.12$ T [in the thermal-vortex-liquid (TVL) phase or the crossover regime between the TVL and QVL phases] and at 0.06 K ($<T_Q$) in 3.24 T (in the QVL phase), respectively, in the *absence* of an applied current I. These ($I=0$) data represent the background contribution due to external noise. The probability distribution of $\delta V(t)$, $P(\delta V)$, is symmetric, as expected usually. We have verified that in the TVL phase, for all the I studied, both $\delta V(t)$ and $P(\delta V)$ are similar to the background data and the substantial $\delta V(t)$ originating from the vortex motion is not detectable [e.g., Fig. 1(b)].

In the QVL phase, in contrast, the contribution of $\delta V(t)$ from the vortex motion is clearly visible. Figure 1(d) shows $P(\delta V)$, as well as $\delta V(t)$, measured at 0.06 K in 3.24 T (in the QVL phase) in the presence of $I = 2.0$ μA ($V_0 = 0.2$ mV). The amplitude of $\delta V(t)$ measured at $V_0 > 0$ is remarkably larger than that at $I = 0$ ($V_0 = 0$) [Figs. 1(a) and 1(c)] and the shape of $P(\delta V)$ is highly asymmetric having a tail which extends to

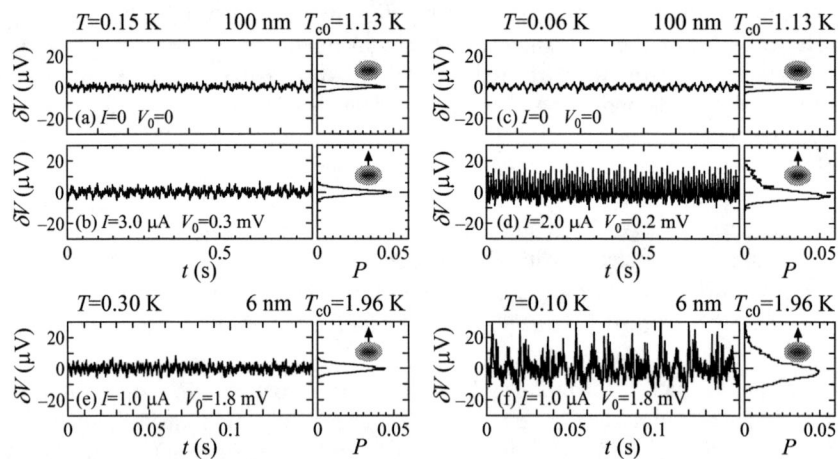

FIGURE 1. $\delta V(t)$ (left) and $P(\delta V)$ (right) for the thick film measured at $T = 0.15$ K in $B = 3.12$ T for (a) $I = 0$ and (b) 3.0 μA, and at 0.06 K in 3.24 T for (c) $I = 0$ and (d) 2.0 μA. (e) The results for the thin film measured (e) at $T = 0.30$ K in $B = 4.83$ T for $I = 1.0$ μA and (f) at $T = 0.10$ K in $B = 5.22$ T for $I = 1.0$ μA are also shown. In each figure (right) the direction of the vortex motion is illustrated.

the direction of vortex motion [$\delta V(t) > 0$]. We have confirmed by changing the polarity of I that the shape of $P(\delta V)$ is determined by the direction of vortex motion. In the T, B, and I regime where $P(\delta V)$ exhibits the unusual asymmetry, large broad-band noise of a Lorentzian type is observed. In the presence of larger I where the film is nearly in the normal state, both $\delta V(t)$ and $P(\delta V)$ are almost identical to the background data. The results obtained in this study are consistent with the view that the physical origin of large $\delta V(t)$ with asymmetric $P(\delta V)$ is due to the anomalous vortex motion in the liquid phase.

Shown in Figs. 2(a)-2(c) are the I dependences of V/I (full circles), the skewness A of the probability distribution $P(\delta V)$ (open circles), and the amplitude of voltage fluctuations $|\delta V(t)|$ (full squares) in the QVL phase (at 0.06 K in 3.24 T), in the crossover regime between the QVL and TVL phases (at 0.10 K in 3.20 T), and in the TVL phase (at 0.50 K in 2.16 T), respectively. Here, $|\delta V(t)|$ is estimated from a base length of $P(\delta V)$. In the high-T regime of the TVL phase (e.g., at 0.50 K in 2.16 T) $A(I) = 0$, namely, P is symmetric for any I studied. In the vicinity of the QVL phase (at 0.10 K in 3.20 T), A and $|\delta V|$ are close to zero at low I, while they take a small broad peak at higher I above which nonlinear $V(I)$ occurs. In the QVL phase (at 0.06 K in 3.24 T) both A and $|\delta V|$ are more pronounced. When we plot the peak value of the skewness, A_p, and that of the amplitude $|\delta V|$, $|\delta V|_p$, extracted from the A vs I and $|\delta V|$ vs I data, respectively, as a function of T, we find that A_p and $|\delta V|_p$ exhibit a rise at around 0.12-0.13 K, upon cooling [9,10]. This temperature is close to or slightly higher than $T_Q \approx 0.1$ K. Upon further cooling, A_p and $|\delta V|_p$ increase gradually. The rise at $T \approx T_Q$ is more dramatic for A_p than for $|\delta V|_p$.

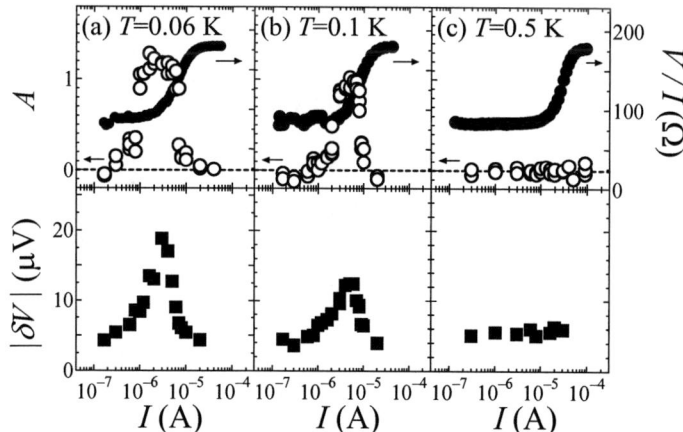

FIGURE 2. (a) I dependences of V/I (full circles), A (open circles), and $|\delta V(t)|$ (full squares) for the thick film (a) in the QVL phase (at 0.06 K in 3.24 T), (b) the crossover regime between the QVL and TVL phases (at 0.10 K in 3.20 T), and (c) the TVL phase (at 0.50 K in 2.16 T).

In the B-T (or QVL) regime where A_p and $|\delta V|_p$ take substantially large values, the vortex motion is not stationary but accompanied by large velocity and/or number fluctuations. Naively, these fluctuations may originate from plastic-flow-like vortex dynamics or random (pinning-)depinning processes dominated by temperature in the presence of I. Once some pinned vortices are depinned randomly in the QVL phase, they are driven by the Lorentz force. After a certain relaxation time τ, the velocity of these vortices reaches the (average) velocity of the underlying flux flow that yields V_0. If viscosity between moving vortices in the QVL phase is smaller than in the TVL phase due to the QVL nature, we may expect that τ in the QVL phase is also smaller than in the TVL phase. This gives rise to large (and narrow) random voltage pulses along the direction of the flux motion, as seen in Fig. 1(d).

Quite recently, it has been suggested [11] that the asymmetric fluctuations observed in our experiment seem to be due to vortex avalanches. This phenomenon has been shown to occur by computer simulations at extremely low T (T=0) in systems containing pinning centers [12]. We note, however, that the proximity of the temperature below which A_p (or $|\delta V|_p$) rises to T_Q below which quantum fluctuations are dominant cannot be explained in any available theory. To explain this fact, we need a theory taking account of quantum-fluctuation effects on vortex dynamics.

Finally, we briefly describe the results for the thin film. For the thin film the similar unusual voltage fluctuation $\delta V(t)$ is observed in nearly the same reduced-temperature regime ($T/T_{c0} \approx 0.1$) as that for the thick film [9,10]. In Figs. 1(e) and 1(f) we representatively show $\delta V(t)$ and $P(\delta V)$ of the thin film measured in the high-T (T=0.30 K, $T/T_{c0} \approx 0.15$) and low-T (T=0.10 K, $T/T_{c0} \approx 0.05$) liquid phases, respectively. We suggest based on all of the results that, although there is the large difference between the static vortex phase diagrams for three and two dimensions, vortex dynamics in the low-T liquid phase of thick and thin films is dominated by common physical mechanisms related to quantum-fluctuation effects.

REFERENCES

1. G. Blatter, B. Ivlev, Y. Kagan, M. Theunissen, Y. Volokitin and P. Kes, *Phys. Rev.* B **50**, 13013 (1994).
2. J. A. Chervenak and J. M. Valles, Jr., *Phys. Rev.* B **54**, R15649 (1996).
3. D. Ephron, A. Yazdani, A. Kapitulnik and M. R. Beasley, *Phys. Rev. Lett.* **76**, 1529 (1996).
4. N. Markovic, A. M. Mack, G. Martinez-Arizala, C. Christiansen and A. M. Goldman, *Phys. Rev. Lett.* **81**, 701 (1998).
5. T. Sasaki, W. Biberacher, K. Neumaier, W. Hehn, K. Andres and T. Fukase, *Phys. Rev.* B **57**, 10889 (1998).
6. T. Shibauchi, L. Krusin-Elbaum, G. Blatter and C. H. Mielke, *Phys. Rev.* B **67**, 064514 (2003).
7. S. Okuma, Y. Imamoto and M. Morita, *Phys. Rev. Lett.* **86**, 3136 (2001).
8. S. Okuma, S. Togo and M. Morita, *Phys. Rev. Lett.* **91**, 067001 (2003).
9. S. Okuma, M. Kobayashi and M. Kamada, *Phys. Rev. Lett.* **94**, 047003 (2005) and references therein.
10. S. Okuma, in *Proceedings of 10th International Vortex State Studies Workshop, Mumbai, 2005*, edited by A.K. Grover, Bangalore, Indian Academy of Sciences, submitted for publication.
11. F. Nori, private communications.
12. S. Field, J. Witt, F. Nori and X. Ling, *Phys. Rev. Lett.* **74**, 1206 (1995).

Study of dendritic avalanches by current noise measurements in High T_c Superconductors

Edvige Celasco*, Roberto Eggenhöffner, Graziano Tolotto† and Marcello Celasco

Department of Physics, LaMIA-INFM and IMEM-CNR, via Dodecaneso 33, 16146 Genova, Italy, University of Genova, Genova, Italy

Abstract. Spectral noise power measurements are reported in a bulk $YBa_2Cu_3O_{7-\delta}$ superconductor at 4.2 K to investigate the flux vortex avalanche processes originated from thermomagnetic instabilitites in an applied magnetic field up to 600 mT and in feeding current close to the critical value. $1/f^\gamma$ behavior is shown at frequencies below 10 Hz, γ values range from 0.7 to 1.7. A sharp peak is observed in applied field and from these results we have obtained a cutoff frequency around 20 Hz for Lorentzian shape knee from which we estimate a mean number of 500 vortices and a mean velocity of about 20 cm/s for the average avalanche.

Keywords: superconductors, 1/f noise, fluxon avalanches, dendritic, SQUID
PACS: 72.70.+m, 74.40.+k, 74.25.Qt, 74.72.Bk

INTRODUCTION

A catastrophic magnetic flux penetration was observed in metallic and ceramic superconductors both in bulk and thin films in the low temperature range (typically T<1/3 T_c). It originates from the thermo-magnetic instabilities triggered by magnetic field, bias current or thermal radiation [1]. These effects, widely studied through the combination of micro Hall probe magnetometry [2-4] and magnetooptical (MO) imaging [5,6], are of deep interest not only for a fundamental point of view but also for their practical outcomes. The magnetic flux avalanches are responsible mainly of the depletion of superconducting properties, i.e. the critical current, limiting, for instance, the applications of superconductor devices. Thus, an extremely noisy behavior of the transport physical properties can be expected in the very low temperature range associated to the avalanche propagation. Surprisingly, current noise spectral power measurements - a remarkable tool to investigate these avalanche topics - was not adopted until recently in ref [7], where a low thermal dissipation was claimed in order to detect appreciable noise in MgB_2 films, a fingerprint of the superconducting properties depletion [8]. In ceramic bulk superconductors we are expecting to exploit the low thermal conductance and the lower dendritic propagation velocity with respect to the thin films.

---

* Present address: DIFIS – Dipartimento di Fisica, Politecnico di Torino, Torino, Italy
† Università degli Studi del Piemonte Orientale *Amedeo Avogadro*, Alessandria, Italy

A critical topic to a full understanding of the avalanche effects in terms of the critical state properties concerns the linear-non linear flux penetration profiles and the different behavior of thin films with respect to bulks and tapes. The flux density profiles in the magnetic field penetration in superconductors were examined also in the recent review by Altshuler and Johansen [1]. In Bean's model, profiles of internal magnetic field are linear, whereas in thin films slight deviations from linearity were observed only in the narrow regions at the edges and the center of the samples. In bulks, we expect that deviations from linearity could dominate in the avalanche processes. Further, MO experiments in thin films mainly resolve avalanches due to sudden exit of the flux lines while direct observation of internal avalanches in thin films and in bulk materials can be observed only by other techniques. We suggest that, among others, noise measurements are one of the most promising tool to investigate these problems. We have thus undertaken these measurements on a $YBa_2Cu_3O_{7-\delta}$ (YBCO) bulk sample.

EXPERIMENTAL DETAILS

YBCO samples were prepared from standard oxides solid state reaction of pressed powders at 920 °C in flowing O_2. The thermal treatment and mechanical powdering was repeated twice to improve chemical uniformity. A 12x2x1 mm slab was cut from the pill and four Pt wires were fixed with a silver epoxy ink on evaporated silver, the distance of voltage wires was 8 mm. The sample was glued on a glass holder placed 40 cm above the position of the SQUID sensor on a copper rod linked to the He-bath. A diode temperature sensor strongly linked to the copper rod was placed 5 cm far from the sample region to avoid any magnetic field influence. Copper wires from the sample to the SQUID sensor were enclosed in a Pb tube thermalized by Al foil. Magnetic field was supplied by seven aligned NdFeB permanent magnets sliding through a screw-operated micromanipulator. SQUID low temperature amplification in flux-looked loop gives an overall amplification of $5.6 \; 10^5$ from DC 5000 Quantum Design control electronics. The overall transfer function of the circuit is flat up to 1 kHz. The voltage output of the SQUID electronics is fed to a HP 3562A spectrum analyzer which performs the analyses of the data and gives in output the spectral results reported. Further details on the apparatus and the experimental procedure were reported elsewhere [7].

RESULTS AND DISCUSSIONS

As discussed above, we report in this work measurements of the noise power in the low frequency range 1-40 Hz at fixed feeding currents of 82 mA, close to the critical current value and in magnetic field up to 600 mT, performed in the YBCO bulk sample at 4.2 K.

Measurements of the spectral power are reported in Fig.1, where a typical $1/f^\gamma$ behavior is shown at frequencies below 10 Hz. Noise increases up to 250 mT, further increases of the field intensity leads to noise reduction. The linear log-log power spectrum below 8-10 Hz was fitted with the $1/f^\gamma$ behavior. The γ dependence is shown in the inset to Fig.1, where the exponent parameter γ increases fastly at increasing the field and it attains the a maximum value of $\gamma =1.7$ at 250 mT. At higher fields, the γ parameter decreases rather slowly.

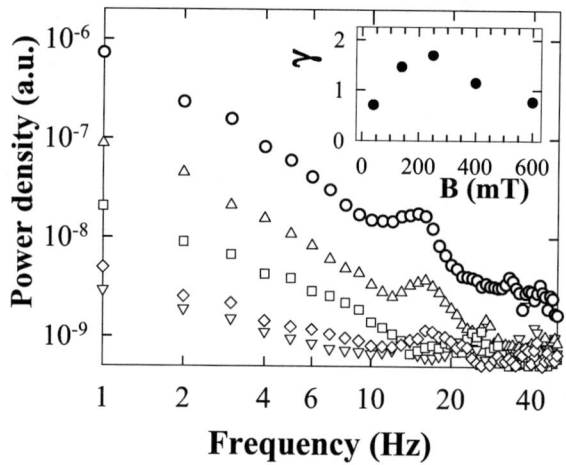

FIGURE 1. Spectral power density at 4.2 K of bulk $YBa_2Cu_3O_{7-\delta}$ at increasing the magnetic field intensity (∇: 40 mT, Δ: 140 mT, O: 250 mT, $\square$: 400 mT, $\Diamond$: 600 mT). A constant feeding current of 82 mA is adopted. In the inset the γ coefficient of the $1/f^\gamma$ behavior is reported vs. the applied magnetic field.

Remarkable peaks are shown around 15 Hz in the spectrum, the higher noise power amplitudes occurs at fields giving the highest exponent γ slope. Peaks in the spectral shapes were observed previously at frequencies above 100 Hz in BSCCO [9] and in NbTi [10] depending on the applied magnetic field intensity. In BSCCO, however, the effect is shown close to the transition temperature. The prominent peak in NbTi tube shaped sample was obtained imposing a magnetic field sweep at 2.9 K (1/4 T_c) at the highest ramp rate. Further, the mesoscopic vortex state in NbGe/NbN double layer shows an oscillatory noise behavior globally slowly decreasing vs. magnetic field [11] as we observe in the γ parameter dependence.

In the NiTi [10] system, Lorentzian like knee is emerging from the power spectrum at 100 Hz. At these frequencies, the system develops around an avalanche instability from a marginally stable state with a SOC-type hallmark. Further, avalanche transitions involving 50-5000 vortices could be traced directly from time evolution of voltage fluctuations. In single crystal YBCO, at 1K, where quantum fluxon tunneling is prevailing, avalanches of 750 fluxons were estimates from Hall probe magnetization measurements [2]. From our measurements in bulk YBCO, the peaks are detected around 15 Hz, a mean cutoff frequency of 20 Hz can be extrapolated for the Lorentzian-like behavior, giving a mean avalanche lifetime of 50 ms. A single flux depinning process was already estimated to occur in 100 μs [12], thus, a mean number of 500 vortices and a mean velocity of 20 cm/s can be estimated for the average avalanche transitions in our bulk YBCO sample. These results are in close agreement with previous slowly-driven observations in NbTi [10] and in YBCO single crystals at 0.2 K [2] and also with SOC-based model calculations [4]. Experiments exploring the propagation speed of these avalanches came across a surprisingly wide range of values from a few cm/s [4] up to 25 km/s [5] exceeding in the latter the sound velocity in the material. The highest velocities were detected at the very early stages immediately after a laser spot while the lower velocities respond to a slow drive like the ramping of the magnetic field. The very wide range occurring in measured avalanche velocities was observed to be influenced by the mechanism of formation, by the 2D/3D conditions as well as by the thickness of the samples, whilst the shape of the dendritic structures is rather similar, irrespectively of the superconducting material.

In conclusions, the main results obtained are: 1) a strong influence of the applied magnetic field in the power density spectrum at low frequencies (in the range 2-40 Hz); 2) a $1/f^\gamma$ behavior obtained at frequencies below 8-10 Hz and 3) the presence of a sharp peak for the intermediate applied magnetic fields at frequencies around 15 Hz.

We have found that in bulk YBCO the very low temperature noise is dominated by the behavior of avalanches. The γ coefficient is strongly influenced by the magnetic field at the adopted feeding current, close to the critical value. Thus, also in a bulk sample we detect the noise signal from the thermal instability which promotes the vortex-antivortex annihilation [7] with avalanche evolutions.

REFERENCES

1. Altshuler E. and Johansen T.H., Rev.Mod.Phys. **76**, 471 (2004)
2. Zieve R.J., Rosenbaum T. F., Jaeger H. M., Seidler G. T., Crabtree G. W. and Welp U., Phys. Rev. **B 53**, 11849 (1996)
3. Nowak E. R., Taylor O. W., Liu L., Jaeger H. M. and Selinder T. I., Phys.Rev. **B 55**, 11702 (1997)
4. Behnia K., Capan C., Mailly D. and Etienne B., Phys.Rev. **B 61**, R 3815 (2000)
5. Bolz U., Biehler B., Schmidt D., Runge B.U.and Leiderer P., Europhys. Lett. **64**, 517 (2003)
6. Leiderer P., Boneberg, J., Brüll P., Bujok V. and Herminghaus S., Phys. Rev. Lett. **71**, 2646 (1993)
7. Eggenhöffner R., Celasco E., Ferrando V. and Celasco M., Appl.Phys.Lett. **86**, 022504 (2005)
8. Choi E.M., Lee H.S., Kim H.J. and Lee S.I., Appl.Phys.Lett. **84**, 82 (2004)
9. Togawa Y, Abiru R., Iwaya K., Kitano H. and Maeda A., Phys.Rev.Lett. **85**, 3716 (2000)
10. Field S., Witt J.and Nori F., Phys.Rev.Lett. **74**, 1206 (1995)
11. Anders S., Smith A.W., Jaeger H.M., Besseling R., Kes P.H. and van der Drift E., Physica **C 332**, 35 (2000)
12. Celasco M., Eggenhöffner R., Gnecco E. and Masoero A., Phys.Rev. **B 58**, 6633 (1998)

Non-Gaussian Fluctuations in Biased Resistor Networks: Size Effects Versus Universal Behavior

C. Pennetta*, E. Alfinito*, L. Reggiani* and S. Ruffo[†]

*Dipartimento di Ingegneria dell'Innovazione and National Nanotechnology Laboratory,
Università di Lecce, Via Arnesano, 73100 Lecce, Italy
[†]CSDC, INFN and Dipartimento di Energetica "Sergio Stecco",
Università di Firenze, Via S. Marta, 3, Firenze, 50139, Italy.

Abstract. We study the distribution of the resistance fluctuations of biased resistor networks in nonequilibrium steady states. The stationary conditions arise from the competition between two stochastic and biased processes of breaking and recovery of the elementary resistors. The fluctuations of the network resistance are calculated by Monte Carlo simulations which are performed for different values of the applied current, for networks of different size and shape and by considering different levels of intrinsic disorder. The distribution of the resistance fluctuations generally exhibits relevant deviations from Gaussianity, in particular when the current approaches the threshold of electrical breakdown. For two-dimensional systems we have shown that this non-Gaussianity is in general related to finite size effects, thus it vanishes in the thermodynamic limit, with the remarkable exception of highly disordered networks. For these systems, close to the critical point of the conductor-insulator transition, non-Gaussianity persists in the large size limit and it is well described by the universal Bramwell-Holdsworth-Pinton distribution. In particular, here we analyze the role of the shape of the network on the distribution of the resistance fluctuations. Precisely, we consider quasi-one-dimensional networks elongated along the direction of the applied current or trasversal to it. A significant anisotropy is found for the properties of the distribution. These results apply to conducting thin films or wires with granular structure stressed by high current densities.

Keywords: Non-Gaussian Fluctuations, Resistor networks, Nonequilibrium stationary states, Disordered materials
PACS: 05.40.-a, 05.70.Ln, 64.60.Fr, 72.80.Ng

INTRODUCTION AND MODEL

Strongly correlated systems usually exhibit non-Gaussian distributions of the fluctuations of global quantities, as a consequence of the violation of the validity conditions of the central-limit theorem. Since correlations become important near the critical points of phase transitions, non-Gaussian fluctuations are usually observed near criticality [1, 2, 3, 4, 5, 6, 7, 8, 9]. In these conditions, the self-similarity of the system over all the scales, from a characteristic microscopic length up to the size of the system, has important implications on the fluctuation distribution [2, 3, 4, 5, 6, 7, 8, 9]. On the other hand, far from criticality, the correlations among different elements of the systems can also be important. This is particularly true for systems in non-equilibrium stationary states, where non-Gaussian fluctuations are frequently present [1, 10, 11, 12, 13, 14, 15]. Therefore the study of non-Gaussian fluctuations and of their link with other features of the system can provide new insights into basic properties of complex systems

Here, we study the distribution of the resistance fluctuations of biased resistor networks in nonequilibrium stationary states [16]. Networks of different size and shape and with different levels of internal disorder are considered. The resistance fluctuations are calculated by Monte Carlo simulations for currents close to the threshold for electrical breakdown. This last phenomenon consists of an irreversible increase of the resistance, occurring in conducting materials stressed by high current densities and it is associated with a conductor-insulator transition [17, 18, 19, 20, 21, 22]. In our study we make use of the Stationary and Biased Resistor Network (SBRN) model [23, 24]. This model provides a good description of many features associated with the electrical instability of composite materials [19, 21, 23] and with the electromigration damage of metallic lines [16, 20], two important classes of breakdown phenomena.

We describe a thin conducting film with granular structure of length L, width W and thickness $t_h \ll W, L$ as a 2D resistor network of rectangular shape and square-lattice structure [16]. The network of resistance R is made by N_L and N_W resistors in the length and width directions respectively. Thus, the total number of resistors in the network (excluding the contacts) is: $N_{tot} = 2N_L N_W + N_L - N_W$. The external bias (here a constant current I), is applied to the network through electrical contacts realized by perfectly conducting bars at the left and right hand sides of the network. The network lies on an insulating substrate at temperature T_0, acting as a thermal bath. Each resistor has two allowed states [20, 25]: (i) regular, corresponding to a resistance $r_{reg,n}(T_n) = r_{ref}[1 + \alpha(T_n - T_{ref})]$ and (ii) broken, corresponding to a resistance $r_{OP} = 10^9 r_{reg,n}(T_0) \equiv 10^9 r_0$ (resistors in this state will be called defects). In the above expression α is the temperature coefficient of the resistance (TCR), r_{ref} and T_{ref} are the reference values for the TCR and T_n is the local temperature. The existence of temperature gradients due to current crowding and Joule heating effects is accounted for by taking the local temperature of the n-th resistor given by the following expression [20]:

$$T_n = T_0 + A[r_n i_n^2 + (3/4 N_{neig}) \sum_{l=1}^{N_{neig}} (r_l i_l^2 - r_n i_n^2)] \quad (1)$$

where, i_n is the current flowing in the n*th* resistor and N_{neig} the number of its nearest neighbors over which the summation is performed. The parameter A represents the thermal resistance of each resistor and sets the importance of Joule heating effects. By taking the above expression for T_n we are assuming an instantaneous thermalization of each resistor at the value T_n [20, 25]. In the initial state of the network (no external bias) we take all the resistors identical (perfect network). We assume that two competing biased processes act to determine the evolution of the network [23, 24]. These two processes consist of stochastic transitions between the two possible states of each resistor and they occur with thermally activated probabilities [25]: $W_{Dn} = \exp[-E_D/k_B T_n]$ and $W_{Rn} = \exp[-E_R/k_B T_n]$, characterized by the two energies, E_D and E_R (where k_B is the Boltzmann constant). The time evolution of the network is obtained by Monte Carlo simulations which update the network resistance after breaking and recovery processes, according to an iterative procedure described in detail in Ref. [23]. The sequence of successive configurations provides a resistance signal, $R(t)$, after an appropriate calibration of the time scale. Depending on the stress conditions (I and T_0) and on the network parameters (size, activation energies and other parameters dependent on the material, like r_{ref}, α and A), the network either reaches a stationary state or undergoes an irreversible

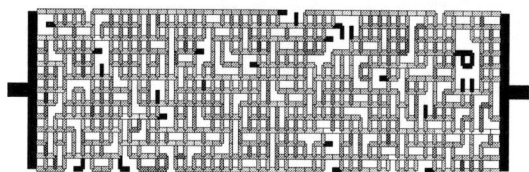

FIGURE 1. Pattern of a network 12×50 stressed by a current density $j = 0.32$ mA. The grey boxes show the backbone of the network, the black ones the "dangling bonds" (branches with zero-current, while the missing boxes correspond to the broken resistors. This pattern has been calculated at $t = 4 \times 10^4$ (time expressed in iteration steps).

electrical failure [16, 23, 24]. This latter possibility is associated with the achievement of the percolation threshold, p_c, for the fraction of broken resistors. Therefore, for a given network at a given temperature, a threshold current value, I_B, exists above which electrical breakdown occurs [23]. For values of the current below this threshold, the steady state of the network is characterized by fluctuations of the fraction of broken resistors, δp, and of the resistance, δR, around their respective average values $<p>$ and $<R>$. In particular, we underline that in the vanishing current limit (random percolation) [26], the ratio $\lambda \equiv (E_D - E_R)/k_B T_0$ determines the average fraction of defects and thus the level of intrinsic disorder inside the network [26]. In the following we analyze the results of simulations performed by considering networks of different size and shape stressed at room temperature, $T_0 = 300$ K, by a current density $j \equiv I/N_W = 0.32$ mA. We have taken: $\alpha = 3.6 \times 10^{-3}$ K^{-1}, $T_{ref} = 273$ K, $r_{ref} = 0.048$ Ω, $A = 2.7 \times 10^8$ K/W, $E_D = 0.41$ eV and $E_R = 0.35$ eV. This choice of the parameters is appropriate to describe the behavior under electromigration of metallic lines of Al-0.5%Cu studied in Ref. [16] and it corresponds to studying a network with an intermediate level of intrinsic disorder.

RESULTS AND CONCLUSIONS

Figure 1 displays the pattern of a network 12×50 calculated at a given time, $t = 4 \times 10^4$, (expressed in iteration steps) in the stationary regime of the network, i.e. for $t > \tau_{rel}$, where $\tau_{rel} \approx 8 \times 10^3$ is the relaxation time for the achievement of the nonequilibrium stationary state. The network in this figure is stressed by a current density ($j = 0.32$ mA) close to the breakdown value, $j_B = I_B/N_W$. The resistance evolution for the same network is reported in Fig. 2. In this figure the grey line shows the average value of the resistance, $<R>$. We note that both the average resistance and the relative variance of the resistance fluctuations, $<(\delta R)^2>/<R>^2$, depend on j. A detailed analysis of the behavior of these two quantities as a function of the current can be found in Refs. [23, 24]. In previous works [13, 14, 15] we have analyzed the effects on the distribution of the resistance fluctuations of the biasing current [13], of the intrinsic disorder and of the size of the network [14, 15], by limiting ourself to discuss square networks. Here, we focus our discussion on shape effects: precisely we analyze the effect on the distribution of the fluctuations of scaling the size of the network separately in the two directions, i.e. of scaling separately the width, N_W and the length, N_L, of the network. Figure 3(a) shows the distributions of the resistance fluctuations obtained for two networks of size 12×50 (big circles) and 50×12 (triangles) stressed by the same current density. In this

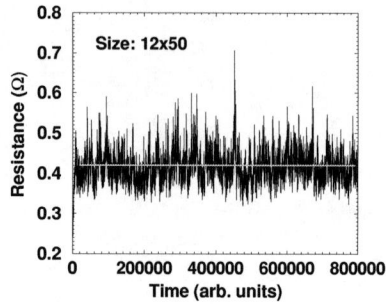

FIGURE 2. Resistance evolution of the network in Fig. 1. The time is expressed in arbitrary units (iteration steps), the resistance in Ohm. The grey line shows the average value of the resistance.

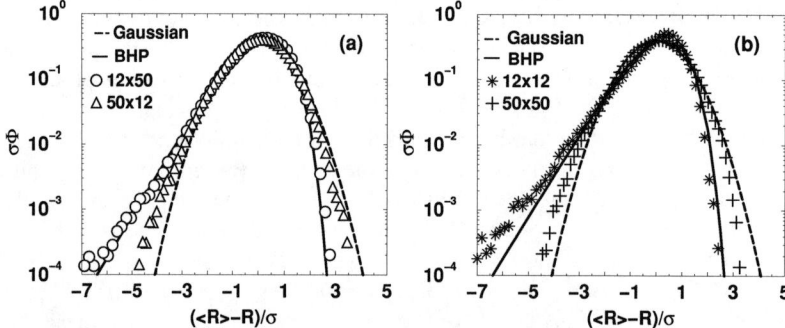

FIGURE 3. Normalized PDF of the resistance fluctuations of networks of different size and stressed by the same current density $j = 0.32$ mA. Precisely, in (a) the size is: 12×50 (big circles) and 50×12 (triangles); in (b): 12×12 (stars), 50×50 (plus). The solid and dashed curves refer to the BHP and Gaussian distributions, respectively.

figure (and in the followings) we denote with Φ the probability density function (PDF) of the distribution and with σ the root mean square deviation from the average value. This normalized representation, by making the distribution independent of its first and second moments, is particularly convenient to explore the functional form of a distribution [3]. A lin-log scale is adopted for convenience to plot the product $\sigma\Phi$ as a function of $(<R> - R)/\sigma$. The PDFs in Fig. 3 and all the others in this paper have been calculated by considering time series containing about 10^6 resistance values. For comparison, in Fig. 3 we also report the Gaussian distribution (dashed curve) and the BHP distribution (continuous curve) [2, 3]. The PDF obtained for the network 12×50 (corresponding to the signal in Fig. 2) exhibits a strong non-Gaussianity, well described by the BHP curve. By contrast, the PDF obtained for the network 50×12 is nearly Gaussian. At a first insight, this result can seem surprising: in fact the two networks are composed by nearly the same number of resistors, moreover the dissipated electric power per unit volume, $RI^2/(LW) \propto j^2$, is the same in both cases. As a consequence, the average fraction of

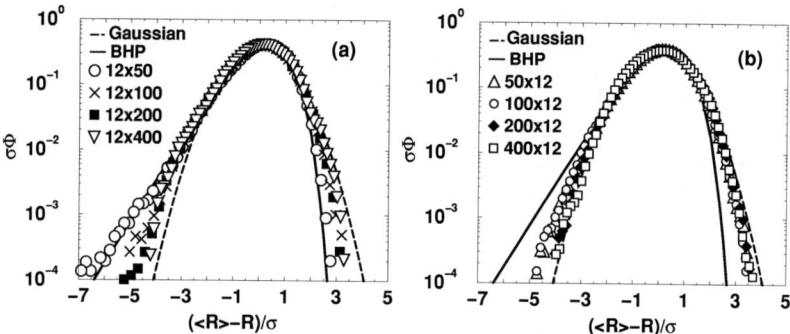

FIGURE 4. Normalized PDF of the resistance fluctuations of networks of different size and stressed by the same current density $j = 0.32$ mA. Precisely, in (a) the size is: 12×50 (big circles), 12×100 (crosses), 12×200 (full squares) and 12×400 (down triangles); in (b) the size is: 50×12 (triangles), 100×12 (small circles), 200×12 (full diamonds) and 400×12 (squares). The solid and dashed curves have the same meaning of Fig. 3.

defects $p \approx 0.19$ is also the same. However, the percolation threshold p_C is different for the two networks [16]. Therefore, the nearly Gaussian distribution of the 50×12 network is due to the higher value of p_C (and thus to the higher stability) of this network [16]. For comparison, we report in Fig. 3(b) the PDFs calculated for two square networks 12×12 and 50×50 biased by the same current density. Again, the dissipated power density and the average fraction of defects are the same for both networks. However, for square $N \times N$ networks the percolation threshold is roughly independent of the size, even for biased percolation [16]. Thus, for these networks the higher instability and the stronger non-Gaussianity for decreasing N is mainly related with the increase in magnitude of the fluctuations associated with the smaller size [13, 14]. The normalized PDFs of the resistance fluctuations calculated for several networks of different size are reported in Fig. 4. Figure 4(a) displays PDFs obtained for networks elongated along the direction of the applied current (precisely networks of a given width, $N_W = 12$, and with increasing length, $N_L = 50 \div 400$), while Fig. 4(b) shows PDFs obtained for networks elongated in a direction trasversal to the current applied (precisely networks of a given length, $N_L = 12$, and with increasing width, $N_W = 50 \div 400$. We can see that for trasversal networks the PDF is rather insensitive to the width and the small non-Gaussianity for small widths completely vanishes already for networks with $N_W = 200$. By contrast, for longitudinal networks, the PDF is sensitive to the length. However, it should be noted that the PDF obtained for $N_L = 200$ practically overlaps with that obtained for $N_L = 400$ and both exhibit non-Gaussian tails. Since the correlation length, ξ, for these networks is estimated to be $\xi < 5$, networks with $N_L = 400$ can be considered as infinitely long. Thus, Fig. 4(a) suggests a persistent non-Gaussianity for longitudinal networks in the limit $N_L \to \infty$, associated with the finite size of the network in the transversal direction. Furthermore, the magnitude of this non-Gaussianity is expecected to be controlled by the level of intrinsic disorder.

In conclusions, we have studied the distribution of the resistance fluctuations of biased resistor networks in nonequilibrium stationary states. We have considered networks biased by currents close to the threshold of electrical breakdown. As a general trend, the distribution of the fluctuations is found to exhibit relevant deviations from Gaussianity, which are in general related to finite size effects [13, 14]. However, for systems close to the critical point of the conductor-insulator transition, the non-Gaussianity persists in the large size limit [13, 14] and it is well described by the universal Bramwell-Holdsworth-Pinton distribution. Furthermore, we have analyzed the role of the shape of the network on the distribution of the resistance fluctuations, by considering quasi-one-dimensional networks elongated along the direction of the applied current or trasversal to it. A significant anisotropy is found for the properties of the distribution. These results apply to conducting thin films or wires with granular structure stressed by high currents.

ACKNOWLEDGMENTS

This work has been performed within the cofin-03 project "Modelli e misure di rumore in nanostrutture" financed by Italian MIUR, the SPOT NOSED project IST-2001-38899 of EC is also acknowledged. P.C.W. Holdsworth and S. Caracciolo are gratefully acknowledged for helpful discussions.

REFERENCES

1. M. B. Weissman, *Rev. Mod. Phys.*, **60**, 537–571 (1988).
2. S. Bramwell, P. Holdsworth, and J. F. Pinton, *Nature*, **396**, 552–554 (1998).
3. S. Bramwell, K. Christensen, J. Y. Fortin, P. C. W. Holdsworth, H. Jensen, S. Lise, J. M. López, M. Nicodemi, J. F. Pinton, and M. Sellitto, *Phys. Rev. Lett.*, **84**, 3744–3747 (2000).
4. S. Bramwell, J. Y. Fortin, P. C. W. Holdsworth, S. Peysson, J. F. Pinton, B. Portelli, and M. Sellitto, *Phys. Rev. E*, **63**, 041106-1–22 (2001).
5. V. Aji, and N. Goldenfeld, *Phys. Rev. Lett.*, **86**, 1107–1010 (2001).
6. T. Antal, M. Droz, G. Györgyi, and Z. Rácz, *Phys. Rev. Lett.*, **87**, 24061-1–4 (2001).
7. K. Dahlstedt, and H. Jensen, *J. Phys. A*, **34**, 11193–11200 (2001).
8. T. Antal, M. Droz, G. Györgyi, and Z. Rácz, *Phys. Rev. E*, **65**, 046140-1–12 (2002).
9. G. Györgyi, P. C. W. Holdsworth, B. Portelli, and Z. Rácz, *Phys. Rev. E*, **68**, 056116-1–14 (2003).
10. N. Vandewalle, M. Ausloos, M. Houssa, P. W. Mertens, and M. M. Heyns, *Appl. Phys. Lett.*, **74**, 1579–1581 (1999).
11. T. Bodineau, and B. Derrida, *Phys. Rev. Lett.*, **92**, 180601-1–4 (2004).
12. S. Kar, A. K. Raychaudhuri, A. Ghosh, H. V. Löhneysen, and G. Weiss, *Phys. Rev. Lett.*, **91**, 216603-1–4 (2003).
13. C. Pennetta, E. Alfinito, L. Reggiani, and S. Ruffo, *Semic. Sci. Techn.*, **19**, S164–S166 (2004).
14. C. Pennetta, E. Alfinito, L. Reggiani, and S. Ruffo, *Physica A*, **340**, 380–387 (2004).
15. C. Pennetta, E. Alfinito, L. Reggiani, and S. Ruffo, , in *Noise in Complex Systems and Stochastic Dynamics*, edited by Z. Gingl, J. M. Sancho, L. Schimansky-Geier, and J. Kertesz, Proceedings of SPIE 5471, Int. Soc. Opt. Eng., Bellingham, 2004, pp. 38–47.
16. C. Pennetta, E. Alfinito, L. Reggiani, F. Fantini, I. D. Munari, and A. Scorzoni, *Phys. Rev. B*, **70**, 174305-1–15 (2004).
17. J. V. Andersen, D. Sornette, and K. Leung, *Phys. Rev. Lett.*, **78**, 2140–2143 (1997).
18. S. Zapperi, P. Ray, H. E. Stanley, and A. Vespignani, *Phys. Rev. Lett.*, **78**, 1408–1411 (1997).
19. C. D. Mukherjee, K. K. Bardhan, and M. B. Heaney, *Phys. Rev. Lett.*, **83**, 1215–1218 (1999).
20. C. Pennetta, L. Reggiani, and G. Trefan, *Phys. Rev. Lett.*, **84**, 5006–5009 (2000).
21. C. D. Mukherjee, and K. K. Bardhan, *Phys. Rev. Lett.*, **91**, 025702-1–4 (2003).
22. C. Pennetta, E. Alfinito, and L. Reggiani, , in *Unsolved Problems of Noise and Fluctuations*, edited by S. M. Bezrukov, 2004, vol. 665 of *AIP Conf. Proc.*, pp. 480–487.
23. C. Pennetta, L. Reggiani, G. Trefan, and E. Alfinito, *Phys. Rev. E*, **65**, 066119-1–10 (2002).
24. C. Pennetta, *Fluct. Noise Lett.*, **2**, R29–49 (2002).
25. Z. Gingl, C. Pennetta, L. B. Kish, and L. Reggiani, *Semic. Sci. Techn.*, **11**, 1770–1775 (1996).
26. C. Pennetta, G. Trefan, and L. Reggiani, *Phys. Rev. Lett.*, **85**, 5238–5241 (2000).

Low-Frequency Current Fluctuations in Post-Hard Breakdown Thin Silicon Oxide Films

Yasuhisa Omura and Kenji Komiya

ORDIST, Kansai University, 3-3-35 Yamate-cho, Suita, Osaka 564-8680, Japan

Abstract. This paper focuses on the current fluctuation of post-hard breakdown thin silicon oxide films with thickness ranging from 3 nm to 5 nm. We characterize the post-degraded structure of silicon oxide films by analyzing current fluctuation spectra after hard breakdown.

Keywords: silicon oxide, hard breakdown, trap, current fluctuation
PACS: 72.80.Sk, 72.20.Jv, 05.40.Ca

INTRODUCTION

It is well known that constant voltage stress generates many kinds of defects inside SiO_2 films. The first stage of degradation is characterized by the emergence of the stress-induced leakage current (SILC), followed by soft breakdown (SBD). Finally, hard breakdown (HBD) occurs. In some cases, the percolation algorithm can provide a qualitative understanding of the degradation mechanisms, but a full analysis and interpretation remain outstanding. Other analyses have been advanced recently to model defect properties of sub-5-nm-thick SiO_2 films because such films are needed to realize many contemporary devices. However, because defect nature has such a complicated dependency on process technology and thickness, no unified interpretation of degradation evolution has been made yet.

In this paper, we focus on the tunneling current characteristics of thin SiO_2 films with thickness ranging from 3 nm to 5 nm. We characterize the post-degraded structure of SiO_2 films by analyzing current fluctuation spectra after HBD.

DEVICE STRUCTURE AND EXPERIMENTAL METHOD

We used the metal-oxide-semiconductor capacitors[1]. The devices were fabricated on n-type (001) Si substrates with resistivity of 4 Ωcm. Thin SiO_2 films with a thickness of 3.3 nm or 5.2 nm were formed by means of rapid thermal oxidation at 950 C. N-type poly-Si film for the gate electrode was formed by an LPCVD technique. The gate electrode was patterned by wet-etching to avoid process-induced damage. Positive gate bias (V_g) was applied to all devices when evaluating device degradation;

both polarities were used in current fluctuation measurements. All measurements were carried out at room temperature[2]. Gate electrode area was 150 μm x 200 μm.

EXPERIMENTAL RESULTS AND DISCUSSION

In the previous paper[3], we stated that a barrier for electrons is present inside post-HBD SiO_2 films as shown in Fig. 1(a). While this assumption has been verified by analyzing dc characteristics, the model is based on a macroscopic view of the phenomenon (Fig. 1(b)), not on a microscopic view. We investigated post-HBD conduction in detail in terms of fluctuation of leakage current under a constant V_g.

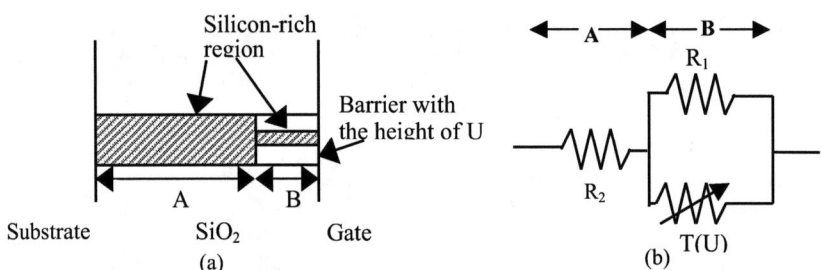

FIGURE 1. Simplified conduction mechanisms of post-hard breakdown oxide films.

Fluctuation characteristics of leakage current are shown in Fig. 2 for various positive V_g. In these measurements, constant V_g was applied to the device for a short time (20 sec.). One of the significant characteristics seen in Fig. 2 is that typical random telegraph noise (RTN) is observed when V_g= 50 mV and 0.2 V, and that semi-analog random noise (SARN) is seen when V_g= 1.5 V. On the other hand, semi-digital random noise (SDRN) is always observed when V_g< 0 V (not shown here). The fluctuation amplitude for V_g< 0 V is one-order smaller than that for V_g> 0 V regardless of V_g. This strongly suggests that the current fluctuation reflects the microscopic aspect of the post-HBD SiO_2 film.

Similar fluctuation characteristics of leakage current were also evaluated for various V_g for a long time (600 sec.). SDRN is observed when V_g= 50 mV, RTN is observed when V_g= 0.2 V, and SARN is seen when V_g= 1.5 V (not shown here). On the other hand, SDRN is observed when V_g= -50 mV and -0.2 V, while RTN occurs when V_g= -1.5 V (not shown here). The fluctuation amplitude for V_g> 0 V was almost identical to that for V_g< 0 V regardless of V_g. Polarity dependence and bias dependence of these noise features suggest that the post-HBD SiO_2 film has an inherently asymmetrical band structure and different microscopic conduction mechanisms.

The numerical calculation results yielded by Fourier transforming the experimental results shown in Fig. 2 are shown in Fig. 3. It should be noted that Lorentzian spectra are observed when V_g= 50 mV and 0.2 V, and that simple 1/f spectra are observed when V_g= 1.5V. Since most electrons go through the narrow-pass region in region B for V_g=50 mV and 0.2 V (<U/e), it is strongly suggested that defects (labeled D_{bp1}) with specific energy level exist inside the narrow-pass region as expected (see Fig. 1); the estimated time constant (τ_{bp1}) ranges from 0.5 sec (Fig. 3(b)) to 2.5 sec (Fig. 3(a)).

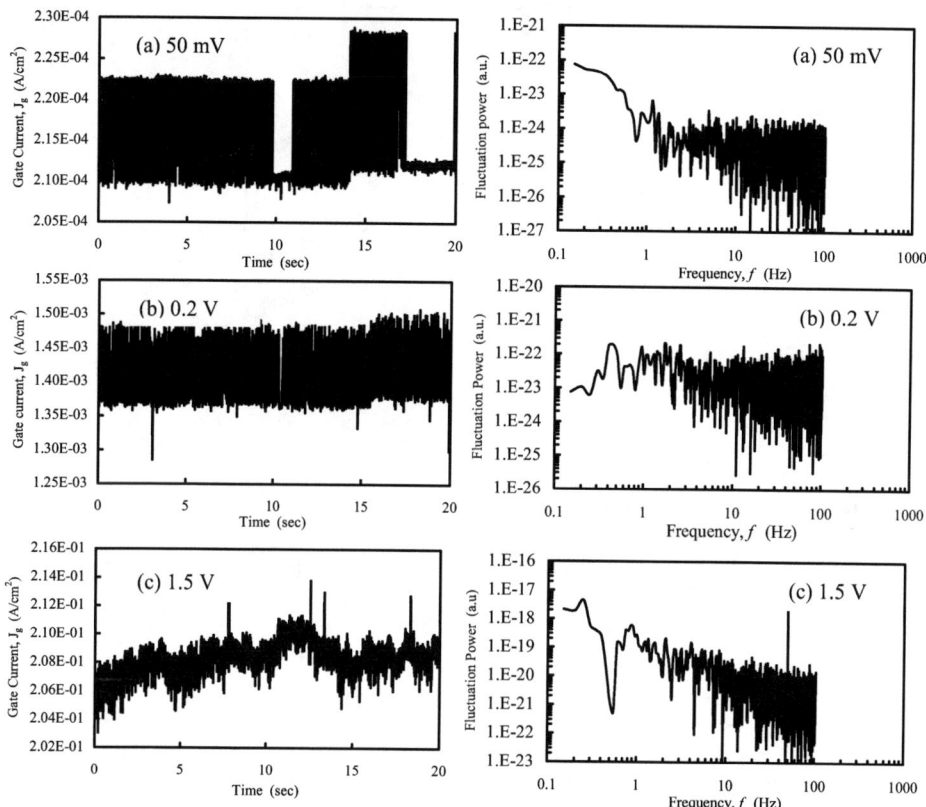

FIGURE 2. Gate current as a function of measurement time (20 sec). A positive gate bias was applied to the device.

FIGURE 3. Fluctuation power of gate current as a function of frequency. A positive gate bias was applied to the device for 20 sec.

For V_g= 1.5 V, most electrons can jump over region B without tunneling, and the electron current mainly reflects the aspect of region A. Since it is anticipated that region A is composed of many silicon-rich clusters, it can be accepted that the 1/f spectra, not the Lorentzian spectra, are typically observed[3]. On the other hand, for V_g< 0 V, it should be noted that clear Lorentzian spectra were observed only when V_g= -0.2 V, and that $1/f^2$ spectra are observed when V_g= -50 mV and -1.5V (not shown here). For V_g= -50 mV, it can be considered that the defects with specific energy level are those aforementioned (D_{bp1}) because most electrons go through the narrow-pass region; the estimated time constant (τ_{bp1}) is 0.3 sec for V_g=-0.2 V. On the other hand, the estimated time constant is larger than 10 sec for V_g= -50 mV and -1.5 V. The sub-linear current characteristic of the gate current (V_g< -0.2 V)[2] suggests a specific generation-recombination process inside the narrow-pass region and region A; this should yield Lorentzian spectra. We consider that the major part of gate current flows through the narrow-pass region at V_g= -0.2 V or -1.5 V. So, it is suggested that the $1/f^2$ spectra for V_g< 0 V correspond to deep level defects inside regions A and B.

The numerical calculation results created by Fourier transforming the experimental

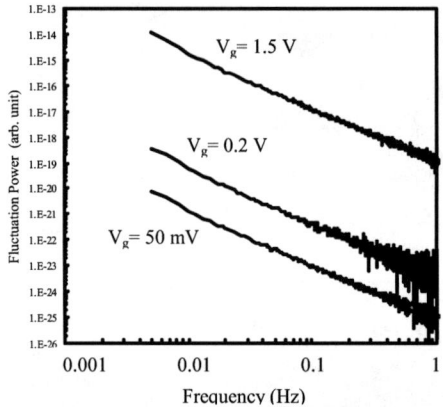

FIGURE 4. Fluctuation power of gate current as a function of frequency. A positive gate bias ranging from 50 mV to 1.5 V was applied to the device for 600 sec.

results for 600-sec measurement are shown in Fig. 4. These analyses are demonstrated for the first time in this paper. In this analysis, it should be noted that $1/f^2$ spectra, not Lorentzian, are observed for all bias conditions. This indicates that high-density defects with a specific time constant exist inside the post-HBD SiO_2 films. Though the time constant (τ_{bp2}) remains unspecified, it is much larger than 100 sec. Most of the defects contributing to current fluctuation lie inside the narrow-pass region for small V_g and inside region A for large V_g; this aspect should be independent of measurement period. Thus, the present experiments indicate that there are two different time constants in the current fluctuation phenomenon; the constants appear to be related to defects with different physical origins.

It is apparent that the present experimental results show phenomena that are different from the past result[4]. Generally speaking, $1/f^2$ spectra mean that the fluctuation source is a group of defects with specific energy level. Since we usually consider that there are various defects with different energy levels before HBD which should show $1/f$ spectra, we have to consider that the density of defects with a specific energy level become dominant after HBD. So, we can thus conclude that a significant and profound relation exists between the SBD event and the HBD event[5].

REFERENCES

1. Omura, Y. and Nakatsuji, H., *Distinct observation of valence-band electron direct-tunneling current in a nanometer-thick silicon oxide film on monocrystalline silicon*, Appl. Phys. Lett., **75**, 513-515 (1999).
2. Omura, Y. and Komiya, K., *Transport characteristics of post hard breakdown thin silicon oxide films and consideration of physical models*, J. Appl. Phys., **91**, 4298-4306 (2002).
3. McWhorter, A. L., *1/f noise and germanium surface properties*, in `Semiconductor Surface Physics` (ed. by R.H. Kingston) as the Proc. The Conf. On the Phys. of Semiconductor Surfaces (Pennsylvania, June, 1956) pp. 207-228.
4. Farmer, K. R., Saletti, R. and Buhrman, R. A., *Current fluctuations and silicon oxide wear-out in metal-oxide-semiconductor tunnel diodes*, Appl. Phys. Lett., **52**, 1749-1751 (1988).
5. Sune, J., Mura, G., and Miranda, E., Are Soft Breakdown and Hard Breakdown of Ultrathin Gate oxides Actually Different Failure Mechanisms?, IEEE Electron Device Lett., **21**, 167-169 (2000).

Noise and Charge Storage in Nb$_2$O$_5$ Thin Films

V. Sedlakova, J. Sikula, L. Grmela, P. Hoeschl[*], Z. Sita[*2], S. Hashiguchi[*3], M. Tacano[*4]

Department of Physics, Brno University of Technology, Technicka 8, 616 00 Brno, Czech Republic
Fax: +425 7261 666, e-mail: sedlaka@feec.vutbr.cz
[*]*Physical Institute, Charles University, Prague, Czech Republic*
[*2]*AVX Czech republic, Dvorakova 328, 563 01 Lanskroun, Czech Republic*
[*3]*Department of Electronics, Yamanashi University, Kofu 400, Japan*
[*4]*Meisei University, Hino, Tokyo, 191-8506 Japan*

Abstract. A low frequency noise and DC leakage current measurements have been performed on the MIS structure NbO - Nb$_2$O$_5$ - MnO$_2$. The mechanism of current flow and current noise sources were determined from these measurements. The insulating layer thickness is 50 to 150 nm, relative permittivity about 38. The charge is accumulated not only on NbO and MnO$_2$ electrodes, but also in the Nb$_2$O$_5$ insulating layer. The charge carrier transport is determined by the Poole-Frenkel mechanism and tunnelling in the normal mode (for NbO electrode positive). In this case g-r noise is dominant for Poole-Frenkel mechanism and 1/f noise is dominant for tunnelling.

Keywords: Noise, Poole-Frenkel transport, Tunnelling, Niobium Pentoxide, MIS structure.

PACS: 71.23.Cq, 72.20.Jv

INTRODUCTION

A charge carrier transport mechanism analysis has been performed on NbO - Nb$_2$O$_5$ - MnO$_2$ structure to determine the mechanism of current flow, charge carriers storage and the noise sources. VA characteristic is exponential in reverse mode (NbO electrode negative) and the charge carrier transport is determined by the Poole-Frenkel mechanism and tunnelling in the normal mode (NbO electrode positive). The charge is accumulated not only on NbO and MnO$_2$ electrodes, but also in the Nb$_2$O$_5$ insulating layer. In this structure the shot noise and g-r noise occurs for the low electric field and 1/f noise is dominant for the high electric field.

VA CHARACTERISTICS

The amorphous insulating oxides including Nb$_2$O$_5$ exhibit current flow, which increases roughly exponentially with applied voltage [1]. To find more information on the current flow processes, the VA characteristics were measured both in the normal and reverse mode. In the reverse mode the VA characteristic (see Fig. 1) can be approximated by the by exponential dependence of current on voltage:

$$I = I_0(\exp(\beta U) - 1) \tag{1}$$

where β = 20 to 25 V^{-1}. This value of parameter β corresponds to high ideality factor n ≥ 2. I_0 is the saturation current from which the Schottky potential barrier was determined. For our samples the potential barrier is 1.0 to 1.4 eV.

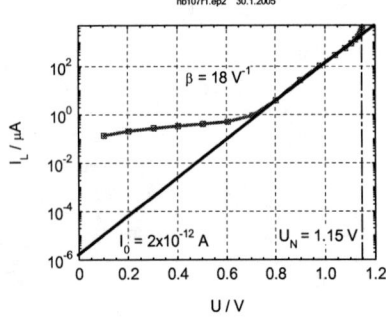

FIGURE 1. Reverse mode VA characteristic   **FIGURE 2.** Normal mode VA characteristic

In the normal mode the Poole-Frenkel mechanism and tunnelling take place (see Fig. 2). For the voltage lover then 7 V Poole-Frenkel mechanisms is dominant and VA characteristic is described by [2,3]:

$$I = G_P U \exp(\beta_P U^{1/2}) \tag{2}$$

Where G_P is conductivity and β_P depends on the isolating layer thickness d and relative permittivity ε_r:

$$\beta_P = (e^3/\pi\varepsilon_0\varepsilon_r d)^{1/2}/kT \tag{3}$$

For our samples β_P is 1 to 2.5V$^{-1/2}$ according to the sample thickness.

Tunnelling is dominant for the voltage higher then 7 V (see Fig. 2). VA characteristic is described by:

$$I = G_T \cdot U \cdot \exp(h/U), \tag{4}$$

where G_T and h are constants.

CAPACITANCE

A typical capacitance-voltage (CV) characteristic for NbO sample is in Fig. 3. In the normal mode the capacitance shows slight decrease with increasing voltage and remains practically constant. In the reverse mode substantial increase of capacitance is observed. We suppose that total capacitance is given by the superposition of C_0 - capacitance component corresponding to the high voltage in normal mode and capacitance C_i - due to existence of depletion region in I - layer:

The capacitance C_i of depletion region in I - layer can be expressed by a formula:

$$C_i = A\left[\frac{e\varepsilon_r\varepsilon_0 N_D}{2(U_D + U)}\right]^{\frac{1}{2}} \tag{5}$$

where N_D is donor concentration, ε_r - I – layer dielectric constant, U_D - diffusion voltage and A – sample area.

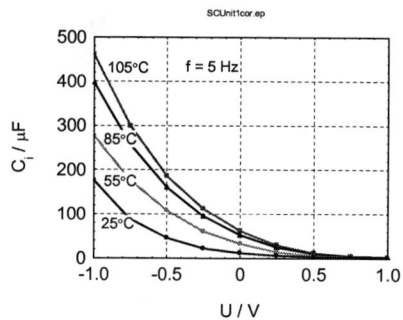
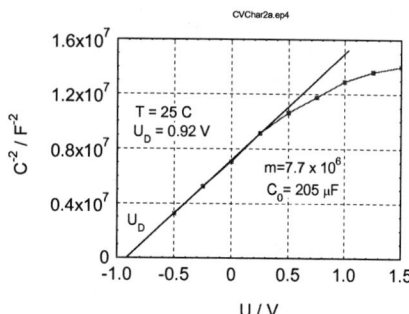

FIGURE 3. C_i vs. U at frequency 5 Hz

FIGURE 4. C_i^{-2} vs. U at frequency 5 Hz

For high value of applied voltage (reverse mode) the $1/C^2$ is a linear function of U as is shown in Fig. 4. For the sample area $A = 300$ cm^2, $\varepsilon_r = 40$ and the slope of $1/C^2 = f(U)$, $m = 7.7 \times 10^6$ we calculate donor concentration $N_D = 10^{18}$ cm^{-3}. For such doping concentration the impurity band can be created.

NOISE

The noise results from the current fluctuations and then the current noise spectral density can be obtained from a measurable quantity – voltage noise spectral density measured on load resistor R_L – from the following relation:

$$S_I = S_U(1 + \omega^2 R_L^2 C^2)/R_L^2 \qquad (6)$$

Where S_U is voltage noise spectral density, R_L is load resistance and C is capacitance of measured capacitor. $\tau = R_L \cdot C$ is the time constant characterizing signal attenuation due to capacitor shunting amplifier input. Measured voltage noise spectral density is shown in Figs. 5 and 6. The voltage noise spectral density is $1/f^a$ type with a in the range 0.8 to 1.2 in the frequency range bellow 1 Hz.

For the frequency above 1 Hz and low applied voltage S_U is proportional to f^2 due to the capacitor shunting of amplifier input. The current noise spectral density calculated from (6) is white noise type (see Fig. 7).

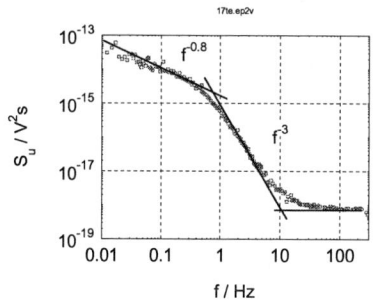
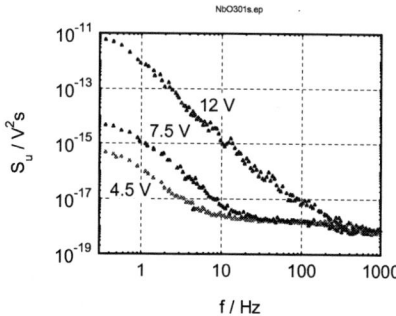

FIGURE 5. Noise spectral density for sample 17t

FIGURE 6. Noise spectral density for NbO301

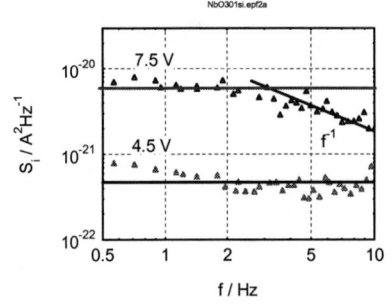
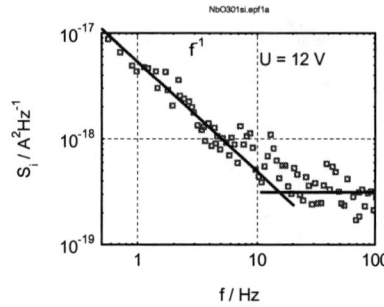

FIGURE 7. Noise spectral density for NbO301

FIGURE 8. Noise spectral density for NbO301

For the frequency above 1 Hz and high applied voltage S_U is proportional to f^3. The current noise spectral density calculated from (6) is 1/f type (see Fig. 8).

CONCLUSION

The low frequency noise in the MIS structure $NbO - Nb_2O_5 - MnO_2$ is given by the superposition of shot noise, g-r noise and 1/f noise. For the normal mode and low electric field the current transport is described by the Poole-Frenkel mechanism and the current noise spectral density is given by the superposition of shot and g-r noise. For the high electric field the current transport is described by tunnelling and the current noise spectral density is 1/f type. Schematically these mechanisms are shown in Fig. 9. Deep energy levels are probably oxygen vacancies in the Nb_2O_5 layer. The leakage current temperature dependence for the low electric field gives the deep level activation energy $E_a = 0.4$ eV.

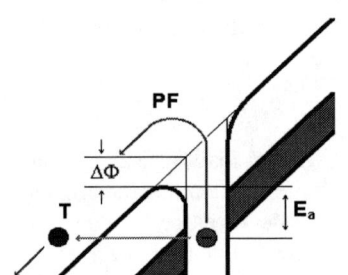

FIGURE 9. Quantum transition diagram

ACKNOWLEDGMENTS

This research has been supported by GACR grant No.102/05/2095 and under the project MSM 0021630503.

REFERENCES

1. C. A. Mead, „Electron transport mechanisms in thin insulating films" Phys. Rev. **128**, 088 (1962)
2. S. M. Sze, Physics of Semiconductor Devices, J.Wiley & Sons New York (1981)
3. J. N. Das, Zsch. f. Phys. **151**, 345 (1958)

1/f Noise In Low Density Two-Dimensional Hole Systems In GaAs

G. Deville[*], R. Leturcq[*], D. L'Hôte[*], R. Tourbot[*], C.J. Mellor[†], M. Henini[†]

[*]*Service de Physique de l'Etat Condensé, DSM, CEA- Saclay, F-91191 Gif-sur-Yvette Cedex, France.*
[†]*School of Physics and Astronomy, University of Nottingham, University Park, Nottingham NG7 2RD, United Kingdom.*

Abstract. Two-dimensional electron or hole systems in semiconductors offer the unique opportunity to investigate the physics of strongly interacting fermions. We have measured the 1/f resistance noise of two-dimensional hole systems in high mobility GaAs quantum wells, at densities below that of the metal-insulator transition (MIT) at zero magnetic field. Two techniques voltage and current fluctuations were used. The normalized noise power S_R/R^2 increases strongly when the hole density or the temperature are decreased. The temperature dependence is steeper at the lowest densities. This contradicts the predictions of the modulation approach in the strong localization hopping transport regime. The hypothesis of a second order phase transition or percolation transition at a density below that of the MIT is thus reinforced.

Keywords: 1/f noise, two-dimensional hole systems, GaAs heterojunction, percolation, hopping.
PACS: 71.30.+h, 71.27.+a, 72.70.+m, 73.21.Fg

INTRODUCTION

The physical nature of electronic systems changes drastically when the density of carriers is reduced. At large densities, the delocalized independent quasi-particle description (Fermi liquid, FL) prevails. At very low density and weak disorder, the ground state of the system is believed to be the Wigner crystal [1,2]. If the disorder is large and the interactions remain weak, decreasing the density leads to the strong localization (SL) regime for independent particles (Anderson insulator) [3]. The challenge for experimentalists is to investigate the strongly correlated systems which should appear as the density is decreased when the disorder is so low that the effect of the interactions should play a major role. Two-dimensional electron or hole systems (2DES or 2DHS) at low temperature offer the possibility of studying such systems as the density is simply tuned by applying a voltage to a metallic gate. The very high mobilities reached in presently available "clean" samples guarantees levels of disorder low enough for such new physics to appear instead of the classical crossover from FL to SL. The ratio which "measures" the magnitude of the interactions between carriers is $r_s = E_{ee}/E_F \propto m^*/p_s^{1/2}$, where E_{ee} and E_F are the interaction and Fermi energies, m^* the effective mass of the carriers, and p_s their areal density. The observation of a metallic behavior for $4 < r_s < 36$, in 2DES or 2DHS in high mobility silicon metal-oxide-semiconductor field effect transistors (Si-MOSFETs) and in GaAs heterostructures has raised the possibility of a new metallic phase due to the interactions [4-9], in contradiction with the scaling theory of localization for independent particles [3]. The metallic behavior is defined by a decrease of the resistivity ρ for decreasing temperature T, for $p_s > p_c$ where p_c is a critical density. When $p_s < p_c$, an insulating behavior occurs (dρ/dT < 0). Explanations of the metallic behavior by corrections to the standard independent particle picture have been put forward [4-9], but the question of the physical nature of the system at large r_s remains open. Several scenarios of the physics at large r_s in clean samples have been proposed. The system may freeze into a glass instead of crystallizing, as found in Si-MOSFETs [10-11]. In GaAs 2DHS, local electrostatic studies [12-13] and transport measurements in a parallel magnetic field [14] suggest the coexistence of two phases. Several calculations predict a spatial separation of a low and a high density phase [15-17], and the transport properties could be due to the percolation of the conducting phase through the insulating one. More generally, the percolation scenario has been put forward by several authors [17-22]. Experimentally, scaling laws observed on the resistance [18,19-22] and its fluctuations [21,22] of 2DHS and 2DES in GaAs favor the percolation scenario. While the "metallic" phase has been widely studied, only few experiments have been performed in the low density insulating phase. In the present study, we use 1/f noise as a tool to investigate the low-density regime in p-GaAs.

EXPERIMENTAL METHOD

Our 2DHS are created in Si modulation doped (311)A high mobility GaAs quantum wells. The metallic gate used to change the density is evaporated onto a 1 μm thick insulating polymide film [21,22]. The experiments were carried out on Hall bars 50 μm wide and 300 μm long. The mobility at a density $p_s = 6 \times 10^{10}$ cm^{-2} and a temperature $T = 100$ mK is 5.5×10^5 cm^2/V.s, a large value which guarantees the very low disorder. The "clean" nature of the 2DHS is confirmed by the temperature dependence of the resistivity ("metallic" behavior: resistivity increases by a factor of almost 2 as T increases [21,22]), and the well defined plateaus and oscillations of the Hall and Shubnikov-de-Haas curves at low density [22]. We performed transport and resistance noise measurements at densities ranging from about 1.0×10^{10} cm^{-2} to 1.6×10^{10} cm^{-2} and temperatures from 50 to 800 mK. Special attention has been paid to minimize parasitic contributions to the noise [22]. Two measurement methods were used: the voltage and its time fluctuations were measured for a fixed injected current I, or the current and its time fluctuations were measured, for a fixed applied voltage V. Both DC and AC techniques were used. The frequency interval in which the spectra were recorded is typically 0.01 Hz < f < 3 Hz. The final result is the normalized resistance noise power S_R/R^2, R being the resistance of the 2DHS in the ohmic region. In the fixed I method, the voltage noise power spectrum S_V was obtained by using the cross-correlation technique [21,22] for the two voltage noise signals measured on opposite sides of a Hall bar to suppress the contribution of the noise due to the contacts, leads and preamplifiers. $[S_V(I) - S_V(0)]/I^2$ did not depend on I, and $S_R = [S_V(I) - S_V(0)]/I^2$. In the fixed V method, the current noise power spectrum S_I was obtained by using the cross-correlation technique for the two voltage signals measured on two resistors in series with the sample. We verified that possible spurious noise sources (e.g. fluctuations of I, T, gate voltage V_G, etc.) did not contribute to S_R [21,22]. Fig. 1 shows that the fixed I or V methods give the same result.

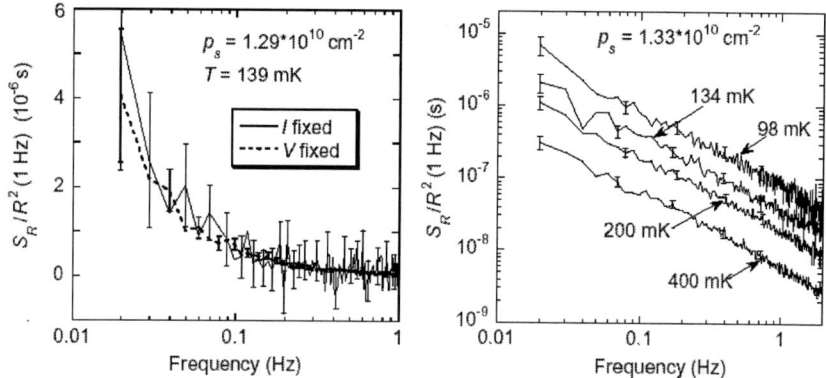

FIGURE 1. Left: an example of comparison between the two spectra obtained by using the fixed I and fixed V method. Right: Typical spectra obtained at four temperatures for $p_s = 1.33 \times 10^{10}$ cm^{-2}.

EXPERIMENTAL RESULTS

The transport measurements give an insulating $\rho(T)$ dependence ($d\rho/dT < 0$) for densities below $p_c \approx 1.46 \times 10^{10}$ cm^{-2}. For low densities and temperatures, the divergence of the resistivity when T goes to zero is described by an activated law [21,22]. Fig. 2 gives the noise power for $0.81 \times p_c < p_s < 0.96 \times p_c$. S_R/R^2 increases strongly when p_s or T decrease. The slope of the T dependence increases when p_s decreases, thus the trend already observed in Ref. [21] for $0.95 \times p_c < p_s < p_c$ continues at lower densities. In Ref. [21], we found that the noise scales with the resistance, which suggests a second order phase transition such as a percolation transition at a critical density $p^* < p_c$. The alternative "standard" description consists in a FL-SL crossover, the transport laws in the SL regime being hopping of carriers between localized sites. Let us examine whether our noise results at low density agree or not with the hopping picture. Experimentally, noise data in the 2D SL regime have been successfully analyzed by Pokrovskii et al. [23], using the modulation approach of Kozub [24]. In this model, $S_R/R^2 \sim 1/N \, (\delta\rho_i/\rho_i)^2$ where N is the number of bonds in the Miller-Abrahams cluster, ρ_i the resistance of the critical hop, and $\delta\rho_i$ its variation due to the electrostatic influence of fluctuating charges outside the cluster. As our transport law indicates simple activation, we would expect nearest neighbor hopping (NNH) in the SL regime. In NNH, N does not depend on the temperature [23,24]. We are thus led to attribute the T dependence of S_R/R^2 to the $\delta\rho_i/\rho_i$ term. Its temperature dependence has been calculated in Ref. [23]: for a size of the dipole modulator larger (resp. smaller) than the distance between the modulator and the critical hop, it is proportional to $T^{-1/3}$ (resp. $T^{1/9}$). This is far from what is shown in Fig. 2 where the power-like dependences indicate T exponents less than -1. We conclude that the T

dependence of the noise contradicts the hopping between localized sites at least in the modulation approach.

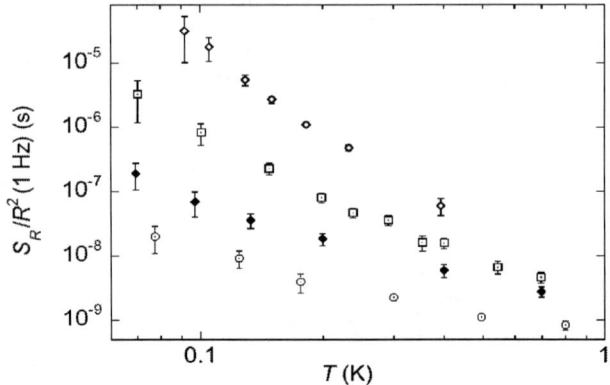

FIGURE 2. Relative resistance noise power at 1 Hz as a function of the temperature for four densities: p_s=1.40×10^{10} cm^{-2} (open circles), 1.33×10^{10} cm^{-2} (closed diamonds), 1.28×10^{10} cm^{-2} (open squares), 1.18×10^{10} cm^{-2} (open diamonds).

REFERENCES

1. Tanatar, B., and Ceperley, D. M., *Phys. Rev. B* **39**, 5005-5016 (1989).
2. Falakshahi, H., and Waintal, X., *Phys. Rev. Lett.* **94**, 046801(4) (2005).
3. Abrahams, E., *et al.*, *Phys. Rev. Lett.* **42**, 673-676 (1979).
4. Abrahams, E., Kravchenko, S. V., and Sarachik, M. P., *Rev. Mod. Phys.* **73**, 251-266 (2001).
5. Altshuler, B. L., Maslov, D. L., and Pudalov, V. M., *Physica (Amsterdam)* **9E**, 209-225 (2001).
6. Kravchenko, S. V., Sarachik, M. P., *Rep. Prog. Phys.* **67**, 1-44 (2004)
7. Pudalov, V. M., Lecture given at 2003 school of Physics "Enrico Fermi" in Varenna, preprint cond-mat/0405315.
8. Pudalov, *et al.*, Chapter 19, Proceedings of the EURESCO conference "Fundamental Problems of Mesoscopic Physics ", Granada, 2003; preprint cond-mat/0401396.
9. Das Sarma, S., Hwang, E. H., preprint cond-mat/0411528.
10. Jaroszyński, J., Popović, D., and Klapwijk, T. M., *Phys. Rev. Lett.* **89**, 276401(4) (2002).
11. Jaroszyński, J., Popović, D., and Klapwijk, T. M., *Phys. Rev. Lett.* **92**, 226403(4) (2004).
12. Ilani, S., Yacoby, A., Mahalu, D., and Shtrikman, H., *Phys. Rev. Lett.* **84**, 3133-3136 (2000).
13. Ilani, S., Yacoby, A., Mahalu, D., and Shtrikman, H., *Science* **292**, 1354-1357 (2001).
14. Gao, X. P. A., *et al.*, *Phys. Rev. Lett.* **89**, 016801(4) (2002).
15. Jamei, R., Kivelson, S., and Spivak, B., preprint cond-mat/0408066.
16. Spivak, B., and Kivelson, S., *Phys. Rev. B* **70**, 155114(8) (2004).
17. Spivak, B., *Phys. Rev. B* **67**, 125205(10) (2003); *Phys. Rev. B* **64**, 085317(6) (2001).
18. Meir, Y., *Phys. Rev. Lett.* **83**, 3506-3509, (1999); *Phys. Rev. B* **61**, 16470-16476 (2000); *Phys. Rev. B* **63**, 073108(4) (2000).
19. Shi, J., and Xie, X. E., *Phys. Rev. Lett.* **88**, 086401(4) (2002).
20. Das Sarma, S., *et al.*, *Phys. Rev. Lett.* **94**, 136401(4) (2005).
21. Leturcq, R., *et al.*, *Phys. Rev. Lett.* **90**, 076402(4) (2003).
22. Leturcq *et al.*, Proceedings of SPIE: Fluctuations and noise in materials, D. Popovic, M.B. Weissman, Z.A. Racz Eds., Vol. 5469, pp. 101-113, Maspalomas, Spain, 2004, preprint cond-mat/0412084.
23. Pokrovskii, V. Ya., *et al.*, *Phys. Rev. B* **64**, 201318(4) (2001).
24. Kozub, V. I., *Solid State Commun.* **97**, 843-846 (1996).

Low Frequency Noise Measurements in $La_{0.7}Sr_{0.3}MnO_3$ Thin Films on (100) $SrTiO_3$

L. Méchin[1], F. Yang[1], S. Mercone[1], J.M. Routoure[1], S. Flament[1],
Ch. Simon[2], R.A. Chakalov[3]

[1]*GREYC (UMR 6072), ENSICAEN & Univ. of Caen, 6 bd Maréchal Juin, 14050 CAEN cedex, France*
[2]*CRISMAT (UMR 6508), ENSICAEN, 6 bd Maréchal Juin, 14050 CAEN cedex, France*
[3]*School of Physics and Astronomy, University of Birmingham, Birmingham B15 2TT, UK*

Abstract. We report measurements of the 1/f noise in 75 nm, 100 nm and 200 nm thick $La_{0.7}Sr_{0.3}MnO_3$ (LSMO) thin films deposited onto (100) $SrTiO_3$ substrates by pulsed laser deposition. The samples were patterned using photolithography into bridges with various widths (20 µm to 150 µm) and lengths (50 µm to 300 µm). The voltage noise spectra $S_V(f)$ clearly showed two regions: excess noise depending on the bias current with a 1/f behavior at low frequency and Johnson noise (4kTR where R is the resistance of the device) at higher frequency. In the investigated 10 Hz – 100 kHz frequency range no deviation from the 1/f behavior was noticed neither as a function of the temperature (300 - 400 K) nor of the bias current. We particularly investigated the validity of the semi-empirical Hooge relation in the 300 – 400 K temperature range, which means across the metal-insulator transition. The voltage noise density S_V in LSMO bridges was found to be proportional to the square of the bias voltage V in the whole temperature range, indicating that noise arises from resistance fluctuations. Unexpectedly the normalized Hooge parameter α_H/n was found not to be volume independent, indicating in a first approach that the noise source localization is not homogeneous. Finally a α_H / n value of 8×10^{-31} m^{-3} have been obtained, which is among the lowest level reported in LSMO thin films and promising for the realization of uncooled bolometers.

Keywords: 1/f noise, manganite, thin films
PACS: 73.50.T, 75.47.Lx

INTRODUCTION

Among the perovskite manganese oxides $La_{0.7}Sr_{0.3}MnO_3$ (LSMO) thin films have attracted big interests because of their high Curie temperature (360 K), thus enabling room-temperature applications [1]. The LSMO epitaxial thin films can be considered as good candidates for the uncooled microbolometer fabrication for example [2]. The influence of the geometry of the devices on the measured electrical fluctuations in the films has not been investigated in detail so far. In the present study we measured the 1/f noise in bridges patterned in a 75 nm, 100 nm and 200 nm thick LSMO epitaxial thin films deposited onto (100) $SrTiO_3$. First, we will very briefly describe the sample preparation and the noise measurement set-up. Results are given and finally compared to literature.

EXPERIMENTAL DETAILS

The LSMO thin films were deposited by pulsed laser deposition from a stoichiometric target onto (100) $SrTiO_3$ single crystal substrate. The deposition conditions were optimized for producing single-crystalline films with smooth surface as judged by X-ray diffraction (XRD) and atomic force microscopy (AFM). The XRD study indicated a highly (100) orientation of the LSMO films. A 500 nm thick gold layer was deposited by the RF sputtering technique to make low resistive four-probe connections. Then the LSMO thin films were patterned by UV photolitography and argon ion etching for forming bridges with various widths (20 µm to 150 µm) and lengths (50 µm to 300 µm). The contacts between the sample holder and the patterned film were finally made using aluminium wires by ultrasonic bonding. The electrical resistance (R) was measured in the 300K – 400K temperature range by the four-probe method. The normalized R(T) curves present the same shape for all the different geometries. The resistivity of the film calculated at 300K ranges in the 15-25 µΩ·m.

The four probe noise measurement set-up consists of a current source made of a low noise voltage source and a high output resistance value (more than ten times the resistance of sample under test) and a low noise differential preamplifier that provides a voltage gain of 10000. The spectral density of the amplified output noise signal was finally obtained using an external spectrum analyzer HP 3562A.

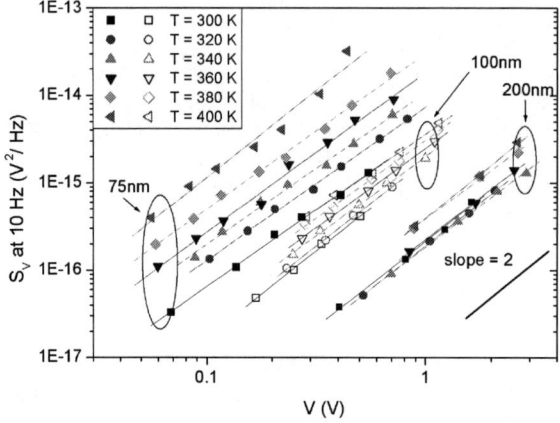

FIGURE 1. Spectral voltage noise density measured at 10 Hz for three LSMO bridges patterned in LSMO films of different thicknesses and geometries (75nm×20µm×300µm; 100nm×150µm×300µm and 200nm×100µm×300µm). For each sample the temperature was increased from 300K to 400K. A line with slope = 2 is shown as a guide for the eye.

RESULTS

Every bridge of each sample presented 1/f noise. No telegraph noise was noticed in the whole temperature range. The Hooge's empirical relation usually describes the 1/f noise in homogeneous sample [3]:

$$\frac{S_V}{V^2} = \frac{\alpha_H}{n} \times \frac{1}{\Omega \times f} \qquad (1)$$

where S_V is the voltage spectral density, R is the resistance of the bridge, V is the bias voltage (i.e. the bias current multiplied by the resistance value), α the Hooge's parameter, n the charge carrier density, Ω the sample volume and f the measuring frequency. α_H / n is defined as the normalized Hooge parameter.

In the following we shall verify the validity of the Hooge relation as a function of its main parameters. As expected in the case of resistance fluctuations figure 1 shows that S_V quadratically varies with V and, in addition, that the slope is independent of the temperature in the 300 – 400 K range. Secondly the normalized Hooge parameter α_H / n was calculated using equation (1) for all the measured bridges. It is plotted as a function of the volume of the bridge in figure 2. Closed symbols represent the 100 nm thick film and open symbols the 200 nm thick film. From the Hooge relation one would expect a volume independent value of α_H / n, which is clearly not the case. A more careful analysis shows actually that α_H / n is length independent but not width independent. Figure 3 shows indeed the evolution of α_H / n for different widths as a function of the temperature in the case of the 100 nm and the 200 nm thick LSMO films. We believe that this dependency can arise from the magnetic microstructures of the films. A study of the magnetic domain size as a function of the width of the bridges is therefore in progress. Figure 3 also remarkably shows that no 1/f noise peak was observed at the transition temperature, thus indicating a quite continuous transition contrary to what has been measured in $La_{0.7}Ca_{0.3}MnO_3$ films [4].

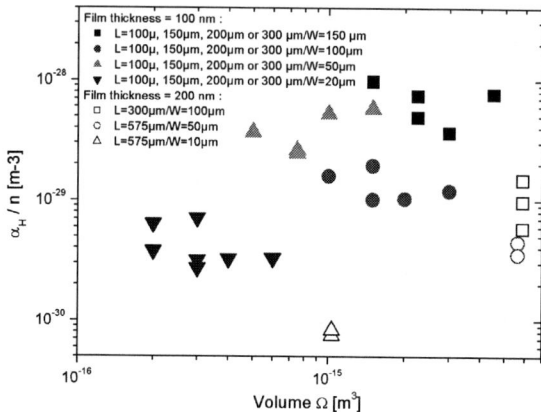

FIGURE 2. Normalized Hooge parameter as a function of the volume Ω of the bridge for the 100 nm thick and the 200 nm thick LSMO films.

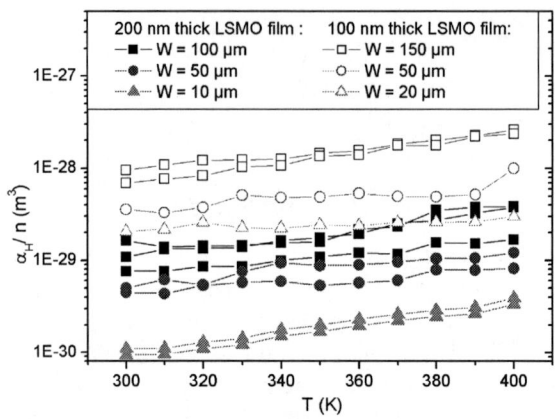

FIGURE 3. Normalized Hooge parameter as a function of the temperature for the 100 nm thick and the 200 nm thick LSMO film.

CONCLUSION

The 1/f noise level measured in the set of three patterned LSMO thin films was very low, even through the metal-insulator transition, which makes this material potential for bolometric and device applications. Our lowest α_H / n value is equal to 8×10^{-31} m^{-3} and was measured in the 10 µm bridge of the 200 nm thick LSMO film, which is among the lowest reported value for LSMO [5, 6]. The normalized Hooge parameter has been found to be length independent but not width independent. Magnetic domain imaging by Kerr microscopy as well as further noise measurements are now in progress in order to study the possible correlation between noise and magnetic domain structures.

REFERENCES

1. Venkatesan T., Rajaswari M., Dong Z., Ogale S. B. and Ramesh R., Rhil. *Trans. R. Soc. Lond. A* **356**, 1661 - 1680 (1998).
2. Yang F., Méchin L., Routoure J.M., Flament, Robbes D., Chakalov R.A., *Proceedings of the 17th ICNF*, edited by J. Sikula, 145-148 (2003).
3. Hooge F.N., *Phys. Lett.*, **29A** (3) 139-140 (1969).
4. Merithew R.D., Weissman, Hess F.M., Sprading P., Nowak E.R., O'Donnell J.O., Exkstein J.N., Tokura Y., Tomiaoka Y., *Phys. Rev. Lett.*, 84 (15) 3442-3445 (2000)
5. Palanisami A., Merithew R.D., Weissman, Warusawithana M.P., Hess F.M., Eckstein J.N., *Phys. Rev. B* **66** 092407 (2002)
6. Raquet B., Coey J.M.D., Wirth S., von Molnar S., *Phys. Rev. B*, **59** 12435-12443 (1999)

Acoustic Noise Spectrum of the Liquid Helium Boiling Process on Superconducting Bolometer Surface

O.V. Pakhomov[(1)], I.A. Khrebtov[(2)]

[(1)]*St.-Petersburg State University of Low Temperature Technology*
Lomonosov str. 9, 195001, St.Petersburg, Russia
Email: oleg_pakhomov@mail.ru

[(2)]*S.I. Vavilov State Optical Institute*
199034, St.Petersburg, Russia
Email: iakhrebtov@yahoo.com

Abstract. As the investigated sample in our experiments we used the film of niobium on the sapphire substrate, and ceramics CTS -19 with the preliminary cooled amplifier was used as the acoustic sensor. Acoustic sensor was fastened to the back side of the substrate. In the experiments we measured the spectrum at the boiling of liquid nitrogen; characters of the frequency dependence of spectrums for helium and nitrogen were identical.

Keywords: noise, liquid helium, cryoelectronic, superconducting films

INTRODUCTION

In physical studies utilizing high-sensitivity and low-noise cryoelectronic devices it is necessary to solve the problem of minimization of both the intrinsic noise of superconducting films [1, 2, 3] and the noise produced by non-stationary thermal processes in the thermostat (gas, liquid, thermal contact with solid body) [1]. When cryogenic liquid is used as thermostat, at low values of scattered power we can observe excess noise caused by equilibrium temperature fluctuations in the liquid phase. However, in the reality, there is always heat flux to liquid helium, resulting in intense surface evaporation. Evaporation, in turn, causes fluctuations of pressure above the liquid and, as a result, the excess noise of bolometer due to the drift of equilibrium boiling point. This problem is usually solved by stabilization of pressure in the cryostat. At higher values of power the bubble boiling begins on the surface of superconducting element. This process is also recorded in the form of excess electrical noise due to the temperature dependence of electrical resistance.

We may suppose that since random process of heat exchange of superconducting film with liquid helium is accompanied by appearance of bubbles and thus produces random acoustic noise, then analysis of changes in the acoustic spectrum may enable us to determine the spectral contribution of random bubble boiling process into the excessive noise of bolometer. That is why our investigation of acoustic noise that accompany boiling of liquid helium was carried out in a wide range of frequencies and levels of scattered power.

TECHNIQUE OF EXPERIMENT AND DISCUSSION

As the investigation sample in our experiments we used niobium film (length – 10 mm, width – 10 μm, thickness – 0.1 μm) on sapphire substrate. Ceramics CTS -19 with the cooled preliminary amplifier was used as the acoustic sensor.

Acoustic sensor was fastened to the back side of the substrate. In our experiments at low frequencies the levels of background acoustic and electrical noise were exceptionally high comparing to level of the wanted signal. That is why the substrate with piezosensor was placed into niobium beaker before being lowered into cryostat with liquid helium. In liquid helium niobium reached superconducting state and thus the substrate with piezosensor were shielded by superconductive niobium screen from the external electromagnetic fields.

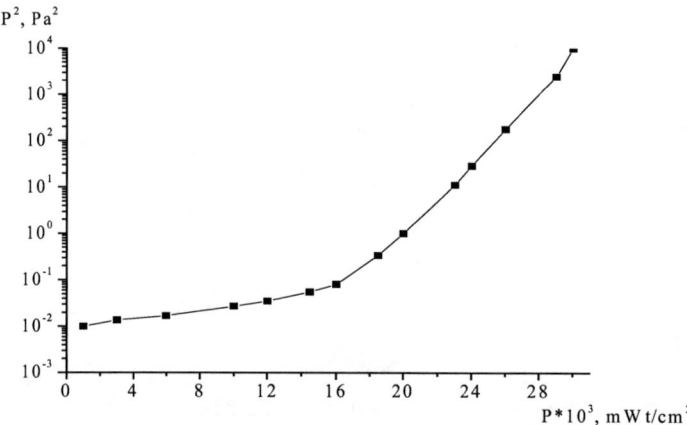

FIGURE 1. Dependence of the level of acoustic noise in the wide band on power dissipating

Cryostat with liquid helium was suspended on elastic cords. Signal from the piezosensor was fed into preamplifier with 20db gain (with gain flatness of 0.2db) and then fed into selective nanovoltmeter and oscillograph. The measured voltage was recalculated into power density of the wanted signal by standard method. Acoustic transfer ratio was measured according to the following procedure. Second CTS-19 piezosensor was attached to the substrate. Alternating voltage of given frequency from the signal generator was applied to this sensor, thus making it an acoustic transmitter. Another piezosensor, attached to the measuring devise worked as acoustic receiver. Frequency dependence of the ratio of output to input signal was accepted as the transfer coefficient of the system film-substrate-sensor.

In order to confirm correctness of the experiment, the measuring devise was tested. For testing purposes, instead of substrate with superconducting film we used polished aluminum disk with a heater. In the center of the disk a calibrated orifice (0.5 in diameter) was made. This orifice served as artificial center for bubble formation while on the rest of polished disk surface boiling did not occur. Experiment demonstrated that at the critical level of scattered power, periodic pressure pulsations were confidently registered with oscillograph. Increase in the level of scattered power resulted in increase of both frequency and amplitude of pulses. Measurements were carried out in the range of frequencies from 3 Hz to 3 kHz in the wide range of power scattered level by the superconducting film. Acoustic background noise was equal to 10^{-4} (Pa^2/Hz) at frequency 10 Hz, and the value of measured noise at the same frequency was from 1 to 100 (Pa^2/Hz). In the experiments we measured the spectrum of boiling of liquid nitrogen. It has to be noted that characteristics of the frequency dependence of spectra for helium and nitrogen were identical. Figure 1 demonstrates the dependence of the acoustic noise level from the level of scattered power in a wide frequency range. It can be seen, that the boiling process begins at some threshold level of heat power.

Figure 2 demonstrates results of measurement of the spectral distribution of pressure fluctuations. Performed measurements show that the form of noise spectrum is constant at low frequencies and has the $1/F^n$ dependence at higher frequencies. Moreover, process of the appearance of pressure fluctuations begins at the specific

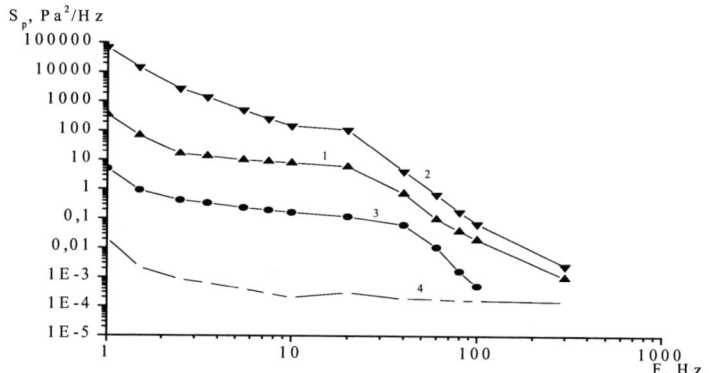

FIGURE 2. Spectrum of pressure fluctuation, 1-He, P=2.5*10^4 mWt/cm^2, 2-He, P=3*10^4 mWt/cm^2, 3-N$_2$, P=5*10^3 mWt/cm^2, 4-He, P=0

level of scattered power on the film. Further increase of the power leads to the decrease of cutoff frequency. Cutoff frequency corresponds to the minimal duration of pressure pulse, or life-time of bubble. In our experiments this value varied from 20 to 50 ms, depending on the level of scattered power. This change of cutoff frequency with increase of power is explained by the reduction of average size of appearing bubbles with increase of heat flux. Average diameter of bubbles for these values of life-time ranges between 20 and 40 μm which corresponds to the results of filming the bubbles. Appearance and growth of each bubble is accompanied by local heat exchange that results in random change of superconducting film temperature under the bubble. Moreover, the energy taken away from the film is used up in the bubble growth for work against forces of surface tension.

Therefore, the experimentally received spectrum of pressure fluctuations can be used to determined the spectrum of random heat exchange between the film and the thermostat. The spectrum of temperature fluctuations of the film is calculated from the solution of heat conduction equation, where the power source is set as a random function with known spectrum. Knowing temperature dependence of the film resistance, we can calculate the spectrum of fluctuations of film's resistance. Furthermore, with known current, the spectrum of fluctuations of voltage on film's surface can be determined. Work [1] presents method and measuring results of the temperature fluctuations spectrum for the bolometer.

Figure 3 presents the comparison between measurements of temperature fluctuations on the bolometer in liquid helium [1] and the dependence calculated from measured spectrum of acoustic pressure fluctuations using solution of heat conduction equation with random source of power.

Comparison of the results shows that in the frequency range up to the cutoff frequency, shapes of the curves are identical, while at higher frequencies measured temperature fluctuations demonstrate flatter slope of frequency dependence, then the calculations. This difference arises probably due to the fact that at higher frequencies the determining role is played by heat flow between different regions of bubble formation on the film. Temperature gradient between these regions arises from the random character of bubble formation.

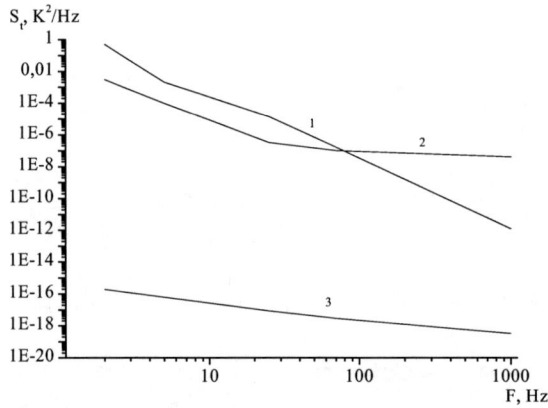

FIGURE 3. Spectrum of temperature fluctuation.
1 – calculated spectrum of temperature fluctuation on film
2 – experimental spectrum of temperature fluctuation on film, T=4.2 K, L-He, P=10^{-4} W

REFERENCES

1. Pakhomov O. V.; Khrebtov, I.A. "An Experimental Setup For Measuring of Low-Frequency Temperature and Pressure Fluctuation Caused by Bubbling of Liquid Helium on Surfaces of Superconducting Niobium Films". *Sixth European Workshop on Low Temperature Electronics WOLTE-6* ESA-ESTEC 23 June 2004, pp.271-277
2. Pankratov N.A. and Khrebtov I.A., " Conversion of the temperature fluctuation spectrum of liquid helium into the noise spectrum of a cryogenic bolometer", *Sov. J. Opt. Technol.* № 6 378-38 (1996).
3. N.A. Pankratov, G.A. Zaitsev, and I.A. Khrebtov, "Low frequency temperature fluctuations of liquid helium in cryostats", Cryogenics 11 № 2 138-140 (1971).

Monte Carlo calculation of diffusion coefficient, noise spectral density and noise temperature in HgCdTe

C. Palermo*, L. Varani*, J.-C. Vaissière*, J.-F. Millithaler*, E. Starikov[†],
P. Shiktorov[†], V. Gružinskis[†] and B. Azaïs**

*Centre d'Electronique et de Micro-optoélectronique de Montpellier
Université Montpellier II — c.c. 084
Place Bataillon — 34095 Montpellier Cedex 5 — France
http://thz.cem2.univ-montp2.fr
[†]Semiconductor Physics Institute
A. Goštauto 11 — 01108 Vilnius — Lithuania
**Centre d'Etudes de Gramat
DGA/DCE/CEG — 46500 Gramat — France

Abstract. Noise in optoelectronic devices is a critical parameter since it can affect dramatically the quality of target discrimination. In the domain of LWIR atmospherical window photodetection, used in the framework of night-vision applications, mercury-cadmium-telluride is a widely used material. We propose the first kinetic calculation of $Hg_{0.8}Cd_{0.2}Te$ noise parameters, that is of diffusion coefficient, spectral density and noise temperature, and investigate the importance of the excess noise due to the presence of impact-ionization processes.

Keywords: HgCdTe, optoelectronics, thermal noise, Auger generation-recombination, Monte Carlo
PACS: 71.55.Gs, 72.70.+m, 02.70.Uu

INTRODUCTION

In the framework of photodetection, mercury-cadmium telluride (HgCdTe) is a widely used material which founds its applications in a wide range of domains, essentially leaded by the strategical field, medical imaging and non-destructive control. Indeed, this II-VI semiconductor material exhibits a band gap which is continuously tunable between 0 and 1.6 eV through the cadmium proportion, which allows it to be calibrated for the detection of the full infrared range wavelengths. In addition, its crystal parameter is practically independent of the band gap. Such features lead to the possibility to use this material for the making of a full set of photodetectors, operating in various atmospherical windows, within the same technological die. Moreover, it makes possible the process of multispectral devices, which represents one of the current technological stakes for optoelectronics progress.

In this work, a particular attention is given to the HgCdTe with a cadmium proportion averaging 20 %, which allows the photodetection in the Long Wavelength InfraRed range (LWIR), an atmospherical window particularly used for night vision applications. At nitrogen temperature, this semiconductor has a narrow gap, smaller than 100 meV, involving interesting physical effects such as impact ionization which can be induced even by small electric fields (that is of the order of few hundreds of V/cm). These Auger processes are responsible for a strong modification of the first order characteristics, that is a multiplication of the carrier concentrations and then of the current density, and, as a consequence, an important influence of these phenomena on the second-order parameters is expected.

In order to evaluate the noise behavior of n-type $Hg_{0.8}Cd_{0.2}Te$, we combine a Monte Carlo kinetic calculation of the thermal noise contribution with a semi-analytical calculation of the impact ionization contribution to the total noise. All calculations are performed at 77 K since this is the standard condition for infrared photodetection.

CALCULATIONS

Due to the difference between velocity relaxation time and Auger lifetime, the processes of electronic scatterings and generation-recombination events can be modeled independently one from the other. Under this assumption, the cross-correlation between electron velocity and number fluctuations is neglected, and the spectral density of current fluctuations can be written as the sum of a diffusion contribution S_D and an excess noise S_{GR}, that is:

$$S_I(f,E) = S_D(f,E) + S_{GR}(f,E) \tag{1}$$

The thermal contribution can be expressed through the relation:

$$S_D(f,E) = \frac{4e^2}{L^2} N_0 D(f,E) \tag{2}$$

where e is the elementary electron charge, N_0 the zero-field electron number, and D the field-dependant diffusion coefficient. This last quantity has been calculated through the autocorrelation function method, as stated in reference [1], by a Monte Carlo Particle simulator which takes into account the main collisions undergone by the electrons in the considered material, that is optical phonon, ionized impurities and alloy scattering, neglecting in this step generation-recombination processes. The obtained results are reported in Fig. 1(a). We can observe that, whatever the electron concentration, the

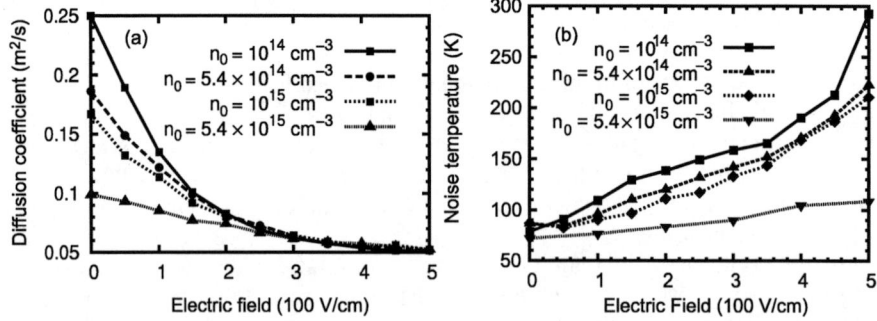

FIGURE 1. Static diffusion coefficient (a) and noise temperature (b) as functions of the electric field for different values of the electron density.

diffusion coefficient decreases with the electric field. For high values of E, D becomes independent of the electron density.

We have calculated the noise temperature T_N as a function of the electric field through the relation:

$$T_N(E) = \frac{e\,D(E)}{k_B\,\mu'} \tag{3}$$

where μ' is the low-frequency differential mobility and k_B the Boltzmann constant, and reported it for different values of the electron density in Fig. 1(b). We can remark that noise temperature increases with the electric field for all the considered values of the electron density.

Fig. 2 represents the spectral density of velocity fluctuations for two values of the electron density, calculated through the autocorrelation functions. At low fields, it exhibits the standard Lorentzian shape. At increasing values of E, we can observe the appearance of a small bump, related to hot-electron effects, around 200 GHz. We remark that, due to the small gap, non-equilibrium conditions appears for electric fields as small as 100 V/cm.

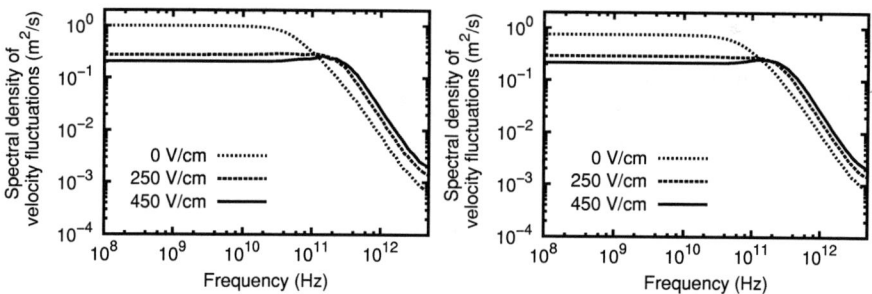

FIGURE 2. Spectral density of velocity fluctuations as a function of frequency for $n_0 = 10^{14}$ cm^{-3} (a) and $n_0 = 5.4 \times 10^{14}$ cm^{-3} (b).

According to McIntyre theory [2], the generation-recombination noise S_{GR} can be written as:

$$S_{GR} = 2eI_D M^2 F(M) \tag{4}$$

where I_D is the dark current, M the multiplication factor and $F(M)$ the noise factor. By neglecting the hole contribution to impact ionization, i.e. by considering a pure electron injection, the noise factor verifies the relation [3]:

$$F(M) = \frac{M}{k} + \left(1 - \frac{1}{k}\right)\left(2 - \frac{1}{M}\right) \tag{5}$$

where $k = \alpha/\beta$ is the ratio between the electron and the hole ionization coefficients respectively. The spectral density of current fluctuation S_I can be rewritten as:

$$S_I/A = 2e^2 n_0 \left(\frac{2D}{L} + v_d M^2 F(M)\right) \tag{6}$$

where v_d is the electron drift velocity, A the cross-section and L the width of the sample. This last has been taken equal to 10 μm. The noise spectral density per unit of surface

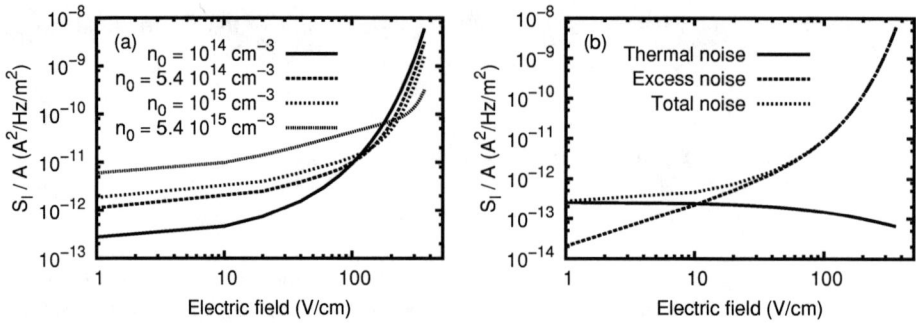

FIGURE 3. Spectral density of total current fluctuations normalized to the device cross-section as a function of the electric field for different values of the zero-field electron density.

has been reported as a function of the electric field for different values of the zero-field electron density on Fig. 3(a) (due to the onset of the ionization process, the electron density increases with respect to the value at zero field). We observe that, for the considered dimensions, the total current noise increases of several orders of magnitudes in the low field region, that is between 10 and 100 V/cm. This increasing is due to the excess noise contribution, as shown in Fig. 3(b), which corresponds to the activation of impact ionization processes.

CONCLUSION

We have calculated through a Monte Carlo approach the diffusion coefficient, the diffusion noise temperature and spectral density of total current fluctuations in the n-type $Hg_{0.8}Cd_{0.2}Te$ at 77 K. Moreover, we have evaluated the excess noise contribution related to the Auger generation-recombination processes and have shown its dramatic effect from very small electric field values corresponding to the activation of impact ionization processes. It is evident that this strong dependence of the noise in the low-field region can greatly affect the sensitivity of HgCdTe-based infrared photodetectors.

ACKNOWLEDGMENTS

This work has been performed in the framework of a French DGA–CNRS thesis grant. The authors would like to thank Professor Lino Reggiani, from the university of Lecce, for helpfull discussions.

REFERENCES

1. L. Varani, and L. Reggiani, "Microscopic theory of electronic noise in semiconductor unipolar structures," in *La rivista del nuovo cimento*, 1994, vol. 17.
2. R. J. McIntyre, *IEEE Trans. Electron Devices*, **ED-13**, 164 (1966).
3. H. M. Menkara, B. K. Wagner, and C. J. Summers, *Appl. Phys. Lett.*, **66**, 1764 (1995).

Noise Properties Of High Resistivity Cl-doped Cadmium Telluride

Ihor S. Virt*[†], Andrzej Kolek[¶], Volodymyr D. Popovych[†], and Igor S. Bilyk[†]

*University of Rzeszow, Rejtana 16A, Rzeszow, 35-959, Poland
[¶]Department of Electronics Fundamentals, Rzeszow University of Technology, W. Pola 2, Rzeszow, 35-959, Poland
[†]Drogobych State Pedagogical University, I.Franko 24, Drogobych, 82100, Ukraine

Abstract. Investigation of noise properties of CdTe:Cl samples with specific resistivity in the range $\rho \approx 10^8 - 10^9$ Ωcm at room temperature were performed. Cadmium telluride single crystals with chlorine concentration in the range $N_{Cl} = 10^{17} - 5 \times 10^{19}$ cm^{-3}) were grown by sublimation traveling heater method (STHM). Noise spectra have been measured over a wide range of frequencies, from below 1 Hz up to 1 MHz, and for voltages applied the samples from 0 up to 200 V. Two type of noise dominates in the range of voltages and frequencies under investigation, namely excess 1/f noise and generation-recombination (G-R) noise. Analyses of short noise were performed taking into consideration the presence of the macrodefects and clusters (inclusions and precipitates) in the crystalline matrix. It is assumed that these types of inhomogeneities affect the current noise spectra especially in highly doped material ($N_{Cl} > 10^{19}$ cm^{-3}).

Keywords: Noise properties, cadmium telluride crystals, macrodefects and clusters
PACS: 05.40.Ca, 72.70.+m, 71.55.Gs

INTRODUCTION

Cadmium telluride is mainly employed in the field of radiation detection, ranging from X-ray up to γ-ray energies. Structural defects and residual impurities create deep levels in the band gap, which degrade the performance of the devices fabricated on the basis of these materials. Photoluminescence (PL), thermally stimulate current (TSC), deep level transient spectroscopy (DLTS), photoinduced current transient spectroscopy (PICTS) are widely used to study these defects. Noise measurements also enable one to investigate deep recombination traps and determine their type (n or p) [1, 2]. It is possible to obtain more realistic model of the defects levels when comparing the results coming from different experimental techniques.

SAMPLE PREPARATION AND MEASURING APPARATUS

Cadmium telluride samples were cut out from the single crystals grown by sublimation traveling heater method (STHM) in vertical configuration [3]. They were chemi-mechanically polished using bromine-ethylenglicole solution to remove damaged layer up to dimensions $5 \times 2 \times 2$ mm^3. Contacts were prepared by soldering of indium. Their ohmic characters were controlled by measuring current-voltage (I-V) characteristics. Resistivities of the samples were in the range 10^8- 5×10^9 Ωcm. All the measurements were performed at room temperature.

Noise measurements were carried out by DC technique using ultra-low noise current amplifier 5182 from EG&G. Output AC noise signal was further amplified in broad-band Unipan 233 amplifier and then was directed into either HP 35660A dynamic signal analyzer or NI-PCI 6111 card supported by Lab View software for signal acquisition and fast Fourier transforming. In both more than 250 transforms were collected to obtain power noise spectral density $S_U(f)$ of the signal. The latter were measured over a wide range of frequencies (from below 0,1 Hz up to1 MHz) at different voltage across the samples.

EXPERIMENTAL RESULTS

At first step of the measurements it was verified that spectra at zero bias are white and agree with the values of thermal noise $S_I = 4k_BT/R$ (see Fig. 1). When biased most samples exhibit monotonous course of dependence which follows the ~ $1/f$ law. Spectra of excess noise were calculated as $S_{Iex}= S_I - S_{I=0}$.

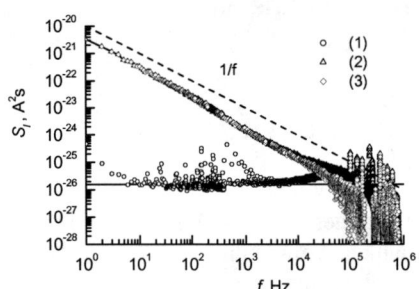

FIGURE 1. Power spectral densities of current fluctuations measured for CdTe:Cl sample ($N_{Cl}= 10^{17}$ cm^{-3}). The most upper spectrum refers to sample biased by the voltage $V = 20$ V. The lowest spectrum was taken with no bias and agrees exactly with thermal noise formula, $S_I = 4k_BT/R$ (solid horizontal line). Spectrum (3) is the difference of the two spectrum (2) and spectrum (1) and shows excess noise. Dashed line shows 1/f dependence.

Power spectral densities of excess noise measured at different bias are shown in Fig. 2. Data measured for two samples which differ in chlorine concentrations are compared. At the limit of low bias S_{Iex} follows the quadratic law versus the dark current $S_{Iex} \sim I^2$. At larger current S_{Iex} decreases in spite that I increases. We explain this effect as follows. It is clear that the larger defect concentration N_{Cl} the larger the excess noise.

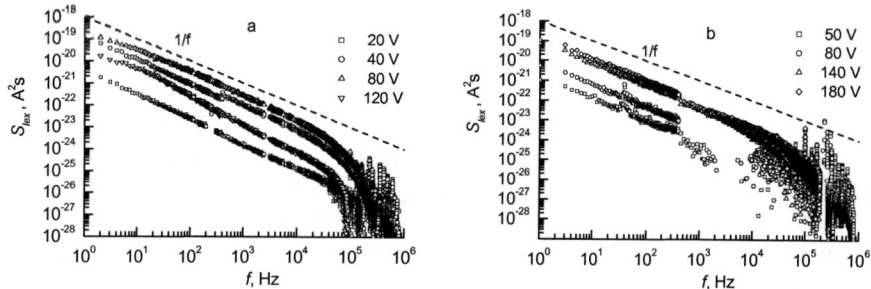

FIGURE 2. Power spectral densities of excess noise measured for CdTe:Cl samples with chlorine concentration of $N_{Cl}= 10^{17}$ cm^{-3} (a) and $N_{Cl}=5 \times 10^{19}$ cm^{-3} (b). Spectra are taken for several bias voltages (from down to up) $V = 0$-180 V.

Not only the magnitude but also the shape of power spectral densities is affected by defects concentration. Consequently there are some peaks on normalized curves $fS_{Iex}(f)$ versus f due to the presence of deep level centers in the band gap of the crystals [4, 5]. This is clearly indicated in high-frequency region.

We suppose that this fact can be explained by the occurring of the second phase clusters due to the chlorine precipitation [3].

It is known that the contribution of any noise source for the sample with resistance R in general case can be expressed as

$$\frac{S_{Iex}(f)}{I^2} = \frac{\alpha}{N_0 f}, \qquad (1)$$

where α is the empirical constant, $N_0 = n_0\Omega$ is the number of free carriers in the semiconductor, n_0 is the free carrier concentration, Ω is crystal volume. Two-phase model were considered to explain experimental results. It was supposed that crystal matrix contains more high resistively domain which create depletion regions with the size equal to the Debye length (L_D). Growth of the local electric field with the bias increasing causes redistribution of the current-carrying regions thus increasing the value of N_0.

Spectral noise density obeys dependence, which is characteristic for generation-recombination (GR) type of noise at the middle-frequency region. Since generation-recombination process in wide band gap materials are determined by the trapping and subsequently recombination of free carriers at the local centers we can use the equation [6]

$$\frac{S_{Iex}(f)}{I^2} = \frac{4N_t}{\Omega N_d^2} \frac{\tau}{1+\omega^2\tau^2}. \qquad (2)$$

where N_t is the trap concentration, τ - relaxation constant. It must be note that the noise spectra analysis of our samples was performed taking into consideration the presence of the inhomogeneities and clusters in crystal matrix. In particular, decreasing of the τ value of G-R lorenzian point out to the occurrence of the additional recombination channel in cluster volume though depletion region can reach up to 100 μm at free carrier concentration of $n_0 = 10^7 - 10^8$ cm^{-3}.

CONCLUSIONS

Noise spectroscopy technique is a very useful method for investigation of the defects in semiconductors, especially in the range where GR- and 1/f-type noises superimpose of spectral dependences of the power noise density reveal different sources of noise in semiinsulating CdTe:Cl single crystals grown by vertical STHM technique. Non-monotonous steep decreasing of noise level and peaks presence on the normalized dependences $S_I(f) \times f \div f$ in the low-frequency range point out to the occurrence of macrodefects - clusters and microinclusions in the volume of these crystals.

REFERENCES

1. Schauer P., Identification of the source of 1/f noise in the CdTe crystals under illumination, Proceedings 17th International Conf. On Noise and Fluctuations, August, 18-22, 2003, Prague, pp. 190-196.
2. Asaad I., Orsal B., Perez J. P., Alabedra R., Shot noise in CdTe resistors: experimental and analytical studies, Proceedings 17th International Conf. On Noise and Fluctuations, August, 18-22, 2003, Prague, pp. 130-136.
3. Tetyorkin V.V., Sukach A.V., Popovych V.D., Popov V. M., Nature of non linearity of I-V characteristics in CdTe:Cl single crystals, Proceeding of SPIE, Epilayers and Heterostructures in Optoelectronics and Semiconductor Technology 5136 (2002), pp. 200-209.
4. Raychaudhuri A.K. *Current Opinion in Solid State and Materials Science* **6**, 67-85 (2002).
5. Cwirko J., Przybysz C., Cwirko R., Kaminski P., Technique of low frequency noise vs. temperature for identification of deep level defects in semiconductor materials Proceeding of SPIE, Epilayers and Heterostructures in Optoelectronics and Semiconductor Technology 4413 (2000), pp. 218-221.
6. Lukyanchikova N.B., Konoval A.A., Sheinkman M.K., *Solid-State Electron*. **18**, 65-70 (1975).

Hot electron noise in n-type semiconductors in crossed electric and magnetic fields

F. Ciccarello and M. Zarcone

Dipartimento di Fisica e Tecnologie Relative Viale delle Scienze, edificio 18, 90128 Palermo (Italy)

Abstract. In this work we have investigated the noise properties of hot electrons in *n*-type GaAs in crossed electric and magnetic fields by single-particle Monte Carlo simulations. The motion of electrons during free flights is treated classically. Nonparabolicity of valleys is taken into account by means of a local parabolic approximation. The velocity auto-correlation function and its spectral density of fluctuations turn out to be strongly affected by the presence of the magnetic field. In particular, significant signatures of nonparabolicity and nonlinearity are observed.

Keywords: Noise, Monte Carlo, electric and magnetic field
PACS: 72.20.Ht, 72.20.My

Up to now the Monte Carlo (MC) determination of hot electron transport properties of *n*-type semiconductors in crossed electric and magnetic fields has been rarely addressed in literature. By using a two-valleys *parabolic* model for the conduction band, Boardman *et al.* [1] used a single-particle MC simulation to deduce the fields-dependence of drift velocity, mean energy, Hall angle and Hall scatering factor in GaAs at room temperature.

On the other hand, the MC method proved to be a powerful tool for deriving the stochastic properties of carriers motion in semiconductors in the presence of an intense electric field (see for example ref. [2] for *n*-type GaAs). However, to our knowledge, no effort was done to generalize the latter study to the case when a magnetic field is simultaneously applied.

In view of this facts, the aim of the current work is to determine by MC single-particle simulations the electron velocity auto-correlation function and the spectral density of its fluctuations in crossed, intense electric and magnetic fields in bulk *n*-type GaAs. Under the latter conditions, neglecting the nonparabolicity of bands is too a rough approximation. However, in this case, since the velocity **v** (and thus the Lorentz force) is not a linear function of the wavevector **k**, the electron classical equations of motion during free flights are analytically unsolvable. To overcome this difficulty, following the idea suggested by Warmenbol *et al.* for InSb at 77 K [3, 4], we expand the Kane dispersion law $\varepsilon(k^2)$ up to the first order in k^2 around $k^2(0)$, that is around the value of the squared wavevector at the beginning of a free flight. According to this local parabolic approximation (LPA), the energy turns out to be a quadratic function of k and the equations of motion are solvable. It can be easily proved that the overall effect of LPA is that in every valley the electron effective mass m^* is renormalized at each free flight to the higher value $m_R^* = m^* \sqrt{1 + 4\alpha \hbar^2 k^2(0)/2m^*}$, where α is the nonparabolicity coefficient of the considered valley. The solutions for **k**(t) obtained in this way differ from those used in ref. [1] (where nonparabolicity of valleys was fully neglected) just for the replacement of m^* with m_R^*. The LPA is physically reasonable in view of the short duration of free flights determining a small variation of k^2 between two next collisions. Furthermore, we checked that the results obtained by a numerical integration of the equations of motion under assumption of the Kane dispersion law show an excellent agreement with those obtained by LPA. Of course, the latter approach is to be preferred to get shorter computational times and to facilitate the understanding of the physical processes involved. With the LPA solutions for $\mathbf{k}(t)$ between two next collisions, a single-particle MC simulation can be performed by including all the relevant scattering mechanisms (scattering with polar optical and acoustic phonons and with impurities and phonon-assisted equivalent and non-equivalent intervalley processes). As suggested in ref.[1], once the total electric field $\mathbf{E_T}$ is chosen as an independent variable and the MC simulation is performed, $\mathbf{E_T}$ is decomposed into two components, parallel and perpendicular to the direction of the drift velocity $\bar{\mathbf{v}}$ (obtained from a temporal average of the electron velocity): the former is interpreted as the applied electric field $\mathbf{E_A}$ while the latter as the Hall field $\mathbf{E_H}$. Due to the exact solubility of the free flights equations of motion under LPA, simple accumulation formula can be derived to calculate $\bar{\mathbf{v}}$ with no need of sampling $\mathbf{v}(t)$ at fixed time steps.

The method adopted to derive the autocorrelation function $C(t)$ and the spectral density of velocity fluctuations $S(f)$ is now illustrated. It is performed a first simulation, using the accumulation formula, to determine $\bar{\mathbf{v}}$ and, in particular, its direction defined by the Hall angle θ_H formed with $\mathbf{E_T}$. Then a second simulation is performed to sample $v_{\theta_H}(t)$ (that is the component of the velocity along the direction of $\mathbf{E_A}$) at fixed time steps dt. The collected sample values $v_{\theta_H}(ndt)$ (with $n = 1,...,N$ with Ndt being the simulation time) are then used to compute the samples $C(jdt)$ of the auto-correlation function as

$$C(jdt) = \frac{1}{N-j+1} \sum_{n=1,N} \delta v_{\theta_H}(ndt) \delta v_{\theta_H}((n-j)dt) \quad (1)$$

with $j = 1,...,N_C dt$ ($N_C dt$ being a time large enough to have $C(N_C dt) \sim 0$) and where $\delta v_{\theta_H}(ndt) = v_{\theta_H}(ndt) - \bar{v}$. Using the Wiener-Kintchine theorem, the autocorrelation function $C(t)$ is finally Fourier-transformed to obtain the spectral density of fluctuations $S(f)$.

Figs. 1 and 2 show the behavior of $C(t)$ and $S(f)$ at room temperature for different values of E_A and for a fixed magnetic field $B=20$ kG[1]. It is important to note that $C(t)$ shows damped oscillations at approximately the cyclotron frequency $f_C = qB/m^*2\pi$ of the Γ valley (~ 888 GHz), as confirmed by the corresponding $S(f)$ curves showing a peak around f_C [2]. As E_A increases, the damping time of $C(t)$ gets shorter (and the linewidth of $S(f)$ around f_C becomes larger). This is mainly due to the fact that scattering processes occurs at a higher rate as electrons are heated. In addition, the more intense

---

[1] To appreciate more clearly some of the features to be discussed, in Fig. 1 not all the $C(t)$ curves corresponding to the S(f) curves of Fig. 2 are shown.
[2] The magnetic field strength $B=20$ kG here considered is low enough not to determine significant magnetic effects on electrons occupying L and X valleys. This is the reason why only the cyclotron resonance of the Γ valley plays a role in the behavior of $C(t)$ and $S(f)$.

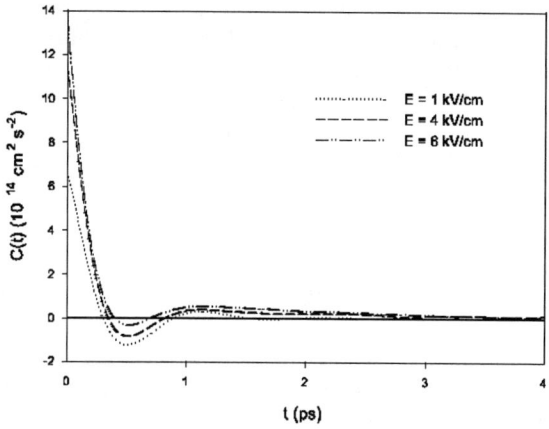

FIGURE 1. Velocity auto-correlation function $C(t)$ (10^{14} cm^2 s^{-2}) at room temperature for B=20 kG and E_A=1, 4, 6 kV/cm.

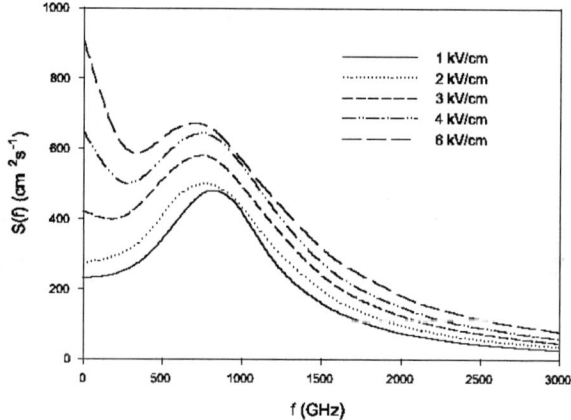

FIGURE 2. Specral density of velocity fluctuations $S(f)$ (cm^{-2} s^{-1}) at room temperature for B=20 kG and E_A=1, 2, 3, 4, 6 kV/cm.

is E_A the less relevant is the influence of the magnetic field. A more careful analysis of Fig. 2 suggests that $S(f)$ is not exactly centered at $f_C = qB/m^* 2\pi$, but at a lower value and this happens more and more markedly as the electric field is increased. The above behavior is a signature of nonparabolicity effects. Indeed, as implied by the Kane dispersion law, as electrons are heated their effective mass gets larger and thus their cyclotron frequency becomes smaller. As it could be expected, this confirms that the overall influence of nonparabolicity effects consists on a smoothing out of the magnetic effects computed on the basis of a parabolic model of valleys. This suggests that any application exploiting the appearance of a cyclotron frequency under the action of an external magnetic field and in the presence of intense electric fields must take this effect into account.

The behavior of $C(t)$ showed in Fig. 1 reveals an interesting property. Indeed, it can be appreciated that, as E_A is increased above ~ 3 kV/cm, the $C(t)$ oscillations starts to exhibit a positive offset. This is shown better in the corresponding $S(f)$ curves that present a maximum at $f = 0$. This behavior can be explained by analyzing the function $\delta v_{\theta_H}(t)$ in a generic free flight. Assuming that $\mathbf{E_T}$ and $\mathbf{B}$ lie along the x and z-axis, respectively, it can be easily proved that

$$\delta v_{\theta_H}(t) = \left(\frac{E_T}{B} + \frac{\hbar k_y(0)}{m_R^*}\right) \sin(\omega t + \theta_H) + \frac{\hbar k_x(0)}{m_R^*} \cos(\omega t + \theta_H) + \bar{v}(r-1) \quad (2)$$

where $k_x(0)$, $k_y(0)$ are the x, y components of $\mathbf{k}(0)$, $\omega = qB/m_R^*$ and where $r = E_H/\bar{v}B$ is the Hall scattering factor. As confirmed by our simulations, below 3 kV/cm (when most of the electrons are in the Γ valley and the response is nearly ohmic) it turns out that $r \simeq 1$, while above this threshold $r > 1$ indicating that the system starts to manifest a nonlinear behavior. This quite well known phenomenon (see for example [1, 5]) implies that, when E_A is high enough, in view of eq. (2), the free flight evolution of $\delta v_{\theta_H}(t)$ exhibits *displaced* oscillations (the term $\bar{v}(r-1)$ being different from zero). The latter gives an explanations of the above discussed offset for $C(t)$.

In conclusion, in this work we have studied the behavior of $C(t)$ and $S(f)$ generalizing the MC analysis of electron transport properties of n-doped GaAs in crossed electric and magnetic fields [1] to the case when nonparabolicity effects are taken into account. It has been shown that nonparabolicity reduces the influence of the magnetic field by decreasing the cyclotron frequency more and more as the applied electric field is increased. Furthermore, it was shown that, as a signature of nonlinearity, the spectrum of velocity fluctuations starts to exhibit a maximum at $f=0$ as E_A is intense enough to determine a nonlinear response of the system.

REFERENCES

1. A. D. Boardman, W. Fawcett, and J. G. Ruch, *Phys. Stat. Sol. (a)*, **4**, 133 (1971)
2. R. Fauquembuergue, J. Zimmerman, A. Kaszynski, E. Constant, and G. J. Microondes, *J. Appl. Phys.*, **51**, 1065 (1980)
3. P. Warmenbol, F. M. Peeters, J. T. Devreese, G. E. Alberga, and R. G. van Welzenis, *Phys. Rev. B*, **31**, 5285 (1985)
4. P. Warmenbol, F. M. Peeters, and J. T. Devreese, *Phys. Rev. B*, **33**, 1213 (1986)
5. R. A. Smith, *Semiconductors*, Cambridge University Press (1978)

Low Frequency Noise Resolution Of Ru-Based Low-Temperature Thick Film Sensors

A. Kolek, P. Ptak, Z. Zawislak, A. W. Stadler, K. Mleczko

Department of Electronics Fundamentals, Rzeszów University of Technology
W. Pola 2, 35-959 Rzeszów, Poland,

Abstract. It is shown that increase of low-frequency noise observed for Ru-based thick film devices in temperatures below few Kelvins results in poorer resolution of temperature measurements when these devices serve as temperature sensors.

Keywords: Low-temperature sensors, RuO_2, Thick film resistors, Low-frequency noise
PACS: 07.07.Df, 07.07.Dt, 07.20.Mc, 72.80.Tm, 72.70.+m

INTRODUCTION

The paper concerns low-frequency (LF) noise of RuO_2+glass thick films in low temperatures. Measurements performed down to 0.35 K have shown that the LF excess noise of these films has 1/f shape of the spectrum and for low biasing voltages, V, scales as $S_V(f) \sim V^2/f$, $S_V(f)$ is the power spectral density (PSD) of $\delta V(t)$. Thus, it can be classified as resistance noise i.e. the noise arising from fluctuations of resistance. The quantity $S \equiv fS_R/R^2 = fS_V/V^2$, $S_R(f)$ is PSD of resistance fluctuations, is the normalized measure of its intensity. This quantity is used on vertical axis of the graph in Fig. 1. It demonstrates the main experimental result of recent investigations, that is the large increase of noise intensity S as temperature T drops below few Kelvins [1].

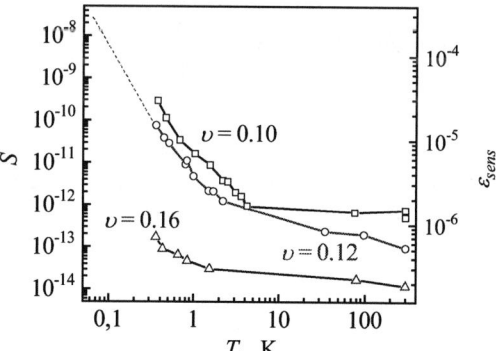

FIGURE 1. Normalized resistance noise S and sensor resolution ε_{sens} plotted as a function of temperature T. Labels $\upsilon = 0.10$, 0.12 and 0.16 refer to various devices used in the experiments (υ is the volume fraction of RuO_2 in the resistive paste used to produce a film).

One of the applications of Ru-based thick films is low-temperature thermometry. This application results from extremely good performance characteristics of these devices. Namely, large temperature sensitivity and very low magnetoresistance. Ru-based temperature sensors are calibrated down to 50 mK. The increase of resistance noise observed for these films as $T \to 0$, limits the resolution of resistance measurements. This means that the resolution of temperature measurement with Ru-based sensors is also limited [2]. The paper discusses this problem in more detail.

RESOLUTION OF TEMPERATURE MEASUREMENT

For the resistance-type sensors the resolution, $\varepsilon_T \equiv \Delta T$, of temperature measurement is limited by the resolution ΔR of resistance measurement according to the expression $\varepsilon_T/T = \varepsilon_r/A$, where $A \equiv d \ln R / d \ln T$ is specific sensitivity and $\varepsilon_r \equiv \Delta R/R$, R is the sensor resistance [3]. Manufacturers of cryogenic equipment define ΔR as rms noise low-pass filtered with $\tau = 3$ s time constant measured on a room temperature resistor. This definition applies to modern cryogenic thermometers which operate with ac current excitation and utilize the benefits of phase sensitive detection (see Fig. 2) [4].

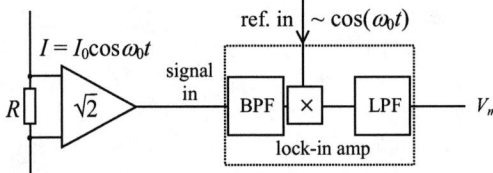

FIGURE 2. Principle of operation of modern AC resistance bridge. LPF and BPF refer to low and band pass filtering, × to signals multiplication. Resistance is calculated as $R_m = V_m/I_{rms}$, $I_{rms} = I_0/\sqrt{2}$

The voltage V_m at the output of the lock-in amplifier in Fig. 2 fluctuates around its average $\langle V_m \rangle = I_{rms} R$ with power spectral density

$$S_{Vm}(f) = |H_{LPF}(f)|^2 (I_{rms}^2 S_R(f) + 4kTR + S_{inst}(f)), \quad (1)$$

where I_{rms} is the excitation current, H_{LPF} is the transmittance of the output low-pas filter, k is Boltzmann constant and $S_{inst}(f)$ is the instrumental noise.

By the virtue of the definition the resistance, which is calculated as the ratio $R_m = V_m/I_{rms}$, is measured with relative resolution

$$\varepsilon_r \equiv \frac{\Delta R}{R} = \frac{1}{\langle V_m \rangle} \left(\int_{1/\theta}^{\infty} S_{Vm}(f) \right)^{1/2} \cong \left(\int_{1/\theta}^{\Delta f} \frac{S_R(f)}{R^2} df + \frac{4kT\Delta f}{I_{rms}^2 R} + \int_{1/\theta}^{\Delta f} S_0(f) df \right)^{1/2}, \quad (2)$$

where $\Delta f = (4\tau)^{-1}$ is the noise bandwidth of the output filter, θ is the observation time, and $S_0(f)$ is the relative PSD of instrumental noise. For the resistance noise-free sensor Eq.(2) simplifies to

$$\varepsilon_r \cong \sqrt{\frac{4kT\Delta f}{I_{rms}^2 R} + \varepsilon_0^2}, \text{ where } \varepsilon_0 \equiv \sqrt{\int_{1/\theta}^{\Delta f} S_0(f) df} \quad (3)$$

is the instrument relative resolution. In Fig. 3 relative resolution ε_r of Eq. (3) is plotted versus excitation current for $R = 135$ kΩ together with the data obtained for Lake

Shore Model 370 AC Resistance Bridge with a wire-wound resistor of resistance R attached to its input. As one can see, the whole system used to measure resistance has more than 5-digit resolution ($\varepsilon_0 = 5 \times 10^{-6}$), which drops to 4.5 digits at low excitation currents. In Fig. 3 the case of noisy sensor is also considered. Eq. (2) then reads

$$\varepsilon_r \cong \sqrt{\frac{4kT\Delta f}{I_{rms}^2 R} + \varepsilon_0^2 + \varepsilon_{sens}^2}, \text{ where } \varepsilon_{sens} \equiv \sqrt{\int_{1/\theta}^{\Delta f} S_R(f)/R^2 df} \qquad (4)$$

is a new parameter. As ε_{sens} is attributed to the sensor rather than to the instrument we shall call it the *sensor resolution*. For the sensor with resistance noise of 1/f type, which is the case of Ru-based sensors being considered, we have $S_R/R^2 = S/f$ and $\varepsilon_{sens} = (S\ln(\theta\Delta f))^{1/2}$. Lines in Fig. 3 are drawn according to Eqs. (3) and (4) assuming room temperature, $R = 135$ kΩ, $\varepsilon_0 = 5 \times 10^{-6}$, and $\varepsilon_{sens} = (3.4 \times 10^{-10}\ln(20))^{1/2}$. This value corresponds to the observation time of $\theta = 20/\Delta f (= 80 \tau)$ and resistance noise intensity of $S = 3.4 \times 10^{-10}$. The next series of experimental points in Fig. 3 refer to carbon resistor of resistance R attached to the instrument (Model 370) input. For this resistor resistance noise of intensity S was identified in an independent experiment (Fig. 4b).

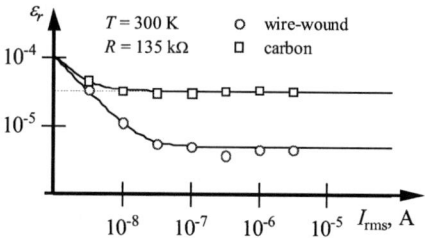

FIGURE 3. Theoretical (lines) and experimental (points) relative resistance resolution ε_r for wire-wound (circles) or carbon (squares) resistors of resistance $R = 135$ kΩ attached to Model 370 R meter.

The experimental data in Fig.3 were obtained in the following way [2]. For the instrument (Model 370) maximum cut-off frequency of $f_{max} = (2\pi\tau_{min})^{-1} = 0.8$ Hz ($\tau_{min} = 0.2$ s) the read-outs of the instrument (R_m) were recorded with the rate of 10 readings per second. Time series of $R_m(t)$ were Fourier transformed to give PSD $S_{Rm}(f)$. Such PSD's were obtained for several excitation currents. After normalization, they are shown in Fig. 4a for wire-wound resistor and Fig. 4b for carbon resistor. In both cases the values of ε_r were calculated by integration PSD's in Fig. 4 over the band [1/θ, Δf], θ is duration of time record used to calculate Fourier transforms of the signal $R_m(t)$.

Concerning Fig. 4a, one may observe that for PSD's taken at low currents the second term in the sum of Eq. (1) prevails over the others and thermal noise describes the shape of the spectra. PSD's taken at large currents overlap each other and reveal the shape of PSD of instrumental noise, $S_0(f)$. For the instrument used in the experiment $S_0(f) = 4.1 \times 10^{-12}/f$ for $f \to 0$. Concerning Fig. 4b, let us note that for noisy sensor and large currents the first term in the sum of Eq. (1) dominates. PSD's $S_{Rm}(f)$ taken in this limit overlap each other and agree with PSD of the sensor's resistance noise, $S_R(f)$. As shown, the relative PSD $S_R(f)/R^2 = 3.4 \times 10^{-10}/f$ obtained in this measurement agrees exactly with relative PSD $S_V(f)/V^2$ obtained in the standard (dc) noise measurement setup. At low currents thermal noise component of Eq.(1) increases and the first two terms of this equation determine the shape of the PSD's.

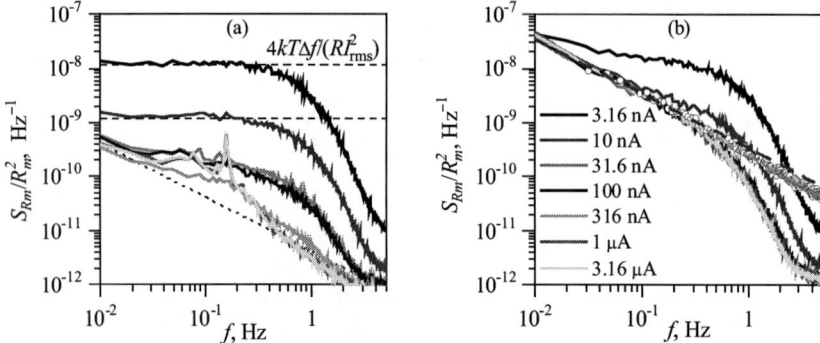

FIGURE 4. Relative PSD's of resistance fluctuations measured for wire-wound (a) and carbon (b) resistors with the instrument Model 370 at several excitation currents I_{rms} (the larger the current the lower the spectrum). In (a) dashed lines are the plots of thermal noise component for respective values of I_{rms}. Dotted line is the plot of the function $S_0(f) = 4.1 \times 10^{-12}/f$. In (b) circles show relative PSD of voltage fluctuations, $S_V(f)/V^2$, measured for carbon resistor in a standard dc measurement setup. Dashed line is the plot of the function $S_R(f)/R^2 = 3.4 \times 10^{-10}/f$.

RESOLUTION OF RU-BASED SENSORS

Once the influence of resistance noise on the resolution of resistance measurements has been proved and sensor resolution has been defined it is possible to calculate this quantity for Ru-based sensor described in Sec. 1. The values of ε_{sens} calculated as defined in Eq. (4) for $\theta \Delta f = 20$ are displayed on the right-hand axis of Fig. 1. One can read out from this figure, for instance for $\upsilon = 0.12$ device, that its resolution, which in 10 K is still of approximately of 10^{-6}, rises (i.e. gets poorer) to $\sim 10^{-5}$ at ~ 0.5 K and will rise to above 10^{-4} if the sensor resistance noise continues to increase with temperature lowering down to 50 mK. This means that the calibration of Ru-based sensors in mK range makes sense only to the resolution of 3 significant digits.

The work was supported by TURz project No U-6759/DS

REFERENCES

1. Kolek A., Ptak P., Zawiślak Z., Mleczko K., and Stadler, A. W., "Low temperature nonlinear 1/f noise in RuO$_2$ based thick film resistors" in *25th Int. Spring Sem. on Electronics Technol., ISSE-2002*, edited by P. Mach and J. Urbanek, Conference Proceedings, Prague: Czech Technical University in Prague, 2002, pp. 142-146.
2. Ptak P., Kolek A., Zawiślak Z., Stadler, A. W. and Mleczko K., *Rev. Sci. Instrum.* **76**, 014901-1-6 (2005).
3. D.S. Holmes, S.S. Courts, „Resolution and accuracy of cryogenic temperature measurements, *Temperature: Its Measurement and Control in Science and Industry*", J.F. Schooley A/P, New York, vol. 6, pp. 1225-30, 1992.
4. LakeShore Model 370 AC Resistance Bridge, *User's Manual*, v. 1.1, 2001, Stanford Research Systems SIM921 AC Resistance Bridge, *User's Manual*, v.1.41, 2003.

Current-driven Large Density Fluctuations of Vortices and Antivortices in the Corbino Disk

S. Okuma and S. Morishima

*Research Center for Low Temperature Physics, Tokyo Institute of Technology
2-12-1, Ohokayama, Meguro-ku, Tokyo 152-8551, Japan*

Abstract. In the Meissner state just below the transition temperature, vortices and antivortices perpendicular to the sample plane move to the opposite directions in the presence of large transport current. In the Corbino disk (CD), where vortices are rotated in concentric circles by a radial current, one vortex can annihilate only by colliding with another vortex with opposite vorticity and, hence, large vortex-density fluctuations are expected. We have found unusual large voltage pulses that oscillate almost periodically in the Meissner state of thick amorphous Mo_xSi_{1-x} films with the CD geometry. Both the period and width of the pulse are much larger than the characteristic time for free vortices to rotate in the circles. This result indicates that the voltage oscillation originates from large density fluctuations of vortices and antivortices.

Keywords: Vortex-antivortex pairs, Meissner phase, Fluctuations, Corbino disk
PACS: 74.40.+k, 74.25.Fy, 74.25.Qt

INTRODUCTION

In the Meissner phase of thick superconductors the linear resistivity (in the limit zero current $I = 0$) is zero, while at large I the nonzero voltage V appears. The I-V characteristics are highly nonlinear, which originates from thermally activated nucleation and subsequent growth of vortex loops [1]. Each grown vortex loop is dissociated into two free vortices with opposite vorticity. Similarly to the case of two-dimensional superconductors, they are driven to the opposite directions by I and produce voltage until they disappear at the sample edges. This is schematically illustrated in Fig. 1(a). We have found that largest I-induced voltage noise appears in the Meissner phase of thick as well as thin amorphous $(a$-$)Mo_xSi_{1-x}$ films, whose origin has been attributed to large density fluctuations of vortices and antivortices [2-4].

Here we focus on the samples with the Corbino-disk (CD) contact geometry, in which vortices move in concentric circles in the presence of the radial current. Dissociated free vortices with opposite vorticity move to the opposite directions by feeling the Lorentz force f_L inversely proportional to the radius r of rotation and continue to rotate without crossing the sample edges [5,6]. Since one vortex can annihilate only by "colliding" with another vortex with opposite vorticity [see Fig. 1(b)], we can expect quite different vortex dynamics, e.g., enhanced vortex-density fluctuations, in CD from that in the usual strip samples. In this paper we report the unusual oscillation of the voltage observed in thick a-Mo_xSi_{1-x} films. The detailed data concerning present work have been shown elsewhere [7-9].

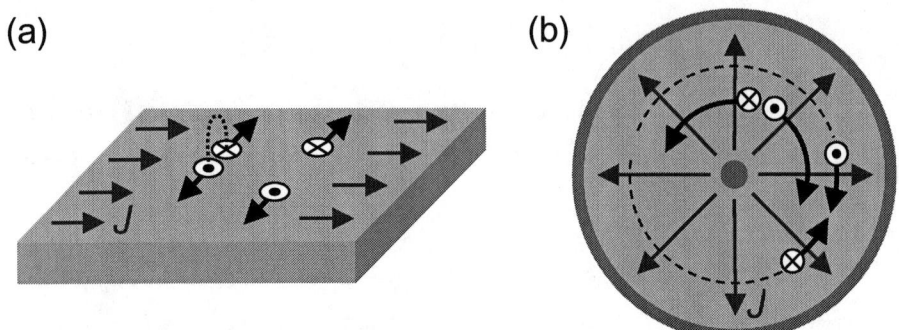

FIGURE 1. The schematic illustration of the vortex flow by the applied current in the (a) strip and (b) CD samples. In the strip sample the current density J and the Lorentz force f_L acting on vortices are uniform, while in CD, both J and f_L are nonuniform, which are inversely proportional to r. "⊙" and "⊗", respectively, denote the vortex and antivortex created and driven by the current.

EXPERIMENTAL

We prepared a thick (100 nm) a-Mo_xSi_{1-x} film with $x=61$ at. % by coevaporation of pure Mo and Si [8]. The transition (zero-resistivity) temperature T_c at zero field ($B=0$) is 3.33 K. The arrangement of the silver electrical contacts is shown in the inset (right) of Fig. 2. The current flows between the contact, +C, of the center and that, -C, of the perimeter of the disk, which produces radial current density J that decays as $1/r$. The voltage was measured using the contacts +V and -V. This contact arrangement was used previously to perform a comparative study of the CD and strip-like geometries on the same sample [6]. The diameter of the CD is ≈ 5.5 mm. The time(t)-dependent voltage $V(t)$ was measured using a four-terminal method [10]. $V(t)$ enhanced with the preamplifier was recorded using a fast-Fourier transform spectrum analyzer. The film was directly immersed in liquid ^{4}He to ensure good thermal contact. Stability of the temperature was as small as ≈ 1 mK, which was maintained by controlling the vapor pressure of ^{4}He electrically.

RESULTS AND DISCUSSION

In the main panel of Fig. 2 we representatively show the time evolution of voltage $V(t)$ for the CD geometry measured at $T =3.23$ K($<T_c$) in the presence of a *constant* current $I =21$ mA. We find that unusual large voltage pulses appear almost periodically on the base line of $V \approx 0$. Such a "spontaneous voltage oscillation" is commonly observed for different currents in the region $I =10$-22 mA. For all the currents at which the $V(t)$ oscillation is observed, both the period t_0 and width t_w of the $V(t)$ pulses are much larger than the characteristic time for free vortices to rotate in the circles. This means that the observed voltage oscillation originates from the collective motion of many vortices, which leads to large density fluctuations of vortices and antivortices in the presence of I, rather than the individual vortex motion rotating in the circles.

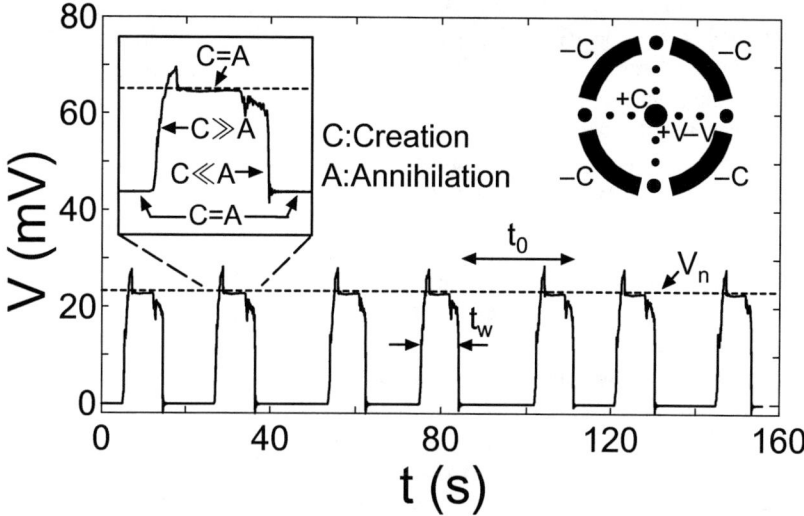

FIGURE 2. $V(t)$ at 3.23 K in the presence of 21 mA. The horizontal dashed lines in the main panel and inset (left) mark the normal-state voltage V_n. Inset: (Left) A voltage pulse is enlarged and shown. "C" and "A" denote, respectively, the creation and annihilation rates of free vortices and antivortices. (Right) Arrangement of the electrical contacts.

The inset (left) of Fig. 2 displays an enlarged view of a voltage pulse. Basically, the shape of the individual voltage pulses is rectangular and the top region of the pulse is nearly flat [8,9]. This implies that there are two "stational" states; a high-V state and a low-V state [11]. In either state where $V(t)$ is nearly constant, a creation rate of free vortices and antivortices, in other words, a dissociation rate of vortex-antivortex pairs, balances with the annihilation rate of vortices and antivortices (as indicated with "C=A" in the inset), while the total number of free vortices in the high-V state is significantly larger than that in the low-V state (see below). In a time duration where $V(t)$ exhibits a steep rise, the creation process dominates the annihilation process ("C>>A"), while in a duration where $V(t)$ falls abruptly, the annihilation process dominates the creation process ("C<<A").

We also notice in Fig. 2 that the height of voltage pulses takes relatively large values. In the inset (left) as well as in the main panel of Fig. 2, the location of the normal-state voltage $V_n (\equiv IR_n$, where R_n is a normal-state resistance) is indicated with a horizontal dashed line. For small currents ($I < 12$ mA) the pulse height is lower than 90 % of V_n, while for currents larger than 15 mA, the top (plateau) region of individual pulses reaches nearly V_n and also, there appears a small peak(s) that exceeds V_n. This result is rather surprising, because usually the voltage larger than V_n is not expected in any superconductor. This result supports the view mentioned earlier that the voltage oscillation we have found does not originate from the individual vortex motion but from the collective motion of many vortices [7-9].

Certainly, the phenomenon presented in this paper is unusual and novel, but experimentally reproducible and systematic. We have observed the essentially same phenomenon in three different samples at temperatures immediate below T_c. In particular, we note a nonmonotonic, though systematic, I dependence of t_0 [8,9]. This behavior was not expected, if the temperature fluctuation or local heating in the sample, which cannot be detected within our experimental resolutions of thermometry, would be a dominant cause of the $V(t)$ oscillation. Unfortunately, there is no available theory to explain the present results comprehensively. Quite recently, Hayashi and Ebisawa have studied vortex nucleation and annihilation in CD theoretically [11]. They have noted the two stational voltage (resistivity) states in our data mentioned earlier and shown that, as the current is increased above a certain critical value, the vortex density increases significantly due to the vortex-vortex interaction effects. This results in large vortex-density fluctuations at large I. They have suggested that the large fluctuations obtained in their calculation may be a key to understand our experimental results.

In summary, we have reported the unusual $V(t)$ oscillation driven by I at $B=0$ in the thick a-Mo$_x$Si$_{1-x}$ film with the CD geometry and interpreted the results in terms of the confined motion of vortices and antivortices within CD. It may be also important to point out that not only the confined geometry but also the nonuniform driving force ($f_L \propto 1/r$) may play a crucial role in generating the $V(t)$ oscillation. To clarify this point, it is necessary to conduct further experiments. For example, we can realize the condition in which the vortex motion is confined while the Lorentz force acting on vortices is *uniform*, using a film deposited on a cylindrical surface [12].

ACKNOWLEDGMENTS

The authors thank M. Hayashi, H. Ebisawa, and Y. Ootuka for useful discussions and M. Kamada for technical assistance. This work was supported by a Grant-in-Aid for Scientific Research from the Ministry of Education, Culture, Sports, Science, and Technology.

REFERENCES

1. D. S. Fisher, M. P. A. Fisher and D. A. Huse, *Phys. Rev. B* **43**, 130 (1991).
2. S. Okuma and N. Kokubo, *Phys. Rev. B* **61**, 671 (2000).
3. S. Okuma and M. Kamada, *Phys. Rev. B* **70**, 014509 (2004).
4. S. Okuma and M. Kamada, *J. Phys. Soc. Jpn.* **73**, 2807 (2004).
5. D. López, W. K. Kwok, H. Safar, R. J. Olsson, A. M. Petrean, L. Paulius and G. W. Crabtree, *Phys. Rev. Lett.* **82**, 1277 (1999).
6. Y. Paltiel, E. Zeldov, Y. Myasoedov, M. L. Rappaport, G. Jung, S. Bhattacharya, M. J. Higgins, Z. L. Xiao, E. Y. Andrei, P. L. Gammel and D. J. Bishop, *Phys. Rev. Lett.* **85**, 3712 (2000).
7. M. Kamada, Y. Watanabe and S. Okuma, in *Proceedings of the 16th International Symposium on Superconductivity, Tsukuba, 2003,* edited by M. Tachiki and M. Nisenoff, *Physica C* **412-414**, 535 (2004).
8. S. Okuma, S. Morishima and M. Kamada, in *Proceedings of the 17th International Symposium on Superconductivity, Niigata, 2004,* edited by K. Kishio, to be published.
9. S. Okuma, S. Morishima and M. Kamada, preprint.
10. S. Okuma, M. Kobayashi and M. Kamada, *Phys. Rev. Lett.* **94**, 047003 (2005).
11. M. Hayashi and H. Ebisawa, *J. Phys. Chem. Solid*, to be published.
12. M. Hayashi, private communications.

Microwave Induced Effects on the Random Telegraph Signal in a MOSFET

Enrico Prati[1], Marco Fanciulli[1], Giorgio Ferrari[2], Marco Sampietro[2], Paolo Fantini[3]

1. Laboratorio Nazionale Materiali e Dispositivi per la Microelettronica, Istituto Nazionale per la Fisica della Materia, Via Olivetti 2, I-20041 Agrate Brianza, Italy
2. Dipartimento di Elettronica e Informazione, Politecnico di Milano, P.za Leonardo da Vinci 32, I-20133 Milano, Italy
3. STMicroelectronics, via Olivetti 2, I-20041, Agrate Brianza, Italy

Abstract. We study the random telegraph signal (RTS) due to defects at the Si/SiO_2 interface of a MOSFET in a microwave field. We observe the change of the characteristic times of the RTS by monitoring the drain current in such device operated under microwave irradiation and the change of the emission/capture time ratio. The random telegraph signal is examined as a function of the microwave power from the temperature of 1.6K to room temperature. We observe a common trend in the RTS modification for all the investigated traps at all temperatures. The effect of increasing the irradiated power is to decrease the emission and capture times, while their ratio may depend on the temperature.

Keywords: Microwave induced Random Telegraph Signal, Interface defects.
PACS: 82.20.Xr, 07.57.Kp, 73.50.Td, 78.70.Gq

The random telegraph signal (RTS) consists in the random switching between two states of the current in a metal-oxide-semiconductor field effect transistor (MOSFET) and it has been observed in several systems. For a Si/SiO_2 MOSFET it is now accepted that, when the Fermi level of the conducting channel is in the proximity of the energy level of a defect at the semiconductor/oxide interface, an electron from the channel can tunnel on and off the trap [1-4]. The single-trap capture/emission phenomena is responsible of such discrete current levels [5] and a complete description based on an extension of the Ramo's theorem was provided as a quantitative prediction of the two-state current levels and characteristic capture and emission times. This effect has also been recognized to be responsible for the 1/f electrical noise in MOSFETs.

Since the RTS is related to a single electron switching between two states, the RTS has been considered as a possible read-out process of qubits in solid state based quantum information devices [6,7]. In these systems the information is managed by the application of microwaves to the device and therefore a detailed study of the interaction of the MOSFET with microwaves is necessary. Here we report our measurements of the RTS in a MOSFET in two standard X-band and Q-band EPR spectrometer resonant cavities, where the structure of the electromagnetic microwave radiation has well known components. We observe the change of the characteristic times of the RTS by monitoring the drain current in such a device operated under microwave irradiation. The random telegraph signal is examined in the temperature range 1.6 K – 293 K as a function of the microwave power. The effect of increasing the irradiated power is to decrease the emission and capture times, while their ratio behaves differently depending on the temperature.

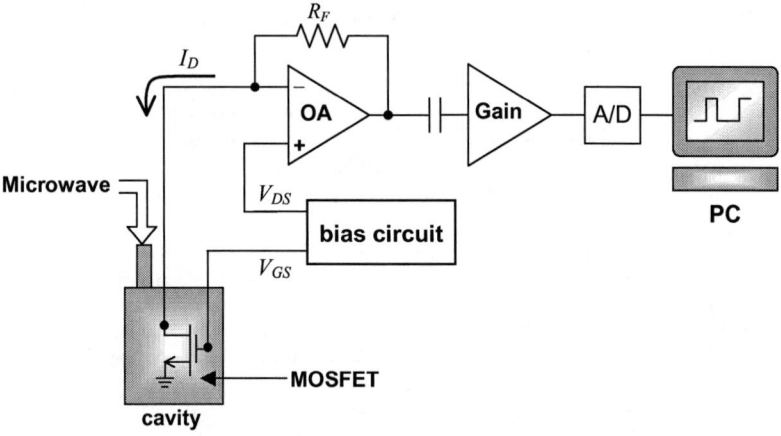

FIGURE 1. Schematic of the set-up for the characterization of the RTS under microwave irradiation.

EXPERIMENTAL SETUP

The experimental setup is depicted in Figure 1. The transistors tested are n-channel MOS devices realized on a p-well, with a length of 0.18 μm and a width of 0.28 μm, a 3.5 nm thick gate oxide, and a threshold voltage V_T variable between 250 mV and 450 mV. The source, drain, gate, and well contacts were directly accessible through the bonding pads and connected to wires to exit the microwave cavity. The current I_{DS} flowing through drain and source is measured by a transimpedance amplifier whose output is sampled and digitized for off-line processing. The bandwidth of the amplifier extends from 1Hz to 100 kHz. The transistor and the electronic are powered by independent batteries to avoid power-line pick-up and interferences. The X-band microwave (TE102) rectangular cavity is excited by a Varian E-101 microwave bridge and tuned to 9.6 GHz while the Q-band (TE012) cylindrical cavity is excited by a Bruker ER051 QG microwave bridge and tuned at 33.9 GHz. In both cases the maximum available microwave power was 200 mW. To safely operate the transistor we never applied more than 100 mW. The phenomena described in this work have been observed in all the investigated samples. We report on the results obtained on two samples (henceforth called n24 and n101) grown on different wafers with threshold voltages respectively of 260 mV and 450 mV.

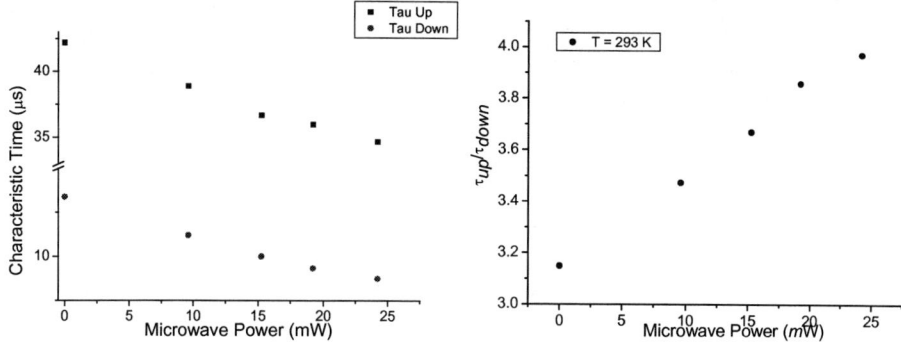

FIGURE 2. Room temperature measurement of the sample n101 in Q band. The MOSFET was operated at $V_g = 600$ mV and $V_d = 582$ mV

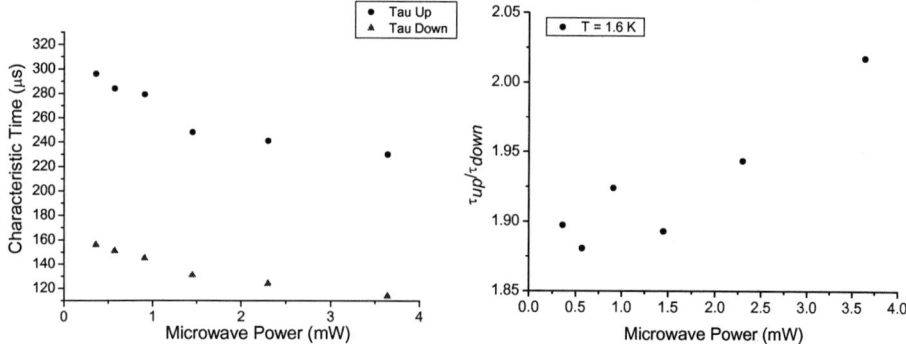

FIGURE 3. Pumped He4 measurement of the sample n101 in Q band. The MOSFET was operated at $V_g = 596$ mV and $V_d = 9.1$ mV

RESULTS

The experimental investigation concerns the determination of the average emission time τ_e and the average capture time τ_c as a function of the microwave power. The reported values of the microwave power refer to nominal values of the klystron. The microwave frequency was tuned to match the resonant frequency of the cavity and a critical coupling was reached. The RTS is reconstructed by eliminating the current fluctuations that overlap the signal and the characteristic times are extracted by an exponential fit of the distribution of permanency times in the two current states. At low temperature the transistor was biased in the ohmic regime with drain-source voltage V_{DS} near to zero volt. This operating condition is advantageous in many experimental investigations on the transport properties of MOSFETs at cryogenic temperatures because DC power dissipation is minimized. The Figures 2 and 3 refer to the sample n101, measured at room and cryogenic (1.7 K) temperatures respectively. All our measurements were made on traps filled in the down current state and empty in the up current state, so $\tau_c = \tau_{up}$. In the full range of investigated temperatures we have observed that the up and down characteristic times decrease when the microwave power is increased. We can therefore exclude that this is a temperature-related effect. Generally the irradiation of microwaves reduces the value ΔI that represents the difference between the up and down state current values. This limits the use of high power to detect and examine the RTS. In the two cases presented, the zero power characteristic time are of the same order of magnitude with τ_{up} longer than τ_{down} and both are of the order of tens and hundreds of microseconds.

At low temperature we observed in the sample n24 a non trivial τ_u / τ_d behavior. This sample was measured in the X band cavity and it was cooled by a coldfinger in thermal equilibrium with an He4 flow. While the trend of the characteristic times does not change, their ratio decreases between 4 K and 15 K, and increases between 15 K and 40 K.

We believe that this effect is related to the microwave induced stationary current previously reported [8] in MOSFETs under microwave irradiation. In our view these effects could be originated by the same mechanism, i.e. a microwave frequency voltage applied to the unavoidable conductors connecting the sample, acting like an antenna under microwave irradiation. Further studies are required on p-MOSFETs and by coupling a microwave signal to the sample by other methods.

REFERENCES

1. K. S. Ralls, W. J. Skocpol, L. D. Jackel, R. E. Howard, L. A. Fetter, R. W. Epworth, and D. M. Tennant *Phys. Rev. Lett.* 52, 228 (1984)
2. M.J. Uren, D. J. Day, and M. J. Kirton, *Appl. Phys. Lett.* 47, 1195 (1985)
3. Amarasinghe NV, Celik-Butler Z, Keshavarz A., *J. Appl. Phys.* 89, 5526 (2001)
4. E. Simoen, C. Claeys, *Mat. Sci. and Eng. B* 91, 136 (2002)
5. K. Kandiah, M.O. Deighton, and F.B. Whiting, *J. Appl. Phys.* 66, 93 (1989)
6. R. Vrijen, E. Yablonovitch, K. Wang, H. W. Jiang, A. Balandin, V. Roychowdhury, T. Mor and D. DiVincenzo, *Phys. Rev. A* 62, 12306 (2000)
7. M. Xiao, I. Martin, E. Yablonovitch, and H. W. Jiang, *Nature*, 430, 435 (2004)
8. G. Ferrari, L. Fumagalli, M. Sampietro, E.Prati, M.Fanciulli, in press on *Journal of Applied Physics*

Low Frequency Noise of the CdTe Crystals

L. Grmela, J. Sikula, J. Zajacek and P. Moravec[*]

Department of Physics, Brno University of Technology, Technicka 8, 616 00 Brno, Czech Republic
Fax: +425 4114 3133, e-mail: grmela@feec.vutbr.cz
[*]*Physical Institute, Charles University, Prague, Czech Republic*

Abstract. Experimental studies of transport and noise characteristics of CdTe crystals have been carried out. There are three different 1/f noise sources; bulk and surface 1/f which is proportional to square of current and contact 1/f noise with current noise spectral density proportional to higher than second power of current. Experimental results are used to characterize contact technology preparation. Contact prepare by deposition of gold from aqueous solution of $AuCl_3$. Low and high ohmic samples reveal contact resistance. This one is dominant noise sources for low ohmic samples, while high ohmic samples are sources 1/f with dominant mobility fluctuation. Hooge constant $\alpha_H = 1.5 \times 10^{-3}$ is very near to theoretical. Noise spectral density of sample with dominant contact noise is in proportion to I^n with n>2. High resistance samples reveal in frequency range f>100 Hz shot noise with spectral density corresponding to theoretical one.

Keywords: Noise, 1/f noise, contact 1/f noise, mobility fluctuation, GR spectra.
PACS: 71.23.Cq, 72.20.Jv

INTRODUCTION

The cadmium telluride (CdTe) is a II–VI semiconductor material useful for the detection of high energy radiations, such as X-rays and gamma rays. The main application of CdTe consists in high-resolution detection of radiation. A fairly wide energy gap $E_g = 1.5$ eV makes room temperature operation of these detectors possible; the detector therefore needs not to be cooled. High atomic number and high stopping power are the main advantage of such a radiation detector.

Many studies have been made in order to compare the electrical noise in CdTe detectors [1]. In this paper Asaad and Orsal, have been suggested a physical model of the current noise spectral density based on Pool-Frenkel effect, we will use in order to interpret these results.

Generally the application of strong electric fields to semiconductors with deep centres yields stimulation of ionisation or capture processes due to: Poole-Frenkel effect, the phonon-assisted tunnelling, and the direct tunnelling. All this effects appear in high ohmic samples.

We have performed transport and noise measurement of CdTe detectors [2], prepared by Physical Institute of Charles University in Prague. The undoped Cadmium Telluride single crystals were grown from 6N-purity Cd and Te elements in the charge by vertical-gradient freezing method. All samples displayed p-type conductivity with the hole concentration in the range from 9×10^{11} to 1.2×10^{15} cm^{-3}. After cutting from the crystal the samples were polished and chemo-mechanically etched. Temperature dependencies of the electrical conductivity and Hall coefficient were measured by a classical six-probe method. Gold electrical contacts were prepared by electrodes deposition from aqueous solution of $AuCl_3$.

The basic material is p-type and features the following parameters:

(i) Sample 1 - F2657B has hole concentration $p = 9 \times 10^{11}$ cm^{-3}, hole mobility $\mu = 74$ $cm^2.V^{-1}.s^{-1}$ and specific resistivity $\rho = 9.3 \times 10^4$ Ωcm.

(ii) Sample 2 - F33B8 has hole concentration $p = 1.2 \times 10^{15}$ cm^{-3}, hole mobility $\mu = 73$ $cm^2.V^{-1}.s^{-1}$, specific resistivity $\rho = 70$ Ωcm. The device dimension was: cross-section area 1.7×2.7 mm^2 and length 11.4 mm.

VA CHARACTERISTICS

VA measurements in the dark were carried out at the room temperature on standard measuring set up. Result of measurements for high ohmic sample CdTe F2657B shows that VA characteristic is linear. Measurement of potential distribution by DisPot shows that exist also contact barrier with negligible contact resistance with comparison with volume resistance. For low ohmic sample as F33B8 the VA characteristic is nonlinear as is shown in Fig. 1.

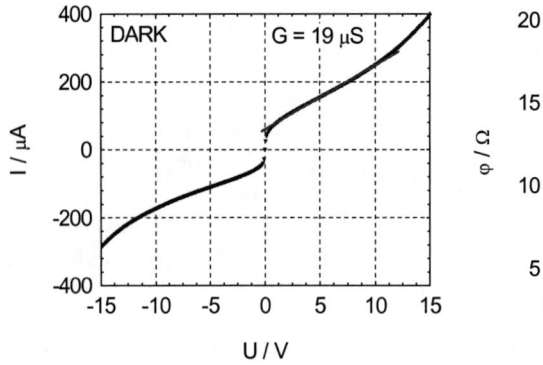
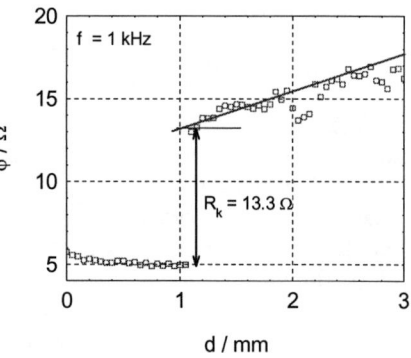

FIGURE 1. VA characteristic of CdTe F33B8   **FIGURE 2.** Contact resistance for CdTe F33B8

Quality of contact technology preparation was analyzed by potential distribution measurement also. Result of this measurement by DisPot for sample CdTe F33B8 is in Fig. 2. The contact resistance on one side is $R_c = 8$ Ω.

NOISE

The standard measuring set-up was used. The sample was fed from dry cells, which proved to exhibit a low own noise, negligible with respect to the background noise of the amplifier. Our set-up allows us to measure the sample current and noise voltage simultaneously without any effect on the noise. The whole set-up was placed in a steel box, which serves to eliminate the electromagnetic smog.

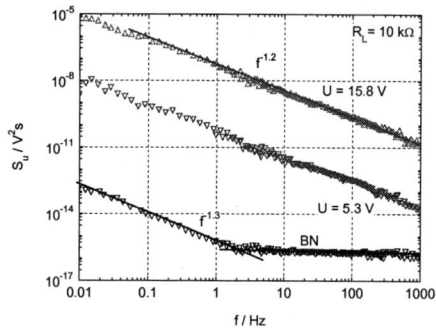
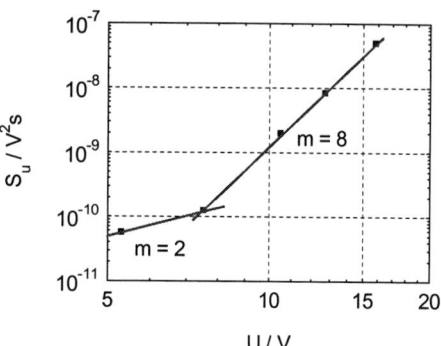

FIGURE 3. Noise spectral density for low ohmic sample CdTe F33B8

FIGURE 4. Voltage noise spectral density vs. applied voltage

The low ohmic sample has 1/f noise spectral density which increases with the square of voltage for low bias. For high bias than 7 volts noise spectral density is proportional to 8 power of voltage. This a result of contact noise generation.

The high ohmic sample CdTe F2657B reveals contact resistance also (see Fig.6) but this one is negligible respect to sample volume resistance R = 5.0 MΩ. Noise spectral density is $1/f^n$ type in low frequency range and shot noise is observed in frequency range about 100 Hz and higher.

In low frequency range noise spectral is proportional to square of voltage of current and than suppose that main source of noise is volume 1/f noise. Noise spectral density at 1 Hz $S_u = 10_{-12}$ V^2Hz^{-1}. Sample length is 11.44 mm, width 2.70 mm and thickness is 1.67 mm. For hole concentration $p = 9 \times 10^{11}$ cm^{-3} we have total number of fluctuators is $N = 4.5 \times 10^{10}$ and

Hooge constant $\alpha_H = 1.5 \times 10^{-3}$.

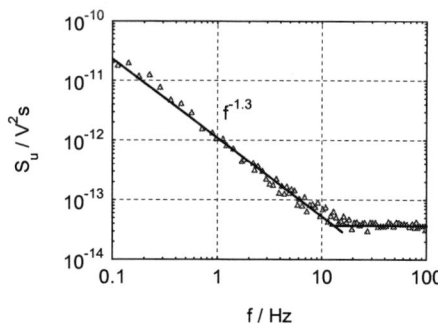
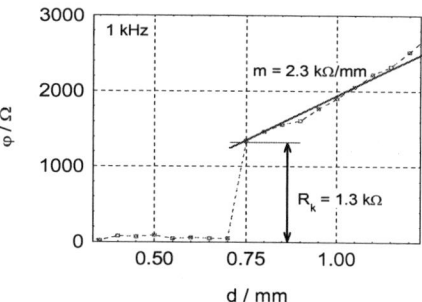

FIGURE 5. Noise spectral density for high ohmic sample CdTe F2657B

FIGURE 6. Contact resistance for high ohmic sample CdTe F2657B

Current noise spectral density given by $S_i = S_u/R^2$ is for frequency 100 Hz

$$S_i = 4 \times 10^{-25} \text{ A}^2\text{Hz}^{-1}.$$

For current $I = 1.1$ μA is this value in good agreement with theoretical value $S_{itheor} = 3.5 \times 10^{-25}$ $A^2 Hz^{-1}$.

CONCLUSION

1. Low and high ohmic samples reveal contact resistance. This one is dominant noise sources for low ohmic samples, while high ohmic samples are sources 1/f with dominant mobility fluctuation. Hooge constant $\alpha_H = 1.5 \times 10^{-3}$ is very near to theoretical value.
2. Noise spectral density of sample with dominant contact noise is proportion to I^n with n>2.
3. High resistance samples reveal in frequency range f>100 Hz shot noise with spectral density corresponding to theoretical one.

ACKNOWLEDGMENTS

This research has been supported by GACR No.102/05/2095 and by the Czech Ministry of Education in the frame of MSM 0021630503 Research Intention MIKROSYN "New Trends in Microelectronic System and Nanotechnologies".

REFERENCES

1. I. Asad at all, „Shot Noise in CdTe Resistors: Experimental and Analytical studies" ICNF 2003 (Prague), 153 (2003)
2. P.Schauer, J.Sikula and P.Moravec, „Transport and Noise Properties of CdTe(Cl) Crystals" Microelectronics Reliability, 431, **41** (2001)
3. S. M. Sze, Physics of Semiconductor Devices, J.Wiley & Sons New York (1981)
4. I. Asad at all, „Physical model of Current Noise Spectral Density versus dark current in CdTe detectors" ICNF 2001 (Gainesville), 339 (2001)

DEVICES

Low Frequency Noise in Si-based MOS Devices

J. Jomaah and G. Ghibaudo

IMEP-ENSERG (CNRS-INPG-UJF), BP. 257, 38016 Grenoble Cedex 1, France

Abstract. In this paper, a review of recent results concerning the low frequency noise in modern Si-based CMOS devices is given. The approaches such as the carrier number and the Hooge mobility fluctuations used for the analysis of the noise sources are presented and illustrated through experimental data obtained on advanced CMOS SOI and Si bulk generations. For SOI devices, a particular attention is paid to the fully-depleted MOSFET's and double-gate structure. The impact of the back gate voltage on the 1/f noise is studied. It is shown that in double-gate devices, due to their thinner silicon film, a volume inversion can be observed which in turn reduces the 1/f noise.

INTRODUCTION

To support the development of conventional and novel architecture Si-based CMOS devices, extensive DC, low and high frequencies electrical characterisation need to be carried out aiming at understanding the device operation, finding the critical parameters limiting the performances of the technologies under development. One of these limitations could come from the low frequency noise (LFN), which is prominent in MOSFETs due to the heterogeneous interface between silicon and silicon dioxide. Indeed, excessive low frequency noise and fluctuations could lead to serious limitation of the functionality of the analog and digital circuits [1-10]. The 1/f noise is also of paramount importance in RF circuit applications where it gives rise to phase noise in oscillators or multiplexors. Therefore, a thorough analysis of LFN in advanced CMOS devices still necessary to identify the main noise sources and to well understand the involved physical mechanisms. Such an analysis allows us to establish an accurate analytical modelling of noise for circuit simulation application.

In this paper, recent issues about the low frequency noise are presented and illustrated through experimental data obtained on advanced CMOS SOI and Si Bulk generations. For SOI devices, the case of Fully Depleted transistors and the double gate structure will be studied, featuring the impact of the thin film or the coupling effects. For Si bulk transistors, the case of advanced generation down to 30nm is analysed with special emphasis on gate dielectric thinning and channel engineering impact.

SOI DEVICES

Partially-depleted, Fully depleted and Double-gate SOI CMOS technologies, processed on Unibond substrates, were used in this study. For PD technology, floating body (FB) and body-contacted (BC) SOI MOS were considered. PD SOI front oxide thickness was T_{ox1} = 4.5nm and 2nm for 0.25µm and 0.12µm technology nodes, respectively. A 2nm front gate oxide was employed and the back oxide thickness was 400nm and 2nm for the fully-depleted and double-gate devices, respectively. Silicon film thickness was T_{Si}=150 nm for PD devices, 15nm for the fully-depleted and 6nm for the double-gate devices. A wide range of channel lengths was considered, from 2 μm down to 50 nm. Noise measurements in both linear and saturation regimes were carried out in the 1 Hz to 100 kHz frequency range using a dynamic signal analyser (HP 35665A) and a BTA noise probe.

Figure 1 shows, in ohmic operation (V_D=50mV), the normalised drain current power spectral density S_{ID}/I_D^2 plotted as a function of the drain current for N and P-channel PD-SOI MOSFETs for different channel lengths. In this plot, the straight line represents the front gate power spectral density S_{VG} multiplied by the ratio $(G_m/I_D)^2$ where G_m stands for the gate transconductance. A good correlation is obtained between these two amounts, confirming results predicted by the Mc-Whorter model which associates the 1/f noise to carrier number fluctuations [3,4].

In this model, the fluctuations of the drain current are due to those of the inversion charge near the silicon-silicon dioxide interface caused by the dynamic trapping and detrapping of free carriers into traps located in the oxide near the interface. The normalized drain current spectral density is given by [4]:

$$\frac{S_{ID}}{I_D^2} = \left(\frac{G_m}{I_D}\right)^2 S_{VG} \qquad (1)$$

where G_m is the gate transconductance and S_{VG} is the equivalent input gate voltage spectral density.

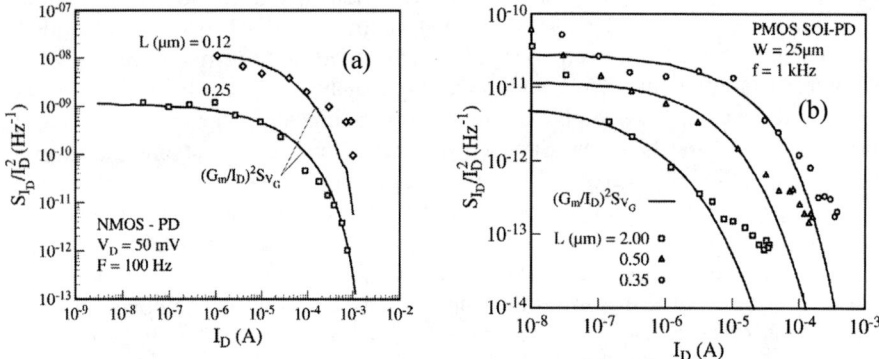

Figure 1: Normalized drain current power spectral density S_{ID}/I_D^2 at V_D=50 mV at different channel lengths for (a) NMOS FB partially-depleted and (b) PMOS FB partially-depleted SOI devices. Full lines: $S_{Vg}.(G_m/I_d)^2$

Moreover, some difference in strong inversion can be observed (case of P-channel) (Fig. 2.b) which is attributed to the correlated mobility fluctuations. Indeed, by taking into account the dependence of the carrier mobility on the insulator charge (Coulomb scattering), the fluctuations of the insulator charge give rise to a supplementary change of the mobility, which induces an extra drain current fluctuation.

Figure 2 shows the normalised drain current power spectral density of a fully depleted device with W/L =25/0.8μm and for back-gate voltages varying from V_{g2}=-

50V to V_{g2}=50V, as a function of the drain current. In this plot the solid line is the the corresponding front gate power spectral density S_{VG} multiplied by $(G_m/I_D)^2$ at V_{g2}=20V.

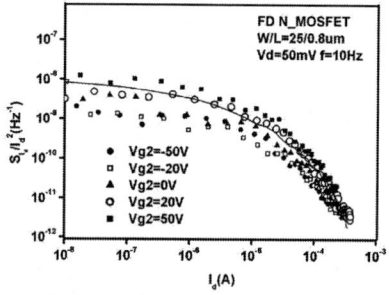 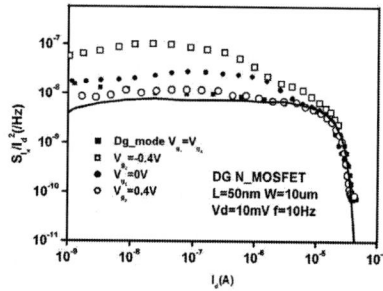

Figure.2. Normalized drain current power spectral density S_{Id}/Id^2 of a W/L=25/0.8um FD N-MOSFET at Vd=50mV and f=10Hz for different back-gate voltages. Solid line: $S_{Vg} \times (Gm/Id)^2$ for V_{g2}=20V [11].

Figure 3. Normalized drain current noise spectral density S_{Id}/Id^2 of a W/L=10/0.05um DG N-MOSFET at Vd=10mV and f=10Hz for different back-gate voltages. Solid line: $S_{Vg} \times (Gm/Id)^2$ for double gate mode [11]

A good correlation exists between S_{ID}/I_D^2 and $(G_m/I_D)^2 \times S_{VG}$, confirming results predicted by the 1/f noise carrier number fluctuations model. By increasing V_{g2}, an increase in the amount of S_{ID}/I_D^2 is observed which is a result of the coupling effect. Indeed, when the back gate is accumulated, the carrier fluctuations induced by the back oxide traps are screened by the accumulation layer, whereas when depleting the back interface, the noise becomes higher due to the coupling effect. Since the silicon film is not thin enough, volume inversion effects have not been observed.

DG-MOSFETs have been characterized in both Single Gate (SG) (back gate voltage controlled separately) and Double Gate (DG) modes (front gate and back gate with the same voltage). The normalized drain current power spectral density S_{ID}/I_D^2 for a DG N_MOSFET, correlates very well with $S_{VG} \times (G_m/I_D)^2$. This fact confirms that in double-gate devices, similar to the partially- and the fully-depleted ones, the noise source is also the carrier number fluctuations. On the other hand, in contrast with the fully-depleted devices studied previously, in the double-gate MOSFETs, the normalized drain current power spectral density decreases while increasing the back-gate voltage, specially at low drain current (weak inversion). An explanation for this behavior is the fact that since the silicon film is very thin (6nm), even when the back interface is in accumulation mode there exists a coupling between the two interfaces and so current flows in both channels giving rise to fluctuations in carrier number. While increasing the back gate voltage, the charge is pushed away from both interfaces (volume inversion) and so there would be a screening effect reducing the number of trapped carriers [11].

BULK DEVICES

Noise measurements have been performed on n-MOSFETs from an advanced 65 nm CMOS technology with 1.2nm gate oxides thickness from STMicroelectronics,

Crolles. In table 1, we report the main technological characteristics of the various technological splits denoted as 45nm and 30nm.

Technology	45nm CMOS	30nm CMOS	
Gate oxides	RTN process (Rapid Thermal Nitridation).	PN process (Plasma Nitridation).	Table 1. Technological characteristics of the studied CMOS technological splits
Polysilicon gate thickness	1500Å	1200Å	
LDD (Lightly Doped Drain)	Arsenic (As)	Arsenic (As)	
Pockets	Boron (B)	Boron fluoride (BF$_2$)	
HDD (High Doped Drain)	Cobalt silicide (CoSi$_2$)	Nickel silicide (NiSi)	

For the two CMOS technologies, the pocket implants allow to obtain an almost constant threshold voltage with gate length but leads to a significant degradation of the low field mobility with gate length [12]. Typical dc drain current versus gate voltage are shown in Figure 4 at V_{DS} = 50 mV, respectively for the shortest and longest geometries of two studied technologies. For the 10x10 µm^2 device from 45nm technology, we can notice a considerable decrease of drain current to negative values due to the gate tunneling current. The MOSFET parameters have been thus extracted by correcting the drain current for the gate-leakage current with the simple following expression [12]:

$$I_{D0} = I_D + \alpha_D \times I_G = I_S - \alpha_S \times I_G \quad (2)$$

where I_{D0} is the intrinsic drain current, I_D the measured drain current, I_S the measured source current, I_G the gate leakage current, α_D the gate-partitioning coefficient at the drain side, and α_S the gate-partitioning coefficient at the source side.

The α_D coefficient is extracted at V_{DS}=0 in strong inversion regime as shown in Figure 5 and is equal to 25%. We also report the drain current, the gate current and the corrected intrinsic drain current with α_D coefficient at V_{DS} = 50 mV. The gate-leakage current is not negligible and is larger than the measured drain current. The corrected drain current for 30nm technology is similar to that of 45nm technology but the gate leakage is much higher. The difference of gate-leakage current between the two studied 1.2nm oxide thickness technologies is due to the difference in the nitridation rate and in turn to capacitive equivalent oxide thickness.

The low-frequency noise characteristics were measured with an original experimental set-up using a programmable biasing amplifier with two entrance bias ports and output trans-impedance amplifier. In Figure 6 we report the evolution of S_{ID} and S_{ID}/I_D^2 with I_D at V_{DS} = 50mV for a large area device for 45nm technology. For all geometries of both studied technologies, drain current spectral density dependencies with gate and drain voltages are well interpreted by carrier number fluctuation noise models [4,13]. In order to obtain an accurate modelling at high gate voltage, drain current is analysed by two noise sources associated to the intrinsic channel and to the access resistances, respectively. In Figure 6, the dotted curve gives the contribution of the access resistances noise to the total noise. In strong inversion, better agreement is obtained using the mobility-correlated number fluctuation model with the access resistances noise (continuous lines). The typical N_t value for both technologies are reported in Figure 7 and are smaller than previous CMOS technologies, indicating a very good dielectric quality despite the use of RTN or PN fabrication processes [14]. We also observe a reduction of N_t when the length decreases being explained by a short channel effect, and an increase of N_t for the shortest length that could be due to process-induced extra defects close to source and drain regions possibly generated during the implantation steps.

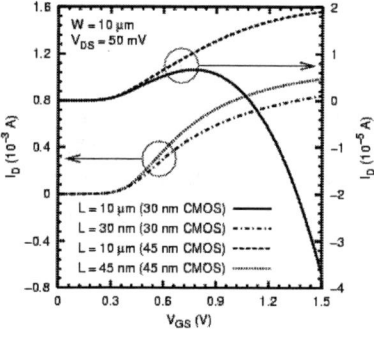

Figure 4. DC measured drain current on 10μm/10μm NMOS for both technologies and 45nm/10μm NMOS for 45nm CMOS technology and 30nm/10μm NMOS for 30nm CMOS technology at V_{DS} = 50 mV [13].

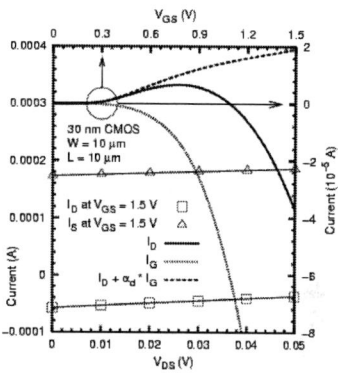

Figure 5. Extraction of α_d and α_s gate partitioning coefficients and DC measured currents for 10μm/10μm NMOS for 30 nm technology [13].

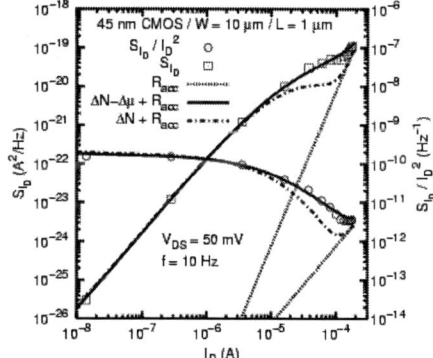

Figure 6. Comparison of measured data and modelling results of drain current noise as a function of drain current at f = 10 Hz for 10μm/1μm NMOS for 45nm technology [13].

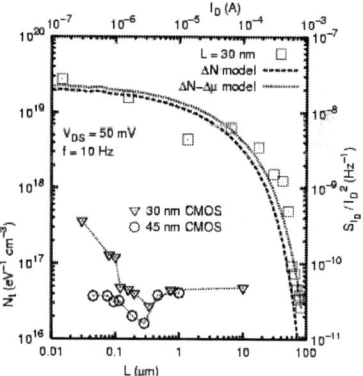

Figure 7. Illustration of good agreement between normalized drain current noise and correlated model for the smallest geometry for both CMOS technologies [13].

The gate current partitioning of equation (2) can also be used to account for the impact of gate current noise S_{Ig} on drain current S_{Id} as [13]:

$$\frac{S_{I_D}}{I_D^2} = \frac{S_{I_{D0}}}{(I_{D0})^2} + (\alpha_D)^2 \times \frac{S_{I_G}}{I_D^2} \quad \text{with} \quad S_{I_{D0}} = \left(1 + \alpha\mu_{eff}C_{ox,eff} \times \frac{I_{D0}}{g_{m0}}\right)^2 \times g_{m0} \times S_{V_{FB}}^D \quad (3)$$

185

where $I_{D0} = I_D + \alpha_D \times I_G$, $g_{m0} = g_m + \alpha_D \times g_G$ and $\dfrac{S_{I_G}}{I_G} = \left(\dfrac{g_G}{I_G}\right)^2 \times S_{V_{FB}}^G$, g_G being the gate conductance ($=dI_G/dV_G$) and $S_{V_{FB}}$ the flat band voltage spectral density associated to the gate current. The sum of both drain and gate current noise contributions enables a very good description of the experimental noise behaviour to obtained showing the impact of the gate-leakage current noise on the output drain current noise for large area device [13].

CONCLUSION

Low frequency noise in both SOI and bulk devices has been investigated. In Partially- and Fully-depleted and Double-gate SOI MOSFETs the noise source was attributed to carrier number fluctuations, while in advanced bulk CMOS technologies the LF noise stems from carrier and correlated mobility fluctuations. No degradation of gate oxide quality as measured by oxide density of traps Nt was observed despite the ultra thin dielectric used.

REFERENCES

1. D. Eggert, P. Huebler, A. Huerrich, H. Kuerck, W. Budde and M. Vorwerk, "A SOI-RF-CMOS technology on high resistivity SIMOX substrates for microwave applications to 5 GHz." IEEE Trans. on Electron Devices, 44, 1981 (1997).
2. O. Rozeau, J. Jomaah, S. Haendler, J. Boussey, and F. Balestra, "SOI Technologies Overview for Low-Power Low-Voltage Radio-Frequency Applications", Analog Integrated Circuits and Signal Processing, 25, 2000, Kluwer Academic Publishers – Special issue of SOI.
3. McWhorter A.L., Semiconductor Surface Physics, University of Pennsylvania Press, Philadelphia 1957.
4. G. Ghibaudo, O. Roux, Ch. Nguyen-Duc, F. Balestra, and J. Brini, "Improved Analysis of Low Frequency Noise in Field-Effect MOS Transistors", Phys. stat. sol. (a) 124, 571 (1991).
5. S. Christensson, I. Lundstrom and C. Svensson, "Low-frequency noise in MOS transistors-I", Sol. State Elec., 11, 797 (1968).
6. R. Kolarova, T. Skotnicki, and J.A. Chroboczek, "Low-frequency noise in thin gate oxide MOSFETs", Mic. Rel., 41, 579, (2000).
7. M. H. Tsai and T.P. Ma, "The impact of device scaling on the current fluctuations in MOSFET's", IEEE-Transactions-on-Electron-Devices., 41, 2061 (1994).
8. C. Jakobson, I. Bloom, Y. Nemirovsky, "1/f noise in CMOS transistors for analog applications from subthreshold to saturation", Solid-State-Electronics, 42, 1807 (1998).
9. E. Simoen and C.Claeys, "On the flicker noise in submicron silicon MOSFET", Solid-State-Electronics, 43, 865 (1999).
10. Y. Nemirovsky, I. Brouk, C.G. Jakobson, "1/f noise in CMOS transistors for analog applications", IEEE-Transactions-on-Electron-Devices, 48, 921 (2001).
11. L. Zafari, F. Daugé, J. Jomaah and G. Ghibaudo, "On the low frequency noise in fully depleted and double-gate SOI transistors", Proc. ULIS'2005, Bologna, April 2005.
12. K. Romanjek, "Characterization and modelling of 50nm and below CMOS transistors technologies", Ph.D. Thesis, INPG, Grenoble, France (2004).
13. T. Contaret, K. Romanjek, T. Boutchacha, G.Ghibaudo and F. Bœuf, "Low Frequency Noise Charaterization and Modelling in Ultrathin Oxide MOSFETs", Proc. ULIS 2005, p. 55.
14. B. Tavel, M. Bidaud et al, "Thin oxynitride solution for digital and mixed-signal 65nm CMOS platform", IEEE-International-Electron-Devices-Meeting-2003, 27.6.1-4, (2003).

The Low-frequency Noise of Strained Silicon n-MOSFETs

E. Simoen, G. Eneman*, P. Verheyen, R. Delhougne*, R. Rooyackers,
R. Loo, W. Vandervorst, K. De Meyer* and C. Claeys*

IMEC, Kapeldreef 75, B-3001 Leuven, Belgium
*also at E.E. Dept., KU Leuven, Kasteelpark Arenberg 10, B-3001 Leuven, Belgium

Abstract. The low-frequency (LF) noise behavior of n-MOSFETs fabricated on strained silicon (SSi) substrates is described and compared with the results obtained on devices in standard silicon wafers. It is demonstrated that a significant lowering (up to a factor 3) of the 1/f noise can be achieved. This improvement is shown correlated with the higher inversion-layer mobility μ_0 and is believed to have a common origin, namely, the biaxial tensile strain in the thin silicon epitaxial layer. This improves the quality of the gate oxide, resulting in lower trapping and Coulombic scattering and, hence, in a better 1/f noise performance.

Keywords: strained silicon substrates, Strain-Relaxed Buffer (SRB) layer, Rapid Thermal Oxidation (RTO)
PACS: 72.20.Jv; 72.20.+m; 73.40.Qv

INTRODUCTION

There exists currently a strong interest in the use of strained silicon (SSi) substrates for the development of high-mobility n-MOSFETs (see e.g., [1],[2]). As an example, the linear characteristics (drain current I_D and transconductance g_m) are compared in Fig. 1 for an L=1 µm long and W=10 µm wide n-MOSFET in a regular silicon and in a SSi substrate, fabricated on a thin Strain-Relaxed Buffer (SRB) SiGe layer [3]. For analog (RF) applications, low-frequency (LF) noise is an important parameter, which should not be degraded by using a tensile-strained silicon layer. Recent reports on the 1/f noise are rather controversial: while some group finds no change in the noise [2], a marked increase has been reported by others [4],[5], that was ascribed to the presence of germanium atoms at the Si-SiO$_2$ interface or in the gate oxide. Also the presence of a threading or misfit dislocation may degrade the noise [2]. Here, the LF noise of SSi n-MOSFETs is compared with the behavior of reference devices fabricated on standard Czochralski (Cz) silicon wafers. It will be shown that a significant reduction in the 1/f noise can be achieved, which appears to be correlated with the static device parameters: the threshold voltage V_T and the effective low-field mobility μ_0. These results will be discussed in terms of the impact of tensile strain on the trap density in the 2 nm SiO$_2$ gate dielectric used in this work.

EXPERIMENTAL

N-channel transistors have been fabricated on 200 mm SSi wafers, whereby the epitaxial layers were deposited before the isolation module, using a novel SRB concept, based on a thin carbon-doped SiGe layer to relax the strain in the 20 % SiGe stack [3]. A 2 nm SiO_2 gate oxide was grown by in-situ steam generated rapid thermal oxidation at 850°C followed by n-type polysilicon gate deposition. LF noise measurements were performed on 10 μm×1 μm n-MOSFETs to reduce the device-to-device scatter, related to scaling [6] and to enhance the chance for observing the effects of threading dislocations on the LF noise. On-wafer measurements have been performed, under control of the NoisePro software from Cadence, in the linear regime, for a drain voltage V_{DS} of 0.1 V.

RESULTS

The spectra in most cases are 1/f-like, with occasionally small GR humps occurring, for both device types. The corresponding drain current noise spectral density is represented in Fig. 2, exhibiting a lower S_I for the SSi component. For low drain currents, an I_D^2 law is approximately followed, while at higher I_D the current exponent tends to reduce to 1. The normalized current spectral density in Fig. 3 is compared with the $(g_m/I_D)^2$ ratio. While there appears to be a good agreement at low I_D, suggesting a number-fluctuations origin, at higher currents, the $(g_m/I_D)^2$ ratio shows a faster reduction. As will be discussed, this could point to a different origin of the 1/f noise in strong inversion. For high gate voltages (V_{GS}), the access resistance starts to play a role, leading to a leveling off of S_I/I_D^2. The input-referred spectral density (S_{VG}) versus the gate overdrive voltage ($V_{GS}-V_T$) in Fig. 4 confirms the improvement by a factor 2 to 3 of the 1/f noise in SSi devices. V_T is the threshold voltage.

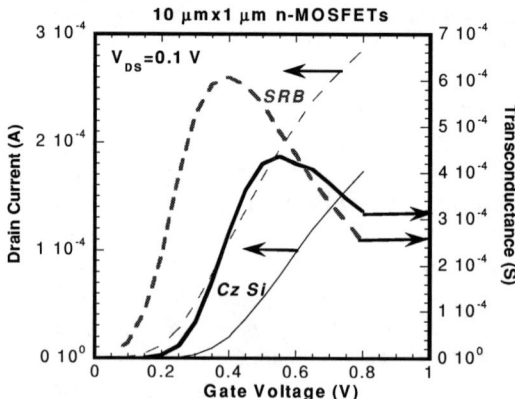

FIGURE 1. Linear characteristics for a SSi (dashes) and a Cz Si (full lines) 10 μm×1 μm n-MOSFET. V_{DS}=0.1 V.

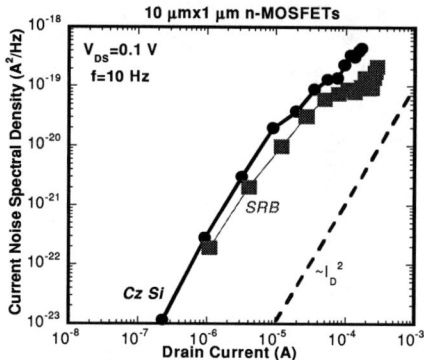

FIGURE 2. Current noise spectral density at f=10 Hz versus drain current, corresponding with the devices of Fig. 1.

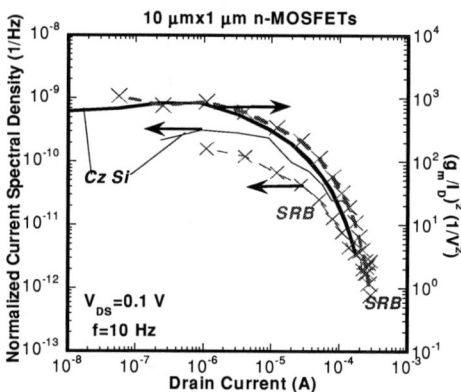

FIGURE 3. Normalised current noise spectral density for a SSi and a reference Cz Si n-MOSFET, compared with the $(g_m/I_D)^2$ ratio, as a function of I_D. V_{DS}=0.1 V and f=10 Hz.

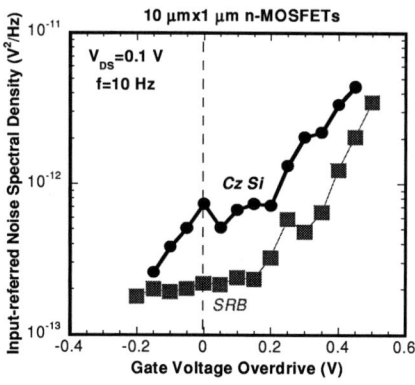

FIGURE 4. Input-referred noise spectral density at 10 Hz, versus gate voltage overdrive for the same transistors.

DISCUSSION AND CONCLUSION

In order to explore the origin of this noise improvement, the correlation was investigated between the S_{VG} at V_T (V_{GS}-V_T=0 V) and V_T and μ_0 extracted for each transistor using the Y-function (=$I_D/g_m^{1/2}$) method proposed by Ghibaudo [7]. It is clear that a lower $S_{VG}(0)$ corresponds to a lower V_T (Fig. 5a) and a higher μ_0 (Fig. 5b). These correlations point to a positive impact of tensile strain on the LF noise. Assuming that $S_{VG}(0)$ can be interpreted in the frame of the number-fluctuations model, we conclude that tensile strain may lead to a gate dielectric with a higher quality (lower density of traps). This is in line with the Electron Spin Resonance observations of Stesmans et al. [8], showing a lower density of dangling bond P_b centers at the Si-SiO$_2$ interface for a gate oxide grown under *in situ* tensile strain, compared with an unstrained reference. The present result suggests that also the density of bulk oxide traps is reduced when grown on a tensile-strained silicon substrate.

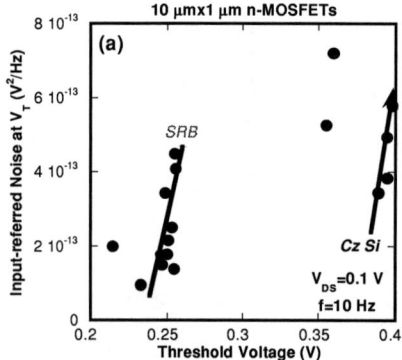 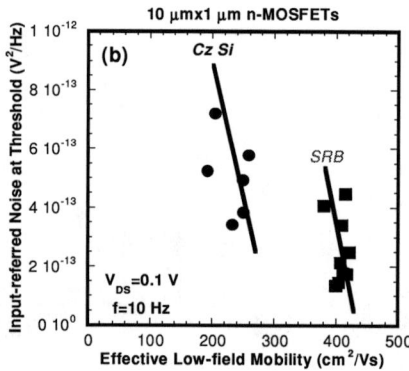

FIGURE 5. (a) Correlation between $S_{VG}(0)$ and V_T for the SSi and reference Si n-MOSFETs investigated. (b) Correlation between $S_{VG}(0)$ and μ_0 for the SSi and reference Cz Si n-MOSFETs studied. The lines are a guide to the eye.

REFERENCES

1. Takagi S. *et al.*, *IEDM Techn. Dig.*, New York: The IEEE, 2003, pp. 03/57-03/60.
2. Lee M.H. *et al.*, *IEDM Techn. Dig.*, New York: The IEEE, 2003, pp. 03/69-03/72.
3. Delhougne R., Eneman G., Caymax M., Loo R., Meunier-Beillard P., Verheyen P., Vandervorst W., and De Meyer K., *Solid State Electron.* **48**, 1307-1316 (2004).
4. Hua W.-C., Lee M.H., Chen P.S., Maikap S., Liu C.W., and Chen K.M., *IEEE Electron Device Lett.* **25**, 693-695 (2004).
5. Tanabe A., Numata T., Mizuno T., Maeda T., and Takagi S., in *Proc. of the First Int. Symp. on SiGe: Materials, Processing, and Devices*, edited by D. Harame *et al.*, Pennington: The Electrochem. Soc., 2004, pp. 483-492.
6. Claeys C, Simoen E., and Mercha A., *J. Electrochem. Soc.* **151**, G307-G318 (2004).
7. Ghibaudo G., *Electron. Lett.* **24**, 543-545 (1988).
8. Stesmans A., Pierreux D., Jaccodine R.J., Lin M.-T., and Delph T.J., *Appl. Phys. Lett.* **82**, 3038-3040 (2003).

Low Frequency Noise sensitivity to technology induced mechanical stress in MOSFETs

Paolo Fantini and Giorgio Ferrari*

STMicroelectronics, via Olivetti 2, Agrate Brianza, Italy, 20041
Politecnico di Milano, Dip. Di Elettronica e Informazione, P.za L.da Vinci 32, Milano, Italy, 200133
E-mail: paolo.fantini@st.com

Abstract. A detailed experimental investigation of 1/f noise in MOSFET devices as a function of technology-induced mechanical stress is presented. Strain in the channel region is obtained by the Shallow Trench Isolation (STI) technique. Both n- and p-MOS have been considered inside the present study. An increasing of 1/f noise intensity with the increasing intensity of mechanical stress has been detected for p-channel transistor. A tentative explanation of this experimental finding has been proposed.

Keywords: 1/f noise, MOSFET, strain, and mechanical stress.
PACS: 73.50.Td Noise processes and phenomena

INTRODUCTION

The possibility to rule the inter-atomic distance in the silicon lattice constituting the "habitat" for the MOSFET channel represents a key element for next generation CMOS technologies. In fact, it has been shown that strained-silicon can significantly improve the MOSFET driving capability. Hereof, today's technology challenge plays on the capability to introduce the desired mechanical stress in the channel of device considering low-cost process modifications. With reference to the present work, a built in biaxial compressive stress is induced on the device active area by the Shallow Trench Isolation (STI) technique that became a must for the CMOS technology scaling down below the 0.25μm node. Hereto, variations in MOSFETs electrical behavior are observed when channel width and length are maintained constant and only the distance from STI oxide is modified. Therefore, the transistor electrical performances show a layout dependence that was ignored in LSI technologies and new layout parameters (Sa and Sb in Fig.1) in addition to W and L must be introduced in order to describe the MOSFET electrical behavior.

The main electrical parameter changing with the mechanical stress of the device active area is the low-field mobility together other ones related with: g_m, I_{DSAT}, R_{ON}...

A dichotomy in the low-field mobility trend as a function of mechanical stress intensity has been observed between N- and P-MOSFETs: it decreases in the first one, while in the second ones it increases with higher stress [2]. The variation ranges observed in the examined technology are around -10% for N-MOS and 25% for P-MOS between the more stressed and more relaxed structure.

Our paper deals with the question if also 1/f noise intensity could show any layout dependence. Some spot studies have just been presented [3]. However, up to date, any complete dedicated work about the layout dependence of 1/f noise is lacked at the knowledge of authors. Later on, we discuss the question with the support of a number of experimental data on both N- and P-channel MOSFETs fabricated in 90nm Flash Memory Technology.

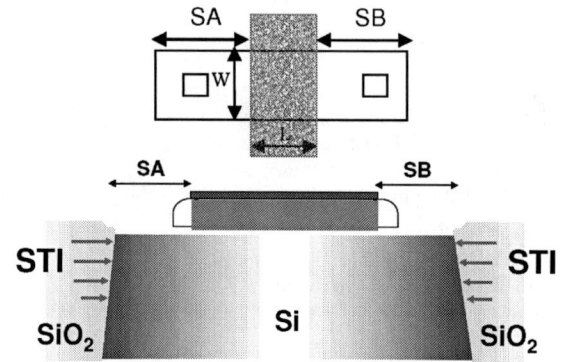

FIGURE 1. Sketch picturing STI induced mechanical stress on the active area device and the new layout parameters: SA and SB.

EXPERIMENTAL DATA

In figure 2, we show the experimental values of the normalized Noise Current Power Spectral Density @ 1Hz as a function of the drain current for P-MOSFET structures with Sa=Sb=0.3µm and Sa=Sb=3.6µm. Measurements of noise drain current power spectral density have been carried out at wafer level by using an EG&G 5182 low noise amplifier and a HP 35665A Dynamical Signal Analyzer, while the dc MOSFETs characterization has been performed by means of a HP 4155 Parameter Analyzer. MOSFETs were biased in linear region with V_D=50mV. The frequency range we have considered was from 1 Hz to 1 KHz and the nominal MOSFET geometry was W=10µm and L=0.18µm. With a so wide device the measured noise spectra normally showed a $1/f^\alpha$ noise spectrum with α in the 0.8-1.2 range. Degradation in low-frequency noise device performance has been found in the overall gate bias range for device with the minimum Sa value (maximum mechanical stress). This can be observed, in particular, in the region close to MOSFET threshold (I_D@V_{TH}) where the correlated number fluctuations are the main low frequency noise source, but also at higher drain current where the correlated mobility fluctuations play an important role. Beneath it has been recently reported a dependence of P_b interface defects density by the applied mechanical stress [4], we have just noticed that in our case no interface traps density variation is induced by mechanical stress since we did not observe any modulation in the subthreshold slope [5]. Also a qualitative analysis

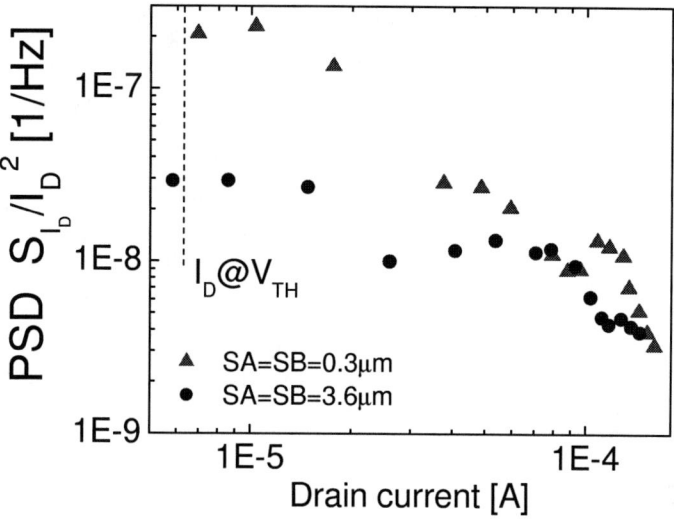

FIGURE 2. Normalized Current Power Spectral Density as a function of drain current for both the P-MOSFETs with minimum active area size (SA=0.3µm) and the larger one (SA=3.6µm).

of C-V curves on transistor arrays with various "Sa" lengths reveals the band structure modification but it did not show any modulation due to a possible increasing of interfacial states density with stress. Therefore, the difference in 1/f noise spectral density cannot be ascribed to a variation of interfacial properties. The same study on N-MOS transistor has been also performed but no appreciable difference in 1/f noise intensity has been found. However, we recall that the variation spread in low-field mobility of N-MOS devices due to STI stress is significantly lower than in P-MOSFETs and it could be too light to see an effect in 1/f noise spectra.

DISCUSSION

Since our experimental data exclude any justification in terms of interfacial effects as reason of the observed difference in 1/f noise intensity, we look for a possible explanation in the modification of the silicon electronic properties as a function of mechanical stress. The lift of the twofold degeneracy in Γ point of the valence band and the energy shift of the bottom of conduction band are the main consequences of mechanical stress. The physical parameter that is responsible for the low-field mobility modulation with strain, at least at small stress intensity, as it occurs in our devices, is the effective mass variation due to the strain-altered band structure [6]. The effective mass variation can affect the noise fluctuations produced by the trapping-detrapping of carriers. To clarify this point, let us recall that a potential fluctuation extends typically in the Debye screening length range, which can be approximated for a 2D-electron gas with the expression [7]:

$$L_S = \sqrt{2}\frac{\varepsilon_s kT}{qQ_S} \sim \frac{1}{m^*}$$

where ε_s is the dielectric constant, k the Boltzmann constant, T the temperature, q the electron charge, Q_S the charge density in the transistor channel and m^* the effective mass.

We note that L_S depends on the electronic properties of material through the static dielectric constant: ε_s. This quantity increases with a hyperbolic trend when m^* decreases. So, we suggest that the observed noise increasing in P-MOS devices with stress could be ascribed to the dielectric properties modification of the p-channel inducing a stronger response of carriers to the trapping-detrapping events. On the contrary, the low-field mobility variation in N-MOS devices is associated to the conduction band repopulation without any energetic dispersion curve modification. This could be a further reason justifying why we did not observe any noise intensity modulation with stress in N-MOS devices.

CONCLUSIONS

We have investigated the effect of mechanical stress on 1/f spectra of MOSFET devices. Mechanical stress modulates the intimate electronic properties of silicon changing its own band structure. An interpretative key that correlates the variations of 1/f noise intensity with the strain-induced electronic band modifications has been proposed. Summarizing, we suggest that the analysis of low frequency noise on MOSFETs, beyond the obvious interaction between carriers and SiO_2 traps, is also linked to the deeper semiconductor properties. In this scenario, the effective mass lowering in next generation strained silicon is expected to produce an increasing contribution to low frequency noise.

REFERENCES

1. G. Scott et al. *IEDM Tech. Dig.* p. 827 (1999).
2. L. R. A. Bianchi et al., *IEDM Tech. Dig.* p. 117 (2002).
3. T. Ohguro et al., *VLSI Tech. Dig. Symp.* p. 37 (2003).
4. A. Stesmans, D. Pierreux, R. J. Jaccondine, M.-T. Lin and T. J. Delph, *Appl. Phys. Lett.* **82**, 3038- (2005).
5. P. Fantini G. Giuga, S. Schippers, G. Ferrari and A. Marmiroli, *ESSDERC Conf.* p. 401 (2004).
6. M. V. Fischetti and S. E. Laux, J. Appl. Phys. **80**, 2234 (1996).
7. E. Simoen, B. Dierickx, C. L. Clayes and G. J. Declerck, *IEEE Trans. Electron Devices* **39**, 422-429 (1992).

Can 1/f noise in MOSFETs be reduced by gate oxide and channel optimization?

M. Marin[1], J.C. Vildeuil[1], B. Tavel[2], B. Duriez[2], F. Arnaud[1], P. Stolk[2] and M. Woo[3]

[1]STMicroelectronics, [2]Philips Semiconductors, [3]Freescale Semiconductor
850 rue Jean Monnet, 38926 Crolles, France.
e-mail: mathieu.marin@st.com

Abstract. This contribution addresses several important topics about 1/f noise in MOSFETs for the 65 nm node. We show that a plasma nitridation technique can significantly improve the 1/f noise performances of the device providing that the insulator is thick enough. This result is explained by correlating the 1/f noise magnitude and the nitrogen concentration profile within the gate oxide. In a second time we investigate the effect of dopant dose and species in the substrate as well as the influence of channel orientation (110) and (100).

Keywords: MOSFET, 1/f noise, nitridation, substrate orientation
PACS: 72.70 m

INTRODUCTION

Many innovations are required to meet the increasingly aggressive ITRS specifications for sub-100 nm MOS technologies. The introduction of high-κ dielectrics or new device architectures are up to the 65 nm node postponed with the use of optimized oxynitride, stress engineering or alternative substrate orientations. Unfortunately some of these technological novelties, mainly used to enhance the "digital" Ion/Ioff figure of merit, are not suited for some analog parameters like Low Frequency Noise (LFN) or even for reliability issues. A good example is the well-known degradation of the 1/f noise and NBTI (Negative Bias Temperature Instability) observed with nitrided oxide devices [1], [2]. In this paper we first discuss the potential benefit in matters of noise of a plasma-based nitridation technique over a conventional thermal method. In a second time we investigate the effect of dopant dose and species in the substrate for both n and p-MOSFETs devices. Finally, we will compare the noise performances of transistors processed with different channel orientations, (110) versus (100).

EXPERIMENTAL

All experiments were performed on n and p-MOS devices issued from a 65 nm CMOS technology. DC characteristics were obtained with a HP4155A semiconductor analyzer and low frequency noise measurements were done directly on wafer using an FFT Dynamic Signal Analyzer (HP35670A) and a low noise current-to-voltage amplifier (EG&G 5182). The Device Under Test (D.U.T) was biased using 12V filtered batteries in order to avoid any external perturbations.

All the measurements were systematically corrected from the noise and gain of the I/V amplifier. Only large geometry devices were investigated (W/L = 10/5 or 10/3 µm) so as not to be disturb with generation-recombination or even RTS noise. Some experimental drain current noise Power Spectral Densities (PSD) are illustrated in FIGURE 1 between 10 Hz and 100 kHz for p-MOS devices with different gate-to-source voltages.

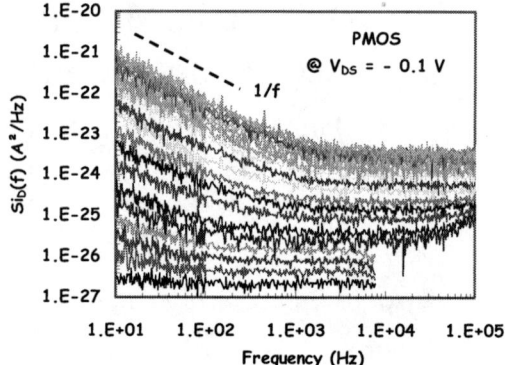

FIGURE 1: Experimental drain current noise spectral densities for p-MOS transistors.

RESULTS AND DISCUSSION

We shall first discuss the gate oxide nitridation influence. Usually, the nitrogen atoms are incorporated within the insulating layer of the MOS device via a thermal anneal under an NO ambient. However, a plasma-based method has recently shown several advantages over the classical thermal technique both in terms of gate leakage current and reliability [3]. We have compared the noise performances of n and p-MOS transistors processed with these two nitridation methods. Their relative 1/f noise level at f = 10 Hz is plotted in FIGURE 2.

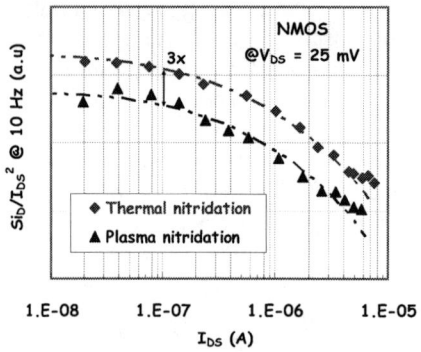 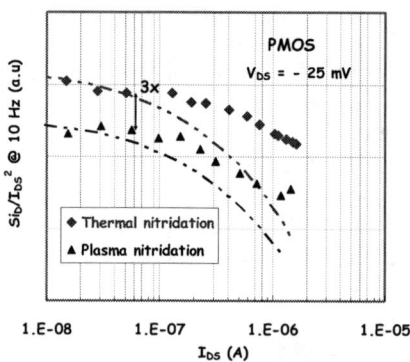

FIGURE 2: Relative 1/f noise (f = 10Hz) for n and p-MOS transistors (t_{ox}=2nm) as a function of I_{DS} – Impact of the gate oxide nitridation technique. The dashed lines stands for the ratio $(g_m/I_{DS})^2$.

We can observe a drastic impact on the 1/f noise level of both n and p-MOS transistors. This result can be explained considering that, on the one hand, nitrided oxide device are reportedly noisier than pure oxide device. Given that the 1/f noise is linked to the oxide trap density near the SiO_2/Si interface, nitrogen must induce extra traps. On the other hand, the nitrogen concentration profiles obtained with the two techniques are quite different [4]. With a plasma process most of the nitrogen is segregated closed to the top interface between the poly and the oxide while in the case of a thermal process N atoms pill up at the SiO_2/Si interface (see **FIGURE 3**, left plot). For a given oxide thickness, a plasma nitridation will result in a lower nitrogen dose at the interface with the channel, and then in a lower 1/f noise level.

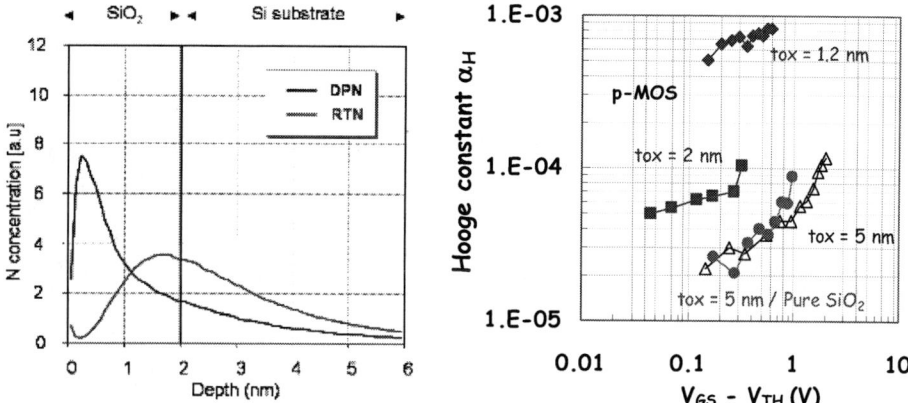

FIGURE 3: Nitrogen profile for plasma (DPN) and thermal (RTN) nitridation [left plot] - Hooge constant for plasma nitrided p-MOS transistors with different oxide thickness [right plot].

The benefit of such technique should then be as much appreciable as the gate oxide is thick since the nitrogen dose at the interface will decay. This assumption is confirmed in **FIGURE 3** (right plot) where we have reported the Hooge constant of plasma nitrided p-MOS transistors with different oxide thickness. The noise obtained with a 5 nm oxide is comparable to what has been measured on a pure SiO_2 device. This result is of particular interest for analog circuits which are designed with the thickest oxide transistors of a given technology platform.

We have also investigated the substrate optimization influence in two ways. First, we have compared the 1/f noise of transistors processed with different dopant dose and species (Indium/Boron for n-MOS and Arsenic/Phosphorus for p-MOS) as described in the next table.

TABLE 1 : Substrate dopant and associated dose ($\times 10^{-12}$ cm^{-2})

n-MOS	p-MOS
Boron 3 / Indium 5	Arsenic 7 / Phosphorus 10
Boron 4 / Indium 3	Arsenic 8 / Phosphorus 6
No Boron / Indium 5	Arsenic 10 / No Phosphorus

In a second time we studied the noise of devices with two distinct channel orientations, (110) and also (100). The latter leads to an enhancement of the drive current (~15%) for p-MOS devices leaving unchanged the n-MOS performances [5]. This particular orientation is obtained by a 45 degrees rotation of the wafer notch from its usual position. All results are reported on the following FIGURE 4 and FIGURE 5 respectively from weak to strong inversion.

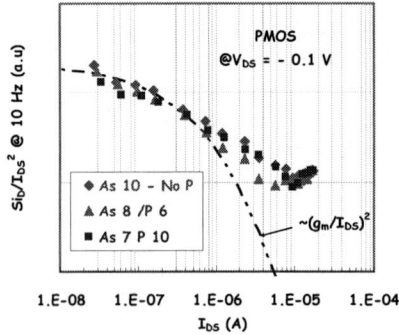 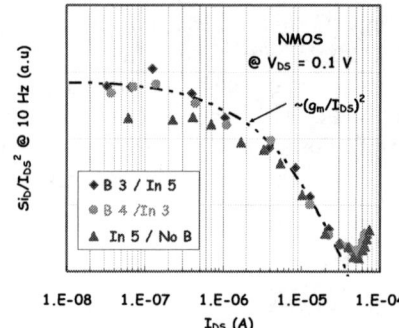

FIGURE 4: Normalized 1/f noise level at f = 10 Hz for n and p-MOS transistors (t_{ox} = 5 nm) with various bulk dopant concentrations

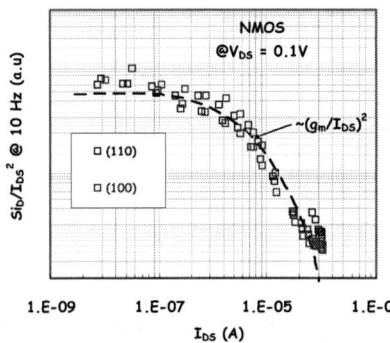 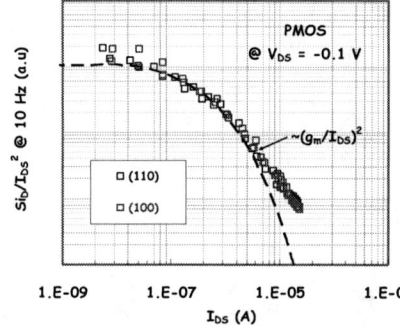

FIGURE 5: Normalized 1/f noise level at f = 10 Hz for n and p-MOS transistors (t_{ox} = 5 nm) with (110) and (100) channel orientation

We can observe that there is no significant impact neither of the dopant dose tuning nor of the channel orientation for both kinds of MOS transistors. Concerning the substrate doping, this result is rather consistent with the number fluctuation model or even with the correlated model. Concerning the results about the channel orientation, one could argue that the interface states should be different. However the density of slow states lying a few nanometers from the interface and which are of interest for 1/f noise, might not vary significantly.

CONCLUSION

We mainly demonstrated that shifting away the nitrogen peak from the Si/SiO$_2$ interface can substantially lower the 1/f noise level of n and p-MOS devices. On the contrary the channel orientation or doping has no appreciable influence.

REFERENCES

[1] P. Morfouli et al., Electron Device Letter, vol. 17, n°8, 1996
[2] D. K. Schroder et al., Journal of Applied Physics, Vol. 94, 2003
[3] B. Tavel et al, Proceedings of VLSI conference, 2003
[4] M. Bidaud et al., Proceedings of ESSDERC conference, 2002
[5] T. Komoda et al., Proceedings of the IEDM Conference, pp 217- 220, 2004

Impact of interface micro-roughness on low frequency noise in (110) and (100) pMOSFETs

P. Gaubert[#], A. Teramoto, T. Hamada[*], M. Yamamoto, K. Nii, H. Akahori, K. Kotani[*] and T. Ohmi

New Industry Creation Hatchery Center, Tohoku University
[*]*Graduate School of Engineering, Tohoku University*
Aza-Aoba 6-6-10, Aramaki, Aoba-ku, Sendai 980-8579, Japan
[#]*gaubert@fff.niche.tohoku.ac.jp*

Abstract. In this paper we describe the evaluation of the dependence of low frequency noise upon the micro-roughness of the surface in pMOSFETs, based on (100) and (110) oriented silicon. For the (110) surface, because RCA cleaning makes the surface rough, we developed a 5 step room temperature cleaning process which does not use alkaline solution. As a result a drop of more than a decade in 1/f noise level was achieved. This low noise level is further reduced by using the process of microwave-excited high-density plasma oxidation of the gate oxide instead of thermal oxidation. This reduction is also observed for a (100) surface if treated in the same way, but the magnitude of the drop is less.

Keywords: Surface micro-roughness, 1/f noise, MOSFET, Surface orientation, Silicon.
PACS: 73.40.Qv, 72.70.+m, 81.65.Cf , 81.65.Mq

INTRODUCTION

Low frequency electronic Flicker noise, or 1/f noise, is still a limiting factor for electronic devices and this is especially true for analog devices. Its reduction or even its complete eradication is a stake of prime importance which could lead the electronic world into a new era of ultra-low power consumption devices.

Improvements in current drivability and noise reduction are two of the most important parameters for the realization of future high-speed analog and digital circuits. The current drivability of pMOS on an Si(110) surface is 3 times larger than that achievable on an Si(100) surface[1,2], but despite this promising feature, with current fabrication techniques, the 1/f noise is still too high for the (110) CMOS to replace the conventional (100) one. In this paper, we report improvements in device fabrication that resulted in a general reduction in noise for both surface orientations. Notably the noise reduction was far more pronounced for the (110) surface, thereby closing the divide between the (110) and (100) pMOS in the race to achieve ultra-low noise devices. These improvements are attributed to the use of an alkali-free 5 step room temperature cleaning process[3], which reduced surface micro-roughness, combined with microwave-excited high-density plasma oxidation technology[4], which gave a more uniform high quality silicon-silicon oxide interface.

EXPERIMENTAL

(110) and (100) pMOS were fabricated using either RCA cleaning or, for comparison, alkali-free room temperature 5 step cleaning for pre-gate oxidation. The pMOS was completed using thermal oxidation at 900°C or, for comparison, microwave-excited high-density plasma oxidation at 400°C to create a 5nm thick gate oxide. Noise measurements were carried out using a vector Signal Analyzer (89410A AGILENT) directly on wafer biased by an Ultra-Low DC Source (PA14A1 SHIBASOKU). The surface micro-roughness was determined using scanning tunneling microscopy (STM).

RESULTS AND DISCUSSION

STM images of (100) surfaces processed using the two cleaning methods are shown in Fig. 1 in which we can see the improvement in surface quality when alkali-free 5 step cleaning is used.

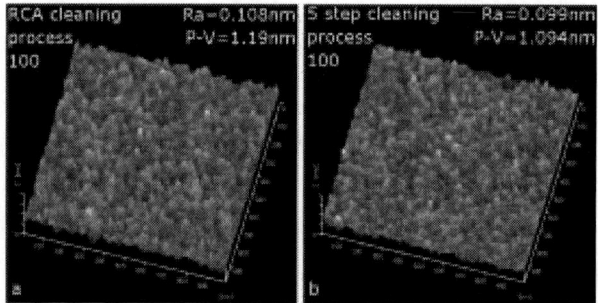

FIGURE 1. Micro-roughness Ra and peak-valley maximum amplitude P-V with the RCA cleaning (a) and the 5 step cleaning (b) for (100) silicon oriented surface (2x1000x1000nm).

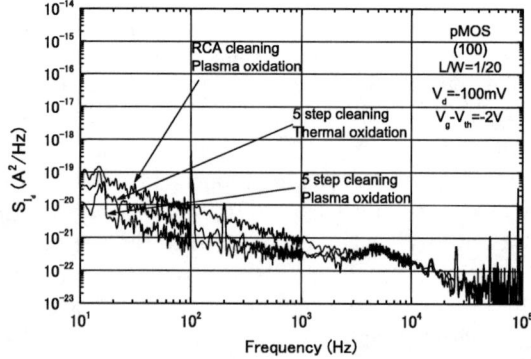

FIGURE 2. Evolution of the spectral density of drain current in a (100) silicon surface oriented pMOS for two different cleaning processes and two different oxidation techniques.

The reduction in the peak-valley height and micro-roughness is found to be around 10%. In addition to the numerous steps and high temperature used in RCA cleaning, a further problem is that it requires the use of alkali solution which leads to anisotropic removal of silicon atoms from the surface. This is because bond cleavage leads to the formation of (111) oriented micro-structured inhomogeneities in the surface which build up over time. On the contrary, the 5 step cleaning process is alkali-free resulting in the formation of a smoother, homogeneous interface.

Typical results of electronic noise measurement on (100) pMOS are shown in Fig. 2. Several measurements were made for all combinations of cleaning and oxidation procedures. The choice of cleaning process has a pronounced impact on the low frequency noise level, which decreases by almost one decade when using the alkali-free 5 step cleaning method, due to the formation of a surface with reduced micro-roughness compared to that obtained with RCA cleaning. Further, it is found that plasma oxidation results in a lower noise level than achieved from thermal oxidation. In fact, the plasma oxidation rate is the same whatever the orientation of the micro-surface upon which it acts whereas thermally induced oxide growth is slower on (100) surfaces than on (110) or (111) surfaces[5].

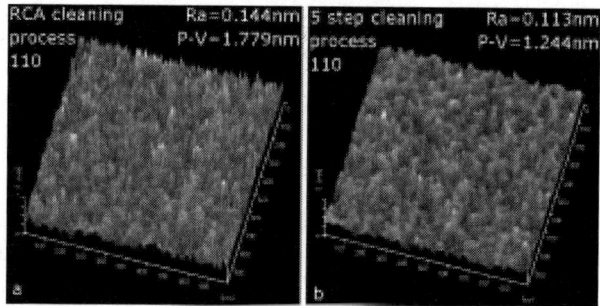

FIGURE 3. Micro-roughness Ra and peak-valley maximum amplitude P-V with the RCA cleaning (a) and the 5 step cleaning (b) for (110) silicon oriented surface (2x1000x1000nm).

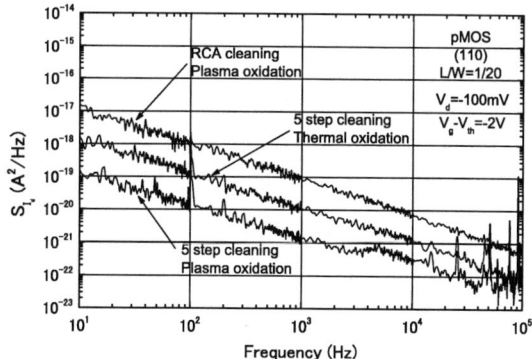

FIGURE 4. Evolution of the spectral density of drain current in a (110) silicon surface oriented pMOS for two different cleaning process and two different oxidation technique.

As a result, the Si-SiO$_2$ interface is less smooth when using the thermal oxidation. STM images obtained for (110) surfaces are shown in Fig. 3. When qualitatively comparing these images with those in Fig. 1, it is obvious that the use of the alkali-free 5 step cleaning again results in improvements in the surface quality. Quantitatively there is a 30% improvement in the micro-roughness with the 5 step cleaning process which is more efficient on (110) than on the conventional (100) orientation. The degradation of the surface resulting from RCA cleaning shown in Fig. 3a is far worse compared to that shown in Fig. 1a. This can be explained by the fact that the alkali solution etching rate on (110) is faster than on (100)[6].

The result of 1/f noise measurements on (110) pMOS, using the same procedure described above, are presented in Fig. 4. In this case, the 1/f noise level is reduced by two decades by simply changing the pre-gate formation cleaning process to the alkali-free 5 step method, while a further one decade decrease in noise results if we use the plasma oxidation method.

This significant change in the noise level has its origin in the cleaning process and the resulting surface quality of the interface and in particular the change in the micro-surface roughness. This can be attributed to the faster degradation of the (110) surface caused by the alkaline solution used during RCA cleaning compared to the degradation of the (100) surface under the same conditions. Thus the choice of cleaning method can have a knock-on effect which may be amplified during the gate formation step and this will be especially true in the case of thermal oxidation.

CONCLUSION

The interface micro-roughness and the 1/f noise in different pMOS have been studied. We successfully demonstrated that a reduction in the micro-roughness and the use of a plasma oxidation process lead to a significant drop in the 1/f noise level due to the formation of a smoother interface and more uniform oxide growth. The advantage of the (100) pMOS over the (110) pMOS has been reduced so much that, considering the improved drivability of (110) pMOS, the (110) CMOS can establish itself as a viable competitor or even a future replacement for the current silicon CMOS technology.

REFERENCES

1. Sato, T., Takeishi, Y., Hara, H., and Okamoto, Y., *Phys. Rev. B*, **4**, 6, 1971, pp. 1950-1960.
2. Ezaki, T., Nakamura, H., Yamamoto, T., Takeuchi, K., and Hane, M., SISPAD, 2004, pp. 53-56.
3. Teramoto, A., Hamada, T., Akahori, H., Nii, K., Suwa, T., Kotani, K., Hirayama, M., Sugawa, S., and Ohmi, T., IEDM Tech. Dig., 2003, pp. 801-804.
4. Ohmi, T., and Hirayama, M., Extend Abstracts of Int. Symp. On Advanced ULSI Tech. – Challenges and Breakthroughs, 1998, pp. 19-22.
5. Sugawa, S., Ohshima, I., Ishino, H., Sato, Y., Hirayama, M., and Ohmi, T., IEDM Tech. Dig., 2001, pp. 817-920.
6. Akahori, H., Nii, K., Teramoto, A., Sugawa, S., and Ohmi, T., Extend Abstracts of the 2003 Int. Conf. SSDM, 2003, pp. 458-459.

Modeling of Suppressed Shot Noise in Stress-Induced Leakage Currents

Giuseppe Iannaccone

Dipartimento di Ingegneria dell'Informazione: Elettronica, Informatica, Telecomunicazioni, Università di Pisa, Via Caruso, I-56122 Pisa, Italy
g.iannaccone@iet.unipi.it

Abstract. We review our approach to model transport and noise in MOS capacitors in the Stress-Induced Leakage Current regime. We show that a model including Pauli and Coulomb interaction among electrons can be implemented in a numerical solver of the coupled Poisson and Schroedinger equations in semiconductor devices, and can reproduce experiments performed on stressed MOS capacitors. We also show that our model suggests a procedure to extract information on the properties of traps from DC and Noise measurements and simulations.

Keywords: Tunneling, Shot Noise, Stress-Induces Leakage Currents, MOS, Trap-assisted-tunneling
PACS: 72.70.+m, 73.25.Hk, 73.40.Gk, 73.40.Sx 73.61.Ng

INTRODUCTION

Stress-Induced Leakage Currents (SILCs) in MOS structures represent one of the main reliability concerns for CMOS non-volatile memory products. At the same time, they are an example of a pressing technological problem that calls the tools of mesoscopic physics to action. Indeed, SILCs are now known to be due to trap-assisted tunneling through single defects created in the oxide by high field stress [1-4]. In recent years, as a result of an experimental and theoretical activity performed in Pisa, it has been shown that noise is a key aspect for understanding the nature of SILCs [4-7].

In this paper, we briefly review our model of shot noise in SILCs, and show that an approach based on the actual device structure can allow to choose the appropriate level of abstraction for the problem at hand, and to include sufficient physical detail to reproduce the experimental DC and noise properties.

It will also be shown that modeling of shot noise and DC properties provides precious insights of the transport mechanisms in the SILC regime and enables one to extract information on the energy distribution of the traps created by stress.

TRANSPORT AND NOISE MODEL IN SILCS

The shot noise behavior in of a MOS capacitor in the SILC regime is based on the following consideration: Pauli exclusion principle and Coulomb repulsion prevent two electrons from occupying the same trap, therefore introducing correlation in the motion of carriers. This causes the suppression of the power spectral density of shot noise of the resulting trap-assisted tunneling current, with respect to the full value $S = 2qI$, obtained for current through fresh oxides, and typical of the uncorrelated motion of electrons.

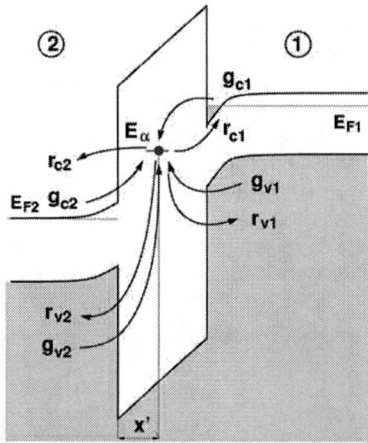

FIGURE 1. Band-edge profile of a MOS capacitors, and indication of the possible transistions of electrons between states in the contacts and a trap state, localized at position x' and energy E_α.

We call "generation rate" the transition rate from an electrode to the nanocrystal, and "recombination rate" the transition rate from the nanocrystal to one electrode [4]. As we can see in Fig. 1 there are eight different generation and recombination rates, with obvious meaning of the subscripts. We dot not enter into the detail of the expressions here, referring the interested reader to [4].

Let us only stress that fact that we allow for inelastic transitions, from a state $|\beta\rangle$ of energy E_β in a contact to a localized state $|\alpha\rangle$ of energy E_α in the oxide layer, by writing the transition rate, according to Fermi's "golden rule", as:

$$v_{\beta \to \alpha} = \frac{2\pi}{\hbar} |M(\beta,\alpha)|^2 h_T (E_\alpha - E_\beta) \tag{1}$$

where

$$h_T(E_\alpha - E_\beta) = \frac{\Gamma/\pi}{(E_\alpha - E_\beta)^2 + \Gamma^2} \tag{2}$$

The larger Γ, the larger degree of inelasticity of the transitions. This assumption corresponds to defining a trap cross section per unit energy as:

$$\sigma_{\alpha,\beta} = k \cdot h_T (E_\alpha - E_\beta) \tag{3}$$

where k is an unknown constant surface, to be determined via a fitting procedure to experiments.

By integrating over all longitudinal and transversal energies, and considering spin degeneracy, we obtain each single generation and recombination rate sketched in Fig. 1. We can define $g_1 \equiv g_{1c} + g_{1v}, g_2 \equiv g_{2c} + g_{2v}, r_1 \equiv r_{1c} + r_{1v}, r_2 \equiv r_{2c} + r_{2v}$. The average occupation factor f of the trap is given by:

$$f' = \frac{g_1 + g_2}{g_1 + g_2 + r_1 + r_2}, \qquad (4)$$

and therefore, in the typical case in which g_2 is neglibile, the average current through the trap I' and the associated current shot noise S' are given by:

$$I' = qr_2 f' \text{ and } S' = 2qI'\left[1 - \frac{g_1 r_2}{(g_1 + r_1 + g_2 + r_2)^2}\right]. \qquad (5)$$

The derivation of S' is very similar to the derivation of the suppressed shot noise expression in resonant tunneling structures, so any detail can be found in Ref. [8].

As can be seen, the Fano factor associated to the tunneling current assisted by a single trap (the term between square parenthesis in (5)) is comprised between 0.5 and 1.

Integrating I' and S' over the distribution of trap energies E_α, in the oxide energy gap, and over x', in the oxide volume, and indicating with η the trap density per unit volume per unit energy, we can obtain the total current density J_{TAT}, the total noise spectral density S_{TAT} and the SILC Fano factor γ_{TAT}:

$$J_{TAT} = \iint I' \eta(E_\alpha, x') dE_\alpha dx' \qquad (6)$$

$$S_{TAT} = \iint S' \eta(E_\alpha, x') dE_\alpha dx' \qquad (7)$$

$$\gamma_{TAT} = \frac{S_{TAT}}{2qJ_{TAT}} = \frac{\iint S' \eta(E_\alpha, x') dE_\alpha dx'}{\iint I' \eta(E_\alpha, x') dE_\alpha dx'}. \qquad (8)$$

By construction γ_{TAT} is therefore in the range 0.5-1. From experiments, the shot noise suppression due to stress-generated traps in the current range where SILC is predominant, found in the range 0.63-0.83.

RESULTS

In the model described above there are many unknowns, mainly related to the density and the distribution of traps in space and energy. One of the objectives of the model was to enable the use of noise measurements to extract properties of those traps that are predominant in the SILC regime. First of all, we verified that a wide and nonuniform energy distribution of traps was absolutely required to fit the experimental I-V characteristics and noise measurements. On the other hand, the trap distribution could be reasonably assumed uniform in real space. We have then implemented the model in a 1D Poisson-Schrödinger solver, that allows us to compute the band profiles taking into account quantum confinement in the 2DEG and the barrier tunneling probabilities required for the computation of the transition rates.

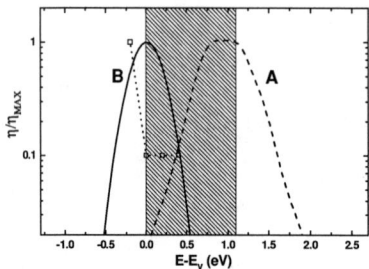
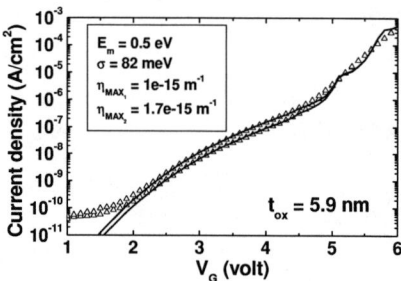

Figure 2. Left: comparison between normalized trap distributions obtained with our model (thick solid line) and those described in [1] (A, dotted line) and in [2] (B, symbols). The patterned region corresponds to the silicon energy gap. Energies on the horizontal axis are referred to the top of the silicon valence band. Right: I-V characteristics of stressed capacitors with 5.9 nm oxide: Symbols represent measurements after two different stress times, and line represents the result of simulations.

We refer to Eq. (6): for any gate voltage, J_{TAT} can be determined by subtracting the fresh current component, i.e. the current not assisted by traps, from the total current and the elementary contribution I' in the right hand side can be numerically computed. Equation (6) is therefore an integral equation in which $\eta(E_\alpha)$ is the unknown. Trying to obtain a closed-form solution is unlikely especially because we do not have any information on the possible analytical form of η, so we need to introduce some approximations. Assuming η to be independent of x', we can transform the integral equation into a linear algebraic system: solving it with the least-square method we extract a distribution that we interpolate with the Gaussian curve plotted in Fig. 2 with thick solid line. Therefore, the trap distribution can be expressed as a function of the position of the energy peak E_m, the corresponding Gaussian distribution peak η_{MAX} and standard deviation σ.

In the same figure our distribution is compared to others (all the distributions are normalized to their maximum value) obtained using models described in [1] and [2] respectively: all the three models agree in locating traps close to the silicon energy gap and, as a consequence, and in considering SILCs due to transitions involving electrons of the cathode valence band. This result is compatible with quantum yield measurements according to which the energy of electrons contributing to SILCs is 1.5 eV lower than that of FN tunneling electrons [3].

The fact that different values for E_m and σ allow to obtain a very good fitting of J-V curves confirm that DC properties are not sufficient to univocally determine the trap energy distribution. For this reason, we obtain further information from the shot noise. Fig. 3 shows numerical results for the Fano factor of the SILC component γ_{TAT}. Curves have been obtained using the trap distribution plotted in Fig. 2 and different values of E_m, and suggest that trap energies closer to the middle of the silicon oxide gap ($E_m \approx 0.5$ eV) provide a better fit with experiments.

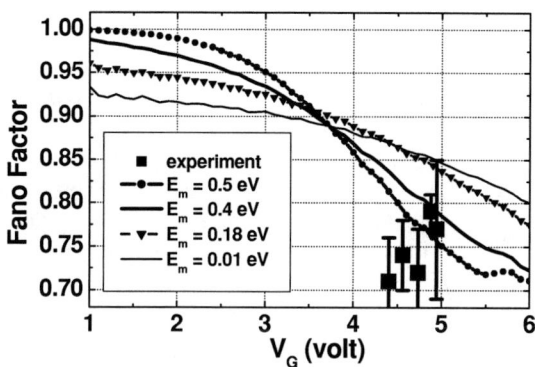

Figure 3. Shot noise suppression for a 5.9 nm oxide, as a function of the applied gate voltage. Symbols and error bars represent measurements on a sample of different MOS capacitors. E_m is the trap energy referred to the silicon valence band edge.

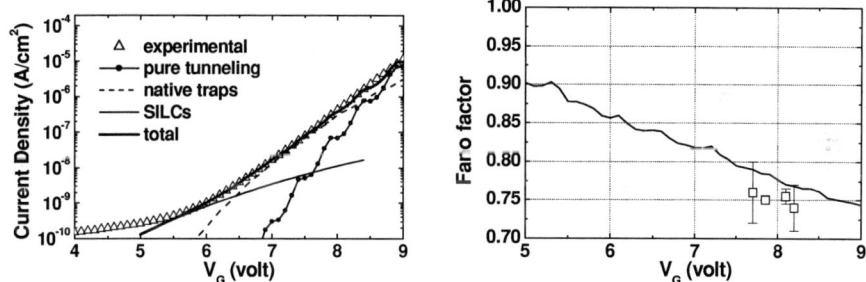

Figure 4. Left: Different current density contributions as a function of the gate voltage for a MOS capacitor with a 10 nm oxide in the SILC regime, compared with experimental results (triangles). Right: Experimental (symbols) and theoretical Fano factor for the 10 nm oxide.

The same procedure can be used for thicker oxides, such as in the case of the MOS capacitor with a 10 nm oxide considered in Ref. [4] and [7]. In that case, even in the case of the fresh oxide a distribution of native traps assisting electron tunneling has to be considered in order to fit the experimental results at very low current levels. In the SILC regime, such traps become predominant even at higher voltages, while traps induced by electric stress are relevant only at very low fields, as can be seen in Fig. 4 (left).

However, also in that case, native traps are responsible for shot noise suppression around 0.75. As can be seen in Fig. 4 (right), experimental results on noise are correctly reproduced by our simulations.

CONCLUSION

We have briefly reviewed our activity on modeling noise in the SILC regime of MOS capacitors. We have seen that mesoscopic transport and noise modeling can have a very practical use here, helping us to extract information on the stress-induced traps that represent the main reliability concern of non-volatile semiconductor memories. This subject is a further exemplification of Rolf Landauer's saying "the noise is the signal" [9], since the measurement of suppressed shot noise was one of the main validations of the hypothesis that a trap-assisted tunneling mechanism was responsible for SILCs. However, let us stress the fact that in this case noise is not telling us anything on the coherence of the tunneling process: in our model we have only assumed that the characteristics times for generation and recombination are much shorter than the inverse maximum frequency of the noise spectrum considered, and that the two barriers are opaque (which is perfectly reasonable for SiO_2 barriers). The two-step tunneling description we have used is only a suitable way of formally treating the problem, and is compatible with both coherent and incoherent tunneling [10].

ACKNOWLEDGMENTS

Support is gratefully acknowledged from the EU through the SINANO Network of Excellence (EU Contract 506844) and from the Italian MIUR through the FIRB Project ``Sistemi miniaturizzati per elettronica e fotonica" (miniaturized systems for electronics and photonics) and through the PRIN Project "Architetture e Modelli Innovativi per nanoMOSFET" (Innovative models and architectures for nanoMOSFETs).

REFERENCES

1. B. Riccò, G. Gozzi, M. Lanzoni, *IEEE Trans. Electron Devices*, **45**, 1554-1560 (1998).
2. D. Ielmini, A.S. Spinelli, M.A. Rigamonti, A. Lacaita, *IEEE Trans. Electron Devices*. **47**, 1258-1272 (2000).
3. S. Takagi, N. Yasuda, A. Toriumi, *IEEE Trans. Electron Devices*, **46**, 335-341 (1999); ibid. **46**, 348-354, 1999.
4. G. Iannaccone, F. Crupi, B. Neri, S. Lombardo, *IEEE Trans. Electron Devices* **50**, 1363-1369 (2003).
5. G. Iannaccone, F. Crupi, B. Neri, S. Lombardo, *Appl. Phys. Lett.*, **77**, 2876-2878 (2000).
6. G. Iannaccone, *Mat. Sci. Tech.* **18**, 736-739 (2002).
7. A. Nannipieri, G. Iannaccone, F. Crupi, *Microelectronics Reliability* **44**, 1497-1501 (2004).
8. G. Iannaccone, M. Macucci, B. Pellegrini, Phys. Rev. B **55**, 4539-4550 (1997).
9. R. Landauer, Nature **392**, 658-659 (1998).
10. G. Iannaccone, B. Pellegrini, Phys. Rev. B **52**, 17406-17411 (1995).

LF Noise and Tunneling Current in Nanometric SiO$_2$ Layers

J. Gurgul,[a] C. Leroux,[b] G. Ghibaudo,[c] and J.A. Chroboczek [b,c]

*[a] Institute of Catalysis and Surface Chemistry,
Polish Academy of Sciences, ul. Niezapominajek 8, 30-239 Krakow, Poland,
[b] Commissariat à l'Energie Atomique, Laboratoire d'Electronique et de Technologie de l'Information
(CEA-LETI), 17 rue des Martyrs, 38045 Grenoble Cedex 9, France,
[c] Ecole Nationale Supérieure d'Electronique et de Radioélectricité de Grenoble,
Institut de Microélectronique, Electromagnétisme et Photonique
(ENSERG-IMEP), 23 rue des Martyrs, BP 257, 38016 Grenoble Cedex 1, France.*

Abstract. We studied current-voltage, I(V) and low frequency noise, LFN, characteristics of SiO$_2$ capacitors with a varying SiO$_2$ thickness, t_{ox}, and the active surface area, A. The tunneling current was found to vary as $\exp(-t_{ox}/\lambda)$, with $\lambda=0.1$nm, λ slightly increasing with the applied voltage. The LFN spectral power density, S_I, was observed to obey the 1/f type dependence and could be characterized in the SPICE formalism $S_I = KI^2/fA$, with 10^{-13}cm^2<K<10^{-12}cm^2. We observed in numerous cases that the layers conductance, $\delta I/\delta V$, and $\sqrt{S_I}$ were linked by a relation close to proportionality in a wide range of bias voltages. Such a relationship results from the flat band potential fluctuations, induced by the charge capture at the dielectric/Si interfaces.

Keywords: 77.55.+f; 05.40.Ca; 03.75.Lm

INTRODUCTION

The downscaling of the Si MOSFET devices requires an appropriate reduction in the gate Si dioxide thickness, t_{ox}, which entails an appearance of undesirable tunneling currents at relatively modest gate biasing voltages. As the gate currents are known to contribute to the low frequency noise, LFN, in the MOSFET drain current, I_d, a systematic study of both, transport and noise in SiO$_2$ layers presents a considerable interest for device physics and technology. We studied those issues on the simplest system: the Si dioxide capacitor. The t_{ox} values, the capacitors dimensions and some fabrication parameters were varied and their impact on the static and LFN characteristics were examined. To our knowledge this paper provides the first systematic study of LFN in the tunneling current of Si/SiO$_2$/poly-Si capacitors.

EXPERIMENTAL RESULTS AND DISCUSSION

Table 1 lists some characteristics of the wafers on which the capacitors used in this work were fabricated (source: LETI/CEA-Grenoble). Their electrical, static characteristics have been previously studied by Leroux et al. [1]. The wafers code is provided in the first line in Table 1, with the nominal t_{ox}, given in the second. Note that for thinner oxides, a compacting annealing step was used (third line: yes). Finally the substrate composition is given in the last line as the substrate layer adjacent to the dielectric plays a predominant role in determining the sample characteristics. The data furnished in Table 1 have been successfully used for I(V) simulation [1]. The capacitors were square and their sides varied from 10μm to 100μm, by steps of 10μm.

TABLE 1. Some characteristics of the capacitors used in this work

Wafer	2	4	6	8	10	18	20	22
t_{ox}	0.6	0.8	1.2	1.2	2.0	1.2	1.2	1.5
Anneal	yes	yes	yes	no	no	yes	no	no
Substrate	p	p	p/p++ Epi: 2,8-3,2 μm	p	p	p+	p+	p+
	7-10 Ωcm $2*10^{15}$ at/cm^3	7-10 Ωcm $2*10^{15}$ at/cm^3	Upper: 10-1 Ωcm $2*10^{15}$/ $5*10^{19}$ at/cm^3	7-10 Ωcm $2*10^{15}$ at/cm^3	7-10 Ωcm $2*10^{15}$ at/cm^3	0.06 Ωcm $5*10^{17}$ at/cm^3	0.06 Ωcm $5*10^{17}$ at/cm^3	0.06 Ωcm $5*10^{17}$ at/cm^3

Noise-Conductance Relationship: Observation and Explanation

One of the salient points of this work was the observation in the experimental data (cf. Fig. 1) that the conductance $g=\delta I(V)/\delta V$ and $\sqrt{S_I(V)}$ are linked by a relation close to a direct proportionality. Similar observations have been made in thin-gate MOSFET structures showing gate dielectric leakage (this Conference [2]). What made us aware of the PSD(V) and I(V) correlation, was the observation of some characteristic features in the near-flat-band range, -1V<V<0, that were seen in both functions. It should be noted that in that biasing range the final states for tunneling are located in the Si band-gap, thus the tunneling must involve intermediate trap states, probably located in the oxide, or at the Si/SiO$_2$ interface. It is worthwhile to remind it here that the stress-induced leakage currents, SILC, in similar structures have been observed in the -1V<V<0 bias range [3]. The SILCs are known to result from the appearance of interface traps. The $g \propto \sqrt{S_I(V)}$ relationship can be explained assuming that the flat band potential on the n-Si side of the capacitor fluctuates by an amount $\delta V_B = \text{const}(-\delta Q_{ox}/C_{ox})$, for the capture/release of a charge δQ_{ox} by the traps. The proportionality constant should contain the DOS of the traps, active, in the case of the tunneling, <u>both</u> in the transport and LFN. As the DOS is not necessarily uniform in the gap, some accidental features may appear in, both, $S_I(V)$ and $\delta I/\delta V$, as observed.

We wish to point out an interesting analogy between the MOSFET's drain current fluctuation PSD, and that of the tunneling current either in the MOSFET's gate dielectric or in the capacitors. Namely, for the former the function proportional to S_{Id} is the square of the drain current **transconductance**, $g_m(V_g)=\delta I_d/\delta V_g$, and for the latter, the square of **conductance** $g_g(V_g) =\delta I_g/\delta V_g$, which is equivalent to g(V) used in this work. Those two derivatives appear in the respective LFN expressions and serve to translate the flat band potential fluctuations, generated by the charge capture/release on near-interface trapping centers, into current fluctuations.

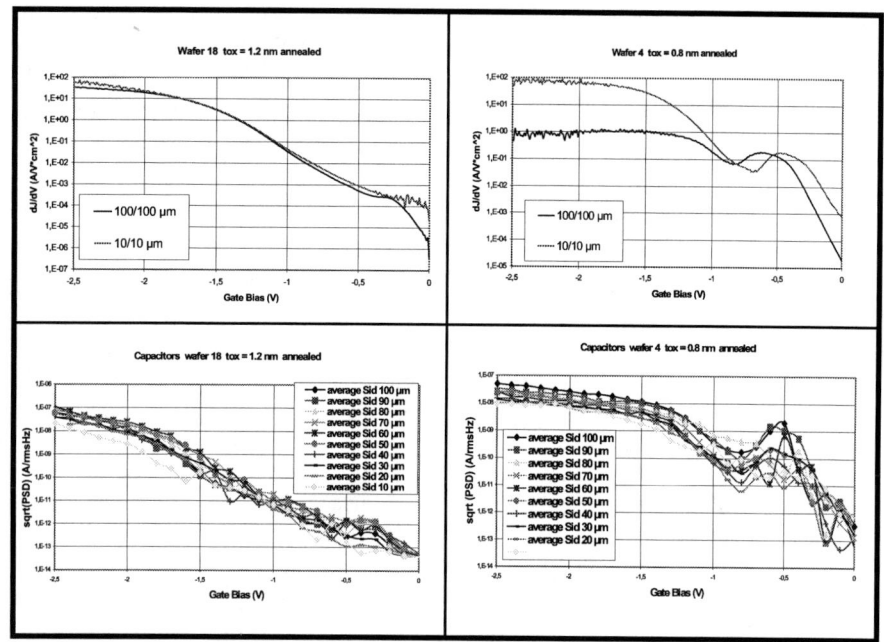

FIGURE 1. Conductance g=δI/δV (upper row of plots) and $\sqrt{S_I}$ vs V taken on two families of capacitors with t_{ox}= 1.3nm (left) and 0.8nm (right). The dispersion in characteristics (just two I(V)'s are shown for clarity) on the right arises from the presence of a series resistance, affecting more severely the characteristics of bigger capacitors with a thinner oxide. For that reason $\sqrt{S_I}$ (V) and g(V) are not always proportional. Note that some characteristic features appearing at -1V <V<0, visible in g(V) appear also in $\sqrt{S_I}$ (V).

Determination of the Tunneling Constant

The current intensity across SiO_2 layers for t_{ox} values listed in Table 1, measured at the bias V equal to -0.8V, -1.0V and -1.2V, revealed that the barrier transparency varied as $\exp(-t_{ox}/\lambda)$, with λ=0.11nm, with a weak augmentation in λ with the increasing V (cf. Fig. 2). We have selected small bias values because the data show less dispersion in that region, as the access resistance has no, or little, effect on the data. Similar results were reported earlier [4].

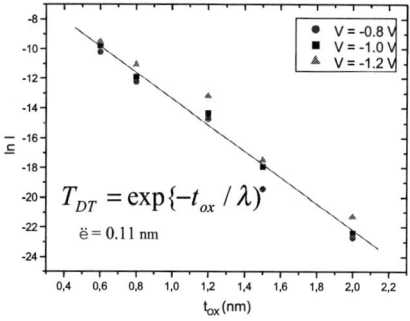

FIGURE 2. Tunneling current dependence on t_{ox} measured in oxide layers listed in Tab. 1. The data points follow the exponential dependence $\exp(-\lambda/t_{ox})$, where the tunneling constant λ is the same at that used in the LFN literature for Si devices.

Note that the value of λ is very close to that extensively used to describe the dominant, number fluctuation, component in the LFN in Si devices. As this type of noise generation involves the tunneling to, or from, the centers localized in the oxide, it is not surprising that the same tunneling constant appears also in the description of the tunneling phenomena in SiO_2 layers.

Determination of the LFN SPICE Parameters

As mentioned, the observed PSD spectra were of the 1/f -type. Figure 3 shows the square root of the PSD data extracted from the spectra at 10Hz, *versus* the tunneling current intensity. The data are seen to scale with the current intensity and (inversely) with the capacitor side dimension. Those observations can be cast in the SPICE formalism as follows: $S_I = K\ I^\alpha / f^\nu A^\beta$, where α=1 and β=ν=1. The value of K is comprised between $10^{-13} cm^2$ and $10^{-12} cm^2$, which is small on the scale of microelectronics devices. This issue will be further examined.

The extraction of the SPICE parameters is not always straightforward, as at small currents the data are limited by the system noise and in the thin-layer capacitors having large surface area, the access resistance noise contributes significantly to the capacitor noise, making the capacitor dimension scaling effect difficult to reveal. We shall address the problem of extracting the access resistance from the static and noise data elsewhere.

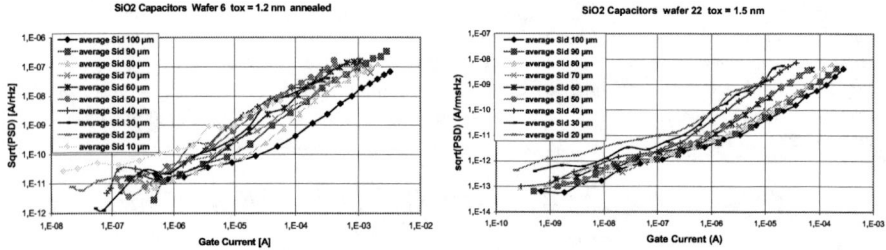

FIGURE 3. Square root of PSD @ 10Hz of the tunneling current fluctuations for two families of capacitors with the intermediate t_{ox} values. The capacitor side dimensions are listed in the figures. The data scale with the current intensity and with the capacitor side (inverse relation).

REFERENCES

1. Leroux, C., Mur, P., Rochat, N., Rouchon, D., Reimbold, G., and Ghibaudo, G., in INFOS Proc, Barcelona, 2003.
2. Contaret, T., Romanjek, K., Ghibaudo, G., Chroboczek, J.A., Bœuf, F., and T. Skotnicki, this Conference.
3. Ghetti A . et al, p. 723I in EDM Techn. Dig.est 1999.
4. Clerc, R., Spinelli, A., Ghibaudo, G., Leroux, C., and Pananakakis, G., Microelectronics and Reliability, **41**, 1027 (2001). Also Clerc, R., et al.50, 13 (13) in WODIM Conf. Proc., Munich, 2000.

Evolution of R.T.S Source Activities in Saturation Range in N-MOSFETs for Different Oxidation Temperatures

C. Leyris[(2)(1)], A. Hoffmann[(1)], M. Valenza[(1)], J.C. Vildeuil[(2)] and F. Roy[(2)]

[(1)] *Centre d'Electronique et de Micro-optoélectronique de Montpellier, CNRS-Université UMR 5507, Place Bataillon,U.M. II, 34095 Montpellier Cedex 5, France.*
[(2)] *ST Microelectronics 850 rue Jean Monnet, 38926 Crolles cedex, France.*

Abstract. Random Telegraph Signal (R.T.S) noise measurements have been performed on N-MOS transistors, in saturation range, from weak to strong inversion. The influence of the gate oxidation temperature on R.T.S noise sources is reported. An oxidation temperature reduction can lead to a decrease of traps-Si/SiO_2 interface distance and to a shift of the trap activation ranges toward higher gate voltage values. A detailed investigation of the evolution of R.T.S source characteristics with the oxidation temperature variation is proposed.

Keywords: N-MOSFETs, R.T.S fluctuations, saturation range, gate oxidation temperature.
PACS: 72.20.Jv

INTRODUCTION

With the down-scaling of CMOS device area the contribution of Random Telegraph Signal (R.T.S) noise sources becomes more pronounced and represents a serious concern as regard to the electrical operation. They may lead to poor signal-to-noise ratio in CMOS compatible analog circuits and to anomalously high R.T.S drain current amplitude in the associated digital circuits. The hold count of these noise sources is necessary to improve performances of the CMOS applications. Up today, many studies present R.T.S evolution in ohmic range mainly in view of design considerations [1,2]. The process flow impact on R.T.S source behavior has not yet been clearly studied for advanced CMOS technologies. Therefore, the detailed characterization of the process flow influences on R.T.S sources are of prime importance. The aim of this paper is to investigate the effect of the gate oxidation temperature on R.T.S sources at high drain voltage. These studied conditions describe the device biases set using in standard advanced CMOS applications.

EXPERIMENT

The R.T.S fluctuations are studied at high drain voltage, 1V (current saturation range), from weak to strong inversion. In order to illustrate the evolution of R.T.S sources with the oxidation temperature, nMOSFETs with W/L = 0.5 µm/0.35 µm and a threshold voltage of 0.65V have been chosen. The studied devices have an oxide thickness of t_{ox}=7 nm. The studied oxidation temperatures are 850°C and 925°C.

RESULTS AND DISCUSSION

From time and frequency measurements, the main characteristic R.T.S parameter values are extracted [3] for each oxidation temperature. A temperature decrease from 925°C to 850°C leads to a strong decrease of the $\Delta I_d/I_d$ ratio. This remains observable from weak to strong inversion in saturation range (Figure 1-a). Each R.T.S results from a trap competition [3]. Three traps are active from weak to strong inversion. They are named I'-II'-III' and I-II-III for oxidation temperatures of 850°C and 925°C, respectively. Corner frequencies (f_c) of each trap are reported versus gate biases on figure 1-b. The decrease of the oxidation temperature induces an f_c increasing pointing out the oxidation temperature impact on R.T.S traps.

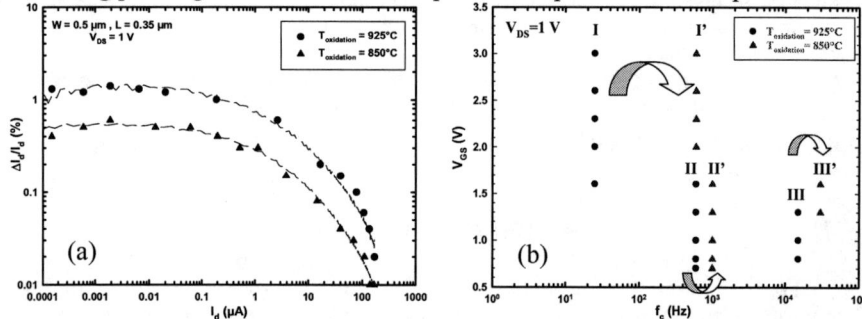

FIGURE 1. (a) Relative drain current fluctuation evolutions and (b) Corner frequency evolutions for an oxidation temperature of 850°C and 925°C.

Classically the $\Delta I_d/I_d$ evolution can be expressed as:

$$\frac{\Delta I_d}{I_d} = \eta \frac{g_m}{I_d} \frac{q}{WLC_{ox}} \left(1 - \frac{x_t}{t_{ox}}\right) \qquad (1)$$

where C_{ox} is the normalized oxide capacitance, g_m the transconductance, η a fit parameter, t_{ox} the oxide thickness, x_t the trap depth with respect to Si/SiO$_2$ interface and W and L the width and the length of the transistor, respectively.

The comparison between the measured $\Delta I_d/I_d$ evolution for the both oxidation temperatures (figure 1-a) shows that the trap- Si/SiO$_2$ interface distance is modified. In the same way, using the McWorther approach [4], a corner frequency (f_c) evolution leads to a variation of the distance between the R.T.S sources and the Si/SiO$_2$ interface (figure 1-b).

R.T.S measurements in the temporal domain allow the determination of the average times, spent at lower current level (τ_0) and at higher current level (τ_1), and their dependence with respect to the gate bias. An acceptor trap is neutral when it is empty and negatively charged when it captures an electron. For all traps, an increase of the gate bias induces a decrease of the time spent at the high level (Figure 2). If the low level of the current tends to be predominant when the gate bias is increasing, the trap tends to be in an ionized state. In this state, an electron can be emitted from the trap. All of the characterized traps have been identified as acceptor type centers.

Knowing emission and capture times the associated trap occupancy probability function ($f = \tau_e / (\tau_c + \tau_e)$) can be calculated (Figure 2-a).

Trap activity remains in a range of few kT around the Fermi level. For N-MOSFETs, a gate bias increase induces an increase of the trap occupancy probability and consequently a decreasing of the τ_c/τ_e ratio.

$$\frac{\tau_c}{\tau_e} = \frac{1-f}{f} = g \exp\left(\frac{E_t - E_F}{kT}\right) \quad (2)$$

where g is the degeneracy factor, which is usually considered as one, E_t-E_F is the trap energy level relative to the Fermi level, T is the absolute temperature and k is the Boltzmann constant.

Thus, the trap activation range is determined. For the both studied oxidation temperatures, each trap activation ranges are found to be identical (Figure 2-b).

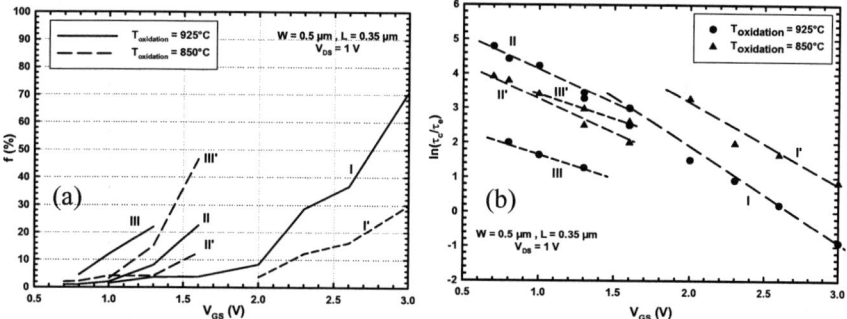

FIGURE 2. (a) Trap occupancy probability function evolutions and (b) Mean emission to capture times ratio evolution for an oxidation temperature of 850°C and 925°C.

Emission and capture times (figure 3-a) or trap occupancy probability function (figure 2-a) studies allow the determination of the trap activation ranges as well as the gate bias values corresponding to the maximum of trap activity ($\tau_e = \tau_c$ or $f_t = 50\%$). It is clearly shown that the decrease of the oxidation temperature leads to a shift of emission and capture times and of the maximum of activities toward higher gate bias values (figure 3-a-b).

Using the Shockley-Read-Hall approach:

$$\frac{\partial \ln(\tau_c/\tau_e)}{\partial V_g} = -\frac{q}{kT}\frac{x_t}{t_{ox}} - \frac{q}{kT}\left(1 - \frac{x_t}{t_{ox}}\right)\frac{\partial \Psi_s}{\partial V_g} \quad (3)$$

where ψ_S is the surface potential.

The extraction of the trap depth in the oxide (x_t) with respect to the Si/SiO$_2$ interface is done using a combination of a study of the τ_c/τ_e ratio with respect to the gate bias and a calculation of the $\frac{\partial \Psi_s}{\partial V_g}$ values. Extracted values are summarized in table 1. The R.T.S traps shift toward the Si/SiO$_2$ interface for a decrease of the oxidation temperature.

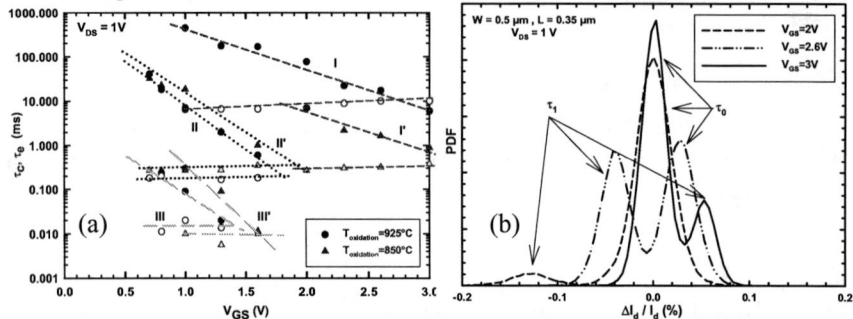

FIGURE 3. (a) Capture (filled symbol) and emission times (open symbol) evolutions of each traps for an oxidation temperature of 850°C and 925°C. (b) Evolution of the Probability Density Function (PDF) showing the high time level current evolution at different gate biases.

TABLE 1. Gate voltage giving the maximum trap activity and traps depth in the oxide (x_t) with respect to the Si/SiO$_2$ interface for $T_{oxidation}$= 925°C and 850°C

Temperature	925°C			850°C		
Trap	I	II	III	I'	II'	III'
Max of activity	2.7 V	1.8 V	1.4 V	> 3 V	1.9 V	1.6V
x_t (nm)	0.43	0.34	0.25	0.34	0.28	0.23

CONCLUSION

The process flow influences on R.T.S sources in advanced CMOS technologies are of prime importance. Our R.T.S measurements performed on N-MOSFETs show greater reduction in the R.T.S magnitude. Without changing the main trap characteristics a shift toward higher gate voltage of the maximum of activity with an oxidation temperature decrease from 925°C to 850°C is observed. The same conclusion can be done as regard to the trap occupancy probability function. For a given bias, a modification of the oxidation temperature provides an evolution of the R.T.S source influences. In this point of view the temperature effect is quite attractive to improve some analog and digital applications.

REFERENCES

1. M.H.Tsai, T.P.MA and T.B.Hook, "Channel Length Dependence of Random Telegraph Signal in Sub-Micron MOSFETs ", IEEE Electron device letters, 15 n°12, pp 504-506,1994
2. T.Boutchacha and G.Ghibaudo, «Electrical noise and RTS fluctuations in advanced CMOS devices», , Microelectronics Reliability, 42, pp 573-582, 2002
3. C. Leyris, A. Hoffmann, M. Valenza, J-C. Vildeuil and F. Roy, "Trap competition inducing R.T.S. noise in saturation range in N-MOSFETs", 3th Conf. Fluctuations and Noise, Texas May 24-26 2005.
4. E.Simoen and C.Claeys, "On the flicker noise in submicron silicon MOSFETs", Solid State Electronics, 43, pp 865-882, 1999

Zero Cross Analysis of RTS Noise

Jan Pavelka[*,¶], Munecazu Tacano[*], Masato Toita[†],
Josef Sikula[¶] and Toshimitsu Musha[#]

[*]AMRC, Meisei University, 2-1-1 Hodokubo, Hino, Tokyo 191-8506, Japan
[†]Asahi Kasei Microsystems, Nobeoka 882-0031, Japan
[¶]FEEC, Brno University of Technology, Technicka 8, Brno 61600, Czech Republic
[#]Brain Functions Lab., KSP, E211, 3-2-1 Sakado, Takatsu-ku, Kawasaki 213-0012, Japan

Abstract. RTS noise of Si MOSFETs and GaN HFET was analysed by means of zero cross method. Noise spectral density of crossing events in 1ms to 100s windows was flat over 10^{-5} to 10^{3} frequency range without apparent 1/f noise. The fluctuation of crossing rate was same in every sample, although other noise components next to RTS noise are quite different in Si and GaN devices. Correlation analysis of pulse length in doublets and triplets of successive pulses didn't reveal any mutual dependence.

Keywords: RTS noise, MOSFET, zero cross
PACS: 72.70.+m, 73.40.Qw, 73.50.Td, 85.30.Tv

INTRODUCTION AND EXPERIMENTAL RESULTS

In small area devices such as submicron MOSFETs, a single defect may be present, which in time domain gives two level switching signals known as random telegraph signal (RTS) noise. An example of RTS noise voltage time dependence is given in Fig.1, measured on 0.35μm Si MOSFET and 2μm GaN/AlGaN HFET. These two levels in drain current are attributed to trapping/detrapping events caused by individual defect in the oxide near the Si-SiO$_2$ interface. The carrier transitions between the oxide trap and channel are governed by the Shockley-Read-Hall (SRH) statistics, which defines an exponential distribution of capture t_c and emission t_e times [1], [2]. When measured over sufficiently long time interval with 15-20 million pulses recorded, in some cases we observed deviation from this statistics (Fig.2b), as well as anomalous capture rate dependence on drain current, and suggested an enhanced two-step capture process model [3]. In the frequency domain, noise spectral density of RTS was always Loretzian (Fig.3a).

In this paper, we discuss another approach based on zero cross analysis. We divided time sequence of measured RTS voltage into series of intervals of constant width w in the range of 1ms to 100s and examined number of pulses corresponding to level up CU or down CD crossing events in each window. This quantity has Gaussian distribution with the dispersion σ inversely proportional to the square root of the window width (Fig.3b,4a). It is interesting to note, that even when the other noise sources next to the RTS noise give different background noise intensity according to the temperature or type of Si or GaN device, as can be seen in Fig.1., the fluctuation of the crossing rate CU characterised by the value of dispersion σ is always same given by the Poisson statistics as the square root of the mean value of number of pulses μ (Fig.4b). The shape of distribution given by the ratio of σ to μ is obviously different in every sample for particular window width because of the various RTS time constant values in the range over 5 decades.

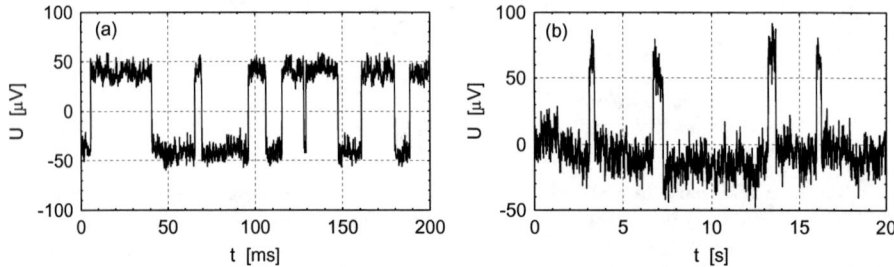

FIGURE 1. Time dependence of RTS voltage on load resistance (a) Si MOSFET, $T=221$K, $R_L=10$kO, $U_G=0.70$V, $U_{DS}=1.27$V, $I_{DS}=1.0\mu$A, (b) GaN HFET, $T=300$K, $R_L=1$kO, $U_G=0$V, $U_{DS}=0.52$V, $I_{DS}=5.8\mu$A.

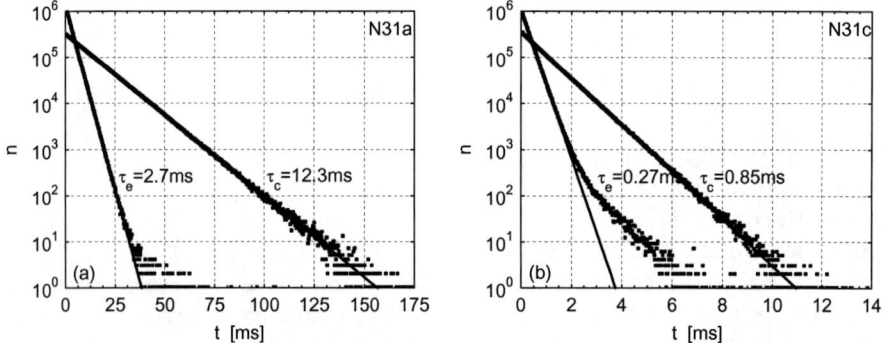

FIGURE 2. Histogram of carrier capture and emission events duration in Si MOSFET, sample N31. (a) $T=234$K, $U_G=0.68$V, $U_{DS}=1.26$V, $I_{DS}=0.83\mu$A, (b) $T=257$K, $U_G=0.70$V, $U_{DS}=1.27$V, $I_{DS}=2.6\mu$A

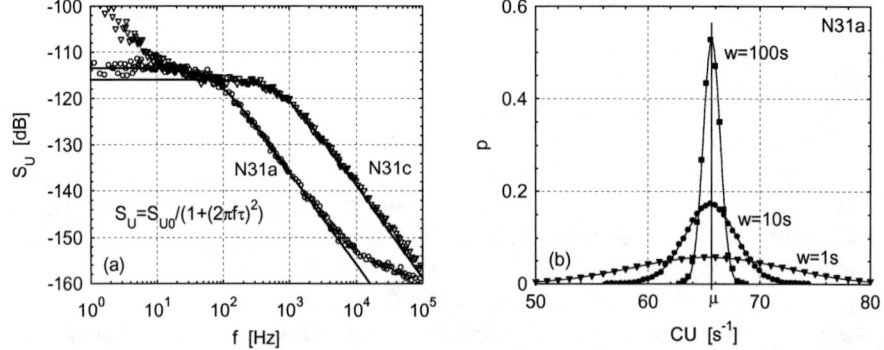

FIGURE 3. (a) Noise spectral density of Si MOSFET N31, biased like in Fig.2. (b) Normalised distribution of crossing up events in windows width $w=1$s, 10s and 100s, Si MOSFET N31 at $T=234$K

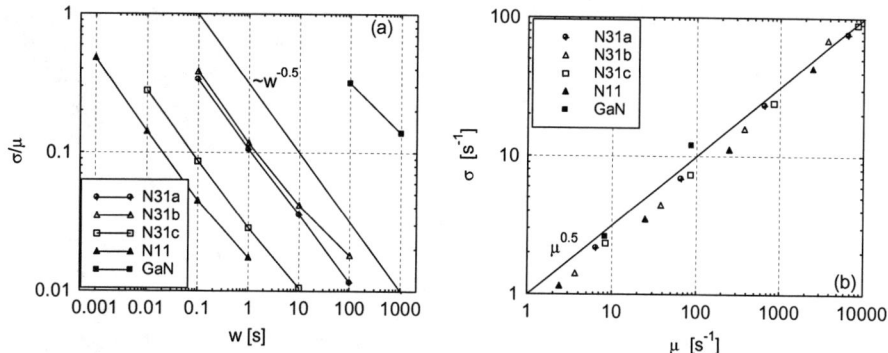

FIGURE 4. Dispersion of crossing rate Gaussian distribution σ as a function of window width w (a) or mean value μ (b). Si MOSFETs samples N31 and N11 and GaN HFET. N31a $T=234$K $U_G=0.68$V, N31b $T=221$K $U_G=0.70$V, N31c $T=257$K $U_G=0.70$V, N11 $T=235$K $U_G=0.695$V, GaN $T=300$K $U_G=0$V

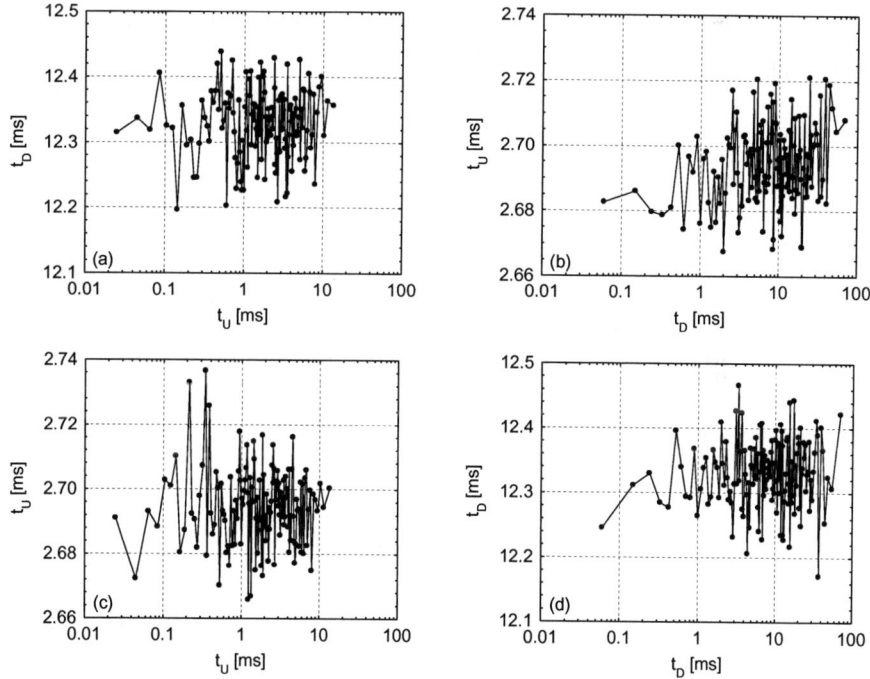

FIGURE 5. Correlation of the successive pulse lengths, Si MOSFET sample N31, $T=234$K, $U_G=0.68$V. (a) lower level duration after upper level, (b) upper after lower level, (c) upper level pulse length and next upper level, (d) lower level and following lower level

The correlation analysis of pulse length in doublets and triplets of successive pulses revealed, that they are completely independent (Fig.5). The FFT analysis of the sequences of crossing up events CU or total time spent on the upper level TU within the series of windows of the duration w gave almost constant noise spectral density in the range from 1kHz down to 10µHz (Fig.5). Although the initial parts of the curves corresponding to the lowest frequencies are elevated in some cases, this is probably mostly effect of the uncertainty of FFT processing in first few coefficients, since it is apparent also in overlapping curves at higher frequency corresponding to shorter window width (Fig.5 MOS N31 at 234K, three points deleted in further curves). There is also some influence of temperature (±0.3K) and gate voltage (±0.5mV) fluctuation.

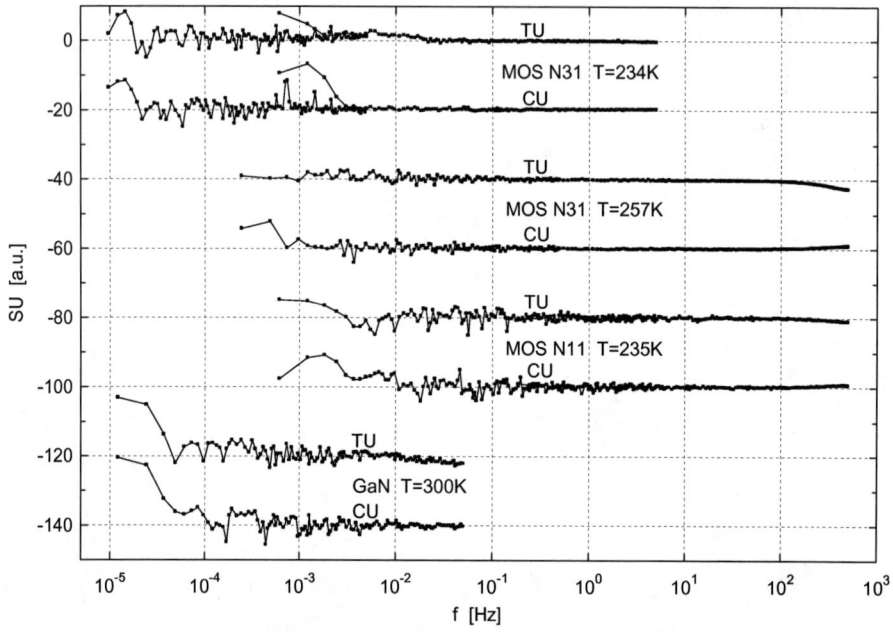

FIGURE 6. Noise spectral density of the signal given as a series of number of crossing up events CU or time spent on the upper level TU within successive windows of width w. Si MOSFETs N31 and N11 and GaN HFET. N31 T=234K, U_G=0.68V, U_{DS}=1.26V, I_{DS}=0.83μA, w=100s and 100ms. N31 T=257K, U_G=0.70V, U_{DS}=1.27V, I_{DS}=2.6μA, w=1s and 1ms. N11 T=235K U_G=0.695V, U_{DS}=1.21V, I_{DS}=1.5μA, w=100ms and 1ms. GaN T=300K U_G=0V, U_{DS}=0.52V, I_{DS}=5.8μA, w=10s

ACKNOWLEDGMENTS

This work was supported by the grant GACR 102/05/2095 and by the project MSM 0021630503. One of the authors (J.P.) gratefully acknowledges the JSPS fellowship.

REFERENCES

1. Kirton, M.J., and Uren, M.J., *Adv. Phys.* **38**, 367-468 (1989).
2. Celik-Butler, Z., and Amarasinghe, N.V., "Random Telegraph Signals in Deep Submicron Metal-Oxide-Semiconductor Field-Effect Transistors," in *Noise and Fluctuations Control in Electronic Devices,* edited by A. A. Balandin, American Scientific Publishers, 2002, pp. 187-199.
3. Sikula, J., Pavelka, J., Sedlakova, V., Tacano, M., Hashiguchi, S., and Toita, M., "RTS in submicron MOSFETs and quantum dots", Proc. of 2nd SPIE Symposium Fluctuation and Noise, Maspalomas, Spain, 2004, pp.64-73.

High Magnetic Field Dependence of Capture/Emission Fluctuations of a Single Defect in Silicon MOSFETs

Enrico Prati[1], Marco Fanciulli[1], Giorgio Ferrari[2], Marco Sampietro[2]

1. Laboratorio Nazionale Materiali e Dispositivi per la Microelettronica, Istituto Nazionale per la Fisica della Materia, Via Olivetti 2, I-20041 Agrate Brianza, Italy

2. Dipartimento di Elettronica e Informazione, Politecnico di Milano, P.za Leonardo da Vinci 32, I-20133 Milano, Italy

Abstract. The random telegraph signal (RTS) due to a defect at the Si/SiO_2 interface in a silicon MOSFET is strongly affected by a static magnetic field. The characteristic capture and emission times vary because of the Zeeman splitting of the defect energy levels. We observe the change of the characteristic times of the RTS by monitoring the dc current in a MOSFET operated at He3 temperature in a static magnetic field up to 12 T parallel to the current flowing from source to drain.

Keywords: Interface defects, high magnetic field, 2DES, random telegraph signal.
PACS: 82.20.Xr, 73.50.Td, 72.25.Dc

INTRODUCTION

Single spin read out is one of the most challenging issue in the physical implementation of spin based qubits [1,2]. Among the several proposed schemes Vrijen et al. [3] suggested to implement a practical field-effect electron spin resonance transistor (ESRT) to demonstrate qubits operation in silicon-germanium heterostructures. In this scheme the random telegraph signal (RTS) is proposed as a mean to perform qubit read-out. Intense magnetic fields and low temperatures, increasing the polarization of the spin system, enhance the effect of the resonance condition. We therefore investigated the random telegraph signal in MOSFETs operated in a static magnetic field up to 12 T and at temperatures below 2 K. The RTS in a MOSFET consists in the random switching of the current between two states. For a Si/SiO_2 MOSFET an electron from the channel can tunnel on and off an interface defect, provided the Fermi level of the conducting channel is in the proximity of the defect energy level [4-8]. Several models [9,10] have been proposed to explain the change of the ratio between the characteristic times (emission and capture) as a function of a static magnetic field without or with additional time dependent electromagnetic fields in the microwave range. The effect of the microwave irradiation is discussed in another contribution [11]. Xiao *et al.* [9] reached a maximum field of 7 T and proposed a model which was found to be inconsistent with the data at low temperatures. More recently the observation, by monitoring the RTS in a MOSFET, of the electron spin resonance due to a single defect in high magnetic fields has been claimed [12].

In this work we present the experimental data of the changes of the RTS when a static magnetic field parallel to the two dimensional electron gas in the channel of a silicon based MOSFET. We demonstrated experimentally that the change in the RTS characteristic times in a n-MOSFET due to static magnetic fields up to 12 T and temperatures in the range from hundreds of mK to 2 K, can reach three orders of magnitude when the field is applied parallel to the electron gas. The observed effect is much stronger than what previously reported by Xiao et al. [9] and in better agreement with the theoretical model presented in that paper.

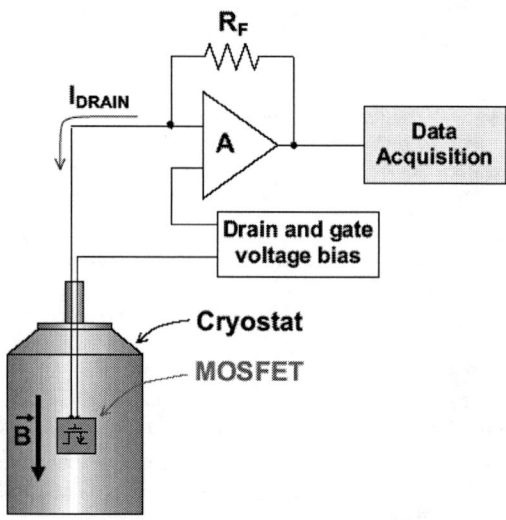

FIGURE 1. Schematic of the set-up used for the characterization of the RTS in a cryo-magnet. Nominal temperature 0.250 K.

EXPERIMENTAL SETUP AND MEASUREMENTS

The experimental setup is depicted in Figure 1. The devices under investigation are n-channel MOSFETs realized on a p-well, with a length of 0.18 μm and a width of 0.28 μm, a 3.5 nm thick gate oxide, and a threshold voltage V_T variable between 200 mV and 500 mV. The source, drain, gate, and well contacts were directly accessible through the bonding pads and connected to wires to exit the cryo-magnetic system. The current flowing through drain and source I_{DS} is measured by a transimpedance amplifier whose output is sampled and digitized for off-line processing. The bandwidth of the amplifier extends from DC to 100 kHz. The transistor and the electronic are powered by independent batteries to avoid power-line pick-up and interferences. The sample was cooled by a copper coldfinger into the He3 cryostat placed into a 12 T superconducting magnet.

The conduction regime is hopping transport. The Figure 2 show an I-V curve at a nominal temperature of 250 mK. The heating effects of the current have to be considered to evaluate the real temperature of the electron gas.

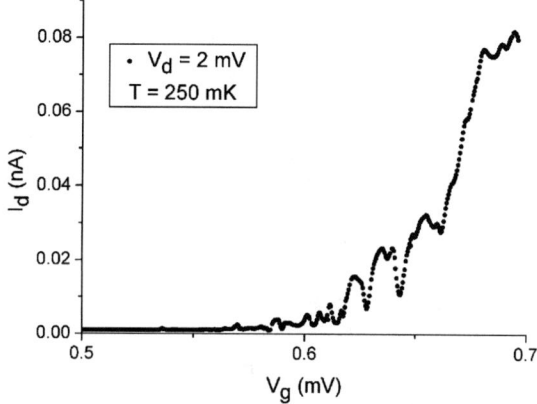

FIGURE 2. I-V curve of the MOSFET at a nominal temperature of 250 mK. The spikes below the threshold voltage are due to transport via quantum state tunnelling. During our measurements the transport was due to finite temperature hopping.

The sample was examined with the channel in plane with the external static magnetic field B, oriented in the direction perpendicular to the electron current flow.

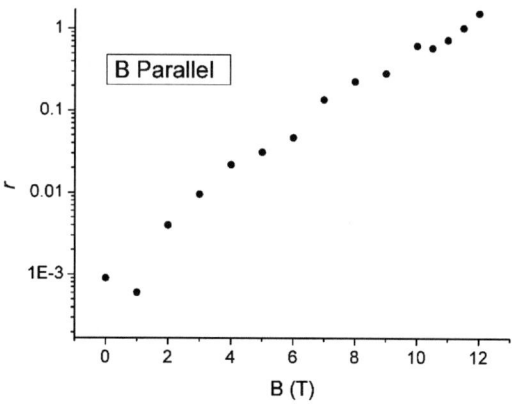

FIGURE 3. Ratio r between the capture and emission times of the trap versus the magnetic field B. The values scale three order of magnitude, twice than what previously reported [9].

The Figure 3 represents the ratio r between the high and the low states of the current:

$$r = \frac{\tau_{high}}{\tau_{low}} \quad (1)$$

at V_g = 560.7 mV and V_d = 15.7 mV, with a current I_d = 37.3 nA. Since the low state occupation increased by increasing the gate voltage (so that the Fermi energy is raised with respect of the trap energy), we deduce that the trap was filled when the current was in the low state.

Despite the better agreement of our data with the model proposed by Xiao et al. [9], a more detailed analytical study is required to fit such experimental data with reasonable parameters.

ACKNOWLEDGMENTS

The authors would like to thank Paolo Fantini and STMicroelectronics for providing the samples. The authors are grateful to Vittorio Pellegrini, Vincenzo Piazza, Pasqualantonio Pingue for the cryomagnet time at the Laboratorio Polvani of NEST-INFM and SNS, Pisa, Italy.

REFERENCES

1. B. E. Kane, *Nature* 393, 133 (1998)
2. D. Loss and D. P. DiVincenzo, *Phys. Rev. A* 57, 120 (1998)
3. R. Vrijen et al., *Phys. Rev. A* 62, 12306 (2000)
4. K. S. Ralls, W. J. Skocpol, L. D. Jackel, R. E. Howard, L. A. Fetter, R. W. Epworth, and D. M. Tennant *Phys. Rev. Lett.* 52, 228 (1984)
5. M.J. Uren, D. J. Day, and M. J. Kirton, *Appl. Phys. Lett.* 47, 1195 (1985)
6. Amarasinghe NV, Celik-Butler Z, Keshavarz A., *J. Appl. Phys.* 89, 5526 (2001)
7. E. Simoen, C. Claeys, *Mat. Sci. and Eng. B* 91, 136 (2002)
8. K. Kandiah, M.O. Deighton, and F.B. Whiting, *J. Appl. Phys.* 66, 93 (1989)
9. M. Xiao, I. Martin, and H. W. Jiang, *Phys. Rev. Lett.* 91, 7, 078301 (2003)
10. I. Martin, D. Mozyrsky, and H.W. Jiang, *Phys. Rev. Lett.* 90, 1, 018301 (2003)
11. E. Prati, G. Ferrari, M. Sampietro, M. Fanciulli and P. Fantini, These Proceedings
12. M. Xiao, I. Martin, E. Yablonovitch, and H. W. Jiang, *Nature*, 430, 435 (2004)

Noise in Si and SiGe MOSFETs with High-k Gate Dielectrics

Martin von Haartman, B. Gunnar Malm, Per-Erik Hellström, and Mikael Östling

KTH (Royal Institute of Technology), Department of Microelectronics and Information Technology, Electrum 229, SE-164 40 Kista, Sweden. Email: mvh@imit.kth.se

Abstract. This paper presents an overview of previous work and new insights on noise in Si-based MOSFETs with high-k gate dielectrics. Results for Al_2O_3, HfO_2, $HfAlO_x$ and composite structures of these materials will be reported and compared. Incorporation of strained SiGe in high-k pMOSFETs in order to enhance hole mobility will be discussed in terms of low-frequency noise. A comparison will be made between devices with a surface Si channel, a surface SiGe channel and a buried SiGe channel. The influence of the gate electrode material and presence of a thin interfacial layer will be investigated. We will discuss noise modeling and highlight important differences compared to CMOS devices with standard gate oxide. Finally, we will discuss possible ways to reduce the $1/f$ noise in high-k MOSFETs. A noise reduction by a factor of two is obtained by forward biasing the substrate.

Keywords: $1/f$ noise, low-frequency noise, high-k dielectrics, MOSFETs, metal-gate, SiGe
PACS: 85.30.Tv; 73.50.Td; 73.61.Ng; 73.40.Qv; 73.40.Ty

INTRODUCTION

Continued downscaling of complementary metal-oxide-semiconductor (CMOS) devices beyond the 65 nm technology node requires, among many new technology features, high-k gate dielectrics to achieve small equivalent oxide thickness (EOT) while maintaining low gate leakage current [1]. Intensive research has been devoted to find and optimize high-k dielectric materials for integration with CMOS technology [2],[3]. Low-frequency noise is a powerful tool to study the quality of the gate dielectric and its interface with the underlying substrate. Most reported results so far on high-k indicate increased noise levels, roughly one order of magnitude, compared to standard CMOS transistors with SiO_2 as gate dielectrics [4]-[20]. It has been shown that the major source of the $1/f$ noise is traps in, or originating from, the high-k layer, but in some cases Hooge mobility fluctuations can give a dominant contribution [7],[16],[20]. Traps in the high-k material, located in its bulk or at interfaces within the gate stack, several nm from the Si-substrate can contribute to the $1/f$ noise [12],[15],[18]. This agrees with the observation of instabilities in the threshold voltage, which has been explained by charging and discharging of traps in the high-k material by tunneling [21]. From noise characterizations, trap densities N_t for the high-k materials in the range 1×10^{18} - 1×10^{20} $cm^{-3}eV^{-1}$ have been extracted. These results are summarized in Fig. 1 together with reported values of the Hooge parameter α_H. Extracted N_ts for nitrided SiO_2 range from 1×10^{16} - 1×10^{18} $cm^{-3}eV^{-1}$ [22]. Trap-density profiles in HfO_2 and Al_2O_3 gate dielectrics derived from various charge-pumping schemes are consistent with the results in Fig. 1 [23],[24]. It has also been reported that the trap densities in SiO_2 increase when high-k materials is deposited on top

[13],[15]. Most values at large EOT in Fig. 1 are for devices with a thick layer of SiO_2 between the high-k layer and the substrate.

In this paper, we will present noise results for Si and SiGe MOSFETs with high-k gate dielectrics. Fabrication and electrical characteristics of the devices used in this study were reported previously in [17],[25],[26] and some $1/f$ noise results was reported in [7],[16],[17],[20].

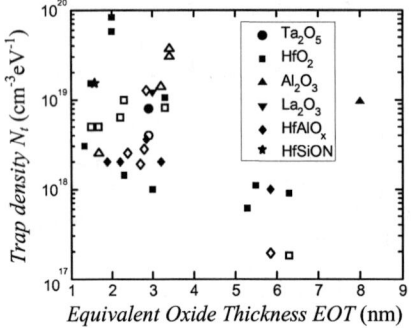 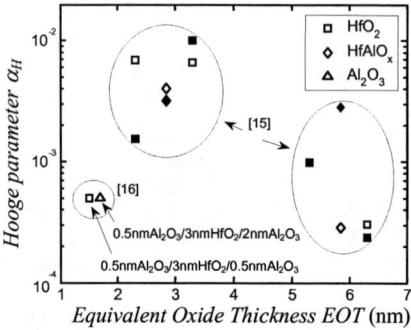

FIGURE 1. A summary of reported trap densites in [3]-[20] plotted versus EOT. Filled symbols denote nMOS, open symbols pMOS.

FIGURE 2. A summary of reported values of α_H plotted versus EOT. Filled symbols denote nMOS and open symbols pMOS, respectively.

LOW-FREQUENCY NOISE RESULTS

On-wafer low-frequency noise measurements were conducted from subthreshold to strong inversion conditions at a constant $V_{DS} = -50$ mV on various pMOSFETs with high-k gate dielectrics. The noise was of the $1/f^\gamma$-type for several decades with γ between 0.9-1.2 for all devices in this study, if not mentioned otherwise. The measured drain current noise S_{ID} was referred to input gate voltage noise S_{VG} by $S_{ID} = S_{VG}/g_m^2$.

Gate Dielectric Material

Figure 3 displays S_{VG} at 10 Hz versus I_{DS}/g_m for pMOSFET with 4-nm Al_2O_3, HfO_2 and $HfAlO_x$ as gate dielectrics, deposited using Atomic Layer Deposition (ALD). The devices have a TiN gate and 0.5-nm Al_2O_3 on both sides of the 4-nm thick high-k layer. Multiplication of S_{VG} with C_{ox}^2 is for normalization purposes in order to compare the different high-k materials. At lower bias, $HfAlO_x$ gives lowest $1/f$ noise, whereas the difference between the high-k materials is small at high bias. Except for the Al_2O_3 devices, the noise level for the 10-μm devices and the 1-μm devices roughly differ by a factor of 10 at all biases. This is in agreement with theoretical predictions for noise sources homogenously distributed under the gate, and also implies that the noise from the source and drain resistance gives a negligible contribution to the measured drain current noise. The reason why the noise level in the Al_2O_3 device does not scale with the gate length at lower bias could be process induced gate edge damage, a problem sometimes observed with high-k gate dielectrics [27].

Using a compressively strained surface SiGe channel in combination with high-k gate dielectrics is advantageous thanks to the higher hole mobility [17],[28]. Although the density of interface states is around 7 times higher in the SiGe devices (4×10^{12} cm^{-2}eV^{-1} compared to 6×10^{11} cm^{-2}eV^{-1}) the $1/f$ noise is not degraded. The SiGe devices show lower $C_{ox}^2 S_{VG}$ by a factor of 2-4 at high gate voltage overdrives, as seen in Fig. 3. Figure 4 show $L \times S_{VG}$ vs I_{DS} for pMOSFETs with HfO$_2$ gate dielectrics. These devices have a poly-Si (SINANO) or poly-SiGe gate (KTH). As observed, the surface SiGe-channel yields some improvement in the noise level compared to Si also in the poly-SiGe gate devices. A buried SiGe channel is often employed to lower the $1/f$ noise in pMOSFETs [29]. However, from our first attempt utilizing a buried SiGe channel in pMOSFETs with HfO$_2$ gate dielectrics (SINANO) no significant reduction in noise level compared to Si could be observed. According to [30], noise originating from trapping/release phenomena in the gate dielectrics need not be reduced although most of the current flows in the buried SiGe channel. If the trap densities in the gate dielectrics are similar one could also expect similar noise levels. Note that most of the SINANO devices showed $1/f^\gamma$-type noise with γ around 0.7 below ~1 kHz, and that high gate leakages were observed. Thus, noise generated in the gate current need to be considered, especially at high gate bias [22].

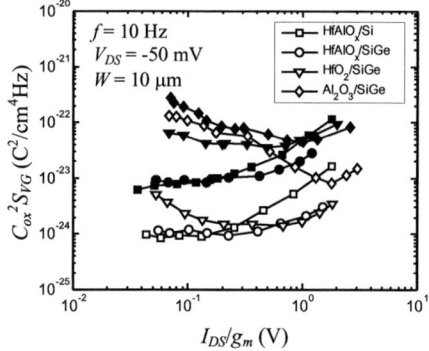 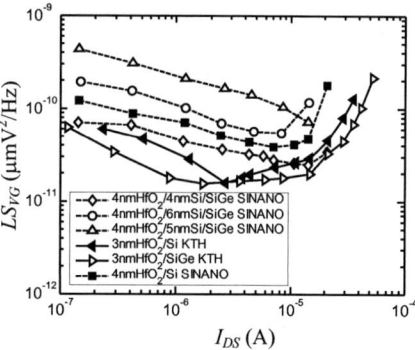

FIGURE 3. Normalized equivalent input gate voltage noise as a function of I_{DS}/g_m for $L = 1$ μm (filled) and $L = 10$ μm (open symbols) pMOSFETs.

FIGURE 4. Normalized equivalent input gate voltage noise as a function of I_{DS} for $L = 1$ μm (SINANO) and $L = 0.9$ μm (KTH) pMOSFETs. $W = 10$ μm, $f = 10$ Hz, $V_{DS} = -50$ mV.

Gate Electrode Material

In Fig. 5 a comparison is made between HfO$_2$ gate dielectric pMOSFETs with a metal gate (TiN) and a poly-SiGe gate. The normalized drain current noise is at the same level for all the HfO$_2$ gate dielectric pMOSFETs at lower bias, whereas the TiN gate devices shows significantly reduced noise in strong inversion. It has been shown that the TiN metal gate electrode is effective in reducing the mobility degradation due to surface optical phonon scattering in the high-k dielectric [31]. Indeed, the hole mobility in the channel is higher for TiN gate compared to poly-SiGe gate, as shown in Fig. 6. The lowering of the noise in our TiN gate MOSFETs may have the same origin, since it has been suggested that only phonon scattering generates mobility fluctuation noise [32]. The $1/f$ noise may also be interpreted with the combined number and mobility fluctuation noise model. In the frame of this model, the reduced noise in strong inversion for the TiN gate devices is explained by a lower value of the Coulomb scattering parameter α.

The interface between the channel and the high-k were prepared differently for the three poly-SiGe gate pMOSFETs labeled A1, B and C in Fig. 5. A 0.5-nm thick ALD Al_2O_3 layer was deposited prior to 3-nm ALD HfO_2 layer for devices A and B, whereas device C has a 2-nm thick Al_2O_3 layer between the HfO_2 and SiGe channel. Moreover, the surface was water-rinsed for device A prior to ALD resulting in a 0.6-0.7-nm thick interfacial layer composed of Al_2O_3 and SiO_2. In contrast, for device B, which was not subjected to water-rinsing, the bottom Al_2O_3 layer was not clearly observable. Device A1 and A2 are from different batches, but were otherwise processed identically. As seen in Fig. 5, the noise properties are very reproducible. For the TiN gate devices a ~1-nm thick interfacial $Si(Ge)O_2$ layer was found to be present. The thicker interfacial layer can play a role in reducing the noise in the TiN gate device. But the different interface preparations for devices A-C do not influence the noise level significantly; therefore we conclude that this is not the main reason. The $1/f$ noise in relation to the interface properties was discussed in more detail in [16].

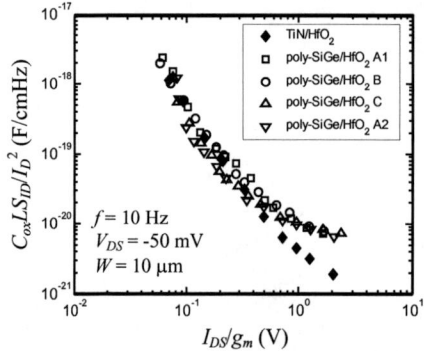
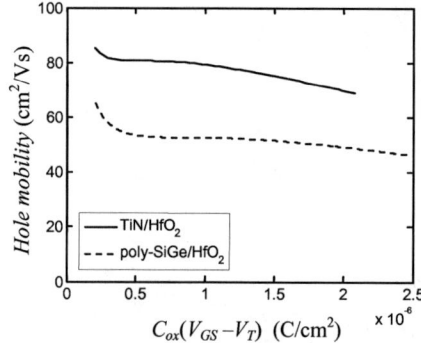

FIGURE 5. Normalized drain current noise versus I_{DS}/g_m for $L = 0.8$-1 μm pMOSFETs.

FIGURE 6. Hole mobility for SiGe surface channel pMOSFETs with poly-SiGe and TiN gate.

Noise Modeling of High-k Gate Dielectric pMOSFETs

The Hooge mobility fluctuations ($\Delta\mu_H$) and the unified noise model (combined number and mobility fluctuations, Δn-$\Delta\mu$) have been considered for modeling of the $1/f$ noise data presented in this work. However, neither of the two models can successfully explain the noise in all our devices at all biases. For the unified noise model, $S_{VG,\Delta n-\Delta\mu} \propto (1 \pm \mu_0 \sqrt{N}/\mu_{C0})^2$, where N is the inversion carrier density, μ_0 is the low-field mobility and μ_{C0} is a Coulomb scattering parameter. The Hooge mobility fluctuation noise model gives $S_{VG,\Delta\mu_H} \propto 1/N \times (I_{DS}/g_m)^2$, and mainly fails below threshold were a flat or weakly bias dependent S_{VG}-curve often is observed (see Fig. 3). However, many devices show a U-shaped S_{VG} curve (see Al_2O_3 and HfO_2 gate dielectric devices in Fig. 3 and 4), which cannot be explained by the Δn-$\Delta\mu$ formalism alone, the curves are often expected to turn up at considerable higher gate voltage overdrives. It is interesting to note that irrespective of the noise level below threshold all devices with the same gate length in Fig. 3 seem to approach the same noise level in strong inversion. Moreover, many of the poly-SiGe gate devices can be well explained only with Hooge mobility fluctuations [16]. Note that an U-shaped S_{VG} curve in the Δn-$\Delta\mu$ model implies that the mobility and number fluctuations are negatively correlated, whereas a positive correlation are almost always reported for SiO_2 gate dielectric MOSFETs.

The Coulomb scattering and correlated mobility fluctuations are further discussed in [20]. The Coulomb scattering parameters (μ_{C0} and α) were extracted using two different methods with the results in good agreement. A maximum α of around 1×10^4 Vs/C was observed in accordance with [33].

Our conclusion is that Δn-type of noise mainly contributes in weak inversion, but decrease rapidly in strong inversion due to increasing N ($S_{ID} \sim 1/N^2$ in strong inversion) and mobility fluctuations negatively correlated to Δn. Hooge mobility fluctuations are important in strong inversion, but may weaken below threshold. The reason for the weakening could be related to the location of the inversion carriers further away from the interface and/or higher influence from Coulomb scattering that can dilute the mobility fluctuation noise. Fig. 7 shows S_{VG} versus I_{DS} for a TiN/HfAlO$_x$/Si device at forward and reverse substrate bias. As seen, the $1/f$ noise decreases in strong inversion when forward biasing the substrate. It is well known that buried channel conduction can give decreased noise. The inversion carriers are located further away from the interface for forward substrate bias leading to reduced $1/f$ noise. Fig. 8 shows S_{VG} at $I_{DS} = 10$ μA versus the average distance of the inversion carriers from the gate dielectric interface, obtained from numerical simulations solving the coupled Poisson's and Schrödinger's equations self-consistently. Note that the correlated mobility fluctuations also depend on the location of the inversion charge [34].

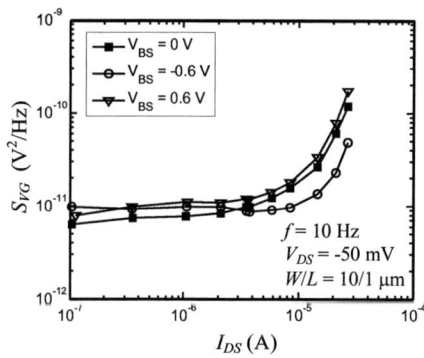 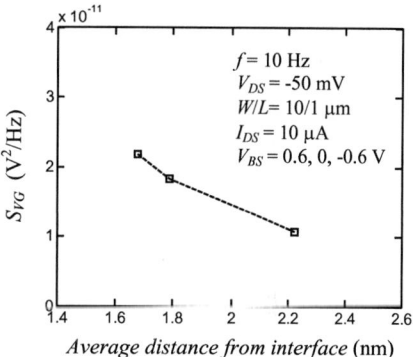

FIGURE 7. Equivalent input gate voltage noise for a $L = 1$ μm TiN/HfAlO$_x$/Si pMOSFET at different substrate biases.

FIGURE 8. S_{VG} at $I_{DS}=10$μA for a TiN/HfAlO$_x$/Si pMOSFET versus the average depth of the inversion carriers under the high-k/Si interface.

CONCLUSION

MOSFETs with high-k gate dielectric show increased noise compared to transistors using SiO$_2$ due to large defect densities originating from the high-k materials. To reduce the $1/f$ noise, TiN metal gate was shown to be advantageous compared to poly-Si, and HfAlO$_x$ often shows lower trap densities N_t than HfO$_2$ and Al$_2$O$_3$ (see Fig. 1). Buried channel devices have been successful lowering the noise in the past. Our results here show that this is not always the case, but the reduced noise at forward substrate bias observed in the present work is encouraging. Lower trap densities have been reported for materials such as HfSi$_x$O$_y$ [23],[35],[36] and Ta incorporated HfO$_2$ [37], possibly they exhibit lower noise as well although initial results for HfSiON indicated higher $1/f$ noise than for HfAlO$_x$ [19]. In conclusion, based on our results the noise properties of high-k gate dielectrics seem promising for future generations of MOSFETs, yet some problems remains to be solved.

ACKNOWLEDGMENTS

The authors would like to thank D. Wu (previously KTH, now Infineon Technologies AG), S.-L. Zhang, and H. H. Radamson for their contribution to this work. The SINANO WP1 sub-project, directed by T. E. Whall, and its participants are greatly acknowledged for providing some samples for this study. Financial support was received from Swedish Foundation for Strategic Research (SSF) through the "high-frequency silicon" program.

REFERENCES

1. International Technology Roadmap for Semiconductors (ITRS), 2003 update.
2. G. D. Wilk et al., *J. Appl. Phys.* **89**, 5243-5275 (2001).
3. E. P. Gusev et al., in *IEDM Technical Digest*, 2001, pp. 451-454.
4. M. Fadlallah et al., *Microelectron. Reliab.* **41**, 1361-1366 (2002).
5. B. Guillaumot et al., in *IEDM Technical Digest*, 2002, pp. 355-358.
6. C. Claeys et al., in *Int. Conf. on Noise and Fluctuations (ICNF)*, 2003, pp. 215-20.
7. M. von Haartman et al., in *Int. Conf. on Noise and Fluctuations (ICNF)*, 2003, pp. 381-384.
8. T. Ishikawa et al., in *Ext. Abstr. Int. Conf. on Solid State Devices and Materials (SSDM)*, 2003, pp. 14-15.
9. E. Simoen et al., in *Physics and Technology of high-k Gate Dielectrics*, Electrochem. Soc. Ser. PV 2003-22, 2003, pp. 319-331.
10. H. Sauddin et al., in *Physics and Technology of high-k Gate Dielectrics*, Electrochem. Soc. Ser. PV 2003-22, 2003, pp. 415-423.
11. H. D. Xiong et al., *Appl. Phys. Lett.* **83, pp.** 5232-5234 (2003).
12. E. Simoen et al., *IEEE Trans. Electron Devices* **51**, pp. 780-784 (2004).
13. T. Horikawa et al., in *Advanced Short-Time Thermal Processing for Si-Based CMOS Devices II*, Electrochem. Soc. Ser. PV 2004-1, 2004, pp. 292-303.
14. E. Simoen et al., *Appl. Phys. Lett.* **85**, pp. 1057-1059 (2004).
15. B. Min et al., *IEEE Trans. Electron Devices* **51**, pp. 1679-1687 (2004).
16. M. von Haartman et al., *Solid-State Electron.* **48**, pp. 2271-2275 (2004).
17. D. Wu et al., *Microelectron. Eng.* **77**, pp. 36-41 (2005).
18. E. Simoen et al., *Solid-State Electron.* **49**, pp. 702-707 (2005).
19. B. Min et al., *Appl. Phys. Lett.* **86**, 082102 (2005).
20. M. von Haartman et al., *Solid-State Electron.* in press (2005).
21. A. Kerber et al., *IEEE Electron Device Lett.*, **24**, pp. 87-89 (2003).
22. M. Valenza et al., *IEE Proc-Circuits Devices Syst.* **151**, pp. 102-110 (2004).
23. C. Leroux et al., in *IEDM Technical Digest*, 2004, pp. 737-740.
24. S. Jakschik et al., *IEEE Trans. Electron Devices* **51**, pp. 2252-2255 (2004).
25. D. Wu et al., *IEEE Electron Device Lett.*, **24**, pp. 171-173 (2003).
26. D. Wu et al., *IEEE Electron Device Lett.*, **25**, pp. 289-291 (2004).
27. H.-H. Tseng et al., in *IEDM Technical Digest*, 2004, pp. 821-824.
28. O. Weber et al., in *Symp. VLSI Tech.*, 2004, pp. 42-43.
29. S. Okhonin et al., *IEEE Trans. Electron Devices* **46**, pp. 1514-1517 (1999).
30. G. Ghibaudo et al., *Solid-State Electron.* **46**, pp. 393-398 (2002).
31. R. Chau et al., *IEEE Electron Device Lett.* **25**, pp. 408-410 (2004).
32. F. N. Hooge, *IEEE Trans. Electron Devices* **41**, pp. 1926-1935 (1994).
33. E. P. Vandamme et al., *IEEE Trans. Electron Devices* **47**, pp. 2146-2152 (2000).
34. A. K. M. Ahsan, et al., *Solid-State Electron.* **49**, pp. 654-662 (2005).
35. E. P. Gusev et al., in *IEDM Technical Digest*, 2004, pp. 79-82.
36. J. Robertson, *Solid-State Electron.* **49**, pp. 283-293 (2005).
37. X. Yu et al., *IEEE Electron Device Lett.* **25** pp. 501-503 (2004).

Impact of Gate Material on Low-frequency Noise of nMOSFETs with 1.5 nm SiON Gate Dielectric: Testing the Limits of the Number Fluctuations Theory

P. Srinivasan[*,#], E. Simoen[*], L. Pantisano[*], C. Claeys[*,∀], and D. Misra[#]

* - IMEC, Kapeldreef, Leuven, Belgium.
\# - Dept. of Electrical Engineering, NJIT, Newark, NJ.
∀ also at Dept. of Electrical Engineering, KU Leuven, Belgium.

Abstract. It is shown that the gate material has a strong impact on the low-frequency (LF) 1/f noise of silicon nMOSFETs with a 1.5 nm SiON gate dielectric. Highest noise is observed for transistors with an n-type polysilicon gate, compared with their counterparts having a metal (TaN) or a fully nickel-silicided polysilicon gate (NiSi). The differences are particularly pronounced in strong inversion (high gate voltage V_{GS}). The observations cannot be explained readily in the frame of the standard correlated-mobility fluctuations theory. They point rather to the impact of the charges/traps at the gate-dielectric interface, which are better screened in case of a metal gate. At the moment, one can only speculate on the origin of the LF fluctuations, giving rise to the higher noise in strong inversion. One hypothesis is that the image charge at the gate induced by a filled oxide trap contributes to excess scattering in the channel.

Keywords: gate material; polysilicon gate; metal gate; fully-silicided gate; nMOSFET; number fluctuations theory
PACS: 72.20.Jv; 72.20.+m; 73.40.Qv

INTRODUCTION

For more than a decade, 1/f noise in silicon MOSFETs has been successfully modeled using the so-called correlated mobility fluctuations theory [1],[2]. This model assumes that the basic noise mechanism is tunneling to and from traps in the gate oxide layer at a distance of 1 to 2 nm from the Si-SiO$_2$ interface. Considering a tunneling parameter of 10^8 cm^{-1}, then the typical tunneling distance is 2 nm for a frequency of 1 Hz (Fig. 1), which is higher than the physical gate oxide thickness of the studied devices.

The observed gate voltage dependence above the threshold voltage V_t is due to mobility fluctuations correlated with a change of trapping-induced charge states in the oxide. CMOS scaling pushes the physical gate oxide thickness t_{ox} within the tunneling distance and whether 1/f noise at low frequencies exists and is adequately explained using accepted theory is a subject of concern. Therefore, the low-frequency noise of nMOSFETs with 1.5 nm SiON gate dielectric with different gate electrode materials has been studied. It is seen that there is a strong impact of the gate material on the 1/f noise, which indicates that screening of the charges at the gate-oxide interface can play an important role in LF channel current fluctuations, most likely through the mobility.

EXPERIMENTAL

The nMOSFETs studied have been fabricated using a 1.5 nm SiON gate oxide and either an n-type polysilicon gate, a TiN/TaN metal gate deposited by Physical Vapour Deposition (PVD) or a fully nickel-silicided gate (NiSi). The device width was W=10 µm and length L= 0.25 or 1 µm. LF noise measurements were performed in the linear regime at a drain voltage V_{ds}=0.05 V and a gate voltage (V_g) ranging from 0 to 1 V.

RESULTS

As seen in Fig. 1, the noise spectra are of $1/f^\gamma$ type with $\gamma \sim 1$. The drain current noise spectral density S_I in most cases follows an I_d^2 dependence for lower currents, as found in Figs 2 and 3. At the first instance, metal or NiSi gate nMOSFETs exhibit similar S_I values in Figs 2 and 3, compared with poly-gate. In order to investigate the underlying fluctuation mechanisms, the normalized current noise spectral density is compared with $(g_m/I_d)^2$ [1]. Figures 4a and 4b exhibit two cases: for the NiSi gate device, both characteristics are parallel for most of the drain current range, while for the poly-gate transistor, there is a strong difference between the two curves. In the NiSi case, the 1/f noise behavior is most likely due to number fluctuations, while for poly, this is not necessarily so, particularly in strong inversion.

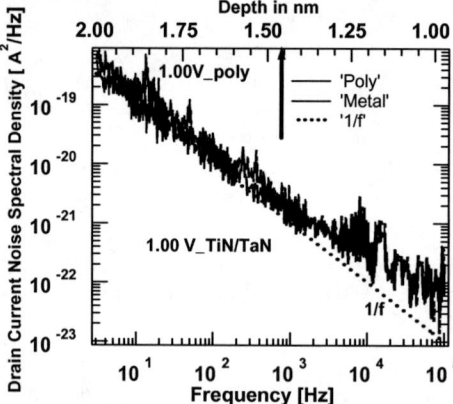

FIGURE 1. Low-frequency noise spectra of 10 µm x1 µm n-MOSFETs at V_g=1 V and V_{ds}=0.05 V for a metal (TiN/TaN) gate and a poly-gate transistor.

DISCUSSION AND CONCLUSION

For the nMOS transistors studied, the foundations of the number fluctuations theory have to be reconsidered, since the physical gate oxide thickness of 1.5 nm corresponds to a frequency of ~1 kHz (Fig. 1). Moreover, the plot of $(LS_{vg})^{0.5}$ versus the gate voltage overdrive (V_g-V_t) in Fig. 5, shows a discernible difference between the three cases studied. The NiSi n-channel transistor exhibits a flat behavior, indicating negligible correlated-mobility fluctuations, while a pronounced V_g-V_t dependence is noticed for the poly-gate nMOSFETs. This suggests that the gate electrode has a strong impact on the 1/f noise characteristics of 1.5 nm SiON devices, both in weak and in strong inversion, which cannot be explained in the frame of the available theory. For deep submicron transistors, the mobility fluctuations due to remote charge and phonon scattering [3],[4] should be considered in the model for adequately explaining the 1/f noise behaviour. The difference in screening efficiency of the fluctuating charges at the gate interface may explain the strong impact of the gate material reported here.

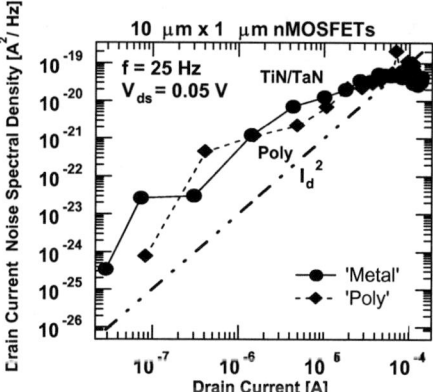

FIGURE 2. Drain current noise spectral density S_I versus I_D for the same transistors as in Fig. 1, for f=25 Hz and V_{ds}=0.05 V.

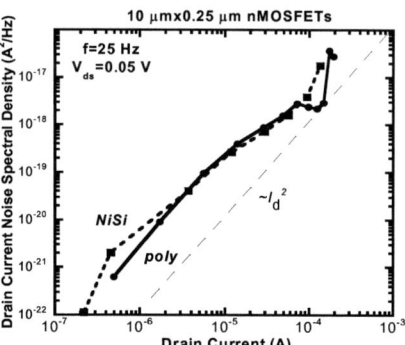

FIGURE 3. S_I versus I_d for a 10 μmx0.25 μm n-MOSFET with fully-silicided (NiSi) and poly gate for 1.5 nm SiON nMOSFET.

The question that remains to be answered is the origin of the LF fluctuations at the gate-oxide interface charges. One possible explanation is the image charge that is induced at the gate electrode by a charge trapped in the oxide. If this is not screened properly, it will induce additional shifts in the threshold (or flat-band) voltage and, hence, number fluctuations. Moreover, this can cause extra remote scattering (mobility fluctuations). The results on the poly-gates seem to pinpoint the latter mechanism as responsible for the higher noise.

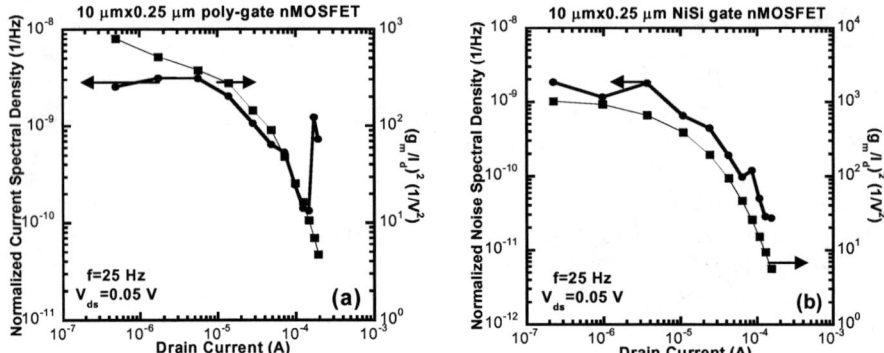

FIGURE 4. (a) Normalized current noise spectral density (f=25 Hz) and $(g_m/I_D)^2$ at V_{ds}=0.05 V for a 10 μm×0.25 μm poly-gate nMOSFET. (b) Normalized current noise spectral density (f=25 Hz) and $(g_m/I_d)^2$ at V_{ds}=0.05 V for a 10 μm×0.25 μm NiSi gate nMOSFET

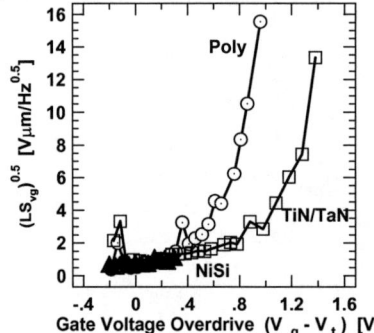

FIGURE 5. Square root of the normalized input-referred noise spectral density versus gate voltage overdrive, for 1.5 nm SiON nMOSFETs with different gate materials.

ACKNOWLEDGMENTS

P. Srinivasan wishes to acknowledge NSF (Award # ECS-0140584) for his financial support.

REFERENCES

1. Ghibaudo G. et al., *Phys. Stat. Sol. (a)* **124**, 571-581 (1991).
2. Hung K.K., Ko P.K., Hu C., and Cheng Y.C., *IEEE Trans. Electron Devices* **37**, 654-665 (1990).
3. Fischetti M.V., *J. Appl. Phys.* **89**, 1232-1250 (2001).
4. Fischetti M.V., Neumayer D.A., and Cartier E.A., *J. Appl. Phys.* **90**, 4587-4608 (2001).

Gate and drain low frequency noise in HfO_2 NMOSFETs

T. Nguyen[1], M. Valenza[1], F. Martinez[1], G. Neau[2], J.C. Vildeuil[2], G. Ribes[2], V. Cosnier[2], T. Skotnicki[2], M. Müller[3]

[1]*CEM2 – UNIVERSITE MONTPELLIER II – UMR CNRS 5507*
Place E. Bataillon, 34095 Montpellier Cedex 5, France
[2]*ST Microelectronics, 850 rue Jean Monnet, F-38926 Crolles Cedex*
[3]*Philips Semiconductors, 850 rue Jean Monnet, F-38926 Crolles Cedex*

Abstract. Gate and drain current noise investigations are performed on nMOS transistors with HfO_2 gate oxides. The drain noise magnitude allows extraction of the slow oxide trap density $N_t(E_F)$ ranging from 3 to 7 10^{19} eV^{-1} cm^{-3}. These values are about 50 times higher than for SiO_2 dielectrics. The 1/f gate current noise component is a quadratic function of the gate leakage current. The gate noise parameter K_{GC} is about 2 10^{-17} m^2, whereas, for SiO_2 dielectrics this gate noise figure of merit is about 10^{-19} m^2.

Keywords: NMOSFETs, high–k, 1/f noise, drain noise, gate noise.
PACS: 72.70.+m ; 73.50.Td

INTRODUCTION

The improvement of speed and chip shrinkage of integrated circuits are achieved by scaling down the thickness of the SiO_2 gate dielectric. However, as oxide thickness is reduced, SiO_2 reaches its ultimate scaling limit such as high tunnelling current and reliability concerns. In order to attenuate this gate current, high-k dielectrics are candidates to replace SiO_2. Various high-k dielectrics have been reported such as HfO_2, ZrO_2, Al_2O_3, Ta_2O_5. HfO_2 gated transistors have shown very encouraging gate leakage reduction, but they present a degraded mobility when non strained Si MOSFETs with a high-k dielectric are achieved [1]. Moreover, instabilities of threshold voltage due to lack of oxygen are observed. The low mobility value is often linked to a high density of traps located in the bulk or at the interface. The interface trap density can be extracted by charge pumping analysis or by 1/f noise measurements. However, the involved traps are not the same. In the first case, fast trap density is extracted, whereas in the second case, slow trap density is extracted. In this contribution, we present gate and drain current noise investigations performed on nMOS transistors with HfO_2 gate oxide and low EOT value.

DEVICES UNDER TEST

The high-K stack consists of a 8Å-thick $HfSiO_x$ capping layer on a 30Å thick HfO_2 deposited on 10Å interfacial oxide giving a 14Å EOT. A NH_3 post deposition anneal at 800°C has been used to suppress the interfacial layer growth during the process. The gate electrode is poly-silicon. The transistors have individual gate and drain electrodes. Devices under test had a width W= 10 µm and a length L between 0.18 to 10 µm.

As shown figure 1, high-k oxide allows reduction of the gate leakage current of 3 decades with respect to the SiO_2 reference with same EOT. However, the mobility is degraded, partly explained by the nitrogen incorporation into the High-K and interfacial oxides layers, the obtained mobility is around 55 $cm^2V^{-1}s^{-1}$, while for the SiO_2 reference the mobility is around 190 $cm^2V^{-1}s^{-1}$.

NOISE INVESTIGATIONS AND DISCUSSION

Gate current and drain current noise measurements have been performed. Concerning drain noise spectra, the main component was 1/f noise, whereas concerning gate noise spectra Lorentzian component associated to Random Telegraph Signal (R.T.S) fluctuations and 1/f component were observed. In this paper we focus our investigations on 1/f noise.

Drain Current noise

To deduce noise figure of merit drain current noise measurements have been performed at V_{DS} = 25 mV and V_{GS} varying up to 1.2 V, i.e. from weak to strong inversion. Figure 2 shows the variations of the normalised drain current noise $W_{eff}L_{eff}S_{I_D}/I_D^2$ at f=1 Hz as a function of I_D for different gate lengths. The normalised noise level is found to level off in weak inversion and decrease as I_D^{-2} in strong inversion. This is in good agreement with carrier number fluctuation theory Δn [2,3]. Moreover, correct scaling of the area normalised noise magnitude is obtained compared with the theoretical variation, that is, $S_{I_D}/I_D^2 \propto (W_{eff}L_{eff})^{-1}$ [2,3]. In figure 2, we have also reported the evolutions of normalised drain current noise obtained in SiO_2 reference n-MOS devices with same EOT. It appears clearly that high-k devices are more noisy. The slow oxide trap density, which is used as figure of merit, has been extracted. We obtain $N_t(E_F)$ between 3 and 7 10^{19} $eV^{-1}cm^{-3}$ which is about 50 times higher than for SiO_2 dielectrics (see figure 3), but in the same range than results found by Simoen et al.[4] in HfO_2 devices with EOT of 2 nm. Simoen et al. [5] have reported a correlation between the increase of trap density and mobility degradation. Our results confirm that the low mobility values obtained for high-K devices is linked to the high magnitude of trap density.

Gate Current noise

Gate current noise measurements have been performed at V_{DS} = 25 mV and 1V for different V_{GS} values. The associated typical spectra are reported figures 4 and 5, respectively. An accurate I-V study shown that at V_{DS}=25 mV, the gate current is predominantly a leakage current flowing between gate and channel, I_{gc}. Whereas, at V_{DS}=1 V, the gate current is predominantly leakage current flowing between drain overlap and gate (I_{gd0}) for V_{GS}<0.6V and gate to channel current for V_{GS}>0.6 V (see figure 6). As shown figure 4, at V_{DS}=25 mV, the low frequency noise is mainly 1/f. Whereas, at V_{DS}=1 V and V_{GS}<0.6V we observe mainly Lorentzian components associated to R.T.S fluctuations (see figure 5).

Concerning gate to channel current noise, the 1/f component is a quadratic function of the gate leakage current following the law: $S_{I_G}(f) = \dfrac{K_{GC}}{f\,W_{eff}\,L_{eff}} I_G^2$. We obtain K_{GC} about 2×10^{-17} m^2 for high-k, whereas for SiO$_2$ reference with same EOT this gate noise figure of merit is about 10^{-19} m^2.

Concerning measurements at V_{DS}=1 V, it appears clearly that the noise associated to off-gate current (V_{GS}=0V) is two decade higher than the one associated to on-gate current (V_{GS}=1.2V) (figure 5). The on-gate current noise is always lower than gate to drain overlap current noise, even when this gate to drain overlap current is lower than on-gate current (figures 5 and 6).

CONCLUSIONS

Gate current noise and drain current noise in HfO$_2$ gated nMOS transistors have been studied.

Drain noise measurements were performed at V_{DS}=25 mV from weak to strong inversion. Noise levels variations agree with the carrier number fluctuation theory. The oxide trap density $N_t(E_F)$ was found 50 times higher than for SiO$_2$ dielectrics.

Gate noise measurements were performed at V_{DS}=25 mV and 1 V. Two paths are well drawn. One between the gate and the channel and one between the gate and the drain region. For gate to channel current 1/f noise is observed, and noise levels variations follow a quadratic current variation inversely proportional to the gate area, characterised by flicker noise parameter K_{GC}. This parameter was found to be in the order of 2×10^{-17} m^2 which is two order of magnitude higher than for SiO$_2$ dielectrics. For gate to drain overlap current mainly R.T.S noise is observed. The gate noise at I_{off} state(V_{GS}=0V, V_{DS}=1V) was higher than the gate noise at I_{on} state (V_{GS}=1V, V_{DS}=1V).

REFERENCES

1. Z. Shi, D. Onsongo, K. Onishi, J.C. Lee and S.K. Banerjee, " Mobility Enhancement in Surface Channel SiGe PMOSFETs with HfO2 Gate Dielectrics" , IEEE Electron Device letters, vol.24, N°1, January 2003.
2. M. Valenza, A. Hoffmann, D. Sodini, A. Laigle, F. Martinez and D. Rigaud, " Overview of the impact of downscaling technology on 1/f noise in p-MOSFETs to 90 nm", IEE Proc-Circuits Devices Syst., vol. 151, No. 2, pp. 102-110, April 2004.
3. M. Marin, M. J. Deen, M. de Murcia and J.C. Vildeuil, " Effects of body biasing on the low frequency noise of MOSFETs from a 130 CMOS technology", IEE Proc-Circuits Devices Syst., vol. 151, No. 2, pp. 95-101, April 2004.
4. E. Simoen, A. Mercha, L. Pantisano, C. Claeys and E. Young; "Low-frequency noise behavior of SiO$_2$-HfO$_2$ dual-layer gate dielectric nMOSFETs with different interfacial oxide thickness", IEEE Trans. On. Electron. Devices, vol.51, N°5, May 2004.

5. E. Simoen, A. Mercha, C. Claeys and E. Young; "Correlation between the 1/f noise parameters and the effective low-field mobility in HfO$_2$ gate dielectric n-channel metal-oxide-semiconductor field-effect transistors", Applied Physics letters, vol. 85, N°6, pp 1057-1059, August 2004.

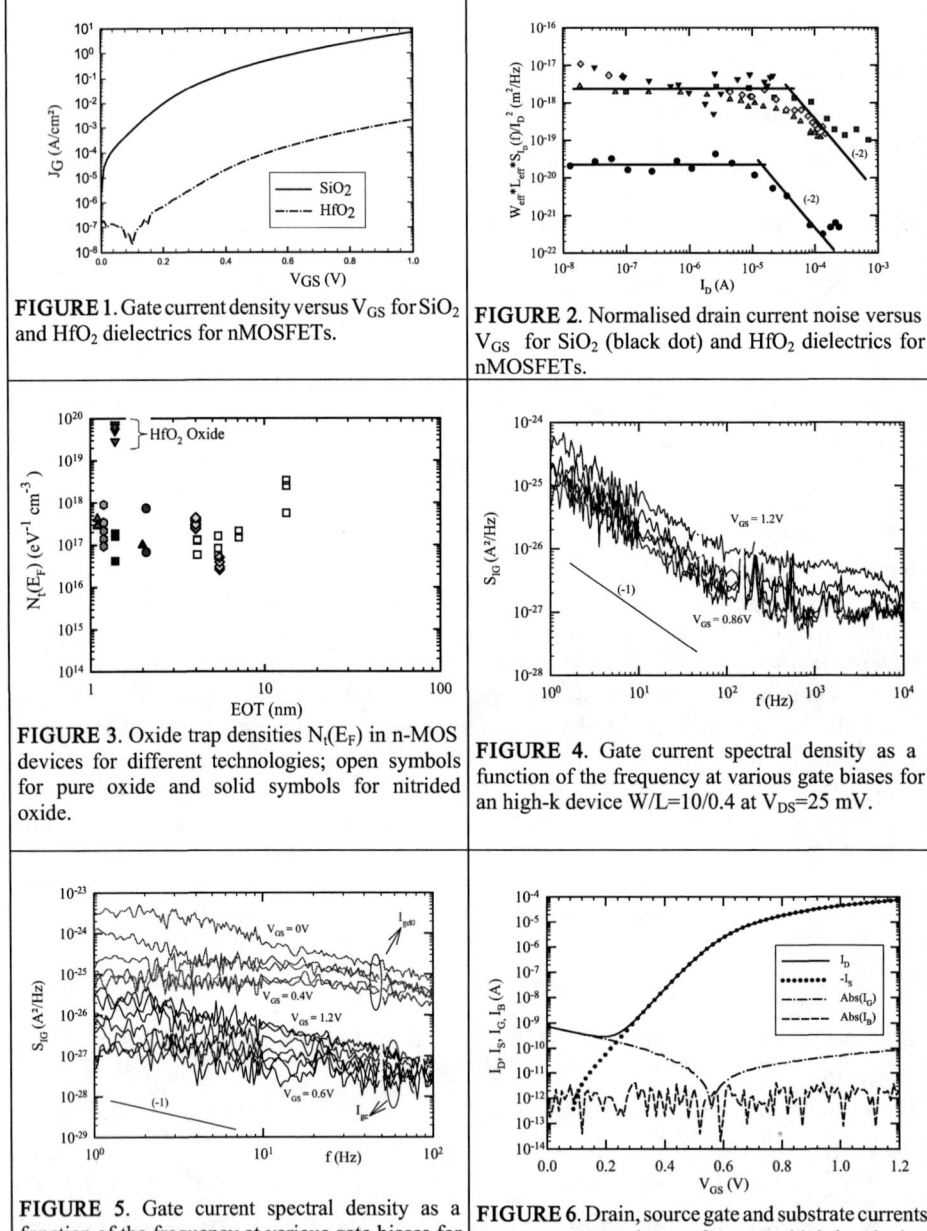

FIGURE 1. Gate current density versus V_{GS} for SiO$_2$ and HfO$_2$ dielectrics for nMOSFETs.

FIGURE 2. Normalised drain current noise versus V_{GS} for SiO$_2$ (black dot) and HfO$_2$ dielectrics for nMOSFETs.

FIGURE 3. Oxide trap densities $N_t(E_F)$ in n-MOS devices for different technologies; open symbols for pure oxide and solid symbols for nitrided oxide.

FIGURE 4. Gate current spectral density as a function of the frequency at various gate biases for an high-k device W/L=10/0.4 at V_{DS}=25 mV.

FIGURE 5. Gate current spectral density as a function of the frequency at various gate biases for an high-k device W/L=10/0.4 at V_{DS}=1 V.

FIGURE 6. Drain, source gate and substrate currents versus gate voltage for an high-k device W/L=10/0.4 at V_{DS}=1V.

Intrinsic fluctuations induced by a high-κ gate dielectric in sub-100 nm Si MOSFETs

A. J. García-Loureiro*, K. Kalna[†] and A. Asenov[†]

*Departamento de Electrónica y Computación, Universidad de Santiago de Compostela, Spain
e-mail: antonio@dec.usc.es
[†]Device Modelling Group, Department of Electronics & Electrical Engineering
University of Glasgow, Glasgow, G12 8LT, United Kingdom
email:kalna@elec.gla.ac.uk

Abstract. We have developed an efficient 3D parallel simulator to study a Si metal-oxide-semiconductor field effect transistor (MOSFET) with a high-κ gate stack. The simulator is employed to study the impact of the intrinsic parameter fluctuations within a high-κ dielectric on the threshold voltage and drive current. We have found that large regions of crystal high-κ dielectrics with a lower dielectric constant than the amorphous high-κ lower the drive current by more than 100% and shifts threshold voltage by 0.2 V. The same effect on the drive current and threshold voltage is observed when high-κ fluctuations follow a Gaussian distribution with a correlation length of 3 nm.

Keywords: high-κ dielectric, MOSFET, fluctuations, 3D simulations
PACS: 85.30.De

INTRODUCTION

A major problem limiting the scaling of the conventional Si metal-oxide-semiconductor field effect transistors (MOSFETs) beyond the 45 nm technology node is the requirement of an extremely thin gate SiO_2 layer which introduces intolerable gate tunnelling [1]. The solution lies in the replacement of SiO_2 or oxinitride by a high-κ dielectric material [2, 3] which provides the same equivalent oxide thickness (EOT) for a much larger physical thickness due to a higher dielectric constant but reduces the gate tunnelling.

The replacement of SiO_2 by a high-κ dielectric material calls also for a replacement of the heavily doped polysilicon gate with a metal gate [1]. Most of the high-κ dielectrics exhibit a thermal instability which leads to the local crystallisation of the initially amorphous high-κ layer. The fabrication of the metal gate can be carried out at a lower temperature thus preserving the amorphous state of the high-κ dielectric material as much as possible. In this work we investigate the impact of the high-κ dielectric structure induced intrinsic fluctuations on the device characteristics using a finite element 3D parallel drift-diffusion (D-D) MOSFET simulator.

Intrinsic fluctuations introduced by the discrete charge [4] and interface roughness [5] create problems when scaling MOSFETs to sub-100 nm dimensions. The introduction of the high-κ dielectric materials into the gate stacks brings also new sources of intrinsic fluctuations as: (i) islands of crystal materials with a different dielectric constant, (ii) trapped and fixed charges occurring at the boundaries of the islands, (iii) an interface roughness between the interfacial layer with a thickness of 1-2 nm and the high-κ dielectric, and (iv) interface roughness between the high-κ dielectric and the metal gate.

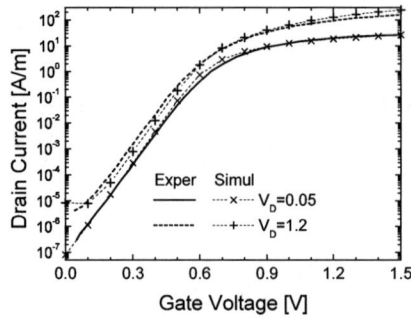

FIGURE 1. I_D-V_G characteristics obtained from the 3D parallel D-D simulator compared with experimental data [8] at low and high drain biases of 0.05 and 1.2 V respectively.

FLUCTUATIONS OF HIGH-κ DIELECTRIC CONSTANT

Our 3D parallel device simulator originally developed for a modelling of heterostructure semiconductor devices [6] was adapted to simulate MOSFET structures while preserving its capabilities to deal with a complex layer geometry. This feature is fully exploited when simulating behaviour of the high-κ dielectric gate stacks since they often consist of different layers and aggregate regions with various material properties. Suitable unstructured tetrahedral element triangulation is employed to describe the complex geometry of the crystal islands in the high-κ dielectric layer and the landscape of the top and bottom interfaces. The whole study is based on a meticulous calibration against a conventional *n*-type 67 nm effective gate length Si MOSFET published by IBM [8]. Figure 1 compares I_D-V_G characteristics obtained from 3D parallel D-D MOSFET simulator against experimental data at a low drain voltage (0.05 V) and at a high drain voltage (1.2 V).

Intrinsic parameter fluctuations introduced due to discreteness of charge and matter [4] and the interface roughness [5] are creating problems in scaling of conventional MOSFETs with the gate SiO_2 layer to sub-100 nm dimensions. When SiO_2 is replaced with a high-κ dielectric its amorphous structure tends to crystallise during a thermal process used to fabricate the metal gate electrode creating localised regions with a smaller dielectric constant [7]. Those regions affect the electron density in the inversion layer as illustrated in Fig. 2. A 1 nm interfacial layer with a lower dielectric constant formatted at the interface between the high-κ dielectric and Si is shown in Fig. 2 as well.

The impact of the high-κ material parameter fluctuations and the interfacial layer on device performance is studied replacing the original 2.2 nm SiO_2 layer with a 11 nm EOT of HfO_2 dielectric. The fluctuations in the dielectric constant are generated using a Gaussian distribution with a correlation length of 3 nm. The resulting distribution of the dielectric constant along a slice of the high-κ dielectric layer is illustrated in Fig. 3. I_D-V_G characteristics at $V_D = 0.05$ V and 1.2 V obtained from simulations using various material compositions in the gate dielectric compared with experimental data are shown in Figs. 4 and 5 respectively. A comparison of the I_D-V_G characteristics obtained when the original 2.2 nm SiO_2 dielectric is used (triangles) reveals that simulated current slightly overestimates the experiments above the threshold due to missing quantum ef-

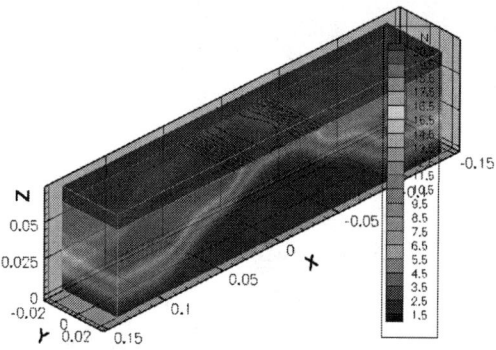

FIGURE 2. Electron density on logarithmic scale in the 67 nm effective gate length Si MOSFET. A contour line below the gate shows the impact of the high-κ dielectric crystal regions on electron density in the inversion layer.

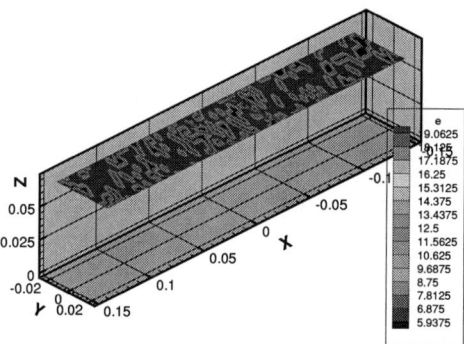

FIGURE 3. Dielectric constant fluctuations in the high-κ layer using a correlation length of 3 nm.

fects in the simulations while it is in an excellent agreement in the sub-threshold region. Then, SiO$_2$ layer is replaced by a 10 nm HfO$_2$ layer ($\kappa_1 = 20$) and a 1 nm interfacial layer of SiO$_2$ ($\kappa_3 = 3.9$). When large crystalline islands with a lower dielectric constant $\kappa_2 = 5$ than the required constant κ_1 are assumed then the drain current significantly drops (diamonds). In addition, a positive shift of the threshold voltage by about 0.2 V is observed. Finally, when the large islands are replaced by fluctuations of high-κ dielectric constant following a Gaussian distribution, the drain current (circles) suffer the same reduction at a low $V_D = 0.05$ V and the positive shift of the threshold is just about 0.16 V. However, a reduction of the island size becomes important at a high $V_D = 1.2$ V since the drain current lowering is then less dramatic [see circles in Fig 5 (a)].

CONCLUSIONS

In this work, we have developed an efficient 3D parallel simulator to study a MOSFET when the original SiO$_2$ layer is replaced with a HfO$_2$ dielectric layer. We have stud-

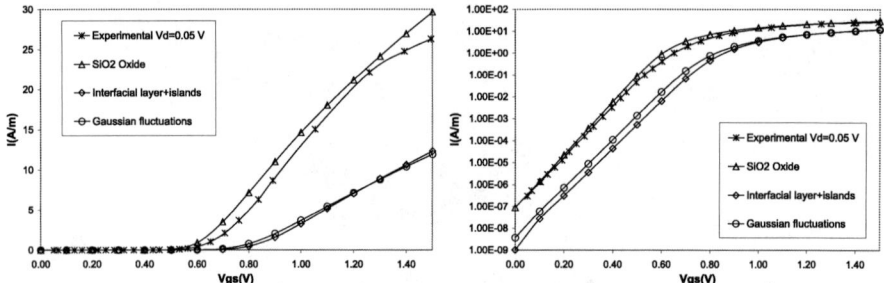

FIGURE 4. I_D-V_G characteristics obtained from the 3D parallel D-D simulator compared with experimental data [8] at a low drain bias of 0.05 V on linear scale (a) and logarithmic scale (b).

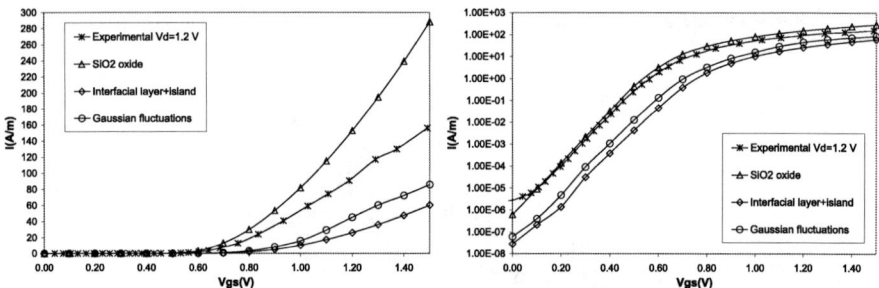

FIGURE 5. I_D-V_G characteristics obtained from the 3D parallel D-D simulator compared with experimental data [8] at a high drain bias of 1.2 V on linear scale (a) and logarithmic scale (b).

ied the impact of intrinsic high-κ dielectric material fluctuations on the drive current and threshold voltage progressively adding sources of mobility and threshold control degradations. The presence of large regions of fluctuating dielectric constants and the interfacial layer leads to a drop in the drain current and the positive shift of the threshold voltage. When high-κ dielectric constant fluctuations are generated using a correlation length of 3 nm the drain current reduction is the same at $V_D = 0.05$ V but at $V_D = 1.2$ V the drain current reduction is much smaller.

Acknowledgements: This work was partly supported by the Spanish Government (MCYT) under the project TIN2004-07797-C02.

REFERENCES

1. H.-S. P. Wong, *IBM J. Res. Dev.* **46**, 133 (2002).
2. S. Datta et al., *IEDM Tech. Dig.*, 28.1.1 (2003).
3. K. Rim et al., *2002 Symp. VLSI Technol., Dig. Tech. Pap.*, 11 (2002).
4. A. Asenov et al., *IEEE Trans. Electron Devices* **50**, 1837 (2003).
5. A. Asenov, S. Kaya, and J. H. Davies, *IEEE Trans. Electron Devices* **49**, 112 (2002).
6. A. J. García-Loureiro, K. Kalna and A. Asenov, *J. Comput. Electron.* **2**, 369 (2003).
7. M. Amrani et al., *J. Phys. D: Appl. Phys* **38**, 596 (2005).
8. K. Rim et al., *2001 Symp. VLSI Technol., Dig. Tech. Pap.*, 59 (2001).

Contributions of Channel Gate and Overlap Gate Currents on 1/f Gate Current Noise for Thin Oxide Gate p-MOSFETs

F. Martinez[1], A. Laigle[1], A. Hoffmann[1], M. Valenza[1],
A. Veloso[2], M. Jurczak[2]

[1]*CEM2 – UNIVERSITE MONTPELLIER II – UMR CNRS 5507
Place E. Bataillon, 34095 Montpellier Cedex 5, France*
[2]*IMEC, Kapeldreef 75, 3001 Leuven, Belgium*

Abstract. Dc Gate current and associated noise sources on a 90 nm CMOS technology are investigated. At high V_{DS} biases two dc components are observed: one due to overlap, the second due to gate to channel tunneling. 1/f noise associated with these two components is studied. The presented results concern PMOST.

Keywords: MOSFETs, 1/f noise, tunneling, gate noise.
PACS: 72.70.+m ; 73.50.Td

INTRODUCTION

Scaling rules in metal-oxide-semiconductor (MOS) impose very thin oxide thickness in the order of 1 nanometer. Several limiting factors associated with ultra-thin gate oxides have been reported; among them, direct tunnelling current increases exponentially with decreasing thickness and affects MOS devices performances. In such devices, gate currents become significant. Depending on the biasing voltages, the gate current affects source and drain currents in different ways. The effect of the gate current is observed on I_{on}-I_{off} characteristics and on the power consumption of CMOS circuits. As a result, the signal-to-noise ratio of analogue circuits is reduced, due to novel noise sources. In this paper, we focus on gate current noise in ultimate MOSFETs. As the gate current is a sum of gate-to-channel current (I_{gc}) and the gate overlap, so-called the edge direct tunnelling (EDT) current (I_{gso} and I_{dso}), we have investigated gate noise sources associated to these two gate current components. It is shown that in both cases 1/f noise can be modelled using a quadratic gate current law, but for each component the characteristic noise parameter has a different value.

DEVICES UNDER TEST

The test devices are processed on 200 mm diameter silicon wafer. The process flow involved have been reported with more precisions in [1], resulting in a RPN processed 1.5 nm EOT gate dielectric

The transistors have individual gate and drain electrodes. The investigated transistors have gate mask width (W) between 100 nm and 10µm, and gate mask length (L) between 90 nm and 10 µm. The I_{on}/I_{off} characteristics have shown that the devices are competitive.

GATE TUNNELLING CURRENTS

A part from gate tunnelling current can be generated in the gate/source- and/or gate/drain-overlap regions. Figure 1 illustrates various gate components for thin gate oxide PMOSFET; the gate-to-channel current (Igc) and gate overlap current I_{gso}, I_{gdo} are shown. Typical dc current variations versus gate voltage are shown in Figure 2 at $V_{DS} = -1$ V. At $V_{GS} = 0$ V, the gate current coincides with the drain current. When $-V_{GS}$ increases the gate current decreases and at $-V_{GS}=0.4$ V the gate current shifts from positive to negative values. This means that for $V_{GS}<0.4$ V the overlap gate current is higher than gate-to-channel current, implying that overlap gate current is the dominant source of the off-state current. Moreover, as $V_S=0V$, $V_D=-1V$ and $V_G=0V$ the overlap current is only due to gate-to-drain overlap region as shown in Figure 2, since $I_D=I_G$ and I_S is negligible. This gate-to drain overlap current is due to the tunnelling of holes from p+ polysilicon gate to drain region. When $-V_{GS}$ increases, the n substrate surface is in inversion and the I_{gc} current is due to tunnelling of holes from channel to the gate, whereas the I_{GDO} decreases.

NOISE INVESTIGATIONS

As shown in the previous section gate tunnelling current limits the performance on the I_{on}/I_{off} characteristics. Moreover, the gate current has potentially a major impact on noise behaviour of the device. The drain noise sources have been already investigated [2]. In this section, we focus on the noise measurements associated with the gate current. As the gate current is combined by two components which can be separated and each of them can be dominant depending on the bias regime. In fact, the I_{off} state current involves the gate-to-drain overlap current, whereas, the I_{on} state current involves the gate-to-channel current. In this paper, we present an investigation of these two associated noise sources, in order to deduce a compact law, and to extract a noise figure of merit for each component.

To investigate the gate-to-drain overlap noise current, the measurements were performed at $V_{GS}=0V$ and V_{DS} varying from $-0.3V$ to -1.2 V, however to investigate the gate-to-channel noise current the measurements were performed at $V_{DS}=-1$ V and V_{GS} varying from -0.5 V to -1.2 V.

A typical gate current noise spectrum for the studied technology shows mainly 1/f noise component and a white noise one. Moreover, for the both cases, of gate-to-drain and gate-to-channel currents spectra we have observed Lorentzian components. In this paper we focus on 1/f noise and white noise.

Concerning the white noise, since gate leakage current is the result of quantum-mechanical direct tunnelling process the expected shot noise level is : $S_{I_{shot}} = 2qI_G$. We have observed a good agreement between experimental and theoretical level.

The evolutions of 1/f gate current noise at f=1 Hz versus the gate current are reported in Figures 3 and 4 for two gate area devices, 10x10 µm², and 10x1 µm², respectively. In these figures we compare for each device, the evolutions of the gate-to-channel current spectral density (V_{DS}=-1 V) and the gate-to-drain current spectral density (V_{GS}=0 V). For both components we observe a quadratic variation. For 10x10 µm² geometry the gate-to-drain current spectral density is higher than the gate-to-channel one, whereas, for 10x1 µm² device the two components follow the same gate current behaviour.

In Figure 5 is reported the normalised gate-to-channel current spectral density for different gate area. We deduce that the gate-to-channel current spectral density is modelled by:

$$S_{I_{GC}}(f) = K_{GC} I_{GC}^2 / W_{eff} L_{eff} f \quad (1)$$

where K_{GC} is the flicker gate-to-channel noise parameter. For this technology $K_{GC} = 2 \cdot 10^{-18}$ m².

Because, the length of the gate-to-drain overlap region is unknown it is not possible to report the area normalised gate-to-drain current spectral density versus the gate current. However, for a given technology the overlap region is the same for different width. So, the Figure 6 shows the width normalised gate-to-drain current spectral density, which is in this case a picture of the gate area normalised representation. We obtain, also a quadratic current evolution with a characteristic flicker gate-to-drain overlap noise parameter K_{GD}:

$$S_{I_{GD}}(f) = K_{GD} I_{GD}^2 / f W_{eff} L_{GDO} \quad (2)$$

As L_{GDO} is unknown it is not possible to extract K_{GD}. Moreover, using eq. (1) and (2) and experimental data, we deduce that:

$$K_{GC} / 1\mu m = K_{GD} / L_{GDO} \quad (3)$$

The data reported in Figure 4, show that K_{GD} is lower than K_{GC}.

In the case where capacitances with identical gate-to-drain overlap regions are available, it is possible from capacitance leakage current noise measurement to extract the K_{GD} parameter and then from MOST gate-to-drain current noise to extract the length of gate-to-drain overlap region.

CONCLUSIONS

Gate current in a 90 nm CMOS technology have been studied. Two paths are well drawn, one between the gate and the channel and one between the gate and the drain region. The gate current noise has been also investigated. The white noise is identified to shot noise. The 1/f for both gate-to-channel and gate-to-drain current noise show a current quadratic variation inversely proportional to the gate area, characterised by two flicker noise parameters K_{GC} and K_{GD}. These parameters are well suited for compact gate current noise modelling for CMOS circuit simulation. The gate-to-drain overlap current noise parameter is lower than the gate-to-channel one. Nevertheless, for a given device with high gate length, the gate noise at I_{off} state can be higher than the gate noise at I_{on} state.

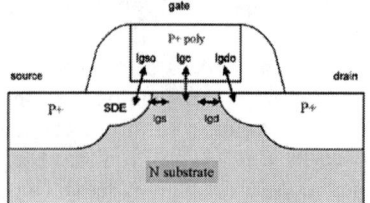

FIGURE 1. Illustration of gate direct tunnelling components of PMOST thin oxide thickness

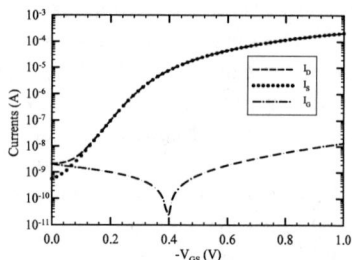

FIGURE 2. Gate current as a function of gate bias for 10 x 1 μm² PMOS device at $V_{DS} = -1$ V.

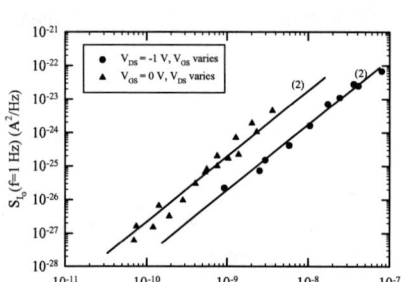

FIGURE 3. Gate current spectral density at f=1Hz versus gate current for 10 x 10 μm² device.

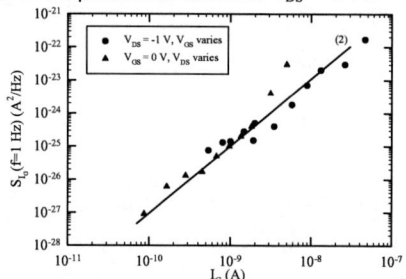

FIGURE 4. Gate current spectral density at f=1Hz versus gate current for 10 x 1 μm² device.

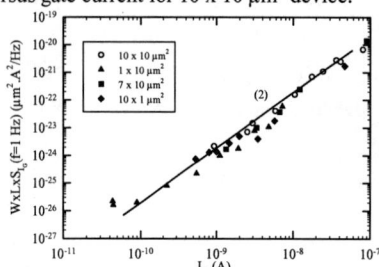

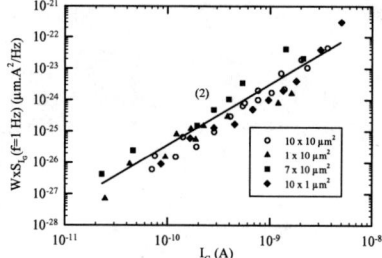

FIGURE 5. Area normalised 1/f gate-to-channel current noise versus gate-to-channel current at f = 1Hz, for various devices.

FIGURE 6. Width normalised 1/f gate-to-drain overlap current noise versus gate-to-drain current at f = 1Hz, for various devices.

REFERENCES

1. A. Veloso and al., "RPN Oxinitride Gate Dielectrics for 90 nm Low Power CMOS Applications", proc. ESSDERC 2002, pp 159-162.
2. M. Valenza and al., "Impact of gate current noise on drain current noise in 90 nm CMOS technology", European Solid-State Device Research, 2003. ESSDERC '03. 33rd Conference on, 16-18 Sept. 2003, pp: 287- 290
3. A.J. Scholten, and al., "Noise modeling for RF CMOS circuit simulation Electron Devices", IEEE Transactions on, Vol.50, Iss.3, March 2003, pp: 618- 632.

RF Noise Modeling in SiGe HBTs

Guofu Niu*, Kejun Xia*, David Sheridan[†] and Susan Sweeney[†]

*Electrical and Computer Engineering Department, 200 Broun Hall
Auburn University, Auburn, AL 36849, USA
Tel: 334 844-1856 / Fax: 334 844-1888 / E-mail: guofu@eng.auburn.edu
[†]IBM Microelectronics, Essex Junction, VT 05452, USA

Abstract. This paper presents RF noise modeling in advanced SiGe HBTs using experimental data from 2 to 26 GHz. Several widely used noise models are evaluated, including the conventional SPICE model, the Van Vliet model, the transport noise model, and a recently developed extraction based model. The connections between various models are investigated.

Keywords: SiGe HBT, noise, bipolar transistors.
PACS: 85.30.De, 85.30.Pq, 72.70.+m

INTRODUCTION

Transistor high-frequency noise is an important issue in wireless and wired communication as it sets receiver sensitivity. For low-noise circuit design, accurate transistor noise models are required. The major RF noise sources in a SiGe HBT include the thermal noise due to the terminal resistances and the RF noise in the base and collector currents (i_b and i_c). The thermal noise behavior is relatively well understood and can be modelled by a noise current with a power spectral density (PSD) of $4kT/R$. In current CAD tools, the i_b and i_c noises are often assumed to be shot like, with a PSD of $S_{ibib*} = 2qI_B$ and $S_{icic*} = 2qI_C$ respectively. Further, they are assumed to be independent of each other. Such an approach is used by SPICE Gummel-Poon, VBIC, Mextram, and Hicum models. Both theoretical analysis and experimental data have shown that such a simplified model is not sufficient for high frequency applications, particularly at higher biasing currents required to achieve high speed operation [1]–[7]. In particular, S_{ibib*} has been shown to increase with frequency, and the correlation between i_b and i_c is significant and cannot be neglected [1]–[9]. This work discusses modeling of RF noise in base and collector currents in SiGe HBTs.

BASE AND COLLECTOR CURRENT RF NOISE MODELS

Van Vliet Model

Van Vliet model is based on microscopic noise analysis, and describes the noise of i_b, i_c and their correlation using the Y-parameters of the intrinsic transistor [10]:

$$S_{ibib*}^{van} = 4kT\Re(Y_{11}) - 2qI_B, \quad S_{icic*}^{van} = 4kT\Re(Y_{22}) + 2qI_C, \quad S_{icib*}^{van} = 2kT(Y_{21} + Y_{12}^* - g_m), \tag{1}$$

where g_m is small signal transconductance, $\Re$ stands for real part, and the superscript * stands for conjugate. The Y-parameters in the above equations are for the intrinsic transistor. Here the frequency dependence of S_{ib} and S_{ic} as well as their correlation S_{icib*} are taken into account through the frequency dependence of the intrinsic transistor Y-parameters. This model serves as the basis of several others models reported in the literature, e.g. [6] [11]–[14]. Observe that the $\Re(Y_{11})$ of the intrinsic transistor is frequency independent for the equivalent circuits used in [13] and [14], which effectively leads to a frequency *independent* S_{ib}. On the other hand, the correlation term is non-zero and frequency dependent, because Y_{21} in the equivalent circuits used is frequency dependent, thus leading to inconsistency.

In typical compact models used in CAD tools, $\Re(Y_{11})$ is frequency independent as well, which prevents a meaningful implementation of the van Vliet model in CAD tools. Even if one includes the frequency dependence of $\Re(Y_{11})$ by adding a non-quasi-static (NQS) resistance r_d in series with the input capacitance [15], the van Vliet model is not accurate for modern SiGe HBTs [16]. Extraction of r_d is also difficult as it is a higher order effect [6].

Transport Noise Model

The transport noise model [1] was shown to work better than the SPICE model [1] [2] [3] [4] [5]. The collector current noise is transported from the electron current shot noise in the emitter-base junction, with a noise transit time τ_n:

$$S^{Tran}_{ibib*} = 2qI_B + 4qI_C[1 - \Re(e^{j\omega\tau_n})], \; S^{Tran}_{icic*} = 2qI_C, \; S^{Tran}_{icib*} = 2qI_C(e^{-j\omega\tau_n} - 1). \quad (2)$$

With only a single dedicated noise model parameter τ_n, simultaneous fitting of measured S_{ib} and S_{icib*} can become difficult. We note that S_{icib*} is a complex number. As a result, τ_n can be chosen to fit only one of S_{ib}, real part of S_{icib*}, imaginary part of S_{icib*}. Simultaneous fitting of NF_{min}, R_n, G_{opt} and B_{opt} is challenging in some cases.

The Noise Extraction Based Model

For accurate modeling of the correlated i_b and i_c noise over a wide current biasing range and a wide frequency range, a semi-empirical model was recently developed based on analysis of experimentally extracted i_b and i_c noise [17] [18]. The i_b and i_c noise, including their correlation, are first modeled as a linear function of ω, the angular frequency, or ω^2. The parameters of the linear functions are then modeled as a function of bias through g_m.

The first step is to extract the noise PSDs of the intrinsic i_b, i_c and their correlation from measured Y-parameters and noise parameters. Fig. 1 shows the small signal equivalent circuit used. The block enclosed in the dash line is the intrinsic transistor. The noise PSDs of intrinsic i_b, i_c and $i_c i_b^*$ are obtained from measured minimum noise figure NF_{min}, noise resistance R_n and optimum noise source admittance $Y_{opt} = G_{opt} + jB_{opt}$ by

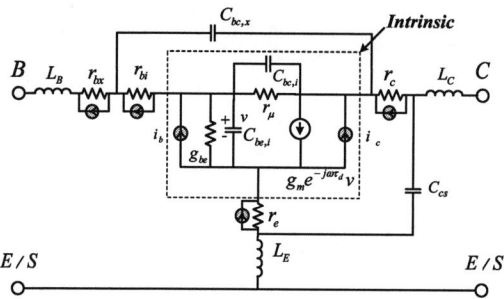

FIGURE 1. Small signal equivalent circuit of SiGe HBTs used.

de-embedding all the components outside the intrinsic transistor. The equivalent circuit parameters are determined from measured S-parameters.

Fig. 2 (a) shows the typical extracted intrinsic noise characteristics for a SiGe HBT from a 50 GHz peak f_T process. The device has an emitter area of $0.24 \times 20 \times 2 \mu m^2$. Note that they are all frequency dependent.

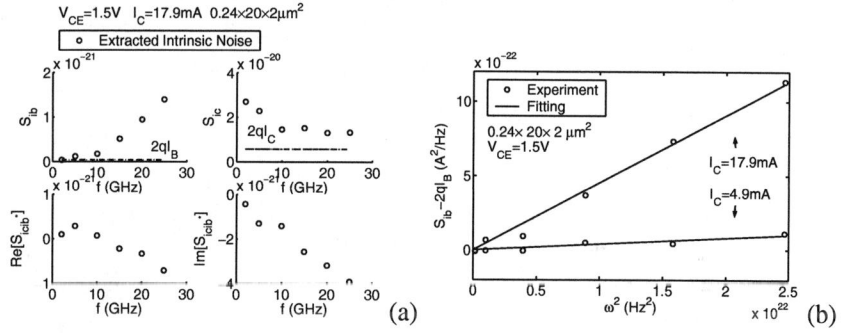

FIGURE 2. (a) Extracted intrinsic noise sources as a function of frequency. (b) $(S_{ib} - 2qI_B)$ versus ω^2 at I_C=4.9 mA and 17.9 mA.

For a given bias, S_{ibib*} is found to be higher than $2qI_B$ by an excess term that is proportional to ω^2, similar to the gate current noise in FETs. An example is given in Fig. 2 (b), which plots the excess base current $(S_{ibib*} - 2qI_B)$ as a function of ω^2 for two representative biases. A linear dependence is observed. Further, the bias dependence of the excess noise can then be modeled using g_m.

The slight frequency dependence of S_{icic*} is modeled through the real part of Y_{21}. The correlation terms, both real and imaginary, are modeled as a function of ω^2 and ω, respectively, with g_m dependent coefficients. The model equations are summarized as follows:

$$S_{ibib*} = 2qI_b + \omega^2(K_{bb*}g_m^{\alpha_{bb*}} + B_{bb*}), \; S_{icic*} = (K_{cc*}g_m^{\alpha_{cc*}} + B_{cc*})\Re(Y_{21}), \quad (3)$$

$$\Re(S_{icib*}) = K_{cb*}^{br}g_m^1 - \omega^2(K_{cb*}^{kr}g_m^2), \; \Im(S_{icib*}) = \omega(K_{cb*}^i g_m^{\alpha_{cb*}^i} + B_{cb*}^i). \quad (4)$$

In general, we found that:

- $\alpha_{bb^*} \approx 2$. $\alpha_{cc^*} \approx 1$. α_{cb^*} is between 1 and 2.
- The $\Re(S_{icib^*})$ is less important than other noise terms as far as its impact on the final noise parameters is concerned.

The model works well up to at least half of the peak f_T prior to high injection f_T roll off, and is scalable [18].

Fig. 3 (a) shows the modeled versus measured S_{ibib^*}, S_{icic^*}, $\Re(S_{icib^*})$, and $\Im(S_{icib^*})$ versus frequency at I_C=16.6 mA. Fig. 3 (b) shows the same parameters versus I_C at 15 GHz. Both the frequency and current dependences are well modeled.

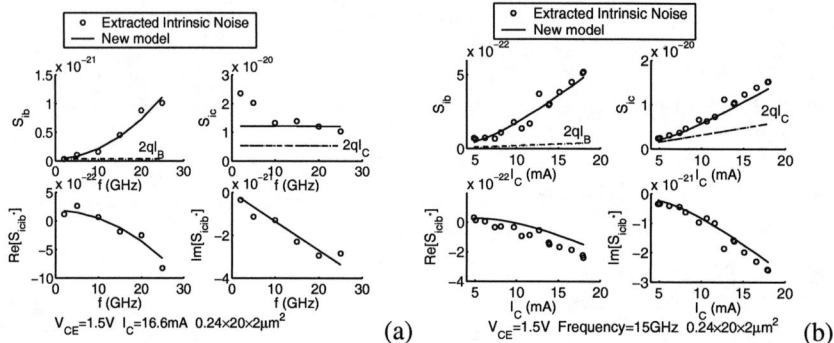

FIGURE 3. (a) Modeled and extracted intrinsic noise sources as a function of frequency. (b) Modeled and extracted intrinsic noise sources as a function of I_C.

Fig. 4 (a) shows the modeled and measured noise parameters versus frequency at I_C=17.9 mA. Fig. 4 (b) shows the noise parameters as functions of I_C at 25 GHz. Excellent fitting is obtained for all of the four noise parameters, at all frequencies and across all biasing currents.

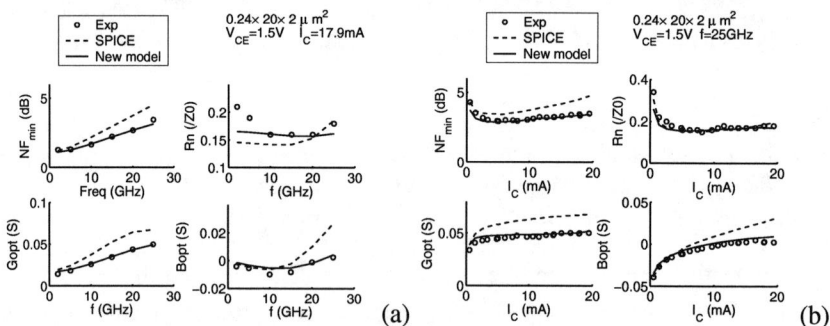

FIGURE 4. (a) Noise parameters versus frequency. I_C=17.9 mA. A_E=0.24 × 20 × $2\mu m^2$. (b) Noise parameters versus collector current at 25 GHz. A_E=0.24 × 20 × $2\mu m^2$.

COMPARISON OF NOISE MODELS

The van Vliet model, the transport model, and the semi-empirical noise extraction based model all describe the excess base current noise and the correlation between i_b and i_c. Despite the seemingly different model equations, the functional form of the frequency dependences of S_{ibib*}, $\Re(S_{icib*})$, and $\Im(S_{icib*})$ is similar for all of the three models, as detailed below. Meaningful use of the van Vliet model requires an input NQS model. A first order input NQS model is to use a NQS delay resistance r_d in series with the base-emitter capacitance [15]. The resulting Y_{11} and Y_{21} are:

$$Y_{11} = g_{be} + \frac{j\omega C_{be}}{1+j\omega C_{be} r_d}, \quad Y_{21} = \frac{g_m e^{-j\omega \tau_d}}{1+j\omega C_{be} r_d}. \tag{5}$$

A Taylor expansion of (5) can be made at $\omega = 0$. Substituting the expansion result into (1), we have:

$$S_{ib}^{van} \approx 2qI_B + 4kTC_{be}^2 r_d \omega^2, \tag{6}$$

$$\Im(S_{icib*}^{van}) \approx -2kT(g_m\tau_d + C_{be}r_d)\omega, \quad \Re(S_{icib*}^{van}) \approx -2kTg_m[(C_{be}r_d)^2 + \tau_d C_{be} r_d + \tau_d^2/2]\omega^2. \tag{7}$$

Interestingly, the van Vliet model also shows a ω^2 dependence for S_{ib}, a ω dependence for $\Im(S_{icib*})$ and a ω^2 dependence for $\Re(S_{icib*})$. These frequency dependences are identical to the extraction based noise model.

Similarly, a second order Taylor expansion of $e^{-j\omega\tau_n}$ in (2) leads to the following approximations for the transport noise model:

$$S_{ibib*}^{tran} \approx 2qI_B + 2qI_C\tau_n^2\omega^2, \quad \Im(S_{icib*}^{tran}) \approx -2qI_C\tau_n\omega, \quad \Re(S_{icib*}^{tran}) \approx -2qI_C\tau_n^2\omega^2. \tag{8}$$

Note that the excess S_{ibib*} is proportional to ω^2, $\Im(S_{icib*})$ is proportional to ω, and $\Re(S_{icib*})$ is proportional to ω^2 in (8). Therefore, the functional form for describing the frequency dependence of S_{ibib*}, $\Re(S_{icib*})$ and $\Im(S_{icib*})$ is identical for all of the three models. The frequency depence of S_{icic*}, however, is *only* present in the extraction based model.

An inspection of the coefficients of ω or ω^2 terms in the relevant model equations will immediately show that the biasing dependence is different for the three models discussed. For example, the excess term in S_{ibib*} is proportional to $C_{be}^2 r_d \omega^2$ for the van Vliet model, and hence approximately propotional to g_m. In the transport noise model, the excess term is proportional to $2qI_C\tau_n^2\omega^2$. In the extraction based noise model, the excess term is proportional to $g_m^{\alpha_{bb*}}\omega^2$. For the SiGe HBTs extracted, α_{bb*} is approximately 2.

CONCLUSION

We have presented an overview of RF noise modeling in bipolar transistors, including the van Vliet model, the transport noise model, and a recently developed semi-empirical noise model that is based on experimental noise extraction. The similarities and differences between various models are investigated. The experimental extraction based model is shown to work well for SiGe HBTs.

REFERENCES

1. G. F. Niu, J. D. Cressler, S. Zhang, W. E. Ansley, C. S. Webster, and D. L. Harame, "A unified approach to RF and Microwave noise parameter modeling in bipolar transistors," *IEEE Trans. Electron Devices*, Vol. 48, No. 11, pp. 2568-2574, Nov. 2001.
2. J. Moller, B. Heinermann, and F. Herzel, "An improved model for high-frequency noise in BJTs and HBTs interpolating between the quasi-thermal approach and the correlated-shot-noise model," *Proceedings of the IEEE BCTM*, pp. 228-231, 2002.
3. P. Sakalas, M. Schroter, P. Zampardi, H. Zirath, and R. Welse, "Microwave noise sources in AlGaAs/GaAs HBTs," in *IEEE MTT-S International Microwave Symposium Digest*, pp. 2117 – 2120, 2002.
4. P. Sakalas, M. Schroter, R. Scholz, H. Jiang, and M. Racanelli, "Analysis of microwave noise sources in 140 ghz SiGe HBTs," in *IEEE RFIC Digest*, 2004.
5. C. Jungemann, B. Neinhus, B. Meinerzhagen, and R. Dutton, "Investigation of compact models for RF noise in SiGe HBTs by hydrodynamic device simulation," *IEEE Transactions on Electron Devices*, Vol. 51, pp. 956-961, 2004.
6. J. C. J. Paasschens, R. J. Havens, and L. F. Tiemeijer, "Modeling the correlation in the high frequency noise of bipolar transistors using charge partitioning," *Proceedings of the IEEE BCTM*, pp. 221-114, 2003.
7. Y. Cui, G. F. Niu, and D. L. Harame, "An examination of bipolar transistor noise modeling and noise physics using microscopic noise simulation," in *Proceedings of the IEEE BCTM*, pp.183-186, 2003.
8. A. van der Ziel, "Theory of shot noise in junction diodes and junction transistors," *Proc. IRE*, vol. 43, pp. 1639–1646, Nov. 1955.
9. A. van der Ziel and G. Bosman, "Accurate expression for the noise temperature of common emitter microwave transistor," *IEEE Trans. Electron Devices*, Vol. 31, No. 9, pp. 1280–1283, 1984.
10. K. M. van Vliet, "General transport theory of noise in pn junction-like devices-I. three-dimensional green's function formulation," *Solid state electronics*, Vol.15, No. 7, pp. 1033-1053, 1972.
11. R. Pucel, T. Daniel, A. Kain, and R. Tayrani, "A bias and temperature dependent noise model of heterojunction bipolar transistors," *Tech. Digest of IEEE MTT-S*, pp. 41-144, 1998.
12. Q. Cai, J. Gerber, U. Rohde, and T. Daniel, "HBT high-frequency modeling and integrated parameter extraction," *IEEE Transactions on Microwave Theory and Techniques*, vol. 45, pp. 2493–2502, Dec. 1997.
13. J. Roux, L. Escotte, R. Plana, J. Graddeuil, S. Delage, and H. Blanck, "Small signal and noise model extraction technique for heterojunction bipolar transistor at microwave frequencies," *IEEE Trans. MTT*, Vol. 43, pp. 293–298, Feb. 1995.
14. T. Daniel, R. Tayrani, "DC and high frequenct models for heterojunction bipolar transistor," *IEEE GaAs IC symposium.*, pp. 299-302, 1996.
15. J. T. Winkel, *Electronic Radio Engineering*, Vol. 36, pp.280-288, 1959.
16. Y. Cui, G. F. Niu, D. L. Harame, "Evaluation of bipolar transistor noise modeling using microscopic noise simulation," to be published.
17. G. F. Niu, K. J. Xia, D.Sheridan, D. L. Harame, "Expreimental extraction and model evaluation of base and collector RF noise in SiGe HBTs," *Tech. Dig. IEEE RFIC*, pp. 615–618, Jun. 2004.
18. K. J. Xia, G. F. Niu, D. Sheridan, S. Sweeney, "Frequency and bias dependent modeling of correlated base and collector current RF noise in SiGe HBTs," to be published.

Impact of Lateral Scaling on Low Frequency Noise of 200 GHz SiGe:C HBTs

Rafael Venegas, Nordin Ouassif, Andreas Piontek,
Stefaan Van Huylenbroeck and Stefaan Decoutere

IMEC VZW, SPDT/ITTO/MSTI, Kapeldreef 75, 3001-Leuven, Belgium

Abstract. We investigate the effect of the lateral scaling of an existing 200GHz single poly SiGe:C HBT quasi self aligned (QSA) BiCMOS technology upon low frequency noise (LFN). The purpose of this lateral scaling is to improve F_{MAX} and to displace the peak f_T and F_{MAX} to lower current per unit length and our process and device simulations predict significant effects on these two quantities. Lateral scaling affects the perimeter design of the device, a region having a strong influence on the LFN behavior of the transistor. We study the impact on LFN of reducing three lateral design parameters, namely the emitter window, the width of the enclosure of the emitter window surrounded by polyemitter and the width of the enclosure of the polyemitter surrounded by the active area. Our results show that the process window, predicted by simulation and tested by DC lateral scaling characteristics, still remains available, maintaining the required HF and LFN good qualities with the scaled architecture and process.

Keywords: Design Rules, Transistor Noise, HBT, Bipolar Transistors, BiCMOS.
PACS: 05.40.Ca, 85.30.(-z., De, Pq), 85.40.Qx

1. INTRODUCTION

State-of-the-art SiGe:C heterojunction bipolar transistors have demonstrated high speed and excellent low frequency noise level, turning them on the best low-cost high-performance choice for RF applications in LNAs, oscillators, PA's and mixers. The last is particularly true when a simple QSA architecture may be applied and optimized. Scaling of lateral dimensions of the device allows higher speeds to be obtained by reduction of key parasitic elements, such as base-collector capacitance and base resistance. At the same time, it is important to insure that lateral scaling does not degrade low frequency noise (LFN) performance, thus leaving enough room for a process window when optimizing geometrical parameters for improved RF performance.

Our purpose is to improve F_{MAX} and simultaneously displace the peak f_T and F_{MAX} towards lower collector current per unit device length to optimize a technology for low power applications. This kind of optimization needs a precise knowledge of the impact that the key dimensions variation has upon the high frequency response and the LFN. Some studies have shown [1] that extrinsic base design parameters may have a strong influence on the observed LFN amplitude. Whereas the high frequency response of the scaled device can be examined by running simulation models of the process and of the resulting device electrical behavior, a study of the LFN response of the HBT to geometrical scaling requires an experimental exploration.

2. EXPERIMENTAL

We study here the noise behavior as a function of scaling parameters of HBT devices from a quasi-self-aligned HBT architecture fabricated using a SiGe:C single poly process described elsewhere[2] and which has already demonstrated 200GHz f_T. The process window for lateral scaling of this technology had been already explored [3] based on the observation of DC parameters as a function of the key dimensions. For the optimization, we performed process and device simulations to explore the HF response of the scaled device. Results of these simulations for an optimal combination of the reduced key dimensions are shown in figure 1. We found that a significant increase of F_{MAX} may be obtained, accompanied by the desired displacement of the curves towards lower normalized currents, as may be verified on the figure.

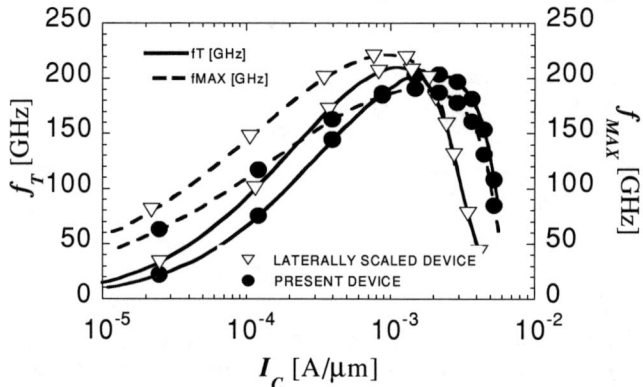

FIGURE 1. f_T-I_c curves for two different technologies, prior to scaling (black circles) and laterally scaled (white triangles).

Measurement of LFN was then performed on a set of devices having three key varying design dimensions, as detailed below, using an automated low frequency noise measurement system (Cadence BTAT-9812B, HP4142 Bias source and HP 8 signal analyser). Measurements were performed under fixed collector-base voltage and swept base-emitter bias, in common-emitter configuration, using a current amplifier and a high input source resistance $R_s \gg R_\pi = dV_{BE}/dI_B$, such that $S_{Ib}=S_{Ic}/\beta^2$. The noise spectral power of the output current presented normally a clean 1/f type noise, sometimes superimposed with one or two lorentzian distributions typically associated to RTS noise, which did not hinder the extraction of S_{Ib} at 1 Hz. The last was extracted by extrapolation of a straight line, fit by least squares, to the longest straight data portions of the log power spectral density. A single slope parameter value was used for each base bias sweep. Curves of area normalized S_{Ib} as a function of each dimension, were obtained by interpolation at selected base bias.

2.1 Scaling Parameters

Lateral scaling was obtained here by the reduction of *three* lateral design parameters (see figure 2). These are: the emitter window width (EW), the width of the enclosure of the EW surrounded by the polyemitter (PE/EW enclosure) and the width of the enclosure of the polyemitter surrounded by the active area (AA/PE enclosure).

All the selected design parameters affect different portions of the perimeter area. This is a region where the fabrication process is most sensitive to the formation of g-r centers responsible of noise generation through various possible mechanisms. For example, effects due to dominant noise sources located at the poly-base epi-base crossover, epitaxial base link region and base emitter depletion region have been identified when varying design parameters related to our PE/EW and AA/PE for a process[3].

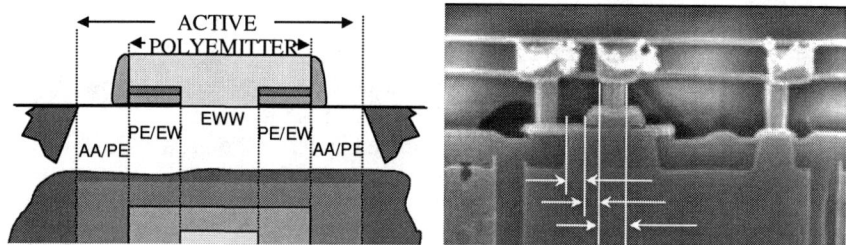

FIGURE 2 a) Transistor profile schematics showing the relevant design dimensions 2b) A SEM view of a typical device using the described architecture and design rule dimensions.

2.2 Noise behavior

Figure 3 shows emitter-area normalized base current noise spectral densities extrapolated at 1Hz as a function of each of the previously indicated design dimension for different magnitudes of the base currents. Variation of the curve shape for most of the curves shown is not significant, with the exception of PE/EW dependency, where noise behavior show strong variations below $0.125\mu m$.

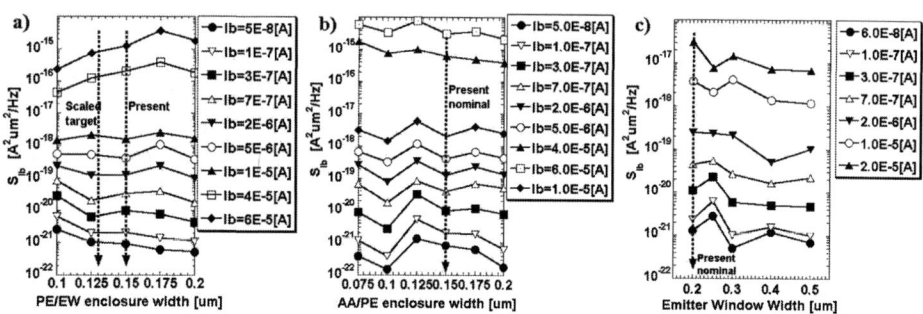

FIGURE 3 Area normalized base current noise spectral density as a function of geometrical design parameter. a) As a function of AA/PE enclosure width. b) As a function of PE/EW enclosure width. c) as a function of emitter window (EW) width.

3. DISCUSSION AND CONCLUSION

The fact that variation of noise level with the three parameters is relatively small for this architecture leaves enough margin for lateral scaling to achieve the predicted HF response.

However, reduction of the PE/EW enclosure is limited, because of strong noise variation below 0.125µm (see figure 3.a). The reduction of this parameter is further limited to 0.13µm, taking into consideration the appearance of a forward tunneling component below this limit, as reported in [3].

On the contrary, LFN results nearly insensitive to variations on the AA/PE parameter (see figure 3.b), indicating that the external base implant sufficiently shields the poly to mono-crystalline transition at the STI edge. Additionally, it was observed that the base collector diode has no increased leakage down to the minimum enclosure considered, implying that the enclosure dimension can be reduced safely to improve F_{MAX}, while holding LFN under control.

Finally, emitter area-normalized noise current spectral density shows only a weak dependence on the emitter window width (see figure 3.c), enabling a further shrink of this parameter with beneficial effects over base resistance reduction.

We conclude that the process window predicted by simulation and tested by the DC lateral scaling results[3] is still available for preserving LFN quality on this architecture and simultaneously achieving high HF performance.

ACKNOWLEDGMENTS

We acknowledge financial support to this work by IWT-Vlaanderen in the frame of a bilateral industrial research project (EXAP-Excellence in Analog Performance, IMEC VZW – NV. PITS -Philips Innovative Technology Solutions).

REFERENCES

1. M. Sanden, B.G. Malcolm, J.V. Grahn and M. Östling, *Solid-State Device Research conference, 2002.* ESSDERC 2002. *Proceeding of the 32th European*, Firenze, Italy, 24-26 September 2002, ESSDERC-2002, pp564-567
2. S. Van Huylenbroeck, A. Sibaja-Hernandez, A. Piontek, , L.J. Choi, M.W. Xu, N. Ouassif, , F. Vleugels, K. Van Wichelen, L. Witters, E. Kunnen, P. Leray, K. Devriendt, X. Shi, R. Loo, S. Decoutere, *Bipolar/BiCMOS Circuits and Technology, Proceedings of the 2004 Meeting 13-14 Sept. 2004* pp. 229 – 232
3. K. Van Wichelen, L. Witters, S. Van Huylenbroeck, P. Leray, D. Laidler,; E. Kunnen, S. Decoutere, *Solid-State Device Research conference, 2004.* ESSDERC 2004. *Proceeding of the 34th European* 21-23 Sept. 2004, pp.333 - 336

Influence of Carbon Incorporation on the Low Frequency Noise of Si/SiGe:C HBTs based on 0.25 µm BiCMOS Technology

P. Benoit[2,1], J. Raoult[1], C. Delseny[1], F. Pascal[1],
J.C. Vildeuil[2], B. Szelag[2], and A. Monroy[2]

[1] *Centre d'Electronique et de Micro-optoélectronique de Montpellier, CNRS-Université UMR 5507, Place Bataillon, U.M. II, 34095 Montpellier Cedex 5, France*
e-mail: benoit@cem2.univ-montp2.fr
[2] *STMicroelectronics 850 rue Jean Monnet, 38926 Crolles cedex, France.*

Abstract. We studied the influence of carbon concentration on the low frequency noise (LFN) of Si/SiGe:C Heterojunction Bipolar Transistors (HBTs). For low and medium concentrations, the same level and evolution of 1/f noise versus bias and geometry is observed. The associated figure of merit, Kb, is closed to $4.10^{-9} \mu m^2$. In this case, noise is mainly generated in the intrinsic transistor at the emitter-base junction. At higher carbon concentration, degradation of the DC and LFN parameter is observed.

Keywords: SiGe:C, Low Frequency Noise, BiCMOS, Carbon
PACS: 72.70.+m : Noise processes and phenomena

INTRODUCTION

The Si/SiGe:C HBTs were supplied by STMicroelectronics Crolles. In this generation of HBTs, carbon is incorporated during deposition of the epitaxial SiGe base layer in order to suppress/minimize boron out-diffusion during emitter process and subsequent thermal treatments (CMOS process). The boron out-diffusion effect, also called transient-enhanced diffusion (TED), is compensated by the incorporation of substitutional carbon atoms[1]. The main advantage is the possibility to process HBTs structures with a very thin and high boron doped base layer. High frequencies operation HBTs can then be achieved together with low base resistances. High performances RF applications (low noise amplifier, mixer, VCO, …) can be addressed using such transistors. The presence of a small content of carbon can additionally reduce the lattice distortion induced by the hetero-epitaxy of the base[1].

The HBTs process consists of a double-polysilicon Quasi-Self-Aligned structure associated with a non selective SiGe:C epitaxy for the base[2]. Transistors with three different carbon concentration were studied: L-type HBT for the lowest concentration, M-type HBT for the medium concentration and H-type HBT for the highest concentration. We have investigated six emitter areas for each type of HBT ranging from 0.32 to 10.24 µm². Gummel-plots and static current gain were investigated from I-V studies. The LFN measurements were performed on-wafer devices in the 1-100kHz frequency range as a function of the base bias current (I_B) and of the emitter area (A_e).

I-V CHARACTERISTICS

Figure 1 shows the Gummel plots for the three types of devices when carbon concentration increases. At low injection ($V_{be} < 0.6V$) for both M and H-Type an increase of the base current was observed, resulting in an increase of the base ideality factor, η_b, (insert in figure 1). This is attributed to the presence of traps in the emitter-base junction which involves a minority carrier lifetime reduction[3] and then an increase of the G-R current component. At high injection, the effect of the carbon concentration on the base current is still present but lower than at low injection.

Whatever carbon concentration and device type, a collector ideality factor close to 1 is observed in figure 1. We also show an increase in the collector current between L and M/H-types devices. This is due to the base width reduction, induced by the presence of the carbon atoms, which ensures an efficiency injection improvement. Nevertheless this effect is not observed between M and H-Types transistors. Hence the increase of the base current involves a static current gain decrease (figure 2).

Thus, our study shows that carbon incorporation in the base leads to an improved static current gain at high injection but its concentration should be properly chosen.

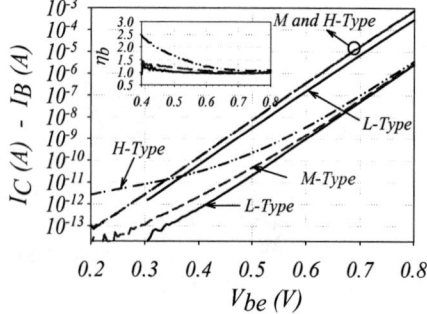

FIGURE 1: Gummel plots and base current ideality factor for the three carbon concentration HBTs.

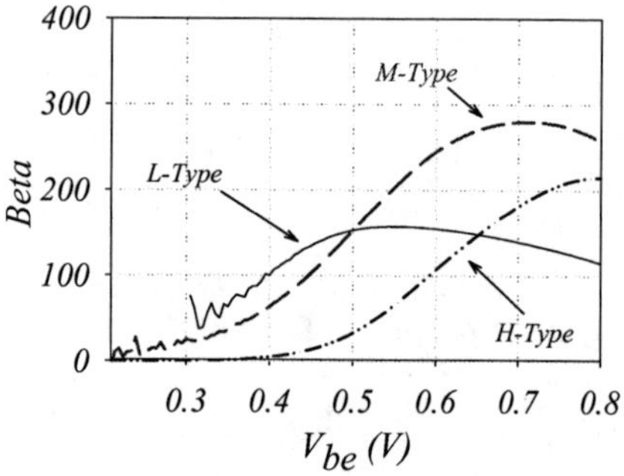

FIGURE 2: Static current gain for the three carbon concentration HBTs.

LOW FREQUENCY MEASUREMENTS

Spectral Analysis

Transistors are biased in the common-emitter configuration. Voltage spectral densities Sv_b and Sv_c are measured simultaneously at the input and at the output of devices in a high impedance configuration. Then, the equivalent current spectral density referred to the input, Si_n, is calculated[4]. The cross-spectrum, Sv_bv_c, is also measured, leading to the coherence function, $\Gamma v_b v_c$. Whatever the transistors, $\Gamma v_b v_c$ is found to be close to 1 in the whole frequency and base current range. Thus, according to our previous works[4], we can directly conclude that the main noise sources are located at the emitter-base junction and that Si_n is equal to the base current spectral density Si_B.

Representative noise spectra of Si_B for the different carbon contents are reported for a large (figure 3) and for a small emitter area (figure 4):

i) For large areas and for low and medium C concentration, spectra are composed of a 1/f component and the white noise, $2qI_B$, is reached at low bias base currents. At high C concentration one or two G-R components appear and the 1/f noise level increases significantly.

ii) For the smallest emitter area the difference in the shape of the spectra is less pronounced but, whatever the transistor, the presence of G-R component(s) is observed. We have already reported[5] such small geometrical effect on SiGe HBTs (i.e. no carbon) and Sanden et al also on advanced Poly-Emitter BJTs.

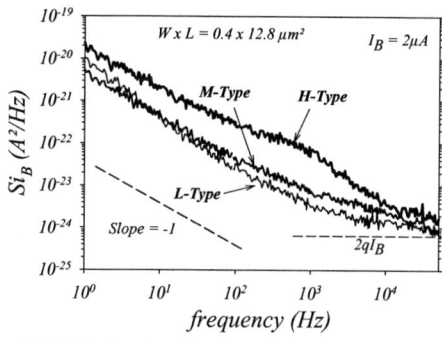
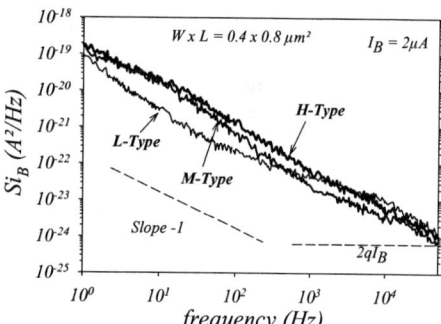

FIGURE 3: Si_B spectra for the L, M and H carbon content HBTs for a large A_e.

FIGURE 4: Si_B spectra for the L, M and H carbon content HBTs for a small A_e.

1/f Noise Level Study

We have studied the evolution of the 1/f noise level (obtained from the decomposition of the Si_B spectra) versus base current. Two representative examples are given in figure 5: for a large emitter area and in the insert for a small area. Firstly, we have systematically (i.e. for all the geometries) found that Si_B evolved quadratically with I_B for L and M-Types HBTs. For H-Type transistors a deviation of this law is sometimes observed. Secondly, we have noticed that high carbon concentration and for largest areas, Si_B increases by a factor of ~10. Unlike L and M-Type devices, reproducibility of LFN results of H-Type is not very good as can be seen in figure 5.

The 1/f noise component of the base current spectral densities can be identified to the SPICE model by: $Si_B = K_f \dfrac{I_B^{A_f}}{f^\gamma}$, with $\gamma \approx 1$.

When $A_f \approx 2$, this model is usually used for a direct comparison of the 1/f noise level by using the unitless parameter K_f.

In order to localize the 1/f noise, we have plotted in figure 6 K_f versus emitter area. For L and M-Type transistors, K_f is found to be inversely proportional to A_e as often reported in the literature for Poly-Emitter BJTs[7] and SiGe HBTs[5]. Hence, for the lowest C concentration, the 1/f noise sources are located in the intrinsic E-B junction. Then, the classical figure-of-merit $K_b = K_f \times A_e$ is deduced and the mean value is around 4.10^{-9} µm². This value is very close to the best results published on no-C SiGe HBTs[5]. For H-Type transistors an erratic evolution is observed and, on the whole, K_f is almost constant with Ae. We can note that the difference in the noise level between H and L/M-Types HBTs is more pronounced for devices with an emitter area larger than 1µm². The large surface effect pointed out in this study could be associated to the fact that noise sources are not homogeneously distributed over the E-B volume or E-B interface. For instance one can suggest the possible influence of carbon precipitate or the increasing role of structural defects (dislocations, growth defects …).

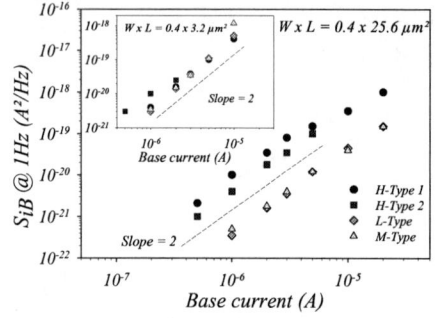
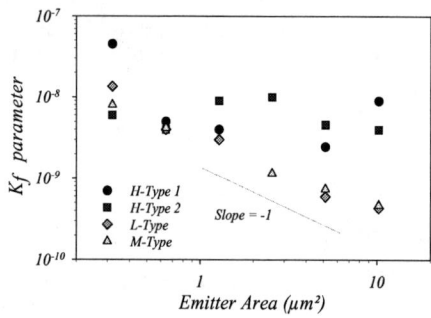

FIGURE 5: S_{iB} versus I_B: 0.4 x 25.6 µm², in insert 0.4 x 3.2µm².

FIGURE 6: K_f parameter versus the emitter area.

CONCLUSION

Carbon incorporation in the epitaxial base layer limits boron diffusion and hence improve the frequency performances of SiGe based HBTs. These additional atoms may create traps in the lattice. A low concentration of carbon atoms significantly improves the static characteristics without LFN degradation. We can notice that high carbon concentration involves an increase of the G-R phenomenon which becomes prejudicial to the low frequency noise level. Concerning the smallest component, the small-geometry effects mask this LFN degradation. To conclude, the carbon incorporation leads to an improvement in the HBT characteristics, but its concentration should be properly chosen.

REFERENCES

1. Lanzerotti, L. D., and Strurn, J. C., Appl. Phys. Letters 70, 1997, pp. 3125-3127.
2. Baudry and al., in the proceedings of BCTM Conference, 2003, pp. 207-210.
3. Ban. I., Öztürk. M. C., Christensen. K., Maher. D. M., Appl. Phys. Letters 68, 1996, pp. 499-501.
4. S. Jarrix, C. Delseny, F. Pascal, G. Lecoy: J. Appl. Phys. 81, 1997, pp 2651-2657.
5. F. Pascal and al., IEE Proc. Circuits Devices Syst, 151,2, 2004, p. 138-146.
6. M. Sanden and al., Fluctuations and Noise Letters, 1, 2001, pp. 251-260.
7. M.J. Deen and F. Pascal, IEE Proc. Circuits Devices Syst, 151,2, 2004, p. 125-137.

Effect of Base/Collector Implant and Emitter-Poly Overlap on Low Frequency Noise in SiGe HBTs

Md Mazhar Ul Hoque[a], Zeynep Çelik-Butler[a], Samuel Martin[b], Chris Knorr[b] and Constantin Bulucea[b]

[a]*University of Texas at Arlington, Electrical Engineering Dept., P. O. Box 19072, Arlington, TX, 76013, USA, E-mail: zbutler@uta.edu*
[b]*National Semiconductor Corporation, 2900 Semiconductor Drive, Mail Stop E-155, Santa Clara, CA 95052, USA*

Abstract. Low frequency noise characteristics of npn SiGe heterojunction bipolar transistors are explored. Transistors with selectively implanted collector and higher doping in the extrinsic base are tested. Variable lengths of the emitter-poly overlap are also studied to investigate their effect on the device noise.

INTRODUCTION

Engineering the dimensional and structural features of the extrinsic regions in SiGe heterojunction bipolar transistors (HBTs) can improve their high speed performance. The doping profiles in the collector and base of the transistors can be altered to take the advantage of a superior performance in analog and RF circuits. However, such aspects of transistor design might deteriorate the device noise. Therefore, the study of the effects of these design features on the low frequency noise in SiGe HBTs is important. The focus of this paper is to investigate the effect of different base/collector doping and varying dimensions of the extrinsic regions on the 1/f noise in SiGe HBTs.

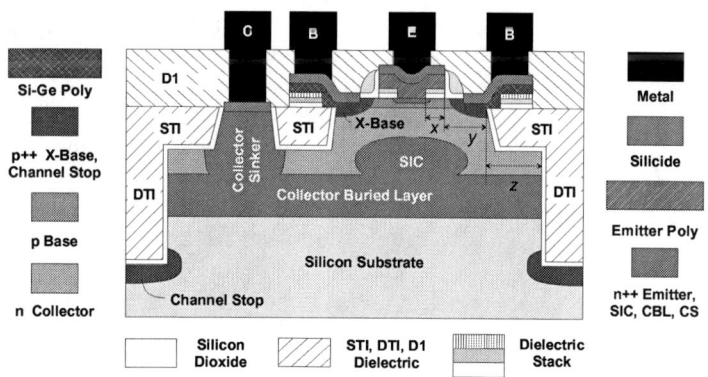

FIGURE 1. Design rules under investigation: x – emitter-poly overlap, y – composite enclosure of poly, z – DTI enclosure of composite.

The transistors studied had non-selective epitaxiallly grown SiGe base, and quasi-self-aligned polysilicon emitters, capable of a peak f_T of 50 GHz. Figure 1 shows a cross section of the SiGe HBT structure with different geometrical design rules. The dimensions x, y and z are the emitter-poly overlap, composite enclosure of the poly, and the DTI (deep trench isolation) enclosure of composite. The transistors are described as "x-y-z" where x, y and z correspond to the dimensions x, y and z in 10 nm units. "SIC:x-y-z" and "HEBI:x-y-z" refer to the transistors with selectively implanted collector and higher extrinsic base implant, respectively.

RESULTS

The voltage noise power spectral densities were measured across the base and collector bias resistances as S_{V_B} and S_{V_C}, respectively. A high base series resistance (~150kΩ) was used to ensure a dominant contribution from the base current noise power spectral density. The base current noise power spectral density (S_{I_B}) is extracted from the two voltage noise power spectral densities measured at 10 Hz [1]. The SPICE noise parameter, AF, is extracted as

$S_{I_B} = KFI_B^{AF}$ at f = 1 Hz.

Figure 2 shows the effect of the selectively implanted collector and the higher extrinsic base implant on the device noise. There is no significant difference in $S_{I_B} f$ (frequency normalized noise) of these three devices. Moreover, the near-identical AF values shown in Fig. 1 suggest similar noise mechanism. The post implantation annealing used for both cases of the selective collector implant and the higher implantation dose in the extrinsic base are believed to have nullified the implantation-induced defects.

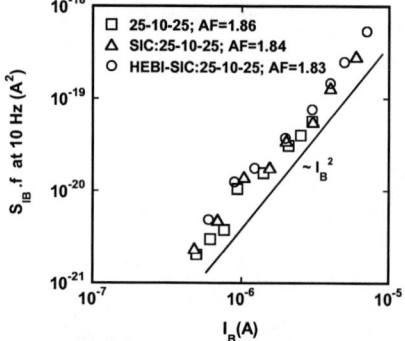

FIGURE 2. $S_{I_B} f$ vs. I_B for 25-10-25, SIC:25-10-25 and HEBI-SIC:25-10-25.

The emitter-poly overlap, referred to as the dimension x, influences both the DC and the noise behavior. Smaller dimension x increases the non-ideality of the base current at lower bias voltage. Figure 3 shows the comparison of the base currents of three transistors with the dimension x of 0.25, 0.2 and 0.1 μm. While the first two show highly ideal base currents, the latter has an ideality factor $\eta_r \approx 3$ for the non-ideal component of I_B. The encroachment of the extrinsic base into the emitter perimeter for smaller dimension x increases the recombination through trap-assisted tunneling to produce such high non-ideality [2].

Figure 4 shows a comparison of $S_{I_B} f$ for the three transistors with different emitter-poly overlap. Although AF values for the larger

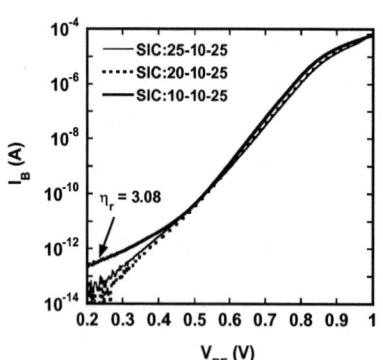

FIGURE 3. I_B vs. V_{BE} for variable emitter-poly overlap (x).

dimension x are closer to 2, it is much lower (1.48) for the transistor with dimension x of 0.1μm. This indicates a different noise mechanism for the latter. Figure 5 shows clean $1/f$ for S_{V_C} measured on SIC:20-10-25 at bias currents of 0.5, 1 and 4 μA, which was the case for all the transistors with larger emitter-poly overlap. However, the devices with smaller x exhibited generation-recombination (g-r) noise at lower bias currents, as shown for a SIC:10-10-25 in Fig. 6 for the bias currents of 0.5, 1 and 4 μA. The g-r noise was overcome by $1/f$ noise which increased at higher bias currents as shown for the base current of 10 μA.

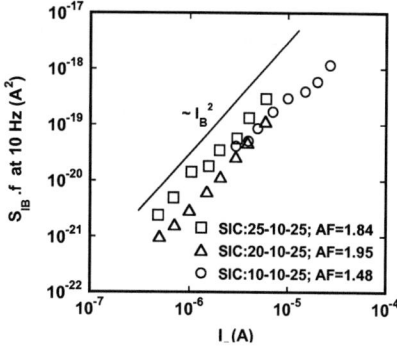

FIGURE 4. $S_{I_B} \cdot f$ vs. I_B for variable emitter-poly overlap (x).

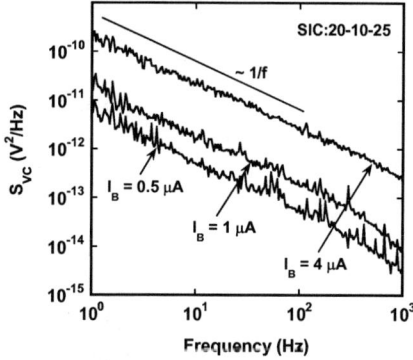

FIGURE 5. S_{V_C} vs. frequency at different base for SIC:20-10-25.

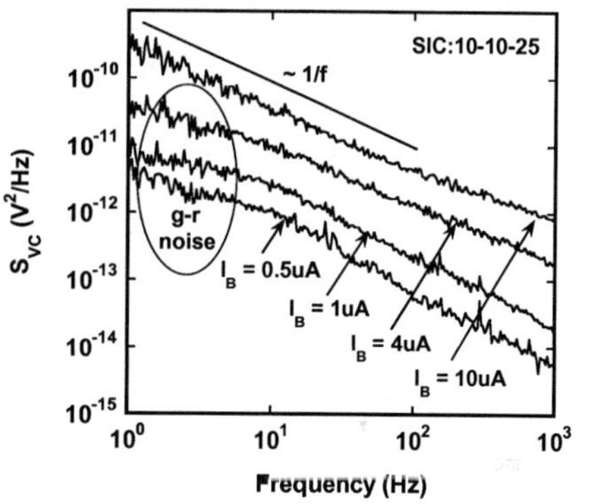

FIGURE 6. S_{V_C} vs. frequency at currents different base currents for SIC:10-10-25.

DISCUSSION

The noise parameter $KF \cdot A_E$ of the transistors with AF =1.83 ~ 1.97 was found to be 2.55×10^{-9} μm² ~ 4.98×10^{-9} μm². Here A_E is the emitter area. An HF clean was used to clean the surface before the polysilicon emitter was deposited. This results in reduced current fluctuations from the oxide at the interface of the polysilicon-monosilicon emitter. Therefore, the quadratic dependence on the base current of S_{I_B} is modeled as a continuous distribution of time constants of the defect centers in the intrinsic emitter-base junction [3].

$$S_{I_B} = K_v \cdot I_B^2 = \sum_{i=1}^{N_v} \frac{\tau_i}{1+(2\pi f \tau_i)^2} I_B^2 \qquad (1)$$

where, N_v (v denotes volume) is the total number of defects in the emitter-base depletion region and τ_i is the characteristic time constant of the ith trap.

For the smaller dimension x, there exists a parasitic transistor through the n+ emitter, p+ SiGe extrinsic base and collector. The n+p+ emitter-base diode produces a thinner depletion region and results in a higher electric field, which increases the recombination through trap-assisted tunneling by the intermediate states within the bandgap introduced by the traps at the oxide interface of the emitter-base junction. The fluctuations in the trap-assisted tunneling current add another component to the current fluctuations in the intrinsic emitter-base junction to reduce AF for the transistors with smaller emitter-poly overlap. Therefore, the transistors with $AF < 2$ can be modeled as [4]:

$$S_{I_B} = K_v \cdot I_{B_v}^2 + K_s \cdot I_{B_s}^2 \qquad (2)$$

where, I_{B_s} (s denotes surface) is the surface recombination component of I_B, I_{B_v} is the volume component of I_B, $K_s = q^4 N_s \lambda / (kTAC_{sc}^2 f)$, k is the Boltzman constant, T is the temperature, q is the electronic charge, A is the area of the surface region, C_{sc} is the base-emitter surface capacitance per unit area, N_s is the oxide slow state volume density and λ is the tunneling attenuation distance. For f=10 Hz, assuming λ =10 Å and C_{sc} =2×10^{-7} F/cm^2, N_s is estimated to be $1\times10^{16} \sim 3\times10^{16}$ (eV.cm^3)$^{-1}$ for the transistors with larger emitter-poly overlap except for the SIC:20-10-25, which has a much lower value of 2×10^{15} (eV.cm^3)$^{-1}$. This indicates the lowest contribution from the trap-assisted tunneling fluctuations (K_s) for SIC:20-10-25.

CONCLUSION

Selectively implanted collectors and higher extrinsic base implants do not deteriorate the noise performance in SiGe HBTs. However, smaller emitter-poly overlaps degrade the ideality of the base current and show an increased contribution from the generation-recombination noise at lower bias currents. Noise originates from the intrinsic emitter-base junction for all the transistors except for the ones with a smaller emitter-poly overlap, for which the most significant contribution comes from the fluctuations in the trap-assisted tunneling current. Hence, there is an inherent trade-off required between noise and speed of the transistors.

ACKNOWLEDGEMENTS

This work is supported in part by SRC under contract No. 2002-NJ-1013 and THECB-ATP Program under Grant No: 003656-0001-2001.

REFERENCES

1. M. M. Hoque, Z. Çelik-Butler, D. Lan, D. Weiser, J. Trogolo and K. Green, *IEEE Trans. Elect. Dev.*, **51**, pp. 1504-1513, 2004.
2. M. Sanden, B. G. Malm, J. V. Grahn and M. Ostling, *Microelect. Reliab.*, **41**, pp. 881-886, 2001.
3. Z. Jin, J. D. Cressler, G. Niu, and A. J. Joseph, *IEEE Trans. Elect. Dev.*, **50**, pp. 676-682, 2003.
4. A. Mounib, G. Ghibaudo, F. Balestra, D. Pogany, A Chantre and J. Chroboczek, *J. Appl. Phys.*, **79**, pp. 3330-3336, 1996.

Low-Frequency Noise in SOI SiGe HBTs Made by Selective Growth of the Si Collector and Non-Selective Growth of SiGe Base

N. Lukyanchikova[1], N. Garbar[1], A. Smolanka[1], M. Lokshin[1], S. Hall[2], O. Buiu[2], I. Z. Mitrovic[2], H. A. W. Mubarek[3], P. Ashburn[3]

[1]*Institute of Semiconductor Physics, 03028, Kiev, Ukraine*
[2]*University of Liverpool, Liverpool, L69 3GJ, UK*
[3]*University of Southampton, Southampton, SO17 1BJ, UK*

Abstract. The Generation-Recombination noise and two components of the 1/f noise of the base current have been identified in SOI SiGe HBTs. It is shown that the GR noise is generated in the emitter depletion layer. One component of the 1/f noise characterized by $S_{IB} \sim (I_B)^{1.7}$ is typical for HBTs made by selective growth of the Si collector and can be due to the mechanical stress created by that technology. Another component is typical for the devices with a polySi emitter being due to fluctuations of the transparency of the interfacial oxide at the polySi/Si interface.

Keywords: SiGe HBTs, 1/f noise, GR noise, SOI, bipolar transistors, polysilicon emitter, stress:
PACS: 72.70,+m; 85.30 De; 85.30.Pq.

INTRODUCTION

While SiGe heterojunction bipolar transistors (HBTs) have been developed for GHz application, their low-frequency noise is of high interest because such noise affects the phase noise of high-frequency circuits where the devices are used and helps to elucidate the physics of different effects in devices. Recently there has been increased interest in SiGe HBTs on SOI for compatibility with SOI CMOS technology [1]. The purpose of this work is to study the low-frequency noise in such devices.

EXPERIMENTAL

The devices were fabricated by selective epitaxial growth (SEG) of the Si collector and non-selective growth of the SiGe base and n-type low doped Si emitter on SOI substrates [1,2]. The doping concentrations were $4 \times 10^{18} cm^{-3}$ for the intrinsic base, $1 \times 10^{17} cm^{-3}$ for the collector and 11.6% for the Ge fraction. Devices with small and large emitter window areas A_{EW} were measured. Noise measurements were carried out on-wafer in the frequency range f = 1 Hz to 200 kHz in a common emitter configuration.

RESULTS AND DISCUSSION

It has been found that the main components of the base current noise are the GR noise at low biases and 1/f noise at high biases (Fig. 1a). Base current GR noise is also observed in the inverse regime of operation (Fig. 1b).

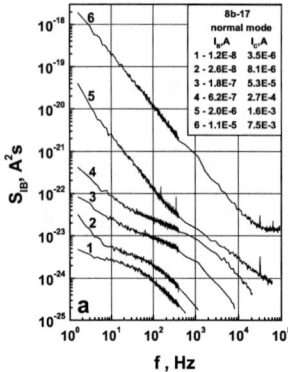 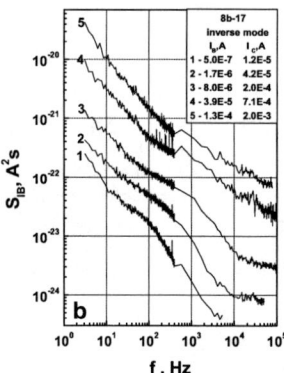

FIGURE 1. Base current noise spectra at different base currents in normal (a) and inverse (b) regimes for an SOI SiGe HBT of $A_{EW}=3\times 3\mu m^2$

Fig. 2a shows a comparison of the behaviour of the GR time constant τ with the collector current I_C in SiGe HBTs and Si control devices for which $I_C(V_{BE})$ are different while $I_B(V_{BE})$ coincide. Fig. 2b shows a comparison of the behaviour of τ with the base current I_B in SiGe HBTs in normal and inverse regimes for which $I_B(V_{BC})$ are different while $I_C(V_{BC})$ coincide where V_{BE} and V_{BC} are the forward base-emitter (normal mode) and base-collector (inverse mode) voltages. These results show that τ does not correlate either with I_B or with I_C.

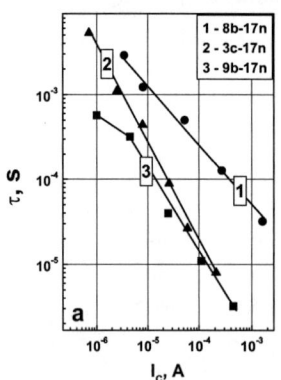 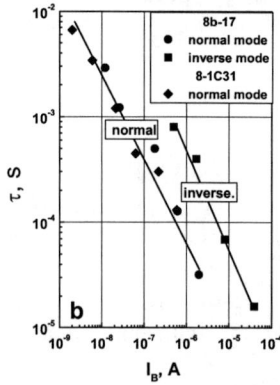

FIGURE 2. Dependences $\tau(I_C)$ for two SOI SiGe HBTs with different concentration of Ge(1, 2) and SOI Si JBT (3) where $I_B(V_{EB})$ coincide while $I_{C1}>I_{C2}>I_{C3}$ at V_{BE}=const (a) and for an SOI SiGe HBT in normal and inverse mode of operation where I_C coincide at $V_{BC}=V_{BE}$ while $(I_B)_{inv}>(I_B)_{norm}$ (b)

At the same time, the dependences $\tau(V_{BE})$ coincide with each other for devices of different types and with the dependences $\tau(V_{BC})$ for the inverse mode if $V_{BC}=V_{BE}$ (Fig. 3a). This suggests that the main factor determining τ is the forward voltage at the

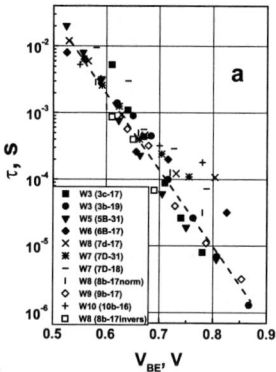 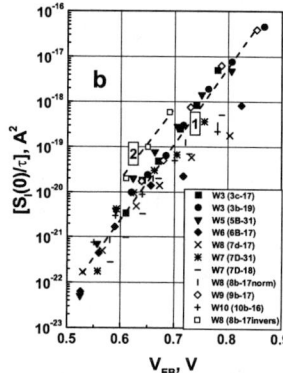

FIGURE 3. Dependences of τ (a) and $[S_{IB}(0)/\tau]$ (b) on V_{BE} for devices of different wafers: wafers #3-#8 are SOI SiGe HBTs, wafer #9 is an SOI Si JBT and wafer #10 is a bulk SiGe HBT

emitter. As is seen from Fig. 3a, $\tau \sim \exp(-qV_{BE}/2kT)$. This is typical for the noise of recombination current I_r flowing through the emitter. In this case [3]

$$[S_I(0)/\tau] \sim I_r^2 N_r / N_d^{1.5} \sim [\exp(qV_{BE}/kT)]N_r^3 / N_d^{2.5} \quad (1)$$

where $I_r \sim (N_r/N_d^{0.5})\exp(qV_{BE}/2kT)$ is the recombination component of the base current, N_r and N_d are the concentrations of recombination centers and donors in the emitter (collector) for the normal (inverted) regimes, q is the electron charge, k is the Boltzmann constant and T is the temperature. The experimental dependences of $[S_I(0)/\tau]$ on V_{BE} presented in Fig. 3b show that $[S_I(0)/\tau]\sim\exp(qV_{BE}/kT)$. This supports the above interpretation. The smaller values of $[S_I(0)/\tau]$ for the normal mode than for the inverse one (Fig. 3b, curves 1 and 2) could be explained by a higher N_d in the emitter than in the collector. We infer that recombination occurs via deep levels arising in the emitter and collector depletion layers from the extrinsic base implantation.

Figure 4a shows the dependences $S_I(I_B)$ typical for the 1/f noise in devices of different A_{EW}. As is seen, $S_I \sim (I_B)^{1.7}$ at all currents investigated for large A_{EW} and at low currents for small A_{EW} while at high I_B the dependence $S_I \sim (I_B)^3$ is observed. It is found that the $I_B^{1.7}$-component is very typical for devices made by SEG of the Si collector [2]. The behaviour of this noise suggests that it accompanies recombination at the oxide/silicon interface where the emitter/base depletion region intersects the oxide surface. The model is considered where the mechanical stress created by the SEG of the Si collector influences significantly this 1/f noise.

As to the $(I_B)^3$-component of the 1/f noise, it should be noted that this component appears at those V_{BE} where $I_C(V_{BE})$ and $I_B(V_{BE})$ turn over, and the smaller A_{EW}, the more severe is the turnover effect. As seen from Fig. 4b, the relation $S_I \sim (I_C)^2$ is valid for this noise component. These results can be explained by the influence of the resistance r_{ox} of the thin interfacial oxide layer between the monocrystalline emitter cap and the poly-Si contact layer as follows. In this case $I_C \sim \exp(qV_E/kT)$ where $V_E=(V_{BE}-I_E r_{ox})$ is the voltage on the emitter junction and I_E is the emitter current.

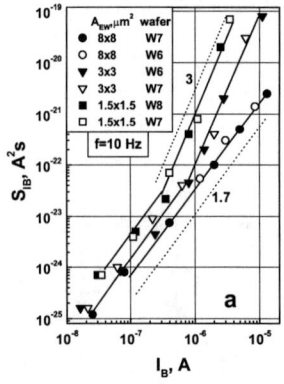 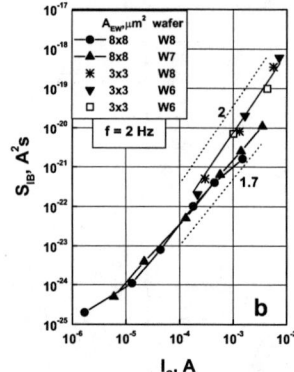

FIGURE 4. Dependences of S_{IB} on the base current (a) and collector current (b) for 1/f noise in small area and large area SOI SiGe HBTs

This means that r_{ox} limits an increase of I_C with increasing V_{BE} giving rise to the turnover effect. For the value of S_{IB} due to 1/f noise of r_{ox} one has [4]:

$$S_{IB} \sim I_{Bp}^2 / A_{EW} \sim A_{EW} \exp(2qV_E/kT) \sim I_C^2 \qquad (2)$$

where $I_{Bp} \sim A_{EW}\exp(qV_E/kT)$ is the base current component due to hole diffusion through the low doped emitter followed by recombination at the interfacial oxide layer. As is seen from Fig 4b, the dependence $S_{IB} \sim (I_C)^2$ has been observed experimentally for the $(I_B)^3$-noise. Moreover, it has been found that $S_{IB} \sim A_{EW}$ at V_E=const. Thus the 1/f noise in devices of small A_{EW} shows the features typical for the noise due to transparency fluctuations of the interfacial oxide layer [15-17] and the turnover effect in curves $I_C(V_{BE})$ can be attributed to the influence of r_{ox} on V_E.

CONCLUSIONS

GR and 1/f noise are observed in SOI SiGe HBTs considered where the GR noise is generated in the emitter junction. The sources of the 1/f noise are located at the SiO_2/Si interface and in the interfacial oxide at the polySi/Si interface.

REFERENCES

1. Hall, S., Lamb, A. C., Bain, M., Armstrong, B. M., Gamble, H., El Mubarek, H. A. W., Ashburn, P., *Microelectronic Eng.* **59**, 449-454 (2001).
2. Lukyanchikova, N., Garbar, N., Petrichuk, M., Schiz, J. F.,W., Ashburn, P., *IEEETrans. Electron Devices,* **48**, 2808-2815 (2001).
3. Lukyanchikova, N. B., *Noise Research in Semiconductor Physics*, London: Gordon&Breach Sci. Publishers, 1996, chapter 5.3.3.
4. Markus, H. A. W., Roche, Ph., Kleinpenning, T. G. M., Solid-St. Electron. **41**, 441-445 (1997).

Noise in SOI MOSFETs and Gate-All Around Transistors

B. Iñiguez[1], A.Lázaro[1], H. A. Hamid[1],
G. Pailloncy[2], G. Dambrine[2] and F. Danneville[2]

[1]*Universitat Rovira i Virgili Department of Electrical, Electronic and Automatic Engineering,
Avinguda dels Països Catalans, 26 43007 Tarragona SPAIN, benjamin.iniguez@urv.net*
[2]*IEMN CNRS UMR 8520, Ave Poincaré, BP 69, 59652 Villeneuve d'Ascq Cédex FRANCE*

Abstract. In this paper, we discuss the RF noise properties of SOI MOSFETs, and we present a suitable model for nanoscale fully-depleted SOI MOSFETs, which is derived from a compact quasi-static SOI MOSFET model by properly extending it to the high frequency regime, using the active line approach and taking into account all the extrinsic parameters. We have used a physically-based noise modeling which takes account diffusion related fluctuations; this allows to study fundamental noise parameters close to the current noise sources and to discuss the downscaling of these noise sources. Finally, we have extended our study to Double-Gate devices, using as a basis a quasi-static model recently presented.

Keywords: Compact noise modeling, SOI MOSFETs, Double-Gate MOSFETs, high-frequency operation
PACS: 85.30.Tv, 85.30.De, 85.40.Qx, 84.40.Lj

INTRODUCTION

SOI MOS transistors are becoming a very promising alternative to conventional bulk CMOS in mixed-mode circuits for wireless applications, because of their better high-frequency performance (due to the lower parasitic junction capacitance, the higher drive current, and the higher transconductance [1]).

The reduced short-channel effects in SOI MOSFETs (in particular in fully-depleted SOI MOSFETs) compared to bulk MOSFETs allow SOI CMOS technology to be scaled down to smaller channel lengths. In order to extend the scalability of SOI CMOS technology, novel SOI MOS structures seem to be necessary, in particular multiple-gate SOI structures: Double-Gate MOSFET, Gate-All Around MOSFET, FinFET,...;they allow the best electrostatic control of the short channel.

Accurate prediction of small-signal equivalent circuit and noise characteristics are a prerequisite for RF-circuit simulation. In this paper, we present a high-frequency physically-based noise model for nanoscale fully-depleted SOI MOSFETs, which is derived from a compact quasi-static SOI MOSFET model by properly extending it to the high frequency regime, using the active line approach and taking into account all the extrinsic parameters. Diffusion related fluctuations are taken into account. The calculated noise performance of the devices has been compared with experimental data. Good agreement has been found. We have also included the influence of a DC tunneling gate leakage current, due to decrease of the oxide thickness. Finally, we have extended our study to Double-Gate devices, using as a basis a quasi-static model recently presented.

NOISE FULLY-DEPLETED SOI MOSFET MODEL

Our noise model is based on the active line approach. The channel is split into n elemental sections, each one is modeled by local equivalent circuit associated to an additional microscopic diffusion noise source (Fig. 1) and quasi-static approximation can be used for each section. The equivalent circuit parameters in each section is derived from a compact quasi-static Fully-Depleted SOI MOSFET [2], which has been validated for channel lengths down to 0.13 µm. The noise model takes into account short-channel effects and the hot carrier effect (a Diffusivity Coefficient is used to define the microscopic noise current sources).

The minimum noise figure is dependent on three noise parameters, which are related to the gate and drain diffusion noise current sources and their correlation by:

$$<i_g^2> = 4.k.T_a.R.\frac{|Y_{11}|^2}{|Y_{21}|}.\Delta f \approx 4.k.T_a.R.\frac{\omega^2.(C_{gsi}+C_{gdi})^2}{G_m}.\Delta f \quad (1)$$

$$<i_d^2> = 4.k.T_a.P.|Y_{21}|.\Delta f = 4.k.T_a.P.G_m.\Delta f \quad (2)$$

$$<i_g.i_d^*> = j.C.\sqrt{<i_g^2>.<i_d^2>} \quad (3)$$

P, R and C are determined using our noise model derived from the active line approach [3]

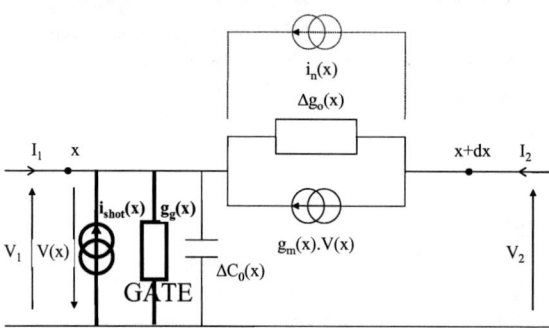

FIGURE 1. Local Small Signal Equivalent Circuit of SOI MOSFET.

By adding extrinsic elements (access resistances, parasitic capacitances) we obtain the main noise parameters for circuit design: F_{min}, R_n (the equivalent noise resistance) and G_{opt} (optimum reflection coefficient).

A good agreement is obtained between theoretical and experimental results (Fig. 2). It has been found that overlap capacitances have a significant influence on NF_{min} magnitude.

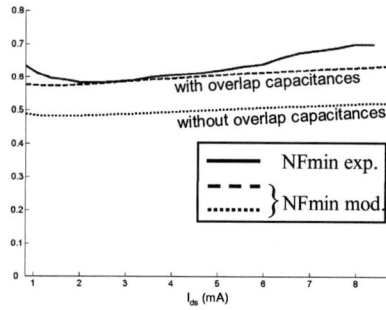

FIGURE 2. Minimum Noise Figure Fmin versus DC bias current Id (Vds = 1.5 V, f = 6 GHz). From [3].

The DC tunneling gate current adds locally a shot noise source (Fig. 1) located across the gate conductance and its contribution to noise is modeled using two gate and drain shot noise sources [4,5]. These gate and drain shot noise sources can be expressed as:

$$<ig_{shot}^2> = 2.q.I_G.\Delta f \qquad (4)$$

$$<id_{shot}^2> = 2.\alpha.q.I_G.\Delta f \qquad (5)$$

with $0.15 < \alpha < 0.3$ depending on the biasing conditions. The correlation between these two noise sources is defined as:

$$C_{shot} = \frac{<ig_{shot}.id_{shot}^*>}{\sqrt{<ig_{shot}^2>.<id_{shot}^2>}} \qquad (6)$$

C_{shot} was found to be purely real, equal to 0.8 over a wide range of DC drain current. The value of C_{shot} and those of α in (5) agrees with previous results published by Van der Ziel in the case of JFETs and MOSFETs [6].

It has been found that as the oxide thickness decreases, the gate shot current noise can reach the same order of magnitude than the gate diffusion noise source, even at microwave frequencies (Fig. 3).

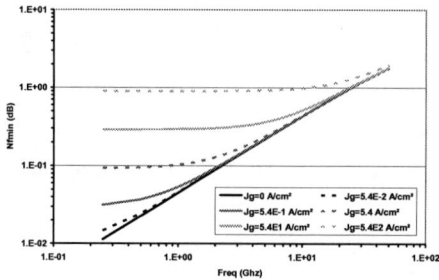

FIGURE 3. Minimum noise figure F_{min} versus frequency for different TGC densities ($V_{GS}=0.5V$, $V_{DS}=0.8V$).

NOISE MODEL FOR DOUBLE-GATE MOSFET

We have also applied the active line approach to develop a noise model for undoped Double-Gate (DG) MOSFET. The assumption of an undoped Si film is reasonable, since for ultra-thin film devices the influence of the doping level is small if the doping is not excessively high; besides, there are practical reasons to use undoped or lightly doped films (higher mobility). The noise model for DG MOSFET has been derived in a similar way as the model for Single-Gate FD SOI MOSFETs. To each channel section we have applied a quasi-static model recently presented for undoped DG MOSFETs [7]. We have included the main short-channel effects in this model.

Fig.4-5 show the frequency dependence of intrinsic and extrinsic noise parameters for two devices with L=1.0 μm L=0.25 μm, respectively. The oxide thickness t_{ox}=1.5 nm and silicon thickness layer t_{si}=5 nm

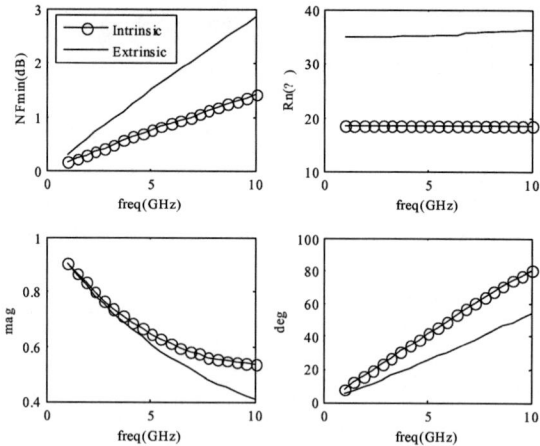

FIGURE 4. Frequency dependence of intrisic and extrinsic noise parameters for DG MOSFET (L=1 μm, W=50 μm, Vgs=1.5V, Vds=1.5V).

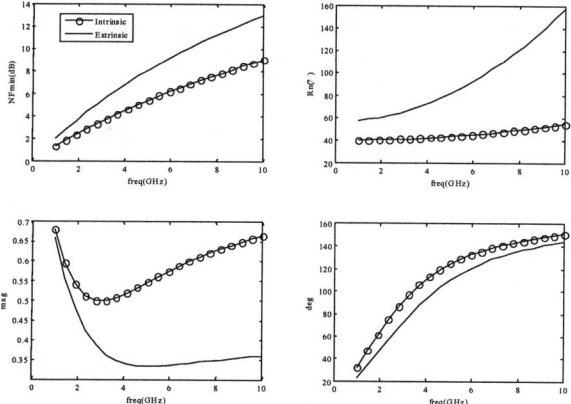

FIGURE 5. Frequency dependence of intrisic and extrinsic noise parameters for DG MOSFET (L=0.25 μm, W=50 μm, Vgs=1.5V, Vds=1.5V).

Similarly, the model can be extended to cylindrical surrounding gate using a recently developed quasi-static model [8].

ACKNOWLEDGMENTS

This work was supported in part by the Ministerio de Ciencia y Tecnología under Project TIC2003-08213-C02-01, and in part by the European Commission under Contract 506844("SINANO") and Contract 506653 ("EUROSOI").

CONCLUSIONS

We have presented a physically-based noise model for nanoscale fully-depleted SOI MOSFETs, which is derived using the active line approach and applying a compact quasi-static SOI MOSFET model to each of the resulting channel sections. Taking into account diffusion related fluctuations, we can discuss fundamental noise parameters related to the current noise sources and the downscaling of these noise sources. Good agreement with measurements has been observed. The effect of the gate tunneling current has also been studied. Finally, we have extended our study to Double-Gate devices, using as a basis a quasi-static model recently presented.

REFERENCES

[1] C.L. Chen, S.J. Spector, R.M. Blumgold, R.A. Neidhard, W.T. Beard, D.-R. Yost, J.M. Knecht, C.K. Chen, M. Fritze, C.L. Cerny, J.A. Cook, P.W. Wyatt and C.L. Keast, "High-performance fully-depleted SOI RF CMOS", *IEEE Electron Device Letters*, vol. 23, no. 1, January 2002.

[2] B. Iñiguez, L. F. Ferreira, B. Gentinne and D. Flandre, "A Physically-Based C_∞-Continuous Fully-Depleted SOI MOSFET Model for Analog Applications," *IEEE Transactions on Electron Devices*, vol. 43, no. 4, p.p., 568-575, April 1996.

[3] G. Pailloncy, B. Iñiguez, G. Dambrine, J. -P. Raskin, F. Danneville, « Noise Modeling in Fully Depleted SOI MOSFETs», *Solid-State Electronics*, vol. 48, no. 5 , pp. 813-825, May 2004.

[4] G. Pailloncy, B. Iñiguez, G. Dambrine and F. Danneville, "Influence of a Tunneling Gate Current on the Noise performance of SOI MOSFETs," *Proc. of the IEEE International SOI Conference*, Charleston, SC (USA), October 2004.

[5] F. Danneville, B. Iñiguez, G. Pailloncy and G. Dambrine, "RF and Noise Properties of SOI MOSFETs, Including the Influence of a Direct Tunneling Gate Current," *Proc. of the 2004 IEEE International Caracas Conference on Devices, Circuits and Systems (ICCDCS'02)*, Punta Cana (Dominican Republic), November 2004 (invited).

[6] A. Van Der Ziel, "Noise in Junction- and MOS-FETs at high temperatures ", Solid-State Electronics, vol. 12, n° 11, pp. 861-866, November 1969.

[7] Y. Taur, X. Liang, W. Wang, and H. Lu, "A continuous, analytic draincurrent model for DG-MOSFETs," *IEEE Electron Device Lett.*, vol. 25, pp. 107–109, Feb. 2004.

[8] D. Jiménez, J. J. Sáenz, B. Iñiguez, J. Suñé, L. F.Marsal, and J. Pallarès, "Modeling of nanoscale gate-all-around MOSFETs," *IEEE Electron Device Lett.*, vol. 25, pp. 314–316, May 2004.

Noise in nanometric s-Si MOSFET for low-power applications

K.Fobelets[*], J.E. Velázquez[†]

[*]*Department of Electrical and Electronic Engineering, Imperial College, Exhibition Road, SW7 2BT London, UK*
[†]*Departamento de Física Aplicada, Pza de la Merced s/n, Universidad de Salamanca, E-37008 Salamanca, Spain*

Abstract. This paper reports on the influence of the gate length reduction on the noise performance of strained-Si surface channel MOSFETs for very low power applications. When the gate length is reduced from 100nm to 20nm an increase of the current gain is achieved that nearly doubles the cut-off frequency of the transistor. This is counterbalanced by a deterioration of the noise figure and the noise resistance of the device for low values of the drain current.

Keywords: Hydrodynamic Model, SiGe, MOSFET, Noise, Fluctuations
PACS: 72.70.+m, 73.50.Td

INTRODUCTION

MOSFET production reached sub-100 nm dimensions in 2001 with Intel shipping 60-nm transistors in the 130-nm bulk CMOS technology node. Recently, the semiconductor industry leaders have shifted the production to the 90-nm node involving even shorter gate lengths. Nevertheless, there is a general consensus in contemplating bulk CMOS as being impractical for future nodes due to heating originated from leakage currents. The introduction of new alternative technologies to bulk CMOS is a pressing issue. The ones based on the system Si/SiGe left the status of promising technology earned in the 90s to enter the production phase in Intel in 90-nm logic circuits. A considerable number of research papers have been published on the possible improvements of n-channel FETs with strained Si channels, both using a buried channel[1] (s-Si MODFET) and a surface channel configuration[2] (s-Si MOSFET). Those technologies have a huge potential for analog and low-power applications. So far minimum noise figures as low as 0.4dB at 2.5GHz and cut-off frequencies (f_T) in excess of 70GHz at 300K have been reported in s-Si MODFET[3]. Recently, we reported encouraging results on the performance of Si/SiGe devices operating in low-power regime when compared with state of the art bulk CMOS[4].

MODEL AND SIMULATED DEVICES

In this paper we report on the calculated noise properties of sub-100nm s-Si MOSFET operating in ultra low-power conditions at room temperature. To this aim we kept the drain to source bias at $V_{DS}=50mV$ and the gate to source bias (V_{GS}) was varied in a voltage range around the threshold voltage (V_{th}). The vertical lay-out of the simulated devices from top to down is: a degenerately doped polysilicon gate, a 6 nm-thick SiO_2 oxide layer, an 8nm-thick quantum well (QW), a 200nm-thick $Si_{0.7}Ge_{0.3}$ set back layer (SBL) grown on top of the graded virtual substrate (VS) and a conventional high-resistivity p-Si wafer. The QW, SBL and VS regions are nominally undoped and the source/drain contacts are assumed self-aligned to the polysilicon gate. Three gate lengths (L_g) have been considered: 100nm (T1), 50 nm (T2) and 20 nm (T3). Two-dimensional simulations have been performed using Synopsys' Taurus/Medici™ implementing an energy balance transport model[6]. In the simulations, impurity de-ionization, Fermi-Dirac statistics and mobility degradation due to both longitudinal and transverse electric field, have been taken into account. The oxide-channel interface and the oxide itself were assumed to be free of traps and fixed charges. Nevertheless, the degradation of the low-field mobility at the channel surface was considered through the use of the Roldan's model[5]. For the simulation of the noise in the MOSFETs we adopted the Impedance Field Method with Langevin stochastics noise sources. We assumed that only thermal (diffusive) noise exists and other noise sources originating from processes such as recombination were not considered in this study.

STEADY-STATE AND DYNAMIC RESULTS

Figure 1 shows the transfer and the transconductance (g_m) characteristics of T1, T2 and T3. There is a noticeable V_{th} roll-off and a degradation of the subthreshold slope (100mV/decade in T1 to over 200mV/decade in T3) as L_g decreases. All the magnitudes presented hereafter are normalized assuming a device's width of 1μm in the non simulated direction.

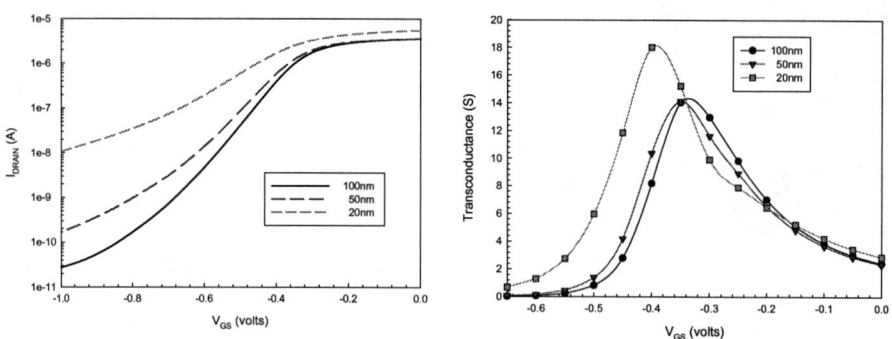

FIGURE 1. I_D-V_{GS} characteristics (left) and transconductance (right) of T1,T2 and T3 ($V_{DS}=50mV$).

The modest increase of the transconductance when shrinking the gate length is a result of several factors: the increased difficulty to keep an adequate carrier confinement in the QW, the mobility degradation due to stronger electric fields in the channel and the fact that the oxide thickness is not scaled down accordingly to the L_g reduction. Nevertheless, maintaining the oxide thickness pays back in terms of the cutoff frequency as can be seen in Fig. 3. As L_g is reduced from 100nm (T1) to 20nm (T3) the maximum of f_T is doubled. This enhancement, not found in g_m, is essentially supported by the gate capacitance reduction.

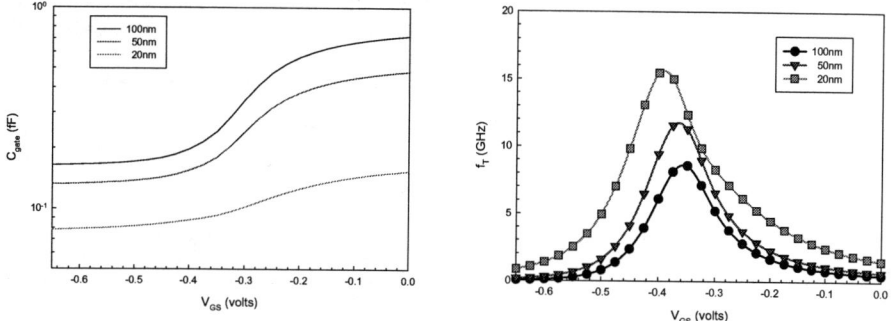

FIGURE 2. Total gate capacitance (left) and cut-off frequency (right) of the simulated transistors.

NOISE ANALYSIS

In this section we present some noise parameters (namely the noise figure and the noise resistance) that are useful to evaluate the noise performance of the device for analog applications. In this study the V_{GS} sweep is limited to a few hundreds of mV around V_{th} in order to maintain the drain current at values sufficiently low to allow for low-power operation (V_{DS} is being kept at 50mV).

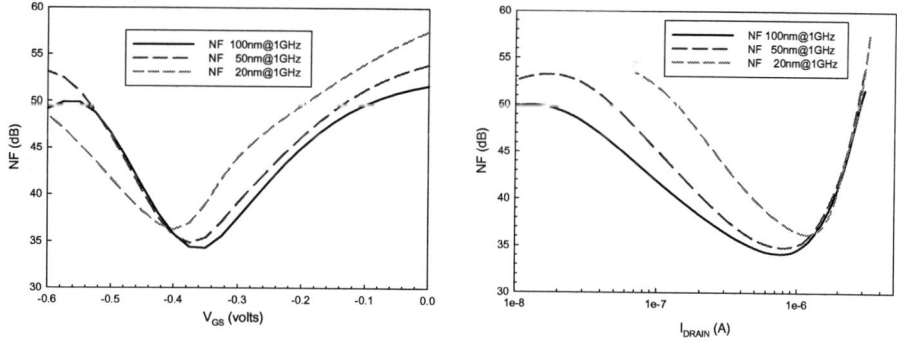

FIGURE 3. Noise Figure of the simulated transistors plotted against the gate voltage (left) and the drain current (right) at 1GHz.

In Fig. 3 we present the noise figure (NF) for the devices under study at a frequency of 1GHz. The minimum value of the NF shifts towards lower values of V_{GS} as L_g decreases (by -60mV as L_g is reduced from 100 to 20nm, Fig. 3, left) at a slower pace than the V_{th} roll-off (by -100mV for the same variation of L_g). As a direct consequence of this the NF steadily grows when the gate length shrinks for all drain current levels under 1µA (Fig. 3, right).

The above behaviour of the noise figure must be understood as a consequence of the combination of the significant resistances across both the intrinsic device and the implanted source and drain regions. Bearing that in mind, we calculated the minimum noise figure (NF_{min}) at the same frequency (Fig. 4, left). Again the shorter gate length transistor (T3) exhibit the largest noise value at very low drain currents. Only for drain current values in excess of $0.8\mu A$ does T3 exhibit the best performance in terms of NF_{min}. Of course, lower values of NF_{min} can be achieved if the source/drain access regions are optimized. In the simulations we maintained the same implantation profile for all devices T1, T2 and T3.

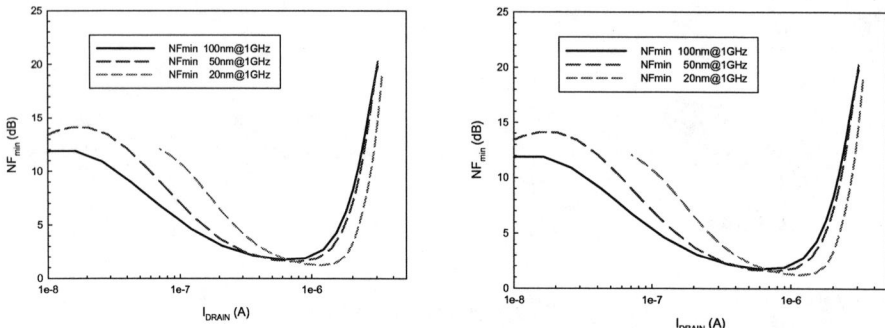

FIGURE 4. Minimum Noise Figure (left) and noise resistance (right) of the simulated transistors plotted against the drain current at 1GHz.

Fig. 4 (right) shows the calculated noise resistance (R_n) at 1GHz. The noise results closely follow the behaviour of NF in Fig. 3 (right). This points out that a significant degradation of the noise performance will follow any gate length reduction when devices operate at very low drain current levels. These effects could be partly mitigated through a careful redesign of the lateral structure.

ACKNOWLEDGMENTS

This work was founded by EPSRC award N. GR/N65844/01, by MCYT/FEDER (TIC2001-1757) and Junta de Castilla y León (SA066/02).

REFERENCES

1. M. Zeuner, T. Hackbarth, M. Enciso-Aguilar, F. Aniel, and H. von Kanel, *Jap. J. Appl. Phys.* **42**, 2363-2366 (2003).
2. K. Rim, J.L. Hoyt, and J.F. Gibbons, *IEEE Tans. Electron Dev.* **47**, 1406-1415 (2000).
   S.H. Olsen, A.G. O'Neill, L.S. Driscoll, K.S.K. Kwa, S. Chattopadhyay, A.M. Waite, Y.T. Tang, A.G.R. Evans, D.J. Norris, A.G. Cullis, D.J. Paul, and D.J. Robbins, *IEEE Trans. Electron Dev.* **50**, 1961-1969 (2003).
3. F. Aniel, M. Enciso-Aguilar, L. Giguerre, P. Crozat, R. Adde, T. Mack, U. Seiler, Th. Hackbarth, H.J. Herzog, U. König, B. Raynor, *Solid-State Electronics* **47**, 283-289 (2003).
4. K. Fobelets, W. Jeamsaksiri, C. Papavasilliou, T. Vilches, V. Gaspari, J.E. Velázquez-Pérez, K. Michelakis, T. Hackbarth, U. König, *Solid-State Electronics* **48**, 1401-1406 (2004).
5. J.B. Roldan, F. Gamiz, P. Cartujo-Cassinello, P. Cartujo, J.E. Carceller, and A. Roldan, *IEEE Tans. Electron Dev.* **50**, 1408-1410 (2003).
6. www.synopsys.com/products/tcad/tcad.html, Synopsys, Inc., 700 East Middlefield Road, Mountain View, CA 94043

Geometry Dependence of 1/f Noise in n- and p-channel MuGFETs

V.Subramanian[1,2], A.Mercha[1], A.Dixit[1,2], K.G.Anil[1], M.Jurczak[1], K. De Meyer[1,2], S.Decoutere[1], H.Maes[1,2], G.Groeseneken[1,2] and W.Sansen[2]

[1]*IMEC, Kapeldreef 75, B-3001 Leuven, Belgium*
[2]*K.U.Leuven ESAT-INSYS, B-3001 Leuven, Belgium*
Email: subraman@imec.be

Abstract. Geometry dependence of 1/f noise in n- and p-channel MuGFETs is investigated for length, width and fin arrays. Based on the geometry and bias dependence of normalized drain current noise spectral density (S_{ID}/I_D^2) and input referred noise (S_{VG}) at low drain bias (V_D=50mV), conclusions are made on the nature and extent to which 1/f noise in these devices follows existing models, and the corresponding noise parameters (oxide trap density, N_t and scattering parameter, α_{SC}) are calculated. It is seen that in general noise follows 1/f behavior and exhibits geometrical scaling. Certain differences are observed between noise in NFETs and PFETs, and also for extremely small devices.

Keywords: FinFET, MuGFET, 1/f noise
PACS: 85.30.De, 85.30.Tv

INTRODUCTION

The **Mu**ltiple **G**ate **FET** (MuGFET), also known as the FinFET, is one of the promising technology solutions for scaling CMOS into the 45nm technology node and beyond. MuGFETs achieve enhanced channel conduction due to contribution from both top surface and side walls of the fin. In addition to meeting the roadmap for digital applications, MuGFETs have a good potential for analog applications [1]. 1/f noise is an important Figure of Merit for analog applications.

This paper focuses on the 1/f noise performance of triple gate MuGFET devices. Trends in 1/f noise as a function of geometrical parameters i.e. gate length (L_G), fin width (W_{FIN}) and fin number (N_{FIN}) are presented, and the results are discussed within the framework of existing theories of 1/f noise.

TECHNOLOGY AND CHARACTERIZATION DETAILS

N and P-channel triple gate MuGFETs have been fabricated on SOI substrates using 193nm optical lithography. A gate stack of 1.8nm nitrided oxide and 100nm poly-Si has been used. The process sequence used for fabricating these devices is described in detail in [2]. Well-shaped fins were obtained, as shown in figure 1(a). A schematic drawing of the MuGFET is shown in figure 1(b).

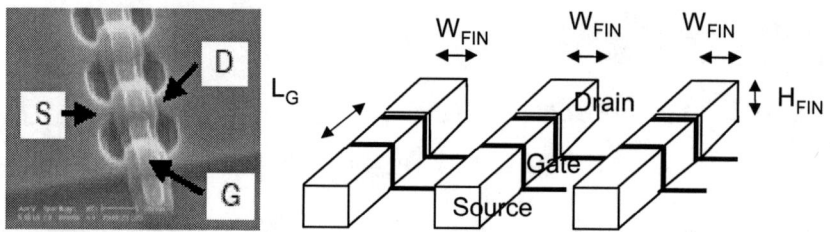

FIGURE 1. (a) SEM picture of a tri-gate MuGFET after spacer etch and (b) Schematic drawing of a MuGFET showing the geometrical parameters L_G, W_{FIN} and H_{FIN}

For 1/f noise characterization, the devices were biased in DC using HP4142 DC Parameter Analyzer, and noise was measured using BTA9812 Noise Analyzer. Gate Voltage (V_G) was swept from 0 to 1.2V for V_D=50mV. At each bias point, the drain current noise spectral density (S_{ID}) was measured for frequencies from 10Hz to 100kHz. The following sets of devices were characterized: (1) Length array for 30-fin NFET (L_G=0.07, 0.2, 0.44 and 1 um); (2) Fin number array for L=70nm NFET (N_{FIN}=1, 5, 15 and 30); (3) Fin width array for L=70nm NFET (W_{FIN}=0.25, 0.5 and 1 um); and (4) Length array for 30-fin PFET (L_G=0.07, 0.2, 0.44 and 1 um).

RESULTS AND DISCUSSION

Frequency spectra (S_{ID} vs. f) of 30-fin NFETs for different gate lengths reveal the 1/f nature of noise in these devices. Hence, the results can be discussed within the framework of 1/f noise theories. A log-log plot of S_{ID}/I_D^2 vs. (V_G-V_T) for different gate lengths (0.07, 0.2, 0.44 and 1 um) is shown in figure 2. A slope of –2 is observed, which is indicative of the number fluctuation (ΔN) theory based on expressions in [3]. To determine whether additional correlated mobility fluctuations are present, S_{VG}.WL is plotted vs. (V_G-V_T) as shown in figure 3.

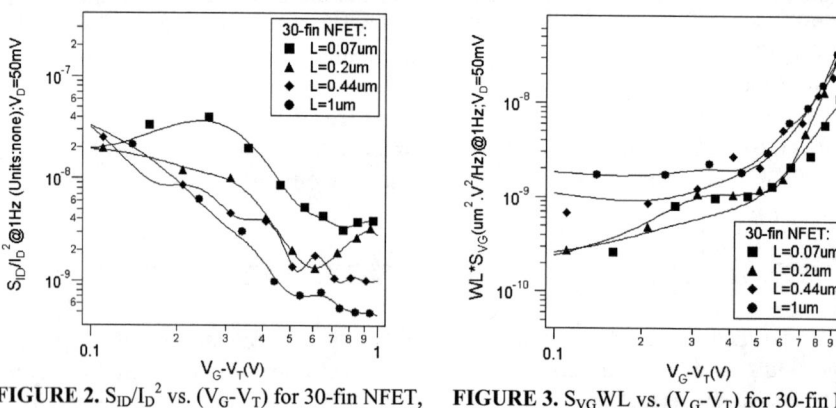

FIGURE 2. S_{ID}/I_D^2 vs. (V_G-V_T) for 30-fin NFET, L_G array

FIGURE 3. S_{VG}WL vs. (V_G-V_T) for 30-fin NFET, L_G array

From this plot, it can be seen that S_{VG} is constant at low (V_G-V_T) as predicted by the ΔN theory, and increases for higher values of (V_G-V_T). This increase in S_{VG} at high gate biases is explained by the unified (ΔN-$\Delta\mu$) theory [4], according to which 1/f noise is the sum of two terms, a bias-independent term due to number fluctuations and a bias-dependent term due to correlated mobility fluctuations, i.e.

$$S_{VG} = [1 + \alpha_{SC}\mu_0 C_{ox}(Vg - Vt)]^2 S_{VFB} \quad (1)$$

where S_{VFB} is the spectral density at flat-band voltage (=$q^2 \lambda kTN_t/WLC_{ox}^2 f$), N_t [cm^{-3}eV^{-1}] is the trap density in the vicinity of the Fermi level, λ is the tunnel attenuation distance (~1A), and α_{SC} is the scattering coefficient. Thus from a plot of S_{VG} vs (V_G-V_T), N_t and α_{SC} can be calculated.

Following this procedure, N_t is calculated to be 4.5×10^{18} cm^{-3}eV^{-1}. This is in agreement with the value of 7×10^{18}cm^{-3}eV^{-1} reported in [5]. From I-V and C-V measurements, the specific capacitance in inversion is calculated to be 1.381×10^{-14} F/um^2, and the electron and hole mobilities are calculated to be 245 cm^2/Vs and 125 cm^2/Vs respectively. With these values, the scattering parameter, α_{SC} for NFETs is calculated to be $\approx 3 \times 10^4$ Vs/C.

The effective width of a triple gate MuGFET is given by W_{EFF}=$N_{FIN} \times (W_{FIN}+2 \times H_{FIN})$. Hence, changing the number of fins is equal to changing device width. A plot of S_{ID}/I_D^2 vs. (V_G-V_T) for different fin numbers (5, 15 and 30) for L_G=70nm is shown in figure 4. Scaling of intrinsic channel noise with fin number is observed down to 5 fin devices. Significant Lorentzian noise is observed in the single-fin FET (spectra not shown). Indeed, in devices with such small dimensions, discrete trapping-detrapping events result in Lorentzian spectra.

In figure 5, S_{ID}/I_D^2 is plotted vs. (V_G-V_T) for different widths of 0.25, 0.5, and 1um for a single-fin, L_G=70nm device. For large (V_G-V_T), 1/f behavior is observed and S_{ID} shows a width dependence, while for lower (V_G-V_T) the spectra are non-1/f in nature (spectra not shown). Significant non-1/f components are also observed for the narrowest device (W_{FIN}=60nm, W_{EFF}=0.18um).

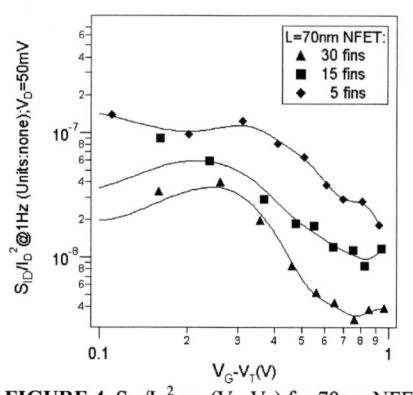
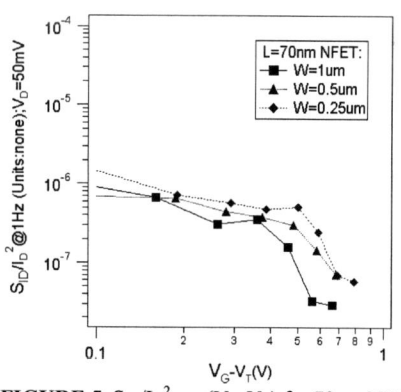

FIGURE 4. S_{ID}/I_D^2 vs. (V_G-V_T) for 70nm NFET, N_{FIN} array

FIGURE 5. S_{ID}/I_D^2 vs. (V_G-V_T) for 70nm NFET, W_{FIN} array

An analysis similar to that for NFETs was performed for PFET devices as well. The noise spectra exhibit 1/f characteristics. Plots of S_{ID}/I_D^2 vs. (V_G-V_T) and $S_{VG}WL$ vs. (V_G-V_T) are shown in figures 6 and 7 respectively. It is seen that S_{ID}/I_D^2 becomes independent of (V_G-V_T) for larger L_G's. This is explained by the fact that since S_{ID}/I_D^2 has an inverse area dependence, the intrinsic noise for large devices is low and the noise arising from the parasitic series resistance becomes visible. Intrinsic noise nevertheless is observed in short channel devices and hence the unified model is applied here and N_t is calculated to be $\approx 1\times10^{18}$ cm^{-3}eV^{-1}, which is lower than the value obtained for NFETs. The scattering parameter α_{SC} is calculated to be 24×10^4 Vs/C, which is higher than obtained for NFETs. From this it can be concluded that there is a much higher contribution to noise arising out of mobility fluctuations for PFETs, compared to NFETs.

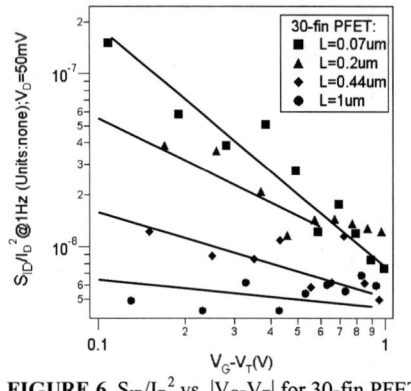
FIGURE 6. S_{ID}/I_D^2 vs. $|V_G$-$V_T|$ for 30-fin PFET, L_G array

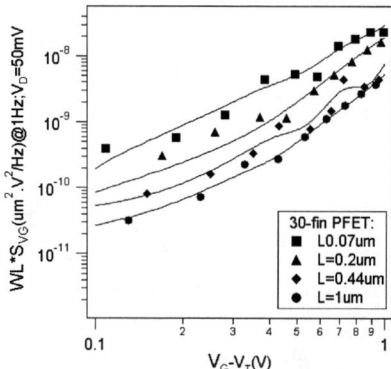
FIGURE 7. $S_{VG}WL$ vs. $|V_G$-$V_T|$ for 30-fin PFET, L_G array

CONCLUSIONS

Length, width and fin number dependence of 1/f noise in n- and p-channel MuGFETs has been investigated. In general, noise is found to scale with L_G, W_{FIN} and N_{FIN}. For devices with very small dimensions, noise spectrum contains significant Lorentzian components. For NFETs, number fluctuations dominate at low V_G-V_T and mobility fluctuations dominate at high V_G-V_T, and the results are explained within the framework of the unified (ΔN-$\Delta \mu$) theory. Trap density (N_t) is calculated to be 4.5×10^{18} cm^{-3}eV^{-1} and is found to be in agreement with values reported in literature. The scattering parameter α_{SC} of the unified model is calculated to be 3×10^4 Vs/C. For PFETs the calculated trap density of 1×10^{18} cm^{-3}eV^{-1} is lower than for NFETs while the scattering parameter is higher. This means that the dominant contribution to noise in PFETs is due to correlated mobility fluctuations.

REFERENCES

1. V.Kilchytska et al, IEEE Trans. Electr. Dev **48**, pp. 65-68 (2004)
2. A.Dixit et al, Proc. of Tech. Papers, Intl. Symposium on VLSI Tech., pp. 112-113 (2005)
3. M.Valenza et al, IEE Proc. Circuits Devices Syst **151**, pp. 102-110 (2004)
4. T.Boutchacha et al, Proc. Intl Conf. on Microelectronic Test Structures **12**, pp. 84-88 (1999)
5. F.Dieudonne et al, IEEE Intl. SOI Conference, pp. 105-106 (2002)

Low Frequency Noise in Si/SiGe HFET

M. Rodriguez[1], N. Zerounian[1], H.-J. Herzog[2], T. Hackbarth[2] and F. Aniel[1]

[1]*Institut d'Electronique Fondamentale, Université Paris-Sud 11, Bt. 220, F-91405 Orsay, France*
E-mail: manuel.rodriguez@ief.u-psud.fr
[2]*Daimler Chrysler Research Center, Wilhem-Runge-Str. 11, D-89081 Ulm, Germany*

Abstract. We report an investigation on low frequency noise behaviour of buried strained Si Channel HFET grown on different strain relaxed virtual substrate which are already very interesting devices for HF and low noise applications. The nature of the low frequency noise, its dependence on gate length, gate width and bias condition are discussed. The Hooge parameter is extracted and compared with other technologies.

Keywords: SiGe HFET, low frequency noise, generation-recombination noise, virtual substrate.
PACS: 85.30.De, 85.30.Tv.

INTRODUCTION

We report an investigation on low frequency noise behaviour of buried and strained Si channel HFETs grown on different strain relaxed buffer (SRB) which are already very interesting devices for low-noise HF applications. SiGe HFETs have high transconductance because both electron and hole mobilities can reach significantly higher values in Si/SiGe heterostructures than in Si devices [1]. These devices can now be considered as reliable. With the same small gate length, SiGe HFET is faster than Si bulk MOSFET. Due to their small active volume, the ultra short gate device are noisy. It is known that noise increases as the size is reducing. This is due of a less efficient fluctuations averaging, when fewer Low Frequency Noise (LFN) sources exist in the device. Naturally, reduction of the noise source density will be a solution, but the coexistence of the Si and SiGe layers for the heterostructure, is a possible source of defects. Previous publication [2] has already pointed out a relatively high level $1/f$ and generation-recombination (g-r) noise like in III-V HEMTs. However recently, encouraging performances in LFN for advanced SiGe n-HFETs devices have been obtained [3]. The authors have demonstrated, thank to correlation measurements that gate and drain noise are originated by different physical mechanisms and that the gate noise does not participate in the phase noise performance.

This work examines the effects of the different strain relaxed buffer (SRB) and gate geometry on the $1/f$ noise. Finally, the $1/f$ noise performance is compared with others technologies, using Hooge parameter as figure of merit, calculated by an original way.

EXPERIMENTAL SET-UP AND DEVICES TECHNOLOGY

The LFN measurements are performed in the frequency range of 10 Hz-100 kHz, using a SR780 FFT analyzer and low noise voltage amplifiers, and the acquisition program corrects measurement to gives the spectral noise density in a band of 1 Hz. Measurements are in linear region of operation, with drain voltage V_{DS} of 50 mV and supplied with a filtered source using a HP4142B, in a room temperature controlled at 293 K. Four Si/SiGe n-HFETs devices with different SRB have been investigated. The Ge content in SRB varies from 30 to 40%. The active devices layers are grown exclusively by MBE technique. Different channel lengths L_G are 0.1, 0.25 and 0.5 μm (mask layout). The channel width W_G is equal to 30, 60 and 100 μm. The device technology is briefly discussed in [4]. The four kind of SRB that have been studied are 1/ low energy plasma enhanced chemical vapour deposition (LEPECVD) [5], 2/ molecular beam epitaxy (MBE) [6], 3/ very low temperature epitaxy (VLTE) [7] and 4/ helium implantation and temperature-assisted relaxation (Helax) [8]. These Si/SiGe n-HFETs devices have been reached outstanding high frequency figures of merit with, f_T of 90 GHz, f_{MAX} of 188 GHz and exhibits a minimum noise figure NF_{min} of 0.3 dB with associated gain G_{ass} of 19 dB at 2.5 GHz [5], [6], [9] but also high dc figure of merit such a 720 mS/mm transconductance and a 570 mA/mm saturation current I_{DSS}.

RESULTS AND DISCUSSION

To facilitate the interpretation and the discussion, we have chosen to present the drain current noise spectral density (S_{ID}) for all devices. Figure 1 shows S_{ID} of several Si/SiGe n-HFETs grown on different SRB for the same drain current I_D = 3 mA/mm (V_{DS} = 50 mV). One can observe 1/f noise and Lorentzian shape noise contribution. The power slope of the 1/f noise is between 0.9 and 1.1 and the Lorentzian spectrum is due to g-r noise involved by traps.

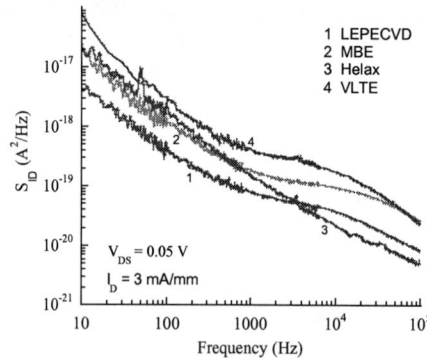 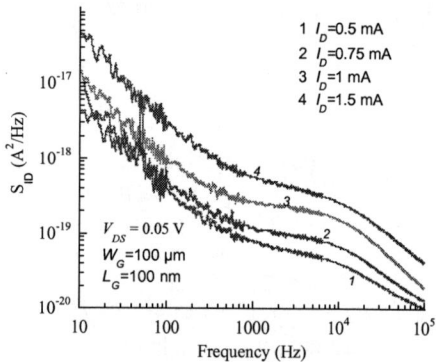

FIGURE 1. Comparison of LFN spectral density for different SRB with W_G / L_G = 100 / 0.1 μm.

FIGURE 2. Comparison of LFN spectral density at different drain current for a L_G = 100 nm LEPECVD Si/SiGe HFET.

Recent measurements of SiGe HFET devices grown on LEPECVD virtual substrate [3], provide power spectral density in the same range and with the same shape.

Between 10 Hz and 1 kHz, S_{ID} is significantly lower in the device grown on LEPECVD virtual substrate. As these devices also exhibit the highest HF performance and particularly the highest transconductance (g_m), we can consider that the low output spectral density is due to a very low noise level in the channel. Besides, it is known that the effective channel electron mobility in tensile surface strained Si channel can decrease due to surface roughness scattering, involving an increase of $1/f$ noise. We can conclude that thick LEPECVD and MBE SRB may show a smaller roughness than thin SRB. On the other hand, the rate of relaxation could also be higher in thick SRB contributing to higher transport properties. Nevertheless, thin SRB are very promising technology and the $1/f$ noise should decrease with maturity.

Figure 2 shows the dependence of the LFN versus I_D for LEPECVD SRB 100 nm gate Si/SiGe HFET. The LFN increases at higher drain current (or gate voltage) due to an increase of both drain current itself and g_m. Figures 3 and 4 provide the normalised drain current noise spectral density (S_{ID}/I_D^2) for several gate length and for several gate width respectively for the LEPECVD SRB HFET. S_{ID}/I_D^2 increases when decreasing the active volume in the device as expected and previously discussed in introduction.

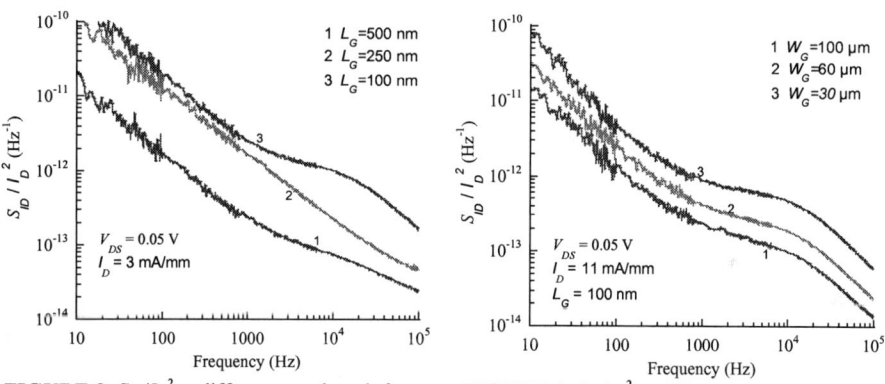

FIGURE 3. S_{ID}/I_D^2 at different gate length for a $W_G = 100$ μm LEPECVD Si/SiGe HFET

FIGURE 4. S_{ID}/I_D^2 at different gate width for a $L_G = 100$ nm LEPECVD Si/SiGe HFET

To evaluate the $1/f$ noise, is useful to use a normalised quantity which take into account the frequency, the bias conditions, the geometric size, etc. As a standard for comparison, in linear regime the Hooge formulation predicts with a relatively good accuracy the $1/f$ noise in the HFET

$$\frac{S_{ID}}{I_D^2} = \frac{\alpha_H}{fN} \qquad (1)$$

where α_H is the Hooge parameter, N is the total number of free carriers controlled by the gate in the channel of the FET. In order to obtain the α_H parameter, N is evaluated taking into account the capacitance in the channel and the parasitic electrostatic capacitances, using HF measurements. Table 1 presents the α_H parameter found in n-HFETs devices studied, with $W_G = 100$ μm and $L_G = 100$ nm, the best performance is reached for the LEPECVD SRB device as already observed on figure 1. The Hooge parameter of all these devices are in the same range as the one of GaAs P-HEMT [10] and slightly higher that state-of-the-art MOSFET [11].

TABLE 1. Hooge parameter in different microwave Si/SiGe n-HFETs for V_{DS} = 50 mV, I_{DS} = 7 mA/mm.

Type of SRB	α_H
LEPECVD	6.1×10^{-6}
MBE	8.5×10^{-6}
HeDI	1.1×10^{-5}
VLTE	1.8×10^{-4}

CONCLUSIONS

We have investigated a large number of Si/SiGe n-HFETs grown on different SRB. The LFN behaviour of these transistors is very close to the III-V HEMT one. Both $1/f$ and g-r noise have been measured and the calculated Hooge parameter are in the same range for SiGe HFETs and for III-V HEMTs. Devices grown on LEPECVD and MBE virtual substrate exhibit a better LFN performance, whereas the devices fabricated on Helax and VLTE SRB contain greater misfit dislocations defects and the rate of relaxation could also be smaller. Dependence of LFN on gate size have been discussed, the presence of $1/f$ noise and of g-r noise increase with downscaling, indicating that the traps plays an important role in the behaviour of the Si/SiGe n-HFETs devices.

REFERENCES

1. F. Aniel et al. "Low temperature analysis of 0.25 µm T-gate strained Si/Si$_{0.55}$Ge$_{0.45}$ n-MODFET's", *IEEE Trans. Electron. Devices* **47**, No. 7, pp.1477-1483, 2000.
2. M. Regis et al. "Noise behavior in SiGe devices", *Solid-State Electron.* **45**, pp. 1891-97, 2001.
3. A. Rennane et al., "Noise behavior of SiGe n-MODFETS", *Material Science in Semicond. Process.* **8**, pp. 383-388, 2005.
4. M. Enciso et al., "HF Noise Performance and Modelling of SiGe HFETs", in the same Proc. of ICNF 2005, and references therein.
5. M. Enciso et al., "DC and high frequency performance of a 0.1 µm n- type Si/Si$_{0.6}$Ge$_{0.4}$ MODFET with f_{MAX} = 188 GHz at 300 K and f_{MAX} = 230 GHz at 50 K", *Elect. Lett.* **39**, 149-151, 2003.
6. M. Enciso et al. "0.3 dB minimum noise figure of 0.13 µm gate-length strained Si/Si$_{0.58}$Ge$_{0.42}$ n-MODFETs". *Elect. Lett.* **37**, pp. 1089-90, 2001.
7. K. Lyutovich et al, "Thin SiGe buffers with high Ge content for n-MOSFETs", *Mat. Sci. and Eng.* **B89**, 341-345, 2002.
8. H.-J Herzog et al., "Si/SiGe n-MODFETs on thin SiGe virtual substrates prepared by means of He implantation", *IEEE Electron Device Lett.* **23**, 485 -487, 2002.
9. F. Aniel et al., "Microwave performances of silicon heterostructure-FETs", *Applied Surface Science* **224**, pp. 370-376, 2003.
10. R. Plana, "Bruit de fond dans les transistors à effet de champ et bipolaires pour micro-ondes", Thétis Université Paul Sabatier, Toulouse, France, 1993.
11. J. Chang et al, "Flicker Noise in CMOS transistors from Subthreshold to strong inversion at various temperatures", *IEEE Trans. Electron Devices* **41**, no. 11, pp. 1965-1971, 1994.

HF Noise Performance and Modelling of SiGe HFETs

M. Enciso[1,3], N. Zerounian[1], P. Crozat[1], T. Hackbarth[2], J.-H. Herzog[2], and F. Aniel[1].

[1]*Institut d'Electronique Fondamentale, Université Paris-Sud 11, Bt. 220, F-91405 Orsay, France*
E-mail: frederic.aniel@ief.u-psud.fr
[2]*DaimlerChrysler Research Center, Wilhem-Runge-Str. 11, D-89081 Ulm, Germany*
[3]*Instituto Politécnico Nacional U. P. Adolfo López Mateos av IPN s/n. Edif. Z 3er piso 07738, México DF*

Abstract. SiGe HFETs have reached in recent years good HF noise performance. We report here an investigation about the contribution of electrostatic parasitics on the noise performance using a PRC electrical model for the intrinsic device. A strong reduction of the capacitance due to gate shape improvement and technological optimization should strongly enhance all the HF performance.

Keywords: SiGe, HFET, noise, electrostatic parasitic capacitances.
PACS: 85.30.De, 85.30.Tv.

INTRODUCTION

The microelectronic market is demanding the development of reliable and cheaper devices with lower power dissipation and low-noise performance. In addition to the development of the Si/SiGe HBT (heterojunction bipolar transistor), it seems that SiGe based hetero-FETs (SiGe HFETs) would satisfy these requirements. The silicon-germanium material heterostructure enables an enhancement of carrier transport properties with a 2DEG (two-dimensional electron gas) or with a 2DHG (two-dimensional hole gas), carrier's mobility exceeding those of bulk Si by far. The 2DEG is generated by the fabrication of a strained Si channel FET (represented in figure 1), and the 2DHG is engendered with a strained SiGe channel FET, both grown on a high quality strain relaxed SiGe buffer layer. In certain cases SiGe FET devices become a serious alternative to high speed III-V devices, while benefiting from the well-established Si technology. Depending on the layer stack configuration, a wide variety of HFET devices can be realized. Recently such kind of devices have demonstrated excellent high-frequency performance that place them as excellent candidates for communication circuit applications like high gain amplifiers, transimpedance amplifiers, ring oscillators and so on. Indeed some circuits' demonstrators have been already manufactured. However to date, microwave noise on such devices has not received much of attention and a limited number of investigations on this subject can be found in the literature.

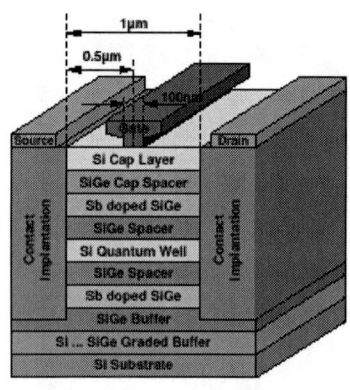

FIGURE 1. Layer stack of the SiGe HFET

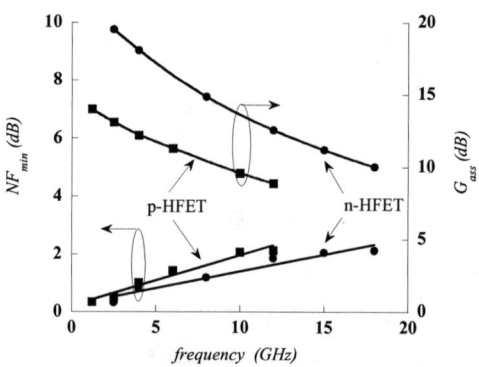

FIGURE 2. NF_{min} and G_{ass} versus frequency of a 130 nm n-HFET (V_{DS} = 1.25 V, I_{DS} = 77 mA/mm), and of a 100 nm p-HFET (I_{DS} = 44 mA/mm, V_{DS} = -0.75 V).

DISCUSSION

After a short review of the noise performances of SiGe HFET, we focus on the impact of parasitic electrostatic capacitances on noise behavior while the contribution of parasitic resistances have been discussed elsewhere [1] and the last part of the article is devoted to noise modelling.

HF Noise performance

The progress of SiGe strained layer heteroepitaxy on virtual buffer substrates has opened up the opportunities for Si-based n- and p-channel HFETs with excellent RF performance. These devices have reached outstanding high frequency figures of merit with f_{MAX} of 188 GHz and 135 GHz, for the n-HFET and the p-HFET, respectively [1]. The first HF noise characterization was performed in 1999 on a 0.1 µm p-type SiGe HFET with a minimum noise figure NF_{min} = 1.1 dB and an associated gain G_{ass} = 18 dB at 3 GHz [2]. Meanwhile the first breakthrough towards experimental noise data concerning the SiGe n-HFET was reported in 2000 for a 0.13 µm gate length device displaying relative high NF_{min} values ranging from 2 to 5 dB between 4 and 26 GHz [3]. These high values were associated with a high leakage current detected in these devices. However, the same year, a SiGe HFET with the same gate length but with a low leakage current shows a very low NF_{min} = 0.3 dB and G_{ass} = 19 dB at 2.5 GHz [4]. Furthermore, noise performance of 0.1 µm pure Ge strained channel p-HFET has been reported with NF_{min} = 0.5 dB and G_{ass} = 13 dB at 2.5 GHz [5]. More recently, IBM labs have reported a SiGe MODFET with L_G = 80 nm presenting a 210 GHz f_{MAX} and an NF_{min} = 1.6 dB at 20 GHz [6]. Figure 2 shows the dependence of NF_{min} and G_{ass} on frequency for the state-of-the-art n and p SiGe HFET of references [4] and [5].

Electrostatic parasitic capacitances

We focuss on the role played by the capacitances on noise performance. The gate-source and the gate-drain capacitances can be separated into a part depending on the gate area and a part which is a residual contribution. These residual capacitances result from the combined effects of two contributions: The electrostatic contribution due to the shape and to the geometry of the gate, involving the coupling between the gate metal edges and the 2DEG, and a fringe contribution arising from the depletion zone within the semiconductor. These parasitic degrade the HF performances of the device by *masking* the strain enhanced transport properties of our SiGe HFETs.

We have measured several HFET technologies for a large range of bias conditions. After de-embedding, the total gate capacitance, i.e. gate-source C_{GS} and gate-drain C_{GD}, capacitance are extracted from S-parameter measurements and subsequent equivalent-circuit parameter extraction for a series of SiGe HFETs with different gate lengths. The extrapolation at $L_G = 0$ of the total gate capacitance $C_{GS} + C_{GD}$ at open channel conditions ($V_{DS} = 1.5$, 2, and 2.5 V at constant $V_{GS}\text{-}V_{TH}$) is depicted in figure 3 for a T shaped gate devices. We do not note any dependence of capacitance versus drain bias. We have investigated several families of devices which essentially differ in the virtual substrate grown technique, doping levels and Ge fraction. We have found residual capacitance at $L_G = 0$ between 130 and 300 fF per mm of gate width depending essentially on doping levels, Ge fraction, and aspect ratio. Higher doping levels are required to achieve lower fringe capacitance, while for a constant doping level, lower Ge fraction yields to lower residual capacitances, because the parasitic MESFET effect increases when the Ge fraction decreases. Such parasitic capacitances are quite high in comparison with data reported in [6]. For devices with $L_G = 0.1$ µm with varying gate width, we have also performed the extrapolation at $W_G = 0$ of the total gate capacitance at open channel conditions and at constant $V_{GS}\text{-}V_{TH}$ (not shown here). In this case the total capacitance exhibits a slight dependence on drain bias. Extrapolated values at $W_G = 0$ are between 30 fF/mm and 55 fF/mm for devices with T gate topology. Again, lower values of residual capacitances were achieved for devices with lower Ge fraction.

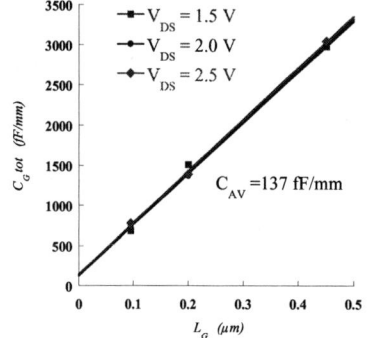
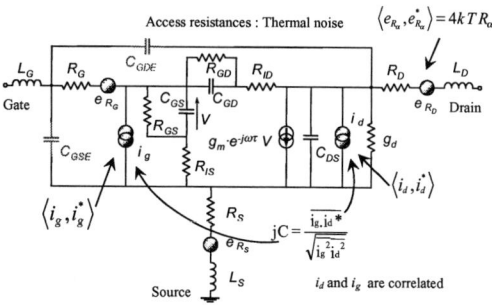

FIGURE 3. Extrapolated zero gate length capacitance from measurements for a SiGe HFET at $V_{DS} = 1.5$, 2, and 2.5 V and at constant $V_{GS}\text{-}V_{TH}$.

FIGURE 4. PRC model for SiGe HFET. Extrinsic elements have been added. For each parasitic resistive element (R_G, R_S and R_D) a thermal noise source is associated.

Small signal Modelling

In order to evaluate the contributions of parasitics on noise performances, we have used the electrical noise model presented in figure 4. The noise of the intrinsic device is modelled using a PRC model and the contribution of the access resistances is included through thermal noise sources. Model and measurements are in good agreement over the 1-18 GHz frequency range. We have taken into account the distributed effects of the residual capacitances controlling values of C_{GS} (C_{GD}) and C_{GSE} (C_{GDE}). Table 1 provides NF_{min} and the equivalent noise resistance R_n at 18 GHz for three configurations. In case #1, C_{GSE} (C_{GDE}) is set to zero and all the gate-source (gate-drain) capacitance is reported in intrinsic C_{GS} (C_{GD}). In case #2, the parasitic electrostatic capacitance is in C_{GSE} (C_{GDE}), the fringe effect which are intrinsic is still in C_{GS} (C_{GD}). It is important to stress that in both cases (#1 and #2) the total capacitance (gate-source plus gate-drain) remains the same. In case #3, we have removed the electrostatic capacitances and let only the value of the intrinsic ones. The distributive effect produces a slight decrease of R_n and NF_{min} with frequency. Thus, the effect of the capacitive coupling between the gate shape and the carrier transport within the semiconductor shall influence the conditions of the device input matching for low noise operation. In fact, the decrease of R_n can be regarded as a better estimation of parasitic capacitances when modelling, outlining the importance to take into account their effects for a proper device noise optimization and for low noise criteria design. The result of case #3 indicates the asymptotic performances one could expect when reducing the parasitic electrostatic capacitances.

TABLE 1. NF_{min} and R_n at 18 GHz for a 100 nm gate Si/SiGe n-HFET.

	#1	#2	#3
NF_{min} @ 18 GHz	3.1 dB	2.95 dB	2.2 dB
R_n @ 18 GHz	45 Ω	43 Ω	49 Ω

CONCLUSION

We have presented low noise SiGe HFET and we discuss the role of electrostatic parasitic capacitances on noise performance. A strong reduction of capacitive parasitics associated with the reduction of gate and source resistance can strongly enhance all the HF figure of merit, including HF noise.

REFERENCES

1. M. Enciso et al., Solid State Electronics, Vol. 48, pp. 1443-1452, 2004.
2. S.J. Koester, Proc. EDMO 1999, pp. 27-32, 1999.
3. M Regis et al., Solid State Electronics, vol. 45 pp. 1891-97, 2001.
4. M. Enciso et al., Elect. Lett., Vol. 37, pp 1089-90, 2001.
5. M. Enciso et al., Elect. Lett, Vol. 37 (24), pp. 1478-1479; 2001.
6. S.J. Koester et al., Proc. ISTDM 2004, Frankfurt, Germany, pp. 125-126, 2004.
7. K. Yazbeck et al., Elect Lett., Vol. 18, No. 19, pp. 1776-1778, 1992.

Model of the 1/f Noise in GaN/AlGaN Heterojunction Field Effect Transistors

M. E. Levinshtein [1], A. P. Dmitriev [1], S. L. Rumyantsev [2]*, and M. S. Shur [2]

[1] *Solid State Electronics Division, The Ioffe Physical-Technical Institute of Russian Academy of Sciences, 194021, St. Petersburg, Russia.*
[2] *Department of Electrical, Computer, and Systems Engineering, CII 9017, Rensselaer Polytechnic Institute, Troy NY 12180-3590*
* *also with Ioffe Physical-Technical Institute of Russian Acad. of Sci. 194021, St. Petersburg, Russia.*

Abstract. The model describing the 1/f noise in GaN/AlGaN Heterojunction Field Effect Transistors (HFETs) has been proposed. This model links the 1/f noise to the phonon assisted tunneling from the two dimensional electron gas in the device channel into the tail states near the conduction band of the GaN layer. The model predicts a fairly weak temperature dependence of the 1/f noise in the temperature interval from 50 K to 600 K with the value of the Hooge parameter α within the range of $10^{-3} - 10^{-5}$. Both these predictions are in agreement with experimental data.

Keywords: 1/f noise, tunneling, GaN/AlGaN, heterostructures.
PACS: 73.40.–c; 73.50.Td

INTRODUCTION

In spite of many efforts, the mechanism of the $1/f$ noise in GaN/AlGaN HFETs has not yet been established. Previous attempts to explain the nature of the $1/f$ noise in AlGaN/GaN HFETs involved three different mechanisms: (i) the occupancy fluctuations of the tail states near the band edges[1], (ii) the fluctuations in the space charge regions surrounding dislocations[2], and (iii) the electron tunneling from the 2D gas into the traps in the adjacent GaN layers[3]. The electron tunneling from the channel into a single Si level in GaN buffer layer was discussed in Ref.[3]. The obtained results allowed us to explain the experimentally observed maximum on the temperature dependence of the low frequency noise. However, the measured $1/f$ noise in GaN/AlGaN HFETs often has a relatively weak temperature dependence in the temperature range from cryogenic temperatures to 600 K. In this paper, we discuss the model of the $1/f$ noise related to the electron tunneling from the 2D channel into the tail states in the GaN buffer and show that this model is qualitative agreement with the experimental results.

THEORY AND DISCUSSION

Figure 1 shows a simplified band diagram of the GaN/Al$_x$Ga$_{1-x}$N heterostructure. Numerical values in Fig. 1 correspond to the typical doped HFET structure with the $x = 0.3$ and with the GaN buffer doped by shallow donors (Si) with the activation energy of $\varepsilon_c - \varepsilon_d = 0.02$ eV. Two-dimensional electron concentration at the GaN/AlGaN interface is $n_0 = 1.3 \times 10^{13}$ cm^{-2}. The density of the tail states in GaN layer, $\rho(\varepsilon)$, decreases exponentially as

$$\rho(\varepsilon) = \rho_0 \exp\left(-\frac{\varepsilon_c - \varepsilon}{\varepsilon_0}\right), \tag{1}$$

where $\rho_0 = \rho(\varepsilon_c)$ is the density of the tail states at the bottom of the conduction band.

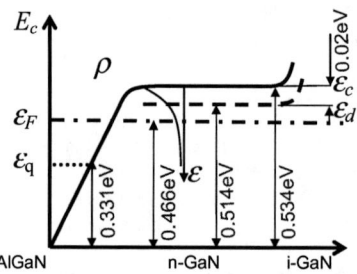

Figure 1. Simplified band diagram of the GaN/AlGaN quantum well. Parameters are calculated for GaN/Al$_{0.3}$Ga$_{0.7}$N heterojunction with doping level of GaN layer 2×10^{17} cm^{-3}. T = 300 K, $n_0 = 1.3 \times 10^{13}$ cm^{-2}.

The 2D electrons in the channel can tunnel into the GaN, where they are captured by the tail states. As seen in Fig. 1, the electrons first tunnel under the bottom of the triangle quantum well and then under the flat part of the conduction band edge. (The electric field at the heterointerface is very high, such the voltage drop across the Bohr radius is comparable or larger then the Bohr energy). The simplest model to capture qualitative feature of electron tunneling in such a system is that of a rectangular potential well that we are using in our calculations below.

The electron tunneling is supposed to be phonon assisted. In this case, electrons within a certain energy interval around the energy level, ε', can be captured by the level in GaN. The probability to be captured is the largest for the electrons with the energy equal to ε' and exponentially decreases with either increase or decrease of the electron energy:

The expression for the noise density of the number of carrier fluctuations (S_n/n_0^2)$_{sl}$ caused by the capture of electrons tunneling on the single level with energy ε' for a triangle quantum well was obtained in Ref.[3]. For a rectangular quantum well it has the form:

$$\left(\frac{S_n}{n_0^2}\right)_{sl} = \frac{4\tau_0 N_e}{WLn_0^2} \int_0^\infty dx \left[\exp(x/x_0) f(\varepsilon')[1-f(\varepsilon')] \frac{1}{1+\omega^2 \tau_0^2 \exp 2(x/x_o)}\right] \tag{2}$$

where N_e is the concentration of the level with energy ε', n_0 is the sheet electron concentration in the channel, W is the gate width, L is source-drain distance, $\omega = 2\pi f$,

$$\tau_0 = \left[A \frac{\sqrt{2m}}{\pi \hbar^2} \sigma_0 \sqrt{\varepsilon'} (\varepsilon_1 + \varepsilon_2) \right]^{-1} \frac{1 + exp[(\varepsilon' - \varepsilon_F)/kT]}{1 + g\, exp[(\varepsilon' - \varepsilon_F)/kT]}, \qquad (3)$$

$f(\varepsilon)$ is the Fermi-Dirac function, and g is the level degeneracy factor.

The expression for the noise spectra caused by tunneling on continuous levels of the tail states can be obtained by integration of Eq. (2) over tail states (see Eq. (1)):

$$\frac{S_n}{n_0^2} = \frac{2\pi \rho_0 x_o}{WLn_0^2 \omega} \int_{-\infty}^{\varepsilon_c} [e^{\frac{\varepsilon-\varepsilon_c}{\varepsilon_0}} f(\varepsilon)(1-f(\varepsilon))]\, d\varepsilon = \frac{2\pi \rho_0 x_o}{WLn_0^2 \omega} \int_{-\infty}^{\varepsilon_c} e^{\frac{\varepsilon-\varepsilon_c}{\varepsilon_0}} \frac{g e^{\frac{\varepsilon-\varepsilon_F}{kT}}}{\left(1 + g e^{\frac{\varepsilon-\varepsilon_F}{kT}}\right)^2} d\varepsilon \quad (4)$$

For two limiting cases $kT \ll \varepsilon_0$ and $kT \gg \varepsilon_0$, the simple analytical expressions for S_n/n_0^2 can be obtained from Eq. (4) [4].

As seen from Eq. (4), this theory predicts the $1/f$ noise spectrum. The level of $1/f$ noise in different semiconductor materials and structures is usually characterized by the dimensionless Hooge parameter, α: $\alpha = S_n/n_0^2 fN$. Here $N = n_0 WL$ is the total number of the conduction electrons in the sample. To estimate the Hooge parameter, we need to know the values of N_t, n_0, ε_0, and temperature dependence of the Fermi level, $\varepsilon_F(T)$.

The N_t and ε_0 values were estimated for Si and GaAs (see Review[5] for more details). For both semiconductors, ε_0 was found to be close to 0.03 eV. The characteristic energy ε_0 should be higher in semiconductors with a larger bandgap and a higher level of crystal imperfection. Absorption coefficient measurements near the bang gap edge yield $\varepsilon_0 \approx 0.1$ eV for GaN. A reasonable estimate for the total concentration of traps in the tail states is 1 to 2 orders of magnitude smaller than the total doping level $N_d + N_a$, where N_d is the shallow donors concentration and $N_a < N_d$ is the concentration of the compensating acceptors. The temperature dependence of the Fermi level was found from the standard neutrality condition.

Solid lines in Figures 2a and 2b show the temperature dependencies of α calculated from Eq. (4) for $n_0 = 1.3 \times 10^{13}$ cm^{-2}, $\varepsilon_0 = 0.1$ eV, $N_d - N_a = 2 \times 10^{17}$ cm^{-3}, $N_t = 0.2(N_d + N_a)$, and different compensation levels N_a/N_d.

As seen, the Hooge parameter α depends only weakly on temperature in the range from 50K to 600K. As expected, the level of the $1/f$ noise increases monotonically with increasing compensation level, while the shape of the temperature dependencies of noise is the same for all compensation levels. The parameter α ranges from $\sim 10^{-5}$ to $\sim 10^{-3}$, in good agreement with experimental data.

The dashed lines in Figures 2a and 2b show the temperature dependencies of the Hooge parameter α for the noise caused by electron tunneling from the 2D channel to the Si doping level ε_d (see Fig. 1). As seen, at low compensation levels, $N_a/N_d \leq 0.5$, the $1/f$ noise from a single Si level dominates in the entire temperature range. At $N_a/N_d = 0.8$, the noise originated from tunneling to the tail states is dominant at $T \geq 400$ K (curves 3, 3' in Fig. 2b), while for $N_a/N_d = 0.98$, the noise from the density of states tail is higher than the noise from the Si level at $T > 100K$ (curves 4, 4' in Fig. 2b).

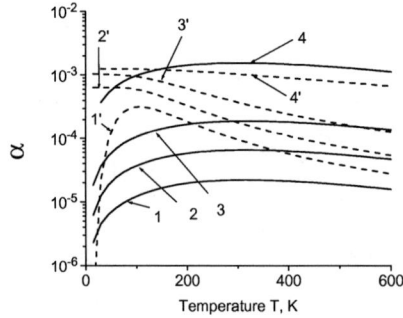

Figure 2. Temperature dependencies of Hooge parameter α for different compensation levels N_a/N_d. Solid lines calculated using Eq. (4) correspond to the noise caused by tunneling on the tail states. Dashed lines represent the noise caused by tunneling on the local Si level. N_a/N_d: 1,1' – 0; 2,2' – 0.5; 3,3' – 0.8; 4,4' – 0.98.

The temperature dependencies of noise for local Si level and for the density of states tail are different. At relatively low temperatures $T < 80K$, the noise from the tail linearly depends on temperature. The temperature dependence of the noise for the local level has a pronounced maximum at low compensation levels, which becomes smaller and disappears at higher compensation levels. At very high compensation levels, this noise only weakly depends on temperature and does not tend to zero at $T \to 0$.

ACKNOWLEDGMENTS

At the Ioffe Physico-Technical Institute this work was supported by Russian Foundation for Basic Research (grants 05-02-17772 and 05-02-17774). The work at RPI has been supported by the National Science Foundation (Project Monitor Dr. Mulpuri Rao).

REFERENCES

1. Levinshtein, M. E., Balandin, A. A., Rumyantsev, S. L., and Shur, M. S., K., "*Low-frequency noise in* GaN-based Field Effect Transistors", edited by A. Balandin, American Scientific Publishers, 2002, pp 49-55.
2. Garrido, J. A., Foutz, B. E., Smart, J. A., Shealy, J. R., Murphy, M. J., Schaff, W. J., and Eastman, L. F., *Appl. Phys. Letters* **76**, 3442-3444 (2000).
3. Wang, R.T., Rumyantsev, S. L., Deng, Y., Borovitskaya, E., Dmitriev, A. Knap, W., Pala, N., Shur, M. S., Levinshtein, M. E., Asif Khan, M., Simin, G., Yang, J., and Hu, X., J. *Appl. Phys.* **92**, 4726-4730 (2002).
4. Dmitriev, A. P., Levinshtein, M. E., Rumyantsev, S. L., Shur, M. S., *Journ. Appl. Phys.*, submitted for publication, (2005).
5. Dyakonova, N. V., Levinshtein, M. E., Rumyantsev, S. L., *Sov. Phys. Semicond.* **25**, 1241-1265 (1991).

Low-frequency Noise Characterization of Hot-electron Degradation in GaN-based HEMTs

S. Jha, J. Gao, C.F. Zhu, E. Jelenkovic, K.Y. Tong, M. Pilkuhn and C. Surya,

Department of Electronic and Information Engineering
The Hong Kong Polytechnic University, Hong Kong

H. Schweizer

Department of Physics
University of Stuttgart, Stuttgart, Germany

Abstract. Hot-electron degradation in MOCVD-grown GaN-based HEMTs, with different gate recess depths, was monitored by flicker noise measurement. Drastic changes were observed in the flicker noise power spectral density, $S_V(f)$ and I-V characteristics when the devices were subjected to voltage stress at V_D=10V for a short stress time of t_S=1 minute. The degradations can be partially reverted by annealing the devices at 100°C for 20 minutes. Further stressing of the devices were performed with V_G=-1.5V and V_D=10V, which results in irreversible degradation in the $S_V(f)$. Detailed analyses of the data suggest that the stressing of the devices, with short t_S, results in the generation of H$^+$ at the AlGaN/GaN interface leading to the observed increase in I_D, and $S_V(f)$. This can be easily annealed as the H$^+$ has relatively low formation energy. The experimental results suggest that the H$^+$ accumulated at the AlGaN/GaN interface may result in a network of percolation paths formed by the depression of the surface potential at the heterointerface. Motion of the H$^+$, arising from the application of a large V_D, leads to the modulation of the percolation paths of the carriers in the 2DEG resulting in large random fluctuations in $S_V(f)$. Further stressing of the devices exhibit substantial increase in $S_V(f)$ due to the generation of traps at the AlGaN/GaN interface. The experimental results demonstrate the significant impact of gate recess depths on the lifetimes of the devices.

Keywords: GaN HEMTs, flicker noise, hot-electron stressing

INTRODUCTION

Gallium nitrides and their related alloys have found important applications in high-temperature and microwave power amplifiers due to their high breakdown fields and high saturation velocities [1]. To achieve better performance in GaN-based HEMTs, gate-recessing technique is used to provide improvements in the transconductance of the devices [2]. Reactive ion etching technique is typically used for the formation of gate recesses. However such processes involve the use of energetic ions and may cause physical damages to the material in recessed regions [3, 4]. This will strongly affect the integrity and the lifetimes of the devices. Flicker noise, being an effective tool for monitoring the integrity of the AlGaN/GaN interface [5], will be utilized for characterizing the hot-electron hardness of the devices.

EXPERIMENT AND RESULTS

GaN-based HEMTs were fabricated using metalorganic chemical vapour deposition-(MOCVD) grown GaN/AlGaN hetero-structures on sapphire substrates. A semi-insulating GaN epitaxial layer, of thickness about 1.5μm, was deposited on a 20 nm-thick low-temperature GaN buffer layer. An undoped AlGaN layer of thickness around 15 nm was deposited on top of the GaN epitaxial layer followed by a 15 nm-thick Si-doped AlGaN layer. A GaN cap layer of thickness around 2.5 nm was deposited on the AlGaN layer. Gate recesses were etched in the AlGaN layer using reactive ion etching technique to enhance the high frequency characteristics of the devices. Recess depths, d_r, of 6 nm (device A) and 18 nm (device B) were used. Also, a device without any gate recess is used as the control device (device C).

The devices were subjected to high dc voltage stressing at V_D = 10V and V_G = 0V and $S_V(f)$ was characterized at V_G = -1.5 V and V_D = 0.22V for different t_S to monitor the degradation of the AlGaN/GaN interface. Significant variations in the device characteristics were observed in the initial phase of hot-electron stressing for t_S =1 minute. Typical results in the I-V characteristics and $S_V(f)$ are shown in Figs. 1 and 2 respectively, which illustrate substantial increase in both I_D and $S_V(f)$. However significant reduction in I_D and $S_V(f)$ were observed after annealing the device at 100°C for 20 minutes. Such phenomenon is observed in all devices, the data exhibited in the figures below are typical results obtained from a device without any gate recess.

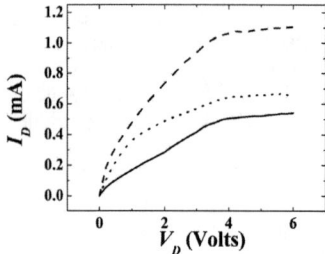

Fig. 1. Measured I-V characteristics of the HEMT device before (solid line) and after (dashed line) 1-minute voltage stress. The dotted line represents the data measured after annealing.

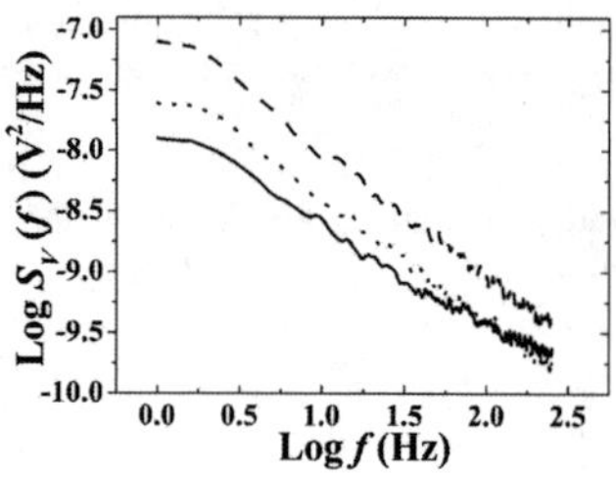

Fig. 2. Experimental $S_V(f)$ of the device before (solid line) and after (dashed line) 1-minute voltage stress. The dotted line represents the data measured after annealing.

Upon further stressing, it is found that the rate of degradation in the $S_V(f)$ is strongly dependent on d_r as shown in Fig. 3. The control device, however, exhibit a gradual increase in the $S_V(f)$ throughout the entire period.

DISCUSSIONS

The significant increase in $S_V(f)$ and I_D for short t_S stipulates substantial increase in the concentration of positively charged states at the GaN/AlGaN interface. Also, the data indicate that the localized states have very low formation energy as shown by the reduction in both I_D and $S_V(f)$ due to a low-temperature anneal at 100°C for just 20 minutes. This shows that the initial degradation cannot arise from the breaking of the Ga-N or Al-N bonds as these have high formation energy and will not be annealed at such low temperatures.

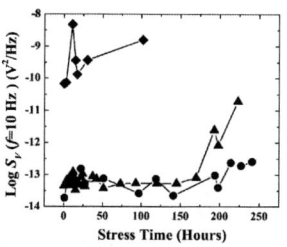

Fig. 3. Experimental $S_V(f)$ as a function of stress time for devices A (solid triangles), B (solid diamonds) and C (solid circles).

The experimental results strongly suggest that the generation of H^+ is responsible for the initial increase in $S_V(f)$ and I_D. It is noted that H^+ are highly mobile in the AlGaN layers. The situation is similar to the incorporation of Na^+ in Si-MOSFETs as reported by Voss [6]. The Na^+ in the SiO_2 layer was found to cause local depressions in the surface potential at the Si-SiO$_2$ interface resulting in the formation of a percolation network. It was reported that a signature V_G dependence of the $S_V(f)$, that reflects the specific arrangement of the Na^+ at the Si-SiO$_2$ interface, can be obtained. The experimental V_G dependencies of $S_V(f)$ are shown in Fig. 4. The results clearly

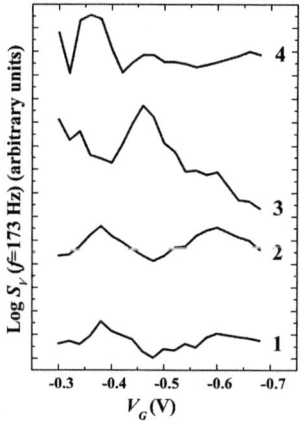

Fig. 4. Experimental variations of $S_V(f = 173$ Hz) vs V_G for measured at 79 K. Lines 1 and 2 were measured consecutively; line 3 was measured after 10 minutes voltage stress at 79K and line 4 was measured after temperature cycling and voltage stress at room temperature.

indicate a signature V_G dependencies for $S_V(f)$, which is repeatable as long as the device is kept at low temperature without being subjected to high voltage bias. For long t_S, the increases in $S_V(f)$ observed in the devices arise from the degradations in the AlGaN/GaN interface. Our experimental data show that d_r has strong influences on the lifetimes of the devices. Device B, with $d_r = 18$ nm, exhibit short lifetime due to hot-electron stressing as indicated by the substantial increase in the flicker noise level after about 50 hours of stress. Device A, with $d_r = 6$ nm, exhibit substantially improved lifetime. Significant increase in $S_V(f)$ for device A was only observed for $t_S >$ 200 hours showing significant improvement in the device lifetime. The control device, with no gate recess at all, did not exhibit any abrupt change in $S_V(f)$ as observed in devices A and B. However, gradual increase in $S_V(f)$ was seen in device C as a function of t_S.

The increase in the magnitudes of $S_V(f)$ indicates the corresponding increase in the trap density at the AlGaN/GaN interface which is related to $S_V(f)$ by

$$S_V(f) = 4 \frac{V^2}{N^2} \int_x \int_y \int_E N_T(x,y,E) \frac{\tau}{1+\omega^2\tau^2} dxdydE. \quad (1)$$

in which V is the dc voltage across the device, N is the total number of electrons in the two-dimensional electron gas (2DEG), $N_T(E)$ is the two-dimensional defect density and τ is the fluctuation time constant. Based on Eq. 1, $N_T(E)$ can be expressed as

$$N_T(E) = \frac{S_V(f) f N^2}{4V^2 AkT},\qquad(2)$$

The computed $N_T(E)$ for device A before and after hot-electron stressing, for $t_S = 200$ hours, is shown in Fig. 5. The results clearly show significant increase in the interface trap density due to the stressing experiment.

CONCLUSION

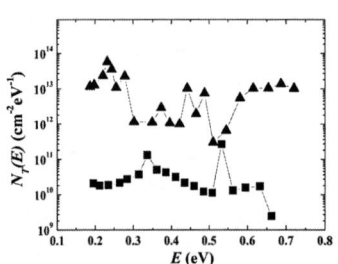

Fig. 5. Computed $N_T(E)$ before (solid squares) and after (solid triangles) the dc voltage stress for device A.

We have investigated the hot-electron hardness of GaN-based HEMTs with various gate-recessed depths fabricated by reactive ion etching technique. Characterization of the flicker noise measured over the two-dimensional electron gas indicated significant increase in the AlGaN/GaN interface traps for devices with deep gate recesses. Devices with gate recesses also indicated substantial shorter lifetimes due to hot-electron injection. Based on our experimental results it is concluded that the hot-electron injection resulted in substantial increase in the interface traps at the AlGaN/GaN heterojunction.

ACKNOWLEDGEMENT

This project is funded by RGC grants (PolyU 1/01C, PolyU 5134/02E and PolyU 5236/04E) and a University Research Grant awarded by The Hong Kong Polytechnic University. Further support is provided by a grant under the Germany/HK Joint Research Scheme (G_HK030/02).

REFERENCES

[1] Burm J, Chu K, Schaff W J, Eastman L F, Khan M A, Chen Q, Yang J W and Shur M S, *IEEE ElectronDevice Lett.*, **18**, 1997, pp. 141.
[2] Hikosaka, Hirachi Y, and Abe M, IEEE *Trans. Electron. Devices*, **33**, 1986, pp. 583.
[3] Pearton S J, Abernathy C R, Vartuli C B, Lee J W, Mackenzie J D, Wilson R G, Shul R J, Ren F and Zavada J M, *J. Vac. Sci. Technol.*, **14**, 1996, pp. 831.
[4] Jones B K, IEEE *Trans. Electron Device*, **41**, 1994, pp. 2188.
[5] Levinshtein M E and Rumyantsev S L, *Semicond. Sci. Technol.*, **9**, 1994, pp. 1183.
[6] R F Voss, *J Phys. C: Solid State Phys.*, **11**, 1978, L923.

Low Frequency Noise Of AlGaN/GaN HEMT Grown On Al₂O₃, Si And SiC Substrates

J.G Tartarin[1-2], G. Soubercaze-Pun[1-2], A. Rennane[1-2], L. Bary[1], S. Delage[3], R. Plana[1-2], J. Graffeuil[1-2]

1- LAAS-CNRS, 7 av. du Colonel Roche, 31077 Toulouse cedex 4, France
2- Université Paul Sabatier, 118 rte de Narbonne, 31062 Toulouse cedex 4, France
3- TRT-TIGER, domaine de Corbeville, 91404 Orsay cedex, France

Abstract. The use of wide bandgap materials for broadcast telecommunication and defense systems allow high power, high efficiency and high integration levels of active devices thanks to their microwave electrical performances. GaN based devices have also demonstrated great potential for high frequency linear low noise applications. However, low frequency noise (LFN) performances characteristics are still under progress as they are related to the material quality and process control. As a consequence, the LFN sources identification and modeling in AlGaN/GaN devices have a twofold stake: on one hand it contributes to the process improvement by the identification of the main noise sources, and on the other hand the non-linear noisy model can be used for CAD of non linear circuits such as low phase noise oscillators. This study focuses on the confrontation of High Electron Mobility Transistors (HEMT) featuring 0.15x2x50μm² gate dimension grown by MOCVD on sapphire (Al₂O₃), silicon (Si) and silicon carbide (SiC) substrates. Each substrate has got its own advantages and drawbacks in terms of cost, wafer size, thermal conductivity and lattice mismatch. This paper deals with the noise mechanisms relative to the use of several substrates: for that purpose, low frequency noise measurements have been performed under different biasing conditions for each substrate. The contributions of the different noise sources (1/f, generation-recombination centers (GR),...) are discussed for each substrate and related to each technological process.

Keywords: Low Frequency Noise, wide band-gap materials, SiC Si and sapphire substrates, G-R centers, noise parameters extraction procedure.
PACS: 05.40.Ca, 73.50.Td, 85.30.De, 85.30.Tv

STATIC MEASUREMENTS

The HEMT devices grown on Silicon, silicon carbide and sapphire substrates have been processed using the same masks set. The Aluminum content in the 2 dimension Electron Gaz (2DEG) layer ranging from 22% to 24% according to the process is however close whatever the substrate. The MOCVD technique has been used for devices grown on sapphire and silicon carbide [1] while the MBE technique has been applied for devices on Silicon [2]. Static and pulsed measurements have been largely performed using some 10 samples for each different gate geometry (length, width) to appreciate the scattering of the electrical performances over the wafer (and so the yielding of each process). The main parameters are reported on Figure 1 for devices featuring 0.15x2x50μm². Devices on SiC exhibit improved performances in terms of higher drain current, higher transconductance gain and lower gate leakage current as well as reduced contact or channel resistance R_{ON}, thus proving a better process maturity. Moreover, a better yielding has been measured for the process on SiC substrate. The following study on low frequency noise performances has been performed on standard devices for each substrate.

TABLE 1. HEMT ($2\times0.15\times50\mu m^2$ gate area) static parameters (mm unit refers to the normalization against the gate width of the devices).

Substrate type	Al_2O_3	Si	SiC
I_{DSS} (saturation drain current, mA/mm)	400	350	1000
V_T (threshold voltage, V)	-3.75	-3.5	-5.5
Gm max (transconductance, mS/mm)	120	100	250
I_G (gate leakage currents, μA)	0.3-0.5	30-80	<0.1
R_{ON} resistance (@V_{GS}=0V, ohmic regime, Ω)	70	85	25
Ft and Fmax (GHz)	30-57	16-37	40-100

LOW FREQUENCY NOISE (LFN)

The experimental setup used is based on the transimpedance amplifier direct measurement technique which allows rapid LFN measurements from 10Hz to 100kHz. The different spectral densities on the drain (S_{ID}) and gate (S_{IG}) accesses can be simultaneously characterized as well as their cross-correlation. From a previous study on S_{ID} versus the gate width and gate length [3], we have found the main noise contribution to occur in the active region (2DEG) under the gate when the device is biased in its saturated $I_{DS}(V_{DS})$ zone. Neither the various complex noise mechanisms revealed by S_{IG} measurements, nor the cross-correlation will be described in this paper. Only the S_{ID} spectral measurements will be discussed in the next paragraphs: the correlation has been found to be null, and so S_{IG} impacts lowly on the overall noise in circuits such as oscillators. Moreover, we have developed an analytical extraction procedure (MatLab) to get the different noise parameters contributions (1/f and generation-recombination G-R centers, noise floor, ...). Figure 1 illustrates the different extracted noise contributions for a HEMT on SiC substrate.

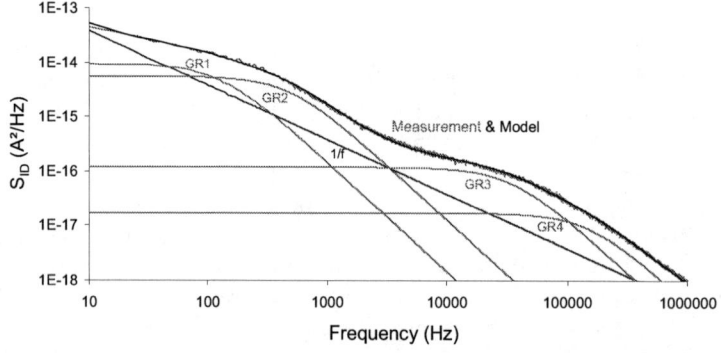

FIGURE 1. Extracted low frequency noise sources ($0.25\times2\times75\mu m^2$ HEMT @ V_{GS}=-2V, V_{DS}=16V)

The robustness of this procedure has been checked with analytical noisy spectra to ensure the uniqueness of the parameters convergence. More than 6 G-R centers, as well as flicker noise and floor noise can be extracted over the considered frequency range. The extraction accuracy is just limited by the noise contribution weight in the overall noise (the weaker the noise contribution, the worse the accuracy).

LFN Measurements - Linear Biasing Operating Mode

This study in the static $I_{DS}(V_{DS})$ Ohmic region allows the location and the quantification of the dominating noise source that take place in the active 2DEG channel or in the contact resistances. The technique proposed by Peransin [4] is used in Figure 2. This method states that only flicker noise is taken into account.

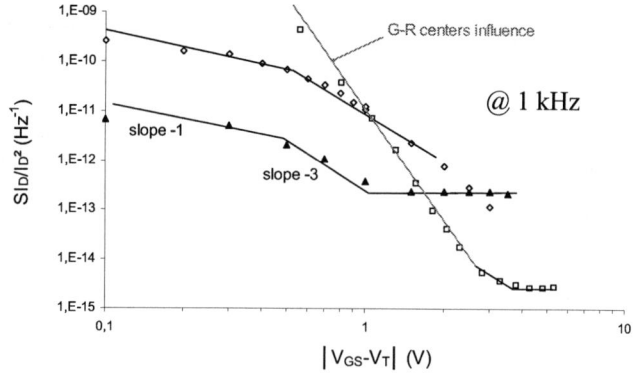

FIGURE 2. LFNmeasured @ 1kHz (Ohmic region) according to Peransin's technique applied to AlGaN/GaN HEMT featuring 0.15x2x50µm²: ▲ Al$_2$O$_3$, □ SiC, ◊ Si substrates

From the -1 slope for devices on Si and Al$_2$O$_3$ substrates (knowing the carrier density and mobility), the Hooge parameters have been estimated respectively at $1,5.10^{-3}$ and 10^{-4}. These results are coherent with the higher dislocations in the 2DEG for device grown on silicon than on sapphire, mainly attributed to higher lattice mismatch and threading dislocations for devices on Si. If only flicker noise contributes to the S_{ID} spectra for the two previous substrates, the devices on SiC substrate suffer from numerous G-R centers. Even after the extraction of the Flicker noise contribution, a non-realistic slope value of -7 already hinder to access the Hooge coefficient for that substrate. The extracted G-R magnitudes also exhibit a -6 slope dependency with the gate biasing voltage for the same biasing range: we assume the presence of another G-R center out of band that raise the S_{ID} spectra, thus raising the apparent extracted 1/f contribution. However, from the extrapolated -3 slope dependency (1/f), the maximum Hooge parameter for devices on SiC still remains below that of devices on Sapphire, indicating a higher degree of structural perfection.

The constant normalized spectral density values near V_{GS}=0V refers to the noise in the access and contact resistances. This plateau does not appear for device on Si, because the noise in the 2DEG channel is still dominating that in the accesses and contacts resistances. The lower value (2 orders of magnitude) for devices on SiC below that of transistors on Al$_2$O$_3$ corroborates the static measurements from table 1. The GaN layer structural perfection and Ohmic contacts appreciations from figure 2 reveal the better process mastership for devices on SiC.

LFN Measurements - Saturated Biasing Operating Mode

Devices on sapphire and silicon with a single 1/f dependency exhibit a proportional variation of S_{ID} versus I_D. But the presence of the G-R centers for devices on SiC is much more difficult to model. From figure 3, we can notice the trapping contribution (4 G-R centers) and the lower 1/f extracted source for devices on SiC. Some of these G-R centers can be attributed to trapping effects due to the presence of hydrogen in the 2DEG (that behaves as an acceptor in n-GaN, and traps the carriers [5]).

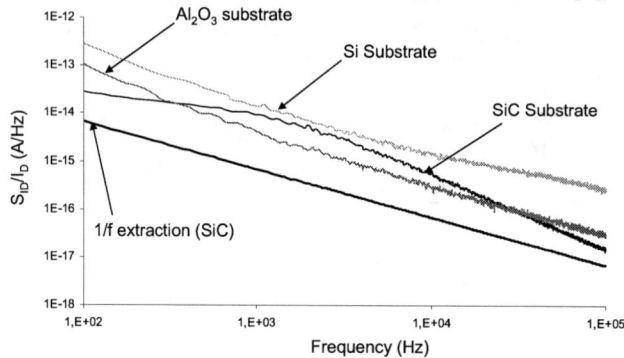

FIGURE 3. Normalized spectral density for a $0.15 \times 2 \times 50 \mu m^2$ HEMT (V_{GS}=-2V, V_{DS}=6V)

CONCLUSIONS

The results presented put forward the differences on the LFN performances of AlGaN/GaN HEMT grown on Al_2O_3, Si, and SiC substrates. Devices on SiC exhibit the best contact process and 2DEG layer quality, in spite of the presence of numerous G-R centers. This technology is still promising, once removing these defects.

ACKNOWLEDGMENTS

The authors would like to thank the French Ministry for Scientific Research and ESA for their financial support (contracts RNRT ANDRO and N°16013/02/NL/CK).

REFERENCES

1. M-A. di Forte-Poisson et Al., "MOCVD growth of group III nitrides for high power, high frequency applications" in Phys. Stat. Sol. (c) 2, No. 3, 2005, pp. 947-955.
2. F. Semond et Al., "Molecular Beam Epitaxy of group-III nitrides on silicon substrates : growth properties and device applications", Phys. Stat. Sol. (a), No 2, 2001, 188, pp. 501-510
3. J.G. Tartarin et Al., "Using low frequency noise characterization of AlGaN/GaN HEMT as a tool for technology assessment and failure prediction", Noise in Devices and Circuits II, Proc. of SPIE, Vol. 5470, Maspalomas, Spain, May 2004, pp.296-306.
4. J.M Peransin et Al., "1/f noise in MODFET's at low drain bias", IEEE Trans. On Electr. Dev., Vol. 37, No. 10, pp. 2250-2253, Oct. 1990.
5. J.M Arroyo et Al., "Effect of Deuterium diffusion on the electrical properties of AlGaN/GaN heterostructures", MRS-symposium E March 2005, San Francisco, USA.

Investigation of Shot Noise Reduction in InGaP HBTs with different Base Thickness

P.Sakalas[$,*], M.Schroter[*,§], P.Zampardi[&]

[$]*Fluctuation Phenomena Lab., Semiconductor Physics Inst., Goštauto 11, Vilnius, Lithuania,*
[*]*CEDIC, Dresden University of Technology, Mommsenstrasse 13, 01062 Dresden,Germany,*
[&]*Skyworks Solutions Inc. 2427 W.Hillcrest Drive Newbury Park, CA, 91320 USA,*
[§]*ECE Dept., University of California, San Diego, USA*

Abstract. DC, AC characteristics and Noise parameters of InGaP/GaAs HBTs with base thicknesses of w_B/nm=90, 70, 50 as well as CCHBTs with w_B=90 nm, were measured and modeled using advanced compact model HICUM. Very good agreement of HICUM versus measured data was observed for AC and DC data. Significant base thickness reduction only slightly increases peak transit frequency f_T/GHz =(45 (90nm), 54 (50nm)) due to reduced base transit time. High speed performance is mainly controlled by nonequilibrium electrons which form minority carrier jam in B/C SCR and thus additional delay. Significant increase of f_T/GHz =60 (90nm) was observed for CCHBTs, which feature lower collector internal resistance and smaller delay in B/C SCR. Therefore measured NF_{min} of different w_B HBTs did not exhibit expected difference, in contrast to CCHBTs, which demonstrated significantly lower NF_{min}. Our analytical noise model clarified that strong shot noise reduction in $A_{III}B_V$ is stemming not only from correlated currents, but also from Coulomb blockade by nonequilibrium electrons.

INTRODUCTION

The physical mechanism for shot noise in p-n junctions is based on thermal fluctuations of minority carriers, which produces a disturbance in the minority carrier distribution, resulting in diffusion current fluctuations [1],[2]. Bipolar transistors, having two p-n junctions, exhibit base and collector current shot noise, which are correlated. The cross-correlation noise of base and collector current is defined by $S_{CB} = 2qI_C(j\omega(\tau_B/3))$, where τ_B is the delay time in the base due to minority carrier drift-diffusion [1]. Thus for HBT noise performance the base transit time is of great importance. The diffusion coefficient of electrons for GaAs is $D_e \cong 200(cm^3/s)$. Therefore, following $\tau_B = w_B^2/(2D_e \cdot \vartheta) + w_B/v_T$ [3], where w_B is base layer thickness, ϑ is a factor, that depends on the built in electric field, and $v_T = \sqrt{((2k_BT))/\pi \cdot m_e^x}$ is the drift velocity with m_e^x as effective mass, the evaluated base transit time is $\tau_B \cong 200 fs$ for 90 nm base HBT, if only diffusion processes considered. If τ_B is the major cause for the E/C delay, then this implies f_T>750 GHz for an ideal transistor, it is not the case in reality. Actually, in $A_{III}B_V$ HBTs electrons are injected across a conduction band barrier, which reduces the base transit time. Drift diffusion simulations, accounting for bandgap discontinuity, show that τ_B reduction due to injection across the barrier is significant only for very narrow base (w_B= 30 nm) HBTs [4]. Velocity saturation in $A_{III}B_V$ HBTs increases τ_B, but not very significantly, (14% for w_B= 50 nm and only 8% for w_B= 100 nm [5]. In the total E/C delay $\tau_{EC} = \tau_E + \tau_B + \tau_C + \tau_{CC}$, where τ_E is emitter charging time, collector transit time $\tau_C = w_{BC}/(2v_S)$ plays an important role. v_S is saturation velocity, and $\tau_{CC} = (R_E + R_C + \eta \cdot k \cdot T/(q \cdot I_C))C_{jC}$ is collector charging time, where C_{jC} is B/C junction capacitance. Electrons in $A_{III}B_V$ HBTs can gain enough energy from electric field and suffer Γ-L intervalley transfer. Higher effective mass and thus reduced mobility and drift velocity in L valley gives rise to τ_C [6],[7]. Since cross-correlation of base and collector shot noise in noise model [1] is described via transit time, experimental high frequency noise investigation of different base thickness and collector composition in $A_{III}B_V$ HBTs can reveal current transfer peculiarities and their impact on DC, scattering and noise parameters. In this work we have investigated an impact of base thickness and collector composition in InGaP HBTs on DC, f_T and noise parameters.

EXPERIMENTAL, RESULTS AND DISCUSSION

DC, AC and Noise parameters were measured on a set of InGaP/GaAs HBTs with the base thicknesses and concentrations of (w_B/nm, N_A/cm^{-3})=(90, 5.5e19), (90, 4.5e19), (70, 5e19), (50, 5.5e19), (50, 5.5e19). High-frequency test transistors with various configurations were realized on a test chip along with adequate de-embedding structures. Composite collector (CCHBTs) with w_B=90 nm were measured as well. Transistors presented in this paper have a CBE contact configuration (i.e. 1 emitter, 1 base and 1 collector) with emitter window areas (W*L) of A_{E0}/ µm^2=2.2*2.2, 2.2*4.4, 2.2*8.8, 2.2*22, 2.2*44 and were fabricated at Skyworks Solutions. Noise and s-parameters were measured with ATN NP5 noise system and HP8510C VNA, using Suss Microtech semiautomatic RF-probestation and "Probebench" software in a 2-26 GHz frequency range. Careful de-embedding of pad parasitics was performed using a 2-step method and, for the noise parameters, employing correlation matrix technique. DC, S- and noise parameters were simulated using advanced compact model HICUM in Agilent ADS2004A and Aplac 8.0 simulator. Noise parameters were also simulated employing our recently derived analytical model (An) [8], [9].

Measured and simulated $J_C(V_{CE})$ and $J_C(V_{BE})$ are in a good agreement for all measured devices. Despite higher but less scattered NF$_{min}$ data, HBTs with larger emitter area A_{E0} =2.2*8.8 [µm^2] have been chosen for the analysis. $J_C(V_{CE})$ of HBT with w_B=90 nm is presented in Fig.1a. HICUM, accounting for selfheating and other physical effects, yields good fit. HBTs with w_B=50 nm, exhibited higher current gain, as it is seen from Fig.1b. Output current density of w_B=50 nm HBT for a given fixed I_B was higher (J_C=0.9 mA/µm^2 @ I_B=80 µA). Composite collector HBTs with w_B=90 nm at I_B=80 µA returned slightly higher J_C=0.46 mA/µm^2 in comparison to conventional w_B=90 nm HBT. Note, that HICUM is in agreement with forward Gummel plot for all devices as well. Base thickness reduction from 90 nm down to 50 nm increases f_T by 18% only (Fig.2a). Simulation of $\tau_B = w_B^2/(2D_e \cdot \vartheta) + w_B/v_T$ shows, that such w_B decrease, even with included velocity saturation term, implies ~70% of f_T change [10]. This means that τ_B itself is not a limiting factor of high frequency behavior for $A_{III}B_V$. Collector transit time $\tau_C = w_{BC}/(2v_S)$ due to saturated drift velocity and impact of Γ-L transfer becomes important. It is obvious from the rapid increase to f_T=59 GHz in CCHBTs, for which collector is designed to avoid the electron jam. HICUM analysis indicates that peak f_T increase towards higher J_C is due to

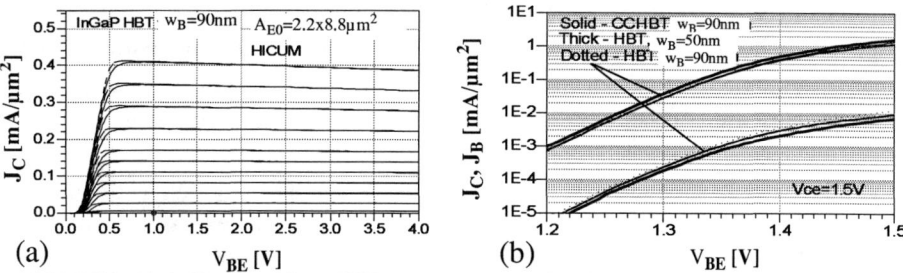

FIGURE 1. (a) $J_C(V_{CE})$ for 90 nm HBT, solid line is measured fixed I_B /μA=1, 5, 10, 20-80, dotted line is HICUM, (b) Measured J_C, $J_B(V_{BE})$ for w_B/ nm=90, 50 and CCHBT w_B= 90 nm.

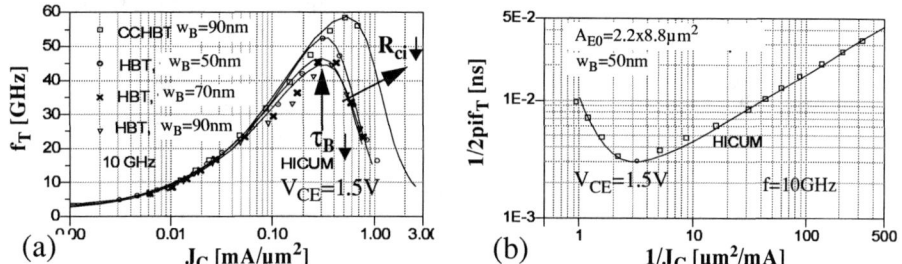

FIGURE 2. (a) Measured (symbols), HICUM (lines) $f_T(J_C)$, (b) $1/2\pi\, f_T(1/J_C)$ for w_B=50 nm.

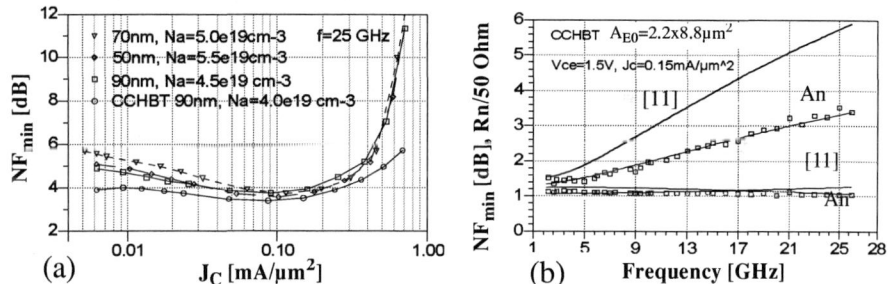

FIGURE 3. (a) Measured NF_{min} (J_C,w_B), (b) measured and simulated NF_{min} with An [8], [9] and with [11].

lower internal collector resistance R_{Ci}. Taking into account that the base resistance increase for w_B=50 nm HBT is negligible [12], the combination of reduced base thickness with composite collector technology should improve device speed and noise performance. The total delay time τ_f can be extracted from the dependence $\tau_f = (1/(2\pi f_T)) - C_{B\upsilon}/g_m$ [13], where $C_{B\upsilon}$ is sum of capacitances at the base node, $g_m = I_C/(m_C \cdot V_T)$ and m_C is non ideality factor, Fig.2b. Actually τ_f is very close to the sum of all EC delay times τ_{EC}. Since τ_E is negligible and $\tau_{CC} \cong 50fs$ is rather small in comparison to τ_B and τ_C, for noise simulations as a delay time we can use

$\tau_f/3$. The comparison of $NF_{min}(J_C)$ (Fig.3a) shows that base thickness reduction only slightly influence noise properties at 25 GHz, while CCHBTs even with w_B=90 nm exhibit an improved $NF_{min}(J_C,f)$. Using HICUM analysis all HBT circuit parameters, including bias dependent, were obtained and used for NF_{min} simulation with analytical noise model [8],[9]. Simulated $NF_{min}(f)$ for CCHBT with An approach (Fig.3b) shows that base and collector cross-correlation with a delay parameter $\tau_f/3$ fairly well fits measurement. High reduction of NF_{min} is observed in a wide J_C. Extracted correlation coefficient at high frequency exceeds "1" indicating an additional noise reduction mechanism, arising in BC SCR. Accumulated charge blocks electron transfer through the base so reducing collector shot noise [14]. Analytical model fairly well accounts for this effect with the additional delay time τ_C.

ACKNOWLEDGEMENTS

We are grateful to R.F.Scholz, H.E.Wulf and F.Korndörfer from IHP Frankfurt (Oder) for the support with noise equipment. Agilent and Aplac are acknowledged for the software donation. Authors are thankful to SINANO for financial support.

REFERENCES

[1] M.J.Buckingham, "Noise in electronic Devices and Systems," ELLIS HORWOOD LIMITED, Publishers Chichester Halsted press: a division of John Willey & Sons, New York, 1983.

[2] A.Blum, "Elektronisches Rauschen", B.G.Teubner Stuttgart, Präzis-Druck, Karlsruhe, Germany, 1996.

[3] M.Lundstrom et al., Fundamentals of Carrier transport, Cambridge Univ. press, UK.

[4] G.Zohar et al., *IEEE Transactions on Electron Devices*, Vol.51, No.5, pp. 658-662, 2004.

[5] K.Suzuki et al., *IEEE Transactions on Electron Devices*, Vol.39, No.3, pp. 623-628, 1992.

[6] Y.M.Hsin et al., *Solid-State Electronics* 49, pp.295-300, 2004.

[7] A.Schwanhäuser et al., *Physical Review B* 70, 085211, pp.1-8, 2004.

[8] P.Sakalas, J.Herricht, M.Schroter, Present. Notes *IEEE CMRF'04*, Montreal, Canada, 2004.

[9] J.Herricht, P.Sakalas, M.Schroter, Proc. *MTT IMS 2005, Long Beach, USA*, 2005, (in press).

[10] M.M.Jahan, A.F.M.Anwar, *Solid-State Electronics*, Vol.49, No.6, pp.941- 948, 1996.

[11] S.Voignescu et al., *Journal of Solid-State Circuits*, Vol.32, No.9, pp.1430-1439, 1997.

[12] M.Kahn, Dr. Th., Universite Paris XI D'Orsay, "Transistor Bipolaire a Heterojonction GaInAs/InP pour circuits ultra-rapides: structure, fabrication et caracterisation", 2004.

[13] M.Schroter, T.-Y. Lee, *IEEE Trans on Electron Devices*, Vol.46, No.2, pp.288-300, 1999.

[14] P.Sakalas et al., Proc. of *SPIE, Noise in Devices and Circuits II*, Vol.5470, pp.151-163, 2004.

Low-frequency Noise in SiGe Channel pMOSFETs on Ultra-Thin Body SOI with Ni-Silicided Source/Drain

M. von Haartman, J. Hållstedt, J. Seger, B. G. Malm, P.-E. Hellström, and M. Östling

KTH (Royal Institute of Technology), Department of Microelectronics and Information Technology, Electrum 229, SE-164 40 Kista, Sweden. Email: mvh@imit.kth.se

Abstract. The low-frequency noise in buried SiGe channel pMOSFETs fabricated on ultra-thin body silicon-on-insulator (SOI) substrates is investigated. The total thickness of the Si/SiGe/Si body structure, which is fully depleted (FD), is 20 nm. The low-frequency noise properties are compared with FD SOI pMOSFETs with a 20 nm Si body. The effect of the Ni-silicide used in the Source/Drain were also studied, especially the case of Schottky-Barrier (SB) MOSFETs when the Ni-silicide is formed at the edges of the channel.

Keywords: $1/f$ noise, low-frequency noise, SiGe, Silicon-on-insulator (SOI), fully depleted (FD), Schottky-Barrier (SB), MOSFETs

INTRODUCTION

Fully depleted (FD) silicon-on-insulator (SOI) technology is very attractive for future generations of ultra-scaled CMOS devices thanks to enhanced performance (high speed, low power consumption) and improved scalability [1]. Ultra-thin body (UTB) SOI offers the possibility of designing buried channel MOSFETs, which can exhibit lower noise and higher mobility compared to surface channel devices [2]. Further enhancements of the mobility, confinement of the carriers towards the middle of the body and noise performance of pMOS transistors are possible by sandwiching a compressively strained SiGe channel between two thin Si layers on top of the buried oxide [3]. The buried SiGe channel forms a quantum well for hole current conduction. A reduction in the low-frequency noise is often observed in buried SiGe channel pMOSFETs thanks to a lower trap density and higher interfacial quality compared to the SiO_2/Si interface [4,5]. However, low-frequency noise studies on SiGe channel SOI pMOSFET are scarce [6]. Results for SiGe pMOSFET on FD SOI have, to the best of our knowledge, never been reported.

In this paper we will investigate the impact of the buried SiGe channel on low-frequency noise properties in FD SOI pMOSFETs. Moreover, we will also study the effect of Ni-silicidation in relation to noise. This is becoming an important concern as Schottky-Barrier (SB) MOSFETs have been suggested as candidates for high-performance radio-frequency devices [7].

DEVICE FABRICATION

100 mm p-type (100) SOI UNIBOND wafers (360 nm Si on 400nm SiO_2) were used as starting material for the device fabrication. The top Si layer was thinned by successive dry oxidation and wet HF etching to a thickness of 10 and 20 nm, respectively. An epitaxial stack of 8 nm $Si_{0.72}Ge_{0.28}$ and 4 nm Si was grown by reduced pressure chemical vapour deposition (RPCVD) on the wafer with 10 nm Si. The gate stack consists of in-situ p+ doped poly-Si and 3 nm thermally grown gate oxide. 8 nm Ni was deposited by an e-beam evaporation procedure and a RTA treatment at 500°C for 30 s induced the solid state reaction between Ni and Si in the gate and S/D areas to form the self aligned silicide (Salicide). Another set of devices was also fabricated where the thickness of the deposited Ni was varied. For thick Ni deposited, the Ni-silicide extends to the lowly doped channel region resulting in a SB MOSFET.

RESULTS AND DISCUSSION

Fig.1 shows hole mobility plotted versus hole density in the channel for a Si and a SiGe FD SOI pMOSFET. As seen, the mobility is enhanced 25-50% in the compressively strained SiGe channel MOSFET. Both the Si and SiGe transistors show an almost ideal subthreshold slope of 62 mV/dec and no evidence of floating body effects except when the back gate is biased in accumulation.

Low-frequency noise was measured from subthreshold to strong inversion at a constant $V_{DS} = -50$ mV as well as at a constant V_{GS} and varying V_{DS} between 0 and -2 V. The low-frequency noise was found to be of the $1/f^\gamma$ type, with $\gamma < 1$ below and around threshold and approaching $\gamma = 1$ at increased bias above threshold. For the Si device, γ between 0.75 and 0.85 was found below threshold, whereas γ around 0.4-0.5 was obtained for SiGe. From the noise spectra we conclude that the low-frequency noise is a sum of $1/f$ noise and a generation-recombination (g-r) noise component.

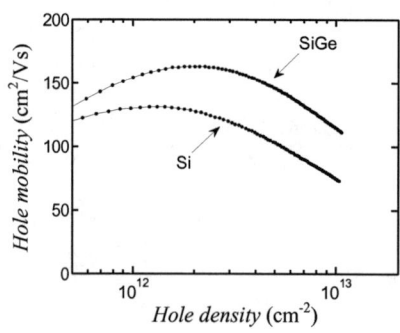

FIGURE 1. Hole mobility vs. hole density in the channel for the Si and SiGe channel FD SOI pMOSFETs.

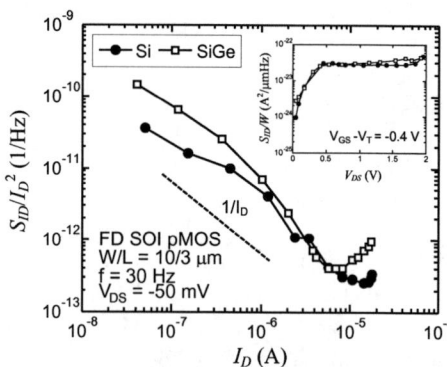

FIGURE 2. Normalized drain current noise vs. drain current for the Si and SiGe channel FD SOI pMOSFETs.

Both the Si and SiGe devices conduct current in buried channels. The Si body has a p-type doping (0.6-1 $\times 10^{15}$ cm^{-3}), thus the devices operate in the accumulation-mode. According to simulations, where the one-dimensional Poisson and Schrödinger equations were solved self-consistently for the Si FD SOI pMOSFET, the average distance of the holes from the top SiO$_2$/Si interface was calculated to range from 7.5 nm at $V_{GS} = V_T$ to 2.9 nm at $V_{GS} - V_T = -1.5$ V with 0 V applied to the back gate. The buried channel conduction results in low noise as evidenced in Fig. 2, where the normalized drain current noise is plotted versus drain current. The minimum value of the Hooge parameter α_H was found to be 6.3×10^{-6} and 8.7×10^{-6} for the SiGe and Si devices, respectively. This can be compared to $\alpha_H = 2.9\times 10^{-5}$ for our bulk surface channel Si pMOSFETs as reported in previous work [5]. The Si and SiGe devices show similar noise level above threshold. The absence of noise reduction by introducing SiGe can be explained by the fact that the Si device also shows buried channel conduction in this case. The SiGe device exhibits higher noise below threshold which could be attributed to a larger g-r noise component. Note that observations of g-r noise in buried channel SOI MOSFETs are not uncommon, which can be caused by traps either at the front or back interface [8]. Excess noise due to charging of the body is a problem in partially depleted (PD) SOI MOSFETs, but arise also in FD devices [9,10]. The inset in Fig. 2 shows drain current noise versus drain-source voltage. No kink-related excess noise can be observed, at least up to $V_{DS} = -2$ V, indicating negligible floating body effects in our FD SOI pMOSFETs.

The drain current noise is dominated by noise from the source and drain access resistances at high drain current, which cause the curves in Fig. 2 to turn up when $I_D \sim$ 10 µA. A combination of low channel noise and poor control of the source/drain resistance makes this noise source considerable also for a long channel device. The importance of the source/drain resistance for the noise performance is highlighted in Fig. 3. The normalized drain current noise is plotted as a function of drain current for a SB MOSFET, where NiSi-Si Schottky barriers are formed at the channel edges, and a reference MOSFET, where the NiSi-Si junctions form in the extension region. The reverse biased Schottky barrier at the Source side controls the drain current [11], which is reflected also on the noise as it increases more than one order of magnitude.

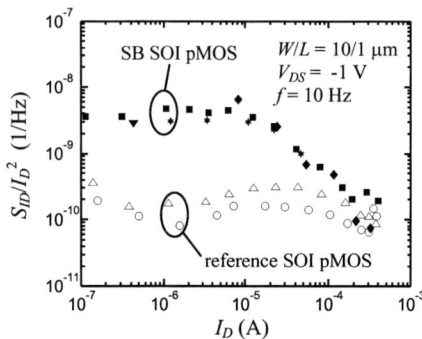
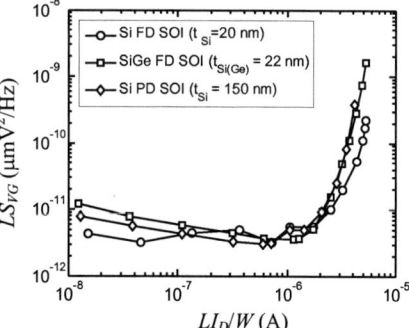

FIGURE 3. Normalized drain current noise for a Schottky-Barrier and a reference SOI pMOSFET.

FIGURE 4. Input gate voltage noise plotted for three different SOI pMOSFETs. The frequency is normalized to 1 Hz. $V_{DS} = -50$ mV, $W = 10$ µm.

Finally, we briefly discuss noise modeling. As seen in Fig. 2, S_{ID}/I_D^2 is roughly proportional to $1/I_D$ for the Si-device and can be successfully modeled using Hooge's empirical mobility fluctuation noise model with $\alpha_H = 8.7\times10^{-6}$. The SiGe device shows a stronger I_D dependence, $S_{ID}/I_D^2 \propto 1/I_D^{1.3}$ which indicates a larger influence from the g-r noise component, as discussed previously. On the other hand, the low-frequency noise can be explained using the number fluctuation noise model as well. Fig. 4 displays the equivalent input gate voltage noise S_{VG} for the SiGe and Si FD SOI pMOSFETs along with a Si PD SOI pMOSFET reference. S_{VG} is almost constant at lower drain current as predicted by the number fluctuation noise model. A trap density $N_t = 5\times10^{16}$ cm^{-3}eV^{-1} is extracted from the flat part of the curve.

CONCLUSIONS

Low-frequency noise results were presented for FD SOI pMOSFETs with a buried SiGe channel. Both the SiGe device and the reference FD SOI pMOSFET show low noise thanks to buried channel conduction and negligible floating body effects, at least up to $V_{DS} = -2$ V. For Schottky-barrier MOSFETs, the drain current is controlled by the reverse biased Schottky-Barrier at the Source at lower currents. In this work it was shown that formation of a NiSi Schottky-Barrier adjacent to the channel lead to a detrimental increase in low-frequency noise.

ACKNOWLEDGMENTS

The authors would like to thank S.-L. Zhang and H. H. Radamson for fruitful discussions. Financial support was received from the Swedish Foundation for Strategic Research (SSF) through the "high-frequency silicon" program.

REFERENCES

1. S. Cristoloveanu, *Solid-State Electron.* **45**, pp. 1403-1411 (2001).
2. M. Matloubian, F. Scholz, and L. Lum, *IEEE Trans. Electron Devices* **41**, pp. 1977-1980 (1994).
3. T. Krishnamohan, C. Jungemann, and K. C. Saraswat, in *IEDM Tech. Dig.*, 2003, pp. 687-690.
4. S. J. Mathew, G. Niu, W. B. Dubbelday, and J. D. Cressler, *IEEE Trans. Electron Devices* **46**, pp. 2323-2332 (1999).
5. M. von Haartman, A.-C. Lindgren, P.-E. Hellström, B. G. Malm, S.-L. Zhang, and M. Östling, *IEEE Trans. Electron Devices* **50**, pp. 2513-2519 (2003).
6. A. Inoue, A. Asai, Y. Kawashima, H. Sorada, Y. Kanzawa, T. Kawashima, H. Hara, and T. Takagi, in *IEEE Int. SOI Conf.*, 2003, pp. 149-150.
7. M. Fritze, C. L. Chen, S. Calawa, D. Yost, B. Wheeler, P. Wyatt, C. L. Keast, J. Snyder, and J. Larson, *IEEE Electron Device Lett.* **25**, pp. 220-222 (2004).
8. N. Lukyanchikova, M. Petrichuk, N. Garbar, E. Simoen, and C. Claeys, *IEEE Trans. Electron Devices* **43**, pp. 417-423 (1996).
9. Y.-C. Tseng, W. M. Huang, M. Mendicino, D. J. Monk, P. J. Welch, and J. C. S. Woo, *IEEE Trans. Electron Devices* **48**, pp. 1428-1437 (2001).
10. F. Dieudonné, S. Haendler, J. Jomaah, F. Balestra, *Microelectron. Reliab.* **43**, pp. 243-248 (2003).
11. B. Winstead, and U. Ravaioli, *IEEE Trans. Electron Devices* **47**, pp. 1241-1246 (2000).

Low Frequency Noise Characteristics of TaSiN/HfO$_2$/SRPO SiO$_2$ MOSFETs

Siva Prasad Devireddy[a], Zeynep Çelik-Butler[a], Hsing-Huang Tseng[b], Philip J. Tobin[b], Fang Wang[c], and Ania Zlotnicka[c]

[a]*University of Texas at Arlington, Electrical Engineering Dept., P. O. Box 19072, Arlington, TX, 76019*
zbutler@uta.edu
[b]*Freescale Semiconductor Inc., 3501 Ed Bluestein Blvd., Austin, TX 78721*
[c]*Freescale Semiconductor Inc., 2100 E. Elliot Rd, MD: EL741, Tempe, AZ 85284*

Abstract. The effect of interfacial SiO$_2$ layer resulting from Stress Relieved Pre-Oxide (SRPO) treatment on the low frequency noise characteristics of TaSiN/HfO$_2$ MOSFETs is studied. Noise comparison with near-equivalent devices having a SiO$_2$ interfacial layer from standard RCA process revealed better overall performance with the new process. Further analysis using the correlated mobility-number fluctuations model yielded close values for the oxide trap density and a considerably different Coulomb scattering coefficient values. In general, SRPO oxide improved the device performance that is further made evident by the mobility values obtained from the split C-V measurements.

INTRODUCTION

High-k materials, especially HfO$_2$, have received considerable focus in the recent years as possible replacements to SiO$_2$ gate dielectric. Although HfO$_2$ is stable with both poly-Si and metal gates, the latter is desired as it eliminates problems such as oxygen and dopant diffusion, depletion effects, extrinsic defect creation at the interface with high-k etc. For ultra-thin MOSFETs, the interfacial layer properties play a significant role in defining the final characteristics of the device and its reliability. While the use of nitrided metal gates like TaSiN offers a good barrier against oxygen diffusion thereby enabling control over SiO$_2$ formation, several other factors like high-k deposition method, pre-surface cleaning, annealing temperatures etc. affect the properties of the interfacial layer and the traps in the entire dielectric. In this paper, the low frequency noise characteristics of TaSiN gate MOSFETs with 70Å of ALD HfO$_2$ on SiO$_2$ of variable thickness and different deposition methods have been studied. One split had 10Å of RCA SiO$_2$ under HfO$_2$. The other two lot-splits had either a 6Å or a 16Å thick SRPO SiO$_2$. Here, SRPO SiO$_2$ was formed after the etch-back of a thick SiO$_2$ layer that is thermally grown above the glass flow temperature for stress relief [1]. Analysis based on unified flicker noise model yielded oxide trap density values that were lower than those previously reported for high-k devices indicating a better overall dielectric quality for this gate stack.

EXPERIMENT AND RESULTS

DC characterization of the devices is done using a HP-4155B semiconductor parameter analyzer in order to obtain parameters like ac conductance ($g_d=\partial I_d/\partial V_d$), transconductance ($g_m=\partial I_d/\partial V_g$), subthreshold slope ($S=\partial log(I_d)/\partial V_g$), threshold voltage ($V_t$) that is used in the subsequent noise expressions and to check for possible drift in device characteristics during and after the noise measurements. Noise measurements were done in a low noise probe station using a custom made DC biasing circuitry, an EG&G PAR113 pre-amplifier and a HP 3562A dynamic signal analyzer. Voltage power spectral density (S_{Vd}) was obtained for varying band-bending conditions in the inversion region at a constant drain bias of 50mV. For each bias point, S_{Vd} was converted into current noise spectral density as $S_{I_d} = S_{V_d}.(g_d)^2$, and the 1Hz value obtained from the straight-line fit is used for further analysis. The flicker noise exponent δ (as in $1/f^\gamma$) was in the range of 0.75-1.2 in these devices and did not exhibit any specific trend with gate bias. Figure1 shows typical noise spectral density plot for 0.165μmx10μm, 6Å SRPO device.

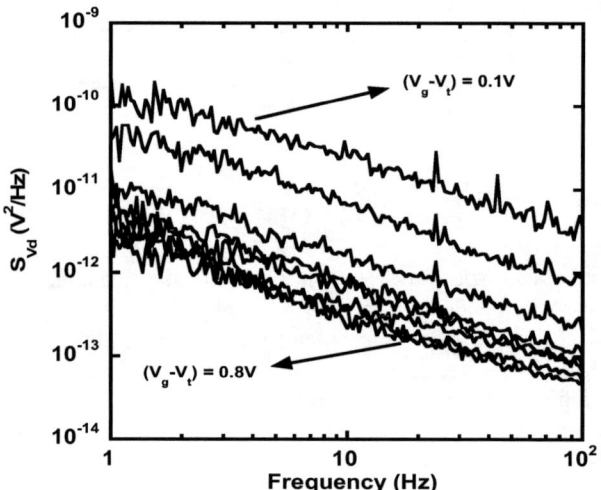

FIGURE 1. Typical voltage power spectral density plot for 6Å SRPO device in the 1-100Hz frequency range for a constant drain bias of 50mV and a gate overdrive varying from 0.1V to 0.8 V.

The observed noise is explained based on the unified flicker noise model [2] that attributes the cause of 1/f noise to the trapping / de-trapping of channel carriers into the oxide traps. The original model is modified according to the observations of Koga et al. [3,4] such that the screened Coulomb scattering coefficient is taken as $\alpha = 1/(\mu_{c0}\sqrt{N})$, and the drain current noise power spectral density is written as

$$S_{Id} = \frac{kTI_d^2}{\gamma f WL}(\frac{1}{N}+\frac{\mu}{\mu_{c0}\sqrt{N}})^2 N_t \qquad (1)$$

where, k is the Boltzmann constant (eVK^{-1}), T is the absolute temperature (K), γ is the tunneling coefficient (cm^{-1}), f is the frequency (Hz), μ is the effective mobility (cm^2/Vs), N_t is the oxide trap density (cm^{-3}eV^{-1}), N is the inversion layer charge (cm^{-2}), μ_{c0} (cm/Vs) is a fitting parameter and WL gives the area of the MOSFET. Here, the Coulomb scattering between the trapped and the channel charges is assumed to be repulsive. The tunneling coefficient value for SiO$_2$ (1x10^8 cm^{-1}) is used in noise parameter extraction. Eq (1) is valid only in strong inversion as it assumes that loss of one charge from the channel results in the gain of the same amount of charge in the dielectric.

The normalized noise data is shown in Fig.2 for the devices under investigation. It can be observed that the noise magnitude for the SRPO devices is lower than that for the RCA devices as the gate bias is increased. The noise parameters extracted using Eq. (1) are shown in Table1.

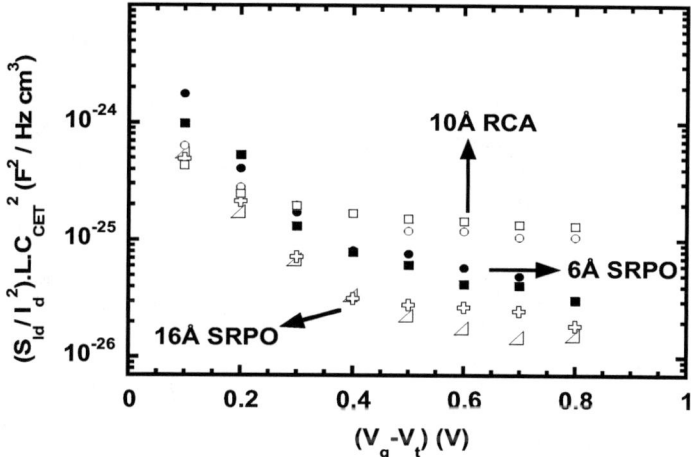

FIGURE 2. Normalized noise data comparison between RCA and SRPO SiO$_2$ devices. A clear branching out of the data can be observed after 0.3V gate overdrive.

TABLE 1. Summary of extracted parameters from noise data. The HfO2 thickness is 70 Å for all devices.

Device	N_t (cm^{-3}eV^{-1})	μ_{c0} (cm/Vs)
0.18μmx10μm, 16Å SPRO SiO$_2$	4.0×10^{17}	1.0×10^9
0.165μmx10μm, 16Å SPRO SiO$_2$	4.4x10^{17}	1.5×10^9
0.18μmx10μm, 6Å SPRO SiO$_2$	4.5×10^{17}	7.0×10^8
0.165μmx10μm, 6Å SPRO SiO$_2$	5.1x10^{17}	9.0×10^8
0.165μmx10μm, 10Å RCA SiO$_2$	4.3×10^{17}	8.0×10^7
0.18μmx10μm, 10Å RCA SiO$_2$	2.5×10^{17}	6.0×10^7

The oxide trap densities in these gate stacks are all within the same order. However, μ_{c0} is comparatively higher for the SRPO devices indicating a lower Coulomb scattering coefficient. In addition, the ratio of the contribution from the number fluctuations to the contribution from the correlated mobility fluctuations to the total noise was higher for the RCA devices, suggesting that mobility limited by Coulomb scattering is more severe in these devices. This is confirmed by the effective mobility obtained from split C-V measurements on 10μm×10μm devices as shown in Fig.3.

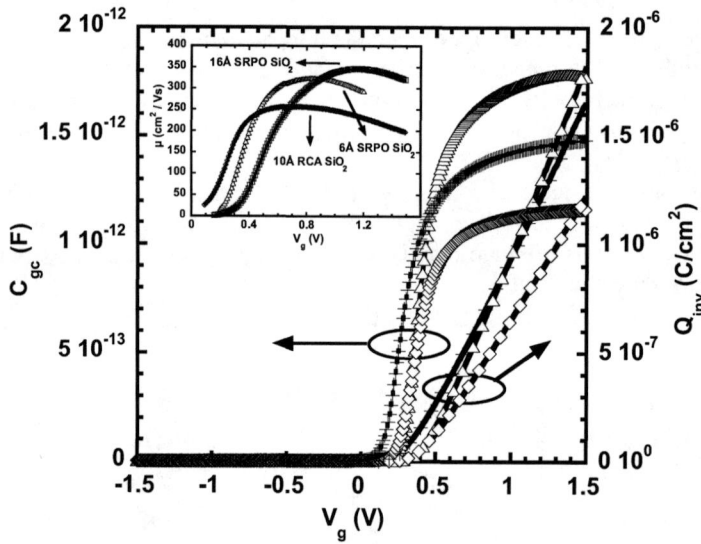

FIGURE 3. Comparison of gate-channel capacitance C_{gc}, inversion charge Q_{inv} and electron channel mobility values obtained from split C-V measurements on 10μm×10μm devices. Δ: 6Å SRPO SiO_2, ◊: 10Å SRPO SiO_2, +: 10Å RCA SiO_2.

In conclusion, SRPO devices were shown to yield better low frequency noise characteristics at strong inversion than RCA devices due to a lower contribution of Coulomb scattering towards the noise component associated with correlated mobility fluctuations. This lower Coulomb scattering is reflected in the effective mobility values.

ACKNOWLEDGMENTS

This material is based upon work supported in part by SRC under contract 2004-VJ-1193 and THECB-ATP under grant: 003656-0001-2001.

REFERENCES

1. H. –H, Tseng et al., *IEDM Tech. Dig.*, pp. 821-824 , 2004.
2. K. K. Hung, P. K. Ko, C. Hu, and Y. C. Cheng, *IEEE Trans. Electron Devices*, **37**, pp.654-665, 1990.
3. J. Koga, S. Takagi, and A. Toriumi, *Proc. Int. Conf. Solid State Devices and Materials, SSDM*, pp. 895-897, 1994.
4. J. Koga, S. Takagi, and A. Toriumi, *IEDM Tech. Dig.*, pp. 475-478, 1994.

Drain and Gate Current LF Noise in Advanced CMOS devices with Ultrathin gate Oxides

T. Contaret[(1)], K. Romanjek[(1)], G. Ghibaudo[(1)], J.A. Chroboczek[(1)], F. Bœuf[(2)] and T. Skotnicki[(2)]

(1) IMEP, UMR CNRS/INPG/UJF, ENSERG, BP 257, 38016 GRENOBLE, FRANCE
(2) STMicroelectronics, BP 16, 38921 CROLLES, FRANCE

Abstract. This paper presents the main results of our study on the low-frequency noise characteristics of ultrathin oxide gate oxide MOSFETs representative of several sub 90 nm CMOS technologies. Simple analytical models for this device are presented and used for interpretation of noise measurements. For the noise data interpretation, a simple drain and gate current noise model based on a flat band voltage fluctuation concept are developed and were shown to account for the overall drain current and gate current noise behavior.

Keywords: Gate-leakage current, 1/f noise, advanced CMOS device, ultrathin gate oxide.
PACS: 85.30.De – Semiconductor-device characterization, design, and modeling

INTRODUCTION

One of the key issues related to the downscaling of advanced CMOS devices is the increasing in the excess low frequency noise, known to vary as a reciprocal gate area. The device scaling requires the gate oxide thickness reduction, which results in the enhanced gate tunnel leakage and the associated gate current noise [1]. Thus, in addition to the intrinsic channel current noise sources, the gate current fluctuations become increasingly important and a study of their impact on the overall MOSFET noise performance is mandatory. We propose a simple noise model, combining, both drain and gate current noise contributions, to interpret the noise data in ultrathin oxide MOSFETs. In the first section, the studied technology and experimental details are presented. The drain current noise characterization and modeling are described in the second section. Finally, we show the experimental results on the gate-leakage current noise measurements and we present a simple noise model based on the flat band voltage fluctuation concept, which is also used for carrier number fluctuation noise models for drain current noise.

EXPERIMENTAL DETAILS

Noise measurements have been performed on n-MOSFETs from an advanced CMOS technology with 1.2 nm gate oxides thickness, fabricated by STMicroelectronics, Crolles. In Table 1, we report the main technological characteristics of two splits, denoted A and B. The 1.2nm thin gate oxides have been fabricated using a RTN (CMOS A) or PN (CMOS B) process [2]. The test structures investigated have variable gate lengths, ranging from 30nm to 10μm with a fixed channel width W=10μm. Prior to the noise analysis, the static characteristics of the devices were investigated and the electrical MOSFET parameters were extracted by an original method which suppresses the effects of the gate leakage perturbations [3]. The low-frequency noise, LFN, characteristics were measured with a set-up comprising a programmable biasing amplifier with two entrance bias ports and output transimpedance amplifier [4].

TABLE 1. Technological characteristics of the studied A and B CMOS technological splits.

Technology	CMOS A	CMOS B
Gate oxides process	Rapid Thermal Nitridation (RTN)	Plasma nitridation (PN)
Polysilicon gate thickness	1500 Å	1200 Å
LDD (Ligthly Doped Drain)	Arsenic (As)	Arsenic (As)
Pockets	Boron (B)	Boron floride (BF_2)
HDD (High Doped Drain)	Colbalt silicide ($CoSi_2$)	Nickel silicide (NiSi)

DRAIN CURRENT NOISE CHARACTERIZATION

Depending on the gate bias, the drain current power spectral density, PSD, shows a clear 1/f noise behavior over which some lorentzian G-R components are often superimposed. In such cases, it has been possible to extract the 1/f noise PSD component by subtracting the G-R components, using an adequate treatment [3]. The drain current PSD dependencies on gate and drain voltages were interpreted by carrier number fluctuation noise models [5]. In particular, oxide trap density N_t and mobility-correlated α were determined for each gate length and for each technology. Typical N_t values are reported in Fig. 1 for the both CMOS technologies. They were found to be smaller than those for p-MOSFETs, or previous CMOS technologies [6], indicating a very good gate dielectric quality, despite the application of RTN or PN fabrication processes.

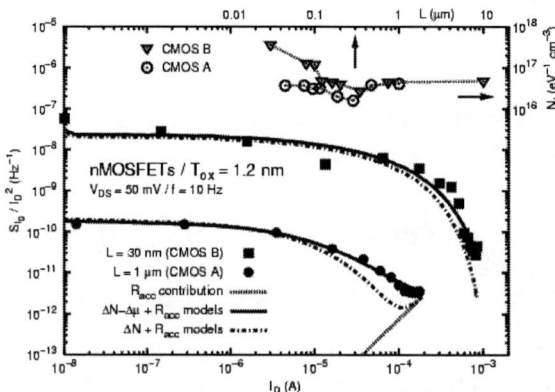

FIGURE 1. Normalized drain current noise PSD and ΔN–Δμ model with R_{acc} noise contribution for long and small geometry and the evolution of N_t with L for the both CMOS technologies studied.

We also observed a reduction of N_t as the length decreases, which we explained by short channel effects, and an increase of N_t for the shortest lengths. We attribute that increase to the process-induced extra defect creation, close to source and drain regions, possibly generated during the implantation steps. In order to obtain an accurate modeling at the high gate voltage, the drain current fluctuations are assumed to be generated by two noise sources, one associated with the intrinsic channel and another to the access resistances. In Fig. 1, the dotted curve gives the contribution of the access resistances noise to the total noise. At strong inversion, a better agreement is obtained using the mobility-correlated number fluctuation model with the access resistances noise (continuous lines in Fig. 1).

GATE-LEAKAGE CURRENT NOISE CHARACTERIZATION

For the gate current noise PSD, with no drain voltage applied, 1/f noise behavior with the shot noise background, and occasionally some trap-related lorentzian features were observed. The evolution of the normalized gate current noise PSD *versus* gate current are reported in Fig. 2, for the largest area geometry for the both technologies. In order to interpret the data, a simple gate current noise model based on the flat band voltage fluctuation concept [7] (already used in the drain current noise modeling) has been developed. We have obtained a relatively good agreement between the data and the model, except for the low gate current region, where system noise perturbs the measurement accuracy.

The drain current PSD in the ohmic regime is reported in Fig. 3, for the largest area device of CMOS B technology. At strong inversion, an important 1/f excess noise is observed, due to the contribution of the gate current noise. In order to justify and explain this particular increase of the 1/f noise, we have plotted the different contributions of the total noise in Fig. 3. The sum of drain and gate current noise contributions is seen to account satisfactorily for the noise behavior. The impact of the gate-leakage current noise on the output drain current noise for the large area device is clearly demonstrated.

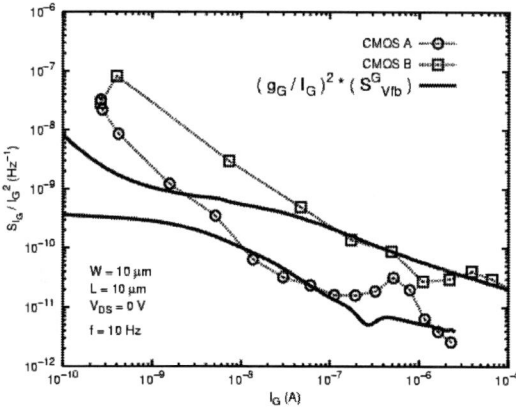

FIGURE 2. Normalized gate current noise versus gate current at drain voltage equal zero for 10μm/10μm geometry for both technologies.

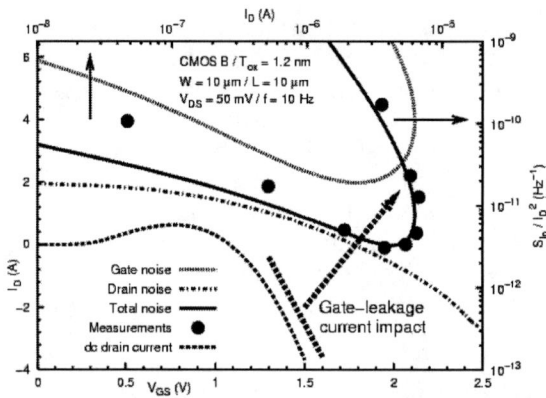

FIGURE 3. Experimental drain noise (points), gate current noise contribution (dotted line), drain current noise contribution (dashed line) and total noise modeling (continuous line) for 10μm/10μm geometry for technology B at V_{DS}=50mV. Drain dc current is also reported.

CONCLUSIONS

Low frequency gate and drain noise PSD in advanced CMOS technologies have been investigated. The 1/f noise data have been well accounted by an original drain and gate current noise models based on the flat band voltage fluctuation concept. Finally, we have explained the huge excess drain current noise appearing at strong inversion by a proper analysis of both drain and gate current noise contributions.

ACKNOWLEDGMENTS

We thank ST-Microelectronics, Crolles for their collaboration and the furnishing of the advanced CMOS devices used in this work.

REFERENCES

1. J. Lee, G. Bosman, K.R. Green, and D. Ladwig, "Noise Model of Gate-Leakage Current in Ultrathin Oxide MOSFETs", IEEE Transactions on Electron Devices, vol.50, 12, pp 2499-2506 (2003).
2. B. Tavel, M. Bidaud & al, "Thin oxynitride solution for digital and mixed-signal 65 nm CMOS platform", IEEE-International-Electron-Devices-Meeting-2003, 27.6.1-4 (2003).
3. K. Romanjek, "Characterization and modeling of 50 nm and below CMOS transistors technologies", Ph.D. Thesis, IMEP/CNRS/INPG, Grenoble, France (2004).
4. J. A. Chroboczek and G. Piantino,1999, Patent No. 15075, France. Registered in November 1999.
5. G. Ghibaudo, O. Roux, C. Nguyen-Duc, F. Balestra, and J. Brini, "Improved analysis of the low frequency noise in field-effect MOS transistors", Physics States Solid(a), vol.124, p157 (1991).
6. M. Marin, "Experimental study of sub-0.18 μm CMOS transistors technologies", Ph.D thesis, CEM2/University of Montpellier II, Montpellier, France (2003).
7. G. Ghibaudo, "A simple derivation of Reimbold's drain current spectrum formula for flicker in MOSFETs", Solid Stat. Electronics, 30, 1037 (1987).

Apparent noise parameter behavior in n-MOS transistors operating from subthreshold to above-threshold regions

A. Boukhenoufa[a], L. Pichon[a], C. Cordier[a], B. Cretu[a], L. Ding[a], R. Carin[a], J-F. Michaud[b], T. Mohammed-Brahim[b]

[a]*Groupe de Recherches en Informatique, Image, Automatique et Instrumentation de Caen (G.R.E.Y.C), CNRS UMR 6072, ENSICAEN-Université de Caen, 6 bd du Maréchal Juin, 14050 Caen, France*
[b]*Groupe de Microélectronique, Institut d'Electronique et de Télécommunication de Rennes (I.E.T.R), CNRS UMR 6164, campus de Beaulieu, bât 11B, 35042 Rennes, France*

Abstract. Low frequency noise is checked in n-channel single crystalline silicon MOS transistors and polysilicon TFTs. The study is performed on devices operating from below to above threshold regions. Apparent noise parameter (α_{app}) is extracted from noise measurements according to the empirical Hooge formula. α_{app} increases in the subthreshold region and decreases above threshold versus the effective gate voltage. For polysilicon TFTs $10^{-2} < \alpha_{app} < 1$ while for MOSFETs $10^{-5} < \alpha_{app} < 10^{-2}$. α_{app} is higher in TFTs due to a higher interface defect density. Distribution of defects in the oxide band gap is also determined for TFTs. Results agree with Mc Whorter model.

Keywords: 1/f noise, noise parameter, McWhorter model, TFTs, MOSFETs

INTRODUCTION

Low frequency (1/f) noise in MOS transistors is attributed to channel conductivity fluctuations and its origin is not clearly stated. Two mechanisms are invoked: i) fluctuations of the carrier number in the channel (ΔN model) or ii) fluctuations of carrier mobility ($\Delta\mu$ model). ΔN model is based on Generation-Recombinaison (G-R) processes of carriers due to defects in the channel region. It is explained as an addition of G-R spectra known as the so-called Mc Whorter model. $\Delta\mu$ model describes mobility fluctuations due to scattering mechanisms of carriers. It was first established by Hooge in homogeneous samples of high quality materials. There has been ample discussion on the 1/f noise origin but they have not been decisive.

Previous studies show that low frequency noise level is related to the active layer and/or to the gate insulator/active layer quality, so that an average oxide and/or grain boundary traps densities can be deduced [1,2]. In this paper we study the 1/f noise level in two types of n-channel MOS transistors: polysilicon Thin Film Transistors (TFTs) and classical single crystalline silicon MOSFETs operating from sub- to above-threshold region. The analysis is based on Mc Whorter model.

RESULTS AND DISCUSSION

In MOS transistors low frequency (1/f) noise can be described by the widely used Hooge empirical relation :

$$\frac{S_{ID}}{I_D^2} = \frac{\alpha_{app}}{fN} \qquad (1)$$

where S_{ID}/I_D^2, N and α_{app} represent the normalized current noise spectra density, the total carriers and the apparent noise parameter respectively. α_{app} can be used as a factor of merit for the noise amplitude. In MOS transistors, for the $\Delta\mu$ model α_{app} stands for a constant value whereas it is gate bias dependent for the ΔN model. This gate bias dependence has been previously experimentally observed in n-channel MOS transistors [3] and explained by assuming that the average total number of carriers in the channel can be approximated by:

$$N \approx \frac{WL}{q} C_{OX} (V_{GS} - V_{FB}) \qquad (2)$$

where W(L), C_{OX} and V_{FB} are the channel width (length), the oxide capacitance per unit area and the flat band voltage respectively. Thus the α_{app} extracted from noise measurements decreases versus the reverse effective gate voltage ($V_G^* = V_{GS} - V_{FB}$). However, and as previously observed [4], this α_{app} behaviour changes by re-considering a more accurate calculation of the average number of carriers in the active layer, in particular for device operating in the subthreshold region, by standing:

$$N \approx \frac{WL}{q} \left[2\varepsilon_{Si} q N_C \exp\left(-\frac{E_G}{2kT}\right) \left(\frac{kT}{q}\right) \left(\exp\left(\frac{qV_s}{kT}\right) - 1\right) - V_s \right]^{1/2} \qquad (3)$$

where N_C is the effective density of states in the conduction band, E_G the silicon band gap, ε_{si} the silicon permittivity, and V_s the surface potential as defined in figure 1. Such N calculation is deduced from activation energy (E_a) of the drain current extracted from temperature measurements. Therefore, as displayed in figure 2, for the two types of MOS transistors α_{app} increases with V_G^* below threshold and decreases above

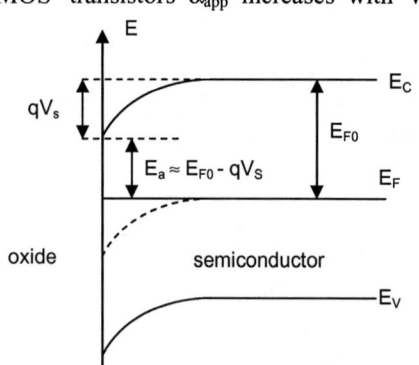

FIGURE 1: Band bending at the interface in a n-channel transistor under a positive gate voltage

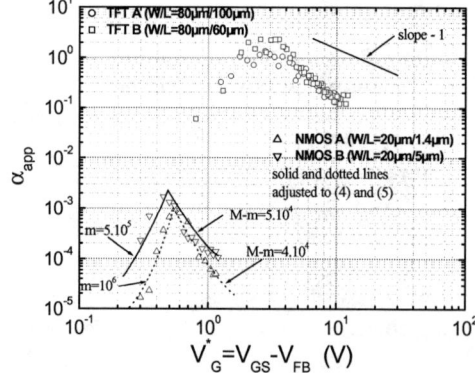

FIGURE 2. Apparent noise parameter versus effective gate voltage

accordingly to the $1/V_G^*$ type behaviour. For polysilicon TFTs $10^{-2} < \alpha_{app} < 1$, whereas for single crystalline silicon MOSFETs $10^{-5} < \alpha_{app} < 10^{-2}$. Such behaviour and corresponding range values of the measured apparent noise parameter are unusually observed.

In order to valid these results, we support our analysis on the Mc Whorter model. Accordingly to Hooge theoretical prediction, concerning the validity of this model in MOS transistors, the noise parameter can be approximated by [5]:

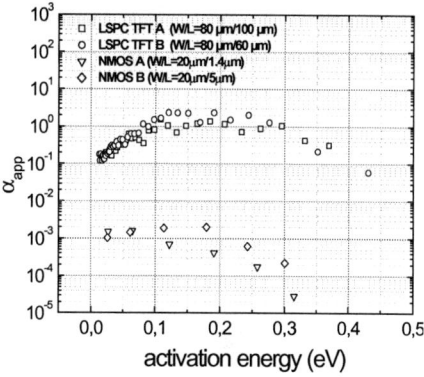

FIGURE 3. Apparent noise parameter versus drain current activation energy

$$\alpha_{app} \approx 4 \frac{N}{m} \frac{\lambda}{z} \frac{1}{f} \qquad \text{for} \quad N < m \qquad (4)$$

$$\alpha_{app} \approx \frac{M-m}{N} \frac{\lambda}{z} \frac{1}{f} \qquad \text{for} \quad N > M - m \qquad (5)$$

where λ is the tunnel attenuation distance of carrier into the oxide (~0.1nm [5]), z the effective oxide thickness (=30nm [5]), m the number of trapped carriers into the oxide close to the interface, and M the number of Mc Whorter traps. Assuming that $E_a \approx E_C - E_F$, thus N<m and N>M-m corresponds respectively to a high activation energy value (subthreshold region) and to a low activation energy value (above threshold region). Reminding that E_a decreases versus V^*_G, therefore these theoretical results predict a maximum value for the $\alpha_{app}(E_a)$ curve: it increases in the above threshold region and it decreases in the subthreshold region. This behaviour is observed by re-plotting the measured α_{app} versus E_a in the two biased regions (figure 3). However, the analysis for single crystalline silicon MOS transistor and polysilicon TFTs is different.

For classical MOS transistors, at low activation energy the drain current is not thermally activated and thus the increase of α_{app} versus E_a cannot be displayed on figure 3. Thermal activation occurs only in the subthreshold region with $E_a \leq 0.3eV$ for a corresponding slight effective gate range $0.3V \leq V^*_G \leq 0.5V$. This situation corresponds to conduction phenomenon described by G-R processes of carriers from trap levels mainly located 0.3eV below the band conduction edge. By adjusting $\alpha_{app}(V^*_G)$ curve with (4) and (5), we deduce m and M values for the two transistors (see figure 2). Considering the total oxide traps defined by $M/(WLzE_G/2)$ we find $\sim 3 \times 10^{18}$ cm^{-3} eV^{-1} which is consistent with previous results [2], deduced from flat band voltage fluctuation following the tunneling theory.

The interface quality in TFTs is poorer due to the higher defect density within the polysilicon active layer and the drain current remains thermally activated in the above threshold region. Thus, we consider that the oxide traps are homogeneously distributed within an energy band of kT around the Fermi level. Therefore, due to the Fermi level shift in the band gap because of the variation of V_{GS} below threshold, the corresponding distribution of defects, close the interface, N_{OX}, is determined from (4) by standing:

$$N_{OX} \approx \frac{m}{WLzkT} \tag{6}$$

and is plotted in the figure 4 for each TFT. These interesting plots are similar to the plots of the density of states (DOS) determined by the field effect conductance method [6]. This latter leads to the DOS related to the active layer defects while in our study it concerns defects into the oxide. Thus, our experimental method is complementary to quantify oxide traps close to the interface in MOS transistors. Moreover, above threshold we find M constant ($\sim 7\times 10^8$) leading to a total oxide traps $\sim 10^{21}$ cm^{-3} eV^{-1} as also deduced accordingly to flat band voltage fluctuation.

Let us remark that $\alpha_{appTFT}/\alpha_{appClass} \sim 10^3$ (fig.2,3). This can be easily justified, in particular above threshold region (5). The corresponding carrier number ratio mainly depends on the channel area ratio and the respective M-m values remain at a constant value for each device. As $(WL)_{TFT}/(WL)_{Class} \sim 10\text{-}10^2$ and $(M-m)_{TFT}/(M-m)_{Class} \sim 10^4\text{-}10^5$ thus α_{app} ratio gives $\sim 10^3$.

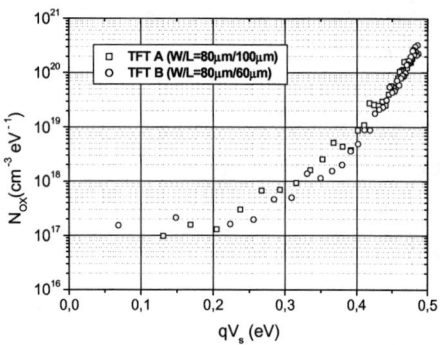

FIGURE 4. Distribution of defects in the band gap into the oxide close to interface in TFTs

CONCLUDING REMARKS

1/f noise is analyzed in MOS transistors operating from sub- to above-threshold regions. A new determination of the apparent noise Hooge parameter α_{app} from noise measurements, based on the Mc Whorter model, is described in particular below threshold. At low effective gate voltages α_{app} increases due to the increase of the free to trapped carrier (into the oxide) numbers ratio. At high effective gate voltages it decreases versus the carrier number as usually described. For equal channel area, α_{app} is higher in TFTs than in classical MOSFETs due to a higher number of oxide traps. The analysis leads to information concerning the oxide quality, in particular to the determination of the oxide band gap defects distribution close to the interface in TFTs.

REFERENCES

1. C. A. Dimitriadis, G. Kamarinos, IEEE Trans. Electron. Dev., 37, 381 (2001).
2. L. Pichon, J. M. Routoure, R. Carin, L. Nze Mekwama, ESSDERC 2003 proceedings, editions frontières, 445 (2003).
3. L. K. J. Vandamme, X. Li, D. Rigaud, IEEE Trans. Electron. Dev., 41, 1936 (1994).
4. A. Boukhenoufa, L. Pichon, C. Cordier, R. Carin, H, El Din Kotb, Noise in Devices and Circuits II, Proceedings of SPIE, 546 (2004).
5. F. N. Hooge, Physica B 336, 236 (2003).
6. G. Fortunato, P. Migliorato, Appl. Phys. Lett., 49, 1025 (1986).

Effect of Oxide Thickness and Nitridation Process on PMOS Gate and Drain Low Frequency Noise

F. Martinez[1], C. Leyris[2], M. Valenza[1], A. Hoffmann[1],
F. Boeuf[2], T. Skotnicki[2], M. Bidaud[2], D. Barge[2], B. Tavel[2]

[1]*CEM2 – UNIVERSITE MONTPELLIER II – UMR CNRS 5507*
Place E. Bataillon, 34095 Montpellier Cedex 5, France
[2]*ST Microelectronics, 850 rue Jean Monnet, F-38926 Crolles Cedex*

Abstract. Gate and drain current 1/f noise parameters have been extracted in p-metal-oxide-semiconductor transistors processed with two nitridation techniques (RTN or DPN). The drain noise magnitude allows extraction of the slow oxide trap density $N_t(E_F)$ in the range of 3 10^{17} eV^{-1} cm^{-3}. We don't observe any improvement of the 1/f noise performance for DPN devices, which is in opposition with the trends observed on thicker oxides. 1/f gate current noise has been also investigated for these devices, and similar results have been obtained concerning the 1/f noise levels. The benefit in changing process for thick oxide doesn't hold any longer in term on 1/f noise performance with an equivalent oxide thickness of 1.2 nm.

Keywords: pMOSFETs, RTN, DPN, 1/f noise, drain noise, gate noise.
PACS: 72.70.+m ; 73.50.Td

INTRODUCTION

Scaling rules in metal-oxide-semiconductor devices impose very thin equivalent oxide thickness (EOT) in the order of 1 nm. Several limiting factors associated to ultra-thin gate oxides have been reported in the literature. As an example, direct tunnelling current increases exponentially with decreasing thickness and affects MOS devices performances. To achieve low EOT with low gate leakage current densities, ultra-thin nitrided oxides have been optimised. This process step is done using Rapid Thermal Nitridation (RTN) or Decoupled Plasma Nitridation (DPN) processes. Up today many works have presented the impact of these nitridation processes on dielectric films thicker than 2 nm [1,2]. It has been shown that the DPN process improves the 1/f noise level about one order of magnitude.

In this paper, we focus on drain and gate 1/f noise in ultra-thin RTN or DPN oxide p-MOSFETs. Noise measurements have been performed from weak to strong inversion in ohmic and saturation range. Flicker noise figure of merit $N_t(E_F)$ is extracted for each nitridation processes. Our data will be compared with others published ones.

DEVICES UNDER TEST

The test devices are processed on 200 mm diameter silicon wafer. The transistors have individual gate and drain electrodes. The investigated transistors have gate mask width (W) between 100 nm and 10 µm, and gate mask length (L) between 90nm and 10µm. The I_{on}/I_{off} characteristics have shown that the devices are competitive.

Noise characterisations have been performed on three targeted EOT values, RTN 1.2 nm, RTN 1.4 nm and DPN 1.2 nm. Since C-V characterisation gives similar EOT for DPN and RTN 1.2 nm devices, 1/f drain and gate noise comparison can be performed between these devices.

NOISE INVESTIGATIONS AND DISCUSSION

Drain Current noise

The drain current power spectral densities of the three technologies are reported on figures 1 and 2. In order to compare the noise performances, the devices have the same width and length. The figure 1 shows measurements at low drain bias (V_{DS}=-25mV) from weak to strong inversion. The 1/f noise levels at 1Hz are the same for the three technologies. From the subthreshold regime and following the ΔN model, we have extracted the slow trap densities. The 1/f noise behaviour in strong inversion can be modelled following the $\Delta \mu$ empirical law [3], allowing the extraction of the Hooge empirical parameter α_H. The mobility fluctuation origin for these devices in strong inversion is confirmed by the slope 3/2 observed in saturation regime. The low nitridation impact on 1/f noise level as measured in ohmic range is obtained. All these parameters are reported in table 1 and are used as support for our discussion.

It has been shown that slow interface trap density involved in subthreshold 1/f noise is an accurate figure of merit for a nitridation process [2]. For the RTN processes, we observe a slight increase with the oxide thickness reduction, which is in accordance with the trend observed for nitrided oxides as shown in figure 4. The defects induced by the nitrogen atoms are responsible of this dramatically increase. In order to preserve the interface, DPN process has been developed, so that nitridation affects only the top of the thermal growth oxide. On thicker oxides than studied ones in this work, it has been shown that DPN process induces less slow interface trap density than RTN process [1,2]. Our measurements show that this behaviour can not be extended to ultra thin oxides and the DPN process on ultrathin dielectrics do not improve the 1/f noise level. For such thin oxides, and following the ΔN model, a 1/f power spectral density is obtained by a trapping mechanism between inversion layer carriers and oxide traps located in a range of 1 nm from the interface [4]. The ratio between this distance and the total oxide thickness is higher for ultrathin oxide so that all defects induced by the DPN process can participate to the 1/f noise. This fact stands as a limitation for the DPN process in term of 1/f noise improvement.

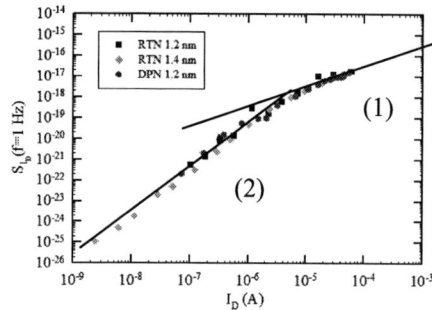
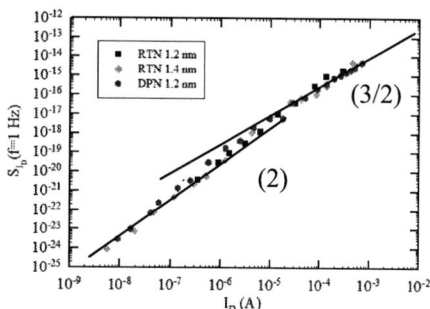

FIGURE 1. Drain power spectral density variations versus drain current at V_{DS}=-25 mV for 10·µm × 200 nm devices from different technologies.

FIGURE 2. Drain current power spectral density variations versus drain current at V_{DS}=-1V for 10 µm × 200 nm devices from different technologies.

Gate Current noise

Gate current 1/f noise has been investigated as a complementary characterisation tool for ultrathin oxide. The gate leakage current is the result of quantum mechanical direct tunnelling process. Several authors have reported that shot noise can be enhanced or suppressed, however we have measured always full shot noise, which means that the tunnelling process through the oxide layer follows Poisson statistics. 1/f gate noise is a trapping related process, which modulates the gate leakage current [4]. In order to evaluate the impact of the nitridation process on the gate current noise, we have used the empirical law for the 1/f gate current noise as follow [5]:

$$S_{I_G}(f) = \frac{K_G}{f W_{eff} L_{eff}} I_G^2 \quad (1)$$

The parameter K_G is used as figure of merit to compare the impact of the nitridation processes. The normalized 1/f gate noise levels at 1 Hz are reported in the figure 3 for the different processes. The parameter K_G is reported in table 1, and as expected from the drain current noise measurements, there is no improvement due to the DPN process. Gate noise power spectral densities exhibit RTS noise, on large area devices, while corresponding drain current power spectral densities don't, as expected for large area devices. It means that single defects can be characterized on large area devices using low frequency gate noise measurements [7]. Gate noise measurement is a very sensitive tool to extract oxide defects and will be useful for future characterizations of ultimate dielectrics.

TABLE 1. Noise parameters for RTN and DPN processes

	RTN 1.4 nm	RTN 1.2 nm	DPN 1.2 nm
$N_t(E_F)$ eV^{-1} cm^{-3}	3 10^{17}	4 10^{17}	4 10^{17}
α_H	9 10^{-4}	10^{-3}	10^{-3}
K_G (m^{-2})	7.5 10^{-20}	1 10^{-19}	2 10^{-19}

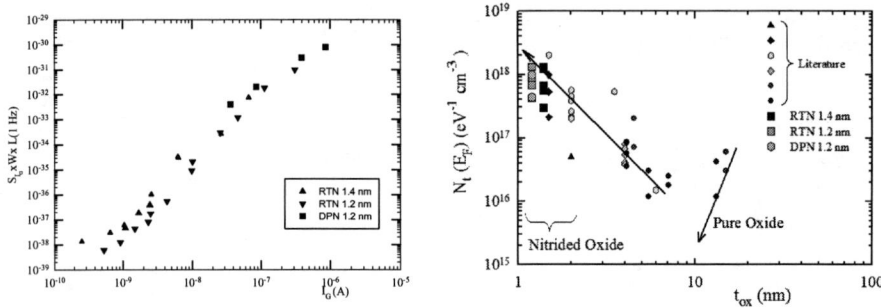

FIGURE 3. Area normalised 1/f gate current noise versus gate-to-channel current at f = 1Hz, for DPN and RTN processes.

FIGURE 4. Evolution of slow interface trap density versus gate oxide thickness for pMOSFETs

CONCLUSIONS

Drain and gate current noise measurements have been performed on ultrathin pMOS transistors in order to evaluate the impact of RTN and DPN processes on 1/f noise levels. The DPN process is as noisy as the RTN one, for both drain and gate current. The benefit in changing process for thick oxide doesn't hold any longer in term on 1/f noise performance with an equivalent oxide thickness of 1.2 nm. Following the ΔN model, the distance between traps and the interface is in the same order of magnitude than the oxide thickness, so that all defects induced by the DPN process can participate to the 1/f noise. Moreover, the same trend has been observed by gate noise measurements.

REFERENCES

1. Da Rold and al., "Impact of gate oxide nitridation process on 1/f noise in 0.18 µm CMOS", Microelectronics reliability, Vol. 41, 2001, pp 1933
2. M. Marin and al.,"Can 1/f noise in MOSFETs be reduced by gate oxide and channel optimization?", published in this preceeding conference.
3. M.Valenza and al., "Overview of the impact of downscaling technology of 1/f noise in p-MOSFETs to 90 nm", IEE Proc-Circuits Devices Syst., Vol 151, No 2, pp 102-110, April 2004
4. E. Simoen and al, " Tunneling 1/f'noise in 5 nm HfO_2/2.1 nm SiO_2 gate stack n-MOSFETs", Solid-State Electronics, vol. 49, pp. 702-707, 2005.
5. J. Lee and al, "Noise model of gate-leakage current in ultrathin oxide MOSFETs", IEEE Trans. On. Electron. Devices, vol.50, N°12, December 2003, pp 2499-2506.
6. M. Valenza and al., "Impact of gate current noise on drain current noise in 90 nm CMOS technology", European Solid-State Device Research, 2003. ESSDERC '03. 33rd Conference on, 16-18 Sept. 2003, pp 287- 290
7. F. Martinez and al., "Oxide trap characterization of 45 nm MOS transistors by gate current R.T.S noise measurements", Insulating Films on Semiconductors (INFOS 2005), June 2005

Characterization of Low-Frequency Noise Sources in Planar Devices Using Cross-Shaped 4-Terminal Devices

Vincent Mosser and Alexandre Kerlain

Itron France, 50 avenue Jean Jaures, F-92120 Montrouge
vincent.mosser@itron.com, +33 146 00 66 74

Abstract. We promote here the use of cross-shaped 4-terminal devices (Hall crosses) to measure LF noise spectra in planar technologies. The implementation of this method is described. When investigating LF noise for the purpose of material or process characterization, such a procedure is more simple and straightforward compared to conventional differential noise measurements based on a Wheatstone bridge or single-ended measurements based on proprietary electronic circuitry. As an example of application, we then use it to extract information on the energetic as well as spatial location of a trap in Metal-Insulator-Semiconductor PHEMT pseudomorphic heterostructures.

Keywords: GaAs, LF-noise, P-HEMT, LF-Noise measurement set-up.

LOW FREQUENCY NOISE IN PLANAR 2DEG STRUCTURES

LF noise in semiconductor devices is known to be due to conductivity fluctuations[1]: the number of free carriers fluctuates, due to trapping/detrapping on localized levels located within a few kT from the Fermi level. These traps can be either interface traps, or bulk traps present in the whole cap layer or buffer layer, only those energetically located a few kT around the Fermi level are active. Morever, in planar 2DEG structure, due to the presence of strong band bending and electric field in either the cap layer or buffer layer, even bulk traps active as noise generator are located in a thin slice a material (cf. Fig. 1a).

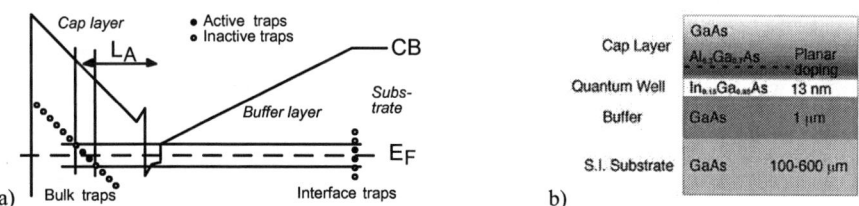

FIGURE 1. a: Band diagram of a P-HEMT-like planar device showing hypothetical bulk and interface traps. Only those in the vicinity of E_F are active. b: Pseudomorphic heterostructure used in the present study

Thus, for a discrete level with sheet density N_T, energy level E_T, mean generation-recombination time τ and occupancy described by the Fermi function $F = F(E_T) = 1/(1 + g^{-1} \exp\frac{E_T - E_F}{kT})$, the level capacitance is $C_T = eN_T/kT \cdot F(1-F)$

and the time constant is $\tau = e_{n0}^{-1} \cdot \exp\frac{E_C - E_T}{kT}(1-F)$. The active trap is located in a plane at a distance L_A from the 2D channel, so that the sheet capacitance between these 2 plates is equal to $C_A = \varepsilon_S / L_A$. Under these assumptions, the noise power spectrum density (PSD) has a Lorentzian behavior which differs from the usual form[1] by the term $(C_T + C_A)^2 / C_A^2$ in the denominator[2]:

$$S_V = \frac{1}{e^4 n_s^4 \mu_n^2} \frac{4kT \tau \, C_T}{\left(\frac{C_T + C_A}{C_A}\right)^2 + \omega^2 \tau^2} \cdot G_N \cdot \frac{I_{DUT}^2}{W^2} \quad (1)$$

Eq. (1) can be rearranged to yield:

$$S_V = S_\rho \cdot G_N \cdot \frac{I_{DUT}^2}{W^2} \quad (2)$$

$$S_\rho = \frac{4kT}{e^4 n_s^4 \mu_n^2} \cdot \frac{C_T C_A}{C_A + C_T} \cdot \frac{\tau'}{1 + \omega^2 \tau'^2} \quad (3)$$

where the apparent time constant τ' is related to the true time constant τ through $\tau' = C_A/(C_A + C_T) \cdot \tau$. The term G_N denotes a geometrical factor, which doesn't depend on the size, but is fully determined by the device type (2 or 4 terminals) and geometry. Apart from the trivial case of a rectangular bar, the geometric factor cannot be calculated by elementary methods, but has to be determined using e.g. Finite Elements calculations[3]. Some useful values are given in Table I.

NOISE MEASUREMENT USING CROSS-SHAPED DEVICES

The standard setup used to measure LF noise spectra are either differential, based on a Wheatstone bridge or single-ended, based on proprietary electronic circuitry[4].
In the proposed method (Fig. 2), a cross-shaped symmetrical 4-terminal device is patterned using a standard process and connected to a commercial current or voltage source and the transverse voltage is amplified by a low noise differential amplifier whose output is fed to a FFT analyzer.

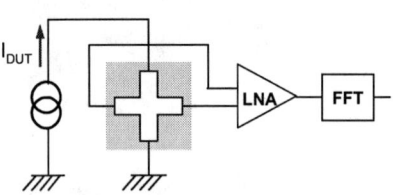

FIGURE 2. Differential measurement of the noise PSD using a cross-shaped device.

TABLE 1. Geometric factor in the noise PSD

Device type	Configuration; # of terminals	Geometrical factor G_N for noise PSD	
		General	Case L/W=4
Rectangular	Wh. bridge; 2T	L/W	4
Greek cross with sharp corners	Wh. bridge; 2T	L/W-0.526	3.474
	Cross; 4T	0.325	0.325
Greek cross with cut corners	Cross; 4T	0.12 to 0.325 depending on shape	

This method is characterized by the following features and presents some advantages compared to standard low frequency noise measurements approaches:

1. The observed noise originates from a small, well defined region of the channel at the intersection of the cross[5].
2. A Hall cross forms by itself an intrinsic, perfectly balanced differential bridge. In particular, the noise of the biasing source is fully rejected, since it is common mode. There is no need for a battery powered or extremely low-noise bias. Moreover the bridge remains perfectly balanced when applying a gate or substrate voltage, or varying the temperature.
3. In addition, the area of the measurement loop can be made extremely small, in contrary to the case of a Wheatstone bridge using external resistors. This makes the measurement much less sensitive to electromagnetic perturbations (external noise).
4. The mobility μ_n and channel carrier density n_S appearing in Eq. (1) or hidden in S_ρ in Eq. (2) can be directly obtained from Hall measurements in the same cross-shaped device.

Application: Noise Source in a PHEMT like Structure

Gated Hall crosses can be used to vertically localize the traps responsible for the excess noise. Indeed, when applying a DC gate voltage, the variation of the channel electron density is a direct image of the variation of the electric field F_A in the cap layer. According to Gauss' theorem:

$$\delta F_A = e/\varepsilon_S \cdot \delta n_S \qquad (4)$$

and the change of the activation energy in the expression of the cutoff frequency is given by

$$\delta(E_T - E_F) = e/\varepsilon_S \cdot \delta n_S \cdot L_A \qquad (5)$$

If the LF noise source is a defect located between channel and surface, its cutoff frequency will shift due to the related energy shift and hence its distance to the channel can be determined. Similarly a backside voltage will allow to probe defects underneath the channel.

This feature has been observed with MIS-Hall crosses patterned using a standard process from heterostructures as shown in Fig. 1-b, with a Ti/Pt/Au gate deposited above the 200 nm thick silicon oxide passivation layer.

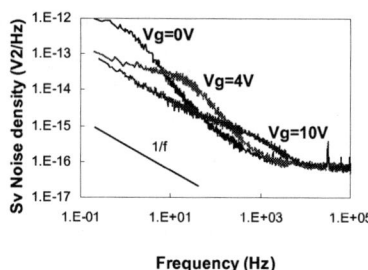
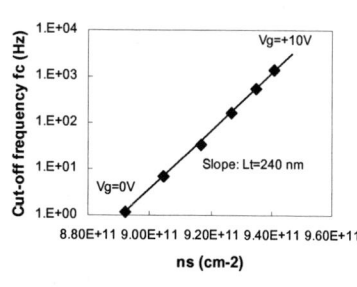

a) b)

FIGURE 3. a: Evolution of the noise spectrum when varying the gate voltage in a cross-shaped MIS device. For the sake of clarity only 3 spectra were drawn. b: Drift of the cutoff frequency of the low-frequency peak f.Sv as a function of the electron concentration in the channel. The slope provides an estimation of the distance between the channel and the trap, according to Eq. (6).

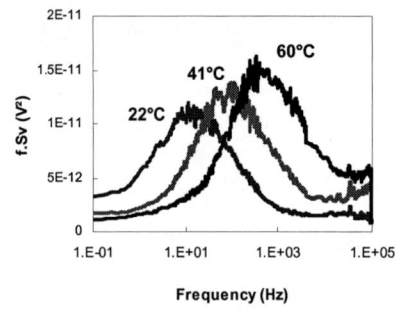 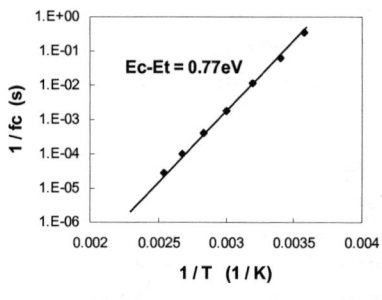

a) b)

FIGURE 4. a: Evolution of the fS_V noise spectrum with temperature in a cross-shaped MIS device. b: Arrhenius plot of $\tau' = 1/f_c$ of traps responsible for the LF noise in the studied structure.

The evolution of the noise spectrum when varying the gate voltage is shown in Fig. 3a. It contains a Lorentzian-type contribution. This behavior is highlighted by plotting $f \cdot S_T(f)$, as in Fig. 4a. The position of the maximum provides the cutoff frequency $f_c = 1/2\pi\,\tau'$ whose evolution as a function of the electron density n_S in the channel (as measured by the Hall effect) is plotted in Fig. 3b. The cutoff frequency shows an exponential dependence over 3 orders of magnitude on the channel electron density:

$$f_c \propto \exp(eL_A \delta n_S / \varepsilon_s kT) \qquad (6)$$

The distance between the related trap and the 2DEG can be estimated to be 240 nm, i.e. the trap is located slightly below the semiconductor surface. This is in agreement with the activation energy of 0.77 eV found from Arrhenius plot (Fig. 4b).

Conclusion

We propose a LF noise measurement method that is more robust, compact and straightforward to implement than standard methods. It is especially well-suited for the material and process characterization of planar devices such as FET and analog devices. This method is also of special interest for the characterization of less-mature materials processing such as AlGaN/GaN or SiC.

ACKNOWLEDGMENTS

We are grateful to Dr. J.L. Reverchon and the technical staff at Thales Research and Technology in Orsay for the photolithography process, to Dr. Daniel Prost, CRIL Technology, for Finite Element calculations and to Mr. Michel Gervais, Akuit, for enlightening discussions.

REFERENCES

[1] M.E. Levinshtein and S.L. Rumyantsev, *Semicond. Sci. Technol.* **9**, 1183-1189 (1994)
[2] Vincent Mosser, Grzegorz Jung, Jacek Przybytek, Miguel Ocio and Youcef Haddab. Fluctuations & Noise 2003, Proc. SPIE Vol.5115, 183-195
[3] Jacek Przybytek, Vincent Mosser and Youcef Haddab, Fluctuations and Noise 2003, Proc. SPIE Vol. 5113, p. 475-483
[4] A. Blaum, O. Pilloud, G. Scalea, J. Victory, F. Sischka, *Proc. IEEE 2001 Int. Conf. on Microelectronic Test Structures*, pp. 125-130, March 2001.
[5] L.K.J. Vandamme and A.H. de Kuijper, *Solid-State Electronics*, **22**, 981-986 (1979)

Low-Frequency Noise Characterization of 90 nm Multiple Gate Oxide CMOS Transistors

N. Lukyanchikova[1], N. Garbar[1], A. Smolanka[1], M. Lokshin[1], S.-C. Lee[2,3], E. Simoen[2], C. Claeys[2,4]

[1]*Institute of Semiconductor Physics, Prospect Nauki 45, 03028 Kiev, Ukraine*
[2]*IMEC, Kapeldreef 75, B-3001 Leuven, Belgium*
[3]*Present address: SanDisk, Yokohama 247-8585, Japan*
[4]*also at E.E. Dept., KU Leuven, Belgium*

Abstract. The $1/f$ noise is investigated in 90 nm MOSFETs with nominal t_{ox}=1.5 and 2 nm prepared with and without a low-energy N or F ion implantation. It is found that a N or F implantation has no impact on the noise. The McWhorter's noise component prevails at not too high currents I and the concentrations of the noisy oxide centers are found to be 4.2×10^{16} cm^{-3}eV^{-1} in nMOSFETs and 5.1×10^{17} cm^{-3}eV^{-1} in pMOSFETs. At high I the noise is due to the device series resistance.

Keywords: $1/f$ noise, McWhorter's noise, MOSFET, System-on-Chip (SoC) technology, selective low-energy ion implantation, surface quantization
PACS: 73.50.Td; 85.30.Tv; 81.16.

INTRODUCTION

The development of so-called System-on-Chip (SoC) technology platforms requires the fabrication of MOS transistors with different gate oxide thickness (t_{ox}) in different parts of a wafer (so-called multiple gate oxide schemes). One viable way to achieve this is by use of selective low-energy nitrogen or fluorine ion implantations (I/I) [1]. In the former case, the oxidation rate will be retarded, resulting in a smaller t_{ox}, while for F I/I a faster oxidation is observed. One issue that should be dealt with is the preservation of the low-frequency noise performance of an I/I gate oxide transistor compared with a standard one with the same thickness. This is particularly true for the thicker oxide(s), used in the analog modules of the platform. Reports in the literature in this respect are a bit controversial: while some reports suggest little change in the $1/f$ noise [2] or even an improvement [3] in the case of N I/I, a strong increase was found for F I/I devices, due to excess GR noise originating from un-annealed implantation damage in the silicon substrate [4].

EXPERIMENTAL

The 90 nm CMOS transistors with a polysilicon gate were fabricated using Shallow Trench Isolation (STI). The dopant activation was achieved by a 1100 °C, 1 s spike anneal. An 8 keV implantation at 5×10^{14} cm^{-2} (F) or 3×10^{15} cm^2 (N) dose was performed in the active regions of different wafers, followed by a 1.5 nm (F I/I) or 2 nm (N I/I) rapid thermal oxidation (RTO). As references, un-implanted 1.5 and 2 nm RTO gate transistors were studied. The electrical gate oxide thickness has been derived from high-frequency C-V measurements on large-area capacitors, in the accumulation regime. The resulting values are: 1.7 nm and 2.2 nm (reference wafers without implant), 2.3 nm (F I/I) and 1.94 nm (N I/I). Noise spectra $S_I(f)$ where S_I is the drain current noise spectral density and f is the frequency were measured on packaged devices in the linear operation regime for a $|V_{DS}|$ =25 mV, with the gate voltage (V_G) varied from weak to strong inversion.

RESULTS AND DISCUSSION

In the frequency range 0.7 Hz≤f≤10 kHz the noise spectra are of $1/f$-type for all devices studied. Figure 1 shows the dependences of S_I on the drain current I measured at f=10 Hz for the devices of different types.

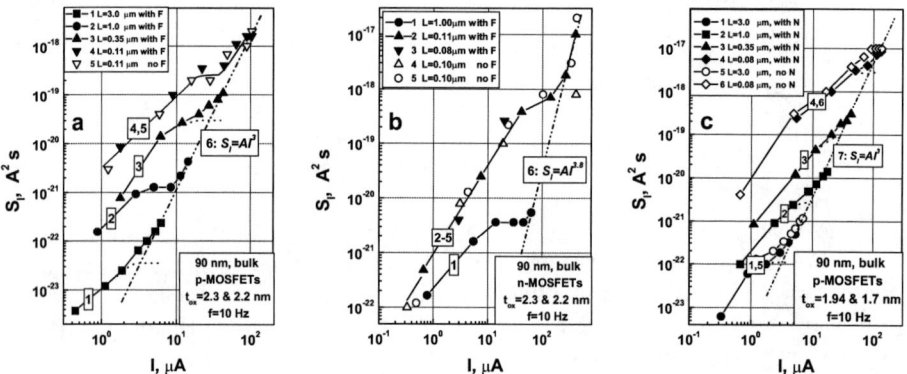

FIGURE 1. Current noise spectral density for different 90 nm pMOSFETs (a, c) and nMOSFETs (b) with and without F I/I (a, b) and with and without N I/I (c)

It is seen that the N I/I or F I/I does not impact the $1/f$-noise considered. It is also seen that the plateau typical for the McWhorter's $1/f$-noise [5] is observed in some curves $S_I(I)$ and S_I decreases with increasing channel length L at low and moderate I. At the same time, the curves $S_I(I)$ for different L fall on one and the same curve at high currents where $S_I(I)$ can be presented as $S_I = AI^n$ where $n \approx 3$. This means that S_I is independent of L at high I and, hence, the corresponding noise is generated out of the channel. The fluctuations of the series resistance r_{ser} [5] can be responsible for the noise described by $S_I = AI^n$. It is easy to show that in such a case

$$S_I = (S_I)_{ch} + (S_I)_{ser}[r_{ser}^2/(r_{ser}+r_{ch})^2] \qquad (1)$$

where $(S_I)_{ch}$ is the contribution of the channel noise into S_I, $(S_I)_{ser} \sim I^2$ is the equivalent current noise generator for the series resistance noise, r_{ch} is the channel resistance. Since $r_{ch} \sim I^{-1}$ in strong inversion, the second term in Eq. (1) has to increase as I^n where

$n \approx 4$ at $r_{ser} < r_{ch}$ and $n \approx 2$ at $r_{ser} > r_{ch}$. Therefore, $2 < n < 4$ can be observed at $r_{ser} \sim r_{ch}$. This explains the values of n observed experimentally and allows to write:

$$S_I = (S_I)_{ch} + AI^n \qquad (2)$$

Then the contribution of $(S_I)_{ch}$ into $S_I(I)$ can be found as $[S_I(I)-AI^n]$ where $S_I(I)$ correspond to experimental curves which are different for different L while the values of AI^n are determined from the curve AI^n vs. I which is the same for different L. For example, the values of $S_I(I)$ for $L=3, 1, 0.35$ and 0.11 μm in Fig. 1a are found from curves 1 to 4 while the values of AI^n for all those values of L are found from curve 6. As a result, the plateaus corresponding to curves $(S_I)_{ch}$ vs. I has been determined for all L's. Those plateaus are shown by dotted curves. It is seen, that the plateau level increases with decreasing L as L^{-3} that has to be observed for the McWhorter's noise.

Figure 2 shows the dependences of the equivalent gate voltage noise spectral density S_{VG} on the overdrive voltage V_G^* for pMOSFETs with F(I/I) where $S_{VG}=S_I/(g_m)^2$, $V_G^*=V_G-V_{th}$, V_{th} is the threshold voltage and g_m is the transconductance.

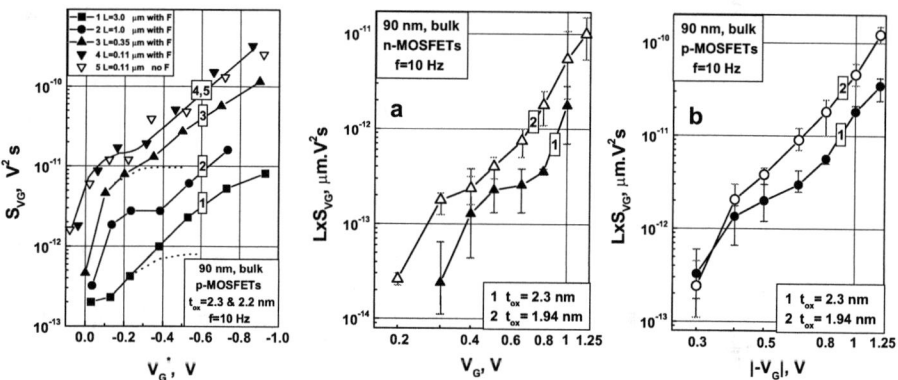

FIGURE 2. S_{VG} vs. V_G^* for 90 nm pMOSFETs of different L with and without F I/I.

FIGURE 3. $(L \times S_{VG})$ vs. V_G for 90 nm nMOSFETs (a) and pMOSFETs (b) of $t_{ox}=2.3$ nm (1) and 1.94 nm (2).

The dotted parts of the curves were calculated by the formula: $S_{VG}=[S_I(I)-AI^n]/(g_m)^2$. As is seen, $S_{VG} \sim L^{-1}$ at the plateaus that is typical for the McWhorter's noise for which:

$$S_{VG} = q^2 kT\lambda N_{ot} / LWC_{ox}^2 f \qquad (3)$$

where q is the electron charge, k is the Boltzmann's constant, T is the temperature, λ is a tunneling parameter of the order of 0.1 nm, N_{ot} is the effective noisy oxide trap density per cm^3 and per eV, W is the channel width and C_{ox} is the gate capacitance per cm^2. Since the surface quantization occurs in the devices considered, C_{ox} has to be replaced by C_{oxeff} where $(C_{oxeff}/C_{ox})^2 \approx 0.6$ and 0.5 for the pMOSFETs and nMOSFETs, respectively [6]. By applying Eq. (3) to the results of Fig. 2, one finds $N_{ot}=5.1 \times 10^{17}$ cm^{-3}eV^{-1}. For the nMOSFETs of $t_{ox}=2.3$ nm, the value $N_{ot}=4.2 \times 10^{16}$ cm^{-3}eV^{-1} was found.

It is seen from Fig. 2 that $S_{VG} \sim L^{-1}$ not only holds for the plateaus, but also at high values of V_G^* where the noise is attributed to the series resistance fluctuations. This does not contradict such an interpretation. Actually, since $S_I(I) \approx AI^3$ at high V_G^* where $I \sim L^{-1}$ for a given V_G^* and $g_m \sim L^{-1}$, one has: $S_{VG}(L) = S_I(I)/(g_m)^2 \sim (L)^{-3}/(L)^{-2} \sim L^{-1}$.

The typical influence of t_{ox} on $L \times S_{VG}$ is demonstrated in Fig. 3. It is seen that the 2.3 nm devices appear to exhibit lower $1/f$ noise than the 1.94 nm ones while Eq. (3) predicts the decrease of $L \times S_{VG}$ with decreasing t_{ox} since $C_{ox} \sim (t_{ox})^{-1}$.

Note that another model based on the increase of capture cross-section σ for the interface traps under conditions of surface quantization could explain the behaviour of $S_{VG}(V_G)$ [6]. In this case, instead of Eq. (3), the following formula has to be used

$$S_{VG} = q^2 kT \lambda N_{ot} (\sigma_{2D}/\sigma_{3D}) / LWC_{oxeff}^2 f \qquad (4)$$

where σ_{2D} and σ_{3D} are the values of σ with and without surface quantization, respectively. However, such a model does not explain our results. Indeed, Eq. (4) does not predict the plateau in curves $S_{VG}(V_G^*)$ for nMOSFETs since $(\sigma_{2D}/\sigma_{3D})$ increases from 3000 to 5000 with increasing V_G corresponding to those plateaus. In the case of pMOSFETs, the value of N_{ot} determined with the help of Eq. (4) decreases from 4.4×10^{14} cm^{-3}eV^{-1} at $|-V_G| \leq 0.65$ V to 3.2×10^{13} cm^{-3}eV^{-1} at $V_G = -1.2$ V, instead of being constant. Moreover, such low values of N_{ot} seem to be unreal.

CONCLUSIONS

1. The use of N or F I/I does not impact on the $1/f$ noise of 90 nm multiple gate oxide devices.
2. The McWhorter's $1/f$-noise manifests itself at low and moderate currents while the series resistance is responsible for the $1/f$-noise observed at high I. The densities of the noisy oxide traps are $N_{ot} = 4.2 \times 10^{16}$ cm^{-3}eV^{-1} and $N_{ot} = 5.1 \times 10^{17}$ cm^{-3}eV^{-1} for nMOSFETs and pMOSFETs, respectively.
3. Reducing t_{ox} below 2.3 nm may lead to a t_{ox} dependence that differs from the one expected for a standard number fluctuations theory.

REFERENCES

1. Adam, L. S., Bowen, C., and Law, M. E., *IEEE Trans. Electron Devices* **50**, 589-600 (2003).
2. D'Sousa, S., Hwang, L.-M., Matloubian, M., Martin, S., Sherman, P., Joshi, A., Wu, H., Bhattacharya, S., and Kempf, P., in *IEDM Techn. Dig.*, The IEEE, New York, 1999, pp. 839.
3. Lee, S.-C., Simoen, E., and Badenes, G., *Solid-State Electron.* **48**, 1687-1690 (2004).
4. Woerlee, P.H., Knitel, M.J., Meyssen, V.M.H., Velghe, R.M.D.A., and Zegers van Duijnhoven, A.T.A., *Proc. ESSDERC 2001*, Eds H. Ryssel, G. Wachutka and H. Grünbacher, Gif-sur-Yvette, Editions Frontières, 2001, pp. 107-110.
5. Lukyanchikova, N., Garbar, N., Petrichuk, M., Simoen, E., and Claeys, C., *Solid-State Electron.* **44**, 1239-1245 (2000).
6. Mercha, A., Simoen, E., and Claeys, C., *IEEE Trans. Electron Devices* **50**, 2520-2527 (2003).

TeraHertz emission from nanometric HEMTs analyzed by noise spectra

J.-F. Millithaler*, L. Varani*, C. Palermo*, J. Mateos[†], T. González[†], S. Perez[†], D. Pardo[†], W. Knap**, J. Lusakowski**, N. Dyakonova**, S. Bollaert[‡] and A. Cappy[‡]

*CEM2 - UMR CNRS 5507 - Université Montpellier II - France
[†]Universidad de Salamanca - 37008 Salamanca - Spain
**GES - UMR CNRS 5650 - Université Montpellier II - France
[‡]IEMN - UMR CNRS 8520 - Avenue Poincaré - 59652 Villeneuve d'Ascq - France

Abstract. TeraHertz emission from High Electron Mobility Transistors has been recently measured from experiments. The experiments show emission spectra with two peaks in the TeraHertz range: one around 1 THz is sensitive to drain and gate voltages, and another one around 5 THz which is fixed. In order to get physical insight into the microscopic mechanism at the basis of the radiation emission we have performed a Monte Carlo simulation of the measured transistors using the current noise spectra as sensitive probes to detect the presence of electrical instabilities. Numerical results are found to be in good agreement with experiments confirming the presence of an oscillatory dynamics in the TeraHertz range.

Keywords: TeraHertz, Monte Carlo, HEMT
PACS: 72.30.+q , 72.70.+m , 02.70.Uu

INTRODUCTION

The development of devices with extremely high operation frequency will not only lead to an increase of the speed of devices but hopefully also to construction of new sources of TeraHertz (THz) radiation which will be useful for different kinds of non-destructive imaging and medical or technical diagnostics. In particular, THz sources based on solid-state devices offer great possibilities of integration with other optoelectronic devices within a single chip [1].

This is the case of InGaAs/InAlAs lattice-matched nanometric High Electron Mobility Transistors (HEMT) where THz emission was observed recently for the first time [2]. It is evident that a detailed analysis of the physical processes characteristics of such devices requires a microscopic approach due to the nanometric dimensions leading to the appearance of hot carriers phenomena, nonstationary transport, ballistic conditions etc.

As a consequence, in order to investigate the observed emission, we have used Monte Carlo simulations to obtain the spectrum of the current noise, which is known to be a quantity very sensitive to different carrier instabilities [3]. While a standard relaxation dynamics gives rise in the noise spectrum to a Lorentzian cutoff frequency related to the inverse of the relaxation time, any carrier instability responsible for an oscillatory dynamics will be evidenced as a peak. This work was carried out by means of a semiclassical two-dimensional Monte Carlo model whose validity was already checked for similar devices [4].

EXPERIMENTS

The study is based on the lattice-mattched InGaAs/AlInAs High Electron Mobility Transistor (HEMT) on InP substrate with a nominal gate length $L_G = 60$ nm, a source-drain distance $L_A = 1.3$ μm and a width $W = 50$ μm (fig.1(a)). The structure was placed, at liquid helium temperature, in a cyclotron emission spectrometer designed to perform a spectral analysis of a weak THz radiation. The gate was short circuited with the source since this boundary condition was supposed to be more favourable for THz emission [2].

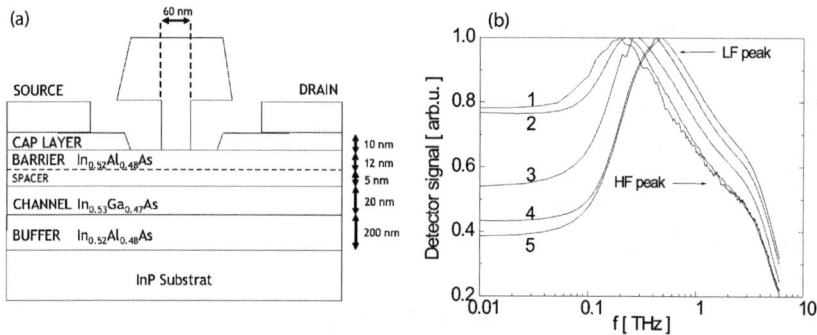

FIGURE 1. (a) The schematic structure of the transistor investigated with a T-shaped gate. The dotted line in the barrier region shows the position of the δ doping (b) HEMT emission spectra for the gate short circuited with the source by a gold wire on the chip. The spectra from 1 to 5 correspond to V_{DS} equal to 0.3, 0.45, 0.6, 0.7 and 0.8 V, respectivlely

The spectrum of THz radiation (fig.1(b)) observed experimentally exhibits two structures: a low frequency peak (lower than 1 THz) which has been found to be sensitive to gate and drain voltages and a high frequency peak (around 5 THz) whose frequency is rather fixed.

MONTE CARLO RESULTS

Figure 2 shows the spatial dependence of the average electron velocity inside the HEMT for different drain voltages V_{ds} and for a constant gate voltage V_{gs}. One can notice that under the gate, the electrons move with a velocity of at least an order of magnitude higher than in the recess and contact parts of the transistor. Such a high velocity is interpreted as a signature of a ballistic regime under the gate. The increase of the velocity under the gate at increasing V_{ds} decreases the electron transit time through the gated part of the transistor. This velocity distribution reflects the observed frequency dependence of the positions of the low frequency peaks in the emission spectra.

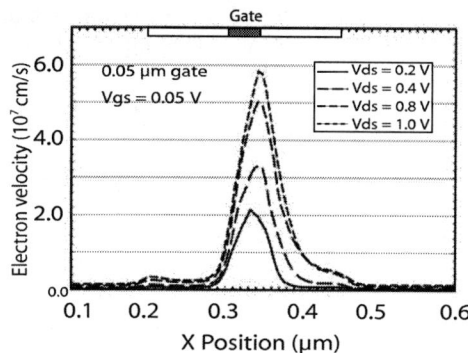

FIGURE 2. Spatial dependence of the average electron velocity in the transistor for a constant V_{gs} and for the indicated values of V_{ds}. The lines at the top of the figure show the spatial extensions of the gate and recess regions.

Fig. 3(a) and (b) show the gate current noise spectra under different operating conditions. The most striking result is that the Monte Carlo calculation of the noise spectra confirms the presence of some oscillatory dynamics in the THz frequency region in good agreement with the experimental results shown in fig. 1(b). Moreover we remark in both cases the presence of two peaks: a low frequency one, around $1-2$ THz, which depends on the drain-to-source as well as on the gate-to-source voltages, and a high frequency one at a fixed frequency around 5 THz.

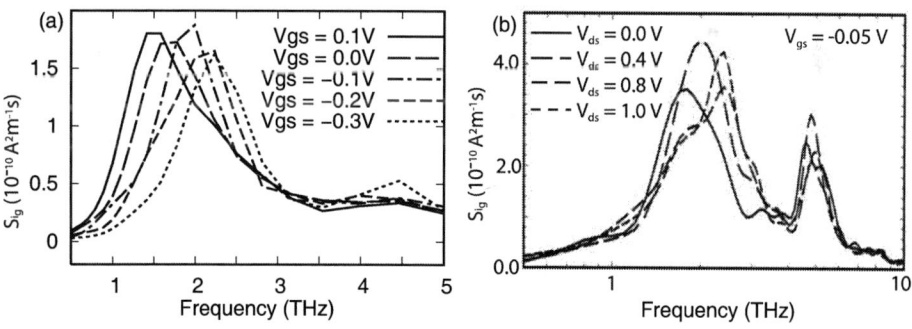

FIGURE 3. Gate current noise spectra for the reported V_{gs} and a constant V_{ds} equal to 0.25 V (a) and for the reported V_{ds} and a constant V_{gs} equal to $-0.05V$ (b).

One possibility to explain these theoretical results relies on the Dyakonov-Shur instability [5] which explains the lower frequency peak as resulting from a plasma instability of the gated two-dimensional electron fluid while the higher frequency one is supposed to result from a plasma instability in the ungated part of the channel [2]. The discrepancies between the experiments and the simulations could be attributed to a peculiar boundary conditions imposed on the resonating transistor cavity in the experiment (short circuiting the gate with the source) that differ from simulations where such a condition is absent. However, additional investigation are necessary to clarify which kind of microscopic instability is at the origin of the observed THz emission.

CONCLUSION

TeraHertz emission from nanometric InAlAs/InGaAs HEMTs has been exprimentally measured. The emission spectrum consists of two peaks around 5 THz and 1 THz. The results of Monte Carlo simulations shows that the electron transport under the gate is ballistic and that the gate current noise spectra evidence the presence of peaks corresponding to the observed frequencies. This analysis shows that the frequency behavior of the emitted radiation can be directly linked to the spectra of current fluctuations. As a consequence, noise spectra are here employed as sensitive indicators of the onset of collective motion of carriers inside the transistor.

REFERENCES

1. R. E. Miles, P. Harrison, and D. Lippens, *Terahertz Sources and Systems*, vol. 27, Kluwer Academic Publishers, 2001.
2. W. Knap, J. Lusakowski, T. Parenty, S. Bollaert, A. Cappy, V. V. Popov, and M. Shur, *Appl. Phys. Lett*, **84**, 2331 (2004).
3. L. Varani, L. Reggiani, T. Kuhn, T. González, and D. Pardo, *IEEE Trans. Electron. Dev.*, **41**, 1916 (1994).
4. J. Mateos, T. González, D. Pardo, S. Bollaert, T. Parenty, and A. Cappy, *IEEE Trans. Electron. Dev.*, **51**, 521 (2004).
5. M. Dyakonov, and M. S. Shur, *Phys. Rev. Lett*, **71**, 2465 (1993).

RTS in Submicron MOSFETs: High Field Effects

J. Pavelka, V. Sedlakova, J. Sikula, J. Havranek, M. Tacano*,
S. Hashiguchi** and M. Toita***

Department of Physics, Brno University of Technology, Technicka 8, 616 00 Brno, Czech Republic
Fax: +425 7261 666, e-mail: sedlaka@feec.vutbr.cz
* Meisei University, Hino, Tokyo, 191-8506 Japan
**Department of Electronics, Yamanashi University, Kofu 400, Japan
***Asahi Kasei, Nobeoka, Miyazaki, Japan

Abstract. The downscaling of electronic devices makes high field transport effects more important. In deep submicron technology high transversal and high lateral electric field exists. Application of drain voltage 1V results in electric field, which exceeds the silicon critical electric field. Electron temperature is then higher than lattice one and field dependent electron mobility must be considered. Due to small gate area we were able to activate one trap only and then in time domain two levels signal was observed. A systematic analysis of two level RTS signal was made to obtain information on capture and emission processes as a function of gate voltage, drain current and temperature for low and high lateral electric field. With increasing drain voltage capture time increases, while dependence on gate voltage is almost the same as for low drain voltage. For constant gate voltage and variable drain voltage emission process is independent on lateral field intensity, while capture time increases with lateral field intensity. For low electric field RTS amplitude is proportional to the current. With increasing lateral field the relative amplitude of RTS pulses also increases. Effect of lateral electric field is temperature dependent. RTS kinetics in high field gives additional information on charge carriers trapping processes in submicron electronic structures.

Keywords: Noise, RTS noise, high electric field, MOSFET, MIS structure.
PACS: 71.23.Cq, 72.20.Jv

INTRODUCTION

In this paper we investigate the high electric field effect on the emission and capture kinetics of random telegraph signals in submicron MOSFETs [1,2]. The downscaling of electronic devices makes high field transport effects more important [3]. In deep submicron technology the thin gate oxide and high channel doping results in high transversal electric field. Application of drain voltage 1V results in high lateral electric field, which exceeds about 5 times the silicon critical field. Electron temperature is then higher than lattice one and field dependent electron mobility must be considered. Due to small gate area and low interface states density we are able to activate one trap only and then in time domain two levels signal is observed.

EXPERIMENTAL

Experiments were carried out for n-channel devices, processed in a 0.35 mm spacer less CMOS technology. The investigated devices have a gate oxide thickness of 6.5 nm, gate with 5 μm, the effective interface area is estimated to be $A = 1.75$ μm^2 and electron low field mobility $\mu_{no} = 370$ cm^2/Vs. The RTS measurements were performed: (i) in the linear operation mode, for low drain bias, where the drain current was changed from weak inversion up to strong inversion by varying the gate voltage U_G, with the source and substrate contact being grounded, (ii) for high gate voltage, and (iii) for the constant gate voltage with the varying drain voltage.

CAPTURE AND EMISSION TIME

The capture time τ_c for the low drain current I_0 (see Fig. 1) is inversely proportional to the square of drain current of the MOSFET for the low and high electric field. For the high drain current value the capture time varies inversely with the drain current.

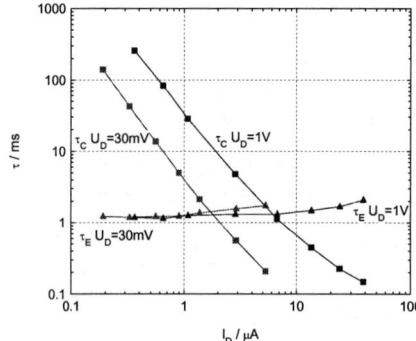 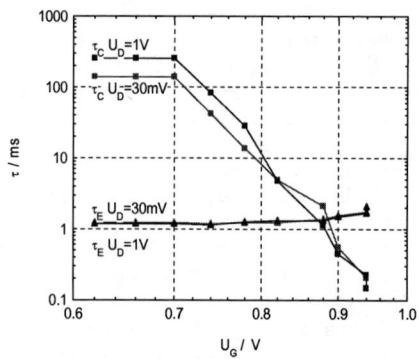

FIGURE 1. Capture and emission time vs. drain current for $U_D = 30$ mV and 1 V

FIGURE 2. Capture and emission time vs. U_G for $U_D = 30$ mV and 1 V

The emission time is independent on the drain voltage value (see Figs. 1 and 2), while capture time increases with the drain voltage for the given value of the drain current. The capture time vs. gate voltage dependence is weak function of the drain voltage (see Fig. 2). We suppose that the capture time coefficient c_n depends on the electric field intensity. There are two variables: $c_n = \sigma(E)v(E)$, where σ and v - capture cross section and charge carrier velocity, respectively – both depend on the electric field.

AMPLITUDE OF RTS PULSES

The RTS current amplitude vs. drain current plot for a submicron MOSFET at low and high drain voltages for N-MOS N51 is in Fig. 3. For low lateral field RTS amplitude ΔI is proportional to drain current, while for high drain current RTS amplitude reach saturation. In this case the quasi Fermi level is near the trap level and dispersion reaches its maximum value. Drain current ΔI amplitude vs. drain current for constant gate voltage and variable load resistance is in Fig. 4.

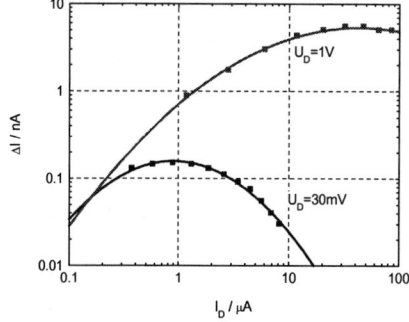 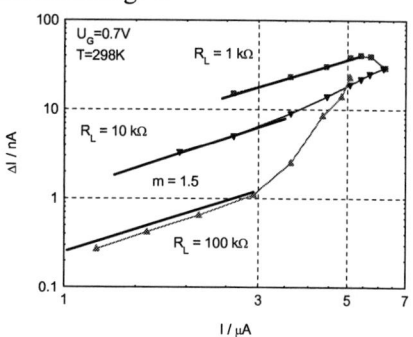

FIGURE 3. The drain current fluctuation amplitude vs. drain current for for $U_D = 30$ mV and 1 V

FIGURE 4. The drain current fluctuation amplitude vs. drain current for constant value of $U_G = 0.7$ V

At small bias the drain current ΔI amplitude is proportional to I^m, where m is approximately 1.5. Exponent m does not depend on load resistance for low lateral field. For the high lateral field saturation appears and all dependencies reach approximately the same value of ΔI.

NOISE SPECTRAL DENSITY

Low frequency noise is g-r type with cut off frequencies from 10 Hz to 1 kHz. The corresponding relaxation time is of the order 0.1 to 10 ms. Results of current noise spectral density vs. frequency are shown in Fig. 5.

Low frequency RTS noise spectral density vs. drain current for $U_D = 30$ mV and 1 V is in Fig. 6. Peak value of the voltage noise spectral density vs. drain current dependence increases with the increasing drain voltage and is shifted to the higher drain current value. Position of these peaks depends on drain current ΔI amplitude vs. drain current and coincidence of quasi Fermi level and trap energy level. These quantities are shifted with temperature.

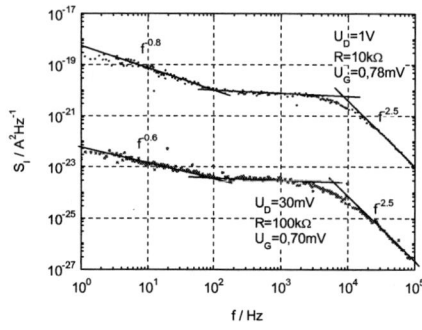 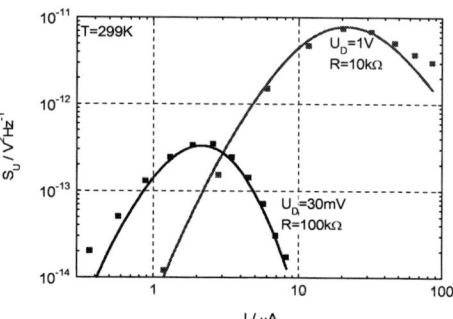

FIGURE 5. RTS current noise spectral density vs. frequency for $U_D = 30$ mV and 1 V

FIGURE 6. Low frequency RTS noise spectral density vs. drain current for $U_D = 30$ mV and 1 V

CONCLUSION

1. The capture time τ_c for the low drain current I_0 is inversely proportional to the square of drain current for both the low and high drain voltage.
2. The capture time for the high drain current is inversely proportional to the first power of the drain current for both the low and high drain voltage.
3. The dependence of emission time on the drain voltage value is very weak.
4. The capture time is almost independent on the gate voltage for both the low and high drain voltage.
5. Peak value of the voltage noise spectral density vs. drain current dependence increases with the increasing drain voltage and is shifted to the higher drain current value.
6. For the constant gate voltage and variable drain voltage: (i) the amplitude of drain current ΔI is proportional to the $I_D^{1.5}$, (ii) the capture time increases with increasing drain current, while the emission time is constant (see Fig. 7). We suppose that the capture time coefficient c_n depends on the electric field intensity.

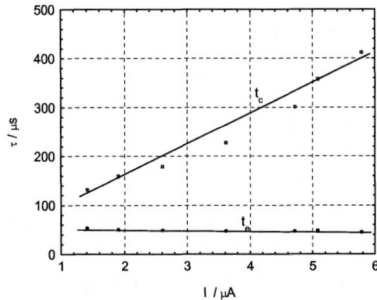

FIGURE 7. Capture and emission time vs. I_D for constant $U_G = 0.7$ V and $T = 298$ K

ACKNOWLEDGMENTS

This research has been supported by GACR No.102/05/2095 and under project MSM 0021630503.

REFERENCES

1. G. Ghibaudo, O. Roux, C. Nguyen-Duc, F. Balestra, J. Brini, Phys. Stat. Sol. (a), 124, 571 (1991).
2. M. J. Kirton and M. J. Uren, Noise in Solid-State Microstructures: A New Perspective on Individual Defects, Interface States, and Low-Frequency Noise, Advances in Phys., 38 (1989) 367
3. Z. Celik – Butler and N.V. Amarasinnghe, Random Telegraph Signals in Deep Submicron Metal–Oxide–Semiconductor Field-Effect Transistors, Noise and Fluctuation Control in Electronic Devices, ed. by A. A. Balandin pp.187 – 199, American Scientific Publisher (2002)

Hooge Noise Parameter of GaN HFETs on SiC

Nobuhisa Tanuma[*], Jan Pavelka[*,¶], Shuichi Yagi[†], Hajime Okumura[†], T. Uemura[*], Munecazu Tacano[*], Sumihisa Hashiguchi[#] and Josef Sikula[¶]

[*]*AMRC, Meisei University, 2-1-1 Hodokubo, Hino, Tokyo 191-8506, Japan*
[¶]*FEEC, Brno University of Technology, Technicka 8, Brno 61600, Czech Republic*
[†]*Power Electronics Research Center, AIST Tsukuba, Ibaraki 305-8568, Japan*
[#]*University of Yamanashi, Kofu 400-8511, Japan*

Abstract. Noise characteristics of epitaxial n-GaN on sapphire layers and GaN/AlGaN on sapphire or SiC HFET structures were investigated in the temperature range of 13K to 300K. Ohmic contacts were made using Ti/Al/Ni/Au and contact noise was found negligible by TLM analysis. The Hooge parameter α_H of epitaxial GaN was 2×10^{-3} at 300K, gradually decreasing to 10^{-4} around 50K. For GaN/AlGaN on sapphire HFET the g-r noise was dominant at almost every temperature, allowing only to determine $\alpha_H = 2\times10^{-4}$ at 22K. The GaN/AlGaN on SiC HFETs were characterized by α_H values of 10^{-4} to 10^{-5}.

Keywords: GaN, 1/f noise, HFET
PACS: 72.70.+m, 72.80.+Ey, 73.40.Kp, 73.50.Td

INTRODUCTION AND SAMPLE CHARACTERISTICS

GaN is wide bandgap compound semiconductor suitable for high-power-density, high-frequency and high-temperature applications. Research efforts are currently focused on the development of GaN/AlGaN heterostructure field effect transistors (HFETs) and optoelectronic devices in the blue and ultraviolet range, such as light emitting diodes, lasers and photodetectors. Device fabrication process technologies are still rather immature and mainly the epitaxial growth of nitrides on the common sapphire substrates usually leads to large defect concentrations due to the significant mismatch in the lattice constants (14%) and thermal expansion coefficients. Better lattice matching in case of SiC substrate is supposed to decrease the number of defects or traps within the epitaxial layers. Low frequency noise is highly sensitive to the presence of crystal imperfections and we used temperature resolved noise spectroscopy to compare the properties of GaN/AlGaN HFETs grown on sapphire and SiC substrates, as well as n-GaN monolayers on sapphire.

In [1] we reported on fabrication of TLM structures with Ti/Al/Ni/Au ohmic contacts on the GaN epitaxial wafer grown by ATMI Ltd. on sapphire substrate with a 0.4μm AlN buffer layer and 2μm Si-doped n-type GaN layer. HFETs were made on POWDEC SiC based substrates and sample structure is given in Fig.1d. Hall measurements of charge carrier density and mobility of GaN layers and both heterostructures are summarized in Fig.1a,b,c. Transistors were dual gated with gate length either 2μm or 4μm and gate width 50μm×2 (Fig.2.). HFET DC characteristics are given in Fig.3. Cut-off frequency was about 6GHz and f_{max} about 16.2GHz.

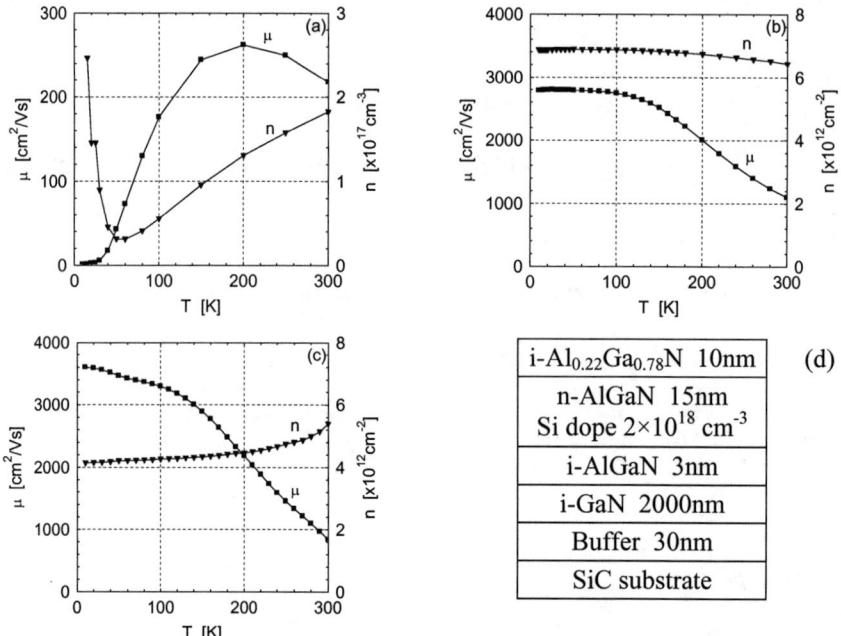

FIGURE 1. Mobility and carrier density: (a) n-GaN on sapphire, (b) GaN/AlGaN HFET on sapphire, (c) GaN/AlGaN HFET on SiC and this sample structure (d)

FIGURE 2. Photography of GaN/AlGaN on SiC structures: Hall device and HFET

EXPERIMENTAL RESULTS AND CONCLUSION

Temperature dependence of noise spectral density was measured from 300K down to 13K using Iwatani helium cryostat, Advantest FFT analyzer, battery power source and Ithaco or NF amplifiers. The influence of contact noise was found negligible by measuring the voltage noise of TLM structures of various lengths L in the range of $L=5\mu m$ to $80\mu m$, where all curves perfectly overlap after normalization to the square of applied voltage and carrier number given by TLM size, both at room temperature and $T=14K$ (see Fig.4a). Noise increases exactly with the square of current (Fig.4b).

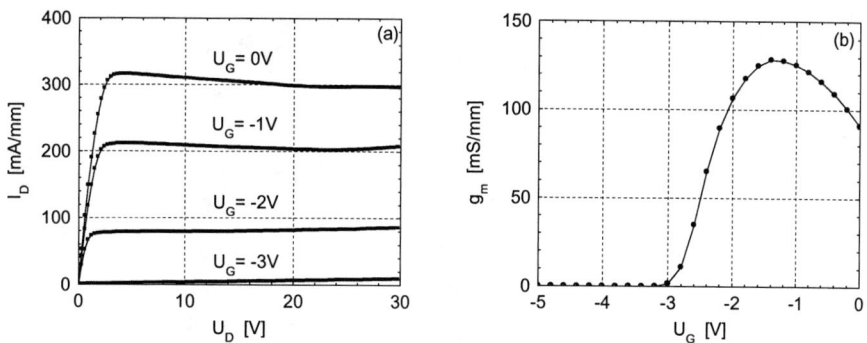

FIGURE 3. DC characteristics of GaN/AlGaN HFET on SiC. (a) Drain current vs. drain voltage, (b) Transconductance vs. gate voltage

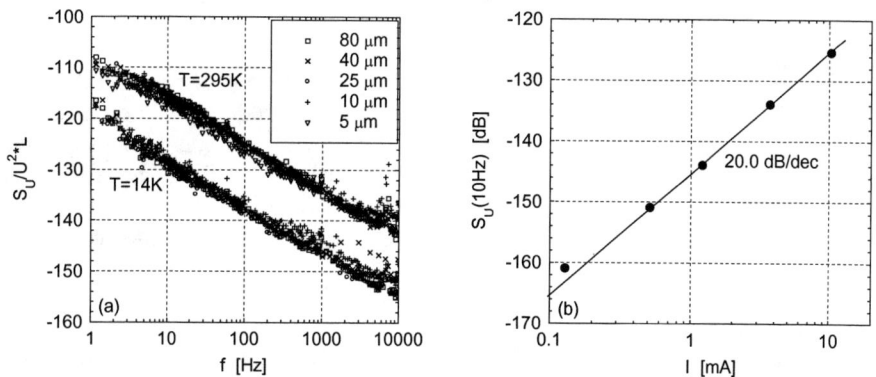

FIGURE 4. (a) Normalised noise spectral density frequency dependence of GaN TLM structures of various length L, measured at temperature 295K or 14K, (b) Noise spectral density of GaN 40µm TLM at frequency 10Hz as a function of sample current

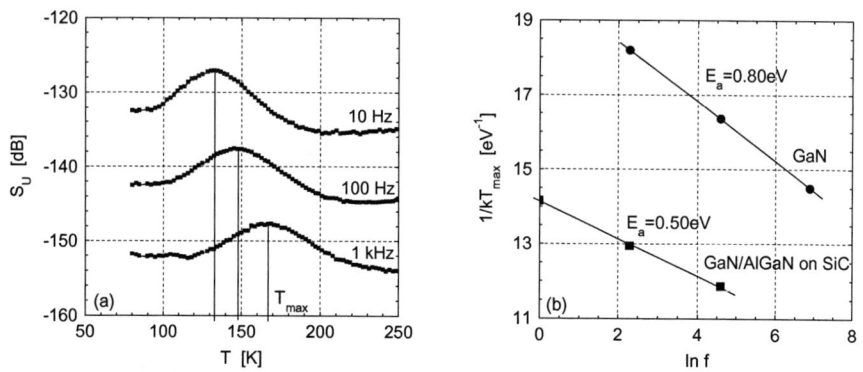

FIGURE 5. (a) Noise spectral density of GaN layer at frequency 10Hz, 100Hz and 1kHz as a function of temperature, (b) T_{max} = temperature corresponding to maximum noise in figure (a) plotted in such coordinates to get activation energy E_a

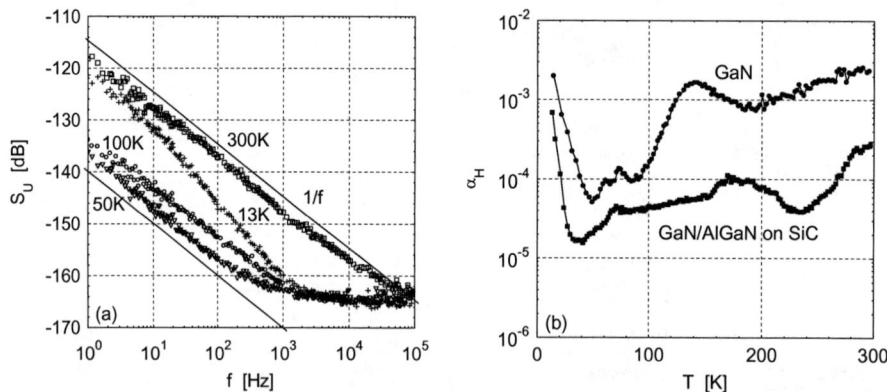

FIGURE 6. (a) Noise spectral density of GaN/AlGaN on SiC HFET measured at various temperatures (b) Hooge parameter temperature dependence for GaN layer and GaN/AlGaN on SiC HFET

Due to the influence of thermally activated traps, the noise spectral density S_U at particular frequency starts to increase at some temperature and then decreases, as the Lorentzian bump in spectra shifts to the high frequencies with increasing temperature. In Fig.5a such behaviour can be observed for S_U measured at frequency 10Hz, 100Hz and 1kHz on GaN layer. Then the temperature T_{max} when noise reaches the maximum can be found and in the Arrhenius plot (Fig.5b) we can determine the activation energy of traps E_A, which is 0.80eV for GaN layer on sapphire and 0.50eV for GaN/AlGaN heterostructure on SiC. In [2] and [3] the values of either 0.42 or 0.85eV for sapphire and 0.20-0.36eV or 0.8-1.0eV for SiC based heterostructures are reported. The noise spectral density was $1/f$ type for GaN/AlGaN on SiC HFET in the whole temperature range except below 30K, as is shown in Fig.6a. Only in one sample out of six we observed g-r noise bump around T=190K with E_a=0.50eV as was mentioned. In Fig.6b the temperature dependence of Hooge parameter is given. In case of sapphire based HFET the g-r noise with $1/f^2$ spectrum is dominant and $1/f$ noise can be observed only around 20K, giving α_H=2×10^{-4}. For the GaN on sapphire layers noise spectral density is mostly $1/f$ type with the exception of strong g-r noise component around T=130K, creating bump in the α_H curve. For temperature below 50K the value of Hooge parameter sharply increases due to the influence of degenerate region with high dislocation density at the GaN/sapphire interface [4].

REFERENCES

1. Tanuma, N., Tacano, M., Pavelka, J., Hashiguchi, S., Sikula, J., and Matsui, T., accepted for *Solid State Electronics*, (2005)
2. Rumyantsev, S.L., Pala, N., Shur, M.S., Borovitskaya, E., Dmitriev, A., Levinshtein, M.E., Gaska, R., Khan, M.A., Yang, J., Hu, X., and Simin, G., *IEEE Trans. Electron Devices* **48**, 530-533 (2001)
3. Levinshtein, M.E., Balandin, A.A., Rumyantsev, S.L., and Shur, M.S., "Low-Frequency Noise in GaN-based Field Effect Transistors," in *Noise and Fluctuations Control in Electronic Devices*, edited by A. A. Balandin, Stevenson Ranch: American Scientific Publishers, 2002, pp. 49-65
4. Look D.C., and Molnar R.J., *Appl. Phys. Lett.* **70**, 3377-3379, (1997)

Fundamental Effects in the Dependence of the 1/f Noise Spectrum on the Bias Current in Semiconductor Diodes

A.V.Yakimov

Nizhni Novgorod State University, Gagarin ave. 23, N.Novgorod 603950, Russia
Fax: +7–8312–656416; E-mail: yakimov@rf.unn.ru

Abstract. Relation by Kleinpenning [1] for 1/f noise in p–n junction is used for all types of semiconductor diodes up to now. It is applied to some diodes with quantum dots and wells but only within restricted range of bias current, and not for all types of diodes, see, e.g. [2]. Possible noise sources in ordinary diode are discussed now. Observed dependences of 1/f noise on the bias current in laser diodes with quantum wells are explained on this basis. It is shown that the main source is the 1/f noise in leakage current; 1/f noise from quantum wells was not detected.

Keywords: 1/F Noise, Laser Diodes, Quantum Wells, Leakage Current
PACS: 72.70.+m; 85.35.Be

INTRODUCTION

Possible 1/f noise sources in modern semiconductor diodes are discussed. Up to now empirical relation by Kleinpenning [1] is widely used. This one was suggested for ordinary p–n junctions having the spectrum of the 1/f noise in the current being proportional to the bias current. This relation may be applied to diodes with quantum dots and wells only within restricted range of bias current, and not for all types of diodes, see, e.g. [2]. In order to solve the existing problem, the noise sources are discussed, which give the possibility to explain observed dependences of the 1/f noise spectrum on the bias current in different types of diodes. Explanation of 1/f noise data in laser diodes with quantum wells is suggested.

1/F NOISE IN ORDINARY DIODES

The total current I through diode may consist of three components, $I = I_d + I_r + I_l$; resistance R_b of the diode base (neutral) region is to be taken into account, see Fig. 1.
Total voltage V_d applied to diode determines the voltage V on p–n junction:

$$V_d = V + R_b \cdot I. \qquad (1)$$

Diffusion current I_d is formed by carriers crossing the depletion region, which recombine in neutral region of the diode. This current and corresponding resistance R_d are determined by following relations:

$$I_d = I_s \cdot [\exp(V/V_T) - 1], \quad R_d = (dI_d/dV)^{-1} = V_T/(I_d + I_s). \qquad (2)$$

Here I_s is the saturation current, $V_T = kT/q_e$ is the thermal potential; the ideality factor of the diffusion current is $\eta_d =1$.

Recombination in the depletion region forms recombination current I_r. This current and corresponding resistance R_r are determined by following relations:

$$I_r = I_{r0} \cdot [\exp(V/\eta_r V_T) - 1], \; R_r = (dI_r/dV)^{-1} = 2\eta_r V_T/(I_r + I_{r0}). \quad (3)$$

Here I_{r0} is characteristic current; the ideality factor is $\eta_r=2$. Recombination in quantum wells and dots produces the same current. Leakage resistance R_l causes current I_l, which is usually nonlinear with large ideality factor $\eta_l \gg 1$.

The main idea is all components of the total current described by resistances in equivalent circuit of the diode, and the base resistance, are subjected to 1/f noise:

$$R_\lambda \Rightarrow (1 + \delta R_\lambda(t)) \cdot R_\lambda, \; \lambda = d, r, l, b . \quad (4)$$

Relative $\delta R_\lambda(t)$ noise in resistances is used here.

The 1/f noise in these equivalent resistances may be caused by mobile defects, see, e.g. [3,4], which are manifested through the random change in structure of different areas of the diode. The semiconductor electronics (see, e.g. [5]) allows considering spectra $S_{\delta R \lambda}$ of relative 1/f noise in resistances be not dependent on the diode current.

The total 1/f voltage noise spectrum S_v is determined by spectra $S_{\delta R \lambda}$ and effect of different components in the total current I, see Fig. 2.

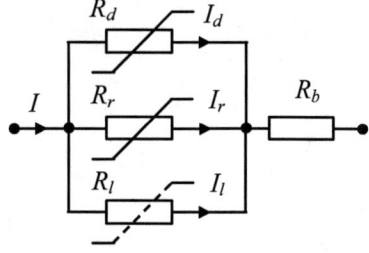

FIGURE 1. Equivalent circuit of the diode with p–n junction.

If the main component is diffusion current ($\lambda=d$) then the 1/f noise is saturated at currents $I \gg I_s$. It was observed in Ge diodes [6] and explained in [7]. Equivalent 1/f current noise source $i_d(t) = I_d \cdot \delta R_d(t)$ yields 1/f voltage noise $v(t) = R_d \cdot i_d(t)$. The increase of the total current $I = I_d$ yields the increase of the current noise source and the decrease of differential resistance R_d determined by eq. (2). At high currents the increase of $i_d(t)$ is compensated by R_d decrease; see line "d" in Fig. 2.

Dependence $S_v \sim I^{-1}$ (line "r" in Fig. 2) was observed in Si and GaAs diodes (see [1]) and explained in [8]. Total current consists of two components, $I=I_d+I_r$, see eqs. (2), (3). Current I_d is the main but noiseless, $I \approx I_d$. Recombination current produces the 1/f current noise $i_r(t) = I_r \cdot \delta R_r(t)$ and voltage noise on diffusion differential resistance $v(t) = R_d \cdot i_r(t)$. As far as $I_r \sim \sqrt{I}$, and $R_d \sim I^{-1}$, we have $S_v \sim I^{-1}$.

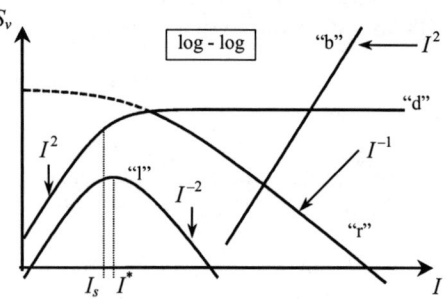

FIGURE 2. Dependences of the voltage noise spectrum on the current through the diode.

Noise in the leakage current, $\delta R_l(t)$, yields the 1/f noise maximization, see line "l" in Fig. 2; it was investigated in [9]. We take into account the noiseless diffusion current, $I \approx I_d$, and the 1/f noise in leakage current $i_l(t) = I_l \cdot \delta R_l(t)$. At small currents, when I-V characteristic of diode is nearly linear, this noise is manifested as in linear resistor, $S_v \sim I^2$. At large currents, when the increase of the voltage on p–n junction is very slow, $V \sim \ln I$, see eq. (2), the increase of leakage noise $i_l(t)$ is also slow. But resistance of the diode is decreased, $R_d \sim I^{-1}$. Thus, after some maximum the 1/f voltage noise spectrum decreases nearly proportionally to I^{-2}. The 1/f noise in the base resistor is pronounced as in all linear resistors but only at large currents; see line "b" in Fig. 2.

1/F NOISE IN LASERS WITH QUANTUM WELLS

The 1/f noise in laser diodes with quantum wells (QW) *InGaAs/GaAs/InGaP* operated in dark and LED modes was investigated.

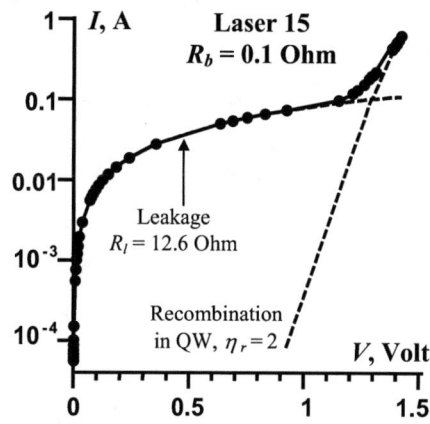
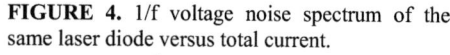

FIGURE 3. I-V characteristic of laser diode with large leakage.

FIGURE 4. 1/f voltage noise spectrum of the same laser diode versus total current.

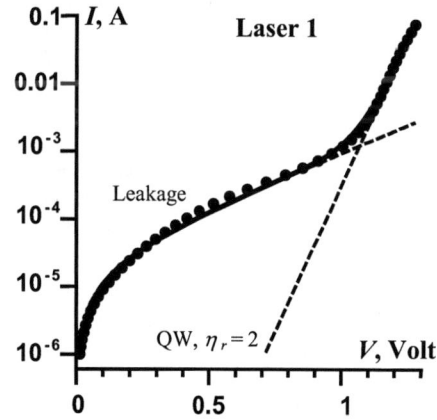
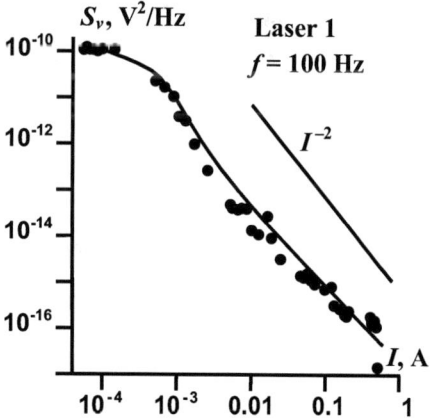

FIGURE 5. I-V characteristic of laser diode with small leakage.

FIGURE 6. 1/f voltage noise spectrum of the same laser diode versus total current.

Laser 15 has large leakage, $R_l = 12.6$ Ohm, and small base resistance $R_b = 0.1$ Ohm. Recombination in QW is described by $\eta_r = 2$, and $I_{r0} = 9 \cdot 10^{-13}$ A, see eq. (3). The I-V characteristic of this diode is shown in Fig. 3. Experimental data for the 1/f noise spectrum S_v, at $f=100$ Hz, versus total current I are shown by dots in Fig. 4; solid line – effect of 1/f noise in leakage current with $S_{\delta Rl} =10^{-12}$ Hz^{-1} at $f=100$ Hz. Thus, experimental data are described by the 1/f noise in leakage current.

An example of diode with small leakage is laser 1. Its I-V characteristic is shown in Fig. 5. Recombination through QW yields $\eta_r=2$, and $I_{r0}=6.5\cdot10^{-13}$ A, that is nearly the same as for laser 15, like $R_b = 0.1$ Ohm. But the leakage current is 10^2 times less and only non-linear component may be detected. In terms of eq. (3), applied to the leakage current, we have $\eta_r => \eta_{ln}=10.5$, and $I_{r0} => I_{ln0}=2.1\cdot10^{-5}$ A. If linear leakage exists it may be described by resistance not less than $R_{ll} = 10^5$ Ohm.

Experimental data for spectrum $S_v(f=100$ Hz) versus I are shown by dots in Fig. 6. This result is explained by additional linear noisy leakage. The total effect at $f=100$ Hz for 1/f noise in linear leakage $R_{ll} = 10^5$ Ohm with $S_{\delta Rll} =9\cdot10^{-7}$ Hz^{-1} and 1/f noise in nonlinear leakage with $S_{\delta Rln} =4\cdot10^{-10}$ Hz^{-1} is shown here by solid line. Again, the 1/f noise in leakage current describes experimental data.

Thus, observed noise data for lasers with quantum wells are explained by 1/f noise in the leakage current. Noise in quantum wells was not detected.

ACKNOWLEDGMENTS

This investigation was supported by the NATO's "Science for Peace" Project SfP-973799, by grants of RFBR 04-02-16708-a, NSh-1729.2003.2 (Scientific School), and Russian Ministry of Education and Science grant No. 4616. Laser diodes were manufactured in Laboratory by Dr. B.Zvonkov at Nizhni Novgorod State University (Russia). All measurements were done by A.Belyakov and M.Perov in Laboratory by Prof. L.K.J.Vandamme at Eindhoven University of Technology (The Netherlands).

REFERENCES

1. T. G. M. Kleinpenning, *Physica* **98B+C**, 289-299 (1980).
2. M. Yu. Perov, N. V. Baidus, A. V. Belyakov, G. A. Maksimov, A. V. Moryashin, S. M. Nekorkin, L. K. J.Vandamme, A. V. Yakimov, and B. N. Zvonkov, "1/f noise in InAs/GaAs quantum dots and InGaAs/GaAs/InGaP quantum well LEDs and in quantum well laser diodes", Proc. 17th Int. Conference "Noise and Fluctuations, ICNF 2003". Prague, Czech Republic, August 18-22, 2003. Ed. by J.Sikula. CNRL s.r.o. pp. 393-396.
3. Sh. M. Kogan, *Sov. Phys. - Usp.* **28(2)** 170 (1985).
4. V. B. Orlov, A. V. Yakimov, *Physica B* **162**, 13-20 (1990).
5. R. P. Nanavati, "Introduction to Semiconductor Electronics", McGraw–Hill, 1963.
6. "Noise in Electron Devices", Ch. 7, edited by K. D. Smullin and A. Haus, Cambridge, MA, MIT Press, 1959.
7. A. N. Malakhov, "Fluctuations in resistance of semiconductor detectors", *Radiotehnika i Elektronika* **3**, 547-551 (1958) (in Russian).
8. A. K. Naryshkin, "Recombination and Generation of Current Carriers as Source of Flicker Noise", in Proceedings of Scientific-Technical Conference of MEI, Section of Radioengineering, Subsection of Receiving-Amplifying Technique, Moscow, Power Institute, 1967, pp. 65-70 (in Russian).
9. E. L. Wall, *Solid-State Electronics* **19(5)** 389-396 (1976).

OPTOELECTRONICS AND PHOTONICS

Photonic Propagation Noise On Long Atmospheric Paths

Paul J Edwards[†], Jeremy Gleeson[‡], Tim Smallhorn[†], Adrian P Whichello[†], David Woodgate[†]

[†]*Centre for Advanced Telecommunications and Quantum Electronics Research, University of Canberra, Canberra, ACT 2601.*
[‡]*School of Civil, Aerospace and Mechanical Engineering, University of New South Wales at ADFA, Campbell, ACT 2612, Australia.*

Abstract. Refractive index fluctuations along the path of an optical beam propagating in the atmosphere give rise to irradiance fluctuations ("scintillations") on millisecond time scales. We present new analyses of concurrent weak (photonic) and bright (photocurrent) scintillation data obtained over atmospheric paths from 4 –40 km in length.

Keywords: photon counting statistics, laser beam scintillation, atmospheric turbulence.
PACS: 42.50.-p, 42.50.Ar, 42.68.Bz

INTRODUCTION

Refractive index fluctuations along the path of an optical beam propagating in the atmosphere cause intensity fluctuations ("scintillation noise") on millisecond time scales. In recent reviews, Milonni et alia [1,2] have presented photon count statistics gathered in the course of operation of a quantum cryptographic key distribution system. They point out that a direct check on the validity of the semi-classical photon count distribution analysis first undertaken by Diament and Teich [3] has not yet been carried out. In the analysis outlined in this paper, measured bright beam scintillation probability density functions, often well described by log-normal statistics, are used to characterise the stochastic intensity modulation imposed by the atmosphere on the laser beam irradiance. We compare the predicted photon count distributions with measurements and discuss the limitations of both.

THEORY

The semi-classical theory of light detection assumes a classical electromagnetic field but a quantum description of the photo-electron excitation. This leads to the Mandel (Poisson Transform) formula [4] for the probability of counting n photons in time T. This is given by the ensemble average over the stochastic variable, E

$$p(n,T) = \langle (E(T)^n / n!) \exp[-E(T)] \rangle \qquad (1)$$

where $E(t,T) = <n(t,T)>$, is the expected number of photons detected in the time intervals, $(t, t+T)$.

In an early interpretation, Diament and Teich [3] viewed the stochastic propagation medium as a linear two port system with stochastic input. The quantum efficiency, the fluctuating path loss, the detector aperture and other geometrical factors are lumped together to describe a time varying modulation transfer function $\alpha(t) = A\, \eta(t')$. This is assumed to be statistically independent of the fluctuating irradiance at the transmitter, $I(t)$. In the absence of path fluctuations, α is fixed, as is the single mode coherent state intensity I in semi-classical theory. The expected photon count, $E(T) = <n(T)>$, is then the same for all realizations in the ensemble. Thus, in the absence of turbulence and extinction variations, the atmospheric photon channel is characterized by a fixed, time independent path loss, and we expect a Poissonian photon count distribution as in the plot labeled $\sigma = 0$ in Fig. 1. This shows Poisson–transformed photon counting distributions calculated from Eqn. 1 for the log-normal statistics commonly encountered over long paths and a range of logarithmic standard deviations, σ [3].

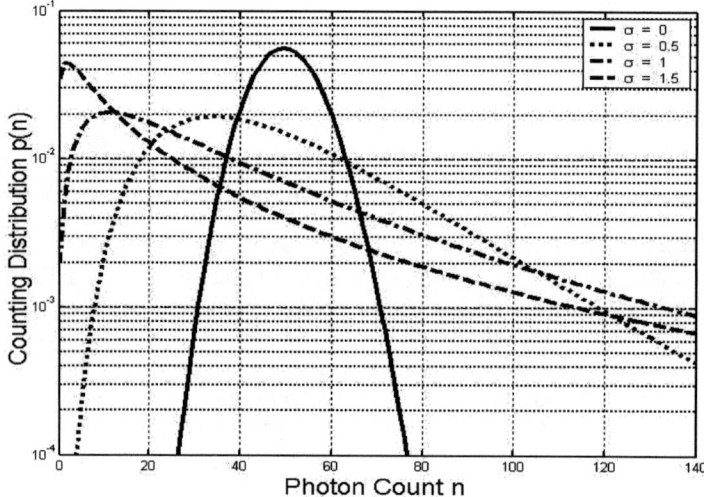

FIGURE 1. Photon count distributions calculated using the Mandel (Poisson Transform) formula. Log-normal bright beam scintillation statistics are assumed with logarithmic standard deviation, σ. [3].

MEASUREMENTS

The typical photocurrent scintillation probability plot of Fig. 2, measured on the University of Canberra quantum key distribution test range [5,6] describes the statistics of the time varying modulation transfer function, $\alpha(t)$. Such a bright beam scintillation distribution is the high photon number limit of the Poisson-transformed photon count distribution and the Poisson count distribution becomes much narrower than the saturated [7] long path log-normal scintillation distribution for which the normalized variance $\sigma_I^2 = exp\,(\sigma^2) - 1 = (<I^2> - <I>^2)\,/\,<I>^2$, typically takes values between 0.2 and 0.6 for long path lengths [8,9,10].

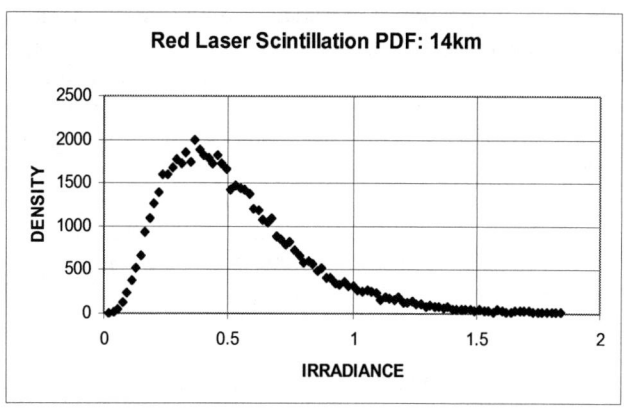

FIGURE 2. Four minute sample of the probability density of the red (630 nm) irradiance fluctuations from a collimated laser diode transmitter observed with a 250mm aperture at a distance of 14.3 km. Normalised variance $\sigma_I^2 = 0.31$.

A comparison between measured and predicted photon count distributions is shown below for 4 ms counts taken over a three minute period. The concurrent bright beam scintillation was measured to be log-normal with scintillation index $\sigma_I^2 = 0.24$.

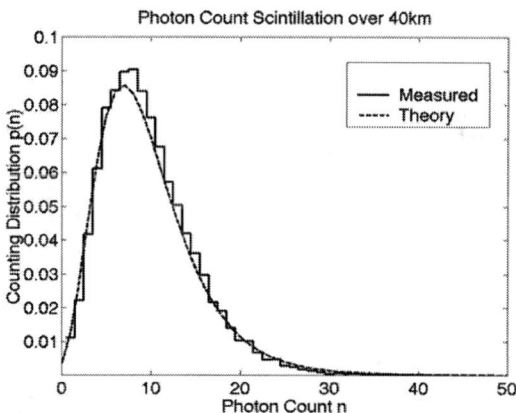

FIGURE 3. Photon count distribution with mean count <n> =9.75 measured with a 200 mm aperture located 40 km from a collimated IR (832 nm) laser transmitter compared with a Poisson transformed log-normal distribution with $\sigma = 0.46$, calculated from concurrent 630 nm bright beam measurements.

DISCUSSION AND CONCLUSIONS

The results appear to confirm the conventional wisdom that the Poisson-transformed bright beam log-normal scintillation probability density function adequately describes the photon counting distribution measured over long atmospheric paths. In the absence of direct measurements Milonni et alia in [1,2] were obliged to assume log-normal bright beam statistics. In this investigation, we have directly confirmed log-normality, measured the scintillation index and demonstrated quantitative agreement with the Diament and Teich photon counting model [3].

Although the results are the same as those calculated from semi-classical theory, phenomenological modeling of the stochastic photon counting distribution following propagation through a stochastic medium can be extended to cover the case of non-classical photon sources and appears to us to be intuitively more appealing. We believe that this is the first definitive measurement and analysis of concurrent bright beam and photon count scintillation and that it is particularly relevant to single-photon cryptographic key exchanges [5] over long atmospheric paths.

ACKNOWLEDGMENTS

We acknowledge financial support from the Australian Research Council and the valuable assistance of Mr A. Blake and Mr P. Zelman with field measurements.

REFERENCES

1. Milonni, P. W., Carter, J. H., Peterson, C. G., and Hughes, R. J., "Effects of propagation in air on photon statistics", in *Fluctuations and Noise in Photonics and Quantum Optics*, edited by Derek Abbott, Jeffrey H. Shapiro and Yoshihisa Yamamoto, Proceedings of SPIE Vol. 5111, Bellingham, WA, 2003, pp.7-11.
2. Milonni, P. W., Carter, J. H., Peterson, C. G., and Hughes, R. J., "Effects of propagation through atmospheric turbulence on photon statistics", *J. Opt. B: Quantum Semiclass. Opt.* **6** S742-S745 (2004).
3. Diament, P. and Teich, M. C., "Photodetection of low-level radiation through the turbulent atmosphere", J. Opt. Soc. Am. **60,** 1489-1494, (1970).
4. Mandel, L. and Wolf, E., "Semiclassical theory of photoelectric detection of light" in *Optical Coherence and Quantum Optics,* Cambridge University Press, New York, 1995.
5. Edwards, P. J., Lynam, P., Cochran, C., and Blake, A., "Simulation of ground-satellite quantum key exchange using a dedicated atmospheric free-space testbed," in *Quantum Communications and Quantum Imaging,* edited by Ronald E.Meyers, Yanhua Shih, Proceedings of SPIE, 5161, Bellingham, WA, pp.152-160, 2004.
6. Edwards, P. J., Blake, A., Cochran, C., Gleeson, J., O'Keeffe, H. B., Woodgate, D., and Zelman, P., "Free space quantum bit transfer over a 26 km urban path", (Abstract only), http://www.ips.gov.au/IPSHosted/NCRS/wars/wars2004, Workshop on Applications of Radio Science, WARS 04, Australian National Committee for Radio Science, Hobart, Feb. 2004.
7. Clifford, S. F., Ochs, G. R., and Lawrence, R. S., "Saturation of optical turbulence by strong turbulence", J. Opt. Soc. Am. **64**, 3, 148-154, (1974).
8. Edwards, P. J., Blake, A., Gleeson, J., Lisle I., and Whichello, A. P., "Optical scintillation noise on long atmospheric paths", http://www.ursi-f2004.com, *Symposium Papers URSI Comm. F* Cairns, June 2004.
9. Edwards, P. J., and Whichello, A. P., "Propagation noise in broadband free-space optical communication systems", Invited Paper 5473-13, 2nd SPIE International Symposium on Fluctuations and Noise, Maspalomas, Gran Canaria, Spain, *SPIE Proceedings FN07*, May 2004.
10. Andrews, L.C., Phillips, R. L., and Young, C. Y., *Laser Beam Scintillation with Applications*, SPIE Press Vol.**PM99**, Bellingham, WA, 2001.

Light transport through Photonic Liquids

L.S. Froufe-Pérez*, S. Albaladejo*, E. Sahagún*, P. García-Mochales*, M. Reufer†, F. Scheffold† and J.J. Sáenz*

*Departamento de Física de la Materia Condensada, Universidad Autónoma de Madrid, Campus de Cantoblanco, E-28049 Madrid, Spain.
†Department of Physics, University of Fribourg, 1700 Fribourg, Switzerland.

Abstract. Colloidal suspensions, in the presence of long-range electrostatic repulsion, have been shown to present an unexpected optical behavior: Structural order enhance the scattering strength while at the same time the total light transmission shows strong wavelength dependence, reminiscent of a photonic crystal. In this work we will present the results of extensive numerical calculations of light transport through *Photonic Liquids* and other systems with strong short range correlations for both two-dimensional and three-dimensional systems. The electromagnetic density of states of those systems is analyzed in detail. As we will show, the interplay between order and disorder and the scattering properties of these systems are strikingly similar to those discussed in an early proposal for strong localization of light.

Keywords: light scattering, diffuse waves, photonic crystals, colloids, random media, liquid metals
PACS: 42.25.Bs, 42.70.Qs, 82.70.Dd

INTRODUCTION

Variations in the dielectric properties of a medium lead to scattering events and typically with increasing amount of scattering the material appears opaque. If such variations can be neatly controlled over macroscopic distances however, totally new, so called photonic properties may appear [1, 2, 3, 4]. At the core of the design of these new materials lies the intelligent way structures are assembled on length scales comparable to the wavelength of light. There are two main promising concepts to achieve lossless guidance and manipulation of light based on seemingly opposite principles: order or disorder. Photonic bandgap materials (PBG) are based on periodic structures predicted to inhibit light propagation completely [1, 2, 3, 4]. In the case of disorder, light cannot propagate in the material due to recurrent interference called strong Anderson localization (SAL) [2, 5].

Attempts to enter these regimes for visible wavelengths often rely on the unique properties of colloidal particles which are chosen both random in size and structure (SAL) or monodisperse and crystalline (PBG) [4, 6, 7, 8]. For colloids typical length scales match the wavelength of visible or near infrared-light. Furthermore colloidal particles self-assemble through Brownian motion, certainly the most efficient route to create large scale structures. Tremendous efforts have been made in engineering random or crystalline configurations [4]. In this article we analyze a different approach to photonic materials. We study the optical properties of amorphous colloidal structures rather than the solid phases usually studied.

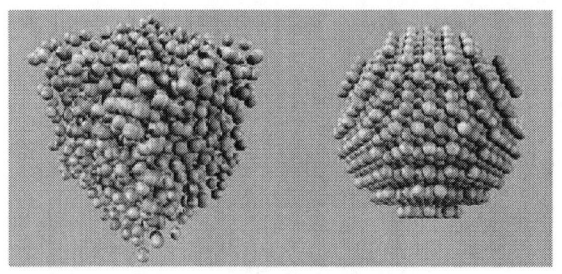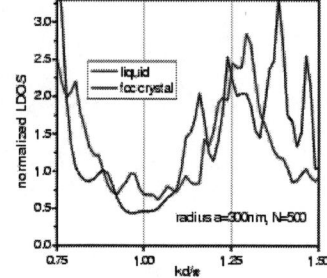

FIGURE 1. Calculated LDOS (normalized to the free space LDOS) of a 500 particle assembly for both correlated liquid structure and FCC crystal structure (a snapshot of the systems shown on the left side).

PHOTONIC LIQUIDS. TUNING THE OPTICAL TRANSMISSION BY LOCAL ORDER

By tuning the interaction potential between the particles it is possible to control the degree of order or disorder and thus explore photonic properties in a completely new regime [9]. In the presence of repulsive interactions, colloidal liquids show fascinating photonic properties despite their overall disorder. Short range structural order enhances the scattering strength at certain configurations while at the same time the total light transmission shows strong wavelength dependence, reminiscent of photonic crystals[9]. On the other hand the interplay between order and disorder and the scattering properties of these systems are strikingly similar to those discussed in the transport of electrons in liquid metals[13, 14]. Close to the Bragg condition the transport cross section becomes anisotropic and the transmission coefficient is reduced.

The observed behavior [9] suggest some similarities between the optical properties of these colloidal liquids and those found in photonic crystals. Recent calculations [15] for Mie-spheres show that the density of "resonant" states of a cluster as small as 32 spheres exhibits a well defined structure similar to the density of electromagnetic states of the infinite photonic crystal. The formation of small ordered clusters could then lead to a significant modification of the density of electromagnetic states in a random collection of monodisperse spheres [15]. As shown below, the analysis of the photon local density of states (LDOS) [16] in strongly correlated colloidal liquids shows that the similarities are much closer than expected, even for an assembly of Rayleigh (non-resonant) particles.

As a first approach, we have numerically calculated the LDOS for an assembly of 500 colloidal particles. For simplicity, we consider each particle as a Rayleigh scatterer with tunable polarizability. The LDOS (at the centre of the assembly) can be obtained from the total power radiated from a dipolar point source [16, 17]. The calculation takes into account all multiple scattering effects. The particle positions for the liquid structure, generated by Brownian Dynamics simulations [18], present a pronounced peak in $S(q)$ at q_{max} $2\pi/d$ similar to that discussed in ref. [9] . Figure 1 shows the calculated LDOS (normalized to the free space LDOS) of a 500 particle assembly for both correlated liquid structure and FCC crystal structure (a snapshot of the system is also included in Figure 1). The strong similarities indicate that amorphous photonic structures might posses similar optical transport properties as compared to some crystal structures. The generalization of these calculations to Mie-spheres and the direct comparison to our experimental results is in progress.

SUMMARY

Interestingly the approach of using local order to manipulate the propagation of light has been ignored almost completely in the past, despite the many studies around photonic materials, light transport and localization [4]. We therefore expect our results to be of substantial use in the interpretation of previous experiments on strongly scattering materials [8, 10]. For example it remains still mysterious why semiconductor powders with comparable optical contrast yield rather different values for $k_0\ell^*$ [11, 12]. Finally we would like to point out that the unusual properties of photonic liquids could also lead to interesting applications. Titanium dioxide nanoparticle based sunscreen lotions for example require efficient blocking of UV light while retaining a high transparency for visible wavelengths. Photonic liquids fulfill this requirement significantly better than random particle assemblies. Since electrostatic interactions are weak in nonaqueous conditions, forces mediated by polymers, surfactants or excluded volume interactions could be used to tailor the desired structural properties. Photonic liquids could also be used as tunable optical filters and switches, for example in windows that change from opaque to clear, provided particle interactions can be controlled externally.

ACKNOWLEDGMENTS

This work has been supported by the Swiss National Science Foundation and the Swiss KTI/TopNano21 initiative (project 5971.2), the Spanish MCyT (Ref. No. BFM2003 01167) and the European Integrated Project "Molecular Imaging" (LSHG-CT-2003-503259)

REFERENCES

1. E. Yablonovitch, Physical Review Letters 58, 2059 (1987).
2. S. John, Physical Review Letters 58, 2486 (1987).
3. J. D. Joannopoulos, R. D. Meade, and J. N. Winn, Photonic crystals : molding the flow of light (Princeton University Press, Princeton, N.J., 1995).
4. C. M. Soukoulis, Photonic crystals and light localization in the 21st century (Kluwer Academic Publishers published in cooperation with NATO Scientific Affairs Division, Dordrecht ; Boston, 2001).
5. P. W. Anderson, Philosophical Magazine B-Physics of Condensed Matter Statistical Mechanics Electronic Optical and Magnetic Properties 52, 505 (1985).
6. D. S. Wiersma, P. Bartolini, A. Lagendijk, et al., Nature 390, 671 (1997).
7. F. Scheffold, R. Lenke, R. Tweer, et al., Nature 398, 206 (1999).
8. J. G. Rivas, R. Sprik, A. Lagendijk, et al., Physical Review E 63, 046613 (2001).
9. L. F. Rojas-Ochoa, J. M. Mendez-Alcaraz, J. J. Saenz, et al., Physical Review Letters 93, 073903 (2004).

10. J. Ballato, J. Dimaio, A. James, et al., Applied Physics Letters 75, 1497 (1999).
    22 H. B. Sun, Y. Xu, J. Y. Ye, et al., Japanese Journal of Applied Physics Part 2-Letters 39, L591 (2000).
11. R. Maynard, B. van Tiggelen, G. Maret, et al., in New Aspects of Electromagnetic and Acoustic Wave Diffusion, 1998), Vol. 144, p. 17.
12. J. G. Rivas, R. Sprik, C. M. Soukoulis, et al., Europhysics Letters 48, 22 (1999).
13. J. M. Ziman, Philosophical Magazine B-Physics of Condensed Matter Statistical Mechanics Electronic Optical and Magnetic Properties 6, 1013 (1961).
14. N. W. Ashcroft and J. Lekner, Physical Review B 145, 83 (1966).
15. A. Yamilov and H. Cao, Physical Review B 68 (2003).
16. K. Joulain, R. Carminati, J. P. Mulet, et al., Physical Review B 68 (2003).
17. R. C. McPhedran, L. C. Botten, J. McOrist, et al., Physical Review E 69 (2004).
18. V. Lobaskin, MPI Mainz, Germany, private commmunication.

About the physical origin of pixel flickering in cooled $Hg_{0.7}Cd_{0.3}Te$ infrared photodetectors

Bernard Orsal*, Jean-Philippe Perez*, Mikhael Myara*, Robert Alabedra*,
Cedric Leyris*, Jean-Philippe Tourrenc* and Philippe Signoret*

*CEM2 Centre d'Electronique et de Micro-optoelectronique de Montpellier, Universite
Montpellier II, 34095 Montpellier Cedex 05, France.

Abstract. We report on electrical noise measurements on both $Hg_{0.7}Cd_{0.3}Te$ test patterns and hybrid 320×256 focal plane array in order to explain the low frequency pixel flickering physical origin. Dark and under infrared illumination test patterns characterization highlights that the detector chip isn't responsible for the flickering phenomenon. Taking into account the silicon readout chip influence when the full IRCMOS infrared detector is investigated, the indium bump based interconnecting system is finally pointed out as a potential excess noise source.

Keywords: HgCdTe, infrared photodetector, pixel, 1/f noise, detectivity, RTS, IRCMOS, indium bump
PACS: 42.79.Pw, 85.60.Bt

INTRODUCTION

With the average composition $x = 0.3$, $Hg_{1-x}Cd_xTe$ material is particularly suitable for MWIR (Medium Wave InfraRed) radiation remote sensing. Nevertheless, these systems usually exhibit some low frequency ($0.1 Hz$ to $1 Hz$) flickering pixels ($\approx 0.05\%$ of the total pixels amount), inconsistent with a convenient photodetection operation. In this paper, we use low frequency noise measurements in order to better understand this flickering phenomenon origin. The results presented consist in a sum-up of several significant measurements carried out on several components of various size. We first perform noise measurements on readout-circuit-less small test patterns in both dark and under illumination configurations. The patterns consists in a $560 \times 600 \mu m^2$ plate containing 16 various areas (A) hybrided photodetectors. The approximate doping profile of the $n^+/n^-/p$ junction-diode array is given in [1]. Then, to complete this step by step study, a full IRCMOS (InfraRed Complementary Metal-Oxyd Semiconductor) detection array is investigated in both time and frequency domains.

NOISE TEST PATTERNS IN DARKNESS

Preliminary dark current measurements exhibit in the $V = -50 mV$ to $V = 0 mV$ voltage range two main conduction regimes as a function of the focal plane temperature [2] : surface leakage current and diffusion current, both separated by the T^* temperature (*Fig. 1-a*). Moreover, these measurements lead to very high dynamic resistance R_d values ($10^{10}\Omega < R_d < 10^{11}\Omega$) leading us to perform noise current measurements. The noise current measurement experimental setup [2] consists in a low noise bias supply for the cryostat enclosed diode, an ultra low noise current-sensitive amplifier ($14 fA/\sqrt{Hz}$) and a Fast Fourier-Transform spectrum analyzer. The noise current data were collected with a $50\,mV$ reverse bias and over a frequency range from $1\,Hz$ to $1\,kHz$.

All the resulting current spectral density measurements exhibited two noise contributions in diffusion regime : a white noise part for the highest frequencies and a $1/f$ contribution in the low frequency part. Both noise shapes are separated by a corner frequency $f_c \in [600\,Hz; 1000\,Hz]$ in the diffusion regime, whereas the surface leakage current regime exhibits much lower f_c, i.e. $f_c < 10\,Hz$ (Fig. 1-b). We show in Fig. 1-c the noise current spectral density evolution as a function of the dark current for several temperatures and at a $50\,mV$ reverse bias. Above $133\,K$, when the diode becomes diffusion-current dominated, the white noise level follows an asymptotic shot noise tendency. Moreover, the low frequency noise current spectral density component $S_i(5\,Hz)$ becomes linearly dependent with I_{dark} in diffusion regime.

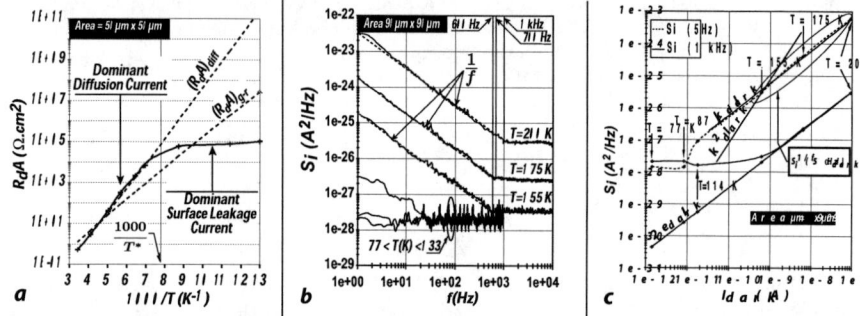

FIGURE 1. Satic and dynamic measurements in darkness : (a) $(R_d \times A)_{-50mV}$, (b) $S_{i_{dark}}$ versus frequency (c) $S_{i_{dark}}$ versus I_{dark}.

Such a behavior in good agreement with theoretical results given by T.G.M. Kleipenning [3] and based on Hooge's $\frac{1}{f}$ noise model, allows to extract the Hooge's α_H parameter considering a diffusion current-dominated reverse-biased n^-/p (or p^+/n) diode. As the base width is close to or lower than the hole diffusion length $L_p = \sqrt{D_p \tau_p}$ in the n-type base, most of the holes cross this one without recombination. Here, $\tau_p = 10^{-7}\,s$, is the typical hole lifetime in the n-region and D_p the hole diffusivity. According to Kleinpenning theory, we assume a mobility originated $1/f$ noise and a shot noise originated flat level. Then, the corner frequency f_c between these two noise sources allows to express the Hooge's parameter as a function of f_c : $\alpha_H = 8\,\tau_p f_c \left(-\frac{k_B T}{eV}\right)$.

For the components under study, this theory leads to α_H between $1.2\ 10^{-4}$ and $2.7\ 10^{-4}$, emphasizing the high maturity degree for this technology.

NOISE TEST PATTERNS UNDER IR ILLUMINATION

The experimental set-up is now completed with an excess-noise-less black body as IR photons source. The black body background temperature T_{BB} is set between $293\,K$ and $308\,K$, and is seen by the detector with a field of view $\approx 30°$. The cutoff wavelenght is $\lambda_{co} = 4.2\,\mu m$ and the magnitude of the background flux is $1.5\ 10^{-5}\,W.cm^{-2}$ @

$T_{BB} = 293K$ and $2.5 \ 10^{-5} W.cm^{-2}$ @ $T_{BB} = 308K$. We observe a lorentzian shape followed by a flat level (*Fig. 2-a*).

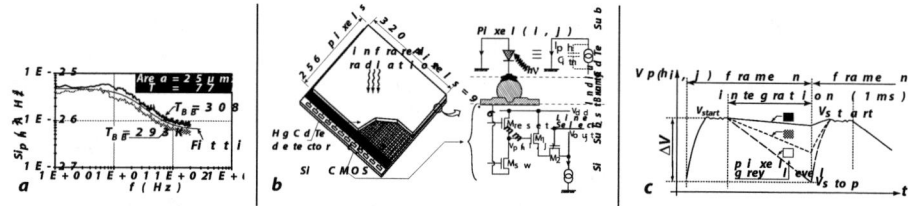

FIGURE 2. Static and dynamic measurements in darkness : (a) $S_{i_{ph}}$ versus frequency, (b) Full IRCMOS architecture (c) Pixel-voltage time-Diagram.

In order to extract a valuable cut-off lorentzian frequency, we fit the experimental data with the following expression : $S_{i_{ph}}(f) = \dfrac{K}{1+\left(\frac{f}{f_{co}}\right)^2} + S_{i_{ph_0}}$ with K the lorentzian contribution amplitude, f_{co} the lorentzian cut-off frequency, and $S_{i_{ph_0}}$ the flat level. Such a noise shape can find its physical origin in background generation-recombination noise [4], due to carrier trapping or fluctuations associated with bandgap states. The resulting f_{co} values belong to the range $6Hz < f_{co} < 22Hz$. Using these experimental noise results under illumination, we can deduce underestimated detectivity D_λ^* values since :

$$D_\lambda^* = \sigma_{4.2} \sqrt{\dfrac{A \Delta f}{\int_{\Delta f} S_{i_{ph}}(f) \, df}} \geq \sigma_{4.2} \sqrt{\dfrac{A}{K}} \text{ with } \Delta f \text{ the analysis bandwith, and } \sigma_{4.2} = 2.5 A/W$$

the mean $HgCdTe$ photodetector sensitivity at $\lambda_{co} = 4.2 \mu m$. The obtained D_λ^* are in the range $2.6 \ 10^{10} cmHz^{1/2}W^{-1} < D_\lambda^* < 1.9 \ 10^{11} cmHz^{1/2}W^{-1}$. The lowest values are very similar to already published measurements and the highest are close to the ideal photovoltaic detector theoretical limit [5].

FULL IRCMOS IR DETECTOR

The studied device is a 320×256 square pixel array, with a $30 \mu m$ pixel pitch (*Fig. 2-b*). The internal $HgCdTe$ photodetector capacitance is charged up as long as both M_{sw} and M_{reset} transistors are in closed state (*Fig. 2-b*). When M_{sw} and M_{reset} are opened, the capacitance is discharged thanks to the photocarriers originated current. Thus, the photodiode voltage decreases with a photocurrent intensity dependent slope (*Fig. 2-c*). Assuming both the capacitance value C_{int} and the photon flux to be constant, we can describe the difference voltage $\Delta V = V_{start} - V_{stop}$ during integration time $T_{int} \approx 1ms$ with the following expression : $\Delta V = \dfrac{\overline{I_{ph}} \times T_{int}}{C_{int}}$.

Time domain noise study : A typical ΔV time dependence is shown in *Fig. 3-a*. In the first cooling cycle, for a flickering pixel, ΔV exhibits a discrete switching between at least two levels like a Random Telegraph Signal (RTS). After a second cooling cycle, we observe a drastic change in the fluctuation nature. Indeed, the same pixel ΔV noise shape is almost flat. This behavior highlights the fact that if the defective pixel number remains about the same from a cooling cycle to another, their spatial distribution strongly fluctuates. We perform measurements with M_{sw} (*Fig. 2-b*) kept in a opened state. The resulting signal is simply the sampling noise of the 14-bit Analog to Digital Converter

(ADC) proving that the readout circuit is not responsible for any background noise when the "follower" (M_1 transistor) does not see any electrical signal.

Frequency domain noise study : We can see in *Fig. 3-b*, the noise current spectral density evolution between two consecutive cooling cycles : three to four orders of magnitude separate the two curves. Moreover, the "safe" pixel exhibits an approximate flat level noise shape, whereas the flickering one displays a $1/f$ noise shape, which is classical either for make-and-break contacts after thermal cycles [6].

Discussion : The available-$CdTe$-substrates thermal-expansion coefficients α_l are much larger than those of the silicon substrates used for the hybrid FPA's readout circuits [7]. The resulting thermal expansion mismatch certainly causes a lateral displacement of the detector and readout substrates when the hybrid FPA is cooled to its operating temperature. On top of that the indium bump contracts on about $0.1\,\mu m$ during a cooling cycle. Because of the resulting stress, broken indium bumps failure may occur leading to make-and-break contacts between the two substrates [8], and then to flickering phenomenons.

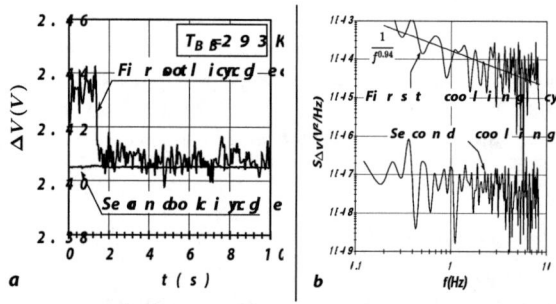

FIGURE 3. Same pixel noise after two cooling cycles : (a) versus time, (b) versus frequency.

CONCLUSION

In this paper, we used low frequency noise measurements in order to explain the pixel flickering physical origin of hybrid $Hg_{0.7}Cd_{0.3}Te$ array with readout Si infrared circuit. The very high technological quality of the detector chip have been proved thanks to very low α_H Hooge parameter and very high D_λ^* detectivity values. Then, it's obvious that both α_H and D_λ^* values rule out the source of flicker as being due to $HgCdTe$ chip. Indium bump bond aspect may be responsible of pixel flickering because of the $CdTe$ and Si different thermal expansion coefficients.

REFERENCES

1. G. Destefanis, *Journal of Crystal Growth*, **86**, 700–722 (1988).
2. J. P. Perez et al., *Fluctuations and Noise Letters*, **3**, 379–388 (2003).
3. T. Kleinpenning, *J. Vac. Sci. Technol.*, **A1**, 176–182 (1985).
4. R. E. Burgess, *Physica*, **20**, 1007–1010 (1954).
5. J. F. Siliquini et al., *IEEE Transactions on Electron Devices*, **42**, 1441–1448 (1995).
6. L. K. J. Vandamme et al., *IEEE Trans. Comp. and Pack. Techno.*, **22**, 446–454 (1999).
7. Y. S. Touloukian et al., *Thermophysical Properties of Matter, vol.13*, IFI/Plenum, Washington, 1977.
8. J. P. Perez et al., *IEEE Transactions on Electron Devices*, **52**, 928–933 (2005).

1/*f* Optical Fluctuations In Quantum Well Laser Diodes

A.V.Belyakov, L.K.J.Vandamme[*], and A.V.Yakimov

Nizhni Novgorod State University, Gagarin ave. 23, N.Novgorod 603950, Russia
Fax: +7–8312–656416; E-mail: beav@yandex.ru; yakimov@rf.unn.ru

[*] *Eindhoven University of Technology, 5600MB Eindhoven, The Netherlands*
Fax: 31–40–2430712; E-mail: L.K.J.Vandamme@tue.nl

Abstract. Electrical and optical intensity noise in quantum well lasers has been investigated. The voltage and optical noise spectra show at low frequencies a 1/*f* noise contribution and a plateau at high frequencies. The coherence function at low frequencies between optical intensity and voltage noise is close to 1. With the coherence function we split optical noise in two parts. One part is fully correlated to electrical noise and has 1/*f* shape. The other part has a flat spectrum, which is somewhat higher than the photo detector shot noise. The 1/*f* optical noise dependence on detector current is in agreement with formerly published experimental results and models.

Keywords: 1/F Noise, Coherence, Correlation, Optical Intensity Noise, Quantum Wells
PACS: 72.70.+m; 78.67.-n

INTRODUCTION

Brophy observed for the first time 1/*f* noise in the optical output of laser diodes [1]. It was noticed that laser diodes with a higher optical 1/*f* noise show a larger spectral width [2]. The spectral properties of the light emission are very important in some applications. This makes the study of the 1/*f* noise in laser diodes an important subject.

We investigated *InGaAs/GaAs/InGaP* quantum well laser diodes manufactured at Physical-Technical Research Institute of Nizhni Novgorod State University (Russia). Optical and electrical noise were sampled simultaneously with an ADS224x48 (Insys, Moscow) AD converter to files with 10^6 readouts per channel.

The main goals of our investigation are: i) the analysis of correlation between electrical and optical noise, and ii) the dependence of the 1/*f* optical intensity noise on the optical power as investigated for the first time in [3]. We measured spectra $S_x(f)$, $S_y(f)$ and cross spectrum $S_{xy}(f)$ of two-channel noise records and calculated the coherence function. The squared coherence function is defined as

$$\Gamma_{xy}^2(f) = \frac{|S_{xy}(f)|^2}{S_x(f)S_y(f)}. \tag{1}$$

The coherence function is the absolute value of the correlation factor between spectral components of $x(t)$ and $y(t)$, which are in our case the electrical and optical laser noise.

ELECTRICAL AND OPTICAL NOISE, COHERENCE FUNCTION

Figure 1 shows a set of voltage noise spectra. Figure 2 shows the dependence on I_d of the $1/f$ part of the current and voltage noise spectra S_V and S_I.

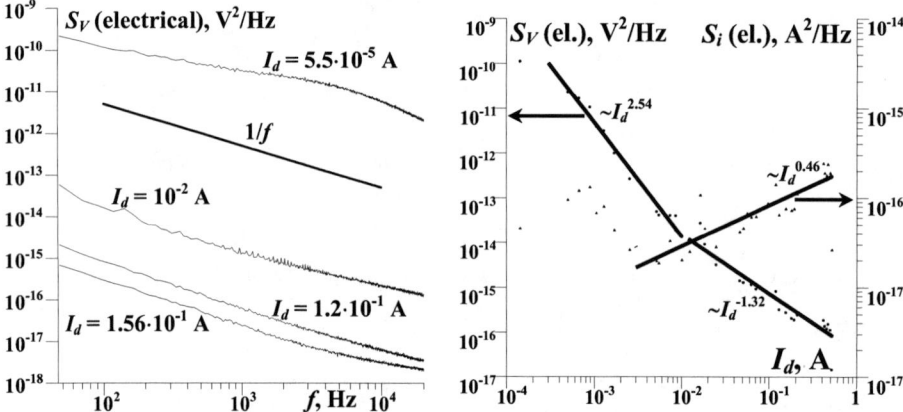

FIGURE 1. Voltage noise spectra

FIGURE 2. Voltage and current noise versus I_d

The optical noise was measured with a photodiode and a low-noise current amplifier. A set of optical noise spectra is shown in Fig. 3 denoted by "RAW". All raw

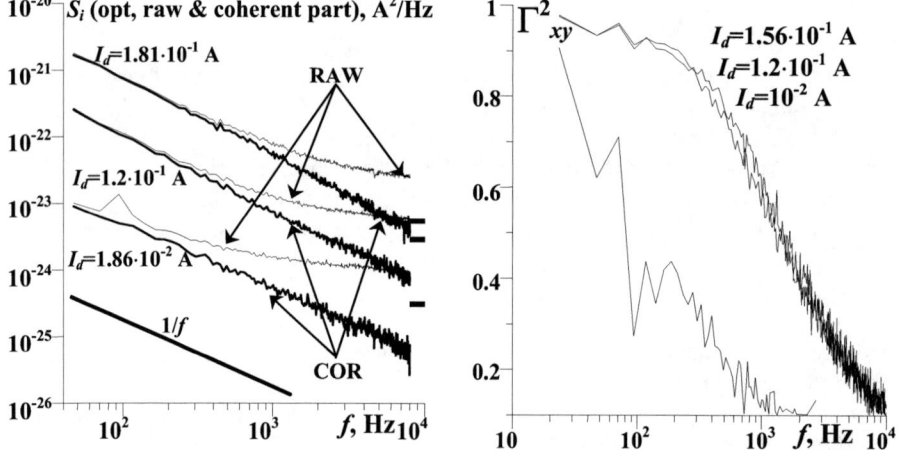

FIGURE 3. Raw optical noise and coherent $1/f$ spectra for different laser currents I_d.

FIGURE 4. Coherence squared versus f.

optical spectra have at low frequencies a $1/f$ contribution and a plateau at $f > 1$ kHz. The photo detector shot noise levels are shown in the same figure by short horizontal lines at the right hand side. The plateau values are higher than the shot noise levels. Fig. 4 shows $(\Gamma_{xy}(f))^2$ versus f. Electrical and optical noise are strongly correlated at low frequencies. The drop in coherence at higher frequencies is due to a white noise contribution in the optical noise (see Fig. 3), which is not present in the electrical noise (see Fig. 1). White noise and $1/f$ noise have different physical origins and are uncorrelated. The coherence at low frequencies between current noise and optical noise is close to 1, even for currents up to stimulated emission threshold.

OPTICAL NOISE DECOMPOSITION AND 1/F NOISE ANALYSIS

The spectrum of optical noise consists of a 1/f and a white noise parts:

$$S_y(f) = S_{1/f}(f) + S_\eta(f), \qquad (2)$$

where $S_{1/f}(f)$ is assumed to be fully correlated with the electrical noise and $S_\eta(f)$ is the non-correlated white noise. Both components are defined as follow:

$$S_{1/f}(f) = S_y(f) \cdot \Gamma_{xy}^2(f), \; S_\eta(f) = S_y(f)\left(1 - \Gamma_{xy}^2(f)\right) \qquad (3)$$

Figure 3 also shows the set of calculated correlated spectra denoted by "COR".

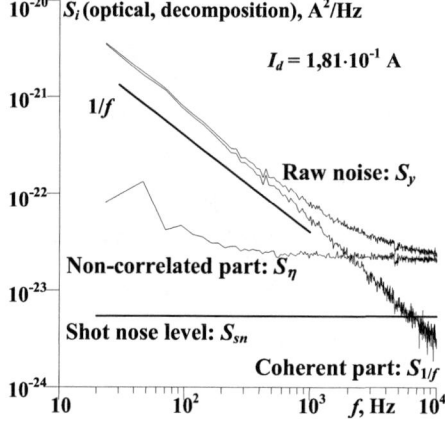

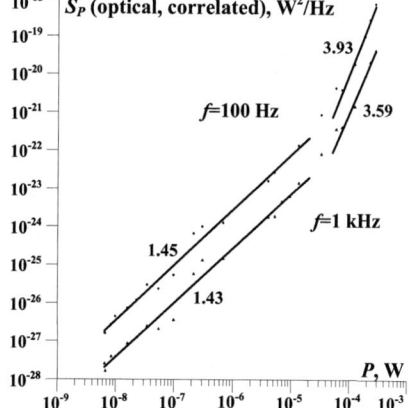

FIGURE 5. Optical noise decomposition.

FIGURE 6. Optical 1/f noise dependence on optical mean power.

Fig. 5 shows an example of decomposition of the observed optical noise spectrum S_y (RAW) for $I_d = 0.181$ A into: the $1/f$ component fully coherent with the electrical noise ($S_{1/f}$) and the uncorrelated flat spectrum (S_η). The photo detector shot noise (S_{sn}), is ten times lower than the corresponding non-correlated component.

The dependence of correlated $1/f$ optical power noise spectrum S_P on the mean light output power P is presented in Fig. 6. Well below stimulated emission threshold we observe the dependence $S_P(P) \sim P^m$, where $m \approx 3/2$, and in the near-threshold region $m \approx 4$. The slopes 2/3 and 4 have been observed and explained in Ref [3] assuming that the $1/f$ noise source is in the absorption coefficient. An uncorrelated $1/f$ fluctuation in the number of electrons and holes also can explain the observed slopes [4, 5].

DISCUSSION AND CONCLUSIONS

The electrical current noise S_I is not proportional to I_d (see Fig. 6) as proposed in the relation $S_I = \alpha q I_d / \tau f$, which was elaborated for homo-junction diodes [7]. We found that such voltage noise spectrum is due to noise in the leakage current.

In the low frequency region, the optical and electrical fluctuations are almost fully correlated as observed earlier [1, 8]. The reduction in coherence at increasing f is explained by the existence of an additional white noise component uncorrelated with the electrical noise. The additional noise component has been investigated from cross spectrum and coherence function analysis.

The calculated $S_\eta(f)$ white noise level is somewhat higher than the expected shot noise level of the detector diode, but it scales with detector current for low currents. Near-threshold the increase of the white noise level above the detector shot noise level becomes outspoken. A possible explanation can be found in [9]. These optical fluctuations are attributed to amplified spontaneous emission and injection current necessary to maintain the condition of almost population inversion for the radiative recombining carriers. A possible onset of generation-recombination noise in the optical output cannot explain our experimental results, because then it also should be observed in the electrical noise and show a strong correlation [8].

The dependence of the correlated $1/f$ noise part in light power on the mean value of this power has been measured. Our experimental results are in agreement with results obtained on gain guided and index guided *GaAlAs* or *GaInAs* based hetero structured laser diodes in [3–6]. If the $1/f$ noise is in the lattice, then the $1/f$ fluctuations in the optical absorption coefficient correlated with electrical noise are plausible.

ACKNOWLEDGMENTS

This investigation was supported by the NATO's "Science for Peace" Project SfP-973799, by grants of RFBR 04-02-16708-a, NSh-1729.2003.2 (Scientific School), and Russian Ministry of Education and Science grant No. 4616.

REFERENCES

1. J.J. Brophy, *J. Appl. Phys.* **38**, 2465-2469 (1967).
2. M.Ohtsu, S.Kotajima, *Japan. J. Appl. Phys.* **23**, 760-764 (1984).
3. L. K. J. Vandamme, and J. R. de Boer, "1/f noise in the light output of laser diodes" in *Noise in physical systems and 1/f noise*, edited by A. D'Amico and P. Mazzetti, Elsevier Science Publishers BV, 1986, pp. 381-384.
4. R. J. Fronen, and L. K. J. Vandamme, "1/f noise in the light output of 0.8 μm and 1.3 μm laser diodes" in *Ninth International Conference on Noise in Physical Systems*, edited by C. M. van Vliet, Singapore: World Scientific, 1987, pp. 187-190.
5. R. J. Fronen, and L. K. J. Vandamme, *IEEE Journal of Quantum Electronics* **24(5)**, 724-736 (1988).
6. S.-L. Jang and J.-Y. Wu, *Solid-State Electronics* **36**, 189-196 (1993).
7. F.N. Hooge, T.G.M. Kleinpenning and L.K.J. Vandamme, *Rep. Prog. Phys.* **44**, 479-532 (1981).
8. S.-L. Jang, K.-Y. Chang and J.-K. Hsu, *Solid-State Electronics* **38**, 1449-1453 (1995).
9. R. Schimpe, *Z. Phys. B Condensed Matter* **52**, 289-294 (1983).

Investigation of the External Feedback Effects on the Relative Intensity Noise Characteristics of AlGaInN Blue Laser Diodes

Jong Chang Yi[*], Jin Yong Kim[†], and Tae Kyung Yoo[¶]

[*]*Department of EEE, Hong Ik University, Seoul, 121-791, Korea; wave@hongik.ac.kr*
[†]*Digital Media Laboratories, LG Electronics, Seoul, 137-724, Korea*
[¶]*EpiValley Co. Ltd., Kyungki, 464-890, Korea*

Abstract. The effect of the external feedback on the relative intensity noise characteristics of 405 nm AlGaInN blue laser diode has been analyzed taking into account the spontaneous emission noise and the high frequency modulation of the injection current. The noise characteristics are modeled after the quantum Langevin noise formalism and the device parameters are extracted by numerical analysis of the multi-band Hamiltonian of the wurtzite multiple quantum well structures.

Keywords: Relative intensity noise, Blue LD, AlGaInN wurtzite crystal, External feedback.
PACS: 42.55.Px; 42.50.Pq; 42.60.Rn; 42.60.Mi.

INTRODUCTION

AlGaInN blue laser diodes are important building blocks for the next-generation DVD or HDD optical data storages. Their data access time and the operation speed are limited in part by the noises of the laser diodes used in the optical pick-ups. In this paper such an intensity noise characteristics known as relative intensity noise has been analyzed for laser diodes having the wurtzite crystal as the active region [1]. The effect of the external feedback into the laser cavity from the reflection by the disc surface as shown in Fig.1 (a) has been investigated using the rate equations. The primary source of the intensity noise in the semiconductor laser diode is the spontaneous emission between the conduction and the valence bands in the active regions. Such quantum noises have been modeled after the quantum Langevin formalism [2]. The required material parameters have also been extracted by using the multiband Hamiltonian for strained wurtzite crystals [3,4]. To show the validity of this method, the simulation results were compared to the analytic solutions when the external feedback effect is ignored, which shows a good agreement. The simulation results of the RIN taking the external feedback effect into account for various feedback strength shows that mere 0.1% external feedback would deteriorate the RIN by more than 20 dB. One way to reduce the external feedback effect is to modulate the injection current to the laser cavity up to a certain high frequency. The optimum current injection frequency on the RIN characteristics has been investigated to reduce the external feedback effect that is inevitable in DVD-ROM or RW pick-up systems.

MODELING OF THE RELATIVE INTENSITY NOISE

The 405 nm laser diode under investigation has typical AlGaInN MQW active structures as shown in Fig. 1 (b). The quantum well regions are 3nm $Al_{0.01}Ga_{0.75}In_{0.24}N$ and the barrier regions are $Al_{0.13}Ga_{0.86}In_{0.01}N$. The SCH regions are $Al_{0.01}Ga_{0.99}N$ and the cladding regions are N- and P-doped $Al_{0.13}Ga_{0.86}In_{0.01}N$ sandwiched by GaN ohmic contact layers. The gain and spontaneous emission spectra of the AlGaInN 3 QW LD are obtained by numerical analysis of the Hamiltonian equations for multiband strained wurtzite multiple quantum well structures [3,4]. Figure 2 shows the calculated gain and spontaneous emission spectra from the AlGaInN 3 quantum well structures when the injection carrier density varies from 10^{12} to 2×10^{13} /cm^2.

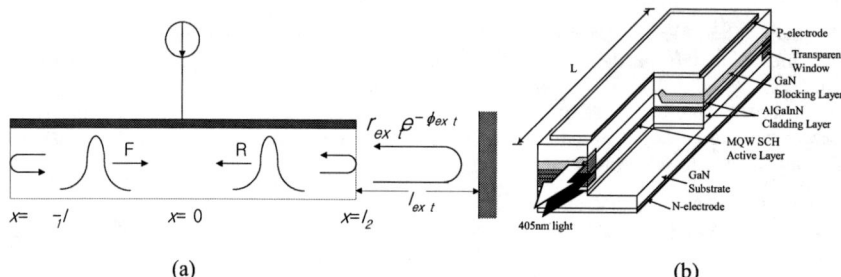

FIGURE 1. (a) Schematic diagram for laser diode cavity with external feedback. (b) Schematic diagram of the 405 nm MQW laser diode structure.

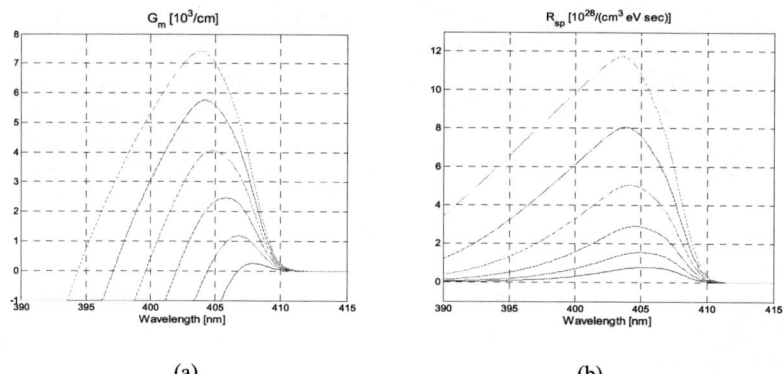

FIGURE 2. Calculated gain and spontaneous emission spectra of AlGaInN MQW structures from the multiband Hamiltonian equations for the strained wurtzite crystals.

Modeling of the Quantum Noise due to the External Feedback

The quantum noise due to the external feedback from the optical disc surface into the laser cavity is modeled as shown in Fig. 1 (a). F and R shown in the figure denote the forward and reverse traveling optical fields, r_{ext}, ϕ_{ext}, l_{ext} the reflection coefficient, phase retardation, and the length of the external cavity, respectively. The primary source of the intensity noise in the semiconductor laser diode is the spontaneous emission between the conduction and the valence bands in the active regions. Such quantum noises have been modeled after the quantum Langevin formalism [2] and the required material parameters have been extracted by using the multiband Hamiltonian as described earlier. The relative intensity noise [RIN] of the laser diode is defined as the ratio of the laser intensity noise $\delta P(t)$ to the average laser power $P_0(t)$, or

$$RIN = \frac{S_P(\omega)}{P^2} = \frac{\int_{-\infty}^{\infty} d\tau \langle \delta P(t+\tau) \delta P(t) \rangle e^{-i\omega\tau}}{P^2} \quad (1)$$

The characteristics of the laser intensity noise can be obtained by analyzing the rate equations [5]

$$\frac{d}{dt}N(t) = \frac{I(t)}{q} - \frac{N(t)}{\tau_e} - G(t)P(t) + F_N(t)$$

$$\frac{d}{dt}P(t) = (G(t)-\gamma)P(t) + R_{sp} + 2\chi P_{fb} + F_P(t) \quad (2)$$

$$\frac{d}{dt}\phi(t) = \frac{1}{2}\alpha(G(t)-\gamma) - \frac{n_m}{n_g}(\omega-\Omega) + \chi\phi_{fb} + F_\phi(t)$$

with

$$P_{fb} = \sqrt{P(t)P(t-\tau)}\cos[\omega\tau + \phi(t) - \phi(t-\tau)]$$

$$\phi_{fb} = \sqrt{\frac{P(t-\tau)}{P(t)}}\sin[\omega\tau + \phi(t) - \phi(t-\tau)] \quad (3)$$

where N denotes the carrier density, P, the photon density, ϕ, the instantaneous phase, F_i, the Langevin noise factors [6], χ, the external feedback coefficient, and τ, the round trip time for the external cavity, respectively.

SIMULATION RESULTS

The simulation results of the carrier and photon density fluctuation after turn-on are shown in Fig. 3. The RIN spectrum shows a good agreement with an analytic solution.

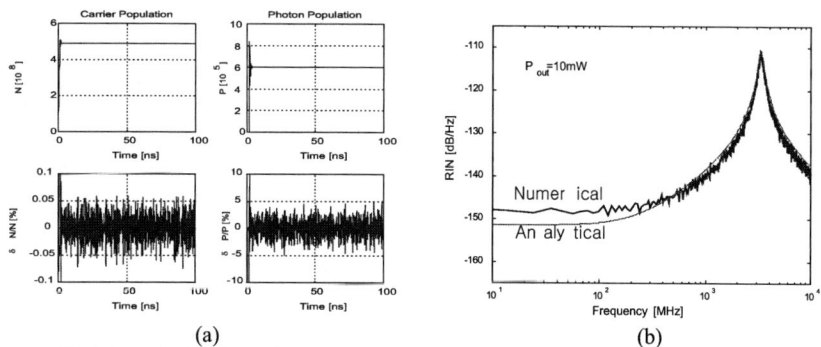

FIGURE 3. (a) Carrier and photon density fluctuation after turn-on. (b) The RIN spectra.

Figure 4 (a) shows the numerical simulation results of the RIN with the external feedback for various feedback coefficients. It shows that 0.1% external feedback would deteriorate the RIN by more than 20 dB. One way to reduce the external feedback effect is to modulate the injection current to the laser cavity up to a certain high frequency [7]. Figure 4 (b) shows the dependence of the RIN characteristics on the current injection frequency for 0.1% external feedback. It shows that by optimizing the injection current levels and its frequency, for example, around 700MHz with a modulation current level at 26.6 mA in this case, one can reduce the external feedback effect inevitable in the optical pick-up systems.

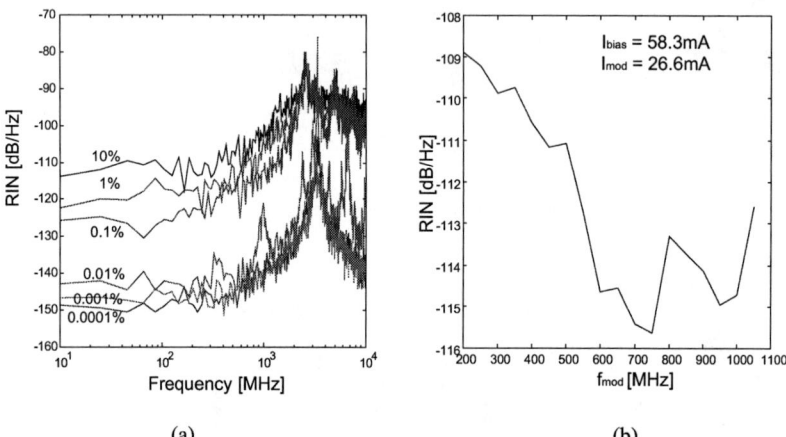

FIGURE 4. (a) The RIN spectra for various external feedbacks. (b) The low-frequency RIN spectra with various modulation frequency of the injection current.

ACKNOWLEDGMENTS

This work was supported partly by Korea Science and Education Fund under ERC program and Institute of Information Technology Assessment under ITRC program.

REFERENCES

1. G. R. Ray, A. T. Ryan, and G. P. Agrawal, *Optical Engineering*, 32, 739-45 (1993).
2. W. W. Chow, S. W. Koch, and M. Sargent III, *Semiconductor Laser Physics*, Springer-Verlag, Berlin, 1994, pp. 337-363.
3. J. C. Yi and N. Dagli, *IEEE J. Quantum Electronics*, 31, pp. 208-18 (1995).
4. J. Piprek, *Semiconductor Optoelectronic Devices*, Academic Press, San Diego, 2003, pp. 32-48.
5. G. P. Agrawal and N. K. Dutta, *Long-wavelength semiconductor lasers*, Van Nostrand Reinhold Co., New York, 1987.
6. P. Gallion, F. Jeremier, and J.-L. Vey, *Optical and Quantum Electronics*, 29, 65-70 (1997).
7. E. C. Gage and S. Beckens, *SPIE* vol.1316, *Optical Data Storage*, 199-204 (1990).

Noise Modelling of Absolute High-T$_c$ Superconducting Measuring Instrument for X-ray Synchrotron Radiation

Igor A. Khrebtov, Konstantin V. Ivanov, Dmitry A. Khokhlov

S.I. Vavilov State Optical Institute, 199034, St. Petersburg, Russia

Abstract. The noise modeling results of absolute high-T$_c$ superconducting measuring instrument for X-ray synchrotron radiation based on *YBaCuO* film bolometer with electrical-substitution are presented. Estimation of the sensitivity, response time and noise equivalent power of bolometer, operating in mode with negative electrothermal feedback and without one is carried out.

Keywords: Synchrotron radiation, superconducting bolometer, noise modeling, electrothermal feedback.
PACS: 41.60.Ap

INTRODUCTION

Synchrotron radiation (*SR*) is a tool useful to conduct metrology measurements in the broad spectral range from infrared to the hard X-ray radiation. One of detection methods for absolute power of incident radiation is transformation of radiant energy to the thermal one with subsequent measurement of the sensitive element temperature [1-2]. The present paper offers the development of an absolute power measuring instrument (or radiometer) cooled by liquid N$_2$ for the absorbed soft X-rays power based on high-T$_c$ superconducting HTS bolometer [3]. The study of noise properties are the most important factors due to both the sensitivity and accuracy of radiometer, that should be < 1 % for measured power <1 µW.

However for these applications HTS bolometer performance is limited by a trade off between speed and sensitivity. Nevertheless, the heat mechanism of bolometer operation allows one to solve this problem using electrothermal feedback (*ETF*) as in passive as in active modes [3-7]. The efficiency of the bolometer operating in constant voltage mode (*CVM*) with use of negative electrothermal feedback *ETF* is due to the loop gain $L_0=P_b\beta/G_b$, here: P_b is Joule power, β is the temperature coefficient of resistance, G_b is the thermal conductance. The negative *FTF* decreases the time constant as $\tau_e=\tau_0/(1+L_0)$, where τ_0 and τ_e is intrinsic and effective thermal time constants. Note, the noise properties of HTS bolometer depends on the *ETF* as well [5,8]. In contrast to passive *CVM* operation, the improvement in the bolometer time response can be obtained using analogue electronic feedback control as well [7].

PRINCIPLE OF OPERATION AND RADIOMETER DESIGN

The absolute radiometer is based on composite HTS bolometer with electrical substitution. The incident *SR* passes through NiCr film heater and is absorbed mainly with thick (50 μm) substrates (Al_2O_3 or $SrTiO_3$) with area size of $3\times3 - 5\times5$ mm^2. The superconducting YBaCuO film meander thermometer is deposited on back side of substrate. Such sensitive element is suspended on four thin Au wires to heat sink. They are used for current bias of YBaCuO and NiCr films and for thermal isolation of bolometer, providing needful thermal conductance G. Teflon gasket and heat Cu sink form the thermal filter that permits to decrease the effect of low frequency temperature fluctuations of N_2 cryostat. Thick substrates provide high absorbed coefficient of *SR* near to 1. However at this case the τ_0 of bolometer is large, so the radiometer should operate without modulation of the *SR* beam. Because of high steepness of superconducting transition, the main limits of accuracy would be temperature fluctuations and drift of N_2 cryostat and 1/f-noise of HTS film. The operation in constant voltage bias mode, when negative *ETF* decreases the response time, will permit to use the modulation and to improve the temperature stability of the HTS bolometer at operating point and its noise parameters and, thus, an accuracy of measurement.

ESTIMATION OF RADIOMETER NOISE PERFORMANCE FOR VARIOUS MODES

The efficiency of the radiometer is due to mainly by noise parameters of HTS bolometer. The theoretical estimation of the influence of the different noise components on the noise equivalent power (NEP) and the detectivity (D*) of bolometers, operating in constant current mode (*CCM*) with positive *ETF* and *CVM* with negative passive *ETF* and active *ETF* at various G_b, T_b, R_b, noise Hooge parameters α_H of YBaCuO films, substrate materials, area size was carried out. The results of estimation, using the theory developed in [4-6,8,9], are shown in Table 1 and on Figures 1,2.

At calculations for *CVM* the value of the sink temperature is equal 80 K, the operation point is selected at temperature with maximum of β, the values of α_H at normal state: 0.09 - for film on Al_2O_3 and 10^{-3} - on $SrTiO_3$ [8]. Table and Figures show that the use of the passive *CVM* permits to decrease time constant τ_e, for example, bolometer on $SrTiO_3$ substrate in ~ 37 times and operate with modulation frequency up to 28 Hz, reaching NEP limited by only phonon noise. Active negative *ETF* provides a large decease of time constant up to 280 times. Comparative calculations show that for supposed operating frequency Fm=10 Hz for the bolometers in active modes calculated NEP can be realized if noise of amplifier has value V_N~(7-8)$\times 10^{-10}$ V/Hz$^{1/2}$ (10 Hz). Note, for passive CVM cooled low noise current measurement (based on SQUID) is also necessarily for practical realized of calculated *SR*-radiometer parameters. Some increase of NEP (10 Hz) for bolometer on Al_2O_3 substrate in active regime is due to 1/f noise of YBaCuO films, but in CCM it is due to large time constant that leads to limit of NEP with Johnson noise of HTS film.

TABLE 1. Calculated parameters of YBaCuO radiometer

Parameter	Bol. on Al_2O_3, 5×5×0,05 mm³			Bol. on $SrTiO_3$, 3×3×0.05 mm³		
	CCM	CVM	Active	CCM	CVM	Active
T_b, K	90.4	84.5	90.4	90.3	89.6	90.3
R_b, kΩ	19.0	0.23	19.0	4.9	1.7	4.9
β, K^{-1}	0,53	1,23	0.53	1.65	2.68	1.65
G, W/K	3×10^{-3}	3×10^{-3}	3×10^{-3}	3×10^{-3}	3×10^{-3}	3×10^{-3}
L_0/L_A	0.3	5.5	5,5/221	0.3	25.7	17/199
τ_0/f_{cut}, s/Hz	0.11/1.4	0.11/1.4	0.11/1.4	0.15/1.1	0.15/1.1	0.15/1.1
τ_e/f_{cut}, s/Hz	0.16/1.0	0.017/9.3	$5\times10^{-4}/314$	0.21/0.74	$5.6\times10^{-3}/28$	$7.5\times10^{-4}/212$
S_V, V/W	141		9.7	95.4		17
S_I, A/W		0.32			0.13	
NEP, W/Hz$^{1/2}$	9.3×10^{-11}	3.68×10^{-11}	8.57×10^{-11}	6.37×10^{-11}	3.67×10^{-11}	3.9×10^{-11}
V_N, V/Hz$^{1/2}$	1.3×10^{-8}		8.3×10^{-10}	6.1×10^{-9}		6.6×10^{-10}
I_N, A/Hz$^{1/2}$		1.18×10^{-11}			4.7×10^{-12}	
D*, cmHz$^{1/2}$W^{-1}	4.6×10^{9}	1.36×10^{10}	5.8×10^{9}	4.7×10^{9}	8.2×10^{9}	7.7×10^{9}

In Table: $L_A=0.5\times L_0\times F\times K$ is the coefficient negative *ETF* for active mode, $F=R_m/(R_m+R_f)$ is the divider, K is the gain of feedback amplifier equal 500, $R_m=R_b$ is the bridge resistance, R_f is the resistance of feedback loop. The values of responsivity S_V, S_I, NEP, V_N and I_N are shown for supposed operation frequency of modulation $F_m=10$ Hz.

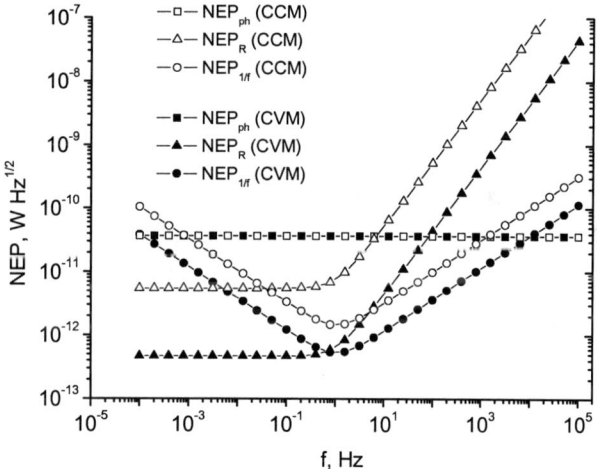

FIGURE 1. The dependencies of the noise equivalent power contributions due to phonon noise, Johnson noise and flicker noise on frequency for various operating regimes of YBaCuO bolometer on $SrTiO_3$ substrate.

Figure 1 shows that negative *ETF* in passive *CVM* depresses Johnson noise and 1/f-noise contributions of NEP. The phonon noise contribution does not depend on frequency and *ETF* does not influence NEP$_{ph}$. The depression of the 1/f-contribution is due to larger value of coefficient β in *CVM* (see Table), the depression of the Johnson noise contribution is due to larger Joule power P_b in *CVM* [8]. The increase of NEP$_{1/f}$ and NEP$_R$ begins at range of f>f$_{cut}$ where responsivity decreases.

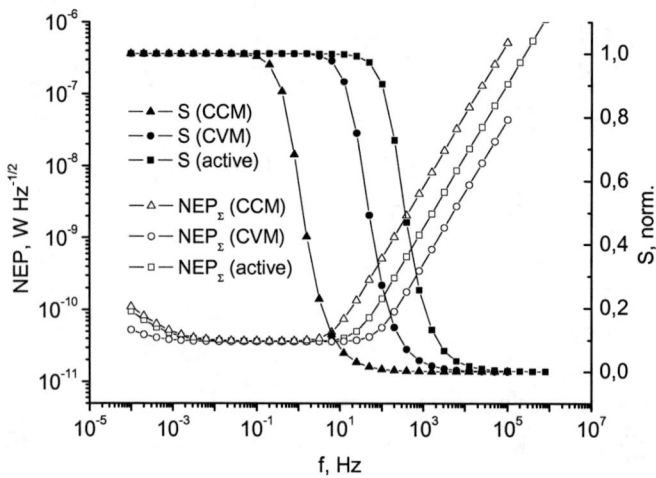

FIGURE 2. The frequency dependencies of the sum noise equivalent power and the responsivity for various operating regimes of YBaCuO bolometer on SrTiO$_3$ substrate.

Figure 2 shows that active negative *ETF* provides more wide frequency interval of uniform responsivity. In range ~10^{-2} - 10 Hz the NEP$_\Sigma$ is due to the phonon noise. On more high frequencies the NEP$_\Sigma$ is due to Johnson noise only. The influence of 1/f noise begins only on frequencies <2×10^{-3} Hz.

ACKNOWLEDGMENTS

This work was carried out in frame of ISTC project, No 2920.

REFERENCES

1. Rabus H., Persch V. and Ulm G., *Appl. Opt.* **36** 5421-5440 (1997).
2. Rice J.P. et al., *Metrologia* **35** 289-293 (1998).
3. Khrebtov I.A., Tkachenko A.D., Ivanov K.V. et al., *J.Phys. IV France* **12** Pr3-137-140 (2002).
4. Neff H., Lima A.M.N., Deep G.S. et al., *Appl. Phys. Lett.* **76** 640-642 (200).
5. Nivelle M. J. M. E. de, Bruijn M. P., Vries R. de et al., *J. Appl. Phys.* **82** 4719-4726 (1997).
6. Lee A.T., Gildemeister J. M., Lee S.-F. and Richards P.L., *IEEE Trans. Appl. Sup.* **7** 2378-2381 (1997).
7. Khrebtov I.A., Tkachenko A.D., Ivanov K.V. and Steinbeiss E ., *J. Opt. Technol.* **68** 290-293 (2001).
8. Khrebtov I.A., *Fluctuation and Noise Letters* **2** No.2 R51-R70 (2002).
9. Ivanov K.V., Khrebtov I.A, Stepanov A.I., *J. Opt.Technol.* **71** 51-54 (2004).

Low and Medium Frequency Noise Levels of fibered very-high-power 1460nm-Pump Laser Diode designed for Raman Amplification

Cedric Chluda*, Mikhael Myara*, Jean-Philippe Perez*, Philippe Signoret* and Bernard Orsal*

*CEM2 Centre d'electronique et de Micro-optoelectronique de Montpellier, Universite Montpellier II, 34095 Montpellier Cedex 05, France.

Abstract. We report low and medium frequency optical noise measurements on very-high-power fibered 1460nm-Pump Lasers ($P > 200mW$) designed for Raman Amplification. These Fabry-Perot Lasers exhibit spectra from a quite monomode situation (Side Mode Suppression Ratio $\approx 10dB$) at near threshold current injection ($I_{threshold} \approx 28mA$) up to very multimode spectra (5nm Full Width at Half Magnitude) at very high biases ($I_{Laser} \approx 1000mA$). In this paper, we have to focus on two main things : on one hand, the low and medium frequency noise behavior of the laser itself and on the other hand the measurement data post-treatment.

Keywords: Pump laser, Noise, Raman Amplification
PACS: 42.60.Mi, 42.55.Px

INTRODUCTION

Fiber optic telecommunication systems are essential for long-distance informations transport. Laser diodes are usually used on one hand as signal light sources, and on the other hand as pump laser diodes designed for optical amplification. In order to increase the distance between two repeaters, the noise study introduced by optical amplifiers becomes essential. Indeed, we have to enhance the signal-to-noise ratio (SNR). We study here the self-noise of a fibered pump laser diode designed for Raman amplification in order to evaluate the contribution of this laser optical noise to the amplifier noise.

OPTICAL STUDY

In order to prevent from any optical feedback [1] and to avoid any measurement set-up saturation, we placed a σ-sensivity, lensed, InGaAs low noise photodiode at a distance d of the optical fiber output. Such a setup reduces the light intensity by a factor M depending on the distance d ($1 < M(d) < 65$). All measurement are actively controlled in temperature at $298K$. Nevertheless, all the exhibited results are absolute, i.d. calculated from the $M(d)$-dependant relative experimental results (*Fig. 1-a*).

A fibered polarizer has been added thereafter in our characterization set-up in order to allow the study of both spontaneous and stimulated emission separately. Surprisingly, the optical power level under threshold ($I_L < 28mA$) is even not in the same order of magnitude with or without polarizer. Above threshold, spontaneous emission magnitude evolution according to injected current is rather chaotic. This kind of instability could originate from the random characteristics of emitted light polarization. Moreover, it does not saturate like in a classical laser. This observed spontaneous emission is thus due to the double heterostructure spacer layers rather than the multi-quantum well gain zone.

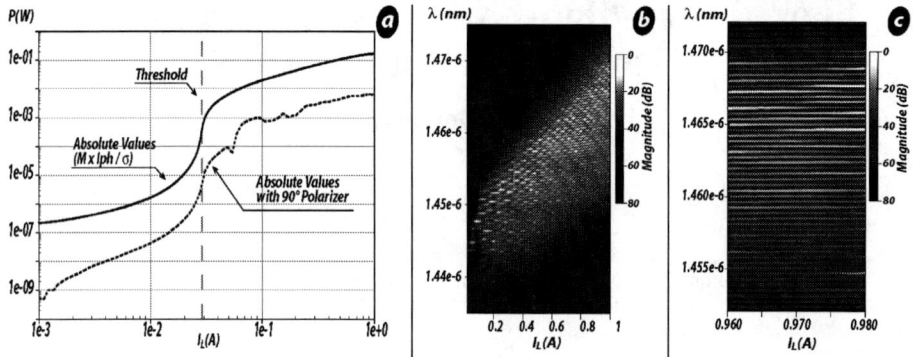

FIGURE 1. (a) Optical power, (b) Spectrum analysis at full study range and (c) Zoom around $960mA$.

SPECTRUM ANALYSIS

General Study : *Figure 1-b* exhibits the laser optical spectrum evolution as a function of the current bias ($298K$). This spectrum strongly evolves with the laser bias and exhibits a very multimode behavior. In addition to the classical widening and global sliding of the spectrum, a broad mode hopping phenomena can be observed. However, its origin still stays unclear. By the way, around $80mA$, the laser is quite monomode.

960mA detailed study : This rather original behavior around $960mA$ (*Fig. 1-c*) is very interesting for the noise study in Raman amplifiers; this current range exhibits optical modes disappearances/creations that will find later in this paper their noise signature.

OPTICAL NOISE STUDY

Relative and absolute noise level : We had to build a low noise current source up to $1000mA$. In order to avoid source parasitic disturbances, we use passive devices : $72V$-batteries with two $1500W$-rheostats. Moreover, the active electronically controlled temperature does not disturb our measurements. Like in the static optical study, we still use here an free space measurement set-up (*fig. 2*) [1]. Variation in distance d between the fiber edge and the lensed photodiode involves an evolution of the attenuation $M(d)$ between laser and photodiode.Such a set-up produces $M(d)$-dependent spectral density noise values : assuming the noise level is homogenous in the laser light (this has preliminarily been checked), we can deduce the absolute photodetection white noise level $S_{ph_{absolute}}$ – as if the photodiode collected the whole laser light – from the observed spectral density noise level $S_{ph_{relative}}$ at the distance d. The absolute white noise level thus can be established thanks to the following formula, considering separately the factor M

and the electrical photodiode shot noise level $2qI_{ph_{relative}}$ effects :

$$S_{ph_{absolute}} = \underbrace{(S_{ph_{relative}} - 2qI_{ph_{relative}}) \times M^2}_{(a)\ Laser\ excess\ noise} + \underbrace{2qI_{ph_{relative}} \times M}_{(b)\ Optical\ power\ induced\ noise} \quad (1)$$

An error calculation allows to predict relative errors $< 10\%$ for noise levels twice as strong as the shot noise level. Nevertheless, the same error calculation process predicts $> 300\%$ for noise levels too close to shot noise and strong M values. However, as if such attenuation values are necessary in order to characterise so high power pump lasers, the expected noise levels are some orders of magnitude over the shot noise level.

All spectral density noise values in this paper are absolute obtained by relative experimental results. Spectra have been fitted using $1/f^k$ low frequency and white noise components.

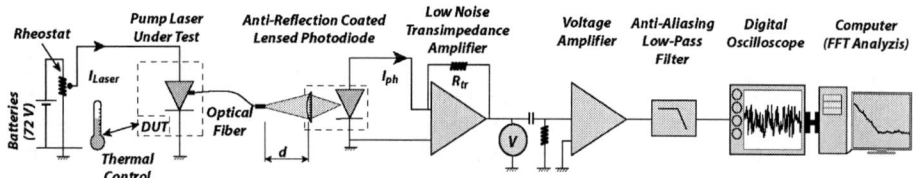

FIGURE 2. Feedback free noise measurement set-up.

Experimental Results : The final result is given in *fig. 3-a*. White noise evolution under threshold is not typical. In this part if the curve, the gap between optical white noise and shot noise may thus find two justifications. One the one hand, this is a measurement error induced by noise level too close to shot noise ($M = 50$). On the other hand, this is really the laser behavior, and this could corroborate that emission under threshold is not only spontaneous in the previous static measurements. For near threshold noise values, we also have to take into account the amplified spontaneous emission phenomenon excess noise.

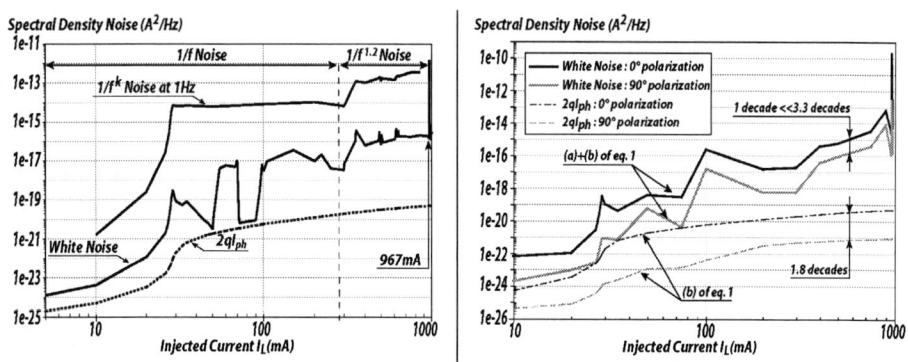

FIGURE 3. Measurements : Noise measurement (a) without and (b) with light polarizer.

White noise evolution around threshold is rather classical : a strong level increase followed by a decrease shows threshold appearance. This behavior is similar to a monomode laser. Above threshold, this rather chaotic behavior is close to previous optical spectrum evolution (*fig. 1-b*). A significant noise decrease around 80*mA* is directly connected to the laser monomode nature in this current range. Moreover, white noise level fluctuates over 3 decades because of the optical spectrum instability at 967*mA* (*fig. 1-c*) [2, 3]. This strong (40*dB* !) and specific fluctuation is perfectly reproducible.

$1/f$ noise evolution is an interesting result. As the theoretical models envisage it [4], we observe a strong saturation from threshold to 300*mA* which shows a quite good gain saturation. Nevertheless, above 300*mA* the low frequency noise shape changes from pure -1 slope to -1.2, with much stronger noise levels (one decade higher). Above this bias, we can not say the laser internal gain is constant. The white noise also varying the same way, these two behaviors probably find their origins into the same physical phenomenon (spectral hole burning for instance) which is responsible for a stronger optical mode competition.

Measurement through Light Polarizer : In *fig. 3-b*, we have two white noise curves : the first with the polarizer angle set to 0° (relative to the stimulated emission direction), and the second with a 90° rotation angle. The result we obtain is at the same time interesting and surprising. On one hand, noise level under threshold is not the same in the two polarization cases, the spontaneous emission following a priviledged polarization direction. On the other hand, the two curves are almost parallel. By a first look at the lower curve, we may think this is stimulated emission attenuated by the polarizer. But the fact is that the shot noise is proportional to the current, and the distance between the two shot noise curves is 1.8 decades apart. Such a distance should be observed, in non shot-noise spectral densities, by a 3.3 decades gap, whereas we only have experimentally a one decade distance. A physical interpretation is hard to set. This result shows the direct link between stimulated optical spectrum and spontaneous emission instabilities.

CONCLUSION

We studied the second order behavior of a high power multimode laser diode. We highlighted the strong link between Intensity Noise and the beam optical spectrum. We also observed correlations between spontaneous emission and stimulated emission noises in competition mode rate. Lastly, it is clearly established that this component is ideal for a pump noise influence study on the noise of a complete Raman amplifier as we can "adjust" its noise level at almost fixed power (around 960*mA*) : this will allow a pragmatic study of the complete amplifier noise evolution according to pump noise.

REFERENCES

1. N. Schunk and K. Petermann, *Electronics Letters*, pp. 63–64 (1989).
2. M. Ohtsu et al., *Applied Physics Letters*, **46**, 108–110 (1985).
3. M. Myara et al., *Fluctuations and Noise Letters*, **3**, L289–L294 (2003).
4. R. J. Fronen and L. K. J. Vandamme, *IEEE J. Quantum Electronics*, **24**, 724–736 (1988).

Transmittances Distributions at the Diffusive-Localized Crossover in Disordered Wave-guides with Absorption

L.S. Froufe-Pérez[1], P. García-Mochales[1], P.A. Serena[2] and J.J. Sáenz[1]

[1] Departamento de Física de la Materia Condensada, Facultad de Ciencias, Universidad Autónoma de Madrid, Campus de Cantoblanco, E-28049 Madrid, Spain
[2] Instituto de la Ciencia de Materiales de Madrid, Consejo Superior de Investigaciones Científicas, Campus de Cantoblanco, E-28049 Madrid, Spain

Abstract. An analysis of the behavior of transmittances distributions with the length in disordered wave-guides with absorption is presented. The study is focused on the crossover between diffusive and localize regimens. Distributions are obtained from numerical calculations from the well-known "tight-binding" model. Absorption in this model produces also an increase of the scattering rate (and a decrease of the mean free path l and localization length ξ. Distributions of the transmission $P(T)$ on the crossover show a smooth crossover between Gaussian and lognormal distribution (unlike the no-absorbing case). The ratio $var(s_a)\xi L$ has in the ballistic regime a value similar to the no-absorbing case, decreases as L rises in the diffusive regime, showing a minimum around ξ increasing again in the crossover. This behavior resembles the results found in microwaves wave-guide experiments.

Keywords: Disordered wave-guides, absorption, transmittances distributions
PACS: 78.67.Lt, 42.25.Bs, 42.25.Db, 71.30.+h

Propagation and localization of waves of various kinds in quasi-one dimensional disordered media have been studied intensively for several decades. Examples of particular interest are electron transport in disordered solid and electromagnetic wave propagation in random dielectric media. In contrast to electronic systems, where the total flux is conserved, in optical systems the transmittance may be affected as a result of absorption.

The conductance (transmittance) distribution in electronic system $P(T)$ is known to evolve from a Gaussian distribution $P(T)$ (deep in the diffusive regime) to a lognormal distribution (deep in the localization regime) [1]. The analysis of the conductance distribution for disordered wires in the diffusive-localization crossover shows a non-obvious transition of the distribution between both regimes [2].

The properties of the transmittance of a multi-channel random wave-guide with absorption have been computed within the Random Matrix framework [3,4]. In the limit of very strong absorption compared with the disorder, statistical properties have been calculated in the diffusive limit (the transmittance distribution is Gaussian) and the localized limit (lognormal distribution) [4]. The understanding of the transmittance properties in the crossover region between these limits and in systems where the disorder strength can be comparable with the absorption may be important in determining the onset of localization in photonic systems [5-8].

In this work disordered wave-guides have been modeled, both with and without absorption. The transport properties (distributions) of the wave-guides have been calculated as their length it is varied, covering the ballistic, diffusive and localization regimens and the crossovers between them. Absorption considered is strong compared with the disorder used but it is far from the limit considered in ref. [4]. In spite of this consideration and the few channel considered in our calculations, our results show a similar behavior to experimental findings of ref. [8].

To calculate the transmittance through disordered wave-guides (with and without absorption) we have used a tight-binding model with a 1s orbital per site and nearest-neighbor interactions. The disordered and the absorption are introduced in a region of size N (width) $\times L$ (length) (in units of the lattice constant a) through complex on-site energies. To model the disorder, the real part of the on-site energy is randomly distributed between $[-W/2, W/2]$ (Anderson disorder); the absorption is introduced using an imaginary part D of the on-site energy (constant and positive) in the disordered zone (the absorption is characterize by mean of the exponential absorption length L_a). This region is connected to two perfectly ordered (complex on-site energy equal to zero) semi-infinite leads of with N. The number of open channels in the system is equal to the width N (Fermi energy used was zero in all the cases considered).

Three types of transmittances can be computed T_{ab}, T_a and T from the $N \times N$ transmission matrix t. The total transmittance T is equivalent of the conductance G on electronic systems, T_a is the total intensity transmitted if the system is illuminated through channel a, and T_{ab} the intensity in channel b when illumination is only through channel a. In order to compare with experimental works, the distributions of the transmittances normalized to their ensemble average values ($s_{ab}=T_{ab}/\langle T_{ab}\rangle$, $s_a=T_a/\langle T_a\rangle$, $s=T/\langle T\rangle$) are calculated.

As in electronic systems, the mean free path l of the system is calculated by a fitting of $\langle T \rangle$ in the ballistic regime ($L<<l,L_a$) to $\langle T\rangle=N(1-L/l)$. The localization length ξ is calculated by $\xi=(N+1)l$ [1]; in absence of absorption, this value is the same one obtained by fitting the linear decay of $\langle \ln T\rangle \approx -2L/\xi$ in the localization regime ($L>>\xi$) [1]. This last method is not possible when absorption is present due to the mixing of the localization and absorption effects, both contributing to the exponential decay of T with the length. The exponential absorption length L_a is obtained from simulation where absorption is present ($D>0$) and there is not disorder ($W=0$), assuming that, in random guides, L_a does not change when the disorder used is weak.

To illustrate our finding, the results of two sets of wave-guides (with and without disorder) will be showed. Both sets have a width of $N=7$ and disorder strength $W=1$. The guides with absorption have $D=0.018$ that correspond to an absorption length of $L_a=39$. The localization lengths are $\xi=108$ for the case of wave-guides with absorption ($L_a<\xi$ but far for the limit considered in ref [4] where L_a/ξ is used as a small parameter) and $\xi=161$ if there is not absorption. Note that a non-zero imaginary component of the on-site energy produces, as well as absorption, an increase of the scattering that results in a decreases of l and ξ.

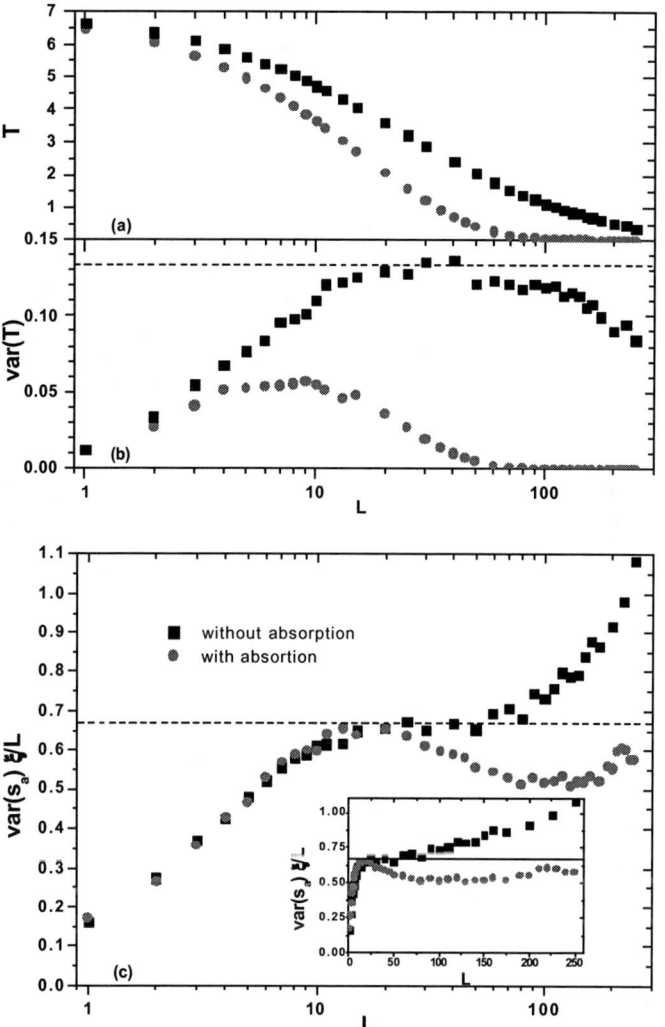

FIGURE 1. (a) $\langle T \rangle$, (b) $var(T)$ and (c) $var(s_a) \cdot \xi/L$ vs. the systems length L for two cases of waveguides simulations: without absorption (black squares) and with a moderate amount of absorption (gray circles). Inset shows the same data of (c) using a linear scale for L. Dashed lines represent predictions of the Random Matrix approach for the diffusive regime: $var(T)=2/15$ and $var(s_a)=2\xi/3L$.

As it is expected, the absorption reduces the transmission $\langle T \rangle$ compared with the no absorbing case. Also $var(T)$ is significantly reduced, showing $P(T)$ for $L \ll \xi$ a Gaussian distribution sharper than guides without losses. In the crossover between diffusive and localized regimes, at difference with the case of no absorbing waveguides [2], the transition of $P(T)$ between Gaussian to log normal distribution is smooth: $P(T)$ can be well fitted with a Gaussian curve when $L \sim \xi$ and then evolves slowly to the log-normal distribution. In figures 1(a) and 1(b) are depicted the evolution with the length L the transmission $\langle T \rangle$ and $var(T)$ for both sets of waveguides, with and without absorption, for comparison.

In figure 1(c) the product of $var(s_a)\xi/L$ is presented versus the system length L. The choice of this representation is due to make easier the comparison with experimental results of ref. [8]. Notice that localization lengths $\xi=(N+1)l$ are different if absorption is considered or not (and that $N+1$ can not be approximated by N due to its low value). The results of both cases are very similar in the semi-ballistic regime, showing an increase with the length. In the diffusive regime results start to diverge. Within this regime, the no absorbing case shows an approximately constant value, close to $var(s_a)=2\xi/3L$ that it is predicted within the Random Matrix framework [9]. As L increases and the system leaves this regime entering into the localized one, $var(s_a) \cdot \xi/L$ increase again showing approximately a parabolic growth.

Wave-guides with absorption behave different after the semi-ballistic regime. At the beginning of the diffusive regime $var(s_a)\xi/L$ reaches the $2/3$ value, then start to decrease as L increases. It presents a minimum (with value $\approx 1/2$) around the localization length ξ, increasing again the ratio $var(s_a)\xi/L$ as the system goes deep into the localization regime. These results cannot be compared with theoretical ones from ref. [4], obtained under the assumption of $L_a/\xi \ll 1$. In spite of this, in that work it is predicted for $var(s_a)\xi/L$ a variation from 2/3 to 1/2 in the diffusive regime, but it assumed that it remains constant after the crossover.

On other hand, a similar behavior to our calculations was found in experiments of ref. [8], both the direct measurements with absorption ($var(s_a)\xi/L$ shows a minimum in that work near the crossover) and processed data where the absorption effect was eliminate ($var(s_a)\xi/L$ shows then a quadratic growth).

ACKNOWLEDGMENTS

P.G.-M. is supported by the Spanish MEC through its "Ramon y Cajal" Program. This work has been supported by the Spanish MEC (ref. BFM2003-01167) and the EU Integrated Project "Molecular Imaging" (EU contract LSHG-CT-2003-503259).

REFERENCES

1. Beenakker, C.W., *Rev. Mod. Phys.* **69** 731 (1997).
2. Froufe-Pérez, L.S., García-Mochales, P., Serena, P.A., Mello, P.A. and Sáenz, J.J., *Phys. Rev. Lett.* **89** 246403-1 (2002).
3. Misirpashaev, T.Sh., Paasschens, J.C.J. and Beenakker, C.W.J., *Physica A* **236** 189 (1997).
4. Brouwer, P.W., *Phys. Rev. B* **57** 10526 (1998).
5. Wiersma, D.S., Bartolini, P., Langendijk, A. and Righini, R., *Nature* **390** 671 (1997).
6. Scheffold, F., Lenke, R., Tweer, R. and Maret, G., *Nature* **398** 206 (1999).
7. Gómez-Rivas, J., Sprik, R., Langendijk, A., Noordam, L.D. and Rella, C.W., *Phys Rev. E* **63** 046613 (2001).
8. Chavanov, A.A., Stoytchev, M., and Genack, A.Z., *Nature* **404** 850 (2000).
9. Kogan, E. and Kaveh, M., *Phys. Rev. B* **52** 3813 (1995).

Rice Representation of Noise processes in Optical PLL's

Arturo Arvizu M., Francisco J. Mendieta J.

Research Center, Km.107 Carretera Tijuana-Ensenada,
Ensenada, B.C., 22880, México, e-mail: arvizu@cicese.mx, jmendiet@cicese.mx

Abstract. The Rice representation is a well-known engineering tool used to describe band-pass electrical signals in radio-frequency and signal processing applications. In previous works, this representation has been used for describing the intensity noise of laser oscillators and optical amplifiers. In this work, we use such tool to describe the amplitude and phase noises in optical phase locked loops.

Keywords: Optical phase locked loops, phase noise, Rice representation.
PACS: 42.79.Sz Optical communication systems, multiplexers, and demultiplexers

INTRODUCTION

Optical Phase Locked Loops (OPLL's) have been studied extensively during the past decades and have found several important applications. They can be used in the implementation of optical carrier synchronizers in coherent communications systems [1]; they can also be used in phase coherent optical-to-microwave frequency chains among others [2]. In the other hand, the Rice representation also known as in-phase/quadrature description is a well-known engineering oriented tool used to describe band-pass electrical signals in radio-frequency and signal processing applications. In previous works [3, 4, 5], the use of such tool to model the noise in semiconductor lasers and optical amplifiers has been shown. In this work we derive the different amplitude and phase noise components appearing in an optical PLL using such representation. This modeling in a simple way is very important for the design of optical communication systems and optical signal processing.

Rice Representation of Noise Signals in OPLL's

A laser oscillator signal can be described by means of a deterministic phasor [3, 4, 5], with normalized amplitude $\bar{E} = \sqrt{\bar{P}_E}$ (where $\bar{P}_E$ is the average power of the optical signal) with an added noise vector $N(t)$ (see figure 1) described by means of the Rice representation as [6]:

$$N(t) = N_I(t)\cos(2\pi v_o t) - N_Q(t)\sin(2\pi v_o t) \quad (1)$$

where $N_I(t)$, $N_Q(t)$ are the so-called in-phase and quadrature components of $N(t)$, and v_0 is the optical frequency. We consider that their respective single-sided power spectrums are equal [5]:

$$S_N(v) = S_I(v) = S_o(v) \qquad (2)$$

For a coherent state the noise vector to take into account is the so-called vacuum fluctuations [7]; therefore the average power (P_N) of the noise vector is:

$$P_N = 2\pi S_N \Delta v \qquad (3)$$

where

$$S_N = \frac{hv}{2} \qquad (4)$$

is the single-sided spectral density, h is the Planck's constant, v is the optical frequency, and Δv is the optical bandwidth of the noise signal respectively.

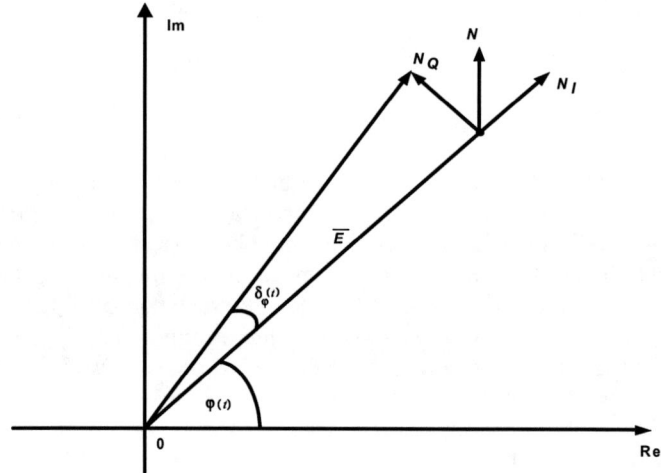

FIGURE 1. Rice representation of a laser oscillator signal

In figure 2 is shown a block diagram of a typical optical PLL. The input signal $E_s(t)$ and the local oscillator $E_{LO}(t)$ can be described, as mentioned above, by means of a deterministic phasor with an added noise vector. The signal $v_p(t) = R_L i_p(t)$ at the output of the photoreceiver stage (R_L: load resistance) is proportional to the total optical power $P_T(t)$ at their input [8] due to the addition of the two optical fields [9]:

$$i_p(t) = \Re P_T(t) \qquad (5)$$

$$P_T(t) = (\overline{E}_T + N_{IT})^2 = \overline{P_T} + \underbrace{2\overline{E}_T N_{IT}}_{\text{fluctuations} = \Delta P_{NIT}} + \underbrace{N_{IT}^2}_{\approx 0} \qquad (6)$$

$$E_T(t) = E_s(t) + E_{LO}(t) \quad (7)$$

$$N_{IT}(t) = N_{IS}(t) + N_{ILO}(t) \quad (8)$$

$$\langle \Delta P_{NIT}^2 \rangle = 4 P_{NIT} \overline{P_T} = 2 P_{NT}.\overline{P_T} \quad (9)$$

where $\Re$ is the photodetector's responsivity, $N_{IT}(t)$ is the total in-phase noise component and $\langle \Delta P_{NIT}^2 \rangle$ is its variance respectively. It can be shown [3], that with the use of the equations (9), (3) and (4) (considering and an equal sharing of the total noise power between the two noise components [5]) and taking into account the integration time of the photodetector, we can obtain, as expected, the usual shot noise $\langle \Delta_i^2 \rangle = 2e\bar{i}\Delta f$ appearing in photodetection. This result shows that the shot noise in the photoreceiver is proportional to the total in-phase noise component $N_{IT}(t)$.

On the other hand, the phase quadrature noise components $N_{QS}(t)$ and $N_{QLO}(t)$ of both optical fields generates a total phase noise signal $\delta\varphi_T(t)$ (and total linewidth $\Delta v_{FWHM\,T}$) proportional, as expected, to the addition of the linewidths Δv_{FWHM} of both optical sources, as can be shown through the use of the equations:

$$\delta\varphi_T(t) = \delta\varphi_S(t) + \delta\varphi_{LO}(t) \approx \frac{N_{QS}(t)}{\overline{E}_S} + \frac{N_{LO}(t)}{\overline{E}_{LO}} \quad (10)$$

$$\langle \delta\varphi_T^2(t) \rangle \approx \frac{P_{QS}}{P_S} + \frac{P_{QLO}}{P_{LO}} = \frac{h\nu B_{0S}}{2\overline{P}_S} + \frac{h\nu B_{0LO}}{2\overline{P}_{LO}} \quad \text{(mean square phase fluctuations)} (11)$$

$$B_0 = \frac{1}{\tau_p} \quad \text{(with } \tau_p \text{ being the photon lifetime)} (12)$$

$$\langle \Delta\varphi_T(\tau)^2 \rangle = \left[\frac{h\nu B_{0S}}{2\overline{P}_S} + \frac{h\nu B_{0LO}}{2\overline{P}_{LO}} \right] \frac{\tau}{\tau_P} = \left[\frac{h\nu B_{0S}^2}{2\overline{P}_S} + \frac{h\nu B_{0LO}^2}{2\overline{P}_{LO}} \right] \tau \quad (13)$$

(mean square phase jitter over a duration τ)

$$\Delta v_{FWHM\,T} = \frac{1}{2\pi} \frac{\langle \Delta\varphi_T(\tau)^2 \rangle}{\tau} = \frac{1}{2\pi} \left[\frac{h\nu B_{0S}^2}{2\overline{P}_S} + \frac{h\nu B_{0LO}^2}{2\overline{P}_{LO}} \right] = \Delta v_{FWHM\,S} + \Delta v_{FWHM\,LO} \quad (14)$$

In addition, the beat signal plus shot noise is filtered to produce the following voltage signal:

$$v_c(t) = [\Re R_L \overline{E}_S \overline{E}_{LO} \cos(\delta\varphi_S(t) - \delta\varphi_{LO}(t) - \phi_c(t)) + n_{shot}(t)] * f_{LPF}(t) \quad (15)$$

is input to the optical VCO, and has two different types of noises; the amplitude noise (proportional to the in-phase components) and the phase noise (proportional to the quadrature components).

Conclusion

In this work, the Rice representation has been used to describe the amplitude and phase noises appearing in an optical phase locked loop when the input signal is not modulated; currently we are working in obtaining the equations for such case. This modeling in a simple way is very important for the design of optical communication systems and optical signal processing.

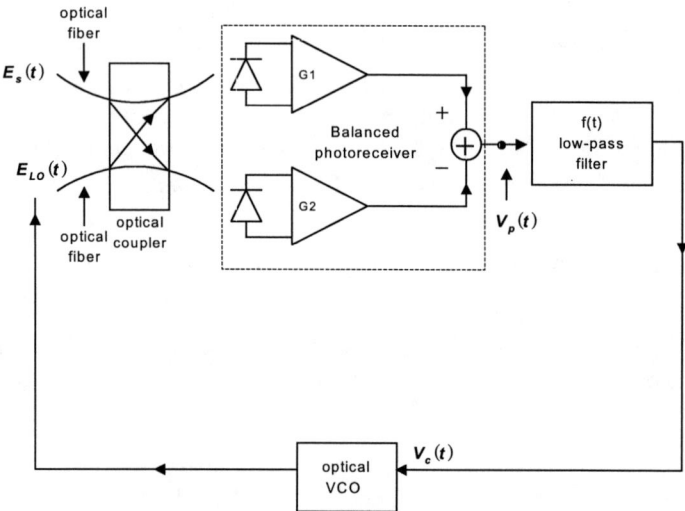

FIGURE 2. Optical Phase Locked Loop (G1, G2: electrical gains)

ACKNOWLEDGMENTS

This work was supported partially by the (CONACYT) - Mexico.

REFERENCES

1. Camatel S., Ferrero V., Gaudino R., and Poggiolini P., Elect. Lett., **40** (6), (2004).
2. Kovacich, R.P., *The Precision of Modern Phase Coherent Optical to Microwave Frequency Chains*, Ph.D. Thesis, The University of Western Australia, 2000.
3. Gallion., P., "A classical corpuscular approach to optical noise", in *Trends in Optics and Photonics, Vol.XXX,* edited by Susumu Kinoshita, Jeffrey C. Livas & M., 1999.
4. Arvizu, A., and Gallion., P., "Rice representation of laser's quantum noise: application to the derivation of Schawlow-Townes relationship", in *IEEE ROC&C'2004*, Conference Proceedings, Acapulco, Gro., Mexico, November, 2004.
5. Gallion., P., "Basics of digital optical communications" in *Undersea communication systems,* edited by José Chesnoy, Academic Press, Harcourt Inc., Boston 2001.
6. Haykin, S., *"Communication systems",* John Wiley and sons, USA., 1978.
7. Yariv, A.,*"Optical electronics in modern communications",* Oxford University Press, USA, 1997.
8. Alexander, S.B., *"Optical communication receiver design",*SPIE, USA, 1997.
9. Kazovsky, Benedetto, S., Willner, A.,*"Optical Fiber Communications",* Artech House, USA, 1996.

Acoustic Emission, Electrical And Light Fluctuations In Optoelectronic Devices

Olexander I. Vlasenko[*], Vitaliy P. Veleschuk[*], Oleg V. Lyashenko[†]

[*]*Institute of Semiconductor Physics of NASU, 03028, 45 Pr. Nauki, Kyiv, Ukraine*
[†]*Kyiv National University of Taras Shevchenko, 03680, 2 Glushkova Str., Kyiv, Ukraine*

Abstract. It is shown, that in heterostructures basic on A^3B^5 acoustic emission occurrence, current and optical noise, evolution electroluminescence spectrums and degradation current-voltage characteristics correlate in time and have the common origin.

Keywords: Acoustic Emission, Fluctuation, Noise, Electroluminescence Spectrum, Device.
PACS: 43.35+d; 43.50+y; 72.70+m; 73.50.TD; 78.60.Fi; 78.66.Fd

INTRODUCTION

The problem of current noise in semiconductor elements is actual. The kind of these noise allows to assume that the source can't have electric nature. Occurrence of electric noise spectrum in semiconductor elements was observed at action of narrow-band acoustic noise [1]. It was an observable correlation in time between electric and acoustic noise [2]. It allows assuming that one of current noises sources may be the transformed internal acoustic noise.

Internal mechanical and induced thermomechanical stresses can result in epitaxial light-emitting diodes occurrence of chaotic acoustic waves - acoustic emission (AE) of materials (at additional superthreshold external action) [3]. There are two kinds of the AE: a high-energy discrete AE, which caused usually by change of a condition ("failure") of 3-dimensional defects or dislocation complexes, and also a low-energy continuous AE, which usually caused by synchronous movement (failure-fastening on a stopper) dislocation segments. Occurrence of the AE means that in crystal local areas there were irreversible changes, there created a new defect or the metastable condition of existing defect has changed. At enough intensive AE such local changes cover a significant part of crystal, are nonreversible and change properties of the crystal - degradation of it's physical parameters is observed.

The complex AE research of optoelectronics devices, connected with mutual transformation and correlation of acoustic (ultrasonic) and electric signals, the change in a electroluminescence spectrum (EL), the change of current-voltage characteristics (CVC) presents are interested. The AE phenomenon under current in semiconductor elements is connected with creation short-term thermomechanical microstresses in semiconductor crystal volume which failure is accompanied by ultrasonic impulses radiation. Quantity of local mechanical stresses before failure on various estimations achieves $10^6 \ldots 10^8$ Pa. Thus, the AE can be considered as a source of the internal acoustic (ultrasonic) vibrations, which comparable with researched in [1] external sources. Arising short-term mechanical stresses can lead to current fluctuations of various quantity, that means change of own noisis level of the device.

EXPERIMENT

Research of correlation in time of noise was performed on experimental setup similar [2, 4-6]. The piezoelectric transducer of specialized acoustic-emissive device AF-15 has registered signals of the AE. In the given experiment 4 independent channels of registration (two devices AF-15), connected with multichannel recorder and computer are used. For recording of EL spectrums was used a monochromator, the signal from photocell of which has been processed by a computer. Current-voltage characteristics (CVC) of structures have been measured simultaneously with recording of EL spectras and registration of the AE.

The specimens were light-emitting epitaxial n^+-*n-p-structures* of A^3B^5 compounds: *GaAs$_{0,15}$P$_{0,85}$:N/GaP, GaP:N/GaP* and *GaP:Zn:O/GaP*. Samples of epitaxial structures had the sizes 400... 450x400... 450 microns. Density of their nominal direct current J_{nom}=4 A/sm^2. Density of a current in range J_i=2...200 A/sm^2, increased steps - in everyone *(i+1)* step J_{i+1}=(2...1,2) J_i. Time between changes of a current has been certain on finished AE (2...15 minutes after increasing J_i), or on absence of the AE during 5 minutes.

Feature of the AE was occurrence of signals not less, than later 10...30 seconds after fast current increasing of the heterojunction was observable. It determines a relative inertance of the AE sources - areas of local microstresses in a crystal. Usually attenuating discrete AE has been registered within the limits of 5...10 minutes as groups of impulses, a time interval between which has achieved 3...5 minutes. During this time the current of a heterojunction only in some cases repeating failures of microstresses. The check by AE method has allowed us to achieve direct current density of these structures up to 30...50 times above declared by the manufacturer.

The AE, electric and "optical" fluctuations (noisis) had been registered simultaneously under direct current of heterojunction. Always correlation was following: to each group of discrete AE signals corresponds the increasing of current noise. AE signals and current noise always correspond to "optical" noise (short-term change of EL intensity). Degradation the EL after intensive AE was observed.

The AE of epitaxial light-emitting structures is accompanied by the displacement of EL strips (fig. 1, a), just as in [4-6]. These displacements for small currents looked as fluctuation (3-5 minutes) [5] or were constant [6]. After intensive AE and formation of EL infrared strip (I-strips) for greater currents this process of EL strips displacement could finished differently. The restoration of initial EL spectrum was in some cases at reduction of current density J_i [5]. There was a nonreversible displacement of a green strip (G-strips) with 2,2 eV up to 2,0 eV and disappearance of a red strip (R-strips) even for small J_i [6].

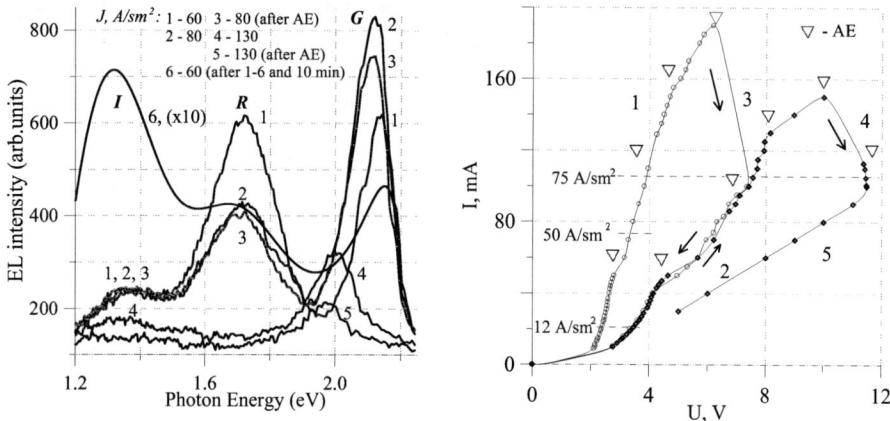

FIGURE 1. Electroluminescence spectrums (a) and current-voltage characteristic (b) n^+-n-p-structures $GaAs_{0.15}P_{0.85}$:N/GaP

Degradation of structures in which was observed AE shows their change of CVC. On fig. 1,b are presented CVC $GaAs_{0.15}P_{0.85}$:N/GaP structures at step cyclic current increasing. The section of a curve 1 corresponds first increasing of J_i. The next reduction of J_i up to zero and increasing of J_i corresponds 2. Sections of curve 2 (in both directions of J_i change) and 5 (at reduction of J_i) are qualitatively similar. 2 and 5 are new CVC (consecutive in time) of degraded structure. The fast degradation during 30-90 seconds (sections of a curve 3 and 4) was accompanied by intensive discrete AE and current noise. This process has finished at $J_i \sim 75$ A/sm^2. The sign ∇ notes values J_i at which carried out a corresponding delay of time for registration of the AE.

DISCUSSIONS

The estimations of probable efficiency of AE signals transformation in current noise have shown that AE signals can't be the initial reason of current noise occurrence. Losses at such transformation are very large and the amplitude of current noise was smaller amplitudes of electric AE signals no more, than in 100 times. Observable conformity in time AE signals, both current and "optical" noise, and degradation the EL are determined by the common nature of these phenomena. The large integrated intensity of current noise is determined by the contribution of mechanisms, that haven't been connected with occurrence of the AE.

The reason of such conformity of these phenomena can be development of microcracks and dislocation in crystal (in the range of a heterojunction) which accompany the AE. Consequences of this are current fluctuations and fluctuation and degradation of EL intensity. Reduction of durability of light-emitting diodes at presence of the AE is connected with the distribution of dispositions in active areas [7]. In GaP diodes in which has been registered AE the greatest density of dislocations and reduction of a light output of diodes have been established.

It is necessary to note that AE at change of a condition only point defects now will not be registered because of too small energy of a separate impulse. Therefore observable features in EL spectra demand additional explanations. It is known that in structures $GaAs_{0.15}P_{0.85}$:N/GaP the emissive recombination in a red strip is determined by Zn-O complexes, which amount in GaP:N/GaP structures insufficient. The absence of an I-strip in GaP:N/GaP structures confirms the assumption [5,6] that the growth of the EL in an I-strip can be connected with the process which causes the fast degradation of structures and disintegration of complexes Zn-O at large J_i. The process is accompanied by a discrete AE. Hence, the current nonequilibrium condition of defect structure of separate crystal local areas defines the EL.

One of primary factors which determine the speed and quantity of degradation of basic parameters of light-emitting diodes and a spectrum and intensity of electroluminescence the direct current of heterojunction is. Hence, received by us for the same samples conformity in occurrence of acoustic, current and optical noise, change of EL spectrum and also occurrence of features in CVC light-emitting diodes – this is consequences of the same reorganization processes of local defect structure.

REFERENCES

1. Kucherov, I.Ya., Lyashenko, O.V., and Perga V.M. *Ukrainian J. of Phys.* **34**, 222-224 (1989)
2. Lyashenko, O.V., and Perga V.M. "Acoustic emission for the diagnostic of semiconductor structures" in *Diagnostics Techniques for Semiconductor Materials Processing II*, edited by S. W. Pang et al., MRS Proceedings 406, Boston, 1996, pp.449-456.
3. Lord, A.E., "Acoustic Emission", in *Physical Acoustics* **XI**, edited by W.P. Mason, New York and London:Academic Press, 1975, pp. 289-353.
4. Veleshchuk, V.P., and Lyashenko, O.V., *Ukrainian J. of Phys.* **48**, 981-985 (2003)
5. Veleshchuk, V.P., Lyashenko, O.V., Myagchenko, Yu.A., and Chuprina R.G., *J .Appl. Spectroscopy* **71**, 553-557 (2004).
6. Lyashenko, O.V., Myagchenko, Yu.A., and Veleshchuk, V.P., "Change of spectrums of an electroluminescence epitaxial lightdiode structures during an acoustic emission" in *"Spectroscopy of Molecules and Crystals"*, edited by G.O. Puchkovska et al., Proceedings of SPIE 5507, Bellingam:SPIE, 2004, pp. 49-52
7. Icoma, T., Ogura, M., and Adachi, Y. "Acoustic emission from single crystals of gallium arsenide" in *"Proceeding Third Acoustic Emission Symposium"*, Tokyo, 1976, pp. 329-341

Local Instabilities In GaAsP Diode PN Junctions

P. Koktavy, B. Koktavy

Brno University of Technology, Department of Physics,
Technicka 8, 616 00 Brno, Czech Republic

Abstract. Currently, the occurrence of microplasma regions in PN junctions is attributed to crystal lattice imperfections. As a rule, these regions feature lower strong-field avalanche ionization breakdown voltages than other homogeneous junction regions. The existence of such regions may lead to local avalanche breakdowns occurring in reverse-biased PN junctions at certain voltages. Macroscopically, these breakdowns are manifested as microplasma noise. Studying the current conductivity bistable mechanism thus may be used as an efficient tool to evaluate the PN junction inhomogeneity.

Keywords: PN Junction, GaAsP diode, microplasma noise, avalanche breakdown.
PACS: 71.55.Eq

INTRODUCTION

In a semiconductor PN junction there are localized regions featuring increased concentration of donor or acceptor impurities or other defects, which cause the PN junction reverse breakdown voltage to be reduced. They can be displayed when picking up the reverse current waveforms at varying reverse voltage or by measuring the U-I characteristics of a PN junction powered from a constant current supply. Below the homogeneous breakdown region, the PN junction reverse current is, in principle, due to the local defect-assisted current conduction only. In regard of very small cross-sectional areas of the regions in question (reported figures range from 10^{-14} m^2 to 10^{-12} m^2), there is a high current density (10^7 A.m^{-2} and even more) in any elevated concentration region, which in turn may result in a strong local heat generation and, consequently, local diffusion or thermal breakdown. These areas are particularly critical for the application of high-power rectifier diodes, which are operated at very high reverse voltages continuously. Our study of the breakdown voltage versus temperature plot makes us suppose that this is an avalanche breakdown in consequence of strong field impact ionization. One of the important avalanche breakdown characteristics is the impact ionization coefficient, which depends strongly on the junction electric field strength. The present paper deals with the GaAs$_{0.60}$P$_{0.40}$ diodes, for which there is no microplasma noise information available in the literature.

MICROPLASMA NOISE VERSUS TIME BEHAVIOUR

The microplasma noise occurs in reverse-biased PN junctions in the form of random two-level or multi-level current impulses. They are usually rectangular in shape (provided that a voltage source is used to supply the circuit, see Fig. 1), however, their shape may be, depending on the outer circuit parameters, approximately triangular with an exponential trailing edge (see Fig. 2, where a current source is used to supply the circuit). The rectangular current impulses feature constant amplitude, random impulse width and separation.

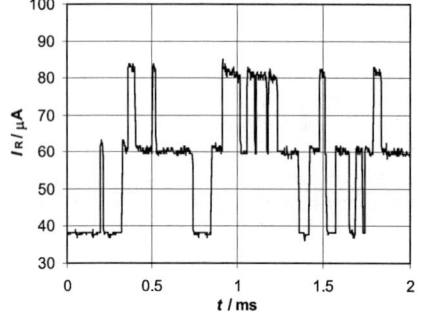

FIGURE 1. Multilevel microplasma current noise for constant voltage power supply, sample M21.

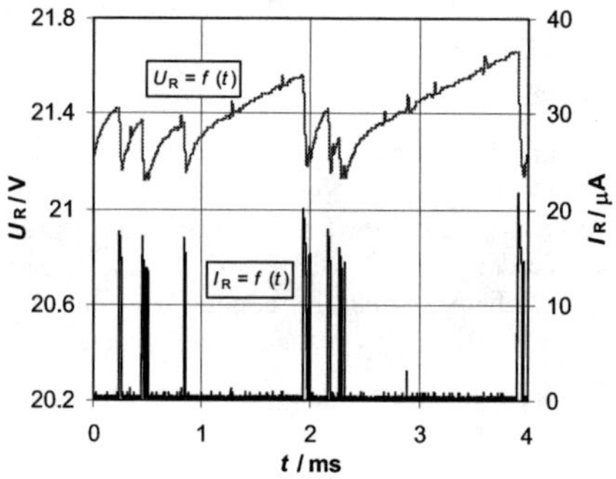

FIGURE 2. Microplasma current noise for constant current power supply, sample M10.

U-I CHARACTERISTICS

The existence of microplasma regions affects the *U-I* characteristics in the reverse state most of all. If a (low-impedance) voltage source is used, there appear abrupt current growth regions in the *U-I* characteristics, being due to local avalanche breakdown in the different microplasma regions, see Fig. 3. On the other hand, if a (high-impedance) current source is used, there occur negative-resistance regions in the *U-I* char., each region corresponding to a single active microplasma zone, see Fig. 4.

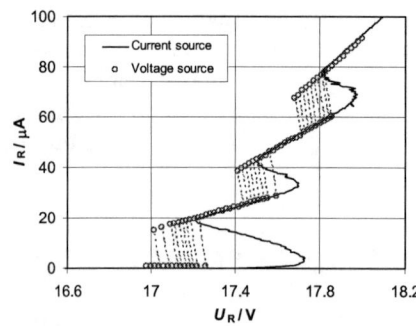
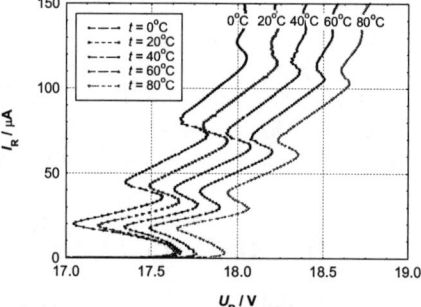

FIGURE 3. *U-I* curves of a diode showing several microplasmas, diode M4, $t = 20$ °C.

FIGURE 4. Partial *U-I* characteristics of the different microplasma regions, reverse biased diode M4, constant current power supply.

Our diode model shows that one and the same process is responsible for both phenomena being observed when measuring the microplasma noise waveforms and U-I characteristics. Therefore, the shape of the U-I characteristics provides us with complete information on the diode bistable behaviour in the instability region, correlating very well with the microplasma noise waveforms measured, see Fig. 3.

MICROPLASMA REGION CURRENT CONDUCTION MODEL

In Fig. 5 a diode showing three microplasma regions is illustrated. The bistable current conduction corresponds to the transitions between two states. If the U-I characteristic linear regions are extrapolated, the intersection points indicate the extrapolated breakdown voltages U_{B1}, U_{B2} and U_{B3}. If the reverse voltage reaches U_{B1}, random impulses corresponding to the first microplasma will start occurring. Their counting rate, amplitude and duration are rather low, corresponding to the low probability that a minority carrier triggers impact ionization of other carriers in the microplasma region and, furthermore, to the high probability that any discharge is quenched in the microplasma region. When the reverse voltage is further increased, the above mentioned quantities are growing, until – at U_{K1} (Fig. 6) – the first microplasma is ionized permanently. The corresponding U-I characteristic shows a continuous straight-line section with a dynamic resistance R_{M1}. Similarly, the current conduction due to the second and other macroplasmas will occur, with the corresponding onset voltages, U_{B3}, ..., U_{Bk}. The onset of the different microplasma transition interval is difficult to discover experimentally, because of their extremely short pulse duration and counting rate. Subsequent superposition of the different microplasma current pulses results in three-level or multi-level pulses. These pulses were however immeasurable for M4 diode (Fig. 5) because of too a large interval between the extrapolated voltages U_{Bk}.

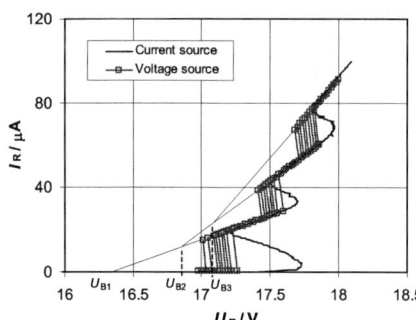
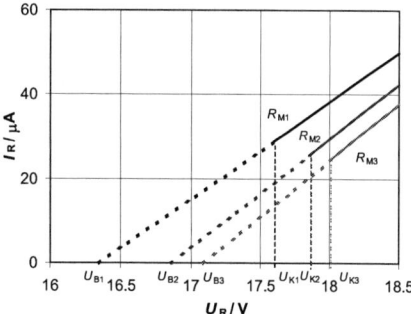

FIGURE 5. U-I characteristics of a multi-microplasma M4 diode, $t = 20$ °C.

FIGURE 6. Partial U-I characteristics of the various microplasma regions, M4 diode, $t = 20$ °C.

Based on the above-mentioned facts, an equivalent circuit of a reverse-biased diode, D, which is connected in series with a load resistor R_L to a power supply of a voltage U, can be accepted in accordance with [1] (Fig. 7).

The above model is based on the assumption that the micro-plasma regions are passing through the entire junction as cylinders whose axes coincide with the current conduction direction, so that they may be segregated from the whole junction. They are characterized by a set of stochastic switches, S_k, which are connected in series with respective voltage sources, U_{Bk}, and series resistors R_{Mk}, $k = 1, \ldots, n$. The S_k switch turn-on and off processes are controlled by a random process, $\xi_k(t)$, whose states are as follows: if $\xi_k(t) = 0$, the switch is off; if $\xi_k(t) = 1$, the switch is on. The statistical characteristics of the $\xi_k(t)$ process versus reverse voltage plots can be measured either for $R_L \rightarrow 0$ (the diode being connected in a low-resistance circuit), or for $R_L \gg R_{Mk}$ (the diode being connected in a high-resistance circuit). The diode denoted D' characterizes an ideal portion of the PN junction, or – for a real diode – the remaining part of the PN junction

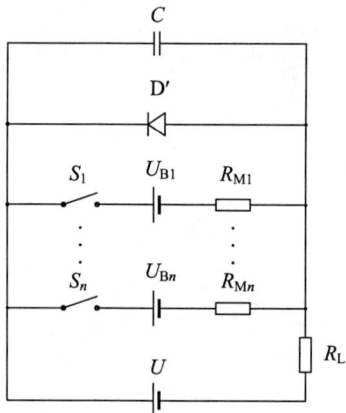

FIGURE 7. Equivalent circuit of the diode.

from which all microplasma regions have been removed. The capacity, C, is made up of the capacity of the PN junction and that of the power supply and measuring instruments.

CONCLUSION

Our study is based on the established fact that the occurrence of microplasma in silicon PN junctions is related to PN junction technology imperfections and that the presence of microplasma leads to the PN junction deterioration or even destruction. This is why we can make use of the methods indicating the occurrence of microplasma in PN junctions and providing their qualitative and quantitative characterization, to assess the quality of PN junction devices.

ACKNOWLEDGMENTS

This paper is based on the research supported by the project MSM 0021630503 and by the projects GACR 102/03/0621 and 102/02/D073.

REFERENCES

1. Haitz, R. H, Model for the Electrical Behaviour of Microplasma. *J. Appl. Phys.*, May 1964, vol. **35**, no. 5, p. 1370-1376.
2. Hazendonk, T. J. and Lodder, A, Noise in Luminiscent GaAs0,60P0,40 Diodes under Non-Uniform Avalanche Conditions – II. *Sol. St. El.*, July – Aug. 1975, vol. **18**, no. 7-8, p. 605-616.
3. Koktavy, P. Local Instabilities in the PN Junction of GaAsP Diodes. Ph.D. Thesis, Brno, 2001. 93 p.

Photocurrent Noise in Quantum Dot Infrared Photodetectors

A. Carbone[*], R. Introzzi[*] and H. C. Liu[†]

[*]*Physics Department and INFM, Politecnico di Torino*
C.so Duca degli Abruzzi, 24 - 10129 - Torino, Italy
[†]*Institute for microstructural Science, National Research Council*
Ottawa - K1A 0R6 - Ontario, Canada

Abstract.
Low-frequency current noise and current-voltage (I-V) characteristics have been studied in InAs/GaAs self-assembled Quantum Dot Infrared Photodetectors in dark conditions and under illumination, at $T = 77K$ and $T = 5K$. The noise behavior is consistent with a generation-recombination fluctuation process mainly related to thermally excited charge carriers at $T = 77K$. At $T = 5K$ the current noise is consistent with a mechanism of fluctuations driven by the electric field, related to tunneling rather than emission-capture of charge carriers from the Quantum Dots. A very effective noise suppression mechanism, related to the tunneling regime, determines a decrease of fluctuation intensity as a function of the voltage. At $T = 5K$, an interesting behavior is observed in the current-voltage and noise power spectra for some of nominally identical QDIP structures in the presence of irradiation. Some devices indeed exhibit (i) a very high photoresponse and (ii) a $1/f$-shaped noise spectrum at low frequencies. The noise suppression mechanism still acts in the presence of radiation, thus reducing the noise intensity proportionally to the photocurrent intensity.

Keywords: QDIP, photocurrent noise
PACS: 73.63.Hs; 72.70.+m

Quantum Dot Infrared Photodetectors (QDIPs) are zero dimensional structures evolved from the Quantum Well Infrared Photodetectors (QWIPs) technology, widely studied in the past decades. QDIPs have attracted more and more attention in the last years since they are expected to reach higher gains and exhibit lower dark current. Furthermore they are sensitive to normally incident infrared radiation not requiring particular optical coupling. QDIPs are similar to QWIPs but the charge confinement is 3-D. The QDIP detection mechanism is based on the intersubband photoexcitation of the charge carriers from confined states in the dots to the continuum [1, 2, 3].

Current noise and current-voltage (I-V) characteristics have been measured in InAs/GaAs self-assembled QDIPs in dark conditions and under illumination, at $T = 77K$ and $T = 5K$. The noise power spectral densities at $77K$ is consistent with a generation-recombination (g-r) fluctuation process related to charge carriers thermally excited from confined to continuum states in the QD layers. The shape of the power spectrum is Lorentzian. The noise gain $g_n = S_I/4eI$ varies as a linear power of I. At $T = 5K$, the noise gain $g_n = S_I/4eI$ varies as I^γ with $\gamma < 1$ in the region of tunneling. This behavior is consistent with current fluctuations driven by the electric field, related to a transport process ruled by tunneling rather than by emission-capture of charge carriers from the Quantum Dots. Since $\gamma < 1$, a noise suppression mechanism of coulombian origin, determines the overall behavior of the average current and its fluctuations.

At $T = 5K$, a complicated behavior is exhibited by the I-V and the noise power spectra in the presence of irradiation for some of nominally identical QDIP structures. Some devices indeed exhibit (i) very high photoresponse and (ii) $1/f$ noise at low frequencies in the presence of infrared radiation. This behavior might be related to the inhomogeneous dopant distribution in the interdot region.

EXPERIMENTAL DETAILS

The devices are made of quantum dot separated by barrier layers, produced by the Stranski-Krastanov technique. The dots are self-assembled and the barriers are thick in order to suppress the dark current between adjacent layers. The dot size and shape determine the electronic shell structure of the bound states. The InAs dots on GaAs substrate are disk-shaped with diameter of about $18nm$ and thickness of about $2.5nm$. This implies a stronger confinement in the growth direction compared to the in-plane one, leading to several energy levels in each dot leading to a broader infrared response compared to QWIPs [1, 2]. Their intersublevel energies are suitable for long wavelength excitations. All layers have been grown on a semi-insulating GaAs substrate: an undoped $300nm$ GaAs buffer layer, a $760nm$ n^+ GaAs bottom contact layer, a $5nm$ GaAs spacer layer, 50 repeats of self-assembled InAs QD layers separated by $30nm$ GaAs barriers and a $400nm$ n^+ GaAs top contact layer. The doping level for QD layers is $1.5 \times 10^{10} cm^{-2}$. The average QD electron density was estimated as 5 electrons per dot. Top and bottom contacts were Silicon doped to $1.2 \times 10^{18} cm^{-3}$, covered with Ni/Ge/Au and annealed. Wet chemical etching was used to define the geometry of the mesa devices with a section area of $240 \times 240 \mu m^2$.

Photoresponse and current noise measurements have been performed in an in-vacuum steady-bath cryostat to reduce all kind of mechanical disturbances, due to boiling and convective motions of cryogenic liquids. The cryostat uses a double vessel and a double thermal shield on the cold plate area. A black-body source (i.e. a white-hot filament) mounted inside the vacuum chamber was used rather than LASER or LED affected by additional noise. A parabolic reflector and baffles served to spot the light on the QDIPs. Thus devices could be exposed directly to the IR source in high vacuum conditions ($10^{-7} mbar$). Appropriate sample-holders properly shone the devices on their 45° polished facet with polarized radiation.

The basic scheme of voltage noise measurements is based on a balanced circuit at the input of a Stanford Research 560 low noise amplifier. The bias was supplied by a low pass filtered dry cell pack. The noise power spectra were obtained as 50 averages of single power spectral densities by a dynamic spectrum analyzer, Hewlett-Packard 3562A via a GPIB interface. The QDIP differential resistance was calculated from the I-V curves measured by a source-measure unit Keithley 236 [4, 5].

RESULTS AND DISCUSSION

The I-V characteristics at $5K$ both in dark (dotted line) and under radiation (solid line) are shown in figure 1 (a) and (b) for two nominally identical devices. A different shape

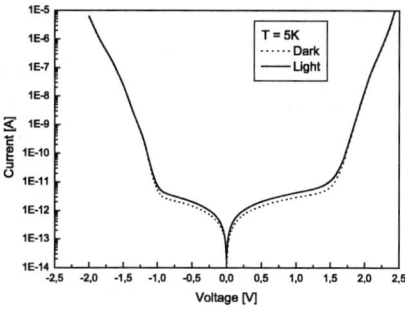 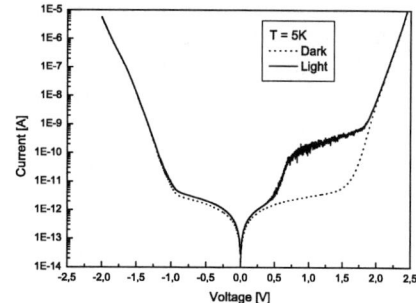

FIGURE 1. Typical behavior of dark (dotted) and photo- (solid) I-V characteristics of InAs/GaAs QDIPs at $5K$.

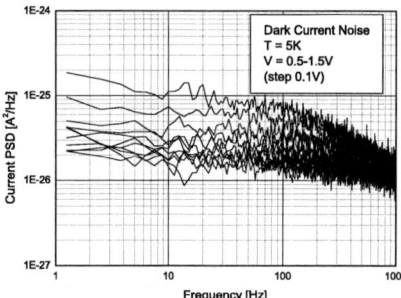 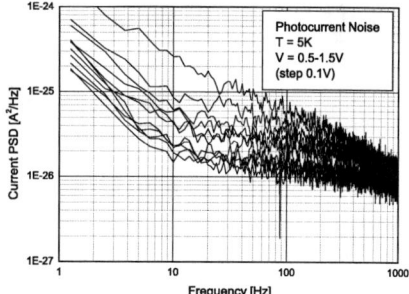

FIGURE 2. Dark (a) and photocurrent (b) noise spectra at $T = 5K$ for the same QDIP of figure 1b. The voltage varies from $0.5V$ to $1.5V$ from top to bottom with step $0.1V$. The roll-off above $100Hz$ is due to the capacitances at the amplifier input (about $25pF$) and along the coaxial cables.

of the photoresponse is observed by comparing the two plots, as already reported by Liu [2].

Current noise power spectral densities are plotted in figure 2 (a) and (b) for dark and irradiated conditions for the sample of figure 1b. The voltage varies from $0.5V$ to $1.5V$ from top to bottom with step $0.1V$. These voltage values corresponding to the tunneling region. The roll-off above $100Hz$ is due to the unavoidable capacitive coupling at the amplifier input (about $25pF$) and along the coaxial cables with the resistance values into play. The I-V characteristics and the current noise spectra measured at $77K$, not shown here, confirm a transport mechanisms where capture-emission processes dominate the average current and its fluctuation. The spectrum amplitudes are proportional to the square of the average current, in agreement with a capture-emission mechanism of the fluctuation process, originated by thermally excited charge carriers, as already observed in QDIPs grown by a different technology in [6, 7].

Current noise power spectral densities are plotted in figure 2 (a) and (b) for dark and irradiated conditions at $5K$ for the sample of figure 1(b). The voltage varies from $0.5V$ to $1.5V$ from top to bottom with step $0.1V$. Considering that the current increases of a factor 5 over the same voltage range, the noise gain g_n decreases with I, as opposed to the behavior observed at $77K$, when the fluctuations are mainly related to thermally excited capture-emission processes. Noise spectra are white in the investigated frequency range, i.e. Lorentzian.

As mentioned above, some samples show a high photogain as can be deduced from the I-V characteristics (figure 1(b)). These detectors exhibit a very high peak of responsivity, $\mathcal{R}_i = I_\phi/(h\nu\phi)$, in the tunneling region of the I-V characteristic [2]. While current noise spectra are white in dark conditions and for samples with very low responsivity, a $1/f$ noise component is observed on the devices with high optical gain. The g_n dependence on the inverse of the average current indicates that the noise suppression mechanism acts in the presence of radiation as well. The strong asymmetry of the I-V plots suggests that the effect should be the result of the combined interactions of the self-assembled quantum dot layer, the wetting layer and the center delta doping as a function of the applied bias and incident radiation.

CONCLUSIONS

Current noise measurements, both in thermal equilibrium and with photo-excitation, have been performed on InAs/GaAs self-assembled QDIPs. The results indicate that a noise suppression effect occurs when the tunneling regime dominates over the thermally emission-capture processes. The photocurrent noise spectral densities are Lorentzian for low optical gain devices. A $1/f$ component is observed for high optical gain detectors under irradiation in the tunneling regime. Remarkably, the noise suppression mechanism is even more effective when the photocurrent flows in the QDs confirming its Coulombian origin.

REFERENCES

1. H. C. Liu, M. Gao, J. McCaffrey, Z. R. Wasilewski and S. Fafard: *Appl. Phys. Letters*, **78**, 79–81 (2000).
2. H. C. Liu, B. Aslan, M. Korusinski, S.J. Cheng and P. Hawrylak: *Inf. Phys. Tech.*, **44**, 503–508 (2003).
3. J.-Y. Duboz, H. C. Liu, Z. R. Wasilewski, M. Byloss and R. Dudek: *J. Appl. Phys.*, **93**, 1320-1322 (2003).
4. A. Carbone, R. Introzzi and H.C. Liu: *App. Phys. Lett.*, **82**, 4292–4294 (2003).
5. A. Carbone, R. Introzzi and H.C. Liu: *Inf. Phys. Tech.*, **46**, (2005).
6. Z. Ye, J. Campbell, Z. Chen, E.T. Kim and A. Madhukar: *Appl. Phys. Lett.*, **83**, 1234–1236 (2003).
7. N. A. Hastas, C. A. Dimitriadis, L. Dozsa, E. Gombia and R. Mosca: *J. Appl. Phys.*, **93**, 5833-5835 (2003).

MESOSCOPICS

Shot Noise in Mesoscopic Conductors: From Schottky to Bell

M. Büttiker*, P. Samuelsson[†] and E. V. Sukhorukov*

Department of Theoretical Physics, University of Geneva, CH-1211 Geneva 4, Switzerland
[†]*Departmenet of Solid State Theory, Lund University, S-233 62 Lund, Sweden*

Abstract. Mesoscopic shot noise is not only probabilistic: it has features which reflect quantum mechanical entanglement. We discuss recent proposals of orbital entanglement generation and detection in mesoscopic coherent conductors. Orbital entanglement avoids the difficulty of manipulating spin and leads to simpler structures. Orbital entanglement schemes invoke two two-particle sources. The index of the source plays the role of a pseudo-spin. The rotation of qubits can be implemented with beam-splitters or even just quantum point contacts. Entanglement is detected via violation of a Bell inequality. The necessary correlations can be extracted from shot noise measurements.

Keywords: Shot noise, two particle Aharonov-Bohm effect, orbital entanglement
PACS: 73.23.-b,05.40.-a,72.70.+m,74.40.+k

INTRODUCTION

For the last two decades the investigation of shot noise in small electrical conductors has been an important frontier in mesoscopic conductors [1]. Shot noise has been a tool to investigate properties of electrical conductors which are not accessible through conductance measurements. Shot noise, investigated by Schottky [2], almost a hundred years ago, is due to thermionic emission from the high-energy Boltzmann-like tail of the Fermi distribution. Schottky's noise is *classical*, with a noise power proportional to the average current,

$$\langle (\Delta I)^2 \rangle_v = 2e|\langle I \rangle| \qquad (1)$$

In contrast, shot noise in mesoscopic conductors is a quantum mechanical phenomenon, a manifestation of the wave-particle duality. An electron incident on a scatterer can be either reflected or transmitted. The final state, however, is in both cases a carrier with a quantized charge, an electron. One the one hand, due to the probabilistic nature of quantum mechanical scattering, the same initial state (the electron approaching the barrier) can have several final states (reflection with amplitude r or transmission with amplitude t). This gives rise to a quantum *partition* noise [3, 4, 5] proportional to $T(1-T)$ where $T = |t|^2$ is the transmission probability.

The scattering approach to electrical transport relates the shot noise to the scattering matrices, $s_{RL} \equiv t$, $s_{RL} \equiv t$, $s_{LR} \equiv t'$, $s_{LL} \equiv r$, and $s_{RR} \equiv r'$ where the indices L,R denote the incident and final state. The shot-noise power is [5]

$$S = 2e\frac{e^2}{h}|eV|Tr(r^\dagger r t^\dagger t) = 2e\frac{e^2}{h}|eV|\sum_n T_n(1-T_n) \qquad (2)$$

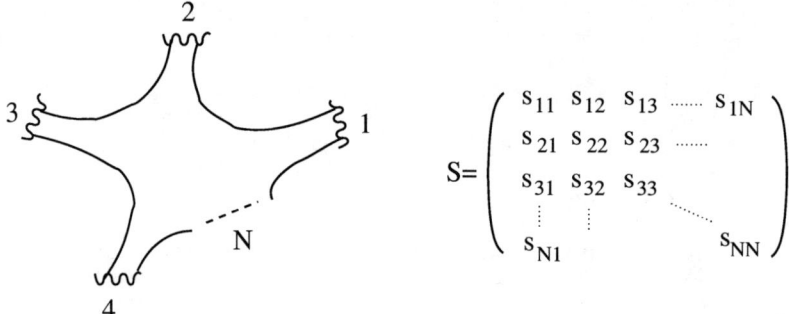

FIGURE 1. Multi-probe conductor and its scattering matrix.

Here the trace Tr is over all transverse modes. The matrix $t^\dagger t$ is hermitian and can be diagonalized with eigenvalues T_n. The shot noise is again a function only of transmission probabilities. The important point which we wish to make, is that a description of shot noise based on transmission probabilities alone, is in general *not possible*.

To highlight this, we now consider a multi-probe conductor (see Fig. 1). The scattering matrix of the conductor is composed of the matrices $s_{\alpha\beta}$ which describe the transmission/reflection amplitudes of carriers incident in contact β and transmitted/reflected into contact α. where N_α is the number of quantum channels in contact α. We are interested in correlations of fluctuating currents with currents measured at different contacts $\alpha \neq \beta$, $S_{\alpha\beta} = 2\int dt \langle \Delta\hat{I}_\alpha(t) \Delta\hat{I}_\beta(0) \rangle$. In the zero-temperature limit, we find [5, 6],

$$S_{\alpha\beta} = 2\frac{e^2}{h} \int dE \operatorname{Tr}\left[B^\dagger_{\alpha\beta} B_{\beta\alpha}\right], \quad B_{\alpha\beta} = \sum_{\gamma=1}^{M} s_{\alpha\gamma} s^\dagger_{\beta\gamma}(f_\gamma - f_0) \quad (3)$$

Here f_γ is the Fermi distribution function of a contact at voltage V_γ and f_0 is the Fermi function of a contact that is grounded. Now if there is only one contact above ground $M = 1$ the correlation function can again be expressed in terms of transmission probabilities only. But if current is incident from two or more contacts, $M \geq 2$, the correlation function contains now products of four scattering matrices in which none of the scattering matrices is the hermitian conjugate of the other. This implies that the correlation function depends on *phases of the scattering matrix elements*. The appearance of phases of scattering matrix elements, emphasized already in Refs. [6], is now central for what follows.

We will show that such phases are connected to the fact that shot noise is a probe of two-particles processes. To illustrate this, we demonstrate first a novel type of Aharonov-Bohm effect [6, 7, 8], which exists only due to two particle scattering processes. These phases are a manifestation of quantum non-locality and subsequently we link them to orbital [9] entanglement.

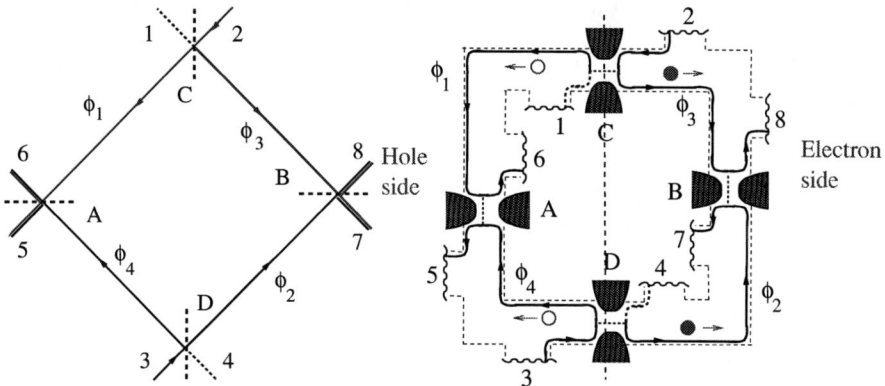

FIGURE 2. Left: Optical intensity interferometer: broken lines represent half-silvered mirrors. Photons incident from 2 and 3 end up in 5,6,7,8. There are no closed path and thus no single photon interference effects. Right: Electrical intensity interferometer: A Corbino disk with four quantum point contacts. Electrons incident from 2 or 3 end in contacts 5,6,7,8. There are no closed path and thus no single electron-interference effects. An Aharonov-Bohm flux penetrates the hole of the disk. After Ref. [7].

TWO-PARTICLE AHARONOV-BOHM EFFECT

To high-light the role of phases of scattering matrix elements we search for a geometry in which the single-particle Aharonov-Bohm (AB) effect is absent and only the two-particle Aharonov-Bohm exists. This implies that none of the conductance matrix elements of such a multi-probe conductor depends on the AB flux, but shot noise correlations are oscillatory functions of flux. In optics, geometries in which time averaged intensities are feature-less but correlation functions exhibit an interference pattern have long been known. Such geometries are known as *intensity* interferometers [10, 11], in contrast to the better known amplitude interferometers, like the Mach-Zehnder or Michelson interferometer. The prime example is the stellar Hanbury Brown Twiss interferometer which was used to measure the angular diameter of stars. In fact the Hanbury Brown Twiss experiments set the stage for the development of quantum optics. A table top version of an optical intensity interferometer [11] is shown in Fig. 2. Photons are injected from 2 and 3. Contacts 5,6 and 7,8 are used to measure intensity correlations. The broken lines represent half-silvered mirrors. Each photon-path is singly connected, there are no single particle interfering alternatives. The time-averaged intensities are not sensitive to the phases ϕ_i accumulated in the traversal from one mirror to the other. However, the intensity correlations do depend on these phases. A topologically equivalent conductor[7], is shown in Fig. 2. It is a Corbino disk [12] with four contacts on the outer perimeter and four contacts on the inner perimeter (not all these contacts are necessary for the demonstration of the two-particle AB effect). The role of mirrors is played by quantum point contacts (QPC) with transmission and reflection probabilities T_i and R_i, $i = A, B, C, D$. The arrows along the sample boundaries are the edge states of a two-dimensional electron gas in the $v = 1$ integral quantum Hall state. An AB flux penetrates the hole of the Corbino disk. Along a path, say along the outer edge from C

to A electrons accumulate a kinetic phase ϕ_1 and due to the AB-flux a phase χ_1. We have $\chi_1 + \chi_3 - \chi_2 - \chi_4 = 2\pi\Phi/\Phi_0$ where $\Phi_0 = h/e$ is the single electron flux quantum. The scattering matrix elements of this conductor contain phases only in the form of a simple multiplicative phase factor. For instance the scattering matrix element for transmission from contact 2 to 5 is, $s_{52} = \sqrt{T_A} e^{i(\phi_1 + \chi_1)} \sqrt{T_C}$. The conductance from contact 2 to 5 is $G_{52} = dI_5/dV_2 = -(e^2/h) T_A T_C$. All other elements of the conductance matrix are similarly independent of the AB-flux. However, if we now evaluate the cross-correlations of currents, the noise-spectra do depend on flux. For example for $T_A = T_B = T_C = T_D = 1/2$ we find [7],

$$S_{58} = -\frac{e^2}{4h}|eV|\left[1 + \cos\left(\phi_1 + \phi_2 - \phi_3 - \phi_4 + 2\pi\frac{\Phi}{\Phi_0}\right)\right] \quad (4)$$

which exhibits a simple oscillation due to the AB-flux. The flux dependence in the current cross-correlation is a consequence of the fact that the correlation probes two-particle processes [7]. Specifically in our interferometer, the contribution to the correlation come from processes in which a particle is emitted from contact 2 towards 5 and from contact 3 towards 8. This process is indistinguishable from the emission of a particle from 2 towards 8 and from 3 towards 5. The amplitudes of these two two-particle processes have to be added. But the two two-particle processes now do encircle the AB-flux!!

ELECTRON-HOLE ENTANGLEMENT IN LOW FLUX LIMIT

We consider the limit of a very asymmetric (tunnel limit) interferometer. We take the reflection probability R_C at C for carriers incident from 2 and the transmission probability T_D for carriers incident form 3 to small, $R_C = T_D = R \ll 1$. As a consequence electrons incident from 2 are very rarely reflected towards B and the missing electrons on the left side of QPC C can be viewed as a hole (see Fig. 2). Similarly electrons from 3 are only very rarely transmitted and through QPC D and the missing electron in the reflected stream can again be viewed as a hole. Thus each reflection event at C and each transmission event at D creates an electron-hole pair. This is similar to the proposal of Beenakker et al. [13] except that here the two sources are spatially separated and thus individually controllable. On the left side of our interferometer excitations are holes in a ground state with Fermi energy $\mu = eV$ and excitations on the right side are electrons in a ground state with Fermi energy 0. (We take the equilibrium Fermi energy to be at $E = 0$). The incident state is a many particle electron state with energies in the range $0 < E < eV$,

$$|\Psi_{in}\rangle = \prod_{0 < E < eV} c_2^\dagger(E) c_3^\dagger(E) |0\rangle \quad (5)$$

We next introduce creation (and annihilation) operators in the out-going regions of the QPC's C and D,

$$c_2^\dagger = t_C c_{2A}^\dagger + r_C c_{2B}^\dagger; \quad c_3^\dagger = r_D c_{3A}^\dagger + t_D c_{3B}^\dagger \quad (6)$$

In terms of these operators we obtain for the out-going state to lowest order in the reflection probability [7],

$$|\Psi\rangle = |\bar{0}\rangle + \sqrt{R} \int_0^{eV} dE \left[c_{3B}^\dagger c_{3A} - c_{2B}^\dagger c_{2A} \right] |\bar{0}\rangle \tag{7}$$

This is an *orbitally* entangled electron-hole state. It is important that here $|\bar{0}\rangle$ is the equilibrium state mentioned above with the voltage drop eV across the QPC's C and D incorporated. It is the introduction of this tunneling ground state [7, 9, 14] which permits to describe the excitations as electron-hole pairs.

BELL INEQUALITY

To analyze the state, we send it onto analyzers A and B which can rotate the state through an angle Θ_A and Θ_B. One reason that orbital entanglement schemes are simpler than spin entanglement is due to the fact that simple beam-splitters rotate orbitally entangled states. As above we consider the tunnel limit and set $\phi_1 + \phi_2 - \phi_3 - \phi_4 + 2\pi\Phi/\Phi_0 = 2\pi$. In the orbital entangler set-up the two detector QPC's A and B "rotate" the qubit. The scattering matrices of these two QPC's are

$$S_{A/B} = \begin{pmatrix} \cos\theta_{A/B} & -\sin\theta_{A/B} \\ \sin\theta_{A/B} & \cos\theta_{A/B} \end{pmatrix}. \tag{8}$$

We now demonstrate that the state Eq. (7) violates a Bell inequality. Originally developed by Bell [15, 16] to separate quantum mechanical non-locality from classical local states, here, the purpose is not to test fundamental aspects of quantum mechanics but simply to use the inequality as a test for entanglement [9, 17, 18, 19]. It is not the only possible criteria but the Bell correlations which enter the inequality are closely related to intensity correlations. We will now show that the electron intensity interferometer [7] in fact contains already all these elements. A Bell test [16] is based on the "equal-time" correlations

$$E(\theta_A, \theta_B) = \frac{\langle (I_{A+} - I_{A-})(I_{B+} - I_{B-}) \rangle}{\langle (I_{A+} + I_{A-})(I_{B+} + I_{B-}) \rangle} \tag{9}$$

where $I_{i\sigma}$ are intensities (currents) at contacts $i\sigma$, $i = A, B$, $\sigma = \pm$. A classical system has

$$S_B = |E(\theta_A, \theta_B) - E(\theta_A', \theta_B) + E(\theta_A, \theta_B') + E(\theta_A', \theta_B')| \leq 2. \tag{10}$$

On the other hand, if there exist angles $\theta_A, \theta_B, \theta_A', \theta_B'$ such that S_B exceeds 2 we know the state is entangled. With these specifications we find for the shot noise current correlations

$$S_{58} = S_{67} = -S_0 P_{++}; \quad S_{57} = S_{68} = -S_0 P_{+-} \tag{11}$$

$$P_{\sigma,\sigma'} = (1 + \sigma\sigma' \cos[2(\theta_A - \theta_B)])/4 \tag{12}$$

where $S_0 = -(4e^2/h)|eV|R$ and $\sigma = \pm, \sigma' = \pm$. Here P has just the form that one would expect in a Bell test of a spin entangled state. Eq. (12) permits to obtain the Bell correlation, $E(\theta_A, \theta_B) = P_{++} + P_{--} - P_{+-} - P_{-+}$. Note, that the Bell test requires an equal-time correlation, whereas the quantity that is presently accessible in a shot noise measurements is the zero-frequency noise. However, in the low flux limit considered here, the cross-correlation is just an equal time measurement run over a long time [9]. Only co-incident current pulses will contribute to the correlation. Such correlated current pulses must necessarily come from the electron-hole pair that was generated in a correlated event. Subsequent pairs are generated with a typical time interval $\tau \sim \hbar/eVR$ (remember $R \ll 1$) which is long compared to a correlation time of an electron-hole pair $\tau \sim \hbar/eV$.

Alternatively we can calculate the joint detection probability Eq. (12) directly using the state Eq. (6). The tunnel limit represents thus a very transparent situation. Since the electron-hole pairs are well separated in time, there is in this limit also the possibility to dynamically manipulate the entangled state during its transfer to the detector. Therefore, the entanglement we have in the tunnel limit is likely "'useful".

Entanglement is not restricted to the tunnel limit considered above but can, like the two-particle AB effect, be found for arbitrary transparency. We conclude by mentioning that the purely stochastic entanglement considered here can be "regulated" by considering the generation of electron-hole pairs through oscillating potentials [20, 21, 22].

ACKNOWLEDGMENTS

This work is supported by the Swiss NSF and the network MaNEP.

REFERENCES

1. Ya. M. Blanter and M. Büttiker, Phys. Rep. **336**, 1-166, (2000).
2. W. Schottky, Ann. Phys. (Leipzig) **57**, 541 -567 (1918).
3. V. A. Khlus, Sov. Phys. JETP **66**, 1243 -1249 (1987).
4. G. B. Lesovik, JETP Lett. **49**, 592 - 594 (1989).
5. M. Büttiker, Phys. Rev. Lett. **65**, 2901 - 2904 (1990).
6. M. Büttiker, Physica B **175**, 199 - 212, (1991).
7. P. Samuelsson, E. V. Sukhorukov, and M. Büttiker, Phys. Rev. Lett. **92**, 026805 (2004).
8. For a wider context, see M. Büttiker, P. Samuelsson and E.V. Sukhorukov, Physica E**20**, 33 - 42 (2003); for a two-particle AB-effect in cotunneling see D. Loss and E. V. Sukhorukov, Phys. Rev. Lett. **84**, 1035 (2000).
9. P. Samuelsson, E.V. Sukhorukov, and M. Büttiker, Phys. Rev. Lett. **91**, 157002 (2003).
10. R. Hanbury Brown and R. Q. Twiss, Nature **178**, 1046 - 1048, (1956).
11. B. Yurke and D. Stoler, Phys. Rev. A **46**, 2229 - 2234, (1992).
12. A similar geometry but for Mach-Zehnder interference has recently been realized, Y. Ji, Y. Chung, D. Sprinzak, M. Heiblum, D. Mahalu, H. Shtrikman, Nature **422**, 415 -418, (2003).
13. C.W.J. Beenakker, et al. Phys. Rev. Lett. **91**, 147901 (2003).
14. P. Samuelsson, E. V. Sukhorukov, and M. Büttiker, (unpublished). cond-mat/0503016
15. J. Bell, Physics **1**, 195 - 200 (1964).
16. J.F. Clauser, M. A. Horne, A. Shimony, and R. A. Holt, Phys. Rev. Lett. **23**, 880 -884, (1969).
17. X. Maître, W. D. Oliver and Y. Yamamoto, ", Physica E **6**, 301 -305, (2000).
18. N.M. Chtchelkatchev, G. B. Lesovik, G. Blatter and T. Martin, Phys. Rev. B **66**, 161320 (2002).
19. A. V. Lebedev, G. B. Lesovik, and G. Blatter, Phys. Rev. B **71**, 045306 (2005).
20. P. Samuelsson and M. Büttiker, Phys. Rev. B (unpublished). cond-mat/0410581
21. C. W. J. Beenakker, M. Titov and B. Trauzettel, cond-mat/0502055
22. A. V. Lebedev, G. B. Lesovik, G. Blatter, cond-mat/0504583

Entanglement in a Noninteracting Mesoscopic Structure

A.V. Lebedev*, G.B. Lesovik* and G. Blatter[†]

L.D. Landau Institute for Theoretical Physics RAS, 117940 Moscow, Russia
[†]*Theoretische Physik, ETH-Hönggerberg, CH-8093 Zürich, Switzerland*

Abstract. We study time dependent electron-electron and electron-hole correlations in a mesoscopic device splitting an incident current of free fermions into two spacially separated particle streams. We analyze the appearance of entanglement as manifested through a Bell inequality test and identify its origin as the Fermi statistics forcing the reservoir to inject spin-singlet pairs. The time window over which the Bell inequality is violated is determined in the tunneling limit and for the general situation with arbitrary transparencies.

Keywords: electronic entanglement, Bell Inequality, shot noise
PACS: 72.70.+m, 03.65.Ud, 03.67.Mn

Quantum entanglement of electronic degrees of freedom in mesoscopic devices has attracted a lot of interest recently. Specific structures have been proposed which generate spacially separated streams of entangled particles. In most cases, the source of entanglement has been identified and measures have been taken to separate entangled pairs in space. One class of devices makes use of a superconducting source emitting Cooper pairs into a normal-metal structure with two leads in a fork geometry: entanglement has its origin in the attractive interaction binding the electrons into Cooper pairs, while the spacial separation of correlated electrons is arranged for by suitable 'filters' [1, 2]. Another class of devices makes use of Coulomb interactions in confined geometries [3]. Besides these proposals for the generation of spatially separated entangled pairs, the implementation of Bell (type) inequality tests probing their entanglement has been discussed in detail [4, 2]. The combination of sources for the creation and methods for testing the correlations of entangled particle streams are first steps towards establishing this quantum resource for solid state based quantum information technology.

An interesting proposal has been made by Beenakker *et al.* [5]: using a two-channel quantum Hall device with a beam splitter they suggest a setup generating two streams of entangled electron-hole pairs and confirm the presence of correlations through a Bell inequality (BI) test. A crucial difference to previous proposals is the absence of interactions generating the entanglement (see also [6]). Here, we investigate a similar setup involving a mesoscopic normal-metal structure arranged in a fork geometry, see Fig. 1, where a constant voltage V is applied to the source lead 's', generating two streams of correlated electrons in the two arms 'u' and 'd' of the fork.

The Bell type setup [7] in Fig. 1(a) measures the current cross-correlators $C^{ee}_{ij}(\mathbf{a},\mathbf{b})$ between the spin-currents $i = 1,3$ (in lead 'u') projected onto the directions $\pm\mathbf{a}$ and their partners $j = 2,4$ (in lead 'd') projected onto $\pm\mathbf{b}$. These correlators enter the Bell

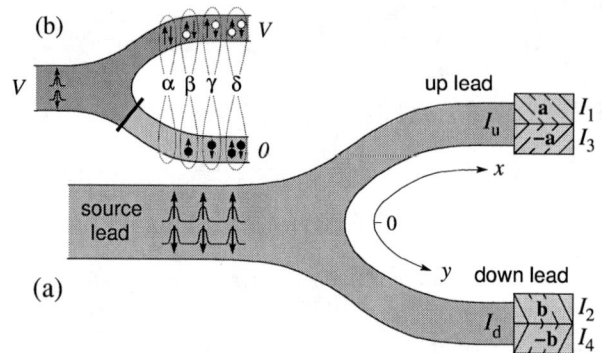

FIGURE 1. (a) Mesoscopic normal-metal structure in a fork geometry generating two streams of spin-correlated electrons in the two arms of the fork. The Bell type setup detects spin-currents I_i, $i = 1,3$, projected onto the directions $\pm \mathbf{a}$ in the upper arm and correlates them with spin-currents I_j, $j = 2,4$, projected onto the directions $\pm \mathbf{b}$ in the lower arm. (b) In the tunneling limit (with a small transparency $T_d \ll 1$ in the 'down' arm) a fraction of spin-entangled electrons is split into two the arms (components β and γ); their correlations can be efficiently measured in a Bell type setup sensitive to hole (electron) currents in the 'up' (down) lead. The component δ accounts for the few processes where both electrons propagate to the 'down' lead; its contribution spoils the maximum violation of the BI and restricts the use of the tunneling limit.

inequality ($\bar{\mathbf{a}}$ and $\bar{\mathbf{b}}$ denote a second set of directions)

$$|E(\mathbf{a},\mathbf{b}) - E(\mathbf{a},\bar{\mathbf{b}}) + E(\bar{\mathbf{a}},\mathbf{b}) + E(\bar{\mathbf{a}},\bar{\mathbf{b}})| \leq 2 \qquad (1)$$

via the current difference correlators

$$E(\mathbf{a},\mathbf{b}) = \frac{\langle [\hat{I}_1^e(\tau) - \hat{I}_3^e(\tau)][\hat{I}_2^e(0) - \hat{I}_4^e(0)]\rangle}{\langle [\hat{I}_1^e(\tau) + \hat{I}_3^e(\tau)][\hat{I}_2^e(0) + \hat{I}_4^e(0)]\rangle} = \frac{C_{12}^{ee} - C_{14}^{ee} - C_{32}^{ee} + C_{34}^{ee} + \Lambda_-^{ee}}{C_{12}^{ee} + C_{14}^{ee} + C_{32}^{ee} + C_{34}^{ee} + \Lambda_+^{ee}}, \qquad (2)$$

with $\Lambda_\pm^{ee} = [\langle \hat{I}_1^e \rangle \pm \langle \hat{I}_3^e \rangle][\langle \hat{I}_2^e \rangle \pm \langle \hat{I}_4^e \rangle]$. The average currents are related via $\langle \hat{I}_1^e \rangle = \langle \hat{I}_3^e \rangle = \langle \hat{I}_u^e \rangle/2$ and $\langle \hat{I}_2^e \rangle = \langle \hat{I}_4^e \rangle = \langle \hat{I}_d^e \rangle/2$ and thus $\Lambda_-^{ee} = 0$, $\Lambda_+^{ee} = \langle \hat{I}_u \rangle \langle \hat{I}_d \rangle$. In the tunneling limit ($T_u \sim 1$ and $T_d \ll 1$, T_u, T_d describe the transmission probabilities into the 'up' and 'down' lead from the source lead 's'), the electronic currents $\hat{I}_i^e$ in the 'up' lead are replaced by hole currents $\hat{I}_i^h$.

First we determine the finite time current cross-correlator between leads 'u' and 'd' for a stream of spinless fermions. Using the standard scattering theory of noise [8] one could split the current cross correlator $C_{x,y}^{ee}(\tau)$ into an equilibrium component $C_{x,y}^{eq}(\tau; V=0)$ and an excess part $C_{x,y}^{ex}(\tau; V)$

$$C_{x,y}^{eq}(\tau) = \frac{e^2 T_{du}}{h^2}[\alpha(\tau+\tau_+,\theta) + \alpha(\tau-\tau_+,\theta)], \qquad (3)$$

$$C_{x,y}^{ex}(\tau) = -\frac{4e^2 T_u T_d}{h^2} \sin^2 \frac{eV(\tau-\tau_-)}{2\hbar} \alpha(\tau-\tau_-,\theta), \qquad (4)$$

with $\alpha(\tau,\theta) = \pi^2 \theta^2/\sinh^2[\pi\theta\tau/\hbar]$, $\tau^\pm = (x \pm y)/v_F$, and θ the temperature of the electronic reservoirs, T_{ud} describes transmission probability from the lead 'u' to the

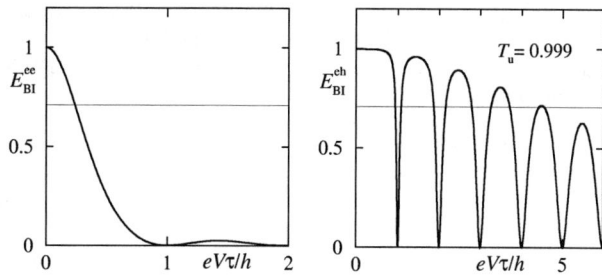

FIGURE 2. Bell inequality test for electron-electron (left) and electron-hole (right) currents. The thin line at $1/\sqrt{2}$ marks the critical value above which the Bell inequality is violated.

lead 'd'. The correlators $C^{ee}_{ij}(\mathbf{a},\mathbf{b})$ relate to the result (3,4) for spinless particles via $C^{ee}_{ij}(\mathbf{a},\mathbf{b};\tau) = |\langle \mathbf{a}_i | \mathbf{b}_j \rangle|^2 C^{ee}_{x,y}(\tau)$ with $\mathbf{a}_{1,3} = \pm \mathbf{a}$ and $\mathbf{b}_{2,4} = \pm \mathbf{b}$. The spin-projections derive from the angle $\theta_{\mathbf{ab}}$ between the directions $\mathbf{a}$ and $\mathbf{b}$ via $\langle \pm \mathbf{a} | \pm \mathbf{b} \rangle = \cos^2(\theta_{\mathbf{ab}}/2)$ and $\langle \pm \mathbf{a} | \mp \mathbf{b} \rangle = \sin^2(\theta_{\mathbf{ab}}/2)$ and the BI (1) assumes the form

$$\left| \frac{C^{ee}_{x,y}(\tau)[\cos\theta_{\mathbf{ab}} - \cos\theta_{\mathbf{a\bar{b}}} + \cos\theta_{\mathbf{\bar{a}b}} + \cos\theta_{\mathbf{\bar{a}\bar{b}}}]}{2C^{ee}_{x,y}(\tau) + \langle \hat{I}^e_u \rangle \langle \hat{I}^e_d \rangle} \right| \leq 1. \quad (5)$$

Its maximum violation is obtained for the standard orientations of the detector polarizations, $\theta_{\mathbf{ab}} = \theta_{\mathbf{\bar{a}b}} = \theta_{\mathbf{\bar{a}\bar{b}}} = \pi/4$, $\theta_{\mathbf{a\bar{b}}} = 3\pi/4$ and the BI reduces to

$$E^{ee}_{BI} \equiv \left| \frac{2C^{ee}_{x,y}(\tau)}{2C^{ee}_{x,y}(\tau) + \langle \hat{I}^e_u \rangle \langle \hat{I}^e_d \rangle} \right| \leq \frac{1}{\sqrt{2}}. \quad (6)$$

In the limit of low temperatures $0 < eV$ and for large distances $x = y \gg \tau_V v_F$, $\tau_V = \hbar/eV$ (allowing us to neglect the equilibrium part in the correlator $C^{ee}_{x,y}$) the above expression (6) reduces to the particularly simple form

$$\frac{2\sin^2(eV\tau/2\hbar)}{\tau^2(eV/\hbar)^2 - 2\sin^2(eV\tau/2\hbar)} \leq \frac{1}{\sqrt{2}}, \quad (7)$$

where we have used that $\langle \hat{I}^e_u \rangle = (2e/h)T_u eV$ and $\langle \hat{I}^e_d \rangle = (2e/h)T_d eV$. We observe that in this limit the Bell inequality is *i)* violated at short times $\tau < \tau_{BI} = \tau_V$, see Fig. 2, *ii)* this violation is independent of the transparencies T_u, T_d and hence universal.

Next, we consider the tunneling limit and determine the outcome of a Bell measurement involving a hole current $\hat{I}^h_u \equiv (2e/h)eV - \hat{I}^e_u$ in the 'up' lead and the electronic current $\hat{I}^e_d$ in the 'down' lead. The calculation proceeds as above but now involves the electron-hole correlator $C^{he}_{x,y}(\tau) \equiv \langle\langle \hat{I}^h_u(\tau,x) \hat{I}^e_d(0,y) \rangle\rangle = -C^{ee}_{x,y}(\tau)$ and the product of the electron and hole currents $\Lambda^{he}_+ = \langle \hat{I}^h_u \rangle \langle \hat{I}^e_d \rangle = (4e^2/h^2)T_d(1-T_u)[eV]^2$ (again, $\Lambda^{he}_- = 0$). The Bell inequality corresponding to (7) now reads

$$E^{eh}_{BI} \equiv \frac{2\sin^2(eV\tau/2\hbar)}{2\sin^2(eV\tau/2\hbar) + T_u^{-1}(1-T_u)\tau^2(eV/\hbar)^2} \leq \frac{1}{\sqrt{2}} \quad (8)$$

and an illustration of this result is given in Fig. 2. The violation of the Bell inequality in the tunneling limit exhibits a much richer structure: *i)* the violation requires a minimum transparency T_u in the upper lead: evaluating E_{BI}^{eh} at $\tau = 0^+$, we obtain the condition $T_u > T_{min} \equiv 2/(\sqrt{2}+1) \approx 0.83$. *ii)* For $T_u > T_{min}$ the Bell inequality is violated during times $\tau < \tau_{BI} - \tau_V \sqrt{2-\sqrt{2}} \sqrt{T_u(1-T_u)^{-1}} \approx \tau_V/\sqrt{T_d}$, where we have assumed $1 - T_u \approx T_d$ as is the case for a splitter with a small back reflection $r_s \ll 1$. *iii)* The BI remains un-violated in narrow intermediate regions separated by the single particle correlation time τ_V and decreases slowly $\propto \tau^{-2}$ with increasing time. The comparison with the electronic result is quite striking: the time interval over which the Bell inequality is violated is extended by a factor $1/\sqrt{T_d} \gg 1$ and the universality (i.e., the independence on the transmissions T_u and T_d) is lost.

The source of entanglement could be found in the reservoir, where the Fermi statistics forces electrons with equal orbits to pair up into singlets which then are injected as a regular stream with time separation τ_V into the source lead. The splitter itself does not contribute to the entanglement of the pair, but redistributes the incoming pairs between outgoing leads. The scattering state describing the propagation of the singlet-pair behind the splitter can be written in the form $|\Psi_{out}^{ee}\rangle = t_{su}^2 |e_{\uparrow\downarrow}^{sg}\rangle_u |0\rangle_d + t_{sd}^2 |0\rangle_u |e_{\uparrow\downarrow}^{sg}\rangle_d + \sqrt{2} t_{su} t_{sd} [|e_\uparrow\rangle_u |e_\downarrow\rangle_d - |e_\downarrow\rangle_u |e_\uparrow\rangle_d]$ where the first two terms describe the propagation of the singlet-pair $|e_{\uparrow\downarrow}^{sg}\rangle_x$ in leads 'x' equal 'u' or 'd', while the last term describes a singlet-pair split between the 'up' and 'down' leads. A coincidence measurement of electrons in leads 'u' and 'd' projects the scattered state $|\Psi_{out}^{ee}\rangle$ onto this spin-entangled component with spatially separated electrons in leads 'u' and 'd'. In the tunneling limit ($T_u \sim 1$ and $T_d \ll 1$) most of the incoming singlet pairs propagate into the well conducting 'up' lead and only rarely split into both leads. The absence of an electron in the 'up' lead then manifests itself as the presence of a hole and it is favorable to go over to a hole representation. The corresponding scattering states of electron-hole pairs are depicted in Fig. 1(b), the most relevant terms β and γ describing the splitting of the singlet electron-hole pair between the two outgoing leads; this electron-hole component is detectable in a coincidence measurement using a hole (particle) detector in the upper (lower) lead.

REFERENCES

1. G. Lesovik, Th. Martin, and G. Blatter, Eur. Phys. J. B **24**, 287 (2001); P. Recher, E.V. Sukhorukov, and D. Loss, Phys. Rev. B **63**, 165314 (2001); C. Bena *et al.*, Phys. Rev. Lett. **89**, 037901 (2002).
2. P. Samuelsson, E.V. Sukhorukov, and M. Büttiker, Phys. Rev. Lett. **91**, 157002 (2003).
3. G. Burkhard, D. Loss, and E.V. Sukhorukov, Phys. Rev. B **61**, 16303 (2000); D. Loss and E.V. Sukhorukov, Phys. Rev. Lett. **84**, 1035 (2000); W.D. Oliver, F. Yamaguchi, and Y. Yamamoto, Phys. Rev. Lett. **88**, 037901 (2002); D.S. Saraga and D. Loss, Phys. Rev. Lett. **90**, 166803 (2003).
4. N.M. Chtchelkatchev *et al.*, Phys. Rev. B **66**, 161320 (2002).
5. C.W.J. Beenakker *et al*, Phys. Rev. Lett. **91**, 147901 (2003) and C.W.J. Beenakker, M. Kindermann, C.M. Marcus, and A. Yacoby, cond-mat/0310199.
6. L. Faoro, F. Taddei, and R. Fazio, cond-mat/0306733, report violation of the Clauser-Horne inequality in a normal three-terminal device.
7. J.F. Clauser, M.A. Horne, A. Shimony, and R.A. Holt, Phys. Rev. Lett. **23**, 880 (1969); A. Aspect, P. Grangier, and G. Roger, Phys. Rev. Lett. **49**, 91 (1982).
8. G.B. Lesovik, Pis'ma Zh. Eksp. Teor. Fiz. **49**, 513 (1989) [JETP Lett. **49**, 592 (1989)] and *ibid* **70**, 209 (1999) [**70**, 208 (1999)]; Y.M. Blanter and M. Büttiker, Phys. Rep. **336**, 1 (2000).

Current Noise Spectrum of Open Charge Qubits

Ramón Aguado*, Neill Lambert† and Tobias Brandes†

*Departamento de Teoría de la Materia Condensada, Instituto de Ciencia de Materiales de Madrid, CSIC, Cantoblanco 28049, Madrid, Spain
†The University of Manchester, School of Physics and Astronomy, P.O. Box 88, Manchester M60 1QD, United Kingdom

Abstract. We study the current noise spectrum of charge qubits under transport conditions (double quantum dots and Cooper pair boxes) in a dissipative bosonic environment. We combine master equations with correlation functions in Laplace-space to derive a noise formula for both weak and strong coupling to the bath.

Keywords: Shot noise, quantum noise, quantum dots, spin-boson model
PACS: 72.70.+m, 03.65.Yz, 73.23.Hk, 73.63.Kv

INTRODUCTION

Electronic current noise has become an important tool for extracting information not available in conventional dc transport experiments. This is in particular true in mesoscopic systems, where energy relaxation and the loss of phase coherence is crucial for the understanding of transport properties. Transport through artificial few-level systems (coupled quantum dots, small superconducting junctions) has received a great deal of attention due to possible applications for quantum information processing [1]. In this contribution, we present a theoretical investigation of current noise in one of the simplest quantum systems, i.e., a quantum two-level system (TLS or qubit) coupled to a dissipative environment and to electron reservoirs [2]. We have developed a formalism in order to calculate the full, frequency dependent charge and current noise spectrum for *arbitrary* coupling to a thermal, dissipative environment.

MODEL

Double quantum dots (DQD) [3, 4] can be tuned into a regime that is governed by an 'open' version of the spin–boson model [5], described by the Hamiltonian $\mathcal{H} = \frac{\varepsilon}{2}\sigma_z + T_c\sigma_x + \frac{1}{2}\sigma_z\sum_\mathbf{Q} g_Q\left(a_{-\mathbf{Q}} + a_\mathbf{Q}^\dagger\right) + \sum_\mathbf{Q} \omega_Q a_\mathbf{Q}^\dagger a_\mathbf{Q}$, where one additional 'transport' electron tunnels between a left (L) and a right (R) dot with energy difference ε and inter–dot coupling T_c, where $\sigma_z = |L\rangle\langle L| - |R\rangle\langle R| \equiv \hat{n}_L - \hat{n}_R$ and $\sigma_x = |L\rangle\langle R| + |R\rangle\langle L| \equiv \hat{p} + \hat{p}^\dagger$. The transport through the DQD is analogous to the Josephson Quasiparticle Cycle in a Cooper pair box [6]. The last two terms in the Hamiltonian appear due to the coupling with the bosonic bath, where ω_Q are the frequencies of bosons, and the g_Q denote interaction constants. The effects of the bath are encapsulated in the spectral density $J(\omega) \equiv \sum_\mathbf{Q}|g_Q|^2\delta(\omega - \omega_Q)$. Although not exactly solvable, the model is quite well understood for closed systems [5]. The coupling to external leads offers the possibility to study its non–equilibrium properties.

Equations of Motion

The dynamics of the open system can be described by a reduced statistical operator $\rho(t)$, allowing for an additional 'empty' state $|0\rangle$ which describes tunneling from a left reservoir at rate Γ_L into the left dot, and from the right dot to the right reservoir at rate. The coupling to the reservoirs within Born and Markov (BM) approximation with respect to $\mathscr{H}_T = \Sigma_{k\alpha\in L,R}(V_k^\alpha c_{k\alpha}^\dagger |0\rangle\langle\alpha| + H.c.)$ yields tunneling rates $\Gamma_\alpha = 2\pi \Sigma_{k\alpha} |V_k^\alpha|^2 \delta(\varepsilon - \varepsilon_{k\alpha})$ (we assume Fermi distributions for the reservoirs $f_L = 1$ and $f_R = 0$; large voltage regime). Then, second order perturbation theory in $\mathscr{H}_T$ becomes exact, and one obtains a set of equations of motion which read in matrix form

$$\langle \mathbf{A}(t)\rangle = \langle \mathbf{A}(0)\rangle + \int_0^t dt' \{M(t-t')\langle \mathbf{A}(t')\rangle + \Gamma\} \qquad (1)$$

with the matrix memory kernel M, the expectation value of the vector $\mathbf{A} \equiv (\hat{n}_L, \hat{n}_R, \hat{p}, \hat{p}^\dagger)$, and $\Gamma \equiv \Gamma_L \mathbf{e}_1$. Eqs. (1) can be solved in Laplace space as

$$\langle \hat{\mathbf{A}}(z)\rangle = [z - z\hat{M}(z)]^{-1}(\langle \mathbf{A}(0)\rangle + \Gamma/z) \qquad (2)$$

and serves as a starting point for the analysis of stationary and non-stationary quantities. The memory kernel has the block structure [2]

$$z\hat{M}(z) = \begin{bmatrix} -\hat{G} & \hat{T}_c \\ \hat{D}_z & \hat{\Sigma}_z \end{bmatrix}, \quad \hat{G} \equiv \begin{pmatrix} \Gamma_L & \Gamma_L \\ 0 & \Gamma_R \end{pmatrix}, \qquad (3)$$

where $\hat{T}_c \equiv -iT_c(1-\sigma_x)$. The blocks $\hat{D}_z$ and $\hat{\Sigma}_z$ are determined by the EOM for the coherences (off–diagonal elements) $\langle \hat{p}\rangle = \langle \hat{p}^\dagger\rangle^*$ and contain the complete information on dephasing of the system. Note that no exact solution of the above model is available and one has to choose between perturbation theory in g_Q (weak coupling, PER), or in T_c in a polaron–transformed frame (strong coupling, POL) [2]. Below, we show results which are obtained using the PER approach.

SPECTRAL DENSITY OF THE CURRENT FLUCTUATIONS

Usually, current noise is described by the power spectral density $\mathscr{S}_I(\omega) \equiv 2\int_{-\infty}^{\infty} d\tau e^{i\omega\tau} \mathscr{S}_I(\tau) = \int_{-\infty}^{\infty} d\tau e^{i\omega\tau} \langle \{\Delta\hat{I}(\tau), \Delta\hat{I}(0)\}\rangle$, with $\Delta\hat{I}(t) \equiv \hat{I}(t) - \langle \hat{I}(t)\rangle$ [7, 8]. The Fano factor $F \equiv \frac{\mathscr{S}_I(0)}{2qI}$, quantifies deviations from the Poissonian noise, $\mathscr{S}_I(0) = 2qI$ (uncorrelated carriers with charge q). To calculate $\mathscr{S}_I(\omega)$, we need to relate the reduced dynamics of the qubit to reservoir operators like the current operator. Note that $S_I(\omega)$ has to be calculated from the autocorrelations of the *total* current I, i.e. particle plus displacement current.[7] Using current conservation together with the Ramo-Shockley theorem, $I = aI_L + bI_R$ (a and b, with $a+b=1$, depend on each junction capacitance [7]), one can express $\mathcal{S}_I(\omega)$ in terms of the spectra of particle currents and the charge noise spectrum $S_Q(\omega)$ [2, 9],

$$S_I(\omega) = aS_{I_L}(\omega) + bS_{I_R}(\omega) - ab\omega^2 S_Q(\omega). \quad (4)$$

Here, $S_Q(\omega) \equiv \lim_{t\to\infty} \int_{-\infty}^{\infty} d\tau e^{i\omega\tau} \langle\{\hat{Q}(t),\hat{Q}(t+\tau)\}\rangle = 2\text{Re}\{\hat{f}(z=i\omega) + \hat{f}(z=-i\omega)\}$, with $\hat{Q} = \hat{n}_L + \hat{n}_R$. $\hat{f}(z)$ is the Laplace transform of $f(\tau) = \sum_{i,j=L,R} \langle\hat{n}_i(t)\hat{n}_j(t+\tau)\rangle$. This is evaluated with the help of the charge correlation functions, $\mathbf{C}_\alpha(\tau) \equiv \langle\hat{n}_\alpha(t)\mathbf{A}(t+\tau)\rangle$. The EOM for $\mathbf{C}_\alpha(\tau)$ can be obtained from the quantum regression theorem [2].

We relate the qubit dynamics with reservoir operators by introducing a counting variable n which represents the number of electrons that have tunneled through the right barrier [10]. This counting statistics scheme can be derived from the EOM of the reduced density operator $\dot{\rho} = L\rho$ by splitting the superoperator $L = L_0 + L_1$ such that L_1 described the 'previous' electron leaving the system. In analogy with the quantum jump approach in quantum optics, one introduces an interaction picture with respect to L_0 and derives a Dyson series in terms of the jump operator L_1 [11]. We define generalized expectation values as $O^{(n)} \equiv \sum_{i=0,L,R} \text{Tr}_{\text{bath}} \langle n,i|\hat{O}\rho(t)|n,i\rangle$. The usual expectation values are recovered as $\langle\hat{O}\rangle = \sum_n O^{(n)}$. This method allows to calculate the particle current and the noise spectrum from $P_n(t) = n_0^{(n)}(t) + n_L^{(n)}(t) + n_R^{(n)}(t)$ which gives the total probability of finding n electrons in the collector by time t. In particular, $I_R(t) = e\sum_n n\dot{P}_n(t)$ and $S_{I_R}(\omega) = 2\omega e^2 \int_0^{\infty} dt \sin(\omega t) \frac{d}{dt}[\langle n^2(t)\rangle - (t\langle I\rangle)^2]$ [12]. Here, $\frac{d}{dt}\langle n^2(t)\rangle = \sum_n n^2 \dot{P}_n(t)$. Alternatively, one can solve the EOMs for the generating functions $G_0(s,t) = \sum_{n=0}^{\infty} s^n n_0^{(n)}(t)$, $G_1(s,t) = \sum_{n=0}^{\infty} s^n n_L^{(n)}(t)$, etc. The EOMs in matrix form $\dot{G}(s,t) = M(s)G(s,t)$ can formally be solved by diagonalising $M(s)$. This second route allows to calculate the generating function, noise and cross-correlations of arbitrarily complex quantum systems described by Master equations [11].

RESULTS

For $a = b = 1/2$, the background noise is half the Poisson value as one expects for a symmetric structure. The Fano factor, F, deviates from this value around $\omega = 0$ where the noise has a peak and at the characteristic frequency of the two level system $\omega = \Delta \equiv \sqrt{\varepsilon^2 + 4T_c^2}$ where the noise is suppressed. The dip in the Fano factor directly reflects the resonance of the subtracted charge noise $S_Q(\omega)$ around Δ, cf. Eq. (4). An increase of ε localizes the qubit and, thus, $F \to 1$ [2]. Moreover, the dip in the high frequency noise at $\omega = \Delta$ is progressively destroyed (reduction of quantum coherence) as ε increases (Fig.1, right inset). A similar reduction of the dip at $\omega = \Delta$ occurs at fixed ε and Γ with increasing dissipation in the weak coupling regime (Fig. 1). This behavior demonstrates that $S_I(\omega)$ reveals the complete internal dissipative dynamics of the TLS. In particular, the *dephasing rate* can be extracted from the half-width of $S_I(\omega)$ around $\omega = \Delta$. For an Ohmic environment, $\gamma_d^b = \gamma_p/2 + 2\pi\alpha(\frac{\varepsilon}{\Delta})^2 k_B T$, with $\gamma_p \equiv 2\pi \frac{T_c^2}{\Delta^2} J(\Delta) \coth(\beta\Delta/2)$, such that the total dephasing rate is $\gamma_d(T=0) = \gamma_d^b + \Gamma/2 = (\gamma_p + \Gamma)/2$ (Fig. 1, arrows denote

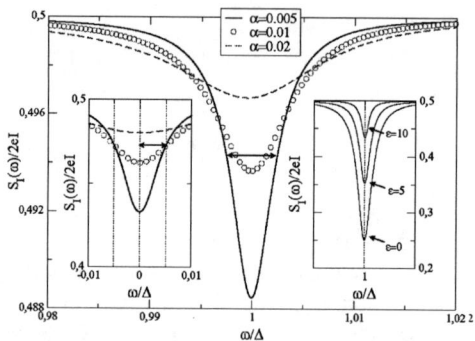

FIGURE 1. Effect of decoherence on current noise near resonance (ohmic dissipation, bias $\varepsilon = 10$, $\Gamma_R = \Gamma_L = \Gamma = 0.01$, and different couplings $\alpha = 0.005, 0.01, 0.02$. Right inset: Current noise ($\alpha = 0$) for increasing ε. Left inset: shot noise limit $\omega \to 0$. Arrows indicate relaxation rate.

full-width, i.e. $2\gamma_d \approx \gamma_p$ as α increases). Close to $\omega = 0$, the peak in $S_I(\omega)$ for $\alpha = 0$ changes into a dip around $\omega = 0$ reflecting incoherent relaxation dynamics for $\alpha \neq 0$. The half-width is now given by the *relaxation rate* such that the full-width of $S_I(\omega)$ around $\omega = 0$ is *twice* that of the high frequency noise (Fig 1, left inset).

ACKNOWLEDGMENTS

This work was supported by EPSRC GR/R44690, DFG BR 1528 and by the MCYT of Spain through the "Ramón y Cajal" program and grant MAT2002-02465 (R. A.).

REFERENCES

1. T. Brandes, *Physics Reports*, **408**, 315-474 (2005).
2. R. Aguado and T. Brandes, *Phys. Rev. Lett.*, **92**, 206601 (2004).
3. T. Fujisawa, T. H. Oosterkamp, W. G. van der Wiel, B. W. Broer, R. Aguado, S. Tarucha, and L. P. Kouwenhoven, *Science*, **282**, 932, (1998).
4. T. Brandes and B. Kramer, *Phys. Rev. Lett.*, **83**, 3021 (1999).
5. A. J. Leggett, S. Chakravarty, A. T. Dorsey, M. P. A. Fisher, A. Garg, and W. Zwerger, *Rev. Mod. Phys.*, **59**, 1 (1987); Ulrich Weiss, *Quantum Dissipative Systems*, volume 2 of *Series of Modern Condensed Matter Physics*, World Scientific, Singapore, (1993).
6. Y. Nakamura, Y. A. Pashkin, and J. S. Tsai, *Nature*, **398**, 786 (1999).
7. Y. M. Blanter and M. Büttiker, *Phys. Rep.*, **336**, 1 (2000).
8. However, quantum noise can be asymmetric in the frequency ω due to the non-commutativity of current operators. It has been recently shown that an asymmetric quantum shot noise spectrum can be detected in situations where a quantum of energy $\hbar\omega$ is transferred from the system to the measurement apparatus. See, R. Aguado and L. P. Kouwenhoven, *Phys. Rev. Lett.*, **84**, 1986 (2000).
9. Note that in symmetric configurations, $a \approx b \approx 0.5$, the charge noise reduces the contribution from particle currents to the noise spectrum in Eq. (4). If, on the other hand, $a \approx 1$ or $b \approx 1$ the main contribution to current noise comes from particle currents.
10. S. A. Gurvitz and Ya. S. Prager, *Phys. Rev. B*, **53**, 15932 (1996).
11. N. Lambert, R. Aguado and T. Brandes, in preparation.
12. D. K. C. MacDonald, *Rep. Progr. Phys.*, **12**, 56 (1948).

Fano factor reduction on the 0.7 structure

P. Roche*, J. Ségala*, D. C. Glattli*,†, J. T. Nicholls**, M. Pepper‡, A. C. Graham‡, K. J. Thomas‡, M. Y. Simmons‡ and D. A. Ritchie‡

*Nanoelectronic group, SPEC, CEA Saclay, F-91191 Gif-sur-Yvette, France
†Laboratoire Pierre Aigrain, ENS 24 rue Lhomond, 75231 Paris cedex 05
**Department of Physics, Royal Holloway, University of London, Egham, Surrey TW20 0EX, UK
‡Cavendish Laboratory, Madingley Road, Cambridge CB3 0HE, UK

Abstract. We have measured the non-equilibrium current noise in a ballistic one-dimensional wire which exhibits an additional conductance plateau at $0.7 \times 2e^2/h$. The Fano factor shows a clear reduction on the 0.7 structure, and eventually vanishes upon applying a strong parallel magnetic field. These results provide the first experimental evidence that the 0.7 structure is associated with two conduction channels which have different transmission probabilities[1].

Keywords: noise reduction 0.7 structure
PACS: 72.70.+m, 73.23.Ad, 05.30.Fk

One of the major advance in the quantum physics of ballistic conductor was the observation of quantized conductance plateaus in 2D electron gas constrictions defined by a Quantum Point Contact (QPC) [2, 3]. These plateaus are quantized in unit of $2G_0$, where $G_0 = e^2/h$ in perfect agreement with the Landauer-Buttiker (LB) description of the balistique transport of non interacting quasi-particles. Surprisingly, since the early observation, an additional plateau appeared at a unexpected value $\sim 0.7 \times 2G_0$, latter called the 0.7 anomaly[4]. Over the last decade, the anomaly has attracted a considerable attention both experimentally and theoretically to find a plausible underlying mechanism. Although extensively studied[4, 5, 6, 7, 8], one main question remains : Does the additional plateau correspond to a perfectly transmitted channel? If it does, the anomaly should result from a perfectly transmitted mode in parallel with a partially reflected one. Here, we provide the first experimental evidence that the 0.7 structure is associated with two modes with different transmission probabilities.

The current noise is used to probe the distribution of the transmission probabilities. It has been shown theoretically [9, 10, 11] and experimentally [12, 13] that the excess noise is suppressed for perfectly transmitted or reflected channels. Whereas the excess noise suppression is rather easy to measure in case of linear systems at low temperatures, the non linear conductance intrinsic to the 0.7 anomaly [6, 8] prevents from performing direct analysis. For channels with energy independent transmission probabilities τ_i, the conductance G is $G = G_0 \times \sum \tau_i$ and the current noise S_I at finite temperature T and bias V is the sum of a thermal noise $S_{I\,Therm} = 4G_0 k_B T \sum \tau_i^2$ and the partition noise $S_{I\,Part} = 2G_0 eV \coth(eV/2k_B T) \sum \tau_i(1-\tau_i)$. Then the excess noise $\Delta S_I = S_I(V) - S_I(0)$ is proportional to the Poissonian noise

$$\Delta S_I = F \times S_{I\,Pois} \text{ with } S_{I\,Pois} = 2eI \coth(eV/2k_B T) - 4k_B T G(0),$$

$$\text{and } F = \sum \tau_i(1-\tau_i)/\sum \tau_i.$$

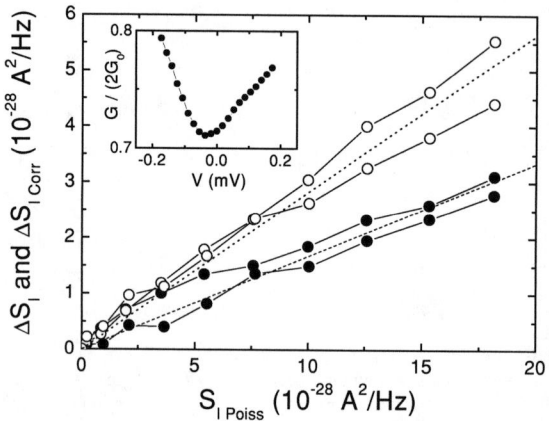

FIGURE 1. Excess noise without (o) and with (•) thermal corrections, as a function of the Poissonian noise at $T = 460$ mK with $G(0) = 0.71 \times 2G_0$. The dotted line fit to the uncorrected data ΔS_I gives an overestimated $F^{++} = 0.28$. The dashed line fit to $\Delta S_{I\,Corr}$ gives $F^+ = 0.17$. The inset shows G as a function of V.

This dependance of the Fano factor F with transmission has been demonstrated in shot noise experiments in the linear regime[12, 13].

In the present case the transmission probabilities are energy dependent[5, 8]. In the analysis of previous noise measurements [14, 15] in the vicinity of the 0.7 anomaly, thermal noise variation due to conductance non-linearities and $1/f$ noise were not properly taken into account. Furthermore, the explored energy range (several meV) have exceeded the energy scale of the 0.7 structure. In the present experiment, we took into account the non linearities in the following way. If F does not vary too strongly with the energy and/or the explored energy range is of the order of few k_BT, the following approximate expressions can be derived:

$$\Delta S_I = F(0) \times S_{I\,Pois} + \Delta S_{I\,Therm},$$

$F(0)$ is the zero bias Fano factor averaged by thermal fluctuations. The second term, which vanishes for linear system is:

$$\Delta S_{I\,Therm} = 2G_0 k_B T \sum (\mathcal{T}_i^2(eV) + \mathcal{T}_i^2(-eV) - 2\mathcal{T}_i^2(0)),$$

$\mathcal{T}_i(\pm eV)$ is the average of $\tau_i(\varepsilon)$ other k_BT around $\pm eV$. It is not possible to determine the exact value of $\mathcal{T}_i(eV)$ as the conductance measurement is proportional to $\sum[\mathcal{T}_i(eV) + \mathcal{T}_i(-eV)]$. When $\tilde{G} = G/G_0 < 2$ we assume that there is only two channels partially transmitted. If these two channels have the same transmission probability at zero bias (which is the case when there is a spin degeneracy of modes) then $\sum_{i=1,2} \mathcal{T}_i^2(0) = \tilde{G}(0)^2/2$ and [1]

$$\Delta S_{I\,Therm} \geq 2G_0 k_B T [\tilde{G}(I)^2 - \tilde{G}(0)^2].$$

---

[1] We used the inequality $\sum_n x_i^2 > (\sum_n x_i)^2/n$ to found a lower bound to $\Delta S_{I\,Therm}$.

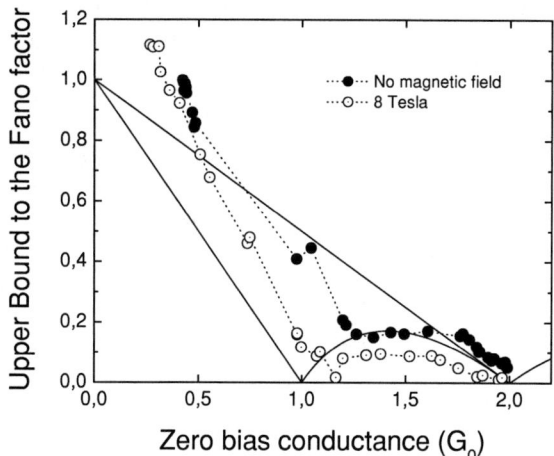

FIGURE 2. Upper bound to the Fano factor F as a function of the zero bias conductance. On the 0.7 structure, the upper bound to F is lower than that we would expect for two channel with same transmission probability. This result prove without ambiguity that there is **two channels with different transmission probability**. The upper bound tends to zero at $G = G_0$ in case of spin splitting under high magnetic field $B = 8$ T.

Therefore, the corrected noise variation $\Delta S_{I\,Corr} = \Delta S_I - 2G_0 k_B T [\tilde{G}(I)^2 - \tilde{G}(0)^2]$ can be fitted with

$$\Delta S_{I\,Corr} = F^+ \times S_{I\,Pois},$$

where F^+ is the fitting parameter which is an upper bound to Fano factor that we would obtain in case of two channels with the same transmission probability ($F(0) = 1 - 2\tilde{G}(0)$).

Details of the experiment are discussed in [1]. In figure 1, we plotted the measured excess noise and the corrected excess noise as a function of the Poissonian noise. The excess noise varies linearly with the Poissonian noise. This result shows that the 0.7 anomaly cannot result from an activated modulation of the conductance. This would give an I^2 dependance of the excess noise. Taking into account the non linearities strongly reduces the upper bound to the Fano factor. If the two channel ($\tilde{G} \leq 2$) have the same transmission probability then F^+ should be larger than $1 - 2\tilde{G}(0)$. In figure 2, F^+ is plotted versus the zero bias conductance. On the 0.7 structure ($\tilde{G} \approx 1.4$), F^+ is below the line defined by $1 - 2\tilde{G}(0)$, definitively demonstrating that the two channel does not have the same transmission: the anomaly is accompanied with a lift of the spin degeneracy of the electronic modes. When applying a magnetic field parallel to the constriction, the anomaly evolves to a conductance plateau at $G = G_0$. At high magnetic field (8 Tesla), the Fano factor vanishes for $G = G_0$ (see figure 2) as expected for a complete splitting.

Although our measurements prove that the underlying mechanism which leads to the anomaly should give 2 distinct modes, the approximation needed for the analysis prevent any claim on the evolution of the transmission probability of these modes with the gate voltage. Nevertheless theoretical approaches relying on the scattering of a plasmon[16] which lead to an activated quasi particle lifetime, on a variation of the compressibility precursor to Wigner critallization[17],on inelastic scattering with a spin wave localized on the QPC [18], on electron-phonon scattering [19] and a recent paper treating the interaction in a 1D wire [20] are not compatible with our experimental result because these theories lead to two channels with the same transmission probability. On the contrary, the mechanisms which lead to either spontaneous spin split of the sub-band [21, 22, 23, 24, 25] or singlet versus triplet pairing [26, 27, 28] with Kondo mechanism for the evolution of the anomaly to $2G_0$ at low temperature [29] are compatible with our result as they give two channels with different transmission probabilities.

It would be interesting to realize the same measurement from low temperature, when the anomaly is not well pronounced, to the high temperature. In the case of spin splitting of the sub band, the Fano would be proportional to $1 - G/G_0$ when $G \leq G_0$ at low temperature. This would allow to distinguish spin splitting approaches from quasi-bound state approaches.

REFERENCES

1. P. Roche, et al., *Phys. Rev. Lett.*, **93**, 116602 (2004).
2. D. A. Wharam, et al., *J. Phys. C*, **21**, L209 (1988).
3. B. J. van Wees, et al., *Phys. Rev. Lett.*, **60**, 848 (1988).
4. K. J. Thomas, et al., *Phys. Rev. Lett.*, **77**, 135 (1996).
5. K. J. Thomas, et al., *Phys. Rev. B*, **58**, 4846 (1998).
6. A. Kristensen, et al., *Phys. Rev. B*, **62**, 10950 (2000).
7. D. J. Reilly, et al., *Phys. Rev. B*, **63**, 121311 (2001).
8. S. M. Cronenwett, et al., *Phys. Rev. Lett.*, **88**, 226805 (2002).
9. G. B. Lesovik, *Pis'ma Zh. Éksp. Teor. Fiz.*, **49**, 513 (1989), [JETP Lett. 49, 592 (1989)].
10. M. Büttiker, *Phys. Rev. Lett.*, **65**, 2901 (1990).
11. T. Martin, and R. Landauer, *Phys. Rev. B*, **45**, 1742 (1992).
12. M. Reznikov, et al., *Phys. Rev. Lett.*, **75**, 3340 (1995).
13. A. Kumar, et al., *Phys. Rev. Lett.*, **76**, 2778 (1996).
14. R. C. Liu, et al., *Nature*, **391**, 263 (1998).
15. N. Y. Kim, et al., *cond-mat/0311435* (2003), W. D. Oliver, Ph.D. thesis, Standford University, (2002).
16. H. Bruus, and K. Flensberg, *Semicond. Sci. Technol.*, **13**, A30 (1998).
17. O. P. Sushkov, *Phys. Rev. B*, **64**, 155319 (2001).
18. Y. Tokura, and A. Khaetskii, *Physica*, **E12**, 711 (2002).
19. G. Seelig, and K. A. Matveev, *Phys. Rev. Lett.*, **90**, 176804 (2003).
20. D. Schmeltzer, et al., *Phys. Rev. B*, **71**, 045429 (2005).
21. C.-K. Wang, and K.-F. Bergreen, *Phys. Rev. B*, **57**, 4552 (1998).
22. S. M. Reimann, M. Koshiken, and M. Manninen, *Phys. Rev. B*, **59**, 1613 (1999).
23. B. Spivak, and F. Zhou, *Phys. Rev. B*, **61**, 16730 (2000).
24. H. Bruus, V. V. Cheianov, and K. Flensberg, *Physica*, **E10**, 97 (2001).
25. A. A. Starikov, I. I. Yakimenko, and K.-F. Berggren, *Phys. Rev. B*, **67**, 235319 (2003).
26. T. Rejec, A. Ramšak, and J. H. Jefferson, *Phys. Rev. B*, **62**, 12985 (2000).
27. V. V. Flambaum, and M. Y. Kuchiev, *Phys. Rev. B*, **61**, R7869 (2000).
28. A. Ramšak, and J. H. Jefferson, *Cond-Mat/0502049* (2005), to appear in Phys. Rev. B Rapid Com.
29. Y. Meir, K. Hirose, and N. S. Wingreen, *Phys. Rev. Lett.*, **89**, 196802 (2002).

Current Noise in Non-Chiral Luttinger Liquids: Appearance of Fractional Charge

Fabrizio Dolcini*, Björn Trauzettel[†], Inès Safi** and Hermann Grabert*

*Physikalisches Institut, Albert-Ludwigs-Universität, 79104 Freiburg, Germany
[†]Instituut-Lorentz, Universiteit Leiden, 2300 RA Leiden, The Netherlands
**Laboratoire de Physique des Solides, Université Paris-Sud, 91405 Orsay, France

Abstract. The current noise of a voltage-biased interacting quantum wire connected to leads is computed in presence of an impurity in the wire. We find that in the weak backscattering limit the Fano factor characterizing the ratio between noise and backscattered current crucially depends on the noise frequency ω relative to the ballistic frequency v_F/gL, where v_F is the Fermi velocity, g the electron interaction strength, and L the length of the wire. In contrast to chiral Luttinger liquids, the noise is not only due to the Poissonian backscattering of fractionally charged quasiparticles at the impurity, but also to Andreev-type reflections at the contacts. The frequency dependence of the noise needs to be analyzed to extract the fractional charge $e^* = eg$ of the bulk excitations.

Keywords: Luttinger Liquid, Shot Noise, quantum wire
PACS: 71.10.Pm, 72.10.-d, 72.70.+m, 73.23.-b

INTRODUCTION

One of the most exciting topic in nowadays mesoscopic physics is represented by shot noise measurements. This is due to the fact that in the Poissonian limit of uncorrelated backscattering of (quasi)particles from a weak impurity, the low frequency current noise is directly proportional to the backscattered charge [1]. This property turns out to be particularly useful in probing the fractional charge of excitations in one-dimensional (1D) electronic systems, where correlation effects destroy the Landau quasiparticle picture and give rise to collective excitations obeying unconventional statistics and carrying a charge different from the charge e of an electron. In particular, for fractional quantum Hall (FQH) edge state devices, which at filling fraction $v = 1/m$ (m odd integer) are usually described by the *chiral* Luttinger liquid (LL) model, it has been theoretically predicted[2] and experimentally verified[4, 3] that shot noise allows for an observation of the fractional charge $e^* = ev$ of backscattered Laughlin quasiparticles.

The question arises whether similar results can be expected also for *non-chiral* LLs, which are believed to be realized in carbon nanotubes [5] and single channel semiconductor quantum wires [6]. Although a non-chiral LL can be modelled through the very same formalism as a pair of chiral LLs, these two kinds of LL systems exhibit important differences. In particular, in chiral LL devices right- and left-moving excitations are spatially separated, so that their chemical potentials can be independently tuned. In contrast, in non-chiral LL systems, right- and left-movers are confined to the same channel, and only the chemical potentials of the reservoirs attached to the 1D wire can be controlled. This implies crucial differences between chiral and non-chiral LLs, for instance, the conductance in the former case depends on the LL parameter $g = \nu$ [7], while in the latter case it is independent of g [8, 9]. Hence, the predictions on shot noise properties of FQH systems are not straightforwardly generalizable to the case of non-chiral LLs, which therefore deserve a specific investigation.

Previous theoretical calculations of the shot noise of non-chiral LL systems have shown that, even in the weak backscattering limit, the zero frequency noise of a finite-size non-chiral LL does not contain any information about the fractional charge backscattered off an impurity [10, 11], but is rather proportional to the charge of an electron. This result, as well as the above mentioned interaction independent DC conductance, prevents easy access to the interaction parameter g. On the other hand, shot noise is never really measured at zero frequency, due to the dominance of $1/f$ noise at low frequencies. This raises the question on the influence of a finite measurement frequency on the shot noise. Previous theoretical analysis on this issue were restricted to a four-terminal FQH edge state geometry (chiral LL) in the case of *infinite system size* [12]. In the present paper, we present instead some results concerning transport properties of a non-chiral quantum wire where the finite length L of the system is explicitly taken into account. We will show that the result of a shot noise measurement crucially depends on the ratio between the noise frequency ω and the ballistic frequency $\omega_L = v_F/gL$ associated with the traversal time of the wire. In particular, we will show that the Fano factor oscillates as a function of frequency, with period $2\pi\omega_L$, and that by averaging over $2\pi\omega_L$, the effective charge $e^* = eg$ can be extracted from noise data.

MODEL AND RESULTS ON THE SHOT NOISE

In order to describe the transport properties of a (non-chiral) quantum wire, we have adopted the inhomogeneous LL (ILL) model [8, 9], which takes explicitly into account the finite length of the interacting wire and the adiabatic coupling to the Fermi liquid leads. The Hamiltonian of this model reads

$$\mathcal{H} = \frac{\hbar v_F}{2}\int_{-\infty}^{\infty} dx \left[\Pi^2 + \frac{1}{g^2(x)}(\partial_x\Phi)^2\right] + \lambda\cos\left[\sqrt{4\pi}\Phi(x_0,t) + 2k_F x_0\right] + \quad (1)$$

$$+ \int_{-\infty}^{\infty} \frac{dx}{\sqrt{\pi}} \mu(x)\, \partial_x\Phi(x,t),$$

where the first term describes the interacting wire, the leads and their contacts, the second accounts for the electron backscattering at the impurity site x_0, and the third represents the coupling to the electrochemical bias applied to the wire.

Here, $\Phi(x,t)$ is the standard Bose field operator in bosonization and $\Pi(x,t)$ its conjugate momentum density [13]. In (1), the interaction parameter $g(x)$ is space-dependent and its value is g in the bulk of the wire ($0 < g < 1$ corresponding to repulsive interactions), and 1 in the bulk of the non-interacting leads. The leads are thus treated as non-interacting 1D systems, and their main effect of applying a bias voltage at the contacts is accounted for by the externally tunable electrochemical bias $\mu(x)$, a piecewise constant $\mu(x < -L/2) = \mu_L$, $\mu(x > L/2) = \mu_R$. The applied voltage is then $V = (\mu_L - \mu_R)/e$.

The current noise is defined as

$$S(x,x;\omega) = \int_{-\infty}^{\infty} dt\, e^{i\omega t} \langle \{\Delta j(x,t), \Delta j(x,0)\} \rangle, \quad (2)$$

where x denotes the measurement point (in the lead), $\{,\}$ the anticommutator and $\Delta j(x,t) = j(x,t) - \langle j(x,t) \rangle$ the current fluctuation operator. In bosonization, the current operator is related to the Bose field Φ through $j(x,t) = -(e/\sqrt{\pi})\partial_t \Phi(x,t)$.

The average current $I \equiv \langle j(x,t) \rangle$ can be expressed as $I = I_0 - I_{BS}$, where $I_0 = (e^2/h)V$ is the current in the absence of an impurity, and $I_{BS} = (e^2/h)\mathscr{R}V$, the backscattering current. Here $\mathscr{R}$ is the effective reflection coefficient that, contrary to a non-interacting electron system, depends on voltage and interaction strength [7, 14]. The fractional charge is expected to emerge only in the limit of weak backscattering ($\mathscr{R} \ll 1$) through the ratio between shot noise and I_{BS}. Let us focus then on the shot noise limit of large applied voltage. Importantly, for temperatures in the window $eV\mathscr{R} \gg k_B T \gg \{\hbar\omega, \hbar\omega_L\}$ the noise (2) can be shown[16, 15] to take the simple form

$$S(x,x;\omega) = 2eF(\omega)I_{BS}, \quad (3)$$

The function $F(\omega)$ represents an effective frequency dependent Fano factor. For simplicity we consider here only the case of an impurity in the middle of the wire ($x_0 = 0$). For this case $F(\omega)$ is plotted in Fig. 1, and it explicitly reads

$$F(\omega) = \frac{2g^2}{1 + g^2 - (1-g^2)\cos(\omega/\omega_L)}. \quad (4)$$

The central result (3) shows that the ratio between the shot noise and the backscattered current crucially depends on the frequency regime one explores. In particular, for $\omega \to 0$, the function F tends to 1, independent of the value of the interaction strength. Therefore, in the regime $\omega \ll \omega_L$ the observed charge is just the electron charge. In contrast, at frequencies comparable to ω_L the behavior of F as a function of ω strongly depends on the LL interaction parameter g, and signatures of LL physics emerge. Averaging over one period, we obtain

$$\langle S(x,x;\omega) \rangle_\omega \equiv \frac{1}{2\pi\omega_L} \int_{-\pi\omega_L}^{\pi\omega_L} S(x,x;\omega) \simeq 2eg I_{BS}. \quad (5)$$

Seemingly, Eq. (5) suggests that quasiparticles with a fractional charge $e^* = eg$ are backscattered off the impurity. The physical origin of the appearance of this fractional charge boils down to the energy resolution that finite frequency allows. Indeed, although the contacts are adiabatic, the mismatch between electronic excitations in the leads and in the wire inhibits the direct penetration of electrons from the leads into the wire; rather a current pulse is decomposed into a sequence of fragments by means of Andreev-type reflections at the contacts [8]. These reflections are governed by the coefficient $\gamma = (1-g)/(1+g)$, which depends on the interaction strength in the wire. The zero frequency noise is only sensitive to the sum of all current fragments, which add up to the initial current pulse carrying the charge e. In contrast, finite frequency noise resolves

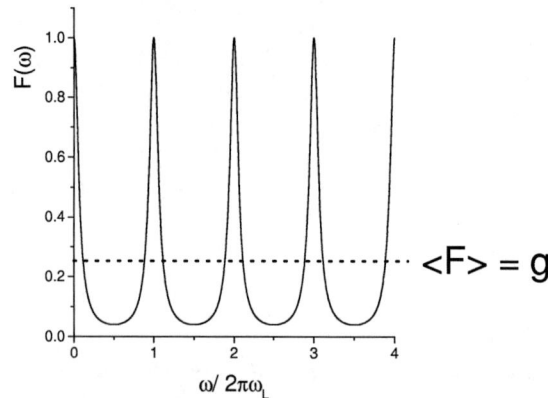

FIGURE 1. The function $F(\omega)$, characterizing the Fano factor, is plotted as a function of $\omega/2\pi\omega_L$ for the case of an impurity at the center of the wire. In the regime $\omega/\omega_L \ll 1$, $F \to 1$, and only the charge of the electron can be observed. In contrast, at finite frequency $F(\omega)$ oscillates, and its average over one period (dotted line) equals the interaction strength g. The fractional charge $e^* = eg$ can thus be extracted.

the current fragmentation at the contacts.

To conclude, we can also estimate the value of ω_L for single-wall carbon nanotubes: since $g \approx 0.25$, $v_F \approx 10^5$ m/s, and $L \sim 1 \div 10\mu$, one has $\pi\omega_L \approx 100$ GHz...1 THz, which seems to be experimentally accessible [18, 19]. Moreover, the requirement $eV \gg \hbar\omega_L$ should be fulfilled in such systems for $eV \approx 10\ldots 50$ meV, a value lying well below the subband energy separation of about 1 eV.

REFERENCES

1. For a recent review on shot noise see: Y. M. Blanter and M. Büttiker, Phys. Rep. **336**, 1 (2000).
2. C. L. Kane and M. P. A. Fisher, Phys. Rev. Lett. **72**, 724 (1994).
3. R. de Picciotto *et al.*, Nature (London) **389**, 162 (1997).
4. L. Saminadayar, D. C. Glattli, Y. Jin, and B. Etienne, Phys. Rev. Lett. **79**, 2526 (1997).
5. M. Bockrath *et al.*, Nature (London) **397**, 598 (1999).
6. A. Yacoby *et al.*, Phys. Rev. Lett. **77**, 4612 (1996).
7. C. L. Kane and M. P. A. Fisher, Phys. Rev. B **46**, 15233 (1992).
8. I. Safi and H. J. Schulz, Phys. Rev. B **52**, R17040 (1995); **59**, 3040 (1999).
9. D. Maslov and M. Stone, Phys. Rev. B **52**, R5539 (1995); V. V. Ponomarenko, *ibid.* **52**, R8666 (1995).
10. V. V. Ponomarenko and N. Nagaosa, Phys. Rev. B **60**, 16865 (1999).
11. B. Trauzettel, R. Egger, and H. Grabert, Phys. Rev. Lett. **88**, 116401 (2002).
12. C. de C. Chamon, D. E. Freed, and X. G. Wen, Phys. Rev. B **53**, 4033 (1996).
13. A. O. Gogolin, A. A. Nersesyan, and A. M. Tsvelik, *Bosonization and Strongly Correlated Systems* (Cambridge University Press, Cambridge, 1998).
14. F. Dolcini, H. Grabert, I. Safi, and B. Trauzettel, Phys. Rev. Lett. **91**, 266402 (2003).
15. F. Dolcini, B. Trauzettel, I. Safi, and H. Grabert, Phys. Rev. B **71**, 165309 (2005).
16. B. Trauzettel, I. Safi, F. Dolcini, and H. Grabert, Phys. Rev. Lett. **92** 226405 (2004)
17. P.-E. Roche *et al.*, Eur. Phys. J. B **28**, 217 (2002).
18. R. J. Schoelkopf *et al.*, Phys. Rev. Lett. **78**, 3370 (1997).
19. R. Deblock, E. Ognac, L. Gurevich, and L. P. Kouwenhoven, Science **301**, 203 (2003).

Comparative Analysis of Sequential and Coherent tunneling Models of Shot Noise in Resonant Diodes

V.Ya. Aleshkin[1] and L. Reggiani[2]

[1] *Institute for Physics of Microstructures RAS, Nizhny Novgorod, GSP-105, 603950, Russia*
[2] *National Nanostructure Laboratory of INFM, Dipartimento di Ingegneria dell' Innovazione, Universita' di Lecce, Via Arnesano s/n, 73100 Lecce, Italy*

Abstract. We compare the sequential and coherent tunneling models for current voltage characteristic and shot noise suppression in symmetric double barrier resonant diodes. Results confirm that the I-V characteristic remains the same for the two models, while at increasing voltage the shot noise power shows significant differences. In the sequential tunneling, shot noise exhibits in general two regions of suppressions with the associated Fano factor never dropping below the value of 0.5. By contrast, in the coherent tunneling shot noise exhibits a single region of suppression with the associated Fano factor having the possibility of dropping below the value of 0.5. We conclude that shot noise suppression below 0.5 of the full shot noise value can be a signature of coherent versus sequential transport model.

Keywords: Quantum transport, resonant tunneling diode, shot noise.
PACS: **72.70+m**, 72.20.-i, 72.30+a, 73.23.Ad

INTRODUCTION

Since its realization [1], the double barrier resonant diode (DBRD) proved to be an electron device of broad physical interest because of its peculiar non Ohmic current voltage (I-V) characteristic. Even the shot noise characteristics of DBRDs are of relevant interest in the sense that suppressed as well as enhanced shot noise with respect to the full Poissonian value has been observed (see the review [2]. These electrical and noise features are controlled by the mechanism of carrier tunneling through the double potential barriers. The microscopic interpretation of these features is found to admit a coherent [3] or a sequential tunneling [4] approach. From the existing literature [5,6] it emerges that both of them are capable to explain the I-V experiments as well as most of the shot noise characteristics. Therefore, to our knowledge there is no way to distinguish between these two transport regimes and the natural question whether tunneling transport is coherent or sequential remains an unsolved one. The aim of this contribution is to carry out a comparative analysis of the sequential and coherent tunneling models for the calculation of the I-V characteristic and of the shot noise suppression in DBRDs at 4.2 K. We anticipate that the main result concerns with the prediction that suppression of shot noise with a Fano factor below 0.5 is found to be an indication of coherent tunneling against sequential tunneling. Such a prediction is confirmed by experimental results [7].

RESONANT DIODE MODEL

The structure here investigated is the standard symmetric double well reported in Fig. 1. The essential features of the structure are described by three parameters: (i) the thickness of each barrier $d_{L,R} = 100$ A, (ii) the energy of the resonant level as measured from the center of the potential well $\varepsilon_r = 50$ meV and, (iii) the partial width of the resonant level due to tunneling through the left and right barriers $\Gamma = \Gamma_L + \Gamma_R = 0.5$ meV. We consider the case when there is only one resonant state and we assume that the resonant tunneling diode has unit square contacts. The double barrier transparency $D(\varepsilon_z)$ is written in the standard form:

$$D = \frac{(\Gamma/2)^2}{(\varepsilon_z - \varepsilon_r + qu)^2 + (\Gamma/2)^2} \quad (1)$$

where ε_z is the energy for electron motion perpendicular to the barriers. A carrier concentration of $n = 5 \times 10^{17}$ cm^{-3} in the emitter and collector regions are considered. In calculations we use everywhere values for the electron effective mass $m = 0.067\, m_0$, m_0 being the free electron mass, and static dielectric constant $\kappa = 12.9$ corresponding to GaAs being this material appropriate for an experimental validation of present results.

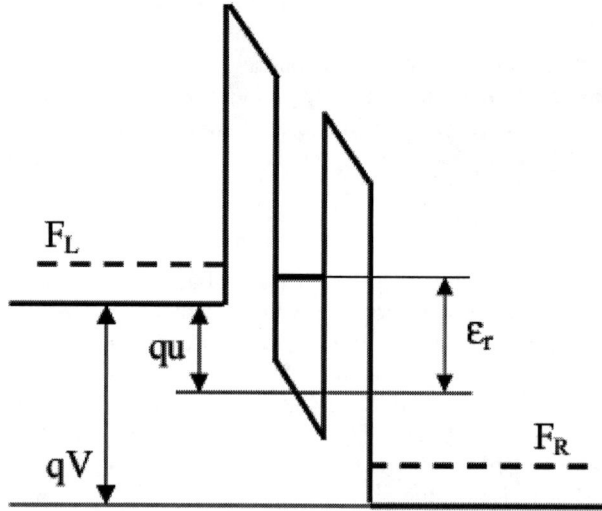

FIGURE 1. Sketch of the band profile of the double barrier structure considered here under typical operation conditions. Here F_L, and F_R are the electrochemical potential in the emitter and the collector regions.

Sequential tunneling approach

In this approach [5,8,9] the resonant tunneling diode (RTD) is considered as a black box connected with two ideal contacts (emitter on the left (L) and collector on the right (R)) acting as ideal thermal reservoirs at given electrochemical potentials $F_{L,R}$. Under steady state, transport and noise are described in terms of the instantaneous current $I(t)$ flowing through the emitter (or the collector) interface which is determined by the probability $P(N)$ of finding $N(t)$ electrons in the device at time t. This probability obeys a master equation, which is governed by four rates, denoted as $g_{L,R}$ and $r_{L,R}$, describing the input or generation (output or recombination) of a carrier from the contacts into (out) the active region of the resonant tunneling diode and providing for the stationary current flowing from the emitter to the collector [9]:

$$I = -q(g_L - r_L) = -q(r_R - g_R) = -\frac{qm}{2\pi^2 \hbar^3} \int_0^\infty D(\varepsilon_z) d\varepsilon_z \int_0^\infty (f_L - f_R) d\varepsilon_\perp \qquad (2)$$

with $f_{L,R} = f_{L,R}(\varepsilon, F_{L,R})$ the Fermi distribution in the emitter (L) and collector (R) contacts, respectively and $\varepsilon = \varepsilon_z + \varepsilon_\perp$ the carrier kinetic energy decomposed in the direction perpendicular and parallel to the barriers. From the above four rates, two differential rates, denoted as $v_{L,R}$, can be constructed. These describe the decay of the carrier number fluctuations in the RTD through the emitter and collector interfaces. We note, that $g_{L,R}$ and $r_{L,R}$ are positive definite quantities, while $v_{L,R}$ can be positive (damping of fluctuations) or negative (enhancement of fluctuations). The spectral density of current fluctuations at low frequency, S_I, takes the form:

$$S_I = 2q^2 \frac{\left[v_L^2(g_R + r_R) + v_R^2(g_L + r_L)\right]}{(v_L + v_R)^2} \qquad (3)$$

The master equation is coupled with the Poisson equation to account for space charge reaction [9].

When the applied voltage is high enough for $qV \gg F_R$ and for neglecting g_R, by defining $\alpha = v_R/(v_L + v_R)$, the Fano factor, $\gamma = S_I/(2qI)$, expressed by Eqs (2) and (3) is conveniently written as:

$$\gamma = 1 - 2\alpha + 2\alpha^2 \frac{I}{I_0} \qquad (4)$$

with $I_0 = qg_L \geq I$ the emitter injecting current. The whole scenario for the possible values of the Fano factor as function of the applied voltage is described by Eq. (4) with $-\infty < \alpha < \infty$. Focusing our interest to the minimum value of the Fano factor, $\gamma_{\min}$, we notice that $\gamma_{\min} = 1 - I/(2I_0)$ is obtained when $\alpha = I/(2I_0)$. Therefore, within this approach shot noise suppression never drops below the minimum value of 0.5.

Coherent tunneling approach

In this approach we use a coherent theory [10]. Again, the RTD is connected with two ideal contacts at given electrochemical potentials. From contacts, carriers impinging on the device are transmitted or reflected at the interfaces with the transmission coefficient of the whole structure given by Eq. (1) and responsible of both conductance and noise (quantum partition noise). Accordingly, the I-V characteristic is found to take the same expression (3), while S_I is given by the sum of three terms as:

$$S_I = S_1 + S_2 + S_3 \tag{5}$$

where

$$S_1 = \frac{q^2 m}{\pi^2 \hbar^3} \int_0^\infty D d\varepsilon_z \int_0^\infty d\varepsilon_\perp \{f_L(1-f_R) + f_R(1-f_L) - D(f_L - f_R)^2\} \tag{6}$$

$$S_2 = -\lambda \frac{q^2 m}{\pi^2 \hbar^3} \int_0^\infty D^2 d\varepsilon_z \int_0^\infty d\varepsilon_\perp [f_L(1-f_L) + f_R(1-f_R)] \tag{7}$$

$$S_3 = \lambda^2 \frac{q^2 m}{4\pi^2 \hbar^3} \int_0^\infty D^2 d\varepsilon_z \int_0^\infty d\varepsilon_\perp \{f_L(1-f_L) + f_L(1-f_R) + f_R(1-f_L)\} +$$
$$+ \lambda^2 \frac{q^2 m}{4\pi^2 \hbar^3} \int_{-qV}^\infty D^2 d\varepsilon_z \int_0^\infty d\varepsilon_\perp f_R(1-f_R) \tag{8}$$

with

$$\lambda = \frac{4\hbar}{\Gamma} \frac{1}{(C_L + C_R + C_{QW})} \frac{\partial I}{\partial u}$$

where $C_{L,R} = \kappa / 4\pi d_{L,R}$ are the capacitance of the left and right barriers, C_{QW} the differential capacitance of the quantum well, λ is a dimensionless parameter describing Coulomb interaction. In case when $\lambda = 0$, Coulomb interaction is negligible, and $S_I = S_1$ [11].

COMPARATIVE ANALYSIS OF THE TWO APPROACHES

A comparative analysis of the sequential and coherent approaches at 4.2 is reported in Fig. 2. We found that the sequential tunneling approach (see Fig. 2a) evidences two regions of suppression, a first starting at low voltages just above the condition $qV > k_B T$, and a second starting in concomitance with the onset of the sharp increase of the current. Here k_B is the Boltzmann constant, T is the temperature. The two suppression regions are a consequence of the repulsion between successive current pulses and are separated by a wide region of full shot noise where current pulses do not interact. The first suppressed region is a consequence of degeneracy (Pauli blockade), while the second suppressed region is due to Coulomb repulsion. This scenario was predicted theoretically [8,12], however it has not yet been confirmed experimentally. By contrast, the coherent tunneling approach (see Fig. 2b) evidences only one region of suppression, which corresponds to the second region of the sequential case. Here both Pauli blockade and Coulomb repulsion act simultaneously. In particular we found that Coulomb repulsion has the net effect to increase the noise for the present symmetric structure [2]. More important, in the coherent approach, there is a dip in the Fano factor near the maximum of the current with a value of 0.48, slightly below 0.5, which is a signature of the coherent tunneling. As a check of the model, we have verified that by increasing the value of Γ and n for a factor of 10, thus moving towards coherence against sequential, a similar dip in the Fano factor takes the value of 0.38, thus confirming that coherent tunneling can be at the origin of a Fano factor drop below 0.5.

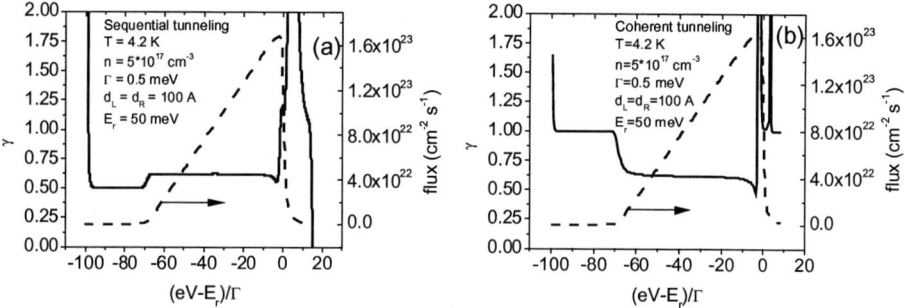

FIGURE 2. Dependence of current and Fano factor on applied voltage in a typical symmetric DBRD at 4.2 K in the presence of Coulomb correlations for a carrier concentration in the contacts $n=5 \; 10^{17}$ cm^{-3}. (a) and (b) refer to the sequential and coherent tunneling approaches, respectively.

In conclusion, we have carried out a comparative analysis of sequential and coherent tunneling models for shot noise suppression in double barrier resonant diodes. By considering a symmetric structure at 4.2 K, we have calculated the current voltage characteristic and the noise power up to voltages around the peak value of the current. Results show that the I-V characteristic remains the same for the two models, as it should be, while the noise power at increasing voltage shows significant differences depending on the sequential or coherent model. In the sequential tunneling, shot noise exhibits in general two regions of suppressions. The first region starts around $qV/k_BT > 1$ and the second around the onset of the sharp current increase. In any case, the associated Fano factor never drops below the value of 0.5. By contrast, in the coherent tunneling shot noise exhibits a single region of suppression in correspondence with the sharp current increase over the Ohmic range. Remarkably, in the coherent tunneling regime the associated Fano factor exhibits the possibility to drop below the value of 0.5. We conclude that within these models shot noise suppression below 0.5 of the full shot noise value is a signature of coherent versus sequential transport model.

ACKNOWLEDGMENTS

Research performed within the project "Noise models and measurements in nanostructures". Partial support by the Italian Ministry of Education, University and Research (MIUR) from the Italian Ministry of Foreign Affairs through the Volta Landau Center (the fellowship of V.Ya. A.), and the NATO Collaborative Linkage Grant PST.EAP.CLG 980629 is gratefully acknowledged.

REFERENCES

1. R. Tsu and L.Esaki, *Appl. Phys. Lett.*, **22**, 562-564 (1973).
2. Y.M.Blanter and M.Buttiker, *Phys. Rep.* **336**, 1-360 (2000).
3. L.L.Chang, L.Esaki and R.Tsui, *Appl. Phys. Lett.*, **24**, 593-595 (1974).
4. S. Luryi, *Appl. Phys. Lett.*, **47**, 490-492 (1985).
5. G. Iannaccone, M. Macucci and B. Pellegrini, *Phys. Rev. B* **55**, 4539-4550 (1997).
6. Y.M. Blanter and M. Buttiker, *Phys. Rev. B* **59**, 10217-10226 (1999).
7. E.R. Brown, *IEEE Trans on Electron Dev.* **39**, 2686-2693(1992).
8. V.Ya. Aleshkin, L. Reggiani and A. Reklaitis, *Phys. Rev. B*, **63**, 085302 (2001).
9. V.Ya. Aleshkin and L. Reggiani, Phys. Rev. B, **64**, 245333 (2001).
10. V.Ya. Aleshkin, L. Reggiani, N.V. Alkeev, V.E. Lyubchenko, C.N. Ironside, J.M.L. Figuiredo and C.R. Stanley, *Phys.Rev. B.* **70**, 115321 (2004).
11. T. Martin and R. Landauer, *Phys. Rev. B*, **45**, 1742-1755 (1992).
12. G. Iannaccone, M. Macucci and B. Pellegrini, *Unsolved problems of noise and fluctuations,* Eds. D. Abbott and L.B. Kish, AIP Conf. Proc. N. 511, 2000 pp. 59-61.

Shot Noise Experiments in Multi-barrier Semiconductor Heterostructures

E. E. Mendez[1], W. Song[1,2], A. K. M. Newaz[1], Y. Lin[3,4], and J. Nitta[3,5]

[1] *State University of New York at Stony Brook, Stony Brook, NY 11794-3800, USA*
[2] *Current address: Korea Research Institute of Standards and Science, Daejeon, Korea*
[3] *NTT Basic Research Laboratories, Atsugi, and CREST-JST, Saitama, Japa*
[4] *Current address: National Tsing Hua University, Hsinchu 300, Taiwan*
[5] *Current address: Tohoku University, Sendai, 980-8579, Japan*

Abstract. We have found strong deviations from Poissonian behavior in the low-temperature noise characteristics of triple-barrier and superlattice heterostructures. Although our results can be explained qualitatively by existing models, a quantitative comparison between experiment and theory suggests an incomplete understanding of shot noise in multi-barrier systems.

Keywords: shot noise, semiconductor heterostructures, resonant tunneling, superlattices
PACS: 73.50Pz, 73.50Td

INTRODUCTION

Until recently our knowledge of electronic tunneling in semiconductor heterostructures has been derived primarily from electrical conductance measurements, but lately the attention has shifted to the measurement of the electrical noise, which can provide additional information about the transport mechanism, especially correlation effects. Thus, when electrons tunnel through a biased single potential barrier between two metallic electrodes (such as in an undoped GaAlAs thin layer between two heavily doped GaAs thick regions), the motion of electrons is uncorrelated and the spectral density of the shot noise, $S_I(\omega)$, is $2eI$, where I, is the tunnel current. In contrast, in double-barrier heterostructures $S_I(\omega)$ can differ significantly from the Poissonian value, $2eI$, because of correlation effects in the transport process: when the correlation is negative noise is reduced while when the correlation is positive the noise is enhanced over $2eI$ [1-3].

In this work we have extended the study of noise to multi-barrier heterostructures, namely, triple-barrier and superlattice structures [4]. We have found experimentally strong deviations from Poissonian behavior, which can be explained qualitatively by existing models. However, a quantitative comparison between experiment and theory suggests an incomplete understanding of shot noise in multi-barrier systems.

EXPERIMENTAL RESULTS

For the triple-barrier experiments, we have used $In_{0.53}Ga_{0.47}As$- $In_{0.52}Al_{0.48}As$ heterostructures, in which the InAlAs regions formed the potential barriers through which electrons tunnel. We focus here on a typical heterostructure, which consisted of two end 100 Å-thick barriers and a 52 Å central barrier, with quantum wells in between having widths of 53 Å and 82 Å. Heavily doped n-type InGaAs electrodes completed the structure. (See Fig. 1) As a reference we used a double-barrier structure identical to that triple-barrier structure except that the central barrier and the 82 Å well had been eliminated.

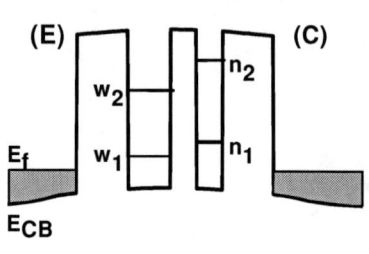
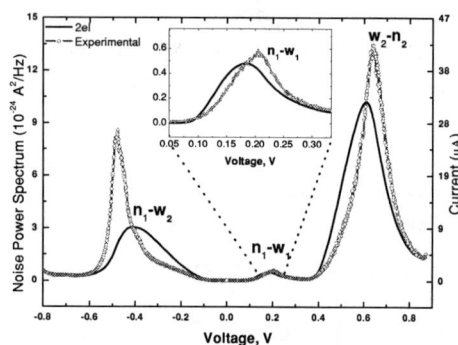

FIGURE 1. Conduction-band profile of an InGaAs-InAlAs triple-barrier structure, showing the quasi-bound quantum-well states. The doped emitter and collector are denoted by (E) and (C), respectively.

FIGURE 2. $2eI$ (Solid line) and noise (circles) characteristics at 4K of the InGaAs-InAlAs structure described in the text. The inset shows the current-noise of the first peak in forward bias. Note that the noise enhancement is much stronger for the reverse bias.

The current-voltage (I-V) characteristic of the triple-barrier structure at T = 4.2K (see Fig. 2) exhibits several regions of negative differential conductance that are the signatures of resonant tunneling involving quantum states. The two current peaks in forward bias correspond to the $w_1 - n_1$ and $w_2 - n_2$ resonances, while the peak in reverse bias is for the $n_1 - w_2$. For each peak, the current rise is gradual (see, e.g., the voltage region between 0.075 and 0.175 V or between -0.1 V and -0.4 V, in Fig. 2), which contrasts with the delta-function-like characteristic expected for electrons tunneling between two quantum wells (each a two-dimensional system), as discussed in detail elsewhere [5].

Figure 2 also shows the measured shot noise characteristic and compares it with the Poissonian value $2eI$. As it is apparent in the figure, the shot noise is reduced below $2eI$ whenever the current rises and is enhanced when the current drops. This behavior, which is qualitatively similar to that found previously in double-barrier structures [1-3] and in our own reference sample, is explained along the same lines as in a double-barrier configuration: while the current is rising the noise is reduced as a consequence of the negative correlation caused by Pauli exclusion principle; when the current starts to drop and the density of states for tunneling becomes small, an electron tunneling into a quantum well creates a potential fluctuation that shifts the quantum state up in energy and in turn induces a positive correlation among the electrons, thus enhancing the shot noise.

There are, however, two aspects of the shot noise in Fig. 2 that should be highlighted. First, the noise reduction approaches 50% of the value $2eI$ for some voltages. Second, the noise enhancement for the $n_1 - w_2$ peak is three times larger than for any of the other two peaks, even though the corresponding minimum in the conductance is much less pronounced for $n_1 - w_2$ than for $w_2 - n_2$. To compare the noise reduction we have observed with theoretical predictions, we calculated the shot noise for a triple-barrier structure using a sequential-tunneling model [6]. The calculated noise reduction for the voltages at which the experimental reduction is the largest is at most 20% of the Poissonian value, that is, much smaller than what is observed experimentally. The discrepancy between the two results may lie on the fact that the calculation did not include Coulomb interaction effects.

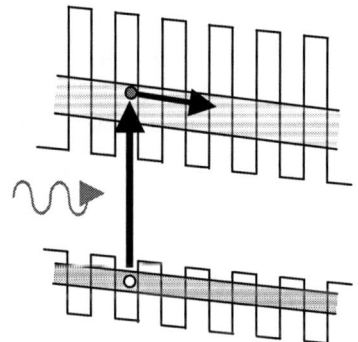

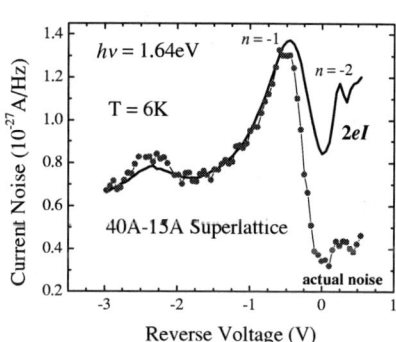

FIGURE 3. Conduction- and valence-band profiles of a superlattice in an electric field. A photon can excite an electron from the superlattice's valence miniband to its conduction miniband and generate a photocurrent. With increasing field, the minibands evolve into Stark ladders, which manifest themselves in the photocurrent.

FIGURE 4. Experimental $2eI$ (solid line) and noise (circles) of the photocurrent in a 40Å/15Å GaAs-GaAlAs superlattice. For voltages larger than −0.5V, the experimental shot noise is much smaller than the Poissonian value ($2eI$), whereas for smaller voltages the two values are practically the same. This result is independent of the wavelength of the excitation, but the details depend on the superlattice configuration.

The large noise enhancement for the $n_1 - w_2$ peak is significant because previous experiments in double-barrier structures have shown a correlation between the strength of the negative differential conductance and the noise enhancement [3], which does not appear in Fig. 2. The reason for this difference between the two cases is not clear, but it may have to do with the charge accumulated in both quantum wells, which depends on the specific quantum states involved in the various current peaks.

The behavior of current noise in superlattices also departs significantly from what a simple multi-quantum-well model predicts [6], as shown by our study of the shot noise in the photocurrent flowing through GaAs-GaAlAs superlattices embedded in the intrinsic regions of *p-i-n* diodes (Fig. 3). We have found experimentally that the noise can be drastically reduced in comparison with the Poissonian value, as shown in Fig. 4, where the measured photocurrent, multiplied by $2e$, of a 40 Å-15 Å superlattice is compared with the experimental shot noise. For small electric fields (of the order of 20 kV/cm), the shot noise is about 1/3 of $2eI$, but at a moderate field of about 60 kV/cm (transition field) the current noise abruptly becomes Poissonian or equal to $2eI$.

This result is independent of the exciting photon's energy, for a given superlattice configuration. The minimum shot noise value, however, depends on the superlattice period (or, consequently, on the interwell coupling). The general trend is for the minimum shot noise to be larger, the larger the period; the transition field depends weakly on the period. For a period of around 100 Å the shot noise is essentially Poissonian, from the low- to the high-field region.

Although the existing multi-barrier model developed for metals explains our results qualitatively [6], it cannot account for either the dependence of noise reduction on superlattice coupling at small fields or the abruptness of the field-induced transition from sub-Poissonian to Poissonian noise that we have observed.

ACKNOWLEDGMENTS

One of us (EEM) has benefited from a stay at NTT Basic Research Labs, where portions of this work were initiated. The work has been sponsored by the National Science Foundation of the US (Grant No. DMR-0305384).

REFERENCES

1. Li, Y. P., Zaslavsky, A., Tsui, D. C., Santos, M., and Shayegan, M., *Phys. Rev. B* **41**, 8388-8391(1990).
2. Iannaccone, G., Lombardi, G., Macucci, M., and Pellegrini, B., *Phys. Rev. Lett.* **80**, 1054-1057 (1998).
3. Kuznetsov, V. V., Mendez, E. E., Bruno, J. D., and Pham, J. T., *Phys. Rev. B* **58**, R10159-10162 (1998).
4. Most of work summarized here is reported in detail in Newaz, A. K. M., Song, W., Mendez, E. E., Lin, Y; and Nitta, J., *Phys. Rev. B* **71**, 195303 (2005).
5. Lin, Y., Nitta, J., Newaz, A. K. M., Song, W., Mendez, E. E., unpublished.
6. De Jong, M. J. M. and Beenaker, C. W. J., *Phys. Rev. B* **51**, 16867-16871 (1995).

Transition between Pauli exclusion and Coulomb interaction in the noise behavior of resonant tunneling devices

I. A. Maione, G. Basso, M. Macucci, G. Iannaccone, B. Pellegrini

Dipartimento di Ingegneria dell'Informazione, Università di Pisa, Via Caruso, I-56122 Pisa, Italy

Abstract. We present results from noise measurements on double barrier resonant tunnel diodes, down to extremely low current bias values. Such measurements were performed for current levels between 0.8 pA and 10 nA, in order to investigate the theoretically predicted deviations from full shot noise due to the interplay of Coulomb repulsion and Pauli exclusion. While Coulomb effects are visible also in the proximity of the current peak, Pauli exclusion is expected to contribute only at very low current levels, which has motivated our effort towards an increase in the sensitivity of the measuring equipment.

Keywords: Shot noise suppression, resonant tunneling devices
PACS: 73.40.Gk, 72.70.+m, 73.23.-b

INTRODUCTION

Theoretical studies on transport in double barrier resonant tunneling diodes and experimental results obtained on such structures show significant deviations from the noise power spectral density that would be expected from a Poisson process (full shot noise). Such deviations are the result of the concurrent effect of Pauli exclusion and Coulomb interaction, which introduce "antibunching" or "bunching" (in the negative differential resistance region) effects among electrons crossing the device.

Some models [1, 2] predict that after the transition from thermal to shot noise (occurring for bias voltage values of the order of kT, and therefore very small at cryogenic temperatures), two separate minima of the Fano factor (ratio of the shot noise power spectral density to that corresponding to full shot noise) appear in the bias region before the current peak, separated by a maximum (still below unity). This is expected to be the result of a transition between shot noise suppression dominated by Pauli exclusion (at lower bias voltages) to a suppression resulting mainly from Coulomb interaction. Such an effect should be more apparent and more easily detectable at low temperatures.

EXPERIMENTAL SETUP

Noise measurements were performed by means of a correlation amplifier, already described in a previous contribution [3], based on a Texas Instruments TLC072 dual operational amplifier, which exhibits a very low input current noise (0.6 fA/$\sqrt{\text{Hz}}$ at 1 kHz) and a voltage noise of 12 nV/$\sqrt{\text{Hz}}$ at 100 Hz; the layout of the correlation amplifier was

FIGURE 1. Air-cushioned suspension system for the liquid helium dewar (a), and non-cryogenic board of the correlation amplifier (b).

redesigned with respect to the previous version, in order to perform reliable measurements of extremely low shot noise levels.

In the current configuration, noise measurements down to liquid helium temperature are possible, while maintaining the operational amplifiers at higher temperature for proper operation. The device under test is introduced in a dewar vessel containing the cryogenic liquid by means of a stainless steel pipe. In order to reduce the thermal noise due to feedback resistors (and as a consequence the overall system noise) they are also kept in the dewar vessel, very close to the helium surface. A minimum distance from it is necessary, in order to avoid the effect of additional noise due to the ebullition of liquid helium.

Although an even higher sensitivity (0.2 pA) had been achieved with a previous setup [4] on a p-n junction at 77 K, reaching 0.8 pA in the present setup, with the liquid helium dewar and longer connecting cables, has been much more challenging, requiring a significantly more complex optimization procedure.

A careful insulation of the system from mechanical vibrations has been obtained, by means of an air cushioned structure which supports the helium dewar with the sample holder and the correlation amplifier (Fig. 1(a)). Furthermore, all coaxial cables connecting the amplifier with the sample and the feedback resistors have been clamped at several locations along the supporting shaft, in order to prevent noise resulting from time-varying deformations.

A dual operational amplifier with a slim outline package was used: this allowed us to further reduce the size of the PCB (Printed Circuit Board) for the amplifier and to obtain a very compact structure suitable for inclusion in the stainless steel insert. The amplifier PCB is shown in Fig. 1(b): jumpers are used to select the different configurations for calibration, measurement and testing. Particularly for very low currents (below 10 pA) the calibration procedure for current measurement is essential, to compensate for the offset due to the amplifier and the contact potentials.

As the frequency range of interest extends from 0.1 Hz to about 200 Hz, time records about 10 s long have to be collected. Besides, since a large number of spectra have to be averaged, in order to take advantage of the cross correlation technique, the measurement process may take a few hours for every bias point.

RESULTS AND DISCUSSION

We have used this setup to investigate the Fano factor at low bias currents (between 0.8 pA and 15 nA) of a resonant tunneling diode with a first AlGaAs barrier of 12.4 nm, a GaAs well of 6.2 nm, and second barrier of 14.1 nm fabricated by the TASC-INFM laboratory. The aluminum mole fraction is 0.36. The experimental data available so far are reported in Fig. 2: starting from the largest bias values and moving to the left we observe a decrease in the Fano factor while moving down the slope leading to the current peak (for this device the height of the resonant peak is 40 nA, at a bias voltage of about 230 mV) until a minimum is reached for a current of about 1 nA (corresponding to a bias of about 170 mV). Below 1 nA the Fano factor increases until it levels off at about 0.7, around 100 pA. For a bias less than 120 mV there is a trend towards a decrease, although much less sharp than what appeared in preliminary data that are not confirmed by the current, more accurate measurements. The leftmost data point is for a bias current of about 0.8 pA, which appears to be the present limit in sensitivity, mainly because of the achievable precision in the bias current measurement. Such a measurement is performed on the basis of the output voltage of one of the channels of the correlation amplifier, with a calibration made by means of an HP4148B picoammeter before and after the noise acquisition for each single data point. There are shifts in time of the calibration curve by a fraction of a picoampere, probably due to varying contact potentials or to a small instability of the amplifier offset.

If we compare the results of Fig. 2 with theoretical predictions obtained with the application of a technique based on the numerical evaluation of the tunneling rates across the barriers by means of a self-consistent solution of the Schrödinger and Poisson equations [5] (shown in the inset), we notice that there is a qualitative agreement with the theoretical curves representing the effect of the Coulomb interaction (dashed line) and the combined action of the Coulomb and Pauli interaction (solid line), although with significant differences in the bias values associated with minima and maxima. Unfortunately, the drop of the Fano factor toward 0.5 that would be evidence of the action of Pauli exclusion is not clearly visible in our current experimental data and, if present, would occur at extremely low current values (below 1 pA), whose shot noise is not easily accessible with the present state of the art. Work is in progress to overcome this limit in sensitivity, with improvements in the current calibration setup and with the development of a cryogenic front end to be placed next to the sample, thus eliminating the connection cables, which represent one of the main sources of interference.

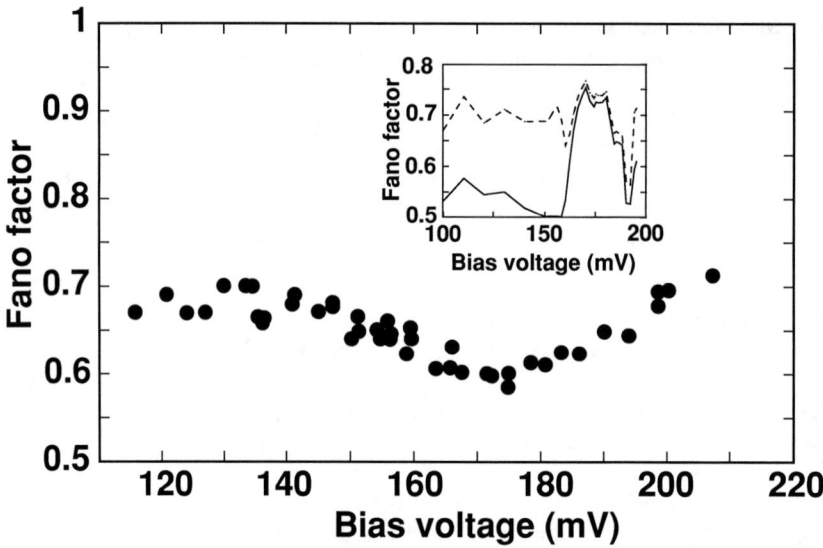

FIGURE 2. Fano factor measured in the low bias region, as a function of the applied voltage; the inset contains a plot of the Fano factor computed with a numerical simulation that includes only Coulomb interaction (dashed line) or both the effect of Coulomb interaction and of Pauli exclusion (solid line).

ACKNOWLEDGMENTS

This work was supported by the Italian Ministry of the Education, University, and Research through the cofunded project "Excess noise in nanoscale devices".

REFERENCES

1. G. Iannaccone, M. Macucci, B. Pellegrini, "Signatures of Electron-Electron Interaction in Nanoelectronic Device Shot Noise," Proceedings of *Unsolved Problems of Noise and Fluctuations*, p. 59, Adelaide, Australia (1999).
2. V. Ya. Aleshkin and L. Reggiani, *Phys. Rev. B* **64**, 245333 (2001).
3. B. Pellegrini, G. Basso, M. Macucci, "Measurement techniques of shot noise in nanostructures," in the Proceedings of the *17th International Conference on Noise and Fluctuations*, p. 693, Prague (Czech Republic), edited by J. Sikula (CNRL, 2003).
4. B. Pellegrini, G. Basso, M. Macucci, "Techniques for high-sensitivity measurements of shot noise in nanostructures," in the Proceedings of the NATO ARW *Advanced Experimental Methods for Noise Research in Nanoscale Electron Devices*, p. 203, edited by J. Sikula and M. Levinshtein (Kluwer Academic Publisher, 2004).
5. G. Iannaccone, M. Macucci, G. Basso, B. Pellegrini, "Concurrent effects of Pauli and Coulomb interaction in resonant tunneling diodes at low bias and low temperature," in the Proceedings of the *17th International Conference on Noise and Fluctuations*, p. 283, Prague (Czech Republic), edited by J. Sikula (CNRL, 2003).

Decoherence and current fluctuations in tunneling through coupled quantum dots

G. Kießlich*, P. Samuelsson[†], A. Wacker[†] and E. Schöll*

*Institut für Theoretische Physik, Technische Universität Berlin, Hardenbergstr. 36, 10623 Berlin, Germany
[†]Department of Physics, University of Lund, Box 118, SE-22100 Lund, Sweden

Abstract. The electronic transport through two coupled quantum dots in series can be described either in a fully coherent approach (e.g. density matrix description or nonequilibrium Green's functions) or in a simple sequential tunneling treatment (Pauli master equation with Fermi's Golden rule for coupling between the quantum dots). It turns out that both descriptions provide the same average current for noninteracting electrons. In contrast, the zero-frequency spectral power density and the skewness are different for intermediate coupling strengths between the quantum dots reflecting their sensitivity on coherence in the tunneling process.

Keywords: skewness, shot noise, full counting statistics
PACS: 72.70.+m, 73.23.-b, 73.63.Kv, 74.40.+k

Current fluctuations allow a detailed insight into the nature of charge transport through mesoscopic conductors. In particular, the concept of full counting statistics adopted from quantum optics [1, 2] provides the full information about the transport process by the knowledge of all its cumulants.

In the full counting statistics the crucial quantity is the distribution function of the number N of transferred charges in the time interval t_0: $P(N,t_0)$ which is related to the characteristic function or cumulant generating function $F(\chi)$ as

$$\exp[-F(\chi)] = \sum_N P(N,t_0)\exp[iN\chi] \qquad (1)$$

From the characteristic function we can obtain the cumulants $C_k = -(-i\partial_\chi)^k F(\chi)|_{\chi=0}$ which are related to the average current $\langle I \rangle = eC_1/t_0$ and to the zero-frequency noise $S_P = 2e^2 C_2/t_0$. The skewness of the distribution of transferred charges is given by the third-order cumulant C_3. Recently, the measurement of the skewness of transport through a tunnel junction was reported in Refs. [3, 4]. For a recent review of the calculation of counting statistics for various mesoscopic conductors see [5].

Our special aim is the investigation of the influence of quantum coherence on the counting statistics. For that purpose, we consider the tunneling through two vertically coupled quantum dots (QDs) [6], experimentally studied regarding the average current e.g. in [7]. Here, first results for the counting statistics of coherent and sequential tunneling and their comparison are presented. The counting statistics for sequential tunneling (incoherent description) in the framework of Pauli master equation is obtained along the lines of [8]. For coherent tunneling we use the density matrix approach starting from [9] which provides the same results for the counting statistics as the scattering matrix approach utilizing Levitov's formula [2].

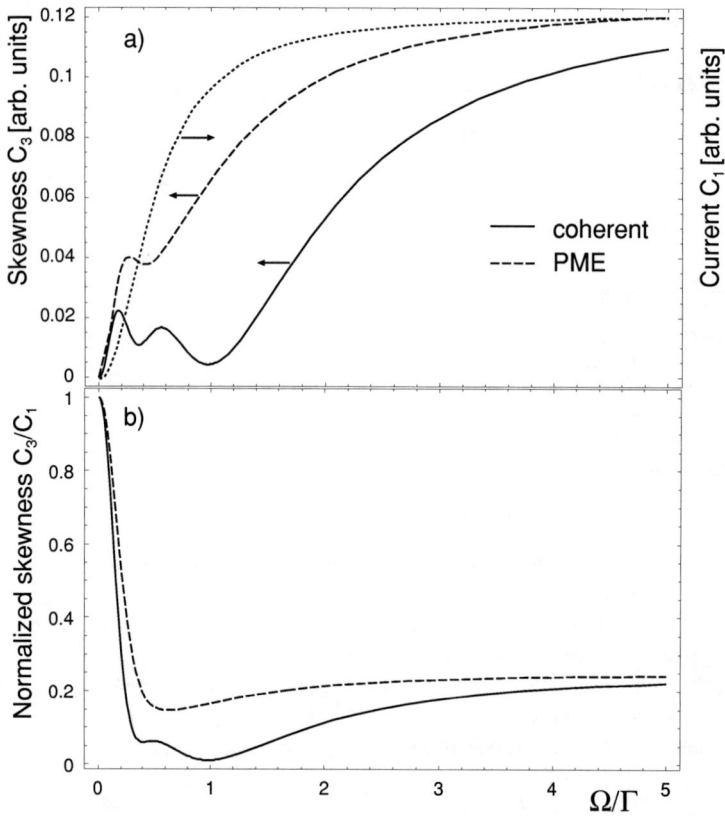

FIGURE 1. a) First-order cumulant C_1 (average current): dotted curve; Third-order cumulant (Skewness) C_3 vs. tunnel-coupling Ω. b) Normalized skewness C_3/C_1. Pauli master equation (PME): dashed curve, Coherent approach: solid curve. Symmetric contact coupling: $\Gamma = \Gamma_e = \Gamma_c$.

By comparing the results for noninteracting electrons it turns out that their first-order cumulants (average current) fully agree (dotted curve in Fig. 1a: C_1 vs. inter-QD coupling Ω), i.e. the average current is not sensitive on coherence during the tunneling process [6]. The second-order cumulants (noise) agree for small and large coupling between the QDs, but deviate for intermediate couplings. The same holds for all higher-order cumulants. As an example, in Fig. 1a the skewness and in Fig. 1b the normalized skewness C_3/C_1 vs. inter-QD coupling Ω for symmetric contact coupling ($\Gamma \equiv \Gamma_e = \Gamma_c$) are shown (dashed and solid curves). There are three distinct extrema in the normalized skewness for coherent tunneling (solid curve in Fig. 1b). The extrema are analytically found to be $\Omega/\Gamma = \{0.40, \frac{1}{2}, 0.98\}$. For $\Omega/\Gamma = 1/2$ we can identify a Poissonian transfer of quarter elementary charges. For the two other extreme points there is not such a particularly "simple" explanation.

By inserting sources of decoherence a continous transition from the coherent approach to the sequential tunneling result can be achieved [10]. This connects both limiting cases discussed here and substantiates the general finding that noise properties are very sensitive towards coherent effects in nanostructure transport.

ACKNOWLEDGMENTS

This work was supported by Deutsche Forschungsgemeinschaft in the framework of Sfb 296.

REFERENCES

1. L. S. Levitov, and G. B. Lesovik, *JETP Lett.*, **58**, 230 (1993).
2. L. S. Levitov, H. W. Lee, and G. B. Lesovik, *J. Math. Phys.*, **37**, 4845 (1996).
3. B. Reulet, J. Senzier, and D. E. Prober, *Phys. Rev. Lett.*, **91**, 196601 (2003).
4. Y. Bomze, G. Gershon, D. Shovkun, L. Levitov, and M. Reznikov, cond-mat/0504382 (unpublished).
5. Y. V. Nazarov, editor, *Quantum Noise in Mesoscopic Physics*, Kluwer Academic Publishers, Dordrecht, Boston, London, 2003.
6. H. Sprekeler, G. Kießlich, A. Wacker, and E. Schöll, *Phys. Rev. B*, **69**, 125328 (2004).
7. M. Borgstrom, T. Bryllert, T. Sass, B.Gustafson, L.-E. Wernersson, W. Seifert, and L. Samuelson, *Appl. Phys. Lett.*, **78**, 3232 (2001).
8. D. A. Bagrets, and Y. V. Nazarov, *Phys. Rev. B*, **67**, 085316 (2003).
9. S. A. Gurvitz, *Phys. Rev. B*, **57**, 6602 (1998).
10. G. Kießlich, P. Samuelsson, A. Wacker, and E. Schöll, in preparation (2005).

Noise and Bistabilities in Quantum Shuttles

Christian Flindt*, Tomáš Novotný[†,**] and Antti-Pekka Jauho*

*NanoDTU, MIC – Department of Micro and Nanotechnology, Technical University of Denmark,
DTU - Building 345 East, DK - 2800 Kongens Lyngby, Denmark
[†]Nano-Science Center, University of Copenhagen - Universitetsparken 5,
DK - 2100 Copenhagen Ø, Denmark
[**]Department of Electronic Structures, Faculty of Mathematics and Physics,
Charles University - Ke Karlovu 5, 121 16 Prague, Czech Republic

Abstract. We present a study of current fluctuations in two models proposed as quantum shuttles. Based on a numerical evaluation of the first three cumulants of the full counting statistics we have recently shown that a giant enhancement of the zero-frequency current noise in a single-dot quantum shuttle can be explained in terms of a bistable switching between two current channels. By applying the same method to a quantum shuttle consisting of a vibrating quantum dot array, we show that the same mechanism is responsible for a giant enhancement of the noise in this model, although arising from very different physics. The interpretation is further supported by a numerical evaluation of the finite-frequency noise. For both models we give numerical results for the effective switching rates.

Keywords: Current noise and fluctuations, bistabilities, quantum shuttles.
PACS: 85.85.+j, 72.70.+m, 73.23.Hk

Introduction – In 1998 Gorelik *et al.* proposed a nano-electromechanical system (NEMS), the charge shuttle, consisting of a movable nanoscopic grain coupled via tunnel barriers to source and drain electrodes [1]. Originally the motion of the grain was modelled using a classical harmonic oscillator. Here we present a study of current fluctuations in two models of (quantum) shuttles, where the oscillator is quantized.

Models – Two models have been proposed as quantum shuttles (the 1-dot shuttle [2] and the 3-dot shuttle [3]). The 1-dot shuttle consists of a single mechanically oscillating quantum dot situated between two leads. In the 3-dot shuttle the mechanically oscillating quantum dot is flanked by two static dots, thus making up an array of dots. Both devices are operated in the strong Coulomb blockade regime, and consequently only one excess electron at a time is allowed in the device. In the 1-dot (3-dot) model the coupling to the leads (the interdot coupling) depends exponentially on the position of the vibrating dot. For detailed descriptions of the models we refer to Refs. [2, 3, 4].

Both models are described using the language of quantum dissipative systems [5]. As the "system" we take in the 1-dot model (3-dot model) the single (three) electronic state(s) of the occupied dot (array) and the unoccupied state plus the quantum harmonic oscillator with natural frequency ω_0. In the limit of a high bias between the leads [6], and assuming that the oscillator is damped due to a weak coupling to a heat bath, the time evolution of the reduced density matrix of the system $\hat{\rho}(t)$ is governed by a Markovian generalized master equation (GME) of the form [2, 3, 4]

$$\dot{\hat{\rho}}(t) = \mathscr{L}\hat{\rho}(t) = (\mathscr{L}_{\text{coh}} + \mathscr{L}_{\text{damp}} + \mathscr{L}_{\text{driv}})\hat{\rho}(t). \tag{1}$$

Here $\mathscr{L}_{\text{coh}}$ describes the internal coherent dynamics of the system, while $\mathscr{L}_{\text{damp}}$ and

$\mathscr{L}_{\text{driv}}$ give the damping and the coupling to the leads, respectively. In the following we consider the stationary state defined by $\dot{\hat{\rho}}^{\text{stat}}(t) = \mathscr{L}\hat{\rho}^{\text{stat}}(t) = 0$. The GME is only valid in the high-bias limit, and hence we cannot use the applied bias as a control parameter. Instead, we vary in the 1-dot model the strength of the damping, denoted γ, and in the 3-dot model the difference between the energy levels corresponding to the outer dots, referred to as the *device bias* and denoted ε_b.

Theory – We have recently developed a systematic theory for the calculation of the n'th cumulant of the current $\langle\langle I^n\rangle\rangle$ for NEMS described by a Markovian GME of the form given in Eq. (1) [7]. In Ref. [7] a numerical evaluation of the first three cumulants showed that the 1-dot model in a certain parameter regime behaves as a bistable system switching slowly between two current channels. The first three cumulants of a bistable system switching slowly (compared to the electron transfer rates) with rates $\Gamma_{1\leftarrow 2}$ and $\Gamma_{2\leftarrow 1}$ between two current channels 1 and 2 with corresponding currents I_1 and I_2, respectively, are [8]

$$\langle\langle I\rangle\rangle = \frac{I_1\Gamma_{1\leftarrow 2}+I_2\Gamma_{2\leftarrow 1}}{\Gamma_{2\leftarrow 1}+\Gamma_{1\leftarrow 2}},$$

$$\langle\langle I^2\rangle\rangle = 2(I_1-I_2)^2\frac{\Gamma_{1\leftarrow 2}\Gamma_{2\leftarrow 1}}{(\Gamma_{1\leftarrow 2}+\Gamma_{2\leftarrow 1})^3}, \quad (2)$$

$$\langle\langle I^3\rangle\rangle = 6(I_1-I_2)^3\frac{\Gamma_{1\leftarrow 2}\Gamma_{2\leftarrow 1}(\Gamma_{2\leftarrow 1}-\Gamma_{1\leftarrow 2})}{(\Gamma_{1\leftarrow 2}+\Gamma_{2\leftarrow 1})^5}.$$

As pointed out by Jordan and Sukhorukov [8, 9] these expressions are very general, *i. e.* they do not depend on the microscopic origin of the rates or the current channels. For the 1-dot model the two current channels were identified from phase space plots of the oscillating dot as a *shuttling* and a *tunneling* channel, respectively, with known analytic expressions for the corresponding two currents [7, 11]. By comparing the numerical results for the first two cumulants with the corresponding analytic expressions given above, the two rates $\Gamma_{1\leftarrow 2}$ and $\Gamma_{2\leftarrow 1}$ could be extracted, and finally a comparison of the numerical results for the third cumulant and the analytic expression given above (with the extracted rates[1] $\Gamma_{1\leftarrow 2}$ and $\Gamma_{2\leftarrow 1}$) confirmed the conjecture about the bistable behavior (see Fig. 1). This in turn explained a giant enhancement of the zero-frequency current noise (the second cumulant) found in Ref. [11].

A similar enhancement of the zero-frequency current noise was found in a study of the 3-dot model [4]. Also in this case, the enhancement was tentatively attributed to a switching behavior, however, neither the number nor the nature of the individual current channels were clarified, and no quantitative explanation could be given. Phase space plots of the oscillating dot seem to indicate the existence of two current channels [4]: One channel, where electrons tunnel *sequentially* through the array of dots, and one channel, where electrons *co-tunnel* between the static dots. The current corresponding to each of the two channels can be read off from the numerical results obtained in Ref. [4]. By proceeding along the lines outlined above, the conjecture that the enhanced noise is due to a slow switching between the sequential and co-tunneling channel can be scrutinized.

---

[1] In a certain limit the rates may even be found analytically, see Ref. [10].

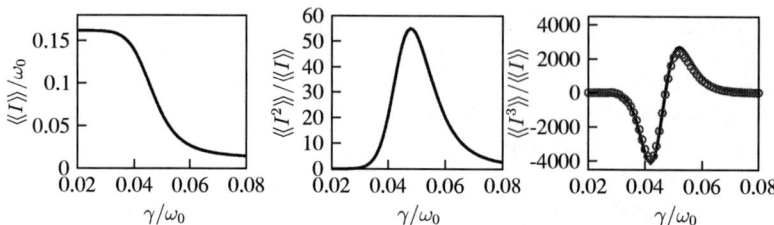

FIGURE 1. First three cumulants for the 1-dot model as a function of the damping strength γ (model parameters correspond to Fig. 3 in Ref. [7]). The shuttling channel current is $I_{\text{shut}} = \omega_0/2\pi$ and the tunneling channel current $I_{\text{tun}} = 0.0082\omega_0$ ($e = 1$). Full lines indicate numerical results, while circles show the (semi-) analytic results for the third cumulant. The central panel shows the giant enhancement of the zero-frequency noise. (Reproduced from Ref. [7]).

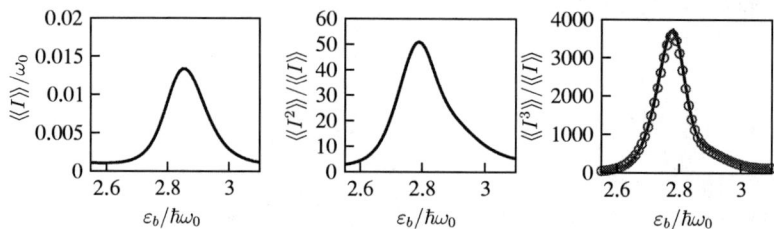

FIGURE 2. First three cumulants for the 3-dot model as a function of the device bias ε_b (model parameters correspond to Fig. 4 in Ref. [4]). The sequential tunneling channel current is $I_{\text{seq}} = 0.043\omega_0$ and the co-tunneling channel current $I_{\text{cot}} = 0.0008\omega_0$ ($e = 1$). Full lines indicate numerical results, while circles show the (semi-) analytic results for the third cumulant. The central panel shows the giant enhancement of the zero-frequency noise. (Left and central panel reproduced from Ref. [4]).

Results – In Figs. 1, 2 we show numerical results for the first three cumulants for the two models together with the analytic expression for the third cumulant of a bistable system with rates extracted from the first two cumulants. We take the agreement between the numerical and (semi-) analytic results as evidence that both models exhibit a bistable behavior. In Ref. [12] this interpretation was further supported by numerical studies of the finite-frequency current noise in the 1-dot model. Correspondingly, we show in Fig. 3 the agreement between the numerical results for the finite-frequency noise in the 3-dot model and semi-analytic results for a slow bistable switching process [12]. In Fig. 4 we show the extracted rates for both models. Most noteworthy is the crossing of the two rates in the 1-dot case, which results in the change of sign of the third cumulant seen in Fig. 1. On each side of the crossing one of the current channels dominates. In the 3-dot case, the two rates close in, however, without crossing each other. Consequently one of the current channels, the sequential tunneling channel, never dominates. It should also be noted that in both models one of the currents is comparable to one of the rates, which implies that some corrections to Eq. 2 are expected [9]. However, we have found that these corrections do not contribute significantly.

Conclusion – We have presented a study of noise in two models of quantum shuttles. By evaluating numerically the first three cumulants of the full counting statistics, we

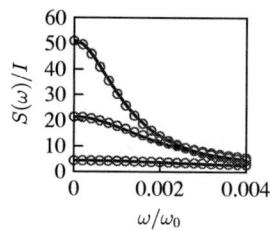

FIGURE 3. Finite-frequency current noise $S(\omega)$ (normalized with respect to the current I) for the 3-dot model. Circles indicate numerical results, while full lines are the corresponding (semi-)analytic results for a slow bistable switching process [12]. The results correspond to Fig. 2 with $\varepsilon_b = 2.60\hbar\omega_0$ (lower curve), $2.70\hbar\omega_0$, $2.79\hbar\omega_0$ (upper curve).

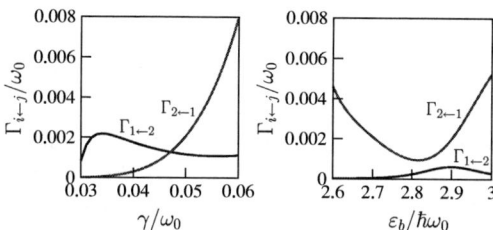

FIGURE 4. Left panel: The two switching rates for the 1-dot model as a function of the damping strength γ. Here the two current channels are the shuttling channel (1) and the tunneling channel (2). The rates correspond to the results shown in Fig. 1. Right panel: The two switching rates for the 3-dot model as a function of the device bias ε_b. Here the two current channels are the sequential tunneling channel (1) and the co-tunneling channel (2). The rates correspond to the results shown in Fig. 2.

have shown that a giant enhancement of the zero-frequency current noise in both models can be explained in terms of a slow bistable switching behavior. For both models, this interpretation is supported further by a numerical evaluation of the finite-frequency current noise. We underline that although the two models behave very differently, it is the same mechanism that is responsible for the giant enhancement of the noise.

REFERENCES

1. L. Y. Gorelik et al., *Phys. Rev. Lett.*, **80**, 4526 (1998).
2. T. Novotný, A. Donarini, and A.-P. Jauho, *Phys. Rev. Lett.*, **90**, 256801 (2003).
3. A. D. Armour, and A. MacKinnon, *Phys. Rev. B*, **66**, 035333 (2002).
4. C. Flindt, T. Novotný, and A.-P. Jauho, *Phys. Rev. B*, **70**, 205334 (2004).
5. U. Weiss, *Quantum Dissipative Systems*, 2nd ed. (World Scientific, Singapore, 1999).
6. S. A. Gurvitz and Ya. S. Prager, *Phys. Rev. B*, **53**, 15392 (1996).
7. C. Flindt, T. Novotný, and A.-P. Jauho, *Europhys. Lett.*, **69**, 475 (2005).
8. A. N. Jordan, and E. V. Sukhorukov, *Phys. Rev. Lett.*, **93**, 260604 (2004).
9. A. N. Jordan, and E. V. Sukhorukov, cond-mat/0503751.
10. A. Donarini, *Dynamics of Shuttle Devices*, PhD thesis, MIC, DTU, cond-mat/0501242 (2004).
11. T. Novotný, A. Donarini, C. Flindt, and A.-P. Jauho, *Phys. Rev. Lett.*, **92**, 248302 (2004).
12. C. Flindt, T. Novotný, and A.-P. Jauho, to appear in *Physica E*, cond-mat/0412425.

Noise Minimization in Quantum Transistors

U. Gavish*, B. Yurke[†] and Y. Imry**

Institute for Theoretical Physics, University of Innsbruck, Innsbruck 6020, Austria
[†]*Bell Laboratories, Lucent Technologies, Murray Hill, NJ 07974*
**Condensed Matter Physics Dept., Weizmann Institute, Rehovot 76100, Israel*

Abstract. We present limits on the minimal noise produced in transistor-type linear amplifiers and a practical procedure for obtaining these limits. The validity of the procedure is verified for a signal which is produced by a set of LC circuits and amplified by a molecular transistor.

Keywords: Quantum noise, Quantum Transistors, Quantum Amplifiers
PACS: 42.50.Lc, 42.65.Yj, 84.30.Le

INTRODUCTION

Consider a current I_{in}, flowing out of a system which has a differential conductance[1] g_s and which is attached to an amplifier input port. The amplified signal coming out of the amplifier, I_{out}, is extracted from a load resistor connected to the amplifier output port and having a differential conductance g_ℓ. We shall consider an amplifier which is impedance-matched[2] to the system and the load, i.e., it has an impedance g_s^{-1} at its input port and g_ℓ^{-1} at its output port. If $I_{out}(t)$ is proportional to $I_{in}(t)$ the amplifier is called linear (and phase insensitive). One can then define the *power gain*, G^2, of the amplifier by the input-output relation $I_{out}(t) = G\left(g_\ell/g_s\right)^{1/2}I_{in}(t)$. Strictly speaking, linear amplifiers do not exist. A linear input-output relation of an actual amplifier has the form

$$I_{out}(t) = G\sqrt{\frac{g_\ell}{g_s}}I_{in}(t) + I_N(t) \tag{1}$$

where $I_{out}(t) = e^{iH_{tot}t}I_{out}(0)e^{-iH_{tot}t}$, $I_{in}(t) = e^{iH_s t}I_{in}(0)e^{-iH_s t}$, $H_{tot} = H_a + H_s + \gamma H_{a,s}$ is the total hamiltonian, H_s is the hamiltonian of the system, H_a is that of the amplifier, and $\gamma H_{a,s}$ is that of the interaction between them. γ is a small dimensionless coupling constant. I_N is called the current noise operator and is a function of operators related to the amplifier degrees of freedom and therefore commutes with I_{in} : $[I_N(t), I_{in}(t)] = 0$. The appearance of I_N is a consequence of Heisenberg's uncertainty principle. For the case of signals which are carried by (non conserved) bosons, it was shown by Caves[1] that a linear amplifier must add a noise power, referred to input, of $\hbar\omega/2$ per unit bandwidth to

---

[1] The differential conductance at frequency ω is the linear coefficient between a small AC field at frequency ω and the AC current which is produced as a response to it. If the AC field is applied on top of an existing DC field, this current is the *additional* current appearing on top of the existing DC one.
[2] The constraints presented below hold for the impedance-matching case. However, the noise minimization procedure which is derived from them holds also in the general case of impedance mismatch.

the output signal. Recently, this result was generalized for arbitrary signals subjected to a linear detection[2][3] and amplification[4]. In particular it was shown in Ref. [4] that the noise operator I_N should satisfy the constraint

$$\Delta I_N^2 \geq (G^2 - 1)\frac{\hbar\omega_0}{2}g_\ell\Delta\nu . \qquad (2)$$

where $\Delta X^2 \equiv \langle X^2 \rangle - \langle X \rangle^2$, $\Delta\nu \equiv \Delta\omega/(2\pi)$ and where $\Delta\omega$ is the narrow bandwidth around ω_0 - the central frequency of the band in which the detection is performed. The main assumptions were that for any (small) value of γ the system and the amplifier are in a stationary state and the differential conductance is constant.

MINIMAL NOISE IN QUANTUM TRANSISTORS

In transistor-type devices the noise current is the form $I_N = I_0 + I_n$, where I_0 is the noise current at $\gamma = 0$, and $I_n \sim \gamma^2$. The considerations leading to Eq.(2) lead also to[5],[6]

$$\Delta I_0(t)\Delta I_n(t) \geq \frac{1}{4}G^2\hbar\omega_0 g_\ell\Delta\nu. \qquad (3)$$

Assuming that I_0 and I_n are uncorrelated, the ideal noise limit is achieved when $\Delta I_N^2 = \Delta I_0^2 + \Delta I_n^2$ is minimal, i.e. when

$$\Delta I_0^2 = \Delta I_n^2 = \frac{1}{4}G^2\hbar\omega_0 g_l\Delta\nu . \qquad (4)$$

NOISE MINIMIZATION PROCEDURE

We now present a practical procedure for achieving the condition Eq.(4). We assume that there are at least two controllable parameters: the transistor source-drain voltage, V, and the coupling constant γ. The latter may be controlled, for example, by varying the transistor gate capacitance. We also assume that $\hbar\omega_0 \ll eV$.

Our procedure consists of two steps, *noise matching* and *gain matching*. The first step, consists of varying the coupling and the voltage until the the initial noise ('shot-noise') ΔI_0^2, and the new noise ('back-action') ΔI_n^2, match:

$$\Delta I_n^2(V,\gamma) = \Delta I_0^2(V) \qquad (5)$$

The matching is enabled by the fact that the shot-noise and the back-action noise dependence on the voltage and the coupling is different. After the noise matching has been achieved, *two* power gains are defined: The first, $G^2(V,\gamma)$, by a direct gain measurement, and the second, the *effective* power gain, $G_{eff}^2(V,\gamma)$, by the relation:

$$\Delta I_0^2(V,\gamma) \equiv G_{eff}^2(V,\gamma)\frac{\hbar\omega_0}{4}g_\ell\Delta\nu. \qquad (6)$$

The second step consists of matching the two gains, that is, varying the bias voltage and the coupling until $G_{eff}^2(V,\gamma) = G^2(V,\gamma)$. This matching should be done while maintaining the condition

$$\gamma G(\gamma, V) = const. \tag{7}$$

Eq.(7) ensures that the gain matching is performed while keeping the shot noise and back-action noise matched as in Eq.(5) and therefore, as a result Eq.(4) is obtained. Eq.(7) is the transistor counterpart of the condition $\gamma Q_p = const$ which is used in the analysis of quantum parametric amplifiers [7], where Q_p is the pump amplitude and γ is the coupling between it and the signal.

EXAMPLE

We now apply our results to a molecular transistor (modelled as a resonant barrier) coupled capacitively to the total charge on the capacitors of a continuous set of LC circuits. The model is similar in many features to that of Ref.[8]. Here,

$$H_a = \Sigma_{i=1,2} \int_0^\infty d\varepsilon \varepsilon b_i^\dagger(\varepsilon) b_i(\varepsilon) + \hbar \omega_A A^\dagger A + \Sigma_{i=1,2} \int_0^\infty d\varepsilon \frac{i\kappa(\varepsilon)}{\sqrt{2\pi}} (b_i^\dagger(\varepsilon) A - A^\dagger b_i(\varepsilon)),$$

$$H_s = \int_B d\omega \hbar \omega a^\dagger(\omega) a(\omega),$$

$$\gamma H_{a,s} = \frac{A^\dagger A e \hat{Q}_s}{C_g}, \tag{8}$$

where $\hat{Q}_s = \Delta Q(\omega_0) \int_B d\omega \frac{1}{\sqrt{\omega}} (a(\omega) + a^\dagger(\omega))$ is the total charge on the capacitors in the LC oscillators and where $B = [\omega_0 - \Delta\omega/2, \omega_0 + \Delta\omega/2]$. The b_i's, A's and $a(\omega)$'s satisfy respectively continuous fermionic, discrete fermionic and continuous bosonic commutation relations. $\kappa^2(\varepsilon)$ which is the resonance width is taken to be wider than eV so that the *second* derivative of the transmission with respect to ε can be neglected. C_g is the gate capacitance of the amplifier and $\Delta Q(\omega_0) = \sqrt{\hbar \omega_0 C/2}$ is the charge fluctuation of an LC oscillator ground state where C is the capacitance in each one of the LC circuits. The coupling constant is taken to be $\gamma \equiv e\Delta Q/(C_g \kappa^2) \ll 1$, and $H_{a,s} = \kappa^2 \hat{Q}_s/\Delta Q$. We assume that bath 1 and 2 have chemical potentials $\mu + eV$ and μ respectively. The current I_{out} is given by $I_{out}(t) = \frac{1}{4}(\dot{Q}_1(t) - \dot{Q}_2(t))$ where $Q_i(t) = e \int_0^\infty d\varepsilon b_i^\dagger(\varepsilon,t) b_i(\varepsilon,t)$. Due to the impedance matching, the current delivered to the load is half of that which would flow if the load was replaced by a short, which is the reason for the factor $1/4$ in I_{out}. Solving the Heisenberg equations of motion to second order in γ we find $I_{out}(t) = I_0(t) + G\sqrt{\frac{g_\ell}{\tilde{g}_s}} \tilde{I}_{in}(t) + I_n(t) + O(\gamma^3)$ where $\tilde{I}_{in} \equiv \frac{1}{2}\omega_0 Q(t)$ and

$$G = \gamma \frac{eV}{\hbar \omega_0} T \sqrt{2(1-T)}. \tag{9}$$

T is the transmission, $g_\ell = Te^2/2\pi\hbar$ and $\tilde{g}_s \equiv \pi \Delta Q^2/\hbar$. $\tilde{g}_s$ is the differential linear response of the "current" $\tilde{I}_{in} = \omega_0 Q/2$. Also the factor $\frac{1}{2}$ in $\tilde{I}_{in}$ is due to the impedance matching: the signal delivered to the amplifier is half of that which would flow out of the system if the amplifier was replaced by a short. The device performs linear amplification of $\tilde{I}_{in}$ instead of $I_{in} = \dot{Q}_s/2$ because of the capacitive coupling. However, this difference is compensated for by the replacement of g_s by $\tilde{g}_s$. One also obtains the expected[9] shot noise spectrum

$$\Delta I_0^2 = T(1-T)\frac{e^3 V}{4\pi \hbar}\Delta v, \tag{10}$$

and the back-action noise spectrum

$$\Delta I_n^2 = \gamma^4 T^5 (1-T)(\frac{eV}{\hbar \omega_0})^2 \frac{e^3 V}{4\pi \hbar}\Delta v. \tag{11}$$

The two types of noise turns out to be uncorrelated $\langle I_n I_0 \rangle = 0$ in accordance with our assumptions. Eqs. (9), (10) and (11) yield

$$\Delta I_0(t)\Delta I_n(t) = \frac{1}{4}G^2 \hbar \omega_0 g_\ell \Delta v, \tag{12}$$

In the present case, the gain matching will automatically be satisfied because Eq.(12) is an equality. To perform the noise matching, we equate the right hand sides of Eqs. (10) and (11). This results in the condition $\gamma^2 \frac{eV}{\hbar \omega_0} T^2 = 1$ which, together with Eq.(9), yields the the ideal ('Heisenberg limit') value

$$G^{(H)} = \frac{1}{\gamma}\frac{\sqrt{2(1-T)}}{T}. \tag{13}$$

In particular, $\gamma G^{(H)} \sim 1$ confirming the arguments that led to Eq.(7).

ACKNOWLEDGMENTS

U. G. and Y. I. are thankful for helpful discussions with Y. Levinson. Research at WIS was supported by the Center of Excellence of the Israel Science Foundation (ISF) and by the German Federal Ministry of Education and Research (BMBF), within the framework of the German Israeli Project Cooperation (DIP).

REFERENCES

1. C. M. Caves, Phys. Rev. D **26**, 1817 (1982).
2. D. V. Averin, cond-mat/0301524
3. A. A. Clerk, S. M. Girvin and A. D. Stone Phys. Rev. B **67**, 165324 (2003).
4. U. Gavish, B. Yurke and Y. Imry, Phys. Rev. Lett. 93, 250601 (2004).
5. U. Gavish, Y. Imry, and B. Yurke, in Proc. Rencontres de Moriond on Quantum Information and Decoherence in Nanosystems, January 2004, Eds. C. Glattli, M. Sanquer and J. Tran Thanh Van.
6. A. A. Clerk, Phys. Rev. B **70**, 245306 (2004). See Eq. (14) which is, under certain assumptions, equivalent to Eq.(3) in the present work.
7. B. Yurke and J. S. Denker, Phys. Rev. A **29**, 1419 (1984).
8. D. Mozyrsky, I. Martin and M. B. Hastings, Phys. Rev. Lett 92, 018303 (2004).
9. V.A. Khlus, JETP 66, 1243 (1987).

Length dependence of the Fano factor in mesoscopic cavities

P. Marconcini, M. Macucci

Dipartimento di Ingegneria dell'Informazione, Università di Pisa, Via Caruso, I-56122 Pisa, Italy

Abstract. The dependence of the shot noise suppression factor on the length of a mesoscopic cavity is investigated using numerical techniques based on the recursive Green's function approach and on the scattering matrix formalism. Our findings show that such a dependence disappears as soon as the magnetic field is raised above a threshold value and that there is no significant effect of the detailed behavior of the potential landscape inside the cavity.

Keywords: Shot noise suppression, mesoscopic cavity
PACS: 72.70.+m, 73.23.-b, 75.75.+a, 75.47.Jn

INTRODUCTION

Cavities defined by means of quantum point contacts along a mesa in a heterostructure represent an interesting model system for the investigation of shot noise suppression in mesoscopic structures: it has been predicted with several approaches that, in the case of a cavity defined by narrow and symmetric constrictions, the Fano factor (ratio of the shot noise power spectral density to that which would be expected for Poissonian noise) has a value equal to 1/4 [1]. This theoretical finding has also received experimental confirmation as a result of the measurements performed by Oberholzer et al. [2]. It is further known that if the constrictions are widened, the Fano factor decreases below 1/4, until it eventually drops to zero when the constrictions are as wide as the cavity (thus forming a noiseless channel with integer transmission). Experimental data [3] are available for the dependence of the Fano factor on the number of modes propagating through the constrictions, and, therefore, on the width of the constriction. In the presence of a magnetic field orthogonal to the structure, a decrease of the Fano factor has been measured [3], resulting from the reduced scattering at the entrance and exit constrictions, as the cyclotron diameter becomes comparable to their width [4].

In this contribution, we focus on the dependence of the Fano factor on the length of the cavity. By means of simulations based on the recursive Green's function method, we find that, in the case of constrictions that are narrow with respect to the width of the cavity, its length does not affect the Fano factor. This is true both in the absence of a magnetic field (with a Fano factor of 1/4), and in the presence of a magnetic field (with a Fano factor below 1/4).

The situation is quite different if the width of the constrictions is not small in comparison with the cavity width. In this case, the Fano factor, starting from a value below 1/4, increases as the length of the cavity is increased, apparently up to an asymptotic value. Most of the theories for the explanation of noise suppression in mesoscopic cavities rely on techniques that were specifically developed for the investigation of chaotic systems,

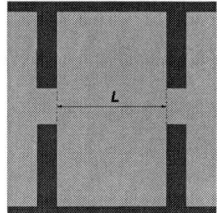 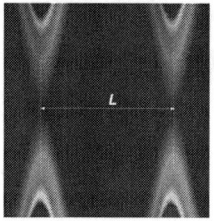 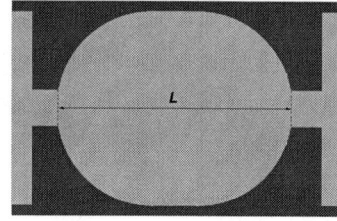

FIGURE 1. Potential landscape defining the considered cavities: rectangular cavity (left), cavity with adiabatic constrictions (center), stadium shaped cavity (right).

moving from the assumption that transport in such structures is in a chaotic regime. These cavities are indeed defined "chaotic cavities" in most of the existing literature, and the fact that chaos is not needed to explain the shot noise suppression down to 1/4 is almost never explicitly stated. We have pointed out in Ref. [4] that only a few simple conditions, not implying a chaotic behavior, are needed to obtain the 1/4 noise suppression: the constrictions must be symmetrical, narrow and with integer conductance, the occupancy of the states in the cavity must be dependent only on their energy.

RESULTS AND DISCUSSION

We start from the analysis of the Fano factor dependence on cavity length for a few relevant cavity shapes in the absence of a magnetic field, with the purpose of determining whether the shape or the characteristics of the confinement potential have any significant effect on such a dependence. The Fermi level is set at 9 meV, a value typical for the 2DEG (2-dimensional electron gas) obtained in a GaAs/AlGaAs heterostructure, and the effective mass of GaAs ($m^* = 0.067$) is used for all calculations.

In Fig. 1 we present a sketch of the three structures that have been considered: a rectangular hard-wall cavity, a cavity defined by adiabatic constrictions, and a stadium-shaped cavity. The width of all cavities in the widest central region is 8 μm and the conventional length L for comparison is assumed as the dimension indicated in Fig. 1. For the potential at the adiabatic constrictions we have adopted the model by Ando [5], given (for each constriction and with reference to its center) by

$$V(x,y) = \frac{E_1}{2}\left[1 + \cos\left(\frac{2\pi x}{L_x}\right)\right] + E_2 \sum_{\pm} \left(\frac{y - y_{\pm}(x)}{\Delta}\right)^2 \theta[\pm(y - y_{\pm}(x))] \quad (1)$$

where

$$y_{\pm}(x) = \pm\frac{L_y}{4}\left[1 - \cos\left(\frac{2\pi x}{L_x}\right)\right]. \quad (2)$$

We have chosen $E_1 = 3.5$ meV, $E_2 = 7$ meV, $L_x = 2.9$ μm, $L_y = 8$ μm, $\Delta = 720$ nm. The function θ is the Heaviside step function, thus equals 0 when the argument is negative and 1 when the argument is positive (notice that here the terms appearing in the argument of the step function are not squared, contrary to what happens in the expression reported

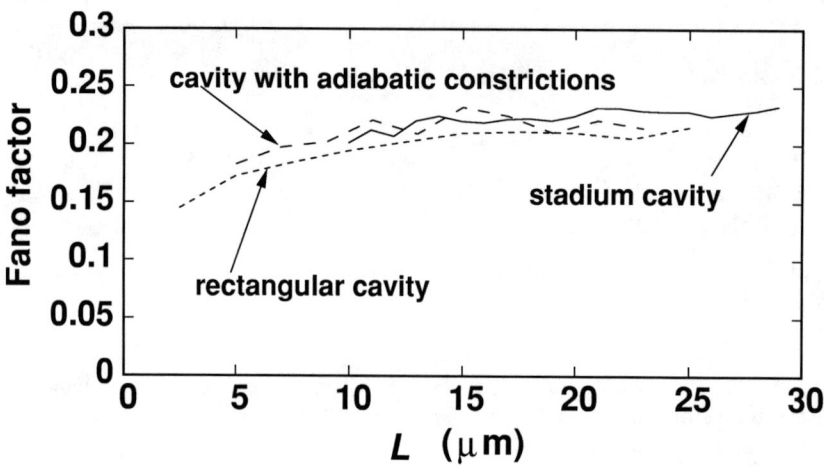

FIGURE 2. Fano factor as a function of the distance L between the input and output constrictions, for the case of a hard-wall cavity (solid line), of a cavity defined by adiabatic constrictions (dashed line), and for a stadium-shaped cavity (dotted line).

in Ref. [5], probably as a result of a typesetting error). The radius of the rounded ends of the stadium equals one half of the maximum cavity width. For all three model cavities, the length is varied by adjusting the longitudinal extension of the central region (8 μm wide), and the number of modes propagating in each constriction is 40.

The results for the Fano factor as a function of cavity length are reported in Fig. 2 with a solid line for the rectangular cavity, a dashed line for the cavity with adiabatic constrictions and a dotted line for the stadium-shaped cavity. It is apparent that there is little difference among the curves, although the potential landscape is quite different in the three cases: this further supports our approach in Ref. [4], where a rectangular model cavity was used to investigate the dependence of the Fano factor on the number of modes propagating through the constrictions, obtaining quite a good agreement with the experimental results of Ref. [3].

A different behavior is observed in the presence of an orthogonal magnetic field that we have studied considering a rectangular hard-wall cavity 8 μm wide, with 1 μm constrictions. Calculations have been performed with a scattering matrix formalism, using a transverse gauge with the only nonzero component along the direction of current propagation and following Tamura and Ando [6] for the calculation of the transverse eigenfunctions.

In the presence of a magnetic field, the number of modes propagating through the constrictions is reduced, as a consequence of the compression of the wave-function against the walls. In Fig. 3 we present results for the Fano factor as a function of cavity length for a perpendicular magnetic field of 0.06 T (empty dots) and 0.12 T (squares), along with the same data, for comparison purposes, in the absence of a magnetic field (solid dots). As the magnetic field B is increased, the Fano factor dependence with length appears to decrease (although some fluctuations are observed for intermediate values of

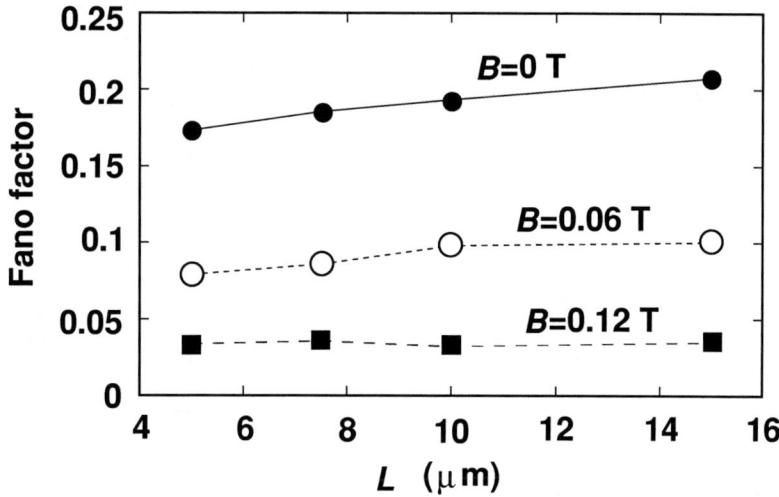

FIGURE 3. Fano factor as a function of the distance L between the input and output constrictions, for the case of a hard-wall cavity in the presence of a perpendicular magnetic field of 0.12 T (squares), of 0.06 T (empty dots), and for no magnetic field (solid dots).

B), until, above about 0.1 T, it vanishes. We can hypothesize a tentative explanation, based on the fact that, as edge states are formed, the actual length of the path going from one constriction to the other becomes much longer than the geometric distance between constrictions, thereby reaching the threshold above which an asymptotic value of the Fano factor is attained.

ACKNOWLEDGMENTS

Financial support from the Italian Ministry of Education, University and Research (MIUR) through the FIRB Project "Nanotechnologies and nanodevices for the information society" and through the cofunded program "Excess noise in nanoscale devices" is gratefully acknowledged.

REFERENCES

1. R. A. Jalabert, J.-L. Pichard, and C. W. J. Beenakker, *Europhys. Lett.* **27**, 255 (1994).
2. S. Oberholzer, E. V. Sukhorukov, C. Strunk, C. Schönenberger, T. Heinzel, and M. Holland, *Phys. Rev. Lett.* **86**, 2114 (2001).
3. S. Oberholzer, E. V. Sukhorukov, and C. Schönenberger, *Nature* **415**, 765 (2002).
4. P. Marconcini, M. Macucci, G. Iannaccone, B. Pellegrini, G. Marola, cond-mat/0411691.
5. T. Ando, *Phys. Rev. B* **44**, 8017 (1991).
6. H. Tamura, T. Ando, *Phys. Rev. B* **44**, 1792 (1991).

Chaotic-to-regular crossover of shot noise in mesoscopic conductors

Stefan Rotter, Florian Aigner, and Joachim Burgdörfer

Institute for Theoretical Physics, Vienna University of Technology, Wiedner Hauptstraße 8-10/136 A-1040 Vienna, Austria, EU

Abstract. We study the shot noise by numerically simulating phase-coherent transport through a quantum dot. The chaotic-to-regular crossover regime of shot noise suppression is investigated explicitly by tuning the disorder potential and the openings of the dot. Employing the Modular Recursive Green's Function Method we obtain results for the Fano factor in regular systems which show a remarkable similarity to the results in chaotic systems. We argue that in the absence of chaotic scattering diffraction at the lead openings is the dominant source of shot noise. Estimates for the shot noise induced by this mechanism are presented, which agree with the numerical data.

Keywords: shot noise, quantum dots, ballistic transport
PACS: 73.23.-b, 05.45.Mt, 73.63.Kv, 72.70.+m

The current noise induced by the discreteness of the electron charge ("shot noise") has attracted attention now already for almost a century [1]. Recently this topic has resurfaced in the field of *mesoscopic physics* [2, 3], where ballistic quantum transport experiments [4, 5] and theoretical advances [6, 7, 8, 9, 10, 11, 12, 13] have mutually stimulated each other. In this context shot noise has been employed to explore the crossover from a deterministic (classical) particle picture of electron motion to a probabilistic (quantum) description, where electrons behave as matter waves. The quantum uncertainty inherent in the latter picture gives rise to noisy transport. In conductance through quantum dots the correlations between electrons in the Fermi sea lead to a suppression of shot noise S relative to the Poissonian value of uncorrelated electrons S_P [14] which is customarily expressed in terms of the Fano factor $F = S/S_P$.

Most investigations to date have focused on quantum dots whose classical dynamics is fully chaotic [6, 8, 9, 10, 11, 12, 14, 15, 16]. In this limit, random matrix theory (RMT) [16] predicts a universal value for the Fano factor, $F = 1/4$. The applicability of this RMT result requires, in addition to the underlying chaotic dynamics, dwell times in the open cavity τ_D which are sufficiently long compared to the Ehrenfest time τ_E. The latter estimates the time for the initially localized quantum wavepackets to spread all over the width d of the cavity (typically $d \approx \sqrt{A}$ with A area of the dot) due to the divergence of classical chaotic trajectories. It can be estimated as [17]

$$\tau_E = \Lambda^{-1} \ln(d/\lambda_F),\qquad(1)$$

where Λ is the Lyapunov exponent ($\Lambda > 0$ for a chaotic cavity), and λ_F is the de Broglie wavelength associated with the wavenumber at the Fermi surface k_F. The limit $\tau_E/\tau_D \ll 1$ corresponds to the quantum (or RMT) regime and $\tau_E/\tau_D \gg 1$ corresponds to the classical limit for which $F = 0$ is expected. For ballistic cavities in the crossover between these two regimes a simple conjecture for F was put forward [6],

$$F = 1/4 \exp(-\tau_E/\tau_D) . \qquad (2)$$

For cavities with a short-ranged disorder potential, an alternative crossover behavior,

$$F = 1/4 \,(1+\tau_Q/\tau_D)^{-1} , \qquad (3)$$

was proposed [4, 11], where τ_Q is a characteristic scattering time within which the wavepacket is scattered into random direction. The quantities τ_Q and τ_E are closely related to another as both denote the characteristic time scale for spreading of the wavepacket by chaotic scattering either at the boundary (τ_E) or the interior (τ_Q) of the cavity. Moreover, for short ranged disorder with a correlation length $l_C < \lambda_F$, τ_Q incorporates, just as τ_E, quantum effects and depends on an effective $\hbar_{\text{eff}}$ of the system. The crossover from the chaotic to the regular regime is therefore predicted to be controlled by a single ratio τ_E/τ_D or τ_Q/τ_D which will be a function of the size of quantum effects ($\hbar_{\text{eff}}$) and the mean rate of irregular (chaotic) scattering, $\langle\Lambda\rangle$. The chaotic-to-regular crossover corresponds to the limit $\langle\Lambda\rangle \to 0$, while the quantum-to-classical limit involves $\hbar_{\text{eff}} \to 0$.

One open question not yet well understood is the behavior of shot noise for motion in a regular rather than chaotic cavity, i.e. in the limit $\Lambda \to 0$. For a mixed system lower values of F have been observed [7] suggesting that for regular systems F may vanish. Taken at face value, Eq. (2) yields $F = 0$ (complete suppression of shot noise) for the case of $\tau_E \to \infty$ or $\Lambda \to 0$ at fixed value of $\hbar_{\text{eff}}$. To investigate this question, we analyze a model system that allows to study the crossover regime from chaotic to regular dynamics, i.e. $\Lambda \to 0$, explicitly. Our scattering system consists of a rectangular cavity to which two leads of width d are attached via tunable shutters with an opening width w (see Fig. 1a). Varying the lead openings and the disorder potential allows to tune the dwell time τ_D and the mean rate of chaotic spreading of the wavepacket, $\langle\Lambda\rangle$, independently. The cavity region of width d and length $2d$ contains a disorder potential V characterized by its mean value $\langle V \rangle = 0$, and the correlation function $\langle V(x)V(x+a)\rangle = \langle V^2\rangle \exp(-a/l_C)$. The correlation length l_C is typically a small fraction of the Fermi wavelength $l_C/\lambda_F \approx 0.12$ and the potential strength $V_0 = \sqrt{\langle V^2\rangle}$ is weak, $V_0/E_F \leq 0.1$. In the limit of vanishing disorder ($V_0 \to 0$) the motion inside the cavity becomes completely regular.

Our quantum calculation proceeds within the framework of the modular recursive Green's function method (MRGM) [18] which allows to treat two-dimensional quantum dots with relatively small λ_F (or small $\hbar_{\text{eff}}$). Details are given in Ref. [13]. We evaluate the transmission amplitudes t_{mn} for an electron injected from the left by projecting the Green's function at the Fermi energy $G(E_F)$ onto all modes $m,n \in [1,\ldots,N]$ in the in- and outgoing lead, respectively. The Fano factor F is then calculated from the N-dimensional transmission matrices t [3],

$$F = \frac{\langle \text{Tr}\, t^\dagger t(\mathbf{1}-t^\dagger t)\rangle}{\langle \text{Tr}\, t^\dagger t\rangle} = \frac{\langle \sum_{n=1}^N T_n(1-T_n)\rangle}{\langle \sum_{n=1}^N T_n\rangle} , \qquad (4)$$

with T_n being the eigenvalues of $t^\dagger t$. The brackets $\langle\ldots\rangle$ indicate that we average over 150 equidistant points in the wavenumber-range $k_F \in [40.1, 40.85] \times \pi/d$, where 40 transverse lead modes are open. Figure 1c displays the Fano factor as a function of the inverse dwell time τ_D^{-1}. For $\tau_D^{-1} \to 0$ (i.e. large dwell times) F approaches the universal value 1/4 irrespective of the strength of the disorder potential V_0, while for shorter dwell times F falls off gradually (Fig. 1c). The steepness of this decrease is clearly dependent on V_0 and thus on the mean scattering rate $\langle\Lambda\rangle$. Most striking is the feature that for $V_0 \to 0$ but long dwell times the shot noise reaches the RMT value even though the dynamics is now entirely regular (see Fig. 1c). This observation suggests that the conjectures [Eq. (2) or Eq. (3)] require modifications to properly account for the shot noise in the regular limit. We argue that the key point is the wavepacket diffraction at the cavity openings which has to be incorporated in the theoretical description of shot noise [19, 20, 21]. Note that this feature is inherent in quantum transport and independent of the underlying regular or chaotic dynamics [18]. Scattering due to *chaotic* dynamics, which lies at the core of RMT, certainly leads to wavepacket spreading but does not constitute the only or, in general, dominant source.

To quantify the amount of diffraction in the cavity we perform a quasi-classical Monte-Carlo transport simulation in which we follow an ensemble of classical trajectories subject to Fraunhofer scattering at the shutter openings [19] and a random Poissonian scattering process in the disorder potential region [22]. For the latter we calculate the transport mean free path ($\tau_S \cdot v_F$) and the differential scattering probability ($P(\theta) \sim d\sigma/d\theta$) in first Born approximation, thus taking into account quantum diffractive scattering (for $l_C \cdot k_F < 1$) along the lines of Refs. [4, 11]. We find that the differential cross section is strongly peaked at small forward scattering angles (see Fig. 1b). The modified Ehrenfest time $\tilde{\tau}_E$ which includes these diffractive corrections is drastically reduced as compared

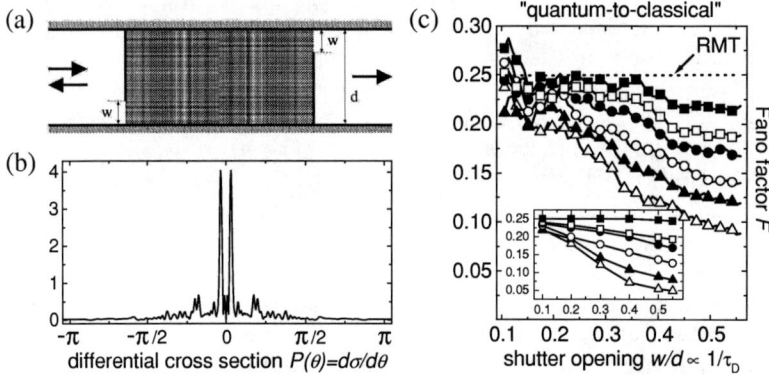

FIGURE 1. (a) Rectangular quantum billiard with tunable shutters and disorder potential (gray shaded area). Tuning the opening of the shutters w, the crossover from quantum-to-classical scattering can be investigated. (b) Normalized differential scattering probability $P(\theta) \sim d\sigma/d\theta$ as calculated in first Born approximation for the employed disorder potential. (c) Fano factor F in the quantum-to-classical crossover regime (numerical data from the full quantum simulation for the geometry depicted in (a)). Curves shown correspond to different strengths of the disorder potential (measured with respect to the Fermi energy E_F): $V_0/E_F = 0.1\,(\blacksquare)$, $0.07\,(\square)$, $0.05\,(\bullet)$, $0.03\,(\circ)$, $0.015\,(\blacktriangle)$, $0\,(\triangle)$. A decrease from the "quantum value" $F = 1/4$ for large τ_D towards the "classical value" $F = 0$ for short τ_D is clearly visible. The inset depicts the theoretical prediction based on a quasiclassical simulation. Note the good agreement with the numerical data from the full quantum calculation.

to the conjecture in Eq. (1). For an improved estimate of the Fano factor F we additionally take into account the exact dwell time distribution $P(t)$, resulting from the quasi-classical simulation for the particular system we study. Following [9] these ingredients determine the Fano factor F as follows:

$$F = 1/4 \left[1 - \int_0^{\widetilde{\tau}_E} P(t)\,dt \right] = 1/4 \int_{\widetilde{\tau}_E}^{\infty} P(t)\,dt. \tag{5}$$

Note that this expression is applicable to chaotic as well as regular systems and valid irrespective of whether the origin of spreading is ballistic scattering at the boundary or diffractive scattering inside the cavity. The estimate according to Eq. (5) (see inset of Fig. 1c) is in very good agreement with the results from the quantum calculations.

To summarize, we have numerically determined the behavior of the Fano factor F in a realistic scattering system with a tunable disorder potential and tunable shutters. We find that diffraction at the lead openings is sufficient to establish the RMT prediction for shot noise suppression ($F = 1/4$), irrespective of regular or chaotic dynamics. The chaotic-to-regular crossover in F can be estimated by a generalization of a previously proposed dependence [9] on the Ehrenfest time $\widetilde{\tau}_E$ [Eq. (5)], provided that the definition of the Ehrenfest time is properly modified to include diffraction.

We thank C. W. J. Beenakker, J. Cserti, V. A. Gopar, F. Libisch, I. Rotter, H. Schomerus, M. Seliger, E. V. Sukhorukov and L. Wirtz for helpful discussions. Support by the Austrian Science Foundation (Grant No. FWF-SFB016 and No. FWF-P17359) is gratefully acknowledged.

REFERENCES

1. W. Schottky, Ann. Phys. (Leipzig) **57**, 541 (1918).
2. C. W. J. Beenakker and Ch. Schönenberger, Physics Today **56**(5), 37 (2003).
3. Ya. M. Blanter and M. Büttiker, Phys. Rep. **336**, 1 (2000).
4. S. Oberholzer, E. V. Sukhorukov and Ch. Schönenberger, Nature **415**, 765 (2002).
5. X. Jehl, M. Sanquer, R. Calemczuk, and D. Mailly, Nature **405**, 50 (2000).
6. O. Agam, I. Aleiner, and A. Larkin, Phys. Rev. Lett. **85**, 3153 (2000).
7. H.-S. Sim and H. Schomerus, Phys. Rev. Lett. **89**, 66801 (2002).
8. R. G. Nazmitdinov, H.-S. Sim, H. Schomerus, and I. Rotter, Phys. Rev. B **66**, R241302 (2002).
9. P. G. Silvestrov, M. C. Goorden, and C. W. J. Beenakker, Phys. Rev. B **67**, 241301(R) (2003).
10. Ph. Jacquod and E. V. Sukhorukov, Phys. Rev. Lett. **92**, 116801 (2004).
11. E. V. Sukhorukov and O. M. Bulashenko, Phys. Rev. Lett. **94**, 116803 (2005).
12. R. S. Whitney and Ph. Jacquod, Phys. Rev. Lett. **94**, 116801 (2005).
13. F. Aigner, S. Rotter, and J. Burgdörfer, cond-mat/0502417 (to be published in Phys. Rev. Lett.).
14. C. W. J. Beenakker and H. van Houten, Phys. Rev. B **43**, R12066 (1991).
15. J. Tworzydło, A. Tajic, H. Schomerus, and C. W. J. Beenakker, Phys. Rev. B **68**, 115313 (2003).
16. H. U. Baranger and P. A. Mello, Phys. Rev. Lett. **73**, 142 (1994); R. A. Jalabert, J.-L. Pichard, and C. W. J. Beenakker, Europhys. Lett. **27**, 255 (1994).
17. G. M. Zaslavsky, Phys. Rep. **80**, 157 (1981).
18. S. Rotter et al., Phys. Rev. B **62**, 1950 (2000); Phys. Rev. B **68**, 165302 (2003).
19. L. Wirtz, J.-Z. Tang, and J. Burgdörfer, Phys. Rev. B **56**, 7589 (1997).
20. L. Wirtz, C. Stampfer, S. Rotter, and J. Burgdörfer, Phys. Rev. E **67**, 016206 (2003).
21. C. Stampfer, S. Rotter, J. Burgdörfer, and L. Wirtz, cond-mat/0504197 (submitted to Phys. Rev. E).
22. J. Burgdörfer and J. Gibbons, Phys. Rev. A, **42**, 1206 (1990).

Excess Noise In Carbon Nanotubes

Shahed Reza, Quyen T. Huynh, Gijs Bosman, Jennifer Sippel* and
Andrew G. Rinzler*

Department of Electrical and Computer Engineering, University of Florida, Gainesville, Florida
**Department of Physics, University of Florida, Gainesville, Florida*

Abstract. The excess noise of single-walled carbon nanotubes is studied over the 77K to 300K temperature range. Lorentzian shaped spectra along with 1/f noise spectra are observed. From the Lorentzian noise components activation energies of 0.21 and 0.46eV for the associated fluctuation mechanisms are obtained.

Keywords: noise, carbon nanotubes.
PACS: 73.63Fg, 72.20Jv, 72.70+m

INTRODUCTION

Carbon nanotubes have attractive mechanical and electrical features[1]. The properties of carbon nanotube single devices, bundles, and films have been studied and many potential applications of these devices have already been proposed[2]. One important figure of merit for electronic and electrical applications is noise characteristics. The shot noise of carbon nanotubes was found to be very low[3]. However, the 1/f noise of carbon nanotubes was found to be unexpectedly high[4].

We report the findings of a study of the excess noise properties of carbon nanotubes over the temperature range 77K to 300K. In addition to a high 1/f noise component[4], Lorentzian noise components were observed. An estimate of the activation energies of the fluctuation processes involved in generating the Lorentzian noise components is presented.

EXPERIMENT

Nanotubes were grown by CVD on silicon chips with a 600 nm thermal oxide layer. The catalyst used was 10μg/ml $Fe(NO_3)_3 \cdot 9H_2O$ dissolved in IPA spun dried onto the oxide/silicon[5]. The nanotubes were grown at 900° C in a 1" tube furnace flowing 200 sccm hydrogen and 200 sccm methane[6]. Optical lithography was used to pattern Cr/Pd (5nm/45nm) electrodes spaced 1 μm apart (the Pd likely overcoats the edge of the Cr layer so that electrical contact to the nanotubes is via the Pd). The length of the electrodes was 500 μm and a total of 126 and 158 nanotubes in parallel spanned the 1 μm spacing between adjacent electrodes in samples A and B respectively. Samples were mounted on a liquid nitrogen cooled finger in a low-pressure cryostat suitable for noise measurements between 77 and 300K. The samples were current biased at 13.9 µA. This bias point was well within the linear range of device operation. The resistance was found to be fairly constant over the temperature range of our experiment. For sample A the resistance changed from 0.98 kΩ at 77K to 1.04 kΩ at 300K and for sample B from 0.83 kΩ at 77K to 0.91 kΩ at 300K.

The noise spectral density was measured with a HP3561A low frequency spectrum analyzer at frequencies between 10Hz and 100kHz. The noise observed is a combination of excess noise and thermal noise. The full shot noise magnitude for the 13.9 µA bias level is 4.5×10^{-24} Amp^2/Hz, far below the excess noise level. A typical excess noise plot measured on sample A at T=300 K is presented in figure 1.

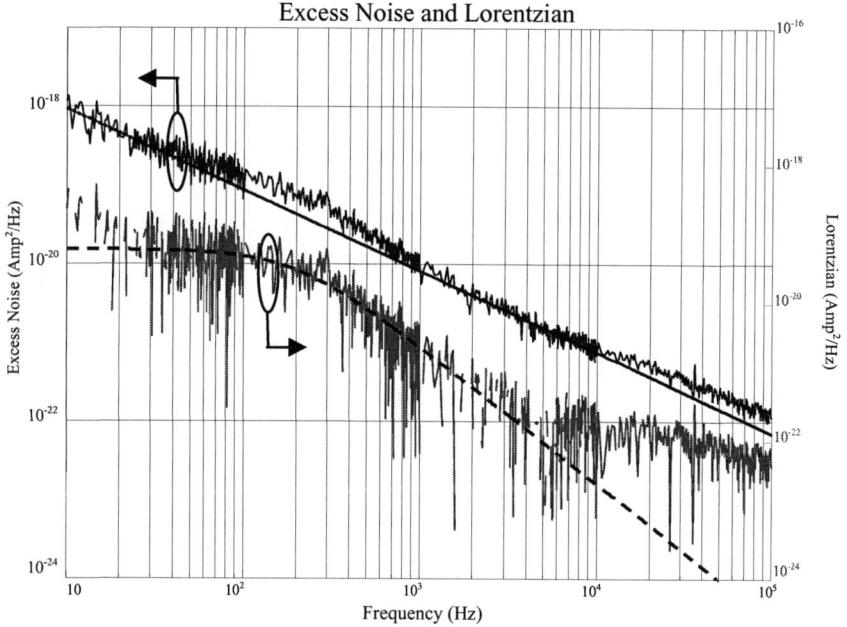

FIGURE 1. (a) A typical excess noise plot (top) with fitted $1/f^\beta$ noise line measured at T=300K. (b) Bottom plot shows the Lorentzian spectrum obtained after subtracting $1/f^\beta$ noise from the excess noise.

DISCUSSION

The measured noise is interpreted as consisting of $1/f^\beta$ noise with Lorentzian components superimposed. The $1/f^\beta$ noise can be expressed by the following equation[4]

$$S_{1/f}(f) = \frac{A \cdot I^2}{f^\beta} \quad (1)$$

where I is the dc current level, f is the frequency, and A and β are constants expressing the relative noise magnitude and frequency exponent. The values of A and β are determined by fitting a line to the experimental data as shown in figure 1. The parameter A can serve as a figure of merit for 1/f noise comparisons of different samples. Collins et al[4], reported that for carbon nanotubes A, normalized to the resistance R, is close to a constant. The measured values of A/R for our samples range from 1.3×10^{-11} to 7.6×10^{-11} between 77 and 300K; within the range of values reported by Collins et el[4] at room temperature. The value of β varies between 0.85 and 1.1 as a function of temperature.

A closer examination of the excess noise plot reveals the presence of Lorentzian noise components superimposed on the $1/f^\beta$ noise. These Lorentzian components arise when random number fluctuations are caused by processes with a single characteristic time and activation energy.

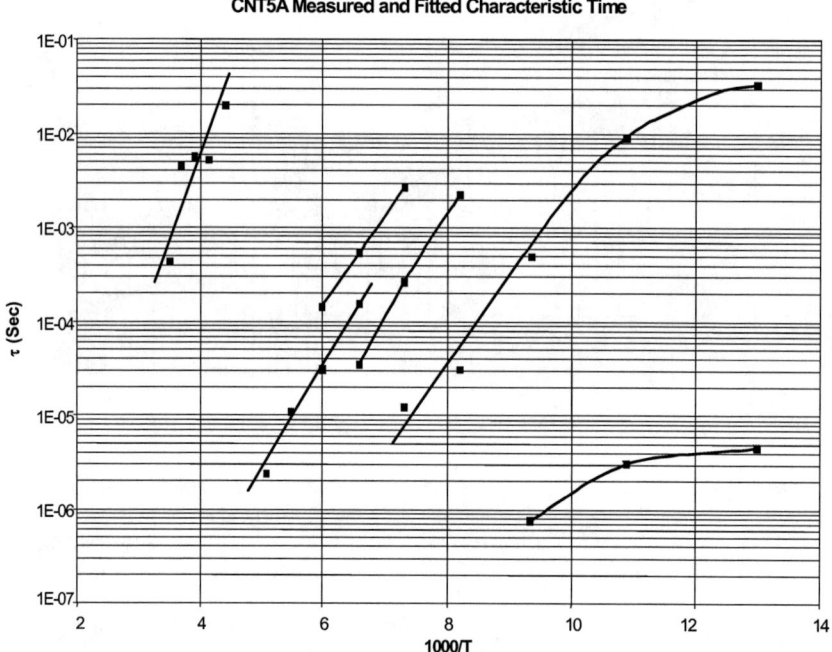

FIGURE 2. Lorentzian characteristic times τ versus 1000/T

The characteristic times calculated from different Lorentzians are shown in figure 2 for sample A. Note that the horizontal axis is 1000/T and the characteristic times are clearly activated i.e.,

$$\tau = \tau_0 \cdot \exp(E/k_B T) \quad (2)$$

where E is the activation energy, k_B is the Boltzmann constant and T is the absolute temperature. The activation energies for the plots shown in figure 2 are 0.21eV ± 0.02eV. The low frequency plateau values of the Lorentzians have a linear relationship with τ, suggesting that the variance of the number of carrier $\langle \Delta N^2 \rangle$ is a constant[7]. The τ values calculated for sample B showed similar trends with an activation energy of 0.46eV ± 0.05eV.

The calculated activation energies for the Lorentzian noise components are within the range of energies of physical processes possibly present in the carbon nanotube samples we studied. For example the fluctuations caused by carrier transitions between one dimensional sub-bands will have activation energies equal to the energy difference between sub-bands. The sub-band energy differences in carbon nanotubes can be close to the activation energies presented here[1].

It may be possible that the noise originates from a local defect center due to adsorption of a chemical species. It has been reported that oxygen chemisorption on the surface of a semiconducting carbon nanotube lowers the energy band gap. For example on a (8,0) SWNT the energy gap decreased from 0.56eV to 0.23eV due to oxidation[8]. Local defects of this nature may produce noise with activation energies in the range reported in this work.

REFERENCES

1. Satio, R., Dresselhaus, G., and Dresselhaus, M.S., *Physical Properties of Carbon Nanotubes*, London: Imperial College Press, 1998, 1st edition.
2. Dresselhaus, M.S., Dresselhaus, G., and Avouris, P., *Carbon Nanotubes: Synthesis, Structure, Properties, and Application*, Berlin: Springer Verlag, 2001, 1st edition.
3. Roche, P. –E, Kociak, M., Ferrier, M., Guéron, S., Kasumov, A., Reulet, B., and Bouchiat, H., *Proc. of SPIE*, **5115**, 104 (2003).
4. Collins, P. G, Fuhrer, M. S., and Zettle, A. *App. Phys. Letters* **76**, 894 (2000).
5. Hafner, J. Cheung, C., Oosterkamp, T., and Lieber, C., *J. Phys. Chem. B* **105**, 743 (2001).
6. Li, Y., Kim, W., Zhang, Y., Rolandi, M., Wang, D., and Dai, H., *J. Phys. Chem. B* **105**, 11424 (2001).
7 Van Rheenen, A. D., Bosman, G., and Zijlstra, R. J. J., *Solid State Electronics*. **30**, 259 (1987).
8 Barone, V., Heyed, J. and Scuseria, G. E., *Chem. Phys. Lett.* **389**, 289 (2004).

Low Frequency Noise in Contacted Single-Wall Carbon Nanotube

S. Soliveres[1], A. Hoffmann[1], F. Pascal[1], C. Delseny[1], A. Salesse[1], M. S. Kabir[2], S. Bengtsson[2], O. Nur[2], M. Willander[2], and J. Deen[3]

[1]*CEM2 CNRS UMR5507, University of Montpellier 2*
Place E. Bataillon CC084
34 095 Montpellier Cedex 5, France
[2]*Dept. of Microtechnology and Nanoscience (MC2), Chalmers University of Technology*
SE-41296, Sweden
[3]*ECE Department, CRL 226, McMaster University, Hamilton, ON L8S 4K1 CANADA*

E-mail: soliveres@cem2.univ-montp2.fr

Abstract. Since their discovery, carbon nanotubes present many interesting electrical properties and can be candidates for future shrinking devices. Electrode realization on nanotubes remains a challenge. The deposited contacts must present interesting electrical parameters: low contact resistance, low series resistance, low parasitic capacity and low excess noise. The understanding of conduction phenomena induced into or near the metal/nanotube contact is necessary to improve the device. In this paper we present characterization of a Au-Ti deposited contact on a single-wall carbon nanotube. Steady-state current-voltage measurements, impedance and noise measurements lead to the frequency device characteristic. From this study a low frequency noise electrical model is presented. Using this model and the low frequency noise measurements, we show that the measured thermal noise has its origin in the electrode/nanotube contacts.

Keywords: carbon nanotube, contact resistance, low frequency noise, impedance, Landauer
PACS: 73.63Fg

INTRODUCTION

Single-Wall Carbon Nanotubes are molecules with a nanometric diameter and with one-dimensional conductor properties i.e. a ballistic conduction at ambient temperature. For these reasons, they are candidates for interconnects or for future nanometric devices as Carbon Nanotube Field Effect Transistor (CNFET) or Schottky Barrier Field Effect Transistor (SBFET) [1]. One challenge for industrial applications is the quality and the reproducibly of the electrical contacts. Electrodes are actually deposited on nanotube by electron lithography process. The studied device is a Single-Wall Carbon Nanotube deposited on an isolating substrate with three electrodes processed on the top of the nanotube to form side-contacts. The device electrodes are composed by 10 nm of Au on 5 nm of Ti. From current-voltage, impedance, and low frequency noise measurements contact contribution on electrical conduction can be clearly pointed-out. The contribution of each contact noise sources to the total output noise is investigated. Variation of the 1/f excess noise level with the current flowing through the device is also characterized.

ELECTRODE-SWCN CONTACT CHARACTERIZATION

Classically [2] the intrinsic contact resistance is due to the one-dimensionality of the nanotube. At low bias voltage only two discrete energy subbands participate to the conduction involving an intrinsic contact resistance $R_{ci}=h/4q^2=6.5$ kΩ [2]. In real device, a Schottky barrier is formed at the interface metal/nanotube due to physical and technological parameters. It implies that the contact conductance is limited by the probability of transmission through the barrier. Consequently, the measured contact resistances (R_c) are higher than 6.5kΩ, usually ranging from 10kΩ to GΩ. If we consider only the intrinsic ballistic resistance of the nanotube, its resistance is close to zero, then negligible compared to the contact resistance.

Our sample is a SWCN with three deposited electrodes (labeled 1, 4, and 2), electrode 3 and the SWCN are not contacted (Fig 1.a). Current voltage measurements, between the three different pairs of electrodes, allow the contact resistance determination. As an example, Fig 1.b presents measurements between electrodes 2 and 4, the resistance is near 10 MΩ ±10%. The comparison between all electrodes combinations allows the extraction of contact resistances values, R1≈430 kΩ, R2≈15 kΩ, and R4≈10 MΩ for electrodes 1, 2 and 4, respectively. The non linear effect can be associated to the metal/nanotube tunneling process. An antagonist variation is obtained when the bias is inversed.

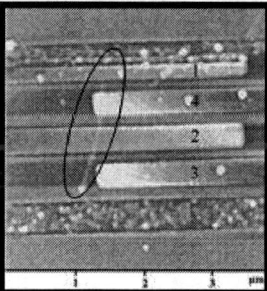 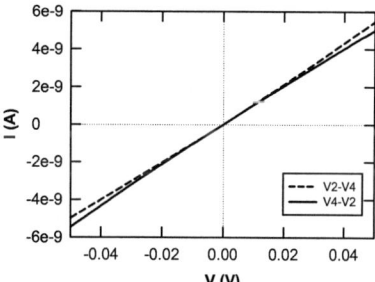

FIGURE 1. (a) Nanotube and labeled electrodes. (b) Current-voltage measurements between electrodes 2 and 4.

The presence of a potential barrier at the contact implies that a potential variation (small AC signal) applied to the metal electrode induces a charge variation on each part of the SWCN-electrode interface. Figure 2a and 2b present measurements of the real and imaginary parts of the impedance without DC supply. Each contact can be modeled by a RC parallel circuit. R is the contact resistance (R_c) and C a capacitance including the potential barrier and depletion effects. These measurements show each contact influence on the electrical transport. The connected nanotube can then be modeled by two RC cells associated to each related contacts and a series resistance R_N. This series resistance includes the intrinsic ballistic resistance of the nanotube and an access resistance (Fig. 4). This electrical model is applicable from DC to the higher measurable frequency. We will use this model to analyze noise measurements.

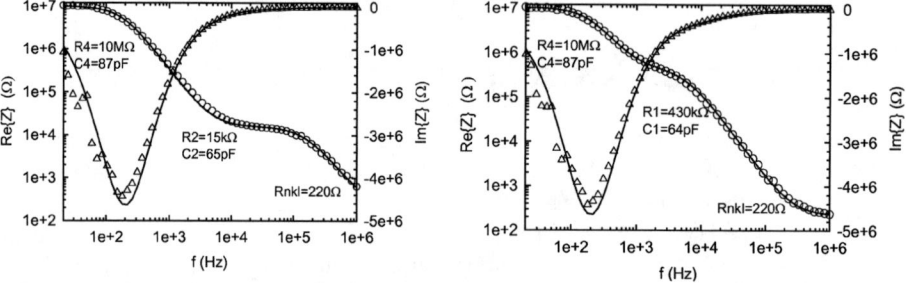

FIGURE 2. (a) Real and imaginary parts of measured impedance (dot) and calculated with the associated model (line) associated to electrodes 4 and 2. (b) Real and imaginary parts of measured impedance (dot) and calculated with the associated model (line) associated to electrodes 1 and 4.

LOW FREQUENCY NOISE CHARACTERIZATION

The current spectral density is measured between each contact at room temperature. As expected the thermal noise, can be associated to the sum of the contact and of the series resistances in accordance with the electrical model. Extracted resistances values being very different, the measured thermal noise correspond to the higher contact resistance noise. As shown by [3], the variation of the excess 1/f noise is proportional to the square of current. Fig.3.a presents the current spectral density between electrodes 1 and 2. In this case, no capacitive effect was observed. To compare the 1/f noise level, the usual equation (1) is used to extract the so-called relative 1/f noise spectral density, K. As shown in Fig 3.b, K is similar for the three studied contact, showing no dependence with the contact resistance.

$$S_I = K \frac{I^2}{f} \quad (1)$$

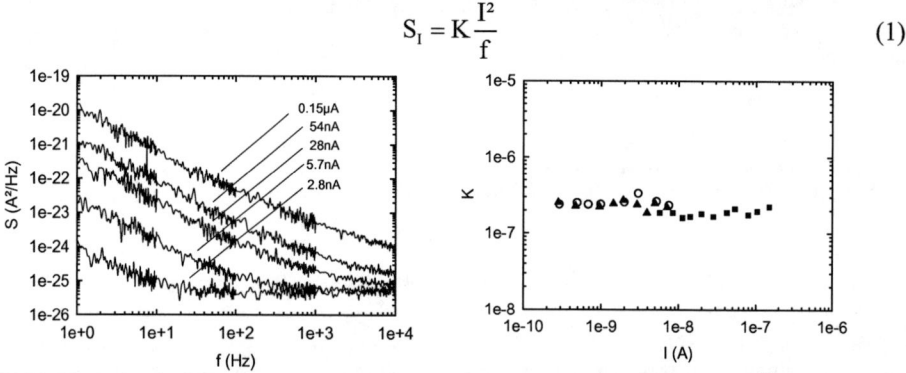

FIGURE 3. (a) Typical current spectral density associated to electrodes 1 and 2. (b) K extracted from measurements between electrodes 1 - 4 (▲), 2 - 4 (o), and 1-2 (■).

As shown previously the noise contribution can be associated to different noise sources associated to the series and contact resistances. Using the proposed electrical model, we can calculate separately each noise source contribution through appropriate transfer functions (Fig 4.a). Without current injection, only thermal noise sources are considered, the output current noise between electrodes k and l can be expressed as:

$$i_{thnoise} = \frac{Z_k\sqrt{\frac{4kT}{R_k}} + \sqrt{4kTR_{nkl}} + Z_l\sqrt{\frac{4kT}{R_l}}}{Z_k + R_{nkl} + Z_l} \quad (2)$$

Where Z_k and Z_l are the impedances associated to the contacts labeled k and l.
A calculation of the total output noise is performed using the extracted resistance and capacitance values obtained from impedance measurements. The calculated spectra are the same as the output total current noise measured (an example is given in Fig 4.b), assuring the model validity. The thermal noise at the lower frequencies is mainly due to the higher resistance value (in our case the contact resistance) in accordance with equation (2).

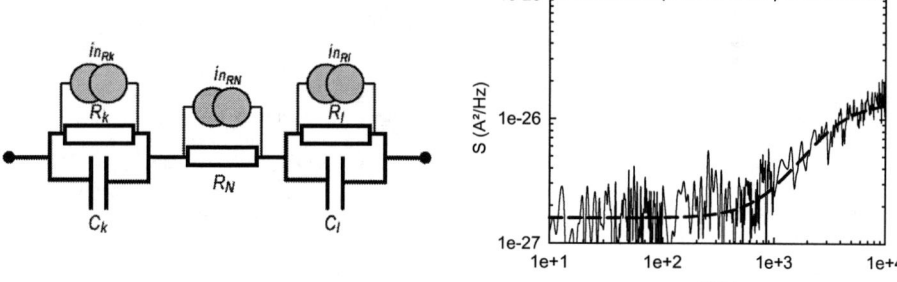

FIGURE 4. (a) Low frequency noise equivalent model. (b) Measured and calculated output thermal noise.

CONCLUSION

A noise model associated to the contacted SWCN is proposed. Using the impedance and the noise measurements we have shown that the main limitation of SWCN use in electrical applications is the contact quality. The noise is dominated by the higher contact resistance value. Moreover, lowering the parasitic capacitance and the contact resistance remains an important challenge to increase the device bandwidth.

'This work has been done in the framework of the GDRE n°2756 'Science and applications of the nanotubes - NANO-E'

REFERENCES

1. S. Heinze, J. Tersoff, et al, *Phys. Rev. Lett.*, **89**, 106801 (2002)
2. S. Datta, *Electronic Transport in Mesoscopic Systems*, Cambridge, U. K.: Cambridge Univ. Press, 1995.
3. P. G. Collins, M. S. Fuhrer, and A. Zettl, *Appl. Phys. Lett.*, **79**, 894-896 (2000)

Hanbury Brown and Twiss Noise Correlations to Probe the Statistics of GHz Photons Emitted by Quantum Conductors

D.C. Glattli*, J. Gabelli[†], L.-H. Reydellet*, G. Fève[†], J.M. Berroir[†], B. Plaçais[†] and P. Roche*

*Service de Physique de l'Etat Condense, CEA Saclay, F-91191 Gif-sur-Yvette, France
[†]Laboratoire Pierre Aigrain, Departement de Physique de l'Ecole Superieure, 24 rue Lhomond, 75231 Paris Cedex 05, France

Abstract. In microwave current noise power measurements, a detector is used which measures the power emitted by the device in the external circuit. This power can be viewed as a flux of (microwave) photons. An interesting question is the nature of the photon statistics associated with the noise power. While for thermal noise one expects classical photon statistics (Bose-Einstein), recently it has been shown theoretically that shot noise may lead to non-classical photon statistics. We propose an original cryogenic experimental scheme to characterize the statistics of photons emitted by conductors at GHz frequency. The experiments, a microwave analog of optical Hanbury Brown Twiss photon correlations, is able to distinguish the statistics of the black-body radiation associated with thermal electronic noise from the statistics of a coherent monochromatic microwave source. The experiments displays the sensitivity required for future investigations of non classical photons emitted by quantum conductors.

Keywords: Electronic Shot Noise, high frequency noise measurements
PACS: 73.23. -b, 42.50.Ar, 73.50.Td

In 1928, Nyquist [1] established that measuring the current noise power $S_I(v) = \overline{(\Delta I)^2}/dv$ of a conductor at frequency v and bandwidth dv is equivalent to monitor the electromagnetic power radiated by the noisy conductor in the measuring circuit. Extending the description of the T.E.M. circuit electromagnetic modes to the quantum regime, he implicitly showed that the photon population N of the modes is proportional to the current noise power. Using notations of Fig.1, conductor of conductance $G = 1/R$ and characteristic circuit impedance Z, the mean power is:

$$\overline{P} = \frac{Z}{(1+GZ)^2} S_I(v) dv = \overline{N} h v dv \tag{1}$$

In this paper we go a step further and consider the low frequency fluctuations of the noise power around its mean value. These fluctuations can equivalently be described as fluctuations ΔN of the T.E.M. photon population. This immediately rises the question of what is the photon statistics associated with a specific type of current noise. For a conductor in equilibrium at finite temperature a black body source photon statistics is expected. The well-known bunching of photons gives rise to large noise and the variance of the photon population N is super-Poissonian: $\overline{(\Delta N)^2} = \overline{N}(1+\overline{N})$, with $\overline{N}$ given by the

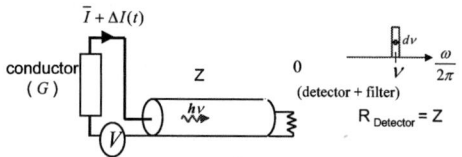

FIGURE 1. A noisy conductor, conductance G, emits photon toward an external circuit made of a transmission line (characteristic impedance Z and a matched detector ($R_{load} = Z$))

Bose-Einstein distribution, and the power fluctuations are:

$$\langle(\Delta P)^2\rangle = 2B\overline{N}(1+\overline{N})(hv)^2 dv \qquad (2)$$

Less obvious is the case of the photon statistics associated with electronic shot noise. Here the noise originates from the granularity of the current : electrons randomly transferred through the conductor generate current noise which at large current I overcomes the thermal noise. Shot noise is best observed for junctions separating two conductors, tunnel barriers or quantum conductors. For conductor biased at voltage V, the detectable finite frequency noise power is the zero frequency noise power reduced by the fraction of electron in the energy range eV having enough energy to emit a quantum hv in the circuit : $S_I(v) = S_I(0)(eV - hv)/eV$. The zero frequency noise is the Schottky noise $S_I(0) = 2eIF$ reduced by a Fano factor $F = (\sum D_n(1-D_n)/\sum D_n) \leq 1$, where D_n are the transmissions of electronic modes in the conductor [3, 4].

For measurement time $\sim 1/B$ much larger than I/e, the current statistics becomes Gaussian like that generated by thermal noise. Following Glauber [5], treating the current classical but the electromagnetic modes quantum, the photon statistics found is that of a black body source whose the temperature is replaced by $S_I/4k_BG$, the equivalent noise temperature. However, the current is no longer a classical observable and the unsymmetrized current noise operator $\hat{S}_I(v,t) = \int_{-\infty}^{+\infty} d\tau \hat{I}(t)\hat{I}(t+\tau)exp(i2\pi v\tau)$ is needed [6]. Nevertheless, the recent approach of Ref.[7] treating quantum both current and electromagnetic modes showed that, for conductors with large numbers of electronic modes or for frequency $v \ll eV/h$, the statistics of photons is indeed a Bose-Einstein distribution with a temperature also given by $S_I/4k_BG$.

More interesting, is the case of the high frequency ($v \leq eV/h$) shot noise of a conductor transmitting very few electronic modes. Implicit in Ref.[7] and subsequently showed by the same authors in Ref.[8], the photon statistics can be *non-classical*. A sub-Poissonian photon population variance was found $(\Delta N)^2 = N(1-\alpha N)$, where $\alpha < 1/N$ depends on electronic mode transmissions. To be more explicit, consider for simplicity a single mode conductor with transmission D, the conductance is $G = De^2/h$ and the current noise $S_I(v) = 2e\frac{e^2}{h}D(1-D)(eV-hv)$ for $hv < eV$ and 0 else. From Eq.(1), the average photon population is $\overline{N} = |S_{12}|^2 \frac{1-D}{2}\frac{eV-hv}{hv}$, where the impedance matching factor $|S_{12}|^2 = 4RZ/(R+Z)^2$ can be view as the probability that a photon emitted by the conductor is absorbed in the measuring circuit. The $1-D$ term is the Fano factor F due to the binomial electron statistics which leads to sub-Poissonian electronic shot noise. The ratio $\frac{eV-hv}{hv}$ represents the number of photons that can be simultaneously emitted by electrons having energy between eV and $eV-hv$ and contributing to the detected current noise. For $eV/2 \leq hv \leq eV$ only a single photon can be emitted and the sub-Poissonian electronic noise leads to sub-Poissonian Photon noise. For example, from Ref.[8], it is shown that for a detection filter with non zero value for $eV/2 \leq hv \leq eV$, the photon variance is $\overline{(\Delta N)^2} = \overline{N}(1 - \frac{2}{3}\overline{N})$

This remarkable result provides a strong motivation to investigate the photon statistics in quantum circuits. Finally, it is worth noting that the variance of the photon number gives a physical meaning to the fourth order moment of the recently derived electron shot noise full counting statistics [9]. One have: $\overline{(\Delta N)^2} = \left(\frac{Ze^2}{h(1+GZ)^2}\right)^2 \times \frac{\overline{(\Delta I)^4} - \left[\overline{(\Delta I)^2}\right]^2}{(e^2 v dv)^2}$

To characterize the qualitative nature of the statistics of photons emitted by conductors is not an easy task as it requires to adapt the photon correlation techniques developed in Quantum Optics to the microwave regime. Here, low frequency is required as quantum conductors are best controlled at temperature and energy below 1 K with eV/h in the 1-20 GHz range. We emphasize that it does not preclude a future extension of this method to the generation optical photons. Nano-conductors such as single wall Carbone nanotube or short molecules remaining quantum at larger energy, several tens of Kelvin, could generate far-infrared sub-Poissonian photons. In this paper we will show that reliable photon correlations can be made at GHz frequency, even in the few photon number regime. We will present the results of the microwave analog [11] of a Hanbury-Brown Twiss optical experiments [10]. The sensitivity is high enough to distinguish between the super-Poissonian statistics of a black-body source from the Poissonian staistics of a monochromatic microwave source. It appears well suited to the study in the future sub-Poissonian photon generation by quantum conductors.

The radio-frequency equivalent of the optical HBT experiment is shown in Fig.2-b. Source (a) is the rf-photon source under test. The emitted TEM photons propagate through a 50 Ohm characteristic impedance coaxial line and are fed to a cryogenic 3dB strip-line power splitter. The splitter scattering matrix, measured with a network analyzer, is identical to that of an optical separatrix (phases included). A built-in 50 Ohms resistor, source (b), plays the role of the vacuum channel of the optical case. Photon vacuum is achieved when its temperature T_0 satisfies $T_0 \ll T_Q$ where $T_Q = hv/k_B$. Outputs (1) and (2) of the power splitter are not immediately detected but a 1-2 GHz linear phase-insensitive amplification chain is inserted before square-law detection (see Fig.2-c). Each chain consists in a microwave circulator followed by an ultra-low noise cryogenic amplifier and room temperature amplifiers. The circulators, at low temperature, ensure that amplifiers do not send back photons towards (a) and (b). The detectors give an output voltage proportional to the photon intensity after amplification $P_{1,2}^{out}$. Their finite 1 μs integration time allows to monitor the low frequency photon intensity fluctuations. A numerical spectrum analyser calculates the autocorrelations $\left\langle (\Delta P_{1,2}^{out})^2 \right\rangle$ and cross-correlations $\langle \Delta P_1^{out} \Delta P_2^{out} \rangle$ in the band 40-200 kHz (B =160 kHz).

In a first series of experiments, A and B, the sensitivity to Bose-Einstein statistics is tested. In a third experiment C, Poisson's statistics is tested using coherent photons. In A, source (a) is a 50 conductor whose temperature T is varied from 20 mK to several K, while source (b) realizes good photon vacuum ($T_0 = 17 - 20$ mK$< T_Q$). A pair of 1.64-1.81 GHz filters select a narrow band frequency around v =1.72 GHz, with $T_Q = 86$ mK and $hdv \ll k_B T$. A quantum amplifier description (see below) predicts the mean

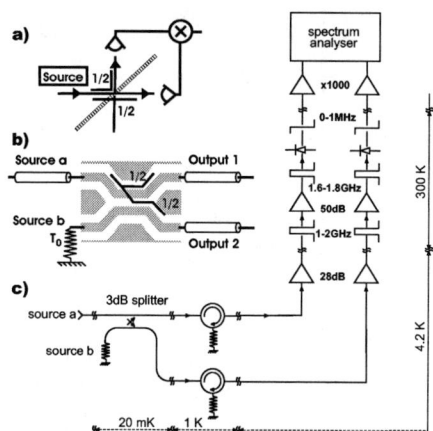

FIGURE 2. Schematics of the HBT experiment (a), of the GHz beam splitter (b), of the amplification and detection chains (c). A detailed description is given in the text.

powers $P_i = P_i^{out}/G_i$, referred to the input, and their fluctuations:

$$\overline{P_i} = (\frac{\overline{N_a}}{2} + \frac{\overline{N_b}}{2} + \frac{k_B T_{N,i}}{h\nu}) h\nu d\nu \qquad i = 1, \ 2 \qquad (3)$$

$$\langle(\Delta P_i)^2\rangle = 2B(\frac{\overline{N_a}}{2} + \frac{\overline{N_b}}{2} + \frac{k_B T_{N,i}}{h\nu}) \times$$

$$\times (\frac{1}{G_i}\frac{\overline{N_a}}{2} + \frac{\overline{N_b}}{2} + \frac{k_B T_{N,i}}{h\nu})(h\nu)^2 d\nu \approx \langle P_i\rangle^2 \frac{2B}{d\nu} \qquad (4)$$

$$\langle\Delta P_1 \Delta P_2\rangle = 2B(\frac{\overline{N_a}}{2} - \frac{\overline{N_b}}{2})^2 (h\nu)^2 d\nu \qquad (5)$$

Here $\overline{N_a}$ and $\overline{N_b}$ are the photon populations of sources (a) and (b) given by Bose-Einstein distribution, with $\overline{N_b} \approx 0$ as $T_0 << T_Q$. Autocorrelations differ from Eq.(2) in two ways: first, the factor 1 in the second parenthesis, revealing the independent particle behavior of thermal photons is replaced by $1/G_i$ (G_i = 80 dB) when refered to the amplifier input; secondly, an extra photon population is added due to amplifier noise expressed in temperature units $T_{N,i}$ ($T_N \approx 15K$ in the experiment A and $T_N \approx 6K$ in experiments B and C). The HBT cross-correlation are unaffected by amplification. Experimental results are shown in Fig.3 . The solid line is the Bose-Einstein theoretical fit of the mean photon power with Eq.(3) taking T_N and G as free parameters. The experimental data reproduce well the quantum crossover at $T_Q/2 \approx 43$ mK. This is the coldest quantum cross-over ever reported for microwave photons. The parameter G is consistent with independent set-up calibration. The experimental scatter, $\delta T \approx 1$ mK in temperature units, corresponds to the resolution $\delta T \approx 2T_N/\sqrt{d\nu \Delta t}$ expected for few seconds acquisition time Δt. The HBT cross-correlations $\langle\Delta P_1 \Delta P_2\rangle$ are expected to vary like $(\overline{N_a})^2 \approx (T/T_Q)^2$ for $T > T_Q$ as $\overline{N_B} = 0$. This is exactly what we observed in Fig.3. This provides the first evidence for Bose-Einstein correlations of photon emitted by

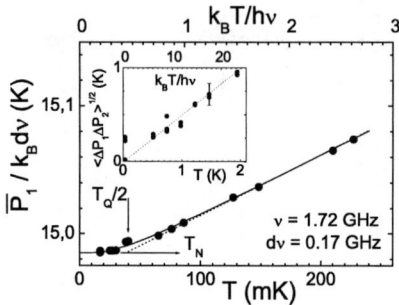

FIGURE 3. Mean photon power $\overline{P_1}$ as function of temperature T, or effective occupation number $k_B T/h\nu$, for a thermal source at $\nu = 1.7$ GHz. The beam splitter temperature is $T_0 \approx 17$ mK and the amplifier noise temperature is $T_N \approx 15$ K. Solid line: Bose-Einstein fit. Dashed: classical limit $P \propto (k_B T - h\nu/2)$. Inset: the positive HBT correlations show the bunching of thermal photons and the absence of residual correlations at $T = 0$ K.

a resistor in the few photon number limit at sub-Kelvin temperature. The correlation resolution is $\delta \langle (\Delta P)^2 \rangle \approx \langle (\Delta P_{1,2})^2 \rangle / \sqrt{B \Delta t}$. This corresponds to detect population fluctuations of $\sqrt{\overline{\Delta N_1 \Delta N_2}} \approx k_B T_N / \left(h\nu \sqrt[4]{B \Delta t} \right) \approx 1.5$ for $\Delta t = 1000$ s.

A second experiment B, not shown here, has provided evidence for vanishing HBT correlations $\langle \Delta P_1 \Delta P_2 \rangle \approx \left(\overline{N_a} - \overline{N_b} \right)^2$ when $\overline{N_a} = \overline{N_b}$. The experiment was performed at higher temperature ($T = 4\text{-}24$ K, $T_0 = 4$ K) while source (b) was no longer in the ground state but at finite temperature: $\overline{N_b}(\nu) \approx k_B T_0 / h\nu \approx 40$. The results of A and B validate the ability of the GHz HBT experiment to characterize super-Poissonian photon statistics.

The third experiment C tests the sensitivity to a different statistics. A microwave source of frequency $\nu_0 = 1.5$ GHz and bandwidth ≈ 100 Hz generates monochromatic photons. Its output is attenuated at cryogenic temperature T_{att}. The average power $\overline{P_a}$ in (a), is chosen comparable to that delivered by the thermal source in experiment B. As for a Laser, the source is expected to generate coherent photon states [5]. The photon statistics being Poissonian, the low frequency power fluctuations of source (a) in bandwidth B are $\langle (\Delta P_a)^2 \rangle = 2Bh\nu_0 \overline{P_a}$. An important question is whether attenuation and amplification before detection change the statistics and if yes how? According to references [12], the commutation rules for bosonic input and output operators imply the addition of an extra bosonic operator describing the amplifier/attenuator noise. Applying this quantum constraint, we find (in units referred to the amplifier chain input):

$$\langle (\Delta P_{1,2})^2 \rangle = 2Bh\nu_0 F_{1,2} \overline{P_{1,2}} \qquad \langle \Delta P_1 \Delta P_2 \rangle = 0 \qquad (6)$$

$$F_{1,2} = 1 + 2k_B \frac{T_{att} + T_{N1,2}}{h\nu_0} \qquad (7)$$

where we have used $T_{att} = T_0$. Again, amplification has no effect on HBT cross-correlations and the absence of cross correlation characterizing Poisson's statistics remains. In the autocorrelations, a Fano factor F appears due to amplification noise. The

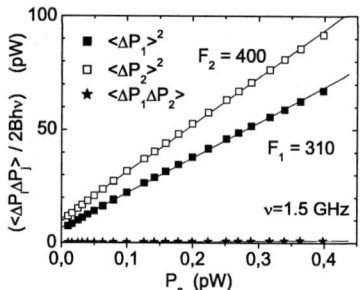

FIGURE 4. Photon noise of an attenuated and amplified monochromatic source showing Poissonnian HBT correlations. Autocorrelations are linear in the input RF power with a large Fano factor F due to the beating of the attenuator/amplifier noise with the source. Cross correlations vanish in accordance with the prediction for a coherent source.

results of experiment C are shown on Fig.4. Cross- and auto-correlations are plotted versus detected power referred to the input. Cross correlations are negligible at the scale of the autocorrelations. Auto-correlations are perfectly linear with power as expected. The large Fano factor $F_{1,2} = 310, 400$ is in accordance with the value of $(T_{att} + T_{N1,2})/h\nu_0$.

The highly sensitive HBT photon correlations performed in the GHz range at sub-Kelvin temperature using phase insensitive LNA and square-law detection easily discriminate between different statistics. In particular the cross-correlations are unaffected by amplification details. The method is simple, versatile and based on currently available electronics. It is suitable to study the photon population statistics of TEM modes emitted by quantum conductors or equivalently the fourth moment of current fluctuations.

REFERENCES

1. H. Nyquist, *Phys. Rev.* **32**, 110 (1928).
2. (throughout this paper, $\overline{\chi}$ denotes the mean value of χ over the full bandwith $(d\nu/2)$ whereas $\langle\chi\rangle$ denotes the estimate of $\overline{\chi}$ in the truncated measurement bandwidth B)
3. G.B. Lesovik, *JETP Lett.* **49**, 592 (1989); Büttiker, *Phys. Rev. Lett.* **65**, 2901 (1990); T. Martin, R. Landauer, *Phys.Rev.B* **45**, 1742 (1992); Y.M. Blanter, M. Buttiker, *Phys. Rep.* **336**, 2 (2000).
4. M. Reznikov, M. Heiblum, H. Shtrikman, D. Mahalu, *Phys. Rev. Lett.* **75**, 3340 (1995); A. Kumar, L. Saminadayar, D.C. Glattli, Y. Jin, B. Etienne, *Phys. Rev. Lett.* **76**, 2778 (1996).
5. R.J. Glauber, *Phys. Rev.*, 2766 (1963).
6. L.S. Levitov, G.B. Lesovik, *JETP Lett.* **58**, 230 (1993); R. Aguado, L.P. Kouvenhoven, *Phys.Rev.Lett.* **84**, 1986 (2000); U. Gavish, Y. Levinson, Y. Imry, *Phys. Rev. B* **62**, R10637 (2000)
7. C.W.J. Beenakker, H. Schomerus, *Phys. Rev. Lett.* **86**, 700 (2001).
8. C.W.J. Beenakker, H. Schomerus, *Phys. Rev. Lett.* **93**, 096801 (2004).
9. L. S. Levitov, H.-W. Lee (MIT), G. B. Lesovik in "Mesoscopic Physics," J. Math. Phys., 37 (10), 4845 (1996).
10. R. Hanbury Brown, R.Q. Twiss, *Nature* **177**, 27 (1956).
11. J. Gabelli *et al. Phys. Rev. Lett.* **93**, 056801 (2004).
12. C.M. Caves, *Phys. Rev. D* **26**, 1817 (1982); Y. Yamamoto, H.A. Haus, *Rev. Mod. Phys.* **58**, 1001 (1986).

Measurements of Correlated Conductances and Noise Fluctuations from 3-Lead Quantum Dots

R.C. Toonen[1], M. Prada[2], H. Qin[1], A.K. Huettel[3], S. Goswami[4], M.A. Eriksson[4], D.W. van der Weide[1], K. Eberl[5], R.H. Blick[1]

[1]*Department of Electrical and Computer Engineering, University of Wisconsin, Madison, Wisconsin 53706-1607* (rctoonen@wisc.edu) [2]*School of Electronic and Electrical Engineering, University of Leeds, LS2 9JT, United Kingdom* [3]*Department Physik, Geschwister-Scholl-Platz 1, Ludwig-Maximilians-Universitaet, 80539 Muenchen, Germany* [4]*Department of Physics, University of Wisconsin, Madison, Wisconsin 53706-1390* [5]*Max-Planck-Institut fuer Festkoerperforschung, 70569 Stuttgart, Germany*

Abstract. We have investigated the conductance properties of a few-electron quantum dot with three terminals. In the regime of strong coupling between the quantum dot and the leads, we have observed the both the integer- and half-integer-spin Kondo effect at zero magnetic field. Within the integer-spin conductance diamond, we find cotunneling spectral lines which correspond to singlet-triplet transitions. We extract the exchange energy from this information and find that the value ($J = 320\ \mu eV$) agrees remarkably well with the theoretical prediction [7]. We believe that spin dependent transport in a three-terminal quantum dot could yield positive cross-correlations between shot noise events on two output channels. To investigate such phenomena, we have designed an analog continuum cross-correlator to analyze the shot noise spectra of our device in the *X*- and *Ku*-bands (*8 to 18 GHz*).

Keywords: Quantum Dots, Three-Terminal, Three-Lead, Spin, Entangler, Entanglement, Flying Qubits, Ebits, Cross-Correlation, Hanbury-Brown and Twiss, Kondo Effect, Noise Fluctuations.
PACS: 72.25.-b, 72.70.+m, 73.21, 73.22.Dj, 73.23.Hk, 73.40.Ei, 73.43.Fj, 73.63.Kv.

There are seven criteria which must be fulfilled in order to physically implement a practical quantum computer [1]. The seventh of these necessary conditions is the ability to transmit *flying qubits* from one location to another. The detection of *flying qubits* will involve measuring bunching and anti-bunching effects in correlated noise spectra [2]. Recently, experimenters have used traditional cross-correlation techniques to measure these quantum statistical effects. Theorists have predicted that three-lead quantum dots can be used as beam splitters for entangling conduction electrons and probing the statistics of dynamic spin interaction [3, 4]. It has also been suggested that by using other three-lead, mesoscopic systems (such as a triple-dot device) spin pairs can be effectively generated and separated into two distinct channels [5]. Our aim is to study the correlation of shot noise events on the output channels of three-terminal quantum dot networks. We have *taken the first step* in this endeavor by fabricating a three-terminal quantum dot which demonstrates spin-dependent transport.

Our quantum dot is formed by laterally constricting a two-dimensional electron gas (2DEG)—located *90 nm* below the surface of a modulation-doped $Al_xGa_{1-x}As/GaAs$

heterostructure—with *36-nm*-high, *NiCrAu*-alloy Schottky split-gates. From *Shubnikov-de-Haas* and *Quantum Hall* measurements, we determined that our heterostructure has sheet density of 1.1×10^{15} m^{-2} and a mobility of 75 $m^2/V\text{-}sec$ when the electron temperature is approximately *1.5 K*. The results reported in this paper were measured in a He-3 / He-4 single-shot cryostat with a base temperature ranging from *230 to 250 mK*. We found that the non-Kondo and non-Fano Coulomb blockade oscillations fit very well to the standard *Beenakker* distribution function (for $h\Gamma \ll k_B T \ll \Delta$), and the temperature of the quantum dot ranged between *800 mK* and *1.5 K*. From the Coulomb diamond contour plot (*Figure 1.c*), we found that the diameter of the quantum dot is approximately *100 nm* which corresponds to a mean energy level spacing of $\Delta \sim 1$ *meV* and a population of less than ten electrons.

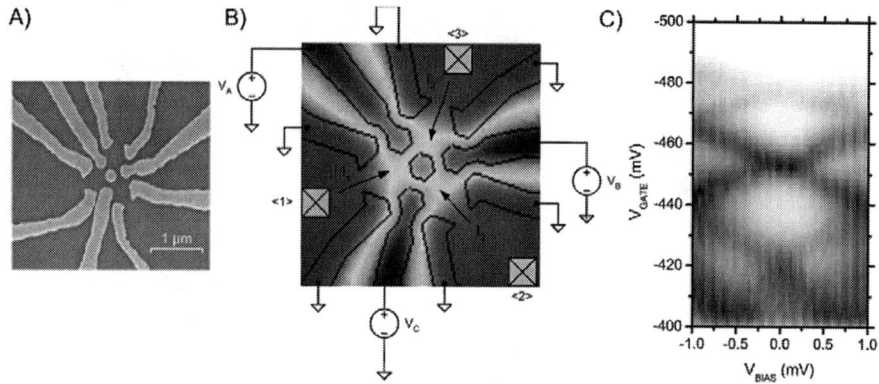

FIGURE 1. (a) SEM micrograph of the nanostructure. (b) FEMLAB simulation of 2DEG constriction (RED areas represent 2DEG regions; BLUE areas represent depleted regions). (c) Coulomb diamond contour plot. The small signal conductance is measured across two terminals with a standard lock-in amplifier measurement. The output lead is shorted to ground so as to emulate an ideal current probe.

Our particular device geometry has three unique features. There is a floating gate positioned directly above the quantum dot. This metal island—having a diameter of *200 nm*—is capacitively coupled to the constriction gates; it helps to ensure that the electrons are centered within the device and aids in flattening the potential profile through the quantum dot (creating a very shallow confinement well). It has been known for some time that shallow well quantum dots are ideal for confining a small number of electrons [6]. The second unique feature of our device is the fact that each of the three constriction gates is shielded in a coplanar fashion by two electrically grounded gates. The advantage to using such a technique is that the 2DEG can come very close to the quantum dot—allowing for a stronger interaction between the dot and lead states. It is possible that this feature accounts for the reason we have detected the integer-spin Kondo effect at such a relatively high electron temperature ($T = 800$ mK). The third unique feature of our system is the fact that there is no single gate dedicated to *plunging* the Fermi level within the dot. Instead, two of the constriction gates are set to a constant value ($V_A = V_B = V_0$) and the third gate is varied around this value ($V_C = V_0 \pm \Delta V_0$). With this technique, we can minimize the size of our device by using fewer active gates. Although it is true that this set-up does not allow us individual control over the strength of the tunneling barriers and the Fermi level of the quantum dot, by tuning the gate voltages appropriately, we were able to control and observe a rich tapestry of tunneling phenomena in different regimes of device operation.

As shown in *Figure 2*, we observed the Kondo conductance signature in the both the odd- and even-numbered electron Coulomb diamond valleys. *Figure 2(d)* shows evidence of the singlet-triplet transition at zero magnetic field. For a quantum dot with a 2DEG sheet density of $n_{2DEG} = 1.1 \times 10^{15} \ m^{-2}$ and a radius of *50 nm*, we calculate the exchange energy $J = \lambda \cdot \Delta = 320 \ \mu eV$—where λ is the *Oreg correction factor*, a function of n_{2DEG} [7]. Our measured value, the distance between the cotunneling conductance peaks within the Coulomb diamond of *Figure 2(d)*, is *320 μeV ± 10 μeV*.

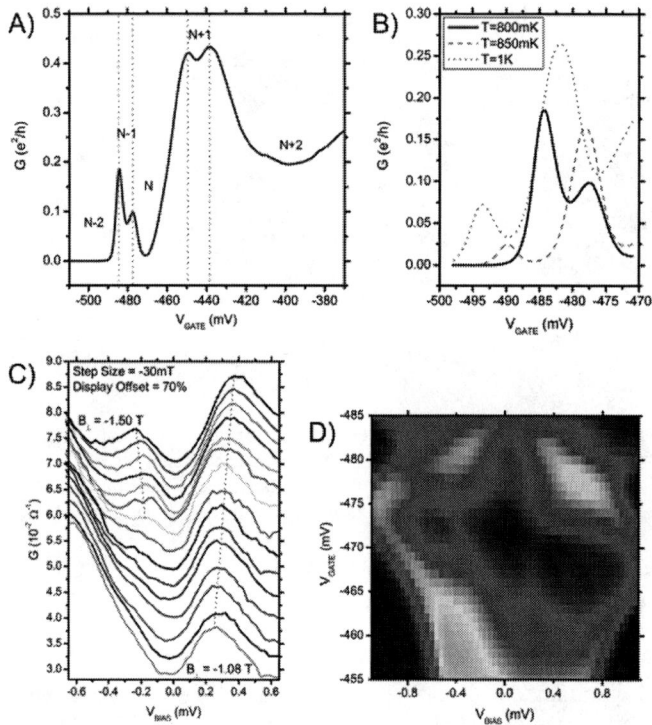

FIGURE 2. (a) Conductance peak trace showing Kondo effect in odd-numbered (*N±1*) electron valleys. (b) Decoherence of Kondo interaction due to temperature. (c) Zeeman splitting of Kondo peak in odd-numbered-electron valley as a function of magnetic field. (d) Coulomb diamond with an even number of electrons. Voltage bias dependent splitting shows singlet-triplet transition ($J = 320 \ \mu eV$).

As previously stated, we are interested in investigating the true nature of electron spin at the single particle level. Until recently, the bulk of measurements performed on mesoscopic systems (Coulomb blockade, electron diffraction, Aharonov-Bohm interference) have had first-order dependence on the electron wave-function. Because these effects are the result of wave-function interference, such measurements can be considered analogous to classical optical effects [8]. We wish to begin probing quantum dots for higher-order effects (analogous to the effects observed in the field of quantum optics). The manifestation of phenomena associated with experiments such as violations of Bell's inequality, quantum non-demolition, and teleportation is directly related to spin entanglement. In order to observe these quantum-optics-like effects, we must measure the correlation of individual shot noise events. We have designed a wideband cross-correlator (shown in Figure 3) based on the topology implemented by *Oliver, et al.* [9] and the microwave regime used in the experiments of *Reznikov, et al* [10]. From the output signals of our circuit ($<\Sigma^2>$ and $<\Delta^2>$), we can compute the normalized cross-covariance estimator $\rho(\tau)$ [9]. We are currently upgrading our cryostat system for wideband measurements. We will present data showing the correlations of shot noise on two channels from three-terminal quantum dot networks at the *18th International Conference on Noise and Fluctuations*.

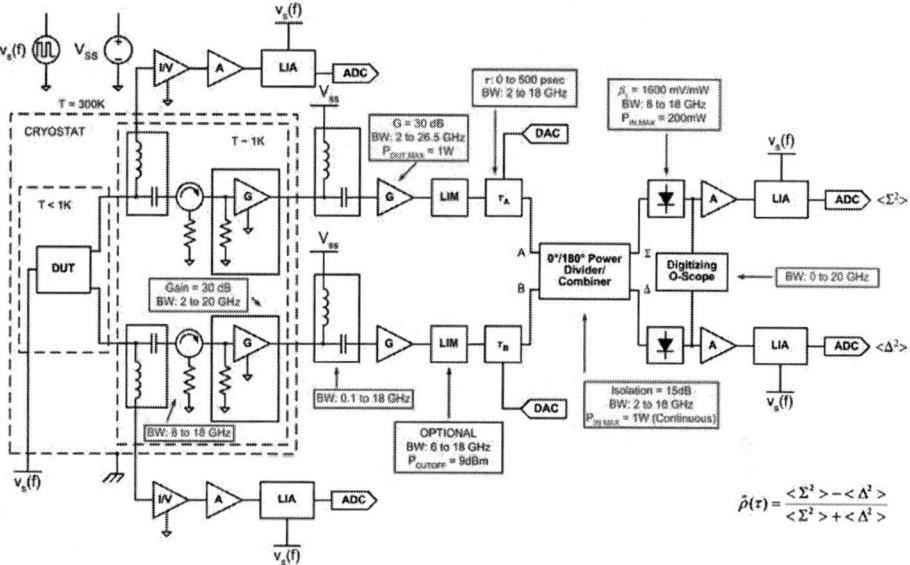

FIGURE 3. Schematic of an analog continuum cross-correlator with a bandwidth covering the *X*- and *Ku*-bands (*8 to 18 GHz*).

REFERENCES

1. D.P. DiVincenzo, *Fortschr. Phys.* **48**, 771 (2000).
2. G. Burkard, et al., *Phys. Rev. B* **61**, 16303 (2000).
3. A.T. Costa & S. Bose, *Phys. Rev. Lett.* **87**, 277901 (2001). W.D. Oliver, et al., *Phys. Rev. Lett* **88**, 037901 (2002).
4. A. Cottet, W. Belzig, & C. Bruder, *Phys. Rev. Lett.* **92**, 206801 (2004). A. Cottet, W. Belzig, & C. Bruder, *Phys. Rev. B* **70**, 115315 (2004). A. Cottet & W. Belzig, *Europhys. Lett.*, **66** (3), 405 (2004).
5. D.S. Saraga and D. Loss, *Phys. Rev. Lett.* **90**, 166803 (2003). P. Zhang, et al., *Phys. Rev. A* **69**, 042307 (2004).
6. M. Ciorga, et al. *Phys. Rev. Lett.* **88**, 256804 (2002).
7. Y. Oreg, et al. cond-mat/0109541. A. Kogan, et al., *Phys. Rev. B* **67**, 113309 (2003).
8. *Quantum Electron Optics and its Applications*, in Quantum Mesoscopic Phenomena and Mesoscopic Devices in Microelectronics, W. D. Oliver, et al., ed. by I. O. Kulik and R. Ellialtioglu, Kluwer, Dordrecht, 2000.
9. W.D. Oliver, et al. *Science* **284**, 299 (1999).
10. M. Reznikov, et al., *Phys. Rev. Lett.* **75**, 3340 (1995).

Observation of giant thermal noise due to multiple Andreev reflection in a ballistic SNS junction with an InGaAs-based heterostructure

T. Akazaki, H. Nakano, J. Nitta, and H. Takayanagi

*NTT Basic Research Laboratories, NTT Corporation,
3-1 Morinosato-Wakamiya, Atsugi-shi, Kanagawa 243-0198, JAPAN*

Abstract. We report on the fabrication of a ballistic SNS junction with a two-dimensional electron gas (2DEG) in an InGaAs-based heterostructure. We have observed the Josephson current as well as the subharmonic energy-gap structures caused by the MAR. Moreover, we experimentally estimated the thermal noise by comparing measured *IV* characteristics with those obtained with an extension of the Ambegaokar and Halperin theory. As a consequence, we have observed giant thermal noise that is much larger than that expected with normal reservoirs. The experimentally obtained "giant" thermal noise can be qualitatively explained by the Martín-Rodero theory that considers both the ballistic transport of the 2DEG and the thermal fluctuation in the coherent multiple Andreev reflection regime.

Keywords: Thermal noise, Multiple Andreev reflection, SNS junction
PACS: 72.70.+m, 73.23.Ad, 74.45.+c

INTRODUCTION

In a mesoscopic superconductor-normal metal-superconductor (SNS) structure, multiple Andreev reflection (MAR) constructs discrete bound states. At a finite temperature the upper level can be thermally populated giving rise to a reverse in the sign of the supercurrent. This switching between positive and negative current caused by thermal fluctuation leads to a huge increase in the thermal noise. In this paper, we report on the fabrication of a ballistic SNS junction with a two-dimensional electron gas (2DEG) in an InGaAs-based heterostructure [1]. This ballistic SNS junction shows clear evidence of enhanced thermal noise. Thermal noise has been experimentally estimated from a fit of current-voltage (*IV*) characteristics, according to an extension of the Ambegaokar and Halperin theory [2, 3]. The experimentally obtained temperature dependence of thermal noise can be qualitatively explained by a theoretical description that considers both the ballistic transport of the 2DEG and the thermal fluctuation in the coherent MAR regime.

EXPERIMENTS AND DISCUSSION

Figure 1(a) shows the schematic structure of the SNS junction formed by 2DEG in an InGaAs-based heterostructure. The InGaAs-based heterostructure was grown by metal-organic chemical vapor deposition (MOCVD) on a semi-insulating (100) InP substrate. The sheet carrier density n_S and mobility μ of the 2DEG at ~ 0.35 K were found to be 2.07 x 10^{12} cm^{-2} and 129,000 cm^2/Vs by Shubnikov-de Haas measurements. These values correspond to a mean free path l of 3.05 µm. The Nb electrodes were coupled by the 2DEG in a 10 nm thick $In_{0.7}Ga_{0.3}As$ channel layer in Fig. 1(a). The critical temperature T_C of the Nb electrodes was about 8.5 K. The coupling length L between the two Nb electrodes was about 0.17 µm. The InGaAs channel width W was about 8.5 µm.

Figure 1(b) shows the IV characteristics of the SNS junction at 0.36 K. A "superficial" critical current I_C^* of about 5 µA can be estimated from the current value at the corners of the IV characteristics ("real" critical current is discussed below). When a magnetic field is applied, the I_C^* follows a Fraunhofer pattern. Moreover, we obtained resistance minima within $|V| \leq 2\Delta_S/e$ as well as dip structures near $|V| = 2\Delta_S/ne$, with n = 1 to 3 or 4. Here, we assumed the Nb superconducting energy gap Δ_S to be 1.2 meV, which is somewhat lower than the typical value of ~ 1.5 meV. A normal resistance R_N of ~33 Ω is obtained in the voltage region above $2\Delta_S$. These dip structures are the subharmonic energy-gap structures caused by the MAR. These results provide clear evidence that our SNS junction formed by 2DEG in an InGaAs-based heterostructure offers ballistic transport in the 2DEG and sufficient transparency between Nb and the 2DEG.

Next, we estimated the thermal fluctuation of our ballistic SNS junction. We can quantitatively evaluate the effect of thermal fluctuation on the SNS junction from a fit of the IV characteristics, by using an extension of the Ambegaokar and Halperin theory including the additional term for phase-dependent dissipative current, namely the $\cos\phi$ term, proposed by Falco et al. [3]. Figure 2(a) shows the IV characteristics of the SNS junction measured at several temperatures. The dashed lines are the results of a parameter fit of the data to the theory in the low-voltage region below ~10 µV. Here, the dimensionless parameter $\gamma \equiv \hbar I_C/ek_B T_{noise}$ is defined as a fitting parameter, where

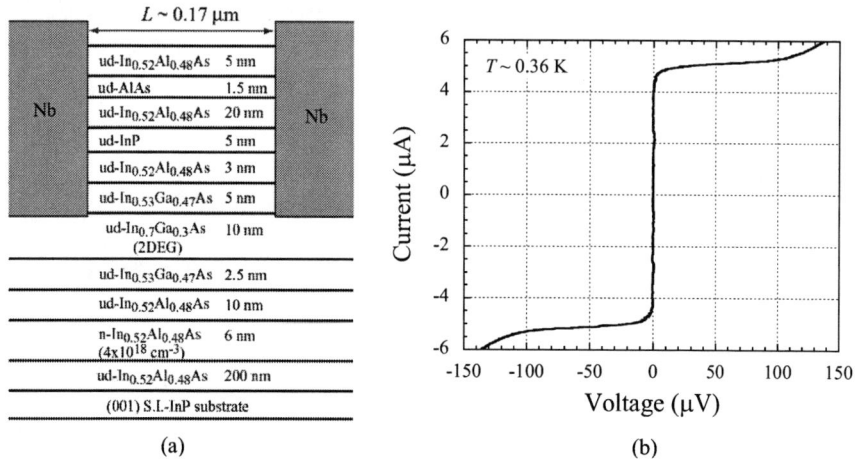

FIGURE 1. (a) Schematic structure of an SNS junction formed by 2DEG in an InGaAs-based heterostructure. (b) IV characteristics of the SNS junction at 0.36 K.

477

I_C is the "real" critical current and T_{noise} is the "effective" noise temperature. By using both an I_C of 5.5 µA and a γ of 70 at the lowest temperature of 0.36 K, we can obtain the best fit between the measured and calculated IV characteristics. In the low temperature region, we can easily determine I_C since it almost the same as I_C^*. However, in the high temperature region, it is hard to determine I_C due to the thermal noise. Therefore, the temperature dependence of I_C has been evaluated theoretically by fitting the I_C of 5.5 µA at 0.36 K to the calculated value. Since our SNS junction can be considered to be ballistic ($L \ll l$) and short [$L \ll \xi_0$, ξ_0: coherence length ($=\hbar v_F/\pi \Delta_S$), v_F: Fermi velocity in the normal region], the temperature dependence of I_C was calculated from the following expression for a ballistic and short SNS junction [4, 5]:

$$I(\phi,T) = \frac{e\Delta_S^2(T)}{2\hbar}\sin\phi \sum_{n=1}^{M} \frac{T_n}{E_n(\phi,T)} \tanh\left(\frac{E_n(\phi,T)}{2k_BT}\right) \quad (1a)$$

$$I_C(T) = \max I(\phi,T) \text{ at each } T. \quad (1b)$$

Here, the energy level of the Andreev bound state $E_n(\phi,T)$ is given by $E_n(\phi,T) = \Delta_S(T)[1 - T_n\sin(\phi/2)]^{1/2}$, where T_n is the transmission probability in the normal region and $\Delta_S(T)$ is the superconducting energy gap. For our SNS junction, we assume that $\Delta_S(0)$ is 1.2 meV and that the temperature dependence Δ_S is in accordance with the BCS theory. M is the number of transmission modes, which is given by $2W/\lambda_F$, [λ_F: Fermi wavelength of 2DEG]. For simplicity, we approximate that all transmission modes are equivalent. The T_n of ~ 0.4 is given from $T_n = R_{sh}/R_N$, [R_{sh}: Sharvin resistance of 2DEG, $R_{sh} = (h/2e^2)(\lambda_F/2W)$]. As shown in Fig. 2(a), we determined the

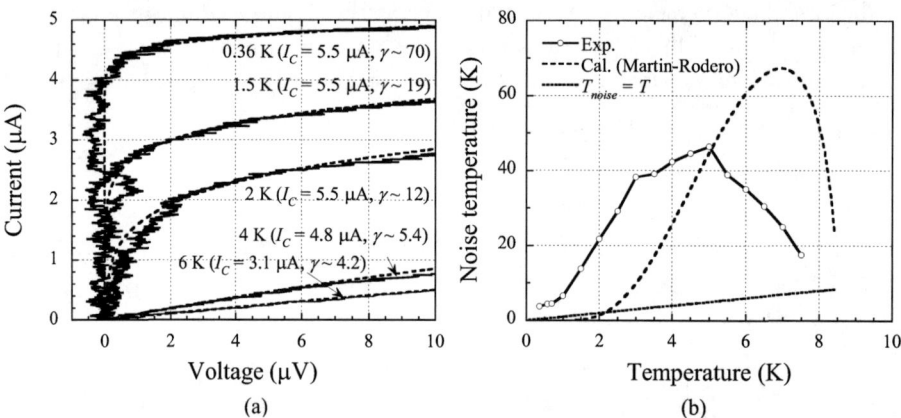

FIGURE 2. (a) The IV characteristics of the SNS junction measured at several temperatures. The solid and lines are experimental and calculated data for the low-voltage region below ~10 µV. (b) The temperature dependence of the noise temperature of the SNS junction. The open circles with the solid line indicate the experimentally obtained noise temperature from the comparison with an extension of the Ambegaokar and Halperin theory. The dashed line represents the calculated temperature dependence of the noise temperature according to the Martín-Rodero theory. The dotted line represents the environment temperature T.

fitting parameter γ using the evaluated I_C. As a consequence, we can evaluate the temperature dependence of T_{noise} from the evaluated I_C and γ. Figure 2(b) shows the temperature dependence of T_{noise} of the SNS junction. We found that T_{noise} is larger than the environment temperature T, represented by the dotted line in Fig. 2(b) as well as an exponential increase with increasing temperature in the sufficiently low temperature regime where the decrease in Δ_S can be disregarded. The theory of thermal noise in SQPC can explain these results as follows [6]. Martín-Rodero et al. have theoretically studied the frequency-dependent current fluctuations in SQPC within the dc transport regime. The zero-frequency thermal noise in SQPC is given by

$$P_{SQPC}(\phi,T) = \frac{2e^2}{h} \frac{\pi}{\eta} \frac{\Delta_S^4(T) T_n^2 \sin^2\phi}{E_n^2(\phi,T)} f(E_n(\phi,T))[1 - f(E_n(\phi,T))]. \quad (2)$$

Here, $f(E)$ is the Fermi distribution function. η is the small energy relaxation rate that takes into account the damping of quasiparticle states due to inelastic processes inside electrodes. A typical estimation of for a traditional superconductor is $\eta/\Delta_S \sim 10^{-2}$ [6]. Thermal noise power P_{SQPC} is directly related to both the noise temperature and the conductance by the fluctuation dissipation theorem [7], $P_{SQPC} = 4k_B T_{noise} G$. We use the estimated T_n of ~0.4 as well as $G = (2e^2/h) \times T_n$. The calculated temperature dependence of T_{noise} is shown by the dashed line in Fig. 2(b). The calculated curve shows the exponential increase with increases in temperature, which agrees qualitatively with the measured temperature dependence of T_{noise}. The experimentally obtained enhanced thermal noise can be qualitatively explained by the Martín-Rodero theory

CONCLUSIONS

We have investigated the superconducting properties of a ballistic SNS junction with a 2DEG in an InGaAs-based heterostructure. We have observed the Josephson current as well as the subharmonic energy-gap structures caused by the MAR. Moreover, we experimentally estimated the thermal noise by comparing measured *IV* characteristics with those obtained with an extension of the Ambegaokar and Halperin theory. As a consequence, we have observed enhanced thermal noise. This enhanced thermal noise can be explained by a theory that considers both the ballistic transport of the 2DEG and the thermal fluctuation in the coherent MAR regime.

REFERENCES

1. Nitta, J., Akazaki, T., and Takayanagi, H., *Phys. Rev. B* **46**, R14286 (1992).
2. Ambegaokar, V., and Halperin, B. I., *Phys. Rev. Lett.* **22**, 1364 (1969).
3. Falco, C. M., Parker, W. H., and Trullinger, S. E., *Phys. Rev. Lett.* **31**, 933 (1973).
4. Beenakker, C. W. J., *Phys. Rev. Lett.* **67**, 3836 (1991).
5. Galaktionov, A. V., and Zaikin, A. D., *Phys. Rev. B* **65**, 184507 (2002).
6. Martín-Rodero, A., Levy Yeyati, A., and García-Vidal, F. J., *Phys. Rev. B* **53**, R8891 (1996).
7. Callen, H. B. and Welton, T. W., *Phys. Rev.* **83**, 34 (1951).

Full Counting Statistics of Mesoscopic Electron Transport

Wolfgang Belzig

Department of Physics and Astronomy, University of Basel, Klingelbergstrasse 82, 4056 Basel, Switzerland

Abstract. Noise in mesoscopic electron transport is usually compared to Schottky's result, which assumes independent charge transfers (i. e. the transfered charge obeys Poisson statistics). Here we discuss an intriguing example in which interactions in a quantum dot lead to a strongly enhanced shot noise. Studying the full counting statistics it is shown that the underlying transport process consists of an infinite number Poisson processes of arbitrary order.

Keywords: Quantum noise, mesoscopic transport, full counting statistics
PACS: 73.23. b, 72.70.+m, 72.25.Rb

Quantum noise of the current in mesoscopic conductors leads directly to the fundamental concept of *full counting statistics* (FCS): during a given time interval a certain number N of charges will pass the conductor, described by a *probability $P(N)$* [1]. FCS is a direct extension of experimental and theoretical studies of current fluctuations in nanoscopic conductors in the last decade [2], which has already led to profound insights in the details of the transport mechanism. In this article we will review some of our recent advances in the field of FCS. Concretely we will address the question how *Coulomb interactions* influence the statistics of the transfered charge.

Interaction effects are particularly prominent in nonlocal current correlations. For noninteracting fermions zero-frequency current crosscorrelations are always negative, but interactions can reverse the sign of crosscorrelations[3]. As example is provided by the statistics of charge transport in a mesoscopic three-terminal device with one superconducting terminal and two normal-metal terminals and calculated the full distribution of transmitted charges into the two symmetrically biased normal terminals[4]. In a wide parameter range, we found large positive cross correlations between the currents in the two normal arms. Next, we have looked at current fluctuations in an interacting three-terminal quantum dot with ferromagnetic leads (but without superconductors)[5]. For appropriately polarized contacts, the transport through the dot turned out to be governed by a dynamical spin-blockade effect, i.e., a spin-dependent bunching of tunneling events not present in the paramagnetic case. As consequence we found positive cross correlations of the currents in the output leads even in the absence of spin accumulation on the dot.

One intriguing question remaining is whether positive cross correlation can be induced by the Coulomb interactions on a quantum dot, i. e. without correlated injection by a superconducting source or spin symmetry breaking by ferromagnetic contacts. That this can indeed be the case has been pointed out in Ref. [7]. One can map for example an internal level structure onto a pseudo spin and then take over the results for the quantum

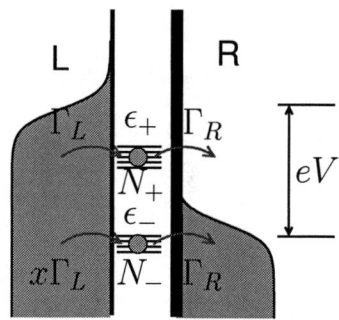

FIGURE 1. Multi-level quantum dot

dot in the presence of Zeeman-split levels due to an external magnetic field [6]. It is important to note that in certain situations positive cross correlations are directly related to super-Poissonian noise. For a symmetric situation (the two output ports, at which the cross correlation are measured have the same properties and are held at the same potentials) and unidirectional transport (no current can go back to the source terminal) one finds

$$F_{12} = \gamma(F-1). \quad (1)$$

Here the generalized Fano factor is defined via the cumulant generating function $S(\chi, \chi_1, \chi_2)$, in which χ is counting field associated with the source and $\chi_{1(2)}$ are counting fields associated with the output terminals 1(2). Currents are obtained as

$$\bar{N} = -i \left.\frac{\partial S}{\partial \chi}\right|_{\chi,\chi_{1(2)}=0}, \quad \bar{N}_{1(2)} = -i \left.\frac{\partial S}{\partial \chi_{1(2)}}\right|_{\chi,\chi_{1(2)}=0}, \quad (2)$$

and the noise correlators are

$$F = -\frac{1}{\bar{N}} \left.\frac{\partial^2 S}{\partial \chi^2}\right|_{\chi,\chi_{1(2)}=0}, \quad F_{12} = -\frac{1}{\bar{N}} \left.\frac{\partial^2 S}{\partial \chi_1 \chi_2}\right|_{\chi,\chi_{1(2)}=0}. \quad (3)$$

Consequently, it is of interest to study also the problem of super-Poissonian noise in quantum dots, from which it is possible to obtain positive cross correlations. The factor $\gamma \leq 1/4$ depends on the details of the setup like e. g. the asymmetry and will play no role in the following. Some examples and further discussion can be found in Refs. [5, 6, 7].

We now turn to the full counting statistics. We consider here a quantum dot which has some internal level structure, see Fig. 1. We assume that the Coulomb energy is so large that only one of the levels can be occupied at a given time. The levels are bunched into two groups, the lower group with N_- levels and the upper group with N_+ levels. We will furthermore assume that the tunneling rate into and from all levels inside a group are the same. Let us introduce $p = N_+/(N_- + N_+)$, which is the relative probability that an electron tunneling from the left electrode jumps on the upper level. The counting statistics (to first order in x) is

$$S(\chi) = -x\Gamma_R t_0 \frac{e^{i\chi}-1}{1-pe^{i\chi}} \equiv -N_{\text{th}}\frac{e^{i\chi}-1}{1-pe^{i\chi}}. \quad (4)$$

This CGF gives the first cumulant $C_1 = N_{\text{th}}/(1-p)$ and the higher cumulants are

$$C_2 = \frac{1+p}{1-p}C_1, C_3 = \frac{1+4p+p^2}{(1-p)^2}C_1, C_4 = \frac{(1+p)(1+10p+p^2)}{(1-p)^3}C_1. \quad (5)$$

We note that the average is larger by the factor $1/(1-p)$ than the estimate by a Poisson statistics of charges defined by C_2/C_1. In other words, higher cumulants obey the relation $C_n > (C_2/C_1)^{n-1}C_1$, which shows the unusual transport characteristics.

We now prove that the counting statistics is indeed stronger correlated than a simple super-Poissonian statistics, i e. it cannot be explained in terms of a Poisson statistics of some effective charge. Let us introduce a set of dimensionless functions

$$c_n(x) = \partial_x^n \frac{e^x-1}{1-pe^x}, \quad (6)$$

from which the cumulants can be obtained as $C_n = N_{\text{th}}c_n(0)$. We write

$$c_2(x) = c_1(x)\frac{1+pe^x}{1-pe^x} \equiv c_1(x)q(x) \quad (7)$$

and find for the effective charge $q \equiv q(0) = (1+p)/(1-p)$. The definition of C_n is

$$C_n = N_{\text{th}}\partial_x^{n-1}c_1(x)|_{x=0} = N_{\text{th}}\partial_x^{n-2}c_1(x)q(x)|_{x=0} \quad (8)$$

It is easy to show that all derivatives of $q(x)$ are positive and it is therefore sufficient to expand $q(x)$ for small x, which yields $q(x) = q(1+2px/(1-p^2))$. Thus,

$$C_n = q\underbrace{\partial_x^{n-2}c_1(x)|_{x=0}}_{C_{n-1}} + (n-2)\partial_x^{n-3}c_1(x)|_{x=0}\frac{2pq}{1-p^2} \geq qC_{n-1}, \quad (9)$$

which holds since the second term in the middle expression is always positive. By induction the desired relation follows

$$C_n \geq q^{n-1}C_1. \quad (10)$$

It turns out that we can decompose the FCS into multiple Poisson processes by the following manipulations

$$S(\chi) = -N_{\text{th}}(e^{i\chi}-1)\sum_{n=0}^{\infty}p^n e^{in\chi} \quad (11)$$

$$= -N_{\text{th}}\left[\frac{1-p}{p}\left(\sum_{n=1}^{\infty}p^n e^{in\chi}\right) - 1\right] \quad (12)$$

$$= -N_{\text{th}}\frac{1-p}{p}\left[\left(\sum_{n=1}^{\infty}p^n e^{in\chi}\right) - p\sum_{n=0}^{\infty}p^n\right]. \quad (13)$$

Combining both terms we finally obtain

$$S(\chi) = -N_{\text{th}}(1-p) \sum_{n=1}^{\infty} p^{n-1} \left(e^{in\chi} - 1\right). \tag{14}$$

The statistics is a sum of Poissonian processes of multiple charges, weighted with probabilities $p^{n-1}(1-p)$. The possible Poissonian processes result from a tunneling event with the small initial rate $x\Gamma_L$. This process is followed by $n-1$ fast cycles, in which an electron tunnels through the dot with probability p. Finally the fast cycle stops with a probability $1-p$. As can be seen from Eq. (14) processes of *all* orders contribute to the transport characteristics. A simplified description in terms of a Poissonian process, in which an effective charge tunnels, can never reproduce the observed cumulants. It is worth to emphasize, that only the complete determination of all cumulants (or equivalently the FCS) can unambiguously identify this transport mechanism.

In this work we have discussed some of the unsusual transport properties of a quantum dot with internal level structure. In certain cases it turns out that the statistics is an infinite sum of Poissonian processes with all possible effective charges. Higher order Poisson processes occur with smaller probabilities, but can in general not be neglected. An experimental study of the full counting statistics or higher order cumulants would be able to reveal these interesting multi-particle correlations.

ACKNOWLEDGMENTS

I acknowledge useful discussion with C. Bruder and A. Cottet. Financial support was provided by the NCCR Nanoscience and the Swiss National Science Foundation.

REFERENCES

1. *Quantum Noise in Mesoscopic Physics*, edited by Yu. V. Nazarov (Kluwer, Dordrecht, 2003).
2. Ya. M. Blanter and M. Büttiker, Phys. Rep. **336**, 1 (2000).
3. M. Büttiker, Phys. Rev. B **46**, 12485 (1992); M. Büttiker, in *Quantum Noise in Mesoscopic Physics*, edited by Yu. V. Nazarov (Kluwer, Dordrecht, 2003).
4. J. Börlin, W. Belzig, and C. Bruder, Phys. Rev. Lett. **88**, 197001 (2002); W. Belzig and P. Samuelsson, Europhys. Lett. **64**, 253 (2003).
5. A. Cottet, W. Belzig, and C. Bruder, Phys. Rev. Lett. **92**, 206901 (2004).
6. A. Cottet and W. Belzig, Europhys. Lett. **66**, 405 (2004).
7. A. Cottet, W. Belzig, and C. Bruder, Phys. Rev. B **70**, 115315 (2004).
8. W. Belzig, Phys. Rev. B **71**, 161301R (2005)

Chaotic transport: from quantum to classical

Robert S. Whitney* and Ph. Jacquod*

*Département de Physique Théorique, Université de Genève, CH-1211 Genève 4, Switzerland

Abstract. We present a semiclassical theory for the scattering matrix $\mathscr{S}$ of a chaotic ballistic cavity at finite Ehrenfest time. Using a phase-space representation we show that the Liouville conservation of phase-space volume decomposes $\mathscr{S}$ as $\mathscr{S} = \mathscr{S}^{\mathrm{cl}} \oplus \mathscr{S}^{\mathrm{qm}}$. The short-time, classical contribution $\mathscr{S}^{\mathrm{cl}}$ generates deterministic transmission eigenvalues $T = 0$ or 1, while quantum ergodicity is recovered within the subspace corresponding to the long-time, stochastic contribution $\mathscr{S}^{\mathrm{qm}}$. This provides a microscopic foundation for the two-phase fluid model, in which the cavity acts like a classical and a quantum cavity in parallel. Our model shows that the Fano factor of the shot-noise power vanishes in this limit, while weak-localization remains universal.

Keywords: noise, conductance, quantum chaos, mesoscopics, semiclassics, random matrix theory
PACS: 73.23.-b, 74.40.+k, 05.45.Mt

Introduction: In recent years it has been possible to make electronic systems clean enough that the electron have a mean free path significantly longer than the size of the potential that confines them[1]. The electrons then move ballistically in these *quantum dots*, in a manner strongly related to *classical* dynamics in the dot. When this classical motion is *chaotic* the transport properties are usually universal and well-captured by random matrix theory (RMT). However once the *Ehrenfest time* becomes a relevant parameter this universality is broken and the transport properties cease to be described by RMT [2]. Elsewhere [3, 4] we explore the role played by the Ehrenfest time in a clean chaotic system connected to two leads in the limit where the Fermi wavelength is much smaller than all system lengthscales (system size, lead widths). Here we define open-cavity Ehrenfest times in such systems and emphasise their relevance to the transport properties. We then show that for finite open-cavity Ehrenfest time the cavity scattering matrix is block diagonal and hence behaves like two cavities in parallel. One of these cavities is classical in nature the other is quantum. We prove that the classical cavity's transmission eigenvalues are all zero or one, so its transport properties are *deterministic* and hence *noiseless*. Meanwhile the quantum cavity is *stochastic*, with its transport properties exhibiting *quantum noise*. Finally we touch on the consequences of this for the Fano factor and the quantum (weak localisation) correction to conductance.

Ehrenfest times : Ehrenfest times are the time-scales on which quantum effects start to become relevant in the evolution of a wavepacket. We consider a chaotic cavity of size L and Lyapunov exponent λ which connected to two leads (Left and Right) with widths W_L, W_R; where L, W_L, W_R are all much larger than the Fermi wavelength, $\hbar/p_F$. There are two *open-cavity* Ehrenfest times [5] associated with modes entering the cavity from the Left (L) lead;

$$\tau_E^{LR} = \lambda^{-1} \ln[\hbar_{\mathrm{eff}}^{-1}(W_L W_R/L^2)], \qquad \tau_E^{LL} = \lambda^{-1} \ln[\hbar_{\mathrm{eff}}^{-1}(W_L^2/L^2)], \qquad (1)$$

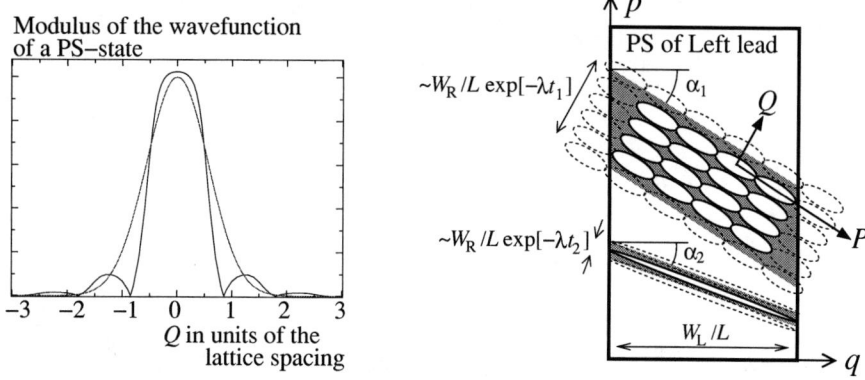

FIGURE 1. On the left is a plot of a PS-state as a function of dimensionless position (dark line); for comparison we plot the wavefunction of the coherent state (grey line). Both wavefunctions have the same shape as a function of P as they do as a function of Q (up to a scaling factor). The PS-state's oscillations make it orthogonal to PS-states centred at finite Q, and its broadened peak (w.r.t. the coherent state) makes it orthogonal to PS-states centred at finite P. **On the right** we show two bands on the Left lead (in grey), with PS-states super-imposed on them (ellipses). The lattice of PS-states has been stretched/rotated to maximise the number of PS-states in each band (solid-edged ellipses) while minimising the number partially in each band (dashed-edged ellipses). Thus the PS-states have the same aspect ratio as the band.

where the dimensionless Planck constant $\hbar_{\text{eff}} = \hbar/(p_F L)$. The first time is for *transmission* (L to R) and the second is for *reflection* (L to L). In addition there is the *closed-cavity* Ehrenfest time [2,6–9], $\tau_E^{\text{cl}} = \lambda^{-1} \ln[\hbar_{\text{eff}}^{-1}]$, unlike those above it is a property of the cavity itself and is independent of the size of the leads.

The three time-scales can be derived as follows. We assume the cavity is a two-dimensional hyperbolic chaotic system. Then the Poincaré surface of section perpendicular to any trajectory is a two-dimensional phase space $(r_\perp, p_\perp)$, which we make dimensionless by writing distances in units of L and momenta in units of p_F. Then the Liouvillian flow on the Poincaré surface of section stretches exponentially, with rate λ in the *unstable* direction, while compressing exponentially in the *stable* direction. The Ehrenfest times are then given by $\lambda^{-1} \ln[\hbar_{\text{eff}}^{-1} XY]$ where X and Y are dimensionless system lengthscales; W_L/L, W_R/L or 1. This is the time for a wavepacket with width X in the *stable* direction (and hence $\hbar_{\text{eff}}/X$ in the *unstable* direction) to spread under the Liouvillian flow to width Y in the *unstable* direction. We note that for all times of relevance here, the evolution of wavepackets inside the system is well approximated by the Liouvillian flow of the *classical* dynamics.

Bands in the classical phase-space (PS): The finiteness of τ_D (the dwell time for trajectories in the cavity) means that classical trajectories injected into a cavity are naturally grouped into PS transmission and reflection bands [12], despite the ergodicity of the associated closed cavity. Each band on the PS cross-section of the L lead (see Fig. 1) consists of a group of classical paths which exit through the same lead after the same number of bounces, τ, (having followed similar paths through the cavity). Because of the chaotic classical dynamics, bands with longer escape times are narrower, having a width (and hence a PS area) scaling like $\propto \exp[-\lambda \tau]$. The Ehrenfest time is the time at which this area becomes smaller than $\hbar_{\text{eff}}$. Thus only for times *shorter* than this can a band carry one (or more) *whole* quantum wavepacket.

Counting classical and quantum modes in the scattering matrix: All trajectories which exit through the R lead at $\tau < \tau_E^{LR}$ will be in bands with phase-space area larger than $\hbar_{\text{eff}}$. We show below that these modes are *classical*. Thus the number of transmitting (reflecting) *classical* PS-states is given by the area of the L lead's phase-space which couples to transmitting (reflecting) trajectories with $\tau < \tau_E^{LR}$ (with $\tau < \tau_E^{LL}$). The total number of classical modes in the L lead is the sum of these two;

$$N_L^{\text{cl}} = [N_L + N_R]^{-1} \left[N_L^2 (1 - e^{-\tau_E^{LL}/\tau_D}) + N_L N_R (1 - e^{-\tau_E^{LR}/\tau_D}) \right] \quad (2)$$

All other modes of the L lead sit over many transmission bands with $\tau > \tau_E^{LR}$ or reflection bands with $\tau > \tau_E^{LL}$, and so they are *quantum* PS-states; thus $N_L^{\text{qm}} = N_L - N_L^{\text{cl}}$. We can do the same for the phase-space of the R lead by swapping L and R throughout.

The phase-space basis: Below we write the scattering matrix $\mathscr{S}$ in a PS-basis, whose construction we now summarize, for details see [4]. We construct the PS-basis by covering all phase-space bands with area bigger than $\hbar_{\text{eff}}$ with a lattice of PS-states of the form shown in Fig. 1. The lattice is stretched and rotated to optimally cover each band (as in Fig. 1). All states on the lattice covering each such band are *orthonormal*, and the basis is *complete* on the parts of phase space which are covered by these bands. The position of the lattice on each band is chosen such that each ingoing PS-state evolves under the cavity dynamics to exit as exactly one outgoing PS-state. Each basis states exits at a time less that τ_E^{LR} (for transmission) or τ_E^{LL} (for reflection). It exit as a single wavepacket at a single time; thus it behave *deterministically*; i.e. like a classical particle with its quantum nature completely hidden.

The remaining phase-space (made of classical bands with phase-space area less than $\hbar_{\text{eff}}$) is covered by states chosen simply to complete the basis. The fact that the basis is already complete on the bands with area larger than $\hbar_{\text{eff}}$, means that each remaining PS-states must sit on many bands in the classical phase space which exit at many different times. Thus these PS-basis states exhibit strongly quantum behaviour.

Scattering matrix in the PS-basis: The transformation from the basis of lead modes to the PS-basis is *unitary* because both bases are complete and orthonormal. Thus this transformation leaves *unchanged* the eigenvalues of the scattering matrix, $\mathscr{S}$, and the transmission matrix $\mathscr{T} = \mathbf{t}^\dagger \mathbf{t}$ (where $\mathbf{t}$ is the L to R transmission block of $\mathscr{S}$). In the PS-basis, the scattering matrix takes the form

$$\mathscr{S} = \mathscr{S}_{\text{cl}} \oplus \mathscr{S}_{\text{qm}} = \begin{pmatrix} \mathscr{S}_{\text{cl}} & 0 \\ 0 & \mathscr{S}_{\text{qm}} \end{pmatrix} \quad (3)$$

The matrix $\mathscr{S}_{\text{cl}}$ is $N^{\text{cl}} \times N^{\text{cl}}$ while the matrix $\mathscr{S}_{\text{qm}}$ is $N^{\text{qm}} \times N^{\text{qm}}$, with $N^{\text{cl}} = N_L^{\text{cl}} + N_R^{\text{cl}}$ and $N^{\text{qm}} = N_L^{\text{qm}} + N_R^{\text{qm}}$. The matrix $\mathscr{S}_{\text{cl}}$ must have only one non-zero element in each row and column. After re-ordering the labels of the modes on L and R, we can write

$$\mathscr{S}_{\text{cl}} \equiv \begin{pmatrix} \mathbf{r}_{\text{cl}} & \mathbf{t}'_{\text{cl}} \\ \mathbf{t}_{\text{cl}} & \mathbf{r}'_{\text{cl}} \end{pmatrix} \qquad \mathbf{t}_{\text{cl}} = \begin{pmatrix} \tilde{\mathbf{t}}_{\text{cl}} & 0 \\ 0 & 0 \end{pmatrix} \qquad \mathbf{r}_{\text{cl}} = \begin{pmatrix} 0 & 0 \\ 0 & \tilde{\mathbf{r}}_{\text{cl}} \end{pmatrix} \quad (4)$$

The matrices $\tilde{\mathbf{t}}_{cl}$ and $\tilde{\mathbf{t}}'_{cl}$ are $n \times n$, where $n = [N_L N_R/(N_L + N_R)] \exp[-\tau_E^{LR}/\tau_D]$ is the number of *classical transmission modes*. The matrix $\tilde{\mathbf{r}}_{cl}$ is $(N_L^{cl} - n) \times (N_L^{cl} - n)$ and $\tilde{\mathbf{r}}'_{cl}$ is $(N_R^{cl} - n) \times (N_R^{cl} - n)$. The matrix $\tilde{\mathbf{t}}_{cl}$ is diagonal with elements given by $\tilde{t}_{ij} = e^{i\Phi_i}\delta_{ij}$ The matrix $\tilde{\mathbf{r}}_{cl}$ has a more complicated structure; it has *exactly one* non-zero element in each row and each column. Thus we have diagonalised N_L^{cl} of the modes of the transmission matrix, $\mathscr{T}$. It has n modes with eigenvalue $T_\alpha = 1$ and $N_L^{cl} - n$ modes with eigenvalue $T_\alpha = 1$. As noise $\propto \sum_\alpha T_\alpha(1 - T_\alpha)$, all these modes are noiseless. In the classical limit the proportion of such classical (noiseless) modes goes to one [10]. The remaining modes (which remain numerous despite their proportion going to zero) are quantum in nature and are unitary *within* their own subspace, $\mathscr{S}_{qm}$ [11].

Average conductance: All *transmitting* quantum and classical modes carry current, so the dimensionless conductance equals $N_L N_R/(N_L + N_R) \propto \hbar^{-1}$.

Zero-frequency noise and the Fano factor : As the classical modes are *noiseless*, all noise is generated by the quantum modes. The number of quantum (noisy) transmission modes is $[N_L N_R/(N_L + N_R)] \exp[-\tau_E^{LR}/\tau_D]$ goes to infinity in the classical limit $\hbar \to 0$. However the Fano factor $\propto$ (noise/average current) scales like $\exp[-\tau_E^{LR}/\tau_D]$, vanishing as $\hbar \to 0$. This fits numerical and experimental [13] observations and has qualitative agreement with the earlier microscopic theory [6].

Weak localisation (WL): Recent numerics [14] have called into question previous microscopic theories [2, 9] which predicted that the WL correction to conductance decays like $\exp[-\tau_E^{cl}/\tau_D]$. We perform a calculation similar to [8, 9] for the quantum modes (the classical modes have no WL correction); for details see [4]. The result in eq. (3), provides a factor of τ_E^{LR} which cancels the τ_E^{cl} in the exponent, leading to the universal (RMT) result even as $\hbar \to 0$ (for well developed chaos, $\lambda \tau_D \gg 1$).

Acknowledgments: We are grateful to E. Sukhorukov for numerous discussions and thank M. Sieber, P. Silvestov, C. Beenakker and İ. Adagideli for helpful comments. This work is supported by the Swiss National Science Foundation.

REFERENCES

1. L.P. Kouwenhoven, C.M. Marcus, P.L. McEuen, S. Tarucha, R.M. Westervelt, and N.S. Wingreen, *Electron Transport in Quantum Dots*, Nato ASI conference proceedings, L.P. Kouwenhoven, G. Schön, and L.L. Sohn Eds. (Kluwer, Dordrecht, 1997).
2. I.L. Aleiner and A.I. Larkin, Phys. Rev. B **54**, 14423 (1996)
3. R.S. Whitney, and Ph. Jacquod, Phys. Rev. Lett. **94**, 116801 (2005)
4. R.S. Whitney, and Ph. Jacquod, *in preparation*.
5. M.G. Vavilov and A.I. Larkin, Phys. Rev. B **67**, 115335 (2003).
6. O. Agam, I. Aleiner and A. Larkin, Phys. Rev. Lett. **85**, 3153 (2000).
7. M. Sieber and K. Richter, Phys. Scr. **T90**, 128 (2001). M. Sieber, J. Phys. A **35**, L613 (2002).
8. K. Richter and M. Sieber, Phys. Rev. Lett. **89**, 206801 (2002).
9. İ. Adagideli, Phys. Rev. B **68**, 233308 (2003).
10. We believe this is the first microscopic proof of this result, despite the fact it was anticipated some time ago; C.W.J. Beenakker and H. van Houten, Phys. Rev. B **43**, R12066 (1991).
11. This gives a microscopic proof of an earlier prediction that the cavity splits into two (one classical, one quantum) [12]; however it remains to be proven that the quantum cavity has RMT behaviour.
12. P.G. Silvestrov, M.C. Goorden, and C.W.J. Beenakker, Phys. Rev. B **67**, 241301(R) (2003).
13. S. Oberholzer, E.V. Sukhorukov, and C. Schönenberger, Nature **415**, 765 (2002)
14. J. Tworzydło, A. Tajic, and C.W.J. Beenakker, Phys. Rev. B **70**, 205324 (2004).

Photo-assisted shot noise in the fractional quantum Hall regime

Adeline Crépieux*, Pierre Devillard† and Thierry Martin*

*Centre de Physique Théorique, Université de la Méditerranée, Case 907, 13288 Marseille, France
†Centre de Physique Théorique, Université de Provence, Case 907, 13288 Marseille, France

Abstract.
The effect of an ac perturbation on the shot noise of a fractional quantum Hall fluid is studied at finite temperature. For a normal metal, it is known that the zero-frequency noise derivative exhibits steps as a function of bias voltage. In contrast, at Laughlin fractions, the backscattering noise exhibits evenly spaced singularities, which are reminiscent of tunneling density-of-states singularities for quasiparticles. The spacing is determined by the quasiparticle charge νe and the ratio of the dc bias with respect to the drive frequency. Photo-assisted transport can thus be considered as a probe for effective charges of the quantum Hall effect.

Keywords: Shot noise, Luttinger liquid, Edge states, Photo-assisted transport.
PACS: 73.43.-f; 73.50.Td; 03.65.Ta.

INTRODUCTION

In mesoscopic systems, the measurement of shot noise makes it possible to probe the effective charges which flow in conductors, and opens the possibility for studying the role of the statistics in stationary quantum transport experiments. This has been illustrated experimentally and theoretically where the interaction between electrons is less important[1, 2, 3, 4, 5, 6] or when it is more relevant[7, 8, 9, 10, 11]. The present work deals with the study of photo-assisted shot noise in a specific one dimensional correlated system: a Hall bar in the fractional quantum Hall regime, for which charge transport occurs via two counter-propagating chiral edges states.

MODEL

We consider the system depicted on Fig. 1 which is described by the Hamiltonian:

$$H = \frac{\hbar v_F}{4\pi} \sum_{r=R,L} \int ds (\partial_s \phi_r(t))^2 + A(t)\Psi_R^\dagger(t)\Psi_L(t) + A^*(t)\Psi_L^\dagger(t)\Psi_R(t) \ . \quad (1)$$

The bosonic fields $\phi_{R(L)}$, which describe the right and left moving chiral excitations along the edge states, are related to the fermionic fields $\Psi_{R(L)}$ through:

$$\Psi_{R(L)}(t) = \frac{F_{R(L)}}{\sqrt{2\pi a}} e^{i\sqrt{\nu}\phi_{R(L)}(t)} \ , \quad (2)$$

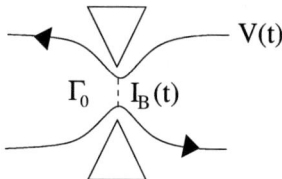

FIGURE 1. Backscattering between edge states in the presence of a bias voltage modulation $V(t)$.

where $F_{R(L)}$ is a Klein factor, a, the short-distance cutoff and v, the filling factor which characterize the charge $e^* = ve$ of the backscattered quasiparticles. The hopping amplitude between the edge states has a time-dependence due to the applied voltage $V(t) = V_0 + V_1 \cos(\omega t)$:

$$A(t) = \Gamma_0 \sum_{n=-\infty}^{+\infty} J_n\left(\frac{\omega_1}{\omega}\right) e^{i(\omega_0 + n\omega)t}, \qquad (3)$$

where we have made an expansion in term of an infinite sum of Bessel functions, which is a signature of photo-assisted processes[12]. It is important to notice that the frequencies ω_0 and ω_1 which appear in Eq. (3) are related to the filling factor v:

$$\omega_0 \equiv veV_0/\hbar, \qquad \omega_1 \equiv veV_1/\hbar. \qquad (4)$$

where $v = 1/(2m+1)$ with m integer.

PHOTO-ASSISTED SHOT NOISE

The symmetrized backscattering current noise correlator is expressed with the help of the Keldysh contour:

$$\begin{aligned} S(t,t') &= \frac{1}{2}\langle I_B(t)I_B(t')\rangle + \frac{1}{2}\langle I_B(t')I_B(t)\rangle - \langle I_B(t)\rangle\langle I_B(t')\rangle \\ &= \frac{1}{2}\sum_{\eta=\pm 1}\langle T_K\{I_B(t^\eta)I_B(t'^{-\eta})e^{-i\int_K dt_1 H_B(t_1)}\}\rangle, \end{aligned} \qquad (5)$$

where H_B is the sum of the second and the third terms in Eq. (1), and:

$$I_B(t) = \frac{ive}{\hbar}A(t)\Psi_R^\dagger(t)\Psi_L(t) - h.c. \qquad (6)$$

We are interested in the Poissonian limit only, so in the weak backscattering case, one collects the second order contribution in the tunnel barrier amplitude $A(t)$, and the product of the average backscattering currents can be dropped. The meaning of the Poissonian limit is that quasiparticles which tunnel from one edge to another do so in an independent manner. Yet by doing so they can absorb or emit n "photon" quanta of

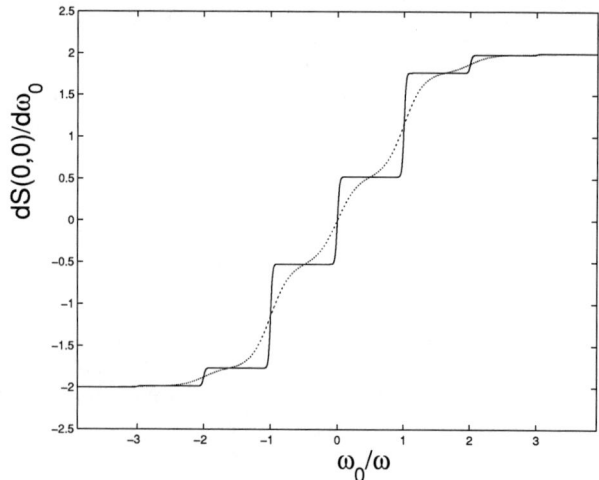

FIGURE 2. Noise derivative for a normal metal at different temperatures: $k_BT/\hbar\omega = 0.01$ (solid line) and $k_BT/\hbar\omega = 0.1$ (dashed line). We take $\omega_1/\omega = eV_1/\hbar\omega = 3/2$.

ω (n integer). The main purpose of this work is to analyze the double Fourier transform of the noise $S(\Omega_1,\Omega_2) \propto \int dt \int dt' S(t,t')\exp(i\Omega_1 t + i\Omega_2 t')$ when both frequencies Ω_1 and Ω_2 are set to zero. Indeed, the presence of the AC perturbation mimics a finite frequency noise measurement. At zero temperature, the shot noise exhibit divergences at each integer value of the ratio ω_0/ω[13]. These divergences are not physical since they appear in a range of frequencies where the perturbative calculation turn out to be no more valid. For this reason, we have performed finite temperature calculations which prevent divergences in the backscattering current and shot noise. At finite temperature, the shot noise reads:

$$S(0,0) = \frac{(e^*)^2 \Gamma_0^2}{2\pi^2 a^2 \overline{\Gamma}(2\nu)} \left(\frac{a}{v_F}\right)^{2\nu} \left(\frac{2\pi}{\beta}\right)^{2\nu-1}$$
$$\times \sum_{n=-\infty}^{+\infty} J_n^2\left(\frac{\omega_1}{\omega}\right) \cosh\left(\frac{(\omega_0+n\omega)\beta}{2}\right) \left|\overline{\Gamma}\left(\nu + i\frac{(\omega_0+n\omega)\beta}{2\pi}\right)\right|^2, \quad (7)$$

where $\overline{\Gamma}$ is the Gamma function and $\beta = 1/k_BT$.

DISCUSSION

We have first test the validity of our result by setting $\nu = 1$, the value which corresponds to non-interacting system (normal metal). The derivative of the shot noise according to the bias voltage exhibits staircase behavior as shown on Fig. 2. Steps occur every time ω_0 is an integer multiple of the ac frequency. This is in complete agreement with the

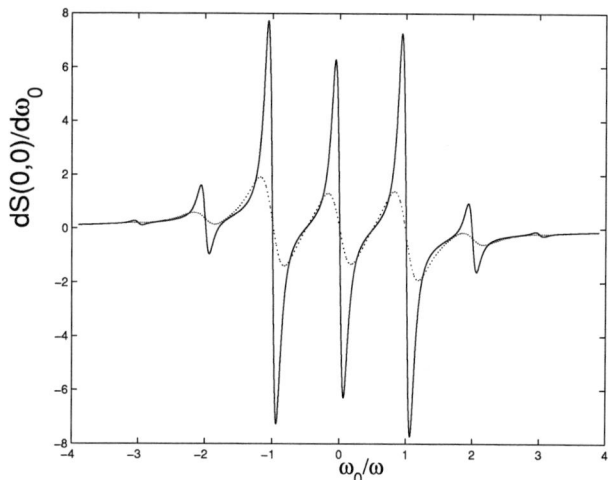

FIGURE 3. Noise derivative in the fractional quantum Hall regime at a filling factor $v = 1/3$ for $k_B T/\hbar\omega = 0.05$ (solid line) and $k_B T/\hbar\omega = 0.15$ (dotted line). We take $\omega_1/\omega = veV_1/\hbar\omega = 3/2$.

results obtained be Lesovik and Levitov for a Fermi liquid[14]. When the temperature increases, the steps are rounded.

For non-integer value of the filling factor ($v = 1/3$, for example), the shot noise derivative exhibits evenly spaced singularities, which are reminiscent of the tunneling density of states singularities for Laughlin quasiparticles. The spacing is determined by the quasiparticle charge ve and the ratio of the bias voltage with respect to the ac frequency, and the amplitude is governed by temperature (see Fig. 3). Photo-assisted transport can thus be considered as a probe for effective charges at such filling factors, and could be used in the study of more complicated fractions of the quantum Hall effect.

REFERENCES

1. M. Reznikov, M. Heiblum, H. Shtrikman, and D. Mahalu, Phys. Rev. Lett. **75**, 3340 (1995).
2. A. Kumar, L. Saminadayar, D. C. Glattli, Y. Jin, and B. Etienne, *ibid.* **76**, 2778 (1996).
3. G. B. Lesovik, JETP Lett. **70**, 208 (1999).
4. M. Büttiker, Phys. Rev. Lett. **65**, 2901 (1990); Phys. Rev. B **45**, 3807 (1992).
5. C.W.J. Beenakker and H. van Houten, Phys. Rev. B **43**, 12066 (1991).
6. T. Martin and R. Landauer, Phys. Rev. B **45**, 1742 (1992).
7. C. L. Kane and M. P. A. Fisher, Phys. Rev. Lett. **72**, 724 (1994).
8. C. de C. Chamon, D. E. Freed, and X. G. Wen, Phys. Rev. B **51**, 2363 (1995).
9. P. Fendley, A. W. W. Ludwig, and H. Saleur, Phys. Rev. Lett. **75**, 2196 (1995).
10. L. Saminadayar, D. C. Glattli, Y. Jin, and B. Etienne, Phys. Rev. Lett. **79**, 2526 (1997); R. de-Picciotto, M. Reznikov, M. Heiblum, V. Umansky, G. Bunin, and D. Mahalu, Nature **389**, 162 (1997).
11. A. Crépieux, R. Guyon, P. Devillard, and T. Martin, Phys. Rev. B **67**, 205408 (2003).
12. M. Büttiker and R. Landauer, Phys. Rev. Lett. **49**, 1739 (1982).
13. A. Crépieux, P. Devillard, and T. Martin, Phys. Rev. B **69**, 205302 (2004).
14. G.B. Lesovik and L.S. Levitov, Phys. Rev. Lett. **72**, 538 (1994).

Observation of $1/f^\alpha$ Noise of GaInP/GaAs Triple Barrier Resonant Tunneling Diodes

Naoya Asaoka, Masakazu Fukumitsu, Michihiko Suhara, and Tsugunori Okumura

Department of Electrical Engineering, Graduate School of Engineering, Tokyo Metropolitan University 1-1 Minami-Ohsawa, Hachioji, Tokyo 192-0397, JAPAN

Abstract. GaInP/GaAs triple barrier resonant tunneling diodes (TBRTDs) were fabricated and temperature dependence of noise measurements were carried out. The $1/f^\alpha$ dependent shaped noise was observed between 100 k and 300 MHz only when TBRTDs were biased in negative differential resistance (NDR) region. The result showed very small biased voltage dependence within the NDR region. The exponential factor α decreased as temperature increased. The possibilities of $1/f^\alpha$ dependent shaped noise were discussed.

Keywords: Triple Barrier Resonant Tunneling Diodes, $1/f^\alpha$ noise
PACS: 73.63.Hs, 81.07.St

INTRODUCTION

Resonant tunneling diodes (RTDs) are promising device for high speed applications such as oscillators, self-oscillating mixers, frequency multipliers, etc., because of their nature of negative differential resistance (NDR). For such kinds of application, it is important to know the noise property of RTDs. Noise measurements on RTDs had been carried out and reported on shot noise [1-10], spectral noise [11-15], and random telegraph noise [12, 16, 17]. Shot noise suppression in the entire positive differential resistance (PDR) region below current peak, and shot noise enhancement in the negative resistance were reported on double barrier RTDs (DBRTDs) [1-8, 10] and triple barrier RTDs (TBRTDs) [9, 10]. Reports on spectral noise and random telegraph noise indicate localized states are responsible for observed low-frequency noise in DBRTDs.

In this paper, temperature dependence of $1/f^\alpha$ dependent noise within NDR region were measured and characterized for first time as TBRTDs in GaInP/GaAs system.

WAFER STRUCTURE AND MEASUREMENTS SETUP

The GaInP/GaAs TBRTDs sample was grown by MOCVD on an n+ GaAs (001) substrate. The thicknesses of the TBRTD structure are 5.3 nm (barrier), 9.3 nm (well), 8.4 nm (barrier), 5.0 nm (well), 5.3 nm (barrier) from substrate to surface respectively. The diameter size of TBRTDs is 20 μm. The more specified wafer structure and fabrication process have been explained in elsewhere [18].

The setup of noise measurements is shown in Fig.1. Fabricated TBRTDs were placed in temperature controlled low temperature prover at various temperatures from 20 K to room temperature. A GSG coplanar prove was placed on the TBRTDs to carry out on-wafer measurements. A bias T was used to measure power spectrum by a spectrum analyzer while applying DC voltage using a parameter analyzer. Band resolution width of spectrum analyzer was set to be 10 kHz. Measured frequency range was from 45 MHz to 13.2 GHz.

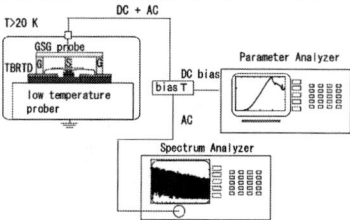

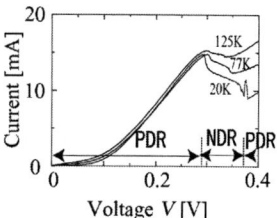

FIGURE 1. Noise measurements setup.

FIGURE 2. Current-voltage characteristics of fabricated TBRTDs. Negative differential (NDR) region and positive differential resistance (PDR) region is also shown.

RESULTS AND DISCUSSION

The measured DC current-voltage (I-V) characteristic of fabricated sample is shown on Fig. 2. The result showed small temperature dependence of peak voltage of NDR region around 0.3 V and peak current of NDR region around 15 mA. The peak to valley (P-V) current ratio of NDR region decreased as temperature increased from 1.64 at 20 K to 1.04 at 125 K.

The $1/f^{\alpha}$ shaped noise was clearly observed only when TBRTDs were biased in NDR region as shown in Fig.3. The noise biased in PDR region is around -83 dBm as shown in Fig. 3 and this was same as background noise level of the measurement setup. These trends were same down to 10 kHz in different measurements with different samples.

The measured noise power spectrum showed very small biased voltage dependence within NDR region at the same temperature as shown in Fig 4 (a). On the other hand, the spectrum showed temperature dependence at the same applied DC voltage as shown in Fig 4 (b). Temperature dependence of exponential factor α is plotted as shown in Fig. 5. The α decreased as temperature increased. The excess signal revealing self-oscillation and/or spurious noise was not observed in the all frequency range up to 13.2 GHz and bias condition.

These kinds of $1/f^{\alpha}$ shaped noise in DBRTDs is due to defect-assisted hopping of electrons from emitter to the quasi-band state in the quantum well that gives rise to fluctuation in the transmission coefficient of the tunnel barrier in pervious work [12-14]. In such case the noise power spectrum has Lorentzian characteristic and, activation energy of the traps is evaluated from the slope and curves. However in this study, we can't clearly identify if frequency dependence of the results has Lorentzian characteristic. Also it is estimated that the low frequency noise mechanism is more complicated than that of DBRTDs since the TBRTDs has extra quantum well.

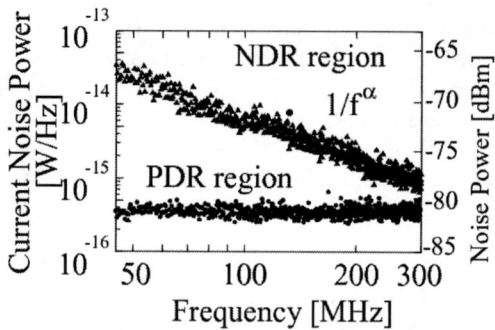

FIGURE 3. Measured current noise power of TBRTD biased in NDR region (T=20K, V=0.305 V) and PDR region (T=20K, V=0.28 V). The exponential factor α in NDR region shown here was evaluated as 1.6.

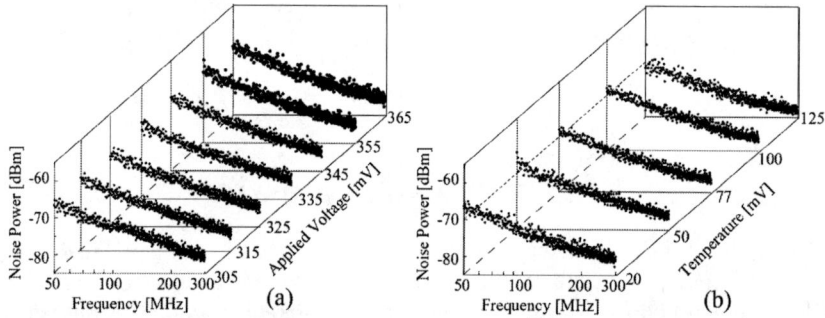

FIGURE 4. Measured noise power spectrum. (a)Applied voltage dependence of measured noise power (T=20 K). (b) Temperature dependence of measure noise power (V=320 mV for T=20 to 100 K and 325mV for T=125 K).

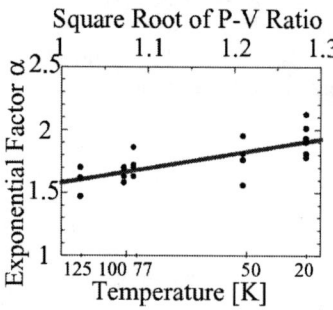

FIGURE 5. Temperature dependence of exponential factor α. Correlation of square root of P-V ratio and the α.

We found that the α showed correlation with square root of the P-V current ratio in the NDR region as shown in Fig. 5. The noise measured in this work has correlation with incoherent tunneling current since peak current of I-V characteristics almost unchanged while valley current of I-V characteristics increased.

The fact that noises exceed background noise of measurements setup only when TBRTDs are biased in NDR region might reveal another possibility that noise comes from non-linearly amplified shot noise and/or amplitude fluctuation of self-oscillating signal due to negative resistance that might lead to $1/f^{\alpha}$ shaped spectrum.

Moreover, noise power spectrum biased in the entire PDR region stayed same in the measured range up to 400 mV.

CONCLUSIONS

GaInP/GaAs triple barrier resonant tunneling diodes (TBRTDs) were fabricated and temperature dependence of noise measurements were carried out. The $1/f^{\alpha}$ dependent shaped noise was observed between 100 kHz and 300 MHz only when TBRTDs were biased in negative differential resistance (NDR) region. The result showed very small biased voltage dependence within NDR region. The exponential factor α decreased as temperature increased. The α changed from 2.12 to 1.47 while peak to valley changed from 1.64 to 1.04.

ACKNOWLEDGMENTS

The authors thank Furukawa Electric Company Ltd. for MOCVD growth of GaInP/GaAs TBRTDs.

REFERENCES

1. Li, Y. P, Zaslavsky, A., Tsui, D. C., Santos, M., and Shayegan, M., *Phys. Rev. B* **41**, 8388-8391 (1990).
2. Van de Roer, T. G., Heyker, H. C., Kwaspen, J. J. M., Joosten, H. P., and Henini, M., *Electron. Lett.* **27** 2158-2160 (1991).
3. Brown, E. R., *IEEE Trans. Electron Devices* **39** 2686-2693 (1992).
4. Kuznetsov, V. V., Mendez, E. E., Bruno, J. D., and Pham., J. T., *Phys. Rev. B* **58**, R10 159-162 (1998).
5. Innaccone, G., Lombardi, G., Macucci, M., and Pellegrini, B., *Phys. Rev. Lett.* **80**, 1054-1057 (1998).
6. Innaccone, G., Lombardi, G., Macucci, M., and Pellegrini, B., *Analog Integrated Circuits and Signal Proceedings* **24**, 73-78 (2000).
7. Song, W., Mendez, E. E., Kuznetsov, V., and Nielsen, B., *Appl. Phys. Lett.* **82**, 1568-1570 (2003).
8. Nauen. A., Hohls, H, Könemann, J., Haug, R. J., *Phys. Rev. B* **69**, 113316-1-113316-3 (2004).
9. Yau, S.-T., Sun, H. B., Edward, P. J., and Lynam, P., *Phys. Rev. B* **55**, 880-883 (1997).
10. Pouyet, V., and Brown, E. R., *IEEE Trans. Electron Devices* **50**, 1063-1068 (2003).
11. Weichold, M. H., Villareal, S. S., and Lux, R. A., *Appl. Phys. Lett.* **55**, 1969-1971 (1989).
12. Surya, C., Ng, S.-H., Brown, E. R., and Maki, P. A., *IEEE Trans. Electron Devices* 41, 2016-2022 (1994).
13. Weichold, M. H., Villareal, S. S., and Lux, R. A., *Appl. Phys. Lett.* **55**, 657-659, (1989)
14. Lin, Y., van Rheenen, A. D., and Chou, Y. C., *Appl. Phys. Lett.* **59**, 1105-1107, (1991)
15. Ng, S.-H., and Surya, C, *J. Appl. Phys.* **73**, 7504-7507
16. Alkeev, N. V., Lyubchenko, V. E., Ironside, C. N., Figueiredo, J. M. L. and Stanley, C. R., *Journal of Communication Technology and Electronics* **47**, 228-231 (2002).
17. Salvino, R. E., and Bout, F. A., *Appl. Phys. Lett.* **63**, 2265-2654 (1993).
18. Asaoka, N., Funato, H., Suhara, M., and Okumura, T., *Appl. Surf. Sci.* **216**, 413-418 (2003).

CIRCUITS AND SYSTEMS

Oscillator Noise Analysis

Alper Demir[1]

Department of Electrical & Electronics Engineering, Koc University, Istanbul, Turkey

Abstract. Oscillators are key components of many kinds of systems, particularly electronic and opto-electronic systems. Undesired perturbations, i.e. noise, that exist in practical systems adversely affect the spectral and timing properties of the signals generated by oscillators resulting in *phase noise* and *timing jitter*. These are key performance limiting factors, being major contributors to bit-error-rate (BER) of RF and optical communication systems, and creating synchronization problems in clocked and sampled-data electronic systems. In noise analysis for oscillators, the key is figuring out how the various disturbances and noise sources in the oscillator end up as phase fluctuations. In doing so, one first computes transfer functions from the noise sources to the oscillator phase, or the sensitivity of the oscillator phase to these noise sources. In this paper, we first provide a discussion explaining the origins and the proper definition of this transfer or sensitivity function, followed by a critical review of the various numerical techniques for its computation that have been proposed by various authors over the past fifteen years.

Keywords: oscillators, noise, Floquet theory, phase noise, timing jitter
PACS: 84.30.Bv,85.40.Qx,84.30.Ng,07.50.Hp

PROBLEM DEFINITION

In oscillator noise analysis, the key problem is the determination of how exactly the various disturbances and noise sources in the oscillator cause phase fluctuations. For the solution of this problem, one defines and computes transfer functions (in a loose sense) or sensitivities, from all of the noise sources to the oscillator phase, that govern and quantify the conversion of the disturbances and noise into phase fluctuations.

PHASE NOISE SENSITIVITIES

In the literature, authors have used various names for the phase noise sensitivities mentioned above, the *Floquet Vector* (FV) [1, 2], the *Impulse Sensitivity Function* (ISF) [3], and the *Perturbation Projection Vector* (PPV) [4]. The terms *Floquet Vector* and *Impulse Sensitivity Function* were inspired by the theory or the techniques the authors have used to define or compute the sensitivities. The term *Perturbation Projection Vector* was chosen considering how the phase noise sensitivities are used in oscillator noise analysis. It can be shown that the definitions for all of the above (FV, ISF and PPV) are, deep down, theoretically equivalent. On the other hand, the (numerical) algorithms that are used for computing the phase noise sensitivities are quite different from each other in the various approaches mentioned above.

FLOQUET VECTOR AND FLOQUET THEORY

From this point on, we will refer to the collection of the phase noise sensitivities as the *Floquet Vector* (FV) to honor the French mathematician G. Floquet for having had developed the *Floquet Theory*, a very simple yet elegant and practically useful theory that forms the basis for the analysis of linear periodically time-varying (LPTV) systems. Without delving into the details and equations, Floquet Theory is essentially an eigen-

---

[1] Sponsored by the Turkish Academy of Sciences GEBIP program and a TUBITAK Career Award.

decomposed solution analysis for LPTV systems, an elegant and simple generalization of the widely utilized eigen-decomposed analysis technique for linear time-invariant (LTI) systems. In eigen-decomposed analysis, the response or the solution of the (linear) system is expressed as a summation of the eigen-modes of the system, from undamped to highly-damped modes corresponding to real or pair of complex-conjugate poles (i.e., eigenvalues) resulting in fast, slow, decaying, growing, persistent, or possibly oscillatory behavior.

ORIGINS OF THE FLOQUET VECTOR

The discussion and definition of a transfer or sensitivity function from the noise sources to the oscillator phase, to the best of our knowledge, appeared first in Kaertner's work [5, 6]. It was Kaertner who first perceptively observed that the disturbances in an oscillator result in both orbital deviations (amplitude deviations in a loose sense) that "eventually" die away and phase fluctuations that persist indefinitely even if the disturbances disappear. He then used the mechanics of the Floquet Theory to decompose the response of an oscillator to disturbances into two sets of eigen-modes, the first containing all of the modes that eventually decay away and the second set comprised of a single mode that persists indefinitely.

Without delving into equations, we now describe how the FV mentioned above was defined and computed by Kaertner in [6]. When a stable nonlinear oscillator is linearized around its periodic steady-state solution (PSSS), one obtains an LPTV system. When an eigen-decomposed analysis for this LPTV system is performed using Floquet Theory, it can be shown that one particular eigen-mode has to be non-decaying. Even though this eigen-mode is persistent, it does *not* grow without bound. The existence of a non-decaying mode, in essence, is equivalent to the steady-state oscillation condition. Any oscillator which supports a periodic, stable, steady-state solution has to have one non-decaying, persistent but non-growing mode associated with the LPTV system obtained by linearization. The *time-derivative* of the periodic steady-state, large-signal solution of the nonlinear oscillator (which is also periodic and persistent) turns out to be the persistent mode of the LPTV system obtained by linearization. Thus, once the large-signal solution for the nonlinear oscillator is computed, the persistent mode of the LPTV system can easily be computed by differentiation in time. In oscillator phase noise analysis, the key is the *decomposition* of the disturbances and noise into their modal components, one component that is *aligned* with the persistent eigen-mode discussed above, and the rest aligned with the eigen-modes which decay away eventually.

The eigen-mode analysis for LPTV (and LTI) systems can be cast as a matrix eigenvalue/eigenvector problem, by using Fourier series representations for the eigen-modes with a harmonic balance formulation, or through time discretization in connection with a finite-difference formulation. Then, the eigen-modes for a linear system become equivalent to the eigenvectors of a matrix and they form a basis set that span the space of all possible solutions. If the eigen-modes are *orthogonal* to each other, one can easily perform the modal decomposition mentioned above using *orthogonal projection* and extract the components of the disturbances aligned with the persistent mode, *without* the need to compute the other decaying eigen-modes. The eigen-modes of a linear system are orthogonal to each other if the system is self-adjoint, or equivalently, when cast as a matrix eigenvalue/eigenvector problem, if the matrix is symmetric. Unfortunately, the LPTV system or circuit obtained by the linearization of an oscillator, in general, is *not* self-adjoint. It turns out, however, that the eigen-modes of the *adjoint system* are *bi*-orthogonal to the eigen-modes of the system itself. The *adjoint* operator for linear systems or circuits is (somewhat loosely) analogous to the transpose operator for matrices. The non-decaying mode of the *adjoint* of the LPTV system obtained by linearization is *orthogonal* to all of the decaying modes of the system itself, and hence, it can be used in an orthogonal projection operation to extract the component of the disturbances and noise that are aligned with the persistent mode for the LPTV system. To extract this component out of a composite noise response, one uses the persistent mode of the *adjoint* system, not the system itself. The composite response is a summation (linear combination) of contributions from all eigen-modes. If one computes the inner product of the persistent *adjoint* mode with the composite response, the result is a quantification of how much of the persistent mode for the system itself is in the composite response. This is so, because the inner product of the adjoint mode with all of the other modes is zero due to bi-orthogonality. Thus, the non-decaying mode of the *adjoint* of the LPTV system obtained by the linearization of the oscillator around the PSSS yields the FV that lies at the heart of oscillator noise analysis.

NUMERICAL TECHNIQUES FOR THE FLOQUET VECTOR

Kaertner in [6] provides the outline of a (numerical) algorithm for computing the FV as the persistent eigen-mode of the adjoint of the LPTV system obtained by a linearization of the system of nonlinear differential equations (DEs) (that describe the dynamics of the oscillator) around the PSSS. In [6], Kaertner uses the shooting method [7, 8] to compute the large-signal PSSS for the oscillator, followed by an eigen-decomposition of the monodromy (fundamental) matrix for the LPTV system. Kaertner observes the fact that the monodromy matrix for the LPTV system is actually computed in the course of the large-signal periodic steady-state analysis using the shooting method. The monodromy matrix for the LPTV system is the Jacobian (plus the identity matrix) of the system of nonlinear algebraic equations solved by the shooting Newton method, with the so-called phase condition removed. The monodromy matrix has one eigenvalue that is equal to exactly 1, corresponding to the persistent oscillatory mode of the oscillator. The shooting method Jacobian matrix is equal to the monodromy matrix minus the identity matrix. Hence, it has an eigenvalue equal to 0, meaning that it is singular. Thus, the eigenvectors corresponding to the persistent mode of the LPTV system and its adjoint lie in the null space of the shooting method Jacobian and its transpose respectively. In a non matrix-implicit implementation of the shooting method, the monodromy matrix (plus the identity matrix) is formed at the last Newton iteration in solving the shooting equations. Kaertner proposes to proceed with a full eigen-decomposition of the monodromy matrix, followed by the identification of the persistent eigen-mode. However, he does not explain how exactly this identification can be performed. The FV is then computed by traversing one period of oscillation, by effectively solving the adjoint LPTV system with the initial condition set to the persistent mode obtained from the eigen-decomposition. As a final note, Kaertner alludes to *numerical problems* that arise when there are other modes of the LPTV system which are very close to being persistent, meaning that they are very slowly decaying. He mentions that this may cause problems in analyzing oscillators with high-Q resonators, even though he does not offer any solutions to these numerical problems that may arise.

Anzill et. al. in [9] present an harmonic balance formulation for the oscillator noise analysis problem. Their formulation is based on linear perturbation analysis, and hence results in a non-physical and incorrect oscillator spectrum that diverges to infinity at the oscillation frequency and its harmonics [9]. The authors are aware of this problem, but they do not question the validity of the linear perturbation analysis. Instead, they portray the divergence problem only as a numerical ill-conditioning issue that causes problems when one would like to compute the noisy oscillator spectrum at frequencies close to the fundamental oscillation frequency. Then, they implicitly use the results of Floquet Theory to compute the noisy oscillator spectrum in an eigen-decomposed form, even though they never explicitly state that they are doing this. In a very clever way, they use eigen-decomposition based computations to solve the problem they have (mistakenly) identified as numerical ill-conditioning. Anzill et. al. in [9] observe that the FV lies in the one-dimensional *null space* of the transpose of the harmonic balance Jacobian matrix evaluated at the PSSS.

Demir et. al. first in [10, 11] and then in [12, 1, 13, 14] describe in detail a time-domain algorithm for computing the FV. Their treatment and the numerical algorithm they present is similar to Kaertner's, founded on Floquet Theory. However, they provide a detailed enough treatment of the numerical methods they propose so that a full-fledged implementation of their numerical techniques can be developed based on their thorough description. Moreover, they address the numerical problems alluded to by Kaertner in [6] regarding the existence of other closely persistent modes for the LPTV system obtained by linearization. In particular, they note the *impossibility* of the identification of the correct oscillatory persistent mode and hence the FV with an inspection of just the eigenvalues of the monodromy matrix. For the genuine persistent mode, the monodromy matrix has an eigenvalue equal to 1. However, Demir et. al. observe in [10, 11] that, for some circuits, numerical eigen-decomposition may produce *many* other eigenvalues close to 1, and that these eigenvalues may be numerically indistinguishable from the genuine eigenvalue that is supposed to be exactly at 1. For this case, Demir et. al. [10, 11] provide a simple recipe grounded on Floquet Theory for the proper resolution of the correct persistent mode, or equivalently the correct eigenvector. They observe that the inner products of the time-derivative of the periodic steady-state solution with all of the near persistent modes of the adjoint LPTV system are theoretically equal to zero, whereas the inner product with the genuine persistent mode has to be nonzero, all due to the bi-orthogonality property discussed before. Of course, due to numerical errors, this property is not exactly satisfied in practical numerical computations, but Demir et. al. [10, 11] claim that, with proper normalization, the comparison of the numerical values of these inner products will reveal the genuine persistent mode.

Hajimiri et. al. in [3] follow a practical and phenomenological approach to define what they call the *Impulse Sensitivity Function* (ISF) in order to quantify the sensitivities between the noise sources and the persisting phase fluctuations in an oscillator. Unlike the approaches described above, their formulation is not founded on Floquet Theory. However, it can be shown that the ISF defined by Hajimiri et. al is deep down, essentially equivalent to the FV defined by the Floquet Theory based approaches. Hajimiri et. al. observe that the disturbances in an oscillator result in both orbital deviations (amplitude deviations in a loose sense) that "eventually" die away and phase fluctuations that persist indefinitely even if the disturbances disappear, an observation originally made by Kaertner. Hajimiri et. al. propose two different numerical algorithms for computing the ISF. In their original proposal, Hajimiri et. al. compute the ISF by using the circuit simulator SPICE directly as a black box. They perturb the oscillator circuit in steady-state with a series of (im)pulses at time points that are swept from a reference time point in the oscillation waveform through the end of the oscillation period. After hitting the circuit with each and every one of these (im)pulses, they simulate the oscillator circuit in SPICE (transient analysis) for a "long enough" duration of time so that all of the amplitude deviations due to the impulse perturbation decay away and only the persistent phase disturbance remains. Then, they measure the amount of the persistent phase shift in the waveforms of the perturbed circuit by comparing its waveforms with an unperturbed version of the same circuit that is simulated concurrently with the perturbed one. They repeat this phase shift computation for every noise source in the circuit and for a number of pulse locations distributed evenly throughout a period of oscillation. Even though this algorithm is not very suitable for an efficient and accurate implementation for the computation of the FV, it *implicitly* defines ISF in such a way that ISF becomes theoretically equivalent to the FV that is defined based on Floquet Theory. There are various aspects (especially the "long" duration of transient analysis needed for the amplitude deviations to decay away after the circuit is hit with an impulsive perturbation) of this algorithm which make it much less efficient and also less accurate compared with algorithms based on Floquet Theory and eigen-decomposed computations. Hajimiri et. al. in [3] propose a simplified, approximate algorithm for computing the ISF based on using the time-derivatives of the periodic steady-state waveforms of the oscillator, possibly motivated by the inefficiency of their original algorithm, and by the desire to obtain a manually executable, analytically calculable, simple and intuitive procedure for the computation of the ISF. However, their simple algorithm for the computation of the ISF does not produce the same results as their original algorithm based on impulsive perturbations or the other algorithms described above. This new algorithm *implicitly* redefines ISF, and no longer produces the sensitivity function that correctly quantifies the conversion of disturbances and noise into persistent phase fluctuations. It turns out, Kaertner, in an earlier work [5], defined a phase sensitivity function equivalent to the one that is implicitly defined by the simplified algorithm of Hajimiri. et. al. However, later in [6], Kaertner realizes that there are problems with his original definition. At the least, the FV or the ISF computed this way becomes dependent on (i.e., changes with) the scaling of the state variables that describe the oscillator. Moreover, this definition *mixes* persistent phase fluctuations with those that die away eventually. Kaertner fixes these problems by properly defining the FV in [6]. The simplified algorithm by Hajimiri et. al. in [3] and the algorithm by Kaertner in [5] for computing the ISF or FV would have been correct and accurate if the LPTV system obtained by the linearization of the oscillator equations at the steady-state is *self-adjoint*. However, this is not the case for almost all practical oscillator circuits. Still, there are some questions that have not yet been fully answered in the literature [15, 16]. Even if the simplified algorithm by Hajimiri et. al. and the earlier algorithm by Kaertner in [5] do not produce exact quantifications for correct phase noise computations, are they "approximately" correct? If yes, is this the case for all kinds of oscillator circuits, and in general? Or for specific types of circuits, in very special cases? A thorough examination of this issue is warranted, given the popularity of the simplified algorithm for computing the ISF in the RF design community.

Demir et. al. in [4] present a novel formulation and efficient numerical method for the computation of the FV. With this new formulation, they address the problem with the correct identification of the genuine persistent eigen-mode. With their new algorithm for the computation of the FV, the identification or selection of the genuine persistent eigen-mode among the other eigen-modes becomes *implicit*. In [4], they observe that the FV lies in the null space of the transpose of the harmonic balance Jacobian matrix evaluated at the periodic steady-state solution. This observation forms the basis for the new formulation and the numerical method they present for the computation of the FV. Their numerical algorithm can be described, in very simple terms with a pure linear algebra perspective, as a specialized technique for the computation of a single vector that spans the one-dimensional null space of a singular matrix. This new technique is not based on eigen-decomposed computations, instead it utilizes an *augmented* non-singular matrix to directly solve for the single vector that spans the null space. The numerical method Demir et. al. propose in [4] relies on the harmonic balance Jacobian having a one-dimensional and well-conditioned null space, unambiguously corresponding to the persistent oscillatory mode. However, as mentioned above, there exist oscillators with many near persistent modes resulting in other near zero eigenvalues for the harmonic balance Jacobian. Moreover, the eigenvalues corresponding to the other near persistent modes may be closer to zero than the eigenvalue corresponding to the genuine persistent mode. These kinds of inaccuracies in numerical computations arise due to the *truncated*

Fourier series that is used to represent the periodic quantities in harmonic balance. Similar inaccuracies also arise in time-domain computations based on shooting method formulations due to the time *discretization* that is used to represent the periodic signals. In fact, using truncated Fourier series in harmonic balance exactly corresponds to using a finite number of equally spaced time samples to represent a periodic waveform.

SUMMARY AND CONCLUSIONS

Based on our discussion of previous work on computational approaches to oscillator noise analysis, we conclude that the ones grounded on Floquet Theory are the most rigorous and most appropriate for the implementation of a general purpose oscillator noise analysis tool in a circuit simulator. All of the approaches based on Floquet Theory somehow try to extract or identify the persistent eigen-mode of the (adjoint) LPTV system corresponding to an eigenvalue 1 (or 0 depending on the flavor of the periodic steady-state analysis algorithm and the specific matrix being eigen-decomposed). However, all of the algorithms discussed above may run into problems when the LPTV system has other near persistent eigen-modes. Eigenvalue based selection of the genuine persistent eigen-mode fails immediately, because the real persistent mode may have an eigenvalue further away from 1 (or 0) compared with the other near persistent modes, due to numerical inaccuracies resulting from time discretization or the truncation of Fourier series. In our opinion, the most effective and robust method currently available for computing the FV is based on performing a full (or partial using iterative Krylov-subspace based techniques) eigen-decomposition for the transpose of either the monodromy matrix, the shooting method Jacobian or the harmonic balance Jacobian followed by an identification of the correct persistent eigen-mode by making use of the bi-orthogonality relationships discussed above. However, even this method may run into problems for some oscillators when other near persistent modes exist.

REFERENCES

1. A. Demir, A. Mehrotra, and J. Roychowdhury. Phase noise in oscillators: A unifying theory and numerical methods for characterisation. *IEEE Transactions on Circuits and Systems-I: Fundamental Theory and Applications*, 47(5):655, May 2000.
2. A. Demir. Phase noise and timing jitter in oscillators with colored noise sources. *IEEE Transactions on Circuits and Systems-I: Fundamental Theory and Applications*, 49(12):1782, December 2002.
3. A. Hajimiri and T.H. Lee. A general theory of phase noise in electrical oscillators. *IEEE Journal of Solid-State Circuits*, February 1998.
4. A. Demir and J. Roychowdhury. A reliable and efficient procedure for oscillator ppv computation, with phase noise macromodelling applications. *IEEE Transactions on Computer-Aided Design of Integrated Circuits and Systems*, February 2003.
5. F. X. Kaertner. Determination of the correlation spectrum of oscillators with low noise. *IEEE Transactions on Microwave Theory and Techniques*, 37(1):90–101, January 1989.
6. F. X. Kaertner. Analysis of white and $f^{-\alpha}$ noise in oscillators. *International Journal of Circuit Theory and Applications*, 18:485–519, 1990.
7. T.J. Aprille and T.N. Trick. Steady-state analysis of nonlinear circuits with periodic inputs. *Proceedings of the IEEE*, 60(1):108–114, Janurary 1972.
8. R. Telichevesky, K.S. Kundert, and J. White. Efficient steady-state analysis based on matrix-free krylov-subspace methods. In *Proc. Design Automation Conference*, June 1995.
9. W. Anzill and P. Russer. Determination of the correlation spectrum of oscillators with low noise. *IEEE Transactions on Microwave Theory and Techniques*, 41(12):2256, December 1993.
10. A. Demir. *Analysis and Simulation of Noise in Nonlinear Electronic Circuits and Systems*. PhD thesis, University of California, Berkeley, May 1997.
11. A. Demir and A. Sangiovanni-Vincentelli. *Analysis and Simulation of Noise in Nonlinear Electronic Circuits and Systems*. Kluwer Academic Publishers, 1998.
12. A. Demir, A. Mehrotra, and J. Roychowdhury. Phase noise in oscillators: A unifying theory and numerical methods for characterisation. In *ACM/IEEE Design Automation Conference*, June 1998.
13. A. Demir. Phase noise in oscillators: DAEs and colored noise sources. In *IEEE/ACM International Conference on CAD*, November 1998.
14. A. Demir. Floquet theory and nonlinear perturbation analysis for oscillators with differential-algebraic equations. *International Journal of Circuit Theory and Applications*, March-April 2000.
15. G.J. Coram. A simple 2-D oscillator to determine the correct decomposition of perturbations into amplitude and phase noise. *IEEE Transactions on Circuits and Systems-I: Fundamental Theory and Applications*, 48(7):896, July 2001.
16. P. Vanassche, G. Gielen, and W. Sansen. On the difference between two widely publicized methods for analyzing oscillator phase behavior. In *ACM/IEEE International Conference on Computer-Aided Design*, November 2002.

Noise Simulation of Continuous-Time ΣΔ Modulators.

J. Arias, L. Quintanilla, D. Bisbal, J. San Pablo,
L. Enriquez, J. Vicente, J. Barbolla

Dpt. de Electricidad y Electrónica, E.T.S.I. Telecomunicación. Universidad de Valladolid. Spain

Abstract. In this work, an approach for the simulation of the effect of noise sources in the performance of continuous-time ΔΣ modulators is presented. Electrical noise including thermal noise, 1/f noise and clock jitter are included in a simulation program and their impact on the system performance is analyzed.

INTRODUCTION

This work is related to the design of a continuous-time (CT) ΔΣ modulator for high-speed data communications. Specifically, a second-order modulator with a 3-bit quantizer and an optimized zero in its noise transfer function, NTF, whose block diagram is shown in Fig.1(a). In this design, integrators are built using transconductors and capacitors. The simple structure of these elements, and the lack of feedback, allows them to operate at very high frequencies, which, together with the low OSR of the modulator, provides a wide signal bandwidth.

The design process for CT modulators requires a detailed simulation in order to analyze the effect of nonidealities on the modulator performance [1]. Circuit-level simulation (Spice-like) are time comsuming and they do not allow the simulation of noise for time-domain analysis. Therefore, a simulation program was developed in order to fill these needs including the effect a broad set of nonidealities specially thermal noise, 1/f noise and clock jitter.

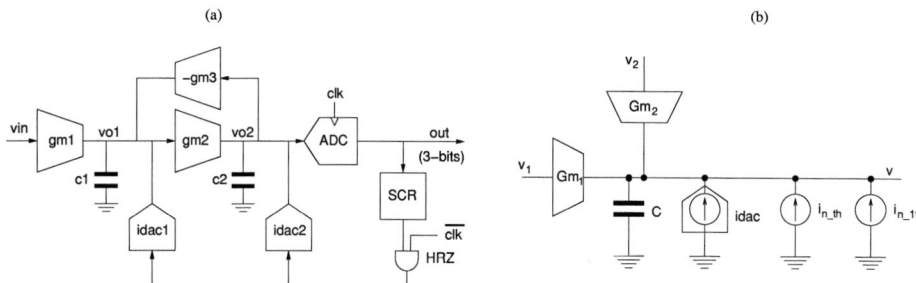

Figure 1. (a) Block diagram of the ΔΣ modulator. (b) Simplified integrator schematic for system simulation. Noise sources are also included.

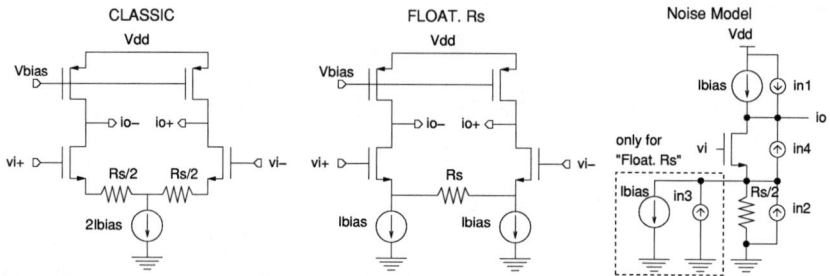

Figure 2. Generic schematic for a source-degeneration transconductor (without and with floating resistor), and its single-ended model including noise sources.

CONTINUOUS-TIME MODULATOR SIMULATION

The simulation of a continuous-time system involves the discretization of time. Time steps must be much smaller than the fastest changing signal, that, in our case, is the clock signal. A simple Euler integration can be performed in each time step. This implies that we assume all signals to be constant during each time step. At each integrator output, two noise sources, i_{n_th} and i_{n_1f}, are attached to include thermal and $1/f$ noise contribution to the integration node (Fig 1(b)). Thermal noise currents are generated by transconductors and DACs. These noise currents are simulated as sources whose value is chosen randomly for each time-step with a standard deviation $\sigma_i = i_n \sqrt{\frac{1}{2\Delta t}}$, where i_n is the current noise density (rms) and Δt is the time-step.

A general structure for transconductors is shown in Fig. 2. The floating-resistor transconductor can be used when the voltage room is small but, as we will see later, it shows a worse noise performance. The equivalent, single-ended, small-signal noise circuit for these transconductors is also shown in Fig. 2. In both cases, the circuit transconductance, G_m is related to the transconductance of the input transistors, g_m, as $g_m = G_m(1+\alpha)$, where $\alpha = g_m R_S/2$ is a design parameter. In Fig. 2, i_{n1} is the current noise of the top biasing source, i_{n2} is due to the degeneration resistor, i_{n3} is the current noise of the bottom biasing source, and finally, i_{n4} is the noise current of the input transistor. If the same overdrive voltage is used for input transistors and biasing currents, their noise will be identical (same transconductance). The resulting output noise current was found to be:

$$i_{no}^2 = 2 \times 4KTG_m(1+\alpha)\left[\frac{\gamma}{(1+\alpha)^2} + \gamma + \frac{\alpha}{(1+\alpha)^2} + \frac{\gamma\alpha^2}{(1+\alpha)^2}\right]$$

Where the last term in brackets is only present for floating-resistor transconductors. $2\times$ stands for the differential output, and γ is a noise parameter for transistors ($i_n^2 = 4KT\gamma g_m$). In both cases, the output noise power is proportional to G_m. Therefore, we can define a noise factor, ϕ_{Gm}, which depends on the transconductor design, so, the output noise is written as $i_n^2 = 2 \times 4KT\phi_{Gm}G_m$. The value of ϕ_{Gm} is plotted as a function of α in Fig. 3. These results were confirmed by transistor-level simulations for a 90nm CMOS technology.

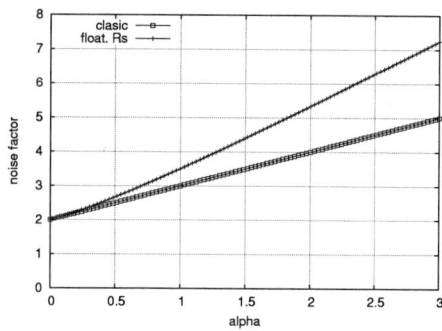

Figure 3. Noise factor for transconductors, ϕ_{Gm}, as a function of the design parameter α. ($\gamma = 1$)

Current-mode DACs are implemented as arrays of 8 identical current sources. The thermal current noise density of DACs is obtained from the transconductance of their current-source transistors, which depends on the current, $I_{DAC}/8$, and their overdrive voltage (V_{ODAC}). The total differential noise current is $i_n^2 = 4.5KT\gamma I_{DAC}/V_{ODAC}$. The noise of feedback DACs is cyclostationary, namely, DAC's noise sources are only turned on during the active phase of DAC (half-return-to-zero waveform, HRZ).

The 1/f noise is a low-frequency noise generated mainly by MOSFETs. Its spectral noise density is $i_{n1f}^2 = K_F/f$. In the program, the algorithm proposed in reference [2] is used. Each 1/f noise source is obtained from 32 white-noise sources followed by sample-and-hold elements with different sampling rates. A very accurate 1/f spectrum is obtained for over 7 decades. The absolute value of 1/f noise depends on the so called "corner frequency", f_{corner}, that is the frequency at which the 1/f noise power density equals the thermal noise power density in a circuit. This gives $K_F = f_{corner} * i_{nth}^2$. In simulations the value of all 1/f noise sources is derived from the corresponding f_{corner} rather than being specified as an absolute value for convenience.

Jitter in clock lines can degrade the modulator performance [1, 3]. This effect consists of the random fluctuation of clock edges. Jitter depends on two contributions: independent jitter and cumulative jitter. In the former, the average edge position remains the same after a large number of clock periods, while phase-noise of the clock oscillator gives rise to a cumulative jitter, and then, after several clock cycles the edge positions can have changed substantially. From a spectral point of view, independent jitter generates a flat noise floor, while cumulative jitter generates skirts on tones [1]. Therefore, two variables are needed to describe the jitter. These variables are the independent clock jitter, "*jitter*" and the phase-noise of clock, "*phn*", that have to be specified for a particular offset frequency from the carrier, "*phnof*". The rms value of the cumulative jitter is calculated from phase noise according to [4]: $\sigma_{T,acc} = phnof \cdot 10^{phn/20} f_{clk}^{-1.5}$. And, the rms fluctuation of the width of DAC's pulses is $\sigma_W = \sqrt{2}\,\sigma_{T,ind}$ where, the $\sqrt{2}$ factor accounts for the two edges of the DAC's pulses. The cumulative jitter will affect the two edges of the DAC pulse in the same way, and then, it is not considered. Finally, the effect of clock jitter is simulated by changing the integration time-step randomly.

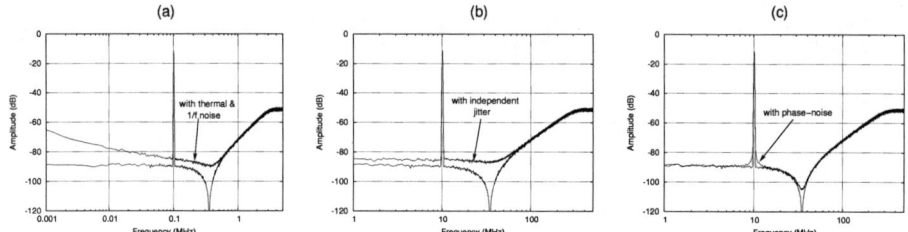

Figure 4. Simulated spectra including (a) thermal and 1/f noise ($f_{corner} = 300 kHz$, $f_{clk} = 10 MHz$). (b) independent clock jitter (3ps, $f_{clk} = 1 GHz$) and (c) phase-noise (-90dBc/Hz @ 1MHz, $f_{clk} = 1 GHz$). The ideal spectrum with only quantization noise is also included as a reference (SNR=52.6dB).

SIMULATION RESULTS

Several simulations were carried out in order to identify and analyze the effect of each noise source in the modulator performance. In these simulations all other system elements are assumed to be ideal. Each clock cycle was divided into 100 time steps for numerical integration. The input signal was a sine-wave with an amplitude of 0.7 relative to full-scale, and its frequency was about 1/100 of clock frequency.

In Fig 4(a) a clock frequency of 10 MHz was selected in order to show clearly the effect of 1/f noise. Thermal noise adds a white noise floor filling in the NTF notch and resulting in a SNR penalty of 2 dB. For 1/f noise a corner frequency of 300kHz, which is a typical value for MOSFET-based transconductors, was assumed. The resulting spectrum shows a 10dB/decade slope due to 1/f noise at low frequencies. The obtained SNR is 48dB, 4dB less than that of an ideal modulator.

For high clock frequencies the effect of jitter is significant. Thus, in these simulations the clock frequency was 1GHz (T=1ns). The effect of clock jitter is shown in Fig. 4(b,c). When only independent jitter is considered the resulting spectrum shows an additional noise floor similar to that obtained with thermal noise. In Fig. 4(b), a clock jitter of 3ps rms (0.3% of T) was simulated resulting in a SNR of 48dB (typical jitter values are about tens of picoseconds for on-chip oscillators). The clock signal can also exhibit phase-noise which results in cumulative jitter. Fig. 4(c) shows the effect of such jitter in the output spectrum. The signal tone shows skirts that degrade slightly the SNR since its contribution to the total integrated noise is small. The phase noise of the clock signal was -90 dBc/Hz at a frequency offset of 1MHz, resulting in a cumulative jitter of 1ps rms, and the calculated SNR is 52 dB, almost that of an ideal modulator (this phase noise value was obtained from measurements of a differential ring oscillator [4]).

REFERENCES

1. J. A. Cherry and W. M. Snelgrove, *Continuous-Time Delta-Sigma Modulators for high speed A/D conversions*, Kluwer Academic Publishers, 2000, ISBN: 0-7923-8625-6.
2. S. J. Orfamidis, *Introduction to signal processing*, Prentice Hall, 1996, ISBN: 0-1324-0334-X.
3. J. A. Cherry, W. M. Snelgrove, P. Schvan, *Electronics Letters*, **33**, 1118–1119 (1997).
4. A. Hajimiri, S. Limotyrakis, T. H. Lee, *Journal of Solid-State Circuits*, **34**, 790–804 (1999).

Noise Performance at Cryogenic Temperature of Microwave SiGeC Low Noise Amplifier using BiCMOS Technology

S. Pruvost[1,2], S. Delcourt[2], F. Danneville[2], I. Telliez[1], G. Dambrine[2], M. Laurens[1], A. Monroy[1]

[1] *STMicroelectronics FTM-CCDS, Crolles, France*
[2] *IEMN UMR CNRS 8520, DHS, USTL Villeneuve d'Ascq, France*

Abstract. This work presents a 1-stage Low Noise Amplifier (LNA) realized using a 0.13µm SiGe:C Heterojunction Bipolar Transistor (HBT). Measured under cryogenic temperature this LNA exhibits a noise figure of 1.8dB at 78K and 40GHz which corresponds to 70% improvement compared with result achieved at room temperature. The associated gain stands higher than 5dB at 78K and 40GHz while it was better than 7.5dB at room temperature. This was explained by the passive components variation with the temperature which shifts the maximum gain around the 25GHz frequency range.

Keywords: BiCMOS, Cryogenic, Heterojunction Bipolar Transistor (HBT), Low Noise Amplifier (LNA), Microwave, Silicon, SiGeC.
PACS: 84.30.Le

INTRODUCTION

The Silicon-Germanium Heterojunction Bipolar Transistor (HBT) has already been reported as suitable for millimeter-wave range applications. The BiCMOS technology also provides competitive results and allows the integration of both digital and radio-frequency applications at room temperature [1], [2]. However, the benefits of SiGe:C HBTs for cooled Low Noise Amplifiers (LNA) have never been demonstrated yet, and this paper focuses on performances of such circuits at cryogenic temperature. Prior to this work, the cooling down temperature benefits on SiGe:C BiCMOS HBT were demonstrated in [3], providing a starting point to the present experimental study to LNA. Feedback simulation is also carried out to analyze the noise figure and more specifically the gain variations.

HBT BEHAVIOR: PRELIMINARY RESULTS

Measurements conditions and de-embedding methodologies were already presented in [3]. The measured SiGeC HBT is a symmetrical interdigitated structure made of 6 quasi-self-aligned base fingers, 3 emitters of 0.17µm width and 7µm length each, and 4 collectors. This transistor topology was designed to obtain good properties in terms of gain and noise characteristics at room temperature. Nevertheless, its behavior along the temperature variation was investigated and results are shown in Fig. 1. A slight increase of the S_{21} is observed while the noise figure decreases when cooling down the temperature (presenting an input impedance of 50Ω). In order to assess the noise performance (minimum noise figure, equivalent noise resistance), an accurate small signal equivalent circuit (SSEC) was extracted and associated which a standard noise model [3]. As shown in Fig. 1, excellent agreement between simulations and experimental data (both on small-signal and noise performance) were obtained for different temperatures.

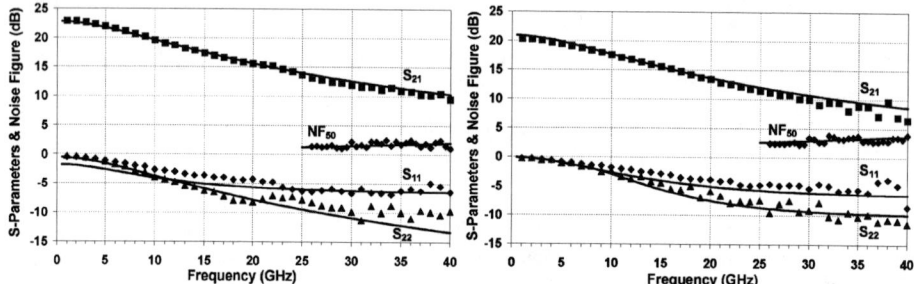

FIGURE 1. HBT noise figure and S-parameters: modeling (solid lines) and measurements (symbols) at 78K on the left hand, and at 173K on the right hand figure (V_{CE} = 1.6V, I_C = 5.3 mA).

This study allowed the HBT noise parameters prediction reported in Table 1.

TABLE 1. Deembedded transistor and modeling values at 40GHz for the HBT.

Temperature	Noise figure : NF_{50}	Min. noise figure : NF_{min}	Equivalent Noise Resistor R_N
296 K	3.6 dB	2.9 dB	31.5 Ω
173 K	2.4 dB	1.7 dB	19.5 Ω
78 K	1.3 dB	1.0 dB	9.4 Ω

Excellent performance were obtained at 78K, with a minimum noise figure as low as 1.0dB@40GHz, close to the NF_{50} value. Due to the availability of the LNA circuit designed at room temperature, the excellent results obtained on single devices when cooling down the temperature strongly motivated the characterization of these LNAs down to 78K, as presented in the next section.

ONE-STAGE COOLED LNA PERFORMANCE

The 1-stage LNA was optimized to operate at room temperature and 40GHz. The LNA design methodology is already described in [2] for a two stage low noise amplifier. The noise figure, shown fig. 2, was measured to be 4.6dB at 40GHz and 296K, which is very close to the state-of-the-art for BiCMOS Technology. Because of the strong improvement of the noise performance reported for the single device level at low temperature, some measurements were carried out for cooled LNAs. Fig. 2 reports the LNA noise figure variation with the temperature. For the 40GHz frequency, a value of 1.8dB at 78K is obtained which is the lowest noise figure value reported in millimeter-wave range for SiGe:C based LNA. Besides these excellent noise figure performance, the associated gain at 40GHz which was about 7.5dB at 296K, stands higher than 5dB for the lowest temperature 123K and 78K. The degradation is explained by the mismatch, because the LNA was optimized to operate at room temperature.

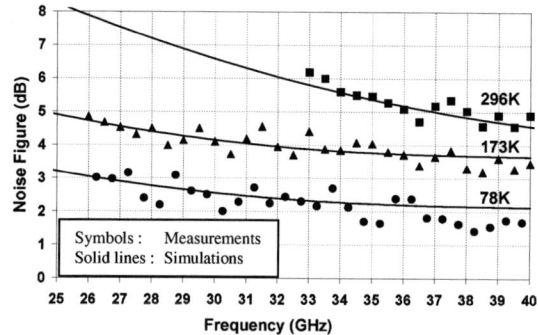

FIGURE 2. 1-stage LNA Noise figure variation with temperature (296K, 173K, and 78K)

The 1-stage LNA S-Parameters comparison between measurement and feedback simulation is presented Fig. 3 for the 173K temperature. Even though the reported noise figure for the lowest temperature seems attractive, the mismatch was too important to present the small-signal parameters at 78K.

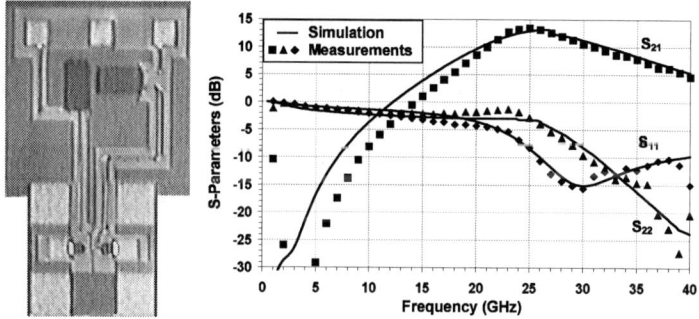

FIGURE 3. 1-stage LNA die photography (on the left hand) and corresponding comparison between measurements (symbols) and simulations (solid lines) of small-signal parameters (on the right hand) versus frequency, at 173K ($V_{CE} = 1.6V$, $V_{BE} = 0.9V$, $I_C = 5.2$ mA).

A good assessment is achieved after feedback simulation. For this purpose, the element values of the transistor SSEC used for the LNA were considered to be those extracted from the de-embedded transistor at 173K (depicted section 1). The first simulation was not so far from measurement but feedback simulation was necessary to account for temperature impact on experimental results reported fig. 3. The transmission lines were modeled using the *multilayer* Advanced Design System tool, for which the conductivity was already taken into account by the simulator along the temperature variation. Nevertheless, the dielectric permittivity versus temperature dependency is not depicted with this model, and its variation was extracted from RF pads capacitance variation with temperature. With decreasing the temperature, the dielectric permittivity value of the Silicon dioxide increases. This element was the main contributor to fit the measurements, shown Fig. 3. The second order variation, more particularly underneath the 20GHz frequency is due to the MIM capacitor, for which it was not possible to extract dielectric permittivity along the temperature (the value at room temperature was set for all simulations). This explains the low frequency difference observed on the gain (S_{21} factor). It turned out from the feedback simulation that passive elements variations when decreasing the temperature explain the discrepancy of the gain in lower temperature range.

CONCLUSION

This paper demonstrates, in addition to results obtained on single devices [3], that cooled LNA is suitable with Silicon technology. Using a BiCMOS technology, the 1-stage LNA exhibits a gain higher than 5dB and an associated noise figure of 3.6dB and less than 2dB, respectively for 173K and 78K. The noise figure value of 1.8dB obtained at 40GHz and 78K is the lowest up-to-date noise figure value reported in the literature, to the authors' knowledge. Through feedback simulations, the work suggests also that an appropriate design considering jointly both the exact transistor and passive elements modeled at 78K would provide comparative noise performance (or better) and an optimized gain at 40GHz. Moreover a dedicated low temperature transistor topology should perform better compromise for low temperature applications at 40GHz or other frequency of interest.

ACKNOWLEDGMENTS

The authors thank the many members involved in the various aspects of this work. First the IEMN laboratory members, especially to D. Vandermoere for the test-set, and secondly the 200mm plant members in STMicroelectronics Crolles.

REFERENCES

1. P. Chevalier, C. Fellous, L. Rubaldo, D. Dutartre, M. Laurens, T. Jagueneau, F. Leverd, S. Bord, C. Richard, D. Lenoble, et al., "230 GHz self-aligned SiGe HBT for 90nm BiCMOS Technology", Proceedings IEEE Bipolar/BiCMOS Circuits and Technology Meeting, pp 225-228, Sept. 2004.
2. S. Pruvost, I. Telliez, F. Danneville, A. Chantre, P. Chevalier, G. Dambrine, S. Lepilliet, *"A Compact Low Noise Amplifier in SiGe:C BiCMOS Technology for 40GHz Wireless Communications"*, Radio Frequency Integrated Circuits, Long Beach, USA, 12-14 June 2005.
3. S. Pruvost, S. Delcourt, I. Telliez, M. Laurens, N.-E. Bourzgui, F. Danneville, A. Monroy, G. Dambrine, "Microwave and Noise Performance of SiGe BiCMOS HBT under cryogenic temperatures", IEEE Electron Device Letters, vol. 26, No. 2, Februray 2005, pp. 105-108.

Nonlinear Noise in SiGe Bipolar Devices and its Impact on Radio-Frequency Amplifier Phase Noise

S. Gribaldo[1], G. Cibiel[2], O. Llopis[1], J. Graffeuil[1]

1 LAAS-CNRS, 7 av. du Colonel Roche, 31077 Toulouse, FRANCE
2 CNES, 18 av. Edouard Belin, 31401 Toulouse, FRANCE

Abstract. The nonlinear behavior of different microwave SiGe bipolar transistors has been studied and models have been extracted. The phase noise of an amplifier is computed, taking into account the microwave additive noise floor and the up-converted 1/f noise. The simulation technique is a combination of different approaches available in a commercial CAD software. Theoretical results are then compared to the experiment.

Keywords: nonlinear modeling, phase noise, SiGe HBT, transistor oscillator
PACS: 84.30.Ng

INTRODUCTION

Amplifier phase noise is the main cause of phase noise in microwave oscillators. Inside the cavity bandwidth, the amplifier phase noise is simply converted into oscillator frequency noise [1]. Two mechanisms are responsible for the amplifier phase fluctuations: the conversion of the transistor low frequency (LF) noise by device nonlinearities, generally of 1/f shape; and addition of high frequency (HF) noise [4]. Improving amplifier phase noise directly improves the oscillator frequency noise. Firstly, amplifier's noise is easier to simulate an open loop circuit than a closed loop circuit. Secondly, the amplifier phase noise measurement is possible from the linear behaviour up to strong compression, whereas the oscillator can only be measured in compression. Studying phase noise on a large input power range is the best way, to check the validity of a modeling approach [2] and thus to understand the noise conversion mechanisms. Moreover an amplifier is free of the phase loop effect.

This paper focus on the phase noise modeling of different commercially available microwave silicon-germanium (SiGe) bipolar transistors (TB1 and TB2). The transistors have first been modeled using a conventional large signal model extraction technique. Then the noise sources have been added to the nonlinear device model and the phase noise has been simulated. The interest of an original two stages amplifier topology to get simultaneously low phase noise and high gain performance is finally pointed out.

NONLINEAR MODELLING AND NOISE MODELLING

Our models are based on the classical Gummel-Poon model [3]. They are extracted from DC characterization and S parameters measurements. Nonlinear validation is performed using output power versus input power data at different harmonic frequencies. The good agreement between the measured and the simulated values allows us to go further in complexity, adding the noise parameters to the model.

Phase noise in a transistor, or an amplifier, is managed by two different processes [4]. The first one is the conversion close to the carrier of the device LF noise. The second one is the addition of the HF noise. These two processes have fundamentally different behaviors. The LF noise conversion is a multiplicative process, which means that the noise level follows the signal level. The HF noise is additive, and has thus a minimum impact on phase noise at high signal level. Both noise processes have to be taken into account in order to accurately simulate an amplifier phase noise. It is therefore essential to be able to model these two phase noise contributions in a device, and also to minimize one or the other noise, according to the application goal. The HF noise addition is probably easier to describe than the LF noise conversion. At low input power, it can be calculated using equation (1).

$$S_\varphi(f) = \frac{FkT}{P_{in}} \quad (1)$$

where F is the amplifier noise figure, k the Boltzmann constant, T the absolute temperature and Pin the amplifier input power. However, some problems appear at high input power. A nonlinear noise figure must be defined because of device compression [4].

Identifying and locating the LF noise sources in a transistor, is a more difficult task. Moreover, the noise source itself can be affected by the RF large signal. In other words, the noise does not depend only on the transistor DC conditions [5]. How to take into account this effect in an equivalent circuit approach is still a debated subject. Each noise source can be associated to a nonlinear element of the equivalent circuit [2], or considered itself as nonlinear [6]. However, an equivalent model is by no means an accurate representation of devices physics which can only be fully described using physical modeling [5] or microscopic models. Unfortunately, these models cannot be used directly to compute the noise in a complex system like an oscillator.

Our approach of the problem is a little different. It uses an extrinsic LF noise source approach, but the dependence of this noise source on the large signal amplitude is taken into account. To this purpose, the device LF noise is measured under large signal conditions and varying the RF power. Even if this approach is not totally rigorous, it has already proven its efficiency [7].Two extrinsic noise sources are generally considered in a classical bipolar transistor model: the base voltage noise source and the base-emitter current noise source, the latter having a very strong contribution on phase noise. However, this contribution can be minimized using a low impedance bias network [7]. The current noise being cancelled in this way, the characterization is made on the base voltage noise source only. The measured data and the LF noise model for a SiGe bipolar device are shown in FIGURE 1. A sudden increase of the noise is observed when the device enters into compression.

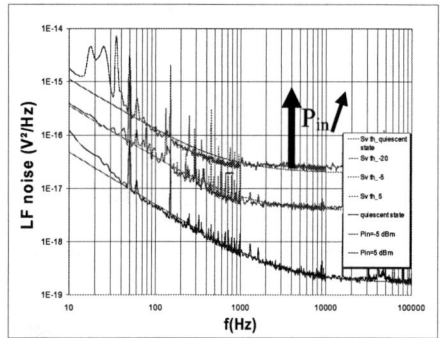 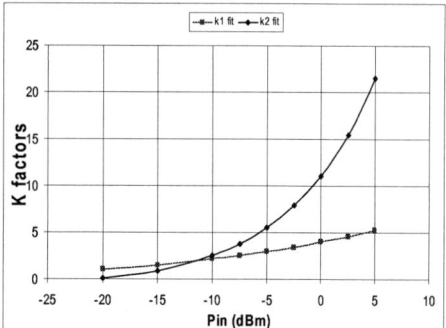

FIGURE 1: LF equivalent input voltage noise spectral density; measured and modeled – TB1

FIGURE 2 : k1 and k2 vs. 3.5 GHz input power (Ic = 10mA, Vce = 2 V) – TB1

The following equation has been used to model this behavior :

$$S_V(P_{in}) = (S_{V\ 1/f} + k_1(P_{in}) * S_{V\ floor}) * (1 + k_2(P_{in})) \quad (2)$$

$S_{V\ 1/f}$ and $S_{V\ floor}$ being the spectral power densities (respectively 1/f and noise floor) measured on the quiescent device and k1(Pin), k2(Pin) two empirical functions of the microwave power P_{in} (FIGURE 2).

Adding this RF power dependent noise source to the transistor nonlinear electrical model allows us to simulate both amplifier and oscillator phase noise at strong compression levels.

PHASE NOISE SIMULATION AND OPTIMIZATION

The above described model is implemented on a commercial software: Agilent ADS. Various approaches may be used on ADS to simulate phase noise. However, many of these tools are restricted to oscillator simulation ("pnmx" and "pnfm") and special techniques must be implemented to simulate amplifier phase noise (particularly the 1/f contribution). A simple but efficient one is the quasi-static perturbation technique, which consists in introducing a small static voltage (or current) shift to evaluate the effect of a LF voltage (or current) noise on the phase of the microwave signal through the amplifier. The conversion noise has been simulated using this technique and the ADS nonlinear noise tool has been used to compute the additive HF noise floor. FIGURE 3a and 3b demonstrate that a good agreement between measurements and simulation has been obtained. The device residual phase noise is measured using previously described techniques [4] [7]. The relative increase of the phase noise floor at low carrier level is typical of an additive noise. The 1/f phase noise is almost constant at high power level, and increases at very low input power. With such a modeling approach, it is possible not only to predict an oscillator phase noise, but also to optimize both the amplifier phase noise and the oscillator phase noise. Considering a 6 dB losses coupling ($Q_L = Q_0/2$), which is an optimum coupling both for additive phase noise [8] or conversion phase noise [7] contributions in a single stage amplifier oscillator, the necessary amplifier small signal gain should be about 9 dB to take into account circuit losses. This gain requirement is easy to fulfill in the low microwave range, but becomes more difficult to reach at higher frequencies (10 GHz). The transistor small signal gain matching is indeed one of the worse loading conditions to get a low amplifier phase noise [9]. Another important application is the one of cryogenic sapphire oscillators. In this case, the resonator is reached through long cables, inducing extra losses, and the use of an amplifier featuring a small signal gain of at least 14 dB becomes mandatory. We found that, both for 10 GHz applications and for cryogenic 5 GHz to 7 GHz applications, it was impossible to get phase noise optimized results staying on a single stage amplifier design. Therefore, a low phase noise two stages amplifier has been designed and features very promising performances [10].

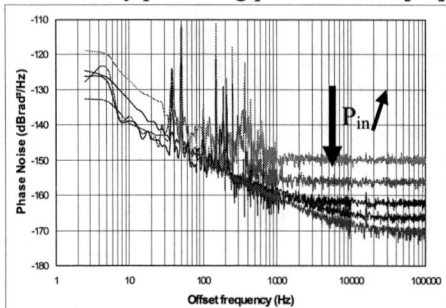 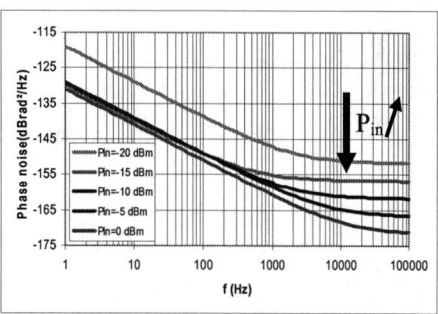

FIGURE 3 **(3A & 3B)** : Measured (left) and simulated (right) @3.5 GHz phase noise of a bipolar transistor (TB2) loaded onto 50 Ω, input power levels from -20 dBm up to 0 dBm

CONCLUSION

A modeling technique, dedicated to microwave amplifier phase noise calculation, has been presented. Different issues to CAD calculation of the two main noise contributors to phase noise in silicon bipolar transistor amplifiers have been presented. The nonlinear effects, which change both the device 1/f converted LF noise and the device HF noise figure are taken into account in our model. This model compares well to the experiment on various single stage microwave amplifiers. It is also used to optimize a two stages amplifier dedicated to cryogenic sapphire oscillator applications.

REFERENCES

[1] D.B. Leeson, Proc. Letters of IEEE, Vol. 54, N° 2, pp. 329-330, 1966.
[2] O. Llopis, et al, Proc. of the IEEE MTT Symposium, pp. 831-834, May 2001
[3] G. Massobrio, P. Antognetti, "Semiconductor Device Modelling with SPICE" McGraw-Hill Ed.
[4] G. Cibiel, L. Escotte, O. Llopis, IEEE Trans. on MTT., vol 52, n° 1, Janvier 2004, pp. 183-190.
[5] F. Bonani et al. IEEE Trans. on Electron. Dev., n° 9, sept 2002, pp. 1640-1647.
[6] H.J. Siweris, B. Schiek, 1986 EuMC Proceedings, pp. 681-686.
[7] G. Cibiel, et al, IEEE Trans. on UFFC, Vol. 51, No. 1, pp. 33-41, January 2004.
[8] J. K. A. Everard, Proc. of the IEEE MTT-Symposium, pp. 1077-1080, 1992
[9] M. Regis, O. Llopis, J. Graffeuil, IEEE Trans. on MTT, Vol. 46, N° 10, pp. 1589-1593, Oct. 1998.
[10]   S. Gribaldo, R. Boudot, G. Cibiel, V. Giordano, O. Llopis, Proceedings of European Frequency and Time Forum 2005, March 2005

A New Method of Minimizing Noise Figure of CMOS LNAs

Dariusz Pienkowski and Georg Boeck

Technical University of Berlin, Microwave Engineering Group, Sekr. HFT 5-1, Einsteinufer 25, 10587 Berlin, Germany

Abstract. This paper shows analysis of a cascode amplifier in terms of input impedance and noise performance. Based on two-port noise theory it is shown, that pad capacitance by decreasing the equivalent noise resistance can improve noise performance of the amplifier.
 Based on the introduced theory an amplifier has been implemented in a 0.13 μm HCMOS technology and shows 0.76 dB noise figure and 12 dB gain at 2.14 GHz for 3.5 mA supply current and 1.2 V supply voltage.

Keywords: CMOS analog integrated circuits, LNA, noise figure, pad capacitance
PACS: 84.30.Le, 84.30.Bv, 85.30.Tv, 85.40.Qx

INTRODUCTION

The design of low noise amplifiers (LNAs) for communication circuits remains a challenging matter due to the serious requirements in terms of low noise, low power consumption, high gain and linearity. Although today leading edge CMOS technologies exhibit very low minimum noise figure (NF_{min}) values it is difficult to meet this values in practical LNA design. The kernel of this paper is the noise performance of a cascode CMOS amplifier using a 0.13 μm HCMOS technology from STMicroelectronics.

State of the art sub-micron RF-MOS-devices show low capacitance values, thus large inductances are needed for commonly used input matching shown in Figure 1a. At low frequencies this large inductance can be realized only by off-chip inductors and input matching circuits changes. Pad capacitance (C_{pad}) appears at the transistor gate, as it is shown in Figure 1b. The analysis of these two amplifiers is performed in the next section.

ANALYSIS

Input impedance

Let us assume, that the frequency of operation is so low that impedance of capacitor C_{pad} compared to 50 Ohm can be neglected. In such a case the input impedance of the amplifier shown in Figure 1a can be expressed as

$$Z_{in} = \frac{g_m L_s}{C_{gs}} + j\left(\frac{\omega^2 (L_g + L_s) C_{gs} - 1}{\omega C_{gs}}\right). \tag{1}$$

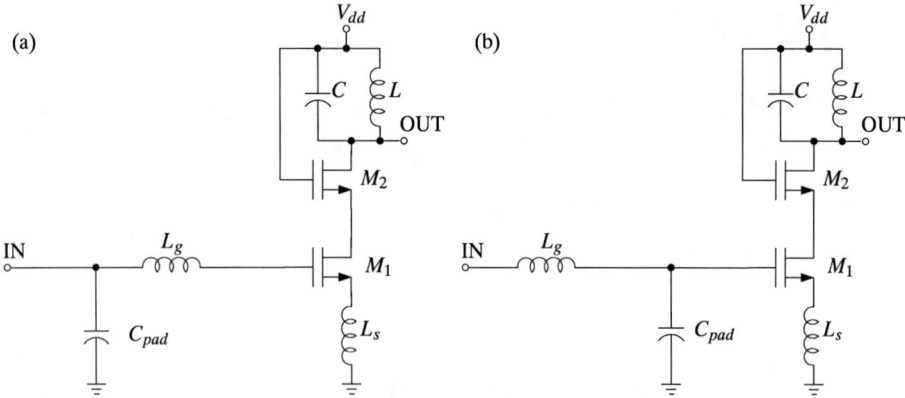

FIGURE 1. Amplifiers with input matching circuits: (a) inductor L_g connected directly to the transistor, (b) pad capacitance C_{pad} connected directly to the transistor.

where L_s and L_g are source and gate inductors, respectively and g_m and C_{gs} denote small signal parameters of transistor M_1 (C_{gd} and g_{ds} are neglected). However, if the influence of C_{pad} can not be neglected only the imaginary part of Z_{in} is affected.

For the amplifier shown in Figure 1b the similar calculation can be performed. In this case C_{pad} is connected parallel to the C_{gs} of the transistor M_1. Since the small transistors are interesting for low noise low current amplifier, these two capacitances are in the similar order. Thus, C_{pad} can not be neglected in this case. The input impedance of such an amplifier can be expressed as

$$Z_{in} = \frac{g_m L_s C_{gs}}{(C_{pad}+C_{gs})^2 + \omega^2 L_s C_{pad} C_{gs}^2 (\omega^2 C_{pad} L_s - 2) + \omega^2 C_{pad}^2 L_s (g_m^2 L_s - 2)}$$
$$+ j\frac{\omega^2(L_g+L_s)C_{gs}^2 - C_{gs} + C_{pad}(g_m^2 L_s \omega^2 (\omega^2 L_g C_{pad} - 1) - (\omega^2 L_s C_{gs} - 1)^2)}{\omega((C_{pad}+C_{gs})^2 + \omega^2 L_s C_{pad} C_{gs}^2 (\omega^2 C_{pad} L_s - 2) + \omega^2 C_{pad}^2 L_s (g_m^2 L_s - 2))}$$
$$+ j\frac{C_{pad}(\omega^4 C_{gs}^2 L_s L_g (\omega^2 L_s C_{pad} - 2) + 2\omega^2 L_g C_{gs}(1 - \omega^2 L_s C_{pad}) + \omega^2 L_g C_{pad})}{\omega((C_{pad}+C_{gs})^2 + \omega^2 L_s C_{pad} C_{gs}^2 (\omega^2 C_{pad} L_s - 2) + \omega^2 C_{pad}^2 L_s (g_m^2 L_s - 2))}.$$
(2)

Contrary to the previous case either the real or imaginary part of input impedance is affected. The real part of input impedance decreases. This decrease requires higher values of inductance L_s. It is rather negative, first and foremost because of lowering the gain, but also because of introducing stability issues and generating noise.

The value of inductance L_g required for matching decreases. In turn, this is positive, since this inductor occupies a large area, particularly when the transistor M_1 (and C_{gs} value) is small. Besides of the area the noise contribution of L_g is limited, because smaller inductor has also lower resistance. In the next section is shown, that C_{pad} connected directly to the transistor has also additional positive influence on noise performance.

Noise calculation

To calculate noise behavior of the amplifiers shown in Figure 1 the noise two-port theory [1] can be used. Let us assume, that the circuit consists of transistors together with inductor L_s, L and capacitor C can be characterized with four noise parameters R_n, B_{opt}, G_{opt}, F_{min} and they are known. Note $Y_{opt} = G_{opt} + jB_{opt}$, B_{opt} is negative and $|G_{opt}| < |B_{opt}|$ for CMOS transistors. We look for the new four noise parameters R'_n, B'_{opt}, G'_{opt}, F'_{min} of the circuits from Figure 1 in relation with elements L_g and C_{pad}. Using this methodology new four noise parameters of the amplifier shown in Figure 1a can be expressed as

$$R'_n = dR_n \quad , \quad G'_{opt} = \frac{G_{opt}}{d} \quad , \quad F'_{min} = F_{min} \tag{3}$$

$$B'_{opt} = \frac{B_{opt}(1 - 2\omega^2 L_g C_{pad}) + \omega L_g |Y_{opt}|^2 - \omega C_{pad}(1 - \omega^2 L_g^2 |Y_{opt}|^2)}{d} \tag{4}$$

$$d = (\omega L_g B_{opt})^2 + 2\omega L_g B_{opt} + (\omega L_g G_{opt})^2 + 1 \tag{5}$$

If we solve (5) for L_g, the minimum achievable d factor is

$$d_{min} = \frac{G_{opt}^2}{G_{opt}^2 + B_{opt}^2} \tag{6}$$

New four noise parameters of the amplifier shown in Figure 1b can be expressed as

$$R'_n = dR_n \quad , \quad G'_{opt} = \frac{G_{opt}}{d} \quad , \quad F'_{min} = F_{min} \tag{7}$$

$$B'_{opt} = \frac{B_{opt}(1 - 2\omega^2 L_g C_{pad}) + \omega L_g |Y_{opt}|^2 - \omega C_{pad}(1 - \omega^2 L_g C_{pad})}{d} \tag{8}$$

$$d = (\omega L_g B_{opt})^2 - 2(\omega^2 L_g C_{pad} - 1)\omega L_g B_{opt} + (\omega L_g G_{opt})^2 + (\omega^2 L_g C_{pad} - 1)^2 \tag{9}$$

Similarly solving (9) for L_g yields to the minimum achievable factor d that can be written as

$$d_{min} = \frac{G_{opt}^2}{G_{opt}^2 + B_{opt}^2 + \omega C_{pad}(\omega C_{pad} - 2B_{opt})} \tag{10}$$

Comparison of (6) and (10) leads to a conclusion that pad capacitance C_{pad} in the amplifier shown in Figure 1b reduce the factor d. Thus, this amplifier shows better noise performance because of reduced R'_n. Interestingly, C_{pad} is mostly treated as a parasitic component that has negative influence on the performance of the amplifier, especially on the noise performance [2].

DESIGN

To prove this conclusion simulations of amplifiers from Figure 1 have been performed. Both transistors M_1 and M_2 have the same widths, supply current is constant and equal

2 mA and analysis frequency is 2.14 GHz. Transistors length is 0.13 μm and C_{pad} value is set to 70 fF. Simulation results are summarized in Table 1.

TABLE 1. Simulation results of amplifiers from Figure 1a and Figure 1b at 2.14 GHz for 2 mA supply current and different transistor sizes.

M_1, M_2 transistor width	24 μm		48 μm		96 μm	
Figure	1a	1b	1a	1b	1a	1b
NF [dB]	1.04	0.25	0.92	0.34	0.54	0.35
Gain [dB]	25.42	15.71	23.08	16.67	19.39	15.69
M (noise measure)	0.27	0.06	0.24	0.08	0.13	0.09

The amplifier with capacitance near the transistor's gate shows better results (noise measure) in all cases. But for 96 μm the difference is not so significant. However, a supply current of 2 mA is too low for such a large transistors and this amplifier exhibit linearity issues.

Based on the given methodology a 0.13 μm CMOS LNA for UMTS has been designed. The characterization of the amplifier, with respect to the noise performance and power consumption confirms our calculations with very good agreement. The designed 2.14 GHz amplifier shows noise figure and gain values of 0.76 and 12 dB, respectively. Supply voltage and current consumption are 1.2 V and 3.5 mA. To the knowledge of the authors, this results belong to the best ever reported in this field [3].

Conclusion

The pad capacitance connected directly to the transistor gate changes the input matching conditions either for real or imaginary part of input impedance and lowers the gain of the amplifier. However, most important is its influence on the noise performance of the amplifier.

Considering as an example a cascode amplifier it is demonstrated based on the two-port noise theory that the pad capacitance can lead to a reduced equivalent noise resistance, and thus improved noise performance. Moreover, it is shown that pad capacitance decreases required values of inductance L_g.

Presented discussion show that it is reasonable to use a capacitor connected directly to transistor's gate. Such a capacitor can be even placed by the designer and optimized together with transistor size to maximize amplifier performance.

REFERENCES

1. J. A. Dobrowolski, *Introduction to Computer Methods for Microwave Circuits Analysis and Design*, Artech House, Boston, 1991.
2. P. Leroux and M. Steyaert, *5 GHz CMOS low-noise amplifier with inductive ESD protection exceeding 3 kV HBM*, in Proc. 30th European Solid-State Circuits Conference ESSCIRC, Sept. 2004, pp. 295–298.
3. B. Floyd and D. Ozis, *Low-noise Amplifier Comparison at 2 GHz in 0.25 μm and 0.18 μm RF-CMOS and SiGe BiCMOS*, in Proc. Radio Frequency Integrated Circuits Symp., June 2004, pp. 185–188.

Analysis and Design of a Frequency Synthesizer with Internal and External Noise Sources

Gurpreet Singh Sangha, Michael H.W. Hoffmann

Department of Microwave Techniques University of Ulm

Abstract: In this paper we demonstrate a rigorous noise analysis of the PLL based synthesizer circuit. In contrast to analyses done previously, we are taking into account a more complete combination of noise sources. A second order PLL is used for this work. In contrast to the standard noise model, we choose an alternative model of describing noise. This leads to the derivation of a very general mathematical description for the output voltage, as well as for the output average power while taking into consideration noise sources at different points in the circuit. This mathematical description shows very clearly how the noise at different points in the circuit is changing the overall behavior of the synthesizer.

Keywords: Synthesizer, PLL, Noise, Modeling, Simulations, Matlab, Simulink
PACS: 01.30.Cc, 05.40.Ca, 02.60.Cb

INTRODUCTION

Today the design of a PLL based synthesizer including component selection and evaluation has become quite challenging. Many parameters must be evaluated. Among the crucial parameters is the phase noise, which is a random modulation of the phase of an oscillator's output due to a variety of internal and external processes.

In this paper we present a thorough analysis of the synthesizer based on a PLL circuit in order to study the effect of electronic noise coming from different sources from within the circuit as well as from the input signals. We represent the theoretical analysis methodology and after that we will follow up with presenting the results from simulations.

Many authors have done valuable studies on the noise behavior of a PLL using different models where they included some effects of noise [1]-[5]. We are continuing their work by doing an analysis of the noise in the synthesizer circuit taking into account the effect of additional noise sources.

NOISE ANALYSIS METHODOLOGY

A second order PLL is used for this work. As usual, it consists of a phase-detector (PD), a control filter (CF), and a voltage controlled oscillator (VCO). The control filter is described by the following transfer function:

$$F(s) = (1 + \tau_2 s)/(b + \tau_1 s) \tag{1}$$

A model for the output of the phase detector is used that does not only make possible to describe the influence of phase fluctuations, θ_i, but also those of amplitude fluctuations, $R(t)$, in the input signal. Similarly, the locally generated VCO-signal will not only be described through its phase fluctuations, θ_o, but it also includes amplitude fluctuations, n'. The detector output voltage is thus described as a function, v_d, that depends not only on the phase-difference, φ, between VCO-signal and input-signal, but also on amplitude fluctuations of its input signals. The novel model shown below is thus more complete as compared to models in use:

$$v_d(t) = g(R(t), n'(t), \varphi(t)) \qquad (2)$$

The noise sources under consideration here are the following:
1. Noise at the input of the PLL circuit, $n(t)$, which can be divided into two parts. A noise source $n_1(t)$ describing the phase variation of input and a noise source $n_2(t)$ included with $R(t)$ describing the amplitude variation of input.
2. The amplitude variations of the VCO signal, $n'(t)$, plays a role as a proportional term at the output of the phase detector, $v_d(t)$. It is designed in a way that it resembles this effect in a synthesizer circuit.
3. Noise source in front of the control filter, $n_3(t)$, is taken to be $1/f$ noise superimposed by white noise. The signal inside the loop is basically a base band signal, which can be expected to be affected by $1/f$ noise that is coming from different electronic devices in the loop.
4. Noise source in front of the VCO, $n_4(t)$, is designed to be a combination of $1/f$ noise and white noise.

The model used for the noise description can be represented by the following random process:

$$u(t) = \sum_{n=-M}^{M} A_n \cos(n\Delta\omega t + \varphi_n) \qquad (3)$$

where $\Delta\omega := 2\pi/T_{ob}$ and T_{ob} is the observation time. This model is different from the standard used model, in the sense that the terms A_n are deterministic and the terms φ_n are random variables which are jointly uniformly distributed in the interval $[0, 2\pi)$. This less frequently used model can be traced back to S.O.Rice [6]. It is mathematically simpler and by virtue of central limit theorem it leads to the same results as the model that is more commonly used, anyway. The nonlinear differential equation of the PLL can be given by:

$$\tau_1 \ddot{\varphi} + \left\{b + K_o \tau_2 \frac{\partial g(R, n', \varphi)}{\partial \varphi}\right\} \dot{\varphi} + K_o g(R, n', \varphi) + K_o \tau_2 \frac{\partial g(R, n', \varphi)}{\partial R} \dot{R} = \tau_1 \ddot{\theta}_i + b\dot{\theta}_i - b\omega_o \qquad (4)$$

Since the mathematical analysis of a nonlinear differential equation is usually mathematically very complex, a novel linearization of the model is used. The phase detector function is linearized in the vicinity of operating point $(R_o, 0, \varphi_\infty)$. Here R_o is the amplitude of the input signal, φ_∞ is the steady state phase error, and φ_c is a constant phase shift of the input signal. The output of the phase detector can be described as:

$$v_d(t) = g(R, n', \varphi) \approx g(R_o, 0, \varphi_\infty) + \frac{\partial g}{\partial \varphi}\bigg|_{(R_o, 0, \varphi_\infty)} (\varphi(t) - \varphi_\infty) + \frac{\partial g}{\partial R}\bigg|_{(R_o, 0, \varphi_\infty)} (R(t) - R_o) + \frac{\partial g}{\partial n'}\bigg|_{(R_o, 0, \varphi_\infty)} n'(t) \quad (5)$$

$$v_d(t) = g(R, n', \varphi) \approx \frac{b\Delta\Omega}{K_o} + K_D(\varphi(t) - \varphi_\infty) + K_A(R(t) - R_o) + K_V n'(t) \quad (6)$$

K_D, K_A and K_V represent phase detector gain, phase detector sensitivity to amplitude variation of input signal, and phase detector sensitivity to amplitude variation of VCO signal respectively. (Note the dependence on amplitude variation of the input and of the VCO). The following equation can be solved for the values of the operating points:

$$g(R_o, 0, \varphi_\infty) = b\Delta\Omega/K_o \quad (7)$$

where K_o is the VCO gain and $\Delta\Omega$ is the initial frequency offset between the input and the VCO. Taking the above discussions into account, the PLL based synthesizer model can be described by the following block diagram.

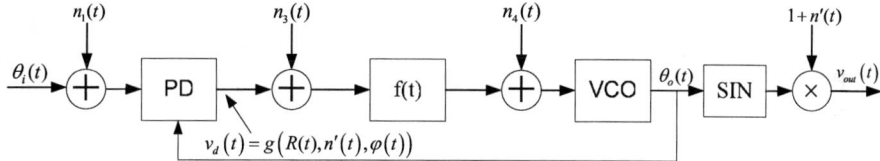

FIGURE 1. Synthesizer model used for the noise analysis.

Using the setup described above, an expression for the output voltage as well as an expression for the average power at the output were derived. Since they are rather complex formulas, they are not being presented in this paper. The calculations were done by taking into account noises at different points in the loop. The spectrum found was the general spectrum having influence of all noise sources described in this paper. The terms A_n from equation (3) allow for designing the noise to have that type of spectrum we want. The spectrum of the output was derived in terms of A_n of each noise source. This makes it a general description which can be designed to have a different type of noise at each point in the circuit.

SIMULATION RESULTS

The calculations were followed by simulations using SIMULINK where the noise sources are designed to represent the noise behavior at different points in a synthesizer circuit. The noise behavior was simulated using the nonlinear model with very satisfying results that show clearly how the noise at different points in the circuit affect the output of the synthesizer. Figure 2 represents the input and the output power spectrum of the synthesizer circuit with noise at different points of the circuit plotted on log-log scale. As it was expected from theoretical results (not shown here) the noise at different points in the circuit affects the system in different ways. Graphs (a), (b), (c) and (d) in Figure 2 show the spectrum of the output affected by amplitude variations of the input and the VCO, input phase noise, noise in front of loop filter, and noise in front of VCO respectively.

In the simulations presented here, the input noise is designed to be white noise that is divided into two components affecting the amplitude and phase respectively. All the random noise sources in this simulation are designed for SNR of -10dB. The angular frequency of the input signal is chosen to be $150\,Krad/\text{sec}$.

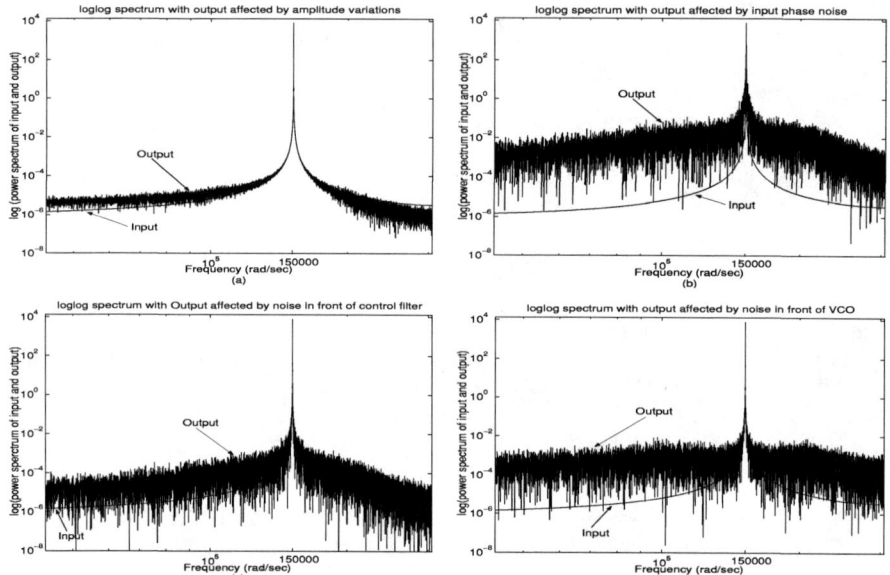

FIGURE 2. Graphs (a), (b), (c) and (d) show the power spectrum of the input and the output with noise present at different points of the circuit in Figure 1.

CONCLUSION

In this paper we have demonstrated the noise analysis of the PLL based synthesizer circuit taking into account a more complete combination of noise sources. A particular model of describing noise is used, which leads to the derivation of a very general mathematical description for the output voltage, as well as for the output average power while taking into consideration noise sources at different points in the circuit. This new setup leads to a deeper understanding of the complex noise phenomenon affecting the output of the synthesizer. It shows how each point in the synthesizer circuit is contributing to the overall noise behavior.

REFERENCES

1. Kundert, K., *Modeling and Simulation of Jitter in PLL Frequency Synthesizers,* Cadence Design Systems, Inc, 2001, www.cadence.com.
2. Hall, R., *Predicting PLL Performance: Phase Noise,* Motorola Logic IC Division.
3. Drucker, E., *Model PLL Dynamics and Phase-Noise Performance,* Microwaves & RF Magazine, February, 2000, http://www.mwrf.com.
4. Kundert, K., *Predicting the Phase Noise and Jitter of PLL-Based Frequency Synthesizers,* http://www.designers-guide.com.
5. Curtin, M., and O'Brien, P., *Phase-Locked Loops for High-Frequency Receivers and Transmitters-Part 2,* Analog Devices, Analog Dialogue 33-5, 1999.
6. Rice, S.O., *Mathematical Analysis of Random Noise,* Bell System Technical Journal Vol. 23, pp. 282-332, 1944 and Vol. 24, pp. 46-156, 1945.

Analytical Calculations For The Influence Of Carrier's Phase Noise On Modulating Signals

D. Kondis, A. Birbas and M. Birbas

Dept. Of Electrical and Computer Engineering
University of Patras, Patras Greece 26500

Abstract. We calculate the exact influence of the phase noise on the quality of modulated signals. Assuming that the noise origin (thermal, shot or flicker) is ergodic and stationary we give quantitative results for the distortion introduced due to the phase noise of the carrier oscillation. We give details for the AM modulation

Keywords: phase noise, oscillators, modulation, and carrier frequency fluctuations.
PACS: 05.40-a, 84.30.Ng

I. INTRODUCTION

The fluctuations of the oscillation frequency of an electrical oscillator are referred as phase noise. Those fluctuations determine a real system and are characterized by parameters such as mean value, variance, stationarity, ergodicity etc. Those small fluctuations might be produced by the presence of thermal, shot even flicker noise.

The here-below analysis of the phase noise is generic and independent of the noise type and includes its change in the time domain as well as its power change as a function of the distance from the central oscillation frequency. The only requirement is the noise signal to be stationary and ergodic. In detail, we apply a coherent mathematic formalism in order to observe the structure of the power spectral density of the phase noise of a specific time modulation scheme in the frequency domain. Then we draw conclusions regarding the shape of the modulated carrier including the phase noise of the carrier (associated with its power spectra after the modulation) convoluted with the modulating signal. We try then to draw conclusions regarding the SNR of the signal regarding the phase noise for every frequency and the aggregated total distortion introduced at the IF stage (phase noise of the local oscillator) for an AM modulation (equally for SSB or NBFM).

II. EXPRESSIONS FOR THE PHASE NOISE AND ANALYSIS

It has been mentioned that the phase noise existing around the central frequency of an oscillator might due to various noise origins. Thus, it is difficult to create a general analytic formulation. If ω_o is the frequency of oscillation of the oscillator then the power spectral density can be expressed by a variable: P ($\omega_o+\Delta\omega$, 1Hz), P is the power of the phase noise per Hz at a distance $+\Delta\omega$ from the frequency ω_0 [1]. We can now refer this power to the total power of the carrier defining:

$$L_{total}\{\Delta\omega\} = 10\log\frac{P_{sideband}(\omega_0+\Delta\omega,1Hz)}{P_{carrier}} \tag{1}$$

In the literature, one can find expressions for P (ω_0 +$\Delta\omega$,1Hz) as a function of $\Delta\omega$ of the type $1/(\Delta\omega)^3$ ($-30db/dec$), then there is a knee and the dependence becomes $1/(\Delta\omega)^2$, thus $-20db/dec$ and finally we have the background noise. When the thermal noise is the cause of the phase noise [2] then the power spectral density is Lorentzian and is given by

$$P = \frac{V_0^2 D}{\Delta\omega^2+D^2} \quad \eta\, P \approx \frac{D}{\Delta\omega^2+D^2} \quad \text{and} \quad L_{total}\{\Delta\omega\} = \frac{D}{\Delta\omega^2} \tag{2}$$

D expresses the increase in entropy. The same dependence is observed in the phase noise stemming from white noise sources, of LC type oscillators [3].

To produce a modulated signal, the signal currying the information could be modulated in the form of, SSB, NBFM, FM, PM, PAM). The modulated signal is not been implied only by the product of the two signals (i.e. the modulating "ω_m" and the carrier signal "ω_c" (this is the case for AM, SSB, PAM), but it stems also from complicated manipulations on the phase of the carrier (NBFM, FM, PM) etc. In most cases in the literature (i.e. AM, NBFM, FM, PM, PAM) [4], as well as in this current work, the modulated signal is expressed in the time domain as a product of exponential and complex factors (functions of the ω_c: $e^{j\omega_c t}$ and the modulating ω_m : $e^{j\omega_m t}$.)

In the case of AM the corresponding expression is: [4]

$$\Phi_{AM}(t) = \text{Re}\left\{Ae^{j\omega_c t}(1+\frac{1}{2}me^{j\omega_m t}+\frac{1}{2}me^{-j\omega_m t})\right\} \tag{3}$$

m is the modulation index ($m \leq 1$).
For NBFM the expression is:

$$\Phi_{NBFM}(t) = \text{Re}\left\{Ae^{j\omega_c t}(1+\frac{1}{2}\beta e^{j\omega_m t}-\frac{1}{2}\beta e^{-j\omega_m t})\right\} \tag{4}$$

where $\beta = a\, k_f/\omega_m$ and $a \sin \omega_m t$ is the modulating signal.
For the FM we have:

$$\Phi_{FM}(t) = \text{Re}\left\{Ae^{j\omega_c t}\sum J_n(\beta)e^{n\omega_m t}\right\} \tag{5}$$

where J_n is the Bessel of nth order, where β is the modulation degree, a k_f/ω_m with a associated with the amplitude of the modulating signal. In all these expressions the carrier signal is considered monochromatic which is rather against the fact that the spectral density of the carrier has constituents (P (ω_0 +$\Delta\omega$, 1Hz)) at a distance $\Delta\omega$.

If we integrate the whole carrier signal (i.e. including its associated phase noise) over all ω we get the total power spectra of the carrier:

$$P = \lim_{T \to \infty} \frac{1}{T} \int_{-T/2}^{+T/2} x^2(t) dt = E\{x^2(t)\} = R_{xx}(0) = \frac{1}{2\pi} \int_{-\infty}^{+\infty} S_x(\omega_c) d\omega_c \qquad (6)$$

and if $A^2(\omega_c) = S_x(\omega_c)$ then :

$$P = \frac{1}{2\pi} \int_{-\infty}^{+\infty} A^2(\omega) d\omega \qquad (7)$$

According to the here above mentioned and by applying the Parseval's theorem [4], $A(\omega)$ should be the coefficient of the Fourier expansion of the stationary and ergodic carrier signal (or part of it):

$$f_{carrier}(t) = \int_{-\infty}^{+\infty} A(\omega_c) e^{j\omega_c t} d\omega_c \qquad (8)$$

As mentioned before thermal noise is the main phase noise producer. Other noise types may co-exist and we can consider those stationary and ergodic as well. This draws as to the conclusions that (3), (4) and (5) before do not include the phase noise factor. Those expressions can be re-expressed if we substitute $e^{j\omega_c t}$ with $f_{carrier}(t) = \int_{-\infty}^{+\infty} A(\omega_c) e^{j\omega_c t} d\omega_c$. Then the relation (3) (representing accurately the AM carrier) becomes:

$$\Phi_{AM}(t) = \operatorname{Re}\left\{ \left(\int_{-\infty}^{+\infty} A(\omega_c) e^{j\omega_c t} d\omega_c\right) \left(1 + \frac{1}{2} m e^{j\omega_m t} + \frac{1}{2} m e^{-j\omega_m t}\right) \right\} \qquad (9)$$

The same expressions can be written for the other cases such as NBFM, and FM. We note here that other than the phase noise factors influencing the carrier spectral density might be more pronounced and of greater importance but the expressions for the phase noise are the accurate ones. Moreover in the case of AM if we further proceed with the calculations, we can derive the correspondence of the time domain with the frequency domain (expressed graphically in Fig. 1):

$$F\left\{ \operatorname{Re}\left[\left(\int_{+\infty}^{-\infty} A(\omega_c) e^{j\omega_c t} d\omega_c\right) \left(1 + \frac{1}{2} m e^{j\omega_m t} + \frac{1}{2} m e^{-j\omega_m t}\right) \right] \right\} =$$

$$= \frac{1}{2\pi} F\left\{ \operatorname{Re}\left[\int_{+\infty}^{-\infty} A(\omega_c) e^{j\omega_c t} d\omega_c \right] \right\} \otimes F\{1 + m\cos\omega_m t\} = \qquad (10)$$

$$= \frac{1}{2\pi} F\left\{ \operatorname{Re}\left[\int_{+\infty}^{-\infty} A(\omega_c) e^{j\omega_c t} d\omega_c \right] \right\} \otimes \{2\pi\delta(\omega) + m\pi\delta(\omega - \omega_m) + m\pi\delta(\omega + \omega_m)\}$$

This results to:

$$A(\omega_c, \omega) + \frac{m}{2} A(\omega_c - \omega_m, \omega) + \frac{m}{2} A(\omega_c + \omega_m, \omega) \qquad (11)$$

If we now calculate the total interfering power as $\omega_{m'}$ takes various values around ω_m (due to phase noise) then the phase noise power is given by the integral:

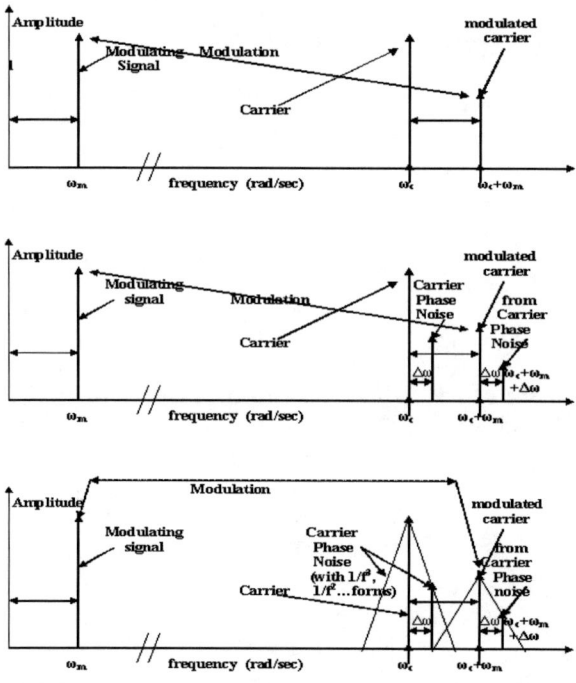

FIGURE 1. The influence of the phase noise of the carrier on the modulated signal through the modulation process

$$PN_{P(AM)}(\omega_m) = \frac{m^2}{4}\int_{\omega_{mmin}}^{\omega_m - \min\Delta\omega} A^2(\omega_{m'}+\omega_c,\omega_m)d\omega_{m'} +$$
$$\frac{m^2}{4}\int_{\omega'_m + \min\Delta\omega}^{\omega_{max}} A^2(\omega_{m'}+\omega_c,\omega_m)d\omega_{m'} \quad (12)$$

As an example, in the case where $P_{SD} = P(\omega_0 + \Delta\omega, 1\,Hz) = \dfrac{a}{(\Delta\omega)^3} = \dfrac{a}{|\omega-\omega_c|^3}$, a being the coefficient indicating the strength of the phase noise then we get:

$$PN_{N/S(AM)1/\Delta\Omega} = \int_{\omega_{min}}^{\omega_m - \min\Delta\omega}\frac{a\,d\omega_{m'}}{(|\omega_{m'}-\omega_m|)^3} + \int_{\omega_m + \min\Delta\omega}^{\omega_{min}}\frac{a\,d\omega_{m'}}{(|\omega_{m'}-\omega_m|)^3} =$$
$$= -\frac{a}{(\omega_m-\omega_{mmin})^3} - \frac{a}{(\omega_{mmax}-\omega_m)^3} + \frac{2a}{(\min\Delta\omega)^2} \cong \frac{2a}{(\min\Delta\omega)^2} = \frac{2a'}{(\min\Delta f)^2} \quad (13)$$

If now one choose min $\Delta f = 20$ Hz (typical carrier's tolerance) then the result turns to be non negligible.

REFERENCES

1. Hajimiri, A. and Lee, T. H., "The Design of Low Noise Oscillators", Boston/Dordrecht/London, *Kluwer Academic Publishers,* 1999.
2. Ham, D. and Hajimiri, A., *IEEE Journal of Solid-State Circuits,* Vol. 38, No. 3, 407-418, (2003).
3. Kouznetsov, K. A. and Meyer, R. G., *IEEE Journal of Solid-State Circuits,* Vol. 35, No. 8, 1244-1248 (2000).
4. Stremler, F. G., "Introduction to Communication Systems", Reading Massachusetts, *Addison-Wesley Publishing Company,* Third Edition, 1990

Observation of noise-induced transitions in radar range tracking systems

Eyad H. Abed* and Sheldon I. Wolk[†]

*Dept. of Electrical and Computer Engineering and the Institute for Systems Research, University of Maryland, College Park, MD 20742 USA
[†]Advanced Techniques Branch, Tactical Electronic Warfare Division, Code 5753, Naval Research Laboratory, Washington, D.C. 20375 USA

Abstract. Noise-induced transitions are studied for nonlinear radar range tracker system models. The paper shows the existence of noise-induced transitions in such models, indicating that the noise level can have a profound effect on the nature of tracker behavior. A scenario is considered in which a centroid tracking radar subtends two physical targets. The perceived track points are studied as a function target distance and noise. The work uses nonlinear models for radar range trackers developed in [1]. The noise-induced transitions found in this work result in unanticipated changes in the bifurcation behavior of the tracking system the presence of noise. The relative impact of noise amplitude and target separation on the noise induced-transitions is studied.

Keywords: Radar tracking systems, nonlinear systems, stability, noise-induced transitions, stochastic systems.
PACS: 05.10.Gg, 05.45.-a, 84.40.Xb

INTRODUCTION

Noise-induced transitions [2] are studied for nonlinear radar range tracker system models. The paper shows the existence of noise-induced transitions in such models, indicating that the noise level can have a profound effect on the nature of tracker behavior. A scenario is considered in which a centroid tracking radar subtends two physical targets. The perceived track points are studied as a function target distance and noise. The work uses nonlinear models for radar range trackers developed in [1] as a tool for analyzing the behavior of centroid tracking-based radar trackers in the presence of multiple competing targets. The equilibrium, bifurcation and stability properties of these models in the case of deterministic target models were studied in [1]. Stochastic extension of the models and stability analysis was carried out in further work of the authors and co-workers.

NONLINEAR RADAR RANGE TRACKER MODELS

In [1], a simple null-tracking system is studied. In this section, we recall from [1] a simplified approximate continuous-time nonlinear model of this gated radar range tracking system. The tracking system is driven by the echo signal E. The main output of the system is the estimated relative slant range ρ.

The model considered here is a system of two first-order ordinary differential equations of the form

$$\dot{\rho}(t) = K\beta^2 e^{-2V(t)} \int_{-T_1}^{T_1} |E(\sigma,t)|^2 w_D(\sigma - \rho(t)) d\sigma \qquad (1)$$

$$\dot{V}(t) = \frac{\beta^2 e^{-2V(t)}}{T_{\text{AGC}}} \int_{-T_1}^{T_1} |E(\sigma,t)|^2 w_S(\sigma - \rho(t)) d\sigma - 1 \qquad (2)$$

where an overdot denotes differentiation with respect to time t.

The model (1),(2) is stochastic if the received signal E is a random process.

Detector laws for a periodic echo signal

It is convenient to introduce notation for the difference and sum detector laws appearing in the deterministic continuous-time model (1)-(2). For any real ρ, denote $\mathscr{D}(\rho) := \int_{-T_1}^{T_1} |E_0(\sigma)|^2 w_D(\sigma - \rho) d\sigma$, $\mathscr{S}(\rho) = \int_{-T_1}^{T_1} |E_0(\sigma)|^2 w_S(\sigma - \rho) d\sigma$. Then $\mathscr{D}(\rho)$ is the difference detector law and $\mathscr{S}(\rho)$ is the sum detector law. Typical difference and sum weighting patterns for split-gate tracking systems are shown in detail in [1]. Note that these weighting patterns have support on a subset of the interval $-T_1 \leq \sigma \leq T_1$. Thus, $w_S(\sigma)$ is a symmetric function, positive on its support, and $w_D(\sigma)$ is an antisymmetric function, and is positive (respectively, negative) for all positive (respectively, negative) values of σ on its support.

To specify the shape of the signal $E_0(\sigma)$, assume the transmitted pulse is rectangular and symmetric. Then, for a point target, the reflected (received) pulse will also be rectangular and symmetric. Assuming an IF filter which is matched to the transmitted pulse, we find that $E_0(\sigma)$ is a symmetric triangle pulse. Note that the pulse width of $E_0(\sigma)$, denoted by T_p, in general differs from T_1 [1].

NOISE-INDUCED TRANSITIONS AND APPLICATION TO RADAR RANGE TRACKER MODEL

Noise-induced transitions were first reported on by Horsthemke and Lefever in the early 1980s, and the basic theory was summarized in their book [2]. They found that noise injected into a dynamical system can lead to transition phenomena not present in the original deterministic system. In a pure noise-induced transition, a system depending on a parameter which originally has a unique equilibrium point for all parameter values undergoes a bifurcation in the presence of noise. In other noise-induced transitions, the nature of a system's bifurcations as a parameter varies is changed by the presence of noise.

We now recall the general calculation method for analyzing noise-induced transitions (see, e.g., [2],[3]). Consider the stochastic differential equation

$$\dot{x} = f(x) + \varepsilon g(x)\xi \qquad (3)$$

where ξ stands for Gaussian white noise with mean $\langle \xi(t) \rangle = 0$ and variance $\langle \xi(t)\xi(t') \rangle = 1$. Any additional parameters are suppressed in (3). Analytical detection of noise-induced transitions for scalar equations of the type (3) is performed by seeking the extrema of the stationary probability density function. The extrema obey the equation $f(\bar{x}) - \frac{\varepsilon^2}{2}g(\bar{x})g'(\bar{x}) = 0$.

In this work, we find that white noise induces a change in the bifurcation behavior of the basic radar range tracking loop model (1)-(2). Since the basic analytical theory described in the previous section applies to scalar stochastic differential equation models (3), it is helpful to approximate the second-order model (1)-(2) by a first-order model. This is done by taking the limit of a fast AGC loop, i.e., by taking the limit as the AGC time constant $T_{AGC} \to 0$.

Two Targets

We consider a two-point-target scenario (impulsive targets), in which one target is noisy and one constant. Let the noisy target have mean amplitude a, with noise perturbation $\varepsilon\xi(t)$ around the mean, where ξ has variance 1. Let the amplitude of the constant target be b. Let the two targets be positioned a distance $2d$ apart in range. Then the return signal $|E(\sigma,t)|^2$ takes the form

$$\begin{aligned}|E(\sigma,t)|^2 &= aE_0(\sigma-d) + bE_0(\sigma+d) \\ &+ \varepsilon a E_0(\sigma-d)\xi(t)\end{aligned} \qquad (4)$$

where the only pure time dependence is subsumed in the noise signal $\xi(t)$, the amplitude of which is scaled by ε. Substituting (4) in (1)-(2) and reducing the system order to one, we obtain the reduced model

$$\begin{aligned}\dot{\rho}(t) &= ae_1(\rho-d) + be_2(\rho+d) \\ &+ \varepsilon a e_1(\rho-d)\xi(t)\end{aligned} \qquad (5)$$

where $e_1(\rho) := \int_{-T_1}^{T_1} E_0(\sigma-d) w_D(\sigma-\rho) d\sigma$ and $e_2(\rho) := \int_{-T_1}^{T_1} E_0(\sigma+d) w_D(\sigma-\rho) d\sigma$.

To be specific, we take the weighting patterns to be $w_D(x) := \frac{2x}{(1+\gamma x^6)}$ and $E_0(x) := e^{-\alpha x^2}$. Note that w_D is an odd-valued function, as required for a difference detector law, and E_0 is even, being the convolution of the (even) sum detector law and the returned rectangular radar pulse signal. For lack of space, detailed calculations are not presented here.

Noise-Induced Transitions in Radar Range Tracker Model

Recall the weighting patterns $w_D(x) = \frac{2x}{(1+\gamma x^6)}$ and $E_0(x) = e^{-\alpha x^2}$. Take $a = 1$, $b = 2$, $T_p = 2T_1 = 2$, $\gamma = 4$ and $\alpha = 1.3863$. The following two scenarios will now be calculated using the system modeling and analysis above:

Scenario 1: The two-target case considered above is taken in the absence of noise in either target, i.e., with $\varepsilon = 0$. The (deterministic) equilibrium points are then calculated as a function of the target separation parameter d.

Scenario 2: The two-target case considered above is taken in the presence of noise in the returned signal from the target with amplitude a, i.e., with $\varepsilon \neq 0$. The stochastic equilibrium points are then calculated as a function of the target separation parameter d for various values of noise amplitude ε.

The calculations for the present example show that there is a single equilibrium point of the noise-free system for $d < d_{cr} = 1.0432$. This is an attracting equilibrium point. At $d = d_{cr}$, two new equilibrium points are born in a saddle-node bifurcation, one attracting and one repelling. The tracking radar is now presented with two possible tracking behaviors, that of tracking the original target and that of tracking the new stable target, which may be a true (physical) target and may be a false (decoy) target.

For the noisy case, we gradually increase the noise amplitude parameter ε and observe the effect on the bifurcation behavior of the stochastic system. With noise, we find that the appearance of the two new track points in a (stochastic) saddle-node bifurcation still occurs but for different d values than the previous critical value of 1.0432. The smaller the value of ε, the closer the critical separation is to that for the deterministic problem. This is no surprise.

What is surprising, however, is that with noise the system supports more than three stochastic equilibria. For example, for d near the deterministic critical value of 1.0432 and for $\varepsilon = 2$, the equilibria corresponding to the deterministic saddle-node bifurcation are now very well defined, and, in addition, the system is exhibits still another pair of stochastic equilibria that is born at a nearby value of d. The result is a total of five (5) stochastic equilibrium points of the system, three of which are attracting and two of which are unstable.

As the noise parameter ε is increased further, the two new track points born at the noise-induced saddle-node bifurcation separate from one another. For larger noise amplitudes, our work shows the presence of up to seven (7) stochastic equilibrium points, four stable and three unstable, depending on target separation.

REFERENCES

[1] E.H. Abed, A.J. Goldberg and R.E. Gover, *IEEE Trans. Aerospace Electron. Syst.* **27**, 68-82 (1991).
[2] W. Horsthemke and R. Lefever, *Noise-Induced Transitions*, Springer-Verlag, Berlin, 1984.
[3] C. Van den Broeck, J.M.R. Parrondo, R. Toral and R. Kawai, *Physical Review E* **55**, 4084-4094 (1997).

BIOLOGICAL AND BIOMEDICAL SYSTEMS

Functional Roles of Noise and Fluctuations in the Human Brain

Yoshiharu Yamamoto*, Rika Soma*, Keiichi Kitajo*[†], Leonid A. Safonov*, Kentaro Yamanaka*, Ichiro Hidaka**, Kyoko Ohashi[‡], Daichi Nozaki[§], Zbigniew R. Struzik*, Lawrence M. Ward[†] and Shin Kwak[¶]

*Educational Physiology Laboratory, Graduate School of Education, The University of Tokyo, 7–3–1 Hongo, Bunkyo-ku, Tokyo 113–0033, Japan
[†]Psychophysics and Cognitive Neuroscience Laboratory, Department of Psychology, The University of British Columbia, 2136 West Mall, Vancouver, BC V6T1Z4, Canada
**National Cardiovascular Center Research Institute, 5–7–1 Fujishirodai, Suita, Osaka 565–8565, Japan
[‡]Developmental Biopsychiatry Research Program, McLean Hospital, 115 Mill St., Belmont, Massachusetts, 02178, USA
[§]Research Institute of National Rehabilitation Center for the Disabled, 4–1 Namiki, Tokorozawa, Saitama 359–8555, Japan
[¶]Department of Neurology, Graduate School of Medicine, The University of Tokyo, 7–3–1 Hongo, Bunkyo-ku, Tokyo 113–0033, Japan

Abstract. In the past decade, it has been recognized that noise can enhance the response of non-linear systems to weak signals, via a mechanism known as stochastic resonance (SR). In this short review, we introduce our experimental demonstration of SR-type behavior in the human brain, and medical (neurological) applications and a possible mechanism of such phenomenon.

INTRODUCTION

The beneficial role of noise and fluctuations in biological systems has been extensively studied over the past decade. One of the fundamental phenomena considered in this respect is that of stochastic resonance (SR) [1, 2], which is generally defined as the optimization of responses of a non-linear system to a weak input signal caused by external noise. For instance, in neural systems SR has been shown to optimize the sensitivity of sensory neurons [3, 4, 5, 6], which may further lead to an enhancement of perception and/or behavior in animals [7] and humans [8, 9, 10].

As a natural extension of these studies, we have recently demonstrated that externally added noise can also enhance input-output coherence within the human brain [11, 12, 13]. This has been achieved by adopting a so-called "double receptor" experimental design, in which noise is injected into one peripheral receptor and a signal is injected into a separate receptor, and the two are only merged within the relevant brain structures.

Here, we first summarize our experimental observations on the double receptor SR at the level of the brain stem, where the neurological noise can help trigger changes in heart rate in response to small changes in blood pressure [11, 12], i.e. the baroreflex response. This will be followed by an introduction to our clinical findings, showing that the noise-enhanced responses of the brain can also be used in the treatment of neurological patients with autonomic failure [14] and Parkinson's disease [15]. Secondly, we introduce another example of the brain's SR at the level of the cortices, including the visual areas, where visual noise can enhance behavioral output [13] and responses in the brain waves to weak visual signals [16]. Finally, we discuss a possible mechanism for this noise-induced sensitization of the human cortex based on a demonstration of noise-induced synchronization in the brain waves [16] and the effects of added noise on adaptive dynamics in coupled non-linear oscillators [17, 18].

SR IN HUMAN BAROREFLEX

In the first biological SR study using the double receptor design, in 2000, we demonstrated experimentally that noise can enhance the homeostatic function in the human blood pressure regulatory system [11]. We studied the human baroreflex system, in which an increase (or a decrease) in blood pressure is automatically compensated for by decreases (or increases) in heart rate and vascular resistance via the autonomic nervous system.

The baroreflex system has two kinds of pressure-sensitive nerve cell receptors. By tilting a subject back and forth on a horizontal table, as described in Ref. [11], we periodically moved blood to the lower part of the body. The draining of blood from the chest area stimulated the cardiopulmonary baroreceptors to fire a weak repeating signal, which the brain interpreted as a drop in (central) venous blood pressure. To create neural noise, we randomly applied and removed mechanical pressure to and from the neck. This caused the arterial baroreceptors to fire randomly, as the artery wall pressure, which normally indicates blood pressure in the arteries, was increasing and decreasing randomly.

The results showed that the compensatory heart rate response to a weak periodic signal introduced at the venous blood pressure receptor was optimized by applying noise to the arterial blood pressure receptor. We thus concluded that this *functional* SR most likely resulted from the interaction between the noise and the signal in the brain stem, where the neuronal inputs from these two different receptors first join together.

Recently, we further conducted a similar experiment, but replaced the mechanical noise added to the arterial baroreceptor with an electrical one applied to the vestibular afferent nerves [12]. With this galvanic vestibular stimulation (GVS), electrical current is delivered transcutaneously to the vestibular afferents through electrodes placed over the mastoid bones, modulating their continuous firing levels [19]. Through the vestibular nuclei in the brain stem, these afferents influence neuronal circuits in the medullary cardiovascular areas [20, 21] and thus the heart rate [22], just like when we stimulate the arterial baroreceptors.

There are two advantages to using GVS. The first is a practical one. As shown in the next section, SR invoked by externally added sensory noise can be used to treat patients with central nervous system dysfunction. While the mechanical device used in our previous studies [11, 14] cannot be used for this purpose, given its large size and immobility, the vestibular afferents can easily be stimulated by using a portable

GVS apparatus. Secondly, the electrical stimulator is capable of applying more precisely controlled input noise than the mechanical actuator. In Ref. [12], we tested whether externally added Gaussian $1/f$ noise, in the frequency range of 0.01–2.0 Hz where spike trains in single medullary neurons exhibiting spontaneous activity show $1/f$-type behavior [23], could more effectively sensitize the medullary cardiovascular areas than Gaussian white noise. In other words, we tested whether the noise mimicking *innate* fluctuations of the brain would be superior to *artificial* white noise.

When we examined the compensatory heart rate response to a weak periodic signal introduced via venous blood pressure receptors while adding $1/f$ or white noise with the same variance to the brain stem through GVS, in both cases, this noisy GVS optimized covariance between the weak input signals and the heart rate responses. However, the optimal level with $1/f$ noise was significantly lower than with white noise, suggesting a functional benefit of $1/f$ noise for neuronal information transfer in the brain.

NEUROLOGICAL APPLICATIONS

The above-mentioned baroreflex system plays a crucial role in maintaining adequate blood perfusion to the brain, and failure can easily lead to fainting, an annoying and potentially dangerous symptom of patients with autonomic failure. While most previous studies on sensory SR have dealt with the noise-induced sensitization in the responses of *normal* sensory neurons to weak input signals [3, 4, 5, 6, 7, 8, 9], we hypothesized that, in neurological diseases with blunted neural responses even to non-weak signals, noise might ameliorate the impaired neural transmission just like the noise-enhanced detection of small signals in normal sensory systems. Thus, we first studied [14] whether adding noise to the arterial baroreceptors might facilitate cardiovascular responses to even strongly hypotensive stimuli in neurological patients with primary autonomic failure (PAF) [24]. PAF is a disorder in which symptoms of autonomic dysfunction, including orthostatic hypotension, are primarily the result of progressive neuronal degeneration of unknown cause.

The patients were tested for their transient responses of heart rate and systolic and diastolic blood pressures to head-up tilt, with and without the continuous application of beat-to-beat mechanical noise to the arterial baroreceptors. The PAF patients exhibited marked drops in blood pressure and a blunted increase in heart rate upon transition from a supine to a head-up position. The addition of noise to the arterial baroreceptors significantly increased the magnitude of the heart rate increment and diminished that of drops in blood pressure. We thus concluded that the addition of external noise to baroreceptor signaling ameliorated the marked postural hypotension seen in patients with PAF.

Recently, we further examined [14], by means of noisy GVS applied by a portable device, whether it might be possible to ameliorate the blunted responsiveness of degenerated neuronal circuits of multiple system atrophy (MSA) [25] patients, the majority of our pool of PAF patients. We evaluated the effect of 24-hour noisy GVS on long-term heart rate dynamics in MSA patients. Short-range or high-frequency fluctuations of heart rate, a reliable marker for cardiac autonomic nervous system activity [26], were significantly increased by the noisy GVS compared with sham stimulation, suggestive of improved autonomic, especially parasympathetic [26], responsiveness.

The vestibular nerves are also known to influence neuronal circuits in the basal ganglia and the limbic system [27, 28] through the cerebellar vermis [20, 29]. This possible vestibular effect on the basal ganglia and the limbic system is intriguing when considering clinical observations in Parkinson's disease (PD) [30] consistent with the lack of a limbic-to-motor link within the basal ganglia, i.e. the disinclination to move by dissociating motivation from executive (spontaneous) motor movement, of which the expressionless face of a PD patient is a common clinical example [31, 32].

Thus, in the same study [14], we further hypothesized that the noisy GVS could alleviate the disinclination to move, or the *bradykinesic* symptom, in PD and MSA patients. We evaluated the effect of 24-hour noisy GVS on daytime trunk activity dynamics in patients with either PD or pharmacologically unresponsive parkinsonism. Long-range anti-persistency of trunk activity patterns probed with an auto-correlation measure [33, 34] increased by the noisy GVS, suggestive of a quickening of bradykinesic rest-to-active transitions. Taken together, we concluded that the noisy GVS is effective in boosting the neuro-degenerative brains of MSA and/or PD patients, including those unresponsive to standard pharmacological therapy, and in improving their autonomic and motor responsiveness.

SR IN HUMAN VISUAL CORTEX

We introduced yet another type of double receptor SR study [13, 35] for human cortical information processing. In Ref. [13], we investigated human subjects' responses, in a sensorimotor integration task, to the slowly changing gray level of an image presented on a computer screen. An important feature of our experiment was that different visual stimuli were separately presented to each eye using a mirror stereoscope; we presented the visual stimulus to the right eye as a signal and investigated how the behavioral response to the weak signal, quantified by handgrip force, would be affected by presenting random visual stimuli to the left eye. Neural inputs from the two eyes first converge in the primary visual cortex, and are integrated via binocular interaction higher in the visual system. Therefore, when signal and noise were presented to separate eyes, changes in sensorimotor integration, if any, took place not at the peripheral level but rather in some higher visual centers.

Such behavioral responses were in fact optimized by presenting randomly changing gray levels separately to the *noise*-side eye. The results indicated that observed behavioral SR was mediated by neural activity within the human brain where the information from both eyes converges.

POSSIBLE MECHANISM FOR SR IN THE BRAIN

The theory of SR has been rigorously studied [2] in dynamical systems, such as over-damped bistable systems or monostable systems with a response threshold—the latter being used as a model for sensory neurons. However, neuronal networks within the brain, especially in the cortices, may not have a definite threshold or bistability, and they rather can be regarded as mutually coupled excitable and/or oscillatory units [36] capable of showing multistable dynamics. Indeed, recent studies have shown that brain-wide, frequency-specific neural synchrony and asynchrony, or the dynamical attractor selection among multiple neural oscillators, seem to be involved in the generation of perceptual, cognitive and behavioral processes in response to external and/or internal stimuli [36]. Thus, we sought a mechanism for SR in the brain, especially for cortical SR, in the context of noise-induced and signal-dependent neural synchronization and/or desynchronization.

Experimentally, we demonstrated that enhanced detection of weak visual signals by the addition of visual noise is accompanied by an increase in phase synchronization of electroencephalograph (EEG) signals across widely-separated areas of the human brain [16]. We again adopted a similar double receptor design, in which observers responded to a weak rectangular gray level signal presented to their right eyes, while randomly changing gray level noise was presented separately to their left eyes. We measured brain electrical activity at the scalp by EEG, calculated the instantaneous phase for each EEG signal, and evaluated the degree of large-scale phase synchronization between pairs of EEG signals. Dynamic synchronization-desynchronization patterns were observed and we found evidence of noise-enhanced large-scale synchronization associated with the detection of changes in the brightness under conditions of noise-enhanced performance. Our results suggest that behavioral SR might arise from noise-enhanced synchronization of neural activities across widespread brain regions.

On the theoretical side, we recently conducted a model study partly to account for this type of phenomenon [17, 18] (also in this volume). We studied the adaptability of coupled non-linear oscillators to external forcing by using a system of globally coupled FitzHugh-Nagumo equations. Each unit was either excitatory or inhibitory. If the numbers of units of both types were in a specific relation (balanced coupling), we observed the presence of multistable oscillatory states with different excitation or *firing* rates. In the presence of noise, there was noise-driven switching between these states. The selection between higher- and lower-frequency oscillations depended on the input, which resulted in increasing coherence between the periodic input and the system's output. We believe that our model can be a simple one for SR-type behavior in the brain.

CONCLUSION

We have shown here that the human brain can work as a *stochastic resonator*. Such a property may be effective in boosting the neuro-degenerative brain by stochastic noise. The theory of SR-type behavior in the brain must, however, still be elaborated on in further research.

ACKNOWLEDGMENTS

This study was in part supported by the Japan Science and Technology Agency and the Toyota Motor Corporation.

REFERENCES

1. Wiesenfeld, K., and Moss, F., *Nature*, **373**, 33–36 (1995).
2. Gammaitoni, L., Hänggi, P., Jung, P., and Marchesoni, F., *Rev. Mod. Phys.*, **70**, 223–287 (1998).
3. Douglass, J. K., Wilkens, L., Pantazelou, E., and Moss, F., *Nature*, **365**, 337–340 (1993).
4. Levin, J. E., and Miller, J. P., *Nature*, **380**, 165–168 (1996).
5. Collins, J. J., Imhoff, T. T., and Grigg, P., *J. Neurophysiol.*, **76**, 642–645 (1996).
6. Nozaki, D., Mar, D. J., Grigg, P., and Collins, J. J., *Phys. Rev. Lett.*, **82**, 2402–2405 (1999).
7. Russell, D. F., Wilkens, L. A., and Moss, F., *Nature*, **402**, 291–294 (1999).
8. Collins, J. J., Imhoff, T. T., and Grigg, P., *Nature*, **383**, 770 (1996).
9. Richardson, K. A., Imhoff, T. T., Grigg, P., and Collins, J. J., *Chaos*, **8**, 599–603 (1998).
10. Simonotto, E., Riani, M., Seife, C., Roberts, M., Twitty, J., and Moss, F., *Phys. Rev. Lett.*, **78**, 1186–1189 (1997).
11. Hidaka, I., Nozaki, D., and Yamamoto, Y., *Phys. Rev. Lett.*, **85**, 3740–3743 (2000).
12. Soma, R., Nozaki, D., Kwak, S., and Yamamoto, Y., *Phys. Rev. Lett.*, **91**, 078101–1–4 (2003).
13. Kitajo, K., Nozaki, D., Ward, L. M., and Yamamoto, Y., *Phys. Rev. Lett.*, **90**, 218104–1–4 (2003).
14. Yamamoto, Y., Hidaka, I., Iso-o, N., Komai, A., Soma, R., and Kwak, S., *Brain Res.*, **945**, 71–78 (2002).
15. Yamamoto, Y., Struzik, Z. R., Soma, R., Ohashi, K., and Kwak, S., *to appear* (2005).
16. Kitajo, K., Yamanaka, K., Nozaki, D., Ward, L. M., and Yamamoto, Y., *Proc. SPIE*, **5467**, 359–269 (2004).
17. Safonov, L. A., and Yamamoto, Y., *Proc. SPIE*, **5467**, 131–138 (2004).
18. Safonov, L. A., and Yamamoto, Y., *to appear* (2005).
19. Fitzpatrick, R. C., and Day, B. L., *J. Appl. Physiol.*, **96**, 2301–2316 (2004).
20. Barmack, H. H., *Brain Res. Bull.*, **60**, 511–541 (2003).
21. Balaban, C. D., and Beryozkin, G., *Exp. Brain Res.*, **98**, 200–212 (1994).
22. Radtke, A., Popov, K., Bronstein, A. M., and Gresty, M. A., *Lancet*, **356**, 736–737 (2000).
23. Lewis, C. D., Gebber, G. L., Larsen, P. D., and Barman, S. M., *J. Neurophysiol.*, **85**, 1614–1622 (2001).
24. Bannister, R., and Mathias, C. J., "Clinical features and evaluation of the primary chronic autonomic failure syndromes," in *Autonomic Failure, A Textbook of Clinical Disorders of the Autonomic Nervous System*, edited by C. J. Mathias and R. Bannister, Oxford University Press, Oxford, 1999, pp. 307–316, 4 edn.
25. Gilman, S., Low, P. A., Quinn, N., Albanese, A., Ben-Shlomo, Y., Fowler, C. J., Kaufmann, H., Klockgether, T., Lang, A. E., Lantos, P. L., Litvan, I., and Mathias, C. J., *J. Neurol. Sci.*, **163**, 4–5 (1999).
26. Task Force of the European Society of Cardiology and the North American Society of Pacing and Electrophysiology, *Circulation*, **93**, 1043–1065 (1996).
27. Albert, T. J., Dempesy, C. W., and Sorenson, C. A., *Biol. Psychiatry*, **20**, 1267–1276 (1985).
28. Anderson, C. M., Polcari, A., Lowen, S. B., Renshaw, P. F., and Teicher, M. H., *Am. J. Psychiatry*, **159**, 1322–1328 (2002).
29. Newlands, S. D., and Perachio, A. A., *Brain Res. Bull.*, **60**, 475–495 (2003).
30. Barbeau, A., "Parkinson's disease: clinical features and etiopathology," in *Handbook of Clinical Neurology*, edited by P. Vinken, G. W. Bruyn, and H. L. Klawans, Elsevier Publishers, Amsterdam, 1986, vol. 49 (Revised Series 5) Extrapyramidal disorders, pp. 87–152, 4th edn.
31. Chaudhuri, A., and Behan, P. O., *J. Neurol. Sci.*, **179**, 34–42 (2000).
32. Joseph, R., *Neuropsychiatry, Neuropsychology, and Clinical Neuroscience*, Williams & Wilkins, Baltimore, 1996, 2nd edn.
33. Ohashi, K., Amaral, L. A. N., Natelson, B. H., and Yamamoto, Y., *Phys. Rev. E*, **68**, 065204(R)–1–4 (2003).
34. Struzik, Z. R., Hayano, J., Sakata, S., Kwak, S., and Yamamoto, Y., *Phys. Rev. E.*, **70**, 050901(R)–1–4 (2004).
35. Mori, T., and Kai, S., *Phys. Rev. Lett.*, **88**, 218101–1–4 (2002).
36. Varela, F., Lachaux, J. P., Rodriguez, E., and Martinerie, J., *Nat. Rev. Neurosci.*, **2**, 229–239 (2001).

Mapping of Synaptic-Neuronal Impairment on the Brain Surface through Fluctuation Analysis

Toshimitsu Musha[1], Takayoshi Kurachi[2], Naohoro Suzuki[2], and Yukio Kosugi[2]

[1] Brain Functions Lab., Inc. and [2] Tokyo Institute of Technology

Abstract. Increase of demented population year by year is becoming a serious social problem to be solved urgently. The most effective way to block this increase is in its early detection by means of an inexpensive, non-invasive, sensitive, reliable and easy-to-operate diagnosis method. We have developed a method satisfying these requirements by using scalp potential fluctuations. We have collected 21ch EEG and SPECT data of 25 very mild Alzheimer's disease (AD) (MMSE=26±1.8), moderately severe AD (MMSE=15.3±6.4) and age-matched normal controls. As AD progresses, local synaptic-neuronal activity becomes abnormal, either more unstable or more inactive than in normal state. Such abnormality is detected in terms of *normalized power variance* (NPV) of a scalp potential recorded with a scalp electrode. The *z*-score is defined by *z* = ((*NPV of a subject*) − (*mean NPV of normal subjects*))/(*standard deviation of NPV of normal subjects*). Correlation of a measured *z*-score map with the mean *z*-score map for AD patients characterizes likelihood to AD, in terms of which AD is discriminated from normal with 75% of true positive and 25% false negative probability. By introducing two thresholds, we have 90% of true positive and 10% of false negative discrimination.

Keywords: Neuronal impairment, EEG, the beta rhythm, functional mapping, brain

INTRODUCTION

One of the most effective ways to reduce the dementia population, which is otherwise increasing every year monotonously, is its early detection. Especially as Alzheimer's disease (AD) is a slow process, the screening of dementia among a large elderly population plays a crucial role in reducing onset of dementia. Earlier detection of AD allows various types of rehabilitation to be more effective. Practically therefore the following are required for a possible early detection tool: it should be *inexpensive, noninvasive*(*without using radioactive material injecting into blood such as in SPECT, PET*), *sensitive, reliable, and easy to handle*.

Electroencephalogram (EEG) is rich in information about neuronal or synaptic-neuronal activity in the brain because EEG is generated by depolarization of neurons in the brain cortex, and hence proper signal processing of EEG signals would allow to realize such a tool. We have already developed a new technique called DIMENSION (Diagnosis Method of Neuronal Dysfunction)[1] which derives the degree of cortical synaptic-neuronal impairment in terms of smoothness of the scalp potential distribution of the alpha rhythm component, especially its mean value and fluctuations around the mean. This method satisfies the four requirements mentioned above and is currently used in some hospitals to monitor the efficacy of treatment of dementia. This method is based on the alpha rhythm, which, however, is not observed in some

CP780, *Noise and Fluctuations: 18th International Conference on Noise and Fluctuations-ICNF 2005*,
edited by T. González, J. Mateos, and D. Pardo
© 2005 American Institute of Physics 0-7354-0267-1/05/$22.50

patients. To overcome this difficulty we have developed another method of detecting synaptic-neuronal impairment in the brain in terms of the beta frequency component (13~30 Hz) which is always observed in any one.

FLUCTUATIONS AND NEURONAL DYSFUNCTION

α-DIMENSION

The scalp potential generated by the alpha rhythm in a normal subject is well approximated by that generated by a properly determined current dipole, and its goodness of fit is called dipolarity.[2] The dipolarity or goodness-of-fit of such a dipole approximation is defined by Eq(1), which characterizes smoothness of the scalp potential. However, the D value fluctuates almost periodically as shown in Fig.1 with

$$D = \sqrt{1 - \frac{\langle (u_{obs} - u_{dip})^2 \rangle_{channel}}{\langle (u_{obs})^2 \rangle_{channel}}} \qquad (1)$$

varying signal-to-noise ratio of the alpha component. We just pick up peak values of D as representing the smoothness of the potential in each period. The left figure refers to a normal subject and the right refers to an AD patient. We introduce the mean values

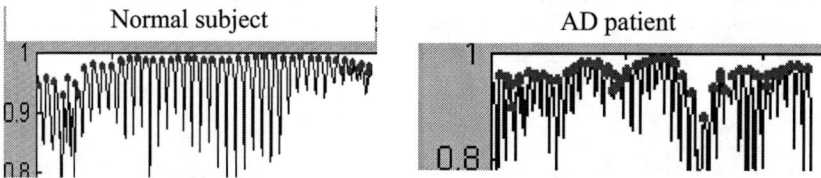

FIGURE 1. Variation of D_α (red points) fluctuates from period to period.

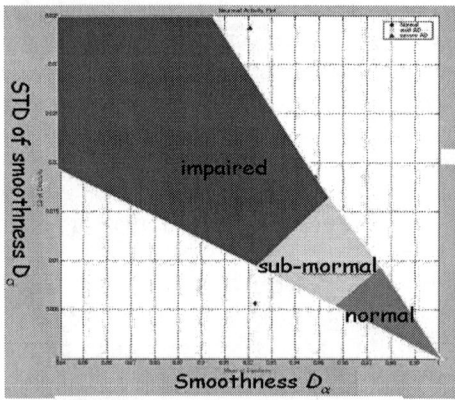

FIGURE 2. Brain activity diagram

of D over 3 min and denote it as D_α. The standard deviation of the peak D values around D_α is denoted as D_σ. In normal subjects D_α mostly stays close to unity, whereas in AD patients it is fluctuating between unity and lower levels. Then the degree of impairment of cortical neurons is characterized by D_α and D_σ. These values of AD patients and age-matched normal controls distribute in a fan-beam area in the brain activity diagram as shown in Fig.2.[1]

This diagram is so sensitive to neuronal impairment that cortical activation in slightly demented people after a proper rehabilitation is clearly observed. This method is named α-DIMENSION. Although it has a high sensitivity to neuronal impairment, it is not applicable to people without the alpha rhythm, and this difficulty has been overcome by the β-DIMENSION which has better discrimination power for AD.

β-DIMENSION

We have observed through behavior of the alpha rhythm that neuronal activity becomes unstable when partially impaired, and we tried DIMENSION with the β-component(13 ~ 30 Hz). Since the β-component spreads over a wider frequency range, its smoothness is not expected even in a normal subject. The degree of instability of a band-limited EEG power is expressed in terms of normalized power variance (NPV_j) of an EEG signal on channel j as

$$NPV_j \equiv \left\langle (u_j^4 - \langle u_j^2 \rangle^2) \right\rangle / \left\langle u_j^2 \right\rangle^2 \qquad (2)$$

where u_j is a potential on channel j and mean $\langle \cdots \rangle$ is taken over several minutes in a resting state. We calculated $\langle NPV_{NL,j} \rangle$ for normal patients where its standard deviation is denoted as $\sigma_{NL,j}$. A deviation of the observed NPV_j from the mean value of normal subjects is scaled in unit of $\sigma_{NL,j}$ as

$$\frac{NPV_j - \langle NPV_{NL,j} \rangle}{\sigma_{NL,j}} \equiv z_j. \qquad (3)$$

By averaging z_j values over all of the moderately severe AD patients in our database, we have obtained a template *z-map* as shown in Fig.3. This is a difference map between normal subjects and moderately severe AD patients. In other words, it is EEG fluctuation map in the β-component of EEG signals. Similarity of a difference map of NPV from the normal template to this difference template map is a measure of likelihood of a subject to AD. The likelihood Z^m of subject m to AD is defined as in Eq.(4) where 21 is the number of scalp electrodes used for EEG recording.

$$Z^m \equiv \frac{\sum_{j=1}^{21}(x_j^m - \overline{x^m})(y_j - \overline{y})}{\sqrt{\sum_{j=1}^{21}(x_j^m - \overline{x^m})^2 \cdot \sum_{j=1}^{21}(y_j - \overline{y})^2}}$$

where   (4)

$$y_j = \langle NPV_j \rangle_{AD} - \langle NPV_j \rangle_{NL}$$

$$\overline{y} = \frac{1}{21}\sum_{k=1}^{21} y_k$$

$$x_j^m = NPV_j^m - \langle NPV_j \rangle_{NL}$$

$$\overline{x}^m = \frac{1}{21}\sum_{j=1}^{21} x_j^m$$

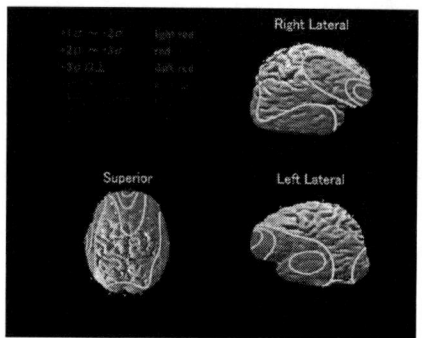

FIGURE 3. The template AD pattern of AD patients

SENSITIVITY OF *NPV* TO ALZHEIMER'S DISEASE

Figure 4 shows sensitivity and specificity to AD, where a severer AD patient has a larger Z-value. A single threshold Z = 0.2 separates 75% of AD patients and 75% of normal subjects to its right and left sides. Hereas two thresholds Z = − 0.1 and 0.35 will reduce both of false positive and false negative probabilities down to 10%. A region limited by these two thresholds is an ambiguous region where no sharp decision should be avoided until the next EEG exam is carried out after, for instance, six months. Unless the Z-value increases in the second exam, he/she is likely to be normal. Such a three-stage classification is more recommended than the binary classification with a single threshold.

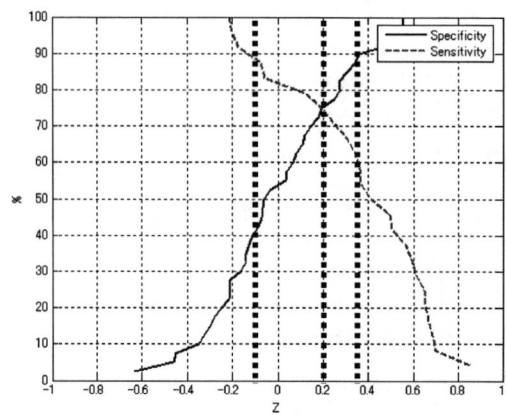

FIGURE 4. Sensitivity and specitivity

SUMMARY

A fluctuation in biological rhythm is essential in maintaining life[34], and absence of fluctuations means nothing but death of a living body. Either too much or too small fluctuations as compared with the normal state is related to abnormality. The degree of fluctuations can be a measure of health. The fluctuation level of cortical neuronal activity is used for diagnosis of dementia. The β-DIMENSION is applicable with no exception and satisfies the four requirements mentioned in INTRODUCTION for easy screening of early AD patients.

REFERENCES

1. T.Musha, T.Asada, F.Yamashita, T.Kinoshita, H.Matsuda, M.Uno, Z.Chen and W.R.Shankle, "A new EEG method for estimating cortical neuronal impairment that is sensitive to early stage Alzheimer's disease," Clinical Neurophysiology, 113 (2002) 1052-1058.
2. T. Musha and Y. Okamoto, "Forward and Inverse Problems of EEG Dipole Localization," Critical Review in Biomedical Engineering, 27(3-5):189-239(1999).
3. "Physics of Living State" ed. by T. Musha and Y. Sawada, Ohm-sha, 1995
4. T. Musha, 1/f Fluctutaions of the Biological Rhythm, Computer Analysis of Cardiovascular Signals, ed. M. di Rienzo, IOS Press, pp.81-94.

Sleep Stage Dependence of Invariance Characteristics in Fluctuations of Healthy Human Heart Rate

Fumiharu Togo*, Ken Kiyono*, Zbigniew R. Struzik* and Yoshiharu Yamamoto*

Educational Physiology Laboratory, Graduate School of Education, The University of Tokyo, 7-3-1, Hongo, Bunkyo-ku, Tokyo 113-0013

Abstract. The outstanding feature of healthy human heart rate is the robust scale invariance in the non-Gaussian probability density function (PDF), which is preserved not only in a quiescent condition, but also in a dynamic state during waking hours [K. Kiyono et al. Phys. Rev. Lett. 93 (2004)]. Together with $1/f$ like scaling, this characteristic is a strong indication of far-from-equilibrium, critical-like dynamics of heart rate regulation. Our results suggest that healthy human heart rate departs from a critical state-like operation during sleeping hours, at a rate which is heterogeneous with respect to sleep stages annotated according to traditional techniques. We study specific contributions of sleep stages to the relative departure from criticality through the analysis of sleep stage dependence of the root mean square of multiscale local energy and the multiscale PDF. There is a possibility that the involvement of cortical activity may be important for a critical state-like operation.

Keywords: Heart rate variability, Humans, Deep sleep, Rapid eye movement sleep
PACS: 87.19.Hh, 87.80.Vt

INTRODUCTION

Fluctuations in a system at a critical point are generally associated with the scale invariance and universal behavior of the scaling function. It has been suggested that the heart rate control system during the waking state and usual daily activity is also in a critical state [1], permanently out of equilibrium, due to the interaction between the activities of the sympathetic and parasympathetic nervous system [2], leading respectively to an increase and decrease in heart rate. However, during the state of sleep, when autonomic nervous system activity is dramatically changed from the waking state, this characteristic is broken [3]. During sleep, especially sleep stages III/IV (deep sleep), when the cerebral cortex activity is thought to be the lowest with decreased cholinergic and noradrenergic neuronal activity [4], bradycardia appears to be due mainly to an increase in parasympathetic nervous system activity, whereas hypotension is primarily attributable to a reduction in sympathetic vasomotor tone [5]. These physiological conditions may possibly be responsible for the breakdown of the critical state. We demonstrate that such fluctuation characteristics in healthy human heart rate change depending on the sleep stage.

METHODS

The subjects were 9 healthy, non-medicated males. Their mean age, height, and weight were 24.5 (range 21–30) yr, 173.5 (range 165.5–183.5) cm, and 68.1 (range 55.6–85.5) kg, respectively. All the subjects were good sleepers and without any history of sleep problems and none were taking any medication at the time of the tests. All gave their informed consent to participating in this institutionally approved study after the test protocol had been fully described.

Each subject underwent three nights of polysomnographic (PSG) sessions in a sound-proof, air-conditioned (22–24 °C), and shaded sleeping room. The first and second night sessions were used for habituation, and the data obtained on the third night were used for analysis. The subjects went to bed at the time they would normally do, and got up voluntarily.

Electroencephalograms (EEG's; P3-A2, C3-A2), bilateral electrooculograms (EOG's; left, right), and a mental electromyogram (EMG) were monitored continuously throughout the night. The sleep stages were manually scored from the PSG recordings every 30 s by two investigators in accordance with the standard criteria [6]. Heart rate was monitored using standard bipolar leads with an electrocardiograph (ECG) (Life Scope6, Nihon-koden). The analog output of ECG meter was differentiated to yield a train of rectangular impulses corresponding to the QRS complex. The impulse train was processed on a real-time basis with a personal computer (PC-9801 BX2, NEC) at a sampling frequency of 1,000 Hz.

In order to characterize the heart rate fluctuations, we use a detrended random walk method, and study (1) the correlation functions of the local energy fluctuations, and (2) the scale dependence of the probability density function (PDF) of detrended increments.

The heart rate data analyzed are measured as sequential heart interbeat intervals $b(i)$, where i is the beat number. We investigate the PDF of heart rate increments at different time scales (in beat number), where the non-stationarity of the data has been eliminated by local detrending [2]. We first integrate the $b(i)$, $B(m) = \sum_{j=1}^{m} b(j)$, and the resultant $B(m)$ is divided into non-overlapping segments $[1+s(k-1), s(k+1)]$ of size $2s$, where k is the index of the subinterval. Then in each segment the best 4th order polynomial is fit to the data. The differences $\Delta_s B(i) = B^*(i+s) - B^*(i)$ at a scale s are obtained by sliding in time over the segments, where $1 + s(k-1) \leq i \leq sk$ and $B^*(i)$ is a deviation from the polynomial fit. By this procedure, the 3rd order polynomial trends are eliminated.

At a scale s, we define the local energy as

$$\sigma_s^2(i) = \frac{1}{s} \sum_{j=1+s(k-1)}^{sk} \Delta_s B(i)^2. \tag{1}$$

We analyze the whole PDF of $\Delta_s B(i)$. For a quantitative comparison, we fit the data to the following function based on Castaing's equation [7]:

$$\overline{P_s}(x) = \int P_L(\frac{x}{\sigma}) \frac{1}{\sigma} G_{s,L}(\ln \sigma) d(\ln \sigma), \tag{2}$$

where P_L is the increment PDF at a large scale $L > s$, and the self-similarity kernel $G_{s,L}$ determines the nature of the cascade-type multiplicative process. Here we assume P_L

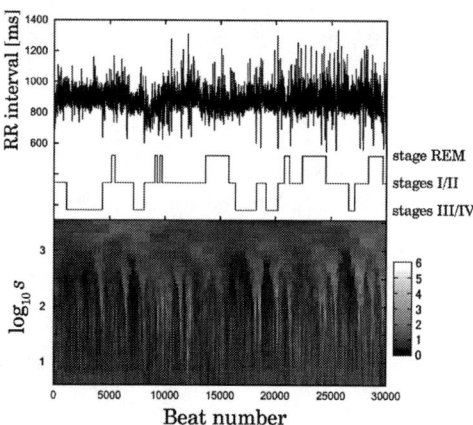

FIGURE 1. A representative recording and root mean square of multiscale (s) local energy of heart rate fluctuations of a healthy subject. REM, rapid eye movement.

and $G_{s,L}$ are both Gaussian,

$$G_{s,L}(\ln \sigma) = \frac{1}{\sqrt{2\pi}\lambda} \exp(-\frac{\ln^2 \sigma}{2\lambda^2}), \qquad (3)$$

and investigate the scale dependence of λ^2.

RESULTS AND DISCUSSION

Figure 1 shows heart rate and the local energy fluctuations as a function of, respectively, the beat number and the scale for one subject. Different characteristics can be observed depending on the sleep stage. Deep sleep is associated with low local energy at $s \approx 200$ as compared to other sleep stages. Recently, we revealed the presence of a non-stationary periodic pattern in the very low-frequency (VLF) range (0.003–0.04 Hz ≈ 25–333 beats) in heart rate variability during deep sleep [8]. These unique VLF oscillations are a likely cause of the low local energy and the breakdown in the invariance characteristics across the scales observed in deep sleep. Neither this breakdown nor the VLF oscillations are observed in REM sleep or the waking state [3], where robust scale invariance has been established.

Indeed, the scale invariance of the non-Gaussian PDF in a range of 70–1000 beats disappears for stages III/IV & stages I/II. In these stages, strongly non-Gaussian fluctuations at a characteristic scale ∼ 200 beats are dominant, as compared to REM sleep & sleep stages I/II (Figure 2). Our results therefore indicate that deep sleep may be responsible for the departure from a critical state-like operation of heart rate during sleep.

The length of data (18000–40300 beats) is limited, so we separate sleep stages into two episodes for studying the PDF. From the point of view of cerebral cortex activity, stages III/IV & stages I/II and stages REM & I/II are categorized as episodes with

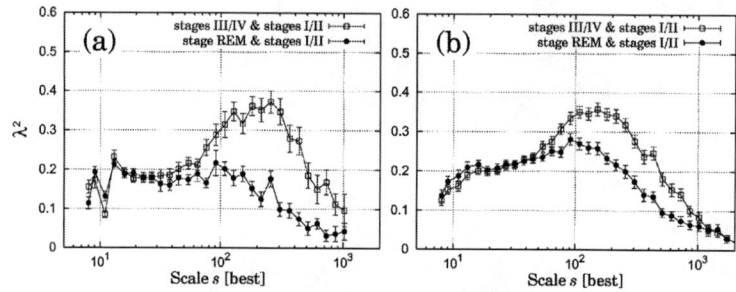

FIGURE 2. Scale dependence of a fitting parameter of Castaing's equation for heart rate fluctuations during nocturnal sleep of a healthy subject, as shown in FIGURE 1 (a), and averaged fitting parameters for 9 subjects (b). The error bars indicate the standard error of the group averages.

relatively weaker and stronger cortical influences on the medullary cardiovascular areas, respectively. Perhaps because of this length limitation and possibly because of the inclusion of stages I/II, the scale invariance in PDF during stages REM/I/II is not as clear as that during rest while awake [3]. A study of large amounts of data for only REM, when the cholinergic neurons are active, would be expected to reveal critical-like scale invariance. This would be required to confirm our hypothesis that cortical activation (during waking and REM) is the key to observing mode-free, critical-like operation.

ACKNOWLEDGMENTS

This study was supported in part by the Japan Science and Technology Agency.

REFERENCES

1. K. Kiyono, Z. R. Struzik, N. Aoyagi, S. Sakata, J. Hayano, and Y. Yamamoto, *Phys Rev Lett*, **93**, 178103 (2004).
2. C. K. Peng, J. Mietus, J. M. Hausdorff, S. Hvlin, H. E. Stanley, and A. L. Goldberger, *Phys Rev Lett*, **70**, 1343 (1993).
3. K. Kiyono, Z. R. Struzik, N. Aoyagi, F. Togo, and Y. Yamamoto, *Phys Rev Lett*, **submitted** (2005).
4. J. A. Hobson, R. Lydic, and H. A. Baghdoyan, *Behav Brain Res*, **9**, 371–448 (1986).
5. G. Baccelli, M. Guazzi, G. Mancia, and A. Zanchetti, *Nature*, **223**, 184–185 (1969).
6. A. Rechtstchaffen, and A. Kales, *A Manual of Standardized Terminology, Techniques and Scoring System for Sleep Stages of Human Subjects*, US Government Printing Office, National Institute of Health Publication, Washington DC, 1968.
7. B. Castaing, Y. Gagne, and E. J. Hopfinger, *Physica D*, **46**, 177–200 (1990).
8. F. Togo, K. Kiyono, Z. R. Struzik, and Y. Yamamoto, *IEEE Trans Biomed Eng*, **submitted** (2005).

Criticality and Universality in Healthy Heart Rate Dynamics

Zbigniew R. Struzik*, Ken Kiyono*, Junichiro Hayano[†], Seiichiro Sakata[†], Shin Kwak** and Yoshiharu Yamamoto*

*The Educational Physiology Laboratory, Graduate School of Education, The University of Tokyo, 7-3-1, Hongo, Bunkyo-ku, Tokyo 113-0013, Japan
[†]Core Laboratory, Nagoya City University Graduate School of Medical Sciences, 1 Kawasumi, Mizuho-cho, Mizuho-ku, Nagoya 467-8601, Japan
**Department of Neurology, Graduate School of Medicine, The University of Tokyo, 7-3-1 Hongo, Bunkyo-ku, Tokyo 113-0033, Japan

Abstract. Methodologies originally developed in the field of statistical physics of complex phenomena have been proven to provide new insights into the modeling, description and understanding of the human heart rate regulatory system. Recent studies have shown the heart rate control system to maintain universality properties characteristic of physical systems exhibiting far-from-equilibrium, critical state-like dynamics [1]. Simultaneously, heart rate regulation has been shown to display correlation properties of antagonist dynamics involving antagonist actors [2], pertinent to some far-from-equilibrium systems. We discuss the range of validity and breakdown scenarios of the universal properties in heart rate regulation leading to the diagnostic capability and also to new challenges for both analysis methods and up-to-date simulation models.

Keywords: Heart Rate Variability, Phase Transitions, Critical Phenomena, 1/f noise
PACS: 87.19.Hh, 87.80.Vt, 89.75.Da, 05.40.-a

INTRODUCTION

In the field of statistical physics of complex phenomena, it has been widely recognized that the analysis of the non-differentiable component, such as noise, fluctuations, and singular or transient changes in the recorded signal, may lead to a fundamentally deeper insight than the study of the smooth, continuous component. Methods have been proposed explicitly to address the fluctuation component separately from the low degree polynomial trend. Structure function analysis, Detrended Fluctuation Analysis (DFA) and wavelet transform based approaches, e.g. WTMM, are the most established. The principle behind these methods is similar - an approximation accounts for the low order trend/baseline and measurement of the residual fluctuations in a window defined by the working resolution. Statistics of the fluctuations are then analyzed. These include the scaling of the average moments or whole PDF shape or, more recently, statistics of two point correlations. This methodology has a remarkable scope of applications in fields as diverse as turbulence [3], financial dynamics [4] and heart rate fluctuations [1].

Since the discovery of 1/f noise in the human heart rate over two decades ago [5], it has been widely recognized that elucidating the origins of complex heart rate dynamics will have profound importance in medical physiology. Yet, a definitive understanding of the mechanism underlying this complexity has as yet not been achieved.

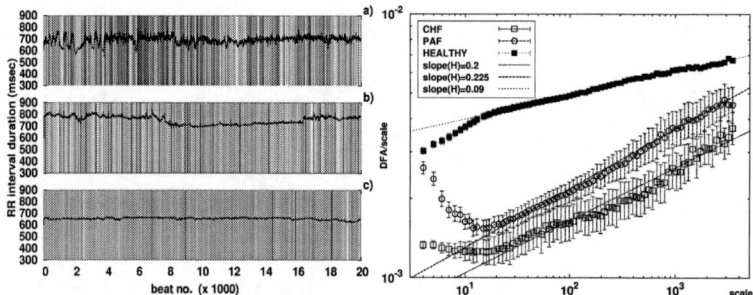

FIGURE 1. Left: Typical traces of daytime heartbeat intervals for: (a) A healthy subject; (b) A PAF (primary autonomic failure) patient; (c) A CHF (congestive heart failure) patient. The grayscale coding used shows the local contribution to multifractality—the spectrum is centered at $h = 0.3$ (mid-gray), with the strongest singularities in black ($h = 0.0$) and the weakest in white ($h = 0.6$). Right: Scale dependency of the mean detrended fluctuation $DFA(s)$ for healthy subjects, PAF patients (neurogenic sympathetic suppression) and CHF patients (neurogenic parasympathetic suppression), showing a near symmetrical response to suppression of either of the antagonistic branches of heart rate control.

ANTAGONIST ACTION IN HEART RATE

Recently, we studied alterations in the scaling of heart rate detrended fluctuations due to the modification of the relative importance of the sympathetic and the parasympathetic branches of the autonomic nervous system [2]. There, we demonstrated that $1/f$ scaling in healthy heart rate requires the existence of and the intricate balance between the antagonistic activity of these two branches, and that affecting this balance leads to a substantial decrease in $1/f$ scaling; see Fig. 1. We further suggested the view of cardiac neuroregulation as a system in a critical state [6], and permanently out of equilibrium, in which concerted interplay of the sympathetic and parasympathetic nervous systems is required for preserving momentary "balance". The criticality hypothesis for the origin of heart rate fluctuation is particularly attractive – the use of concepts from the theory of phase transitions and critical phenomena in a broad range of non-equilibrium systems has proven to be a fruitful approach in the physics community. Our recent studies provide more direct evidence for the criticality hypothesis through an analysis of the whole PDF statistics [1] and two-point correlations [7].

THE CRITICALITY SCENARIO

A characteristic feature at a critical point of a second order phase transition is the divergence of the relaxation time and strongly correlated fluctuations with $1/f^\beta$ ($\beta \sim 1$) scaling in the power spectrum. Healthy human heart rate has long been known to show this type of scaling behavior [5, 8]. However, the existence of robust universal scaling functions at criticality, required to support the criticality hypothesis, has been confirmed only recently. In Kiyono *et al.* [1], we demonstrate that heart rate (increment) PDF's retain robust scale invariance of the non-Gaussian increment probability density function (PDF) during usual daily activity and constant routine protocol; see Fig. 2.

These findings also suggest that an alternative *cascade* paradigm may not be adequate to explain the origin of complex fluctuations in heart beat. A common feature of cascade-type multifractal models - proven in modeling energy (in turbulence) and information (in market regulation) cascades – is an evolution in the shape of the PDF of the increments from stretched exponential at smaller scales to Gaussian at large scales [3].

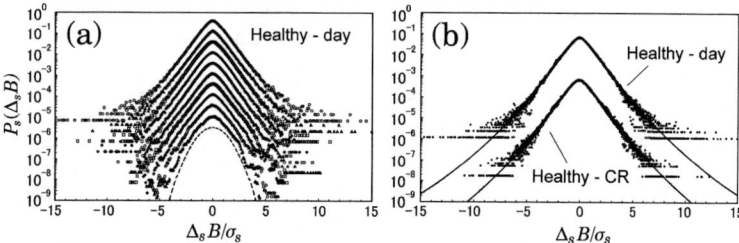

FIGURE 2. Deformation of increment PDF's across scales. (a) Standardized PDF's (in logarithmic scale) for different time scales, from 8 beats (top) to 1024 beats (bottom). The PDF's are estimated from daytime heart rate time series from 50 healthy subjects (b) Superposition of PDF's at different scales from healthy subjects during normal daily life (day) and during constant routine protocol (CR).

In further support of the criticality hypothesis, we recently provided evidence that heart rate fluctuations in healthy individuals remain at criticality within a narrow transition range between distinct, stable 'phases' [7]. Within these phases — experimentally controlled behavioral states of sleep (see also Togo *et al.* in this volume) and prolonged strenuous exercise (Aoyagi *et al.* in this volume) — healthy human heartbeat fluctuations undergo a dramatic breakdown of critical characteristics, such as PDF scale invariance and long-range correlations (see Fig. 3 for the plots of the PDF in the states of sleep and exercise, which shows convergence to Gaussian). Through the direct observation of phase transition-like phenomenon in heart rate dynamics, these results provide a definitive step towards an understanding of heart rate complexity and establishing its basis as a critical phenomenon at the boundary of distinct phases of heart rate dynamics.

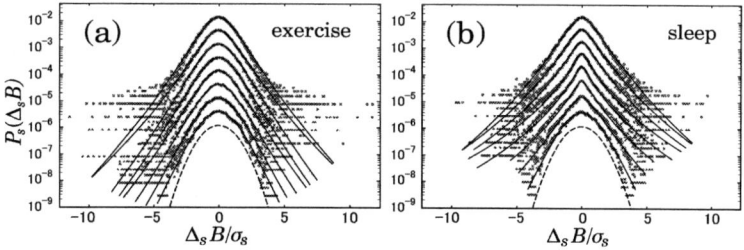

FIGURE 3. Standardized PDF's (in logarithmic scale) for different time scales, from 8 beats (top) to 1024 beats (bottom). The PDF's are estimated in: (a) Constant exercise; (b) Sleep. The dashed line is a Gaussian PDF for comparison - convergence to Gaussian can be observed for both conditions.

CONCLUSIONS

The two scenarios, that of criticality and that of the multifractal cascade, have been particularly successful as possible explanations of the underlying dynamics of heart rate and stock market regulation, respectively. A relevant question to ask is whether they are necessarily mutually exclusive. Perhaps not - they remain valid as the closest paradigms available, but a dramatic change in behavior has been discovered outside their respective applicability ranges, pointing to a more complex picture. As demonstrated by Kiyono et al. [7] (see also Togo et al. and Aoyagi et al. in this volume), the PDF scale invariance is not observed in sleep or strenuous activity, and is limited to daily activity records. This suggests that criticality-like characteristics may not be as omnipresent in heart rate as one would expect from works focusing on $1/f$-like scaling analysis. Interestingly, the cascading characteristics may also not be fully adequate to explain the market regulation - the criticality-like behavior has recently been revealed in the Standard & Poor 500 (S&P500) records during the dramatic crash of '87 [9]. Adequate answers are thus far from being found, and the search for models should likely move towards dynamical models with an explicit temporal axis. Surprisingly few such models, capable of displaying the invariant characteristics of the type discussed in this paper, are known. The first principles heart-rate model [10] representing the state of-the-art in modeling the neuro-regulatory mechanism in both health and neurogenic selective suppression inducing disease, remains to be verified with respect to PDF functional invariance. The recently proposed extension to the self-organized criticality model of Bak et al. by Kiyono et al. (in this volume) is at the other, parsimonious end of the specialization spectrum. This model of time dependent cascades may be particularly attractive, as it has been shown to display both PDF functional invariance criticality as well as phase-transition-like behavior. We believe that further study of the criticality and universality characteristics of heart rate will contribute to elucidating the mechanism of complex heart rate dynamics.

ACKNOWLEDGMENTS

This study was supported in part by the Japan Science and Technology Agency.

REFERENCES

1. K. Kiyono, Z. R. Struzik, N. Aoyagi, S. Sakata, J. Hayano, and Y. Yamamoto, *Phys. Rev. Lett.*, **93**, 178103 (2004).
2. Z. R. Struzik, J. Hayano, S. Sakata, S. Kwak, and Y. Yamamoto, *Phys. Rev. E*, **70**, 050901(R) (2004).
3. B. Castaing, Y. Gagne, and E. J. Hopfinger, *Physica D*, **46**, 177–200 (1990).
4. S. Ghashghaie, W. Breymann, J. Peinke, P. Talkner, and Y. Dodge, *Nature*, **381**, 767–770 (1996).
5. M. Kobayashi, and T. Musha, *IEEE Trans. Biomed. Eng.*, **BME-29**, 456–457 (1982).
6. P. Bak, C. Tang, and K. Wiesenfeld, *Phys. Rev. Lett.*, **59**, 381–384 (1987).
7. K. Kiyono, Z. Struzik, N. Aoyagi, F. Togo, and Y. Yamamoto, *to appear* (2005).
8. N. Aoyagi, K. Ohashi, and Y. Yamamoto, *Am. J. Physiol.*, **285**, R171–R176 (2003).
9. K. Kiyono, Z. Struzik, and Y. Yamamoto, *to appear* (2005).
10. K. Kotani, Z. Struzik, K. Takamasu, E. Stanley, and Y. Yamamoto, *to appear* (2005).

Fluctuations of the Single Photon Response in Visual Transduction

Lixin Shen*, Daniele Andreucci[†], Heidi E. Hamm** and Emmanuele DiBenedetto[‡]

*Biomathematics Study Group, Department of Pharmacology, Vanderbilt University Medical Center, Nashville, Tennessee 37232, USA
[†]Dipartimento di Metodi e Modelli Matematici, Università di Roma La Sapienza, 00161 Rome, Italy
**Department of Pharmacology, Vanderbilt University Medical Center, Nashville, Tennessee 37232, USA
[‡]Biomathematics Study Group, Department of Mathematics, Stevenson Center, Vanderbilt University, Nashville, Tennessee 37240, USA

Abstract. In phototransduction, there is experimental evidence of considerable fluctuations in the current as a response of the rod to single photon events. We propose that the location of the activation site is one of the components responsible for such a fluctuation. We have created a mathematical model of this system which is capable of detecting activation sites and tracing the dependence of the current profiles on such locations. Using such a mathematical and computational model, we have analyzed the variation in the global current drop, and then compared it to the variation of the existing experimental data. The overall conclusions are that the location of the activation site is, at least partially, responsible for the fluctuations of the current response.

Keywords: Fluctuation, Mathematical Model, Signal Transduction
PACS: 87.15.Ya

INTRODUCTION

Phototransduction is the process by which photons of light get converted into electrical pulses in the rod outer segments(ROS) or in cones, thus initiating vision[4]. In phototransduction, there is experimental evidence of considerable fluctuations in the current as a response of the rod to single photon events. Several reasons for this have been proposed, e.g., fluctuations in the shutting off of Rhodopsin, a G-protein coupled receptor protein initiating phototransduction. We propose that at least one phenomenon is one of the components responsible for such a fluctuation, e.g., the location of the activation site. To our knowledge it is experimentally impossible to control the precise activation site on the activated disc and, as a consequence, current drops cannot be directly related to such a geometrical location. It is however accepted and believed that the same phenomenon with the same parameters, but with different activation sites, yields different current drop profiles.

We have created a mathematical model of this system which is capable of detecting activation sites and tracing the dependence of the current profiles on such locations [1] [2]. Using such a mathematical and computational model, we have analyzed the variation in the global current drop, and then compared it to the variation of the existing experimental data. The overall conclusions are that the location of the activation site is, at least partially, responsible for the fluctuations of the current response.

ANALYSIS OF THE EXPERIMENTAL DATA

Here and below, we refer to the experimental data of Rieke for the Salamander ROS, whose parameters are shown in Table 1 in [1]. The global circulating current in the ROS were recorded. Denote by $J_{exp,i}$, the global current relative to the i-th experiment, which we understand as a piecewise constant function of time. The experimental raw data have been normalized by the dark current J_{dark}. Set

$$f_{exp,i}(t) = 1 - \frac{J_{exp,i}(t)}{J_{dark}}, \quad \text{and its average} \quad f_{exp,av}(t) = \frac{1}{N}\sum_{i=1}^{N} f_{exp,i}(t). \quad (1)$$

The variance of these data, at each time $0 < t < T$, is

$$\sigma_{exp}^2(t) = \frac{1}{N}\sum_{i=1}^{N}[f_{exp,i}(t) - f_{exp,av}(t)]^2, \quad 0 < t < T. \quad (2)$$

There are several ways to quantify the variation of the experimental data over the time interval (T_1, T_2) for $T_1 = 0$ and $T_2 > T_1$. We have chosen the L^2-norm criteria

$$V_{exp} = \left(\frac{\frac{1}{T_2-T_1}\int_{T_1}^{T_2}\sigma_{exp}^2(t)dt}{\frac{1}{T_2-T_1}\int_{T_1}^{T_2}f_{exp,av}^2(t)dt}\right)^{1/2} = \left(\frac{\int_{T_1}^{T_2}\sigma_{exp}^2(t)dt}{\int_{T_1}^{T_2}f_{exp,av}^2(t)dt}\right)^{1/2}. \quad (3)$$

All these integrals reduce to finite sums, since the integrands are piecewise constant functions. Using the experimental data of Rieke [2], we compute $V_{exp} = 0.268977$.

NUMERICAL SIMULATIONS

The data arising from numerical simulations have been dealt with essentially in the same way, after a preliminary calibration of the constants against the averaged response $f_{exp,av}(t)$. Let R be the radius of the disc in the ROS, for the Salamander about $5.5\mu m$. We generated $(m+1)$ simulations corresponding to the activation sites located at

$$r = r_n = \frac{n}{m}R, \quad n = 0, 1, \ldots, m, \quad \text{at the activation disc.} \quad (4)$$

We computed the probabilistic average of these outputs by the formula

$$f_{av}(t) = \frac{2m}{(m+1)R^2}\sum_{n=1}^{m} f_n(t;r_n)r_n\frac{R}{m} \quad (5)$$

corresponding to an approximate discretized evaluation of the integral

$$\frac{2}{R^2} \int_0^R f(t;r) r\,dr. \tag{6}$$

By this procedure we use $(m+1)$ simulations to predict the expectation of a random response. The variance of such a random response is

$$\sigma^2(t) = \frac{2m}{(m+1)R^2} \sum_{n=1}^{m} [f_n(t;r_n) - f_{av}(t)]^2 r_n \frac{R}{m}. \tag{7}$$

Paralleling the same treatment of the experimental data we introduce the criteria of variation

$$V = \left(\frac{\frac{1}{T_2 - T_1} \int_{T_1}^{T_2} \sigma^2(t)\,dt}{\frac{1}{T_2 - T_1} \int_{T_1}^{T_2} f_{av}^2(t)\,dt} \right)^{1/2} = \left(\frac{\int_{T_1}^{T_2} \sigma^2(t)\,dt}{\int_{T_1}^{T_2} f_{av}^2(t)\,dt} \right)^{1/2}. \tag{8}$$

The simulated value for V is 0.057203.

COMPARISON OF EXPERIMENTAL AND SIMULATED VARIATION FOR THE ROS OF THE SALAMANDER

We see that the relative variation of the simulated response at the different activation sites is of the order of 5.7203%, whereas the corresponding variation in experimental data is 26.8977%. So it does appear that the variation generated by the simulation is about 20% of the experimental one. This is not surprising since the mathematical model and the relative simulations do not reflect experimental variabilities due to instrumentation, temperature, and controls and mostly Rhodopsin shut off. So, in conclusion, we believe that this can be used as evidence to argue that location of the activation site is a significant factor in the variability of the response, while perhaps not the only one. Figure 1(a) shows seven of experimental raw data against their average. Figure 1(b) exhibits the seven simulated responses for 7 different activation sites.

An interesting by-product of the analysis carried out in this note, is the conclusion that, at least in the range of parameters we considered, the mean value of the response essentially coincides with its value at $2R/3$. The expected value of the random variable

$$X = \text{distance of the activation site from the center of the disc.} \tag{9}$$

is computed from (uniform distribution is assumed)

$$\frac{1}{area(disc)} \int_{disc} X\,dx\,dy = \frac{2}{R^2} \int_0^R r(r\,dr) = \frac{2}{3}R. \tag{10}$$

Assume now a random variable $Y = F(X)$ is affine in X, i.e., $Y = aX + b$, with a and b constants. Then the expected value of Y is

$$\frac{1}{area(disc)} \int_{disc} F(X)\,dx\,dy = \frac{2}{R^2} \int_0^R (ar+b)(r\,dr) = \frac{2}{3}aR + b = F\left(\frac{2}{3}R\right). \tag{11}$$

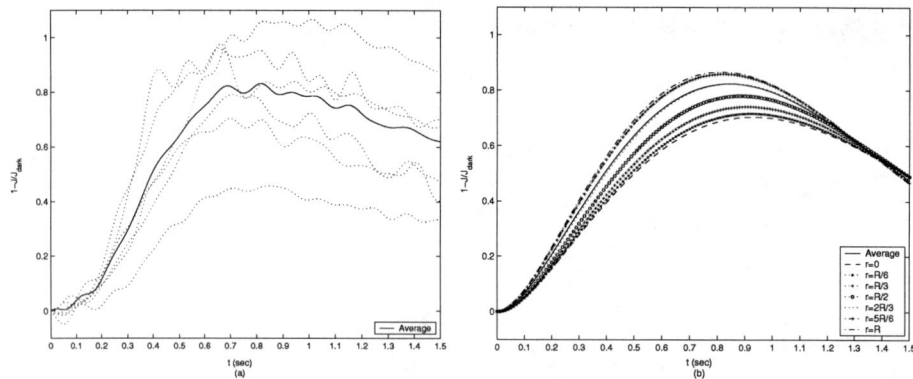

FIGURE 1. (a)Normalized response by experiment; (b)Normalized response by simulation

We assumed that the response of the rod to the photon event, attains its expected value at $r = 2R/3$. Actually, this turns out to be the case, at least for the numerically simulated response, and with good approximation, in the following sense. For a fixed set of parameters, we computed the responses for different locations of the activation site (as in (4) above), and averaged them (as in (5)). This is indicated by the quantity

$$\tilde{V} = \left(\frac{\int_{T_1}^{T_2} \left(f\left(t; \frac{2}{3}R\right) - f_{av}(t) \right)^2 dt}{\int_{T_1}^{T_2} f_{av}^2(t) dt} \right)^{1/2}, \quad (12)$$

whose value is 0.008862.

ACKNOWLEDGMENTS

The work was supported by NIH grant NIH-1-R01-GM 68953-01. We thank F. Rieke of the University of Washington, Seattle, USA, for sharing unpublished data.

REFERENCES

1. D. Andreucci, P. Bisegna, G. Caruso, H. E. Hamm, and E. DiBenedetto, *Biophysical Journal*, **85**(3), 1385–1376 (2003).
2. H. Khananl, V. Alexiades, G. Caruso, F. Rieke, H. E. Hamm, and E. DiBenedetto, preprint, (2004).
3. T. D. Lamb, and E. N. Pugh, *Journal of Physiology*, **449**, 719–758 (1992).
4. E. N. Pugh and T. D. Lamb, *Handbook of Biological Physics, Vol. 3, Chap. 5*, Elsevier Science, St. Louis, 2000, pp. 1–75.
5. H. Khanal, V. Alexiades, E. DiBenedetto and H. Hamm, "Numerical Simulations of Diffusion of Second Messengers $cGMP$ and Ca^{2+} in Rod Photoreceptors Outer Segment of Vertebrates," in *Unsolved Problems of Noise and Fluctuations* AIP Conference Proceedings 665, American Institute of Physics, Washington DC, 2002, pp. 165-172.

Spontaneous Movements of Mechanosensory Hair Bundles

Björn Nadrowski*, Pascal Martin[†] and Frank Jülicher*

*Max Planck Institute for the Physics of Complex Systems, Nöthnitzerstr. 38, 01187 Dresden, Germany
[†]Physicochimie Curie, Institut Curie, 26 rue d'Ulm, 75248 Paris Cedex 05, France

Abstract. The ear relies on nonlinear amplification to enhance its sensitivity and frequency selectivity. It has been suggested that this active process results from dynamical systems which oscillate spontaneously. In the bullfrog sacculus, hair bundles, which are the mechanosensitive elements of sensory hair cells display noisy oscillations. These oscillations can be described in a simple model which takes into account the properties of mechanosensitive ion channels coupled to motor proteins which are regulated by inflowing Ca^{2+} ions. The role of fluctuations can be studied by adding random forcing terms with characteristic amplitudes that result from the number and properties of ion channels and motor molecules. This description can account quantitatively for the experimentally measured linear and nonlinear response functions and reveals the relevance of fluctuations for signal detection.

Keywords: Hair cells, oscillations, hearing
PACS: 87.16.-b,87.19.Bb,05.40.-a

INTRODUCTION

Detecting sounds from the outside world imposes stringent demands on the design of the inner ear, where acoustic stimuli are transduced to electrical signals[1]. The vertebrate ear contains highly specialized cells called hair cells, which act as mechanosensors[2]. Each of these cells is responsive to a particular frequency component of the auditory input. In order to achieve high sensitivity and frequency selectivity, non-linear amplification is necessary. Because of the viscous damping at microscopic scales, the familiar resonant gain of a passive system is far from sufficient for the required demands[3].

The ear relies on active systems to achieve exquisite sensitivity and sharp frequency selectivity[3, 4, 5]. The most striking evidence for active behaviors in the ear are so-called otoacoustic emissions which are sounds emitted from the ears of mammals, birds and amphibians[6]. It has recently been proposed that the ear contains active dynamical systems which are close to an oscillating instability or Hopf-bifurcation [7, 8, 9]. This concept can explain many of the observed features of the ear, in particular, the nonlinearities that are generally observed at resonance conditions[10], the interference effects of multiple frequencies[11] as well as the occurrence of spontaneous emissions.

Although the cellular mechanisms that mediate this active process have remained elusive, *in vitro* [12, 13, 14] as well as *in vivo* [15] experiments have revealed that the mechanosensory organelle of the hair cell - the hair bundle - can generate active oscillatory movements that might underlie frequency-selective amplification. The observed noisy oscillatory system exhibits the signature of a system near a Hopf bifurcation: The

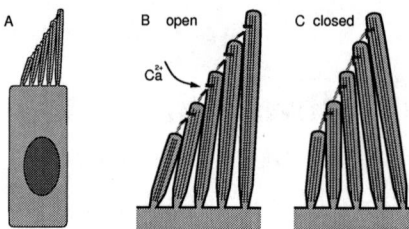

FIGURE 1. (A) Schematic representation of a hair cell with nucleus (grey) and a hair bundle of stereocila which is the sensitive element of mechanosensory hair cells. It consists of 30-300 rod-like stereocilia which have a length of 1-10 μm and are connected by fine filaments, called tip-links (green). The stereocilia contain mechanosensitive ion channels which open as a result of mechanical stimulation (blue). Myosin motor molecules control the tension in the tip-links (red). (B) open and (C) closed states of a hair bundle.

response to periodic stimuli was nonlinear for sufficiently large amplitudes near the oscillation frequency[16]. For small amplitudes there was a linear response regime which exhibited stable behaviors. Comparing the linear response to the autocorrelation function it was shown that the fluctuation dissipation theorem is violated, indicating the activity of the system[13].

The hair bundles of vertebrate hair cells consist of about 50 stereocilia which are stiff, rod-like extensions of the cell with a length of several micron and a diameter of about 300nm, see Fig. 1 [5]. The stereocilia merge at the tip and are grouped in a bundle. Fine filaments, so-called tip-links form bridges between neighboring stereocilia [17]. The micromechanical properties of hair bundles in living hair cells can be rich and range from adaptive movements in response to abrupt force steps with both fast and slow components[18], to spontaneous oscillations[5, 12, 16, 19].

Oscillatory instabilities of the hair bundle can be generated by at least three active mechanisms. First, a collection of molecular motors can generate spontaneous oscillations [20] and such motors occur in the hair bundle [21]. Second, the Ca-dependence of ion channel opening [18, 22, 23, 24] can give rise to oscillations [7, 25]. Finally, the interplay between negative hair-bundle stiffness and the Ca^{2+}-dependent activity of the adaptation motors can generate oscillations [26, 27]. This third mechanism provides the most convincing description of the hair-bundle oscillations observed in the bullfrog's sacculus [28].

ACTIVE HAIR BUNDLE MECHANICS

The dynamic behavior of the hair bundle can be described by three coupled equations:

$$\lambda \dot{X} = -K_{gs}(X - X_a - DP_o) - K_{sp}X + F_{ext} + \eta \quad , \tag{1}$$
$$\lambda_a \dot{X}_a = K_{gs}(X - X_a - DP_o) - \gamma N_a f p(C) + \eta_a \quad , \tag{2}$$
$$\tau \dot{C} = C_0 - C + C_M P_o + \delta c \quad . \tag{3}$$

Eq. **1** describes the dynamics of the hair-bundle position X subjected to an external force F_{ext}. The hair bundle is subjected to friction, characterized by the coefficient λ, as well as to the elastic forces $-K_{sp}X$ and $-K_{gs}Y$, where K_{sp} and K_{gs} are the stiffnesses of stereociliary pivots and of the tip-links, respectively, with $Y = X - X_a - DP_o$. Active hair-bundle movements result from forces exerted by a collection of N_a molecular motors within the hair bundle. The variable X_a can be interpreted as the position of the motor collection. At stall, these motors produce an average force $F_0 = N_a \gamma f p$ that is proportional to the force f generated by a single motor and to the probability p that a motor is bound to an actin filament, where $\gamma \simeq 1/7$ is a geometric projection factor. Because adaptation depends on the Ca^{2+} concentration C, we set $p = p_0 + p_1 C$. Active force production by the motors corresponds to motors climbing up the stereocilia, i.e. $dX_a/dt < 0$, which tends to increase the extension of the gating springs and to open transduction channels. In a two-state model for channel gating [29], the open probability can be written as

$$P_o = \frac{1}{1 + A e^{-(X - X_a)/\delta}}, \quad (4)$$

where $A = \exp([\Delta G + (K_{gs}D^2)/(2N)]/k_B T)$ accounts for the intrinsic energy difference ΔG between the open and the closed states of a transduction channel and $\delta = N k_B T / (K_{gs} D)$.

STATE DIAGRAM

To explore the dynamic behaviors of the system described by Eqns. **1-3**, we first ignore the effects of fluctuations and assume $F_{\text{ext}} = 0$. Linear stability analysis of these steady states reveals conditions for stability as well as for oscillating instabilities that lead to spontaneous oscillations via a Hopf bifurcation [30]. The state diagram is a function of two parameters: the maximal force $f_{\max} = N_a f p_0$ produced by adaptation motors along their axis of movement, and the dimensionless feedback strength S of the Ca^{2+} regulation.

The state diagram exhibits different regimes (Fig. 2). If the force $f_{\max}$ is small, the motors are not strong enough to pull transduction channels open. In this case, the system is monostable with most of the channels closed. Increasing $f_{\max}$ leads to channel opening. For intermediate forces and weak Ca^{2+} feedbacks, the system is bistable, i.e. open and closed channels coexist. For strong Ca^{2+} feedbacks, however, the motors can't sustain the forces required to maintain the channels open. Spontaneous oscillations occur in a region of both intermediate forces and feedback strengths. Note that there is no oscillation in the absence of Ca^{2+} feedback, i.e. for $S = 0$.

EFFECTS OF FLUCTUATIONS

Spontaneous hair-bundle oscillations are noisy [13]. Noise terms η, η_a and δc in Eqns. **1-3** formally take into account the effects of various sources of fluctuations that destroy the phase coherence of hair-bundle movements. The stochastic forces η and η_a act respectively on X and X_a. The consequences of these forces have been analyzed for

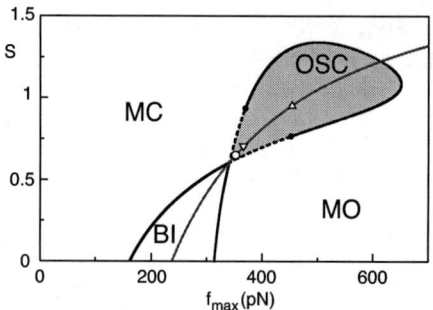

FIGURE 2. State diagram of the active hair bundle as a function of two control parameter, the maximal motor force f_{max} in the absence of Ca^{2+} and the strength S of the Ca-feedback. The diagram reveals several regions: MO monostable with open channels; MC monostable with closed channels; BI bistable and oscillations (OSC). Along the red line, the open probability of ion channels is $P_o = 0.5$.

non-oscillating hair bundles [32]. The fluctuations δc of the Ca^{2+} concentration in the stereocilia result from stochastic transitions between open and closed states of the transduction channels [33].

Assuming that the motors are deactivated ($f = 0$), we first discuss thermal contributions to the noise. The noise term η in Eq. **1** then results from brownian motion of fluid molecules which collide with the hair bundle and from thermal transitions between open and closed states of the transduction channels. By changing the gating-spring extension, this channel clatter generates fluctuating forces on the stereocilia. The fluctuation-dissipation theorem implies that $<\eta(t)\eta(0)>= 2k_B T \lambda \delta(t)$. The friction coefficient $\lambda = \lambda_h + \lambda_c$ results from two contributions: $\lambda_h \simeq 1.3 \; 10^{-7} \text{N·s·m}^{-1}$ accounts for hydrodynamic friction, which depends on bundle geometry and fluid viscosity [34, 18], whereas λ_c results from channel clatter. The contribution λ_c can be estimated from the autocorrelation function of the force η_c that results from stochastic opening and closing of N transduction channels

$$<\eta_c(t)\eta_c(0)> \simeq D^2 K_{gs}^2 P_o(1-P_o)N^{-1} e^{-|t|/\tau_c} \simeq 2D^2 K_{gs}^2 P_o(1-P_o)N^{-1}\tau_c\delta(t) \quad . \quad (5)$$

Assuming that $<\eta_c(t)\eta_c(0)> \simeq 2k_B T \lambda_c \delta(t)$, we define a hair bundle friction λ_c which is associated to channel opening and closing. Using Eq. 5, we estimate

$$\lambda_c \simeq \frac{K_{gs}^2 D^2 P_o(1-P_o)\tau_c}{Nk_B T} \quad . \quad (6)$$

Using typical parameter values our estimate reveals that channel clatter dominates friction and $\lambda \simeq 3 \; 10^{-6} \text{N·s·m}^{-1}$.

The noise strength resulting from stochastic motor action can also be estimated [28]. This noise strength can be described by introducing an effective temperature T_m defined by $<\eta_m(t)\eta_m(0)> \simeq 2k_B T_m \lambda_a \delta(t)$. With $f \simeq 1$pN, $\tau_a \simeq 10$ms and $p \simeq 0.05$, we find $T_m/T \simeq N_a \gamma^2 p(1-p) f^2 \tau_a/(k_B T \lambda_a) \simeq 0.5$. Writing $<\eta_a(t)\eta_a(0)> = 2k_B T_a \lambda_a \delta(t)$, we thus get $T_a \simeq 1.5T$. Finally, we can show that fluctuations in the Ca-concentration can be neglected [28].

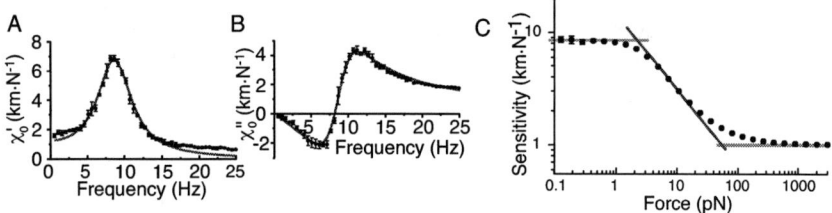

FIGURE 3. Response functions calculated from numerical simulation of the model equations with noise terms in presence of a periodic stimulus force. (A) Real part χ' of the linear response function. (B) Imaginary part χ'' of the linear response. (C) nonlinear response function at fixed frequency 8Hz.

The calculated linear response function χ_0 as a function of frequency agrees quantitatively with the experimental observations [13]. At the characteristic frequency of the spontaneous oscillations, the sensitivity $|\chi|$ of the system to mechanical stimulation exhibits the three regimes observed experimentally [14] as a function of the stimulus amplitude $|F_1|$ (Fig. 3C): a linear regime of maximal sensitivity $|\hat{\chi}_0| = 8.5 \text{km} \cdot \text{N}^{-1}$ at $\omega = \omega_0$ for small stimuli, a compressive nonlinearity for intermediate stimuli and a linear behavior of low sensitivity for large stimuli. The maximal sensitivity as well as the breadth of the nonlinear region are in quantitative agreement with experiments. An important parameter that influenced the system's maximal sensitivity is the stiffness of the load to which the hair bundle is coupled. For $f_{\max} \simeq 352\text{pN}$, power spectra of spontaneous oscillations and response functions were not significantly affected by varying P_o in the range 0.2-0.8. Agreement between simulations and experiments thus did not qualify a particular value of P_o.

DISCUSSION

We have presented a physical description of active hair-bundle motility that emphasizes the role played by fluctuations. The mechanical properties of oscillatory hair bundles in the presence of a periodic stimulus force can be described quantitatively only if fluctuations are taken into account. Fluctuations arise in part from brownian motion of fluid molecules and from the stochastic gating of transduction channels. By consuming energy, the motors power frequency-selective amplification but also generate non-thermal fluctuations that add to the inevitable thermal fluctuations. We find, however, that the magnitude of fluctuations due to active processes remain below the level of thermal noise.

The ability of a single hair bundle to detect oscillatory stimuli using critical oscillations is limited by fluctuations which conceal the critical point. This limitation could be overcome if an ensemble of hair cells with similar characteristic frequencies were mechanically coupled. Coupled noisy oscillators could approach the ideal case of a critical oscillator near a Hopf bifurcation. In an intact mammalian cochlea, the gain that characterizes amplification of basilar-membrane motion is up to 10^3 [10], which can be compared to a gain of only about 10 for a single hair bundle in the bullfrog's sacculus.

This suggests that in the cochlea the effects of fluctuations of individual hair cells could be reduced by the cooperative action of many oscillatory cells, whether the oscillations are provided by active hair-bundle motility or by a different mechanism.

ACKNOWLEDGMENTS

We thank Thomas Duke, Martin Göpfert, Jim Hudspeth, and Jaques Prost for stimulating discussions.

REFERENCES

1. Hudspeth, A.J.: Nature (London) **341**, 397-404 (1989).
2. Dallos, P., Popper A.N., and Fay R.R., *The cochlea* (Springer, New York) (1996).
3. Gold, T.: Proc. R. Soc. London Ser B **135** (1948).
4. Dallos, P.: J. Neurosci. **12**, 4575-4585 (1992).
5. Hudspeth A.J.: Curr. Opin. Neurobiol. **7**, 480-486 (1997).
6. Probst, R.: Adv. Otorhinolaryngol. **44**, 1-91 (1990).
7. Choe, Y., Magnasco, M., and Hudspeth A.J.: Proc. Natl. Acad. Sci. **95**, 15321-15326 (1998).
8. Camalet, S., Duke, T., Jülicher F., and Prost J.: Proc. Natl. Acad. Sci. **97**, 3183-3188 (2000).
9. Eguiluz, V.M., Ospeck, M., Choe Y., Hudspeth. A.J., and Magnasco, M.: Phys. Rev. Lett **84**, 5232-5235 (2000).
10. Ruggero, M.A., Rich, N.C., Recio, A., Narayan, S.S., and Robles, L.: J. Acoust. Soc. Am. **101**, 2151-2163 (1997).
11. Jülicher, F., Andor, D., and Duke, T.: Proc. Natl. Acad. Sci. **98**, 9080-9085 (2001).
12. Martin, P., and Hudspeth, A.J.: Proc. Natl. Acad. Sci. **96**, 14306-14311 (1999).
13. Martin, P., and Hudspeth A.J., and F. Jülicher: Proc. Natl. Acad. Sci. **98**, 14380-14385 (2001).
14. Martin, P., and Hudspeth, A. J.: Proc. Natl. Acad. Sci. **98**, 14386–14391 (2001).
15. Manley, G. A., Kirk, D. L., Koppl, C., and Yates, G. K.: Proc. Natl. Acad. Sci. **98**, 2826–2831 (2001).
16. Martin, P., and Hudspeth A.J.: Proc. Natl. Acad. Sci. **98**, 14386–14391 (2001).
17. Kachar, B., Parakkal, M., Kurc, M., Zhao, Y., and Gillespie, P.G. : Proc. Natl. Acad. Sci. **97**, 13336-13341 (2000).
18. Howard, J., and Hudspeth, A.J.: Neuron **1**, 189-199 (1988).
19. Fettiplace., R., Ricci, A.J., and Hackney, C.M.: Trends in Neurosci. **24**, 169-175 (2001).
20. Jülicher, F., and Prost, J.: Phys. Rev. Lett. **75**, 2618–2621 (1995).
21. Hudspeth, A. J., and Gillespie, P. G.: Neuron **12**, 1–9 (1994).
22. Benser, M. E., Marquis, R. E., and Hudspeth, A. J.: J. Neurosci. **16**, 5629–5643 (1996).
23. Wu, Y. C., Ricci, A. J., and Fettiplace, R.: J. Neurophysiol. **82**, 2171–2181 (1999).
24. Ricci, A. J., Crawford, A. C., and Fettiplace, R.: J. Neurosci. **20**, 7131–7142 (2000).
25. Vilfan, A., and Duke, T.: Biophys. J. **85**, 191–203 (2003).
26. Martin, P., Mehta, A., and Hudspeth, A.J.: Proc. Natl. Acad. Sci. **97**, 12026-12031 (2000).
27. Martin, P., Bozovic, D., Choe, Y., and Hudspeth, A. J.: J. Neurosci. **23**, 4533–4548 (2003).
28. Nadrowski, B., Martin, P., Jülicher, F.: Proc. Natl. Acad. Sci. **101**, 12195–12200 (2004).
29. Markin, V. S., and Hudspeth, A. J.: Annu. Rev. Biophys. Biomol. Struct. **24**, 59–83 (1995).
30. Strogatz, S. T.: (1997) *Nonlinear Dynamics and Chaos*. (Addison-Wesley, Reading, MA), 7th edition.
31. Lumpkin, E. A., and Hudspeth, A. J.: J. Neurosci. **18**, 6300–6318 (1998).
32. Frank, J. E., Markin, V., and Jaramillo, F.: Biophys. J. **83**, 3188–201 (2002).
33. van Netten, S. M., Dinklo, T., Marcotti, W., and Kros, C. J.: Proc. Natl. Acad. Sci. **100**, 15510–15515 (2003).
34. Denk, W., Webb, W. W., and Hudspeth, A. J. Proc. Natl. Acad. Sci. **86**, 5371–5375 (1989).
35. Hacohen, N., Assad, J. A., Smith, W. J., and Corey, D. P.: J. Neurosci. **9**, 3988–3997 (1989).
36. Duke, T., and Jülicher, F.: Phys. Rev. Lett. **90**, 158101 (2003).

Brownian dynamics simulations of ionic current through an open channel

R. Tindjong*, R.S. Eisenberg[†], I. Kaufman**, D. G. Luchinsky* and P.V.E. McClintock*

*Department of Physics, Lancaster University, Lancaster LA1 4YB, UK.
[†]Department of Molecular Biophysics and Physiology, Rush Medical college,
1750 West Harrison, Chicago, IL 60612, USA.
**VNII for Metrological Service, Gosstandart, Moscow, 119361, Russia.

Abstract. Ionic motion through an open ion channel is analyzed within the framework of self-consistent Brownian dynamics. A novel conceptual model in which the ions motion is coupled to the vibrations of the pore walls is introduced. The model allows to include into simulations an important additional mechanism of energy dissipation and the effects of self-induced strong modulation of the channel conductivity.

Keywords: ion channels, Poisson equation, Langevin equation, self-consistent approach
PACS: 87.16.Uv

INTRODUCTION

Ion channels in the cell membrane are natural nanotubes that control a vast range of physiological activity [1]. Understanding their structure-function relationship is the main challenge. Much progress has been made in that direction in recent years. Theories ranging from Poisson Nernst Planck (PNP) theory , Brownian dynamics (BD) simulations and Molecular Dynamics (MD) have been developed, each of them presenting its own particular strengths and weaknesses [2]. In our model, we use a one dimensional approximation of the ion dynamics inside the channel. To this end, we neglect the off-axis motion of ions in the channel to be able to include in an effective way interaction with the channel wall. One of the goals of the simulations is to infer the diffusion coefficient for ionic motion in the channel, rather than assuming it to be known. To this end we therefore integrate out fast degrees of freedom related to the ion motion perpendicular to the channel axis. This effectively reduces the problem to that of one-dimensional motion along the reaction coordinate, which is in our case the channel axis. And the reaction is the transition of ion from the left to the right (or vice versa). This approach is standard in chemical kinetics, allowing self-consistent treatment of ionic diffusion in an open protein pore at the level of the Brownian motion approximation. Our approach allows us to introduce, in addition another very important degree of freedom related to the wall motion. The latter is a source of both self-induced enhancement of transition probabilities and a real source of dissipation.

MATERIALS AND METHODS

We start our analysis by considering the Langevin equation for the i-th ion when the channel walls are fixed, which takes the form:

$$m_i \dot{\mathbf{v}}_i = -m_i \gamma_i \mathbf{v}_i + \mathbf{f}_i + \sqrt{2\gamma_i m_i k_B T}\, \xi_i(t), \qquad (1)$$

where m_i and $\mathbf{v}_i$ are the mass and the velocity of the ion. The friction force is proportional to $\mathbf{v}_i$ with a coefficient of proportionality $m_i \gamma_i$. The equation for a j-th different type of ion is written in the same way, but with different mass and friction coefficient. The force acting upon the ion has a number of components

$$\mathbf{f}_i = \mathbf{f}_D + \mathbf{f}_E + \mathbf{f}_R, \qquad (2)$$

where $\mathbf{f}_D$ is the dielectric force due to the channel walls, fixed charge, and external applied potential, $\mathbf{f}_E$ is the electrostatic force due to the Coulomb interaction between moving ions, finally $\mathbf{f}_R$ is a short range repulsive force. We accommodate the short range repulsive potential through the following expression [3]

$$U_{sr}(r) = \frac{F_0}{9} \frac{(r_1 + r_2)^{10}}{r^9}, \qquad (3)$$

where r_1 and r_2 are the Pauling radii of the ions, r is the ion-ion distance, and F_0 is a short-range force constant taken to be $F_0 = 2 \times 10^{-10} N$.

Eqs. (3) can be integrated simultaneously in the bulk solution. A snapshot of ions near the channel mouth is shown in Fig.1

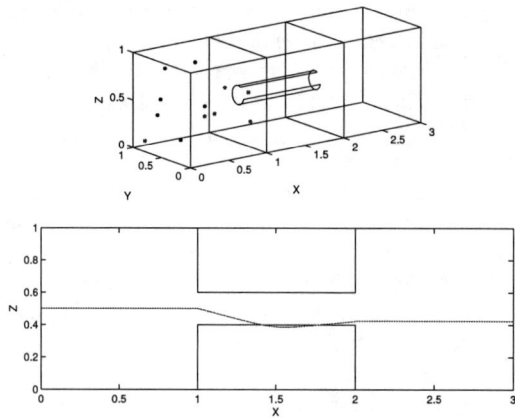

FIGURE 1. A snapshot of ions in the bulk solution obtained by simultaneous solution of Eqs. (1).

The line in the bottom figure is a representation of the one-dimensional approximation of the potential across the channel. The shift between the two ends is the potential drop due to the applied voltage. As soon as an ion enters the channel we assume that its motion is one dimensional along the axis of the channel. In particular this means that the multiple interactions of the ion with the wall during its diffusion within the channel are included in the unknown diffusion coefficient, which depends in turn on the position of the ion in the channel. We note that such an approximation is traditional in chemical physics where reaction in a system with many degrees of freedom is often considered as a one-dimensional dynamics along the so-called "reaction coordinate". In the present case, reaction is the transition of an ion from the left volume to the right (or vice versa), and the reaction coordinate is the x-coordinate along the channel axis.

For the ion in the channel the main driving force is f_D. Since we have assumed symmetry of the system about the channel axis, we calculate the corresponding force by integrating the Poisson equation using the Finite Volume Method in the system shown in the Fig.1, with only the ions within the channel taken into account.

In the bulk reservoir, we consider a symmetric concentrations of $NaCl$. We used a stochastic boundary condition. Each time that an ion cross the channel, it is re-injected back into a random location with zero velocity. This way, we maintain fixed the specified concentrations in the reservoirs. The transmembrane potential which is generated by the applied electric potential across the membrane far away from the channel is explicitly taken into account in the solution of Poisson's equation.

RESULTS AND DISCUSSION

Before using the Poisson's equation solution, we make sure that the latter is sufficiently robust by comparing the numerical potential with a simple electrostatic problem. For example, the program reproduces accurately the potential created by one charge in a space divided in two with two different dielectric coefficients. After this check, we feel confident in using the numerical simulation of the Poisson equation to estimate the energy faced by an ion entering the channel. A Na^+ entering the channel is trapped in the energy well created by a negatively charged ring located at the middle of the channel, as can be seen on Fig.2

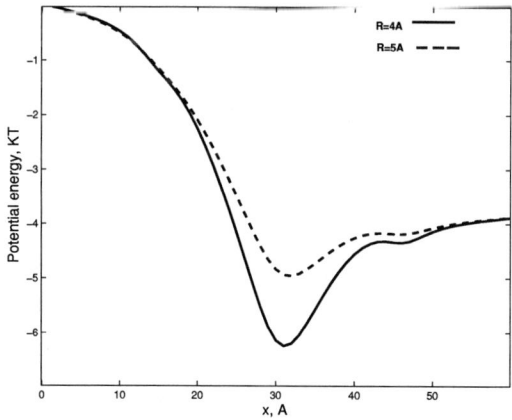

FIGURE 2. Profile of the Energy faced by a N_a^+ ion.

For each applied potential, the Langevin equation is solved. The trajectories of ions traversing the channel are tracked. By counting all the transitions of ions crossing the channel during a given period of time, the current can be calculated for each applied voltage. The current-voltage characteristic of the channel can thus be obtained. This can then be compared with experiments. The position-dependent diffusion coefficient of the ion as it moves across the channel has also to be calculated. As can be seen in Fig.2, the variation of the channel radius due to its interaction with the ion moving in the channel has a huge effect on the potential energy. The ion will face a smaller barrier for larger channel radius. The equations of motion for the moving wall and for the single ion in the channel can be written as:

$$m_i \ddot{x}_i = -m_i \gamma_i \dot{x}_i + \mathbf{f}_i + \frac{q_1 q_2 |(x-L/2)|}{4\pi\varepsilon\varepsilon_0 \tilde{R}^3} + \sqrt{2\gamma_i m_i k_B T}\, \xi_i(t), \quad (4)$$

$$M\ddot{\delta\tilde{r}} + M\tilde{\Gamma}\dot{\delta\tilde{r}} + k\delta\tilde{r} = -\frac{q_1 q_2 \delta\tilde{r}}{4\pi\varepsilon\varepsilon_0 \tilde{R}^3} + \sqrt{2M\tilde{\Gamma} k_B T}\, \tilde{\eta}(\tilde{t}), \quad (5)$$

Here $\tilde{R} = \sqrt{(x-L/2)^2 + \delta\tilde{r}^2}$ is the distance between moving ion on the channel axis and charged vibrating segment of the wall, and q_1 and q_2 are respectively the charge of the moving ion inside the channel and the charged ring located at the middle of the channel.

CONCLUDING REMARKS

We have presented a self-consistent method for determination of the current-voltage characteristic of the ion channel. So far, we have successfully calculated the energy barrier faced by a Na^+ ion traversing the channel. Studies of the bi-ionic solution and the effect of the flexible protein wall, as well as the space dependant diffusion coefficient, are in progress. As results, we expect the coupling of ionic motion with vibration of the wall to introduce energy dissipation and self-induced modulation of the transition probability through the ion channel.

ACKNOWLEDGMENTS

The work was supported by the Engineering and Physical Sciences Research Council (UK), The Russian Foundation for Basic Research, INTAS, and ESF.

REFERENCES

1. B. Hille, *Ionic Channel Of Excitable Membranes*, Sinauer Associates, Sunderland, MA, 1992.
2. Gennady V. Miloshesvky and Peter C. Jordan, "Permeation in ion channels: the interplay of structure and theory " *TRENDS in Meurosciences* **27** (6),pp. 308–314, 2004
3. G. Moy, B. Corry, S. Kuyucak, and S.-H. Chung, "Tests of continuum theories as models of ion channels. I. Poisson-Boltzmann theory versus Brownian dynamics," *Biophys. J.* **78**(5), pp. 2349–2363, 2000.

Stochastic Resonance of Artificial Ion Channels inserted in Small Membrane Patches

Robert H. Blick, Si-Young Choi, Hyun S. Kim, Sujatha Ramachandran, and Daniel W. van der Weide

University of Wisconsin-Madison, Laboratory for Molecular Scale Engineering, Electrical & Computer Engineering, 1415 Engineering Drive, Madison, WI 53705, USA.

Abstract. We investigate stochastic resonance in artificial ion channels inserted in bilipid membranes with varying membrane diameters. Stochastic resonance (SR) is fundamental for sensory reception of organisms, cell systems, single cells and bundles of ion channels. This indicates the importance of SR for applications in systems, such as nano scale single molecular biosensors. For Alamethicin, a standard ion channel model system, SR was found for large membrane patches. It was suggested that noise stemming from individual ion channels may affect collective properties of the whole ensemble. From the data obtained we find that a finite membrane size is required for SR.

Keywords: Ion channels, stochastic resonance, micromachining
PACS: 87.80.Jg;87.80.Mj;87.80.Tq

INTRODUCTION

Stochastic resonance (SR) addresses the fact that the addition of noise to a weak signal can improve the overall sensitivity of the system. This is of particular importance in biology when it comes to cell networks and ion channels embedded in cell membranes [1]. The fundamental question arising is, if cellular systems and in particular single ion channels as the fundamental gating mechanism of cells are inherently contributing to the generation of noise. Hence, we are aiming at clarifying whether SR is a cooperative phenomenon or can be traced down to single ion channels. In the case of the peptide Alamethicin – an artificial ion channel – SR was demonstrated for large membranes [2]. In recent theoretical work it was suggested that noise stemming from single ion channels will contribute to the signal transduction efficiency of cellular networks [3]. Hence, it was concluded that the membrane patch size in which ion channels are inserted plays a crucial role.

MOTIVATION

Here, we apply the on-chip patch clamping technique [4] to study SR on Alamethicin inserted in membrane patches of different diameters. In contrast to earlier work where large so-called black lipid films (membranes) were used [2], we use glass chips with drilled pores of different sizes. The typical glass chip has a size of 20 x 20 mm^2 and a thickness 150 μm. Apertures are drilled with openings ranging between 50 to 200 μm. The glass itself is hydrophilic and was pretreated by a mixture of dichloro-dimethyl-silane to change the hydrophilic characteristics around the orifice. This strongly supports formation of the bilayer.

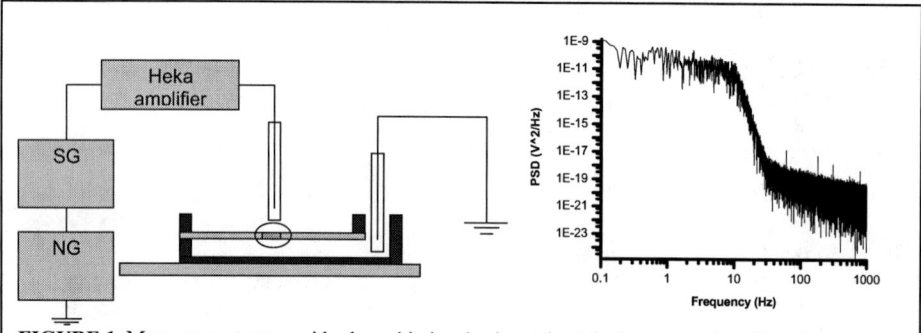

FIGURE 1. Measurement setup with glass chip in a horizontal patch clamp chamber. Signal and noise generators (SG & NG) connecting to the amplifier. The Ag/AgCl-electrodes are dipped into the solution. Right panel gives the noise output power of the NG.

For the measurements we designed a low-frequency noise source and built two different setups, a lateral (see Fig. 1) and a vertical (not shown) patch clamp chamber. The noise generator signal has a white noise spectrum in the range 0 – 10 Hz, which is superimposed onto the sub-threshold small sine wave input signal and then applied to the Alamethicin channels. For the current recordings we made use of an Axopatch 200B and a HEKA EPC10 amplifier. The channel currents were filtered at 1 kHz by a 4-pole Bessel filter and recorded by an A/D-D/A converter.

The setup is shown in Fig. 1: the chamber is separated into two compartments, filled with solutions of 1M KCl and 10mM HEPES. The trans-membrane current is measured using Ag/AgCl electrodes immersed into a PE tube containing 1% agar and 0.5 M KCl. A sine wave voltage from an HP 3325B signal generator (SG) with the noise generator (NG) added, was finally connected to the external trigger input of the

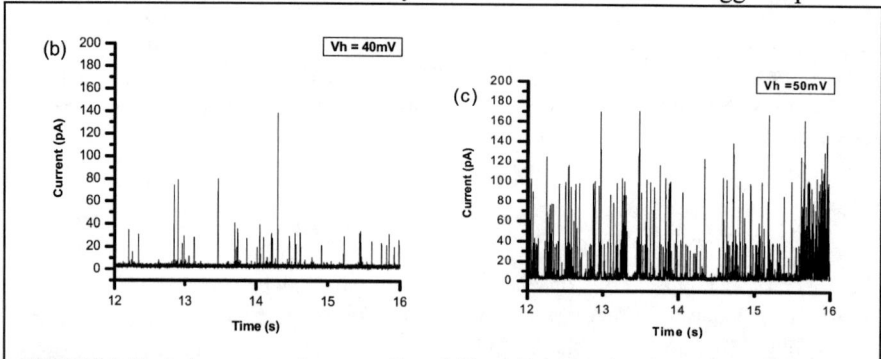

FIGURE 2. Typical current vs. time recording of Alamethicin forming channels in a bilipid membrane. Left panel shows channel insertion at a holding potential of 40 mV, while the rhs panel is taken at 60 mV. The discrete nature of the current steps can be identified.

Axopatch 200B or HEKA EPC10. Great care was taken in building the noise generator with an output characteristic shown in the right panel of Fig. 1. For these experiments the 3dB corner was designed to be at 10 Hz. A characteristic measurement is shown in Fig. 2 where a current vs. time trace is represented. The current spikes indicate membrane openings by the peptide. Depending on the number of Alamethicin peptides forming a channel the current amplitude can be as large as 200 pA. Applying a bias potential inserts the rod-like peptides into the membrane, due to their dipole moment of about 70 Debye. As seen in the base line a certain noise level is already present. Electronic background noise and microphonic noise can be suppressed by placing the setup in a Faraday cage. The recordings shown were taken at a bias voltage (holding potential) of 40 and 60 mV. All measurements were performed for a number of holding potentials from 20 to 60 mV with a 4 mV r.m.s. sine signal at 0.5 Hz.

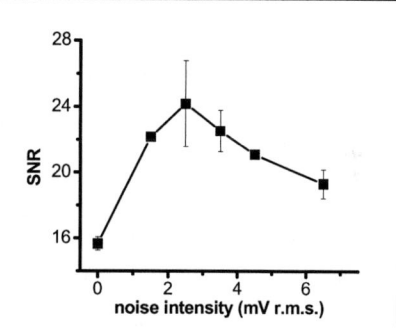

FIGURE 3. Stochastic resonance signal-to-noise ratio (SNR) vs. noise intensity for a membrane patch of 200 μm diameter.

RESULTS

We observe standard SR where the output signal-to-noise ratio (SNR) increases at an optimum level of input noise and ion channel gating-voltage as shown in Fig. 3 & 4. The signal-to-noise ratio (SNR) is commonly used to quantify SR. Fig. 3 & 4 show the SR signature for which the SNR increases with enhanced noise intensity until it reaches a certain threshold after which it rolls off. The straight forward approach to measure SNR is to determine the power spectrum density. The SNR is then quantified as the signal peak maximum compared to the background noise averaged over a certain bandwidth according to

$$SNR = \frac{[S(fs) - N(fs + \Delta f)]}{dN(fs \pm \Delta f)}. \quad (1)$$

Fig. 3 shows the SNR indicating SR for a 200 μm membrane, while Fig. 4 shows SR for a 125 micron membrane. In further measurements we found SR also in 75 μm membrane patches, but not below this value. We found that the optimum value of membrane diameters lies around 125 μm for measurements with the horizontal and vertical setups. This leads us to the assumption that noise is inherently generated by membrane fluctuations in conjunction with bundles of interacting Alamethicin peptides. The results are summarized in Table 1 shown below. We varied membrane patch diameters as well as holding potential strengths and Alamethicin concentration. For all membranes except below 75 μm SR occurs. Membranes with diameters around 50 μm could still be formed, whereas 25 μm we could not realize with the current setup.

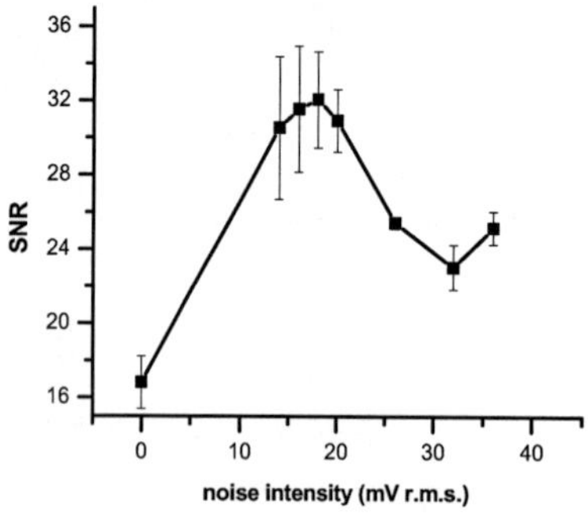

FIGURE 4. Stochastic resonance signal-to-noise ratio (SNR) vs. noise intensity for a membrane patch of 125 micron diameter.

SUMMARY

In summary we have validated the occurrence of SR for Alamethicin embedded in bilipid membranes. We have shown that an on-chip patch clamp setup can be successfully applied to probe SR for different membrane patch diameters. Towards smaller membrane diameters we find that SR is not pronounced below 75 µm membrane diameters. We can conclude that this is a strong indication for SR being induced by the collective interaction of ion channels with membrane patches of finite size.

TABLE 1. SR for a number of membrane diameters.

Membrane patch diameter (microns)	SR	Alamethicin concentration (µM/ml)	Capacitance (pF)	Holding potential (-mV)
200	Yes	1 and 50	110 and 140	10 and 10
125	Yes	0.5, 0.5, 0.5, 3.5	85, 85, 85, 100	30, 40, 50, 40
75	Yes	0.5	60	35
50	No	--	--	--

ACKNOWLEDGMENTS

The authors like to thank for support from DARPA within the MOLDICE program. RHB likes to thank Meyer Jackson for detailed discussion.

REFERENCES

[1] Peter Hanggi, ChemPhysChem **3**, 285 (2002).
[2] M. Bezrukov and I. Vodyanoy, Nature **378**, 362 (1995).
[3] G. Schmid, I. Goychuk and P. Hanggi, Europhysics Letters **56**, 22 (2001).
[4] N. Fertig, R.H. Blick, and J.C. Behrends, Biophysical Journal **82**, 3056 (2002).
[5] S.M. Bezrukov and I. Vodyanoy, Biophyscial Journal **73**, 2456 (1997).

Starting and Stopping a Bistable Pacemaker: Stochastic Stimulation Identifies Critical Perturbations

David Paydarfar[1,4], Daniel B. Forger[2,4], and John R. Clay[3,4]

[1]*University of Massachusetts Medical School, Worcester, MA, USA (david.paydarfar@umassmed.edu);*
[2]*New York University, NY, USA (forgerd@courant.nyu); [3]National Institutes of Health, Bethesda MD, USA (jrclay@helix.nih.gov); [4]Marine Biological Laboratory, Woods Hole, MA, USA*

Abstract. Bistable pacemakers exhibit highly nonlinear properties, such as abrupt transitions between repetitive firing and quiescence in response to small perturbing stimuli. We describe a search method for estimating optimum stimulus shapes and intensities for starting or stopping the repetitive firing of a bistable pacemaker. A large library of randomly generated stimuli is used to perturb the pacemaker, and a library of responses is recorded. From these two libraries, a rank order of desirability of the stimulus is generated to arrive at an estimate of the optimum stimulus shape. The search method was validated by calculus of variations applied to the Bonhoeffer-van der Pol (Fitzhugh-Nagumo) model of a bistable pacemaker. We found that the optimum stimulus for inducing a switch from one stable attractor to the other is a critically timed oscillatory stimulus. While the optimum stimulus shape for stopping the oscillator is similar to that for starting the oscillator, they differ in that stopping the oscillator requires that the stimulus is in antiphase to the natural rhythm, while the optimum stimulus for starting the oscillator is in phase with the natural rhythm. These theoretical predictions can be tested in real biological pacemakers, such as a recently described squid giant axon preparation that exhibits membrane bistability. Elucidation of optimum stimulus shapes may be useful for studying many periodic phenomena in biology and medicine. Our findings also suggest a novel approach to understanding how bistable membranes encode information over long time scales using fast noisy transients.

Keywords: neuron, oscillator, noise, optimum, action potential, information processing
PACS: 87.19.La, 87.17.Nn, 87.80.Vt, 87.19.Hh, 87.19.Dd

INTRODUCTION

Bistable pacemakers exhibits two stable behaviors for the same set of experimental conditions, for example stable repetitive firing and stable quiescence. A brief duration current pulse can switch the preparation from quiescence to repetitive firing. Moreover, a similar pulse, appropriately timed, can annihilate the repetitive firing thereby resulting in quiescence. Switching between these two states may be an important control mechanism which governs a variety of normal and pathological behaviors involving excitable cells [1]. We describe a search method for estimating optimum stimulus shapes and intensities for starting or stopping the repetitive firing of a bistable pacemaker.

NOISE-BASED SEARCH METHOD

Many mathematical tools have been developed that can be applied towards finding optimum stimuli for inducing an oscillator to achieve a certain behavior [2]. While such methods can yield exact solutions, they have limited applicability in biology because they require a mathematical description of the oscillator. Another way to find optimum stimuli would be to try every possible stimulus, and choose the best among them. The mean of the multiple stimuli which precede these events could be an optimum stimulus, but only if the system is linear [3]. We recently proposed a noise-based method that can be used to find optimum stimulus shapes for starting, stopping, or resetting nonlinear biological oscillators [2]. The method is summarized as follows:

1. <u>Choose stimuli that are biologically plausible</u>. Consider the spectrum of stimuli the system naturally responds to. For example, most perturbations that a neuron receives are combinations of post-synaptic potentials (PSPs). The rise and decay time constants and statistical properties of these perturbations may be estimated from experimental data and could guide the range of polysynaptic PSP shapes that are considered candidate stimuli for finding the optimum shape of a neural PSP that induces a certain behavior in the neuron. Filtering the noise in this way greatly enhances the number of test stimuli that are close to the desired optimum stimulus shape.

2. <u>Perturb the system using a random combination of biologically plausible stimuli</u>. A random combination of plausible stimuli is an efficient method of generating a broad enough set of stimuli that includes optimum stimuli. Random stimuli can be easily generated since once an algorithm is developed for choosing one random stimulus, the same algorithm can be used to choose all others.

3. <u>Choose the duration and intensity of the noisy stimulus so that detection of the desired event is reliable and unbiased</u>. The protocol would use a library of random biologically plausible stimuli, and a library of responses is generated. From the two libraries, the "best" sub-set of stimuli can be selected. Continuous noisy perturbations may allow the system to remain near one or the other steady states but intermittently jump from one to the other. This requires the intensity of stimulation (e.g., L2 norm) to be adjusted carefully. The ideal response is to cause the oscillator to sporadically flip from one state to the other, *but not so frequently that more than one event occurs within the time scale of the longest stimulus fluctuations that might be considered as an optimum*. In this way, no events will be excluded and the library of stimuli and responses can be ranked solely on the basis of stimulus and response desirability. Exclusion of events can create a large error in estimating optimum stimulus profiles.

4) <u>Find the optimum stimuli</u>. Once a library of identified events and their associated test stimuli has been generated, the next step is to select a subgroup of stimuli with optimum properties that cause the desired behavior. One convenient way of doing this is to plot the stimulus desirability versus the outcome desirability. From this plot, a subgroup of optimum stimuli can be selected. The final step in the protocol is to send candidate optimum stimuli back into the system and re-test to confirm that the stimuli achieve the desired response. Check for experimental error that may have overestimated the desirability of the outcome of some stimuli. Any average should be checked to see if it captures the relevant signal. Check the amplitude of the signals.

Perhaps a lower amplitude might still cause the desired response and thereby further optimize the stimulus. Try parts of the stimulus. Maybe the first or last half gives most of the effect. Also compare the best stimulus against a randomly chosen control.

OPTIMUM STIMULI FOR STARTING AND STOPPING A BISTABLE OSCILLATOR: THEORY AND EXPERIMENT

The Bonhoeffer-van der Pol (BvP) model of the electrical activity of a nerve cell [4,5] is given by

$$\frac{dx}{dt} = c\left(y + x - \frac{x^3}{3} - r\right) + z$$

$$\frac{dy}{dt} = -(x - a + by)/c$$

Here, -z is a mathematical representation of an applied current and a = 0.7, b = 0.8 and c = 3.0. Choosing r = 0.342 results in two stable states [6]: oscillation (repetitive firing with a period of 12.8 msec), and the quiescent resting state. We consider two separate events for the bistable BvP oscillator. 1) Starting with repetitive firing (RF), what is the shape of the stimulus having the lowest intensity that annihilates RF? 2) Starting at the resting state, what is the shape of the stimulus having the lowest intensity that induces RF? We measure intensity of stimulation using the root-mean-square (i.e., L2 norm). The relevant input stimuli are chosen as follows. A neuron typically receives combinations of short bursts of current induced by post-synaptic potentials (PSPs) each of which can be mathematically represented as:

$$\pm\left(1 - e^{-\lambda_1 t}\right)e^{-\lambda_2 t}$$

Simulating a neuron in a network with fast synapses, we choose λ_1 as 1/(0.25 mSec) and λ_2 as 1/(1 msec). This acts to filter out any behavior that occurs on a time scale faster than 1 msec. Simulating a neuron in an active network, we choose each positive (or negative) PSP to be timed randomly with a Poisson rate so that, on average, one excitatory and one inhibitory PSP occurs every 0.1 msec. The BvP model can sporadically switch between two identifiable behaviors during noisy stimulation [6]: 1) quiescence (Q) during which there is low amplitude subthreshold activity near steady state, and 2) large oscillations characteristic of repetitive firing (RF). The intensity of the noisy stimulus was adjusted to avoid a switch between subthreshold activity and RF more often than once per 100 msec. When a transition from one state to the other is observed, we record the stimulus during the 60 msec period around the time of the switch (i.e. 30 msec before and 30 msec after the time of the peak action potential closest to the Q state). We computed the stimuli with minimum intensity (L2 norm) that caused the BvP model to switch between Q and RF. The stimulus shapes computed from the calculus of variations was similar to the shape found using our noise-based method [2]. We found that optimum stimuli were sinusoids whose frequency was close to the frequency of RF. The optimum sinusoidal stimulus that induces a transition from Q to RF was in phase with RF, whereas the optimum sinusoidal stimulus that induces a switch from RF to Q is in antiphase to the RF cycle.

Notably, we found that sinusoidal stimuli are much more effective than rectangular pulses for RF-to-Q or Q-to-RF transitions [2].

Our theoretical analysis of optimum stimuli can be tested in an experimental preparation of a bistable pacemaker neuron [7]. The preparation is of isolated squid giant axons, with membrane potential recordings using the intracellular perfusion-axial wire technique. The axon is perfused with an alkaline solution of potassium glutamate, and the bath temperature is kept constant (13-23°C). We have completed experiments in seven axons. Membrane bistability was demonstrated by inducing stable quiescence after releasing a voltage clamp near the equilibrium potential, and inducing stable repetitive firing after a brief current pulse to the quiescent membrane. Stochastically varying current without offset was administered to the axon for 10 sec periods. We have validated this method for noise induced noise transitions between RF and Q [8] (Figure 1). This preparation allows for a critical test of the noise based method for finding optimum stimuli that start and stop repetitive firing of a bistable cellular pacemaker.

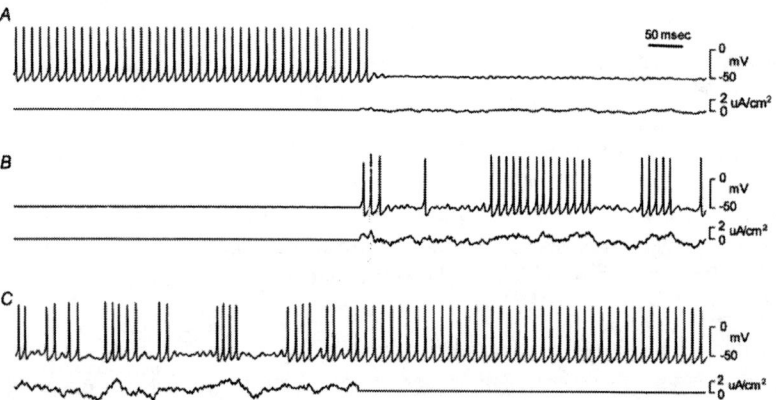

FIGURE 1. Membrane potential recording of alkalinized squid giant axon with membrane bistability. *A*, shows stable repetitive firing switched to quiescence by small stochastic currents administered to the cell through an axial wire. *B*, shows stable membrane quiescence switched to sporadic bursts of repetitive firing by more intense stimulation. *C*, shows switch from sporadic bursting to stable repetitive firing upon cessation of stochastic stimulation.

REFERENCES

1. Winfree, A.T. *The Geometry of Biological Time*. Springer, Berlin (1980).
2. Forger, D.B. and Paydarfar, D. *J. Theor. Biol.* **230**, 521-532 (2004).
3. Bryant, H.L. and Segundo, J.P. *J. Physiol. (London)* **260**, 279-314 (1976).
4. Fitzhugh, R. *Biophys. J.* **1**, 445-466 (1961).
5. Nagumo, J., Arimoto, S., and Yoshizawa, S. *Proc. IRE* **50**, 2061-2070 (1962).
6. Paydarfar, D., and Buerkel, D.M. *Chaos* **5**, 18-29 (1995).
7. Clay, J.R., and Shrier, A. *J. Membrane Biol.* **187**, 213-223 (2002).
8. Paydarfar, D., Forger, D.B., and Clay, J.R. *J. Physiol. (London)* **536**, 120p (2001).

Nanoscale electronic noise measurements

Laura Fumagalli[1], Ignacio Casuso[2], Giorgio Ferrari[1], Gabriel Gomila[2], Marco Sampietro[1], Josep Samitier[2]

1. Dipartimento di Elettronica e Informazione, Politecnico di Milano, P.za Leonardo da Vinci 32, I-20133 Milano, Italy
2. Centre de Referènica en Bioenginyeria de Catalunya (CREBEC) and Departament d'Electrónica, Universitat de Barcelona, C/ Martí i Franquès 1, 08028 Barcelona, Spain.

Abstract: Electronic noise measurements were performed at nanoscale by means of Atomic Force Microscopy (AFM). To this aim, a new AFM electrical setup has been developed, which combines the contact-mode AFM operation with complete custom-made low noise electronics in a Correlation Spectrum Analyser (CSA) scheme. This new electrical AFM operation mode allows noise measurements with nanometer spatial resolution simultaneously to topographic imaging of materials. Nanoscale noise measurements are particularly suited for biological sample characterisation as an alternative method for resistance extraction at nanoscale, since no electrical bias is required. Noise measurements were tested and results are discussed.

Keywords: Noise measurement, Correlation Spectrum Analyser, Atomic Force Microscopy (AFM)
PACS: 05.40.Ca, 74.78.Na, 39.25.+k

INTRODUCTION

An increasing interest in electrical characterisation of nanoscale systems has recently emerged. The motivation comes from, among the other, the development of new nanoscale electronic devices (i.e. molecular electronic devices) and the need to electrically characterize biological systems at the nanoscale (i.e. ionic channels, single proteins, cell membranes, etc.). The aim is to improve the basic knowledge of these systems and to help the development of new electronic nano-bio-devices (i.e. nanobiosensors [1]).

Atomic Force Microscopy (AFM) constitutes one of the more versatile techniques to electrically characterise materials with nanoscale resolution, especially for biological samples since it allows performing measurements under physiological conditions [2]. In particular, to measure the electrical resistance of materials at the nanoscale, there exists a variety of electrical characterization methods based on AFM, namely scanning spreading resistance and current sensing AFM. For a review of the different electrical modes see for instance [3].

Electrical noise measurement is a very interesting characterisation method, since it allows investigation of the electrical behaviour through the whole frequency spectrum. However, to our knowledge it has not been performed yet at nanoscale by means of AFM. In addition, existent AFM techniques for resistance extraction require the biasing of the sample under measurement, what, together with the small radius of the probe tip, can induce unnatural electrical stress of the samples due to high electrical field and sample heating through high current densities. Noise measurement is an alternative to extract electrical resistance with no bias applied, thanks to the Nyquist's relationship, thus being specially suited for the characterization of molecular and biological samples.

To precisely address this issue, we have developed a new AFM electrical operation mode to perform electronic noise measurements at the nanoscale. A commercially available AFM with a conductive tip has been connected to custom-made electronics of a Correlation Spectrum Analyser (CSA). By combining the contact-mode AFM operation, typically used for measuring current–voltage characteristics, with correlation noise measurement, it is possible to perform very sensitive noise measurement with nanometer spatial resolution, simultaneously to the topographic imaging. This new setup required the design of very low noise/high sensitivity electronics, particularly the input amplifiers. In addition, great effort was necessary to minimise spurious noise, partially generated by the commercial AFM electronics itself and picked up by electrical connection between the nanoscale system and the macroscopic instrumentation. In this paper, the developed setup of the new AFM mode is described and nanoscale noise measurement are demonstrated by first test measurements.

EXPERIMENTAL SETUP

Noise measurements were performed by a commercial AFM (Nanotec Electrónica S.L.) using a conductive cantilever as a first electrode for two-terminal current detection. The sample under test is mounted on conductive substrate that works as the second electrode. By operating the AFM in contact-mode, the sample is electrically probed by the cantilever tip with a nanometer spatial resolution, set by the tip radius that are typically in the order of 10nm.

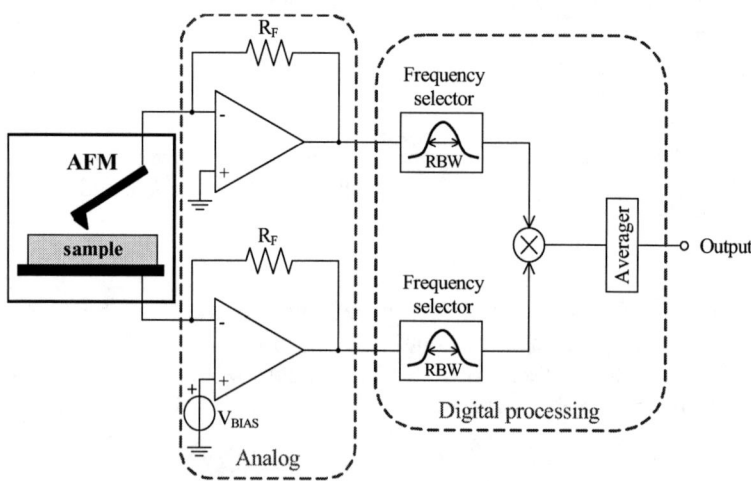

FIGURE 1. Simplified schematic of the electronic setup for correlation noise measurement with AFM.

In order to measure the noise of the sample with very high sensitivity, a complete custom-made electronics was designed such that the AFM acts as the probes of a Correlation Spectrum Analyser (CSA).

As sketched in figure 1, the two AFM electrodes (the cantilever and the sample substrate) were connected to two independent channels. Thus, as in a typical CSA scheme [4], in order to minimise effect of the instrumental noise, the sample noise is fed to two distinct input amplifiers operated in parallel, followed by a frequency selector circuit and a correlation stage that multiplies each component of the two channels and that averages out the result in time . The sample noise is, therefore, processed in phase by the two channels and multiplied frequency by frequency, thus giving at the output the noise power density of the sample. On the contrary, since the instrumental noise of the two channels is uncorrelated one to each other, it adds at the output of the multiplier a contribution with zero mean value and with standard deviation that can be strongly reduced by increasing the averaging time.

Specifically designed current-to-voltage amplifier were used as input amplifiers in order to limit the instrumental noise ($7fA/\sqrt{Hz}$ up to a few kHz). While input amplifiers were implemented in analog components, the rest of CSA functions are digitally performed. The outputs of the two amplifiers are sampled by high-speed acquisition board (10MS/s 12bit ADC) and processed by fully custom-developed software. Thus, the new implemented instrumentation provides the state-of-the art sensitivity in noise measurement (up to an noise equivalent resistance of few $G\Omega$) combined with nanometer spatially resolution.

Connection cables from the input amplifiers to the AFM electrodes have a fundamental role for the instrumentation performance. Since the noise of input amplifiers increases proportionally with input parasitic capacitance, the input amplifier were installed as near as possible to the AFM head. Work is in progress to further minimise the input parasitic capacitance, by direct mounting the front-end amplifiers onto the AFM tip holder and substrate.

RESULTS AND DISCUSSION

Noise measurements by AFM were demonstrated by first tests. An example of test noise spectra is shown in figure 2. The sample under test was a discrete resistor of well-known value ($1M\Omega$, $10M\Omega$, $100M\Omega$) connected in series of a gold substrate. Noise spectra are measured by the AFM setup in single-point contact mode. They show current power density level corresponding to resistor under test, as expressed by the Nyquist's formula $4kT/R$ (indicated by continuous lines).

The stability of the electrical contact between tip and surface is a crucial issue for conductive AFM measurement [5]. Particularly, noise measurements require a long-term stable contact during the whole averaging time. To this aim, a hard diamond-coated tip was used ((DDESP Veeco, force constant of 20N/m and resonant frequency of about 260kHz), to avoid rapid wear of the conductive tip. Moreover, for the electrical connections were used thin unshielded cables to reduce mechanical vibration. This is the reason for the high level pick-ups of 50Hz and its harmonics shown in figure 2.

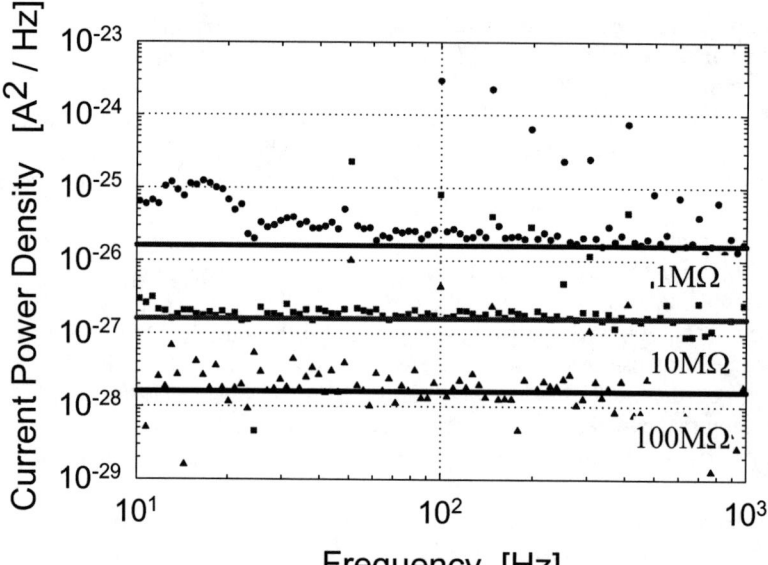

FIGURE 2. Test noise measurements performed by AFM of discrete resistor of well-known value (1MΩ, 10MΩ, 100MΩ) connected in series of a gold substrate. The continuous lines indicate the theoretical thermal noise.

CONCLUSION

The proposed experimental setup allows to perform noise measurement with nanometer spatial resolution. The possibility to probe by the AFM tip any nanometric sample combined with the high sensitivity given by the correlation spectrum analyzer opens new fields of investigation for biological systems and molecular devices.

REFERENCES

1. "Single protein nanobiosensro grid array", www.nanobiolab.pcb.ub.es/projectes/spotnosed
2. F. Rico, X. Fernández-Busquets and G. Gomila, "Applications of atomic force microscopy to biology", in *Scanning Probe Microscopy: Basic Concepts and Applications* (Phantoms, Madrid, 2003), pp. 81-89.
3. K. L. Sorokina and A. L. Tolstikhina, "Atomic Force Microscopy modified for studying electrical properties of thin films and crystals. Review", *Crystallography Reports* 49, 476 (2004).
4. G. Ferrari and M. Sampietro, *Rev. Sci. Instrum.* 73, 2717 (2002).
5. T.Trenkler et al. "Evaluating probes for "electrical" atomic force microscopy", *J.Vav. Sci.Technol*, B18 (1) Jan/Feb 2000.

What Information Is Hidden in Chaotic Signals of Biological Systems?

S.F. Timashev, *G.V. Vstovsky, **A.Ya. Kaplan, *A.B. Solovieva

L.Ya. Karpov Institute of Physical Chemistry, Moscow, Russia
**N.N. Semenov Institute of Physical Chemistry, Russian Academy of Science, Moscow*
****M.V. Lomonosov Moscow State University, Biological Department, Moscow, Russia*

Abstract. Applications of the Flicker-Noise Spectroscopy (FNS) to analysis of electroencephalograms (EEG) are demonstrated. We present the double correlation function for the EEG measured in the C4 and O2 points for two patients – a healthy ("normal") child and a sick Schizophrenia child. The drastic differences in the behavior of the two-correlators manifest the information meaning of the similar dependences. We conclude that the FNS approach could be considered as a new instrument to early diagnostics of various brain diseases.

Keywords: Flicker-Noise Spectroscopy, informative essence of chaotic signals, electroencephalograms, double correlation function.

INTRODUCTION

Chaotic time series $V(t)$ (t is a time) which is given at an interval T, obtained under studying dynamics of biological and medical systems (ECG and EEG data, pulse and variation of blood pressure, etc.), among other chaotic signals, contain much information. What type of information is hidden within a chaotic signal? Is it possible to propose an algorithm for taking out so much the hidden information as one need for solving problems under consideration? In this paper we demonstrate that the practical problems related to revealing the informative essence of various chaotic signals could be resolved by introducing a new image of information hidden in chaotic signals. This image is presented in Flicker-Noise Spectroscopy (**FNS**) [1-3]. According to this phenomenological approach, the main information hidden in chaotic signals is provided by sequences of distinguishing types of irregularities – spikes, jumps, and discontinuities of derivatives of different orders, at all space-time hierarchical levels of systems. FNS method with getting information by analyzing the power spectra $S(f)$ and difference moments of the second order $\Phi^{(2)}(\tau)$ can be used to display of redistribution dynamics in distributed systems by analysis of dynamic correlations in the chaotic signals measured simultaneously at spaced locations.

The behavior of the real $S(f)$ and $\Phi^{(p)}(\tau)$ dependences is very specific and individual to each study case. These dependences can be considered as "characteristic passport patterns" of the state of systems under study, especially, as these dependencies have a very clear physical sense. For raising the pattern specificity we split the considered signal into low frequency $V_R(t)$ and high frequency $V_F(t)$ components. The introduced decomposition $V_G(t) = V_R(t) + V_F(t)$ makes it possible to find new characteristic features of the studied signals. The $V_R(t)$ term is obtain by using a relaxing ("diffusive") procedure [2-3] for the total set of the initial time series readings. By this smoothing procedure, the "evolution of the dynamical variable" of the chaotic series is a realization of minimum high frequency information in the $V_R(t)$ component. This means that the high frequency information is contained in the function $V_F(t) \equiv V(t) - V_R(t)$. Now, we can calculate $S_J(f)$ and $\Phi_J^{(2)}(\tau)$ for each of the functions $V_J(t)$ ($J = R$ or F) or G), where the subscripts R and F and G refer to $V_R(t)$, $V_F(t)$, respectively.

DOUBLE CORRELATION FUNCTIONS

In this paper applications of the FNS method to analysis of electroencephalograms (EEG) are presented. The tentative investigations have shown a high individuality of local parameters obtained by analysis of EEG of different patients, definite sensitivity of the parameters to the change in external conditions (the acquisition of new information or new task setting, sound and light signals, emotional effects, etc.). The signals were obtained during EEG registration from different electrodes using a 16-channel system which measured electric potentials for the standard sampling interval 128 seconds with sampling digital frequencies 128 Hz and 256 Hz at the standard "points" C3, C4, Cz, F3, F4, F8, O1, O2, P3, P4, Pz, T3, T4, T5, T6. The analysis carried out has shown a high sensitivity of local EEG signals and excitation flows in the cortex to the change in patient state. It is very important phenomenon because of main informational processes in the brain based on interractive collaborations between different brain areas.

We have analyzed EEG of 11 healthy children and 13 children suffering schizophrenia. Every EEG set demonstrates a very high individual features. A high level of specificity was revealed for the $S_F(f)$ and $\Phi_F^{(2)}(\tau)$ patterns, which were obtained by analyzing the C4 and O2 electrode signals measured for the two groups (healthy children, 27SW, and suffering schizophrenia, 545W). The initial signals and the corresponding $S_F(f)$ patterns are presented in Fig. 1. The investigations carried out so far show that FNS method with getting information by analyzing the $S(f)$ and $\Phi^{(2)}(\tau)$ dependences can be used to reveal the redistribution dynamics in distributed systems by analysis of dynamic correlations in the chaotic signals measured simultaneously at spaced locations. In the frame of FNS a new type of multi-point correlation functions is introduced to obtain spatial-temporal maps of flows of different nature on different scales of spatial-temporal system arrangement. For example, the double correlation function $q_{ij}(\tau, \theta_{ij})$ for dynamic variable $V_i(t)$ and $V_j(t)$ measured in the ith and jth points, which are formed exclusively by jumps-irregularities of the dynamic variables, are presented as:

$$q_{ij}(\tau;\theta_{ij}) = \left\langle \left[\frac{V_i(t) - V_i(t+\tau)}{\sqrt{2}\sigma_i} \right] \times \left[\frac{V_j(t+\theta_{ij}) - V_j(t+\theta_{ij}+\tau)}{\sqrt{2}\sigma_j} \right] \right\rangle,$$

where τ is a delay; θ_{ij} is a parameter; σ_i is the variance of the measured dynamical variable or "normalized dispersion"

$$\sigma_i^2(\tau) = \left\langle [V_i(t) - V_i(t+\tau)]^2 \right\rangle^{1/2}.$$

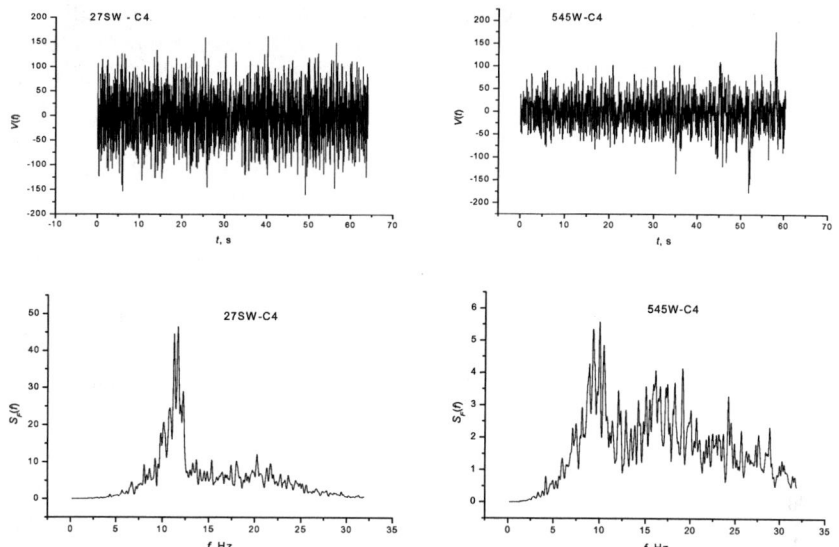

FIGURE 1. Dynamics of the C4 EEG electric potentials (mcV) for the "normal" (27SW) and "sick" (545W) patients, and the corresponding $S_F(f)$ dependences ("patterns").

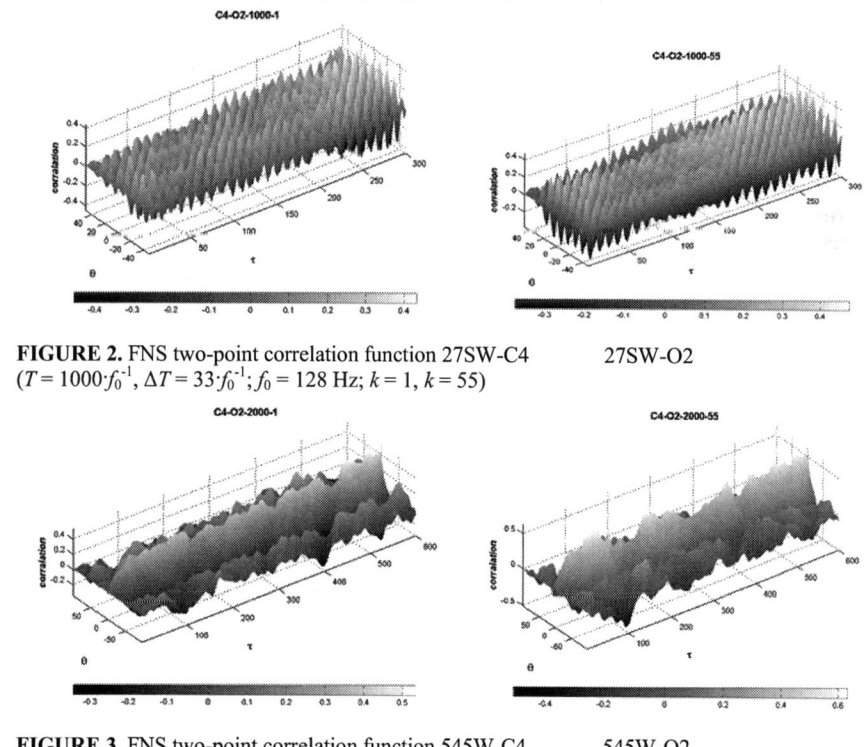

FIGURE 2. FNS two-point correlation function 27SW-C4 ($T = 1000 \cdot f_0^{-1}$, $\Delta T = 33 \cdot f_0^{-1}$; $f_0 = 128$ Hz; $k = 1$, $k = 55$)

FIGURE 3. FNS two-point correlation function 545W-C4 ($T = 1000 \cdot f_0^{-1}$, $\Delta T = 33 \cdot f_0^{-1}$; $f_0 = 128$ Hz; $k = 1$, $k = 55$))

The θ_{ij} parameter is found from the demand that $q_{ij}(\tau, \theta_{ij})$ is a positive and has a maximum value among possible ones. The sign and value of this parameter determine the velocity and direction of the flow between the ith and jth points. It is easy to understand that the similar analysis with introduction of the double, triple and other multi-point correlation functions gives a possibility of obtaining the needed spatial-temporal maps of the studying flows in the system under consideration.

APPLICATION TO ANALYSIS OF EEG DATA

Here we present (Figs. 2 and 3) the double correlation function $q_{ij}(\tau, \theta_{ij})$ for the EEG measured in the C4 ("i") and O2 ("j") points (see Fig. 1) for a healthy ("normal", 27SW) and suffering schizophrenia (545W) children. We analyzed the time variation of the two-point correlators, that are calculated within the averaging interval [(k – 1)·ΔT, t_k] of duration T, where k = 1, 2, 3 ... and $t_k = T + (k-1)\cdot\Delta T$, shifting discretely through the entire observation period T_{tot} = 128 sec by steps ΔT. The time intervals T = 1000·Δt_1 = 2000·Δt_2 and ΔT = 33·Δt_1 = 66·Δt_2 were chosen. The two-point correlators C4(27SW) – O2(27SW) and C4(545-1W) – O2(545-1W) in the cases k = 1 and 102 for parameters τ and $\theta_{ij} \equiv \theta$ which are varied in the intervals [0, 0.3T] and [- 0.05T, + 0.05T] correspondingly, are presented in the Figure. Note, that the interval 0.3T (0.05T) corresponds to k = 300 (45) and k = 600 (90) in the "normal" and "sick" cases. The drastic differences in the behavior of the two-correlators manifest the information meaning of the similar dependences. Every EEG pattern set (among 11 healthy persons and 13 children suffering schizophrenia children) demonstrates a very high individual features. "Intermediate" cases were observed where the schizophrenia related EEG feature appeared at several T intervals whereas the correlators demonstrated "normal" behavior at other intervals. We conclude that the FNS approach could be considered as an instrument to early diagnostics of brain diseases.

Introduction of new parameterization to characterize the dynamics of cortex excitations on the base of EEG signals together with other standard indexes of individual EEG activity will enable to introduce a classification of the functional states of the whole organism, to develop a base for establishing "normal" and "pathological" states of different types, to discover the new approaches to early diagnostics and revelation of the features of dynamics of a number of diseases due to violation of functional interrelations between the cortex regions – Schizophrenia, Alzgeimer and Parkinson diseases and others.

ACKNOWLEDGMENTS

The work was supported by the ISTC grant (2280).

REFERENCES

1. Timashev S.F., and Vstovsky G.V. This issue (R019).
2. Timashev S.F., and Vstovsky G.V. *Russian J. Electrochemistry*, **39**, 156-169 (2003).
3. Telesca L., Lapenna V., Timashev S., Vstovsky G. and Martinelli G. *Physics and Chemistry of the Earth*, **29**, 389-395 (2004).

Information Transfer Analysis of Spontaneous Low-frequency Fluctuations in Cerebral Hemodynamics and Cardiovascular Dynamics

Takusige Katura[*,1], Naoki Tanaka[1,2], Akiko Obata[1]
Hiroki Sato[1] and Atsushi Maki[1]

[1]*Advanced Research Laboratory, HITACHI Ltd.*
[2]*Depertment of Physical Electronics Graduate School of Science and Engineering, Tokyo Institute of Technology*

Abstract. In this study, from the information-theoretic viewpoint, we analyzed the interrelation between the spontaneous low-frequency fluctuations around 0.1Hz in the hemoglobin concentration in the cerebral cortex, mean arterial blood pressure and the heart rate. For this analysis, as measures of information transfer, we used transfer entropy (TE) proposed for two-factor systems by Schreiber and intrinsic transfer entropy (ITE) introduced for further analysis of three-factor systems by extending the original TE. In our analysis, information transfer analysis based on both TE and ITE suggests the systemic cardiovascular fluctuations alone cannot account for the cerebrovascular fluctuations, that is, the regulation of the regional cerebral energetic metabolism is important as a candidate of its generation mechanism Such an information transfer analysis seems useful to reveal the interrelation between the elements regulated each other in a complex manner.

Keywords: Fluctuation, Transfer entropy, NIRS
PACS: 24.60.Ky, 24.60.-k, 87.57.-s, 87.64.-t

INTRODUCTION

Spontaneous low-frequency fluctuations in cerebral hemodynamics [1], which have often been observed by the functional magnetic resonance imaging and the near-infrared spectroscopy, are very complex. The complexity of these physiological fluctuations seems to come from the interaction of the systemic regulation of cardiovascular and respiratory systems and that of regional cerebral energetic metabolism. Indeed, similar low-frequency fluctuations have been observed also in the mean arterial blood pressure (ABP) and the heart rate (HR), which are typical indices of the cardiovascular system. Clarifying the interrelation between low-frequency fluctuations in the cerebral hemodynamics and the cardiovascular dynamics is not only interesting in itself but also essential to extracting the true functional information of brain from the observed signals.

In this study, we concentrate on the information-theoretic analysis of the interrelation between the low-frequency fluctuations around 0.1Hz in the hemoglobin concentration (HbCC) in the cerebral cortex, ABP and HR. For this analysis, as measures of information transfer, we used transfer entropy (TE) proposed for two-factor systems by Schreiber [2] and intrinsic transfer entropy (ITE) introduced for further analysis of three-factor systems by extending the original TE.

MEASUREMENT

Twelve subjects participated in the experiment. The data for analysis were obtained by simultaneous measurement of cerebral hemodynamics with near-infrared optical topography (ETG-100, Hitachi Medical Corporation, Japan) [3, 4] and by beat to beat finger arterial blood pressure measurement with infrared finger plethysmography (Finometer, FMS Finapres Medical Systems BV, The Netherlands), which also measures the heart rate, for 15 minutes under rest condition of subjects. We applied band-pass filter between 0.04 Hz and 0.15 Hz to the data to extract the low-frequency fluctuations.

ANALYSIS

The point of our analysis is quantifying the contribution of the two remainders to each of the three variables with respect to the information carried with the variables based on both TE and ITE. This is a so-called causality analysis from an information-theoretic viewpoint.

We used the transfer entropy (TE) proposed by Schreiber (Schreiber, 2000) as a measure of information transfer. Here, the TE (bits) from variable X to variable Y with delay τ, $TE(X, Y, \tau)$, is given by

$$TE(X,Y,\tau) = \left\langle \log_2 \frac{p_{Y|YZX}(y(t+\tau)|y(t),x(t))}{p_{Y|YZ}(y(t+\tau)|y(t))} \right\rangle_t, \quad (1)$$

where $p_{Y|YZX}(y(t+\tau)|y(t),x(t))$ denotes the conditional probability density function (PDF) that Y will fall within interval $[y(t+\tau), y(t+\tau)+dy]$ at time $t+\tau$ when Y and X fall within intervals $[y(t), y(t)+dy]$ and $[x(t), x(t)+dx]$ at time t, respectively. $\langle \cdot \rangle_t$ stands for the average over time. The numerical procedures to estimate conditional PDFs can be found elsewhere [1, 4].

The system in this study had thee variables, HbCC, ABP, and HR. The X and Y corresponded to any two variables for these three. To analyze the system, we modified the original TE and defined intrinsic transfer entropy (ITE) (bits) from X to Y with delay τ in a system consisting of X, Y, and Z as

$$ITE(X,Y,\tau;Z) = \left\langle \log_2 \frac{p_{Y|YZX}(y(t+\tau)|y(t),z(t),x(t))}{p_{Y|YZ}(y(t+\tau)|y(t),z(t))} \right\rangle_t, \quad (2)$$

where it differs from Eq. 1 in that variable Z (its realization z) has been added to the argument variables of the conditional PDF at the right hand side. By the modification, we can exclude the common information transferred from X and Z to Y. We can interpret delay τ as the time required to transfer information from X to Y.

Because we did not know the pathways for information transfer in detail, we used the maximum value of TE (ITE) over an adequate range of τ as a measure of information transfer. This value corresponded to the amount of information transferred through the most plausible pathway. Using this measure, we can regard X as a causal variable and Y as a resultant variable if we can find

$$\max_{\tau}\{TE(X,Y,\tau)\} > \max_{\tau}\{TE(Y,X,\tau)\}, \quad (3)$$

and *vice versa*. This is the principle behind the analysis we applied to the causality between LFOs in HbCC, HR, and ABP. The same principle can be applied to ITE. We set the range of delay τ from 0 to 5 seconds corresponding to half the main period of the LFOs.

We estimated the joint probability functions with precision r using the naive kernel method [2, 5].

$$\hat{p}_{YYX}^{r}(y(t+\tau),y(t),x(t)) = \frac{1}{N-\tau}\sum_{t'=1}^{N-\tau}\Theta\left(r - \left\|\begin{array}{c}y(t'+\tau)-y(t+\tau)\\ y(t')-y(t)\\ x(t')-x(t)\end{array}\right\|_{\max}\right), \quad (4)$$

where the Θ stands for the step kernel

$$\Theta(z) = \begin{cases} 1 & \text{for } z \geq 0 \\ 0 & \text{for } z < 0 \end{cases} \quad (5)$$

and the norm $\|\cdot\|_{\max}$ is the maximum distance.

RESULTS AND DISCUSSION

FIGURE 1.(a) and (b) show the delay dependence of the inter-subject mean ratio of TE (upper) and ITE (lower) between HbCC and HR (left), HR and ABP (middle), ABP and HbCC (right), respectively. The values for TE and ITE stand for those normalized by amount of the information carried with the destination variables.

From these results, we found that the peak values for the dependence of ITE on delay are smaller than those for TE, which indicates that a considerable part of the information transferred from two of the three variables to

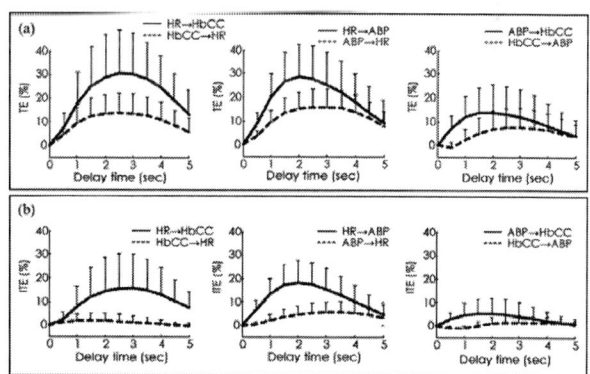

FIGURE 1. Dependence of the mean TE (a) and ITE (b) averaged across all subjects on delay. One-sided error bars stand for standard deviations (n= 9). TE for (HbCC, HR), (HR, ABP), and (ABP, HbCC) are in upper, middle, and lower rows.

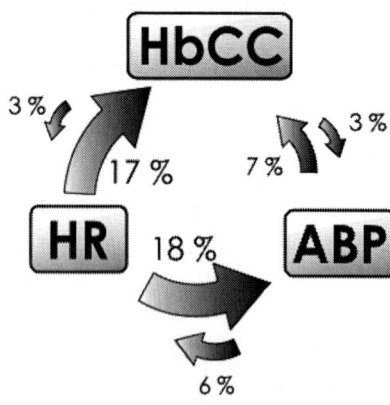

FIGURE 2. Schematic diagrams of interrelation between HbCC, HR, and ABP. Directions of arrows indicate information transfer direction and their sizes indicate amount of information transferred, based on results of maximum mean ITE.

the remaining one is common.

There are two main streams of information in LFOs around 0.1 Hz. The first is from HR to ABP and the second is from HR to HbCC. This suggests the major information source is HR in the HbCC-HR-ABP system. However the contribution of HR to HbCC is only about 17%. In addition, the contribution of ABP to HbCC is about 7% and the common contribution of HR and ABP to HbCC is about 10%. From these, HR and ABP cannot account for more than a half of information carried with variable HbCC, which suggests the origin of the low-frequency fluctuations in cerebral hemodynamics may be the regulation of the regional cerebral blood flow and energetic metabolism rather than the systemic regulation of the cardiovascular system.

In conclusion, the information transfer analysis based on both TE and ITE suggests that the systemic cardiovascular fluctuations alone cannot account for the cerebrovascular fluctuations, that is, the regulation of the regional cerebral energetic metabolism is important as a candidate of its generation mechanism. Such an information transfer analysis seems useful to reveal the interrelation between the elements regulated each other in a complex manner

ACKNOWLEDGMENTS

This research was supported by the Network Human Interface Project Foundation of the Ministry of Internal Affairs and Communications of the Japanese Government.

REFERENCES

[1] Obrig H, Neufang M, Wenzel R, Kohl M, Steinbrink J, Einhaupl K, Villringer A. 2000. "Spontaneous low frequency oscillations of cerebral hemodynamics and metabolism in human adults." Neuroimage 12(6), 623-639.
[2] Schreiber T. 2000. "Measuring information transfer." Phys Rev Lett 85(2), 461-464.
[3] Maki A, Yamashita Y, Ito Y, Watanabe E, Mayanagi Y, Koizumi H. 1995. "Spatial and temporal analysis of human motor activity using noninvasive NIR topography." Med Phys 22(12), 1997-2005.
[4] Koizumi H, Yamashita Y, Maki A, Yamamoto T, Ito Y, Itagaki H, Kennan R. 1999. "Higher-Order Brain Function Analysis by Trans-Cranial Dynamic Near-Infrared Spectroscopy Imaging." Journal of Biomedical Optics 4(4), 403.
[5] Silverman BW. 1986. "Density Estimation for Statistics and Data Analysis." Chapman&Hall /CRC, London.

Stochastic resonance of Na, K-ion pumps on the red cell membrane

Cheng-Hung Chang[1,2] and Tian Yow Tsong[3,4]

1 Physics Division, National Center for Theoretical Sciences, Hsinchu 300, Taiwan
2 Institute of Physics, National Chiao Tung university, Hsinchu 300, Taiwan
3 Institute of Physics, Academy of Sciences, Taipei 115, Taiwan
4 University of Minnesota, College of Biological Science, St. Paul, Minnesota 55108

Abstract. The ion transport of the Na, K-ion pump through the cell membrane can be driven by an oscillating electric field. Noise in this field is not necessarily destructive to the transport efficiency. The recent experimental result showed constructive effect of noise in this system [1]. This phenomenon was later on confirmed theoretically in numerical simulations [2]. The theoretical results not only catch the main feature of stochastic resonance in the experiment but also predict unknown transport behavior beyond the experimentally studied parameter region.

Keywords: Stochastic resonance, Na, K-ATPase ion pump.
PACS: 82.39.Fk, 82.39.-k, 87.15.Aa

INTRODUCTION

The Na, K-ATPase is a transmembrane enzyme, which can transport Na^+, K^+ ions against the concentration gradient across the membrane. The transport behavior of the pump in the living cell can be either generated by the ATP (adenosine triphosphate) hydrolysis or by the transmembrane potential fluctuation [3,4]. The latter driving mechanism is especially of fundamental interest from the point of view of nano-bio technology and its behavior under signal and noise has been investigated recently from two different ways: In the theoretical study [5], the idea of intrinsic noise is introduced, which is induced by the ambient ion channels and coupled to the external signal. In the experiment [1], independent external noise and signal are imposed to drive the system. Both studies lead to stochastic resonance phenomena.

Figure 1 depicts the kinetic model of the Na, K-ATPase, which has four conformations with certain kinetic rate constants k_i's, where the ligand L can be Na^+ or K^+. Suppose k_i's are time-independent and the enzyme, which opens to the left (right) hand side of the membrane, prefers to adapt the ligand (release the ligand) under certain $k_{\pm 1}$ and $k_{\pm 3}$. Then the enzyme concentrations in Fig. 1 tend to flow from E_1 (E_2L) to E_1L (E_2) and finally converge to an equilibrium state, with more concentration distributed on E_1L and E_2. However, the equilibrium can be broken, when a force is applied to change the rate constants and activates the processes from E_1L (E_2) to E_2L (E_1), which shifts the ligand across the membrane. After the enzyme releases the ligand (adapts a new ligand) on the right (left) hand side, the force is lifted, which drives the enzyme conformations back to the original equilibrium state again. A proper force oscillation then generates a clockwise flow, which transports the ligand from left to right.

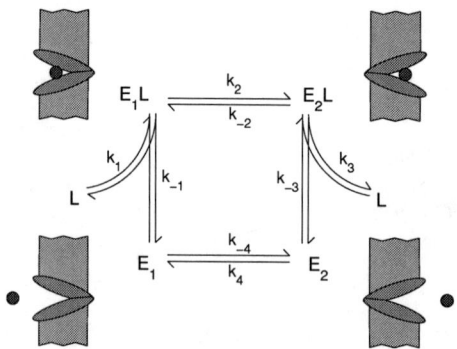

FIGURE 1. The four-state kinetic model of a transmembrane enzyme motor. The enzyme is depicted by the pacmans with two different opening orientations. The ligand is depicted by the red ball.

The kinetic equation for the concentrations of the four conformations in Fig. 1 is a four dimensional non-autonomous linear dynamical system $dV(t)/dt=M(t)V(t)$ with

$$V(t) = \begin{pmatrix} [E_2L] \\ [E_1L] \\ [E_2] \\ [E_1] \end{pmatrix}, \quad M(t) = \begin{pmatrix} -k_3-k_{-2} & k_2 & k_{-3}[L_3] & 0 \\ k_{-2} & -k_{-1}-k_2 & 0 & k_1[L_1] \\ k_3 & 0 & -k_4-k_{-3}[L_3] & k_{-4} \\ 0 & k_{-1} & k_4 & -k_{-4}-k_1[L_1] \end{pmatrix}$$

The values $[L_1]$ and $[L_3]$ in $M(t)$ denote the concentrations of L on the left respectively right hand side of the membrane, which can be regarded as constants. The rate constants $k_i=h_i \exp(q_i \varphi(t) a_i / RT)$, for $i=\pm1, \pm2, \pm3, \pm4$, consist of the gas constant R, the temperature T, the effective charge q_i of different enzyme conformations, the transmembrane potential $\varphi(t)$, the rate constant h_i in zero potential $\varphi(t)=0$, and the apportionment constant a_i. The values k_i's are positive, time-dependent, and can be simplified to the form $k_i =h_i \exp(d_i\Psi(t))$, characterized by parameters h_i, d_i and the fluctuation $\Psi(t)$. The instantaneous ligand flux is $j(t)=k_3[E_2L]-k_{-3}[E_2][L_3]$, which yields the transported amount $S(t)= \int_0^t j(t') dt'$ and the averaged flux $J=\lim_{t\to\infty}S(t)/t$.

Figure 2(a) shows the concentration evolutions and the net transported amount $S(t)$ under parameter values:

i	1	-1	2	-2	3	-3	4	-4
d_i [cm/V]	0	0	-2	-3	0	0	4	-2
h_i [1/s]	40	60	25700	12000	70	200	20	10

and $[L_1]=[L_2]=1$ cm^{-3} (h_1 and h_{-3} have dimension cm^3/s). The values $d_{\pm1}$ and $d_{\pm3}$ are set to zero, because the ligand association with and dissociation from the enzyme are less affected by the fluctuation $\Psi(t)$. A concrete fluctuation $\Psi(t)=A\sin(\omega t)+\xi(t)$ is taken, with a dominating signal $A\sin(\omega t)$ and a secondary noise $\xi(t)$, where A denotes the signal amplitude and the dichotomous noise with values $\xi\in\{\pm1\}$ and frequency ω_n. The noise strength is in the scale of noise level $\eta=\eta_{rms}/A$ with the root mean square $\eta_{rms}=(\int_0^t \xi(t)^2 \, dt/\tau)^{1/2}$, where $\tau=1/\omega_n$ denotes the period of the noise.

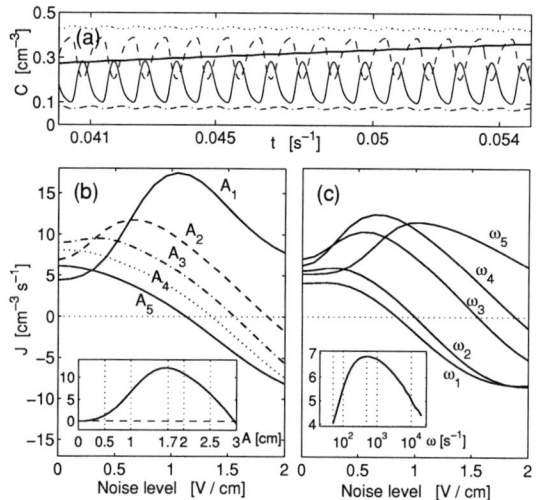

FIGURE 2. (a) The oscillations of $[E_1]$ (dotted), $[E_2]$ (dash-dotted), $[E_1L]$ (thin solid), $[E_2L]$ (dashed), induced by $\Psi(t)=A\sin(\omega t)$, and the positive transported amount S(t) across the membrane (thick solid). (b) The flux J versus the noise level η under five different signal amplitudes with signal frequency $\omega=10^3$ and $\omega_n/\omega=10^3$, where $A_1=0.5$, $A_2=1$, $A_3=1.4$, $A_4=1.7$, and $A_5=2$. The inset shows the amplitude dependence of J without noise. (c) The flux J versus the noise level η under five different signal frequencies ω's with signal amplitude $A=1$ and $\omega_n/\omega=10^3$, where $\omega_1=50$, $\omega_2=10^2$, $\omega_3=500$, $\omega_4=10^3$, and $\omega_5=10^4$. The inset shows the frequency dependence of J without noise.

The inset in Fig. 2(b) shows the amplitude dependence of the flux J without noise. The optimal signal amplitude A_{op} for the transport is around $A=1.7$. If the signal amplitude is too large, e.g., $A>3$, J can be negative and current reversal can occur. Taking five different A's from the inset, the curves of J v.s. η in Fig. 2(b) indicate the destructive influence of the noise on the transport in large noise regime, such as $\eta>1.5$. This destructive role of the noise coincides with our experience in the macro world. However, this is not always true for small noise, e.g., $\eta<0.5$, in which the noise is constructive for $A<A_{op}$ and destructive for $A>A_{op}$. The inset in Fig. 2(c) shows the frequency dependence of J without noise. The apparent peak indicates the natural frequency $\omega_{op}=500$, which is the optimal frequency of the system. Taking five ω_n from the inset, J decreases monotonically for small frequency like $\omega=50$. Other larger frequencies exhibit a bell shape and imply the stochastic resonance, which can be recognized after replotting the figure to a ratio of J/η_{rms} versus η_{rms}, as that in Ref. [5]. Similar to the plot in Fig. 2(b), the transport efficiencies in Fig. 2(c) decrease for strong noise. However, different from the plot in Fig. 2(b), the noise in Fig. 2(c) is constructive for both $\omega<\omega_{op}$ and $\omega>\omega_{op}$ in the weak noise regime, up to the slow signal like $\omega=50$.

Similar features as in Fig. 2(b) and 2(c) can be found in Fig. 3, which are measured in the experiment of the Rb^+ transport across the human red cell membrane [1]. The plot from random-telegraph-fluctuation (RTF) in Fig. 3 (a) and (b) have similar bell shapes as those in the insets of Fig. 2(b) and 2(c) under sinusoidal fluctuation. The dotted, dashed, and dash-dotted curve families in Fig. 3 (c) are the same as the features of the dotted, dashed, and dash-dotted curves in Fig. 2(b) in the small noise strength regime, e.g., $\eta<0.6$. The decreasing transport efficiency after A_{op} predicted in Fig. 2(b), as well as the curves in the main plot of Fig. 2(c) are beyond the current experimental data and suggest further experimental investigation.

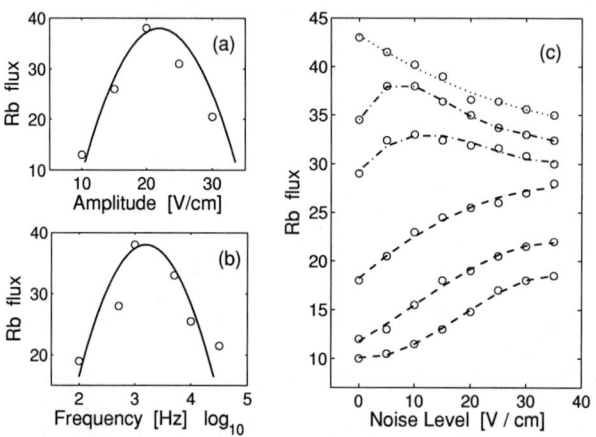

FIGURE 3. Electric field induced Rb^+ pumping in Ref. [1]: flux [amole/hour] versus (a) amplitude of the RTF with mean frequency 1 kHz; (b) amplitude of the RTF with amplitude 20 V/cm; (c) noise level η under sinusoidal signal with amplitude 20 V/cm and frequency 1 kHz. The curves have signal amplitudes 20, 17.5, 15, 12.5, 10, and 7.5 V/cm from top to bottom.

Both the experimental observation and the numerical results calculated conclude the same fact that the noise is destructive, when the signal magnitude A has performed the optimal transport efficiency. But a noise can enhance this efficiency and induce stochastic resonance phenomenon when the signal is apart from this optimal magnitude.

REFERENCES

1. Chang, C.-H. and Tsong, T.Y., *Phys. Rev. E* **69**, 021914 (2004).
2. Tsong, T.Y. and Xie, T.D., *Appl. Phys. A* **75**, 345 (2002).
3. Tsong, T.Y. and Astumian, R.D., *Bioelectrochem. Bioenerg.* **15**, 457 (1986); *Prog. Biophys. Mol. Biol.* 50:1-45 (1987).
4. Astumian, R.D., Chock, P.B., Tsong, T.Y., and Westerhoff, H.V., *Phys. Rev. A* **39**, 6416 (1989).
5. Fulinski, A., *Phys. Rev. Lett.* **79**, 4926 (1997); *Chaos* **8**, 549 (1998).

Contraction of Information on Brain Wave Fluctuations by Information Geometrical Approach

Hidetoshi Konno

Department of Risk Engineering, Faculty of Systems and Information Engineering, University of Tsukuba, Tsukuba, Ibaraki 305-8573 Japan

Abstract. We will first propose a method of EEG signal identification with the use of the stochastic complex Ginzburg-Landau (CGL) equation having complex coefficients with the aid of the method of information geometrical approach to determine the system parameters. After the contracting information on the natures of fluctuations of amplitude and phase in the EEG signals on human scalp, we combine the information with other information such as complex measures like Higuchi's fractal dimension and multi-scale entropies. A new theory of unification of the information is also proposed. To exhibit the potentiality of our new method, we show the result of application of the theory and method to practical EEG data from elderly sound and demented people.

Keywords: Space-Dependency, EEG, Stochastic GL Equation, Information Geometrical Approach
PACS: 05.40.+j, 05.20.-y

INTRODUCTION

Many works have been done for characterizing and understanding the EEG signals in steady state and non-steady state situations with classical standard techniques like correlation functions, power spectral densities. "Chaos analysis" by using fractal dimensions, lyapunov exponents and other complex measures have been also done for understanding the physical meaning of the EEG signals, and to extract information on the symptoms of aging and/or diseases. The mutual information measures and the random matrix theory have been further examined. However, there are still various uncertainties when one apply simply these mathematical measures to the EEG and MEG signals. To step up the present situation, one must take more physical and/or physiological approach in combined with a sophisticated physical model of EEG fluctuations.

STOCHASTIC CGL EQUATION

To exhibit our information geometrical approach, we utilize here the stochastic complex Ginzburg-Landau (CGL) equation [1]:

$$\frac{d}{dt}A = (i\omega_0 + a)A - (b + ic)|A|^2 A + A F_p(t) \tag{1}$$

where A is the dynamical variable of complex number, ω_0, a, $b > 0$ and c are assumed to be real constants, $F_p(t)$ represents parametric noise source of complex number with the Gaussian-white nature (assume that the real and the imaginary parts of them are independent and strengths of them are identical for simplicity). To reduce the number of unknown parameters, let us assume that $\langle F_p(t)F_p(t')^*\rangle = 2D\delta(t-t')$.

Without $F_p(t)$, the model represents the feature of nonlinear dynamics near the onset point of an limit cycle oscillation. The positive definite parameter b is required to keep the global stability of the system. When $a < 0$, $A = 0$ is the stable fixed point and the system stays in this point without perturbation. On the other hand, when $a > 0$, $A = 0$ becomes the unstable fixed point. In this case, one obtains the oscillating solution $A_0(t) = \sqrt{\frac{a}{b}}\exp\{i(\omega_0 - cR_0^2)t\}$. The imaginary part of nonlinear term $c|A|^2A$ expresses the phenomenon of nonlinear pulling.

INFORMATION GEOMETRICAL APPROACH

The relevant amplitude and phase equation for Eq.(1) become

$$\frac{dR}{dt} = aR - bR^3 + RF_R(t) \quad \text{and} \quad \frac{d\Phi}{dt} = \omega_0 - cR^2 + F_\Phi(t), \tag{2}$$

where F_R and F_Φ are random forces. The amplitude motion is independent of the phase motion, which makes it possible to have a generalized Rayleigh distribution of amplitude R as

$$P_{amp}(R) = \exp\left\{\theta_1 \log R - \theta_2 R^2 - \varphi(\theta_1, \theta_2)\right\}, \tag{3}$$

where $\varphi(\theta_1, \theta_2) = \log\Gamma(\frac{\theta_1}{2}+1) - \log 2 - (\frac{\theta_1}{2}+1)\log\theta_2$, $\theta_1 = \frac{a}{D}$ and $\theta_2 = \frac{b}{2D}$. The theory of information geometry (TIG) [2] teaches us that the dual coordinates of the model (a generalized fluctuation-dissipation theorem (FDT)) are obtained as $\eta_1 = \frac{\partial \varphi}{\partial \theta_1} = \langle \log R \rangle$ and $\eta_2 = \frac{\partial \varphi}{\partial \theta_2} = -\langle R^2 \rangle$. This means that the information of the two statistical measures are given, the two system parameters θ_1 and θ_2 can be determined. The relevant relations are obtained as $\langle \log R \rangle = \psi(\frac{\gamma+1}{2}) - \frac{1}{2}\log\beta$ and $\langle R^2 \rangle = \frac{\gamma+1}{2\beta}$.

Estimation of the parameters θ_1, θ_2 can be possible by using the Newton method since the di-gamma function ψ is a nonlinear function of γ. If one adopt the higher order statistical quantity $\langle R^2 \log R \rangle = \langle R^2 \rangle \langle \log R \rangle + \frac{1}{2\beta}$, the parameters are estimated by simple algebraic manipulation. Thus one can escape from estimation error. [3]

On the other hand, the phase Φ motion in eq.(2) is influenced by the random movement of the amplitude R when $c \neq 0$. Letting $\Phi = \omega_0 t + \phi$ and eliminating the amplitude, we obtain the SDE of phase velocity $v_\phi (= \dot{\phi})$ as

$$\dot{v}_\phi = 2\{[a+D] + F_R(t) - \frac{2b}{c}F_\Phi(t)\}v_\phi + \frac{2b}{c}v_\phi^2 + F_{add}(t), \tag{4}$$

where $F_{add}(t) = -\dot{F}_\Phi(t) + 2(a+D)F_\Phi(t) - \frac{2b}{c}F_\Phi(t)^2 + 2F_R(t)F_\Phi(t)$. This is a SDE with the parametric and additive noises are incorporated. So the fat-tailed distribution is expected. The physical origin of unstable behavior comes from the frequency-modulation due to the parametric noise in eq.(2). As shown by Konno, Kanemoto and Takeuchi

[4], the corresponding pdf of phase-velocity $v_\phi = \dot{\phi}$ takes in many cases the truncated stretched exponential (SE). $P_{pha}(v_\phi) = P_0 \exp(-|v_\phi - v_{\phi_0}|^\alpha)$, where P_0 is the normalization constant, α and v_{ϕ_0} are the fitting parameters in the range $1 < \alpha \leq 2$ and $-\pi \leq v_\phi \leq \pi$. Since the SDE for the phase velocity $v_\phi(t)$ is explicitly obtained in eq.(4), we obtain a generalized Cauchy distribution

$$P_{pha}(v_\phi) = (v_\phi^2 + 2v_1 v_\phi + v_0^2)^{-\alpha} \exp(-\beta v_\phi + \gamma \arctan[\delta v_\phi + \varepsilon] - \varphi(\vec{\theta})), \qquad (5)$$

where $\varphi(\vec{\theta})$ denotes the IGP in this pdf. Note here that The form of the pdf can be derived provided that $c < 0$. The unknown constants $\vec{\theta}$ (v_1, v_0, α, β, γ, δ and ε) are the fitting parameters in the range $1 < \alpha \leq 2$ and $-\pi \leq v_\phi \leq \pi$. Interestingly, in the actual experimental data analysis, the pdf with taking $v_1 = \beta = \gamma = 0$ is enough in many cases. The set of these parameters or the variance $\sigma_{v_\phi}^2$ and the kurtosis $\gamma_4 (= m_4/\sigma_{v_\phi}^4)$ become statistical mechanical quantities characterizing the intermittent bursting of phase slip motion. In some reactors, the crossover phenomenon between the Gaussian and the SE profile is observed. In the fat-tailed distributions, the divergence of the theoretical variance might be problematic. Since $-\pi \leq v_\phi \leq \pi$, the variance of v_ϕ does not diverge.

APPLICATION TO FLUCTUATION OF EEG

Figure 1 shows (a) the feature of phase, (b) of phase-velocity (PV) and (c) of the pdf of PV for EEG data of a healthy old person and a demented one. The phase-velocity exhibits bursting (phase-slip) motion (Fig.1(b)). Consequently, the pdf becomes a generalized Cauchy distribution (Fig.1(c)) in many cases for EEGs in stead of Eq.(5) (cf. the crossover with Gaussian PDF [4])

$$P_{ph}(v) = \frac{a^{2b-1}}{B(b-1/2, 1/2)} \frac{1}{(v^2 + a^2)^b} , \quad (-\pi < v < \pi), \qquad (6)$$

where $B(x,y)$ is the Beta function. As expected in these, the variance (the kurtosis) tends to have larger (smaller) values for healthy people. The corresponding PV maps for EEG are compared with the fractal maps in Fig.2. To give accurate discrimination method between healthy and demented one, one can utilize the entropy measures.

SUMMARY AND REMARKS

We have demonstrated a new nodal approach to the space-dependent analysis of fluctuation. By taking the method combined with the TIG, we extract successfully the space-dependency of EEGs via the PV map, the fractal map. Physically, we have found first (i) the existence of nonlinear pulling (NLP) $c|A|^2$ term; (ii) the fat-tailed pdf of v_ϕ is closely related to the existence of this NLP (the PV-PDF has Gaussian nature when $c = 0$).

For the detailed analysis of inhomogeneous structures for demented people, one must account for a class of space-dependent SGL equation under the influence of parametric excitations. [3] Application of the present IG approach can be extended to infer the local diffusion constants at local spatial position, which will be published elsewhere.

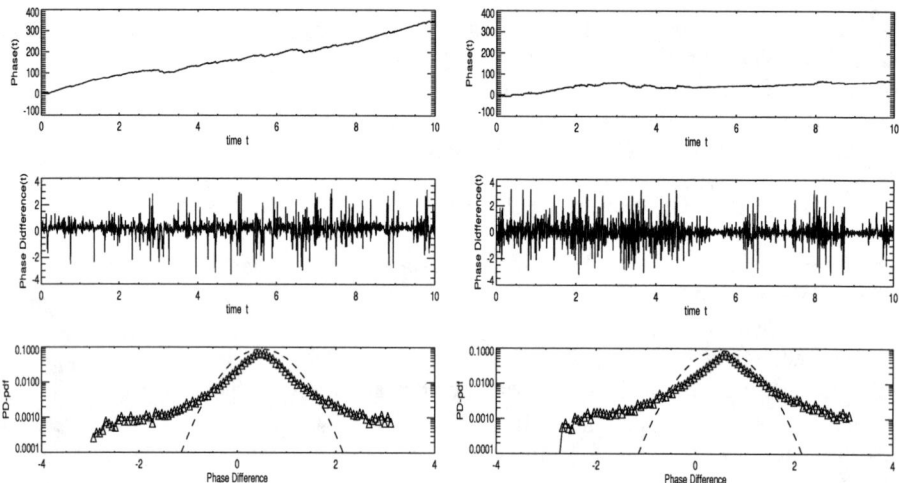

FIGURE 1. Left : (a) Phase, (b) PV , (c) PV-PDF for a healthy subject; Right : (a) Phase, (b) PV, (c) PV-PDF for a demented subject. The profile of Gaussian PDF is shown by dashed line in (c).

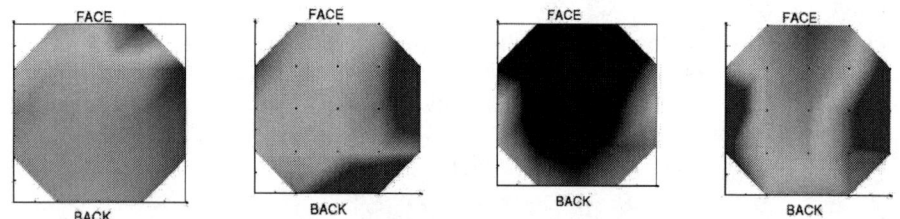

FIGURE 2. From left: (a) Fractal map and (b) PV map for a healthy subject; (c) Fractal map and (d) PV map for a demented subject. Lattice mapping is adopted for the international ten-twenty EEG system.

ACKNOWLEDGMENTS

This work is partially supported by the Japanese Society for the Promotion of Science (JSPS) through KIBAN 16500169.

REFERENCES

1. H. Konno, and P. Lomdahl, *J. Phys. Soc. Jpn*, **73**, 573–579 (2004).
2. S. Amari, and H. Nagaoka, *Methods of Information Geometry*, AMS, NY, 2000.
3. F. Watanabe, H. Konno, and S. Kanemoto, *Ann. Nucl. Energy*, **31**, 375–397 (2004).
4. H. Konno, S. Kanemoto, and Y. Takeuchi, *Progr. Nucl. Energy*, **43**, 201–207 (1999).

Fluctuations in Neuronal Activity: Clues to Brain Function

José L. Pérez Velazquez[1], Ramón Guevara[2], Jason Belkas[1], Richard Wennberg[3], Goran Senjanoviè[2], Luis García Dominguez[1]

[1]*Hospital for Sick Children, Brain and Behaviour Programme and Division of Neurology, 555 University Avenue, Toronto, Ontario M5G1X8, Canada*
[2]*Abdus Salam ICTP, Strada Costiera 11, 34014, Trieste, Italy*
[3]*Toronto Western Hospital, Krembil Neuroscience Centre, Toronto, Ontario M5T2S8, Canada.*

Abstract. Recordings from neuronal preparations, either in vitro or in the intact brain, are characterized by fluctuations, what is commonly considered as "noise". Due to the current recording and analysis methods, it is not feasible to separate what we term noise, from the "meaningful" neuronal activity. We propose that fluctuations serve to maintain brain activity in an optimal state for cognitive processing, not allowing it to fall into long-term periodic behaviour. We have studied fluctuations in magnetoencephalographic (MEG) recordings from normal subjects and epileptic patients, in electroencephalographic (EEG) recordings from children with impact injury, as well as in intracerebral electrophysiological recordings in freely moving rats. Specifically, we have determined phase locking patterns between brain areas from these recordings, which display fluctuations at different scales. We submit the idea that the variability in phase synchronization affords a more complete search of all possible phase differences in a hypothetical phase-locking state space that contributes to brain information processing. In brain pathologies, like epileptiform activity here studied, different levels of fluctuations in phase synchrony may favour the generation of stable synchronized states that characterize epileptic seizures. While the border between noise and high-dimensional dynamics is fuzzy, the scrutiny of neuronal fluctuations at different levels will provide important insights to the unravelling of the relation between brain and behaviour.

Keywords: neuroscience; phase synchrony; magnetoencephalography; electroencephalography.
PACS: 05.45.Xt; 87.19.Nn; 87.18.Hf

INTRODUCTION

The description of brain function involves the analyses of electrical recordings of brain activity, which are normally "noisy". This noise is normally removed and almost never considered, as we do not attribute any meaning to it, for, after all, it is not the "signal". However, the essence of brain function is reflected, to some extent, in that "noise", and some have considered the question of the brain as a "noisy processor" [1]. Recent evidence suggests that spontaneous, ongoing and apparently random neuronal activity, may be similar to "evoked" activity [2,3] and has a role in the determination of evoked responses and even in behavioural performance [4]. One of the central questions in neuroscience is whether synchronization of cellular activity is a mechanism for linking distributed cell networks that contributes to brain information processing. We propose that consideration of the fluctuations in cellular synchronization patterns, or phase locking, will reveal features of brain function that may help explain some characteristics of normal and pathological brain activity. We thus studied phase locking in two pathological conditions: epileptiform activity and traumatic brain injury, as well as during some cognitive tasks in healthy subjects. We present evidence for fluctuations in phase synchrony, even during apparently highly synchronous epileptic seizures.

METHODS

Magnetoencephalographic (MEG) recordings and selection of subjects.

MEG recordings were obtained from four patients with generalized epileptic seizures, as well as from four non-epileptic individuals. Electroencephalographic (EEG) recordings were obtained from brain trauma injured patients.

Phase synchronization analysis

For each patient and control subject, 16 two-minute segments of recordings were analyzed. Each of these segments was initially bandpassed using a FIRLCS filter [5] with a band pass of ± 2 Hz around a central frequency. Fifteen equally spaced frequencies (central frequencies) were studied from 3 to 55 Hz. The Hilbert Transform was applied and successive values of instantaneous phases were derived from the corresponding analytic signal. These phase series were then analyzed using sliding windows of 2.5 seconds. The Mean Phase Coherence was extracted as described in [6]. The number of channels in each MEG recording was 148, and 19 in EEG.

FLUCTUATIONS IN PHASE SYNCHRONIZATION

Phase synchrony patterns during epileptic seizures revealed enhanced local synchronization (within neighbouring neocortical areas), but the distant phase locking was not different from that observed during interictal periods (between seizures) and in normal subjects. We observed prominent fluctuations in phase locking during seizures between two specific MEG channels, as shown in figure 1. When comparing the time course of phase differences between two channels, a desynchronization was observed before and after the seizure (figure 2).

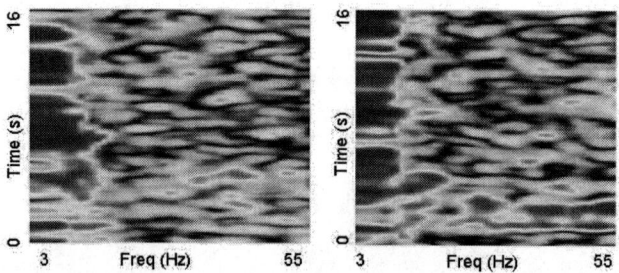

FIGURE 1. Phase synchrony between two neighbouring neocortical areas during two different seizures in an epileptic patient with generalized seizures. More synchronization is observed in the low frequency range (2-15 Hz), but note the fluctuations in synchrony during the seizure, when close cortical areas are supposed to be highly synchronized. Synchronization is colour-coded, from blue to red (minimum to maximum)

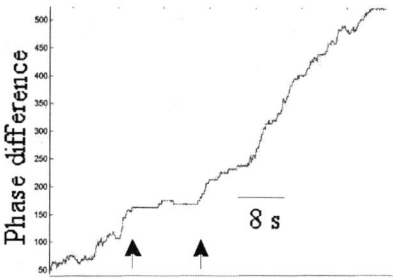

FIGURE 2. Phase difference computed between two MEG recording channels that recorded activity in neighbouring regions of the left frontal cortex in an epileptic patient. The seizure occurs between the arrows. Notice the steep slope preceding the flat, phase-locked region during the seizure, as well as that immediately after the ictal event. The slope indicates the phase slips, therefore increasing slope means less phase locking, and no slope indicates perfect phase locking.

Determination of phase synchronization in patients with traumatic brain injury and in control subjects revealed desynchronization immediately preceding periods of phase locking, similar to that observed with the epileptic patients aforementioned, though shorter in these cases as compared with the longer phase locking during seizures in the epileptic patients. In addition, less noise is observed in the raw electrographic traces during the seizure (figure 3B), and estimation of the instantaneous frequency of the recordings revealed a decrease in the fluctuations of instantaneous frequency as the seizure approaches (figure 3A).

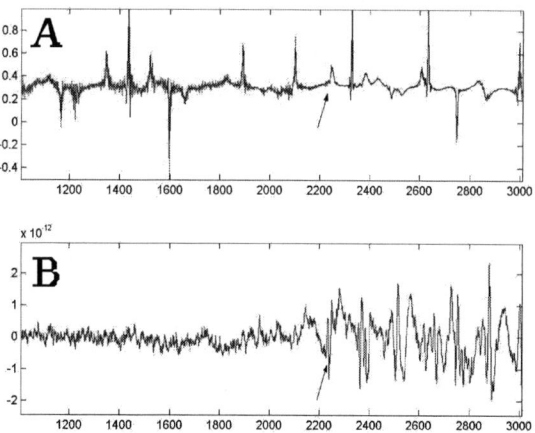

FIGURE 3. Instantaneous frequency (**A**) corresponding to the MEG trace in (**B**), depicting the start of a spike-and-wave seizure (arrow). Notice that the noise level in the instantaneous frequency (red trace) diminishes as the seizure approaches (arrow). Black line in (**A**) represents the wavelet-denoised signal. Subtraction of this signal from the original (red trace) provides an estimate of the noise, or fluctuations, in the instantaneous frequency.

To further examine the decrease in the instantaneous frequency fluctuations observed in the biological time series (figure 3), we used a computer simulation study to couple two Rossler oscillators, with a fixed level of noise, and increased coupling coefficient, and the instantaneous frequency of the computer-generated traces was determined as we did for the MEG traces shown in figure 3. There was a decrease in noise in the instantaneous frequency of the coupled Rossler oscillators as the level of coupling was increased. This noise was estimated as the residues of the difference between the wavelet-denoised skeleton of the signal and the signal itself. Thus, as we know exactly how much noise was present in this system (a parameter in the simulation), the reduced noise observed with increased coupling indicates that this reduction is not due to a recording effect, but maybe to a "real" diminution of the fluctuations when more synchronized activity is observed.

DISCUSSION

The description of complex systems with noise is best achieved with probabilistic approaches. Synchronization, in the presence of noise, can certainly occur but only in short time intervals [7]; hence, we can expect that fluctuations in phase locking in the activity of nervous systems will be most prominent. The enhanced desynchronization immediately before and after a period of phase locking suggests that desynchronization may favour, sometimes, frequency locking. We speculate that distinct "noise" levels may control the fall into a synchronization manifold. The description of synchronization in terms of invariant manifolds has been proposed [8], and it will be of interest to determine the characteristics of the manifolds that may be obtained from time series recordings, such as MEG or EEG. In summary, we propose that fluctuations and noise in neuronal activity should be studied in detail, rather than being discarded by averaging methods.

ACKNOWLEDGMENTS

Our work is supported by a Discovery grant from the National Science and Engineering Research Council of Canada (NSERC).

REFERENCES

1. Adey, W.R. *Intern. J. Neuroscience* **3**, 271-284 (1972).
2. Tsodyks, M., Kenet, T., Grinvald, A. and Arieli, A. *Science* **286**, 1943-1946 (1999).
3. Kenet, T., Bibitchkov, D., Tsodyks, M., Grinvald A. and Arieli, A. *Nature* **425**, 954-956 (2003).
4. Linkenkaer-Hansen, K., Nikulin, V.V., Palva, S., Ilmoniemi, R.J. and Matias Palva, J. *J. Neurosci.* **24**, 10186-10190 (2004).
5. Rosenblum, M.G., Pikovsky, A.S. and Kurths, J. *Phys. Rev. Lett.* **76**, 1804 (1996).
6. Mormann, F., Lehnertz, K., David, P. and Elger, C.E. *Physica D* **144**, 358-369 (2000).
7. Stratonovich, R.L. "Selected problems of fluctuation theory in radiotechnics", in Sov. Radio, Moscow 1961
8. Josic, K. *Phys. Rev. Lett.* **80**, 3053-3056 (1998)

Changes in the Hurst Exponent of Heart Rate Variability during Physical Activity

Naoko Aoyagi[*†], Ken Kiyono[*], Zbigniew R. Struzik[*], and Yoshiharu Yamamoto[*]

[*]*Educational Physiology Laboratory, Graduate School of Education, University of Tokyo, 7--3--1 Hongo, Bunkyo-ku, Tokyo 113-0033, Japan*
[†]*The Institute for Science of Labour, 2-8-14 Sugao, Miyamae-ku, Kawasaki, Kanagawa 216-8501, Japan*

Abstract. We examine fractal scaling properties of heart rate variability using detrended fluctuation analysis (DFA), during physical activity in healthy subjects. We analyze 11 records of healthy subjects, which include both usual daily activity and experimental exercise. The subjects were asked to ride on a bicycle ergometer for 2.5 hours, and maintained a heartbeat interval of 500-600 ms. In order to estimate the long-range correlation in the series of heartbeat intervals during controlled physical activity, we apply DFA to the data set with the third-order polynomial trend removed. For all records during exercise, we observe a characteristic crossover phenomenon at $\approx$ 300 beats. The scaling exponent in the range > 300 beats (> 3 minutes) during exercise decreases and tends to be closer to white noise ($\approx$ 0.5), which corresponds to uncorrelated behavior. The long-range scaling exponent during exercise is significantly lower than that during daily activity in this range. Contrary to the currently held view, our results indicate a breakdown in long-range correlations and $1/f$-like scaling, rather than the increase in the Hurst exponent characteristic of a (congestive) increase in afterload and observed, e.g., in congestive heart failure (CHF) patients. Further, our results suggest an increased load imbalance induced departure from critical-like behavior, which has recently been reported in healthy human heart rate during daily activity [1].

INTRODUCTION

Healthy human heart rate has long been known to exhibit $1/f$-type fluctuations [2–5], and has recently also been attributed properties of multifractal scaling [6]. This complex dynamics, resembling non-equilibrium [7] and/or multiscale [8] dynamics in physics, has been demonstrated to be independent of human behavior — the statistical properties of the heart rate remain unaltered even after eliminating known behavioral modifiers [9, 10] — suggesting that, during usual daily activity, the origin of heart rate complexity lies in the intrinsic dynamics of the physiological regulatory system.

Recently the view has been suggested of cardiac neuroregulation as a system in a critical state, and permanently out of equilibrium, in which concerted interplay of the sympathetic and parasympathetic nervous systems is required for preserving momentary "balance" [11]. In Kiyono *et al.* [1], we demonstrate that the heart rate retains robust scale invariance of the non-Gaussian increment probability density function (PDF) during usual daily activity and constant routine protocol. One possible hypothesis is that the criticality in healthy human heart rate is linked to the optimality of its functional performance. To confirm this, it is important to study the relation between the breakdown of the optimal control from a physiological point of view and characteristics of the critical state-like fluctuations, such as the presence of $1/f$ long-range correlations. The question arises as to whether such $1/f$-type long-range correlation properties are observed in conditions with consistently higher heart rate during physical exercise, when the regulatory system is continuously forced.

Some recent studies of heart rate variability during exercise report significant scaling differences between rest and exercise in healthy individuals, and an increase in the *short*-range scaling exponent during physical activity [12-14]. However these studies have severe limitations: 1) The analyses are based solely on short-term recordings, which are not long enough accurately to determine the long-range correlation properties of heart rate variability; 2) They are performed with low intensity load, which results in only a slight increase in the heart rate. To account for this, in the present study, we conducted experimental exercise in which the subjects exercised constantly until sufficiently large amounts of data were collected. Then, by studying changes in the correlation properties of heart rate during: 1) Normal daily activity; and 2) Constant experimental exercise, we seek to identify conditions in which correlations in long-range heart rate variability hold and break down.

METHODS

Eleven non-smoking healthy subjects participated in this study (9 male and 2 female; age: 21-35 yr). Each subject gave his or her informed consent to participating in this institutionally approved study after the test protocol had been fully described.

The subjects were instructed to keep to their regular sleep schedules and to refrain from vigorous exercise or alcohol consumption during three days prior to the experiment. All the subjects reported to the laboratory at 09.30 h. After the placement of electrodes for the electrocardiogram (ECG), the heart rate data collection commenced at 10:00~11:00 h. The experimental exercise was performed on a bicycle ergometer (Aerobike 75XL, COMBI-Wellness Corporation, Japan) between 14:00-18:30 h. The subjects were familiarized with the cycling protocol prior to performing the exercise test. They were asked to perform the experimental exercise for 2.5 hours, at a pedal speed of 50-80 rpm, with a workload of 50-90 watts, and maintaining a heart rate of 100-120 beats/min (corresponding to heartbeat intervals of 500-600 ms). They were required to follow their normal daily routine, without taking any daytime nap and without consuming alcohol, until 18:00 h the next day.

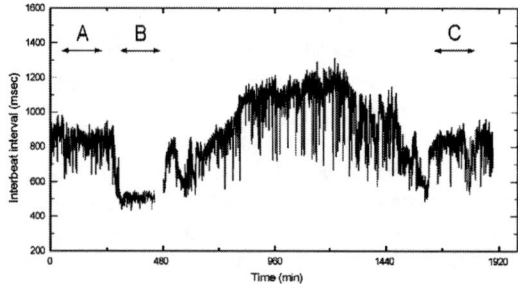

FIGURE 1. A representative example of heartbeat intervals for a healthy subject measured over 30 hours. We extracted subintervals during: (A) Daily activity before the exercise; (B) Experimental exercise; (C) Daily activity the next day.

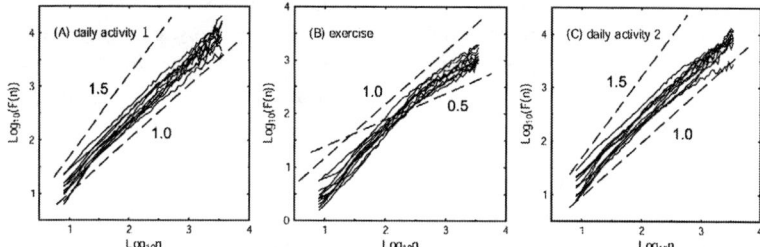

FIGURE 2. DFA scaling exponents of 11 healthy subjects. The line of log $F(n)$ vs log n for data from heartbeat interval time series during: (A) Usual daily activity; (B) Experimental exercise; (C) Daily activity the next day.

During the experiments, beat-to-beat RR intervals (RRI; from the standard V_5 lead of the ECG) were measured using a portable heart rate monitor (AMX, Nihon Koden Wellness Corporation, Japan). Any abnormal RRI's, caused either by body movements or occasional extrasystoli, were corrected by either omitting beats (for those < 300 ms) or inserting beats (for those double or triple the length of the preceding intervals).

In the present study, we examine the long-range scaling properties of heart rate variability, i.e., the interval fluctuations, by using detrended fluctuation analysis (DFA) [15]. For fractal signals, a power law relation is found between the average magnitude of the fluctuations $F(n)$ and the number of points n: $F(n) \sim n^\alpha$, where the scaling exponent α quantifies the degree of the correlations. Uncorrelated time series yield $\alpha = 0.5$, long-range anti-correlations result in $\alpha < 0.5$, and long-range positive correlations result in $\alpha > 0.5$. In order to estimate the long-range correlation in the series of heartbeat intervals during usual daily activity and exercise (14,000 beats for all conditions, Fig. 1), we apply DFA with the third-order polynomial trend removed.

RESULTS AND DISCUSSION

A representative example of heart rate variability time series during the experiment is shown in Fig. 1. The heartbeat intervals decrease consistently during exercise (tachycardia). During the experimental exercise protocol, the mean heartbeat interval (522 ms) was successfully maintained at a target value of 500-600 ms (Table 1).

We extract subintervals during: (A) Usual daily activity before the exercise; (B) Experimental exercise; (C) Daily activity the next day. These are next analyzed for all eleven subjects using the DFA; see Fig. 2. For daily activity 1 and 2, we observe similar behavior with consistent scaling, spanning both short- and long-range regimes. On the other hand, during the experimental exercise, we observe a characteristic crossover phenomenon at around 300 beats (about 3 min) for all

Table 1. Heart rate variability measures during daily activity 1, exercise and daily activity 2.

	daily activity 1	exercise	daily activity 2
heartbeat interval (ms)	839.5±130.4	522.0±21.7	869.6±129.7
α_1	1.07± 0.1	1.20± 0.2	1.09± 0.11
α_2	1.06± 0.1	0.73± 0.1 *†	0.91± 0.16

Values are means ± SD. α_1, short-term fractal scaling exponent; α_2, long-term fractal scaling exponent.
*$p < 0.01$ daily activity 1 vs exercise, and †$p < 0.01$ daily activity 2 vs exercise.

individuals, where there is a change in the correlation behavior between short- and long-range regimes. The scaling exponents are estimated for the short and long range, excluding the crossover region. We denote the average values of scaling exponents in the short-range as α_1 (estimated for scales $50 < n < 200$) and the scaling exponent of the long regime as α_2 (estimated for $500 < n < 2000$) in Table 1. Please note that our definition of α_1 and α_2 does not aim to reproduce that in Ref. [15].

We observe a significant difference in the α_2 of both daily activity 1 and 2 in comparison with the α_2 of experimental exercise (by paired t-test, $p < 0.001$); see Tab. 1. During exercise, α_2 decreases and tends to be globally closer to white noise (≈ 0.5), which corresponds to uncorrelated behavior. In previous studies, using only short data lengths, the scaling exponent during physical activity is shown to increase [14], and its value to correspond to correlated behavior [12, 13]. In contrast, we have demonstrated a breakdown of long-range correlations during strenuous exercise, as opposed to the $1/f$ scaling in a waking, non-exercise state.

The higher heart rate is observed both during exercise and in CHF patients. It is characterized pathologically by an increased afterload with an unchanged preload in CHF. On the contrary, in the experimental exercise, there is an increased preload of the heart from the working muscles (i.e., venous return) and a decreased afterload by vasodilation. In this study, we have demonstrated that the long-range correlation is lost in the exercise phase. It indicates that under too much loading, the heart control system is unable to maintain long-range correlation, and suggests the importance of "flow-limited" balance for long-range correlation.

In some transport systems, the critical point corresponds to the phase transition point from an "uncrowded" state to a "congested" state in the transportation routes [16-18]. It has also been shown that a functional advantage of the system being at or near the critical point is that maximum efficiency of transportation is realized [17]. The criticality hypothesis in healthy human heart rate argues that the central neuroregulation continually brings the cardiovascular system to a critical state in order to maximize its functional ability. The observed breakdown of the long-range correlations suggests a departure from the optimal functional ability regime, likely caused by the imbalance between the cardiac preload and afterload during strenuous exercise.

REFERENCES

1. K. Kiyono et al, *Phys. Rev. Lett.*, **93**, 178103 (2004).
2. M. Kobayashi and T. Musha, *IEEE Trans. Biomed. Eng.*, **29**, 456 (1982).
3. C. K. Peng et al., *Phys. Rev. Lett.*, **70**, 1343 (1993).
4. J. P. Saul et al., *Comput. Cardiol.*, **14**, 419 (1987).
5. Y. Yamamoto and R. L. Hughson, *Am. J. Physiol.*, **266**, R40 (1994).
6. P. C. Ivanov et al., *Nature (London)*, **399**, 461 (1999).
7. C. K. Peng et al., *Integr Physiol. Behav. Sci.*, **29**, 283 (1994).
8. D. C. Lin and R. L. Hughson, *Phys. Rev. Lett.*, **86**, 1650 (2001).
9. L. A. N. Amaral et al., *Phys. Rev. Lett.*, **86**, 6026 (2001).
10. N. Aoyagi, K. Ohashi, and Y. Yamamoto, *Am. J. Physiol.*, **285**, R171 (2003).
11. Z.R. Struzik et al., *Phys. Rev. E*, **70**, 050901 (2004).
12. R. Karasik et al., *Phys. Rev. E*, **66**, 062902 (2002).
13. M. Martinis et al., *Phys. Rev. E*, **70**, 012903 (2004).
14. M.P. Tulppo et al., *Am J Physiol.*, **280**, H1081 (2001).
15. C. K. Peng et al., *Chaos*, **5**, 82 (1995).
16. D. Chowdhury, L. Santen, and A. Schadschneider, *Phys. Rep.*, **329**, 199 (2000).
17. M. Takayasu, H. Takayasu, and K. Fukuda, *Physica A*, **277**, 248 (2000).
18. M. Hou et al., *Phys. Rev. Lett.*, **91**, 204301 (2003).

Spatial asymmetric retrieval states in binary attractor neural network

Kostadin Koroutchev[*,†] and Elka Korutcheva[**,‡]

*Escuela Politecnica Superior, Universidad Autónoma de Madrid, Spain
†Inst. for Personal Computing and Communication Systems, Bulgarian Academy of Sciences, 1113 Sofia, Bulgaria
**Depto. de Fisica Fundamental, UNED, Madrid, Spain
‡G.Nadjakov Inst. Solid State Physics, Bulgarian Academy of Sciences, 1784 Sofia, Bulgaria

Abstract. In this paper we show that during the retrieval process in a binary Hebb attractor neural network, spatial localized states can be observed when the connectivity of the network is distance-dependent and there is an asymmetry between the retrieval and the learning states.

Keywords: attractor neural networks, bump formations, replica formalism
PACS: 64.60.Cn, 84.35.+i, 89.75.-k, 89.75.Fb

INTRODUCTION

In a very recent publication [1] it was shown that using linear-threshold model neurons, Hebb learning rule, sparse coding and distance-dependent asymmetric connectivity, spatial asymmetric retrieval states (SAS) can be observed. Similar results have been reported in the case of Hebb binary model for associative neural network [2]. This asymmetric states are characterized by a spatial localization of the activity of the neurons, described by the formation of local bumps.

In this paper we show that the major factor to observe spatial asymmetric activity is the distance-dependent connectivity of the neurons in the network and the different level of activity in the retrieval and learning states.

Our model consists of an attractor neural network (NN) model of Hebbian type, formed by N binary neurons $\{S_i\}, S_i \in \{-1,1\}, i=1,...,N$, storing p binary patterns $\eta_i^\mu, \mu \in \{1...p\}$, and we assume a symmetric connectivity between the neurons $c_{ij} = c_{ji} \in \{0,1\}, c_{ii} = 0$. $c_{ij} = 1$ means that neurons i and j are connected. We regard only connectivities in which the fluctuations between the individual connectivity are small, e.g. $\forall_i \sum_j c_{ij} \approx cN$, where c is the mean connectivity.

The learned patterns are drawn from the distribution:

$$P(\eta_i^\mu) = \frac{1+a}{2}\delta(\eta_i^\mu - 1) + \frac{1-a}{2}\delta(\eta_i^\mu + 1),$$

where the parameter a is the sparsity of the code. Further on we will work in terms of variables $\xi_i^\mu = \eta_i^\mu + a$.

To impose a condition on the mean activity, in order to introduce asymmetry between the learning and the retrieval states, we add an extra term H_a to the Hamiltonian

$$H_a = NR \sum_i S_i/N.$$

This term actually favors states with lower total activity $\sum_i S_i$ that is equivalent to decrease the number of active neurons, creating asymmetry between the leaning and the retrieval states. When $H_a = 0$ we have equal mean activity of the learned pattern and the the retrieval state.

ANALYTICAL ANALYSIS

For the analytical analysis of the SAS states, we consider the decomposition of the connectivity matrix c_{ij} by its eigenvectors $a_i^{(k)}$:

$$c_{ij} = \sum_k \lambda_k a_i^{(k)} a_j^{(k)}, \quad \sum_i a_i^{(k)} a_i^{(l)} = \delta_{kl}, \tag{1}$$

where λ_k are the corresponding (positive) eigenvalues.

Following the classical analysis of Amit et al. [3], we study binary Hopfield model [4]

$$H = -\frac{1}{cN} \sum_{ij\mu} S_i \xi_i^\mu c_{ij} \xi_j^\mu S_j - \sum_{v=1}^s h^v \sum_i \xi_i^v S_i + NR\overline{S_i b_i^0}, \tag{2}$$

where we have assumed Hebb's rule of learning [5].

By using the "replica formalism" [6], for the averaged free energy per neuron we get:

$$f = \lim_{n \to 0} \lim_{N \to \infty} \frac{-1}{\beta nN} (\langle\langle Z^n \rangle\rangle - 1), \tag{3}$$

where $\langle\langle ... \rangle\rangle$ stands for the average over the pattern distribution $P(\xi_i^\mu)$, n is the number of the replicas, which later are taken to zero and β is the inverse temperature.

The saddle-point method [3], gives the following final form for the free energy per neuron:

$$\begin{aligned} f = &\frac{1}{2c}\alpha + \frac{1}{2}\sum_k (m_k)^2 + \frac{\alpha\beta(1-a^2)}{2}\sum_k r_k q_k + \frac{\alpha\beta(1-a^2)}{2}\sum_k \mu_k^2 r_k + \\ &+ \frac{\alpha}{2\beta}\sum_k [\ln(1-\beta(1-a^2)\mu_k + \beta(1-a^2)q_k) - \\ &- \beta(1-a^2)q_k(1-\beta(1-a^2)\mu_k + \beta(1-a^2)q_k)^{-1}] - \\ &- \frac{1}{\beta}\int \frac{dz e^{-\frac{z^2}{2}}}{\sqrt{2\pi}} \ln 2\cosh\beta \left(z\sqrt{\alpha(1-a^2)\sum_l r_l b_i^l b_i^l + \sum_l m_l \xi_i b_i^l + Rb_i^0} \right). \end{aligned} \tag{4}$$

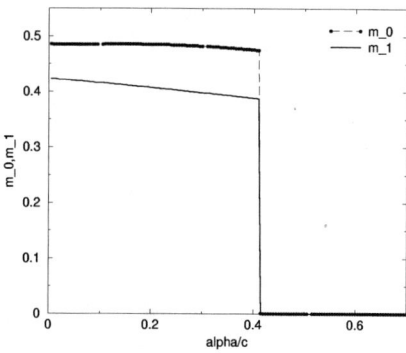 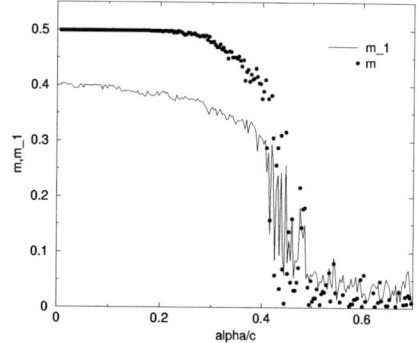

FIGURE 1. Left: The order parameteres m_0, m_1 for $a = 0.2$ and $R = 0.57$. The result of the simulations is represented on the right side.

In the last expression we have introduced the variables $\mu_k = \lambda_k/cN$ and $b_i^k \equiv a_i^{(k)}\sqrt{\lambda_k/c} \sim 1$ and we have used the fact that the average over a finite number of patters ξ^ν can be self-averaged [3].

The different order parameters, present in the above expression are the following: the overlap between patterns and neurons $m_k = \overline{(\eta - a)b_i^k S_i}$, where the k-mode corresponds to the k-th eigenvalue of the connectivity matrix and the over line is the space average, the neural activity, $q_k = \overline{\sum_i (b_i^k)^2 S_i^2}$ and r_k is the parameter conjugate to q_k. Finally, $\alpha = p/N$ is the storage capacity of the network.

RESULTS AND DISCUSSION

The analysis of the equations for the corresponding order parameters at zero temperature is shown in Fig. 1 (left). The sharp bound is due to the limitation of the equations only to the first two terms of the OP m, r, C, as well as to the finite scale effects.

In order to compare the results, we also performed computer simulations. To this aim we chose the network's topology to be a circular ring, with a distance measure

$$|i - j| \equiv \min(i - j + N \bmod N, j - i + N \bmod N)$$

and used the same connectivity as in Ref.[1] with typical connectivity distance $\sigma_x N$:

$$P(c_{ij} = 1) = c\left[\frac{1}{\sqrt{2\pi\sigma_x N}} e^{-(|i-j|/N)^2/2\sigma_x^2} + p_0\right].$$

The corresponding behavior of the order parameters m_0, m_1 in the zero-temperature case, obtained by the simulations, is represented in Fig. 1 (right). As can be seen, there is a good correspondence between the numerical solution of the analytical results and the results obtained by simulation.

The behavior of the information capacity, presented in the left panel of Fig.2, is non trivial and shows a well pronounced maximum for intermediate values of the sparsity of

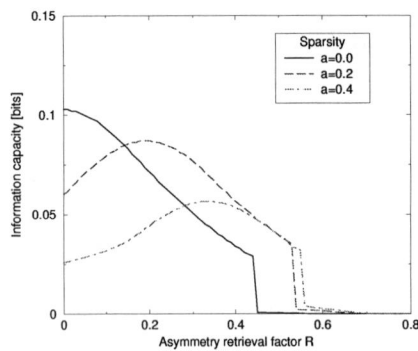 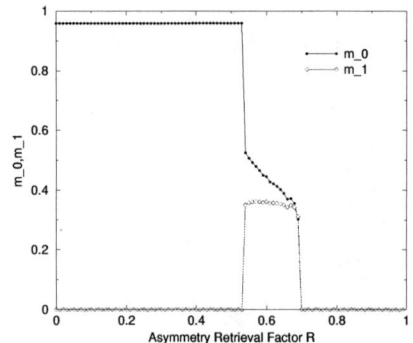

FIGURE 2. Left: the information capacity versus the asymmetry retrieval factor R for different values of the sparsity a. Right: m_0, m_1 for $a = 0.2$.

the code a. There is no maximum of the mutual information for $a = 0$. The corresponding behavior of the order parameters m_0 and m_1, presented in the right panel of Fig. 2, shows that inside the SAS region, their values are significantly different form zero, but the information capacity is very low although non zero.

The analysis of the corresponding phase diagram [7], shows that there exist states with $a = 0$ and spatial asymmetry retrieval states. Therefore, the sparsity of the code is not a necessary condition for SAS, but it makes the effect of SAS better pronounced.

On the other hand there is no state with SAS and $H_a = 0$, and therefore the asymmetry between the retrieval activity and the learned patterns is essential for the observation of the phenomenon.

In conclusion, we have shown that the presence of the term H_a, which introduces asymmetry between retrieval and learning states is sufficient for the existence of the spatially asymmetric retrieval states. The drop of the information capacity of the network is very significant in the phase transition point between symmetric and asymmetric retrieval states.

Acknowledgments

This work is financial supported by Departamento de Física Fundamental, UNED and by Spanish Grants CICyT, TIC 01–572, TIN 2004–07676, DGI.M.CyT.BFM2001-291-C02-01 and "Promoción de la Investigación UNED'02".

REFERENCES

1. Y.Roudi and A.Treves, *JSTAT* P07010 (2004).
2. K.Koroutchev and E.Korutcheva, *Preprint ICTP*, Trieste, Italy, IC/2004/91.
3. D.Amit, H.Gutfreund and H.Sompolinsky, *Ann.Phys.*, **173**, 30 (1987).
4. J.Hopfield, *Proc.Natl.Acad.Sci.USA*, **79** 2554 (1982).
5. D.Hebb, *The Organization of Behavior: A Neurophysiological Theory*, Wiley, New York, 1949.
6. M.Mézard, G.Parisi and M.-A.Virasoro, *Spin-glass theory and beyond*, World Scientific, 1987.
7. K.Koroutchev and E.Korutcheva, Central European Journal of Physics, 2005, in press.

Relationships among One-Minute Oscillations in Oxygen Saturation Level of Blood and Hemoglobin Volume in Calf Muscular Tissue and One-Minute Wave in Body Fluid Volume Change during Upright Standing in Humans

Kinsaku Inamura*, Tadaaki Mano**, and Satoshi Iwase ***

*Department of Physical and Health Education, Faculty of Education,
Shizuoka University, Shizuoka 422-8529, Japan
** Tokai Central Hospital, Kakamigahara 504-8601, Japan
*** Department of Physiology, Aichi Medical University, Nagakute 480-1195, Japan

Abstract. One-minute oscillations in the oxygen saturation level of blood and the hemoglobin volume in calf muscular tissue were found during upright standing in humans. Spectral analyses indicated that one source of the one-minute wave in body fluid volume change is the spontaneous constriction of blood vessels triggered by an elevation of transmural pressure when blood pooling is evoked.

Keywords: one-minute oscillation, blood volume, postural sway

INTRODUCTION

When a human stands up, gravity causes about 500 ml of venous blood to shift to the lower part of the body. If the compensatory mechanism for the blood shift acted insufficiently, humans would suffer from syncope or fainting due to insufficient cerebrovascular circulation during prolonged upright standing. Major compensatory measures for this blood shift in the physiological system are vagal withdrawal and subsequent sympathetic activation due to the unloading of baroreceptors.

Our previous study showed that "minute rhythms" in postural sway and cardiovascular circulation, as well as sympathetic nerve activation, contribute to the compensatory action. In postural sway during quiet standing in humans, there is a frequency component of approximately one oscillation per minute. Muscle contractions in the lower legs and abdomen enhance the one-minute wave in body fluid volume change (1-MW in FV), which pumps venous blood to the heart by upward propagation. Our previous reports suggested that the origin of the 1-MW in FV might be attributed, in part, to the autonomic constriction of vessels because transmural pressure is elevated when venous blood pooling is evoked. However, there have been no reports of data obtained by directly measuring blood volume change.

The purpose of the present study was to find the one-minute oscillation (1-MO) in blood volume change by near-infrared spectroscopy (NIRS) and to clarify the mechanism for the action of the 1-MO of the blood parameters on 1-MW in FV.

METHODS

Subjects were 11 healthy male volunteers 18-22 years old. Their cardiovascular values at the initial stage of standing were as follows: 123.7 ± 3.4 mmHg systolic blood pressure (BP), 77.7 ± 2.1 mmHg diastolic BP, and 73.9 ± 3.2 bpm heart rate (HR). Each subject stood upright on a force plate for 40 minutes. The oxygen saturation level of the blood (StO_2) and the hemoglobin (Hb) volume in the tissue of the calf muscle (soleus) were measured by near-infrared spectroscopy (Biomedical Science PSA-III N) with the wavelength 700-830 nm, and the distances between the transmitter and two receivers were set at 10 mm and 25mm. The total blood volume was calculated from the sum of $Hb-O_2$ and $Hb-CO_2$. The stroke volume (SV) of the heart, cardiac output (CO), blood flow, and mean BP in the lower leg, as well as fluid volumes of the calf and thorax, were calculated from impedance plethysmography (Nihon Koden AI-601G and ED-601G). Vascular resistance was calculated from the mean BP and CO. HR and surface electromyograms were measured using a bioelectric amplifier (Nihon Koden AM-601G). Peripheral BP at the heart level was measured with Finapres (Ohmeda 2300). Body circumferences (BC) at 14 parts of the body were measured by mercury-infused rubber strain gauge plethysmography. Fluid volume in the abdomen was calculated from BC. Displacement of the foot pressure center (FPC), indicating postural sway, was measured by platform stabilometry using a gravicorder (Patella K-105s). All measurements were carried out simultaneously while the subject was standing. Analog signals were converted into digital data at a sampling frequency of 1 kHz and entered into a personal computer. The 1-MO was detected by means of auto-power spectral analysis. Cross-power spectral density functions and the coherency between two 1-MOs were calculated from the time delay of fluctuations.

RESULTS

A direct current trend and very low frequency components in the oscillation of BC indicated a shift in body fluid. During prolonged upright standing, the BC of the upper body had an overall tendency to decrease, whereas that of the lower body had an overall tendency to increase, indicating venous blood pooling in the calf and abdomen.

Auto-power spectral analysis revealed a 1-MO of 0.0147 ± 0.0045 Hz (period: 68.0 s ± 1.9 s) in BC and FPC. Cross-power spectral analysis on BC with reference to the 1-MO in FPC revealed that the 1-MO at the calf propagated upward and disappeared at the lower abdomen. The 1-MO at the abdomen propagated to the chest as a 1-MW in FV. A clear 1-MO was found in the StO_2 and the Hb volume in the muscular tissue of the soleus. The same 1-MO was found in the other parameters. The mean frequency was almost the same as that in BC: 0.0146 ± 0.0047 (period: 68.5 s ± 2.1 s).

Phase differences of 1-MOs between FPC and each variable, obtained from cross-power spectral analysis, were used for a time-series analysis of these variables. The order of the increase phase of the 1-MOs in one cycle of 1-MW in FV was as follows: 1) fluid volume in the lower abdomen, 2) mean BP in the lower leg, 3) calf fluid volume, 4) HR, 5) Hb volume in the muscular tissue of the soleus, 6) blood flow in the lower leg, 7) StO_2 in the muscular tissue of the soleus, 8) mean BP at the heart level, 9) total peripheral vascular resistance, 10) FPC, 11) electromyogram of the soleus, 12) vascular resistance in the lower leg, 13) SV of the heart, 14) electromyogram of the abdomen, 15) thoracic fluid volume, and 16) CO (Figure 1).

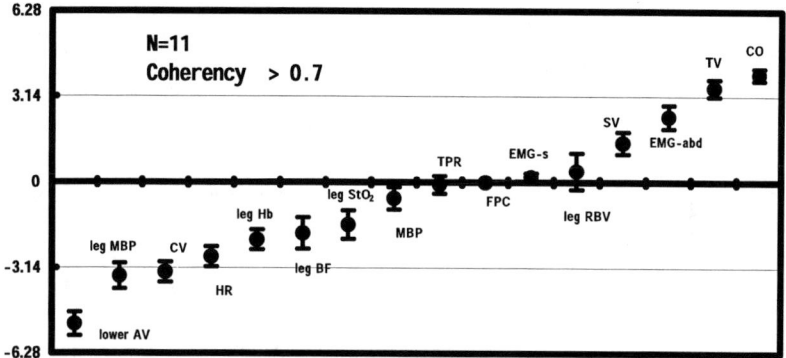

FIGURE 1. Phase difference of one-minute oscillation refers to that in FPC. Ordinates exhibit phase difference (rad). Plots are increase phase. Vertical line denotes ± 1 SE. Plus value denotes a phase delay from the phase of FPC. Minus value denotes an advance phase from the phase of FPC. Lower AV: fluid volume in the lower abdomen, leg MBP: mean BP in the lower leg, CV: calf fluid volume, leg Hb: Hb volume in the muscular tissue of the soleus, leg BF: blood flow in the lower leg, leg StO_2: oxygen saturation level of the blood in the muscular tissue of the soleus, MBP: mean BP at heart level, TPR: total peripheral vascular resistance, EMG-s: electromyogram of the soleus, leg RBV: vascular resistance in the lower leg, EMG-abd: electromyogram of the abdomen, TV: thoracic fluid volume.

DISCUSSION

Near-infrared spectroscopy (NIRS, PSA-III N) revealed the StO_2 and Hb volumes in micro-vascular circulation within the muscular tissue. The oscillation of Hb volume directly indicates the blood volume in the tissue. In the previous study, the volume shift between lymph volume and blood volume could not be discriminated. In the present study, 1-MOs were found in the Hb volume and in the StO_2, important evidence for the origin of a 1-MW in FV. These results indicate that the origin of the 1-MW in FV is caused, in part, by a spontaneous constriction of the blood vessels triggered by an elevation of transmural pressure when pooling is evoked. In addition, the phase order obtained from the time-series analysis of the valuables explains the functioning mechanism of the 1-MW in FV more clearly than previous reports have. The phase order in the 1-MO in the variables can be classified as follows.

Phase 1. When a human stands up, gravity causes venous blood to be pooled initially in the abdomen due to the valveless structure of the abdomen. Simultaneously, calf venous blood volume increases, leg BP rises, and FPC is placed backward on the sole of the foot. HR increases just after the reaction to standing. In this phase, thoracic blood volume is in the increase phase of advanced 1-MW in FV. Therefore, the increase in HR is mainly caused by a sympathetic nerve response from the carotid sinus reflex. An increase in blood flow and Hb volume is delayed slightly because of the perfusion time to the capillary. Then StO_2 is elevated by an increase in Hb.

Phase 2. When the pooling in the calf muscle exceeds a threshold level, the smooth muscles of the blood vessels begin to contract automatically by the stretching stimulus. Venous blood is pressed toward the heart in cooperation with the sympathetic nerve response. The FPC is shifted forward by the rising center of mass associated with the blood shift because the center of mass is placed in front of the supporting point (the ankle joint) in the human body during upright standing.

Phase 3. The righting reflex caused by the vestibular postural reflex and the stretch reflex of the calf muscle is activated by a forward inclination of the body. The righting reflex brings out a stronger contraction of the soleus, a rise in vascular resistance in the lower leg, and a rise in total peripheral vascular resistance. Contractions of the calf and other muscles pump up venous blood, and an increase in SV is derived from an increase in venous return.

Phase 4. During phases 2 and 3, the initial calf blood volume is shifted to the abdomen. When the body is inclined backward by the righting reflex, the abdominal muscles contract. Muscle and abdominal pumping increase the blood volume in the chest, increasing the thoracic blood volume. In this phase, HR is in the increase phase of the next 1-MW in FV. Therefore, increases in the thoracic blood volume and in HR bring about an increase in CO. The subsequent increased loading into baroreceptors in the thorax and carotid arteries inactivates the sympathetic nerve responses. Thus, a cycle of 1-MW in FV is completed, and the next cycle begins with a downward blood shift by continuous hydrostatic pressure.

In conclusion, the human body possesses a cyclic rhythm of approximately one minute, which facilitates the blood volume shift against blood pooling due to gravity. This mechanism may contribute to cardiovascular homeostasis and help to maintain sufficient cerebral blood flow in upright standing. The results of this study strongly indicate that the origin of 1-MW in FV is attributed, in part, to the spontaneous constriction of the blood vessels triggered by an elevation of transmural pressure when pooling is evoked.

This work was supported by a Grant-in-Aid for Scientific research (C) (No. 11680020) from the Ministry of Education, Science, Sports, and Culture of Japan.

REFERENCES

1. K. Inamura, T. Mano, S. Iwase, Y. Amagishi and S. Inamura, J. Appl. Physiol. **81**(1), 459-469 (1996).
2. K. Inamura, T. Mano, S. Iwase, Y. Amagishi and K. Aoki, "Fluctuation of body sway which has about 1 minute period and muscle pumping in the lower legs during static standing in humans." In: Noise in Physical Systems and 1/f Fluctuations, edited by T. Musha et al., ICNF 1991 Proceedings, Kyoto: Ohmusha, Ltd., 1991, pp. 727-730.

Modelization of Thermal Fluctuations in G Protein-Coupled Receptors

C. Pennetta*, V. Akimov*, E. Alfinito*, L. Reggiani*, G. Gomila[†], G. Ferrari**, L. Fumagalli** and M. Sampietro**

*Dipartimento Ingegneria dell'Innovazione and National Nanotechnology Laboratory, Università di Lecce, Via Arnesano, 73100 Lecce, Italy
[†]Department d'Electronica and Research Centre for Bioelectronics and Nanobioscience, Universitat de Barcelona, C/ Josep Samitier 1-5, 08028 Barcelona, Spain
**Dipartimento di Elettronica ed Informazione, Politecnico di Milano, P.zza Leonardo da Vinci 32, 20133 Milano, Italy

Abstract. We simulate the electrical properties of a device realized by a G protein coupled receptor (GPCR), embedded in its membrane and in contact with two metallic electrodes through which an external voltage is applied. To this purpose, recently, we have proposed a model based on a coarse graining description, which describes the protein as a network of elementary impedances. The network is built from the knowledge of the positions of the C_α atoms of the amino acids, which represent the nodes of the network. Since the elementary impedances are taken depending of the inter-nodes distance, the conformational change of the receptor induced by the capture of the ligand results in a variation of the network impedance. On the other hand, the fluctuations of the atomic positions due to thermal motion imply an impedance noise, whose level is crucial to the purpose of an electrical detection of the ligand capture by the GPCR. Here, in particular, we address this issue by presenting a computational study of the impedance noise due to thermal fluctuations of the atomic positions within a rhodopsin molecule. In our model, the C_α atoms are treated as independent, isotropic, harmonic oscillators, with amplitude depending on the temperature and on the position within the protein (α-helix or loop). The relative fluctuation of the impedance is then calculated for different temperatures.

Keywords: Thermal fluctuations, proteins, G protein coupled receptors, complex networks
PACS: 87.15Ya, 87.15.Aa, 87.14Ee, 89.75.-k

MODEL AND RESULTS

G protein-coupled receptors (GPCRs) constitute the largest family of trans-membrane receptors, with functions going from revealing light and smells to the individuation of drug and virus intruders [1]. For this reason, many efforts are devoted to the study of their properties [1]. In particular, we are interested to develop electronic nanobiosensors based on GPCRs. Actually, the detection of an electrical signal from hybrid nanodevices based on a single or few receptors and associated with the capture of the ligands, is a challenging goal, rich of potential applications [2, 3]. Here, our aim concerns with the calculation of the electrical properties of a device realized by a G protein coupled receptor [1], embedded in its membrane and in contact with two metallic electrodes through which an external AC voltage is applied [4]. To this purpose, recently, we have proposed a model based on a coarse graining [5] description, which describes the protein as a network of elementary impedances [4]. The network is built from the knowledge of the position of the C_α atoms of the amino acids [6], which are taken as the nodes of the network [7]. At least in the case of rhodopsin (photonic receptor) these positions are known by X-rays diffraction experiments [6] for both, the basic and the most stable excited state (metarhodopsin) [6]. Though these positions are generally unknown for the other receptors, a coarse graining, complex network approach offers the possibility of taking advantage of the common topology of the GPCR family [1]. In fact, all GPCRs share a seven-helices trans-membrane structure, where the seven α-helices are interconnected by extracellular and intracellular loops [1, 6]. Additionally, there are two terminal chains: an extracellular chain (N terminus) and an intracellular chain (C terminus) [1, 6]. We assume that the amino acids interact electrically among them and that charge transfers between neighboring residues and/or changes of their electronic polarization [8] affect these interactions [4]. Accordingly, a link is drawn between any pair of nodes neighboring in space within a given a distance, $d = 2R_a$ (R_a electrical interaction radius) [5, 7] and an elementary impedance is associated with each link [4]. Moreover, two extra nodes can be introduced in the network, associated with the electrodes, which are linked to a given set of amino acids, depending on the particular geometry of the contacts in the real device. The elementary impedance is taken as the impedance of a RC parallel circuit (the most usual equivalent passive AC circuit) [4]. Precisely, by denoting with $Z_{i,j}$ the impedance associated with the link between the i-th and j-th nodes, separated by a distance $l_{i,j}$, we take [4]:

$$Z_{i,j} = \frac{l_{i,j}}{\pi(R_a^2 - l_{i,j}^2/4)} \frac{1}{(\rho^{-1} + i\epsilon_{i,j}\epsilon_0\omega)} \quad (1)$$

where ω is the frequency of the external voltage, ρ the resistivity of the resistor, ϵ_0 the vacuum permittivity and $\epsilon_{i,j}$ the relative dielectric constant of the capacitor [4] expressed in terms of the intrinsic polarizabilities α_i and α_j of the corresponding amino acids [8]. By taking the values $R_a = 12.5$ Å and $\rho = 10^9$ Ωm, we have found [4] that the conformational change of the receptor induced by the capture of the ligand (i.e. the transition rhodopsin $\to$ metarhodopsin) implies a significant variation of the impedance at all frequencies, and in particular we have found a variation of about 20 % in the static value of $Re[Z]$ [4]. On the other hand, the fluctuations of the atomic positions due to the thermal motion [5, 9, 10] imply an impedance noise, whose level, in comparison with the impedance change due to variation of conformation and with the electrode/amplifier noise, is crucial to the purpose of an electrical detection of the ligand capture by the GPCR. Therefore, here we consider the effect of the thermal atomic motion on the electrical response to an external field of a rhodopsin molecule. To this purpose, we allow the nodes of the network (C_α atoms) to fluctuate around their equilibrium positions. For the sake of simplicity and to get a qualitative estimation, we describe the system of coupled oscillators as a set of independent, isotropic, harmonic oscillators. When the oscillators are in their ground state, their positions, $\vec{r}$, referred to the equilibrium ones, are distributed with a probability density:

$$|\psi(\vec{r})|^2 = (\frac{M\omega_0}{\pi\hbar})^{3/2} \exp[-\frac{M\omega_0}{\hbar}r^2] = \frac{1}{(2\pi <x^2>)^{3/2}} \exp[-\frac{3}{2}\frac{x^2}{<x^2>}] \quad (2)$$

where M is the average mass of the amino acids, $\omega_0 = \sqrt{\gamma/M}$ the oscillator frequency, γ the elastic constant and $<x^2> = 1/3 <r^2>$ is the mean square displacement of the oscillator from its equilibrium position along the x-direction. If each oscillator is in

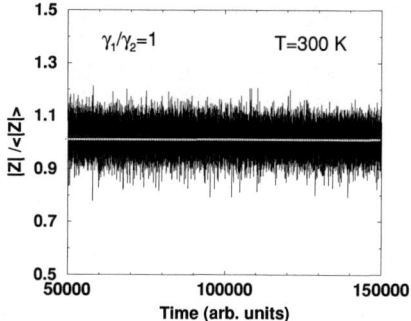

FIGURE 1. Simulation of the modulus of the network impedance, $|Z|$, versus time at $T = 300$ K. The time is expressed in simulation steps while $|Z|$ has been normalized to its average value (shown by the gray line).

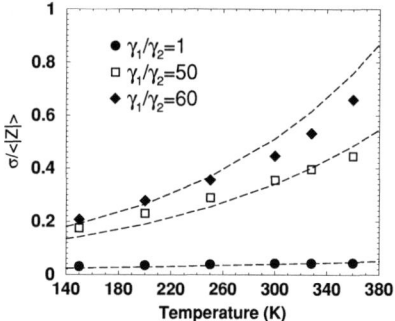

FIGURE 2. Relative root mean square fluctuation of the impedance modulus as a function of the temperature (in K). The three set of data refer to simulations performed wit the ratio of the helix and loop elastic constants, γ_1/γ_2, equal, respectively, to 1 (full circles), 50 (open squares) and 60 (full diamonds). The dashed curves show the best-fit with an exponential function.

contact with a thermal bath at temperature T, the value of $<x^2>$ at the equilibrium is:

$$<x^2> = \frac{1}{2}\frac{k_B\theta}{\gamma} + \frac{k_B\theta}{\gamma}\frac{1}{\exp[\theta/T]-1} \qquad (3)$$

where $\theta = 3\hbar\omega_0/k_B$. When $T \gg \theta$, Eq. (3) simplifies in:

$$<x^2> \approx k_B T/\gamma \qquad (4)$$

Of course, for an arbitrary temperature $T \neq 0$ the wave function of the oscillator is a superposition of several excited states and the density probability cannot be expressed in the simple form of Eq. (2). However, again for simplicity, we keep this expression and we account for the effect of the temperature by assuming in Eq. (2) $<x^2>$ given by Eq. (4). Moreover, as a first approximation, we take the mass M and the elastic constant γ equal for all the oscillators [10], and precisely: $M = \bar{M} = 100$ Dalton, $\gamma = 2.5$ KJ mole^{-1}

Å^{-2} [5]. This choice provides $\theta = 36$ K. Thus, the condition $T \gg \theta$ is satisfied at room temperature. Figure 1 shows the results of simulations at 300 K of the modulus of the network impedance, $|Z|$, versus time (the modulus has been normalized to its average value). On the other hand, it is well known [1, 9, 11] that loops and terminals are very flexible structures compared with the quite rigid α-helices. Therefore, to overcome the crude approximation of a unique elastic constant for all the oscillators, we consider two different elastic constants, γ_1 and γ_2, for oscillators belonging to the α-helices and to loops/terminals, respectively, with $\gamma_1 > \gamma_2$. Figure 2 shows the relative root mean square fluctuation of the modulus of the impedance, i.e. the root mean square fluctuation, σ, normalized to the average value of $|Z|$, as a function of the bath temperature. The three sets of data report the results of simulations performed by taking the ratio, $F = \gamma_1/\gamma_2$, equal 1 (full circles), 50 (open squares) and 60 (full diamonds). In all cases the values of γ_1 and γ_2 have been chosen to keep constant the average value of the elastic constant, $\bar{\gamma} = (1/N)\sum \gamma_i$, where the sum is performed over the whole number N of considered oscillators. The dashed curves in this figure represent the best-fit with exponential functions. We can see that once the higher flexibility of loops and terminals is accounted for, the relative fluctuation of the impedance increases significantly and it becomes strongly sensitive to the temperature. From this study we can conclude that a careful modelization of thermal motion [9, 11] is necessary to provide reliable estimates of the relative fluctuation of the impedance.

ACKNOWLEDGMENTS

This work has been performed within the SPOT NOSED project IST-2001-38899 of EC. Partial support from the cofin-03 project "Modelli e misure di rumore in nanostrutture" financed by Italian MIUR is also acknowledged. Authors thank E. Pajot-Augy, R. Salesse and J. Minic (INRA, Jouy en Josas, France) for helpful discussions.

REFERENCES

1. R. J. Lefkowitz, *Nature Cell Biology*, **2**, E133–E136 (2000).
2. C. Joachim, J. Gimzewski, and A.Aviram, *Nature*, **408**, 541–548 (2000).
3. F. Patolsky, G. Zheng, O. Hayden, M. Lakadamyali, X. Zhuang, and C. M. Lieber, *PNAS*, **101**, 14017–14022 (2004).
4. C. Pennetta, V. Akimov, E. Alfinito, L. Reggiani, and G. Gomila, "Fluctuations of Complex Networks: Electrical Properties of Single Protein Nanodevices," in *Noise and Information in Nanoelectronics, Sensors and Standards II*, edited by J. M. Smulko, Y. Blanter, M. I. Dykman, and L. B. Kish, Proceedings of SPIE 5472, Int. Soc. Opt. Eng., Bellingham, 2004, pp. 172–182.
5. A. R. Atilgan, S. R. Durell, R. L. Jernigan, M. C. Demirel, O. Keskin, and I. Bahar, *Biophys. J.*, **80**, 505–515 (2001).
6. R. C. for Structural Bioinformatics, *Protein data bank*, State University of New Jersey, http://www.rcsb.org/pdb, 1.
7. R. Albert, and A. L. Barabasi, *Rev. Mod. Phys.*, **74**, 47–97 (2002).
8. X. Song, *J. Chem. Phys.*, **116**, 9359–9383 (2002).
9. F. G. Parak, *Rep. Prog. Phys.*, **66**, 103–129 (2003).
10. M. M. Tirion, *Phys. Rev. Letl.*, **77**, 1905–1908 (1996).
11. P. W. Fenimore, H. Frauenfelder, and R. D. Young, "Proteins as Paradigms Complex Systems," in *Fluctuations and Noise in Biological, Biophysical and Biomedical Systems*, edited by S. M. Bezrukov, H. Frauenfelder, and F. Moss, Proceedings of SPIE 5110, Int. Soc. Opt. Eng., Bellingham, 2003, pp. 1–9.

Noise-driven switching between limit cycles and adaptability in a small-dimensional excitable network with balanced coupling

Leonid A. Safonov and Yoshiharu Yamamoto*

Educational Physiology Laboratory, Graduate School of Education, The University of Tokyo, 7-3-1 Hongo, Bunkyo-ku, Tokyo 113–0033, Japan

Abstract. We study the adaptability of coupled non-linear oscillators to external forcing by using a system of globally coupled FitzHugh-Nagumo equations. Each unit is either excitatory or inhibitory. If the numbers of units of both types are in a specific relation (balanced coupling), we observe the presence of multistable oscillatory states with different excitation or *firing* rates. In the presence of noise, there is noise-driven switching between these states. The selection between higher- and lower-frequency oscillations depends on the input, which results in increasing coherence between the periodic input and the system's output.

Keywords: stochastic resonance, chaotic itinerancy, balanced excitation/inhibition
PACS: 87.10.+e, 84.35.+i, 05.45.-a

Biological oscillators are generally stable. This is the main reason why they are usually modeled by limit cycles [1, 2]. Limit cycles show *stability* in their "activity", when evaluated by both the amplitude and frequency, over a wide range of control parameters, including external force. For example, in neural models such as FitzHugh-Nagumo (FHN) [3] and Hodgkin-Huxley [4] equations, the activity does not change much when they are forced by inputs above their "firing" threshold.

In contrast, networking limit cycles [1, 5] or general non-linear elements [6] are known to produce diverse spatio-temporal (synchronized) patterns with different frequencies and amplitudes, giving rise to variable activity levels over the whole network. Each pattern corresponds (in a broad sense) to a stable attractor, and the existence of multistable attractors with different activity levels accounts for *diversity* or *variability* in real-world oscillators, including biological ones. Here we consider the *adaptability* issue, which is one of the unique characteristics of living organisms.

We study a system of globally coupled suprathreshold FHN equations with external forcing and stochastic noise. Each unit is either excitatory or inhibitory. If the numbers of units of both types are in a specific ratio (balanced coupling), we observe the presence of multistable oscillatory states with different excitation (firing) rates, and switches between high- and low-frequency oscillatory states can occur easily in the presence of noise. In the presence of an optimal level of noise, in particular, the system demonstrates considerably improved coherence between the external forcing and the mean firing rate, similar to what is known as the stochastic resonance (SR) phenomenon [7, 8], indicating that the selection of multistable attractors with different activity levels is statistically more "ordered" in response to the changes in the external forcing. We conclude that the coexistence of inhibitory and excitatory connections and the noise effect is essential for this novel phenomenon leading to network adaptability due to the statistically ordered selection/deselection of multistable attractors with different intrinsic frequencies.

We consider a system of coupled units whose dynamics is described by the FHN equations:

$$\varepsilon \dot{v}_i = v_i(v_i - a)(1 - v_i) - w_i + I + S(t) + \frac{1}{N}\sum_{j=1}^{N} k_{ij}(v_j - v_i) + \xi_i(t), \quad (1)$$
$$\dot{w}_i = v_i - w_i - b,$$

where $\varepsilon = 0.005$, $a = 0.5$, $b = 0.15$, I is a constant input, $S(t)$ is a time-dependent input, $i, j = 1, \ldots, N$, $\xi_i(t)$ is Gaussian white noise with $\langle \xi_i(t)\xi_j(s) \rangle = 2D\delta_{ij}\delta(t-s)$, where $\langle \ldots \rangle$ denotes the ensemble average, and $2D \cdot \Delta t = \langle \xi_i^2(t) \rangle = \sigma^2$ in the discrete case. $k_{ij} > 0$ if the jth unit is excitatory and $k_{ij} < 0$ if it is inhibitory.

For a single uncoupled unit, there is a stable stationary solution for small I. At $I_0 \approx 0.11$ a Hopf bifurcation produces a stable limit cycle, which corresponds to the excited (or *firing*) state of the unit.

The adaptability of excitation or *firing* rates in the suprathreshold regime to changing levels of I can best be understood by considering the system's dynamics for $N = 3$ with four possible choices of the number of excitatory units (N_E).

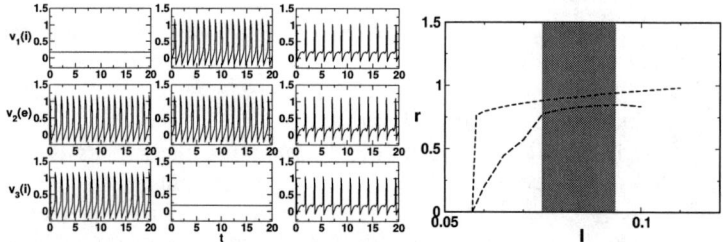

FIGURE 1. Firing regimes for cycles *1* (left), *2* (center) and *3* (right). $N = 3$, $N_E = 1$, $I = 0.07$. The second unit is excitatory (**e**), the others are inhibitory (**i**).

FIGURE 2. Dependence of the firing rate on the input I for $N = 3$ and $N_E = 1$. Upper line - high-frequency regimes (cycles *1* and *2*). Lower line - low-frequency regime (cycle *3*).

At $I_1 \approx 0.057$ there is a double resonant Hopf bifurcation, at which two stable limit cycles emerge, which have two firing units and one silent. The plots of v-variables of these cycles are shown in Fig. 1 (left and central columns). Along with the two cycles, there is a third limit cycle emerging at or near the double Hopf bifurcation. In the following, we refer to these three cycles as cycles *1*, *2* and *3* respectively.

With increasing I, we observe that the firing rate of cycles *1* and *2* grows rapidly over a short parameter interval, after which it changes only insignificantly (Fig. 2). In contrast, the firing rate of cycle *3* grows slowly with I for a long interval, after which it stabilizes.

At $I_0 \approx 0.11$ (as in the single unit case) there is an ordinary Hopf bifurcation, where the stationary solution becomes globally unstable and a new limit cycle emerges. In the present study we consider values of I between I_1 and I_0 (shown by a shaded area in Fig. 2).

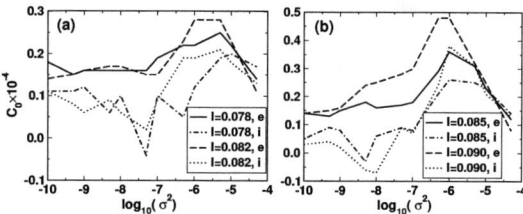

FIGURE 3. Stochastic resonance in a system with $N = 3$, $N_E = 1$ for excitatory and inhibitory units. $S = 0.003\sin(2\pi t/30)$. Results are averaged over 200 trials. Total time for each trial $T = 600$.

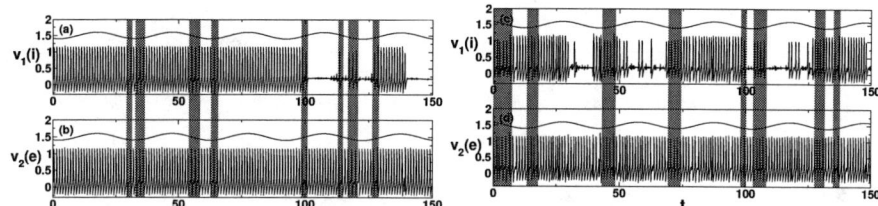

FIGURE 4. Time series of excitatory and inhibitory units with a periodic input. $\sigma^2 = 10^{-8}$: (a) inhibitory unit, (b) excitatory unit; $\sigma^2 = 5 \cdot 10^{-7}$: (c) inhibitory unit, (d) excitatory unit. $N = 3$, $N_E = 1$, $I = 0.09$, $S(t) = 0.003\sin(2\pi t/30)$. Shaded areas denote visits to cycle 3. Sinusoid lines schematically represent the input.

In the presence of weak noise, the system performs random transitions between domains of attraction of the three cycles. This is similar to a phenomenon known as chaotic itinerancy (CI) [6].

The system also demonstrates the phenomenon of SR in rate coding. We apply periodic input $S(t)$ to all units and plot the covariance measure

$$C_0 = \overline{S(t)R(t)},$$

(where $R(t)$ is the instantaneous firing rate (IFR) [9] and overbar denotes the time-average) against the noise intensity. Fig. 3 shows the dependence of C_0 on the noise intensity for different baseline input and amplitude of periodic input. We can observe the bell-shape curve, especially for the IFR of excitatory units, characteristic of SR. This phenomenon resembles suprathreshold SR studied previously [10], but has a different dynamical nature, as explained below.

Fig. 4 shows sample spike trains of an excitatory and an inhibitory unit for weak and moderate noise intensities. Although for most of the time the system stays near cycles *1* or *2*, occasionally it jumps to cycle *3*, and the probability of such a transition is higher for lower input than for higher input.

With increasing noise the transition probability becomes higher, and the time spent by the orbit near cycle *3* is greater. Therefore the resulting IFR of an excitatory unit is more coherent with the input for moderate noise [Fig. 4(d)] than for weak noise [Fig. 4(b)]. Thus, an increase of noise from weak to moderate values leads to an increase of the ability of excitatory units to encode the incoming signal (Fig. 3). When the noise becomes too strong, the coding ability deteriorates, giving the bell-shape curve of C_0, characteristic of SR.

Unlike excitatory units, inhibitory ones have three possible firing regimes – a high-frequency and a silent one at cycles *1* and *2*, and a low-frequency one at cycle *3*. Therefore, the IFR of an excitatory unit coincides with that of an inhibitory unit only for part of the time, otherwise it is zero. This results in a worse coding ability compared to that of an excitatory unit, which is shown by the lower SR-curve in Fig. 3.

For $N_E = 0$ all limit cycles are all globally stable and no noise-driven switching is observed. For $N_E = 2$ and $N_E = 3$ there are no bifurcations for $I < I_0$, hence no stable attracting set other than the stationary state exists in that interval.

Therefore the case of $N_E = 1$ has special significance. For $N_E = 1$ we observe CI-like dynamics with small noise and improved coding ability with increased levels of noise, while for lower and higher N_E the system's dynamics is much simpler and does not qualitatively depend on noise. Our simulations show that this is the case for $N > 3$ as well. In order for the switching to be observed, it must hold that $N_E = N_E^0 = \lceil N/2 \rceil - 1$. This relation gives the definition of balanced coupling, for which the system has specific properties described above. It has been observed that the balancing of excitation and inhibition can bring about complex dynamics in neuronal models [11], which is related to variability of cortical neuronal dynamics.

Collections of non-linear oscillators encountered in biological organisms are frequently quite adaptable to changes in the external environment. Recent studies have shown that brain-wide, frequency-specific neural synchrony and asynchrony, or the dynamical attractor selection among multiple neural oscillators, seem to be involved in the generation of perceptual, cognitive and behavioral processes in response to external and/or internal stimuli [12]. We believe that ours can be a simple model of this sort of emergent adaptability by using small dimensional non-linear oscillators.

REFERENCES

1. A.T. Winfree, *The Geometry of Biological Time*, Springer, Berlin, 1980.
2. L. Glass, *Nature*, **410**, 277–284 (2001).
3. R. FitzHugh, *Byophys. J.*, **1**, 445–466(1961); J. Nagumo, S. Arimoto, and S. Yoshizawa, *Proc. IRE*, **50**, 2061–2070 (1960).
4. A.L. Hodgkin and A.F. Huxley, *J. Physiol.*, **117**, 500–544 (1952).
5. Y. Kuramoto, *Chemical Oscillations, Waves, and Turbulence*, Springer, Berlin, 1984.
6. K. Ikeda, K. Matsumoto, and K. Otsuka, *Prog. Theor. Phys. Suppl.*, **99**, 295–324 (1989); K. Kaneko, *Physica D*, **41**, 137–172 (1990); K. Kaneko and I. Tsuda, *Chaos*, **13**, 926–936 (2003).
7. K. Wiesenfeld and F. Moss, *Nature*, **373**, 33–36 (1995).
8. L. Gammaitoni, P. Hänggi, P. Jung, and F. Marchesoni, *Rev. Mod. Phys.*, **70**, 223–287 (1998).
9. Hanning window filtered IFR is defined as follows. If t_i are firing times, then the spike train is $s(t) = \sum \delta(t - t_i)$, and the IFR is $r(t) = \int_{-T/2}^{T/2} \cos[4\pi(t-\tau)/T]s(\tau)d\tau$, where T is the window width, $T=10$.
10. N.G. Stocks, *Phys. Rev. E*, **63**, 041114 (2001); D. Rousseau, F. Duan, and F. Chapeau-Blondeau, *Phys. Rev. E*, **68**, 031107 (2003).
11. C. van Vreeswijk and H. Sompolinsky, *Science*, **274**, 1724–1726 (1996); *Neural Computation*, **10**, 1321–1371 (1998).
12. F. Varela, J.-P. Lachaux, E. Rodriguez, and J. Martinerie, *Nat. Rev. Neurosci.*, **2**, 229–239 (2001).

Stochastic Nonlinear Evolutional Model of the Large-Scaled Neuronal Population and Dynamic Neural Coding Subject to Stimulation

Rubin Wang[1] and Wei Yu

Institute for Brain Information Processing and Cognitive Neural Dynamic System,
East China University of Science and Technology,130 Meilong Road, P.R.China

Abstract. In this paper, we investigate how the population of neuronal oscillators deals with information and the dynamic evolution of neural coding when the external stimulation acts on it. Numerically computing method is used to describe the evolution process of neural coding in three-dimensioned space. The numerical result proves that only the suitable stimulation can change the coupling structure and plasticity of neurons.

Keywords: External Stimulation, Nonlinear Coupling, FPK Equation, Average Number Density
PACS: 84.35.+i

INTRODUCTION

Neuronal impulse aroused by external stimulation can form new synapse and change the density of the postsynaptic membrane through the coupling of synapse, thus the information is stored for a long time, that is to say the long-term memory is formed. Reference [1] proved that the stimulation's intensity is one of the important factors that transform the short-term memory to long-term memory. In this paper, we suppose the duration of stimulation coincides with the coupling of neurons when we numerically calculate. In order to obtain the result that neural stimulation would exert some effects on neural coding, based on our previous research [2-7], this paper takes into account of the effect of external stimulation on the neural coding when doing numerical analysis. The expected result of this numerical analysis would be that only appropriate intensity of stimulation could change the neuronal coupling structure. Weak stimulation can only have some impacts on the coupling structure, whereas strong stimulation would damage neurons, and even destroy one's cognition.

DERIVATION OF STOCHASTIC MODEL EQUATION

Setting the amplitude and the phase of N oscillators under the random noise independently are r_j, ψ_j (j=1,2,......, N). The phase and the amplitude dynamics obey the following evolution equations,

---
[1] Email address: rbwang@163.com or rbwang@dhu.edu.cn

$$\dot{\psi}_j = \Omega + \frac{1}{N}\sum_{k=1}^{N} M(\psi_j - \psi_k, r_j, r_k) + S(\psi_j, r_j) + F_{j_1}(t) \quad (1)$$

$$\dot{r}_j = g(r_j) + F_{j_2}(t) \qquad (j = 1, \cdots, N) \quad (2)$$

We assume that in equation (1) all oscillators have the same eigenfrequency Ω, and the oscillators' mutual interactions are modeled by the term $M(\psi_j - \psi_k, r_j, r_k)$ that model the impact of the kth on the jth oscillator. $\psi_j - \psi_k$ is the difference of their phase. $S(\psi_j, r_j)$ is taken into account stimulation of oscillators of the amplitude r_j. $g(r_j)$ is a nonlinear function of the amplitude. For the sake of simplicity the random force, $F_{j_i}(t)$ (i=1,2) is modeled by Gaussian white noise.

Introduce average number density

$$n(\psi, R, t) = \int_0^{2\pi}\cdots\int_0^{2\pi} d\psi_l \int_0^{\infty}\cdots\int_0^{\infty} \frac{1}{N}\sum_{k=1}^{N}\delta(\psi - \psi_k)\delta(R - r_k) f dr_l \quad (3)$$

We get the Fokker-Planck equation of averaging number density according to equations (1), (2) and (3).

$$\frac{\partial f}{\partial t} = \frac{Q_1}{2}\sum_{j=1}^{N}\frac{\partial^2 f}{\partial \Psi_j^2} + \frac{Q_2}{2}\sum_{j=1}^{N}\frac{\partial^2 f}{\partial \Psi_j^2} - \sum_{j=1}^{N}\frac{\partial}{\partial r_j}(g(r_j)f) - \sum_{j=1}^{N}\frac{\partial}{\partial \Psi_j}[\frac{1}{N}\sum_{k=1}^{N}\Gamma(r_j, r_k, \Psi_j, \Psi_k) f] \quad (4)$$

Where R denotes the probability when the amplitude r_j of every oscillator equals R, ψ denotes the probability when the phase Ψ_j.

And
$$\Gamma(r_j, r_k, \Psi_j, \Psi_k) = \Omega + M(\psi_j - \psi_k, r_j, r_k) + S(\psi_j, r_j). \quad (5)$$

And $M(\psi - \psi', R, R')$ can be expanded as the sum of progression by Fourier transform for it is a 2π-perioded function. We define the nonlinear function of amplitude and the mutual interaction term as follows,

$$M(\psi_j - \psi_k, r_j, r_k) = -\sum_{m=1}^{4} r_j^m r_k^m (K_m \sin m(\psi_j - \psi_k) + C_m \cos m(\psi_j - \psi_k)) \quad (6)$$

$$S(\psi_j, r_j) = \sum_{m=1}^{4} I_m r_j^m \cos(m\psi_j + \gamma) \quad (7)$$

We assume $C_m = 0$ and $\gamma = 0$, K_m denotes the mth coupling coefficient among neurons, I_m denotes the mth stimulation. The first term of average number density which has been Fourier transformed is given by

$$\hat{n}(0, R, t) = B_1 e^{\frac{2}{Q_2}\int_0^R g(x)dx} \int_0^R e^{-\frac{2}{Q_2}\int_0^x g(r)dr} dx \quad (8)$$

where B_1 fills $\int_0^{R_0} \hat{n}(0, R, t) dR = \frac{1}{2\pi}$, and R_0 equals 2.

THE NUMERICAL RESULT OF EQUATION

We consider one-mode coupling structure, namely, $K_1 = 1$ and other coupling coefficients are zero. The same initial condition is chosen as

$n(\psi,R,0) = \bar{n}(0,R,0)(1+0.05\sin\psi)$. The four stimulation intensities are set in [0,7] and increase in turn. Equation (4) was computed numerically and Fig.1 and Fig.2 are obtained as follows.

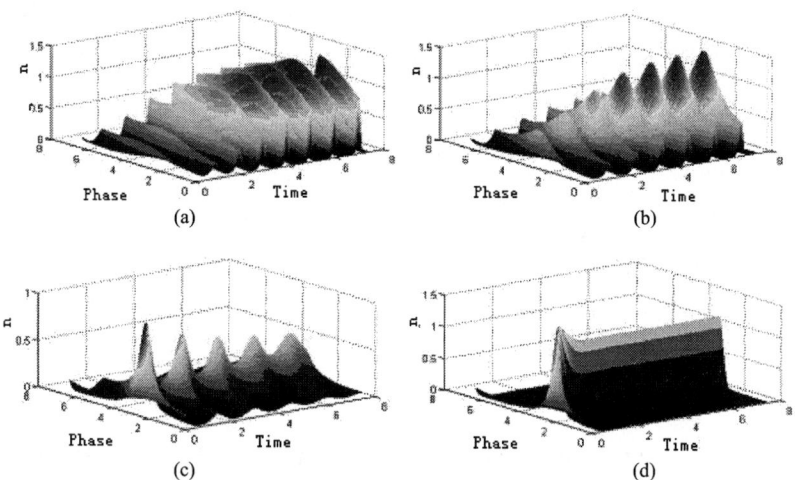

FIGURE 1. Average number density's evolution along with time (a) $I_1 = 0.5$ (b) $I_1 = 3$ (c) $I_1 = 5$ (d) $I_1 = 7$

From function (7), one knows that its value is minimum at $\psi = \pi$ for any amplitude R, then the value of function (5) is minimum, so that the distributing probability of neuronal population's action potential at $\psi = \pi$ is bigger than others phases. For this reason one can obtain the average number density's evolution figure about phase as Fig.1.

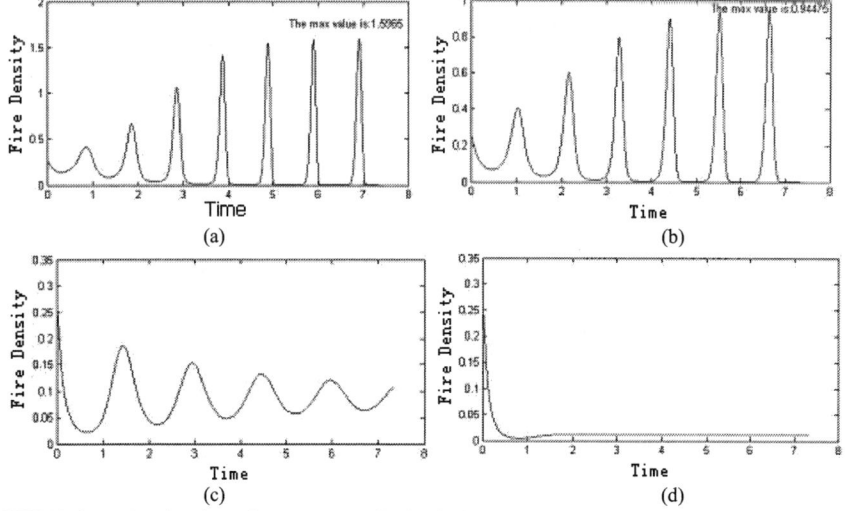

FIGURE 2. Fire density of neurons on limit circle R=1:(a) $I_1 = 0.5$ (b) $I_1 = 3$ (c) $I_1 = 5$ (d) $I_1 = 7$ $p(t) = n(0,1,t)$

According to the numerical result and theory analysis, one can obtain that there is a critical stimulation intensity between 3 and 5 from Fig.2.(b) and Fig.2.(c). New neural coding can be formed if the stimulation intensity is less than the critical intensity, otherwise the new neural coding cannot be formed. When stimulation intensity I_1 equals 0.5 or 3, which is smaller than critical intensity, the average number density of the neuronal fire changes stochastically along with time on phase. The waves of fire density possesses period when they are steady state, namely, the new neural coding is formed. When I_1 equals 5 which exceeds the critical value, one can obtain that the amplitude of the wave is attenuated incessantly from Fig.2.(c), this is a coding transitional process, the fire density will be a constant at last according the trend of the evolution, the wave does not oscillate again, here strong external stimulation make the distribution of neurons' amplitude and phase to be a constant, new neural coding cannot be formed. When the stimulation intensity I_1 equals 7, the fire density is faster to be a constant because too strong stimulation has injured some neurons.

CONCLUSION

The space coding of the average number density is mainly dominated by the stimulation intensity that is larger than the coupling intensities K_1 and K_2. It indicates that the neural structure does not dominate the average number density's evolution. It incarnates well the effect of learning and memory because it changes completely the coupling intensity and structure among neurons. The numerical analysis result shows that for the population's neural coding, the population's amplitude does not take part in the neural information processing. The conclusion is of signification because the experiment could not prove it.

ACKNOWLEDGMENTS

Project (30270339) supported by National Natural Science Foundation of China (NSFC)

REFERENCES

1. Barbara, F., Roozendaal, and B., McGaugh, J. L., *Society of Biological Psychiatry* **46**, 1140-1152 (1999).
2. Wang, R. B., and Zhang, Z. K., *Neurocomputing* **51C**, 401-411 (2003).
3. Wang, R. B., Hayashi, H., and Zhang, Z. K., *Proceedings of 9th International Conference on Neural information Processing* .**5**, 2497-2501 (2002).
4. Wang, R. B., and Zhang, Z. K., *International Journal of nonlinear science and numerical simulation* **4**, 203-208 (2003).
5. Wang, R. B., and Zhang, Z. K., Proc. ICNF. World Scientific, 408-411 (2002).
6. Wang, R. B., Proc. IEEE, Vol.1, 139-143 (2003).

CHEMISTRY AND ELECTROCHEMISTRY

The Electrochemical Noise Technique – Applications in Corrosion Research

Florian Mansfeld

Corrosion and Environmental Effects Laboratory (CEEL)
Department of Materials Science and Engineering
University of Southern California
Los Angeles, CA 90089-0241, USA

Abstract. Some of the applications of the electrochemical noise (ECN) technique in corrosion research as well as some of the promises and realities concerning the use of electrochemical noise analysis (ENA) are discussed. Perceived advantages of ENA include the ease of data collection and analysis. It is often assumed that the noise resistance R_n equals the polarization resistance R_p and that certain parameters such as the localization index LI are indicative of the prevailing corrosion mechanism. It has also been postulated that the slopes of the power spectral density (PSD) plots contain mechanistic information. Based on the author's evaluation of some of these assumptions it is concluded that experimental difficulties such as drift and asymmetry complicate the analysis of the ECN data. It has been shown that R_n equals R_p only in those cases for which the impedance has reached its dc limit in the bandwidth of the ECN measurement (ECNM). LI is mainly an indicator of the degree of asymmetry of the two electrodes used for the ECNM and not an indicator of corrosion mechanisms. The slopes of the potential and current PSD plots depend on the bandwidth of the ECNM and are in general not related to a specific corrosion mechanism

Keywords: electrochemical noise data, electrochemical noise analysis, PSD plots, localization index, noise resistance.
PACS: 01.50Kw, 01.50Pa

INTRODUCTION

The ECN technique has become quite popular in recent years – most likely due to some of its perceived advantages over other electrochemical techniques. These advantages include the low cost of the equipment and the ease of data collection. It is sometimes assumed that valid ECNMs can be carried out in a much shorter time than that required for the measurement of an impedance spectrum. In many cases it is expected that the noise resistance R_n is equal to the polarization resistance R_p. Furthermore, it has been suggested that certain parameters derived from statistical analysis of ECN data such as the localization index LI provide mechanistic information. Likewise it has been assumed that the slope of a PSD plot is an indicator of the corrosion mechanism. It is becoming increasingly clear however that some of these advantages do not exist [1]. In addition, experimental problems such as drift and electrode asymmetry can cause difficulties in the analysis and interpretation of ECN data [2-4]. The promises and realities of the ECN technique will be discussed in the following using experimental and theoretical data obtained by the author and his co-workers.

RESULTS AND DISCUSSION

ECN data are usually determined by measuring the current fluctuations ΔI between two electrodes of the same material that are connected by a ZRA and the potential fluctuations ΔV between the two coupled electrodes and a stable reference electrode [5]. For corrosion monitoring the reference electrode might be replaced by a third electrode of the same material. The bandwidth $\Delta f = f_{max} - f_{min}$ of the ECNM is given by the sampling frequency f_s that determines the maximum frequency f_{max} and the measuring time T that determines the minimum frequency f_{min}. In this laboratory a sampling frequency of 2 points/sec and T = 1024 sec are commonly used [6,7]. The bandwidth for ECNMs is limited at high frequencies by instrumentation noise, while the time needed for the collection of ECN data at very low frequencies limits f_{min}.

ENA consists of the analysis of the ECN data in the time and frequency domains. The parameters determined by statistical analysis in the time domain include the mean value of the potential of the coupled electrodes E_{coup} and the mean value of the coupling current I_{coup}, the rms value of the current fluctuations I_{rms}, the standard deviations of the current (σI) and potential (σV) fluctuations, the noise resistance $R_n = \sigma V/\sigma I$, skewness of current (I_{skew}) and potential (E_{skew}) fluctuations and the corresponding kurtosis values E_{kurt} and I_{kurt} (Table I). The latter parameters are also referred to occasionally as "moments". For a normal distribution of the ECN data the skewness has values of zero, while kurtosis has values of 3 [2,3]. The localization index LI, which has values between 0 and 1, has been suggested as a parameter that can be used to distinguish between cases of uniform and localized corrosion with values of 1 supposedly being characteristic of localized corrosion [8,9].

ENA in the frequency domain results in PSD plots for potential (V) and current (I) fluctuations that can be described by:

$$\log PSD_i = m_i \log f + b_i \tag{1}$$

where i=V or I and b is the intercept of the log PSD-log f plot at the frequency f=1 Hz. Fast Fourier Transform (FFT) has been used to convert ECN data from the time domain to the frequency domain. The spectral noise plot or noise impedance $R_{sn}(f)$ is defined as the ratio of the FFT functions of the potential and current fluctuations [8,10] or the square root of the corresponding PSD plots [11,12]. It has been shown for many different systems that the spectral noise plots are identical to impedance plots [1, 2, 7, 8, 10, 13].

The frequency dependence of the spectral noise plot is given by:

$$\log R_{sn} = m_{rsn} \log f + b_{rsn} \tag{2}$$

The spectral noise resistance $R°_{sn}$ is defined as the dc limit of the spectral noise plot.

Experimental Limitations

1.Drift. The main experimental problems in ECNMs are drift or trend of the signal and asymmetry of the two electrodes used for the experiment. A theoretical and experimental analysis of the effects of trends in the potential and current noise fluctuations on the slopes of the PSD plots as well as some of the other parameters listed in Table I has been reported elsewhere [2]. This subject was also discussed recently by Bertocci et al [4], who evaluated several methods of drift removal.

TABLE 1. Statistical analysis of ECN data.

Mean	$\bar{x} = \dfrac{1}{N}\sum_{i=1}^{N} x_i$	Standard deviation	$\sigma = \sqrt{\dfrac{1}{N}\sum_{i=1}^{N}(x_i - \bar{x})^2}$
Root mean square	$rms = \sqrt{\dfrac{1}{N}\sum_{i=1}^{N} x_i^2}$	Localization index	$LI = \dfrac{\sigma I}{Irms}$
Noise resistance	$R_n = \dfrac{\sigma V}{\sigma I}$	Skewness	$S = \dfrac{\dfrac{1}{N}\sum_{i=1}^{N}(x_i - \bar{x})^3}{\sigma^3}$
Kurtosis	$K = \dfrac{\dfrac{1}{N}\sum_{i=1}^{N}(x_i - \bar{x})^4}{\sigma^4}$		

Mansfeld et al have discussed the effects of trends on the analysis of ECN data for mild steel exposed to 0.5 N NaCl (open to air) [2]. The trends in the experimental data were removed by assuming that the trends followed a linear expression such as:

$$V = V_o + at \qquad (3a)$$
and
$$I = I_o + bt \qquad (3b)$$

From Eq. 3 it follows that a plot of the potential fluctuations vs. the current fluctuations should results in a straight line:

$$V = V_o + a\,(I - I_o)/b \qquad (4)$$

I = 0 will be observed for: $\quad V = V_o - aI_o/b \qquad (5)$

Fig. 1 is a plot of the potential fluctuations vs. the current fluctuations for the raw ECN data obtained after 24 h. As predicted by Eq. 4 a straight line was observed with a slope of 1811 ohm which is close to the value of R_n – 1821 ohm. However this result is a coincidence and the observed slope in Fig. 1 is equal to a/b = 1825 ohm according to Eq. 4 as calculated from the linear equations used for trend removal, where a = -5.27×10^{-6} V and b = -2.89×10^{-9} A [2]. According to Eq. 5, I = 0 occurs at -0.771 V in agreement with the fit equation ($V_o = -0.759$ V and $I_o = 6.59 \times 10^{-6}$ A). A plot similar to that shown in Fig. 1 of the ECN data after trend removal showed a random distribution of the potential and current data (Fig. 2).

2. Asymmetry. The experimental approach described above for ECNM assumes identical behavior of the two coupled electrodes. However asymmetry can occur during an ECNM especially for cases of very low corrosion rates, localized corrosion and polymer coated metals. The latter case has been discussed in detail by Mansfeld et al [14] who evaluated the effect of asymmetric electrodes on the analysis of electrochemical impedance and noise data. When the polymer coating on one electrode develops defects, while the coating on the other electrode remains perfect, the EIS and ECN data are dominated by the impedance of the undamaged coating. This situation poses severe problems in corrosion monitoring since coating damage remains undetected. Bautista et al [15] as well as Cottis [16] have discussed the influence of electrode asymmetry on the interpretation of ECN data in detail.

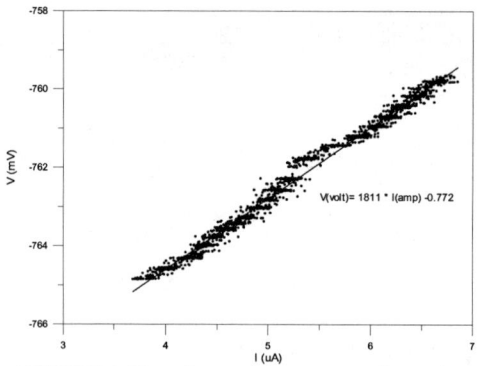

FIGURE 1. Plot of potential vs. current fluctuations.

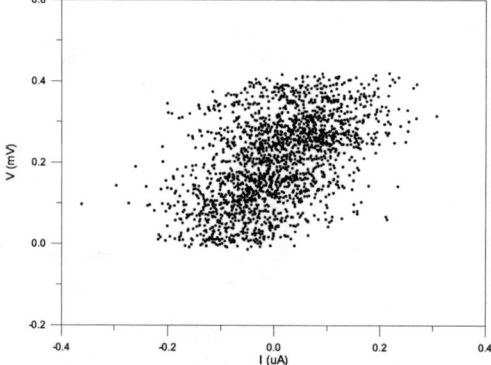

FIGURE 2. Plot of potential vs. current fluctuations after linear trend removal.

Limitations of Data Interpretation

1. <u>The relationship between the noise resistance and the polarization resistance</u>. While it has often been assumed that R_n equals R_p for all cases, it has been show repeatedly that this equality exists only for those systems for which the impedance has reached its dc limit in the Δf range used for the ECNM. Mansfeld et al have performed ENA for several active and passive systems in chloride media [1]. It is obvious from these data that R_n has no relationship to the impedance properties of a passive system such as the SS316L/Ringer's solution system. In general, an agreement of R_n and R_p can only be expected for the case where f_{max} is lower than the breakpoint frequency f_b, which is defined as the frequency for which the phase angle Φ is equal to $-45°$. Similar problems were found for polymer coated steel samples exposed to artificial seawater [13]. Mansfeld and Lee [17] have discussed the frequency dependence of R_n for polymer coated metals. For very protective polymer coatings with capacitive spectra within Δf R_n was found to depend on f and therefore no definite relationship to any particular coating or steel corrosion property could be expected. On the other hand, for severely degraded coatings for which the impedance had reached its dc limit within Δf R_n was equal to R_p with the meaning of these parameters depending on the equivalent circuit appropriate for a given metal/coating system [13].

2. The significance of the localization index LI. Mansfeld and Sun [6] have evaluated the significance of LI obtained from ECNMs. The experimental LI data for mild steel exposed to 0.5 M NaCl were 0.07 for an exposure time of 24 h and 0.03 for 48 h, while for another test the LI values were 0.16 and 0.17, respectively. For Ti-6Al-4V exposed to Ringer's solution for 24 h LI was 0.94. For Al 2024 exposed to 0.5 M NaCl LI changed from 0.79 after 2h to 0.16 after 48 h, while for Al 6016 LI changed from 0.99 after 2h to 0.07 after 48 h [1]. Since uniform corrosion occurs for the mild steel/NaCl system and the Ti alloy is passive in Ringer's solution, the question arises why LI has value close to 0 and 1, respectively for these two systems. Also, active pitting still occurred for the two Al alloys after 48 h, but LI was close to zero [1].

The answer to the questions arising for the experimental observation summarized above can be found by inspecting the definition of LI given in Table I, which shows that LI = 1 should be observed for ideal systems in which the two electrodes of the same material have identical corrosion kinetics, i. e. $\bar{x} = 0$. This situation apparently occurred for the Ti alloy where fairly large current fluctuations occurred around the mean value I_{coup}, which was very small. For mild steel the individual data points showed only small fluctuations around I_{coup} which was quite large [2]. For the Al alloys it is quite likely that asymmetry of the two electrodes developed with increasing exposure time producing large values of I_{coup}.

Based on the results discussed here it has been concluded that LI should be considered as a measure of the deviation of a system from ideal behavior for which LI equals 1 and not as an indicator of corrosion mechanisms [1, 2].

3. The significance of the slopes of PSD plots. In 1988 Searson and Dawson based on ECNMs for mild steel in 0.5 N H_2SO_4 suggested that the slope of a potential PSD plot is indicative of the corrosion mechanism with a slope of –1 being characteristic of pitting corrosion, while a slope of –2 was expected for uniform corrosion [18]. Since then this concept has often been applied towards the interpretation of ECN data despite the lack of a theoretical derivation.

Due to the limited bandwidth of the ECNM only a small portion of an impedance spectrum can be compared with the spectral noise plot calculated from the potential and current PSD plots. For passive metals the spectral noise plot will fall into the capacitive region, where $m_Z = -1$, which means that $m_V = -2$ and $m_I = 0$ as observed for the SS316L/Ringer's solution system [1]. For the mild steel/NaCl system the noise impedance plot fell into the transition region of the Bode-plot with $m_{rsn} = -0.46$ [2]. For the brass/NaCl+BTA system similar results were obtained with $m_{rsn} = -0.60$ [3]. For the Al 2024/NaCl system the noise impedance plot was located in the transmission line region of the Bode plot which is characteristic of pitting [19]. The slope of this plot was close to –0.5 which also is the slope that is observed for diffusion processes. It is clear that it is not possible to determine that localized corrosion has occurred in this case based solely on the information provided by the noise impedance plot.

SUMMARY AND CONCLUSIONS

It has been demonstrated that while ECNMs can be carried out in a simple manner with low-price equipment there are experimental problems that can severely affect the results obtained by ENA. Drift of the ECN data during the measurement is frequently observed especially in the early stages of the ECNM. Linear drift produces erroneous slopes of the PSD ($m_V = m_I = -2$) as a result of which the spectral noise plot will have a slope of zero. Drifts can be successfully removed using linear or polynomial trend removal methods. Asymmetry of the two electrodes used for the ECNM is a more difficult problem that cannot be solved by mathematical methods. It has been observed mainly for cases of localized corrosion and polymer coated metals. In the latter case the ECN and EIS data are dominated by the electrode with the more protective coating. Asymmetry produces values of LI close to zero. For this and other reasons discussed above LI can not be considered as an indicator of corrosion mechanism.

R_n has often been assumed to be equivalent with R_p. However, inspection of ECNMs for different systems has shown that often R_n has no relationship with R_p as determined as the dc limit of an impedance spectrum. It is often possible to determine R_p by fitting of the experimental impedance data and extrapolation to $f = 0$. However, such procedures are not available in ENA. The experimental value of R_n depends in general on Δf and $R_n = R_p$ is only observed for cases where $f_{max} < f_b$.

Many experimental results have shown that traditional ac impedance spectra and spectral noise plots or noise impedance spectra are identical in the absence of experimental artifacts such as drift. The question of whether the slopes of potential and or current PSD plots contain mechanistic information has to be considered based on this fact. Contrary to EIS data, ECNM can only be obtained in a limited Δf. The slopes of the PSD plots depend therefore to some extent on Δf. For cases for which Δf lies entirely in the capacitive region of a Bode plot $m_V - m_I = -2$, while for cases for which Δf lies in the frequency -independent region of the Bode plot at very low f, $m_V = m_I$. In many cases, Δf will fall into the transition region between the capacitive region and the dc limit and the slopes of the PSD plots will be between -2 and zero.

REFERENCES

1. F. Mansfeld, Z. Sun and C. H. Hsu, Electrochim. Acta 46, 3651 (2001).
2. F. Mansfeld, Z. Sun, C. H. Hsu and A. Nagiub, Corrosion Science 43, 341 (2001).
3. A. Nagiub and F. Mansfeld, Corrosion Science 43, 2147 (2001).
4. U. Bertocci, F. Huet, R. P. Nogueira and P. Rousseau, Corrosion 58, 337 (2002).
5. D. A. Eden, K. Hladky, D. G. John and J. L. Dawson, paper No. 274, Corrosion/86, (1986), NACE.
6. F. Mansfeld and Z. Sun, Corrosion 55, 915 (1999).
7. F. Mansfeld, L. T. Han, C. C. Lee, C. Chen, G. Zhang and H. Xiao, Corrosion Science 39, 225 (1997).
8. D. A. Eden, paper No. 386, Corrosion/98, (1998), NACE.
9. F. Mansfeld and Z. Sun, Corrosion 55, 915 (1999).
10. H. Xiao, L. T. Han, C. C. Lee and F. Mansfeld, Corrosion 53, 1025 (1997).
11. U. Bertocci, C. Gabrielli, F. Huet, and M. Keddam, J. Electrochem. Soc. 144, 31 (1997).
12. U. Bertocci, C. Gabrielli, F. Huet, M. Keddam and P. Rousseau, J. Electrochem. Soc. 144, 37 (1997).
13. F. Mansfeld, L. T. Han, C. C. Lee and G. Zhang, Electrochim. Acta 43, 2933 (1998).
14. F. Mansfeld, C. Chen, C. C. Lee and H. Xiao, Corrosion Science 38, 497 (1996).
15. A. Bautista, U. Bertocci and F. Huet, J. Electrochem. Soc. 148, B412 (2001).
16. R. A. Cottis, Corrosion 57, 265 (2001).
17. F. Mansfeld and C. C. Lee, J. Electrochem. Soc. 144, 2068 (1997).
18. P. C. Searson and J. L.Dawson, J. Electrochem. Soc. 135, 1908 (1988).
19. F. Mansfeld, C. H. Tsai and H. Shih, ASTM STP 1154, 186 (1992).

Chaos-Order Transitions During Electrochemical Corrosion of Silicon

Vitali Parkhutik, Eugenia Matveyeva

R&D Centre "Materials and Technologies of Microfabrication", Technical University of Valencia, Cami de Vera s/n 46022 Valencia, Spain

Abstract. In a series of recent works performed at the R&D Centre MTM we have registered and analysed quite intriguing effects of oscillating current and voltage during silicon electrochemical treatments in fluoride-containing electrolytes. Depending on the experimental conditions the oscillations can exhibit an amplitude from milliVolts to 20-60 Volts and periods from 0.2 to 200 seconds. They are quite sensitive to the temperature of solution, electrolyte stirring, concentration of acid, sample history, periodic external impact, etc. In each particular case there is a coupling between different parallel electrochemical reactions that gives rise to the oscillatory behaviour.

We discuss general conditions for the occurrence of the oscillations of current or voltage in a system Si/electrolyte and show that this case is described by the same phenomenological laws as other cases of chaos-order transitions (f.i. Belousov-Zabotinsky reactions) though in the present case reaction products and variables are very different from other existing oscillatory systems. Through studying the oscillatory processes at corroding Si electrode we can get a general information which is useful in studying other chemical and physical processes that yield similar behaviour.

Keywords: Silicon, corrosion, chaos-order transitions
PACS: 82.40.Bj, 82.45.Vp

INTRODUCTION

Oscillatory processes are the universal property of physical, chemical, biological and socioeconomic sys¬tems related with non-linear interactions between their components. Studies of temporal and spatial oscillatory phenomena are now in progress due to their importance for the physics (f.i. oscillations of magnetic susceptibility in layered structures) in chemistry (one of the most beautiful reactions in chemistry is the oscillatory Belousov-Zabotinsky reaction), in chemical technology (chem¬ical reactors with oscillatory regime, oscillatory thermo-capillary convection etc.), biology (oscillatory neural networks, sociology, to mention but a few.

Basically, the macroscopic oscillations in complex systems are due to the effect of stochastic resonance in a multi-component medium with interacting parts. This resonance can be reasonably well described using some general theoretical formalism (like logistic equation [1], etc.). In each case, however, the nature of this interaction may be different though the kinetics of oscillatory behaviour can be surprisingly very similar.

Recently, rather curious effect of the electrochemical oscillations when silicon is corroding in aqueous electrolytes under the application of positive potential to the Si electrode or even without polarization [2-8]. The nature of these oscillating phenomena is presently disputed and different mechanisms are proposed as responsible for the oscillatory behaviour.

Sasano and Ogata [9] have discovered that immersion plating of pSi with copper yields oscillations of Open Circuit Potential (OCP). At first stage the OCP value experiences a transient aperiodic change associated with copper deposition at the surface of the sample. Further on, the OCP value starts to fluctuate or oscillate depending on the properties of the porous silicon layer and composition of the electrolyte. In the course of these oscillations a deposited Cu layer lifts off exhibiting bare Si surface beneath.

EXPERIMENTAL PROCEDURE AND RESULTS

We have used both n- and p-type monocrystalline Si samples with (100) crystallography and various doping levels (0.01-20 $\Omega \cdot cm$) and porous silicon layers grown on these substrates. The electrolyte solution used in the work included 0.01M $CuSO_4$ and small amounts of HF added into the base electrolyte. The experimental set-up has been described in previous papers [16,18]. The OCP transients were registered using HOKUTO DENKI HZ-3000 and PAR 273A potentialgalvanostats operating at sampling rate of 10 measurements per second. Electrolytic cell was equipped with Pt or Cu reference electrode) and Cu counter electrode. Stirred and unstirred operation modes were employed.

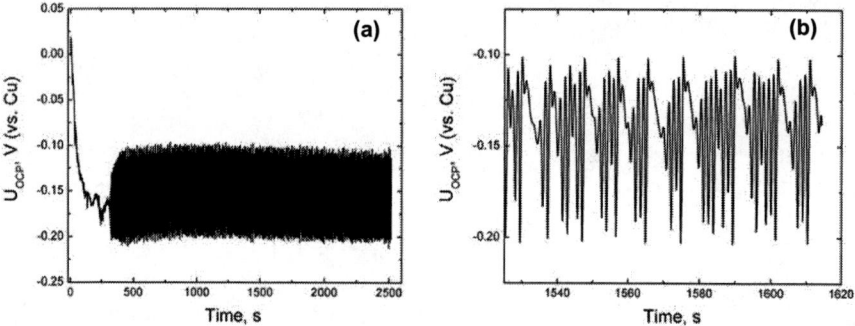

FIGURE 1. Oscillations of Open Circuit Potential during immersing of porous silicon layer formed on 10 $\Omega \cdot cm$ p-Si in $CuSO_4$ /HF solution: kinetics of the immersion process with initial transient (a), zoomed portion of steady-state oscillatory process (b).

Fig. 1 shows the kinetics of the OCP variation during immersion plating of porous silicon in the solution of $CuSO_4$/HF. During initial 300 seconds we observe OCP transient which corresponds to rapid deposition of Cu layer at the surface of porous silicon sample. Further on, oscillations of the OCP value emerge and simultaneously the deposited Cu layer detaches from the surface of the sample.

The oscillations are irregular (Fig. 1,b). Phase map of the oscillatory process (Fig. 1,c) indicates that the oscillations have a chaotic nature and a tendency to form a strange attractor is seen. This case therefore corresponds to a chaos in a system Si/Cu/electrolyte.

Further studies have shown that the presence of porous silicon layer on top of Si wafer is not a key factor to obtain the oscillations of the OCP value. They are much better shaped and long lasting in the case then copper immerse-plated at a surface of monocrystalline silicon. Fig.2 shows the kinetics of the OCP oscillations in the case of p- Si (10 Ohm·cm). Here we see shorter initial transient of the OPC value after which the oscillatory process starts developing. Similarly to the case of the deposition onto PSi, deposited Cu layer peals-off from the surface with time. The oscillations are much more regular in the case of monocrystalline Si then in the case of the pSi - this is seen from the nearly round shaped phase trajectory of the oscillatory process (Fig. 2,c).

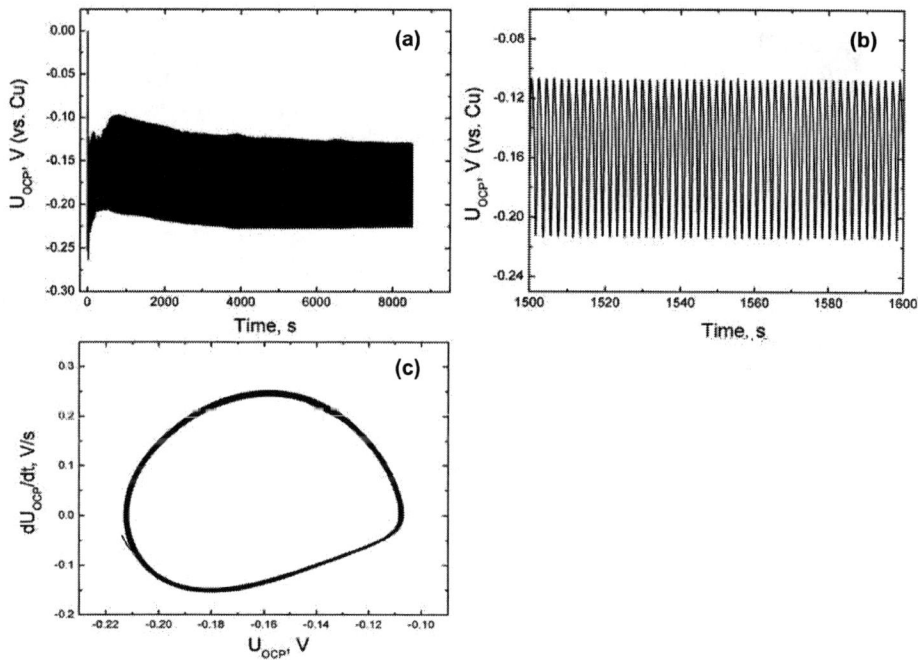

FIGURE 2. Kinetics of the OCP oscillations during immersing of p-Si in $CuSO_4$/HF solution (a), zoomed view of the oscillatory process (b) and its phase map (c).

DISCUSSION

Immersion plating process on any surface will take place if the electrolyte and sample contain components in solution which allow for at least two processes 1) reduction and 2) oxidation in a way that they compensate each other to maintain the electroneutrality of the system (as the electrical circuit is broken up). In case of Si surface immersed in $CuSO_4$ solution the overall immersion plating reaction is the chemical redox process:

$$Si + 2Cu^{2+} + 2H_2O \rightarrow SiO_2 + 2Cu + 4H^+ \qquad (1)$$

As a result of reaction (1) copper is deposited at the Si electrode which is simultaneously oxidized. At the same time electrolyte solution at the vicinity of the Si electrode is acidified by H^+ ions. Most important to note here is that the reactions of Cu deposition and Si oxidation can not occur in a steady-state manner. Indeed, to ensure continuous growth of Cu at the Si surface permanent supply of electrons is necessary. But the growth of SiO_2 film which possesses good insulating properties inhibits the transport of the electrons to the electrolyte/electrode interface. Therefore the presence of HF is necessary first to dissolve the native SiO_2 layer present at the Si surface and start the reaction (1) and then – to dissolve the SiO_2 produced in the process (2).

$$SiO_2 + 6HF \rightarrow H_2SiF_6 + 2H_2O \qquad (2)$$

The role of HF is to activate the Si electrode and to maintain the process (1). But in its turn, the reaction depends not only on the presence of HF and SiO_2 in the system but also on the conditions of the Cu layer deposited on top of the Si electrode. Copper layer should possess essentially porous morphology to ensure a flow of HF to the SiO_2 layer beneath it.

ACKNOWLEDGEMENTS

The work on oscillatory electrochemical dissolution of silicon was partially financed through R&D projects MAT 2001-3203 of the Ministry of Science and Technology of Spain and project GROO-153 of the Valencian Autonomous Community (Spain).

REFERENCES

1. S.Lundqvist, N. March, M.P. Tosi, *Order and Chaos in Non-Linear Physical Systems*, Plenum Press, New York, 1988.
2. V.Parkhutik, *Electrochim. Acta,* **36,** 762 (1991).
3. J.Rappich, V.Y. Timoshenko, T. Dittrich, *J. Electrochem. Soc*, **144,**493 (1997).
4. J. Carstensen, R. Prange, H. Foil, *J. Electrochem. Soc.* **146,** 1134 (1999)
5. H. Lewerenz, *J. Phys. Chem. B* **101,** 2421 (1997).
6. V. Parkhutik, E. Matveeva, *Electrochem. Solid State Lett.* **2,** 347 (1999).
7. S. Langa, J. Carstensen, et al., *Electrochem. Solid State Lett.* **4,** G50 (2001).
8. V.Parkhutik *Solid State Electronics* **45,**1451(2001).
9. J.Sasano, Y.Ogata, *(private communication)* 2002.

Chaos and Resonance in the Model for Current Oscillations at the Si/Electrolyte Contact

J. Grzanna, H. Jungblut, H. J. Lewerenz *

Hahn-Meitner-Institut Berlin GmbH, Glienicker Str. 100, 14109 Berlin, Germany
* Department of Materials Science and Engineering
North Carolina State University, Raleigh, N.C. 27695, USA

Abstract. The transition from damped current oscillations (chaos) to sustained current oscillations (resonance) and reverse path is described for single crystal silicon electrodes in aqueous fluoride containing solutions using a model based on morphological aspects.

Keywords: Oscillation, Silicon, Silicon Oxide, Stress, Cracks, Pores, Etching, Synchronization
PACS: 05.45.xt , 62.20.Mk , 68.47.Jn , 82.40.Bj

INTRODUCTION

The transition from damped current oscillations (chaos) to sustained current oscillations (resonance) at single crystal silicon electrodes in aqueous fluoride containing solutions is considered. Oscillations are observed at higher anodic potentials (>3V vs. SCE) and they are connected with a cyclic increase and decrease of the mean oxide thickness [1]. The transition from damped to sustained oscillations (see 1 below) and the reverse behavior (see 2 below) is observed for two cases:
1) The composition of the solution is constant (for instance pH 4 and 0.1 M NaF). The potential is increased from smaller (<3V vs. SCE) to larger (>3V vs. SCE) values.
2) The potential is constant (for instance 6V vs. SCE). The etching rate of silicon oxide is increased from lower (for instance pH 4 and 0.1 M NH_4F) to larger pH values by solution composition adjustment.

THE MODEL

The assumptions of the model during a cycle which consists of the oxidation of a silicon layer and the following etching of this layer are:
(i) the initial oxide growth rate is larger than the oxide etching rate (bare silicon surface); (ii) the oxidation process causes a current proportional to the increase in oxide thickness; (iii) the oxide growth rate decreases in dependence on the growing oxide thickness; (iv) the oxidation process is terminated after a potential dependent maximum oxide thickness is reached. The electrode is passivated and no further oxide is built during this cycle; (v) subsequently, the oxide is removed by etching; (vi) after the oxide is removed (not necessary completely) a new cycle starts.

The process of cyclic oxidation and the following removal of the oxide is not a uniform process. Local effects such as the existence of cracks and pores in the grown oxide lead to a dynamical behavior. One consequence is that at different positions on the silicon electrode, the process is in different stages (or cycles). This leads to further features of the model:

(vii) the lattice mismatch between silicon and its oxide leads to stress (see ref. [2]) and stress leads to cracks and pores in the oxide.

(viii) earlier grown oxide (type I) in a silicon layer "feels" little stress (no oxide in the neighborhood) resulting in less cracks and pores with only two-dimensional etching perpendicular to the oxide surface. This results in a larger life time of this oxide labeled I.

(ix) later (or last) grown oxide (type II) on a silicon layer experiences more stress from the already grown oxide I islands in its neighborhood leading to more cracks and pores and more three-dimensional etching (additionally in pores) which is known to be faster. This results in a shorter life time of the later grown oxide labeled II.

The existence of oxide I and II inside one layer leads to a synchronization because the later grown oxide II (a little bit desynchronized or delayed in the process) has a shorter overall lifetime so that the delay vanishes or is compensated at the end of a cycle.

For the mathematical description of the model (see ref. [3-4]), we introduce the so-called synchronization states $p_i(t)$ (i-cycle number) and the probability distribution q for the periods of the (oxide) thickness oscillators linked by the convolution $p_i = p_{i-1} \circ q$ creating a Markov Process. For the derivation of p_i and q the phase $\Phi(t,x,y)$ of the thickness oscillators is considered. The phase grows uniformly in time (at each period from the time $t_{i-1}(x,y)$ to $t_i(x,y)$) and gains 2π (see ref. [5]). Consequently Φ is a piecewise linear time function (see Fig. 1) at each location (x,y).

$$\Phi(t,x,y) = (i-1)\,2\pi + \frac{t - t_{i-1}(x,y)}{t_i(x,y) - t_{i-1}(x,y)}\,2\pi, \quad \text{for } t_{i-1}(x,y) < t \le t_i(x,y),\ i = 1,2,3,...$$

$t_i(x,y)$ defines the time at which at the location (x,y) the oxide thickness is zero (or minimal) during the i-th cycle. Hence in the phase space yields $\Phi(t_i(x,y),x,y) = i*2\pi$.

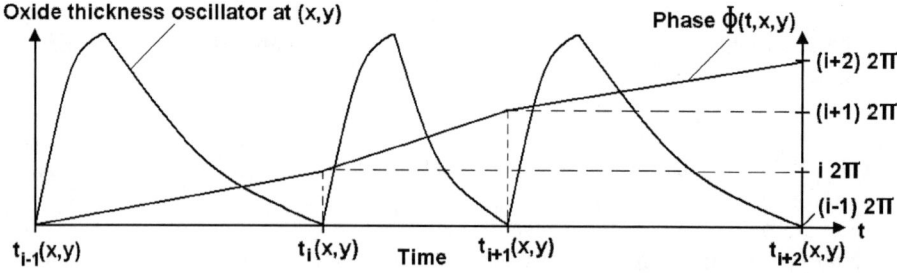

FIGURE 1. The phase $\Phi(t,x,y)$ of a thickness oscillator at the location (x,y) of the silicon electrode.

The passing of the phase through the (i 2π)-planes is considered whereas the (i 2π)-planes are planes of a constant phase (here i 2π) in the phase space. The passing is a dynamical process therefore a stroboscopic map of the phase is introduced at these (i 2π)-planes. The image of the map is the so called synchronization state $p_i(t)$ which depends on the gradient of Φ at the points of contact (intersection) of Φ and the (i 2π)-plane at the time t. Synchronization (sustained current oscillations) occurs when the p_i does not vary its shape after some cycles (Fig. 3).

q results from the evolution of the so called characteristic domain which is the spatio-temporal mean of the evolution of real domains. Real domains are defined as domains of silicon oxide arranged in a layer concentric around cracks and pores. Consequently real domains located in type I oxide have a larger diameter than real domains in oxide of type II. The characteristic domain is non-local but it can be shown to reflect important details of real domains such as the diameter and hence the mean pore distance. The feedback mechanism is reflected by a life time shortening of the later grown oxide II, equivalent to a successive shortening of the base interval of q.

A convolution of $p_i(t)$ with the elementary current peak (resulting from the evolution of the characteristic domain during one cycle) leads to the current contribution of the i-th cycle and the summation of the contributions of all cycles results in the current of the whole electrochemical system.

FROM DAMPED TO SUSTAINED OSCILLATIONS

Based on 1) the transition from damped to sustained current oscillations is investigated. Here, the potential plays an important role. A linear correlation between the potential and the maximum oxide thickness is given in ref. [6]. An increasing potential leads to an increasing integral oxide thickness and hence to an increasing thickness of the silicon layer which is consumed by the oxidation during one cycle. In context with the developed model, the fundamental force stress can act only in the silicon layer in the presence of a sufficiently thick oxide. The stress produces two types of oxide leading to a synchronization in the oxidation and etching process. In this case the formation of oxide I and II acts during a short time leading to the known current peaks in the case of sustained current oscillations (see ref. [4] Fig. 5).

In the case of a small potential (<3V vs. SCE) the built oxide layer is to thin to produce stress and hence only one type of oxide exists. The consequence is that the number of cracks and pores per area unit does not vary during one cycle. The corresponding real domains and finally the characteristic domain together with q do not vary. In this case the convolution of $p_{i-1} \circ q$ leads to a more damped shape of p_i resulting in a sequence of more and more damped synchronization states (see Fig. 2). The sum of all synchronization states p(t) is a constant function after some cycles.

If the potential is increased (>3V vs. SCE) the built oxide layer is thick enough to produce stress to the neighborhood in the considered layer due to lattice mismatch between silicon and its oxide. The result is the existence of two types of oxide inside one layer. In this case (see Fig. 3) the base of q at the start of the cycle (for instance at the time s') is shortened to a smaller base at the end (for instance at the time s"). The resulting p_i of the convolution $p_{i-1} \circ q$ is very similarly to p_{i-1} that means sustained oscillations occur.

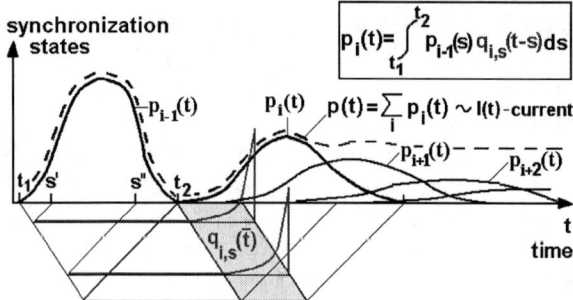

FIGURE 2. The synchronization states $p_i(t)$ in the case of damped current oscillations. The convolution of p_{i-1} with a constant q leads to a damped p_i.

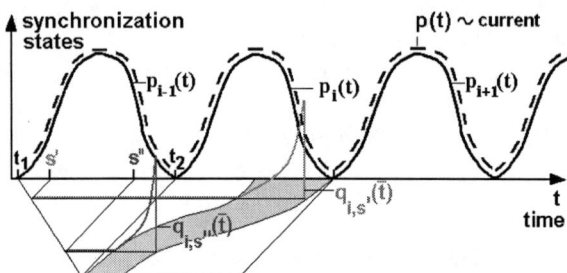

FIGURE 3. The synchronization states $p_i(t)$ in the case of sustained current oscillations. The convolution of p_{i-1} with a variable (shortened) q leads to an nearly unchanged shape of p_i.

FROM SUSTAINED TO DAMPED OSCILLATIONS

Based on 2) the transition from sustained to damped current oscillations is investigated. If the etching rate is increased at a potential of 6V the initial maximum oxide thickness is not reachable because during the oxide growth process more and more oxide is etched concurrently. The result is a smaller maximum oxide thickness. If the etching rate is large enough, the oxide thickness is too small to generate stress. Hence only one type of oxide is built during one cycle leading to a constant q and to damped oscillations (Fig. 2). It is a similar situation as in the case of a small potential with the difference that the current peak frequency, the consumed silicon during one cycle and hence the flown charge during the oxidation time is increased.

REFERENCES

1. Lewerenz, H. J., *J. Phys. Chem. B* **101**, 2421-2425 (1997).
2. Lehmann, V., *J. Electrochem. Soc.* **143**, 1313-1318 (1996).
3. Grzanna, J., Jungblut, H., and Lewerenz, H. J., *J. of Electroanal. Chem.* **486**, 181-189 (2000).
4. Grzanna, J., Jungblut, H., and Lewerenz, H. J., *J. of Electroanal. Chem.* **486**, 190-203 (2000).
5. Pikovsky, A., Rosenblum, M., and Kurths, J., "Synchronization" Cambridge University Press 2001.
6. Da Fonseca, C., Ozanam, F., and Chazalviel, J.-N., *Surface Science* **365**, 1-14 (1996).

Stochastic resonances and highly selective separation methods: application to the detection of DNA mutations

A. Estévez Torres*, L. Jullien* and A. Lemarchand[†]

*Ecole Normale Supérieure, Département de Chimie, CNRS UMR 8640, 24 rue Lhomond, 75231 Paris Cedex 05, France
[†]Université Pierre et Marie Curie, Laboratoire de Physique Théorique de la matière condensée, CNRS UMR 7600, 4 place Jussieu, case courrier 121, 75252 Paris Cedex 05, France

Abstract. The application of a time-periodic field on a chemical reacting system dramatically increases the dispersion coefficient of one reactant, provided that the periods of the field and the chemical reaction are in resonant conditions. We show how this principle can be used to perform highly selective separation and we illustrate the building-up of an experimental setup for detecting DNA mutations.

Keywords: Stochastic resonance, molecular sorting, diffusion, single nucleotide polymorphism
PACS: 05.40.-a, 07.10.Cm, 87.83.+a, 82.39.-k, 82.80.-d, 82.45.Rr, 82.80.Bg,

OPTIMIZED DIFFUSION OF A REACTANT IN A TIME-PERIODIC FIELD

We wish to separate a given chemical species, C_I, from a mixture of compounds C_i with close properties, that target species P according to the reaction:

$$C_i + P \underset{k_2^i}{\overset{k_1^i}{\rightleftharpoons}} Q_i \qquad (1)$$

and that differ by their kinetic properties, i.e. the values of the rate constants k_1^i and k_2^i. To this goal, we prepare a chromatographic medium filled with species P in excess, spot the mixture in the middle, and impose a uniform, time-periodic, electric field. We assume that species C_i and Q_i have different mobilities μ_{C_i} and μ_{Q_i}. At the macroscopic level, the system is governed by partial differential equations for the concentrations C_i and Q_i, that can be viewed as distribution functions for the position of the two species. The variance of the position increases linearly with time with a dispersion coefficient that reduces after a short transient, for species i, to:

$$D_{disp}^i = [a(\mu_{C_i} - \mu_{Q_i})]^2 \frac{k_1^i P k_2^i}{2(k_1^i P + k_2^i)\left[(k_1^i P + k_2^i)^2 + \omega^2\right]} \qquad (2)$$

for a sinusoidal field of pulsation ω and amplitude a, sufficiently large to neglect the intrinsic contribution of diffusion D_{diff}^i [1, 2]. Considering the dispersion coefficient

D_{disp} as a function of $\kappa_1 = k_1 P$ and k_2, we find that it is maximum for species I verifying

$$\kappa_1 = k_1^I P = k_2^I = \omega/2 \qquad (3)$$

that can be seen as stochastic resonance conditions [3]. Thanks to reaction (1), the molecules undergo random jumps between a more mobile state C and a less mobile state Q. When the jumps between the two states occur with the same frequency as the changes in direction of the field, the distance covered during a given half-period, π/ω, by the molecules in state C, of average lifetime $1/\kappa_1$, is not retraced when the field is reversed because they are in the less mobile state Q, of average lifetime $1/k_2$. Hence, the couple (C_I, Q_I) whose rate constants obey relations (3) visits a more extended domain than any other couples. This enhancement of the effective diffusion of a couple (C_I, Q_I) with specific kinetic properties with regard to the reaction with a target P allows us to define a simple protocol for extracting it from a mixture. Fixing the target concentration at $P_I = k_2^I / k_1^I$ and the field pulsation at $\omega_I = 2k_2^I$, we collect the desired species in the tails of the distribution. We recently validated this procedure on the example of the association of dyes C_i with α-cyclodextrin P [4, 5].

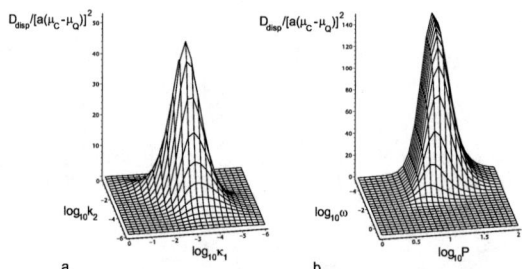

FIGURE 1. Theoretical variation in the normalized dispersive coefficient $D_{disp}/[a(\mu_C - \mu_Q)]^2$ of species $\{C, Q\}$ submitted to a sinusoidal periodic field ($E(t) = a\sin(\omega t)$) in logarithmic units with **a:** $\kappa_1 = k_1 P$ and k_2 ($\omega = 0.002$); **b:** ω and P ($k_1 = 0.001$ and $k_2 = 0.001$).

The main advantages of the procedure described are that: i) the selected species is always separated the first, ii) selectivity relies on two different parameters, k_1, k_2, instead of a single one as is the general case in chromathography and iii) separation is at the fastest because the minimal time is of order of the chemical reaction relaxation time $\tau = (\kappa_1 + k_2)^{-1}$.

APPLICATION TO THE DETECTION OF SINGLE NUCLEOTIDE POLYMORPHISMS

In routine genomic analysis it is needed to analyze many different DNA sequencies, related to diseases or to the identification of harmful microorganisms. The most common mutation in human DNA is Single Nucleotide Polymorphism, SNP, where the normal basepair in a given position, i.e. CG, has been changed, i.e. by AT. Many different methods exist nowadays for the screening of SNPs [6]. The most common one uses the technology of microarrayed DNA chips that relies on the association between complementary DNA single strands. It has the advantage of allowing high-throughput but the drawbacks of being a time-consuming technique —about 2 hours for an experiment— and relying on one single parameter for selectivity —the association constant between DNA single strands—. We would like to apply the preceeding separation protocol to considerably improve both duration and selectivity of the DNA identification process.

Hybridization of a single stranded DNA with its complementary strand follows the pattern of reaction (1). The presence of a SNP changes k_2 by two to three orders of magnitude. Thus, two sets of values for the pulsation, ω_I, and the target concentration, P_I, can be defined which maximize the dispersion coefficient of either the mutated or the normal DNA. The idea is to track the concentration profile of fluorescent single stranded DNA, species C, over a typical length $l = \sqrt{D\tau}$, given by the maximal distance covered by dispersion during the relaxation time. At 298 K, typical values for short DNA strands (13mer) at resonant conditions (3) are $\kappa_1 = k_2 = 10^{-1} - 1 \ s^{-1}$, thus $\tau = (\kappa_1 + k_2)^{-1} \sim 10 - 1 \ s$. With $D^i_{diff} \ 10^{-10} \ m^2 s^{-1}$, a, the amplitude of the electric field, could be reasonably set to enhance the effective diffusion coefficient of the resonant species by a factor of ten, leading to a diffusion length $l \sim 30 - 100 \ \mu m$.

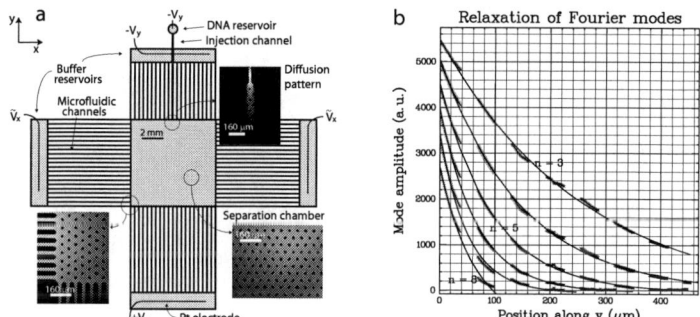

FIGURE 2. a Scheme of the microfabricated device. Fluorescently labeled DNA is introduced vertically in the chamber by migration due to the application of a time-constant voltage drop along the y axis. The time-periodic electric field is applied along x axis. An array of microfluidic channels ensures the electrical connection between the buffer reservoirs, where platinum electrodes are placed, and the separation chamber to ensure a homogeneous electric field [7]. The chamber is 10 μm height and an array of posts separated by 80 μm is needed to maintain mechanic rigidity [8]. The action of migration along the y axis and diffusion along the x axis creates a stationary diffusion pattern. b The diffusion pattern is analyzed by Fourier transform along the x axis. Relaxation of the Fourier modes, n, along the y axis fit to a diffusion-like relaxation law of type $e^{-D/vq_x^2 y}$, v being the migration velocity along the y axis, $q_x = 2\pi n/L_x$, where L_x is the length of the image along the x axis and n the Fourier index. Dashes represent experimental points and lines the fit. We extract, for out of resonance conditions, $D = 1.6 \times 10^{-10} \pm 0.1 m^2 s^{-1}$ in agreement with the result obtained by other techniques.

A VERSATILE MICROFABRICATED SEPARATION CHAMBER

With the idea of detecting SNPs fast and with high selectivity, we have constructed a thermostated microfabricated laboratory that is compatible with the experimental and theoretical constraints of the present approach. It allows to apply electric fields in different directions, time periodic and eventually with spatial patterns, to visualize the concentration profile of a fluorescent species.

Our present device is displayed in figure 2. Imaging of a concentration profile over $\sim 100 \mu m$ by fluorescence microscopy in a microfabricated device of $\sim 10 \mu m$ height allows us to set a range of electric fields of amplitude 10^2 to $10^4 V m^{-1}$ in a constant or time-periodic way in any direction of the xy plane. The largest values are required to observe the enhanced diffusion phenomenon. The application of such a strong electric field gives rise to large Joule heating. This could be critical when kinetic constants greatly vary with temperature (one order of magnitude every five degrees celsius for k_2 in DNA hybridization). Reducing the device height to $10 \mu m$ reduces the heat flux to $\sim 1 W cm^{-2}$ that can be dissipated by coupling a Peltier module to the microfabricated device.

ACKNOWLEDGMENTS

This work relies on the contribution of many people: Damien Alcor, Jean-François Allemand, Anne Bourdoncle, Vincent Croquette, Charlie Gosse, Laurent Lacroix and Jean-Louis Mergny. Hervé Lemarchand and David Bensimon are gratefully acknowledged for fruitful discussions. This work was supported by a special grant from the French ministry of research in the frame of the Action Concertée Incitative "Nouvelles méthodologies analytiques et capteurs".

REFERENCES

1. L. Jullien, A. Lemarchand, H. Lemarchand, "Procédé de séparation d'un composé chimique ou biologique dans un mélange de composés similaires par diffusion dans un milieu tel qu'un gel", n^0 FR 99 133 66, 26/10/1999 ; n^0 PCT/FR 00/02974, 25/10/2000.
2. L. Jullien, A. Lemarchand, H. Lemarchand, *J. Chem. Phys.*, **112**, 8293–8301 (2000).
3. C. van den Broeck, *Physica A*, **168**, 677–696 (1990); I. Claes, C. van den Broeck, *Phys. Rev. A*, **44**, 4970–4977 (1991).
4. D. Alcor, V. Croquette, L. Jullien, A. Lemarchand, *Proc. Nat. Acad. Sci. USA*, **101**, 8276–8280 (2004).
5. D. Alcor, J.-F. Allemand, E. Cogné-Laage, V. Croquette, F. Ferrage, L. Jullien, A. Kononov, A. Lemarchand, *J. Phys. Chem. B*, **109**, 1318–1328 (2005).
6. P.Y. Kwok, *Annu. Rev. Genomics Hum. Genet.*, **2**, 235–258 (2001).
7. L.R. Huang, J.O. Tegenfeldt, J.J Kraeft, J.C. Sturm, R.H. Austin, E.C. Cox, *IEDM Technical Digest*, 363–366 (2001).
8. C.-X. Zhang, A. Manz, *Anal. Chem.*, **75**, 5759–5766 (2003).

Baseline Noise in High-Performance Liquid Chromatography with Electrochemical Detection

Akira Kotani[*], Yuzuru Hayashi[†], Rieko Matsuda[†], and Fumiyo Kusu[*]

[*] *School of Pharmacy, Tokyo University of Pharmacy and Life Science, Hachioji, Tokyo 192-0329, Japan*
[†] *National Institute of Health Sciences, Setagaya, Tokyo 156-8511, Japan*

Abstract. The chromatographic baseline was converted to power spectrum by Fourier transform to show the frequency dependence of noise. The spectral analysis of chromatographic baseline was made under various experimental conditions for high-performance liquid chromatography with electrochemical detection (HPLC-ECD). From the spectral analysis, it was found easily that a system using a small stroke pump and a wall jet type cell with a plastic formed carbon working electrode had a low level of the baseline noise. The optimization of HPLC-ECD system was performed by evaluation of the detection limit based on a chemometric tool, called the Function of Mutual Information (FUMI) theory, which treats the chromatographic baseline fluctuation as a $1/f$ fluctuation model. We found that the observed relative standard deviation (RSD) of peak area by repetitive measurements ($n = 5$) of chromatograms was parallel to the predicted RSD curve based on the FUMI theory. The result indicates that the FUMI theory can provide a measure for estimating the precise and reliable detection limit from a single measurement of noise and signal in HPLC-ECD. The present spectral analysis of baseline noise and the prediction of RSD based on the FUMI theory are an efficient optimization strategy for determining catechins using HPLC-ECD.

Keywords: HPLC; Electrochemical detection; FUMI theory; Power spectrum; catechins
PACS: 82.80.Fk (Electrochemical analysis)

INTRODUCTION

High-Performance liquid chromatography with electrochemical detection (HPLC-ECD) is a sensitive and selective method for the determination of redox compounds. However, ECD has sometimes low reproducibility, because current signals and noise are susceptible to the characteristics of interface between an electrode and solution in the electrochemical flow cell. If suitable conditions of HPLC-ECD were selected for keeping stable interface, the above drawback would be overcome. For selecting and describing the conditions of HPLC-ECD, repetitive measurements on the same samples have been required to obtain a relative standard deviation (RSD) of measurements, thus a lot of experimental time and effort are usually required.

In this study, we proved that an optimization strategy for determining catechin by HPLC-ECD using the spectral analysis of chromatographic baseline noise. Moreover, the optimization of HPLC-ECD system was performed by chemometric tool based on the $1/f$ fluctuation model which is made up of white noise and a Markov process, called the Function of Mutual Information (FUMI) theory [1,2]. The above methods were an attempt to save considerable amounts of chemicals and experimental time for the optimization of HPLC-ECD systems.

EXPERIMENTAL SECTION

The HPLC-ECD system consisted of a pump, an ODS column (250 mm x 4 mm i.d. 5 µm), a mobile phase (acetonitrile: 0.1 mol/L phosphate buffer, pH 2.5, 15:85, v/v), and an electrochemical detector. Jasco PU-880 (a dual series of piston pump, 80 µL/stroke, pump A) or Tosoh DP-8020 (a dual parallel piston pump 6 µL/stroke x 2, pump B) was used. Figure 1 shows the composition of the electrochemical detectors of Jasco EC-840 (abbreviated as detector A, Fig. 1A) and Kotaki EDP-1 (abbreviated as detector B. The detection potential for determining (-)-epicatechin was +600 mV vs. SCE. The analog data of a chromatogram from the electrochemical detector were amplified by an amplifier, converted to digital chromatogram data by an A/D converter, and the digital data were recorded by a personal computer at sampling intervals of 0.2 s/point.

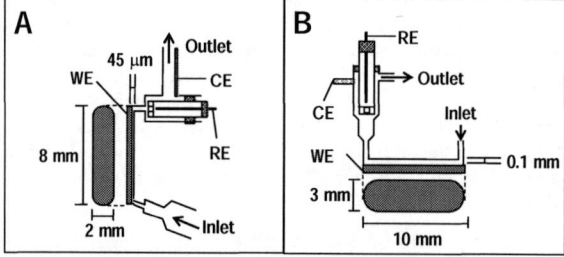

FIGURE 1. Electrochemical cells and dimensions of working electrodes for Jasco EC-840 (A) and Kotaki EC-840 (B). WE, working electrode; RE, reference electrode (saturated calomel electrode, SCE); CE, counter electrode (stainless). (A) Glassy carbon (GC) was used as WE. (B) GC or plastic formed carbon (PFC) was used as WE.

RESULTS AND DISCUSSION

The baseline noise of the chromatogram in HPLC-ECD was converted to power spectrum by Fourier transform. The power density of the low frequency was larger than that of the high frequency, and it was apparent that the baseline noise contained $1/f$ noise. These data demonstrate that the reduction of the low frequency noise in $1/f$ fluctuation improves sensitivity and precision on HPLC-ECD analysis. To trace the source of low frequency noise, and to select a suitable pump, electrochemical flow cell, and working electrode material for the cell, the analysis of power spectra of chromatographic baselines was made.

To show the effect of pulsation on chromatographic baselines in HPLC-ECD, the time series and power spectra of baselines were examined. Fig. 2A illustrates the chromatographic baseline obtained using pump A and detector A at a flow rate of 0.45 mL/min, but it is difficult to recognize the effect of pulsation on the noise by visual inspection. However, two major bands were clearly spotted in the power spectra in Fig. 2B. At a flow rate of 0.45 mL/min, the rotation speed of pump A (dual piston series type, 80 μL/stroke) was 0.09 Hz, when the flow delivery was ideal. This frequency was identical with the low frequency of the one band in Fig. 2B. The higher frequency of another band was twice the low frequency. Hereinafter, the low and high frequencies are called the fundamental tone and harmonic, respectively, from acoustics terminology. The frequencies of the fundamental tone and harmonic were proportional to the flow rate. Clearly, the cause of the excess fluctuation was the reciprocating motion of the pistons.

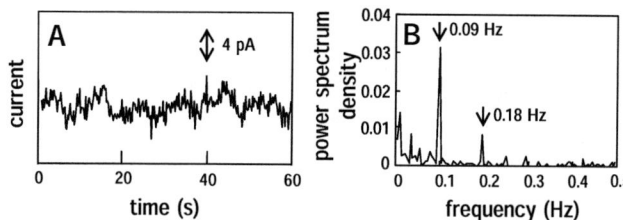

FIGURE 2. Baseline of chromatogram (A) and power spectrum (B). Baseline was obtained using the HPLC-ECD system composed of pump A (PU-880) and detector A (EC-840) at a flow rate of 0.45 mL/min. The baselines of chromatograms were converted to power spectra by Fourier transform. The arrows indicate the noises derived from the pulsation of the pump.

Fig. 3 shows the power spectra of chromatographic baselines obtained using various pumps and detectors. With the pulse damping system in HPLC-ECD, the bands of pulsation were remarkably decreased (Fig. 3(a); compare Fig. 2A). The power spectrum shown in Fig. 3(b) was obtained using the HPLC system using pump B (dual piston parallel type) and detector A. The power density for pump B (Fig. 3(b)) was considerably smaller than that for pump A (Fig. 2B). Fluctuations in flow rate caused by piston pump mechanics often plague high-sensitive determination using HPLC-ECD. However, the frequencies of the fundamental tone and harmonic for pump B were difficult to recognize these noises at the frequencies on the power spectrum. Pulse damping systems and small stroke pump are useful for decreasing pump pulsation. A spectral analytical comparison was made of the effects of the electrochemical cell structure and working electrode material (GC or PFC) on chromatographic baseline noise. These spectra were obtained using detector B with the GC electrode (Fig. 3(c)) and the PFC working electrode (Fig. 3(d)). In the former case, bands at 0.09 and 0.18 Hz in the power spectrum in Fig. 2B were no longer present in Fig. 3(c). Detector B was a wall jet type cell whose inlet line was perpendicular to the working electrode though the inlet line of detector A and was oblique; the power density for the detector B was considerably less than that of the detector A. The wall jet type cell thus reduces the noise created by pump pulsation. With detector B with the PFC working electrode, power density was found to be very small, as seen in Fig. 3(d). When the wall jet type cell and PFC working electrode were used in HPLC-ECD, the noise level of baseline was very low.

Moreover, the detection limit for determining (-)-epicatechin by HPLC-ECD was obtained by the FUMI theory [1,2]. When the main cause of the uncertainty is the chromatographic baseline noise, the FUMI theory can predict measurement RSD and detection limits using noise and signal from a single measurement of a chromatogram. In the case of HPLC-ECD, the predicted RSD based on the FUMI theory was parallel to the observed RSD by repetitive measurements ($n = 5$), indicating the FUMI theory is useful for predicting the measurement precision and detection limit in HPLC-ECD without repetitive measurements of chromatograms. The detection limits at which 33% RSD was given for (-)-epicatechin at various combinations of the pump, the pulse damping system, electrochemical detector, and working electrode are shown in Table 1. Using the FUMI theory, the optimization of HPLC-ECD systems is easily carried out in the shortest possible time.

We conclude that the present power spectral analysis of baseline noise and the detection limit based on the FUMI theory are thus an efficient optimization strategy for determining catechins using HPLC-ECD.

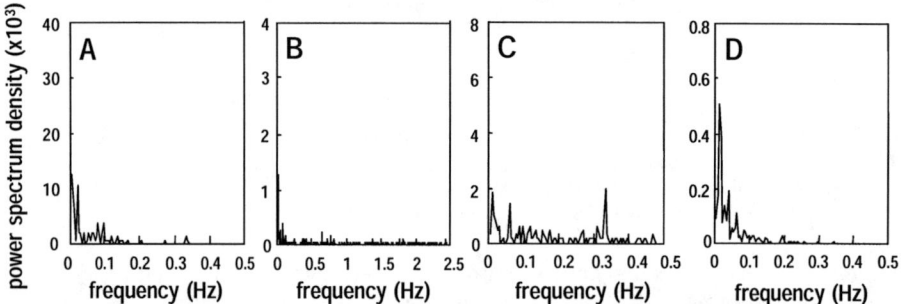

FIGURE 3. Power spectra of baselines of chromatograms using HPLC systems of (A) pump A, detector A and pulse damper system, (B) pump B and detector A, (C) pump A and detector B using GC working electrode, and (D) pump A and detector B using PFC working electrode.

TABLE 1. Detection limit of (-)-epicatechin determination, calculated by FUMI theory.

Pump		Detector	WE	Detection limit (fmol)
A		A	GC	96.0
A	+ Damper	A	GC	38.4
B		A	GC	26.1
A		B	GC	74.5
A		B	PFC	25.7

ACKNOWLEDGMENTS

This work was supported in part by a grant from the Japan Health Science Foundation.

REFERENCES

1. Matsuda, R., Hayashi, Y., Sasaki, K., Saito, Y., Iwaki, K., Harakawa, H., Satoh, M., Ishizuki, Y., and Kato, *Anal. Chem.* **70**, 319-327 (1998).
2. Hayashi, Y., and Matsuda, R., *Anal. Chem.* **66**, 2874-2881 (1994).

The Noise Diagnostics of Organic Electrolytes for Rechargeable Lithium Batteries

Leonid S. Kanevskii, Boris M. Grafov, and Mikhail G. Astafiev

*A.N. Frumkin Instutute of Electrochemistry of Russian Academy of Sciences,
31 Leninskii prospect, Moscow 119071, Russia*

Abstract. We investigated the electrochemical noise of lithium electrode under the current control. Different aprotic organic electrolytes containing different solvents and lithium salts were used as well as anodic and cathodic currents. It was found that level of electrochemical noises depends strongly on the kind of electrolyte. The electrolyte systems with high lithium cycling efficiency and without the dendrite forming posses the low level of noise. This means the electrochemical noises method can be used for express screening organic electrolytes for rechargeable batteries with negative electrode of metal lithium.

INTRODUCTION

The method associated with fluctuation electrochemical phenomena (noises), which was elaborated more than 35 years ago [1, 2], at present is widely used in studies of complex electrochemical and corrosion processes of passivation type. It is advantageous to apply this method for a study of electrode processes in Li batteries. It is known that the surface of Li electrode in an aprotic electrolyte is coated with a passive insulating film consisting of the products of Li interaction with the electrolyte [3]. Thus, it may be said that Li is in the quasi-passive state, which is characterized by a certain level of electrochemical noises. The aim of this work is to study the character of electrochemical noises on the Li electrode in the steady state and under the polarization, i.e. under actual conditions of rechargeable battery operation. Particular attention is given to the dependence of noise phenomena on the kind of aprotic organic electrolyte.

RESULTS AND DISCUSSION

The experimental approaches which are common for corrosion studies using the method of electrochemical noises [4, 5], were applied. The following electrolyte systems containing different Li salts and organic solvents were used: 1 M $LiClO_4$ in 1,3-dioxolane (El1), 1 M $LiN(CF_3SO_2)_2$ in 1,3-dioxolane (El2), and 1 M $LiPF_6$ in the ethylene carbonate–diethyl carbonate equivolume mixture (El3). The electrochemical measurements were performed using a low-noise galvanopotentiostat – Solartron Electrochemical Interface 1286. The electrode potentials were recorded with a sampling rate of 1 sec^{-1}. The fluctuations of potential difference were measured between two smooth Li plates equalled in sizes and similarly prepared [6].

On results the measurement, the time rows for noise of Li electrode in each electrolyte were calculated and corresponding correlative functions were elaborated.

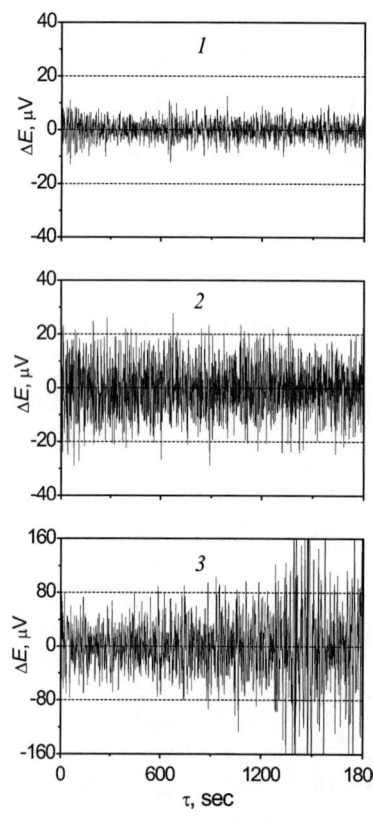

FIGURE 1. Electrochemical noises of Li electrodes under the cathodic polarization (j = 1000 µA/cm^2) in electrolytes El1 (1), El2 (2), El3 (3).

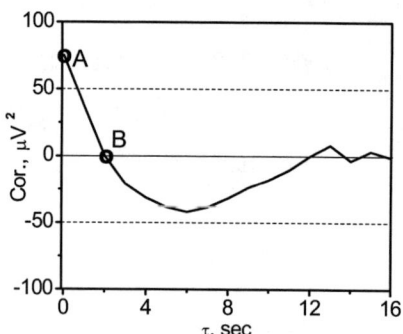

FIGURE 2. The typical view of the correlative function.

Examples of initial data for investigated electrolyte presented on Fig. 1. Fig. 2 demonstrates the typical correlative function. In Table are given the averaged over set experiments values of correlative functions maxima (the point A on Fig. 2) and characteristic times of noise process (the point B on Fig. 2). The value of correlative function maxima is characteristic of average noise density. As may be seen from the Table, the steady state of electrodes under open-circuit condition in all electrolytes under studying is characterized by low density of potential fluctuations (5–8 µV^2)

TABLE. Parameters A and B of correlative functions

Electrolyte kind	Polarization mode	Maxima A, µV^2	Parameter B, sec
1 M LiClO$_4$ in the 1,3-dioxolane (El1)	No polarization	4.7	1.7
	Cathodic	20.4	3.0
	Anodic	15.8	2.0
1 M LiN(CF$_3$SO$_2$)$_2$ in the 1,3-dioxolane (El2)	No polarization	3.9	1.5
	Cathodic	64.0	1.6
	Anodic	65.0	1.5
1 M LiPF$_6$ in the ethylene carbonate–diethyl carbonate mixture (El3)	No polarization	8.6	1.8
	Cathodic	370	1.8
	Anodic	310	1.9

reached quickly after immersing the electrodes into the electrolyte. The intensity of noises remained virtually constant on long-term staying the electrodes in the solution indicating that stable passive film exists on the Li surface.

Polarization of Li electrode by small currents (~50 $\mu A/cm^2$) causes a certain change in the Li surface state manifested in a slight increase in the potential fluctuations density (average value is 15–20 μV^2). In the process the noises of anodically and cathodically polarized electrodes are similar in character and intensity. This permits to consider the problem about localization of anodic and cathodic processes. The fact that Li anodic dissolution is localized immediately at the Li/passive film interface is beyond this problem. However, there is some disagreement on the localization of Li cathodic deposition. Along with the concept of Li deposition at the aforementioned interface, there is an opinion that Li forms on the film surface or in the its bulk [7]. If this were so, potential fluctuations under the cathodic and anodic polarization would differ. The results of this study indicate that Li cathodic deposition is localized under the passive film.

The electrode polarization with large currents (1000 $\mu A/cm^2$, this corresponds to the standard electrode loading under cycling of Li batteries) significantly changes Li state. This is reflected in the character of electrochemical noises. First of all, the noise amplitudes increase significantly pointing to the working electrode activation and destruction of passive film on its surface. The intensity and character of potential fluctuations under forced polarization mode depend significantly on the electrolyte composition. In the perchlorate electrolyte (El1), the electrochemical noises of loaded Li electrodes only slightly differ from those under the polarization with small currents. The intensity of potential fluctuations in electrolyte system El2 is considerably stronger. The maximum amplitude of fluctuations was recorded in electrolyte system El3 (Fig. 1).

The dynamics of electrochemical noises of polarizable Li electrode in the aprotic electrolyte can be explained as follows.

Under the anodic polarization, Li dissolution proceeds under the film, through which metal ions are transferred to the electrolyte, and void cavities form under the film. In the case of sufficiently flexible film, it is pressed to the metal by outer pressure; more rigid film can crack. In the latter case, the activated surface area increases and it becomes more heterogeneous leading to an increase in the noise intensity. In parallel, the exposed surface is passivated, and the dynamic equilibrium is reached. In the case of rapid Li passivation, metal surface which becomes exposed as a result of film cracking is intensively passivated reducing the heterogeneity. In the case of slow passivation, the heterogeneity increases. In the first case, an insignificant intensity of potential fluctuations is recorded, and in the latter case, the potential fluctuations are more intensive.

Under the cathodic polarization, Li deposition on the metal surface proceeds under the passive film. Deposition of a large amount of Li on the negative electrode is accompanied by the film perforation by growing metal crystals. The freshly formed active Li surface, as well as in the case of film cracking under the anodic polarization of electrode, raises the electrode's heterogeneity and noise intensity. As well as under the anodic polarization, in the case of rapid passivation, Li heterogeneity decreases in time, whereas in the case of slow passivation Li heterogeneity does not decrease.

According to literature data and our experimental results, behavior of Li electrodes in used electrolytes is very different. Li surface in the electrolyte El1 is coated with a thin flexible homogeneous passive film [8] which is not perforated under the cathodic polarization. Li, which is deposited on the electrode, is rapidly coated with this film. As the result, the dendrite formation on battery charge is hampered providing a high Li cycling efficiency [9, 10]. In the electrolyte El2 containing imide, which is aggressive with respect to Li [10], the metal passivation rate is slightly lower than in the perchlorate electrolyte based on 1,3-dioxolane. This makes possible the dendrites growth on active Li surface areas. In this case, the electrode surface is more heterogeneous than in the perchlorate electrolyte, as is evidenced by an increase in the intensity of potential fluctuations. The study of Li electrode cycling showed that the dendrite-free service life of the electrode in this electrolyte is 200 –300 charge-discharge cycles.

Based on $LiPF_6$ electrolyte El3 contains a considerable amount of acidic impurities [11] strongly activating Li surface. Here, dendrite passivation rate decreases, and dendrites grow rapidly. Li surface heterogeneity increases significantly, i.e. highly active areas form along with the passivated metal areas. The experimental results and their explanation are in agreement with the data on the dendrite formation and cycling of Li electrode in this electrolyte. Cycling of the Li electrode in electrolyte El3 is unsatisfactory: noticeable dendrites form on the electrode even after several cycles.

It is particularly remarkable that despite the very high scatter of noise densities in different electrolytes, the characteristic times of noise process are only slightly dependent on electrolyte composition. This suggests a single mechanism of the phisico-chemical processes of Li passivation-activation in aprotic organic electrolytes.

The established correlation between the intensity of Li electrode potential fluctuation and stability of this electrode on cycling (the higher noise intensity, the poorer cycling) enables to use the method of electrochemical noises for express screening organic electrolytes for rechargeable Li batteries with negative electrode of metal Li.

ACKNOWLEDGMENTS

The authors are grateful to Russian Foundation for Basic Research for financial support of this study (Grant No 05-03-32294).

REFERENCES

1. Sh. Kogan, *Electronic Noise and Fluctuations in Solid*, Cambridge University Press, 1966.
2. V.A. Tyagai, *Soviet Electrochemistry*, **10**, 1–18 (1967).
3. E. Peled, "Lithium Stability and Film Formation in Aprotic Electrolytes for Lithium Battery Systems", in *Lithium Batteries*, J.-P. Gabano (Ed.), London, Academic Press, 1983, pp. 43–72.
4. A. Legat and V. Dolecek, *J. Electrochem. Soc.*, **142**, 1851–1858 (1995).
5. J.C. Uruchurtu and J.L. Dawson, *Corrosion*, **43**, 19–26 (1987).
6. L.S. Kanevskii, A.M. Skundin, *Elektrokhimicheskaja Energetika*, **2**, 140–143 (2002) (in Russian).
7. A. Kedrinskii, V.E. Dmitrenko and I.I. Grudyanov, *Litievie istochniki toka*, Energoatomizdat, Moskwa, 1992, pp. 29-30 (in Russian).
8. D. Aurbach, O. Yongman, Y. Gofer and A. Meitav, *Electrochim. Acta.* **35**, 625–638 (1990).
9. K.M. Abraham, and S.B. Brummer, "Secondary Lithium Cells", in *Lithium Batteries*, J.-P. Gabano (Ed.), London, Academic Press, 1983, pp. 371– 406.
10. D. Aurbach, I. Weissmann, A. Zaban and O. Chusid, *Electrochim. Acta*, **39**, 51–71 (1994).
11. C.G. Barlow, *Electrochem. Solid-State Lett.* **2**, 362–364 (1999).

MEASUREMENT TECHNIQUES

Novel Transimpedance amplifier for Noise Measurements on Bio-Electronic devices

Giorgio Ferrari and Marco Sampietro

Politecnico di Milano, Dip. Di Elettronica e Informazione, P.za L.da Vinci 32, Milano, Italy, 200133
E-mail: ferrari@elet.polimi.it

Abstract. The paper introduces a novel transimpedance amplifier based on the sequence of an integrator and a differentiator stage with a feedback loop that allow the system to discharge the standing current from the device under test. The proposed solution maintains signal amplification over a large bandwidth while ensuring a timeless measuring opportunity. The amplifier is ideal for noise spectral measurements of nanodevices and biomolecules covering a spectral range from few Hz to few MHz, and processing DC leakage currents as high as few hundred nanoAmperes. The circuit architecture can be easily transferred an integrated solution, opening a wide range of applications in miniaturized systems.

Keywords: Noise measurements, Transimpedance amplifier, Current converter.
PACS: 05.40.Ca, 72.70.+m, 73.43.Fj

INTRODUCTION

The electrical characterisation of molecular and biological systems may profit from being performed with very low signal levels to avoid mechanical stress on the samples due to high electrical field or to heat them with high current density. Noise spectroscopy, measuring the signal produced by the nano-bio device without external bias, offers a unique possibility to study these systems in the equilibrium and near equilibrium conditions [1]. Due to the small size of the nano-bio devices, the noise levels are vanishing low and efforts must consequently be put in the design of very high sensitivity front-end preamplifiers to perform these noise measurements [2].

Noise current measurements on nanostructures or on molecular devices impose not only high sensitivity (which may even be enhanced by using correlation techniques on two channels [3,4]) but also wide bandwidth and large dinamic range front-end amplifiers. Wide bandwidth gives the possibility to investigate all details of "colored" features in the noise spectra or to extract the thermal noise level also when the low frequency spectra is dominated by the 1/f noise. Large dinamic range enable the detection of a small noise on top of a large standing bias. Most of available electronic instruments fulfill these three requirements one at a time [5,6]; the few very best two at a time [7,8].

The classical configuration for noise measurement using a transimpedance amplifier, for example, trades sensitivity for bandwidth and can not be generally adopted in this contest. In order to minimize noise, the feedback resistance should be chosen as big as possible but technological considerations limit its value to few tens of Ohm in discrete components. The measuring bandwidth of these high sensitivity transimpedance amplifiers is fixed by the unavoidable parasitic capacitance of the resistance to a very small value (e.g. if R_f=1Gohm and C_f=1pF, the bandwidth extends only to about 160Hz) insufficient in all cases in which large noise bandwidth analysis is desired. Note that the maximum feedback resistance that can be obtained in an integrated solution is much lower and consequently, the corresponding thermal noise ($4kT/R_f$) is so high to fade the signal.

The alternative circuit is the integrator-differentiator amplifier, that reaches the highest possible sensitivity over a large bandwidth. In this case the conversion of the DUT current into a voltage is made by a capacitance, a truly noiseless component in

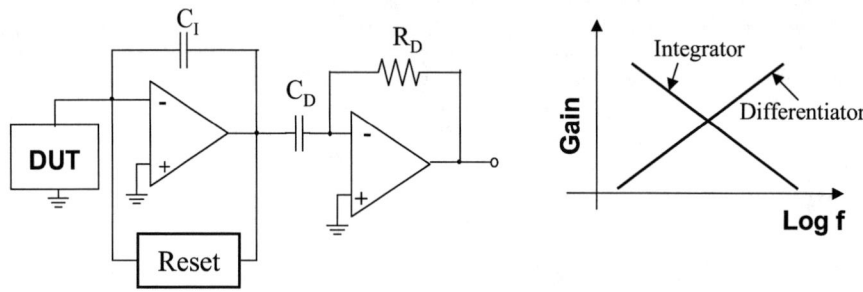

FIGURE 1. Integrator-differentiator scheme (left) and corresponding frequency bandwidth (right). The reset circuit in the scheme is necessary to discharge the feedback capacitance C_I conveniently.

most cases. The integrator stage is followed by a differentiatior stage in order to obtain the overall desired conversion from current to voltage constant for all the frequencies, as sketched in Fig.1. The integrator-differentiator configuration not only reduces the instrumental noise, but the overall bandwidth of the system can be extended to the gain-bandwidth product of the used amplifiers. The disadvantage of this class of instruments is the saturation of the integrator stage due to standing current (leakage or sample bias), thus limiting the available measuring time.

LOW-NOISE CONTINOUS CHARGE RESET

To overcome the disadvantage of saturation of the integrator stage due to the standing current, a novel circuit is introduced that provides a DC path to ground for the input standing current, leaving free the signal current to go through the "noiseless" amplifying stage. Figure 2 reports a schematic of this continuous charge reset realisation. The network has high gain from point A to B for low frequencies signals (from DC to a value set by C_H and R_{H2}, typically less than 1Hz). In this way, as soon as the standing leakage current from the DUT starts to be integrated on C_I and moves A, it is swept away through R_F thanks to the corresponding large voltage change in B. Conversely, the gain from A to B is very low (ideally equal to zero, in the schematic of Fig.2 set by the ratio R_{H2}/R_{H1}) for frequencies higher than 1Hz so that B is not

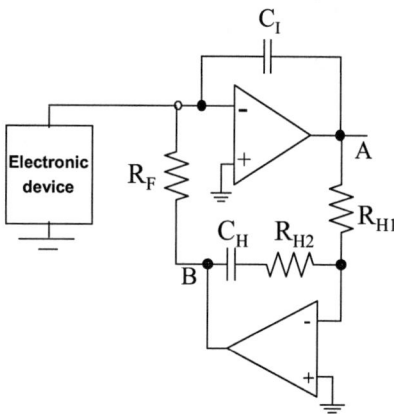

FIGURE 2. Schematic of the continuous reset circuit around the main integrator stage of the transimpedance amplifier. The feedback network has a frequency dependent voltage gain from point A to point B to ensure transfer of signal through C_I and leakage current through R_F.

moving when signals of interest from the bioelectronic device are processed. The result is a system that maintains signal amplification over a large bandwidth while ensuring a timeless measuring opportunity.

From the noise point of view, the largest source of noise is the resistor R_F, which therefore should be chosen as high as possible, limited by the maximum leakage current that is foreseen from the DUT. It is important to note that this high value resistor does not affect in any matter the bandwidth of the circuit for the signals of interest [9].

An important point to be emphasized is that none of the components forming the feedback loop need to be of a well defined value nor linear. The resistor R_F and R_H may therefore be obtained from MOSFETs operating in subthreshold ohmic regime without the bandwith of the full transimpedance amplifier be affected at all. Only attention is to be paid to the stability of the feedback loop. This imply that the integration of the amplifier in a single chip is feasible without altering the proposed schematic.

EXPERIMENTAL DATA

Figure 3 (left graph) shows the frequency response of the prototype (continuous line) and, for comparison, the frequency response of a traditional transimpedance amplifier (dashed line) with the same instrumental noise. Note that frequencies as high as few MHz may be perfectly reached with the new amplifier thus extending considerably the spectrum over which to perform noise analysis.

In the right graph of Fig.3 we report the equivalent input current noise spectrum of the amplifier. The flat part is due to the current noise of the resistor R_F which becomes dominated, at higher frequencies, by the input series noise of the operational amplifier over the input capacitance.

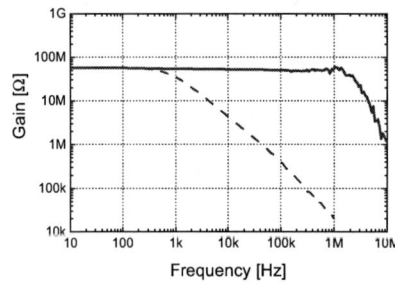 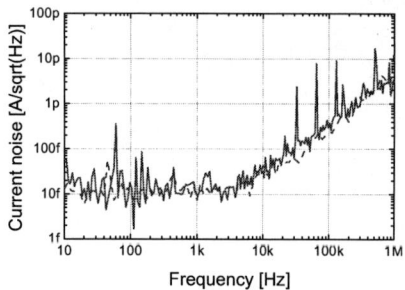

FIGURE 3. (Left) Experimental frequency behaviour of the amplifier (continuous line) as compared to a standard transimpedance amplifier (dashed line) with the same instrumental noise. Lower bandwidth limit is defined by the choice of C_H and R_H components (see Fig.2) while highest bandwidth limit is set by the bandwidth of the operational amplifier. (Right) Equivalent input current noise of the amplifier over the full spectrum

CONCLUSIONS

The proposed circuit senses input currents with the highest possible sensitivity and on a bandwidth as large as few MegaHerts, with the ability to discharge a DUT leakage current as large as hundred of nanoAmperes. It is consequently ideal for applications in the field of nano or molecular device characterisation. Emphasis should be given to the fact that the proposed solution has the fundamental advantage to be easily integrated in standard CMOS technologies, thus matching bio-electronic (lab-on-chip) applications at its best.

REFERENCES

1. C. Pennetta, V. Akimov, E. Alfinito, L. Reggiani, G. Gomila, Proceedings of SPIE -- Volume 5472, Noise and Information in Nanoelectronics, Sensors, and Standards II, J. M. Smulko, Y. Blanter, M. I. Dykman, L. B. Kish, Editors, May 2004, pp. 172-182
2. J. Bylander, T. Duty, P. Delsing, Nature 434, 361-364 (2005)
3. M. Sampietro, L. Fasoli, and G. Ferrari, Rev. Sci. Instrum. 70, 2520 (1999).
4. G. Ferrari and M. Sampietro, Rev. Sci. Instrum. 73, 2717 (2002).
5. Keithley sub FemtoAmpere SourceMeter model 6430, Keithley Instruments Inc., USA.
6. HP4142BSource-monitor Unit, Agilent Inc. USA
7. Femto LCA-1-10T, FEMTO Messtechnik GmbH, Germany.
8. Axopatch 200B patch-clamp amplifier, Axon Instruments Inc., USA.
9. M. Sampietro, G. Ferrari, D. Natali, European patent PCT/IB2004/004080.

Suppression of Offset and Drift in a dc Amplifier by Combination of Multiple Amplifiers

S. Hashiguchi, Y. Takemoto, M. Ohki, M. Tacano*, J.Sikula**

University of Yamanashi, Japan
*Meisei University, Japan
**Brno University of Technology, Czech

Abstract. Offset voltage and its drift were suppressed by generating the weighted sum of multiple dc amplifiers. The weighting factors were determined so as to make the total gain and the signal-to-noise ratio as large as possible.

Keywords: dc amplifier, offset, drift, parallel operation, multi-amp
PACS: 84.30.Le, 07.50.Hp,

INTRODUCTION

Offset voltage and its drift in a low-noise head amplifier are the limiting factors for noise measurements, especially in very low frequency region such as 0.001Hz. Insertion of coupling capacitors is not a reasonable solution because it causes the increase of noise at lower frequency. The offset originates from the dispersion of circuit parameters, and the drift originates from power dissipation of transistors and from the variation of ambient temperature. Practically these origins are not controllable so as to make offset and drift sufficiently small.

The offset voltage from integrated amplifiers scatters around zero, the weighted sum can be made zero if the weighting factors are properly selected. The drift is the variation of the offset due to the variation of the operating condition such as ambient temperature. Under various operating conditions the weighted sum of the offset can be made zero, accordingly the drift is suppressed in a certain extent of operating conditions.

BASIC CONCEPT

We consider a set of m identical unit amplifiers. The weighted sum of the offset voltage $V_{\text{off}}(i, j)$ of i-th amplifier at the j-th operating condition can be made zero for non-trivial set of weighting factors $k(i)$ if the number of the unit amplifiers is less than the number of the operating conditions as

$$\sum_{i}^{m} k(i) V_{\text{off}}(i, j) = 0 \text{ for } j < m. \qquad (1)$$

The total signal gain A_T is given as

$$A_T = A_U \sum_{i}^{m} k(i), \qquad (2)$$

where A_U is the gain of the unit amplifier. The total gain A_T should be as high as possible.

The total noise output voltage $<e_{\text{noT}}^2>$ is given as

$$<e_{\text{noT}}^2> = <e_{\text{noU}}^2> \sum_{i}^{m} \{k(i)\}^2 = A_U^2 <e_{\text{niU}}^2> \sum_{i}^{m} \{k(i)\}^2, \qquad (3)$$

where $<e_{\text{noU}}^2>$ is the noise output of the unit amplifier, and $<e_{\text{niU}}^2>$ is the noise referred to the input. The total noise $<e_{\text{noT}}^2>$ should be made as low as possible.

In order to satisfy these two additional requirements the number of the operating conditions, where the offset voltage is zero, must be m-2.

METHOD

Figure 1 shows the configuration of the present method. The system consists of m identical dc amplifiers AMP1-AMPm, two adding amplifiers ADDp and ADDn, preceded by resistors $pk1$-pkm, $nk1$-nkm for weighting factors, and a difference amplifier DIFF.

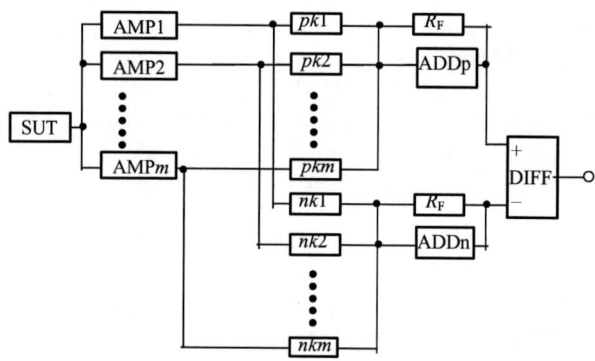

FIGURE 1. Configuration of Drift-free Amplifier

Calibration

At first the offset voltage $V_{\text{off}}(i, j)$ $(1 \leq i \leq m)$ at j-th operating condition are measured simultaneously. The measurements at m-2 different operating conditions give an array of (m, m-2) offset voltage. From this array the weighting factors $k(i)$ are obtained by solving Eqs.(1)-(3). The values of resistors *pki* and *nki* in Fig.1 are set according to the sign of the weighting factors $k(i)$ as

$$pki = \frac{R_F}{k(i)}, \quad nki = \infty, \quad \text{if } k(i) > 0,$$

$$pki = \infty, \quad nki = \frac{R_F}{-k(i)}, \quad \text{if } k(i) < 0, \quad (4)$$

$$pki = \infty, \quad nki = \infty, \quad \text{if } k(i) = 0,$$

where R_F is the value of the feedback resistors attached to ADDp and ADDn.

EXPERIMENTS

A system including 32 unit amplifiers (LM741) was assembled. The voltage gain of each unit amplifier was set at 100. In the preliminary measurements it was found that the power supply voltage dependence of offset was similar for all unit amplifiers. The selected operating conditions are the six points of ambient temperature, i.e., 0, 10, 25, 40, 50 C, and the room temperature around 20 C. The temperature dependence of the offset ranges from 20 µV/K to 1.4 mV/K. Eight unit amplifiers were selected to make total offset zero at six points. We obtained three groups of eight amplifiers out of 32 unit amplifiers.

Figure 2 shows the temperature dependence of the offset voltage of the first group. The plot "Total" shows the performance of the weighted sum, and the others show the characteristics of unit amplifiers. The offset 1.2 mV of "Total" is the least of all and its temperature dependence 68 µV/K is the second best. The total gain factor

$$K_A = \frac{A_T}{A_U} = \sum_i^m k(i) \quad (5)$$

was 3.2 and the total noise factor

$$K_N^2 = \frac{<e_{noT}^2>}{A_U^2 <e_{niU}^2>} = \sum_i^m \{k(i)\}^2 \quad (6)$$

was 3.6. The noise voltage referred to the input (K_N/K_A) is reduced to 0.6 times of the unit amplifier.

The second group gave the values of 2.3 mV at 25 C, 42 µV/K, K_A=2.5, and K_N=3.6, and the third group gave 8.0 mV at 25 C, 21 µV/K, K_A=2.0, and K_N=3.8, respectively.

The combination of three groups gave 11 mV at 25 C, 130 µV/K, K_A=7.8, and K_N=10.9. This means that the total gain A_T is 7800, noise voltage referred to the input (K_N/K_A) is 0.4 times of the unit amplifier. The offset and it temperature dependence referred to the input are 1.4 µV and 170 nV/K, respectively.

Figure 3 shows the noise (referred to the input) of the first group. The upper trace shows the noise spectrum of one of the unit amplifiers, and the lower shows that of the combined system. The total noise is about 5db lower than that of the unit amplifier, which corresponds to the reduction of noise by 0.6(-4.4 db) as stated above.

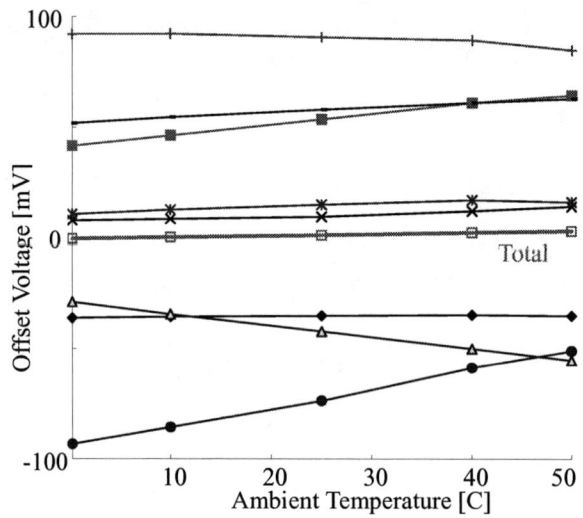

FIGURE 2. Temperature Dependence of Offset Voltage

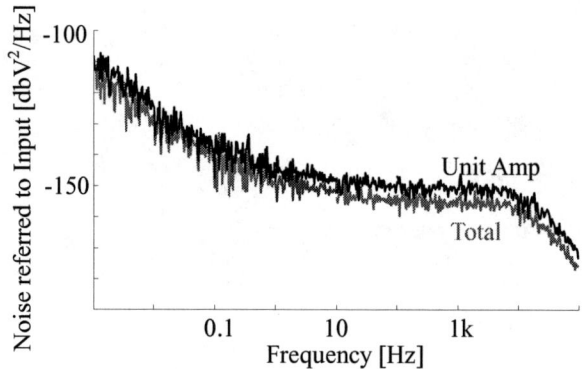

FIGURE 3. Noise Spectra of a Unit Amp and the Combined System

CONCLUSION

It is demonstrated that the weighted sum of the multiple dc amplifiers results in the reduced offset at various operating conditions. The combination of multiple amplifiers improves the noise performance. Further investigations will be needed to optimize the weighting factors.

Experimental Study Of Hysteretic Josephson Junctions As Threshold Detectors Of Shot Noise

M. Meschke*, T. E. Nieminen* and J. P. Pekola*

Low Temperature Laboratory, Helsinki University of Technology, P.O. Box 3500, FIN-02015 TKK, Finland, E-mail: meschke@boojum.hut.fi

Abstract.
We present an experimental study of Josephson Junctions as threshold detectors for an investigation of shot noise. The main advantage of this technique is that the detector and the noise source can be located in close vicinity on a single chip, avoiding natural problems of usual approaches, like filter and bandwidth requirements on connecting wires between samples in a cryogenic environment and measurement apparatus at room temperature. Our first experimental results demonstrate the potential of the proposed technique by exhibiting sensitivity to the shot noise and they allow an estimation of the feasibility of further experiments to characterize mesoscopic conductors.

Keywords: Shot noise, non-Gaussian noise, higher moments, full counting statistics, Josephson Junction
PACS: 72.70.+m, 73.23.-b, 85.35.-p, 85.25.Cp

INTRODUCTION

Noise measurements provide significant information about transport mechanisms in mesoscopic conductors. Intensive theoretical investigations were performed in this field and even first measurements of the third moment of voltage fluctuations across a tunnel junction were recently reported [1, 2]. On the other hand, experiments to observe these higher moments of current or voltage remain difficult and time consuming, because they are normally characterized by very weak signals. What is more, filter and bandwidth requirements for the experiments are hard to fulfill.

Therefore, new experimental approaches are needed and were recently proposed [3, 4]: Josephson Junctions (JJ) may be used as threshold detectors to detect higher moments or even the full counting statistics, but no experimental results were published until now. To overcome this somewhat unsatisfying situation, we studied experimentally the properties of a JJ as a threshold detector to investigate shot noise.

EXPERIMENTAL SETUP: JJ AS A THRESHOLD DETECTOR

We describe our detector with the common RCSJ (Resistively and Capacitively Shunted Junction) model, which provides the escape probability of a JJ for a current pulse of length τ and amplitude I_P as $P = 1 - e^{\Gamma_{esc}(I_P)\tau}$. At low enough temperature ($T \leq 50$ mK), the dominating escape mechanism (Γ_{esc}) is the MQT (Macroscopic Quantum Tunneling)

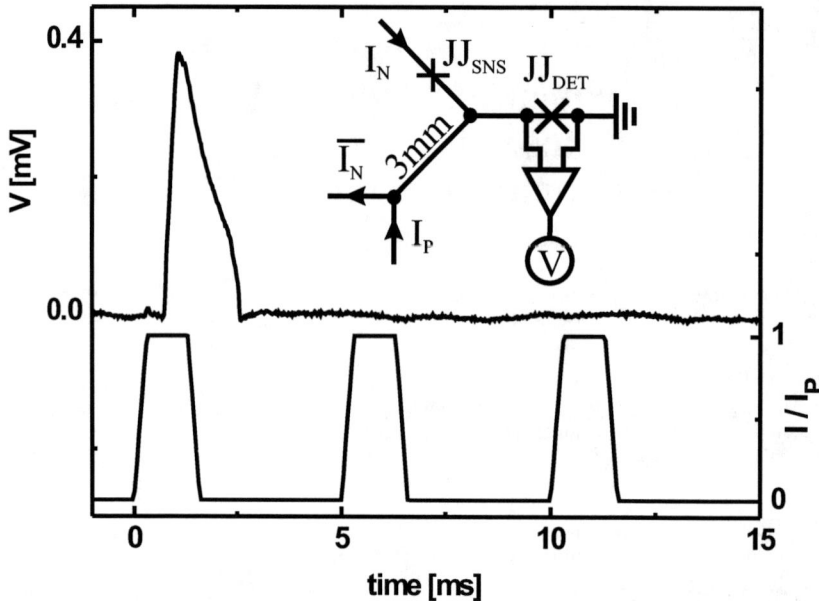

FIGURE 1. Current pulses (lower graph, right scale) and resulting voltage across the detector (upper graph, left scale). In this particular example, only the first current pulse leads to an escape of the detector junction from the superconducting to the normal state, which corresponds to a value of twice the superconducting gap ($\simeq 0.39$ mV for aluminum) and is counted as an escape event. After an escape event, the detector needs at the end of the current pulse additional time to return to its initial state (re-trapping time). The inset shows the experimental setup: I_P is applied to the detector (JJ_{DET}) through a long (3 mm) superconducting line on the chip. A room temperature amplifier allows the observation of the resulting voltage across JJ_{DET}. I_N creates a current running through a 2nd Josephson Junction JJ_{SNS}, acting as our shot noise source (see text).

effect [5].

We use a second, normal conducting tunnel junction as a noise source, which generates a close to Poissonian charge distribution [6]. Both detector and the noise source are fabricated by standard electron beam lithography and two angle shadow evaporation of aluminum. The noise source is in our case a second Josephson Junction with a critical current ($I_C = 20$ nA) well below the detector value ($I_C = 700$ nA). The inset of Fig. 1 shows the experimental setup: noise source and detector are placed on a silicon chip, which is cooled below 50 mK with a dilution refrigerator. An additional long (3 mm) superconducting line transports the current pulses (I_P) to the detector (JJ_{DET}) and compensates with $-I_N$ the DC component of the shot noise current (I_N) through the detector. On the contrary, the high frequency current noise components of I_N pass through the detector. This special setup should allow the investigation of higher odd moments [4], like the third moment of current $\langle \delta I^3 \rangle$: any asymmetry with respect to the mean current value results in a detectable change of the escape probability of JJ_{DET}, when the polarity of I_N is changed.

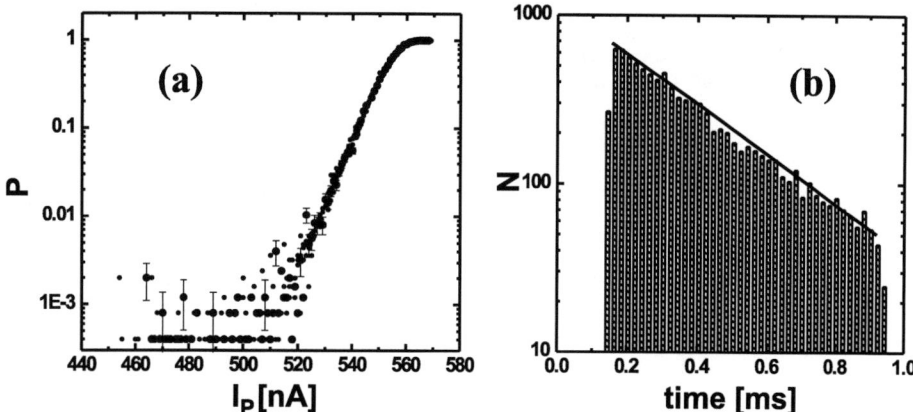

FIGURE 2. (a) Experimentally obtained escape probability (P) histogram for a Josephson Junction at $T \leq 50$ mK, measured with $N = 2500$ pulses for each point. The graph shows four series of measurements taken under equal conditions to demonstrate the reproducibility of the measured values. Selected error bars (see Eq. (1)) show, that the noise for low escape probabilities arises from the statistical error, when only few escape events are detected. (b) Number of escape events for an interval of 20 μs as a function of time with respect to the current pulse. The plot shows data for a current which generates a 94 % escape probability for one full pulse (duration 800 μs plus 100 μs rise and fall time). The line follows the expected behavior according to Eq. (2).

To measure the escape probability for a given amplitude of I_P, a certain number of pulses (N_{PULSE}) is applied. By counting the number of detected escape events N_{ESCAPE} (see Fig. 1) given by voltage pulses of the amplifier, the escape probability is then determined as $P = N_{ESCAPE}/N_{PULSE}$.

Typically, few hundreds to few thousands of pulses are measured to reach a reasonable number of escape events in order to reduce the statistical error, which reads:

$$\Delta P = \sqrt{\frac{P(1-P)}{N}}. \tag{1}$$

A typical measurement time for one point of the full escape histogram (Fig. 2a) is on the order of seconds, even if more than thousand pulses are applied, as the typical pulse lengths plus the detector re-trapping times do not exceed few milliseconds. The logarithmic plot emphasizes the low noise floor, which can be further reduced with an increasing number of pulses. We found within reasonable numbers of applied pulses (up to 10000), that the observed noise is still limited by the statistical error as described by Eq. (1). The threshold detector reaches its optimal sensitivity for an escape probability of ~ 70 %, where it is in the typical range of below 1‰ of I_C.

Special care was taken to assure the stability of the detector: it should show a constant escape probability (P) during the whole duration of the pulse. The number of escapes (Z) for the ith time interval with length Δt during the pulse is described by

$$Z_i = P\Delta t N(1 - P\Delta t)^i \tag{2}$$

663

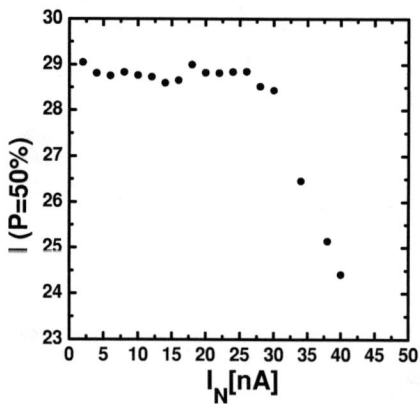

FIGURE 3. Current value corresponding to the 50 % escape probability for a detector with $I_C = 75$ nA as a function of noise current I_N through JJ_{SNS} with $I_C = 27$ nA at $T \leq 50$ mK.

and corresponds well to the experimental observations, as shown in Fig. 2b. Such an ideal result can be obtained with optimized pulse forms (see Fig. 1): it is necessary to assure slowly rising and falling current edges to avoid escape events due to circuit resonances. In any case, the edges of the pulse do not contribute noticeably to the escape rates, because the escape probability depends exponentially on the current: even for equally long edge and pulse durations, the edge would generate less than one escape event within 30000 pulses.

CONCLUSION

The detector has a sufficient current sensitivity and reliability with respect to time stability to investigate noise sources. Figure 3 shows one example: as soon as I_N exceeds the critical current of JJ_{SNS} and the latter switches from the noise free, superconducting to the normal state, the observed current corresponding to 50 % escape probability of the detector is significantly reduced. A detailed quantitative analysis of the response of the detector to noise would exceed the availible space of this contribution and is published elsewere [7].

REFERENCES

1. B. Reulet, J. Senzier and D. E. Prober, *Phys. Rev. Lett.* **91**, 196601 (2003).
2. Yu. Bomzea, G. Gershona, D. Shovkuna, L. S. Levitov, and M. Reznikov *cond-mat/0504382* (2005).
3. J. Tobiska and Yu. V. Nazarov, *Phys. Rev. Lett.* **93**, 106801 (2004).
4. J. P. Pekola, *Phys. Rev. Lett.* **93**, 206601 (2004).
5. A. O. Caldeira and A. J. Leggett, *Phys. Rev. Lett.* **46**, 211 (1981).
6. Ya. M. Blanter and M. Büttiker, *Phys. Rep.* **336**, 1 (2000).
7. J. P. Pekola, T. E. Nieminen, M. Meschke, J. M. Kivioja, A. O. Niskanen and J. J. Vartiainen *cond-mat/0502446* (2005).

Accuracy of 1/f Noise Parameter Extraction in the Presence of Background Noise

Ilmars Slaidins and Maris Zeltins

Faculty of Electronics and Telecommunications, Riga Technical University,
Azenes 12, LV-1048, Riga, Latvia

Abstract. In this work the methodology of measurement data processing and noise component identification is presented leading to extraction of $1/f$ noise parameters on the background of RTS and G-R noise. For identification of RTS noise presence deviation from the Gaussian probability density distribution and matched optimum filtering is used. Only after the identification and subtraction of RTS and R-G components from the power spectral density $1/f$ noise parameter extraction with appropriate accuracy is possible.

Keywords: $1/f$ noise, parameter extraction.
PACS: 72.70.+m , 85.30., 85.40.Qx

INTRODUCTION

Investigation of $1/f$ noise features is based on experiments and role of up-to-date methodology and technology in success of research is increasing. Extraction of pure $1/f$ noise parameters from the measurement data is a complex task because other noise sources contribute to the output noise of the device being tested. Parameter extraction method [1] for $1/f$ noise in the case when other background noises are thermal noise and shot noise was proposed and robust measurement and parameter extraction system elaborated [2] for the same case. Thermal noise and shot noise components are always present and their parameters could be calculated from.

There are also G-R noise and RTS noise components in low frequency region of noise spectra. These noise components are not always present and their parameters could not be calculated but in most cases must be determined from a measurement data. When determining $1/f$ noise parameters contribution of G-R noise and RTS noise also must be subtracted but first these noise components must be identified using different temporal, frequency (spectral) or/and temperature dependence. Presence of these components could also serve as an indicator of low reliability of particular device [3].

In this work the methodology of measurement data processing and noise component identification is presented leading to extraction of $1/f$ noise parameters.

DISCRIMINATION OF LOW FREQUENCY NOISE COMPONENTS

Low frequency noise power spectral density (PSD) consisting of $1/f$ noise component and one or several additional $1/f^2$ spectral components is typical. In general such power spectral density $S(f)$ could be described with equation:

$$S(f) = 4kTG + 2eI + KF\frac{I^{AF}}{f^\gamma} + \frac{B}{f_o\left[1+(f/f_o)^2\right]} \qquad (1)$$

For identification of RTS noise component deviation from Gaussian distribution could be used. All other noise components except RTS have Gaussian probability distribution. If background noise has Gaussian probability distribution with zero mean value and dispersion σ^2 then for RTS noise with pulse height A and probability of the pulse P_i probability density function for total noise voltage $w(u)$ could be described by equation:

$$w(u) = (1-P_i)\frac{1}{\sigma\sqrt{2\pi}}\exp\left(-\frac{u^2}{2\sigma^2}\right) + P_i\frac{1}{\sigma\sqrt{2\pi}}\exp\left(-\frac{(u-A)^2}{2\sigma^2}\right) \qquad (2)$$

In Fig. 1 is presented RTS noise voltage as a time function $u(t)$ and an estimate of probability density $p(u)$. Due to RTS pulses resulting distribution is formed from two Gaussian distributions and could have well observable bimodality at definite values of parameters.

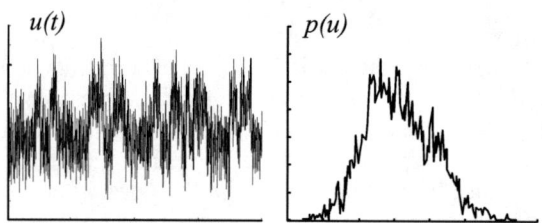

FIGURE 1. RTS noise detection using deviation from Gaussian probability density.

Deviation from Gaussian probability density could be determined by Pierson test, kurtosis and skewness coefficients. Sensitivity of these tests is evaluated using as a criteria ratio $a = A/\sigma$. Analysis and simulation show that for pulse probability values round $P_i = 0.5$ the coefficient of skewness becomes insensitive because noise waveform is symmetrical to mean value. Coefficient of kurtosis becomes insensitive at pulse probability values $P_i = 0.2$ and $P_i = 0.8$ because at these values the coefficient becomes equal to 0 (changing from $-$ to $+$ values). Pierson test appears to be most sensitive for pulse probability values round $P_i = 0.5$ and shows good sensitivity in large range of values $P_i = 0.1-0.9$. In this range of values RTS noise component could be identified using Pierson test with high accuracy if $a > 0.5$. Identification accuracy decreases for very short pulses (low probability). For $P_i = 0.001$ identification of RTS noise component is possible just for $a > 2.9$ using Pierson test and for $a > 2.0$ using the coefficient of kurtosis.

Large RTS or burst noise pulses with a low probability of appearance could be easy detected using a matched optimum filtering with subsequent decision taking. For the background noise with characteristic corner frequency ω_c consisting of $1/f$ noise, thermal noise and shot noise components analytical expression for PSD is presented by equation:

$$S(\omega) = S_0 (1 + \frac{\omega_c}{|\omega|}) \qquad (3)$$

The matched optimum filter for RTS with characteristic pulse duration τ_p will have characteristic presented by following equation:

$$K_0(\omega) = \frac{cA}{S_0} \cdot \frac{1}{\left(1 + \omega_c/|\omega|\right)} \left[1 - \exp(-j\omega \tau_p)\right] \qquad (4)$$

In practice filter could be implemented as near-optimum matched filter shown in Fig.2. It consists of low pass filter, delay with inversion allowing to subtract this delayed signal from the main signal.

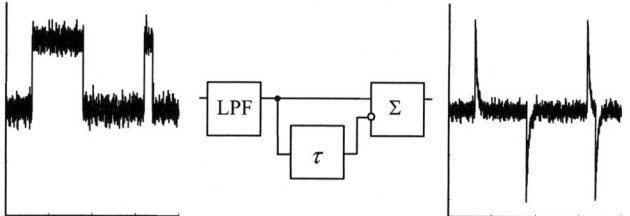

FIGURE 2. RTS noise detection using matched near-optimum filtering.

If RTS noise is present then positive and negative spikes in the output signal will represent accordingly beginning and end of pulse. This output signal is used to take decision on RTS noise presence. Filter is maximising signal to noise ration making identification possible even for pulses with small height.

DATA PROCESSING

Data processing must include: a) the filtering of measured noise power spectral density to avoid such interfering components as 50 Hz with harmonics and other, b) subtraction of unwanted noise spectra components, c) determination of the specific parameters in the noise model, such as *KF*, *AF* and γ for $1/f$ noise.

Manual processing of PSD measured at each current value is important. As we can see in Fig.3a G-R or RTS noise may appear at some particular current and other parameter values. In such case manual introduction of initial processing parameters can substantially increase a speed and accuracy of processing. In Fig.3b is presented example on how to use different initial processing parameters and equations in particular frequency ranges of PSD. Boundaries of these typical frequencies are marked with f_1, f_2, f_3 and f_4. In each frequency range, similar as in [1], it could be recommended to introduce different initial parameters and equations.

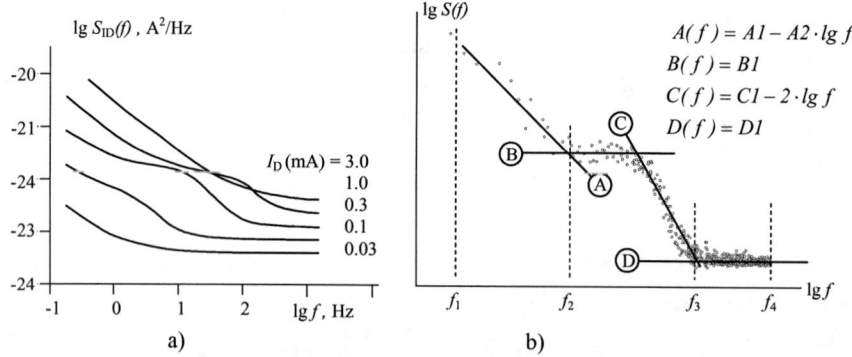

a)   b)

FIGURE 3. FET drain current noise source PSD at several drain current values (a) and example of noise spectra (b) with "white", $1/f$ and G-R components, manually entered data - frequencies and lines for further computing using Least Square Method.

After subtracting unwanted noise components could be made final extraction of $1/f$ noise parameters. For this procedure following equations are used:

$$lg\, S(f,I) = KF + AF \cdot lg\, I - \gamma\, lg\, f \qquad (5)$$

$$lg\, S(f,I) = const + lgI \qquad (6)$$

It is important to evaluate accuracy in all procedures else it can lead to the false interpretation of results.

CONCLUSIONS

For identification of RTS noise presence a deviation from Gaussian probability density distribution and matched optimum filtering is used. For RTS noise pulse probability range $P_i = 0.1 - 0.9$ Pierson test provides the highest accuracy in noise component identification. Large RTS or burst noise pulses with low probability of appearance ($P_i < 0.01$) could be better identified with the matched optimum filtering.

Only after identification and subtraction of RTS and R-G components from the power spectral density $1/f$ noise parameter extraction with appropriate accuracy is possible.

REFERENCES

1. Costa J. C, Ngo D., Jackson R., Camilleri N., and Jaffee J., *IEEE Trans. Electron Devices*, 41, 1992-1999 (1994).
2. Blaum A., Pilloud O., Scalea G., Victory J., and Sischka F., "A New Robust On-Wafer $1/f$ Noise Measurement and Characterization System" in *Proc. IEEE 2001 Int. Conference on Microelectronic Test Structures*, **14**, 125-130 (2001).
3. Slaidins I., "Measurement of low frequency noise as a quality and reliability indicator" in *Noise and Reliability of Semiconductor Devices*, Proc. Int. NODITO Workshop, July 18-20, 1995, Brno, Czech Republic, 1995, pp.188-192.

RTS Noise in Optoelectronic Coupled Devices

Alicja Konczakowska, Jacek Cichosz, Barbara Stawarz[1]

Gdańsk University of Technology, Faculty of Electronics, Telecommunications and Informatics, Department of Metrology and Electronic Systems, ul. G. Narutowicza 11/12, 80-952 Gdańsk, Poland

Abstract. The low frequency noise of optoelectronic coupled devices (OCDs) was measured in the system designed and constructed by the authors. The RTS noise was observed in some devices. The analysis of RTS noise in time and frequency domains is presented. The values of f_{RTS} were found from spectrum and from observed RTS noise on the base of t_{up} and t_{down} evaluation. The results of evaluations were compared. The dependencies of t_{up}, t_{down} and f_{RTS} on values of I_d were described.

Keywords: noise measurement, RTS noise.
PACS: 05.40Ca,71.55.-i.

INTRODUCTION

The Random Telegraph Signal (RTS) noise observed in noise signal of investigated devices is an indicator of their quality. The analysis of RTS noise can be useful to obtain information about the physical location of the defect centre (trap). The two level fluctuations are generally attributed to the capture and emission time of a trap.

CONDITIONS OF NOISE MEASUREMENTS

The investigations were carried out for the optoelectronic coupled devices (OCDs) CNY 17 type. The CNY 17 is a pair consisting of a Gallium Arsenide infrared emitting diode optically coupled to a silicon npn phototransistor. The low frequency noise of OCDs was measured at DC diodes currents I_d = 1, 2, 5, 10 mA. The measurement system, designed and constructed by the authors, consists of the noise measurement circuit of OCD, low noise preamplifier, antialiasing filter, A/D converter and computer [1]. The OCD under test and preamplifier were biased by batteries and carefully shielded. The voltage output noise of OCD was filtered, sampled, processed and stored under LabVIEW control. The data were acquired within the frequency range 10Hz – 3kHz. The corner frequency of the antialiasing filter and the f_p sampling frequency were equal to 2.85 kHz and 6,14 kHz, respectively. The low frequency noise of OCDs, at mentioned above I_d currents, was analyzed in time and frequency domains. The $N=10^6$ data points of each results of noise measurements were used for time domain analysis. In frequency domain the spectrum $S_v(f)$ was calculated on the

---

[1] Ph D Student

base of 3072 data points with 100 number of averaging. The RTS noise was observed in some devices.

TIME DOMAIN ANALYSIS

The results of measurements with RTS noise were taken into account. The histograms were calculated for all results of measurements.

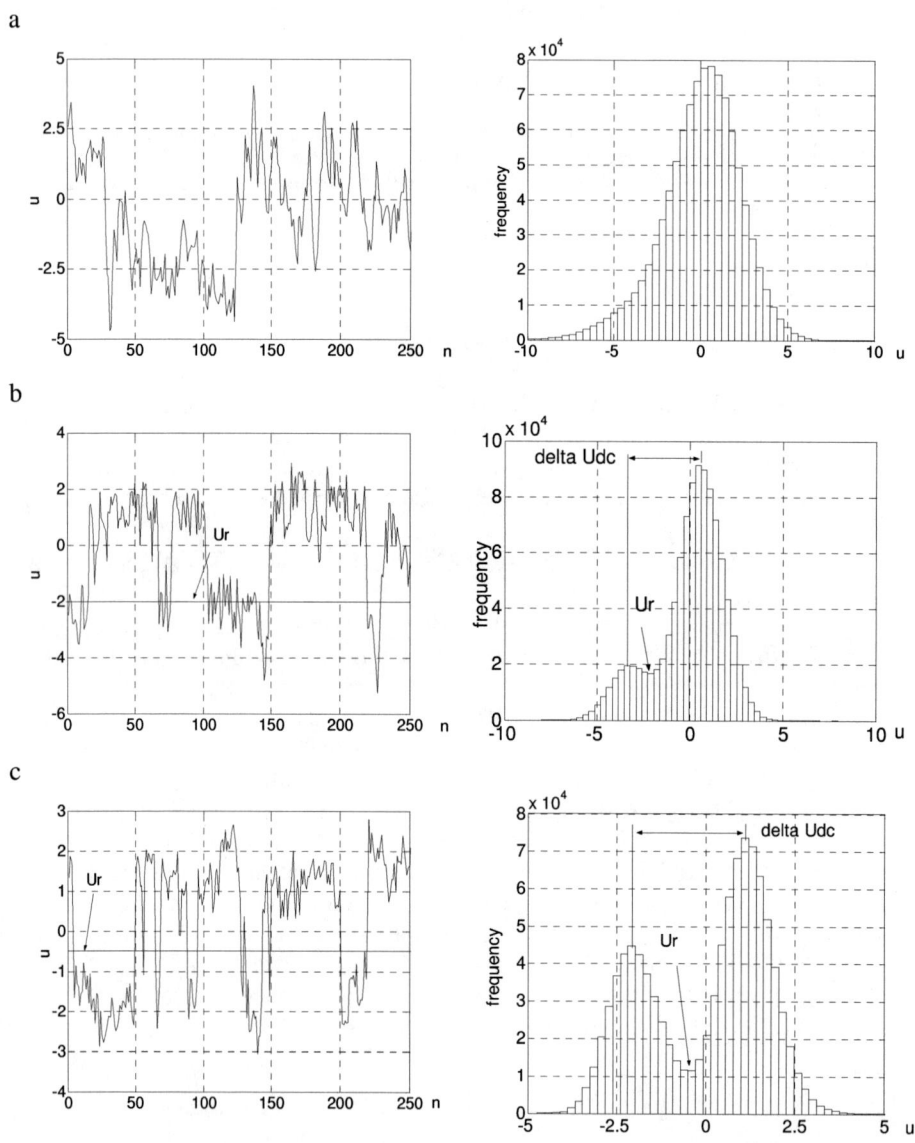

d

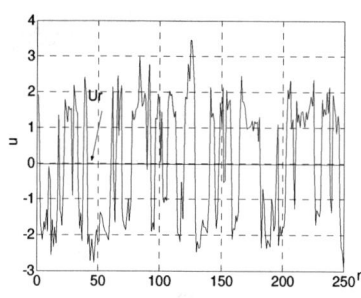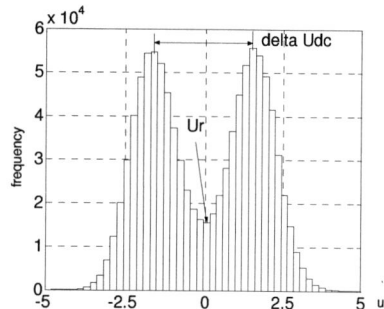

FIGURE 1. The observed noise and histograms for device No.8,
a - I_d = 1mA, b - I_d = 2mA c - I_d = 5 mA, d - I_d = 10 mA.

In Fig. 1 the examples of RTS noise of No. 8 device in the form of waveform and of histograms are presented. In Fig. 1 the view of noise, contains of n = 250 data points, is given, where the current time is equal to t = n/f_p. The reference voltage level between the maximum and minimum levels in noise versus time was defined and was noted as U_r. Every data point above reference level was classified as a "1" and every data point equal to or below this level was classified as a "0". The number of "1" and the number of "0" was multiplied by the $\Delta t=1/f_p$ sampling rate and separately averaged. The mean value of "1" was noted as t_{up}, and the mean value of "0" was noted as t_{down} [2]. The amplitude of RTS noise was noted as ΔU_{dc}. The values of U_r and of ΔU_{dc} were evaluated from histograms (Fig. 1). The values of t_{up} and t_{down} were evaluated from noise for devices with RTS.

FREQUENCY DOMAIN ANALYSIS

The voltage power spectral density $S_v(f)$ of an RTS noise of the measured OCDs can be expressed by [2, 3]:

$$S_v(f) = \frac{A \cdot (\Delta U_{dc})^2}{1+(f/f_{RTS})^2} \qquad (1)$$

where: A is a constant and f_{RTS} is frequency of RTS noise.

The spectrum at f = f_{RTS} has a characteristic maximum which some times is not well visible. Because of this the relation f·$S_v(f)$ is evaluated. The spectra $S_v(f)$ and relations f·$S_v(f)$ were calculated in virtual instrumentations of LabVIEW for all the measured devices. The values of f_{RTS} were evaluated from relations f·$S_v(f)$ for devices with RTS. The results of measurements for No. 8 device are presented in Fig.2.

In Table 1 the parameters of RTS noise for No.8 device are collected.

TABLE 1. The parameters of RTS noise for No. 8 device

I_d [mA]	U_r [V]	t_{up} [ms]	t_{down} [ms]	f_{RTS} [Hz] TD	f_{RTS} [Hz] FD
2	- 2,00	1,09	6,02	140	110
5	-0,66	1,82	3,12	202	210
10	0,10	1,17	0,77	512,5	600

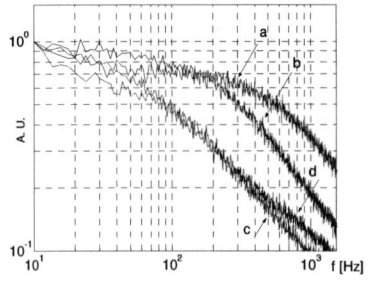 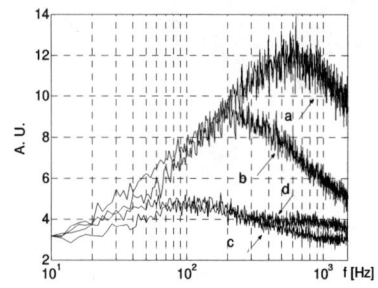

FIGURE 2. The spectra and relations f·S$_v$(f) for No.8 device in arbitrary unit (A.U.),
a - I$_d$ = 10mA, b - I$_d$ = 5mA, c - I$_d$ = 2 mA, d - I$_d$ = 1 mA.

RESULTS DISCUSSION

From the presented results of the low frequency noise measurements of OCDs, one can see, that it is more easier to evaluated the RTS noise parameters on the base of time domain analysis than of frequency domain. The presented spectra are 100 times averaged, but they are not sufficiently smooth. We can obtain better results of noise measurements (more smooth) but we have to increase the average time. It means the time of collecting data points have to be longer.

The f$_{RTS}$ frequency decreases when values of I$_d$ decrease. The evaluated t$_{up}$ and t$_{down}$ are proportional to capture and emission time constants. The value of t$_{up}$ for all DC diode currents I$_d$ is similar and it is near 1 ms, but the values of t$_{down}$ are different, what can be expected for bipolar transistors with RTS noise.

The OCDs with RTS noise are of poor quality. The easy method of an recognizing the RTS noise is presented in [4].

ACKNOWLEDGMENTS

This work was supported by Ministry of Science and Information Society Technologies - the project No. 3 T10C 026 28.

REFERENCES

1. A. Konczakowska, J. Cichosz, S. Galla, "Noise of optoelectronic coupled devices" in *Noise and Fluctuactions,* edited by J. Sikula Conference Proceedings, Prague, Czech Noise Research Laboratory, Brno University of Technology, 2003, pp. 703-706.
2. Z. Celik-Butler, "Measurement and analysis methods for Random Telegraph Signals" in *Advanced Experimental Methods for Noise Research in Nanoscale Electronic Devices,* edited by J. Sikula and M. Levinshtein, NATO Science Series, vol.151, 2003, pp.219-226.
3. C. Claeys, E. Simoen, *J. Electrochem Soc.*, **145**, 2058-2067 (1998).
4. J. Cichosz, A. Szatkowski, "Noise scattering patterns methods for recognition of RTS noise in semiconductor components", this issue.

Noise Scattering Patterns Method for Recognition of RTS Noise in Semiconductor Components

Jacek A. Cichosz*, Andrzej Szatkowski[†]

*Gdansk University of Technology, Department of Metrology and Electronic Systems,
11/12 G. Narutowicza str., 80-952 Gdansk, Poland, jcichosz@eti.pg.gda.pl

[†]Pomeranian Pedagogical University, Department of Computer Science and Statistics,
8 Slowianska str., 76-200 Slupsk, Poland, andrzej-szat@wp.pl

Abstract. A new method for visualization of RTS noise is described. The method is useful in quick selection of semiconductor components for further RTS noise examination. The results of median filtering of RTS noise are also presented.

Keywords: RTS noise, stochastic analysis, semiconductor components.
PACS: 72.70.+m, 05.10.Gg

INTRODUCTION

There are several methods of obtaining RTS (Random Telegraph Signal) noise parameters of semiconductor components [1, 2, 3]. They are base mostly on storing and processing the digital sampled data obtained as a result of measuring the low frequency noise semiconductor devices. All the methods are troublesome and time consuming.

The main problem is that RTS noise is being observed only in small parts of the examined populations of semiconductor components. Noise scattering patterns (NSP) method enables a quick selection of the measured semiconductor devices which exhibit RTS noise from among a very large number of the examined devices.

PRINCIPLES OF THE METHOD AND EXPERIMENTAL RESULTS

The NSP method for recognition of RTS noise is based on the analysis of time domain data obtained from a digitizing low frequency noise of semiconductor devices. As it is shown in Fig.1, the sequence $(z[n])_{n=1,2,\cdots,N}$ of digital samples had been divided into two subsequences $(x[i])_{i=1,2,\cdots,N/2}$ and $(y[j])_{j=1,2,\cdots,N/2}$.

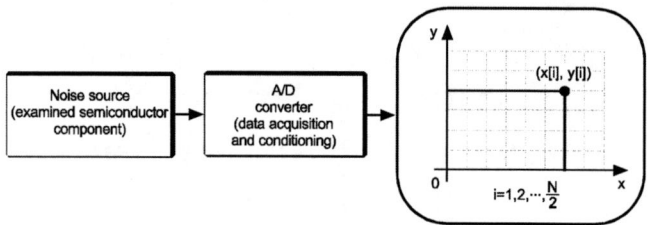

FIGURE 1. The NSP method schematic diagram.

Using the noise digital samples one can obtain the pictures which contain points whose coordinates are $(x[i], y[i])_{i=1,2,\cdots,N/2}$, as it is shown in Figs 2 to 5.

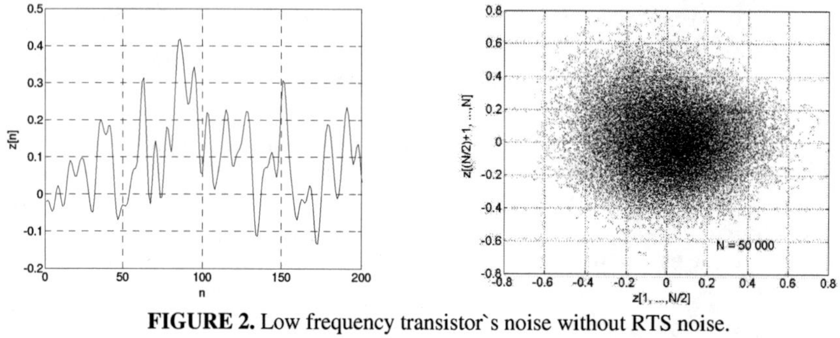

FIGURE 2. Low frequency transistor's noise without RTS noise.

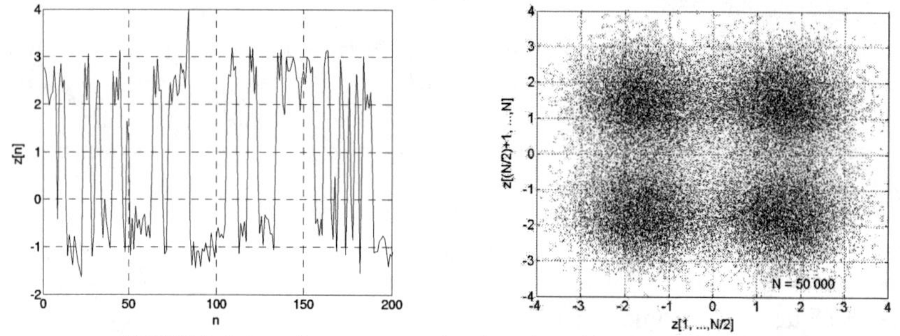

FIGURE 3. Low frequency transistor's noise with two-level RTS noise.

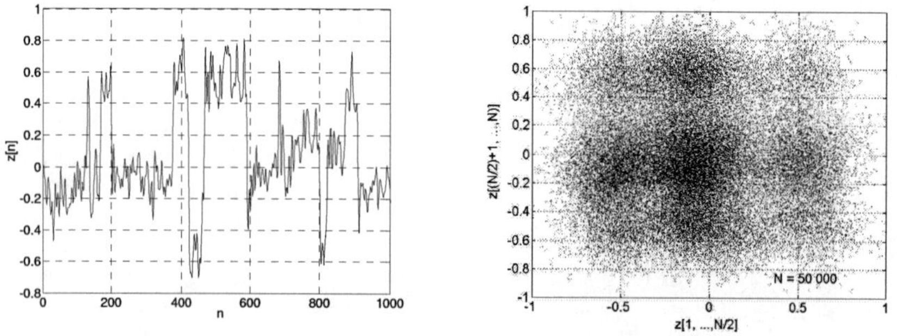

FIGURE 4. Low frequency transistor's noise with three-level RTS noise.

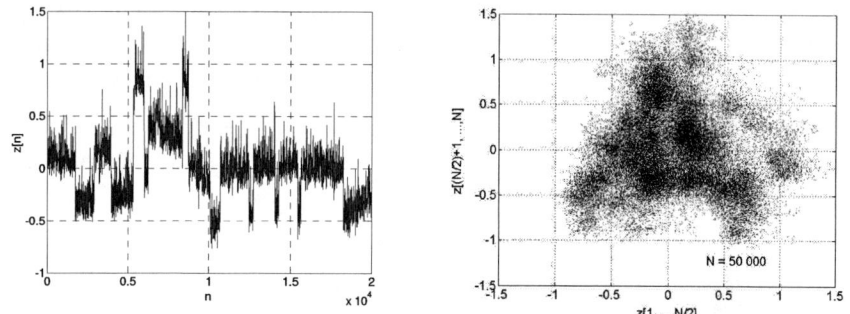
FIGURE 5. Low frequency transistor's noise with multi-level RTS noise.

From the presented patterns it is easy to recognize RTS noise and to decide whether RTS noise exists or not in the noise of the examined transistors.

IMPROVEMENTS

A signal $z(\cdot)$ being measured is considered, by assumption, to be the sum $z(\cdot) = v(\cdot) + s(\cdot)$, where $v(\cdot)$ is assumed to be the RTS component of $z(\cdot)$ and, $s(\cdot)$ is the remainder noise component of $z(\cdot)$. Analogically, there is considered the decomposition $z[\cdot] = v[\cdot] + s[\cdot]$ of the sequence $(z[n])_{n=1,2,\cdots,N}$ of discrete samples of the signal $z(\cdot)$ being measured, where it has been assumed for $v[\cdot]$ to be the RTS component of $z[\cdot]$, and $s[\cdot]$ is the remainder noise component of $z[\cdot]$.

We have applied the median filtering in order to filter the remainder noise components from the sequences of discrete samples of the measured signals.

The median map, denoted by $med(\cdot)$, is defined on finite sets of real numbers. The median of a finite set of real numbers is calculated by first sorting all the numbers of the set into numerical order and then assigning the middle value of the obtained sequence to the considered set of numbers (if the set under considerations contains an even number of elements, the average of the two middle elements is used).

Let $(z[n])_{n=1,2,\cdots,N}$ be a given sequence of real numbers, $N \geq 1$ being a natural number. In this text $(z[n])_{n=1,2,\cdots,N}$ is a sequence of discrete samples of a signal $z = z(t)$, where t is time. The domain of $z(\cdot)$ is a subset of the time axis $\boldsymbol{R}$.

The median filter of order l, l being a number in the set $\{0, 1, 2, \cdots\}$, is the map $F_l(\cdot)$ which assigns to the given sequence $(z[n])_{n=1,2,\cdots,N}$ a sequence $(\hat{z}[n])_{n=1,2,\cdots,N}$ according to: $\hat{z}[n] = med\{z[n_1], z[n_1+1], \cdots, z[n_2]\}$, for $n = 1, 2, \cdots, N$, where: $n_1 = \max\{1, n-l\}$, and $n_2 = \min\{n+l, N\}$.

There is considered the median filter $F_l(\cdot)$. Let the sequence $(z[n])_{n=1,2,\cdots,N}$ be given to be a discrete rectangle wave in $N \times \boldsymbol{R}$, N being the set of natural numbers. That is, for each $1 < n < N$ $z[n-1] = z[n]$, or $z[n+1] = z[n]$.

Let us consider the following problem, which concerns distortions of discrete rectangle waves which one could observe in result of the median filtering. The problem is formulated as follows. Let the sequence $(z[n])_{n=1,2,\cdots,N}$ be given to be a discrete rectangle wave in $N \times R$. Under what conditions one obtains $\hat{z}[n] = z[n]$, for each $n = 1, 2, \cdots, N$, where $(\hat{z}[n])_{n=1,2,\cdots,N}$ is the sequence $F_l((z[n])_{n=1,2,\cdots,N})$ of median values of the sequence $(z[n])_{n=1,2,\cdots,N}$.

Let μ denote the minimal number of the neighbouring elements in $(z[n])_{n=1,2,\cdots,N}$ which have the same values. Assume that the image set of the considered rectangle wave in $N \times R$ contains at least two values. Then $(\hat{z}[n])_{n=1,2,\cdots,N}$ and $(z[n])_{n=1,2,\cdots,N}$ are the same sequences if, and only if, $l \leq \mu - 1$.

The RTSs are rectangle waves. Hence, we have assumed that the respective relation given above is satisfied in order to avoid unexpected distortions of the sets of discrete samples of the RTSs, which could be observed in result of the median filtering of the RTSs with noise.

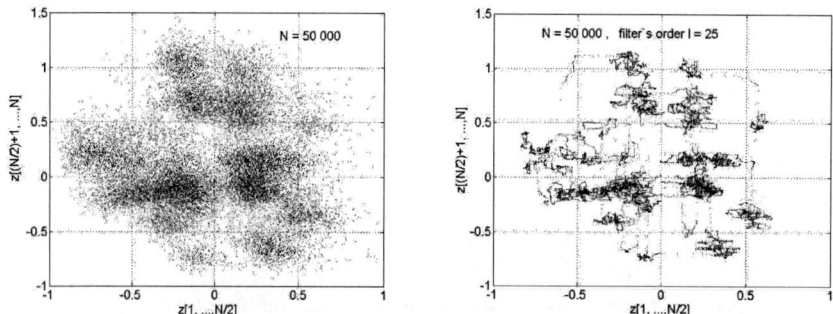

FIGURE 6. An example of median filtering of RTS with noise.

ACKNOWLEDGMENTS

This work was supported by the Ministry of Science and Information Society Technologies, project No. 3 T10C 026 28.

REFERENCES

1. Celik-Butler, Z., *Measurement and Analysis Method for Random Telegraph Signals*, Advanced Experimental Methods for Noise Research in Nanoscale Electronic Devices, edited by J. Sikula and M. Levinshtein, Dordrecht, Boston, London: Kluwer Academic Publishers, 2004, pp. 219-226.
2. Wel, A.P. van der, Klumperink, E.A.M., Kolhatkar, J.S., Hoekstra, E., Nauta, B., *Visualisation Techniques for Random Telegraph Signals in MOSFETs*, ProRisc, Veldhoven, Proceedings and Poster, http://icd.el.utwente.nl/publications/index.php .
3. Kolhatkar, J.S., Vandamme, L.K.J., Salm, C., Wallinga, H., *Separation of Random Telegraph Signals from 1/f Noise in MOSFETs under Constant and Switched Bias Conditions*, Proceedings of the 33rd European Solid State Device Research Conference ESSDERC'03, IEEE, 2003, pp. 549-552.

A Method Of Two-Terminal Excess Noise Measurement With A Reduction Of Measurement System And Contact Noise

Arkadiusz Szewczyk, Ludwik Spiralski, Lech Hasse

Gdansk University of Technology, Faculty of Electronics, Telecommunications and Informatics
Dep. of Metrology and Electronic Systems, G. Narutowicza str. 11/12, 80-952 Gdańsk Poland

Abstract. The method and the system for excess noise measurement of electrical two-terminal components with the significant reduction of the system and contact noise influence is presented. The proposed method can be applied for noise measurements either on devices with terminals or on the wafer level, where as a reference noise the noise of resistive structure fabricated directly on the wafer in the vicinity of the tested component can be used.

Keywords: noise measurement, wafer level measurement
PACS: 84.37.+q

INTRODUCTION

Noise level of nowadays fabricated electronic components, particularly microelectronic devices, is very low. In these elements, the noise intensity is often comparable and even lower than the inherent noise of the measurement instrumentation and the noise of contacts between a device and measurement setup.

The authors experience shows, that the conducting connections between the point probe and a contact pad of a device being tested are a significant source of unwanted signals. Disturbances originating in the contact region are usually caused by the contact resistance fluctuations and by the fluctuation effects of physical and chemical degradation of surfaces being in contact [1, 2].

The contact resistance fluctuations that originate from vibrations and shocks contain usually stochastic and deterministic components. In order to minimize disturbances emerging in the contact area due to vibrations and shocks, the measurement instrumentation is usually placed on an anti-vibration stone table using shock-absorbing pillows. Unwanted signals having periodic character can be eliminated in a relatively simple way using the algorithm of periodic and stochastic components separation in a measurement signal [3,4].

The presented measurement method allows the elimination of unwanted signals, both stochastic and periodic ones, originated in the vicinity of contacts and in the measurement system [5].

MEASUREMENT PRINCIPLE

The simplified block diagram of the system for excess noise measurement of electrical two-terminal components with the reduction of the system and contacts noise is shown in Fig. 1.

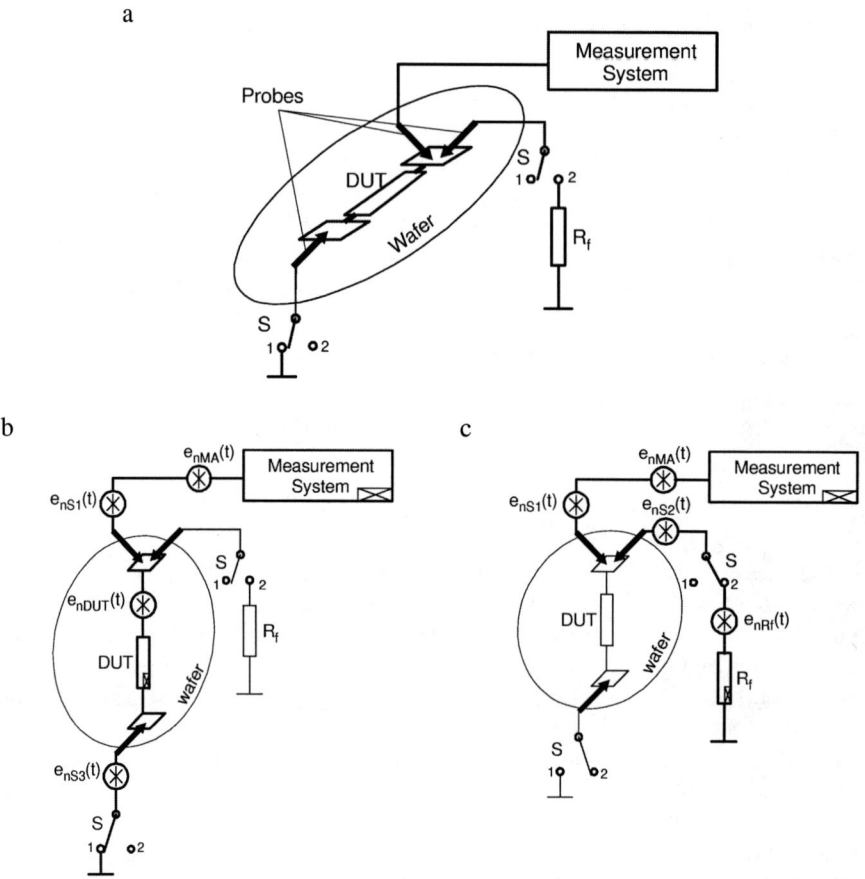

FIGURE 1. Noise measurement of electrical two-terminal component with the reduction of the system and contacts noise: a – measurement setup; b, c – schematic circuits with equivalent noise sources in the first and the second measurement steps, respectively; R_f - reference resistance.

The measurement is carried out in two steps. First (switch S in position 1), the reference resistor R_f is disconnected from the measurement system and the total noise of the Device Under Test (DUT; here the structure of two-terminal electrical component), the contacts and the measurement system is measured. In the next step (switch S in position 2) the DUT is disconnected from the system ground and the reference resistance is connected to the system. Now only the noise of the R_f, contacts and system is measured.

As the noise sources in the measurement circuit are uncorrelated, therefore the root mean square (RMS) value of the voltage at the input of the measurement system in the first step is given by:

$$E_{n1} = \sqrt{E_{nMA}^2 + E_{nS1}^2 + E_{nDUT}^2 + E_{nS3}^2} \qquad (1)$$

and in the second one by:

$$E_{n2} = \sqrt{E_{nMA}^2 + E_{nS1}^2 + E_{nS2}^2 + E_{nRf}^2} \qquad (2)$$

where E_n denotes the RMS value of the voltage noise of: $E_{nMA} = (\overline{e_{nMA}^2(t)})^{1/2}$ – measurement system, $E_{nS1} = (\overline{e_{nS1}^2(t)})^{1/2}, E_{nS2} = (\overline{e_{nS2}^2(t)})^{1/2}, E_{nS3} = (\overline{e_{nS3}^2(t)})^{1/2}$ – contacts of first, second and third probe with contact pads of the DUT, $E_{nDUT} = (\overline{e_{nDUT}^2(t)})^{1/2}$ – DUT, $E_{nRf} = (\overline{e_{nRf}^2(t)})^{1/2}$ –reference resistance (with known noise level).

In the case of narrow band measurements, when the power spectral density is measured, its value for the signal at the input of the measurement system is given in the first ($S_{n1}(f)$) and in the second ($S_{n2}(f)$) measurement step, respectively, by:

$$S_{n1}(f) = S_{nMA}(f) + S_{nS1}(f) + S_{nDUT}(f) + S_{nS3}(f) \qquad (3)$$

$$S_{n2}(f) = S_{nMA}(f) + S_{nS1}(f) + S_{nS2}(f) + S_{nRf}(f) \qquad (4)$$

For stable measurement conditions, the intensity of the noise sources located in the point probe – DUT's pad contact area has usually the same value for all the probes. One can then calculate the RMS value of the voltage or power spectral density of DUT's noise (knowing the noise level of the reference resistance) using equations (1) and (2) or (3) and (4), respectively as:

$$E_{nDUT} = \sqrt{E_{n1}^2 - E_{n2}^2 + E_{nRf}^2} \qquad (5)$$

$$S_{nDUT}(f) = S_{n1}(f) - S_{n2}(f) + S_{nRf}(f) \qquad (6)$$

In detailed solutions as a reference resistance can be used the wire low-noise resistor or the resistive structure fabricated on the wafer in the immediate vicinity of the DUT's structure. In the second case, one of the DUT terminals and one of the reference resistive structure terminals can be connected to the same contact pad and then the measurement setup from the Fig. 1 is modified to that one shown in Fig. 2. The common contact pad for the DUT and for the reference resistance can be connected directly through the one of probes to the ground (Fig. 2a) or to the measurement system (Fig. 2b). The advantage of the second configuration is that the changes of the input load of the measurement system during switching between the DUT's and reference resistance are limited and the system is more immune to the disturbances.

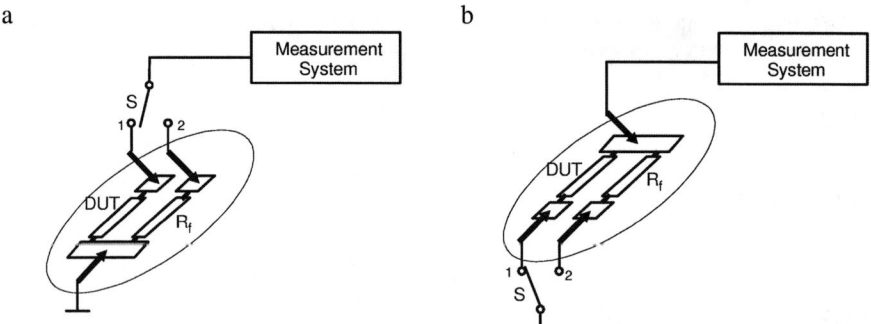

FIGURE 2. System configuration with the resistive reference structure; the common pad for the DUT and R_f is connected: a – to the ground; b – to the measurement system.

The measurement procedure for the setup shown in Fig. 2 is the same as for the one in Fig. 1, beside the necessity of using the double switch.

CONCLUSIONS

The noise measurement of microelectronic elements usually requires additional procedures for reduction of the disturbances influence, including disturbances originating in the contact area and in the measurement system, on measurement results. That involve the necessity of additional costs incurring for the measurement setup development. Proposed measurement method does not require significant modifications in standard measurement system and the cost of its implementation is relatively low

All proposed solutions related to the measurement system can be automated by applying the digitally controlled switches and digital data acquisition with effective computer procedures for data processing.

REFERENCES

1. Szewczyk, A., Spiralski, L. "Noise in Semiconductor Structures Probe Measurements" (in Polish) in *Elektronizacja* No 2, 2003 pp. 12-13.
2. Szewczyk, A., Spiralski, L., "The method of disturbances reduction in low frequency noise measurements on wafer-level" in Proc. Polish Metrology Congress KKM 2001, 24-27 June 2001, Warsaw, vol. 2 pp. 321-324.
3. Pałczyńska, B., Spiralski, L. and Turczyński, J., "The new method of interference assessment in low-voltage power supply lines" in Proc. 11th IMEKO TC-4 Symposium, 13-14 Sept. 2001, Lisbon, pp. 130-134.
4. Pałczyńska, B., Turczyński, J., Konczakowska, A. and Spiralski, L. "Method of parasitic signal measurement in low voltage supply lines in the low and high frequency range, particularly in ship lines" Polish Patent Office, Patent Application No. 140152 (2000)
5. Szewczyk, A., Spiralski, L. and Hasse, L. "Circuit for excess noise measurement in electrical two-port structures", Polish Patent Office, Patent Pending No. P 368669 (2004).

Progress toward "optical beam induced noise" measurement set-up

Jean-Marc Routoure, Laurence Méchin and Stéphane Flament

GREYC, CNRS UMR 6072, ENSICAEN, University of Caen 6, Bd du maréchal Juin 14050 Caen Cedex. France. routoure@greyc.ensicaen.fr

Abstract. The paper presents an experimental set-up dedicated to noise measurements of samples under laser spot illumination. The laser is scanned on the surface which locally heats. The noise level is recorded at each positions. The final purpose of the system is to detect spatial inhomogeneities in the microscopic noise sources of resistive samples ($La_{0.7}Sr_{0.3}MnO_3$ for example).

Keywords: noise, defects, optical beam induced
PACS: 73.50.Td,79.20.Ds

INTRODUCTION

The optical beam induced techniques have been used for a long time in order to detect defaults in integrated circuits or devices [1, 2]. This paper describes an extension of these techniques based on noise measurements instead of current or voltage measurements done during OBIC (Optical Beam Induced Current) or OBIV (Optical Beam Induced voltage) techniques respectively. This technique may be promising due to the high sensitivity of defects on noise level [3, 4].

PRINCIPLE AND SENSITIVITY ANALYSIS

Principle

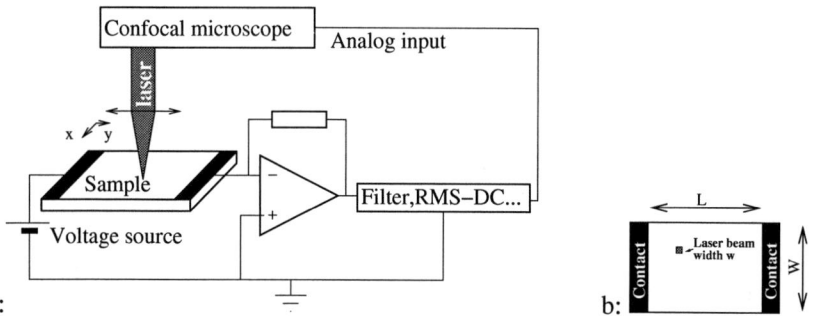

FIGURE 1. a: schematic diagram of the experimental set-up. b : definition of the sample dimensions.

A confocal system (Zeiss LSM 310) was used (cf. figure 1a) to control the laser position on the sample and to convert the output signal of the measurement system into gray-level as a function of the laser position. A "noise image" of the sample could then be built.

The devices were biased using a voltage source. The noise was measured using a I->V converter as a first stage, a high-order band-pass filter, some rejection filters in order to eliminate the 50Hz and harmonic perburtations, a voltage amplifier and finally a RMS-> DC converter. OBIC images can also be obtained using the output voltage of the I->V converter.

From the current noise spectral density i_n of the device, the output voltage V_{OUT} writes : $V_{OUT} = K \cdot \int_{f_1}^{f_2} i_n \cdot df$ where f_1 and f_2 are the characteristic frequencies of the band-pass filter and K a trans-impedance factor. If we assume that white noise is due to thermal noise and that the sample exhibits 1/f noise at frequencies smaller than the corner frequency f_C. Depending on the choice of f_1 and f_2, one can have access to two kinds of information. If f_1 and f_2 are much higher than f_C, a "white noise" image is obtained, leading to an image of the resistance variation of the device whereas if f_1 and f_2 are much smaller than f_C a "1/f" noise image is obtained which can inform on the spatial localization of the microscopic 1/f noise sources.

Sensitivity analysis

Since the confocal system uses a 8 bits Analog to Digital Converter, an analysis of the sensitivity is important in order to estimate if some contrast could be obtained on the images. As described in figure (1b), we assume that the device has a width W, an electrical resistance R between two contacts and a temperature coefficient $\beta = (1/R) \cdot \left(\frac{\delta R}{\delta T}\right)$. We also assume that the laser beam (of width w) increases the device temperature of a quantity ΔT at the position of the laser beam. Using these assumptions, when the laser is applied at a position (x,y), the electrical resistance R' is given by :

$$R'(x,y) = R \cdot \left(1 + \frac{w}{W} \cdot \beta(x,y) \cdot \Delta T\right)$$

For white noise ($i_n = 4 \cdot k \cdot T / R'(x,y)$), assuming that $\Delta T \ll T$ and $\frac{w}{W} \cdot \beta(x,y) \cdot \Delta T \ll 1$, the output voltage is easily derived:

$$V'_{OUT} = K \cdot \sqrt{\frac{4 \cdot k \cdot T}{R'(x,y)}} \cdot (f_2 - f_1) \cong K \cdot \sqrt{\frac{4 \cdot k \cdot T}{R}} \cdot (f_2 - f_1) \cdot \left(1 - \frac{w}{2 \cdot W} \cdot \beta(x,y) \cdot \Delta T\right)$$

Finally, in order to obtain a contrast of at least one gray level in the image, the variations of $\frac{w}{2 \cdot W} \cdot \beta(x,y) \cdot \Delta T$ must be higher than $4 \cdot 10^{-3} (\cong 1/2^8)$ for all the positions (x,y) of the laser on the sample. To respect this condition, we have to choose devices of small width and try to use a laser that heats as much as possible.

The same analysis could be led on 1/f noise if the temperature dependency could be known but it is clear that the smallest width devices will exbihits highest sensitivity.

RESULTS

Two kinds of samples have been investigated. The first one is a silicon educational process with simple PMOS transistors. The second one consists in patterned LSMO thin films[5, 6]. Some OBIC and noise images have been obtained and are discussed in the following.

Silicon devices

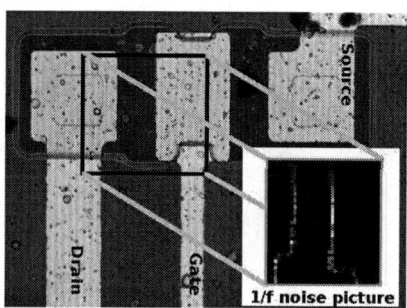

FIGURE 2. image obtained for 1/f noise on a basic pMOS transistor. An effect on the 1/f noise is obtained when the spot heats the limit of the aluminum region.

Contrasted noise images have been obtained on a MOS device biased in the linear regime (using another DC source to bias the gate). The result is shown in figure (2). An increase of the noise level is obtained at the limit of the aluminum region. This result needs to be interpreted but it clearly indicates that contrasted images can actually be provided by the system.

LSMO thin films

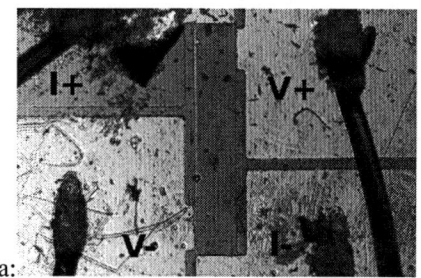 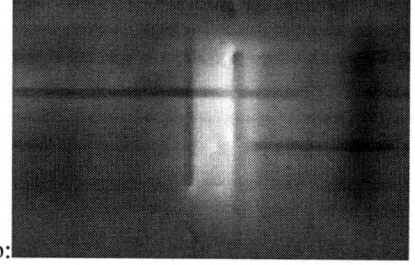

FIGURE 3. a: optical image of the sample showing the four probe V+, V-, I+ and I-. b: an OBIV image obtained for a LSMO patterned thin films. A contrast is obtained at the proximity of the voltage probes.

Only OBIV images have been obtained on four probe patterned LSMO samples as shown in the figure (3). In this case, the bias system is different since a current is applied between the I+ and I- contacts and the voltage across the sample is measured between the V+ and V- contact. Some contrast has been obtained at the proximity of the voltage contacts as well as in the layer. It may be interpreted as spatial inhomogeneities in the current lines. The horizontal black line obtained is related to the horizontal nature of the laser spot scanning.

CONCLUSION AND PERSPECTIVES

Further inverstigations and improvements are needed in order to obtain valuable results. These first results are promising and may allow to localize using a non destructive technique the main microscopic noise sources. The limitations of this technique will be the spatial resolution due to the laser beam width, the laser wavelength and the heat diffusion in the sample, and the time needed to obtain one image.

REFERENCES

1. E. Cole, P. Tangyunyong, and D. Barton, *Microelectronics Reliability*, **39**, 681–693 (1999).
2. N.-C. Park, A. Abbate, F. Palme, and P. Das, *Microelectronic Engineering*, **31**, 227–234 (1996).
3. L. Vandamme, *IEEE transactions on electron devices*, **41**, 2176–2186 (1994).
4. I. Lartigau, J.-M. Routoure, R. Carin, A. Mercha, E. Simoen, and C. Claeys, "Low temperature spectroscopy of $0.1 \mu m$ partially depleted silicon on insulator mosfets," in *Proceedings of the 17th International conference, Noise and fluctuations, ICNF 2003*, edited by J. Sikula, 2003, pp. 763–766.
5. F. Yang, L. Mechin, J.-M. Routoure, S. Flament, D. Robbes, and R. Chakalov, "low 1/f noise in $La_{0.7}Sr_{0.3}MnO_3$ Thin film on (100) $SrTiO_3$," in *Proceedings of the 17th International conference, Noise and fluctuations, ICNF 2003*, edited by J. Sikula, 2003, pp. 145–148.
6. L. Mechin, F. Yang, S. Mercone, J. Routoure, S. Flament, C. Simon, and R. Chakalov, *This proceeding* (2005).

Finite Bandwidth Related Errors in Noise Parameter Determination of PHEMTs

Wojciech Wiatr[1], David Adamson[2]

[1] *Warsaw University of Technology, Nowowiejska 15/19, 00-665 Warszawa, Poland*
[2] *National Physical Laboratory, Hampton Road, Teddington, Middlesex, TW11 0LW, United Kingdom*

Abstract. We analyze errors in the determination of the four noise parameters due to finite measurement bandwidth and the delay time in the source circuit. The errors are especially large when characterizing low-noise microwave transistors at low microwave frequencies. They result from the spectral noise density variation across the measuring receiver band, due to resonant interaction of the highly mismatched transistor input with the source termination. We show also effects of virtual de-correlation of transistor's noise waves due to finite delay time at the input.

Keywords: noise, measurement, parameter estimation, errors, microwave field-effect transistors
PACS: 05.40.Ca, 06.20.Dk, 06.60.Mr, 07.05.Tp, 07.57.Ty, 84.40.-x, 85.40.Qx

INTRODUCTION

Noise characterization relies on determining two-port parameters from a series of noise power measurements performed for distinct impedances and noise temperatures of the source. This approach is called the noise source-pull technique and requires the use of an adequate noise model to describe the measurements analytically. The parameters are then determined by fitting [1], as a solution of an inverse problem.

There are various noise models utilized for this purpose [1],[2]. All assume that the source reflection coefficient and the parameters of the device under test (DUT) are invariant across the receiver's bandwidth. This requirement may not be, however, fulfilled when testing highly mismatched low-noise transistors, since, due to resonant interaction between a mismatched source and the transistor input, one observes large variations of the noise power density [3]. Though the variation may cause severe errors in the DUT noise characterization, this issue has not been addressed yet.

In this paper, we analyze for the first time errors in the determination of the four noise parameters, which arise due to finite measurement bandwidth and the delay time in the DUT input circuit. We show that the errors are especially significant at low microwave frequencies when characterizing low-noise microwave transistors like pseudomorphic high-electron mobility transistors (PHEMTs) or RF CMOS transistors.

MODEL

The dependence of the effective input noise temperature T_e of a two-port on the source (generator) impedance can be described analytically using four real quantities. In our analysis, we employ the source reflection coefficient Γ_g and the following formula:

$$T_e(\Gamma_g) = T_{min} + T_N \frac{|\Gamma_g - \Gamma_{opt}|^2}{(1-|\Gamma_g|^2)(1-|\Gamma_{opt}|^2)} , \qquad (1)$$

where T_{min} is the minimum value of $T_e(\Gamma_g)$, complex Γ_{opt} is the optimum source reflection coefficient for which the minimum occurs and T_N determines how rapidly $T_e(\Gamma_g)$ increases when moving away from that minimum.

These four noise parameters (two being $\mathrm{Re}\Gamma_{opt}$ and $\mathrm{Im}\Gamma_{opt}$) are determined from noise power measured at the two-port's output in a band which is narrow enough to assume that the noise power density is constant. Generally, the dimensionless power meter indication p is a complex function of Γ_g [2]:

$$p = G \frac{(T_g + T_{min})(1-|\Gamma_g|^2) + T_q|\Gamma_g - \Gamma_{opt}|^2}{T_r|1-\Gamma_i\Gamma_g|^2} , \qquad (2)$$

where T_g is the source's noise temperature, $T_q = T_N(1-|\Gamma_{opt}|^2)^{-1}$, T_r denotes a receiver's constant expressed in temperature units and two other quantities, the power gain G and the complex input reflection coefficient Γ_i, represent the measured two-port. Thus, the total set of parameters in (2) comprises seven real quantities: the four noise parameters and three others, T_r, $\mathrm{Re}\Gamma_i$ and $\mathrm{Im}\Gamma_i$, related to the gain variation of the two-port.

All the parameters of (2) can be determined using the well-known cold source measurement technique [1] and a constrained least-squares method based on the eight-term linear noise model [2]. This approach is very effective in the characterization of low-noise semiconductor devices using the multi-state radiometer [4], a novel noise instrument measuring simultaneously the noise and complex scattering parameters [5].

The assumption of a "white" noise spectrum may, however, be violated when characterizing contemporary low-sized semiconductor devices like PHEMTs. We simulated the noise measurements of a modern PHEMT having a gate of 0.2 μm long and 4×40 μm wide at 2 GHz. We used the parameters: T_{eo} =18.3 K, T_N =36.0 K, Γ_{opt} =0.874 ∠11.8°, Γ_i =0.977 ∠-21.3° and T_r =47.5 K for the bias I_D =20 mA and V_{DS} =3 V. Our calculations showed, the output noise level vary rapidly as a function of Γ_g. The output noise peaks at the reflection coefficient $\Gamma_g \approx \Gamma_i^*$ and drops down abruptly to minimum close to $\Gamma_g \approx \Gamma_{opt}$, undergoing over 25 dB change. Such a large variation may cause serious problems in the noise characterization if the reflection coefficient Γ_g either has been measured with errors [6] or alters in frequency across the measurement band of the receiver. This latter problem is discussed here for the first time.

ERRORS DUE TO FINITE MEASUREMENT BANDWIDTH

The errors analyzed here arise when the DUT is placed at a distance from an impedance tuner employed for the source-pull noise characterization. This is a frequent situation in on-wafer measurement systems that use a cable interconnecting the tuner with the microwave probe at the DUT input. Due to the delay of the cable and tuner, one observes fast phase variation of the generator reflection coefficient

$$\Gamma_g(\omega) \approx \Gamma_{gc} e^{-j2\omega\tau}, \qquad (3)$$

where Γ_{gc} is the reflection coefficient measured at the center frequency f_c in the receiver's pass band $[f_c\text{-}B/2, f_c\text{+}B/2]$, $\omega=2\pi(f\text{-}f_c)$ represents the angular frequency within the bandwidth B and τ is a delay time. The approximation sign in (3) refers to the assumption of constant magnitude of Γ_g across the bandwidth. In the case of a PHEMT, the phase variation of Γ_g may easily convert into significant changes of the output noise spectrum and affect the power level readings. Such noise level errors may cause in turn errors in the determination of the two-port parameters. We analyze them both here.

Our error analysis accounts for two basic factors, the delay time τ of the reflection coefficient Γ_g and the finite receiver bandwidth B. For the analysis, we simulate noise measurements of the PHEMT as would be done using the well-known cold source technique [1]. In the simulations, we mimic measurement conditions met in a real source-pulled noise measurement system composed of an impedance tuner, the DUT and a noise figure analyzer (NFA) measuring the output noise power. We account for the real frequency characteristics, measured for the analyzer's IF section and the variation of the source reflection coefficient with frequency. From our point of view, the DUT and NFA constitute one device to be characterized.

With our simulations, we have observed that real noise power measurements significantly deviate from the levels predicted by conventional noise models. The deviations increase rapidly with the magnitude of the source reflection coefficient, and the product of measurement bandwidth and the delay time. They range from few tenths of a dB to about 6 dB for 4 MHz bandwidth of the NFA and delay times from 1 ns to 6 ns.

We analyzed the errors of the complex noise characterization using the eight-term linear model [2] and our procedures for the constrained least-squares calculations. Since they depended to some extent on particular distribution of the measurement points in the complex reflection coefficient plane Γ_g, we decided to study regular and evenly spaced distributions that can be easily realized using precise automatic mechanical tuners. Such distributions are usually confined inside of a centered circle with the radius $\Gamma_{g\,max}$.

We have found out, similar to [7], that the power level errors affect much more the noise parameters than Γ_i and T_r. Moreover, the noise parameter errors are very sensitive to the way the measurements are weighted in the procedures. Since simple weighting strategies turned out to be not so effective as the complex one proposed in the full least-square method [8], which we employed for our calculations.

When changing $\Gamma_{g\,max}$ gradually from 0.4 to 1.0, we observed very good fits of our model as long as $\Gamma_{g\,max}$ did not exceed 0.7. Then, the fits worsen quickly for higher values of $\Gamma_{g\,max}$, due to the approach of the region of the highest noise level variation. Thus, it is not beneficial in practice to arrange the measurements in this region.

The results of our analyses are shown in Fig. 1. Fig. 1. a.) shows that T_{min} and T_N generally rise as the delay time increases with the only exception being that T_{min} decreases at higher $\Gamma_{g\,max}$ as explained above. This is accompanied with decreasing magnitude of Γ_{min} shown in Fig. 1 b.). We attribute this phenomenon to de-correlation of the reflected noise wave due to its delay. To prove this, we calculated and show in Fig. 1 b.) the correlation coefficient ρ of the noise waves at the two-port's input. We interpret its decreasing magnitude as the effect of the de-correlation. Generally, all these results demonstrate the importance of keeping the product of the delay time and bandwidth in real measurement systems low.

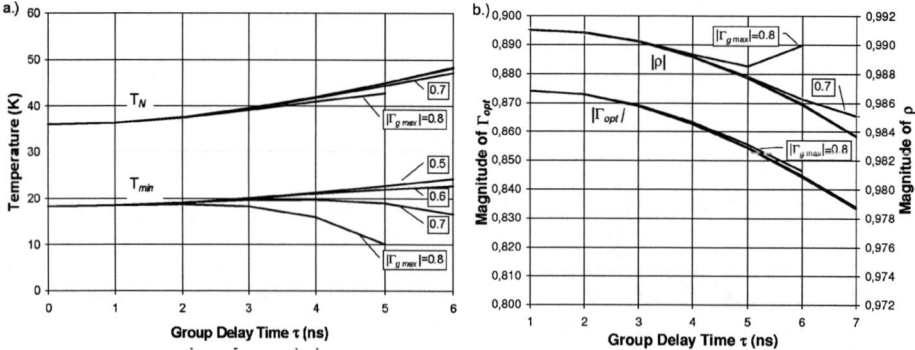

FIGURE 1. Dependence of the Noise Parameters on the Delay Time with Γ_{gmax} as a parameter.

CONCLUSIONS

We have analyzed errors in the determination of the four noise parameters, which arise due to finite measurement bandwidth and the delay time in the source circuit. The errors are especially significant at low microwave frequencies when characterizing noise of low-noise microwave transistors, e.g. PHEMTs or RF CMOS transistors. We have assessed the optimum range of the source reflection coefficient for the characterization and shown errors due to de-correlation of the noise waves in systems with high product of the bandwidth and the delay time.

ACKNOWLEDGMENTS

This work was supported in part by the Polish Ministry of Science and Information Society Technology under the Grant 0804/T11/2002/23.

REFERENCES

1. Adamian, V., and Uhlir, A., *IEEE Trans. Instrum. Meas.*, **22**, No.2, 181-182, (1973).
2. Wiatr, W., *IEEE Trans. Instrum. Meas.*, **44**, 343-346, (1995).
3. Valk, E. C., et all, *IEEE Trans. Instrum. Meas.*, **42**, 983-989, (1993).
4. Wiatr W. and Schmidt-Szalowski M., *IEEE Trans. Instrum. Meas.*, **46**, 486-489, (1997).
5. Wiatr W., and Schmidt-Szalowski M., "The Multistate Radiometer: A Novel Means For Broadband Noise and Small-Signal Characterization of Microwave Semiconductor Devices". 49th ARFTG Conference Digest, New York: Institute of Electrical and Electronics Engineers, 1997, pp. 171-180.
6. Wiatr W., and Walker D. K., *IEEE Trans. Instrum. Meas.*, **54**, 696-700, (2005).
7. Schmidt-Szalowski M., and Wiatr W., *Journ. of Telecommunications and Information Technol.*, 1, 34-38, (2002)
8. D'Antona G., *IEEE Trans. Instrum. Meas.*, **52**, 189-196, (2003)

Uncertainty In Measuring Noise Parameters Of a Communication Receiver

Karol Korcz, Beata Palczynska and Ludwik Spiralski

*Department of Marine Radio Electronics, Gdynia Maritime University,
Morska 81-87, 81-225 Gdynia, Poland*

Abstract. The paper presents the method of assessing uncertainty in measuring the usable sensitivity E_s of communication receiver. The influence of partial uncertainties of measuring the noise factor F and the energy pass band of the receiver Δf on the combined standard uncertainty level is analyzed. The method to assess the uncertainty in measuring the noise factor on the basis of the systematic component of uncertainty, assuming that the main source of measurement uncertainty is the hardware of the measuring system, is proposed. The assessment of uncertainty in measuring the pass band of the receiver is determined with the assumption that input quantities of the measurement equation are not correlated. They are successive, discrete values of the spectral power density of the noise on the output of receiver. The results of the analyses of particular uncertainties components of measuring the sensitivity, which were carried out for a typical communication receiver, are presented.

Keywords: Measurement uncertainty, noise parameter, communication receiver.
PACS: 06.20.Dk

INTRODUCTION

Communication receiving equipment, in respect to the noise, is described with the help of the noise factor F and its sensitivity E_s. When a low level useful signal is received the dominant unwanted waveforms of the output of the professional receiver are its own noise. There is a strict connection between the noise factor F and the sensitivity E_s. It is expressed by the following equation [1]:

$$E_s = \left(R_s T n_0 \Delta f F\right)^{\frac{1}{2}} \quad (1)$$

where Δf is the energy (noise) pass band of the receiver, R_s is the signal source (antenna) resistance, n_o is the ratio of the signal power to the noise power of the output of receiver in the Δf pass band.

In order to calculate the sensitivity of a receiver E_s according to the formula (1), it is necessary to measure the noise factor F and the energy pass band Δf (for assumed standard values: $R_s = 50\ \Omega$, $T_0 = 290K$; $n_o = 20$ dB).

THE METHOD OF MEASURING NOISE PARAMETERS OF A COMMUNICATION RECEIVER

The measurement of the receiver noise parameters F and Δf was carried out with the help of the digital measuring system [1], where as an input signal source the standard noise generator has been used. The measurement of F is carried out using the method of a constant relative output power [1], however the indirect method of measuring the energy pass band Δf consists in measuring spectral power density of the noise signal $G_n(k)$ on the output of receiver and then calculating the energy pass band Δf from the dependence [1]:

$$\Delta f = \frac{\sum_{k=0}^{\frac{K}{2}} G_n(k) \cdot \frac{1}{K\Delta t}}{G_n(k)_{max}} \qquad (2)$$

where $G_n(k)_{max}$ is the maximal value of spectral power density of the noise signal, K is the number of a single set of samples, Δt is a sampling period of a virtual instrument.

UNCERTAINTY OF MEASURING USABLE SENSITIVITY

In order to determine combined standard uncertainty of measuring the usable sensitivity of the receiver $u_c(E_s)$, the measurement function (1) was approximated with a 1st-order Taylor series assuming as input variables the noise factor F and the energy pass band Δf of the receiver,

$$u_c^2(E_s) = (\frac{R_s T n_o F}{2E_s})^2 u^2(\Delta f) + (\frac{R_s T n_o \Delta f}{2E_s})^2 u^2(F) + \frac{R_s^2 T^2 n_o^2 F \Delta f}{2E_s^2} r(\Delta f, F) u(\Delta f) u(F) \qquad (3)$$

where $u(\Delta f)$ and $u(F)$ are designate partial standard uncertainties in measuring the noise factor F and the energy pass band Δf respectively, $r(\Delta f, F)$ is the correlation coefficient between these values.

It was assumed in the considerations that the noise factor F and the energy pass band Δf of the receiver are independent input variables, so $r(\Delta f, F) = 0$.

The participation of particular uncertainty components in measuring respectively the noise factor F and the energy pass band Δf in the combined standard uncertainty, depends inversely on the value of the usable sensitivity.

Standard Uncertainty in Measuring the Noise Factor

The assessment of uncertainty of measuring the noise factor $u(F)$ was carried out on the basis of the systematic component of uncertainty (B-type) assuming that the main source of uncertainty of the measurement is a hardware part of the measurement system [2]. It was assumed that uncertainty depends on the accuracy of the reading from the noise generator Δ_S, which has rectangular distribution of the probability of this error. Then standard uncertainty $u(F)$ is determined as [2]:

$$u(F) = \frac{\Delta_S}{\sqrt{3}}. \qquad (4)$$

Standard Uncertainty in Measuring the Receiver Energy Pass Band

In order to assess uncertainty in measuring the energy pass band $u(\Delta f)$ it was assumed that input variables of the measurement equation (2), which are successive, determined with the help of the program, discrete values of the spectral power of the noise signal on the receiver output $G_n(k)$, are not correlated. Thus, uncertainty of measuring the energy pass bandwidth $u(\Delta f)$ is expressed by the following dependence:

$$u^2(\Delta f) = \sum_{k=0}^{\frac{K}{2}} (\frac{1}{K\Delta t G_n(k)_{max}})^2 u^2(G_n(k)) + (\frac{-\Delta f}{G_n(k)_{max}})^2 u^2(G_n(k)_{max}) \qquad (5)$$

where $u(G_n(k))$ is partial uncertainty in measuring spectral power density determined for the frequency f_k and $u(G_n(k)_{max})$ is partial uncertainty in measuring the maximal spectral power density.

Standard uncertainty of the measurement of spectral power density $u(G_n(k))$ depends inversely on the total number of samples of the analysed waveform and the spectral analysis resolution; on the other hand it depends directly on the value $G_n(k)$ and the uncertainty connected with the registration of a single sample of the measurement waveform in the digital measurement system [3].

THE ANALYSIS OF SELECTED MEASUREMENT RESULTS

The assessment of standard uncertainty in measuring the usable sensitivity of the communication receiver $u_c(E_s)$ for various frequencies of its operation was carried out (table 1). Relative uncertainty in measuring the usable sensitivity in all the analysed cases assumes similar values in spite of the fact that uncertainty components connected with measuring the pass bandwidth $u(\Delta f)$ differ from one another significantly. This is a result of the dominant participation of partial uncertainty in measuring the noise factor $u(F)$ in combined uncertainty of measuring the usable sensitivity of the receiver $u_c(E_s)$ (fig. 1).

The assessment of partial uncertainty in measuring the pass bandwidth $u(\Delta f)$ was carried out on the basis of the presented above dependencies describing uncertainties in measuring spectral power density resulting from the measurement equation (2). Uncertainty $u(\Delta f)$ can be also assessed on the basis of a series of the measurement results (their dispersion) as the standard deviation of the probability distribution of occurrence these results - the A-type method [2]. The obtained results of the analyses of this uncertainty component, determined by both methods, are comparable and show its similar participation in combined uncertainty in measuring the usable sensitivity.

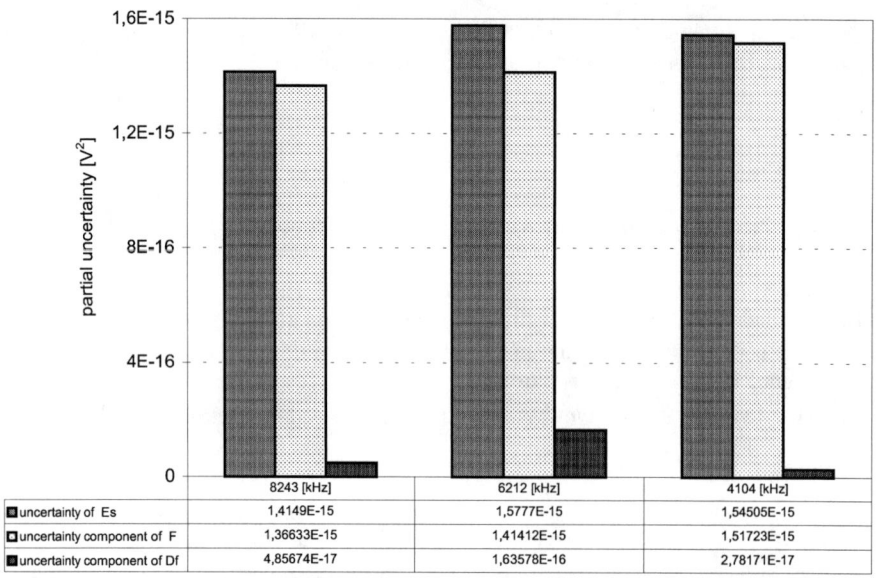

FIGURE 1. Partial standard uncertainties for various frequencies of communication receiver operation.

TABLE 1. Measured values of noise parameters and their uncertainties.

The frequency of receiver operation	The values of the parameter and their uncertainties								
	E_s [μV]	F [dB]	Δf [Hz]	$u_c(E_s)$ [μV]	$u(F)$	$u(\Delta f)$ [Hz]	$u_c(E_s)/E_s$ [%]	$u(F)/F$ [%]	$u(\Delta f)/\Delta f$ [%]
8243 [kHz]	1,28	9,6	4069,9	3,76E-02	0,62	52,05	3,45	6,78	1,28
6212 [kHz]	1,06	9,4	4022,7	3,97E-02	0,62	97,18	3,75	6,78	2,42
4104 [kHz]	1,02	9,1	4027,9	3,93E-02	0,62	41,51	3,84	6,78	1,03

CONCLUSIONS

The dominant component of uncertainty in measuring the usable sensitivity of the receiver $u_c(E_s)$ is partial uncertainty in measuring the noise factor $u(F)$. Uncertainty $u(F)$ depends mainly on the accuracy of the reading of the standard noise generator. The influence of component uncertainty connected with a/d processing and the propagation of this uncertainty in the algorithm of digital signal processing (DSP) on combined uncertainty in measuring the usable sensitivity of a communication receiver is negligibly small.

REFERENCES

1. Korcz K., Palczynska B., Spiralski L., *Digital measurement of noise parameters of communication receiver*, 12th IMEKO TC4 International Symposium, Zagreb, Croatia, pp.506-511, 2002.
2. ISO – *Guide to the Expression of Uncertainty in Measurement*, 1995.
3. Palczynska B., Spiralski L., *Uncertainty in measuring the power spectrum density of a random signal*, 14th IMEKO TC4 Symposium - New Technologies in Measurement and Instrumentation, Jurata – Poland 2005.

-190 dBV^2/Hz Preamplifier for Low Frequency Noise Measurements

Saburo Yokokura[1], Nobuhisa Tanuma[1], Munecazu Tacano[1], Sumihisa Hashiguchi[2], Josef Sikula[3], Yoko Kajiwara[4], Masami Hirasita[5]

[1]*Meisei University, Tokyo, 191-8506, Japan*, [2]*Yamanashi University, Kofu, Japan*, [3]*Brno University of Technology, Brno, Czech Republic*, [4]*Bunkyo University, Saitama, Japan*, [5]*Kinjyou University, Kanazawa, Japan*
E-mail:yokokura@ee.meisei-u.ac.jp

Abstract. Low cost highly sensitive preamplifier was made by using the commercially available operational amplifier AD797 which has the input noise equivalent power of -190 dBV^2/Hz with 80 dB amplification.

Keywords: Noise Measurements, Low Noise Preamplifier
PACS: 84.30.Le

INTRODUCTION

The AD797 of Analog Devices Co. has the low input noise of 0.9 nV/(Hz)$^{1/2}$ and low total harmonic distortion of -120 dB at audio frequencies. In order to improve the input sensitivity of the preamplifier, we used 16 AD797 in parallel to the feed back resistance of 10/100 Ω for each input stage followed by the same single AD797 2nd stage amplifier. Each stage amplifies 40 dB, and the input noise power must be reduced to 1/4 for 16 parallel amplifiers, 0.23 nV/(Hz)$^{1/2}$ (-193 dBV^2/Hz). Each device requires DC current of 30 mA at ±12 V, 12W in total, and fairly large heat sink is necessary to reduce the heating up of the device. The highest sensitivity of about -190 dBV^2/Hz is realized between 100 Hz and 10 kHz. The input impedance of this amplifier is 1.5 kΩ with the capacitance of 320 pF.

HIGHLY SENSITIVE PREAMPLIFIER

The preamplifier is the key device in noise measurements and the highly sensitive wide range reliable preamplifier is necessitated to improve the measuring technology. Typical preamplifier #113 of PAR/EG&G has the input sensitivity of 1nV/(Hz)$^{1/2}$ below 100 Hz [1], but we may need further sensitivity to detect more small noise levels in a wider frequency range. The extension to the lower frequency limit is required especially in measuring small level RTS noises, and to the higher frequency limit in assigning the shallow trap levels at elevated temperatures.

We used parallel connection of 16 AD797 operational amplifiers to reduce the input equivalent noise level [2], [3], followed by a single summing amplifier together with the zero balance circuit and 80 dB gain, as is shown in Fig. 1. The amplifier AD797 made by Analog Devices has the sensitivity of 0.9 nV/(Hz)$^{1/2}$ with the low total harmonic distortion of -120 dB at audio frequency. Low feedback resistances (10 Ω, 1 kΩ) were used in the non-inverting amplifier circuit to reduce the input noise and obtain 40 dB gain. Input noise equivalent power is expected to decrease to 1/16$^{1/2}$, 0.25 nV/(Hz)$^{1/2}$.

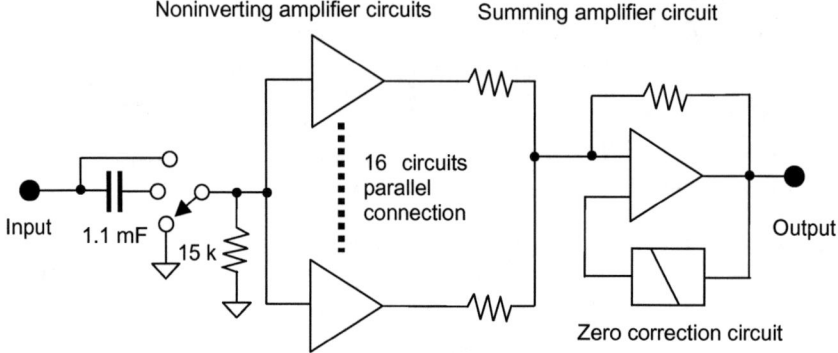

FIGURE 1. Preamplifier Block Diagram.

The coupling capacitor in the input terminal is set 1.1 mF and the resistance 15 kΩ.

The first stage output voltage includes DC voltages induced by the offset voltages and drifts [4], as well as by the leakage currents within the coupling capacitors and the circuit boards. These DC voltages are eliminated by the zero correction feedback circuit in the second stage. In order to reduce the thermal fluctuations appearing in the low frequency region below 1 Hz, 16 operational amplifiers are enclosed in the 8-Lead Standard Small Outline Package (SOIC) and further effectively cooled down by the heat sink via a gel sheet, as shown in Fig. 2. The amplifiers are powered using batteries at ±6V. Figure 3 shows the printed circuit board of the amplifier.

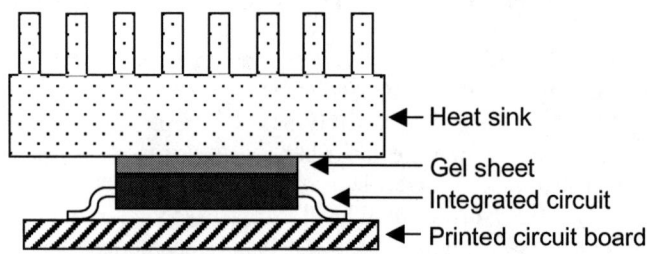

FIGURE 2. Attachment of AD797 to heat sink.

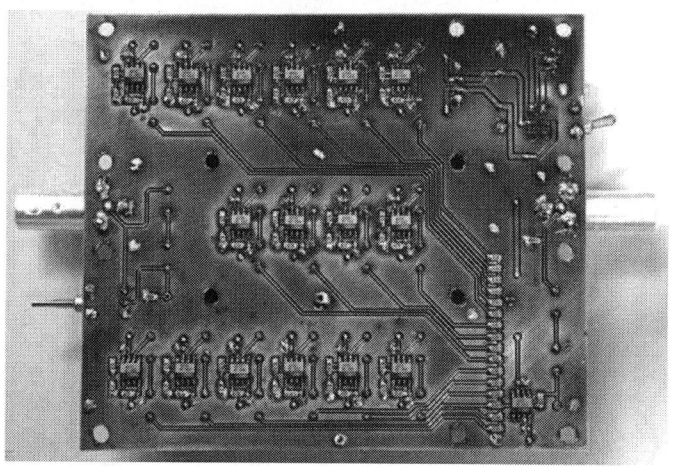

FIGURE 3. Printed circuit board of highly sensitive preamplifier.

RESULTS

The noise power spectrum density of our preamplifier was measured by a dynamic signal analyzer HP35665A. The output noise power spectrum densities are shown in Fig. 4 for 1. shorted input, 2. input resistance 10 Ω, 3. 10 Ω with coupling capacitor 1.1 mF, 4. 10 Ω with 11mF. The noise equivalent power of the shorted input indicates -190 dBV^2/Hz above 100 Hz, while the equivalent input resistance changes by frequency between 10 Ω and 1.5 kΩ in the presence of the coupling capacitors

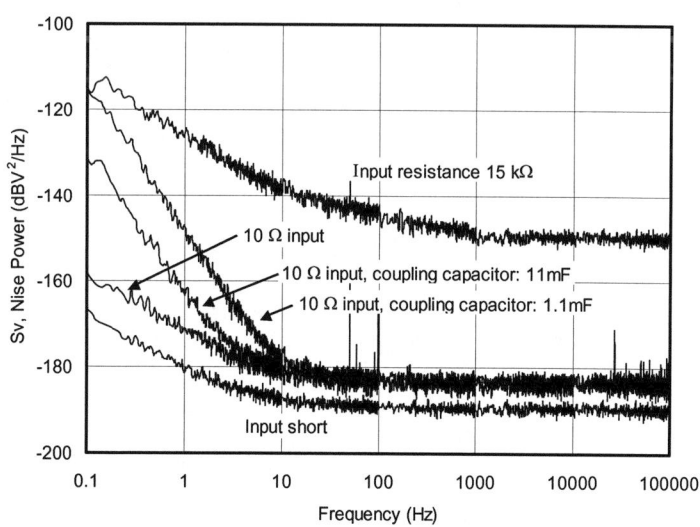

FIGURE 4. Noise power spectrum density of preamplifier

The resistor thermal noises are shown for various resistors in Fig. 5. The thermal noise for 10 Ω resistor, −188 dBV^2/Hz, is clearly discriminated from the shorted input noise. In the low frequencies below 10 Hz, the noise increased by the increase of offset voltage for higher input resistance. The maximum input voltage to the preamplifier was 100 μV rms.

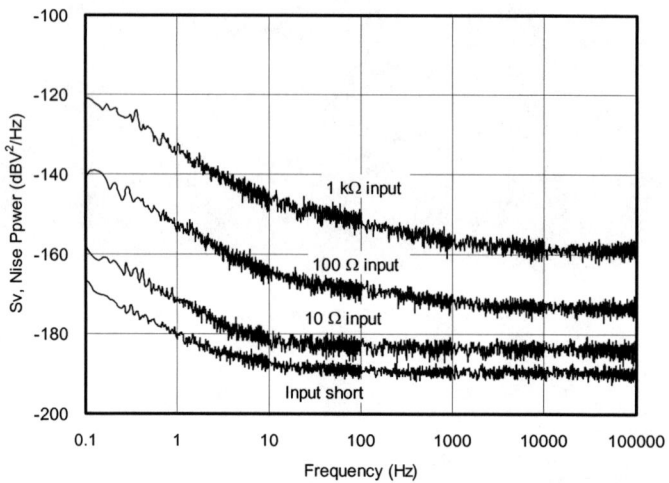

FIGURE 5. Noise power spectrum density of resistance noise (10 Ω, 100 Ω, 1 kΩ)

CONCLUSION

The most sensitive preamplifier ever reported was made by AD797. This preamplifier has the noise equivalent power of -190 dBV^2/Hz, and the thermal noise of 10 Ω resistor is apparently discriminated from the background. Further extension to the low frequency below 10 Hz will definitely improve the noise measurement technology.

REFERENCES

1. Scofield, J.H., *Rev. Sci. Instrum.* **58**, 985 (1987)
2. Yokokura, S., Tanizaki, H., Tanuma, N., and Tacano, M., "Improved instrumentation for low frequency noise measurements", Proc. of ICNF 2003, edited by J. Sikula, CNRL, Prague, 2003, pp. 847-850
3. Sikula, J., Tacano M., Yokokura, S., and Hashiguchi, S., "New tools for fast and sensitive noise measurements", Proc. of NATO ARW "Advanced Experimental Methods for Noise Research in Nanoscale Electronic Devices", edited by J. Sikula and M. Levinshtein, Kluwer, Dordrecht, 2003, pp. 345-354
4. Hashiguchi, S., Takemoto, Y., Ohki, M., Tacano, M., and Sikula, J., "Suppression of offset and drift in a DC amplifier by combination of multiple amplifiers", this conference proceedings

Analysis of Error Sources in On-Wafer Noise Characterization of RF CMOS Transistors

Wojciech Wiatr

Warsaw University of Technology, Nowowiejska 15/19, 00-665 Warszawa, Poland

Abstract. This paper analyzes how erroneous source impedance measurement due to residual errors within a VNA affect the four-noise parameter determination based on the cold-source noise measurement procedure and the eight-term linear model. It shows that although the errors disturb the complex noise characterization of a CMOS transistor at RF, mismatch and finite bandwidth errors seem to be more significant.

Keywords: noise, measurement, parameter estimation, errors, radio frequency CMOS transistors
PACS: 05.40.Ca, 06.20.Dk, 06.60.Mr, 07.05.Tp, 84.40.-x, 85.40.Qx

INTRODUCTION

For years, the noise characterization of semiconductor devices has been a very difficult metrology task due to many error sources affecting the noise measurements. In many instances the measurement errors go undiscovered and then heavily deteriorate the noise parameter determination making all the measurement effort worthless. To prevent such a situation, one may simulate the laborious measurement process on computer getting instantly a thorough insight into its potential errors. Such a simulation becomes now an important element in the experiment design that precedes real noise measurements.

Difficulties of the noise characterization of RF CMOS transistors stem mostly from their small sizes. The gate width shrinks steadily and the noise decreases as a result of the CMOS technology progress. Having high input and output impedances at RF, CMOS transistors hardly couple to typical measurement systems equipped with standard 50 Ω lines. Moreover, this high mismatch promotes resonant interacting between the device and the system interconnected with usually long cables. Since such an interaction deteriorates the noise spectrum to be measured, a thorough experiment planning should preclude this to avoid large measurement errors [1].

Another problems occur when measuring CMOS transistors on wafer. The contacts realized with microwave probe tips on aluminum pads are highly unstable and worsen quickly in time due to oxidation. Unfortunately, the noise characterization at RF using the source-pull technique is time consuming, so drifts caused by the contact instability can totally kill the noise characterization. Although, the drift effects can be mitigated with the use of right probes and shortening the measurement time, a complete solution of the problem is vital to precisely characterize noise of the CMOS transistors.

The measurement of the four noise parameters with the source-pull noise technique relies on varying the impedance and noise temperature of the source termination, and measuring resultant noise power at the output. This technique allows determining the noise parameters from a minimal necessary number of output noise powers using the popular cold-source procedure [2]. It employs several 'cold' source terminations (passive loads at ambient temperature) and one termination having a different (non-ambient) noise temperature than the others to establish a reference for the temperature scale.

The noise parameter determination is based on an assumption of the exactly known impedance and noise temperature of each termination. Though the cold-source approach eliminates inconsistencies in the terminations' noise temperatures, but does not exile the exposition to errors of the source impedance measurement. Especially, the systematic ones may destroy the noise parameter determination [3]. They arise from residual errors within a vector network analyzer (VNA) due to imperfections of the calibration standards, random measurement errors during the calibration and any change in the measurement conditions after the calibration.

On-wafer measurements are especially prone to such errors due to the use of coplanar waveguide (CPW) as the transmission line. Beside mechanical imperfections of the CPW standards, differences in probe-tip-to-line geometry and substrate permittivity when switching from off-wafer calibration to the device measurement on wafer are another important error factor [4]. Moreover, some less known errors arise due to propagation of various modes in the CPW [5]. All these result in an uncertainty with which any electrical device is characterized on wafer. This uncertainty needs to be considered when interpreting the noise measurement results.

This work analyzes how residual measurement errors within a VNA affect the four-noise parameter determination of a RF CMOS transistor. The analysis regards the cold-source noise measurement procedure [2] and the eight-term linear model [6] applied to the determination. It utilizes the approach developed in [3] to explain difficulties faced when characterizing noise of small-sized RF CMOS transistors.

METHOD

The eight-term linear model [6] describes the dependence of the output noise power on the source reflection coefficient Γ_g. It is based on the known formula for the effective input noise temperature T_e of a two-port:

$$T_e(\Gamma_g) = T_{min} + T_N \frac{|\Gamma_g - \Gamma_{opt}|^2}{(1-|\Gamma_g|^2)(1-|\Gamma_{opt}|^2)}, \qquad (1)$$

where T_{min} is the minimum value of $T_e(\Gamma_g)$, complex Γ_{opt} is the optimum source reflection coefficient for which the minimum occurs and T_N determines how rapidly $T_e(\Gamma_g)$ increases when moving away from that minimum. Beside the noise parameters, the model accounts also for the input reflection coefficient Γ_i and the gain of the two-port.

Due to the residual errors, the measured reflection coefficient Γ'_g usually differs from its truth value Γ_g and this can be expressed with the bilinear formula

$$\Gamma'_g = \frac{e_1 \Gamma_g + e_2}{1 - e_3 \Gamma_g} \qquad (2)$$

where e_i (i = 1, 2, 3.) represent residual VNA errors. After a valid calibration $\Gamma'_g \approx \Gamma_g$, and the errors are approximately $|e_1| \approx 1$, $|e_2| \approx 0$ and $|e_3| \approx 0$.

The transform (2) is typically depicted with an error box that can be then decomposed into two cascaded two-ports, from which one comprises a fully-resistive two-port and the other is lossless [3]. Each two-port contributes to transformation of Γ_g with its own fractional transform, the lossless and lossy one:

$$\Gamma''_g = \frac{\Gamma_g - e_3^*}{1 - e_3 \Gamma_g} e^{j\psi},$$
$$\Gamma'_g = r\Gamma''_g + c \qquad (3)$$

where Γ''_g is an intermediate reflection coefficient, and

$$r = \frac{|e_1 + e_2 e_3|}{1 - |e_3|^2}, \qquad c = \frac{e_2 + e_2 e_3^*}{1 - |e_3|^2} \qquad \text{and} \qquad \psi = \arg(e_1 + e_2 e_3).$$

In the complex Γ'_g plane, the parameters r and c are interpreted as the radius and center of the $|\Gamma_g| = |\Gamma''_g| = 1$ circle, respectively.

These two linear fractional transforms (3) enable one to decompose the residual VNA errors into two relevant factor sets and then analyze how each affects the determined parameters [3]. Results of such an analysis published in [3] showed a very high sensitivity of the noise parameters when characterizing noise of a PHEMT.

ERROR ANALYSIS

Results presented here refer to a RF transistor with the gate width 20×5 μm realized using the 0.35 μm gate-length CMOS technology. The transistor structure was shielded beneath with a grounded metal layer in order to mitigate substrate effects on pads and interconnects, and thus reduce losses and the device noise.

The analysis simulated measurement conditions in a real system by utilizing the following two-port's parameters: T_{min} = 89.2 K (the noise factor NF_{min} = 1.17 dB), T_N = 151.6 K, Γ_{min} = 0.883∠7.27° and Γ_i = 0.995∠-11.18°. They were determined in the system at 1 GHz for the transistor biased at V_{DS} = 2 V and V_{GS} = 1.5 V (I_D = 15.1 mA).

This analysis followed the approach introduced in [3] by considering each transform of (3) in turn. Since the effects of the first transform (3a) are usually negligible (T_{min} and T_N are invariant to lossless cascade embedding) [3], it was focused just on effects of the second transform (3b) resulting from variation of c and r.

Exemplary results of this analysis are shown in Fig. 1. The graphs evidence effects of the VNA residual errors on the four noise parameters. The largest changes are observed for T_{min} that swings regularly at ±7.5 K versus the phase of c and its average value increases with r. A similar nature reveals in T_N though at smaller amplitude. Changes of Γ_{min} are almost insignificant. Generally, the errors become vital at higher magnitudes of c and variations of r. Because of higher noise of the RF CMOS transistor, they are not, however, so destructive to the noise parameters as in the case of a PHEMT presented in [3].

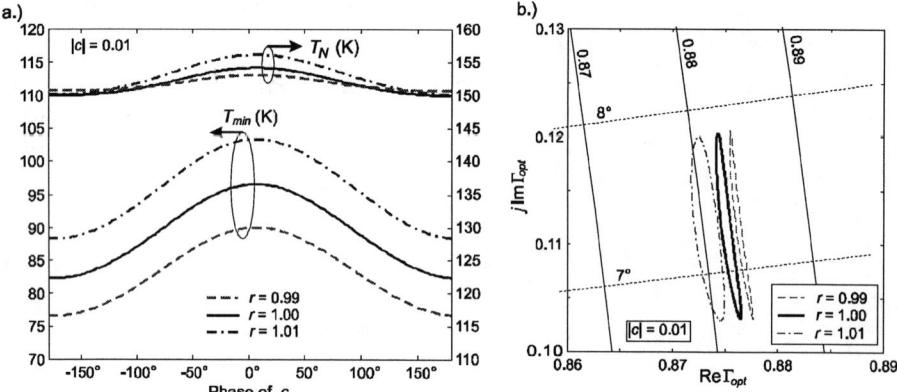

FIGURE 1. Effects of the Residual Errors in the Transform (3b) Shown for $|c|=0.01$ and r as a Parameter for the Noise Temperatures T_{min} and T_N Versus the Phase Angle of c in (1a) and as Loci of c in the Complex Γ_{min} Plane in (1b).

CONCLUSIONS

The analysis of systematic measurement errors in the determination of the four noise parameters caused by VNA residual errors in the source impedance measurement has been presented for the first time for a RF CMOS transistor. It shows that the errors rise with the residual errors to a significant level, but do not disturb the noise characterization to such an extent as observed for the PHEMT in [3]. Thus, mismatch and finite bandwidth seem to be more critical error factors in the noise measurement of such devices.

ACKNOWLEDGMENTS

The author would like to thank Dr. Zbigniew Nosal from the Institute of Electronic Systems of Warsaw University of Technology who designed the RF CMOS transistors for this search. This work was supported by the Polish Ministry of Science and Information Society Technology under the Grant 4 T11B 040 23.

REFERENCES

1. Wiatr, W., and Adamson, D., "Finite Bandwidth Related Errors in Noise Parameter Determination of PHEMTs", AIP Conference Proceedings ICNF, 2005, in this issue.
2. Adamian, V., and Uhlir, A. *IEEE Trans. Instrum. Meas.*, **22**, 181-182, (1973).
3. Wiatr, W., Walker, D. K., *IEEE Trans. Instrum. Meas.*, **54**, 696-700, (2005).
4. Carchon, G., et al, *Electronic Letters*, **35**, 1087-1088, (1999).
5. Lewandowski A., Wiatr, W., *Journ. of Telecommunications and Information Technol.*, 2, in print, (2005).
6. Wiatr, W., *IEEE Trans. Instrum. Meas.*, **44**, 343-346, (1995).

RELIABILITY

Low Frequency Noise Considerations for CMOS Analog Circuit Design

Ralf Brederlow[1], Jeongwook Koh[2], Gilson I. Wirth[3], Roberto da Silva[4], Marc Tiebout[1], and Roland Thewes[1]

[1] *Corporate Research, Infineon Technologies AG, Otto-Hahn-Ring 6, D-81730, Munich, Germany, Phone: +49 (89) 234-50575, Fax: +49 (89) 234-9555341, Email: ralf.brederlow@infineon.com*
[2] *Now with SAIT, Samsung Electronics Co. Ltd., Kiehung, Republic of Korea,*
[3] *State Univiversity of Rio Grande do Sul-UERGS, Rio Grande do Sul, Brazil,*
[4] *Informatics Institute, Federal University RGS-UFRGS, Porto Alegre, RS, Brazil*

Abstract: This paper gives an overview on 1/f-noise issues relevant for today's CMOS analog circuit design. The device-to-circuit relation of noise and the relevant operating conditions are reviewed. Modeling of the biasing dependence of 1/f-noise amplitude including large signal and statistical effects are discussed. The noise corner frequency[1] is shown to increase with CMOS technology scaling, and statistical effects are shown to even scale worse compared to the 1/f-noise. Moreover circuit design measures against noise are investigated. Finally, reliability issues concerning 1/f-noise in analog circuits are reviewed.

Keywords: 1/f-noise, flicker noise, low noise circuit design, CMOS low noise
PACS: 85.40.Qx

INTRODUCTION

At low frequencies the noise of CMOS integrated circuits is dominated by 1/f-noise or even by the noise of individual traps (see Figure 1a). This noise increasingly threads design of low noise analog, mixed-signal and radio frequency circuits in increasingly larger frequency bandwidths. The noise corner frequency[1] is roughly 10 MHz for 90nm technology and will further increase in future CMOS generations (see Figure 1b). Careful design for low noise is mandatory in many cases. In this paper we review the 1/f-noise properties of CMOS devices at circuit-relevant operating conditions, their modeling, and discuss strategies to minimize noise in circuits.

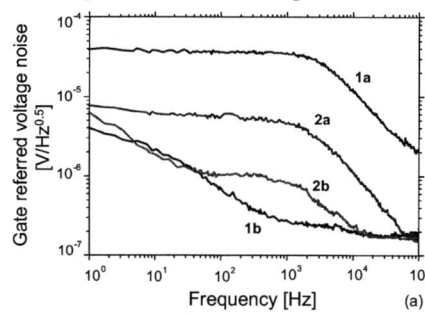
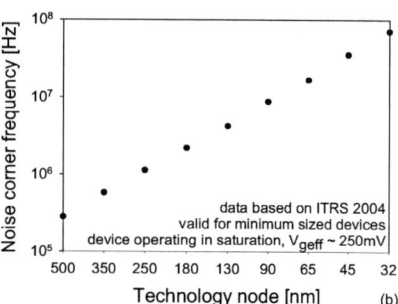

FIGURE 1(a): Gate referred voltage noise of two different W=0.16µm/L=0.13µm n-MOS transistors: first device biased at V_g-V_t =0.5V, with V_d=0.15V (1a) and V_d=1.0V (1b), second device, biased at V_d=1.0V, with V_g-V_t =0.5V (2a) and V_g-V_t =0.2V (2b). **(b):** Noise corner frequency of minimum sized CMOS devices versus technology node calculated from ITRS 2004 data [1].

---
[1] Intercept point where amplitudes of 1/f-noise and thermal noise are equal.

1/F-NOISE IN CIRCUITS

To optimize the noise properties of CMOS analog circuits, noise propagation within the circuit and the device noise behavior in the circuit at the respective operating point needs to be considered. This is detailed for two examples and two classes of circuits.

The circuit in Figure 2a) shows a differential stage as it is frequently used for amplification of small signals or for buffering of signals. This circuit is an example for a small signal type circuit with almost constant biasing conditions. All devices are operated in saturation with gate voltages a few hundred millivolts above threshold (typically 100–500mV). Circuit noise calculation is performed considering that each noise voltage propagates like an uncorrelated signal in a small signal equivalent circuit.

The circuit in Figure 2b) shows a voltage controlled oscillator as it is frequently used for tunable frequency references. This circuit is an example for another class of circuits where the operating points of the devices undergo almost every value between maximum and minimum supply voltage for both gate and drain. As a consequence operating point dependent signal transfer functions including non-linear effects (e.g. frequency-mixing) have to be taken into account. A simple estimation of the noise from extrapolation of one operating point is insufficient. Especially here, circuit simulation tools are very useful. Since large-signal circuits often show periodic behavior they can either be described by a transient simulation over one period or by Fourier transformation techniques. As a rule of thumb, the most critical operating points for noise propagation are the points of equal current flow on both branches, when the output voltages (and therefore both drain and source voltages) are roughly half the supply voltage.

From a circuit-designers point of view, noise optimization is done by circuit sizing and choosing appropriate circuit architecture. Circuit simulation together with simple but sufficiently accurate noise models helps to reach an optimum solution in a reasonable amount of time. Moreover, a good understanding of the device physics behind 1/f-noise helps to identify possible bottlenecks for scaling the circuit architectures to new CMOS process generations.

1/F-NOISE COMPACT MODELING

1/f-noise of MOSFETs originates from trapping and de-trapping of charged carriers at the interface and oxide traps in the MOS structure [2] (Figure 3a). By variation of the number and mobility of free carriers in the MOS channel, the traps directly influence the low frequency noise behavior. Single traps produce a Lorentzian shaped spec-

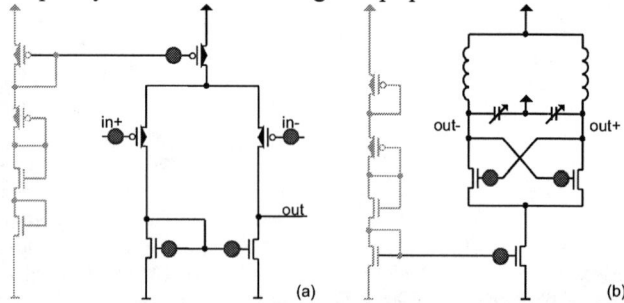

FIGURE 2(a): The picture shows a differential stage as example of a typical CMOS linear analog circuit for small signal gain. **(b)**: Voltage controlled oscillator as example for a noise sensitive circuit with large signal output. The devices depicted in light gray color represent simple biasing branches.

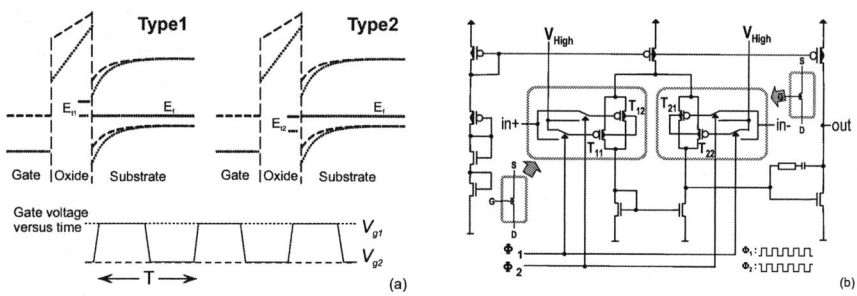

FIGURE 3(a): Energetic situation of two noise relevant traps at two different gate voltages as shown in the lower left. **(b)**: Test circuit used for measuring the 1/f-noise reduction of switched MOSFETs [10].

trum, an ensemble of traps produces a noise spectrum roughly inversely proportional to the frequency [2,3,4] (Figure 4a). For small signal circuits using devices operated in saturation it is useful to describe the 1/f-noise S_{Vg} as referred to their gate voltage V_g:

$$S_{Vg}(V_g, f) = S(V_g) \cdot \frac{N_{t,f}}{W \cdot L} \cdot \frac{1}{f} \qquad (1)$$

In this representation the Coulomb-equivalent charge of these trapped oxide charges well describes the noise behavior almost independent of the biasing [5]. $N_{t,f}$ is the trap density close to the Fermi level which is mostly relevant for noise, and W and L are the device width and length. The biasing relation is described by the function $S(V_g)$. It depends on the relative influence of number and mobility fluctuations on the noise amplitude [3,6], but the general influence on the noise amplitude over the range of relevant gate voltages is relatively weak for modern technologies.

1/f-noise amplitude is reduced under certain non-equilibrium biasing conditions [7,8,9]. The physics behind the noise reduction is understood when considering two types of traps and two voltages alternately applied to the gate of a device (see Figure 3a) [9]. The densities of those two types differ with respect to their energetic position and the corresponding alternating Fermi-level or gate voltage. If the period of oscillation between the two gate voltages is faster than the average capture and emission time of the trap levels, the traps cannot follow the fast oscillation. They often remain in the

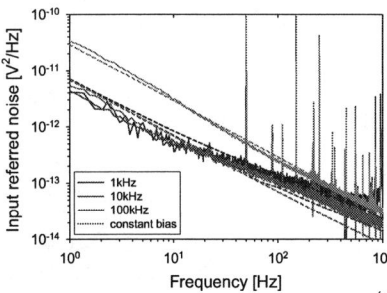 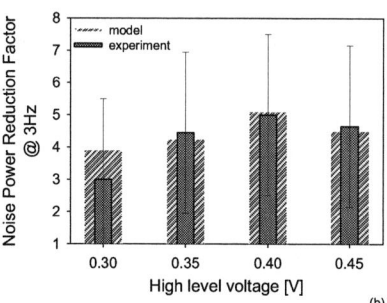

FIGURE 4(a): Experimental (solid lines) and calculated (dashed lines) gate referred voltage noise versus frequency of a W/L=12μm/0.6μm p-MOSFET with a rectangular gate-source voltage (V_{g1}=-500±30mV, V_{g2}=300±30mV, see Figure 3a) at switching frequencies indicated in the graph. Drain voltage is 800±100mV. **(b)**: Measured (thin bars) and calculated (thick bars) noise reduction versus higher gate-source voltage ($V_{g2} = V_{high}$ in Figure 3b) for a W/L=12μm/1μm p-MOSFET at a lower gate-source voltage V_{g2} of -500±30mV, a switching frequency of 1kHz, and a drain-source voltage of 800±100mV.

energetic position where one of the two gate voltages drives them into a defined steady state where they do not contribute to the noise. This 'memory' reduces their trapping and de-trapping activity and their contribution to the noise at the other gate voltage where they normally are active is reduced as well. However, it also enhances their noise activity in the next half period when driven back into a defined state. Only if there is an imbalance in the trap densities at the two energetic levels corresponding to the two types of traps, this finally results in a reduction of the total 1/f-noise [9]. If one of the two voltage levels is close to the equivalent Fermi level of mid-gap, where the trap density is lowest, the highest noise reduction effect is observed.

Today's standard compact models are not capable of simulating the effect. However the effects of noise reduction under periodic large signal excitation can be analytically described for compact modeling using a stepwise constant approximation for the gate voltage:

$$S_{V_g}(V_g(t),T,f) = \sum_{i=1}^{k}\left(S(V_{gi}) \cdot A_{\omega}^i(V_g(t),T,f)\right) \text{ with } V_g(t) = \begin{cases} V_{g1} \text{ for } n \cdot T < t \leq (n+1/k) \cdot T \\ V_{gj} \text{ for } (n+(j-1)/k) \cdot T < t \leq (n+1) \cdot T \\ \vdots \end{cases} \quad (2)$$

Here, A_{ω}^i is the Fourier transform of the average autocorrelation function of the noise relevant traps corresponding to the fraction of the period where the gate voltage $V_g(t)$ is equal to V_{gi}. This is detailed in [9]. T is the oscillation period, and j is an integer accounting for the k different constant voltage values, each valid for equal time fractions of the gate voltage oscillation period (repeated for the n-th time).

Figure 3b shows a test circuit for measuring the 1/f-noise of p-MOSFETs under periodic large signal excitation [9,10]. Results of those measurements and of the model (Eq. (4)) are shown in Figure 4a) and b). The model agrees with the experimental observed dependencies for the noise-frequency, the oscillation period and the biasing. For circuit design the effect may be used to reduce 1/f-noise below the constant biasing noise values [8,10] as in the circuit shown in Figure 3b).

STATISTICAL EFFECTS AND 1/F-NOISE

For modern CMOS technologies the amount of traps generating 1/f-noise is relatively small and for the smallest devices even single traps may dominate certain frequency bands of the low frequency noise (see Figure 1a). Compared to small signal gate voltage related changes in the gate referred 1/f-noise, device and intra-die varia-

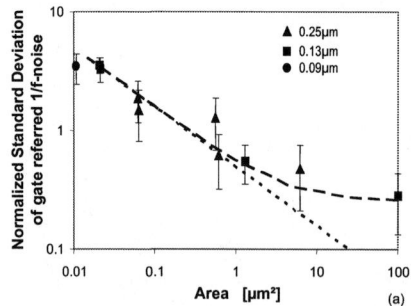
(a)

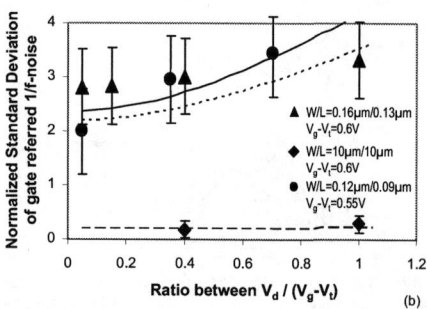
(b)

FIGURE 5(a): Standard deviation of 1/f-noise amplitude for several CMOS technologies from 0.25µm to 90nm versus area at $V_g-V_t=0.6V$, $V_d=1V$ and **(b)** versus ratio of gate to drain voltage. The dashed and the dotted lines in (a) show results of a model for the 0.13µm data which includes and excludes global variations. The lines in (b) are results predicted by the same model.

tions are relatively large, at least for the most relevant operating conditions for analog circuits. A circuit however, has to be designed to yield in production. To cope with this effect, worst case statistics for the noise amplitude needs to be introduced for noise modeling and circuit design [6,11,12]. Since the trap distribution follows a Poisson statistic also the 1/f-noise of an ensemble of different devices approximately follows a similar statistics [6]. For hand calculations a simple worst case function for the 1/f-noise of a single device under 3σ conditions $S_{Vg,3\sigma}$ [12] is given by:

$$S_{Vg,3\sigma}(f,V_g) = S_{Vg}(f,V_g) \cdot \left(1 + 3\sqrt{\frac{\sigma_{N+\mu}}{W \cdot L \cdot N_{t,f}}}\right) \qquad (3)$$

Here $\sigma_{N+\mu}$ describes the noise variability of traps for both number and mobility related effects and is in the order of one [6]. In addition to the local variations also global variations due to manufacturing imperfections have to be considered [12]. Figure 5a) shows the 1/f-noise standard deviation versus device area. In this figure both local (left, small area devices) and global (right, large area devices) variations are seen. In Figure 5b) the standard deviation versus operating conditions is shown. Both graphs show good agreement between model and experiment.

Statistical parameter fluctuations in the total 1/f-noise of analog circuits are often larger than noise variations due to voltage or environmental effects. For accurate estimation of noise related yield in larger circuits with many different noise contributors, Monte-Carlo approaches for circuit noise estimation can give more area efficient results than simple worst case approximations.

1/F-NOISE AND RELIABILITY

For analog circuits the two most important reliability aspects are Negative Bias Temperature Instabilities (NBTI) and Hot Carrier (HC) induced degradation.

NBTI today is the most important thread to the reliability of analog circuits [13, 14]. However no report on the impact of NBTI on 1/f-noise does exist so far. The most likely physical origin of NBTI degradation is the Si-H interface bond breaking and movement of hydrogen ions within the gate-oxide [15]. This mechanism involves interface trap formation. Since the energy of the interface states must be close to the Fermi level to have 1/f-noise relevance, NBTI related 1/f-noise degradation is not necessarily expected. Nevertheless 1/f-noise after NBTI degradation should be monitored, especially when new gate-stack materials are introduced into the CMOS technology.

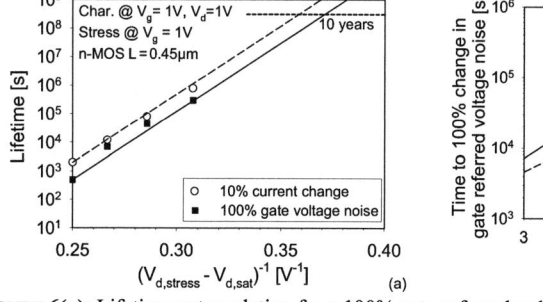 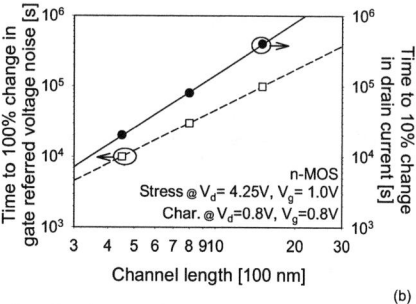

FIGURE 6(a): Lifetime extrapolation for a 100% gate referred voltage noise and a 10% drain current failure criterion. **(b)**: Changes in low frequency noise and drain current of n-MOS transistors after hot carrier stress versus channel length. Process details see [17].

HC damage has frequently been observed in the 1/f-noise characteristics in the past [e.g. 16]. This is also true for operating and stress conditions relevant for analog applications [17]. Hot carrier stress for both p- and n-MOS devices today is mainly caused by interface trap generation between pinch-off and drain. As soon as traps are generated in the oxide close to channel regions where the current density is close to the interface, additional 1/f-noise is resulting. For analog operation in saturation this happens when stress damage generation moves into the region of the pinch-off point [17]. The noise related change in 1/f-noise amplitude $\Delta S_{Vg}(t)/S_{Vg}$ over stress time t approximately follows an effective stress voltage $(V_d - V_{d,sat})$ related power law (Figure 6a):

$$\frac{\Delta S_{Vg}(t)}{S_{Vg}(t=0)} \approx \left(\frac{I_{d,stress}}{W} \cdot \exp\left[\frac{V^*}{V_d - V_{d,sat}}\right]\right)^m \cdot \left[a \cdot t^m + b \cdot t^{n+m}\right] \quad (4)$$

Here $I_{d,stress}$ is the stress drain current, V_d and $V_{d,sat}$ the drain and saturation drain voltage. The other parameters are physics related fit constants [17].

Since hot carrier generation needs a minimum energy, in modern devices with decreasing operating voltages, HC damage is becoming less problematic. However I/O and precision analog devices with higher voltages are incorporated in those technologies. For these devices, HC related 1/f-noise degradation is a concern for reliable analog circuit design. The most important measure to enhance device lifetime in such circuits is to increase the channel length (Figure 6b), but also certain bias conditions [17] may enhance circuit lifetime.

CONCLUSION

We have reviewed the 1/f-noise properties of CMOS devices with emphasis on their application in analog circuits. It is shown that the corner frequency and statistical effects in the 1/f-noise increase for each new technology node. Compact models for the correct description of the statistical effects and the non-equilibrium behavior of the 1/f-noise are discussed. Reliability issues for 1/f-noise in analog designs need to be monitored but are expected to remain non-critical as long as no severe material changes in the device structure are implemented in new CMOS process generations.

REFERENCES

1. H. S. Bennett, et al., Circuits and Devices Magazine, Nov/Dec 2004, pp. 39-51
2. S. Christenson et al., Solid-State Electronics 11, pp. 791-812, 1968
3. R. Jayaraman et al., IEEE Transaction on Electron Devices 36, pp. 1773-82, 1989
4. K.K. Hung et al., IEEE Transaction on Electron Devices 37, pp. 654-65, 1990
5. J. Chang et al., IEEE Transaction on Electron Devices, 41, pp. 1965-71, 1994
6. G. Wirth et al., accepted for publication, IEEE Transaction on Electron Devices
7. I. Bloom et al., Applied Physics Letters 58, pp. 1664-6, 1991
8. E.A.M. Klumperink et al., Journal of Solid-State Circuits 35, pp. 994-8, 2000
9. R. Brederlow et al., submitted to ESSDERC 2005
10. J. Koh et al., Proc. VLSI Circuits Symposium 2004, pp. 222-5
11. G. Ghibaudo et al., Physica Status Solidi (a) 132, pp. 501-7, 1992
12. R. Brederlow et al., IEDM 1999 Tech. Dig., pp. 159-162
13. R. Thewes et al., IEDM 1999 Tech. Dig., pp. 81-84
14. C. Schlünder et al., Microelectronics Reliability, vol. 45, pp. 39-46, 2005
15. S. Ogawa et al., Physical Review B, p. 4218- 30, 1995
16. Z.H. Fang et al., IEEE Electron Device Letters EDL-7, pp. 371-3, 1986
17. R. Brederlow et al., IEEE Transaction on Electron Devices, 49, pp. 1588-1596, 2002

Accelerated Aging of GaN Light Emitting Diodes Studied by 1/f and RTS Noise

S. Bychikhin[1], L. K. J. Vandamme[2], J. Kuzmik[1,3], G. Meneghesso[4], S. Levada[4], E. Zanoni[4], D. Pogany[1]

[1]Institute for Solid State Electronics, TU Vienna, Floragasse 7, A-1040 Vienna, Austria,
[2]Eindhoven University of Technology, The Netherlands,
[3]on leave from IEE, Slovak Academy of Sciences, Bratislava,
[4]Dept. of Information Engineering, University of Padova, 35131 Padova, Italy,
e-mail of corresponding author: dionyz.pogany@tuwien.ac.at

Abstract. We study variations of $1/f$ and random telegraph signal (RTS) noise with accelerated aging in blue GaN light emitting diodes. RTS noise is dominant at low currents and its parameters vary with the aging. $1/f$ noise has a junction component dominant at medium currents and a series resistance component dominant at elevated currents. While the junction noise is relatively non-sensitive to aging, the series resistance contact noise is strongly aging-dependent. The contact noise increases by the factor of 60-800 after aging and is found to be a good reliability indicator. The physical origin of degradation is discussed.

Keywords: light emitting diode, gallium nitride, $1/f$ noise, RTS noise, reliability assessment
PACS: 72.70+m, 78.30.Fs, 78.60.Fi

INTRODUCTION

Because of their ultra brightness, GaN-based light emitting diodes (LEDs) becomes promising light sources [1]. The growing market requires high reliability and long lifetime. Degradation mechanisms due to accelerated aging in GaN-based LEDs have previously been investigated by capacitance spectroscopy or light emission mapping [2]. Low frequency noise (LFN) characterization, in particular $1/f$ and random telegraph signal (RTS) noise, has previously been found as a suitable indicator of physical degradation of materials, electronic and opto-electronic devices during different kinds of accelerated aging and for reliability estimation [3-5]. In this paper low frequency noise (LFN) is used as a diagnostic tool to study degradation mechanisms in GaN-based LEDs submitted to accelerated aging.

DEVICES AND EXPERIMENTAL TECHNIQUE

Devices are blue LEDs grown on n-SiC substrate with an InGaN quantum well as an active layer [2]. A semitransparent platinum film of 300μm × 300μm and 5nm thick deposited on p-doped GaN is used as a top contact. LFN spectra were measured from open circuit voltage noise $S_V(f)$ using a low noise preamplifier and FFT analyzer and then recalculated to current noise $S_I(f)$ according to $S_I = S_V / r_d^2$, where r_d is the experimentally obtained current-dependent differential resistance of the forward biased LED. The RTS noise was studied in time domain using a current-voltage preamplifier and a digital oscilloscope.

Two kinds of accelerated long duration stresses are applied: thermal aging 300h at 240°C (sample THS) and forward electrical stress 2500h at 50mA (ELS). A short duration current stress (minutes, up to 100mA) was also applied. Unstressed sample is denoted by NSS.

RESULTS AND DISCUSSION

Figure 1 shows the forward current characteristics of unstressed and aged samples. A weak and strong increase of current can be seen in THS and ELS device, respectively, for the region $I < 100\mu A$. For $I > 100\mu A$, where the effect of series resistance dominates, the aging causes a higher series resistance increase in THS ($\approx$ 90%) compared to ELS device ($\approx$13%). The series resistance increase has previously been correlated with the decrease in the LED optical power [2].

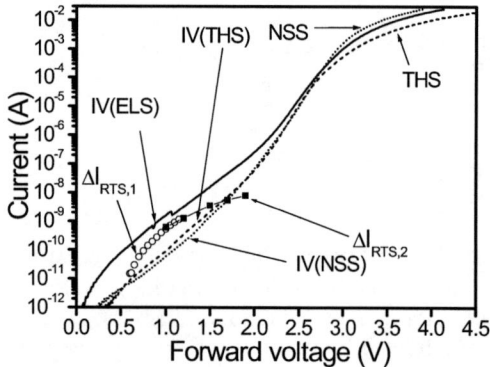

FIGURE 1. Forward IV characteristics in unstressed and aged LEDs. $\Delta I_{RTS}(V)$ characteristics are also shown for the ELS device (curve $\Delta I_{RTS,1}$, corresponding to IV(ELS)) and for ELS device stressed by an additional short duration stress of 15min@50mA (curve $\Delta I_{RTS,2}$).

At low currents ($I < 100nA$), two level or multilevel RTS fluctuations can be observed in both bias directions in both unstressed and stressed samples. The aging causes either (a) a creation of new RTS or (b) changes the RTS parameters or (c) leaves RTS parameters unaffected or (d) vanishing RTS fluctuations. Curve $\Delta I_{RTS,1}$ in Fig. 1 represents typical RTS amplitude – forward voltage dependence in an aged LED (ELS). The $\Delta I_{RTS}(V)$ dependence first steeply increases with voltage, while at higher voltages it tends to level-off. This behavior can be explained by a movement of the edge of the space charge region with increasing forward bias towards the RTS controlling defect [5]. The relative RTS amplitude $\Delta I_{RTS}/I$ in the aged devices reaches a value up to 50% (see $\Delta I_{RTS,1}/I$(ELS) in Fig. 1), which is much higher than that observed in unstressed devices (<5%). The large $\Delta I_{RTS}/I$ indicates a trap located at a strategic position in the space charge region.

The RTS mean pulse width in the upper $<t>_+$ (lower $<t>_-$) RTS state is typically decreasing (unchanged) with increasing voltage (see Fig. 2a). This behavior is attributed to changes in quasi-Fermi level relative to the energy position of the RTS controlling trap [6].

We have also applied short duration current stress to study variations in RTS characteristics. The sample ELS was additionally stressed by I_{stress}=50mA for 15 minutes. The RTS mean pulse widths $<t>_+$ increased relatively to the original value (see Fig. 2b), which indicates changes in trap parameters (e.g. carrier capture cross sections, ionization energy,...) with stress. Due to long $<t>_+$ for $V < 0.95V$, the $\Delta I_{RTS}(V)$ characteristics (curve $\Delta I_{RTS,2}$ in Fig. 1) were only measurable for $V > 0.95V$. The dependence $\Delta I_{RTS,2}(V)$ follows the shape of $\Delta I_{RTS,1}(V)$, which indicates that the RTS fluctuations before and after the short stress are due to the same defect – a current constriction in the space charge region (e.g. an extended defect as dislocation).

For currents $I > 1\mu A$ the noise is dominated by $1/f$ fluctuations. The $S_I(I)$ dependence exhibits the following behavior (see Fig. 3): For currents $I < I_C$, where I_C is a critical current, one finds $S_I \propto I$, which is attributed to quantum well junction noise. A typical value of extracted noise (Hooge [3]) parameter $\alpha = 3\times 10^{-3}$ is almost independent on the stress, which indicates a good crystalline quality of the junction.

For $I > I_C$ a leveling-off region is found which continues with a $S_I \propto I^3$ dependence at higher currents. This region is attributed to series resistance noise [4]. This noise component increases by a factor 60 to 800 for the electrical (ELS) and thermal stress (THS), respectively (see Fig. 3). As the value of series resistance increases with aging at most by a factor of 2 (see Fig.1), the series resistance noise is a very sensitive reliability indicator for this type of accelerated aging.

We assume that the strong increase in the series resistance noise indicates the degradation of a part of contact resistance, where a noisy contribution is in a series with a noise-free contribution. The problem is similar to formation of multi-spot contact giving rise to a current crowding effect resulting in an enhanced noise [4,7]. We suppose that the degradation takes place in the semi-transparent platinum top layer. Both thermal stress (in THS) and electromigration (in ELS) can accelerate the structural changes in this layer.

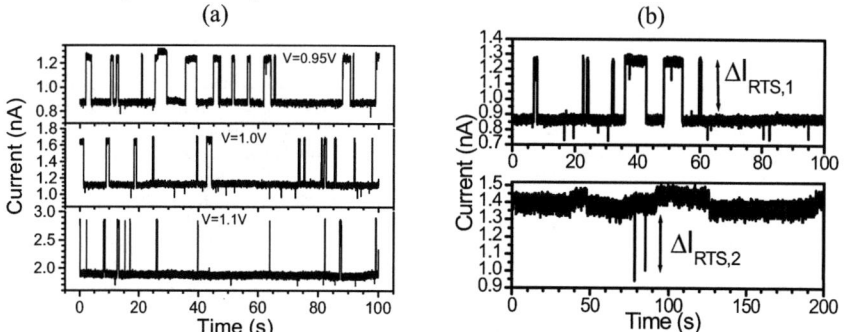

FIGURE 2. (a) RTS waveforms measured in ELS device at different forward biases. (b) RTS waveform at $V = 0.95V$ in ELS device before (top) and after (bottom) the additional short duration electrical stress of 15min@50mA. See Fig. 1 for $\Delta I_{RTS,1}(V)$ and $\Delta I_{RTS,2}(V)$ curves.

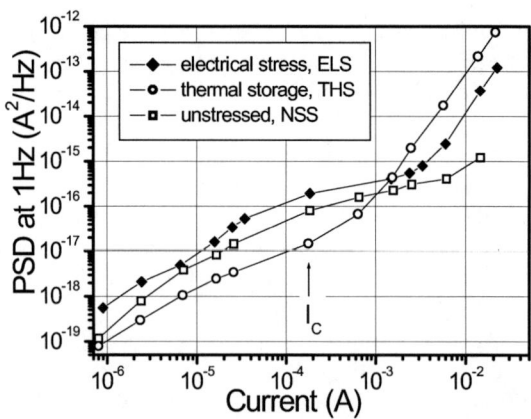

FIGURE 3. $S_I(I)$ dependencies calculated from the measured $S_V(I)$ characteristics, for the unstressed and stressed devices. The critical current I_C is indicated for THS device.

CONCLUSIONS

The accelerated aging of GaN LEDs causes modifications of RTS noise parameters and a strong increase in series resistance noise (up to factor of 800). The modification in RTS characteristics by stress is attributed to defects modification in the junction. The increase in the series resistance noise, attributed to platinum contact degradation and current crowding effect, is a good reliability indicator.

ACKNOWLEDGMENTS

The work was partially supported by PRIN 2002 project of Italian Ministry of University and Research (MIUR). Prof. Gornik is acknowledged for the support.

REFERENCES

1. H. Morkoç, *Nitride Semiconductors and Devices*, Springer-Verlag, (1999).
2. G. Meneghesso, S. Levada, R. Pierobon, F. Rampazzo, E. Zanoni, A. Cavallini, A. Castaldini, G. Scamarcio, S. Du, I. Eliashevich, "Degradation mechanisms of GaN-based LEDs after accelerated DC current aging", *International Electron Device Meeting (IEDM) Technical Digest*, San Francisco, IEEE, 2002, pp.103-106.
3. L.K.J. Vandamme, "Noise as a diagnostic tool for quality and reliability of electronics devices", *IEEE Trans. Electron. Dev.*, **41**, 2176-2187 (1994).
4. L.K.J. Vandamme, "Opportunities and limitations to use low-frequency noise as a diagnostic tool for device quality", *Proc. Int. Conf. on Noise and Fluctuations (ICNF)*, Edited by J. Sikula, Prague, 2003, pp.735-748.
5. D. Pogany, A. Chantre, J.A. Chroboczek, G. Ghibaudo, "Origin of large amplitude random telegraph signal in silicon bipolar junction transistors after hot carrier degradation", *Appl. Phys. Lett.* **68**, 541-543 (1996).
6. S.T. Hsu, R. J. Whittier, and C. A. Mead, "Physical model for burst noise in semiconductor devices", *Solid-St. Electron.*, **13**, 1055-1071 (1970).
7. L.K.J. Vandamme, M.G. Perichaud, E. Noguera, Y. Danto and U. Behner, "1/f noise as a diagnostic tool to investigate the quality of isotropic conductive adhesive bonds", *IEEE Trans. Components and Packaging Technology*, **22**, 446-454 (1999).

Influence of Small Doses of Gamma Irradiation on Transport and Noise Properties of SiC MESFETs

S.A.Vitusevich[1], M.V.Petrychuk[2], A.M.Kurakin[1], S.V.Danylyuk[1], A.E.Belyaev[3], H.-Y.Cha[4], M.G.Spencer[4], L.F.Eastman[4] and N.Klein[1]

[1] *Insitut für Schichten und Grenzflächen and CNI-Center of Nanoelectronic Systems for Information Technology, Forschungszentrum Jülich, 52425 Jülich, Germany*
[2] *Taras Shevchenko National University, Kiev 01033, Ukraine*
[3] *V.Lashkaryov Institute of Semiconductor Physics, NASU ,03028 Kiev, Ukraine*
[4] *School of Electrical Engineering, Cornell University, Ithaca, New York 14853, USA*

Abstract. Steady-state characteristics and low-frequency noise spectra of SiC-based metal-semiconductor field-effect transistors (MESFETs) before and after small doses (1×10^6 rad) of gamma radiation treatment are studied. The structural ordering of non-controllable impurities with radiation leads to an increase in threshold voltage, decrease of the channel's resistance and reduces the number of G-R components observed in the total noise spectra of the devices.

Keywords: SiC, MESFET, gamma radiation treatment, low-frequency noise.
PACS: 72.70.+m, 61.80.Ed

In recent years special attention has been paid to the research and development of silicon carbide metal-semiconductor field-effect transistors (MESFETs), which are presently being developed to meet the demands of high-power electronic [1] and radiation-resistant [2,3] systems. Despite the remarkable results demonstrated by many groups in exploiting the superior properties of SiC-based devices, the various trapping effects [4] causing a current collapse in the I-V characteristics, frequency dispersion of transconductance and output resistance, drain- and gate-lag transients are present in these devices and deteriorate the performance and reliability of the microwave devices. Therefore, developing SiC-based devices that are stable in operation and have improved performance is of current importance. In this communication, we study the effect of gamma-radiation treatment on transport and noise characteristics of MESFETs. Improvement of the MESFET devices performance under small irradiation doses was observed and analysed.

The investigated samples were epitaxially grown on vanadium-doped semi-insulating 4H-SiC substrate. P-type buffer, grown on the substrate, has a thickness of 250nm and an acceptor concentration of less than 5×10^{15} cm^{-3}. An N-type channel layer was formed by the film with a thickness of 260nm and a donor concentration equal to 2×10^{17} cm^{-3}. To improve the ohmic behaviour of the contact regions of the film under the source and drain, the contacts were implanted with phosphorus. The channel region between the source and drain was etched 60 nm deep resulting in a 200 nm-thick layer. The total length of the channel L_{SD} was equal to 2.45 µm, of which the source-gate distance L_{SG} is 0.5 µm, and the gate length L_G is about 0.45 µm. The channel width, W, of the MESFETs was varied from 100 µm to 500 µm.

Current-voltage (I-V) and low-frequency noise characteristics of the devices were measured in a wide temperature range of T=60-300 K. The devices were irradiated at room temperature by ^{60}Co gamma rays by a dose equal to 10^6 rad with a flux of 10^2 rad/s. Typical I-V characteristics of the MESFET measured at different gate voltages are shown in Fig.1a. It can be seen that the characteristics are linear up to drain–source voltages equal to 0.5V, except for some voltage ranges measured at high bias on gate. To compare devices with different channel widths, we calculated the channel resistance normalized to its width. The measured I-V dependences of channel current on source drain voltage demonstrated a 6 - 17% decrease in the channel's resistance after a 10^6 rad gamma-irradiation dose. The observed device performance exhibited limited variation and all devices showed similar responses to gamma radiation.

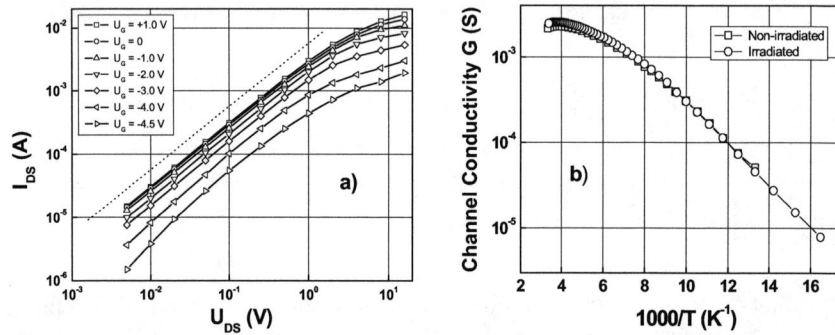

FIGURE 1. a) I_{DS}-U_{DS} characteristics of the MESFET with a channel width of 100 µm, measured before gamma irradiation at T = 300K. b) Temperature dependence of conductivity of the MESFET before and after irradiation.

The data of the channel resistance R_{SD} obtained from ten MESFET samples before and after gamma radiation exposure with a total dose of 1x10^6 rad at T =300K are summarized in Table 1.

TABLE 1. Average value of normalized channel resistance for ten MESFETs before and after treatment with gamma radiation.

W, µm	$R_{SD}\cdot W/100$, Ω, non-irradiated	$R_{SD}\cdot W/100$, Ω, irradiated
100,200,300,500	354±11%	317±2%

A variation of the channel resistance values can be evaluated from the drain-gate characteristics of the devices. Before irradiation a variation of the drain current at the same voltages, U_{DS} and U_G, was in the order of 30%, while after irradiation the characteristics practically completely coincided and the threshold voltage increased from U_{th} = -(3.9÷5)V to U_{th} = -5.5 V. Temperature dependences of the channel's conductivity normalized to U_{DS} = 100 mV before and after irradiation are shown in Fig.1b on a logarithmic scale. At low temperatures both curves coincide and demonstrate exponential dependence with activation energy E_a = 0.064 eV. At temperatures higher then 100 K the conductivity increase decelerates and eventually a small decrease of conductivity is observed at higher temperatures (T > 270 K). Starting from $T \approx 100$ K, the curves for irradiated and non-irradiated samples do not coincide anymore, and the conductivity of the irradiated sample is slightly higher. This fact is in agreement with results of room-temperature conductivity measurements performed on a large set of samples.

The peculiarities of the steady-state characteristics of the SiC MESFETs are reflected in low-frequency noise spectra. The temperature dependence of the low-frequency current noise of the devices are presented in Fig.2a. As a noise parameter we used S_I/I^2 quantity, where S_I – spectral density of current noise and I – channel's current. To analyse generation-recombination (GR) components of noise we shall use a model developed in [5], where GR noise is considered to be dominant in the space charge region since the free carriers appearing are quickly dragged away from the depleted region. Using the results of this work and taking into account the effective mass value from [6] and the temperature dependence of the band gap from [7] it can be shown that in our case the time constant of the GR process is described by the equation:

$$\lg(\tau T^2) = -0.4343 \frac{E_{ti}}{kT} - \lg\left(3.95 \cdot 10^{15} \sigma_n \sqrt{3k/0.36 m_0}\right), \qquad (1)$$

where σ_n – electron trapping cross section, E_{ti} – energy level of GR centres, k – Boltzmann constant, T – absolute temperature, m_0 - free electron mass.

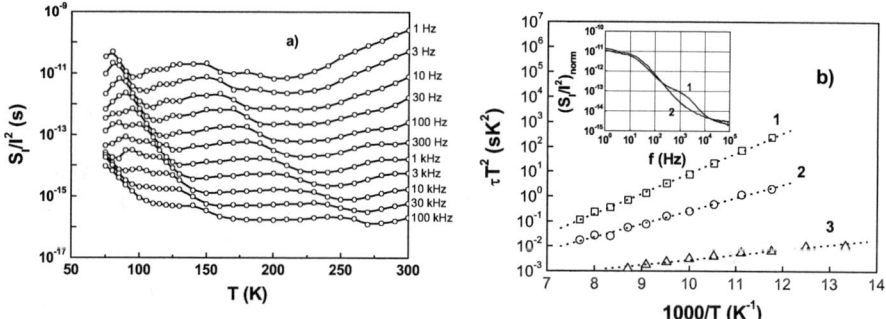

FIGURE 2. a) Temperature dependences of current noise spectra for non-irradiated device ($W = 100$ μm), measured at $U_G = -1.5$ V, $U_{SD} = 0.25$ V. b) Dependences of GR time constants on temperature for the three most pronounced GR processes. Inset: Normalized noise spectra of MESFET before (1) and after (2) small-dose gamma irradiation.

This equation allows us to evaluate the energies of GR traps, E_{ti}, and their cross sections, σ_n, by plotting experimental dependence $\tau T^2 = f(1/T)$. These dependences for the three most pronounced GR processes observed at temperatures $T = (75 \div 130)K$ are shown in Fig.2b. As we can see, the characteristic time constants of the different processes differ strongly and we shall analyse them as independent effects. The energies and cross sections obtained are summarized in Table 2.

TABLE 2. Parameters of GR centres, obtained from Fig.2b.

N of traps	E_t, eV	σ_n, cm^2
1	-0.16	$3.6 \cdot 10^{-13}$
2	-0.10	$1.6 \cdot 10^{-14}$
3	-0.04	$8.2 \cdot 10^{-16}$

Traps 2 and 3 (E= - 0.10 eV and E= - 0.04 eV) can be identified as donor nitrogen centres (N). Their energy levels are close to those observed in the literature [8]: E= - 0.092 eV – nitrogen in Si site and E= - 0.052 eV – nitrogen in C site. Trap 1 with energy E= - 0.16 eV could not be reliably identified. Presumably, it is a level of titanium (Ti) [8]. Another trap with a close energy level (around 0.15 eV) was reported in [9] and ascribed to a nitrogen-defect complex. The estimated cross section value σ_n= 5x10^{-15} cm^2, however, is two orders of magnitude lower than in our case.

Small-dose irradiation causes changes in the noise spectra of one of the components (inset of Fig.2b). For an irradiated sample one of the GR components observed at frequencies $f = (0.3 \div 10)$ kHz in the non-irradiated spectrum vanishes. At the same time, GR noise components connected with dopant levels are immutable. This is reflected in the conductivity characteristics of the device (Fig.1b), in particular in the coincidence of curves for irradiated and non-irradiated samples in the exponential part.

After gamma irradiation, the noise connected with trap 2 (nitrogen in Si site) is reduced by one order of magnitude. This means that the concentration of this type of defect decreases. Most likely this centre is metastable and breaks upon irradiation. Therefore, small-dose irradiation of SiC MESFETs leads to reorganization of the defect structure of SiC, in particular, to a relaxation of the metastable states. Apparently, this is a reason for the reduced variance of drain-gate characteristics and increase of the threshold voltage. The latter is caused by removing negatively charged centres located under the gate region of the device.

In conclusion, the investigation of low-frequency noise of SiC MESFETs revealed three local trap levels near the conduction band edge of SiC. Their energy positions and cross sections were evaluated and compared with values reported in the literature. Analysis of changes in noise behaviour caused by gamma irradiation leads to the conclusion that one of the centres, namely nitrogen in the Si site, is metastable and relaxes under small-dose irradiation. Such an ordering of non-controllable impurities enhances the stability of the devices and improves their steady-state characteristics, in particular the threshold voltage.

This work is supported by the Deutsche Forschungsgemeinschaft (Project KL 1342).

References

1. A. Agarwal, M. Das, S. Krishnaswami, J. Palmour, J. Richmond and R. Sei-Hyung, *Mat.Res.Soc.Symp.Proc.* **815**, J1.1 (2004).
2. K.K. Lee, T. Ohshima and H. Itoh, *IEEE Trans. on Nuclear Science* **50**, 194-200 (2003).
3. C. Brisset, O. Noblanc, C. Picard, F. Joffre and C. Brylinski, *IEEE Trans. on Nuclear Science* **47**, 598-603 (2000).
4. S.C. Binari, P.B. Klein and T.E. Kazior, *IEEE MTT-S International Microwave Symposium Digest* **3**, 1823-1826 (2002).
5. C.T. Sah, Theory of Low Frequency Generation Noise in Junction Gate FET, *Proceedings of IEEE* **52**, 795-814 (1964).
6. N.T.Son , W.M. Chen, O. Kordina, A.O. Konstantinov, B. Monemar, E. Janzen, D.M. Hofman, D. Volm, M. Drechsler. and B.K. Meyer. Electron effective masses in 4H SiC. *Appl. Phys. Lett.* **66**, 1074-1076 (1995).
7. Yu. Goldberg, M.E. Levinshtein and S.L. Rumyantsev in *Properties of Advanced SemiconductorMaterials GaN, AlN, SiC, BN, SiC, SiGe,* edited by Levinshtein M.E., Rumyantsev S.L., Shur M.S., John Wiley & Sons, Inc., New York, 2001, pp.93-148.
8. A.A. Lebedev, *Semiconductors* **33**, 2, 107-130 (1999).
9. S. Mitra, M.V.Rao, N. Papanicolaou, K.A. Jones, M. Derenge, O.W. Holland, R.D. Vispute and S.R. Wilson, *J.Appl.Phys.* **95**, 69-75 (2004).

Effects of oxygen-related traps in silicon on the generation-recombination noise

J.A. Jiménez Tejada, J.A. López Villanueva, A. Godoy,
J.E. Carceller, F. M. Gómez-Campos and S. Rodríguez-Bolívar

*Departamento de Electrónica y Tecnología de Computadores. Facultad de Ciencias.
Universidad de Granada. Avenida Fuentenueva s/n. 18071 Granada, Spain.
Phone: +34 958243386; Fax: +34 958243230; E-mail: tejada@ugr.es*

Abstract. This work shows the effects of oxygen related traps in silicon on the generation-recombination noise and presents a procedure to determine their capture cross-sections and densities. It compares the current noise spectral density measured in $p-n$ junctions fabricated on Czochralski-grown silicon (Cz-Si) with an analytical expression proposed by us. In those systems where two or more levels are present in the bandgap additional electrical measurements are necessary in order to discern which level is the origin of the noise. Multiple oxygen related traps can be found in the literature. A thorough study of their associated levels, and of different techniques employed to characterize them is made in this paper. We have found that one of these levels behaves both as a minority or majority trap, depending on the temperature. The parameters proposed for this level can link different results found in the literature.

Keywords: Generation-recombination noise, oxygen traps, $p-n$ junctions, Cz-Si
PACS: 72.20.Jv, 72.70.+m, 71.55-i

INTRODUCTION

In a previous work [1] we proposed an analytical expression for the current noise spectral density associated with carrier generation and recombination through a trap level E_T with concentration $N_T(x)$ in the space-charge region of a $p-n$ junction:

$$S_{\Delta I_{rec}} = 2q \int_{x_n}^{x_p} \frac{I_{eq}^L(n_s, p_s, c_n, c_p, x)}{1 + \omega^2 \tau_t^2} dx, \quad (1)$$

$$\tau_t = \frac{1}{c_n(n_1 + n_s) + c_p(p_1 + p_s)}, \quad (2)$$

where x_p and x_n are the limits of the depletion region and I_{eq}^L is a function that depends on n_s and p_s, the steady-state concentrations of electrons and holes, respectively, and on c_n and c_p, the capture coefficients for electrons and holes, respectively. τ_t is the trap time constant, and $n_1 = N_C exp[(E_T - E_C)/KT]$ and $p_1 = N_V exp[(E_V - E_T)/KT]$ are the standard Shockley-Read quantities.

The main goal of this paper is to use this expression as a way to determine capture cross-sections, $\sigma_{n,p}$, and densities of oxygen related traps in silicon. This expression has been compared with experimental low-frequency noise measurements [2], taken in $p-n$ junctions fabricated on Czochralski-grown silicon (Cz-Si). We choose this system

TABLE 1. Samples with their trap specifications

	$N_{T1}(cm^{-3})$*	$N_{T2}(cm^{-3})$†	$\sigma_{n1}(cm^2)$	$\sigma_{p1}(cm^2)$	$\sigma_{n2}(cm^2)$	$\sigma_{p2}(cm^2)$
a	0	$\sim 10^{12} \times F(x)$	—	—	10^{-23}	10^{-14}
b	$\sim 10^{12} \times F(x)$	$\sim 10^{12} \times F(x)$	10^{-8}	10^{-8}	10^{-23}	10^{-14}
c	$\sim 10^{12} \times F(x)$	$\sim 10^{12} \times F(x)$	10^{-23}	10^{-14}	10^{-13}	10^{-12}
d	$\sim 10^{12}$	$\sim 10^{12}$	10^{-23}	10^{-14}	10^{-13}	10^{-12}

* $E_C - E_{T1} = 0.17$ eV
† $E_C - E_{T2} = 0.43$ eV

for two reasons. First, because multiple oxygen related traps are present, making the characterization method a more difficult task. Second, because different values of the parameters of these traps can be found in the literature, depending on the voltage and temperature conditions of the characterization technique. In order to interpret different electrical measurements in the Cz:Si system, traps with different nature can be found in the literature: capacitance transients at low temperatures have been explained by electron traps [3, 4], generation-recombination (g-r) noise in forward biased junctions at room temperature by hole traps with $\sigma_p \gg \sigma_n$ [2], leakage current transients by hole traps [5, 6], and steady leakage current by the migration of electrons from the valence band to the conduction band via a shallow Coulomb center, supported by a combined phonon-assisted tunneling and Poole-Frenkel mechanism [7].

EXTRACTION OF TRAP PARAMETERS

We focus our attention on two of the levels present in Cz-Si that are most commonly detected by different authors: $E_C - E_{T1} = 0.17$ eV and $E_C - E_{T2} = 0.43$ eV. Analyzing multiple traps means that the number of parameters to be determined increases. Moreover, when two or more levels are simultaneously present in the semiconductor any of them could be the origin of the measured current noise. Actually, we have reproduced experimental measurements of the current noise density in forward-biased $n^+ - p$ junctions with different contents of oxygen [2], assuming that the origin of the noise lies in either of the two levels these authors detected by DLTS. Fig. 1 shows some of our comparisons with experimental data taken in forward biased $n^+ - p$ junctions with high content of oxygen. The values of the capture cross-sections and concentrations for the levels used in these fittings are summarized in Table 1. The doping profile used in all these cases is shown in solid line in Fig. 2a. This profile allows us to calculate the actual distribution of electrons and holes necessary to evaluate (1). It was calculated by SUPREM using the fabrication data [2, 3, 4]. The resulting profile corresponds to an $n^+ - p$ junction. We could have adopted the usual approximation of an abrupt, asymmetrical junction. However, as we show below, the main electrical activity of the junction is confined in the n^+-side. The trap profile used in samples a-c is also depicted in this figure (dashed line). A non uniform profile is considered, reflecting an increase towards an oxygen-precipitated region inside the wafer ($x < 0$) [4]. Sample d considers a uniform trap profile.

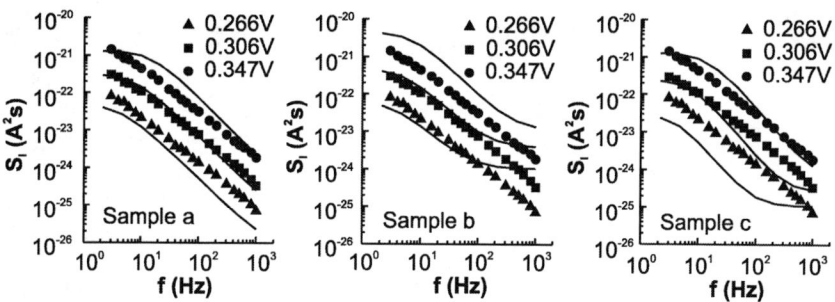

FIGURE 1. Current noise spectral density in a forward-biased $n^+ - p$ junction with high oxygen content. Experimental results from Ref. [2] are shown with symbols and our calculations with solid lines. In samples a and b the origin of the noise is E_{T2}; in sample c the noise comes from E_{T1}.

Samples a and b assume that the noise stems from level E_{T2}, and samples c and d from level E_{T1}. To prove this we have represented in Fig. 2b the characteristic time for these two pairs of samples: $\tau(E_{T1})$ in samples a-b and $\tau(E_{T2})$ in samples c-d (Fig. 2b). Both contribute with the same values and at the same regions of the junction, confirming the similarity among the spectra of Fig. 1. No differences are found because, in both cases, $\tau \approx 1/(c_p(p_s + p_1)) \approx 1/(c_p p_s)$. As can be seen in the representation of the hole concentration in Fig. 2b, a non-negligible tail of holes is present in the n^+-region, making $p_s > p_1$. The frequency range represented in Fig. 1 corresponds to characteristic-time values between $10^{-4} - 10^{-2}$ s, a range wholly included in the n^+-side. The small differences among these spectra lie on the contribution to the current-noise density of E_{T1} in sample b and of E_{T2} in sample c, whose characteristic times are not represented in Fig. 2b. The capture cross-sections of these former levels are not arbitrary. They have been chosen in order to fit current-voltage curves measured in the same set of devices [2], and biased in a region where the g-r component is high (Fig. 2c). This fitting help us to distinguish which trap is responsible for the g-r noise, and which one is responsible for the generation current. Samples a-c are represented in Fig 2c. A constant value for the trap concentration was also used (sample d). A value of 7×10^{12} cm^{-3} was necessary to obtain a fitting with noise measurements similar to sample c in Fig. 1c. However, in order to reproduce $I - V$ measurements the value of the trap concentration should be enhanced by a factor of 4. This means that the regions that contribute to the g-r current are farther from the diode surface than the region that contributes to the g-r noise. This validates the fact of considering an increasing trap profile towards the p-side. As can be deduced from Figs. 1 and 2c, sample c is the one that simultaneously fits current-noise and current measurements.

Once the values of the capture cross sections of both levels are determined they can be compared to other authors' results. Nevertheless, with these results in mind, and seeking to find the link between all of them, the following model is proposed for E_{T1}: a coulomb trap with activation energy 0.17 eV whose transitions with the conduction band are temperature and electric field dependent according to the Poole-Frenkel theory. The rest of parameters are the ones defined for sample c.

At first sight, this model might be considered one more among the different results

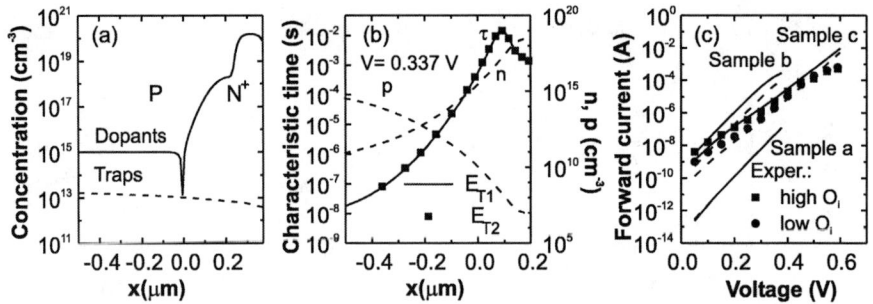

FIGURE 2. a) Doping and trap profile in the $n^+ - p$ junction considered in this work. b) Distribution of the characteristic time for levels E_{T1} in samples *a-b* and E_{T2} in samples *c-d*, and electron and hole densities along the junction. c) Experimental (symbols) and fitted (solid and dashed lines) diode current versus forward voltage for $n^+ - p$ junctions with different starting interstitial oxygen content.

described in the literature. However, to analyze the effects of this model, the transient response of sample *c* to voltage pulses at different temperatures was numerically calculated [8]. The objective was to make a complete electrical study of the junction with these levels, under different voltage and temperature conditions, and to reproduce the response of a Cz-Si diode to other characterization methods, such as DLTS. The evolution of the occupation factors of levels E_{T1} and E_{T2} and the capacitance of the junction during the transient response were studied, and both led to the conclusion that the E_{T1} level can act as a minority trap at low temperatures and high electric fields, and as a majority trap at room temperature and low electric fields, thus connecting different studies found in the literature. Another conclusion is that care must be taken when the results of an isolated technique are analyzed apart. The combination of several techniques is necessary when the complexity of a system increases.

ACKNOWLEDGMENTS

This work has been carried out within the framework of research project No. TEC2004-02612/MIC supported by the Ministerio de Educación y Ciencia and FEDER.

REFERENCES

1. J. A. Jiménez Tejada, A. Godoy, A. Palma, and P. Cartujo, *J. Appl. Phys.*, **90**, 3998–4006 (2001).
2. F. C. Hou, G. Bosman, E. Simoen, J. Vanhellemont, and C. Claeys, *IEEE Trans. Electron Devices*, **45**, 2528–2536 (1998).
3. J. Vanhellemont, E. Simoen, A. Kaniava, M. Libezny, and C. Claeys, *J. Appl. Phys.*, **77**, 5669–5676 (1995).
4. A. Kaniava, J. Vanhellemont, E. Simoen, and C. Claeys, *Semicond. Sci. Technol.*, **9**, 1474–1479 (1994).
5. Y. Murakami, and T. Shingyouji, *Appl. Phys. Lett.*, **65**, 2591–2593 (1994).
6. Y. Murakami, and T. Shingyouji, *J. Appl. Phys.*, **75**, 3548–3552 (1994).
7. M. Tsuchiaki, H. Fujimori, T. Iinuma, and A. Kawasaki, *J. Appl. Phys.*, **85**, 8255–8266 (1999).
8. J. Jiménez Tejada, A. Godoy, J. Carceller, and J. López Villanueva, *J. Appl. Phys.*, **95**, 561–570 (2004).

Noise and I-V Characteristic as Characterization Tools for GaSb based Laser Diodes

Z. Chobola*, J. Vaněk*, J. Kazelle*, E. Hulicius**, T. Šimeček**

*Department of Physics, Faculty of Civil Engineering, University of Technology Brno
**Institute of Physics, Academy of Sciences, Czech Republic, Praque

Abstract. Transport and noise characteristic of forward biased 2.3 µm CW GaSb laser diodes were measured in order to evaluate new technology. From the measurement results it follows that noise spectral density related to defects is of 1/f type and its magnitude was found to be proportional to the square of DC forward current at low injection levels.

Keywords: Noise, Laser diodes

INTRODUCION

Noise has been used for a long time as a diagnostic tool in device research [1]

Low frequency noise is a sensitive tool for degradation phenomena like electro-migration and short breakdown [2]. All types of noise, thermal, shot, generation-recombination and 1/f noise play different role in reliability analysis. The correlation between noise in a device, its reliability and the question why conduction noise, especially 1/f noise, is a quality indicator for devices is indicated in [3, 4]. Opportunities and limitations of the use of low/frequency noise as a diagnostic tool for device quality is discussed in Wandame [5].

We have studded a set of eight samples:97/21-2. 97/21-6, 97/21-7, 97/23-6, 97/23-8, 97/25-9, 97/25-10 and 97/25-11.

EXPERIMENTAL RESULTS AND DISCUSSION

Fig. 1 shows I-V characteristic of forward-biased samples Nos.97/25-9 and 97/25-11. We can see excess current around $U_F = 0,1 - 0,6$ V. The exponent β in $I = I_0\, e^{\beta U}$ plot equal between $\beta = 17,9$ V^{-1} to $\beta = 19,3$ V^{-1}.

The I-V characteristic give evidence of poor contact quality, with the contact resistance ranging from $R_S = 2.3\ \Omega$ for up to $R_S = 5.4\ \Omega$.

Figs. 2 and 3 shows the noise voltage power spectral density versus DC voltage plots. The noise voltage was measured across load resistance $R_L = 1$ kΩ, 10 kΩ and 100 kΩ respectively. The pass band central frequency was 1kHz, the band-width being equal to 20 Hz. A marked excess current component can be observed at higher

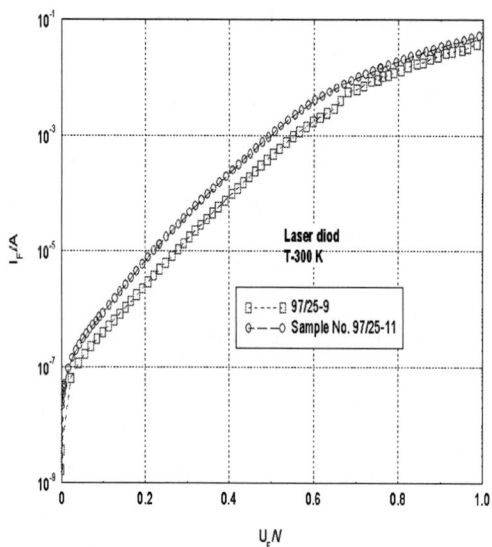

FIGURE 1. I-V characteristic for laser diodes Nos. 97/25-9 and 97/25-11.

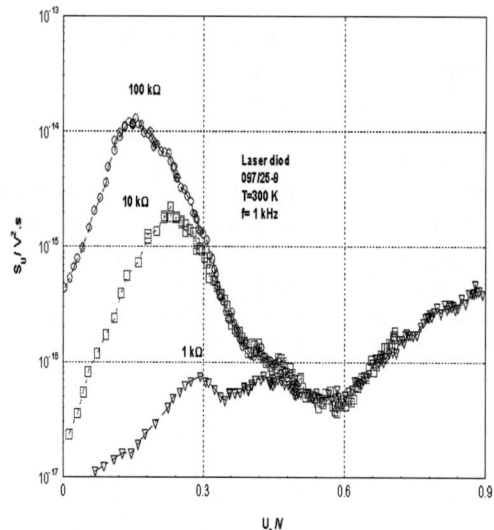

FIGURE 2. The noise spectral density as a function of forward voltage.

voltages, reaching maximum at the power match point, where the PN junction dynamic resistance equals the load resistance. This peak is shifting toward lower DC voltage if the load resistance is increased. When the load resistance is $R_L = 1\ k\Omega$, we can see the second maximum. It may be deduced that at least two separate structure defect related noise sources are present in the PN junction region.

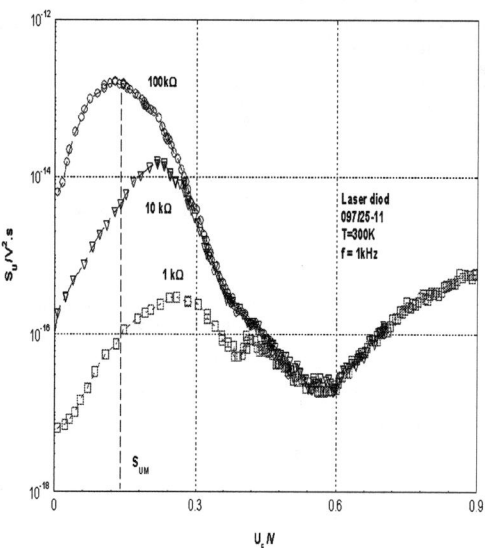

FIGURE 3. The noise spectral density as a function of forward voltage.

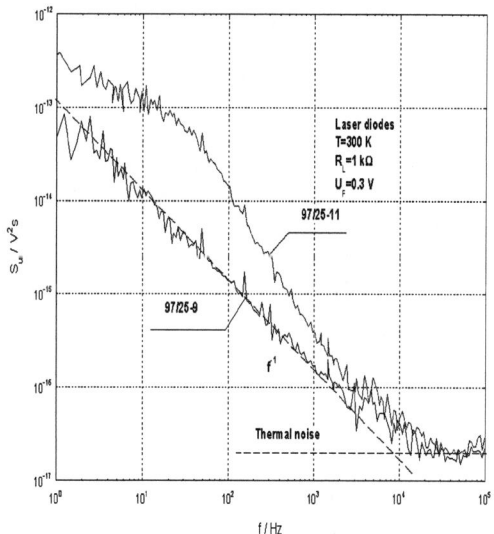

FIGURE 4. The noise spectral density versus frequency.

At forward voltages exceeding $U_F = 0.6$ V, all samples show an increase in the excess noise component, which is characteristic of imperfect contacts, where resistance ranges from 2 Ω to 6 Ω.

The greatest excess noise $S_{UM} = 2.10^{-13}$ V²s we can see for sample No. 97/25-11 when $R_L = 100$ kΩ and voltage $U_F = 0.15$ V

Fig. 4 shows the noise voltage power spectral density versus frequency plots for the laser diode No. 97/25-9 and 97/25-11.7675. The noise voltage is measured across the load resistance $R_L = 1$ Ω, the DC forward bias voltage being $U_F = 0.3$ V, i.e. in the power match region. It may be seen that the excess noise component is of $1/f^\alpha$ type when α is very close to unity. This noise component is masked be thermal noise at higher frequencies.

CONCLUSION

Our experimental studies of GaSb laser diodes have shown that these diodes are of rather "inferior quality", which concerns both the PN junctions and the contacts as compared to wide-gap GaAs based lasers. The I-V characteristics have marked excess current component occurring at DC forward voltages below $U_F = 0.1$ V to 0.6 V. Also the contact resistance $R_S = 2$ Ω to 6 Ω should be improved.

These assumptions concerning the imperfections being present in the samples under study have been fully confirmed by our noise measurements, indicating $1/f^\alpha$ - type excess noise components for both the PN junction and the contact region. Moreover samples contains at least two different types of defects, which were manifested by two separate local maxima in the spectral density versus frequency plots.

ACKNOWLEDGMENTS

This work is a part of the research programs of the AS CR: Scientific Goal No. AVOZ 010 914 and Research Project No. K 1010104 and is also supported by EC Project IST-2004-35178 GLADIS.

REFERENCES

1. Gusta, M.S., *Applications of Electrical Noise*, Proceedings od IEEE, vol. 63, no 7, 1975, pp. 996 – 1010.
2. Cioti, C., and Nevi, B., *Low – frequency noise measurements as a characterisation tool for degradation phenomena in solid-state devices*. Journal of Physics – applied physics, vol. 33, no. 21, 2000, pp. R199 – R216.
3. Vandamme, L. K. L., *Noise as a Diagnostic Tool for Quality and Reliability of Electronic Devices*, IEEE Transaction on Electron Devices, vol. 41, no. 11, 1994, pp. 2176 – 2187, Now.
4. Chobola, Z., *Noise as a tool for non-destructive testing of single-crystal silicon solar-cells*, Microelectronics Reliability, vol. 41, no. 12, 2001, pp. 1947 – 1952, Dec.
5. Vandamme, L. K. L., *Opportunities and limitations to use low-frequency noise as a diagnostic tool for device quality*, Proceedings of the 17th International conference ICNF 2003, 2003, pp. 735 – 748, Prague.

Foil Capacitors Reliability Prediction Based On Fluctuation And Non-Linear Phenomena

Lech Hasse[1], Ludwik Spiralski[1,2], Janusz Turczyński[3]

[1] *Gdańsk University of Technology, Gdańsk, Poland;* [2] *Gdynia Maritime University, Poland*
[3] *Industrial Institute of Electronics, Warsaw, Poland*

Abstract. Non-linearity and noise properties of a capacitor have been analysed to established criteria for selection of interference suppressor capacitors for different endurance (sustainability) and reliability groups. It can improve the process of quality estimation of high reliability capacitors. Selected experimental results of measurements for capacitors produced by MIFLEX (Poland) have been presented. The implementation of techniques for non-linearity and noise measurements in the system for production testing of high reliability interference suppressor capacitors is described.

Keywords: capacitor noise, third harmonic measurement, technical diagnostics.
PACS: 84.37.+q; 84.32.Tt; 05.40.Ca; 81.70.-q.

INTRODUCTION

One can distinguish two approaches to reliability problems in electronics: traditional (measurement results or times to failure are the basis to statistical calculations of reliability indicators) and contemporary one (parameters are measured before a failure enabling to built-in reliability to elements by means of the proper modification of the technological process.

Quality and endurance tests are required for all interference suppressor capacitors during their production. Such tests are carried out mainly using destructive and long-lasting environmental (endurance) trials for selected samples. Therefore a simple, non-destructive and fast approach is needed for this purposes.

The problem can be solved by implementation in the production testing system some direct nonlinearity or/and noise measurement tasks. A dependence between these two phenomena was identified in the case of other electronic elements, and furthermore an explanation of the observed relation on, for example, geometrical factors, has been found to apply with the similar success to the both magnitudes. The choice of third harmonic index (THI), noise parameter (method and electrical circumstances of their measurement) and rules of classification into reliability group gives a possibility to predict a reliability of tested capacitors individually.

Increased level of THI in capacitors is mainly caused by instability of contacts, improper adhesion, electrodes and dielectric heterogeneity, weak contact between an electrode and a terminal, ferric oxide existence in dielectric particles, slow processes of insulating layer degradation or mechanical instability of a capacitor.

MEASUREMENT OF NONLINEARITY AND FLUCTUATIONS

Interference suppression capacitors WXPC X2 **0.1 µF**, WXPC X2 **0.68 µF**, WXP X2 **0.22 µF** and **0.33 µF** 275 V~ produced by MIFLEX [1] were investigated. They have been made using a 7.5 µm thickness and 13.5 or 21 mm width polypropylene or polyester foil with a 2 mm margin. The dielectric constant and resistivity of metallic layer are equal ε = 2.2 and r = 7.5 $\Omega/\square$ ± 30%, respectively. The foil is souped up near the edges (3 mm width) by means of metallization layer up to 4 $\Omega/\square$.

The basic causes of capacitor failures are: improper silver adhesion (flicker of capacitance – 1/f noise), inhomogeneities and microcracks of foil (1/f noise), electrodes and dielectric heterogeneity, higher temperature of plates, improper terminal construction (impulse, burst noise), silver migration, dielectric aging.

It was established the correlation between THI and long term reliability (in foil capacitors it can be related to gas bubbles as defects). The THI is proportional to the extent of elementary nonlinearity. It comprises a built-in component (its level is to be considered as a nonlinearity mean value), excessive component (due to high contact resistance of any junction affecting the U-I curve, physical properties of the base material, defects and inhomogeneities in the material structure or interaction with the environment; equal zero at no defect present) and capacitor instability in time.

The main modules of the instrument for THI measurement can be used to measure also the voltage fluctuations on the load resistance $R_{Li}, i = 1,2$ when the tested capacitor C_x is stimulated by a large sine wave signal E_s (Fig. 1). It is specially useful

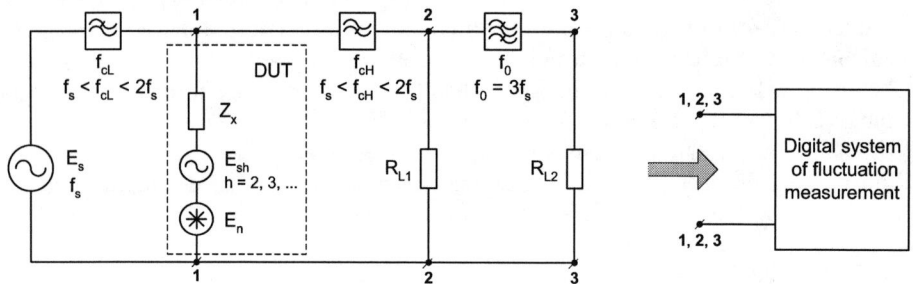

FIGURE 1. The simplified diagram for complex THI and fluctuation measurements of C_x

instead of $1/\Delta f$ noise measurement when the appropriate resolution of noise power spectral density is extremely hard for obtaining at large amplitudes of E_s. Due to noise and nonlinearities the electromotive forces of the harmonics E_{sh} and noise E_n are generated when the signal current is flowing through a capacitor. The E_{sh} can be measured on the load R_{L1} and the third harmonic f_0 ($3f_s$) at the output of the selective filter. The correlation between the stimulus and response signals can be analysed by the measurement of the coherence function of the voltage fluctuations on the components C_x, R_{L1} and R_{L2}. A residual inherent system nonlinearity should be no higher than –140 dB.

It can be proved that TH is dependent on the signal rms value by the third order and on the second order of the foil thickness. Therefore the dependence of the foil thickness is a good tool to find capacitors with weak spots. In case of polystyrene capacitors the utmost stability is required and it is fulfilled at the THI measurement.

The THI value of capacitors (stemming from the same batch) with nominally the same capacitance should have the Gaussian distribution around the mean value. However, usually a few of components exhibit a higher level of THI due to defects or deviations in material composition. Exposing the batch to an accelerated life trial, the components having a higher value of THI will also be prone to exhibit inferior reliability. The actual value of THI in a good component should be found experimentally. The mean values of $U_{3h}[\mu V]$ versus applied $U_1[V]$ for several batches of MIFLEX capacitor samples (100 pieces each) and corresponding mean values is shown in Fig. 2. Third harmonic distortion is proportional to the mth order of the applied voltage U_1. Experimentally measured mean values of the exponent m are equal 1.75 to 2.1 for measured capacitors.

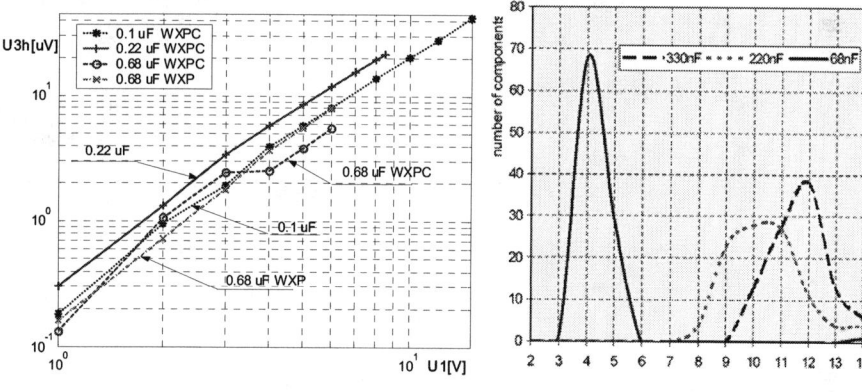

FIGURE 2. $U_{3h}[\mu V]$ versus applied $U_1[V]$ for some type of MIFLEX capacitors

FIGURE 3. Statistical distribution of the THI for selected single batches of foil capacitors

The actual distribution of the third harmonic level within the selected populations (stemming from the same batch samples 100 pieces each) is shown in Fig. 3.

The shape of the TH distribution is very important as a basis for the criterion of capacitors classification on the reliability classes. The analysis of the achieved results confirmed the relation that if the value of capacitance increases the mean value of TH distribution for every batch also increases.

Standard endurance test are performed in the test chamber. Class X capacitors are submitted to an endurance test of 1000h at upper category temperature and with 1.25 Voltage Rating (once every hour the voltage is increased to 1000 V_{rms} for 0.1s).

Partial discharges (PD) was measured using the bridge method in the TETTEX 9120 system for effective noise suppression. For all capacitors 0.68 µF WXPC the striking voltage was equal 230–255 V and was exactly 10V higher than the extinction voltage. More immune to electromagnetic noise is PD detection based on acoustic emission.

SYSTEM FOR PRODUCTION TESTING

To assure a high flexibility of manufacturing and cooperation with a general management in the quality domain, the system for testing capacitors [2] should fulfil high functional requirements with possibilities of its easy reconfiguration (Fig.4) enabling to predict reliability of an every tested capacitor individually.

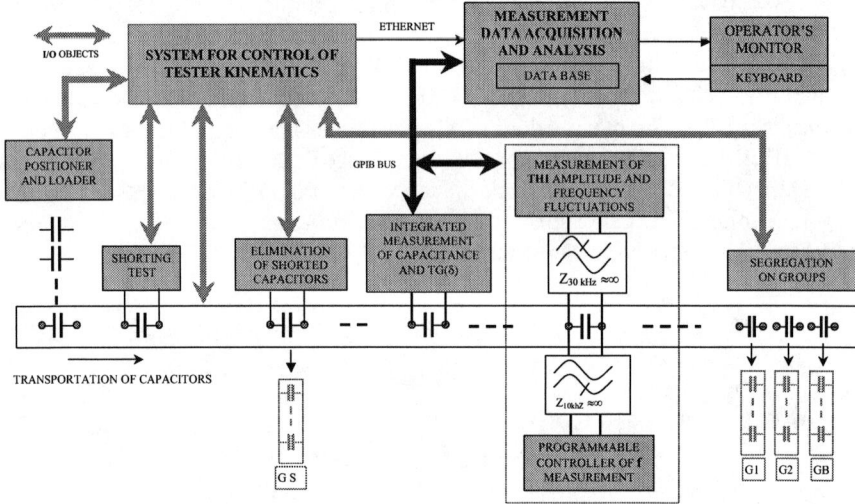

FIGURE 4. General block diagram of the capacitor testing system during manufacturing

CONCLUSIONS

The idea of quality and reliability control by non-linearity and noise testing has been already proved for other electronic components. The implementation of techniques for nonlinearity and noise measurements in the system for production testing of high reliability interference suppressor capacitors gives a possibility of individual testing of an every produced element for accepted criteria of classification.

ACKNOWLEDGMENTS

This work was partially supported by the Polish State Committee for Scientific Research (KBN) project No. 3 TC10 C 041 27.

REFERENCES

1. http://www.miflex.com.pl/eng/capacitors/interference.htm
2. Hasse, L., Rogala, K., Spiralski, L. and Turczyński, J.,*Factors determining the production testing of high reliability interference suppressor capacitors* J., in Proc. XIII IMEKO TC-4 Int. Symp. On Measurements for Research and Industry Applications, Athens 2004, ed. E. Kayafas and V. Loumos, pp. 116-121.

Morphology of Nanostructured Semiconductors Studied Using Atomic Force Microscopy Combined with Stochastic Signal Spectroscopy

Vitali Parkhutik, Yuri Makushok, Eugenia Matveyeva

R&D Centre "Materials and Technologies of Microfabrication", Technical University of Valencia, Cami de Vera s/n 46022 Valencia Spain

Abstract. Nano-structured semiconductors obtained by porosifying monocrystalline materials in electrochemical bath are the subject of intensive research due to their specific quantum-mechanical properties and emerging applications. To characterize their morphology, electron microscopy (SEM and TEM) and scanning probe microscopy (STM and AFM) have been used, providing qualitative rather than quantitative knowledge on the porosity, pore wall size and pore alignment.

We have applied space series analysis which has been developed within the framework of Stochastic Signal Spectroscopy (3S) to the digitalized images of surface and cross sections of nano-structured inorganic (porous silicon) and organic (polyaniline) semiconductors. We show that SPM/3S study yields quantitative information about alignment of the pores, depth-dependent pore structure and other issues which were not considered before adequately to their importance.

Keywords: Porous silicon, polyaniline, AFM, roughness analysis
PACS: 05.40.Ca, 82.47.–a, 68.37.–d

EXPERIMENTAL SET-UP AND RESULTS

In this work we have studied two types of nano-structured semiconductors: one inorganic (porous silicon) and organic (polyaniline) semiconductors. Porous silicon is a synthetic material which is formed by etching (electrochemical or chemical) of pores in monocrystalline Si wafers. Porous silicon (hereafter pSi) is now a subject of very intensive research and development in view of its applications in electronics, optics, biology and medicine [1].

We have formed thick (70-110 μm) porous silicon layers of medium porosity by polarizing p+ silicon (0.01 Ω·cm) in 24%HF water/ethanol solution at 50 mA/cm^2 during 20 min. Then the samples were cleaved and their cross section was sandwiched between two flat glass wafers to form a sample with area and flatness sufficient for safe positioning of tip of AFM analyzer (Nanoscope-3) at predetermined areas of the pSi cross section.

At every sample two characteristic areas (each 10x10 μm^2) were analyzed. First area was that corresponding to the internal interface of pSi with monocrystalline Si substrates (Fig.1, left), while second area was chosen at the centre of the pSi layer (Fig.2, left).

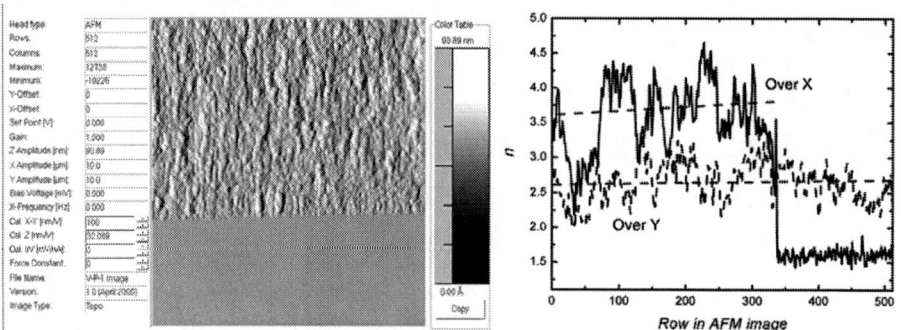

FIGURE 1. AFM image of the interface between porous silicon layer and underlying Si substrate (left) and asymmetry coefficient n calculated for OX and OY scan directions (right)

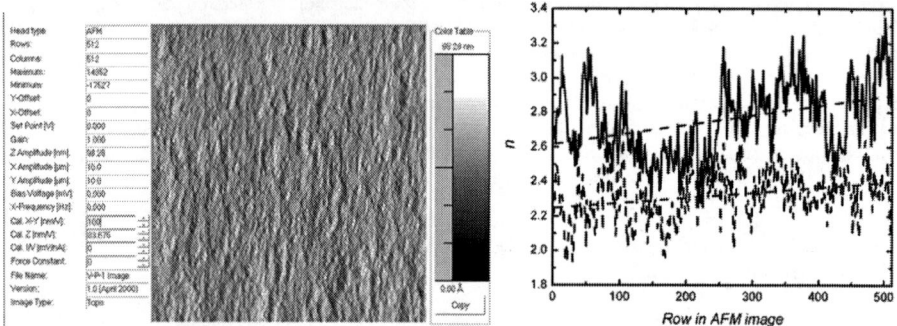

FIGURE 2. AFM image of the middle part of porous silicon layer shown in Fig.1 (left) and asymmetry coefficient n calculated for OX and OY scan directions (right)

To characterize the features of roughness profiles and distinguish between cases shown in Fig. 1 and 2, the AFM files were processed using software utilities which have been developed at the Technical University of Valencia. Using this software the original Nanoscope-3 files were transformed into a table of roughness data (512 rows, each consisting of 512 data points) and further processed using a package of statistical analysis developed by the authors.

For each line of this table we have calculated power spectra of autocorrelation function (ACF) in order to reveal the existence of correlation links in a sequence of stochastic data [2]. Autocorrelation function (ACF) of stochastic signal $V(x)$ is determined by equation

$$\psi(\chi) = \frac{1}{\sigma^2} \left\langle (V(x) - \langle V \rangle)(V(x - \chi) - \langle V \rangle) \right\rangle \tag{1}$$

where χ is the space shift (if the coordinate is variable). For practical purposes instead of ACF its Fourrier transform (power spectrum) is used

$$S(f) = \frac{1}{2\pi} \int_{-\infty}^{\infty} \exp(i 2\pi f \chi) \cdot \psi(\chi) d\chi \tag{2}$$

Power spectrum indicates with what amplitude different frequencies contribute into stochastic events. In many cases function $S(f)$ can be approximated by hyperbolic law $S(f) \approx f^{-n}$ with power coefficient $n \geq 1$. When $n = 1$, so-called flicker-noise takes place. The higher is the value of n, less correlated are the events in stochastic series [2]. After calculating power spectra according to Eq.(2) for each row of AFM image matrix this function was further approximated with a function $S(f) \approx f^{-n}$ to obtain the value of the parameter n.

Processing of the image of Fig.1 using noise analysis (calculating power spectra of autocorrelation function and difference moments of second order) for directions OX (parallel to the pores) and OY (perpendicular to them) allows to show that there exists an asymmetry of noise parameters as it is shown on Fig.1 (right) and Fig.2 (right). For example non-linearity parameter n is 3.6 for OX direction and only 2.6 for OY one for the image shown on Fig.1, thus indicating specifics of the mechanism of the pore growth in silicon samples of different doping level and crystalline orientation. The difference in values of n_{OX} and n_{OY} means that the growth of the pore is more correlated in the direction of their propagation then the growth of adjacent pores.

Other interesting feature revealed by AFM analysis is depth-dependent pore structure in porous silicon: the cross section of the sample at the middle of its thickness (Fig.2) shows rather different noise parameters ($n_{OX}=2.7$, $n_{OY}=2.3$) than that close to the interface with Si layer (Fig.1) ($n_{OX}=3.5$, $n_{OY}=2.6$). First attempt of using this approach was done while analyzing the surface of porous silicon samples with a lateral non-homogeneity of porosity [3]. Here we expand this analysis to cross-sectional mode of the AFM analysis, when the roughness of cleaved pSi is analyzed rather than the roughness of pSi surface.

Another type of samples were thin films of polyaniline (PANI) deposited onto glass substrates. The experimental procedure used to deposit the PANI films was described earlier in [5]. Characteristic feature of these films is their intrinsic crystallinity introduced by polymerization of PANI in the presence of hydro-quinone [4] - the ordering of the polymer fiber s is clearly seen in Fig.3, left. Higher resolutions images of these samples (Fig.4, left) shows the ordering of the polymer molecules extends to the level of several tens of nanometers.

We have applied the described method to the analysis of surface topography PANI samples. AFM image of such sample is shown in Fig.3 and in Fig.4 (with higher resolution). Statistical analysis of the image shown in Fig.3 (left) indicates a pronounced asymmetry of the values of the coefficient n ($n_{OX}=2.2$, $n_{OY}=0.7$ as follows from Fig.3, right). It is quite natural result provided apparent anisotropy of the roughness profile.

While analysing the smaller area of the same sample (Fig.4, left) we can see that the sample has two different zones: one apparently ordered with characteristic width of ordered fibres of about 50 nm and another rather disordered. While performing the 3S analysis on this image we obtain apparently different results for these two regions: within a cluster of ordered PANI fibers we obtain strong asymmetry ($n_{OX}=3.0$, $n_{OY}=2.0$) while disordered area yields ($n_{OX}=n_{OY}=1.9$).

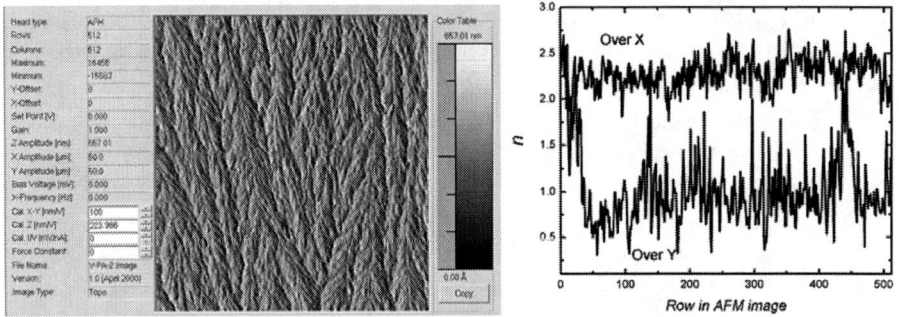

FIGURE 3. AFM image (50x50 μm) of the surface of crystalline PANI sample (left) and asymmetry coefficient n calculated OX and OY scan directions (right)

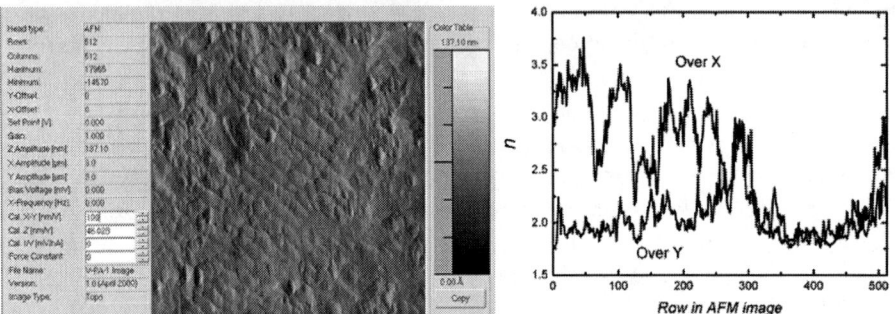

FIGURE 4. AFM image (3x3 μm) of the surface of crystalline PANI sample (the same as in Fig.3) (left) and asymmetry coefficient n calculated OX and OY scan directions (right)

In order to be sure that the observed anisotropy of the asymmetry coefficient n is not the artefact of the AFM analyser (due to possible asymmetry of displacement of piezo-crystal used for scanning of the surface) we have performed similar study of the surface of pSi sample with apparently symmetrical distribution of the pores at the surface. The result (not shown here) have confirmed the symmetry of the image in what concerns the value of n.

In sum, statistical analysis of AFM images acquired on cross-section and surface of pSi samples shows its suitability for obtaining quantitative information on the pore growth mechanisms, particularly to resolve depth-dependences of pore morphology, anisotropy of pore shape, etc. This is new and promising tool in studies of morphology of porous semiconductors.

REFERENCES

1. L.T.Canham, *Properties of Porous Silicon*, INSPEC, London, 1997.
2. R.Carmona, W.Young, *Practical Time-Frecuency Analysis*, Academic Press, N-Y., 1998.
3. V.Parkhutik, B.Collins, M.Saylor, et al. *Phys.Stat.Solidi* (a), **197**, 88(2003).
4. V.Parkhutik, E.Matveyeva, *J.Phys. Chem. B*, 102, 1549-1555(1998).

MODELING

Modeling and characterization of noise in 90-nm RF CMOS technology

A.J. Scholten, L.F. Tiemeijer, A.T.A. Zegers-van Duijnhoven, R.J. Havens, R. de Kort, R. van Langevelde, D.B.M. Klaassen*, W. Jeamsaksiri[†] and R.M.D.A. Velghe**

Philips Research Laboratories, Prof. Holstlaan 4, 5656AA, Eindhoven, The Netherlands
[†]*IMEC, Kapeldreef 75, B-3001, Leuven, Belgium*
**Philips Research Leuven, Kapeldreef 75, B-3001, Leuven, Belgium*

INTRODUCTION

Recently, the suitability of 90-nm CMOS technology for RF circuit design has been demonstrated at the device level [1, 2, 3], as well as at the circuit level [4, 5] In order to facilitate RF CMOS circuit design, the availability of RF CMOS compact models is essential. Noise is one of the key ingredients of these compact models. Here, accurate modeling of both low- and high-frequency noise are on the designers' wish list.

In this paper, we will discuss the modelling of both low- and high-frequency noise using new data taken on a 90-nm RF CMOS technology. This technology features an equivalent oxide thickness of ~ 2 nm and a minimum drawn gate length of L=100 nm. Note that this corresponds to an actual polysilicon gate length of roughly 80 nm.

1/F NOISE

Low-frequency noise measurements in the frequency range from 100 Hz to 100 kHz were carried out on-wafer with a BTA 9812A noise measurement system. Several n- and p-channel geometries have been investigated. For a number of them, we also investigated the spread of the measured noise across the wafer.

In most compact MOS models [6, 7], the so-called unified 1/f noise model [8] has been adopted, which attributes the noise to trapping/de-trapping of charge carriers, causing both number- and mobility fluctuations in the conducting channel. As shown in Fig. 1, this model is indeed capable of describing the measured low-frequency noise as a function of bias and geometry, both for NMOS and PMOS.

For small-area devices, the shape of the low-frequency noise spectra deviates strongly from the $1/f$-like behavior commonly observed in larger-area devices, see Fig. 2. Instead, the spectra show structures that can be identified as Lorentzian noise spectra. Moreover, the noise spectra of these devices differs strongly from one device to another. Nevertheless, as shown in Fig. 2, the sum of a large number of such devices takes the $1/f$-like shape that we know form larger-area devices. Also the magnitude of the average, when multiplied with the effective device area A_{eff}, compares well with mea-

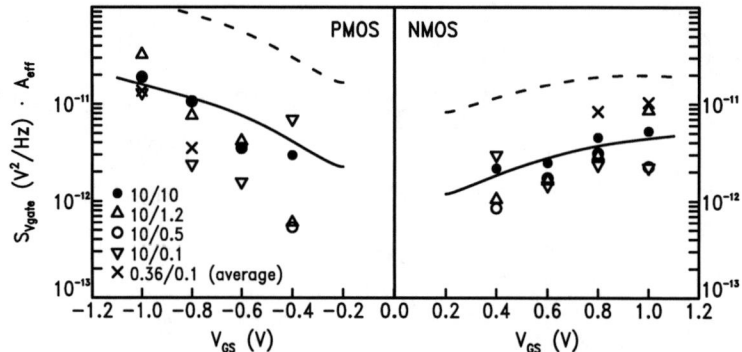

FIGURE 1. Input-referred voltage noise spectral density at $f = 100$ Hz, multiplied by effective area A_{eff}, as a function of gate-source voltage for PMOS (*left*) and NMOS (*right*) devices. Different markers represent measurements on different geometries, as indicated in the figure. The drain-source bias is 1 V and -1 V for NMOS and PMOS respectively. The solid lines are model calculations for the 10/10 μm devices. Model curves for the other geometries are very similar and have been omitted for clarity's sake.

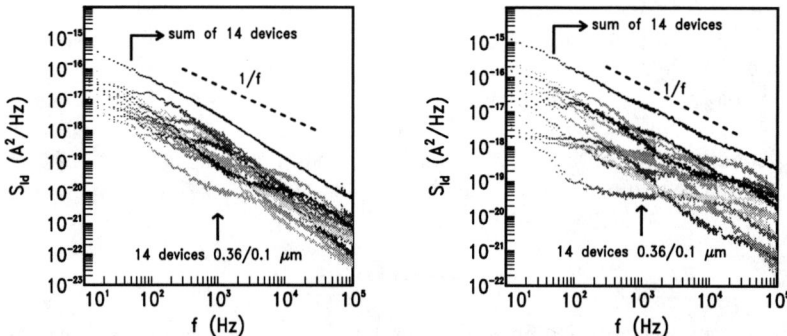

FIGURE 2. Low-frequency drain current noise spectra for 14 different 0.36/0.1 μm devices located on different positions on the wafer (grey lines), for PMOS (*left*) and NMOS (*right*). The bias condition is $V_{\text{GS}} = V_{\text{GS}} = 1$ V for NMOS and -1 V for PMOS. The sum of these spectra is represented by the solid line and shows $1/f$-like behavior.

surements on larger geometries (crosses in Fig. 1). This result is important because (*i*) it shows that it still makes sense to describe the (average) spectrum of these small-area devices as $1/f$ or $1/f^\alpha$, and (*ii*) it confirms one of the basic assumptions of the unified $1/f$ noise model, i.e. that the $1/f$-like noise in MOSFETs is built up from Lorentzian spectra associated with individual traps. Note that similar results have been obtained earlier on other CMOS technology nodes, see Refs. [9, 10].

In the unified $1/f$ noise model, the parameter n_t is the oxide trap density per unit of energy and per unit of volume. From this, we can estimate the effective number of traps as follows. First, we multiply by the effective area A_{eff}. Next, we multiply $k_B T$ because the traps that contribute to the noise are situated in an energy band $k_B T$ around the Fermi level. Similarly, one can show that the traps located in a band $\pi \lambda$ (in the direction perpendicular to the interface) contribute the most to the noise. Here, λ is the attenuation depth of the electron or hole wave function into the oxide. Thus, the effective number of traps is estimated as $A_{\text{eff}} \pi \lambda k_B T n_t$, and assuming Poisson statistics the relative device-to-device spread is expected to be:

$$\sigma_{rel., S_{V_{\text{gate}}}} = \frac{C}{\sqrt{A_{\text{eff}}}} \qquad (1)$$

where the proportionality the constant C is estimated as:

$$C \approx 1/\sqrt{\pi \lambda k_B T n_t} . \qquad (2)$$

Our experiments, see Fig. 3, are in agreement with Eq. (1), although the observed device-to-device spread is seen to level off at $\sim 30\%$ for large device areas. Most probably, this is due to experimental limitations, because only a small part of this $\sim 30\%$ can be attributed to oxide-thickness fluctuations. Thus, we fit the data of Fig. 3 with the expression

$$\sigma_{rel., S_{V_{\text{gate}}}} = \sqrt{\sigma_1^2 + \frac{C^2}{A_{\text{eff}}}} \qquad (3)$$

where σ_1 represents the experimental uncertainty. The constant C is found to be $C = 20 \pm 4 \% \ \mu\text{m}$ and $C = 15 \pm 3 \% \ \mu\text{m}$ for NMOS and PMOS, respectively. These values are of the same order of magnitude as observed in Ref. [9]. The value of C can be translated into a value of λn_t, which, irrespective of the value for λ, determines the magnitude of the $1/f$ noise, since all other model parameters such as C_{ox} are accurately known. Using these values for λn_t we obtain the dashed curves in Fig. 1, which are a factor of 5–10 above the measured data. Given the uncertainty in our determination of C, as well as the approximate nature of Eq. (2), the disagreement is not even that bad. Nevertheless, more elaborate statistical experiments on $1/f$ noise as well as more sophisticated theories than Eq. (2) are called for. In this context, it is worth noting that alternative versions of the unified noise model exist, e.g. based on thermally activated trapping at interface states instead of tunneling to oxide states [11], or a combination of those [12]. It would be interesting to compare statistical predictions based on such models with experiments.

Modelling and characterization of $1/f$ noise device-to-device spread has received surprisingly little attention in the literature so far. However, this topic will become increasingly important for the industrial application of minimum-area devices in upcoming CMOS technologies [11]. In addition, further study of this phenomenon is important because it can help to discriminate between different models for $1/f$ noise in MOSFETs.

THERMAL NOISE

It is well known that CMOS downscaling leads to an increase in important RF figures-of-merit such as f_T and f_{max}, as well as to a steady decrease in the minimum noise figure [14]. In order to obtain the best possible minimum noise figures, it is necessary

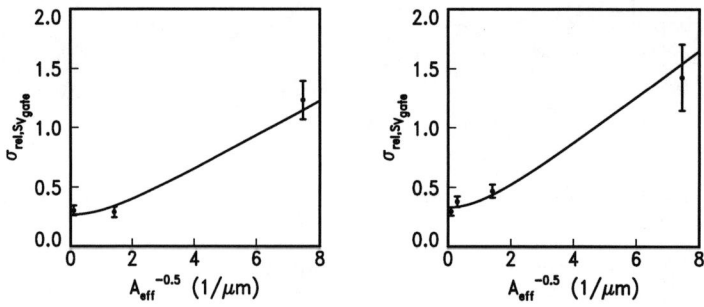

FIGURE 3. Relative spread in input-referred noise spectral density at f=1 kHz versus the square root of the inverse effective area for PMOS (*left*) and NMOS (*right*). Markers are measurements, and solid line is a fit of Eq. (3) to the data.

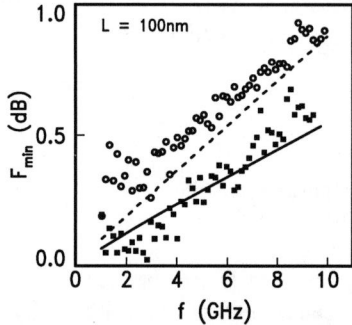

FIGURE 4. Minimum noise figure as a function of frequency for two 100-nm devices with different layouts. Markers are measured data and lines represent our model. The open markers and dashed line represent a device with 24 fingers which are 5 μm wide and contacted from one side. Solid markers and solid line represent a device with 96 fingers of 4 μm width and double-sided contacting.

to carefully consider the device layout [10, 15]. This is illustrated by the experiment in Fig. 4. Here, two 100-nm devices are compared. The device with double-sidedly contacted 4 μm wide fingers shows a significantly better minimum noise figure than its less-well designed counterpart which has single gate contacts and 5 μm wide fingers. Obviously, the effective gate resistance, which is four times larger in case of single-sided gate contacting, is the culprit.

The modelling of thermal noise has attracted considerable attention lately, mainly triggered by the fear that hot-carrier or nonequilibrium effects cause excessive noise in deep-submicron CMOS. We have developed a thermal-noise model [16], in which the local noise source is given by $S_I = 4k_B T g(x)/\Delta x$, where T is the lattice temperature, and $g(x)$ is the local conductance $W \mu_{eff} Q_{inv}$. Here, μ_{eff} is the effective mobility, including velocity saturation effects. The transfer function to drain and gate terminals was calculated using the Langevin method. Integrating the contributions of all channel segments from source to drain leads to the so-called modified Klaassen-Prins equation [16], which has successfully predicted the thermal noise in drain current and gate current, as well as their

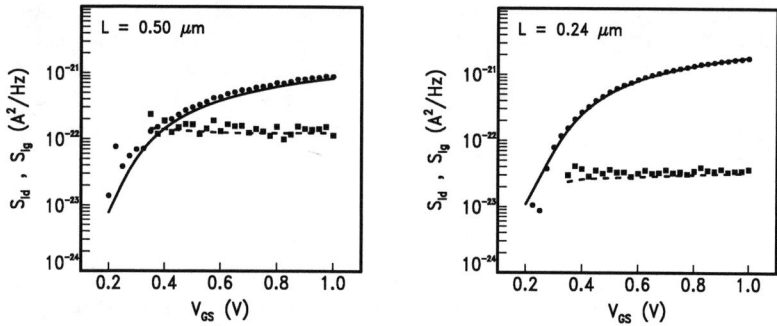

FIGURE 5. Drain current and gate current noise against gate voltage for n-channel devices with $L = 0.50$ μm (*left*) and $L = 0.24$ μm (*right*). Both devices consist of 24 single-sidedly contacted fingers of 5 μm width. The frequency is 10 GHz and the drain-source bias is 1 V. Markers and lines represent measurements and model, respectively.

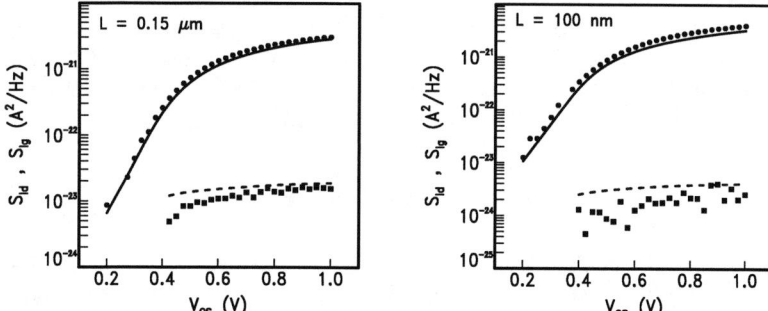

FIGURE 6. Same as Fig. 5, but now for $L = 0.15$ μm (*left*) and $L = 100$ nm (*right*). The 0.15 μm device consist of 24 single-sidedly contacted fingers of 5 μm width, the 100-nm device has 60 double-sidedly contacted fingers of 2 μm width.

correlation in 0.18 μm and 0.13 μm technologies. The model has been implemented in MOS Model 11, level 1102 [7].

Here, we test this model using new data taken on our 90-nm technology. The measurement procedure is extensively described in Ref. [17]. The results for drain and gate current noise, as a function of gate-source voltage, are displayed in Figs. 5 and 6. As expected, the model predicts the drain current noise for relatively long channel lengths ($L = 0.50$, 0.24 μm, and 0.15 μm) very well, see Fig. 5. For these channel lengths, the deviation between model and measurements is typically less than 10 %. For the shortest channel length, $L=100$ nm, the deviation is somewhat higher and the predicted noise is typically 20 % below the measured values. In terms of the often-used white-noise γ value [10], we measure $\gamma = 1.67$ and model $\gamma = 1.35$. Note that these γ values are still fairly low compared to values reported in older literature, such as [18]. The measured γ value is very close to the value predicted by Ref. [19] for a 100-nm device, based on hydrodynamic and Monte Carlo simulations [19]. Therefore, it would be interesting to see whether the incorporation of their expression for the local noise source into our model can cure the discrepancy we find for the shortest channel in our 100-nm technology.

For all channel lengths, the gate current noise is predicted fairly well in view of the measurement accuracy. For the short-channel, the gate current noise is less sensitive to the thermal noise in the channel, because it is dominated by the gate resistance even for a well-designed device layout [10, 15].

CONCLUSION

In this paper, we have discussed the modelling of both $1/f$ noise and thermal noise in a 90-nm CMOS technology. As far as $1/f$ noise is concerned, we have shown that the unified $1/f$ noise model explains both its bias and geometry dependence. Device-to-device variations in the $1/f$ noise will become increasingly important as devices are miniaturized further [11]. Modelling and characterization of this phenomenon is not only important for the application of downscaled CMOS technologies, but is also essential in the verification of $1/f$ noise models.

For thermal noise, we have shown that our previously developed model [16] predicts the measured data very well, with the exception of the shortest channel for which the noise was underestimated by $\sim$20 %. Results from device simulations [19] may explain this effect, which is likely to become more prominent in upcoming CMOS technologies.

ACKNOWLEDGMENTS

The authors would like to acknowledge the support of the EU IMPACT IST-2000-30016 project. We also thank H.P. Tuinhout (Philips Research) for stimulating discussions.

REFERENCES

1. V.C. Venezia et al., ESSDERC 2002, pp. 491–494.
2. L.F. Tiemeijer et al., IEDM Tech. Dig., pp. 155–158 (2004).
3. W. Jeamsaksiri et al., VLSI technology symposium 2005.
4. D. Linten et al., CICC2004, pp.701–704.
5. J. Ramos et al., ESSDERC 2004, pp.329–333.
6. http://www-device.eecs.berkeley.edu
7. http://www.semiconductors.philips.com/Philips_models
8. K.K. Hung et al., IEEE Trans. El. Dev., Vol. 37, pp. 654–665 (1990); ibid., pp. 1323-1332 (1990).
9. R. Brederlow et al., IEDM Tech. Dig., pp. 159–162 (1999).
10. A.J. Scholten et al., IEEE Trans. El. Dev., Vol. 50, pp. 618–632 (2003).
11. G. Ghibaudo, Proceedings of SPIE Vol. 5113 Noise in devices and circuits, pp. 16–28 (2003).
12. G. Bosman et al., Proceedings of SPIE Vol. 5113 Noise in devices and circuits, pp. 232–236 (2003).
13. P. Dutta and P.M. Horn, Rev. Mod. Phys., 53, pp.497–516 (1981).
14. P.H. Woerlee et al., IEEE Trans. El. Dev., Vol. 48, pp. 1776-1782 (2001).
15. A.J. Scholten et al., IEE Proc.-Circuits Devices Syst., Vol. 151, No. 2, pp. 167–174 (2004).
16. R. van Langevelde et al., IEDM Tech. Dig., pp. 867–870 (2003).
17. L.F. Tiemeijer et al., to be published in IEEE Trans. on microwave theory and techniques, 2005.
18. A.A. Abidi, IEEE Trans. El. Dev.,Vol. ED-33, 1986, pp. 1801-1805.
19. T. Oh, C. Jungemann, and R.W. Dutton, Proceedings of SISPAD, pp. 87–90 (2003).

An Analytical Thermal Noise Model of the MOS Transistor Valid in All Modes of Operation

A. S. Roy* and C. C. Enz*,†

*Ecole Polytechnique Fédérale de Lausanne (EPFL)
Electronics Lab, CH-1015 Lausanne, Switzerland
†Swiss Center for Electronics and Microtechnology (CSEM)
Av. Jaquet-Droz 1, CH-2007 Neuchâtel, Switzerland

Abstract. Although some of the recently proposed compact models for thermal noise in MOS Transistor exhibit good match with experimental data, we believe most of the existing compact models suffer from incorrect physical assumptions or modeling (e.g absence of carrier heating, incorrect modeling of velocity saturation effect, wrong modeling of diffusivity etc.). This work presents a new completely analytical thermal noise model based on consistent physical assumptions and valid in all modes of inversion.

Keywords: MOSFET, Thermal noise, weak inversion, velocity saturation
PACS: 05.40.Ca

INTRODUCTION

We have already presented a new physics based analytical model developed by using an expression of carrier temperature which is consistent with the mobility model. The model uses a new methodology to calculate the so called "impedance field" for any arbitrary velocity-field relation [1]. However, the model was limited to strong inversion. In this article we present an extension of our modelling methodology to weak inversion. The final result of the model is a closed form analytical solution of noise conductance valid in any inversion mode. Although there are noise models valid in all region of operation, [2] we believe they suffer from inconsistent physical assumptions [1]. Since short-channel effects degrade the ratio of noise conductance to gate transconductance [3] (perhaps the most important factor for a circuit designer), it will be beneficial to work in moderate inversion. So this modeling work can be of great interest for low-voltage and low-power circuit design in deep-submicron CMOS technologies.

MODEL DERIVATION

In case of a MOSFET we can identify the fundamental noise source as a current noise source between channel slice x and $x + \Delta x$ (The axis is directed along source to drain, $x=0$ being the source). Let the power spectral density (PSD) of the elementary noise current be $S_{\delta i_n^2}$, the resistance across the channel slice (between x and $x + \Delta x$) be Δr, the conductance that noise element sees looking at the source be g_s and the conductance looking at the drain be g_d. Then the PSD of the current fluctuation at the drain terminal

will be
$$S_{\delta I_D^2} = S_{\delta i_n^2} \cdot (\Delta r)^2 \cdot g_{eq}^2, \qquad (1)$$

where g_{eq} is the channel conductance given by $g_{eq}^{-1} = g_s^{-1} + g_d^{-1}$.

In this work we use the same methodology developed in [1] to calculate g_{eq}, Δr and $S_{\delta i_n^2}$. However, as this model is extend to weak inversion we need to distinguish between channel potential V (difference between electron and hole quasi fermi level) and the surface potential Ψ_s and their derivatives $\tilde{E} = dV/dx$ and $E = d\Psi/dx$. We start by noticing that the total current (including drift and diffusion) is given by

$$I_D = I = W \cdot (-Q_i) \cdot \mu_{eff} \cdot \frac{dV}{dx} = -W \cdot (-Q_i) \cdot \mu_{eff} \tilde{E}, \qquad (2)$$

Now admittance between x and a, looking at a, g_{xa} is given by $g_{xa} = \frac{dI}{dV_{ch}}|_{V_x}$, V_x being the channel potential at variable point x. We must remember that μ_{eff} is a function of electric field E but as the current I is constant along the channel we can write μ_{eff} as a function of V and I. Therefore

$$\frac{dI}{dV} \cdot (x-a) = \frac{\partial}{\partial V}\left(\int_{V_a}^{V} W \cdot (-Q_i) \cdot \mu_{eff}(V,I) \cdot dV\right) + \left(\int_{V_a}^{V} W \cdot (-Q_i) \cdot \frac{\partial \mu_{eff}(I,V)}{\partial I} \cdot dV\right)\frac{dI}{dV}. \qquad (3)$$

In order to evaluate $\frac{\partial \mu_{eff}(I,V)}{\partial I}$ we notice that

$$\frac{\partial \mu_{eff}(I,V)}{\partial I} = \frac{\partial \mu_{eff}(I,V)}{\partial E} \cdot \frac{\partial E}{\partial \tilde{E}} \cdot \frac{\partial \tilde{E}}{\partial I}. \qquad (4)$$

In order to calculate $\frac{\partial E}{\partial \tilde{E}}$ we will use the following definitions

$$Q_i = n \cdot C_{ox} \cdot (\Psi_p - \Psi_s) \text{ and } 2 \cdot \frac{Q_i}{2nC_{ox}U_T} + \ln\left(\frac{Q_i}{2nC_{ox}U_T}\right) = \frac{V_P - V_{ch}}{U_T},$$

where Ψ_p and V_P are the pinch of value of surface potential and channel potential respectively, and n is the slope factor [4].

From the above relations one can easily obtain the relation between E and $\tilde{E}$ as

$$E = \tilde{E} \cdot \frac{2 \cdot Q_i}{2nC_{ox}U_T + 2 \cdot Q_i}. \qquad (5)$$

Therefore combining (3), (4) and (5) the final expression for g_{xa} becomes

$$g_{xa} = \frac{W \cdot (-Q_i) \cdot \mu_{eff}}{(x-a) + \int_{V_a}^{V} \frac{\mu'_{eff} \cdot \frac{\partial E}{\partial \tilde{E}}}{\mu_{eff} + \mu'_{eff} \cdot E} \cdot dV}, \qquad (6)$$

where μ'_{eff} is the derivative of μ with respect to E.

We can now easily find out the g_s, g_d and g_{eq} from (6) at a position x resulting

$$g_{eq} = \frac{W \cdot (-Q_i) \cdot \mu_{eff}}{L_{eff} + \int_{V_s}^{V_{deff}} \frac{\mu'_{eff} \cdot \frac{\partial E}{\partial \tilde{E}}}{\mu_{eff} + \mu'_{eff} \cdot E} \cdot dV}, \qquad (7)$$

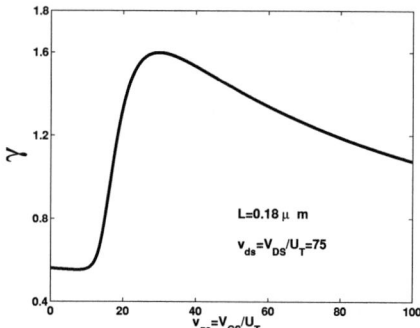

FIGURE 1. γ versus v_{gs} plot. model.

where V_s and V_{deff} are the voltages at the source and at the end of the velocity non-saturated region respectively, and L_{eff} is the length of velocity non-saturated region.

In order to calculate Δr we use the fact the the voltage difference between x and $x+\Delta x$ is $\Delta V = -\tilde{E} \cdot \Delta x$. Using this with (6) we obtain

$$\Delta g = \frac{1}{\Delta r} = \frac{W \cdot (-Q_i) \cdot (\mu_{eff} + \mu'_{eff} \cdot \frac{\partial E}{\partial \tilde{E}} \cdot \tilde{E})}{\Delta x} = \frac{W \cdot (-Q_i) \cdot (\mu_{eff} + \mu'_{eff} \cdot E)}{\Delta x}. \quad (8)$$

$S_{\delta i_n^2}$ in non equilibrium can be expressed as [1]

$$S_{\delta i_n^2} = 4 \cdot q \cdot W \cdot (-Q_i) \cdot \frac{D}{\Delta x} = 4 \cdot k \cdot T_C \cdot \Delta g = 4 \cdot k \cdot T_L \cdot \Delta g \cdot \left(\frac{\mu_0}{\mu_{eff}}\right)^2, \quad (9)$$

where T_C and T_L are the carrier and lattice temperatures respectively and μ_0 is the low lateral field mobility.

From the above discussions we can write down the contribution of the elementary current noise source between x and $x+\Delta x$ to the total drain current PSD and since the elementary current noise sources at different positions are uncorrelated one can obtain the total drain current noise PSD as an integral over x

$$S_{I_D^2} = 4 \cdot k \cdot T_L \cdot \frac{W}{L_{eff}^2 \left(1 + \frac{1}{L_{eff}} \int_{V_s}^{V_{deff}} \frac{\mu'_{eff} \cdot \frac{\partial E}{\partial \tilde{E}}}{\mu_{eff} + \mu'_{eff} \cdot E} \cdot dV\right)^2} \times \int_0^{L_{eff}} (-Q_i) \cdot \frac{\mu_0^2}{\mu_{eff} + \mu'_{eff} E} \cdot dx. \quad (10)$$

If we compare this expression of $S_{I_D^2}$ to our previously derived strong inversion model [1] we find that in strong inversion $Q_i \gg 2nC_{ox}U_T$ resulting $E \approx \tilde{E}$ and (10) reduces to eqn. 14 in [1].

Although the model holds for any arbitrary mobility field relation, in order to obtain a closed form expression we will choose $\mu_{eff} = \frac{\mu_0}{1+E/E_c}$, E_c being the critical field.

It can be shown that the exact analytical solution under the assumed mobility model is given by

$$S_{I_D^2} = \frac{4kT_L G_{spec}}{(1+\lambda_c \cdot (q_s-q_d))^2} \left(\frac{2}{i_d}\right) \left(\frac{1}{3}(q_s^3-q_d^3) + (\frac{\lambda_c i_d}{2})^2 \cdot (q_s-q_d) + \frac{1}{2} \cdot (\frac{\lambda_c i_d}{2} + \frac{1}{2}) \cdot (q_s^2-q_d^2) \right.$$
$$\left. + (\frac{\lambda_c i_d}{2} - \frac{1}{2}) \cdot (\frac{\lambda_c i_d}{2})^2 \cdot \ln(\frac{q_s+\frac{1}{2}-\frac{\lambda_c i_d}{2}}{q_d+\frac{1}{2}-\frac{\lambda_c i_d}{2}}) \right), \quad (11)$$

where, $i_d = \frac{I}{I_{spec}}$, $q_s = \frac{Q(0)}{Q_{spec}}$, $q_d = \frac{Q(L_{eff})}{Q_{spec}}$, $I_{spec} \triangleq 2n \cdot \mu_0 \cdot C_{ox} \cdot \frac{W}{L} \cdot U_T^2$, $Q_{spec} \triangleq 2n \cdot C_{ox} \cdot U_T$, $\lambda_c \triangleq \frac{2U_T}{E_c \cdot L}$, $G_{spec} \triangleq 2n \cdot \mu_0 \cdot C_{ox} \cdot \frac{W}{L} \cdot U_T^2$.

RESULTS AND DISCUSSIONS

Fig. 3 shows γ (defined as $S_{I_D^2} = \gamma \cdot 4 \cdot k \cdot T_L \cdot g_{dso}$, where g_{dso} is the output conductance at zero drain bias) versus $v_{gs} = V_{GS}/U_T$ (U_T is the thermal voltage) plot in saturation. From the plot we see in strong inversion, as expected, we get back the recently observed experimental tend [5, 6]. In weak inversion we find the γ has a value of 0.5. This indicates that the noise in weak inversion is pure shot noise. Indeed, from (11) we can see that in weak inversion and saturation ($q_s \to 0$ and $q_s >> q_d$), $S_{I_D^2} = 2qI$. However, as on average carrier velocity can not exceed thermal velocity, for very short channel device the condition $q_s >> q_d$ is not satisfied anymore. It can be shown that because of this γ increases by a factor $\frac{1+2\lambda_c}{1+\lambda_c}$ in weak inversion but for all practical purpose this value lies close to 1.

CONCLUSION

In this work we have presented a physics based analytical expression of thermal noise valid from strong to weak inversion using an extension of the methodology developed in [1]. Although, the γ degrades in strong inversion (from ideal value 2/3 to a value of 1.5-2), the value in weak inversion remains close to it's ideal value 0.5. This can be understood by considering the fact the reasons for the thermal noise degradation like carrier heating, channel length modulation and mobility degradation are absent in weak inversion.

REFERENCES

1. A. S. Roy, and C. Enz, *IEEE Trans. Electron Devices*, **52**, 611–614 (2005).
2. A. J. Scholten, L. F. Tiemeijer, R. v. Langevelde, R. J. Havens, A. T. A. Z. v. Duijnhoven, and V. C. Venezia, *IEEE Trans. Electron Devices*, **50**, 618–632 (2003).
3. A. S. Roy, and C. C. Enz, "Compact modeling of thermal noise in the MOS transistor," in *MIXDES*, 2004, pp. 71–78.
4. J. M. Sallese, M. Bucher, F. Krummenacher, and P. Fazan, *Solid-State Electronics*, **47**, 677–683 (2003).
5. C.-H. Chen, and M. J. Deen, *IEEE Trans. Electron Devices*, **49**, 1484–1487 (2002).
6. K. Han, H. Shin, and K. Lee, *IEEE Trans. Electron Devices*, **51**, 261–269 (2004).

On the high-frequency noise figures of merit and microscopic channel noise sources in fabricated 90 nm PD SOI MOSFETs

Raúl Rengel[1*], María J. Martín[1], Guillaume Pailloncy[2], Gilles Dambrine[2] and François Danneville[2]

[1]*Departamento de Física Aplicada, Universidad de Salamanca, Spain*
2*Département Hyperfréquences et Semiconducteurs, I.E.M.N., France*
* *e-mail : raulr@usal.es; Phone : +34 923 294436 ; Fax : +34 923 294584*

Abstract. A Monte Carlo investigation of the high-frequency noise performance of 90 nm Partially-Depleted Silicon-On-Insulator MOSFETs is presented. The good agreement of the simulation results to the experimental measurements (using only the surface scattering and gate workfunction as fitting parameters) confirms the reliability of the simulator as a global tool for predicting the performance of such devices. Velocity fluctuations are investigated in order to analyze the noise behavior of the diffusive and velocity overshoot sections of the channel; the results show that the velocity overshoot region presents a much stronger spectral intensity of longitudinal velocity fluctuations. Moreover, Monte Carlo simulation shows that electrons populating the transversal X valleys are the major contributors to the local noise source in the channel.

Keywords: Partially-Depleted SOI, Monte Carlo simulation, High-frequency noise.
PACS: 72.70.+m, 85.30.Tv, 85.30.De, 05.10.Ln

INTRODUCTION

Silicon-On-Insulator (SOI) MOSFET technologies are becoming a new contender in the RF design field due to their improved transconductance, cut-off frequency, higher soft-error immunity, excellent isolation and absence of latch-up as compared to bulk devices [1]. From the point of view of industrial applications, Partially-Depleted (PD) SOI transistors are an interesting and widespread used alternative.

In this work, we have employed a 2D ensemble Monte Carlo (MC) simulator to analyze the high-frequency figures of merit and the microscopic velocity fluctuations within the channel of a 90 nm effective gate-length body contacted PD SOI n-MOSFET. The simulated structure replicates the main features of the topology of the experimental device: the effective gate length of the transistors was considered to be 90 nm, the oxide thickness is 2 nm, and the active layer is p-doped ($2 \cdot 10^{18}$ cm^{-3}) and 150 nm thick. The doping of the n$^+$ regions is 10^{19} cm^{-3}. To reproduce the experimental *I-V* characteristics, a 20% of diffusive reflections were considered in our surface scattering model. Acoustic and optical phonons and impurity scattering are also incorporated. To mimic the body contact an additional contact was placed at the back interface of the active layer, thus allowing the flux of holes in the substrate.

STATIC AND HIGH-FREQUENCY PERFORMANCE

The MC simulator was able to reproduce adequately the main static characteristics and the high-frequency dynamic and noise figures of merit of the fabricated device by using only two parameters (surface scattering and metal gate workfunction) that were fitted through the I_D-V_{DS} curves. The modeling procedure employed was already successfully used for the analysis of the static and high-frequency dynamic and noise properties of fabricated Fully-Depleted transistors [2,3].

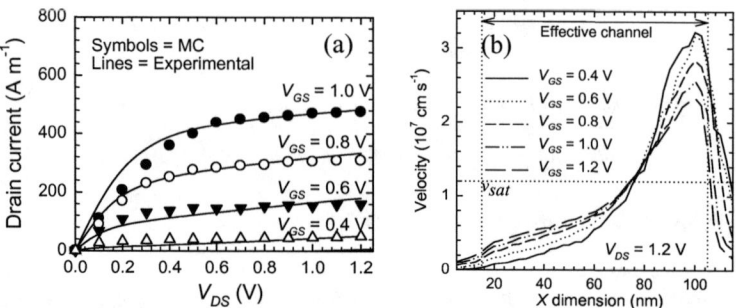

FIGURE 1. I_D-V_{DS} curves (a) and longitudinal drift velocity of electrons in the channel (b).

Figure 1(a) shows the I_D-V_{DS} curves for several values of V_{GS}. A slight underestimation of the drain current exists at high V_{GS} in the triode region; additional simulations using ATLAS® showed an excellent agreement with our MC results, what suggests that the discrepancy with the experimental data is due to some feature of the experimental topology for the contacts unknown to us. However, in saturation conditions the agreement to experimental data is fairly good (the noise and dynamic measurements to be shown correspond to V_{DS} = 1.2 V). Figure 1(b) shows the values of longitudinal electron velocity as a function of the X position. The velocity overshoot can be clearly appreciated (the maximum values are well over the saturation velocity in bulk Silicon, $1.2 \cdot 10^7$ cm s^{-1}), a phenomenon already described in bulk and FD SOI MOSFETs [2,4]. It is significant the reduction of the maximum peak (from 3.2 to 2.2×10^7 cm^{-3}) when increasing V_{GS}, due to the reduction in the drain-to-gate electric field. Simultaneously, the velocity in the diffusive part of the channel (where $v < v_{SAT}$) is readily augmented. From this graph, the channel can be split into two different sections for the study of velocity fluctuations: the diffusive region ($v < v_{sat}$) and the overshoot region ($v > v_{sat}$).

Fig. 2(a) shows the MC results (obtained through the calculation of Y parameters) and the experimental measurements for the two most relevant dynamic figures of merit, the transconductance and the cut-off frequency. g_m shows a maximum value of 900 Sm^{-1} for V_{GS} = 0.8 V and V_{DS} = 1.2 V and f_T = 90 GHz for those same bias conditions. From the spectral densities of current fluctuations at terminals (given in a natural way by MC simulation) and considering the previously determined MC Y parameters and the parasitic resistances provided by the experimental extraction procedure method, the main noise parameters were calculated. As an example, in Figure 2(b) we plot the MC and experimental values for the minimum noise figure and the associated gain at 6 GHz. As it can be observed, the agreement between simulation and measurements is quite reasonable. The remarkably reduced NF_{min} (0.4 dB) observed at low V_{GS} confirms the suitability of the device for analog high-frequency applications.

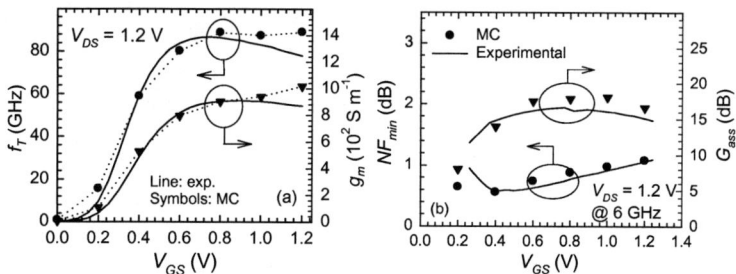

FIGURE 2. Cut-off frequency f_T and transconductance g_m (a) and minimum noise figure (b) at 6 GHz.

ANALYSIS OF VELOCITY FLUCTUATIONS

In deterministic approaches, the contribution to noise of carriers in the velocity overshoot region is frequently neglected basing on impedance field reasoning, while the noisy nature of carriers in this area and the values of the spectral density of velocity fluctuations (S_v) are barely discussed. The features of the MC method allow addressing the investigation of this quantity in the previously defined sections of the cannel, the diffusive and velocity overshoot regions. We have employed the following procedure: the instant values of longitudinal velocity for all particles inside one region are added in the absence of fluctuations of the self-consistent electric field. From the record of all accumulative velocities and calculating instantaneous fluctuations, the correlation function is determined; Fourier transformation of this quantity gives the spectral density of velocity fluctuations, which is properly normalized by the total number of particles in order to get *per electron* values. S_v (that is related to the diffusion coefficient, $S_v = 4D$) gives the local microscopic current noise source through [5]:

$$S_{in}(x,f) = q^2 N(x) S_v(x,f) W / dx$$

where q is the electron charge and N the charge sheet concentration. Figure 3(a) shows the results for S_V in the diffusive and overshoot regions as a function of V_{GS} for V_{DS} = 1.2 V. The value of S_V obtained from the low-field value of the bulk diffusion coefficient corresponding to the active-layer doping concentration is also plotted ($4D$). The results correspond to the 100 MHz-10GHz range, where S_V was found to be constant. In the gate bias range considered the overshoot region shows a much stronger spectral density of velocity fluctuations as compared to the diffusive part of the channel, especially at low gate voltage. As V_{GS} is raised, the differences between both channel sections are reduced: S_V is decreased in the overshoot area and slightly increased at the diffusive region. The analysis of the average inner quantities of interest (energy, velocity, electric field, scattering mechanisms, etc.) and the velocity fluctuations and correlation functions showed that in the diffusive region the amplitude of the fluctuations is mainly raised due to the increase of the local electric field V_{GS}. In the overshoot region, the proportional increase of isotropic scattering mechanisms with V_{GS} (phonons and surface scattering) reduces the time decay of the correlation and consequently the total S_V. Electrons suffer approximately 10 scattering mechanisms in the overshoot area (which is one order of magnitude less than the average number of collisions in the diffusive section) and the mean free path is near three times longer in that region. In

Figure 3 it can be also observed how the prediction of S_v given by the low-field bulk diffusion coefficient yields and underestimation as compared to the actual value given by the spectroscopy of electrons within the channel. Regarding the values of the local current noise source, the much elevated sheet concentration in the diffusive area gives higher values of the q^2NS_v factor than in the overshoot area for medium and high gate voltages (over 0.7 V). However, it is necessary to mention that to evaluate the final influence of the overshoot section on the current noise at the drain terminal it would be necessary to determine the local impedance field.

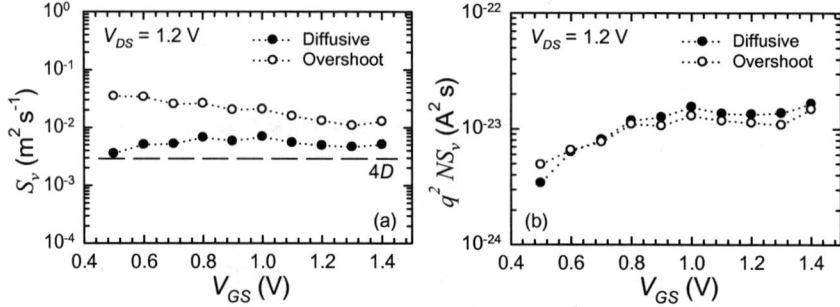

FIGURE 3. Values of S_v as a function of the gate voltage (a) and $q^2\,N\,S_v$ (b).

We also checked separately the spectra of velocity fluctuations in longitudinal and transversal valleys. The results evidence that the weight of electrons on transversal valleys represent near 90% of the total local current noise at high frequencies, a fact that is directly related to the reduced effective mass of carriers in this case (0.19 m_0).

CONCLUSIONS

A detailed investigation of the static, dynamic and high-frequency noise properties of 90 nm effective gate-length PD SOI MOSFETs has been carried out by means of a MC simulator. By using only the gate workfunction and the surface scattering as fitting parameters for *I-V* curves, a good general agreement is found for the main static dynamic and noise figures of merit. The devices show an exceptional behavior in terms of elevated g_m, f_T and low NF_{min}. The study of the spectral density of velocity fluctuations showed that electrons within the overshoot area evidence a stronger intensity of fluctuations than electrons in the diffusive region. Electrons in transversal *X* valleys are the main contributors to velocity fluctuations in the whole channel.

REFERENCES

1. Colinge, J. P., Silicon-on-Insulator Technology: Materials to VLSI, Norwell, Kluwer, 1997.
2. Rengel, R. *et al.*, *Semicond. Sci. Technol.* **17** pp. 1149-1156 (2002).
3. Rengel, R. *et al.*, "Microscopic analysis of the high-frequency noise behaviour of Fully-Depleted Silicon-On-Insulator MOSFETs" in *Proceedings of the 17th ICNF*, Prague: CNRL, 2003, pp. 585-588.
4. Laux, S. E. and Fischetti M. V., *IEEE Electron Device Lett.* **9** pp. 467-469 (1988).
5. Cappy, A. and Heinrich, W., *IEEE Trans. Electron Dev.* **36** pp. 403-409 (1989).

3D Monte Carlo Study of Thermal Noise in DG-MOSFET

P. Dollfus[1], A. Bournel[1], J.E. Velázquez[2]

[1]*Institut d'Electronique Fondamentale, CNRS UMR 8622, Université Paris-Sud 11,
F-91405 Orsay, France*
[2]*Departamento de Física Aplicada, Pza de la Merced s/n, Universidad de Salamanca,
E-37008 Salamanca, Spain*

Abstract. By means of self-consistent 3D Monte Carlo simulation we study steady-state and noise characteristics of Single (SG) and Double Gate (DG) MOSFET based on unintentionally doped ultra-thin silicon film. The effect of possible residual discrete impurity in the channel is considered. Preliminary results suggest that DG MOSFETs can offer improved noise performance as compared to SG ones at high frequencies.

Keywords: Monte Carlo simulation, Atomistic simulation, Noise, Fluctuations
PACS: 72.70.+m, 73.50.Td

INTRODUCTION

The history of CMOS technology is governed by aggressive scaling of the feature sizes to increase speed and packing density. Until recently it has been very conservative in the rate of introduction of non-bulk MOSFET configurations. However, in the sub-35 nm gate length area, new multi-gate architectures with ultra thin silicon film become more and more attractive. They may be the unique option likely to overcome short channel effects and to provide the possibility of downscaling CMOS devices into the nanometer regime if problems related to heat dissipation and interconnection density are solved [1]. As gate length is reduced, electron noise appears as a severe limit for both analog and digital applications. Thermal noise, generated by carrier velocity fluctuations in the channel, is considered as the most important noise source in FET amplifiers operating at RF and microwave frequencies. It can be fully and easily described using Monte Carlo device simulation. In nano-scale FETs, the inescapable fluctuations in the number and position of dopants in the channel are known to induce spreading in overall device performance and should be reflected on noise characteristics [2]. It makes necessary the use of a transport model including the atomistic nature of ionized impurities and 3D Poisson solver. The present work deals with the study of thermal noise in Single-Gate Fully-Depleted SOI (SG) and Double-Gate (DG) MOSFET using self-consistent 3D Monte Carlo simulation including the effect of possible discrete impurities in the channel. Such a simulation tool is used here for the first time for noise analysis of realistic devices.

MODEL AND SIMULATED DEVICES

Details on the analytical band structure and scattering parameters used in the simulator for the calculation of electron transport in Si may be found elsewhere [3]. The atomistic approach to describing the electron-ion interaction in the presence of discrete impurities was presented more recently. This model has been successfully used to compute the electron mobility in Si resistors versus average doping and to study the effect of impurity position in the channel of 50-nm bulk-MOSFETs [4]. Lately, the influence of atomistic doping on carrier transport and noise in nano-resistors has been tackled [5]. The simulated devices (Fig. 1) consist of ultra-thin SG and DG transistors with 17 nm-long unintentionally doped channel (denoted as SG1 and DG1, respectively). They operate with midgap gate material under power supply $V_{DD} = 1$ V. Source/channel and channel/drain junctions are assumed to be abrupt with continuous doping of 5×10^{19} cm^{-3} in S/D regions. However, we consider the possible presence of a single residual P-type impurity in the channel of a DGMOS.

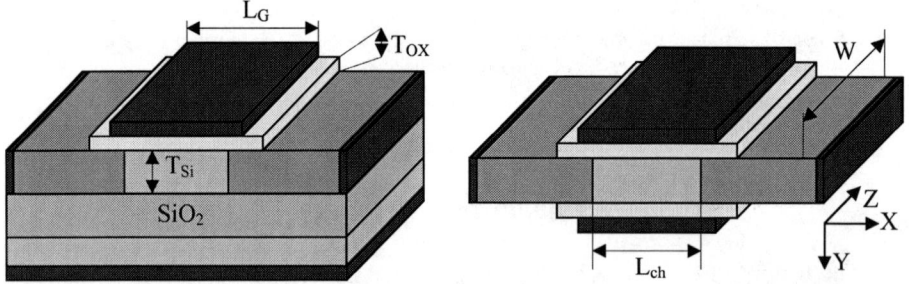

FIGURE 1. Schematic picture of the simulated SGMOS (left) and DGMOS (right) transistors. The set of values $L_G = 25$ nm, $L_{ch} = 17$ nm, $T_{Si} = 5$ nm, $T_{OX} = 1.1$ nm, and $W = 15$ nm apply for both structures.

The position (X_I, Y_I, Z_I) of the single channel impurity is specified as follows. Along source-drain direction, $X = 0$ corresponds to the source/channel junction and $X = 17$ nm corresponds to the channel/drain junction. The vertical position $Y = 0$ corresponds to the top SiO$_2$/channel interface. We study the effect of the position X_I between 0 and 17 nm for an impurity in the plane ($Y_I = 2.5$ nm, $Z_I = 7.5$ nm).

The time step between two solution of Poisson's equation is $\Delta T = 0.2$ fs. The instantaneous terminal currents are calculated on the basis of the Ramo-Shockley theorem using the technique described by Babiker et al. [6].

STEADY-STATE RESULTS

Figure 2 shows the typical effect of the presence of a single residual impurity on the transfer characteristics of both SG and DG MOSFET. We compare the case of channel without impurity (solid line) to the case of channel with one impurity in $X_I = 5.25$ nm (dashed line). In the case without impurity, the I_{on} current in DG1 (26 µA) is only 40% higher than in SG1 ($I_{on} = 18.5$ µA). This apparently good result for SG is only due to strong short-channel effects which induce high output conductance and also poor subthreshold behavior (not shown here).

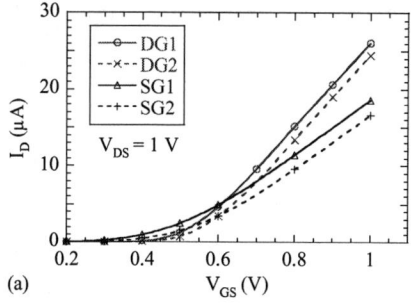 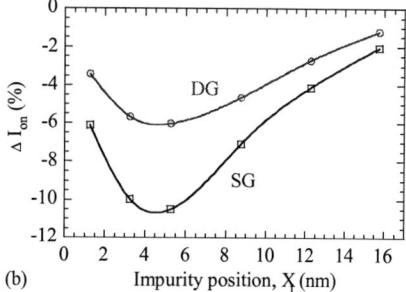

FIGURE 2. (a) I_D-V_{GS} characteristics of SG and DG MOS at $V_{DS}=1$V. We consider both cases of channel free of impurity (SG1, DG1) and channel with one impurity in $X_I=5.25$ nm (SG2, DG2); (b) I_{on} versus lateral position X_I of a single impurity in the channel with respect to SG1 and DG1.

The presence of a single impurity in the channel (dashed lines) perceptibly shifts the transfer characteristics and degrades I_{on}. This effect seems stronger in SG than in DG which is confirmed by Fig. 2(b) where the relative variation of I_{on} is plotted as a function of impurity position between X=0 and X=17 nm. The I_{on} degradation reaches 6% for DG and 10.5% for SG if the impurity is located near the source-end. On the contrary, an impurity located near the drain-end of the channel, i.e. in the high-field region, has a much smaller impact.

NOISE ANALYSIS

The results presented in this work are necessarily limited as a direct consequence of the vast CPU time needed for the calculations. Therefore we will limit ourselves to present the analysis for a point in the saturation regime. In order to evaluate the noise performance of both SG and DG transistors we first studied the current fluctuations for the same gate overdrive ($V_{GS}=V_{DS}=V_{DD}$). We considered three structures: SG1, DG1 and a DG with one P-type residual impurity located in $X_I=5.25$ nm (DG2 from now on). From Fig. 3(a) it follows that for the same gate overdrive the spectral density of the gate current ($S_{igig}(f)$) is considerably larger in SG1 than in both DG1 and DG2. In all the structures $S_{igig}(f)$ exhibits the typical f^2 dependence and the slower increase with frequency for the DG structures must be attributed to a partial cancellation of the fluctuations due to the double gate topology. On the contrary, the SG exhibits the lowest value of the spectral density of the drain current ($S_{idid}(f)$) (Fig. 3(b)).

For the sake of comparison, in Fig. 3 (c) and (d) we present $S_{igig}(f)$ and $S_{idid}(f)$ for the same value of the drain current (to this aim simulations were carried out biasing the gates of the DG1 structure to $V_{GS}=0.862$ V). Results are qualitatively similar to the ones previously described.

From the above gate and drain current noise spectral densities and the Y-parameters obtained from time-series as described in [6] we performed the calculation of the minimum noise figure (NF_{min}) and the noise resistance (R_n). At high-frequencies (namely, 20GHz), for the same gate overdrive SG1 exhibited a value of R_n of 150Ω that is considerably higher than in DG (R_n=105Ω in DG1 and R_n=95Ω in DG2). This difference increases for operation at the same current (R_n=75Ω in DG1 for V_{GS}=0.862V). Nevertheles, it turns out that the three structures exhibit similar values of NF_{min} (close to 1dB).. Further work is needed in order to extend the calculations to other regimes and fully establish if DG can effectively guarantee lower noise operation than SG ones.

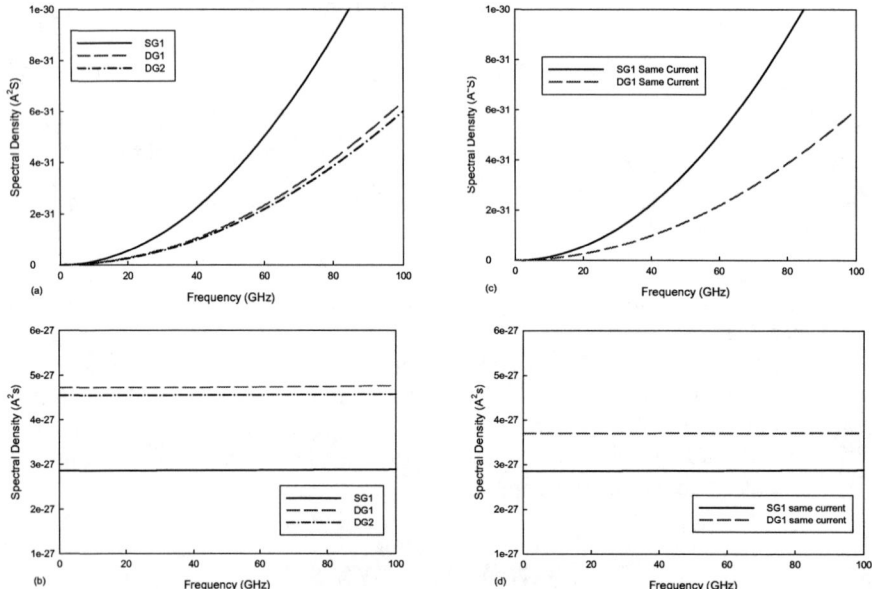

FIGURE 3. $S_{igig}(f)$ (a) and $S_{idid}(f)$) (b) at the same value of the gate overdrive. $S_{igig}(f)$ (c) and $S_{idid}(f)$) (d) at the same value of the drain current.

ACKNOWLEDGMENTS

This work has been partially supported by the European Community 6th FP under contracts IST-507857 (IP NANOCMOS) and IST-506844 (NoE SINANO).

REFERENCES

1. Taur,Y., *IBM Journal of Research and Development* **46**, 213-222 (2002).
2. Asenov, A., Balasubramaniam, R., Brown, A.R., and Davies, J.H., *IEEE Trans. Electron Devices* **50**, 839-845 (2003).
3. Dollfus, P., *J. Appl. Phys.* **82**, 3911-3916 (1997)
4. Dollfus, P., Bournel, A., Galdin-Retailleau, S., Barraud, S., and Hesto, P., IEEE Trans. Electron Devices **51**, 749-756 (2004).
5. P. Dollfus, J.E. Velázquez, A. Bournel, S. Galdin-Retailleau, Int. Workshop on Computational Electronics (IWCE-10), West Lafayette, USA, Oct. 2004, to be published in *J. Comput. Electron*.
6. Babiker, S., Asenov, A., Cameron, N., Beaumont, S.P., and Barker, J.R., *IEEE Trans. Electron Devices* **45**, 1644-1652 (1998).

Non-linear noise in nanometric Schottky-barrier diodes

S. Pérez[1], T. González[1], P. Shiktorov[2], E. Starikov[2], V. Gruzinskis[2], L. Reggiani[3], L. Varani[4], and J. C. Vaissière[4]

[1]*Departamento de Física Aplicada, Unversidad de Salamanca,*
Plaza de la Merced s/n, 37008 Salamanca, Spain
[2]*Semiconductor Physics Institute, A. Gostauto 11, 2600 Vilnius, Lithuania*
[3]*INFM National Nanotechnology Laboratory and Dipartimento di Ingegneria dell'*
Innovazione, Università di Lecce, Via Arnesano s/n, 73100 Lecce, Italy
[4]*Centre d'Electronique et de Micro-optoélectronique de Montpellier (CNRS UMR 5507),*
Université Montpellier II, 34095 Montpellier, France

Abstract. We present a microscopic analysis of current fluctuations in a heavily doped nanometric GaAs Schottky-barrier diode operating in series with a parallel resonant circuit under the action of a high-frequency large-signal voltage. An ensemble Monte Carlo simulation is used for the calculations. Important modifications of the noise spectra are found in the THz region with respect to the unloaded diode. An analytical model based on the static I-V and C-V characteristics of the diode is developed to explain the spectra obtained by Monte Carlo simulations.

Keywords: Schottky-barrier diode, noise, Monte Carlo simulation, THz operation
PACS: 72.70.+m, 73.40.Kp, 05.40.-a, 05.10.

INTRODUCTION

The increasing importance of broadband high-speed telecommunications demands new high power generators working at room temperature within the THz frequency region.

Schottky-barrier diodes (SBDs) are commonly used as multipliers and mixers of time-varying signals because of their strongly nonlinear current-voltage (I-V) and capacitance-voltage (C-V) characteristics. However, the use of these devices as THz generators requires high technology and accurate design. In particular, SBDs for THz generation involve extremely short sizes, usually in the nano-scale range or beyond, and very high dopings at the limit of degeneracy [1]. The improvement in the fabrication processes during the last years has made possible the production of SBDs with active layer thickness below 50 nm, making these structures promising devices for THz generators [2].

To extract the high-frequency signal, the SBD must be loaded by a given external resonant circuit, and the optimal conditions for signal extraction are achieved by a proper choice of the parameters of the resonant circuit. An additional complexity in the design of these devices lies in the fact that the same microscopic mechanisms are responsible for both the nonlinearities (which provide mixing and/or high-order harmonic generation) and the intrinsic noise (which masks harmonics and limits their extraction from the noise level). Furthermore, under large-signal THz operation, the regular and noise responses are superimposed in the whole frequency range of interest. Therefore a theoretical investigation of the intrinsic noise spectrum of the SBD loaded by an external resonant circuit becomes a key issue.

The noise properties and the intensities of generated harmonics in nano-scale SBDs operating under cyclostationary conditions have been recently studied [3]. The current noise spectra were found to be similar to those obtained under static excitation. However, in the case of circuit operation one can expect that, due to feedback, the external circuit should modify the intrinsic noise spectrum of the unloaded SBD, especially under large-signal operation inherent to frequency multiplication.

The aim of this work is to study such an influence by Monte Carlo (MC) simulations of the circuit performance of a nanometric GaAs n^+-n-metal SBD. We also present an analytical model able to identify the influence of the different circuit elements on the modifications found in the noise spectra with respect to the unloaded case.

MODEL AND RESONANT CIRCUIT

We consider room temperature operation of a heavily doped GaAs n^+-n-metal structure similar to that reported in Ref. 2, favourable for THz applications. The main parameters of this diode are the following: $n^+ = 8 \times 10^{18}$ cm^{-3}, $n = 1.1 \times 10^{18}$ cm^{-3}, n^+ region length $l_{n+} = 0.02$ μm and n region length $l_n = 0.03$ μm. l_{n+} has been reduced to 0.02 μm in order to optimize the computation time. The semiconductor-metal barrier height at equilibrium is assumed to be 1.03 V. The cross-sectional area A of the structure is 7.1×10^{-14} m^2.

In experiments, to obtain frequency mixing or multiplication, the SBD is seriesly connected with some kind of resonant system tuned at the frequency of interest, and a microwave voltage $U(t) = U_0 + U_1 \sin(2\pi v_0 t)$ of frequency v_0 is applied to the whole system. Since the noise characteristics of SBDs operating under static and cyclostationary conditions have been studied in previous works [3, 4], in this paper we will focus on the influence of the resonant system on the noise performance. We will investigate, by the MC technique, the large-signal operation of the SBD connected in series with a resonator whose equivalent circuit is represented by the parallel connection of a resistance R, an inductance L and a capacitance C_l. This circuit is shown schematically in Fig. 1. R is also the load resistance (assumed to be noisless) where the output power generated by the whole system at the frequency of a given high-order harmonic is extracted. No additional losses apart from those of the load resistance are considered. L and C_l take values of 1.02×10^{-23} Hm^2 and $0.5 C_g$, respectively, with $C_g = \varepsilon \varepsilon_0 /(l_{n+} + l_n)$ the geometrical capacitance times unit surface of the diode. These values correspond to resonant operation optimized to obtain the third harmonic of the applied frequency v_0 (600 GHz in our calculations with $v_0 = 200$ GHz).

Since the output power at a given harmonic (mainly the third one in our case) is extracted from the load resistance, we will evaluate the current (or voltage)

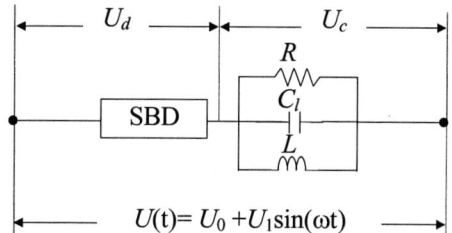

FIGURE 1. Scheme of the SBD connected to the resonant circuit.

fluctuations through (at the terminals of) this resistance.

MONTE CARLO SIMULATIONS

For the calculations, the Boltzmann transport equation self consistently coupled with the Poisson equation is solved by means of a MC procedure together with the external circuit equations in a similar way as that reported in Ref. 5.

As follows from the scheme shown in Fig. 1, the circuit equations can be written as:

$$C_l \frac{dU_c}{dt} + j_L + \frac{U_c}{R} = j_d,$$

$$L \frac{dj_L}{dt} - U_c = 0, \quad (1)$$

$$U_d + U_c = U(t),$$

where U_d and U_c are the voltage drops between the terminals of the SBD and of the external circuit, respectively; j_L is the current density flowing through the inductance L, and j_d is the total current density flowing through the SBD, given by:

$$j_d(U_d,t) = j_d^{RS} + C_g \frac{dU_d}{dt}, \quad (2)$$

with $j_d^{RS}(t) = \frac{q}{(l_{n+} + l_n)} \sum_{i=1}^{N(t)} v_i(t)$ the drift current directly obtained from the MC simulations by using the Ramo-Shockley theorem, where $v_i(t)$ is the instantaneous velocity of the i-th quasiparticle, $N(t)$ the number of quasiparticles inside the SBD at the time moment t, and q the unit charge of the quasiparticle. We remark that the microscopic noise source associated with scattering events is naturally included in $j_d^{RS}(t)$ which is a stochastic quantity.

Figure 2(a) shows the spectral density of current fluctuations through the load resistance $S_{\delta j \delta j}^R$ calculated by using the finite Fourier transform of the fluctuating current density $j_R(t) = U_c/R$ for different values of $R = 10^{-12}$, 10^{-10}, 10^{-9} and 10^{-8} Ωm^2. Only the noise part of the spectrum is shown, i.e. the vertical spikes corresponding to regular contributions of the fundamental and high-order harmonics are omitted. The solid line shows the spectral density of current fluctuations for the

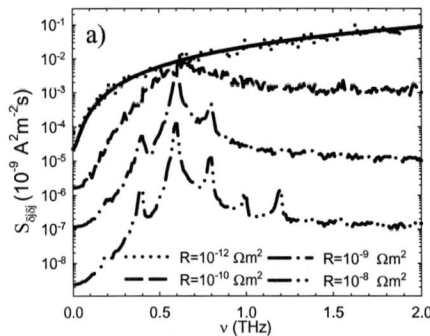 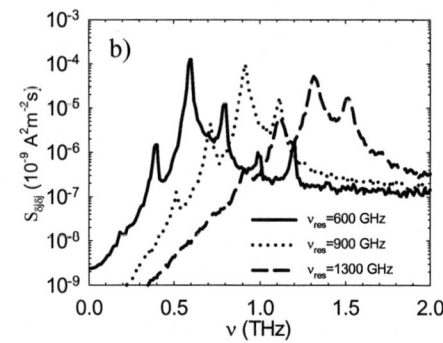

FIGURE 2. Spectral density of current fluctuations through the load resistance R calculated at U_0=0.6V, U_1=0.2V, v_0=200GHz for different values of: (a) the load resistance when the parallel circuit has a resonance at the third harmonic and (b) the resonant frequency of the parallel circuit v_{res} when R=$10^{-8}\Omega m^2$.

unloaded SBD under the same biasing conditions, $U_0 = 0.6\,V$, $U_1 = 0.2\,V$, v_0= 200 GHz, plotted for comparison.

When the load resistance is considerably lower than the SBD differential resistance $Z_d(0)$ (about $10^{-10}\,\Omega m^2$ in our diode), $S^R_{\delta j \delta j}(v)$ practically coincides with the current spectral density of the unloaded SBD (short-circuit operation) [curve for R=$10^{-12}\,\Omega m^2$ and solid line in Fig. 2(a)]. With the increase of R, a suppression of the low- and high-frequency wings of the spectrum centred on the frequency of the third harmonic (v= 600 GHz in this case) is observed (see curve for R=$10^{-10}\,\Omega m^2$). With a further increase of R, an additional suppression of the spectrum wings (and, in part, of the spectrum centre) takes place, leading to the appearance of a peak centred around the frequency of the third harmonic. Moreover, a series of additional peaks centred around other harmonics emerge in the spectrum. The higher the load resistance, the higher the number of satellite peaks appearing in the spectrum (curves for R=10^{-9}, $10^{-8}\,\Omega m^2$).

Figure 2(b) shows the influence of the tuning frequency of the resonant circuit v_{res} on $S^R_{\delta j \delta j}(v)$. The spectrum keeps the same structure for the different values of v_{res}, with the peaks shifting to higher frequencies following the main resonance v_{res} and obeying the same series at frequencies $v_{res} \pm n v_0$ ($n = 1,2,3,\ldots$).

ANALYTICAL MODEL

To provide a more direct physical interpretation of the noise spectra obtained by MC simulations, we have developed a simple analytical model for the SBD circuit operation. In this model, the average current flowing through the SBD is approximated as:

$$j_d(U_d) = j_d^s(U_d) + \frac{d}{dt} Q^s(U_d), \tag{3}$$

where $j_d^s(U_d)$ is the drift component, obtained from the static current-voltage relation, and $\frac{d}{dt}Q^s(U_d)$ takes into account the variations of total charge of free carriers in the diode due to the oscillatory character of U_d. This latter contribution is calculated from the static total charge $Q^s(U_d)$ evaluated from MC simulations, which provides the SBD varactor capacitance $C_v(U_d) = dQ^s(U_d)/dU_d$. Linearizing equations (1) to (3) and working in the frequency domain, one obtains the following implicit expression for the voltage fluctuations at the terminals of the SBD

$$\delta U_d(\omega) = -Z(\omega)\left[\delta j_d(\omega) + i\omega \sum_{n\neq 0}^{\pm\infty} C_n^v \delta U_d(\omega - \omega_n)\right], \quad (4)$$

where $Z(\omega) = \dfrac{i\omega}{\overline{C}(\omega_{res}^2 - \omega^2 + i\omega\ \nu_{RC})}$ is the net small-signal impedance of the considered system (SBD+resonant circuit), $\overline{C} = C_l + C_0^v$, $\omega_{res}^2 = 1/\overline{C}L$, $\nu_{RC} = 1/\overline{C}\overline{R}$, $\overline{R} = RR_d^0/(R + R_d^0)$, R_d^0 is the SBD average resistance and C_n^v the Fourier coefficients of the varactor capacitance C_v.

In the square brackets of the r.h.s. of Eq. (4) the first term corresponds to the noise source related to fluctuations of the drift current flowing through the unloaded SBD. The second term describes the influence of the C-V nonlinearities on the fluctuations of U_d when the SBD operates in the external resonant circuit, and it can be considered as an additional noise source induced by the harmonic modulation of the varactor capacitance. Note that this second term involves frequency mixing that leads to an implicit dependence between $\delta U_d(\omega)$ and $\delta j_d(\omega)$.

Such an implicit dependence in Eq. (4) can be resolved by using sequential iterations with respect to C_n^v. In the zero-order approximation, by omitting the second term in square brackets of Eq. (4), the spectral density of voltage fluctuations takes the form:

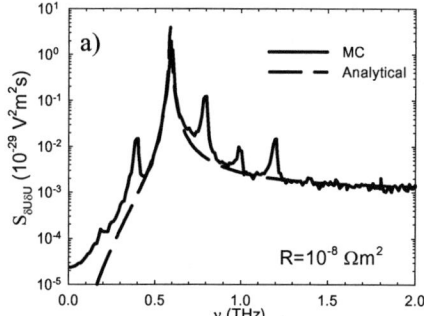
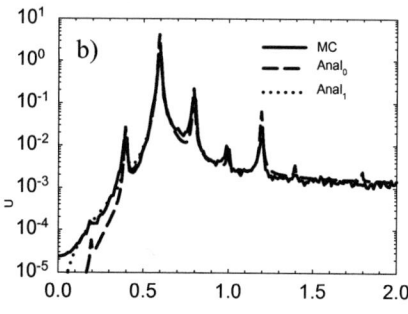

FIGURE 3. Spectral density of voltage-drop fluctuations at the load resistor obtained by Monte Carlo simulations (solid-line) and by the analytical approach for (a) zero-order and (b) first-order approximations.

$$S^0_{U_d U_d}(\omega) = |Z(\omega)|^2 S^d_{jj}(\omega), \qquad (5)$$

where $S^d_{jj}(\omega)$ is the spectral density of the current noise source in the unloaded SBD.

Figure 3(a) compares the result of the zero-order approximation with the MC spectrum for $R=10^{-8}$ Ωm^2. As observed, the zero-order approximation provides a qualitative description of the $S_{U_d U_d}(\omega)$ spectrum in practically the whole frequency range of interest, including the resonant frequency $\nu_{res} = 3\nu_0$ at which the external circuit is tuned. However, it does not describe the extra resonance-like noise contributions appearing around frequencies $\nu_{res} \pm n\nu_0$. To reproduce these satellite peaks, the influence of the C-V nonlinearities on the spectrum of fluctuations must be taken into account. In the first-order approximation one obtains:

$$S^1_{U_d U_d}(\omega) = |Z(\omega)|^2 \left[S^d_{jj}(\omega) + \omega^2 \sum_{n \neq 0}^{\pm\infty} |C^v_n|^2 |Z(\omega - \omega_n)|^2 S^d_{jj}(\omega - \omega_n) \right], \qquad (6)$$

where the second term can be considered as an additional source of fluctuations of the current flowing through the SBD induced by the harmonic modulation of the varactor capacitance. The modified total spectrum including the first-order iteration is reported in Fig. 3(b) and compared with direct MC simulations. The first-order iteration describes qualitatively and quantitatively the origin of the extra noise at the harmonics of the applied signal in the whole frequency region of interest excluding the lowest range, where the noise remains underestimated with respect to the MC results. The origin of this discrepancy is the high value of the average current $\bar{j}_0$ flowing through the diode when R increases, which is not accounted for in the noise source of the unloaded SBD. By adding the corresponding shot-noise value we have recalculated the circuit noise. The result, plotted in Fig. 3(b) by the dotted line, is found to improve significantly the agreement with the MC curve. Therefore, the analytical model is fully validated by MC simulations.

ACKNOWLEDGMENTS

Financial support from the MEC (Spain) and FEDER through the project TEC2004-05231 and from NATO grant PST.EAP.CLG 980629 are gratefully acknowledged.

REFERENCES

1. Jelenski, A., Grub, A., Krozer, V., and Hartnagel, H.L., *IEEE Trans. Microwave Theory Techn.* **41**, 549-556 (1993).
2. Gelmont, B., Woolard, D., Hesler, J., and Crowe, T., *IEEE Trans. Electron Devices* **45**, 2521–2527 (1998).
3. Shiktorov, P., Starikov, E., Gruzinskis, V., Pérez, S., González, T., Reggiani, L., Varani, L., and Vaissiere, J. C., *IEEE Electr. Dev. Lett.* **25**, 1–3 (2004).
4. González, T., Pardo, D., Varani, L., and Reggiani, L., *J. Appl. Phys.* **82**, 2349-2358 (1997).
5. Mitin, V., Gruzinskis, V., Starikov, E., and Shiktorov, P., *J. Appl. Phys.* **75**, 931-941 (1994).

Analytical Model of Noise Spectrum in Schottky-Barrier Diodes

P. Shiktorov*, E. Starikov*, V. Gružinskis*, L. Reggiani[†], L. Varani** and J.C. Vaissière**

Semiconductor Physics Institute, A. Goštauto 11, LT 01108 Vilnius, Lithuania
[†]*INFM - National Nanotechnology Laboratory, Dipartimento di Ingegneria dell' Innovazione, Università di Lecce, Via Arnesano s/n, 73100 Lecce, Italy*
**CEM2 - Centre d'Electronique et de Micro-optoelectronique de Montpellier, (CNRS UMR 5507) Université Montpellier II, 34095 Montpellier Cedex 5, France*

Abstract. We develop an analytical model for the high-frequency current noise spectral density of n^+n-metal Schottky-barrier diodes (SBDs). The model is validated by comparison with Monte Carlo simulations of GaAs SBDs operating from barrier-limited to flat-band conditions. The high-frequency spectrum is shown to be governed by the self-consistent collective motion of carriers in the SBD.

Keywords: High-frequency noise, Schottky-barrier diodes
PACS: 72.20.Ht, 72.30.+q, 72.70.+m

INTRODUCTION

Terahertz (THz) applications of Schottky-barrier diodes (SBDs) demand an estimation of the intrinsic high-frequency noise. Usually, such a task is solved in the framework of Monte Carlo Particle (MCP) method (see, e.g., [1] and references therein). Indeed, the existing analytical approach [2], being based on the independent motion of carriers in the static electric field of the depletion region only, presents some drawbacks in the description of the resonances appearing in the high-frequency region of the spectral density of current fluctuations. For example, such an approach: (i) does not give the dependences of these resonances on the SBD parameters and applied voltage, (ii) predicts a unphysical modulation of the returning carrier resonance, etc. By noticing that the low-frequency noise of SBDs is well described by the universal shot-noise law, the main aim of this work is to develop an analytical model able to describe the behaviour of the high-frequency current spectral density of a SBD operating from barrier-limited to flat-band regimes which overcomes these drawbacks.

THEORETICAL MODEL

We consider carriers moving in the self-consistent electric field of the whole n^+n-metal SBD structure, as sketched in Fig. 1 (a). Here we assume that: (i) the change of free carrier concentration in the n^+n-homojunction and at the edge of the depletion region is abrupt, (ii) all the carriers in n-region are reflected when hitting the Schottky barrier so that in the presence of a depletion region no static current is flowing in the structure (i.e.

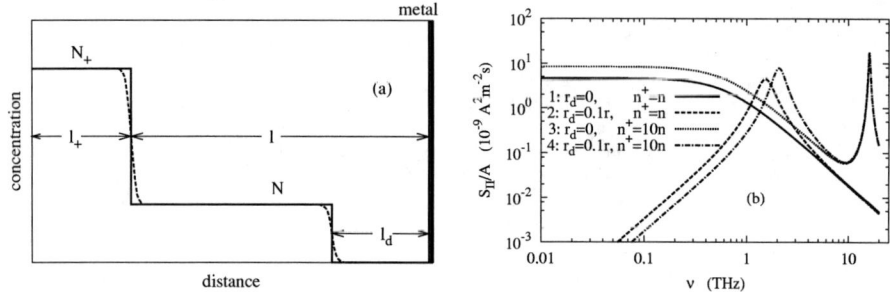

FIGURE 1. (a) Schematic representation of the n^+n–metal SBD with the depletion region near to the metal contact, and (b) four typical solutions for the noise spectrum given by Eq. (5) of the analytical model.

the shot noise is not considered), and (iii) the potential drop at the n^+n-homojunction is negligible so that the electric field is zero outside the depletion region. Accordingly, we take $E(x) = 0$ for $x < L - l_d$ and $E(x) = (e/\varepsilon\varepsilon_0)N(x - L + l_d)$, for $L - l_d < x < L$, where $L = l_+ + l$ is the total length of the SBD and the length of the depletion region, $l_d = \sqrt{2\varepsilon\varepsilon_0(\phi_0 - U)/eN}$, is determined by the barrier height ϕ_0 at the semiconductor-metal boundary and the voltage U applied to the SBD.

Under constant voltage conditions, the instantaneous response to any local perturbation originated by a scattering event of a free carrier leads to the appearance inside the whole SBD of an homogeneous component of the electric field fluctuation, $\delta E(x) = \Delta E$, which results in a collective motion (spatial shift) of all the free carriers in the SBD volume. By assuming that at a certain stage of the perturbation evolution the collective shifts with respect to the donor centers of free carriers in the n^+ and n regions, labeled below as $i = +, -$, are equal to δx_+ and δx_-, respectively, the field fluctuation $\delta E(x)$ corresponding to these shifts is:

$$\delta E(x) = \begin{cases} \Delta E - (e/\varepsilon\varepsilon_0)N_+\delta x_+ & , 0 < x < l_+ \\ \Delta E - (e/\varepsilon\varepsilon_0)N\delta x_- & , l_+ < x < l_+ + l_- \\ \Delta E & , l_+ + l_- < x < L \end{cases} \quad (1)$$

where $l_- = l - l_d$ is the length of the neutral part of the n region, $\Delta E = (e/\varepsilon\varepsilon_0 L)[N_+l_+\delta x_+ + Nl_-\delta x_-]$ the homogeneous part of the self-consistent electric field fluctuation. To describe free carrier fluctuations in the n^+ and n regions, we shall take advantage of the Langevin equations for the collective carrier motion in these regions:

$$\frac{d^2}{dt^2}\delta x_i - v_i \frac{d}{dt}\delta x_i = \frac{e}{m}\delta E_i + f_i \quad (2)$$

where

$$\delta E_i = \frac{e}{\varepsilon\varepsilon_0} \begin{cases} -(1-r_+)N_+\delta x_+ + r_- N\delta x_- & , i = + \\ r_+ N_+\delta x_+ - (1-r_-)N\delta x_- & , i = - \end{cases} \quad (3)$$

is the electric field perturbation, v_i the carrier momentum relaxation rate, f_i the Langevin force describing thermal fluctuations in n^+ and n regions, $r_i = l_i/L$ the relative length of

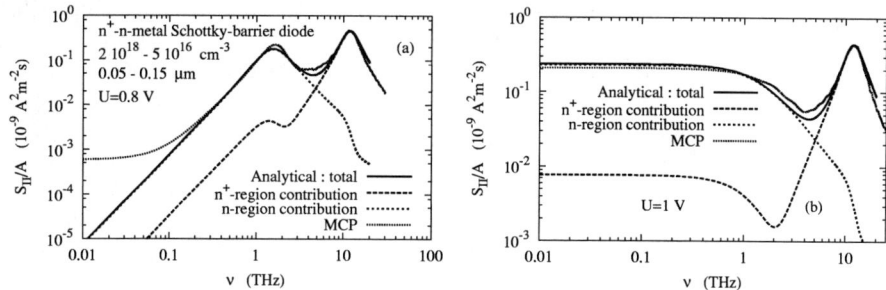

FIGURE 2. Spectra of current fluctuations per unit surface (A) in GaAs $0.05 - 0.15 \, \mu m \, n^+n$–metal SBDs at room temperature calculated under (a) barrier-limited and (b) flat-band conditions.

the i-region. To be consistent with Nyquist theorem, the spectral density of the thermal Langevin force, $S^i_{ff}(\omega) = \frac{1}{AN_i l_i} 4 k_B T \frac{v_i}{m}$, must be normalized to the total number of free carriers inside the i-region.

By solving Eq. (2) in the spectral representation and taking into account that the fluctuation of the current in the external circuit is given by $\delta I = i\omega \varepsilon \varepsilon_0 \Delta E$, one obtains the spectral density of thermal fluctuations of the SBD current as:

$$S_{II}(\omega) = 4k_B T \frac{e^2}{m} \frac{A}{L^2} \left[v_+ N_+ l_+ \left| \frac{\omega}{\Delta(\omega)}(a_{22} - a_{12}) \right|^2 + v_- N l_- \left| \frac{\omega}{\Delta(\omega)}(a_{11} - a_{21}) \right|^2 \right] \quad (4)$$

where $\Delta(\omega) = a_{11}a_{22} - a_{12}a_{21}$, $a_{12} = r_-\omega_-^2$, $a_{21} = r_+\omega_+^2$, $a_{11} = \omega^2 - \omega_+^2 - i\omega v_+ + a_{21}$, $a_{22} = \omega^2 - \omega_-^2 - i\omega v_- + a_{12}$, $\omega_i^2 = (e^2/\varepsilon \varepsilon_0 m) N_i$ is the plasma frequency of the free carriers in the i-region.

By assuming that the carrier momentum relaxation rates are the same in the n^+ and n regions, i.e. $v_+ = v_- = v$, the fluctuation spectrum in Eq. (4) takes the form

$$S_{II}(\omega) = 4k_B T v C \omega^2 \frac{r_+ \omega_+^2 o(\omega_-) + r_- \omega_-^2 o(\omega_+)}{o(\Omega_1) o(\Omega_2)} \quad (5)$$

where $C = \varepsilon \varepsilon_0 A/L$ is the geometrical capacitance of the SBD, $o(\omega_i) = (\omega^2 - \omega_i^2)^2 + v^2 \omega^2$ and Ω_i ($i = 1, 2$) are the resonant frequencies determined by the roots of the characteristic equation $\Delta(\omega) = 0$ at $v_i = 0$. For $l_d < l$, one obtains the following approximation for the resonant frequencies: $\Omega_1^2 = \omega_h^2 - \beta \omega_-^2 r_d$, $\Omega_2^2 = (1+\beta) \omega_-^2 r_d$ where $\omega_h^2 = \omega_+^2(r_- + r_d) + \omega_-^2 r_+$, $\beta = r_+(\omega_+^2 - \omega_-^2)/\omega_h^2$, $r_d = l_d/L$.

Figure 1 (b) illustrates four typical spectra of current fluctuations given by Eq. (5) for a SBD with $l_+ = l = 0.1 \, \mu m$ and $v/\omega_- = 0.1$. In the simplest case of the homogeneous SBD ($\omega_+ = \omega_- \equiv \omega_0$, curves 1 and 2) only the second resonance $\Omega_2 = \omega_0 \sqrt{r_d}$, related to the presence of the depletion region is physically meaningful for current fluctuations:

$$S_{II}(\omega) = 4k_B T C \frac{v\omega^2}{(\omega^2 - \Omega_2^2)^2 + v^2 \omega^2} (\omega_0^2 - \Omega_2^2) \quad (6)$$

while the plasma oscillations of free carriers described by $\Omega_1 = \omega_0$ give no contribution to $S_{II}(\omega)$ (see curve 2). Such a case gives only evidence of the resonance associated with the reflection of carriers by the barrier. For $r_d \to 0$, the depletion region disappears, $\Omega_2 \to 0$ and Eq. (6) recovers the Lorenzian spectrum typical of thermal fluctuations of an homogeneous resistor: $S_{II}(\omega) \sim \nu/(\omega^2 + \nu^2)$ (see curve 1). When the SBD contains the n^+n-homojunction (see curves 3 and 4), $\omega_+/\omega_- = 3.16$ and the noise spectrum exhibits a resonant peak at the frequency of the hybrid plasma resonance, $\Omega_1^2 = \omega_+^2(r_- + r_d) + \omega_-^2 r_+$, in the highest-frequency region of the spectrum. The position and height of this peak is practically independent of the presence or less of the depletion region.

It is worthwhile to note that the variance of current fluctuations, $<\delta I^2> = k_B T C[r_+ \omega_+^2 + r_- \omega_-^2] = k_B T e^2 N_0/(L^2 m)$, obtained from Eq. (5) independs of the presence and position of resonances Ω_i of collective motions in the SBD and similarly to the case of a uniform resistor it is determined by the temperature and a total number of carriers N_0 inside the SBD. Therefore, the resonant enhancement of the noise power at frequencies $\Omega_i \neq 0$ (see Fig. 1(b)) is merely a spectral redistribution of the thermal noise power under the conservation of the integral value given by $<\delta I^2> = \int_0^\infty S_{II}(\omega) d\omega/2\pi$. Such a redistribution can be treated as the upconversion of the low-frequency thermal noise with Lorentzian spectrum into the high-frequency regions at Ω_i due to collective motions excited in the SBD.

Figure 2 reports the comparison of the spectra of current fluctuations for a GaAs n^+n SBD with $n = 5 \times 10^{16}$ cm^{-3}, $n^+ = 2 \times 10^{18}$ cm^{-3}, $l = 0.15$ μm and $l_+ = 0.05$ μm calculated at 300 K with the MCP technique and with the analytical model described above. For completeness, the figure also presents the separate contributions coming from n^+ and n-regions obtained from the analytical model (long- and short-dashed curves, respectively). The results refer to the two operation regimes corresponding to barrier-limited Fig. 2(a) and flat-band Fig. 2(b) conditions. Analytical calculations are performed in accordance with Eq. (4) by taking $\nu_+ = 4 \times 10^{13}$ and $\nu_- = 10^{13}$ s^{-1} as estimated from MC simulations of bulk GaAs with the same carrier concentrations. The good agreement found between analytical and MCP results in the frequency range of interest fully validates the analytical model proposed here.

ACKNOWLEDGMENTS

This work is supported by NATO Collaborative Linkage under Grant PST.EAP.CLG 980629 and project "Noise models and measurements in nanostructures".

REFERENCES

1. P. Shiktorov, et al, IEEE Electr. Dev. Lett. **26**, 2 (2005).
2. M. Trippe, G. Bosman, and A. van der Ziel, IEEE Trans. Microwave Theory Techn. **MTT-34**, 1183 (1986).

Monte Carlo Simulation of Quantum Noise

X. Oriols

Departament d'Enginyeria Electrònica, Universitat Autònoma de Barcelona, SPAIN
E.mail: xavier.oriols@uab.es

Abstract. The experimental current measured in quantum-based devices fluctuates around average values due to the quantum mechanical wave-particle duality. An alternative quantum noise approach, based on the de Broglie-Bohm interpretation of the quantum theory, is presented. Since this formalism directly deals with the wave and particle nature of electrons, it provides an excellent framework to study quantum noise. In fact, most of the numerical techniques used in classical Monte Carlo simulations can be directly adapted to our quantum formalism. Simulated noise results for interacting electrons tunnelling through double barriers are presented in agreement with experimental results. The present approach opens a new path for studying electron transport and quantum noise in mesoscopic systems.

Keywords: Noise, Bohm trajectories, Quantum transport, Monte Carlo simulation.
PACS: 05.40.Ca, 72.70.+m, 73.63.-b, 05.60.Gg,

INTRODUCTION

The classical Monte Carlo (MC) method for the simulation of transport properties in semicondcutor devices is based on the direct solution of the Boltzmann equation by numerically simulating the motion of individual electrons. Each electron path is described via semi-classical transport equations where its initial conditions are selected randomly. Such microscopic description of the electron transport has become the most versatile and powerful technique to describe average values, and current fluctuations, in classical devices.

In quantum scenarios, it is mandatory to overcome the previous semi-classical assumptions and to take into account the wave-like nature of electrons [1]. In general, quantum devices simulators (with few exceptions) emphasize the wave nature of electrons, but neglect its complementary particle-like behavior. However, an approach that allows the discussion of both aspects of the electron is mandatory to correctly model the current fluctuations. For example, in tunneling barriers, the electron probability of being reflected or transmitted is determined by the wave nature of electrons (the transmission coefficient) and the fact that the electron can only be transmitted or reflected (but not both!) shows its particle-like behavior.

The Buttiker formalism [1,2] has become the standard procedure for noise predictions of quantum-based devices. It contains the essential wave and particle description of the electron. Buttiker generalized the Landauer scattering approach via the second quantization formalism. The basic elements of his approach are creation and annihilation operators. Due to the commuting algebra of these operators, the simultaneous presence of one electron at different contacts is not possible [2]. The Buttiker formalism has been successfully applied in many experimental situations [1,3,4] and it has the additional advantage that it directly includes the fermionic/bosonic symmetries when describing many particle systems [1,2].

Contrarily to classical MC simulations (where all electron dynamics are coupled by the Poisson equation), most quantum simulators are based on a picture of non-interacting electrons (the Fermi liquid). The standard procedure to include the coulomb interaction among electrons in quantum scenarios is the mean field theory. It focuses on a single particle Hamiltonian replacing the Coulomb interaction of all other electrons by a mean field potential. In spite of its unquestionable success, such approximation cannot describe the Coulomb correlation in the motions of different electrons.

Here, we present an alternative quantum transport approach based on the de Broglie-Bohm (dBB) interpretation of the quantum theory [3-7]. Each electron is described by a well-defined quantum trajectory guided by a wave function. Since the dBB formalism directly deals with the wave and particle nature of electrons, it provides an excellent framework to study quantum noise. In addition, the dBB approach allows the introduction of the Coulomb interaction with a procedure similar to those of the classical MC. In fact, most of the numerical techniques used in classical MC, which are related with the microscopic description of electron trajectories [8], can be directly adapted to our dBB formalism [9-12].

OUR APPROACH

The fundamental difference between the classical MC and our approach is the computation of the electron velocity. In the former, it is directly related to the electric field (solution of the Poisson equation), while in our approach it is determined from the wave-nature of the electron. We consider the presence of N electrons in a device active region described by the many-particle Schrödinger equation within the approximation of the effective mass, m, for non-relativistic electrons:

$$i\hbar \frac{\partial \Psi(\vec{r}_1,...,\vec{r}_N,t)}{\partial t} = \left\{ \sum_{k=1}^{N} \left(-\frac{\hbar^2}{2m} \nabla_k^2 \right) + U(\vec{r}_1,...,\vec{r}_N,t) \right\} \Psi(\vec{r}_1,...,\vec{r}_N,t) \quad (1)$$

where $\vec{r}_k = (x_k, y_k, z_k)$ is the particle position, $U(\vec{r}_1,...,\vec{r}_N,t)$ is the potential energy that takes into account the external applied voltages and the electron-electron Coulomb interaction and $\Psi(\vec{r}_1,...,\vec{r}_N,t) = R(\vec{r}_1,...,\vec{r}_N,t) \cdot \exp(iS(\vec{r}_1,...,\vec{r}_N,t)/\hbar)$ is the many-particle wave-function rewritten, for convenience, in a polar form. In order to emphasize the computation of Bohm trajectories, let us notice that an intuitive Newton-like equation can be obtained from the previous Schrödinger equation:

$$\frac{d^2 \vec{r}_k}{dt^2} = -\nabla_k \left(U(\vec{r}_1,...,\vec{r}_N,t) + Q(\vec{r}_1,...,\vec{r}_N,t) \right) \bigg|_{\vec{r}_i=\vec{r}_i(t), i \neq k} \quad ; \quad k=1,...,N \quad (2)$$

where, in addition to the classical term, a quantum potential appears which is responsible for all non-local quantum features:

$$Q(\vec{r}_1,...,\vec{r}_N,t) = -\left\{ \sum_{k=1}^{N} \frac{1}{R(\vec{r}_1,...,\vec{r}_N,t)} \left(\frac{\hbar^2 \nabla_k^2 R(\vec{r}_1,...,\vec{r}_N,t)}{2m} \right) \right\} \quad (3)$$

The computation of the electron trajectory $\vec{r}_i(t)$ is coupled to all other trajectories by the potentials. The initial values of the Bohm trajectories are selected randomly according to the initial wave-function uncertainty. By construction, all standard (Copenhagen) probabilistic results obtained from the wave function are exactly reproduced by an appropriate ensemble of Bohm trajectories [5,7-9]. In fact, such equivalence explains the limited success of Bohmian mechanics in solving practical quantum problems. In general, since the computation of Bohm trajectories needs the detailed knowledge of the wave function, the extra effort needed for the trajectories seems unnecessary. However, the re-formulation of a system of (non-relativistic) interacting electrons in terms of Bohm trajectories provides and advantaged numerical framework to study Coulomb correlations. The details are explained somewhere else [13]. As a numerical example, we study current noise in resonant tunneling diodes (RTD). Experimental results [14,15] shows noise enhancement at the region of negative conductance due to Coulomb correlation.

NUMERICAL RESULTS

We study the role of the electron-electron Coulomb interaction in the transport properties of a typical RTD with a quantum well of GaAs with a length of 6 nm between two barrier of AlGaAs of 2nm. Numerical details of the simulation can be found in reference 12. The role of the Coulomb interaction as a source of electron correlation is shown on the noise properties of the RTD depicted in figures 1 and 2 in terms of the Fano factor. The power of the current fluctuations at low frequency, $S(0)$, is normalized to the well-known Poissonian power spectral density, $S(0) = 2 \cdot q \cdot I$ with I the average current value and q the electron charge. Therefore, the Fano factor is defined as $F = S(0)/2 \cdot q \cdot I$. We have used room temperature, 300 K, for all numerical simulations.

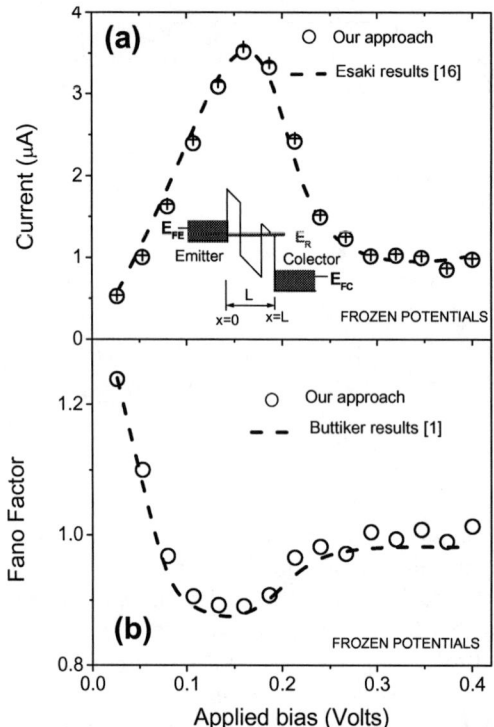

FIGURE 1. (a) Current voltage characteristic for the standard RTD diode of the text. The inset shows the static (frozen) potential profile. In dashed line, current results obtained from the Esaki expression [16]. (b) In circles, numerical values of F for each bias points computed using our Quantum Monte Carlo approach. In dashed line, Buttiker formalism [1].

In order to be able to distinguish the effects originated by the electron-electron interaction, first in figure 1, we consider frozen potentials (i.e without using self-consistence between the electron charge and the potential energy). As seen in the inset of figure 1, a hand-made linear approximation is used for defining the potential profile for different bias. For this simple system, with independent electrons, the excellent agreements between our numerical values and the theoretical predictions for the current (obtained from reference [16]) and the noise (obtained from reference [1]), provides a satisfactory test of the numerical implementation of our approach.

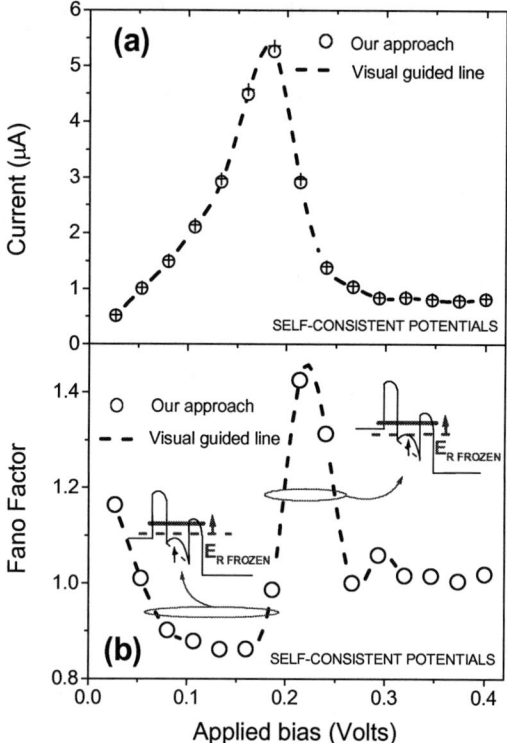

FIGURE 2. (a) Current-voltage characteristic for the RTD diode of fig. 1 when the Poisson equation is solved self-consistently with the time-dependent Schrödinger equation. (b) Fano factor for each bias point computed with our approach. The insets show the effect of time-dependent variations of the potential energy profile (due to the presence of an electron at the quantum well) on the transmission of other electrons, before and after the resonant voltage.

The results depicted in figure 2 are obtained for Coulomb interacting electrons. The Poisson and the Schrödinger equations for the same RTD considered in figure 1 are solved self-consistently. The instantaneous charge is computed from the Bohm trajectories. The effect of the Coulomb correlation between electrons is clearly manifested in the Fano factor depicted in figure 2b. Just after the resonant voltage, in the negative differential conductance region, (when the resonant energy, E_R, is below the conduction band at the emitter contact) the presence of one electron inside the quantum well raises the value E_R (see inset in fig. 2b). Thus, the transmission for the next electron is highly enhanced at this voltage. Roughly speaking, the Coulomb interaction affects the electron dynamics by trying to regroup the electrons and providing a Fano factor higher than one. The noise shows a process of bunching of electrons (i.e. super-Poissonian noise). On the contrary, just before the resonant voltage, the presence of an electron inside the quantum well obstructs the transport of the second one. Therefore, the noise presents a process of anti-bunching of electrons (i.e. sub-Poissonian noise). These results are in good agreement with experimental data [14-15]

CONCLUSIONS

In conclusion, we show that quantum noise can be naturally understood within the dBB interpretation of the quantum theory. Most of the numerical techniques used in classical MC simulations can be directly adapted to our quantum MC formalism. Simulated noise results for non-interacting and interacting electrons through double tunnelling barriers are presented, in agreement with experimental results. The present approach opens a new path for studying average values and current fluctuations in either static [9-11] or time-dependent [17] quantum scenarios.

ACKNOWLEDGMENTS

The author is really grateful to Javier Mateos, Tomás González, Emilio Méndez and Giuseppe Iannaccone for fruitful discussions. This work has been partially supported by the Ministerio de Ciencia y Tecnología and FEDER through project TIC2003-08213-C02-02.

REFERENCES

1. Blanter, Y.M. and Büttiker, M., *Phys. Rep.*, 336, 1, (2000) ; De Jong, M.J.M. and Bennakker, C.W. J., NATO ASI Series E, 345 Kluwer Academic Press (1997)
2. Büttiker, M., *Phys. Rev. Lett.* 65, 2901-2904 (1990); *Phys. Rev. B* 46(19), 12485-12507 (1992).
3. Bell, J.S., *Speakable and Unspeakable in Quantum Mechanics*, Cambridge U.P., England (1997).
4. Bohm, D., *Phys. Rev.* 85, 166-179, (1952).
5. Suñé, J. and Oriols, X., *Phys. Rev. Lett.*85, 894 (2000).
6. Oriols, X., Martín, F. and Suñé, J., *Phys. Rev. A* 54, 2594-2604, (1996).
7. Oriols, X. *Physical Review A* 71, 017801 (2005).
8. Jacoboni, C., and Reggiani, L., *Rev. Mod. Phys.*55, 645-700, (1983).
9. Oriols, X., Martín, F., and Suñé, J., *Appl. Phys. Lett.*79, 1703-1705, (2001).
10. Oriols, X., Martín, F., and Suñé, J., *Appl. Phys. Lett.*80, 4048-4050 (2002).
11. Oriols, X., *IEEE Trans. on Electron Devices*50(9), 1830 (2003).
12. Oriols, X., Trois, A., and Blouin., G., *Appl. Phys. Lett.* 85(16), 3596-3598 (2004).
13. Oriols, X., Submitted.
14. V.Ya Aleshkin *et al.*, Semicond. Sci and Technol, 18, L35 (2003); and references therein.
15. Iannaccone, G., Macucci, M., and Pellegrini, B., *Phys. Rev. B*55(7), 4539-4550 (1997); Iannaccone, G., Lombardi, G., Macucci M., and Pellegrini, B., *Phys. Rev. Lett* 80(5), 1054-1057(1998); Song, W., Méndez, E.E., Kuznetsov V., and Nielsen,B., *Appl. Phys. Lett.* 82(10), 1568 (2003).
16. Esaki, L., and Chang, L.L., *Phys. Rev. Lett.* 33, 495-498, (1974).
17. Oriols, X., Alarcón, A., and Fernández-Díaz, E., *Physical Review B* in press (2005)

Noise and diffusion in superlattices within the Wannier-Stark approach

Marcello Rosini* and Lino Reggiani*,†

*Dipartimento di Ingegneria dell'Innvazione, Università di Lecce, via Arnesano, 73100 Lecce Italy
†National Nanotechnology Laboratory NNL-INFM

Abstract. Hopping transport in a superlattice is studied by a Monte Carlo method in the Wannier-Stark picture. Velocity fluctuations are determined by the diffusion coefficient. Einstein relation is found to be fulfilled at moderate fields even in the presence of negative differential mobility. By contrast, at the highest fields a more complete description is needed for the diffusion coefficient. This includes the effects of space quantization and the presence of a complex set of hopping transitions.

Keywords: diffusion, electron transport, noise, superlattices, Wannier-Stark
PACS: 05.40.Ca 02.70.Uu 68.65.Cd 73.63.-b 72.70.+m

INTRODUCTION

In recent years, superlattices have received a relevant scientific and technological interest, owing to their peculiar electrical and optical properties. In particular, superlattices exhibit a strong negative differential conductivity (NDC) regime [1]. In the NDC regime, charge transport can be described in terms of hopping between Wannier-Stark (WS) states [2] thus being a good model for studying hopping conduction in semiconductor structures.

The aim of this paper is to analyze velocity fluctuations through the calculation of the longitudinal diffusion coefficient. To this purpose, we have used an ensemble Monte Carlo simulator of independent carriers. In this case the low frequency spectral density of velocity fluctuations is related to the diffusion coefficient by the simple relation $S_v(0) = 4D$.

Under thermal equilibrium the diffusion coefficient is given by the Einstein relation $D = \frac{2}{3e}\mu\langle\varepsilon\rangle$ where μ is the carrier mobility and $\langle\varepsilon\rangle$ their mean energy which for thermal electrons is given by $\langle\varepsilon\rangle = \frac{3}{2}K_B T$. Under far from equilibrium conditions the Einstein relation is in general no longer valid and in the case of WS-hopping transport another formulation for the diffusion coefficient has been proposed in the literature [3]. In this paper we will investigate to which extent the diffusion coefficient in a superlattice compares with the Einstein and WS-hopping models.

MODEL AND RESULTS

The physical system is a Si/SiO_2 superlattice with period $L = 3.1$nm, where silicon is grown along the (100) direction. Within the envelope-function and effective-mass approximations, it is possible to reduce the electronic problem to the following

Schrödinger equation for the envelope function:

$$\left[-\frac{\hbar^2}{2}\nabla_i\left(\frac{1}{m(z)}\right)_{ij}\nabla_j + W_{SL}(z) - eEz\right]\mathscr{F}(\mathbf{r}) = \varepsilon\mathscr{F}(\mathbf{r}), \quad (1)$$

where $\left(\frac{1}{m(z)}\right)_{ij}$ is the effective mass tensor, $\mathscr{F}(\mathbf{r})$ is the envelope wave function, $W_{SL}(z)$ is the Kronig-Penney potential, which describes the superlattice along the z direction, and E is the applied electric field. The solution of the problem is found to be [4]

$$\mathscr{F}_{n\mathbf{k}_\parallel}^\nu(\mathbf{r}) = ce^{ik_xx}e^{ik_yy}\Phi_n(z-\nu L) \quad (2)$$

$$E_n^\nu(\mathbf{k}_\parallel) = -\nu eFL + \bar{\varepsilon}_n + \frac{\hbar^2}{2m_x}k_x^2 + \frac{\hbar^2}{2m_y}k_y^2, \quad (3)$$

where: ν is the spatial index of the well, n is the level index, $\bar{\varepsilon}_n$ is the energy offset relative to the n-th level, and $\Phi_n(z-\nu L) = \Phi_n^\nu(z)$ is the Wannier-Stark eigenfunction for the n-th level and centered in the ν-th well.

In the model we have considered only optical phonons, in particular the Si deformation potential optical mode and the SiO_2 polar and deformation potential optical modes. Moreover, we have used the scheme of confined optical phonons [5]: in layered systems, the frequencies of the optical phonons [6] of the two materials do not match in general, so we assume that the optical phonon of one layer does not propagate into the other one.

Scattering mechanisms are introduced through the Fermi golden rule, obtaining the following probability per unit time for the transition from a state $(\nu n\mathbf{k}_\parallel)$ to a state $(\nu'n')$ with any $\mathbf{k}'_\parallel$

$$P(\nu'n',\nu n\mathbf{k}_\parallel) = \frac{\sqrt{m_xm_y}}{(2\pi)^5\hbar^3}\left[\begin{array}{c}n_q\\n_q+1\end{array}\right]\int d\theta\int dq_z c^2(\mathbf{q})\left|\Phi_{n'}^{*\nu'}(z)e^{\mp iq_zz}\Phi_n^\nu(z)\right|^2. \quad (4)$$

In the finite difference formulation the drift velocity v_d and the longitudinal diffusion coefficient D_z are estimated, from the Monte Carlo simulation, as

$$v_d(t) = \frac{1}{N}\sum_{i=1}^N\frac{\delta z_i(t)}{\delta t} = \frac{L}{N\delta t}\sum_{i=1}^N\delta\nu_i(t). \quad (5)$$

$$D_z = \frac{1}{2}\frac{\delta\langle(\Delta z)^2\rangle}{\delta t}. \quad (6)$$

with $\Delta z = z - \langle z\rangle$ the instantaneous fluctuation of the carrier position along the field direction.

The drift velocity versus high electric fields is reported in Fig. 1. At intermediate fields we observe a remarkable decrease of the drift velocity and thus the presence of an NDC regime. Here NDC is due to localization effect driven by the electric field. At the highest fields we find a substantial increase of the drift velocity. Looking at the figure, where different contributions to drift velocity are reported, we explain the sudden increase in

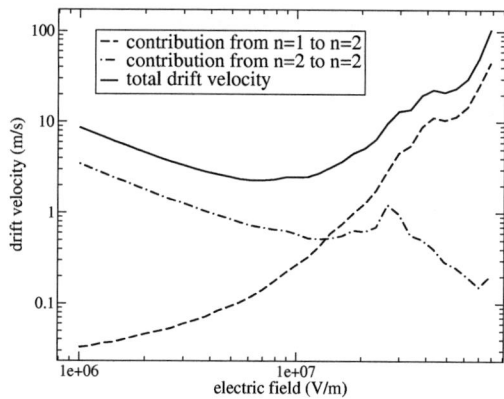

FIGURE 1. Electron drift velocity in the superlattice. The total drift velocity is the sum of different contributions: the two curves report the intrasubband and intersubband contributions respectively.

terms of two kind of phenomena. The first one is the presence of a resonance between the energy difference of two adjacent wells and the Si phonon energy $\hbar\omega_{op}$ (which is the most important scattering mechanism) that occurs when $eEL = \hbar\omega_{op}$; this resonance, occurring at $E \simeq 2.5 \cdot 10^7 \text{V/m}$, is clearly visible in the velocity arising from intrasubband hopping. The second phenomenon is the intersubband hopping, that becomes significant only at very high electric fields.

The diffusion coefficient obtained from the simulation is reported in Fig. 2, together with those obtained from the Einstein relation and the WS-hopping model. We observe, that the diffusion coefficient obtained from the Einstein relation with the mobility calculated as v_d/E from the data of Fig. 1, well reproduces the first part of the simulated curve, corresponding to the NDC regime. The second part at the highest fields is substantially underestimated. To overcome this underestimation, we used an analytic form calculated from the Wannier-Stark formulation of transport [3]:

$$D_z = \frac{v_d L}{2} \coth \frac{L}{L_E} \qquad (7)$$

where $L_E = \frac{2K_B T}{eE}$ is a thermal field length. In the limit $L/L_E < 1$ the above formula reproduces the Einstein relation and thus agrees with the simulations. However, at the highest fields, when $L/L_E > 1$, even if this formula gets closer to the simulated value, it still underestimates the diffusion coefficient for a factor 3-4 (see Fig. 2). The main reasons for the remaining disagreement is that eq.(7) is calculated in the approximation of nearest-neighbor coupling between the wells of the superlattice, and that it does not take into account the contribution to diffusion of the intersubband transitions. This conjecture has been tested by complementary simulations which neglect the above features and prove the good agreement between the numerical results and Eq. (7). Finally, we want to point out that the crossover between the two diffusion regimes

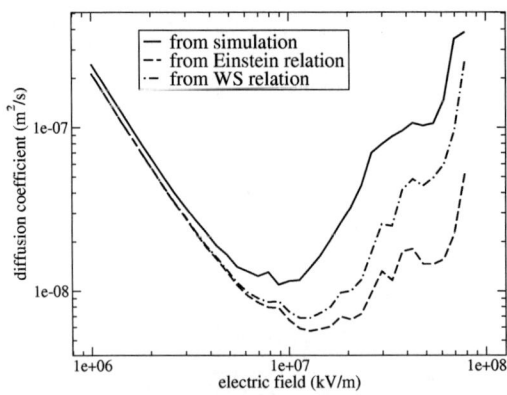

FIGURE 2. Diffusion coefficient in the superlattice. The simulated curve (continuous line) is compared with those obtained from the Einstein relation (dashed line) and from the relation of Ref.[3] (dash-dotted line)

occurs at a field for which $L = L_E$, and that, in correspondence of this crossover, the mean energy of the carriers is found to start increasing with the electric field.

CONCLUSIONS

The high-field behavior of the velocity fluctuation noise calculated through the diffusion coefficient is found to exhibit a two regime behavior. In the NDM region it is well described by the Einstein relation with a field dependent mobility. In the highest field region, where the drift velocity starts increasing again with the electric field, the results of the simulations are only qualitatively described by the WS-hopping model which is found to represent a too simple model in the present formulation. Remarkably, the superlattice period represents an intrinsic size parameter for the structure which dominates the behavior of longitudinal diffusion at high fields. The crossover between the Einstein and the hopping expressions is determined by the relation $L = L_E$.

Support by Italian Ministry of Education, University and Research (MIUR) under the project "Noise models and measurements in nanostructures" is gratefully acknowledged.

REFERENCES

1. H. T. Grahn, K. von Klitzing, K. Ploog, and G. H. Döhler, *Phys. Rev. B*, **43**, 12094 (1991).
2. A. Wacker, and A.-P. Jauho, *Phys. Rev. Lett.*, **80**, 369 (1998).
3. V. V. Bryskin, and P. Kleinert, *J. Phys.: Condens. Matter*, **15**, 1415 (2003).
4. G. H. Wannier, *Phys. Rev.*, **117**, 432 (1960).
5. H. Rücker, E. Molinari, and P. Lugli, *Phys. Rev. B*, **45**, 6747 (1992).
6. M. Rosini, C. Jacoboni, and S. Ossicini, *Phys. Rev. B*, **66**, 155332 (2002).

Coulomb Suppression of Avalanche Noise in Double-Drift IMPATT Diodes

Antanas Reklaitis[1] and Lino Reggiani[2]

(1) Semiconductor Physics Institute, Goshtauto 11, 2600 Vilnius, Lithuania
(2) INFM-National Nanotechnology Laboratory and Dipartimento di Ingegneria dell' Innovazione, Università di Lecce, Via Arnesano s/n, 73100 Lecce, Italy

Abstract. Current noise in GaN double-drift impact avalanche diodes is investigated by Monte Carlo simulations. For values of the current multiplication factor greater than ten avalanche noise is suppressed down to three order of magnitude with respect to the standard excess noise factor. Negative feedback between fluctuations in space charge and in number of generated electron-hole pairs is proven to be the responsible of this giant suppression

Keywords: avalanche-noise, GaN, IMPATT, Monte Carlo, shot-noise
PACS: 72.70.+m, 72.20.-i, 72.30+a, 73.23.Ad

INTRODUCTION

In this paper we report on a giant suppression of avalanche noise which occur at increasing current because of long range Coulomb interaction. To this purpose we investigate current voltage (I-V) and noise performance of an avalanche GaN double-drift diode starting from a microscopic model of the steady current and using a self-consistent Monte Carlo (MC) simulation [1]. By comparing the results of simulations performed with a static and a dynamic Poisson solver we determine unambiguously the role played by space charge fluctuations on avalanche noise.

MODEL AND RESULTS

The structure of the GaN diode is of the type $p^{++}/p/p^{++}/n^-/n^{++}/n^+/n/n^{++}$ with length in nanometer of $|30|30|10|10|50|60|$ and doping concentration in 10^{17} cm^{-3} of $500|3|7|400|1|400|3|1|100$. Besides phonon scattering, we consider ionized impurity scattering, band-band impact ionization initiated by electrons and holes and electron-hole pairs created through Zener tunneling [3]. Thus, the resulting current fluctuations account for all the noise sources, namely: velocity, number and their cross-correlations.

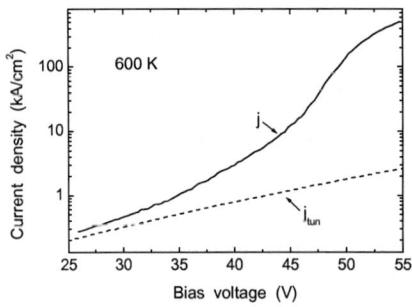 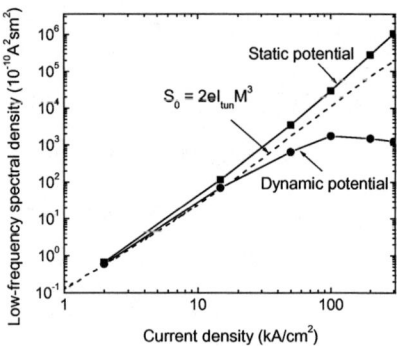

FIGURE 1. Steady-state I-V characteristic of the IMPATT diode at T=600 K. Solid curve shows the I-V characteristic calculated for the full model of the device. The dashed curve is obtained when carrier multiplication by impact ionization is disregarded, and current is induced by Zener tunneling process.

FIGURE 2. Low-frequency value of spectral density of current fluctuations in the IMPATT diode as a function of the current density. Open (solid) circles refer to the case in which the instantaneous fluctuations of the self-consistent potential distribution are considered (neglected) in the simulation. Dashed curve shows the results of Eq. (1).

Figure 1 reports the I-V characteristic calculated taking into account carrier multiplication by impact ionization (j and continuous curve) and when carrier multiplication is neglected (j_{tun} and dashed curve) so that current is controlled by Zener tunneling process. From the results reported in Fig. 1, the average current multiplication factor M is estimated by $M=j/j_{tun}$. At low voltages around 30-40 V, the sharp increase of j is caused by the exponential increase of the electron and hole ionization rates in the avalanche zone which is located between the p^+ and n^+ layers. At high voltages around 50 V, due to the space charge of generated carriers, the electric field is distributed enough uniformly throughout the diode. When the bias voltage increases above about 50 V, the electric field increases faster in the drift regions than in the avalanche zone. As a consequence, the rate of increase of carrier multiplication is reduced at further increasing voltages until the diode breakdown occurs at about U = 80 V.

The fluctuations of the potential distribution do not modify the I-V characteristics. However, they can play a very significant role in the determination of the noise level. Accordingly, we have studied the influence of the potential fluctuations on the noise performance of the diode by comparing the noise properties obtained with the fluctuating (dynamic) and non-fluctuating (static) potential distributions. In MC simulations, the static potential distribution is obtained from the self-consistent potential distribution averaged over a sufficiently long time. Figure 2 reports the low frequency spectral density as a function of the current density. The kinds of noise present in the device are: shot noise due to carrier multiplication, thermal noise due to fluctuations in carrier velocity, and cross-correlation noise between the above two sources. As expected, in the avalanche region considered here the shot noise associated with fluctuations in carrier number is the dominant noise source. Under static potential conditions, the results of MC simulations (full circles in the figure) exhibit a cubic increase with current of the noise level, in general agreement with the theoretical prediction [3]

$$S_0 = 2eI_{tun}M^3 \tag{1}$$

where e is the unit charge, I_{tun} the primary leakage current (Zener in this case) before the multiplication. By contrast, under dynamic potential conditions (open circles in Fig. 2) the noise level at high current densities is found to saturate being reduced down to about three orders of magnitude with respect to what predicted by Eq. (1). The comparison between static and dynamic simulations is detailed in Figs. 3 and 4 where the variance of current fluctuations and the characteristic correlation times are shown as a function of current. Both quantities are found to be reduced significantly by the presence of Coulomb correlations at the highest currents. In particular, the fact that the reduction occurs when the dielectric relaxation time

$$\tau_d = \frac{dV}{dj} \frac{\varepsilon \varepsilon_0}{L} \tag{2}$$

(with ε_0 the static dielectric constant, ε the vacuum permittivity and L the total length of the diode) becomes the shortest time scale (see Fig. 4) confirms that the suppression of the noise level is associated with the role played by long range Coulomb interactions. Indeed, the effect of potential fluctuations on the avalanche noise becomes of importance at current values above about 10 kA/cm² when the dielectric relaxation time is found to become comparable with or shorter than the dynamic correlation time associated with the decay of the correlation function of current fluctuations. Coulomb interaction provides the mechanism of suppression which is explained as follows. Electrons and holes generated by impact ionization drift in opposite directions and the associated space charge reduces the built-in electric field in the avalanche zone. Now, in case the number of generated electron-hole pairs exhibits a positive fluctuation, the electric field in the avalanche zone is reduced and the successive carrier generation rate is suppressed. Hence, a positive fluctuation in the created pair number (and, consequently, in current) is damped. In the opposite case that the number of generated pairs exhibits a negative fluctuation, the electric field in the avalanche zone is enhanced and the successive carrier generation rate is also enhanced. Again, a negative fluctuation in the created pair number is damped.

CONCLUSIONS

We have investigated transport and noise characteristics of a GaN homojunction IMPATT diode by performing microscopic Monte Carlo simulations. Giant suppression of avalanche noise is found at increasing current density because of a negative feedback between fluctuations in the number of electron-hole pair generated

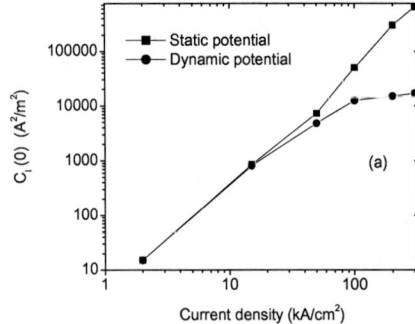 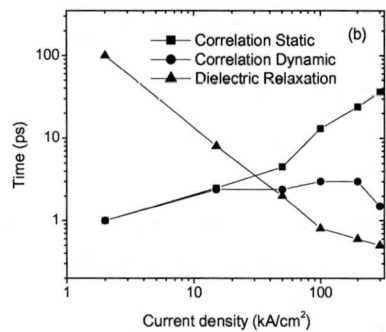

FIGURE 3. Amplitude of the current correlation functions obtained from the simulations under static (full square) and dynamic (full circles) conditions as a function of the steady current for the IMPATT diode under test.

FIGURE 4. Different times scales relevant to current fluctuations as a function of the steady current for the IMPATT diode under test. Full squares refer to the static correlation time, full circle to the dynamic correlation time and full triangles to the differential dielectric relaxation time of the diode. Curves are guide to the eyes.

by impact ionization and fluctuations of the self consistent potential. This suppression mechanism, which is reminiscent of that occurring in vacuum diodes before current saturation conditions, can be of valuable help to design low noise photodetectors [4].

ACKNOWLEDGMENTS

Partial support by Italian Ministry of Education University and Research under the project "Noise models and measurements in nanostructures" is gratefully acknowledged.

REFERENCES

1. Reklaitis A. and Reggiani L., Phys Rev. **B60** 11683-11693 (1999).
2. Reklaitis A. and Reggiani L., J. Appl. Phys., **95**, 7925-7935 (2004).
3. McIntyre R.J., IEEE Trans. Electron Devices vol. **ED-13**, 164-168 (1966).
4. Carbone A., Introzzi R. and Liu H.C., Appl. Phys.Lett., **82**, 4292-4294 (2003).

A Frequency Domain Spherical Harmonics Solver for the Langevin Boltzmann Equation

Christoph Jungemann* and Bernd Meinerzhagen*

*NST, Postfach 33 29, TU Braunschweig, 38023 Braunschweig, Germany

Abstract. A new solver for the Langevin-Boltzmann equation (LBE) is presented, which in contrast to the Monte Carlo (MC) method calculates noise directly in the frequency domain by means of a spherical harmonics expansion of the LBE. This new noise simulation approach is orders of magnitude more CPU efficient than MC and allows to solve technically relevant problems for which the MC method fails. This is demonstrated for generation/recombination noise, noise in cyclostationary systems, and devices. In the latter case it is shown that hot electrons in a velocity saturation region actually produce less noise than cold electrons.

Keywords: Noise, Langevin Boltzmann equation, spherical harmonics expansion, silicon
PACS: 72.10.Bg,72.20.Ht,72.30.+q

INTRODUCTION

The rapid proliferation of wireless and broadband applications has caused an increased demand for physics-based noise modeling [1]. The most fundamental models in the framework of semiclassical transport theory are based on the Boltzmann equation (BE) [2], which is usually solved by the Monte Carlo (MC) method [3]. MC simulations inherently contain noise, but they are performed in the time domain and the CPU time is at least inversely proportional to the minimum frequency investigated. In Si devices already frequencies below 100GHz require excessive CPU times [4, 5]. On the other hand, RF noise (noise in the lower GHz range) and low frequency noise are the most important types of noise for practical applications. Thus, other methods than MC are required to solve the BE and they should be based in the frequency domain.

A numerical solver for the BE in the frequency domain based on a first order spherical harmonics expansion (SHE) was demonstrated in [6] for bulk systems. In [7] the first frequency domain based solver utilizing an n-th order SHE for the Langevin-type BE (LBE) was presented, which like the MC method solves the full BE without any approximations. Albeit it is numerically more challenging than the MC method, this new SHE solver has many distinct advantages over the MC method: **1)** It can handle the full frequency range from arbitrarily high frequencies down to zero frequency, where the CPU time is almost independent of the frequency; **2)** A full small-signal analysis is possible; **3)** It can be combined with the harmonic balance technique to calculate higher harmonics and intermodulation products; **4)** It can simulate rare events at low computational cost; **5)** It gives more physical insight than MC, because the Green's functions are available; **6)** It is orders of magnitude more CPU efficient than MC. This is demonstrated below.

THEORY

For the sake of brevity only the case of a stationary and spatially homogeneous system is discussed in this section, for which the real space coordinates can be neglected and the LBE is linear. It reads in this case [2]

$$\left\{\frac{\partial}{\partial t} - \frac{q}{\hbar}\vec{E}\nabla\right\}p = \hat{W}\{p\} + \xi . \tag{1}$$

$p(\vec{k},t)$ is the single-particle probability density, which is fluctuating due to the Langevin force $\xi(\vec{k},t)$, and $\hat{W}$ the scattering integral

$$\hat{W}\{p\} = \frac{\Omega}{(2\pi)^3}\int W(\vec{k}|\vec{k}'')p(\vec{k}'',t) - W(\vec{k}''|\vec{k})p(\vec{k},t)\mathrm{d}^3k'' , \tag{2}$$

where $\vec{E}$ is the electric field, $W(\vec{k}|\vec{k}'')$ the transition rate, and Ω the system volume [3]. The Langevin force has zero mean and is delta-correlated in time

$$\mathrm{E}\left\{\xi(\vec{k},t)\xi(\vec{k}',t')\right\} = S_{\xi\xi}(\vec{k},\vec{k}')\delta(t-t') . \tag{3}$$

The white and symmetric power spectral density (PSD) of the Langevin force is given by [2]

$$\begin{aligned}S_{\xi\xi}(\vec{k},\vec{k}') &= \frac{2\Omega}{(2\pi)^3}\left[\int W(\vec{k}''|\vec{k})p(\vec{k}) + W(\vec{k}|\vec{k}'')p(\vec{k}'')\mathrm{d}^3k''\,\delta(\vec{k}-\vec{k}')\right.\\ &\quad \left.-W(\vec{k}|\vec{k}')p(\vec{k}') - W(\vec{k}'|\vec{k})p(\vec{k})\right] \\ &= S_{\xi\xi}(\vec{k}',\vec{k}) ,\end{aligned} \tag{4}$$

where $p(\vec{k})$ is the noiseless stationary probability density [8]. The corresponding Green's functions $G(\vec{k},\vec{k}',\omega)$ are the solutions of a modified BE in the frequency domain

$$\left\{i\omega - \frac{q}{\hbar}\vec{E}\nabla\right\}G = \hat{W}\{G\} + \delta(\vec{k}-\vec{k}') , \tag{5}$$

where ω is the angular frequency. With the Wiener-Lee theorem [9] the PSD of the single-particle probability density reads

$$S_{pp}(\vec{k},\vec{k}',\omega) = S_{pp}^*(\vec{k}',\vec{k},\omega) = \int\int G(\vec{k},\vec{k}_1,\omega)S_{\xi\xi}(\vec{k}_1,\vec{k}_1')G^*(\vec{k}',\vec{k}_1',\omega)\mathrm{d}^3k_1\mathrm{d}^3k_1' . \tag{6}$$

The asterisk indicates the complex conjugate. The PSD of two microscopic variables $x(\vec{k})$, $y(\vec{k})$ (e.g. velocity and energy) is given by

$$S_{xy}(\omega) = \int\int x(\vec{k})S_{pp}(\vec{k},\vec{k}',\omega)y(\vec{k}')\mathrm{d}^3k\mathrm{d}^3k' = S_{yx}^*(\omega) . \tag{7}$$

Green's functions are defined for the microscopic quantities [10, 6]

$$G_x(\vec{k}',\omega) = \int x(\vec{k})G(\vec{k},\vec{k}',\omega)d^3k, \tag{8}$$

which are direct solutions of the adjoint of (5). Since $G_x(\vec{k}',\omega)$ depends only on one wave-vector argument, it can be evaluated much more CPU efficiently than $G(\vec{k},\vec{k}',\omega)$, which depends on two arguments. The final expression for S_{xy} reads

$$S_{xy}(\omega) = \int\int G_x(\vec{k},\omega)S_{\xi\xi}(\vec{k},\vec{k}')G_y^*(\vec{k}',\omega)d^3kd^3k'. \tag{9}$$

This formulation has the additional advantage that the source of the fluctuations, the PSD of the Langevin force, and their propagation by the Green's functions are separated. This gives more physical insight than solutions of the BE, where this separation is not possible.

Electron transport is described with the Modena group's aniosotropic and non-parabolic six valley model [11] including a Brooks-Herring like impurity scattering model with an empirical correction for high doping concentrations [8]. The generation/recombination (GR) process is modeled similar to [12]. In contrast to most previous works (e.g.: [13, 14]), which are based on the isotropic approximation for the band structure, in this work the full anisotropy is retained. The LBE is expanded up to the nth order with Legendre polynomials (LP) and discretized in energy and real space by finite differences [13, 7]. The cyclostationary case is simulated with a new harmonic balance technique, which is a generalization of the method developed for momentum-based noise models [15]. For the calculation of the Green's functions the CPU efficient adjoint method is used [16]. In the case of devices the LBE and Poisson equation are solved self-consistently by the Newton-Raphson method. The resultant large sparse matrix systems are solved using a new numerical software package [17].

RESULTS

In Fig. 1 (a) the first three LP coefficients of the electron distribution function (EDF) in bulk Si are shown for the valleys with the heavy mass parallel and perpendicular to the electric field in $\langle 100 \rangle$ direction. Without problems the EDF is evaluated over many orders of magnitude by SHE. In the case of MC simulations this is only possible if statistical enhancement is used, which biases the statistics [18] and makes it impossible to calculate noise. Therefore, only SHE is able to simulate rare events and noise at the same time.

The convergence of SHE is demonstrated in Fig. 1 (b) by comparison with consistent MC results for velocity fluctuations in a bulk system. Good agreement of both methods is found, where already a third order LP expansion is sufficient and the SHE CPU time is more than two orders of magnitude smaller than MC.

This CPU-time advantage becomes even larger if GR processes involving traps are present. Due to their relatively long life times these processes have a strong impact on noise in the GHz range and below. This is shown in Fig. 2 (a) for a process with a life

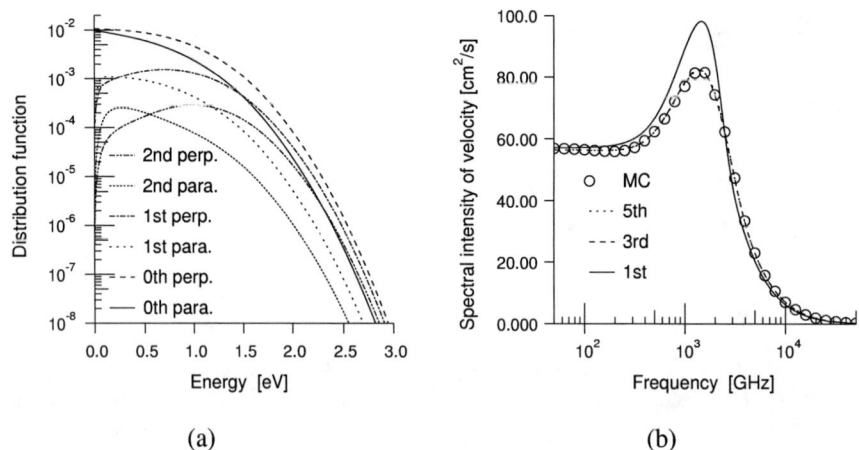

FIGURE 1. First three LP coefficients of the electron distribution function (left) for the two sets of valleys with the heavy mass parallel (para.) and perpendicular (perp.) to the electric field of 300kV/cm and the PSD of the longitudinal velocity (right) for an electric field of 30kV/cm based on different orders of SHE and MC results for bulk Si at room temperature.

time of 5ns. No MC results are shown, because the CPU time would be excessively long. The noise shows a strong dependence on the electric field at low frequencies because GR noise is proportional to the square of the drift velocity. In Fig. 2 (b) the corresponding side band correlation matrix (SCM) of the current fluctuations is shown for a cyclostationary electric field. The single-tone simulations include the DC component and eight harmonics. Upconversion of noise is clearly visible at the fundamental frequency of 1GHz. In the case of the larger electric field, which varies between 0 and 60kV/cm and results in velocity saturation, noise upconversion is also found at multiples of the fundamental frequency because of the strong nonlinearity. In contrast to Ref. [12], where similar simulations were performed by solving a severely simplified BE by the MC method, here the full LBE is solved and a technically relevant fundamental frequency of 1GHz is used instead of the 500GHz of Ref. [12], which were necessary to make the MC simulations feasible. Despite this very high fundamental frequency the MC simulations of Ref. [12] were extremely CPU intensive. Thus, already MC bulk simulations for technically relevant frequencies can entail prohibitively long CPU times.

In Fig. 3 results are presented for a 1D N^+NN^+ structure. At a bias of 1.0V the drift velocity saturates in the low electron density region of the device, where also the mean energy has its peak. In addition, the local contribution to the terminal current noise is shown. At equilibrium the distribution is symmetric and roughly inversely proportional to the conductivity. Thus it peaks in the low density region. For an applied bias of 1.0V the peak of the spatial noise distribution does no longer coincide with the minimum of the electron density, but moves to the LHS of the low density region, where the electrons have a relatively low mean energy. The hot electrons on the RHS on the other hand produce less noise than under equilibrium conditions. This result, which cannot be obtained by MC simulation, clearly verifies previous findings based on the impedance

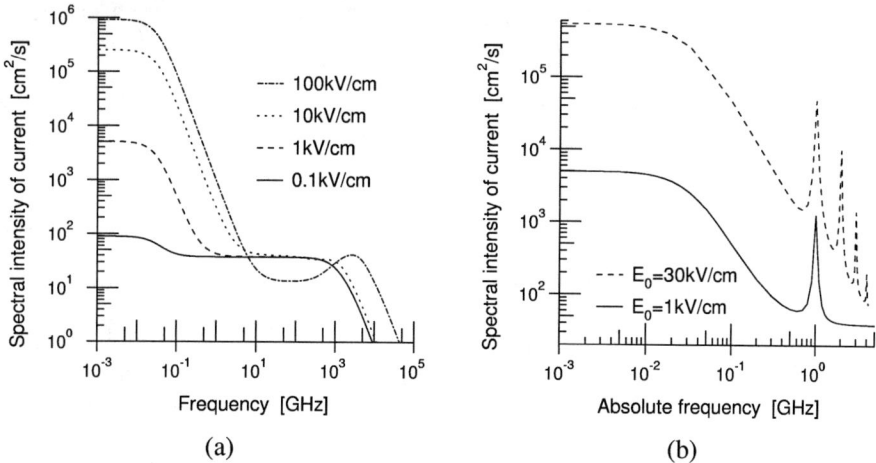

FIGURE 2. PSD of the normalized current fluctuations for n-Si with a donor concentration of $10^{17}/\text{cm}^3$ at room temperature including a GR process with a life time of 5ns and the corresponding central diagonal element of the SCM for a periodic electric field ($E = E_0(1+\cos(2\pi f_0 t))$, $f_0 = 1\text{GHz}$).

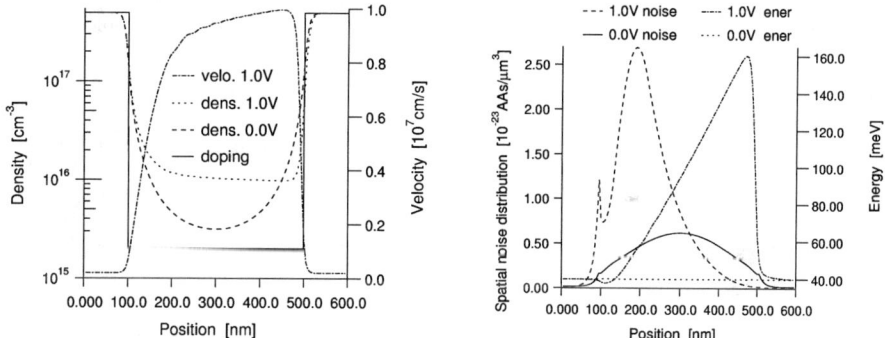

FIGURE 3. Doping concentration, electron density, drift velocity, mean energy, and local contribution to the terminal current noise for the Si-N^+NN^+ structure biased at 0.0 and 1.0V for room temperature.

field method [19, 20]. One reason for this decrease are the Green's functions, which propagate the fluctuations to the device terminals. Due to velocity saturation they decay with increasing mean energy.

CONCLUSIONS

We have presented the first frequency-domain based LBE solver for cyclostationary systems and devices including GR processes, which can handle arbitrarily low frequencies without prohibitive CPU times. It is much faster than the MC method for technically relevant problems and gives more physical insight. It has been demonstrated that hot electrons in a velocity-saturation region produce less and not more noise than cold electrons.

REFERENCES

1. T. H. Lee, *The Design of CMOS Radio Frequency Integrated Circuits*, Cambridge University Press, Cambridge, UK, 1998, 1st edn.
2. S. Kogan, *Electronic Noise and Fluctuations in Solids*, Cambridge University Press, Cambridge, New York, Melbourne, 1996.
3. C. Jacoboni, and P. Lugli, *The Monte Carlo Method for Semiconductor Device Simulation*, Springer, Wien, 1989.
4. C. Jungemann, B. Neinhüs, S. Decker, and B. Meinerzhagen, *IEEE Trans. Electron Devices*, **49**, 1258–1264 (2002).
5. T. Gonzalez, J. Mateos, M. J. Martin-Martinez, S. Perez, R. Rengel, B. G. Vasallo, and D. Pardo, *Proceedings of the 3rd International Conference on Unsolved Problems of Noise*, pp. 496–5038 (2003).
6. C. E. Korman, and I. D. Mayergoyz, *Phys. Rev. B*, **54**, 17620–17627 (1996).
7. C. Jungemann, and B. Meinerzhagen, "A Legendre Polynomial Solver for the Langevin Boltzmann Equation," in *Proc. IWCE*, 2004, pp. 22–23.
8. C. Jungemann, and B. Meinerzhagen, *Hierarchical Device Simulation*, Computational Microelectronics, Springer, Wien, New York, 2003.
9. A. Papoulis, *Probability, Random Variables, and Stochastic Processes*, McGraw–Hill, 2001, 4th edn.
10. K. M. van Vliet, *J. Math. Phys.*, **12**, 1981–1998 (1971).
11. R. Brunetti, C. Jacoboni, F. Nava, L. Reggiani, G. Bosman, and R. J. J. Zijlstra, *J. Appl. Phys.*, **52**, 6713–6722 (1981).
12. S. Pérez, T. González, S. Delage, and J. Obregon, *J. Appl. Phys.*, **88**, 800–807 (2000).
13. N. Goldsman, L. Henrickson, and J. Frey, *Solid–State Electron.*, **34**, 389–396 (1991).
14. A. Gnudi, D. Ventura, G. Baccarani, and F. Odeh, *Solid–State Electron.*, **36**, 575–581 (1993).
15. F. Bonani, S. D. Guerrieri, and G. Ghione, *IEEE Trans. Electron Devices*, **50**, 633–644 (2003).
16. F. H. Branin, *IEEE Transactions on circuit theory*, **20**, 285–288 (1973).
17. M. Bollhöfer, and Y. Saad, ILUPACK — preconditioning software package, release 1.1 available online at www.math.tu-berlin.de/ilupack/ (2004).
18. M. G. Gray, T. E. Booth, T. J. T. Kwan, and C. M. Snell, *IEEE Trans. Electron Devices*, **45**, 918–924 (1998).
19. J.-S. Goo, C.-H. Choi, F. Danneville, E. Morifuji, H. S. Momose, Z. Yu, H. Iwai, T. H. Lee, and R. W. Dutton, *IEEE Trans. Electron Devices*, **47**, 2410–2419 (2000).
20. C. Jungemann, B. Neinhüs, S. Decker, and B. Meinerzhagen, "Hierarchical 2D RF Noise Simulation of Si and SiGe Devices by Langevin-type DD and HD Models based on MC Generated Noise Parameters," in *IEDM Tech. Dig.*, Washington (USA), 2001, pp. 481–484.

Two-dimensional physics-based low-frequency noise modeling of bipolar semiconductor devices in small- and large-signal operation

F. Bertazzi, G. Conte, S. Donati Guerrieri, F. Bonani, G. Ghione

Dipartimento di Elettronica, Politecnico di Torino,
Corso Duca degli Abruzzi, I-10129 Torino, ITALY

Abstract. The paper presents an efficient, multidimensional noise simulator accounting for trap-assisted generation-recombination (GR) noise, both in small- and forced large-signal conditions. As a case study, trap-assisted GR noise in a bipolar diode is considered, highlighting a complex conversion behaviour.

Keywords: Semiconductor device noise, Circuit noise, Semiconductor device modeling, Microwave devices, Nonlinear systems
PACS: 85.40.Qx, 07.05.Tp

INTRODUCTION

State-of-the-art analog RF devices are affected by variable amounts of colored, low-frequency small-signal noise in addition to the white, high-frequency noise floor related to diffusion noise: low-frequency noise can be, in many cases, traced back to trap-assisted generation-recombination (GR) phenomena. Due to device nonlinearities, up-conversion of low-frequency noise can be a limiting factor for wireless TX/RX links. Therefore, accurate physics-based GR noise models not only in small-signal, but above all in large-signal conditions are called for. Finally, GR noise modelling could be a paradigm for the treatment of $1/f$ noise, since flicker noise can be modelled, on a certain frequency band, through a proper superposition of noninteracting GR spectra [1]: a proper use of GR noise sources could be a viable, though possibly empirical, substitute for a $1/f$ fundamental noise source.

In this work, we have implemented a two-dimensional bipolar device simulator, accounting also for the presence in the device of an arbitrary number of (position-dependent) trap levels. Besides the standard DC and small-signal analyses, this simulator is able to evaluate the large-signal (LS) device working point in forced periodic operation through the Harmonic Balance (HB) solution of the coupled drift-diffusion and trap rate equations. Concerning noise, both the small-signal, i.e. stationary, and LS, i.e. cyclostationary, cases are treated with the Green's function approach presented in [2]. A similar implementation was presented in [3]. We will show results on the cyclostationary GR noise behaviour for a bipolar diode, wherein, besides a bulk trap uniformly distributed in the device, surface traps are also accounted for in a region close to the catode device contact. This example was chosen to model the simultaneous effect of bulk traps in the device substrate and of the damage possibly introduced on the device

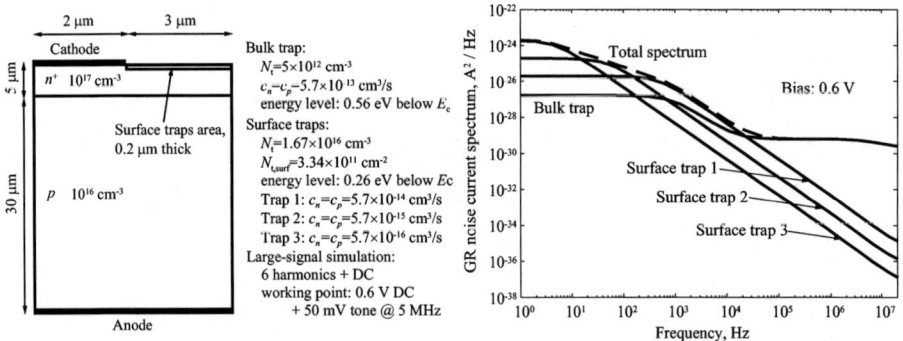

FIGURE 1. Structure and trap parameters for the n^+p Si junction (left). Small-signal stationary GR spectra for the n^+p abrupt junction (right). The surface traps yield a $1/f$-like spectrum.

surface during fabrication.

MODEL IMPLEMENTATION

A unified Green's function approach to trap-assisted GR noise modeling for physical, PDE-based simulators amenable for implementation both in small- and large-signal device operation was presented in [4]. The total partial differential equations (PDE) system is made of the standard drift-diffusion model plus one rate equation for each trap included in the simulation. The model main advantage is the use of stationary white microscopic noise sources [5, 6], allowing for a uniquely defined modulation scheme in cyclostationary conditions [7].

A simulator implementing the above model in two dimensions has been developed. The LS device working point is evaluated in the frequency domain with the HB approach, while Green's functions required for noise analysis, both in the stationary and cyclostationary case, are efficiently evaluated using the generalized adjoint approach [5, 2]. Due to the large number of discretization nodes necessary to accurately simulate a 2D device, to solve the HB nonlinear system through the Newton method, an iterative approach for the solution of the linearized system is required: we have exploited the GMRES linear solver, using the preconditioner originally proposed in [8]. The GMRES algorithm is also used for the determination of the Green's functions required for cyclostationary noise analysis: in this case, an original implementation has been devised to improve the efficiency of the simulation [9].

TEST CASE

As a test case we have simulated a 2D Si abrupt n^+p diode, with cross section and simulation parameters shown in Fig. 1 (left). We included a bulk trap and three surface traps modelled as bulk traps present in a thin surface volume. The surface trap parameters were chosen to yield logarithmically spaced lifetimes, in order to possibly recover

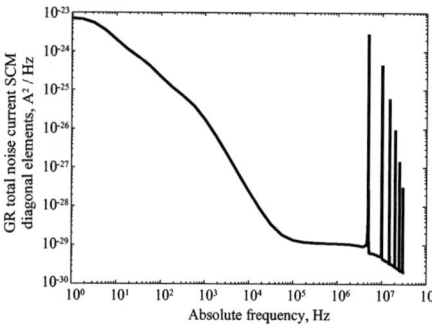 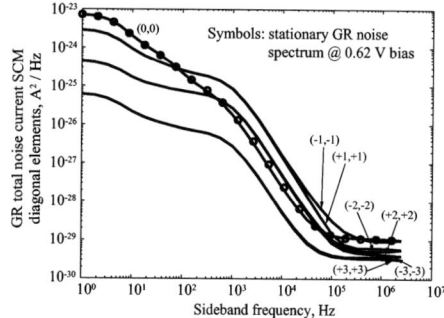

FIGURE 2. Absolute (left) and sideband (right) frequency dependence of the diagonal SCM elements for the total GR noise current. Symbols in the right figure are the stationary total GR noise current spectrum for a DC bias yielding the DC component of the LS current. On the left figure, 6 noise sidebands are shown, while 3 only are represented on the right one.

a $1/f$-like stationary noise spectrum [1]. The device was discretized with 4100 nodes, while the LS working point in forward operation required 6 harmonics plus DC for an accurate HB simulation.

The stationary noise current spectrum due to GR noise only is shown in Fig. 1 (right) as a function of frequency, for a forward bias of 0.6 V. The contributions of the four trap levels are shown separately, pointing out that a $1/f$-like behaviour is actually obtained for a limited frequency range, as a result of the superposition of the lorentzian spectra due to the surface traps.

In the LS case, we have applied a 50 mV input tone at 5 MHz, superimposed to a 0.6 V DC bias component. The simulation results for what concerns the first 3 noise sidebands are shown in Fig. 2 (left) and Fig. 2 (right), for the absolute and sideband frequency dependence, respectively. In those figures, we have plotted the diagonal elements of the total GR noise current sideband correlation matrix (SCM), i.e. we have summed up the contributions of the bulk and surface traps. A strong frequency conversion effect from baseband to upper sidebands is clearly observed. A more detailed analysis, based on the right part of Fig. 2, points out that the baseband total spectrum (element $(0,0)$ of the SCM) is basically equal to the stationary spectrum evaluated for a bias point (0.62 V) yielding a current equal to the DC component of the LS device current. Furthermore, the frequency conversion effect is intricated, since the shape of the upper sidebands significantly differs from the baseband one. This behaviour can be traced back to the relative contributions of the four traps to GR noise, and to their different frequency conversion. To gain further insight, we show in Fig. 3 the sideband frequency dependence of the diagonal SCM elements for the bulk and, as an example, for surface trap 3. While for the surface trap the frequency conversion exhibits a decreasing value of the low (sideband) frequency plateau as the sideband order increases, for the bulk trap the trend is different. From Fig. 1 (right), one observes that the plateau of the trap 3 contribution is three orders of magnitude higher than the bulk trap. Fig. 3 shows that the same ratio is maintained in the baseband, while a much lower ratio is present for the upper sidebands. This proves that, although the GR spectra corresponding to the various

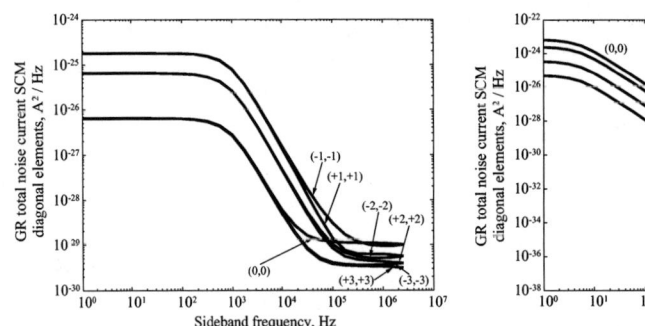

FIGURE 3. Sideband frequency dependence of the diagonal SCM elements for the GR noise current due to the bulk trap (left) and surface trap 3 (right).

traps are converted maintaining the same sideband frequency dependence, the shape of the total GR spectrum at the various sidebands is different due to the relative importance of the single trap contributions.

This result suggests that a simple terminal level modulation of the stationary spectrum [10], often exploited in circuit-level cyclostationary noise simulations, is not in general an accurate approach. Furthermore, care must be exerted when a $1/f$-like stationary spectrum is modelled as a superposition of trap spectra, since the exact conversion of such a frequency dependence towards upper sidebands should not be taken for granted.

ACKNOWLEDGMENTS

This work was partially supported by the Italian Minister of University and Research through the PRIN 2003 project "Non-linear noise models and design of low-phase noise oscillators for high performance communication systems". Support from the Azione Integrata 2005 "Analisi cinetica e PDE del rumore nonlineare in dispositivi RF e a microonde" funded by the Italian and Spanish governments is also acknowledged.

REFERENCES

1. F. N. Hooge, *IEEE Trans. El. Dev.* **41,** No. 11, pp. 1926–1935, Nov. 1994.
2. F. Bonani et al., *IEEE Trans. El. Dev.*, Vol. ED-48, No. 5, pp. 966–977, May 2001.
3. J. E. Sanchez et al., *IEDM Technical Digest*, pp. 477–480, Washington, 2001. J. E. Sanchez, PhD Thesis, Univ. of Florida, 2002.
4. S. Donati Guerrieri et al., *FaN04* pp. 37–48, May 2004.
5. F. Bonani, G. Ghione, *Noise in semiconductor devices: modelling and simulation*, Springer Verlag: Heidelberg, 2001.
6. F. Hou, PhD Thesis, University of Florida, 2002.
7. F. Bonani et al., *IEEE Trans. El. Dev.*, Vol. ED-49, No. 9, pp. 1640–1647, September 2002.
8. B. Troyanovsky et al., *Computer Methods in Applied Mechanics and Engineering*, vol.181, No. 4, pp. 467–482, 2000.
9. F. Bertazzi et al., submitted to *Modeling and Simulation of Electron Devices*, Pisa, Italy, July 2005.
10. F. Bonani et al., *IEDM Technical Digest*, pp. 133–136, San Francisco, December 2002.

Governing Equations of the Green's Functions for the Short-Circuit Terminal Noise Currents in Semiconductor Devices

Sung-min Hong*, Chan Hyeong Park[†], Hong Shick Min* and Young-June Park*

*School of Electrical Engineering and Nano-Systems Institute (NSI-NCRC), Seoul National University, Korea
[†]School of Electronics Engineering, Kwangwoon University, Korea

Abstract. We derive the governing equations of the Green's functions for the terminal noise currents of semiconductor devices. Using the properties of the newly derived governing equations, we prove the Nyquist theorem using the transport equations under the drift-diffusion (DD) scheme at arbitrary frequencies for the first time.

INTRODUCTION

The Green's functions in the semiconductor devices, which are the linear responses of the transport equations to impulsive perturbations, play crucial roles in the noise calculation [1]. Unfortunately, from a numerical standpoint, the computational burden for evaluating the Green's functions through a brute-force method is fairly high.

Ghione applied the adjoint approach to the unipolar devices for numerical efficiency [2]. In this approach, the "discretized" adjoint system - a linear system whose unknowns are the Green's functions for the terminal voltages - can be identified directly from the principle of the circuit theory. The desired Green's functions are obtained by solving the adjoint system just once, therefore the computational cost is greatly reduced. However, it was found that the adjoint approach was not suitable to the bipolar devices, since the electron and hole concentrations are coupled in Poisson equation. Bonani solved this difficulty in the bipolar device through the generalized adjoint approach [3], which employs the "discretized" semiconductor equations. Although this approach is clearly convenient for a computational standpoint [1] [2], we propose another equivalent method where the "continuous" adjoint system of a bipolar device can be explicitly identified. The knowledge of the governing equations of the Green's functions in the new approach allows deeper insight into the noise transfer mechanism inside a device and is expected to provide a valuable tool for the noise analysis as shown in the last section.

DERIVATION OF THE GOVERNING EQUATIONS

In the DD model, the small-signal semiconductor equations with the Langevin noise sources are given as follows [1]:

$$\nabla \cdot \delta \mathbf{J}_\psi = j\omega q (\delta p - \delta n) + s_\psi, \tag{1}$$

$$\nabla \cdot \delta \mathbf{J}_n = j\omega q \delta n + s_n, \tag{2}$$

$$\nabla \cdot \delta \mathbf{J}_p = -j\omega q \delta p + s_p. \tag{3}$$

Here the generation-recombination processes are neglected for the simplicity of derivation and all symbols have their conventional meanings. s_α's for $\alpha = \psi, n, p$ are the Langevin noise sources.

We consider the test functions G_α^k ($\alpha = \psi, n, p$) for the k-th terminal [4] [5], which satisfy the following boundary conditions:

$$G_\alpha^k = \delta_{kl} \quad \text{at the } l\text{-th terminal,} \tag{4}$$

$$\mathbf{n} \cdot \nabla G_\alpha^k = 0 \quad \text{on the free surfaces.} \tag{5}$$

Using these boundary conditions, the outward short-circuit noise current for the k-th terminal, δI^k, can be written as

$$\delta I^k = \int_\Omega d^3 r \nabla \cdot \left(G_\psi^k \delta \mathbf{J}_\psi + G_n^k \delta \mathbf{J}_n + G_p^k \mathbf{J}_p \right). \tag{6}$$

We impose the following requirement on G_α^k for $\alpha = \psi, n, p$ so that G_α^k's become the Green's functions for the k-th terminal short-circuit current (the current Green's functions for the k-th terminal):

$$\delta I^k = \int_\Omega d^3 r \left(G_\psi^k s_\psi + G_n^k s_n + G_p^k s_p \right). \tag{7}$$

Note that (6) and (7) are two different expressions for the same scalar quantity δI^k. After equating two equations followed by replacing $(\nabla \cdot \delta \mathbf{J}_\alpha - s_\alpha)$ for $\alpha = \psi, n, p$ by $j\omega q (\delta p - \delta n), -j\omega q \delta p, j\omega q \delta n$, respectively, we get

$$\int_\Omega d^3 r \left[\nabla \cdot \left\{ (j\omega\varepsilon)\nabla G_\psi^k + (q\mu_n n_s)\nabla G_n^k + (q\mu_p p_s)\nabla G_p^k \right\} \right] \delta\psi$$
$$- q \int_\Omega d^3 r \left[j\omega(G_\psi^k - G_n^k) + \mu_n \nabla \psi_s \cdot \nabla G_n^k + \nabla \cdot \{D_n \nabla G_n^k\} \right] \delta n \tag{8}$$
$$+ q \int_\Omega d^3 r \left[j\omega(G_\psi^k - G_p^k) - \mu_p \nabla \psi_s \cdot \nabla G_p^k + \nabla \cdot \{D_p \nabla G_p^k\} \right] \delta p = 0,$$

where the auxiliary equations for the current densities and boundary conditions are used in deriving (8).

Mathematically, three unknown functions can be determined by three appropriate governing equations. Therefore, G_ψ^k, G_n^k, G_p^k can be determined by the following three equations.

$$\nabla \cdot \left\{ (j\omega\varepsilon)\nabla G_\psi^k + (q\mu_n n_s)\nabla G_n^k + (q\mu_p p_s)\nabla G_p^k \right\} = 0, \tag{9}$$

$$(j\omega)G_\psi^k - (j\omega)G_n^k + \mu_n \nabla \psi_s \cdot \nabla G_n^k + \nabla \cdot \{D_n \nabla G_n^k\} = 0, \tag{10}$$

$$(j\omega)G_\psi^k - (j\omega)G_p^k - \mu_p \nabla \psi_s \cdot \nabla G_p^k + \nabla \cdot \{D_p \nabla G_p^k\} = 0. \tag{11}$$

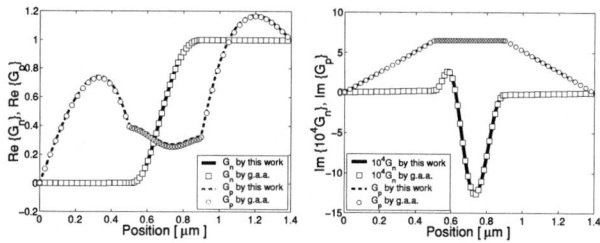

FIGURE 1. The electron and hole current Green's functions for a 1-D n^+-n-n^+ resistor obtained from both the governing equations (this work) and the generalized adjoint approach (g.a.a.). $f = 1$ GHz.

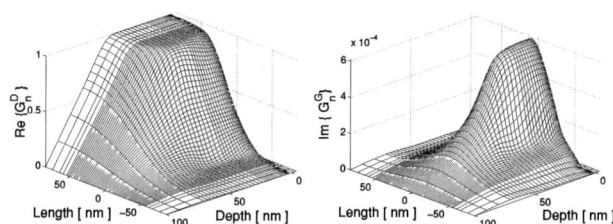

FIGURE 2. Real part of the drain electron current Green's function G_n^D and imaginary part of the gate electron current Green's function G_n^G for an NMOSFET obtained from both the governing equations and the generalized adjoint approach. We find no difference between the two results. The MOSFET has $L = 90 nm, t_{ox} = 4.0 nm$. $f = 1$ GHz.

When (9)-(11) hold, we find from (8) that (7) also holds, which implies that the solutions of (9)-(11) with the boundary conditions (4), (5) satisfy the requirement of the current Green's functions. Thus, the governing equations with the boundary conditions constitute the generalized continuous adjoint system [2].

Numerical simulation are also carried out to confirm the correctness of the governing equations by comparing our results with those of the conventional generalized adjoint approach [1] [3]. Figure 1 shows the electron and hole current Green's function for a 1-D n^+-n-n^+ resistor. Figure 2 shows the drain electron current Green's function G_n^D and the gate electron current Green's function G_n^G for the NMOSFET. We find no difference between our results and those of the conventional generalized adjoint approach.

PROOF OF THE NYQUIST THEOREM

Under dc zero-bias condition the proof of the Nyquist theorem in the DD model at zero frequency limit can be found in the literatures [4]. However, the proof at arbitrary frequencies is still lacking, because the relations between Green's functions in semiconductor devices and the small-signal admittances are not clearly known.

Through the comparison between the governing equations (9)-(11) and the small-signal transport equations of the semiconductor device under the dc zero-bias condition, we find that G_ψ^k, G_n^k, G_p^k become equal at arbitrary frequencies to $\delta\psi^k, \delta f_n^k, \delta f_p^k$, respectively, which are the ac responses of the electrostatic potential, the electron quasi-Fermi potential and the hole quasi-Fermi potential to an unit ac voltage excitation at the k-th terminal.

At the zero dc bias, only the diffusion noise sources are considered. The cross power spectral density for the k-th and l-th terminal noise currents is obtained from (7),

$$S_{\delta I^k \delta I^l} = \int_\Omega d^3 r (\nabla G_n^{k*} \cdot \nabla G_n^l) 4q^2 D_n n_s + \int_\Omega d^3 r (\nabla G_p^{k*} \cdot \nabla G_p^l) 4q^2 D_p p_s. \quad (12)$$

Here we use the relation of $S_{\eta_n}(\mathbf{r}, \mathbf{r}', \omega) = 4q^2 D_n n_s \delta(\mathbf{r} - \mathbf{r}'), S_{\eta_p}(\mathbf{r}, \mathbf{r}', \omega) = 4q^2 D_p p_s \delta(\mathbf{r} - \mathbf{r}')$, where $S_{\eta_n}(\mathbf{r}, \mathbf{r}', \omega), S_{\eta_p}(\mathbf{r}, \mathbf{r}', \omega)$ are the power spectral densities of the electron diffusion noise source and the hole diffusion noise source, respectively.

A simple manipulation shows that the first integral in the r.h.s. of (12) gives

$$\int_\Omega d^3 r (\nabla G_n^{k*} \cdot \nabla G_n^l) 4q^2 D_n n_s = -2k_B T \left[\delta I_n^{k,l} + \delta I_n^{l,k*} - \int_\Omega d^3 r j\omega q (G_\psi^{k*} \delta n^l - G_\psi^l \delta n^{k*}) \right] \quad (13)$$

where $\delta I_n^{k,l}$ and δn^l are the outward electron ac current for the k-th terminal and the ac response of the electron concentration, respectively, when the unit ac voltage is applied only to the l-th terminal. Similar relation holds for the second integral in the r.h.s. of (12). We can show that the last term in the r.h.s. of (13) and the similar term obtained from the second integral in (12) gives $(-2k_B T \delta I_\psi^{k,l} - 2k_B T \delta I_\psi^{l,k*})$, where $\delta I_\psi^{k,l}$ is the outward displacement ac current for the k-th terminal, when the unit ac voltage is applied only to the l-th terminal. Then we can show that summation of the two integrals in (12) yields

$$S_{\delta I^k \delta I^l} = 2k_B T (Y_{kl} + Y_{lk}^*), \quad (14)$$

at arbitrary frequencies, where Y_{kl}'s are the small-signal admittances.

ACKNOWLEDGMENTS

This work was supported by the National Core Research Center program of the Korea Science and Engineering Foundation (KOSEF) through the NANO Systems Institute at Seoul National University. C. H. Park's work was supported by the National Program for Tera-Level Nanodevices of the Ministry of Science and Technology as one of the 21st century Frontier Programs.

REFERENCES

1. F. Bonani and G. Ghione, *Noise in Semiconductor Devices. Modeling and Simulation.* Heidelberg, Germany: Springer-Verlag, 2001.
2. G. Ghione et al., *IEEE Trans. CAD*, vol. 12, pp. 425-438, 1993.
3. F. Bonani et al., *IEEE Trans. ED*, vol. 45, pp. 261-269, 1998.
4. C. Jungemann and B. Meinerzhagen, *Hierarchical Device Simulation. The Monte-Carlo Perspective.* Wien, Austria: Spinger-Verlag, 2003.
5. P. D. Yoder et al., *J. Appl. Plys.*, vol. 79, pp. 1951-1954, 1996

Noise Enhancement as Indicator of Instability Onset in Semiconductor Structures

E. Starikov*, P. Shiktorov*, V. Gružinskis*, L. Reggiani[†], L. Varani** and J.C. Vaissière**

*Semiconductor Physics Institute, A. Goštauto 11, LT 01108 Vilnius, Lithuania
[†]INFM - National Nanotechnology Laboratory, Dipartimento di Ingegneria dell' Innovazione, Università di Lecce, Via Arnesano s/n, 73100 Lecce, Italy
**CEM2 - Centre d'Electronique et de Micro-optoelectronique de Montpellier (CNRS UMR 5507), Université Montpellier II, 34095 Montpellier Cedex 5, France

Abstract. The spectral density of current fluctuations and the impedance spectrum in short n^+nn^+ structures made of InN and GaInAs is calculated by the Monte Carlo particle method under conditions favourable for the instability onset in the THz frequency range. Main attention is paid to situations when the instability is related with: (i) quasi-ballistic motion of carriers in the n-region, (ii) spatially-coherent optical phonon emissions in the n-region, and (iii) electron transfer to upper valleys. Results show that the noise investigations in the frequency range below the n^+n hybrid plasma resonance allows, from one hand, to look for and individuate the microscopic mechanism of such instabilities and, from another hand, to evaluate the conditions necessary to realize a microwave generator driven by these instabilities.

Keywords: THz radiation generation, High-frequency noise, Monte Carlo simulation
PACS: 72.20.Ht, 72.30.+q, 72.70.+m

INTRODUCTION

It is well known, that when the applied voltage approaches the threshold for the exponential evolution of an instability in a two-terminal semiconductor device a sharp enhancement of the noise takes place in the frequency region characteristic of the instability. Above this threshold, the noise can transform into an harmonic or quasi-harmonic signal, i.e. the generation process originated by the instability starts working. By considering the significant case of n^+nn^+ structures, we stress that the upper frequency limit of the various instabilities associated with the plasma oscillations of free carriers in the n region is restricted by a hybrid plasma resonance originated by the n^+n-homojunction (see, e.g. [1] and references therein). This hybrid plasma resonance has been recently observed at frequencies around $6-7$ THz [2]. Thus, the appearance of a resonant-like enhancement of the noise at frequencies below that of the n^+n hybrid plasma resonance should be considered as an indicator of the onset of an instability which can transform into generation. As a consequence, the recent development of experimental techniques and setups able to perform noise measurements in the wide sub- and near-THz frequency range [2,3] opens a new approach to the experimental investigation of instabilities. This, in turn, stimulates more detailed theoretical analysis of such an approach.

The aim of this communication is to verify and justify such an approach by applying the Monte Carlo Particle (MCP) simulation of noise to various n^+nn^+ structures.

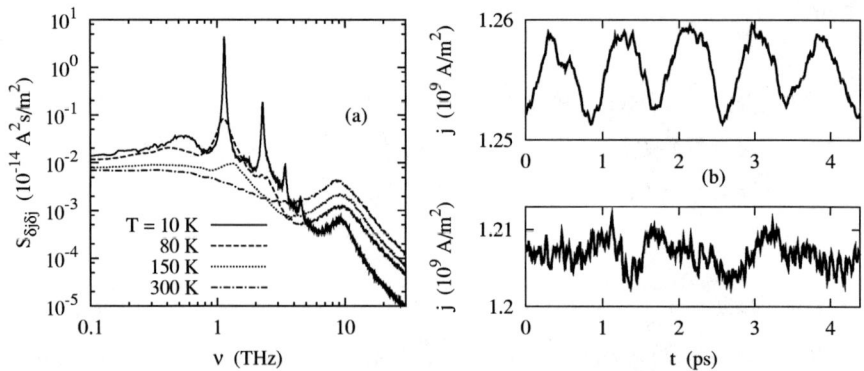

FIGURE 1. (a) Spectral density of current fluctuations at different temperatures T and (b) time dependence of current at $T = 10$ and 80 K (upper and lower cases, respectively) calculated by the MCP approach for 0.02-0.4-0.02 μm n^+nn^+ InN structure with $n^+ = 2 \times 10^{18}$ cm^{-3}, $n = 4 \times 10^{16}$ cm^{-3} at U = 0.15 V.

NUMERICAL RESULTS AND DISCUSSION

Below we shall present three examples of the noise behavior of n^+nn^+ structures under constant voltage operation when the structures move from a passive state to an active one. The latter is characterized by the appearance of a frequency region where the real part of the small-signal impedance of the structure $Z(\omega)$ becomes negative, $ReZ(\omega) < 0$, so that the structure can amplify and/or generate microwave radiation. Here two main scenarios are possible. The first refers to the case when the imaginary part of $Z(\omega)$ crosses the zero axis in this frequency range. Then, when $ImZ(\omega) = 0$, the regular component of the current exhibits an oscillatory behavior (the so-called self-oscillation regime). The second refers to the case when $ImZ(\omega) \neq 0$. Then, the oscillatory behavior of the current at the frequency ω_0 is possible only when the structure is loaded by an external circuit with an impedance $Z_e(\omega)$ which, at $Re[Z(\omega_0) + Z_e(\omega_0)] < 0$, satisfies the resonance condition: $Im[Z(\omega_0) + Z_e(\omega_0)] = 0$. In the absence of the external resonant circuit (i.e. $Z_e(\omega) = 0$) the regular component of the current will not exhibit the oscillatory behavior, however, there will appear a resonant enhancement of the current noise at frequencies where $|Z(\omega)|^2$ reaches minimum values, thus indicating the approaching of instability conditions [1].

To illustrate the first scenario, let us consider first the streaming instability associated with a near ballistic electron motion in the *n*-region of an InN submicron n^+nn^+ structure at low lattice temperatures. The instability is based on the excitation of coherent electron plasma oscillations driven by a single optical phonon emission during the near ballistic carrier flight across the *n*-region [4,5]. The effect is most pronounced when the frequency associated with the transit-time through the *n*-region is close to the plasma frequency proper of the *n*-region. This property is a signature to identify the instability (see, e.g., [4,5] and references therein). The transition from the stable state of the n^+nn^+ structure to the oscillator state due to the instability onset is illustrated in Figs. 1 (a) and (b). Here, we report the frequency behavior of the current noise and the time

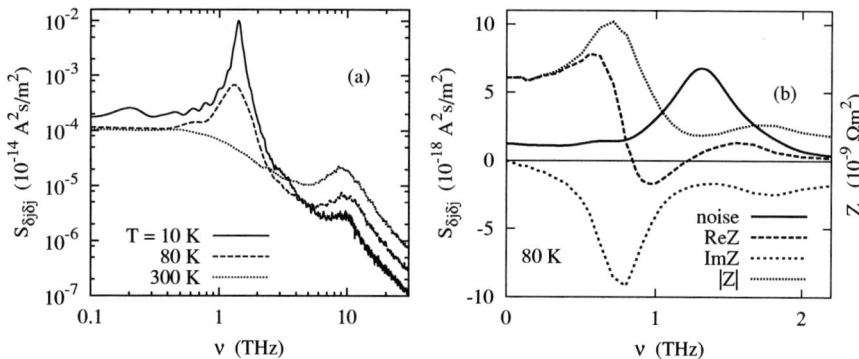

FIGURE 2. (a) General behavior of noise spectra and (b) frequency dependence of spectral density of current fluctuations and impedance in the near-THz region calculated by the MCP method for $0.02-3.0-0.02\ \mu m\ n^+nn^+$ InN structure with $n^+ = 2 \times 10^{18}$ and $n = 2.2 \times 10^{16}\ cm^{-3}$ at U=1.2 V.

behavior of the current flowing through the structure calculated by MCP simulations of the electron transport at different lattice temperatures T. At room temperature the spectral density of current fluctuations, $S_{\delta j \delta j}(\nu)$, shows the standard behavior consisting of a low-frequency plateau and a high-frequency peak at 10 THz caused by the hybrid plasma resonance originated by the n^+n-homojunctions [1]. With the decrease of the temperature, the conditions for a ballistic motion through the n-region are improved. As a consequence, at $T = 80$ K there appears an additional peak of $S_{\delta j \delta j}(\nu)$ at $\nu = 1.14$ THz. The amplitude of the peak sharply increases at $T = 10$ K. The peak and its harmonics are caused by the streaming plasma instability, as directly evidenced by Fig. 1 (b). Indeed, at $T = 10$ K the instantaneous current, $j(t)$, exhibits a near-periodic noisy behavior, while at $T = 80$ K a more random behavior becomes evident. Nevertheless, the current noise spectrum (see Fig. 1 (a)) evidences the peak caused by the instability formation already at $T = 80$ K.

Two next examples illustrate the second scenario when, to obtain an oscillatory behavior, an external resonant circuit is necessary. In analogy with the first example, we consider the case when a spatial grating of carrier concentration is formed inside the n-region due to the coherent repetition of optical phonon emission. As shown in [6], at low temperatures and when the n-region plasma frequency approaches the frequency between successive optical phonon emissions, the n^+nn^+ structure can enter the active regime where microwave power generation is possible. The current noise and impedance spectra corresponding to this situation are shown in Fig. 2. Here, at decreasing temperatures $S_{\delta j \delta j}(\nu)$ exhibits an additional peak in the near THz region (see Fig. 2 (a)). As follows from Fig. 2 (b), the formation of the noise peak is directly related to the appearance of negative values of $ReZ(\nu)$ and a minimum of $|Z(\nu)|$. As shown by MCP simulations [6], in the frequency range where $ReZ(\nu) < 0$, microwave power generation can be obtained by placing the structure into an appropriate resonant circuit.

As last example, we consider an n^+nn^+ GaInAs structure similar to that used for an experimental investigation of enhanced noise [2,3]. Figure 3 (a) shows the general

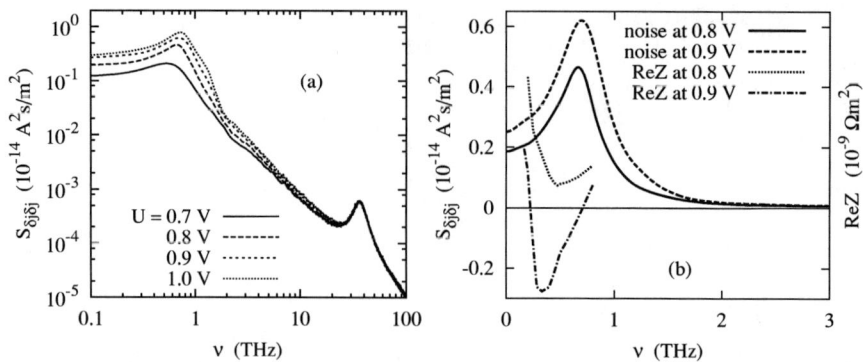

FIGURE 3. (a) General behavior of noise spectra and (b) frequency dependence of spectral density of current fluctuations and real part of the impedance in the near-THz region calculated by the MCP method for $0.005 - 0.5 - 0.005$ μm n^+nn^+ GaInAs structure with $n^+ = 10^{19}$ and $n = 10^{17}$ cm^{-3} at $T = 10$ K.

behavior of the current noise spectrum with the increase of the applied voltage U. Similarly to the previous cases, the spectrum exhibits a peak growing with the bias in the near THz region before the hybrid plasma resonance here located around 35 THz. Figure 3 (b) compares the frequency behavior of $S_{\delta j \delta j}(\nu)$ and $ReZ(\nu)$ in the frequency region centred around this resonant-like noise enhancement. Again, the noise enhancement indicates the approaching of an instability. Indeed, the structure is found to become active in the sub-THz region where $ReZ(\nu)$ is negative (see curve for $U = 0.9$ V). As confirmed by MCP simulations, the instability is caused by electron transfer into upper valleys and negative values of $ReZ(\nu)$ are obtained even at room temperature and above. Direct MCP simulations of the structure performance in external resonant circuits confirm a microwave power generation in this case too.

In conclusion, from the examples considered above, we have proven that the resonant enhancement of noise below the hybrid plasma resonance can be used as a reliable indicator for the formation of instabilities in semiconductor n^+nn^+ structures.

ACKNOWLEDGMENTS

This work is supported by NATO Collaborative Linkage under Grant PST.EAP.CLG 980629 and project "Noise models and measurements in nanostructures".

REFERENCES

1. E. Starikov et al, J. Appl. Phys. **79**, 242 (1996).
2. J. Lusakowski et al, J. Appl. Phys. **97**, 064307 (2005).
3. W. Knap et al., Appl. Phys. Lett., **84**, 2331 (2004).
4. N. Bannov et al, Solid State Electron., **29**, 1207 (1986).
5. E. Starikov et al., phys. stat. sol. (a), **190**, 287 (2002).
6. V. Gružinskis et al., Semicond. Sci. Technol., **19**, S173 (2004).

Fluctuation of the Electron Scattering Probability in One-dimensional Atomic Chain Having Plural Degree of Freedom

Hideo Akabane and Masahiro Agu*

Department of Media and Telecommunications Engineering,
Faculty of Engineering, Ibaraki University,
Nakanarusawa 4-12-1, 316-8511 Hitachi, Japan
E-mail: akabane@mx.ibaraki.ac.jp
*Fukushima National College of Technology,
970-8034 Iwaki, Japan
E-mail: agu@fukushima-nct.ac.jp

Abstract. One-dimensional atomic chain having plural degree of freedom was investigated by computer simulation. In this system, the fluctuation of scattering probability shows $1/f^\alpha$ ($\alpha \sim 1$) spectrum in the low frequency range.

Keywords: $1/f^\alpha$ fluctuation, One-dimensional atomic chain, Scattering Probability.
PACS: 05.40.Ca:

INTRODUCTION

The generation mechanism of $1/f$ resistance fluctuation has not been fully clarified. $1/f$ resistance fluctuation in thermal equilibrium sample was observed by Voss and Clarke.[1] Kogan and Nagaev reported that the origin of $1/f$ conductance fluctuation in metals is the mobility fluctuation.[2] On the other hand, Vandamme et al. reported that the origin of $1/f$ conductance fluctuation is the carrier number fluctuation in n-type metal oxide semiconductor (n-MOS) devices and it is the mobility fluctuation in p-MOS devices.[3] Musha et al. measured the fluctuation of Brillouin scattering from quartz and they reported that the origin of mobility $1/f$ fluctuation is the number fluctuation of thermally excited phonon.[4] We have performed the simulation and have reported that the scattering probability shows $1/f^\alpha$ ($\alpha \sim 1.2$) spectrum by choosing the shape of the anharmonic potential around the atoms. The fluctuation of scattering probability had the same correlation property as that of the $1/f$ fluctuation generated by 1/2-order integration of white noise.[5]

In this study, one-dimensional atomic chain having plural degree of freedom is investigated by computer simulation. If the degree of freedom is plural, phonon mode interaction can be caused in one-dimensional atomic chain that has harmonic potential and it was found that the fluctuation of electron scattering probability shows $1/f^\alpha$ ($\alpha \sim 1$) spectrum in the low frequency range.

SIMULATION

Fluctuation of the electron scattering probability is investigated in one-dimensional atomic chain. In case of general one-dimensional system, the movements of the atoms are limited along the longitudinal axis, but in our system, they have plural degree of freedom as shown in Fig. 1.

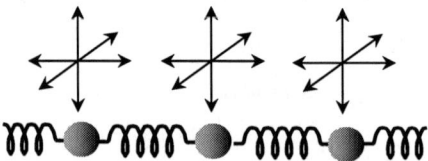

FIGURE 1. One-dimensional atomic chain having plural degree of freedom.

The reversion force of an atom is assumed as

$$M\frac{d^2}{dt^2}r(i,t) = \alpha[r(i-1,t) - 2r(i,t) + r(i+1,t)]. \tag{1}$$

Here, M is the atomic mass, $r(i,t)$ the three-dimensional place of the ith atom at time t and α the coefficient. Firstly, we have considered the non-linear reversion force of the atoms,[5] but after, it was found that the plural degree of freedom works instead of the nonlinearity. By the effect of plural degree of freedom, phonon mode interaction is caused. For every calculation step, spatial Fourier transform is given to the atomic displacements along the chain. These Fourier amplitudes can be regarded as the amplitudes of phonon modes and they are fluctuating. The fluctuation of electron scattering probability $S(t)$ by the acoustic phonon is calculated from Fourier amplitudes using deformation potential model and it can be written as

$$S(t) = \left[\sum_k kV\sqrt{\frac{\hbar n(k,t)}{2NM\omega(k)}}\right]^2. \tag{2}$$

Here, V is the constant related to deformation potential, N the number of atom, M the atomic mass, $\omega(k)$ the angular frequency of the phonon mode and $n(k,t)$ the phonon number whose wave number is k. In eq. (2), the temperature of the system is high and $n(k,t) \gg 1$ is assumed. The number of atom is 32. Only minimum wave number mode is excited as an initial condition. Figure 2 shows the energy distribution of each phonon mode. The horizontal axis means wave number(phonon mode) and vertical axis means energy in logarithm scale. Here, r_0 is the lattice constant. The result after 2^8 steps from the simulation beginning is indicated by triangle and that of 2^{20} steps is indicated by circle. The open mark indicate the result of longitudinal mode, the closed mark indicate that of transverse mode. At first, the energy is distributed to the modes whose wave number is small and which are close to initially excited mode, but after 2^{20} steps, every phonon mode gets almost the same energy. Thus, we considered that equipartition of the energy is materialized.

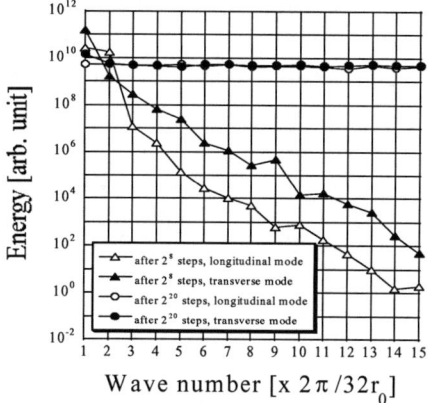

FIGURE 2. Energy distribution to each phonon mode.

Figure 3 shows the power spectral density (PSD) of the scattering probability fluctuation. The analysis of the time series is started after 2^{20} calculation steps have been passed. The solid line shows the result of two degrees of freedom, and the dotted line shows that of three degrees of freedom. In this figure, low frequency part of both solid line and dotted line show $1/f^\alpha$ ($\alpha \sim 1$) spectra. The fluctuation of three degrees of freedom is larger than that of two degrees of freedom. Even if the temperature is the same, the total energy in three degree of freedom system is larger than that in two degree of freedom system. This fact might be the cause of above difference.

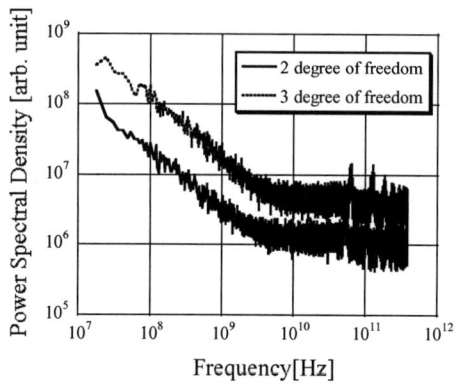

FIGURE 3. PSD of the electron scattering probability fluctuation.

The peaks seen in high frequency area are the characteristic frequencies of the phonon modes. In this simulation, the amplitude of phonon mode is estimated only by the displacement of atoms and kinetic energy is not used. Thus, the calculated amplitude of the phonon mode is fluctuating on its characteristic frequency.

In one-dimensional atomic chain with harmonic potential, the scattering probability of the electron shows $1/f^\alpha$ ($\alpha \sim 1$) fluctuation by the existence of plural degree of freedom. If the movements of the atoms are limited along the longitudinal axis, achievement of energy equipartition and the generation of $1/f^\alpha$ ($\alpha \sim 1$) fluctuation by the phonon mode interaction are not easily caused. The reason why $1/f^\alpha$ ($\alpha \sim 1$) type fluctuation is easily generated by the existence of plural degree of freedom might be related to the easiness of interaction between longitudinal phonon mode and transverse phonon mode. This kind of interaction complicates the energy transfer. Further, it might be related to the very low characteristic frequency of transverse mode. This situation becomes remarkable in one-dimensional or two-dimensional systems.

CONCLUSION

One-dimensional atomic chain having plural degree of freedom was investigated by computer simulation. It was found that the plural degree of freedom take place of the nonlinearity and it was also found that the fluctuation of electron scattering probability shows $1/f^\alpha$ ($\alpha \sim 1$) spectrum in the low frequency range. The generation of $1/f$ fluctuation might be facilitated by the existence of the plural degree of freedom.

REFERENCES

1. R. F. Voss and J. Clarke: Phys. Rev. B **13**, 556-573, (1976).
2. M. Kogan and K. E. Nagaev: Solid State Commun. **49**, 387 (1984).
3. L. K. J. Vandamme, X. Li and D. Rigaud: IEEE Trans. Electron Devices **41**, 1936-1945 (1994).
4. T. Musha, G. Borbely and M. Shoji: Phys. Rev. Lett. **64**, 2394-2397 (1990).
5. H. Akabane and M. Agu: Noise in Physical Systems and 1/f Fluctuations 2003, pp. 667-670.

On the High Frequency Limit of the Impedance Field Method for Si

Christoph Jungemann* and Bernd Meinerzhagen*

NST, Postfach 33 29, TU Braunschweig, 38023 Braunschweig, Germany

Abstract. In the Impedance Field method, which is based on the drift-diffusion approximation, acceleration effects are neglected by considering the derivative with respect to time only in the continuity equation. This also implies white noise sources. The high frequency limit of this approximation is investigated for the first time based on solutions of the Langevin Boltzmann equation for stationary and cyclostationary systems. It is found that the approximation holds up to about 100GHz for Si under small and large signal conditions.

Keywords: Noise, Impedance Field method, Langevin Boltzmann equation, silicon
PACS: 72.10.Bg,72.20.Ht,72.30.+q

INTRODUCTION

SiGe HBTs with cutoff and maximum oscillation frequencies of over 200GHz have been demonstrated[1]. With such transistors circuits can be operated at 100GHz. If the circuit operates under large signal conditions, higher harmonics occur at multiples of the fundamental frequency. These high frequencies call into question some of the approximations of the drift-diffusion model (DDM) [2], which is the basis of the Impedance Field method [3]. This is investigated for the first time for Si devices including cyclostationary excitations based on solutions of the full Langevin Boltzmann equation (LBE), which is solved for transport and noise by numerical means [4].

THEORY

The DDM consists of the continuity equation for the particle density (ignoring generation/recombination phenomena)

$$\nabla \cdot \vec{j} + \frac{\partial n}{\partial t} = 0 \qquad (1)$$

and the constitutive equation for the particle current density

$$\vec{j} = -\mu(n\vec{E} + U_T \nabla n) + \vec{\xi}_{\vec{j}}, \qquad (2)$$

where $\vec{j}$ is the particle current density, n the particle density, $\vec{E}$ the electric field, μ the mobility, U_T the thermal voltage, and $\vec{\xi}_{\vec{j}}$ the Langevin force of the current fluctuations.

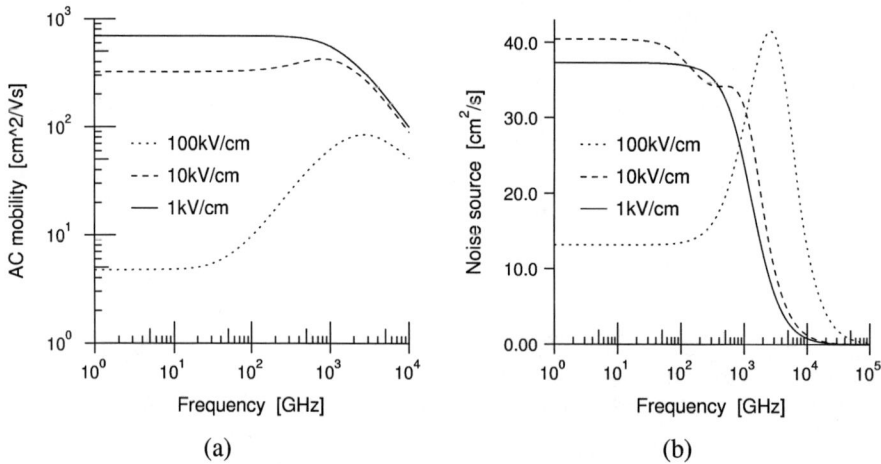

FIGURE 1. Absolute value of the AC mobility ($\mu = dv/dE$) (left) and the local noise source (right) of the DDM for n-Si doped with 10^{17}/cm^3 at room temperature.

While Eq. (1) is exact, Eq. (2) contains several approximations. With respect to high frequencies these are:

1. The disregarding of the acceleration term $\tau_v \partial \vec{j}/\partial t$, where τ_v is the macroscopic velocity relaxation time. This implies a relation between the particle density, electric field, and mobility, which is local in time.
2. The assumption that the noise source (power spectral density of the equivalent bulk velocity fluctuations) is white.

These two assumptions, which are also used in the derivation of most hydrodynamic models, are investigated by solving the LBE with and without these approximations. Because the acceleration term is neglected in Eq. (2) this approximation is also called quasistationary. It has the advantage that plasma oscillation are suppressed in the solutions of the DDM, which allows to use time steps in transient simulations, which can be many orders of magnitude larger than without this approximation. In addition, the numerical stability is improved. Assumption 2 is consistent with assumption 1.

Transport is described with the electron model of the Modena group [5] and the LBE is solved using a spherical harmonics expansion [4].

RESULTS

Most of the consequences of assumptions 1,2 can be studied under homogeneous real space conditions neglecting generation/recombination processes. In Figs. 1 (a), (b) the AC mobility and local noise source of the DDM are shown for bulk Si as obtained by the LBE. Up to about 100GHz both quantities do not depend on the frequency and assumptions 1,2 appear to be well justified.

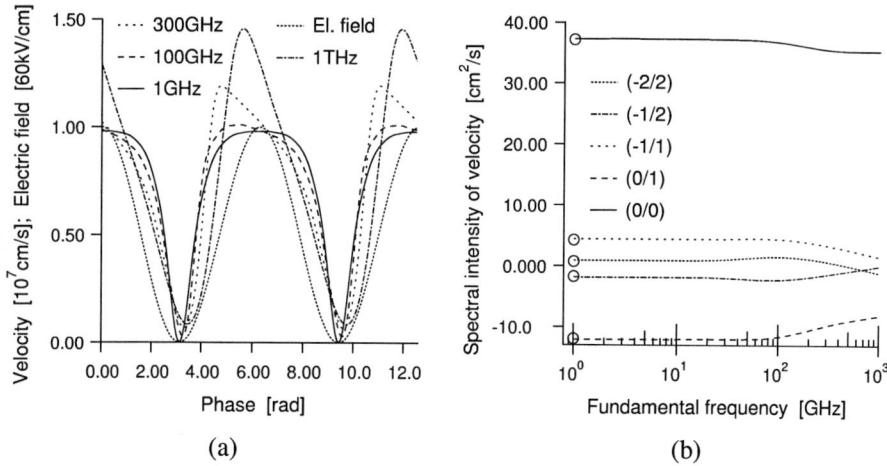

FIGURE 2. Velocity (left) for two cycles versus phase for a periodic electric field ($E = 30\text{kV/cm} \cdot (1 + \cos(2\pi f_0 t))$) in $\langle 100 \rangle$ direction at different fundamental frequencies and the real part of the sideband correlation matrix (right) for the longitudinal velocity fluctuations including the quasistationary approximation (symbols) for undoped Si at room temperature.

In Fig. 2 (a) the electron velocity is shown evaluated by solving the LBE for periodic electric fields of different frequencies. The result for 1GHz corresponds to the quasistationary result of the DDM. Non-quasistationary effects occur only at higher frequencies. At 100GHz a small velocity overshoot is seen, which becomes much larger at higher frequencies. For small and large signal transport, assumptions 1,2 seem to be valid up to nearly 100GHz. In Fig. 2 (b) the local noise source of the cyclostationary Impedance Field method, the real part of the sideband correlation matrix of the velocity fluctuations [6] is shown, where the imaginary part is negligible. Again the results of the full LBE are frequency independent up to about 100GHz and the quasistationary approximation yields good results up to this frequency. The reason for this is that the velocity and energy relaxation times of silicon are smaller than one picosecond at room temperature.

For a 1D N^+NN^+ structure, which consists of a 400nm long region with a doping concentration of $2 \cdot 10^{15}/\text{cm}^3$, to which on both sides 100nm long regions of $5 \cdot 10^{17}/\text{cm}^3$ are attached, the LBE has been solved and the current flows into the $\langle 100 \rangle$ direction. In Fig. 3 the small-signal admittance and terminal current noise are shown calculated with and without assumptions 1,2. Again good results are obtained up to 100GHz. Above 100GHz the peak due to the plasma resonance is missing if the acceleration term is neglected.

CONCLUSIONS

We have investigated for the first time the quasistationary approximation, which is used in the Impedance Field method, for transport and noise including cyclostationary sys-

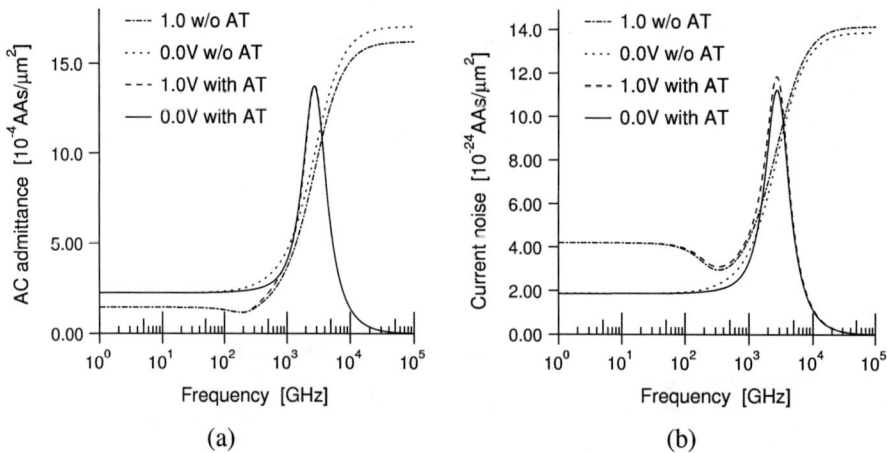

FIGURE 3. Real part of the small-signal admittance (left) and the spectral intensity of the terminal current fluctuations (right) of an N^+NN^+ structure calculated with and without the acceleration term (AT) for two applied voltages at room temperature.

tems based on the LBE. It turns out that up to 100GHz the quasistationary approximation is valid under small and large signal conditions in Si due to the small relaxation times of the velocity and energy.

REFERENCES

1. B. Jagannathan, M. Khater, F. Pagette, J. Rieh, D. Angel, H. Chen, J.Florkey, F. Golan, D. Greenberg, R. Groves, S. Jeng, J. Johnson, E. Mengistu, K. Schonenberg, C. Schnabel, P. Smith, A. Stricker, D. Ahlgren, G. Freeman, K. Stein, and S. Subbanna, *IEEE Electron Device Lett.*, **23**, 258–260 (2002).
2. W. van Roosbroeck, *Bell System Technical Journal*, pp. 561–607 (1950).
3. W. Shockley, J. A. Copeland, and R. P. James, "The Impedance Field Method of Noise Calculation in Active Semiconductor Devices," in *Quantum theory of atoms, molecules and solid state*, edited by P. O. Lowdin, Academic Press, 1966, pp. 537–563.
4. C. Jungemann, and B. Meinerzhagen, "A Legendre Polynomial Solver for the Langevin Boltzmann Equation," in *Proc. IWCE*, 2004, pp. 22–23.
5. R. Brunetti, C. Jacoboni, F. Nava, L. Reggiani, G. Bosman, and R. J. J. Zijlstra, *J. Appl. Phys.*, **52**, 6713–6722 (1981).
6. F. Bonani, S. D. Guerrieri, and G. Ghione, *IEEE Trans. Electron Devices*, **50**, 633–644 (2003).

Properties of Some Deterministic Noises

Juraj Kumičák

Department of Thermodynamics, Technical University, Vysokoškolská 4, 042 00 Košice, Slovakia

Abstract. This contribution studies a family of generalized baker maps. Mapping the unit square onto itself, the maps are partly contracting and partly expanding, but they preserve the global measure of the definition domain. Stochastic properties of the maps are analyzed in terms of noise they generate. The analysis shows that one can link properties of noise with the parameters characterizing the contractivity of the model. It appears that the crucial role is played, in this respect, by the presence of attractor/repellor pair in the phase space of the system.

Keywords: Noise, Numerical simulation
PACS: 05.40.Ca, 05.45.Pq, 72.70.+m

INTRODUCTION

There are deterministic processes which generate apparently random behavior in one dimension. It is therefore appropriate to denote the latter as deterministic noise. The study of such processes is interesting mainly due to interrelations existing between statistical properties of the noises and their generating laws. This contribution is devoted to the analysis of one such model system, the generalized baker map.

GENERALIZED BAKER MAP

The generalized baker map (or GBM for short) is defined for any $w > 1$ and is reversibly mapping the points $\gamma \equiv (x,y)$ of the unit square $E = [0,1] \times [0,1]$. For the points to the left of the "dividing line" $x = (w-1)/w$ its action is:

$$B_w(x,y) = \left(\frac{w}{w-1}x, \frac{y}{w}\right) \qquad (1)$$

and for the remaining ones $(w-1)/w < x \leq 1$:

$$B_w(x,y) = \left(w(x-1)+1, \frac{w-1}{w}y + \frac{1}{w}\right). \qquad (2)$$

The properties of the map can be examined by viewing the author's animation [1] showing all essential details of the map.

The expansion caused by B_w in the x-direction and contraction in the y-direction are characterized by local logarithmic rates l_x and l_y

$$l_x = \ln\frac{\partial B_w(x,y)}{\partial x} \quad \text{and} \quad l_y = \ln\frac{\partial B_w(x,y)}{\partial y}. \qquad (3)$$

Since the action of B_w is different for points lying to the left and to the right of the dividing line, we will have two rates in the x-direction

$$l_x^L = \ln \frac{w}{w-1} \quad \text{and} \quad l_x^R(w) = \ln w \tag{4}$$

and two in the y-direction

$$l_y^L = \ln \frac{1}{w} \quad \text{and} \quad l_y^R(w) = \ln \frac{w-1}{w}. \tag{5}$$

Averaging the above logarithmic expansion rates over typical trajectories in E, one obtains what is usually called Lyapunov, or time-averaged, exponents — the positive one

$$\lambda_1 = \frac{w-1}{w} \ln \frac{w}{w-1} + \frac{1}{w} \ln w = \frac{1}{w} \ln \frac{w^w}{(w-1)^{w-1}} \tag{6}$$

and the negative one

$$\lambda_2 = \frac{w-1}{w} \ln \frac{1}{w} + \frac{1}{w} \ln \frac{w-1}{w} = \frac{1}{w} \ln \frac{w-1}{w^w}. \tag{7}$$

One sees that with growing $w > 2$ both exponents are monotonously decreasing, and their limit behavior for $w \to \infty$ is $\lambda_1(w) \to 0$ and $\lambda_2(w) \to -\infty$.

The existence of positive Lyapunov exponent suggests that one should expect chaotic behavior of the iterates of $B_w^n \gamma_0$ (for almost every $\gamma_0 \in E$), and the existence of the negative one the existence of a (strange) attractor [2]. Both are observed when one iterates B_w beginning with almost any starting point — see the animation in [1]. The attractor is a self-similar fractal object consisting of an infinite set of lines parallel to the x-coordinate. Reversibility implies also the existence of a repellor — an object from which all phase points depart — consisting of lines orthogonal to the lines of the attractor.

One of the most important characteristics of the attractor is its information dimension which is calculated, for general w, to be:

$$D_1(w) = 1 - \frac{\ln(w^w(w-1)^{1-w})}{\ln((w-1)w^{-w})}. \tag{8}$$

Information dimension can be viewed as the measure of attractor inhomogeneity so its approach, with growing w, to the value of 1 suggests that the iterated points tend to accumulate (condense) on smaller subsets of E.

The analysis [3] of this model system has shown that the model is strictly reversible but approaches different limit structures in the future and in the past — the attractor and the repellor, respectively (which are symmetric with respect to the diagonal $y = 1 - x$). One can moreover introduce the well defined notion of "age" that can be applied along the trajectory. The age grows monotonously with iterations and is negative for the past and positive for the future. In this sense the model exhibits what is sometimes called the "arrow of time".

The map also possesses, for any w, a rich structure of periodic orbits of any period. The number of different cycles grows exponentially with the cycle period p, amounting to a total of $(2^p - r - 2)/p$, with $r < p$. The cycles of B_w exhibit a peculiar property: there are trajectories, which unwind from a cycle in the past and approach a cycle with different period in the future. This demonstrates that there exist two different limits in the behavior of B_w, separated by intermediate states, thus confirming that there is a difference between the past and the future. This may be regarded as an independent confirmation of the arrow of time in the system.

DETERMINISTIC NOISES

The statistical properties of GBM are evidently dominated by chaotic approach to attractor (and departure from repellor) but the zero measure of starting points approaches limit cycles as well. The study of properties of both types of behavior [3] yielded several interesting observations.

The motion of a typical point under the action of B_w in the unit square E has white power spectrum. However, the projection of this motion onto the second diagonal $y = 1 - x$, has a spectrum with maximum around $f = 0$ and decaying in its neighborhood as $1/f^k$. At a greater distance from $f = 0$ the spectrum approaches that of white noise.

This specific behavior seems to be based on the fact that at very high measure of contraction (high w), the behavior of trajectories simulates burst noise whose spectrum is known to approach $1/f^2$ at higher frequencies. With growing value of w also the correlations of consecutive points seem to be growing, since the higher the value of w, the more points are being contracted by the same factor: one observes a kind of condensation in the phase space E.

One can take still another approach to the explanation of the spectrum observed in the system under consideration. As already mentioned, one can always find, for arbitrary integers $p_1, p_2 > 1$, trajectories which unwind from a p_1-cycle in the past and approach a p_2-cycle in the future. Such trajectories will evidently have spectrum with two peaks corresponding to the two periods. This becomes less trivial when we consider extremely long periods. In this case the approach to a periodic cycle will take a long time so that the frequencies will be the less sharply pronounced, the longer the periods and they will be obviously very low. This suggests a possible way how to approach the understanding of the predominance of higher amplitudes at lower frequencies.

Projections of the trajectories onto a straight line of any angle (within the unit square) generate what can be denoted as scaled noises. Since the projections onto horizontal and vertical lines were shown (and confirmed in simulation) to have white and brown spectrum, respectively [4], one obtains a "combination" of white and brown noises by changing the angle. In view of the above mentioned asymmetry between the past and the future, one will therefore observe, on the second diagonal $y = 1 - x$, noises with white past and brown future, or *vice versa*. This seems to be rather intriguing phenomenon deserving further investigation since the transition from white to brown noises might be indicative of a kind of selforganization.

The results given here can be expressed also using the language of ergodic theory as applied to strange attractors: the GBM system approaches the ergodic strange attractor with multifractal structure, and this in its turn imposes restrictions on the recurrent trajectories related to periodic orbits, which results in observed types of noises [5].

CONCLUSIONS

The generalized baker map was studied primarily to demonstrate the interrelation between reversibility and irreversibility. At first sight it might seem that this interrelation is not related to the problem of noise, but the contrary is true. Irreversibility is found in systems with sufficiently chaotic dynamics and the latter is at the same time a necessary condition for observing noise in deterministic systems. Preliminary results seem to point to the conclusion that the joint treatment of irreversibility and noise represents the correct direction in which to go to arrive at the solution of both problems: the origin and essence of irreversibility, and the characterization of noise, appearing in deterministic microscopically reversible systems with apparent irreversible macroscopic behavior.

ACKNOWLEDGMENTS

This work was partly supported by the Scientific Grant Agency (VEGA) of the Slovak Academy of Science and Ministry of Education of the Slovak Republic under the Grant No. 1/0428/03. The support is herewith greatly acknowledged.

REFERENCES

1. J. Kumičák, *Animation of generalized baker map*, see at: http://www.sjf.tuke.sk/ket/kumicak/BakAnima.html, 2004.
2. H. G. Schuster, *Deterministic Chaos: An Introduction*, VCH, Weinheim, 1988.
3. J. Kumičák, *Phys. Rev. E*, **71**, 016115 (2005).
4. J. Kumičák, "Stochastic Properties of Deterministic Systems," in *Proc. 16th Internat. Conf. "Noise in Physical Systems and $1/f$ Fluctuations" (ICNF 2001)*, edited by G. Bosman, World Scientific, Singapore, 2001, pp. 606–609.
5. J. Kumičák, "Simple Deterministic Model of Noise," in *Proc. "17th Internat. Conf. on Noise and Fluctuations" (ICNF 2003)*, edited by J. Šikula, CNRL, Brno, 2003, pp. 675–678.

Phase Transition and $1/f$ Noise in a Modified Bak-Tang-Wiesenfeld Sand Pile Model with Time-dependent Avalanche Propagation

Ken Kiyono*, Zbigniew R. Struzik* and Yoshiharu Yamamoto*

Graduate School of Education, The University of Tokyo, Japan

Abstract. We study numerically the statistical properties of a modified Bak-Tang-Wiesenfeld sand pile model by introducing a non-zero driving rate, and time-dependent avalanche propagation on a two-dimensional lattice. When we consider a very small driving rate, this model approaches the original BTW model [P. Bak, C. Tang, and K. Wiesenfeld, Phys. Rev. Lett. 59, 381 (1987)]. As the driving rate increases, the model exhibits a continuous phase transition. At the critical point, the variation in the sand flux, the amount of moving grains, exhibits long-range temporal correlations with a $1/f$ power spectrum. Moreover, the critical fluctuations show scale invariance of the non-Gaussian probability density function. We will discuss how the characteristics in this system depend on the driving rate.

Keywords: Sand pile model, Critical point phenomenon, Long-range temporal correlation
PACS: 05.40.-a, 05.70.Jk

INTRODUCTION

Temporal fluctuations with a long-range power-law correlation, indicating a $1/f$ power spectrum (also known as $1/f$ noise), have been observed in a variety of fields, including economics, physics, and biology. In order to explain the widespread occurrence of $1/f$ noise, Bak, Tang, and Wiesenfeld (BTW) introduced the concept of self-organized criticality (SOC) [1], which proposed that, without fine tuning of external parameters, certain non-equilibrium systems spontaneously evolve to a critical state, characterized by power law event size distributions and a $1/f^\alpha$ power spectrum. However, the original BTW sand pile model of SOC displays temporal fluctuations with a $1/f^2$ scaling of the power spectrum [2]. Thus, the reason why the exponent α close to one is ubiquitous is still not clear, although some SOC models display $1/f$-like noise [3]. In addition, it has been pointed out by several authors that the driving rate is a parameter that has to be fine-tuned to zero in order to observe criticality in SOC systems [4]. In the SOC models, a separation of time scales between the external driving and the avalanche propagation has been generally assumed. The assumption of an infinitesimal driving rate is unlikely to be valid in many non-equilibrium (far from equilibrium) systems.

In a more general situation of non-equilibrium phase transitions, it has been demonstrated that certain systems at a critical point exhibit $1/f$ noise [5, 6]. Recently, the functional advantage of the critical state, such as maximum efficiency of transportation [7, 8], has been demonstrated. If criticality possesses properties of optimality and efficiency, it is natural to expect that, especially in biological systems, the state of operation will be tuned toward its critical point [9].

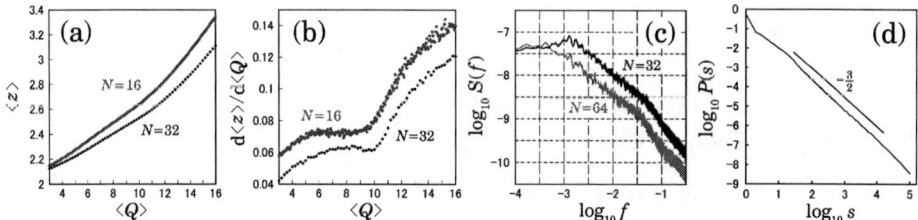

FIGURE 1. (a) Average height $\langle z(t) \rangle$ of the pile of grains as an order parameter vs. average inflow of sand, $\langle Q \rangle \equiv pN^2$; (b) Differential coefficient of the order parameter. A singular point is observed at $\langle Q \rangle \approx 10$; (c) Power spectrum of the flux fluctuations at the critical value, $\langle Q \rangle = 10$; (d) The corresponding avalanche size distribution.

At present there is no general theory of non-equilibrium phase transitions. So the relation between criticality and $1/f$ noise is still not clearly understood. In order to address this issue, here we study numerically a modified BTW sand pile model with a non-zero driving rate as a control parameter and time-dependent avalanche propagation, in which a separation of time scales is not assumed. We discuss this system as a simple model of a non-equilibrium phase transition. As the driving rate increases, our model exhibits a continuous phase transition. At the critical point, the variation in the flux exhibits long-range temporal correlations with a $1/f$ power spectrum. Moreover, the critical fluctuations show non-Gaussian behavior and scale invariance of the probability density function (PDF). In the following sections, we will discuss how the characteristics of this system depend on the control parameter.

THE MODEL

In a stochastic cellular automaton model inspired by the BTW sand pile model [1], we assume the following:

1. At each time step t, a grain is added with probability p to each site (x,y) on a two-dimensional lattice of size $N \times N$:

$$z(x,y;t) = z(x,y;t-1) + 1, \quad (1)$$

where $z(x,y)$ represents the number of grains of sand (or height) at site (x,y).

2. If $z(x,y;t)$ exceeds a prescribed threshold z_c, where we set $z_c = 4$, the site (x,y) topples at the next time step, and the grains are transferred to the nearest neighbors (NN):

$$z(x,y;t+1) = 0, \quad \text{if } z(x,y;t) \geq 4, \quad (2)$$
$$z_{NN}(t+1) = z_{NN}(t) + 1 + \xi_{NN}, \quad (3)$$

where the integer values of ξ are randomly assigned under the condition $\sum \xi_{NN} = z(x,y;t) - 4$ (conservation of mass).

3. The above processes 1 and 2 progress simultaneously.

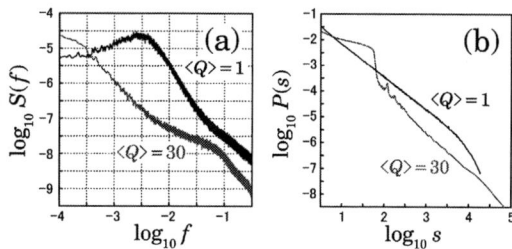

FIGURE 2. (a) Power spectrum of the flux fluctuations in the off-critical region; (b) The corresponding avalanche size distribution. $N = 64$.

4. Grains leave the system when a toppling site is adjacent to the boundary.

Instead of Eqs. (2) and (3), we can assume the same toppling rule as the original BTW model: $z(x,y;t+1) = z(x,y;t) - 4$, $z_{NN}(t+1) = z_{NN}(t) + 1$. In this case, the maximum transfer rate at each site is limited to 4. Thus, the average height $\langle z \rangle$ rapidly grows, as the p increases, and diverges if the p exceeds a certain threshold. To avoid this behavior, we assume Eqs. (2) and (3). Nevertheless, in the vicinity of the phase transition point, no significant difference in the scaling properties is observed.

RESULTS AND DISCUSSION

In this section, we present numerical results. In our model, the driving rate $\langle Q \rangle \equiv pN^2$ is the control parameter. After some transient period, the system reaches a stationary state with an average height $\langle z \rangle$. Although the $\langle z \rangle$ is monotonically increasing with $\langle Q \rangle$ [Fig. 1 (a)], the differential coefficient of the $\langle z \rangle$ exhibits singular behavior at $\langle Q \rangle = Q_c \approx 10$ [Fig. 1 (b)] — an indication of a critical point. At the critical point, temporal fluctuations of the sand flux $J(t)$, the amount of moving grains, exhibit long-range temporal correlations with the $1/f$ power spectrum [Fig. 1 (b)]. In addition, the distribution of the avalanche size induced by a single addition follows a power law with an exponent close to $-3/2$. This fact suggests an analogy with a critical branching process, which creates $P(s) \sim s^{-3/2}$ for large s. All of these features demonstrate the existence of a dynamical phase transition. Note that this criticality cannot be associated with self-organized criticality. If the driving rate $\langle Q \rangle$ departs from the critical value, a breakdown of the critical properties occurs, as shown in Fig. 2.

In addition to the $1/f$ scaling, our model shares some features of heart rate fluctuations in healthy humans. To date, healthy human heart rate has been confirmed robustly to show $1/f$-type fluctuations under the conditions of daily life [10]. In addition, non-Gaussian behavior and scale invariance of the PDF have also been reported [9]. Our model can reproduce scale-invariant properties of the non-Gaussian PDF, as shown in Fig. 3(c) and (d). In the off-critical region, the PDF's converge to a Gaussian [Fig. 3(a) and (b)], as the scale n increases. On the other hand, at the critical point, over a wide range of scales the PDF's do not show any convergence to the Gaussian. Thus, our model may describe universal behavior in a certain class of non-equilibrium phase transitions,

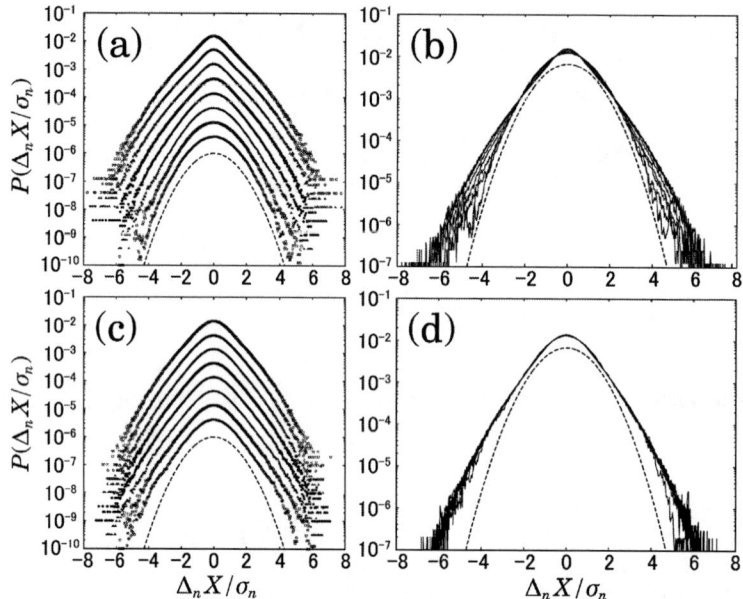

FIGURE 3. Deformation of increment PDF across scales, where $\Delta_n X$ is the sand flux over a time scale n (normalized to zero mean and unit variance) and σ_s is the standard deviation of $\Delta_n X$: (a)(b) In the off-critical region ($\langle Q \rangle = 1$, $N = 64$); (c)(d) at the critical value ($\langle Q \rangle = 10$, $N = 64$); (a)(c) Standardized PDF's on scales (from top to bottom) $s = 8, 16, 32, 64, 128, 256, 512, 1024$; (b)(d) The PDF's at different scales, shown in (a) and (c), are superimposed. The dashed line is a Gaussian PDF for comparison.

including $1/f$ noise and scale invariance of the non-Gaussian PDF, although further studies are needed in order to understand the mechanism generating the critical fluctuations.

REFERENCES

1. P. Bak, C. Tang, and K. Wiesenfeld, *Phys. Rev. Lett.*, **59**, 381–384 (1987).
2. H. J. Jensen, K. Christensen, and H. C. Fogedby, *Phys. Rev. B*, **40**, 7425–7427 (1989).
3. K. Christensen, Z. Olami, and P. Bak, *Phys. Rev. Lett.*, **68**, 2417–2420 (1992).
4. D. Sornette, A. Johansen, and I. Dornic, *J. Phys. I*, **5**, 325–335 (1995).
5. M. Takayasu, H. Takayasu, and T. Sato, *Physica A*, **233**, 824–834 (1996).
6. S. Chen, and D. Huang, *Phys. Rev. E*, **63**, 036110 (2001).
7. M. Takayasu, H. Takayasu, and K. Fukuda, *Physica A*, **277**, 248–255 (2000).
8. S. Valverdea, and R. V. Solé, *Physica A*, **312**, 636–648 (2002).
9. K. Kiyono, Z. R. Struzik, N. Aoyagi, S. Sakata, J. Hayano, and Y. Yamamoto, *Phys. Rev. Lett.*, **93**, 178103 (2004).
10. N. Aoyagi, K. Ohashi, and Y. Yamamoto, *Am. J. Physiol.*, **285**, R171–R176 (2003).

Effect Of Noise In The Estimation Of Magnitudes With Spatial Dependence: A Spatial Statistics Technique Based On Kriging

Luis Miguel Sanchez-Brea, Eusebio Bernabeu

Universidad Complutense de Madrid. Departamento de Optica. Facultad de Ciencias Físicas.
Ciudad Universitaria s.n., 28040, Madrid (Spain)
sanchezbrea@fis.ucm.es

Abstract. Kriging is a family of linear methods for the estimation of physical quantities with spatial dependence which are optimal in the squared minima sense. To perform the interpolation, kriging considers, in addition to the value and location of the observations, the spatial correlation of the quantity by means of variogram, the random fluctuations of the measured magnitude and the resolution of the measuring devices. The traditional way kriging equations are solved involves the resolution of inverse of great matrices, so that it is normally quite time consuming. Comparing the uncertainty obtained with kriging (for magnitudes with spatial dependence) with standard techniques for uncertainty estimation, we have seen that for the case of regular sampling, the uncertainty estimation can be computed as a convolution.

Keywords: Kriging, uncertainty as a convolution.
PACS: 05.40.Ca, 42.79.Pw

ESTIMATION AND UNCERTAINTY OF QUANTITIES WITH SPATIAL DEPENDENCE USING KRIGING

Kriging is a technique to estimate a quantity with spatial dependence and its uncertainty that explicitly considers the spatial correlation [1-3]. For this it performs a best linear unbiased estimation in the minimal squared sense. Kriging is widely used in geostatistics and other experimental sciences such as geology, mining, biology, medicine, etc. where few data are available, are disposed irregularly, and present strong random fluctuations. Kriging has also been applied to image processing [4,5]. However, in its general form, kriging equations involves inverse of matrices, which can be quite time consuming [6]. When the locations of the measuring devices are regularly disposed, Kriging equations are simplified resulting that the interpolation can be applied as a convolution [7]. With this assumption, kriging technique can be applied to image processing, including the finite size of the pixels in the design of the interpolator [8], and it has been proven to obtain better results than adaptive Wiener filter [9].

However, uncertainty estimation still needs to be solved by the conventional matrix form, even for the case of regular sampling. When the standard kriging approach is compared to the conventional statistical technique for uncertainty estimation it results that they do not coincide for the limit case where all the observations are placed at the same location. As a consequence, Sanchez Brea and Bernabeu [7] modified kriging equations to solve this inconvenience. It results that both equations present a similar structure.

Comparing both procedures of uncertainty estimation, in this work we propose, a convolutional method for the uncertainty estimation of magnitudes with spatial dependence. For this, we assign to each observation a function, which we have named Distributed Measurement (*DM*) that informs us how many experimental observations should be obtained, without considering the spatial dependence, for decreasing the uncertainty as kriging (which considers the spatial dependence) does. *DM* function only depends on the variogram and the resolution of the measuring devices. As a result, the uncertainty of magnitudes with spatial dependence can be easily computed, without the need of performing inverse of matrices.

UNCERTAINTY AS CONVOLUTION

The traditional way that kriging performs the uncertainty estimation [1-3] does not agree, in the limit case where all observations are performed at the same location, with the standard technique for estimation the uncertainty of a magnitude [10, 11]

$$u^2 = I^2 + \frac{s^2}{N}, \tag{1}$$

being I the resolution of the measuring device, s the standard deviation of noise, and N the number of observations. For that reason Sanchez Brea et. Bernabeu [7] have modified the equations for kriging estimation. Let us consider that we know the functional dependence of the variogram, which can be obtained using

$$\gamma(h) = \left\langle \left[Z(x+h) - Z(h) \right]^2 \right\rangle. \tag{2}$$

Normally experimental variogram is fitted to a theoretical model of variogram, such as a Gaussian variogram. With these modifications, the proposed uncertainty estimated using kriging when N observations are performed at the same location, $\mathbf{x}_0$, results

$$u_K^2(\mathbf{x}) = I^2 + \frac{s^2}{N} + 2\left[\gamma(\mathbf{x} - \mathbf{x}_0) - s^2 \right]. \tag{3}$$

which coincides with the definition of the uncertainty for the limit case $\mathbf{x} \to \mathbf{x}_0$. Comparing Eq. (3) with Eq. (1), we propose that the uncertainty of magnitudes with spatial dependence be described as

$$u^2(\mathbf{x}) = I^2 + \frac{s^2}{N_{EQ}(\mathbf{x})}. \tag{4}$$

Solving Eq. (4) it is obtained

$$N_{EQ}(\mathbf{x}) = \frac{s^2}{u^2(\mathbf{x}) - I^2}. \tag{5}$$

To determine $N_{EQ}(\mathbf{x})$ let us use the known uncertainty for just one observation. Introducing the uncertainty estimated using Eq. (3) for $N=1$ it results

$$N_{EQ}(\mathbf{x})\big|_{N=1} \to DM(\mathbf{x}) = \frac{s^2}{2\gamma(\mathbf{x}-\mathbf{x}_0)-s^2}. \qquad (6)$$

We have called this function *Distributed Measurement*. Since the standard deviation of a magnitude with spatial dependence can be obtained also from the variogram [12] by using $s^2 = \lim_{h\to 0}\gamma(h)$, then to determine the uncertainty only the knowledge of the variogram and the resolution of the measuring devices is required. As a result, *DM* function can be written as

$$DM(\mathbf{x}) = \frac{\gamma(0)}{2\gamma(\mathbf{x}-\mathbf{x}_0)-\gamma(0)}. \qquad (7)$$

DM function presents a maximum at $x=x_0$ and decreases when increasing the distance to the observation place, provided that the spatial correlation decreases with the distance.

We will assume that when several observations are obtained at regular locations, $N_{EQ}(\mathbf{x})$ is a linear process so it can be obtained as a sum of individual $DM(\mathbf{x})$ functions assigned to each observation.

$$N_{EQ}(\mathbf{x})\big|_{N=1} = DM(\mathbf{x})*\text{Ш}(\mathbf{x}), \qquad (8)$$

where $\text{Ш}(\mathbf{x}) = \sum_i \delta(\mathbf{x}-\mathbf{x}_i)$.

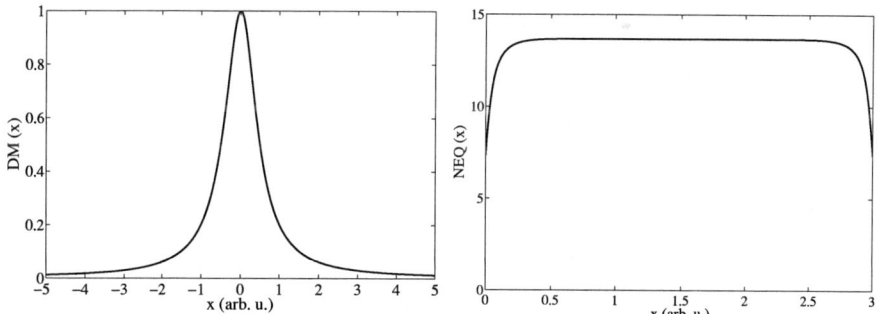

FIGURE 1. DM(x) function and NEQ(x) function for the case of a sinusoidal signal $f(x)=sin(2\pi x)$ with a gaussian random noise of $s=0.5$ arb.u.. The sampling frequency for determining NEQ(x) was $v=100$ arb.u.$^{-1}$.

As an example of the technique, in Figure 1 we show *DM(x)* and *NEQ(x)* functions obtained for the case of a sinusoidal signal $f(x)=sin(2\pi x)$ which presents a Gaussian random noise $s=0.5$ arb.u. and a resolution of $I=0.05$ arb.u. In Figure 2 we show the uncertainty estimation obtained with the proposed technique, and the uncertainty obtained when the kriging estimation is compared to $f(x)$.

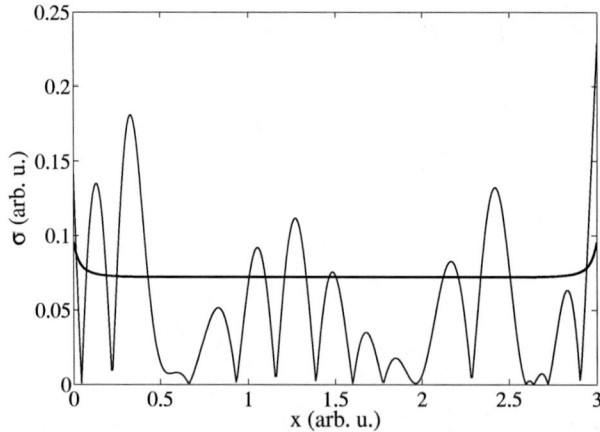

FIGURE 2. Estimation of error estimated with kriging (thin) and with Eq. (4). In this case $I=0.5$ arb.u and the standard deviation of noise (Gaussian noise) is $s=0.05$ arb.u..

ACKNOWLEDGMENTS

Sanchez-Brea is currently contracted by the Universidad Complutense de Madrid under the "Ramón y Cajal" research program.

REFERENCES

1. R. Christiensen, *Linear Models for Multivariate, Time Series, and Spatial Data*, Springer-Verlag, Berlin, 1985.
2. N. Cressie, *Statistics for Spatial Data*, John Wiley & Sons, New York, 1991.
3. J.P. Chilès, P. Delfiner, Geostatistics: *Modeling Spatial Uncertainty*, John Wiley & Sons, New York, 1999.
4. D. Mainy, J.P. Nectoux, D. Renard, *Mat. Charact.*, **36**, 327-334 (1996).
5. E. Bernabeu, I. Serroukh, L.M. Sanchez-Brea *Opt. Eng.*, **38**, 1319-1325 (1999).
6. W. H. Press, S. A. Teukolski, W. T. Vetterling, B.P. Flannery, *Numerical Recipes in C*, Cambridge University Press, New York, 1992.
7. L.M. Sanchez-Brea, E. Bernabeu, *Appl. Opt*, (in press)
8. W.Y.V. Leung, P.J. Bones, R.G. Lane, *Opt. Eng.*, **40,** 547-553 (2001).
9. T.D. Pham, M. Wagner, *Int. J. Pattern Recogn.*, **14**, 1025-1038, (2000)
10. *"Guide to the Expression of the Uncertainty in Measurement"*, International Standardisation Organisation (ISO), Geneva, 1995.
11. P. Bevington, *Data Reduction and Error Analysis for the Physical Sciences*, McGraw-Hill, New York (1969).
12. L.M. Sanchez-Brea, E. Bernabeu, *J. Electron. Imaging*, **11**, 121-126, (2002).

MISCELLANEOUS

Density correlations in ultracold atomic Fermi gases

Wolfgang Belzig[*], Christian Schroll[*] and Christoph Bruder[*]

[*]*Department of Physics and Astronomy, University of Basel, Klingelbergstrasse 82, 4056 Basel, Switzerland*

Abstract. We investigate density fluctuations in a coherent ensemble of interacting fermionic atoms around the BEC-BCS crossover. Adapting the concept of full counting statistics, well-known from quantum optics and mesoscopic electron transport, we study second-order as well as higher-order correlators of density fluctuations. This method is applied to the crossover from a molecular BEC state to a fermionic BCS state and yields a transition from Poissonian statistics to binomial statistics. The statistics can thus be used as an experimental tool to gain information on the many-body ground state in the crossover regime.

Keywords: Quantum noise, full counting statistics, Bose Einstein condensation, BCS-BEC crossover
PACS: 03.75.Ss,03.75.Hh,05.30.Fk

Following the successful creation of Bose-Einstein condensates (BECs) in ultracold atomic clouds [3], recently ultracold Fermionic clouds have been produced [4]. This has attracted a lot of attention both theoretically and experimentally, especially due to the ability to tune the mutual interaction between atoms via a Fano-Feshbach resonance. The unique opportunity to study the crossover from weak attractive to strong attractive interactions in one and the same system makes this interesting from a fundamental many-body point of view. For weakly interacting Fermions a superconducting state occurs, as described by the BCS theory, and the crossover to a molecular BEC state can be studied by tuning the magnetic field [5].

In this work we propose to measure the statistics of density fluctuations as a tool to gain access to the many-body nature of the ground state in the BEC-BCS crossover regime [7]. Our main result is a general expression for the particle number statistics for the BCS wave function valid in the full crossover regime. In the limiting cases, the statistics allows a straightforward interpretation. Deep in the molecular BEC limit the statistics is Poissonian, i.e. that of independent pairs of atoms. In the opposite limit, on the BCS side of the crossover, the fluctuations are strongly suppressed and reflect the particle-hole symmetry.

Our goal is to find the number statistics for the BCS theory. The BCS wave function is given by [8]

$$|\text{BCS}\rangle = \prod_k \left(u_k + v_k c_{k\uparrow}^\dagger c_{-k\downarrow}^\dagger \right) |0\rangle . \qquad (1)$$

The variational procedure yields $v_k^2 = 1 - u_k^2 = (1 - (\varepsilon_k - \mu)/E_k)/2$, where $E_k^2 = (\varepsilon_k - \mu)^2 + \Delta^2$ is the energy of quasiparticle excitations. The order parameter Δ and the chemical potential are fixed by the self-consistency equations

$$\Delta = -\lambda \sum_k u_k v_k \quad, \quad \bar{N} = 2\sum_k v_k^2, \tag{2}$$

where λ is the BCS coupling constant. After renormalization of the coupling constant λ and considering only the low-energy limit, the gap equation can be related to the two-particle scattering amplitude [9, 10, 11].

To calculate the statistics we note that the product form of the BCS wave function greatly simplifies the calculation, since different k states can be treated separately. For a single pair of states $(k\uparrow, -k\downarrow)$ the sum over all possible configurations can be easily performed

$$e^{-S_k(\chi)} = \sum_{n_{k\uparrow},n_{-k\downarrow}=\pm 1} \langle \text{BCS}|e^{i(\hat{n}_{k\uparrow}+\hat{n}_{-k\downarrow})\chi}|\text{BCS}\rangle = u_k^2 + v_k^2 e^{i2\chi}. \tag{3}$$

The sum of all states yields the result

$$S(\chi) = -\sum_k \ln\left[1 + v_k^2(e^{2i\chi} - 1)\right]. \tag{4}$$

This is one of the main results of our paper. We would like to stress that it is valid for a pairing state in any kind of trapping potential (e.g. box or harmonic) and for any dimension. In the following we will concentrate on a box potential in two and three dimensions.

We now discuss some limiting cases in which analytical expressions can be obtained. On the BEC side, $\mu < 0$ and $\Delta \ll |\mu|$ leads to $v_k^2 \ll 1$ for all energies and allows to expand the logarithm in Eq. (4). The result is

$$S(\chi) = -\sum |v_k|^2 \left(e^{i2\chi} - 1\right) = -\frac{\bar{N}}{2}\left(e^{i2\chi} - 1\right), \tag{5}$$

which corresponds to a Poissonian number statistics of *pairs of atoms*. This supports the picture of strongly bound pairs, which act like independent particles of a classical gas. Note, that the factor of 2 in the exponent leads to exponentially growing cumulants, viz., $C_n = 2^n \bar{N}/2$. We therefore expect very strong fluctuations.

We now evaluate the CGF in the BCS limit $\Delta \ll \mu = \varepsilon_F$. We start with (using $v^2(-\xi) = 1 - v^2(\xi)$, where $\xi = \varepsilon_k - \mu$)

$$\begin{aligned}
S(\chi) &= -\sum_k \ln\left[1 + v_k^2(e^{i2\chi} - 1)\right] \tag{6}\\
&= -N_0 \int_{-\varepsilon_F}^{\infty} d\xi \underbrace{\ln\left[1 + v^2(\xi)(e^{i2\chi} - 1)\right]}_{s(\xi,\chi)} \\
&= -N_0 \int_0^{\infty} d\xi\, s(\xi,\chi) - N_0 \int_0^{\varepsilon_F} d\xi\, s(-\xi,\chi) \\
&= -i\chi\bar{N} - N_0 \int_0^{\infty} [s(\xi,\chi) + s(\xi,-\chi)] \\
&= -i\chi\bar{N} - \frac{N_0}{2} \int_{-\infty}^{\infty} \ln\left[1 + 2v^2(1-v^2)(\cos(2\chi) - 1)\right].
\end{aligned}$$

Here, we have used the following property of the binomial distribution

$$\ln(1+T(e^{i\chi}-1)) = i\chi + \ln(1+(1-T)(e^{-i\chi}-1)) \qquad (7)$$

and assumed that $\Delta \ll E_F$, so that we can replace the density of states by its value at the Fermi energy N_0. Furthermore, in the third line of Eq. (6), we have replace the upper limit of the second integral by ∞, since $s(\varepsilon_F, \chi) = 0$.

To evaluate the energy integral, we define

$$I(q) = \frac{d}{dq}\int_{-\infty}^{\infty} d\xi \ln\left[1+2v^2(1-v^2)q\right] = \int_{-\infty}^{\infty} d\xi \frac{2v^2(1-v^2)}{1+2v^2(1-v^2)q}. \qquad (8)$$

Using that $2v^2(1-v^2) = \frac{1}{2}\frac{\Delta^2}{\xi^2+\Delta^2}$, we can evaluate the integral

$$I(q) = \int_{-\infty}^{\infty} d\xi \frac{\Delta^2}{\xi^2+\Delta^2(1+q/2)} = \frac{\pi\Delta}{\sqrt{1+q/2}} \qquad (9)$$

and therefore

$$S(\chi) = -i\bar{N}\chi + 2\pi\Delta N_0 \left(\sqrt{\frac{1+\cos(2\chi)}{2}} - 1\right) = -i\bar{N}\chi + 2\pi\Delta N_0 \left(|\cos(\chi)|-1\right). \qquad (10)$$

Note, that we have to take the absolute value here to account for the π-periodicity. Thus, the CGF is very different from the BEC side. We observe that the first term is dominant but contributes only to the first cumulant. The fluctuations come from the second term in Eq. (10) which is smaller by a factor Δ/ε_F. Furthermore, similar to the degenerate Fermi gas, the odd cumulants C_n for $n \geq 3$ vanish, which can be traced back to the particle-hole symmetry present in the BCS limit.

We would like to address briefly the experimental observability of the statistics of density fluctuations. Real atomic clouds consist of a fixed number of atoms. However, in a homogeneous system with fixed atom number it is possible to consider a small subsystem as connected to a particle reservoir and treat this subsystem in the grand-canonical ensemble. Therefore our considerations apply to the particle number statistics measured in a small subsystem (which should however still contain a large number of atoms). In recent experiments [6] the spatial structure of an atomic cloud has been directly observed (without the expansion used in most other experiments). This makes it possible to determine the density fluctuations either by repeating the experiment many times or by taking densities at different positions in a homogeneous system to extract the statistics. Even more recently spatial correlations have been measured experimentally [12, 13] in ultrcold atomic clouds.

The number statistics in the vicinity of the BEC-BCS crossover displays interesting features which reveal the nature of the many-body ground state[7]. Poissonian fluctuations of a molecular condensate on the BEC side are strongly suppressed on the BCS side. The size of the fluctuations in the BCS limit are a direct measure of the pairing potential. We have also discussed the BEC-BCS crossover of the third cumulant and the temperature dependence of the second cumulant. These quantities can be accessed experimentally and provide additional information on the many-body ground state in the crossover regime. The concept of counting statistics in ultracold gases opens interesting possibilities to study the interplay between coherence and correlation in quantum many-body systems - a direction in which further work is clearly needed.

ACKNOWLEDGMENTS

We acknowledge useful discussions with T. Esslinger, J. Schmiedmayer, and W. Zwerger. Financial support is from the Swiss National Science foundation.

REFERENCES

1. *Quantum Noise in Mesoscopic Physics*, edited by Yu. V. Nazarov (Kluwer, Dordrecht, 2003).
2. Ya. M. Blanter and M. Büttiker, Phys. Rep. **336**, 1 (2000).
3. M. H. Anderson, J. R. Ensher, M. R. Matthews, C. E. Wieman, and E. A. Cornell, Science **269**, 198 (1995); K. B. Davis, M.-O. Mewes, M. R. Andrews, N. J. van Druten, D. S. Durfee, D. M. Kurn, and W. Ketterle, Phys. Rev. Lett. **75**, 3969 (1995).
4. B. DeMarco and D. S. Jin, Science **285**, 1703 (1999); A. G. Truscott, K. E. Strecker, W. I. McAlexander, G. B. Partridge, R. G. Hulet, Science **291**, 2570 (2001); F. Schreck, L. Khaykovich, and K. L. Corwin, G. Ferrari, T. Bourdel, J. Cubizolles, and C. Salomon, Phys. Rev. Lett. **87**, 080403 (2001); S. R. Granade, M. E. Gehm, K. M. O'Hara, and J. E. Thomas, Phys. Rev. Lett. **88**, 120405 (2002); K. M. O'Hara, S. L. Hemmer, M. E. Gehm, S. R. Granade, J. E. Thomas, Science **298**, 2179 (2002).
5. M. Bartenstein, A. Altmeyer, S. Jochim, C. Chin, J. Hecker Denschlag, and R. Grimm, Phys. Rev. Lett. **92**, 203201 (2004) J. Kinast, S. L. Hemmer, M. E. Gehm, A. Turlapov, and J. E. Thomas, Phys. Rev. Lett. **92**, 150402 (2004); C. A. Regal, M. Greiner, and D. S. Jin, Phys. Rev. Lett. **92**, 040403 (2004); M. W. Zwierlein, C. A. Stan, C. H. Schunk, S. M. F. Paupach, A. J. Kerman, and W. Ketterle, Phys. Rev. Lett. **92**, 120403 (2004); M. Greiner, C. A. Regal, C. Ticknor, J. L. Bohn, and D. S. Jin, Phys. Rev. Lett. **92**, 150405 (2004).
6. C. Chin, M. Bartenstein, A. Altmeyer, S. Riedl, S. Jochim, J. Hecker Denschlag, and R. Grimm, Science **305**, 1128 (2004).
7. W. Belzig, C. Schroll and C. Bruder, cond-mat/0412269.
8. J. Bardeen, L. N. Cooper, and J. R. Schrieffer, Phys. Rev. **108**, 1175 (1957).
9. D. M. Eagles, Phys. Rev. **186**, 456 (1969).
10. A. J. Leggett, in *Modern Trends in the Theory of Condensed Matter*, edited by A. Pedalski and J. Przystawa (Springer, Berlin, 1980).
11. M. Randeria, J.-M. Duan, and L. Y. Shieh, Phys. Rev. B **41**, 327 (1990).
12. M. Greiner, C. A. Regal, J. T. Stewart, and D. S. Jin, Phys. Rev. Lett. **94**, 110401 (2005).
13. S. Fölling, F. Gerbier, A. Widera, O. Mandel, T. Gericke, and I. Bloch, Nature **434**, 481 (2005).

Spatial and angular intensity correlations of waves in disordered media

Luis S. Froufe-Pérez[*], Antonio García-Martín[†], Pedro García-Mochales[*], Gabriel Cwilich[**] and Juan José Sáenz[*]

[*]*Departamento de Física de la Materia Condensada, Universidad Autónoma de Madrid, Campus de Cantoblanco, E-28049 Madrid, Spain.*
[†]*Instituto de Microelectrónica de Madrid, Consejo Superior de Investigaciones Científicas, Isaac Newton 8, Tres Cantos, E-28760 Madrid, Spain.*
[**]*Department of Physics, Yeshiva University, 500 W 185th Street, New York, NY 10033 USA.*

Abstract. Spatial intensity correlations between waves transmitted through random media are analyzed within the framework of the random matrix theory of transport. Assuming that the statistical distribution of transfer matrices is isotropic, we found that the the spatial correlation function of the normalized intensity can be expressed as the sum of three terms, with distinctive spatial dependences. This result, that coincides with the one obtained from microscopic perturbative calculations valid in the diffusive regime, holds all the way from quasi-ballistic transport to localization. While correlations are positive in the diffusive regime, we predict a transition to *negative* correlations for both angular and spatial correlations as the length of the system decreases.

Keywords: Waves in disordered media, Mesoscopic systems, Correlations, Universal Conductance Fluctuations, Random Matrix Theory
PACS: 71.30.+h,72.15.Rn, 42.25.Dd

INTRODUCTION

When a wave propagates coherently through a random medium important correlations emerge between the different propagating paths, which manifest themselves as correlations in the intensity speckle pattern. They have been the subject of great interest over the last decade for the case of temporal, angular, and frequency correlations [1, 2, 3, 4, 5, 6, 7, 8, 9]. Recently, the direct observation of spatial correlations in the intensity speckle pattern [10, 11] and in the polarization [12] of electromagnetic waves transmitted through a random medium has led to a renewed interest in this problem [13].

One of the theoretical approaches followed in the study this problem involved a microscopic diagrammatic calculation [5, 4, 3]. This was also the approach in the recent work that showed that the spatial correlation function of the normalized intensity can be expressed as the sum of three terms, which differ in their spatial dependence [10], and in the work finding an equivalent structure for the correlations in the polarized radiation [12]. While the diagrammatic approach has the appealing advantage of illuminating explicitly the nature of the correlations in terms of the microscopic trajectories of the underlying paths, its application is strictly limited to the diffusive regime. An alternative approach, macroscopic in nature, has been applied successfully to study angular correlations in random media [6, 8, 9] ; it considers the correlations between the transport coefficients in the scattering matrix describing the system in the framework of Random

Matrix Theory (RMT) [7, 14, 15]. Most of the work based on RMT has been focused on the study of angular or channel-channel correlations. It shows that only the assumption of isotropy of the transfer matrix distribution, discussed below, determines the structure of the correlations as a function of channel indices. It is the purpose of this work to apply the RMT approach to study the spatial intensity correlation functions.

SCATTERING MATRIX

We will consider a wave propagating in the z-direction in a constrained geometry. The input and output faces of the cavity are the planes $z = 0$ and $z = L$ respectively. The transverse coordinates in the system are describe by $\vec{\rho}$. We will only discuss the case of scalar waves in this work, neglecting polarization effects. The eigenfunctions of the cavity (in the absence of disorder) naturally separate into a longitudinal and a transverse part,

$$\phi_n^{\pm}(\vec{r}) = \frac{1}{\sqrt{k_n}} \psi_n(\vec{\rho}) \exp(\pm i k_n z) \tag{1}$$

The integer $n = 1, 2, \ldots, N$ labels the propagating modes, also referred to as scattering channels. Mode n has a real wave number $k_n = \sqrt{k^2 - k_{\vec{\rho}}^2}$, where k is the wavenumber of the incident radiation, and $k_{\vec{\rho}}$ is the momentum associated with the normalized transverse wave function $\psi_n(\vec{\rho})$. The normalization of the total wave function ϕ_n is chosen to carry unit current.

The scattering matrix $\mathbf{S}$ relates the asymptotic propagating outgoing waves $(\mathbf{o}^L, \mathbf{o}^R)$ to the incoming ones $(\mathbf{i}^L, \mathbf{i}^R)$

$$\begin{pmatrix} \mathbf{o}^L \\ \mathbf{o}^R \end{pmatrix} = \begin{pmatrix} \mathbf{r} & \tilde{\mathbf{t}} \\ \mathbf{t} & \tilde{\mathbf{r}} \end{pmatrix} \begin{pmatrix} \mathbf{i}^L \\ \mathbf{i}^R \end{pmatrix}. \tag{2}$$

where the matrix elements r_{ba} and t_{ja} denote the reflected amplitude in channel b and the transmitted amplitude in channel j when there is a unit flux incident from the left in channel a; $\tilde{r}_{ji}$ and $\tilde{t}_{bi}$ have an analogous meaning when the incident flux in channel i comes from the right. Flux conservation implies that the matrix $\mathbf{S}$ is unitary. Calling $\mathcal{T}_1, \mathcal{T}_2, \ldots, \mathcal{T}_N$ the set of N common transmission eigenvalues of the four Hermitian matrices $\mathbf{tt}^{\dagger}$, $\tilde{\mathbf{t}}\tilde{\mathbf{t}}^{\dagger}$, $\mathbf{rr}^{\dagger}$ and $\tilde{\mathbf{r}}\tilde{\mathbf{r}}^{\dagger}$, it can be shown [14, 15] that the matrix S can be written in terms of the $\mathcal{T}_n$'s by means of the polar decomposition

$$\mathbf{S} = \begin{pmatrix} \mathbf{u}^{(1)} & 0 \\ 0 & \mathbf{u}^{(2)} \end{pmatrix} \begin{pmatrix} -\sqrt{1-\tau} & \sqrt{\tau} \\ \sqrt{\tau} & \sqrt{1-\tau} \end{pmatrix} \begin{pmatrix} \mathbf{u}^{(3)} & 0 \\ 0 & \mathbf{u}^{(4)} \end{pmatrix} \tag{3}$$

where $\mathbf{u}^{(i)}$ are unitary matrices and $\mathcal{T} = \text{diag}(\mathcal{T}_1, \mathcal{T}_2, \ldots, \mathcal{T}_N)$ is a $N \times N$ diagonal matrix with the transmission eigenvalues on the diagonal.

We will consider from here on that our system satisfies reciprocity, which requires that the matrix $\mathbf{S}$ should be symmetric. From reciprocity it immediately follows that the matrices $\mathbf{u}$ satisfy that $\mathbf{u}^{(1)} = (\mathbf{u}^{(3)})^T$ and $\mathbf{u}^{(2)} = (\mathbf{u}^{(4)})^T$.

The transmission and reflection matrices can, then, be written as

$$r_{ba} = -\sum_n u^{(1)}_{bn}\left(\sqrt{1-\mathcal{T}_n}\right)u^{(1)}_{an} \quad ; \quad t_{ja} = \sum_n u^{(2)}_{jn}\left(\sqrt{\mathcal{T}_n}\right)u^{(1)}_{an} \tag{4}$$

CORRELATIONS AND THE "ISOTROPY ANSATZ"

Let us consider two point sources at $\vec{r} = \vec{r}_A$ and $\vec{r} = \vec{r}_B$ ($z_{A,B} < 0$). The incoming field is then proportional to the Green function of the clean waveguide

$$G^+_0(\vec{r}_A,\vec{r}) = \sum_a a_a \phi^+_a(\vec{r}) = \frac{i}{2}\sum_a \phi^{+*}_a(\vec{r}_A)\phi^+_a(\vec{r}) \quad (z > z_A), \tag{5}$$

We also consider two point "detectors" at $\vec{r} = \vec{r}_1$ and $\vec{r} = \vec{r}_2$ ($z_{1,2} > L$). The spatial correlation function is defined in this system as $C(a,1;b,2) \equiv \langle I(a,1)I(b,2)\rangle - \langle I(a,1)\rangle\langle I(b,2)\rangle$ and the first term in the left hand side of this expression can be evaluated as

$$\langle I(a,1)I(b,2)\rangle = \sum_{aa'bb'}\sum_{ii'jj'}\{(a_a a^*_{a'} b_b b^*_{b'}) \times$$
$$\left(\phi^+_j(\vec{r}_1)\phi^{+*}_{j'}(\vec{r}_1)\phi^+_i(\vec{r}_2)\phi^{+*}_{i'}(\vec{r}_2)\right)\langle t_{ja}t^*_{j'a'}t_{ib}t^*_{i'b'}\rangle\} \tag{6}$$

One of the key assumptions in the macroscopic approach is the hypothesis of *isotropy* [14, 15, 7]. Under this hypothesis the statistical distribution of the transmission eigenvalues $\{\mathcal{T}_n\}$ is independent of the unitary matrices $\mathbf{u}^{(i)}$, and the calculation of the statistical averages in (6) factorizes. Moreover, $\mathbf{u}^{(1)}$ and $\mathbf{u}^{(2)}$ are statistically independent from each other, each being distributed according the invariant measure of the unitary group. By using the averages over the unitary group $\langle(u_{jn})(u_{j'n'})^*\rangle$ and $\langle(u_{jn}u_{im})(u_{j'n'}u_{i'm'})^*\rangle$ (evaluated by Mello in ref. [17]), after some algebra, we found

$$\frac{\langle I(a,1)I(b,2)\rangle}{\langle I(a,1)\rangle\langle I(b,2)\rangle} - 1 =$$
$$C_1\left(|F(\vec{r}_a,\vec{r}_b)|^2|F(\vec{r}_1,\vec{r}_2)|^2\right) + C_2\left(|F(\vec{r}_1,\vec{r}_2)|^2 + |F(\vec{r}_a,\vec{r}_b)|^2\right) + C_3 \tag{7}$$

where the correlation coefficients C_1, C_2 and C_3 are *exactly the same* as those appearing in the well known channel-channel correlation function C_{jaib} [5, 6]

$$\frac{\langle T_{ja}T_{ib}\rangle}{\langle T_{ja}\rangle\langle T_{ib}\rangle} - 1 = C_1\left(\delta_{ij}\delta_{ab}\right) + C_2\left(\delta_{ab} + \delta_{ij}\right) + C_3. \tag{8}$$

$|F(\vec{r}_1,\vec{r}_2)|^2$ is the square of the field-field correlation function and is defined in terms of the Green function (5):

$$|F(\vec{r}_1,\vec{r}_2)|^2 = \frac{|\Im\{G^+_0(\vec{r}_1,\vec{r}_2)\}|^2}{\Im\{G^+_0(\vec{r}_1,\vec{r}_1)\}\Im\{G^+_0(\vec{r}_2,\vec{r}_2)\}} \tag{9}$$

SUMMARY

The structure of the spatial correlations is equivalent to that obtained for channel correlations with the angular "δ_{ab}" functions replaced by the spatial functions $|F(a,b)|^2$. It should be noted that our expression (7) is consistent, after a slight reordering of the terms, with the expression (5) in reference [10], which was obtained through a perturbative diagrammatic expansion. The equivalence with polarization correlations (equation (2) of reference [12]) is also evident. However, while the diagrammatic expansions are strictly valid in the diffusive regime, our result **does not depend at all on the transport regime**, as a direct consequence of the isotropy hypothesis. Only the relative *size* of C_1, C_2 and C_3 will depend on the length of the system L and the mean free path ℓ through the distribution of transmission eigenvalues $P(\{\mathcal{T}_n\},s)$, with $s = L/\ell$. Their values, obtained from the Monte Carlo solution of the Dorokhov, Mello, Pereyra, and Kumar (DMPK)[14, 16, 18] scaling equation, are in full agreement with microscopic numerical calculations of bulk disordered wires. While correlations are positive in the diffusive regime, we predict a transition to *negative* correlations for both angular [9] and spatial correlations as the length of the system decreases.

ACKNOWLEDGMENTS

This work has been supported by the Spanish MCyT (Ref. No. BFM2003-01167) and the European Integrated Project "Molecular Imaging" (LSHG-CT-2003-503259)

REFERENCES

1. *Waves and Imaging through Complex Media*, ed. by P. Sebbah,(Kluwer, Dordrecht 2001).
2. *Wave Scattering in Complex Media: From Theory to Applications*, ed. by B. van Tiggelen and S. Skipetrov, (Kluwer, Dordrecht 2003).
3. M. J. Stephen and G.A. Cwilich, Phys. Rev. Lett. **59**,285 (1987)
4. J. Freund, M. Rosenbluh and S. Feng, Phys. Rev. Lett. **61**, 2238 (1987)
5. S. Feng, C. Kane, P. A. Lee and A.D. Stone, Phys. Rev. Lett. **61**, 834 (1988)
6. P.Mello, E. Akkermans and B. Shapiro, Phys. Rev. Lett. **61**, 459 (1988)
7. P.A. Mello and A. D. Stone, Phys. Rev. **B44**, 3559 (1991).
8. E. Bascones, M.J. Calderón, D. Castelo, T. López and J.J. Sáenz, Phys. Rev. B **55**, R11911 (1997).
9. A. García-Martín, F. Scheffold, M. Nieto-Vesperinas and J.J. Sáenz, Phys. Rev. Lett. **88**, 143901 (2002); J.J. Sáenz, L.S. Froufe-Pérez and A. García-Martín, in reference [2], page 175.
10. P. Sebbah, B. Hu, A.Z. Genack, R. Pnini and B. Shapiro, Phys. Rev. Lett **88**, 123901 (2002)
11. V. Emiliani, F. Intonti. M. Cazayous, D.S. Wiersma, M. Colocci, F. Aliev and A. Lagendijk, Phys. Rev. Lett. **90**, 250801 (2003).
12. A.A. Chabanov, N.P. Trégourès, B.A. Van Tiggelen and A.Z. Genack, Phys. Rev. Lett **92** 173901 (2004).
13. Y.H. Kim et al., Phys. Rev. Lett. **94**, 036804 (2005); S. E. Skipetrov, *ibid.* **93**, 233901 (2004); A. A. Chabanov et al., *ibid.* **93**, 123901 (2004); V. M. Apalkov et al., *ibid.* **92**, 253902 (2004).
14. P.A. Mello, P. Pereyra, and N. Kumar, Ann. Phys. (N.Y.) **181**, 290 (1988).
15. C.W.J. Beenakker, Rev. Mod. Phys. **69**, 731 (1997).
16. O.N. Dorokhov, Solid State Commun. **51**, 381 (1984)
17. P.A. Mello, J. Phys. **A 23**, 4061 (1990).
18. L.S. Froufe-Pérez, P. García Mochales, P.A. Serena, P.A. Mello and J.J. Sáenz, Phys. Rev. Lett.**89**, 246403 (2002).

Seismic Pattern Recognition By Wavelet Based-Higher Order Statistics

S.S. Kharintsev, M.Kh. Salakhov

Physics Department, Kazan State University, Kremlevskaya str., 16
Kazan, Russia, 420008
email: red@ksu.ru, msalakh@ksu.ru

Abstract. In this work we develop an approach for detecting nonlinearity in chaotic dynamical systems using the higher order statistics and wavelet analysis. A special attention is paid to the consideration of three-wave interaction in a quadratically coupled medium. The knowledge of nonlinearities allows one to extract order parameters both for reconstruction of a dynamical system and for the study of transient processes between oscillatory regimes. To demonstrate a power of this approach we verify the latter for real data coming from the seismology.

Keywords: wavelet, higher order statistics, bicoherence, tricoherence, earthquake, artificial explosion.
PACS: 05.45.–a, 05.40.Ca, 02.50.–r

INTRODUCTION

The dynamical future of evolution of complex system is unpredictable. This is caused by the fact that characteristic temporal scale is absent in a fractal structure. However we are interested to predict bifurcations rather not the knowledge of prospective evolution. It is important to note that apparent and/or possibly real non-determinancy of natural processes prior to anomalous events, masks if not a reason than anyway precursors of the events. The precursors of critical events may be deciphered until we approach to this region.

In this study we show that the higher order statistics can be used effectively in combination with the wavelet transform to analyze a bound frequency composition of the wave process in the vicinity of the critical point. We intend to demonstrate that high frequency part of the wave evolution spectrum is determined by the higher harmonics of the dominate waves. This means a particular phase relationship between each high-frequency mode and the dominant frequency mode. Thus, the bicoherence of any pair of frequency modes becomes nonzero. The wavelet-based higher order statistics can be successfully used for pattern recognition and fault events prediction in nonlinear dynamical systems.

THEORY

Over the last two decades, the higher order statistics has been successfully used by many researchers for studying nonlinear effects in the different applications [1]. This technique allows one to look beyond the conventional power spectrum estimation to extract information regarding, in the first turn, phase relations. Certainly, higher order statistics technique has more wide possibilities. In particular, it provides to suppress additive colored Gaussian noise of unknown power spectrum, extract information due to deviations from Gaussinity and detect and characterize nonlinear properties in signals as well as identify nonlinear systems.

Here we will consider particular cases of higher order spectra – the third-order spectrum, so called bispectrum (by definition, this is the Fourier tansformation of the third-order statistics) and the trispectrum (the fourth-order statistics) of a stationary signal. Traditionally, when the higher order statistics is used in signal processing the emphasis is placed on the second, third and fourth moments and/or cumulants and their respective Fourier transform (power spectrum, bispectrum and trispectrum) [2-4]. Without of loss of generality we will deal mainly with the bispectral estimation, bearing in mind that basic results obtained below may be generalized for moments and polyspectra when $n > 3$.

The response P of a dynamical system to external field can be represented by the following way:

$$\left\langle \vec{P}(t) \right\rangle_T = \varepsilon_1 B(\omega) + \varepsilon_{12} B(\omega_1, \omega_2) + \varepsilon_{123} B(\omega_1, \omega_2, \omega_3) + \ldots \quad , \qquad (1)$$

where $\varepsilon_1, \varepsilon_{12}, \varepsilon_{123}, \ldots$ are constants. $B(\omega) = \left\langle W(t, \omega) W^*(t, \omega) \right\rangle_T$ stands for squared wavelet spectrum (the brackets stands for the averaging over temporal interval T). The second term in Eq.() may be interpreted as a bispectrum of a signal

$$B(\omega_1, \omega_2) = \left\langle W(t, \omega_1) W(t, \omega_2) W^*(t, \omega_1 + \omega_2) \right\rangle_T +$$
$$+ \sum_{\Delta \neq 0} \left\langle W(t, \omega_1) W(t, \omega_2) W^*(t, \omega_1 + \omega_2 - \Delta) W^*(t, \omega_1 + \omega_2 + \Delta) \right\rangle_T \quad . \qquad (2)$$

The interpretation of the bispectrum defined by Eq.(2) is shown in Fig.1.

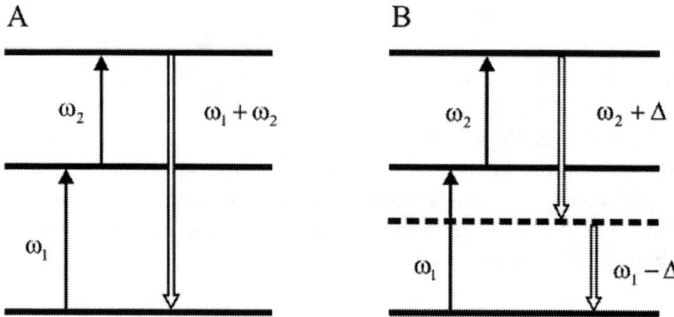

FIGURE 1. Energy diagrams of quadratically coupled oscillators. A – one-photon decay, B – two-photon decay.

It is important to emphasize that only phase coupled components contribute to the third-order cumulant sequence of a process is what makes the bispectrum a useful tool for detecting quadratic phase coupling and discriminating phase components from those that are not.

A near zero value of $|B(\omega_1,\omega_2)|$ at harmonically related frequency pairs will suggest an absence of phase coupling. Certainly, one of the advantages of using the conventional approach to bicoherence index estimation is its ability to serve as a good quantifier by providing good estimates of degree phase coupling at harmonically related frequency pairs.

By analogy, the trispectrum can be used to resolve cubic phase coupling and that if a signal contains components due to both quadratic and cubic phase coupling, the bvispectrum of that signal is blind to the cubucally coupled component and can resolve the quadratically-coupled components, whereas the trispectrum of that signal blind to the quadratically coupled components and can resolve the cubically-coupled components. Hence, higher order statisticscan resolve different types of nonlinearities.

EARTHQUAKE AND ARTIFICIAL EXPLOSION

Here we apply the bi- and tri-spectral wavelet transformation for analyzing the real seismic data. We are trying to solve the following problems: 1) strong earthquake forecasting, 2) differentiation of small earthquake from technogenic explosion.

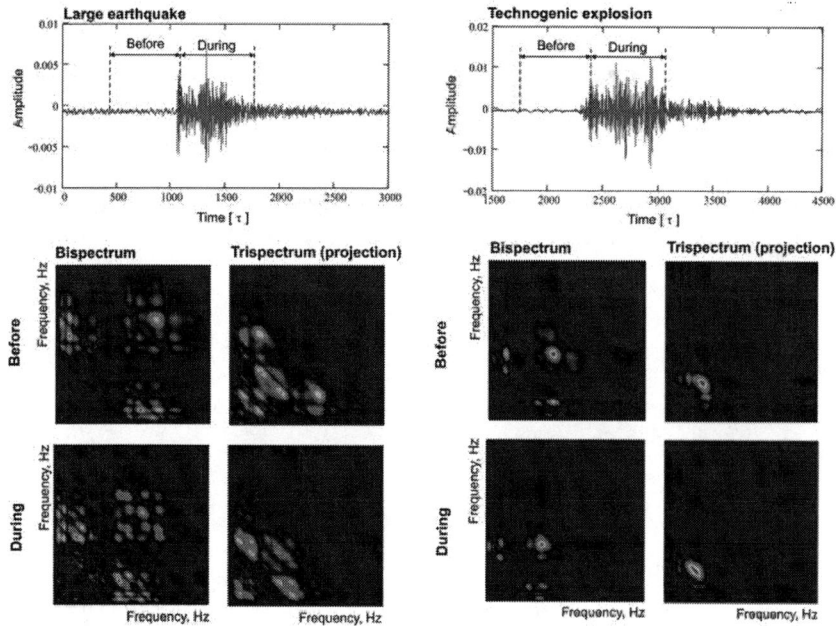

FIGURE 2. Bispectrum and trispectrum for large earthquake and man-made explosion.

For this purpose we analyzed two parts of the real seismogram [5]: 1) before the event (earthquake and explosion) and during the event. A typical seismogram contains 3000 registered points. The complete analysis includes the following information: 1) the complexity parameter dynamics, 2) wavelet analysis, 3) bi- and tri spectral wavelet estimation.

We used two types of available experimental data courteously given by Laboratory of Geophysics and Seismology (Amman, Jordan) for the following seismic phenomena: strong earthquake in Turkey (case I) (Summer 1999) and small earthquake in Jordan (case II) (Summer 1998). As a man made explosion we used the local underground explosion (case III).

Al the data correspond to transverse seismic displacements. The real temporal step of digitization t between registered points of seismic activity has the following values: $t = 0.02$ sec for the case I and $t = 0.01$ sec for the case II and case III.

As follows from the results obtained bi- and tri-spectral wavelet transformation allows one to reveal quadratic and cubic coupling as before as during the seismic event for case I, whereas for the technogenic explosion this approach shows the lack of local skewness and kurtosis. In the case of small earthquake we observe the presence of skewness during the event only. This behavior is caused by the local long-range memory effects arising in natural conditions.

Finally we can conclude that the wavelet based higher order statistics is a powerful instrument for small earthquake and man made explosion and large earthquake forecasting.

ACKNOWLEDGMENT

This research has been supported by the CRDF (REC 007). This work has been supported by the CRDF (REC 007). We are very grateful to Prof. Dr. A.I. Fishman (Faculty of Physics, Kazan State University) and Dr. A.I. Skvortsov (Faculty of Physics, Kazan State University) for fruitful discussion that has noticeably improved the manuscript. We thank to Prof. R.R.Nigmatullin (Faculty of Physics, Kazan State University) for giving the experimental data.

REFERENCE

1. A. Gallego, C. Urdials, D.P. Ruiz, *Nonlinear Dynamics* **19**, 273-294 (1999).
2. A. A. Koronovskii, A. E. Khramov, *Plasma Physics Reports* **28**, 666-682, (2002).
3. L. Persson, *Computers & Geosciences* **19**, 243-250 (1993).
4. T.Dudok de Wit and V.V. Krasnosel'skikh, *Phys. Plasma* **2**, 4307-4312 (1995).
5. R.Yulmetyev, F. Gafarov, P. Hangii, R. Nigmatullin, S. Kayumov, *Phys. Rev. E* **64**, 2201-2216 (2001).

Author Index

A

Abed, E. H., 529
Adamson, D., 685
Agu, M., 795
Aguado, R., 413
Aigner, F., 454
Akabane, H., 795
Akahori, H., 199
Akazaki, T., 476
Akimov, V., 611
Alabedra, R., 361
Albaladejo, S., 357
Aleshkin, V. Y., 425
Alfinito, E., 125, 611
Amann, A., 37, 41
Andreucci, D., 553
Aniel, F., 283, 287
Anil, K. G., 279
Aoyagi, N., 599
Arias, J., 505
Arnaud, F., 195
Aroutiounian, V. M., 87
Arvizu M., A., 385
Asaoka, N., 492
Asenov, A., 239
Ashburn, P., 265
Asriyan, H. V., 87
Astafiev, M. G., 647
Azaïs, B., 151

B

B, M., 403
Balanov, A. G., 37, 41
Barbolla, J., 505
Barge, D., 323
Bary, L., 299
Basso, G., 435
Belkas, J., 595
Belyaev, A. E., 713
Belyakov, A. V., 365
Belyi, V., 63
Belzig, W., 480, 817
Bengtsson, S., 462
Benoit, P., 257

Bernabeu, E., 811
Berroir, J. M., 466
Bertazzi, F., 777
Bidaud, M., 323
Bilyk, I. S., 155
Birbas, A., 525
Birbas, M., 525
Bisbal, D., 505
Bisquert, J., 59
Bœuf, F., 315, 323
Blatter, G., 409
Blick, R. H., 472, 567
Boeck, G., 517
Bollaert, S., 335
Bonani, F., 777
Bosman, G., 458
Boukhenoufa, A., 319
Bournel, A., 749
Brandes, T., 413
Brederlow, R., 703
Bruder, C., 817
Buiu, O., 265
Bulucea, C., 261
Burgd, J., 454
Bychikhin, S., 709

C

Cappy, A., 335
Carbone, A., 397
Carceller, J. E., 717
Carin, R., 319
Casuso, I., 575
Celasco, E., 121
Celasco, M., 121
Çelik-Butler, Z., 261, 311
Cha, H.-Y., 713
Chakalov, R. A., 143
Chang, C.-H., 587
Chluda, C., 377
Chobola, Z., 721
Choi, S. Y., 567
Choi, W., 109
Chong, F., 95
Chovet, A., 109
Chroboczek, J., 315

Chroboczek, J. A., 209
Cibiel, G., 513
Ciccarello, F., 159
Cichosz, J., 669
Cichosz, J. A., 673
Claeys, C., 187, 231, 331
Clay, J. R., 571
Contaret, T., 315
Conte, G., 777
Cordier, C., 319
Cosnier, V., 235
Crépieux, A., 488
Cretu, B., 319
Crozat, P., 287
Cwilich, G., 821

D

Dambrine, G., 269, 509, 745
Danneville, F., 269, 509, 745
Danylyuk, S. V., 713
da Silva, R., 703
Decoutere, S., 253, 279
Deen, J., 462
Deen, M. J., 3
de Kort, R., 735
Delage, S., 299
Delcourt, S., 509
Delhougne, R., 187
del Rio, E., 21
Delseny, C., 257, 462
De Meyer,, 279
De Meyer, K., 187
Demir, A., 499
Devillard, P., 488
Deville, G., 139
Devireddy, S. P., 311
DiBenedetto, E., 553
Ding, L., 319
Dinís, L., 15
Dixit, A., 279
Dmitriev, A. P., 291
Dolcini, F., 421
Dollfus, P., 749
Donati Guerrieri, S., 777
Dubkov, A. A., 25
Duriez, B., 195
Dyakonova, N., 335

E

Eastman, L. F., 713
Ebeling, W., 21
Eberl, K., 472
Edwards, P. J., 353
Eggenh, R., 121
Eisenberg, R. S., 563
Enciso, M., 287
Eneman, G., 187
Enriquez, L., 505
Enz, C. C., 741
Eriksson, M. A., 472
Estévez-Torres, A., 639
Evans, D., 99

F

Fanciulli, M., 171, 221
Fantini, P., 171, 191
Ferrari, G., 171, 191, 221, 575, 653
Ferrrari, G., 611
Fève, G., 466
Flament, S., 143, 681
Flindt, C., 442
Fobelets, K., 275
Forger, D. B., 571
Froufe-Pérez, L. S., 357, 381, 821
Fukumitsu, M., 492
Fumagalli, L., 575, 611

G

Gabelli, J., 466
Gao, J., 295
Garbar, N., 265, 331
García-Dominguez, L., 595
García-Loureiro, A. J., 239
García-Martín, A., 821
García-Mochales, P., 357, 381, 821
Gasparyan, F. V., 87
Gaubert, P., 199
Gavish, U., 446
Ghibaudo,, 181
Ghibaudo, G., 209, 315
Ghione, G., 777
Glattli, D. C., 417, 466
Gleeson, J., 353

Godoy, A., 717
Golub, A., 51
Gómez-Campos, F. M., 717
Gomila, G., 575, 611
González, T., 335, 753
Goswami, S., 472
Grabert, H., 421
Graffeuil, J., 299, 513
Grafov, B. M., 647
Graham, A. C., 417
Gribaldo, S., 513
Grmela, L., 135, 175
Groeseneken, G., 279
Gruzinskis, V., 753
Gružinskis, V., 759, 791
Gružinskis, V., 151
Grzanna, J., 635
Guevara, R., 595
Gurgul, J., 209

H

Hackbarth, T., 283, 287
Hall, S., 265
Hållstedt, J., 307
Hamada, T., 199
Hamid, H. A., 269
Hamm, H. E., 553
Hänggi, P., 45
Hashiguchi, S., 135, 339, 343, 657, 693
Hasse, L., 677, 725
Havens, R. J., 735
Havranek, J., 339
Hayano, J., 549
Hayashi, Y., 643
Hellstr, P. E., 225
Hellstr, P.-E., 307
Henini, M., 139
Herzog, H. J., 283
Herzog, J.-H., 287
Hidaka, I., 535
Higuchi, H., 75
Hirasita, M., 693
Hizanidis, J., 41
Hoeschl, P., 135
Hoffmann, A., 213, 243, 323, 462
Hoffmann, M. H. W., 521
Hong, S.-M., 787
Hoque, M. M. U., 261

Howard, R. M., 83
Huettel, A. K., 472
Hulicius, E., 721
Huynh, Q. T., 458

I

Iannaccone, G., 203, 435
Imry, Y., 446
Inamura, K., 607
Iñiguez, B., 269
Introzzi, R., 397
Ivanov, K. V., 373
Iwase, S., 607
Izpura, J. I., 113

J

J, F., 557
Jacquod, P., 484
Jauho, A. P., 442
Jeamsaksiri, W., 735
Jelenkovic, E., 295
Jha, S., 295
Jiménez Tejada, J. A., 717
Jomaah and G., J., 181
Jullien, L., 639
Jungblut, H., 635
Jungemann, C., 777, 795
Jurczak, M., 243, 279

K

Kabir, M. S., 462
Kainuma, K., 117
Kajiwara, Y., 693
Kalna, K., 239
Kanevskii, L. S., 647
Kaplan, A. Y., 579
Katilius, R., 29
Katura, T., 583
Kaufman, I., 563
Kaulakys, B., 91
Kazelle, J., 721
Kerlain, A., 327
Kharintsev, S. S., 825
Khokhlov, D. A., 373

Khrebtov, I. A., 147, 373
Kießlich, G., 439
Kim, H. S., 567
Kim, J. Y., 369
Kishimoto, T., 117
Kitajo, K., 535
Kiyono, K., 545, 549, 599, 807
Klaassen, D. B. M., 735
Klein, N., 713
Knap, W., 335
Knorr, C., 261
Koh, J., 703
Kohler, S., 45
Koktavy, B., 393
Koktavy, P., 393
Kolek, A., 155, 163
Komiya, K., 131
Konczakowska, A., 669
Kondis, D., 525
Konno, H., 591
Korcz, K., 689
Koroutchev, K., 603
Korutcheva, E., 603
Kosugi, Y., 541
Kotani, A., 643
Kotani, K., 199
Kumicák, J., 803
Kurachi, T., 541
Kurakin, A. M., 713
Kurin, V. V., 33
Kusu, F., 643
Kuzmik, J., 709
Kwak, S., 535, 549

L

L, A., 269
Laigle, A., 243
Lambert, N., 413
Langevelde, R. van, 735
Laurens, M., 509
Lebedev, A.V., 409
Lee, J., 109
Lee, S.-C., 331
Lemarchand, A., 639
Leroux, C., 209
Lesovik, G. B., 409
Leturcq, R., 139
Levada, S., 709

Levinshtein, M. E., 291
Lewerenz, H. J., 635
Leyris, C., 213, 323, 361
L'Hôte, D., 139
Liberis, J., 105
Lin, Y., 431
Liu, H.C., 397
Llopis, O., 513
Lokshin, M., 265, 331
Loo, R., 187
López-Villanueva, J.A., 717
Luchinsky, D. G., 563
Lukyanchikova, N., 265, 331
Lusakowski, J., 335
Lyashenko, O. V., 389

M

M, M., 235
Macucci, M., 435, 450
Maes, H., 279
Maione, I. A., 435
Maki, A., 583
Makushok, Y., 729
Malm, B. G., 225, 307
Malo, J., 113
Mano, T., 607
Mansfeld, F., 625
Marconcini, P., 450
Marin, M., 195
Marinov, O., 3
Martín, M. J., 745
Martin, P., 557
Martin, S., 261
Martin, T., 488
Martinez, F., 235, 243, 323
Mateos, J., 335
Matsuda, R., 643
Matulionis, A., 105
Matveyeva, E., 631
McClintock, P. V. E., 563
Méchin , L., 143
Méchin, L., 681
Meinerzhagen, B., 777, 795
Melkonyan, S. V., 87
Mellor, C. J., 139
Mendez, E. E., 431
Mendieta J., F. J., 385
Meneghesso, G., 709

Mercha, A., 279
Mercone, S., 143
Meschke, M., 661
Meškauskas, T., 91
Metveyeva, E., 729
Michaud, J.-F., 319
Millithaler, J. F., 151
Millithaler, J.-F., 335
Min, H. S., 787
Misra, D., 231
Mitrovic, I. Z., 265
Mleczko, K., 163
Mohammed-Brahim, T., 319
Monroy, A., 257, 509
Moravec, P., 175
Morishima, S., 167
Mosser, V., 327
Mubarek, H. A. W., 265
Munakata, T., 79
Musha, T., 217, 541
Myara, M., 361, 377

N

Nadrowski, B., 557
Nakamura, S., 75
Nakano, H., 476
Nam, H., 109
Neau, G., 235
Newaz, A. K. M., 431
Nguyen, T., 235
Nicholls, J. T., 417
Nieminen, T. E., 661
Nii, K., 199
Nitta, J., 431, 476
Niu, G., 247
Novotný, T., 442
Nozaki, D., 535
Nur, O., 462

O

Obata, A., 583
Ochi, S., 75
Ohashi, K., 535
Ohki, M., 657
Ohmi, T., 199
Okuma, S., 117, 167

Okumura, H., 343
Okumura, T., 492
Omura, Y., 131
Oriols, X., 763
Orsal, B., 361, 377
Östling, M., 225, 307
Ouassif, N., 253

P

Pailloncy, G., 269, 745
Pakhomov, O. V., 147
Palczynska, B., 689
Palermo, C., 151, 335
Pantisano, L., 231
Pardo, D., 335
Park, C. H., 787
Park, Y.-J., 787
Parkhutik, V., 631, 729
Parrondo, J. M. R., 15
Pascal, F., 257, 462
Pavelka, J., 217, 339, 343
Paydarfar, D., 571
Pei-Rong, W., 95
Pekola, J. P., 661
Pellegrini, B., 435
Pennetta, C., 125, 611
Pepper, M., 417
Perez, J. P., 361
Perez, J.-P., 377
Pérez, S., 335, 753
Pérez-Velazquez, J. L., 595
Petrychuk, M. V., 713
Pichon, L., 319
Pienkowski, D., 517
Pilkuhn, M., 295
Pimenov, I. V., 33
Piontek, A., 253
Plaçais, B., 466
Plana, R., 299
Pogany, D., 709
Popovych, V. D., 155
Prada, M., 472
Prati, 221
Prati, E., 171
Pruvost, S., 509
Ptak, P., 163

Q

Qin, H., 472
Quintanilla, L., 505

R

Ramachandran, S., 567
Ramonas, M., 105
Raoult, J., 257
Reggiani, L., 125, 425, 611, 753, 759, 769, 773, 791
Reggiani, S., 29
Reklaitis, A., 773
Rengel, R., 745
Rennane, A., 299
Reufer, M., 357
Rey, M., 45
Reydellet, L.-H., 466
Reza, S., 458
Ribes, G., 235
Rinzler, A. G., 458
Ritchie, D. A., 417
Roche, P., 417, 466
Rodriguez, M., 283
Rodríguez-Bolívar, S., 717
Romanjek, K., 315
Rooyackers, R., 187
Rosini, M., 769
Rotter, S., 454
Routoure, J. M., 143, 681
Roy, A. S., 741
Roy, F., 213
Rudan, M., 29
Ruffo, S., 125
Rumyantsev, S. L., 291

S

Sáenz, J. J., 357, 381, 821
Safi, I., 421
Safonov, L. A., 535, 615
Sahagún, E., 357
Sakalas, P., 303
Sakata, S., 549
Salakhov, M. Kh., 825
Salesse, A., 462
Samitier, J., 575

Sampietro, M., 171, 221, 575, 611, 653
Samuelsson, P., 403, 439
Sanchez-Brea, L. M., 811
Sangha, G. S., 521
San Pablo, J., 505
Sansen, W., 279
Sato, H., 583
Sch, E., 37, 41, 439
Scheffold, F., 357
Scholten, A. J., 735
Schroll, C., 817
Schroter, M., 303
Schwarz, W., 79
Schweizer, H., 295
Sedlakova, V., 135, 339
Ségala, J., 417
Seger, J., 307
Senjanoviè, G., 595
Serena, P. A., 381
Shen, L., 553
Sheridan, D., 247
Shiktorov, P., 151, 753, 759, 791
Shur, M. S., 291
Signoret, P., 361, 377
Sikula, J., 135, 175, 217, 339, 343, 657, 693
Simecek, T., 721
Simmons, M. Y., 417
Simoen, E., 187, 231, 331
Simon, C., 143
Sippel, J., 458
Sita, Z., 135
Skotnicki, T., 235, 315, 323
Slaidins, I., 665
Smallhorn, T., 353
Smolanka, A., 265, 331
Soliveres, S., 462
Solovieva, A. B., 579
Sols, F., 45
Soma, R., 535
Song, J., 109
Song, W., 431
Soubercaze-Pun, G., 299
Spagnolo, B., 25
Spencer, M. G., 713
Spiralski, L., 677, 689, 725
Srinivasan, P., 231
Standler, A. W., 163
Starikov, E., 151, 753, 759, 791
Stawarz, B., 669

Stegemann, G., 37
Stolk, P., 195
Strass, M., 45
Struzik, Z. R., 535, 545, 549, 599, 807
Subramanian, V., 279
Suhara, M., 492
Sukhorukov, E. V., 403
Surya, C., 295
Suzuki, N., 541
Sweeney, S., 247
Szatkowski, A., 673
Szelag, B., 257
Szewczyk, A., 677

T

Tacano, M., 135, 217, 339, 343, 657, 693
Takayanagi, H., 476
Takemoto, Y., 657
Tanaka, N., 583
Tanuma, N., 343, 693
Tartarin, J. G., 299
Tavel, B., 195, 323
Telliez, I., 509
Teramoto, A., 199
Thewes, R., 703
Thomas, K. J., 417
Tiebout, M., 703
Tiemeijer, L. F., 735
Timashev, S. F., 67, 579
Tindjong, R., 563
Tobin, P. J., 311
Togo, F., 545
Toita, M., 217, 339
Tolotto, G., 121
Tong, K. Y., 295
Toonen, R. C., 472
Tourbot, R., 139
Tourrenc, J.-P., 361
Trauzettel, B., 421
Tseng, H. H., 311
Tsong, T. Y., 587
Turczyński, J., 725

U

Uemura, T., 343

V

Vaissière, J. C., 151, 753, 759, 791
Valenza, M., 213, 235, 243, 323
Vandamme, L. K. J., 365, 709
Vandervorst, W., 187
van der Weide, D. W., 472, 567
Vaněk, J., 721
Van Huylenbroeck, S., 253
Varani, L., 151, 335, 753, 759, 791
Velarde, M. G., 21
Velázquez, J. E., 275
Velázquez, J.E., 749
Veleschuk, V. P., 389
Velghe, R. D. M. A., 735
Veloso, A., 243
Venegas, R., 253
Verheyen, P., 187
Vicente, J., 505
Vildeuil, J. C., 195, 213, 235, 257
Virt, I. S., 155
Vitusevich, S. A., 713
Vlasenko, O. I., 389
von Haartman, M., 225, 307
Vstovsky, G. V., 579
Vstovsky, G. V., B. M., 67

W

Wacker, A., 439
Wang, F., 311
Wang, R., 619
Ward, L. M., 535
Wei- Yong, Z., 95
Wennberg, R., 595
Whichello, A. P., 353
Whitney, R. S., 484
Wiatr, W., 685, 697
Willander, M., 462
Wio, H. S., 55
Wirth, G. I., 703
Wolk, S. I., 529
Woo, M., 195
Woodgate, D., 353

X

Xia, K., 247

Y

Yagi, S., 343
Yakimov, A. V., 347, 365
Yamamoto, M., 199
Yamamoto, Y., 535, 545, 549, 599, 615, 807
Yamanaka, K., 535
Yang, F., 143
Yang, H., 109
Yi, J. C., 369
Yokokura, S., 693
Yoo, T. K., 369
Yu, B., 109
Yu, W., 619

Yurke, B., 446

Z

Zajacek, J., 175
Zampardi, P., 303
Zanoni, E., 709
Zarcone, M., 159
Zawislak, Z., 163
Zegers-van Duijnhoven, A. T. A., 735
Zeltins, M., 665
Zerounian, N., 283, 287
Zhe, X., 95
Zhu, C. F., 295
Zlotnicka, A., 311